D1179562

BRITISH MEDICAL ASSOCIATION
0831568

Biotechnology

Biotechnology

Second Edition

David P. Clark
Department of Microbiology
Southern Illinois University
Carbondale, Illinois, USA

Nanette J. Pazdernik
Washington University School of Medicine
St. Louis, Missouri, USA

ELSEVIER

AMSTERDAM • BOSTON • HEIDELBERG • LONDON
NEW YORK • OXFORD • PARIS • SAN DIEGO
SAN FRANCISCO • SINGAPORE • SYDNEY • TOKYO
Academic Cell is an imprint of Elsevier

Academic Cell is an imprint of Elsevier
125 London Wall, London EC2Y 5AS, UK
525 B Street, Suite 1800, San Diego, CA 92101-4495, USA
225 Wyman Street, Waltham, MA 02451, USA
The Boulevard, Langford Lane, Kidlington, Oxford OX5 1GB, UK

ISBN: 978-0-12-385015-7

British Library Cataloguing-in-Publication Data

A catalogue record for this book is available from the British Library

Library of Congress Cataloging-in-Publication Data

A catalog record for this book is available from the Library of Congress

For information on all Academic Cell publications
visit our website at http://store.elsevier.com/

Typeset by TNQ Books and Journals
www.tnq.co.in

Printed and bound in China

An Online Study Guide is now available with your textbook, containing a relevant journal article with a case study to focus understanding and discussion about each chapter.

---

1. To access the Online Study Guide, as well as other online resources for the book, please visit: http://booksite.elsevier.com/9780123850157

2. For instructor-only materials, please visit: http://textbooks.elsevier.com/web/Manuals.aspx?isbn=9780123850157

This book is dedicated to Donna. —DPC

This book is dedicated to my children and husband. Their patience and understanding have given me the time and inspiration to research and write this text. —NJP

HOW WE GOT HERE

In speaking with professors across the biological sciences and going to conferences, we, the editors at Academic Press and Cell Press, saw how often journal content was being incorporated in the classroom. We understood the benefits students were receiving by being exposed to journal articles early: to add perspective, improve analytical skills, and bring the most current content into the classroom. We also learned how much additional preparation time was required on the part of instructors finding the articles, then obtaining the images for presentations and providing additional assessment.

So we collaborated to offer instructors and students a solution, and *Academic Cell* was born. We offer the benefits of a traditional textbook (to serve as a reference to students and a framework to instructors), but we also offer much more. With the purchase of every copy of an *Academic Cell* book, students can access an online study guide containing relevant, recent Cell Press articles and providing bridge material in the form of a case study to help ease them into the articles. In addition, the images from the articles are available as zipped .jpg files and we have optional test bank questions.

We plan to expand this initiative, as future editions will be further integrated with unique pedagogical features incorporating current research from the pages of Cell Press journals into the textbook itself.

CONTENTS

PREFACE

From the simple acts of brewing beer and baking bread has emerged a field now known as biotechnology. Over the ages the meaning of the word biotechnology has evolved along with our growing technical knowledge. Biotechnology began by using cultured microorganisms to create a variety of food and drinks, despite its early practitioners not even knowing of the existence of the microbial world. Today, biotechnology is still defined as any application of living organisms or bioprocesses to create new products. Although the underlying idea is unchanged, the use of genetic engineering and other modern scientific techniques has revolutionized the area.

The fields of genetics, molecular biology, microbiology, and biochemistry are merging their respective discoveries into the expanding applied field of biotechnology, and advances are occurring at a record pace. Two or three years of research can dramatically alter the approaches that are of practical use. For example, the simple discovery that double-stranded RNA can block expression of any gene with a matching sequence has revolutionized how we study and apply genetic interactions in less than a ten-year period.

This rapid increase in knowledge is very hard to incorporate into a textbook format, and often instructors who teach advanced molecular biology classes rely on the primary research to teach students novel concepts and applications. This type of teaching is difficult and requires many hours to plan and organize.

The new partnership between Academic Press and Cell Press has adopted a solution to teaching advanced molecular biology and biotechnology courses. The partnership combines years of textbook publishing experience with the most relevant and high impact research. What has emerged is a new teaching paradigm. In *Biotechnology*, the basic ideas and methodologies are explained using very clear and concise language. These techniques are supplemented with a wide variety of diagrams and illustrations to simplify the complex biotechnology processes.

These basics are then supported with a *Biotechnology* online study guide that not only tests the student's knowledge of the textbook chapter, but also contains primary research articles. The articles are chosen from the Cell Press family of journals, which includes such high-impact journals as *Cell, Molecular Cell,* and *Current Biology.* The articles expand upon a topic presented in each chapter or provide an exemplary research paper for that particular chapter. The entire full-color research article is included online.

In addition to the article itself, the *Biotechnology* study guide includes a synopsis of the research paper. The synopsis includes a thorough discussion of the relevant background information. This material is often absent from primary research articles because their authors assume that readers are also experts. Then each synopsis breaks the paper into sections, explaining each individual experiment separately. Each experiment is explained by defining the underlying hypothesis or question, the methods used to study the question, and the results. The final section of the synopsis provides the overall conclusions for the paper. This approach reinforces the basic scientific method. The instructor does not have to find an article, create a presentation on the background, and then work with the student to explain each of the methods and results. The study guide synopsis provides all of this information already.

The online format ensures that only the most recent papers are associated with the chapter. The combination of the online study guide with the newest relevant research and a solid basic textbook provides the instructor with the best of both worlds. You can teach students the basic concepts using the textbook, and then use the relevant research paper to stretch the student's knowledge of current research in the field of biotechnology.

ACKNOWLEDGMENTS

We would like to thank the following individuals for their help in providing information, suggestions for improvement, and encouragement: Laurie Achenbach, Rubina Ahsan, Phil Cunningham, Donna Mueller, Dan Nickrent, Holly Simmonds, and Dave Pazdernik. Special thanks go to Marshall Spector for helping us understand bioethics, to Michelle McGehee for writing the questions and online supplements and to Karen Fiorino for creating most of the original artwork for the first edition. Alex Berezow was responsible for writing a major part of the following chapters: Chapter 16, Transgenic animals, Chapter 22, Biowarfare and bioterrorism, and Chapter 24, Bioethics in biotechnology.

INTRODUCTION

MODERN BIOTECHNOLOGY RELIES ON ADVANCES IN MOLECULAR BIOLOGY AND COMPUTER TECHNOLOGY

Traditional biotechnology goes back thousands of years. It includes the selective breeding of livestock and crop plants as well as the invention of alcoholic beverages, dairy products, paper, silk, and other natural products. Only in the past couple of centuries has genetics emerged as a field of scientific study. Recent rapid advances in this area have in turn allowed the breeding of crops and livestock by deliberate genetic manipulation rather than trial and error. The so-called green revolution of the period from 1960 to 1980 applied genetic knowledge to natural breeding and had a massive impact on crop productivity in particular. Today, plants and animals are being directly altered by genetic engineering.

New varieties of several plants and animals have already been made, and some are in agricultural use. Animals and plants used as human food sources are being engineered to adapt them to conditions that were previously unfavorable. Farm animals that are resistant to disease and crop plants that are resistant to pests are being developed in order to increase yields and reduce costs. The impact of these genetically modified organisms on other species and on the environment is presently a controversial issue.

Modern biotechnology applies not only modern genetics but also advances in other sciences. For example, dealing with vast amounts of genetic information depends on advances in computing power. Indeed, the sequencing of the human genome would have been impossible without the development of ever more sophisticated computers and software. It is sometimes claimed that we are in the middle of two scientific revolutions, one in information technology and the other in molecular biology. Both involve handling large amounts of encoded information. In one case the information is human made, or at any rate man-encoded, and the mechanisms are artificial; the other case deals with the genetic information that underlies life.

However, there is a third revolution that is just emerging—nanotechnology. The development of techniques to visualize and manipulate atoms individually or in small clusters is opening the way to an ever-finer analysis of living systems. Nanoscale techniques are now beginning to play significant roles in many areas of biotechnology.

This raises the question of what exactly defines biotechnology. To this there is no real answer. A generation ago, brewing and baking would have been viewed as biotechnology. Today, the application of modern genetics or other equivalent modern technology is usually seen as necessary for a process to count as "biotechnology." Thus, the definition of *biotechnology* has become partly a matter of fashion. In this book, we regard (modern) biotechnology as resulting in a broad manner from the merger of classical biotechnology with modern genetics, molecular biology, computer technology, and nanotechnology.

The resulting field is of necessity large and poorly defined. It includes more than just agriculture: it also affects many aspects of human health and medicine, such as vaccine development and gene therapy. We have attempted to provide a unified approach that is based on genetic information, while at the same time indicate how biotechnology has begun to sprawl, often rather erratically, into many related fields of human endeavor.

CHAPTER 1

Basics of Biotechnology

Biotechnology

http://dx.doi.org/10.1016/B978-0-12-385015-7.00001-6

ADVENT OF THE BIOTECHNOLOGY REVOLUTION

Biotechnology involves the use of living organisms in industrial processes—particularly in agriculture, food processing, and medicine. Biotechnology has been around ever since humans began manipulating the natural environment to improve their food supply, housing, and health. Biotechnology is not limited to humankind. Beavers cut up trees to build homes. Elephants deliberately drink fermented fruit to get an alcohol buzz. People have been making wine, beer, cheese, and bread for centuries (Fig. 1.1). For wine, the earliest evidence of wine production has been dated to c. 6000 BC. All these processes rely on microorganisms to modify the original ingredients. Ever since the beginning of human civilization, farmers have chosen higher yielding crops by trial and error, so that many modern crop plants have much larger fruit or seeds than their ancestors (Fig. 1.2).

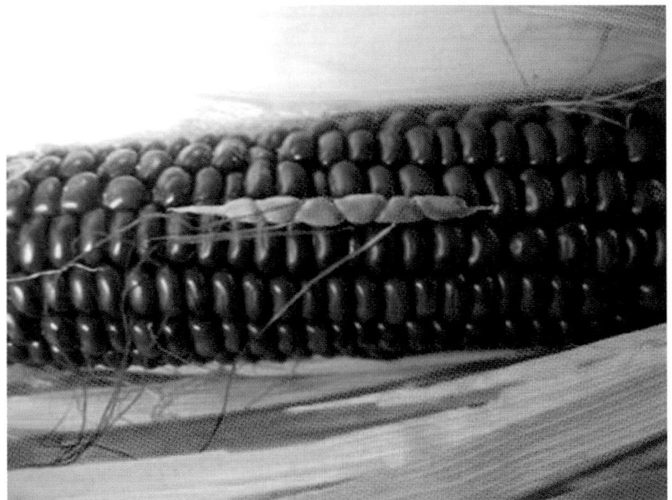

FIGURE 1.1
Traditional Biotechnology Products
Bread, cheese, wine, and beer have been made worldwide using microorganisms such as yeast. Photo taken by Karen Fiorino, Clay Lick Creek Pottery, IL, USA.

FIGURE 1.2
Teosinte versus Modern Corn
Since early civilization, people have improved many plants for higher yields. Teosinte (smaller cob and green seeds) is considered the ancestor of commercial corn (larger cob; a blue-seeded variety is shown). Courtesy of Wayne Campbell, Hila Science Camp.

We think of biotechnology as modern because of recent advances in molecular biology and genetic engineering. Huge strides have been made in our understanding of microorganisms, plants, livestock, as well as the human body and the natural environment. This has caused an explosion in the number and variety of biotechnology products. Face creams contain antioxidants—supposedly to fight the aging process. Genetically modified plants have genes inserted to protect them from insects, thus increasing the crop yield while decreasing the amount of insecticides used. Medicines are becoming more specific and compatible with our physiology. For example, insulin for diabetics is now genuine human insulin, although produced by genetically modified bacteria. Almost everyone has been affected by the recent advances in genetics and biochemistry.

Mendel's early work that described how genetic characteristics are inherited from one generation to the next was the beginning of modern genetics (see Box 1.1). Next came the discovery of the chemical material of which genes are made—**DNA** (**deoxyribonucleic acid**). This in turn led to the central dogma of genetics: the concept that genes made of DNA are expressed as an **RNA** (**ribonucleic acid**) intermediary that is then decoded to make proteins. These three steps are universal, applying to every type of living organism on earth. Yet these three steps are so malleable that life is found in almost every available niche on our planet.

Biotechnology affects all of our lives and has altered everything we encounter in life.

Box 1.1 Gregor Johann Mendel (1822–1884): Founder of Modern Genetics

As a young man, Mendel spent his time doing genetics research and teaching math, physics, and Greek to high school children in Brno (now in the Czech Republic). Mendel studied the inheritance of various traits of the common garden pea, *Pisum sativum*, because he was able to raise two generations a year. He studied many different physical traits of the pea, such as flower color, flower position, seed color and shape, and pod color and shape. Mendel grew different plants next to each other, looking for traits that mixed together. Luckily, the traits he studied were each due to a single gene that was either dominant or recessive, although he did not know this at the time. Consequently, he never saw them "mix." For example, when he grew yellow peas next to green peas, the offspring looked exactly like their parents. This showed that traits do not blend in the offspring, which was a common theory at the time.

Next Mendel moved pollen from one plant to another with different traits. He counted the number of offspring that inherited each trait and found that they were inherited in specific ratios. For example, when he cross-pollinated the yellow and green pea plants, their offspring, the F_1 generation, was all yellow. Thus, the yellow trait must dominate or mask the green trait. He then let the F_1 plants produce offspring, and grew all of the seeds. These, the F_2 generation, segregated into 3/4 yellow and 1/4 green. When green seeds reappeared after skipping a generation, Mendel concluded that a "factor" for the trait—what we call a gene today—must have been present in the parent, even though the trait was not actually displayed.

Mendel demonstrated many principles that form the basis of modern genetics. First, units or factors (now called genes) for each trait are passed on to successive generations. Each parent has two copies of each gene but contributes only one copy of the gene to each offspring. This is called the **principle of segregation**. Second, the **principle of independent assortment** states that different offspring from the same parents can get separate sets of genes. The same phenotype (the observable physical traits) can be represented by different genotypes (combinations of genes). In other words, although a gene is present, the corresponding trait may not be seen in each generation. When Mendel began these experiments, he used pure-bred pea plants; that is, each trait always appeared the same in each generation. So when he first crossed a yellow pea with a green pea, each parent had two identical copies or alleles of each gene. The green pea had two green alleles, and the yellow pea had two yellow alleles. Consequently, each F_1 offspring received one yellow allele and one green allele. Despite this, the F_1 plants all had yellow peas. Thus, yellow is dominant to green. Finally, when the F_1 generation was self-pollinated, the F_2 plants included some that inherited two recessive green alleles and had a green phenotype (Fig. A).

Mendel published these results, but no one recognized the significance of his research until after his death. Later in life he became the abbot of a monastery and did not pursue his genetics research.

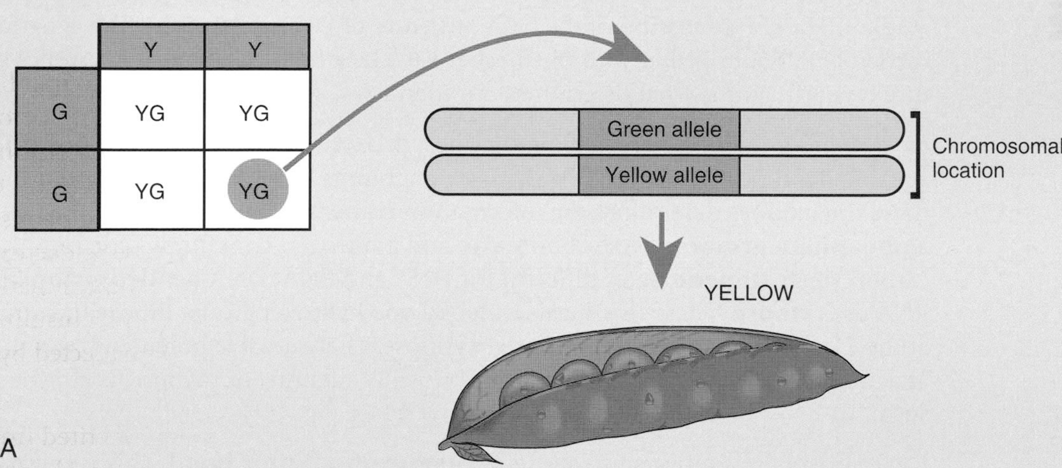

FIGURE A Relationship of Genotype and Phenotype
(A) Each parent has two alleles, either two yellow or two green. Any offspring will be heterozygous, each having a yellow and a green allele. Since the yellow allele is dominant, the peas look yellow. (B) When the heterozygous F_1 offspring self-fertilize, the green phenotype re-emerges in one-fourth of the F_2 generation.

(*Continued*)

Box 1.1 Gregor Johann Mendel (1822–1884): Founder of Modern Genetics—cont'd

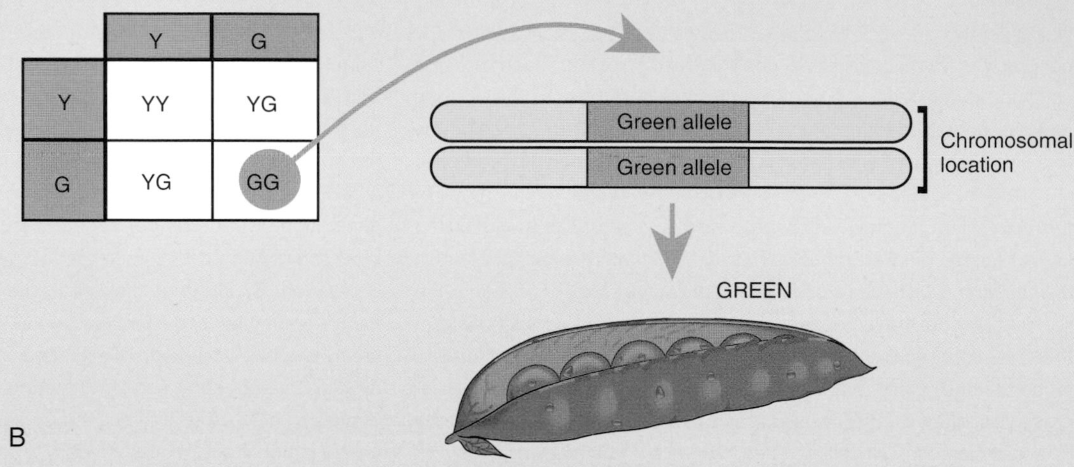

FIGURE A Relationship of Genotype and Phenotype—cont'd
(B) When the heterozygous F_1 offspring self-fertilize, the green phenotype re-emerges in one-fourth of the F_2 generation.

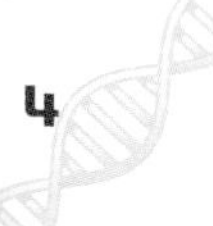

CHEMICAL STRUCTURE OF NUCLEIC ACIDS

The upcoming discussions introduce the organisms used extensively in molecular biology and genetics research. Each of these has genes made of DNA that can be manipulated and studied. Thus, a discussion of the basic structure of DNA is essential. The genetic information carried by DNA, together with the mechanisms by which it is expressed, unifies every creature on earth and is what determines our identity.

Nucleic acids include two related molecules: deoxyribonucleic acid (DNA) and ribonucleic acid (RNA). DNA and RNA are polymers of subunits called **nucleotides**, and the order of these nucleotides determines the information content. Nucleotides have three components: a **phosphate group**, a five-carbon sugar, and a nitrogen-containing base (Fig. 1.3). The five-carbon sugar, or **pentose**, is different for DNA and RNA. DNA has **deoxyribose**, whereas RNA uses **ribose**. These two sugars differ by one hydroxyl group. Ribose has a hydroxyl at the 2′ position that is missing in deoxyribose. There are five potential bases that can be attached to the sugar. In DNA, guanine, cytosine, adenine, or thymine is attached to the sugar. In RNA, thymine is replaced with uracil (see Fig. 1.3).

Each phosphate connects two sugars via a **phosphodiester bond**. This connects the nucleotides into a chain that runs in a 5′ to 3′ direction. The 5′-OH of the sugar of one nucleotide is linked via oxygen to the phosphate group. The 3′-OH of the sugar of the following nucleotide is linked to the other side of the phosphate.

The nucleic acid bases jut out from the sugar phosphate backbone and are free to form connections with other molecules. The most stable structure occurs when another single strand of nucleotides aligns with the first to form a double-stranded molecule, as seen in the DNA **double helix**. Each base forms hydrogen bonds to a base in the other strand. The two strands are **antiparallel**; that is, they run in opposite directions with the 5′ end of the first strand opposite the 3′ end of its partner and vice versa.

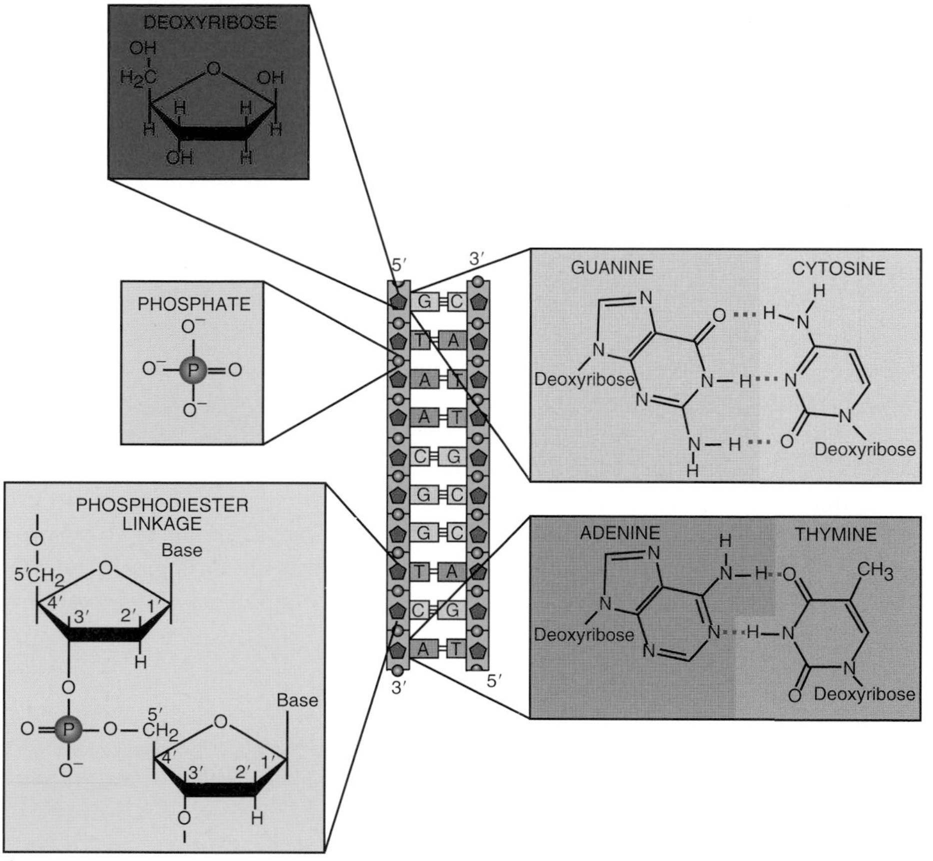

FIGURE 1.3 Nucleic Acid Structure

(A) DNA has two strands antiparallel to each other. The structure of the subcomponents is shown to the sides. (B) RNA is usually single-stranded and has two chemical differences from DNA. First, an extra hydroxyl group (-OH) is found at the 2′ position of ribose, and second, thymine is replaced by uracil.

The bases are of two types: **purines** (guanine and adenine) and **pyrimidines** (cytosine and thymine). Each base pair consists of one purine connected to a pyrimidine via hydrogen bonds. Guanine pairs only with cytosine (G-C) via three hydrogen bonds. Adenine pairs only with thymine (A-T) in DNA or uracil (A-U) in RNA. Because an adenine–thymine (A-T) or adenine–uracil (A-U) base pair is held together with only two hydrogen bonds, it requires less energy to break the connection between the bases than in a G-C pair.

Double-stranded DNA takes the three-dimensional shape that has the lowest energy constraints. The most stable shape is a double-stranded helix. The helix turns around a central axis in a clockwise manner and is considered a **right-handed helix**. One complete turn is 34 Å in length and has about 10 base pairs. DNA is not static but can alter its conformation in response to various environmental changes. The typical conformation just described is the **B-form** of DNA and is most prevalent in aqueous environments with low salt concentrations. When DNA is in a high-salt environment, the helix alters, making an **A-form** that has closer to 11 base pairs per turn. Another conformation of DNA is the **Z-form**, which has a left-handed helix with 12 base pairs per turn. This form occurs when certain proteins bind to the DNA in regions around genes and induce the change in shape. In this form, the phosphate backbone has a zigzag conformation. These forms are biologically relevant under certain conditions, but the exact role the shape of DNA plays in cellular function is still under investigation.

DNA and RNA are both structures with alternating phosphate and sugar residues linked to form a backbone. Base residues attach to the sugar and stick out from the backbone. These bases can base-pair with another strand to form double-stranded helices.

PACKAGING OF NUCLEIC ACIDS

Most bacteria have just a few thousand genes, each approximately 1000 nucleotides long. These are carried on a chromosome that is a single giant circular molecule of DNA, although there are exceptions. A single DNA double helix with this many genes is about 1000 times too long to fit inside a bacterial cell without being condensed somehow in order to take up less space.

In bacteria, the double helix undergoes **supercoiling** to condense it. Supercoiling is induced by the enzyme **DNA gyrase**, which twists the DNA in a left-handed direction so that about 200 nucleotides are found in one supercoil. The twisting causes the DNA to condense. Extra supercoils are removed by **topoisomerase I**. The supercoiled DNA forms loops that connect to a protein scaffold (see Fig. 1.4).

In humans and plants, much more DNA must be packaged, so just adding supercoils is not sufficient. Eukaryotic DNA is wound around proteins called **histones** first. Histones have a positive charge to them, and this neutralizes the negatively charged phosphate backbone. DNA plus histones looks like beads on a string and is called **chromatin**. Each bead, or **nucleosome**, has about 200 base pairs of DNA and nine histones—two H2A, two H2B, two H3, two H4, and one H1. All the histones form the "bead" except for H1, which connects the beads by holding the DNA in the linker region. The histones are highly conserved proteins that are found in all eukaryotes and, in simplified form, in archaebacteria. Histone tails stick out from the nucleosome and are important in regulation. In regions of DNA that are expressed, the histones are loose, allowing regulatory proteins and enzymes access to the DNA. In regions that are not expressed, the histones

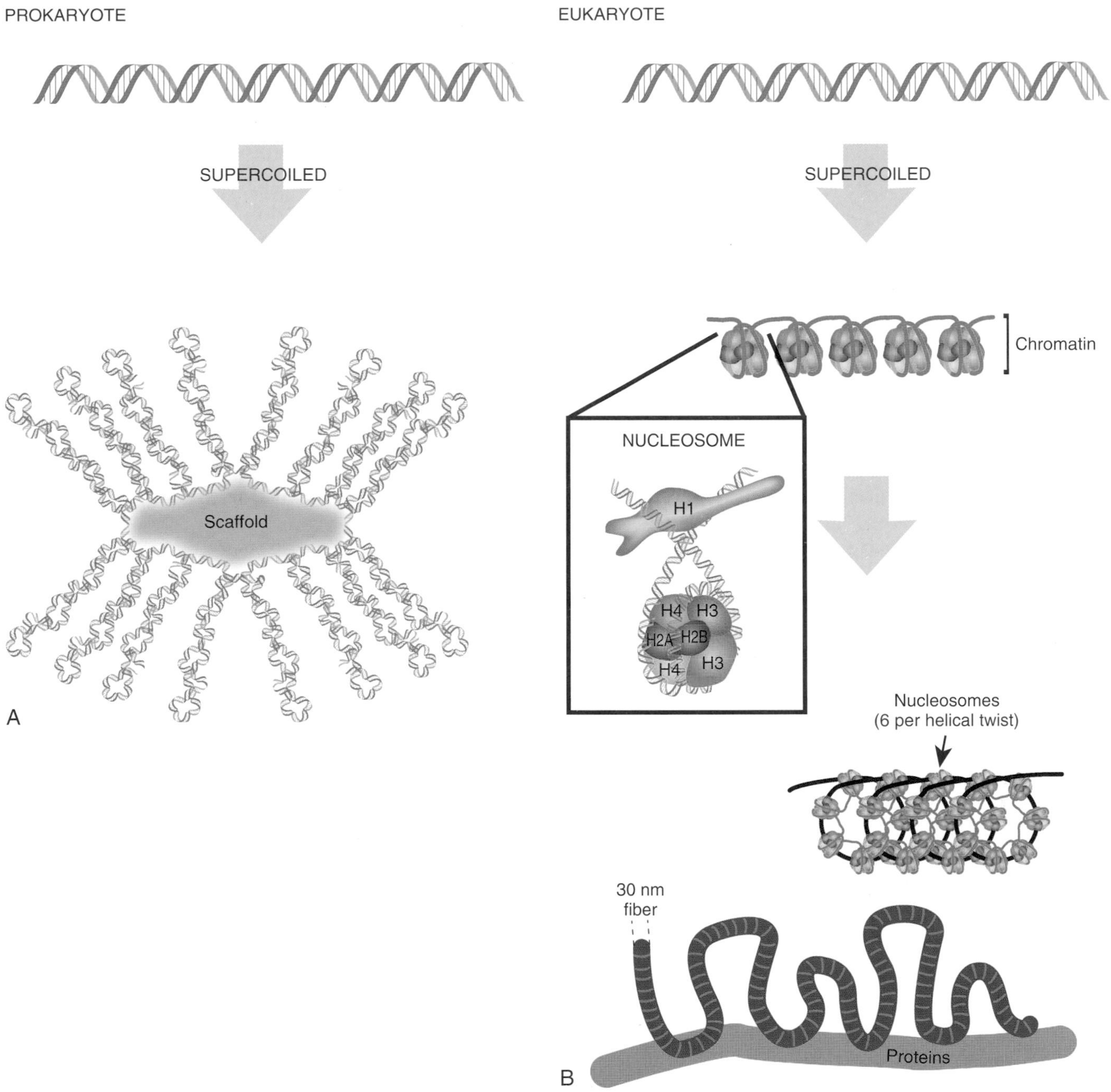

FIGURE 1.4 Packaging of DNA in Bacteria and Eukaryotes
(A) Bacterial DNA is supercoiled and attached to a scaffold to condense its size to fit inside the cell. (B) Eukaryotic DNA is wrapped around histones to form a nucleosome. Nucleosomes are further condensed into a 30-nm fiber attached to proteins at MAR sites.

are condensed, preventing other proteins from accessing the DNA (this structure is called **heterochromatin**).

Chromatin is not condensed enough to fit the entire eukaryotic DNA genome into the nucleus. It is coiled into a helical structure, the **30-nanometer fiber**, which has about six nucleosomes per turn. These fibers loop back and forth, and the ends of the loops are attached to a protein scaffold or chromosome axis. These attachments occur at **matrix attachment regions (MAR)** and are mediated by **MAR proteins**. These sites are 200–1000 base pairs in length and have 70% A/T. The structure of A/T-rich DNA is slightly bent, and these bends promote the connection between proteins in the matrix and the DNA. Often, enhancer and regulatory elements are also found at these regions, suggesting that the structure here may favor the binding of protein activators or repressors. This structure refers to chromosomes during normal cellular growth. When a eukaryotic chromosome readies for mitosis and cell division, it condenses even more. The nature of this condensation is still uncertain.

DNA must be condensed by supercoiling and wrapping around nucleosomes to form chromatin, and finally attached to protein scaffolds in order to fit into the nucleus.

BACTERIA AS THE WORKHORSES OF BIOTECHNOLOGY

DNA is the common thread of life. DNA is found in every living organism on Earth (and even in some entities that are not considered living such as viruses—see later discussion). Only a tiny selection of living organisms has been studied in the molecular biology laboratory. These few chosen species have special traits or features that make them easy to grow, study, and manipulate genetically. Each of the model organisms has had its entire genome sequenced. The model organisms are used both as a guide to understand other related organisms not investigated in detail and for various more practical biotechnological purposes.

Bacteria are the workhorse of model organisms. Bacteria live everywhere on the planet and are an amazing part of the ecosystem. There are an estimated 5×10^{30} bacteria on the Earth, with about 90% of these living in the soil and the ocean subsurface. If this estimate is accurate, then about 50% of all living matter is microbial. Bacteria have been found in every environmental niche. Some bacteria live in icy lakes of Antarctica that only thaw a few months each year. Others live in extremely hot environments such as hot sulfur springs or the thermal vents at the bottom of the ocean (Fig. 1.5).

There has been great interest in these extreme microbes because of their physiological differences. For example, *Thermus aquaticus*, a bacterium from hot springs, can survive at temperatures near boiling point and at a pH near 1. Like others, this bacterium replicates its DNA using the enzyme **DNA polymerase**. The difference is that *T. aquaticus* DNA polymerase has to function at high temperatures and is therefore considered **thermostable**. Molecular biologists have exploited this enzyme for procedures like **polymerase chain reaction** or **PCR** (see Chapter 4), which is carried out at high temperatures. Other bacteria from extreme environments provide interesting proteins and enzymes that may be used for new biotechnological applications. Hydrothermal vents found on the ocean floor have revealed a fascinating array of novel organisms (see Fig. 1.5). Water temperatures in different vents range from 25°C to 450°C.

FIGURE 1.5
Hydrothermal Vent
Mineral-rich fluid is escaping from an opening in the bottom of the ocean along the East Pacific Rise, which has temperatures as high as 403°C. Surprisingly, bacteria are able to survive in this high-heat environment. The vent base is covered with a bed of tube worms, and a probe surrounds the vent. Photo courtesy of NOAA PMEL EOI program and obtained from http://www.pmel.noaa.gov/eoi/gallery/.

Bacteria are highly evolved into every niche of the planet and provide researchers with many unique properties.

ESCHERICHIA COLI IS THE MODEL BACTERIUM

Although extreme bacteria are interesting and useful, more typical bacteria are the routine workhorses for research in molecular biology and biotechnology. The most widely used is *Escherichia coli*, a rod-shaped bacterium about 1 by 2.5 microns in size. *E. coli* normally inhabits the colon of mammals including humans (Fig. 1.6). *E. coli* is a Gram-negative bacterium that has an outer membrane, a thin cell wall, and a cytoplasmic membrane surrounding the cellular components. Like all prokaryotes, *E. coli* does not have a nucleus or nuclear membrane, and its chromosome is free in the cytoplasm. The outer surface of *E. coli* carries about 10 flagella that propel the bacteria to different locations, and thousands of pili that allow the cells to attach to surfaces.

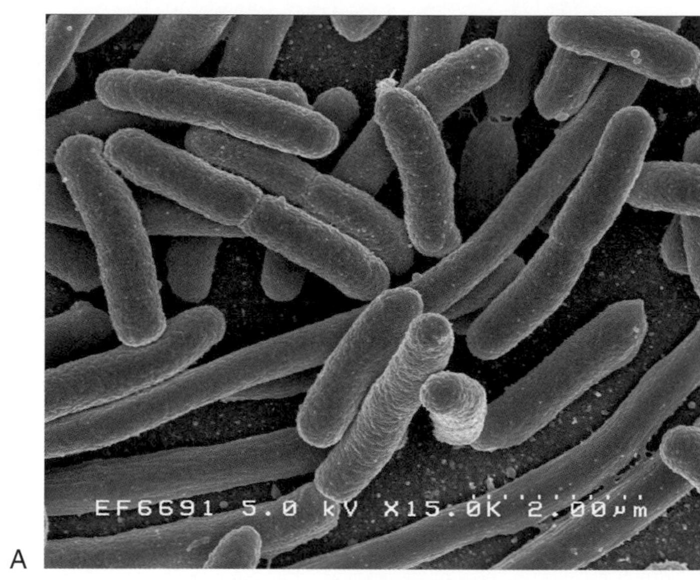

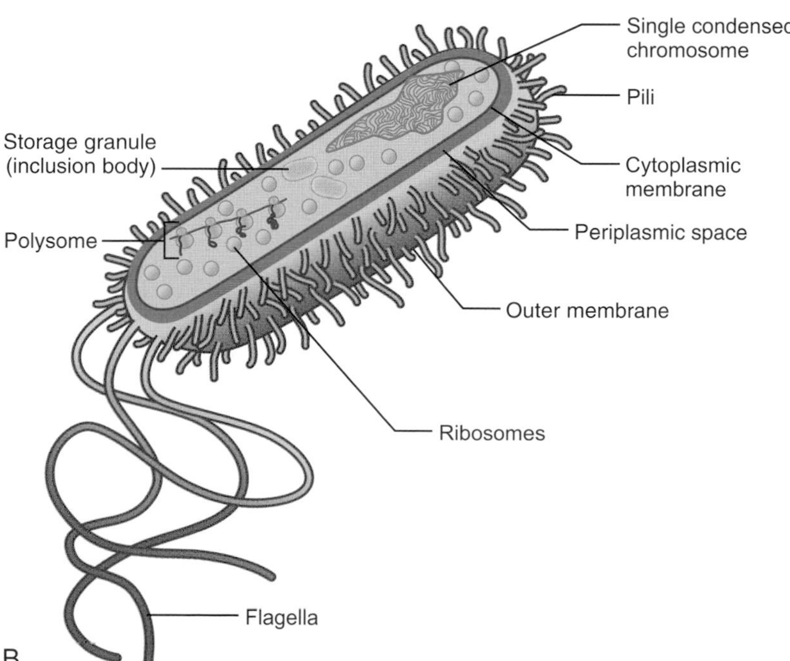

FIGURE 1.6 Subcellular Structure of *Escherichia coli*
(A) Scanning electron micrograph of *E. coli*. The rod-shaped bacteria are approximately 0.6 microns by 1–2 microns. Courtesy of Rocky Mountain Laboratories, NIAID, NIH. (B) Gram-negative bacteria have three structural layers surrounding the cytoplasm. The outer membrane and cytoplasmic membrane are lipid bilayers, and the cell wall is made of peptidoglycan. Unlike eukaryotes, no membrane surrounds the chromosome, leaving the DNA readily accessible to the cytoplasm.

Although the media often report about *E. coli*-contaminated food, *E. coli* is usually harmless. However, occasional strains of *E. coli* are pathogenic and secrete toxins that cause diarrhea by damaging the intestinal wall. This results in fluid being released into the colon rather than being extracted. *E. coli* O157:H7 is a particularly potent pathogenic strain of *E. coli* with two toxin genes that can cause bloody diarrhea. It is especially dangerous to young children, the elderly, and those with compromised immune systems.

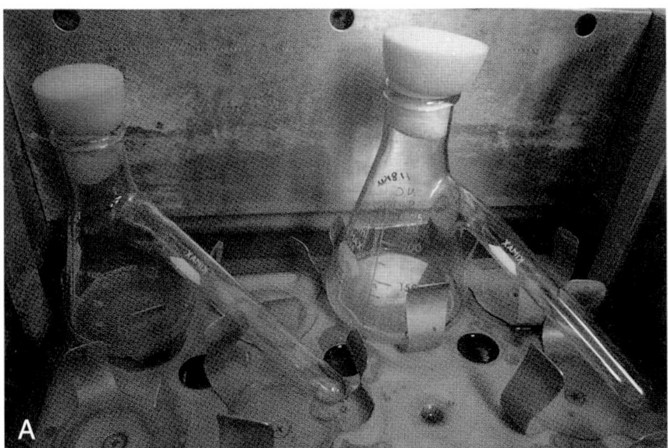

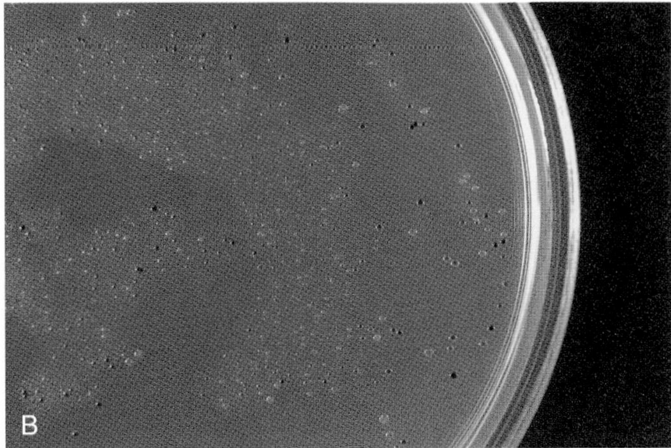

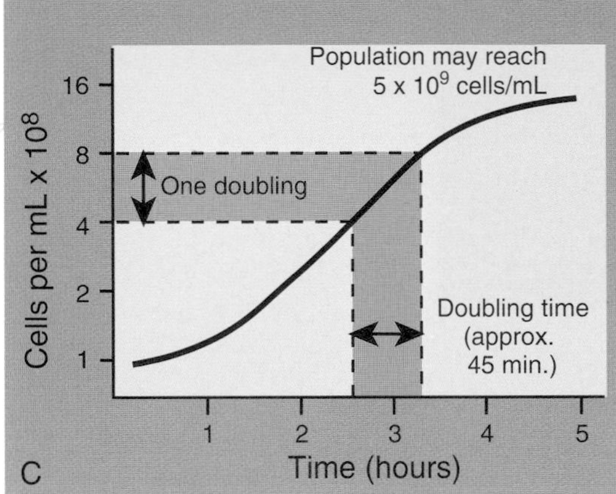

FIGURE 1.7 Bacteria Are Easy to Grow
(A) Bacteria growing in liquid culture. (B) Bacteria growing on agar. This photo shows a mixture of bacterial colonies from the blue/white method for screening plasmid insertions. (C) Fast-growing bacteria can double in numbers in short periods. Here, the number of bacteria double after approximately 45 minutes and reach a density of 5×10^9 cells/mL in about 5 hours.

Bacteria provide many advantages for research. Bacteria have growth characteristics that are very useful when large numbers of identical cells are needed. A culture of bacteria can be grown in a few hours and can contain up to 109 bacterial cells per milliliter. Growth can be strictly controlled; that is, the amount and types of nutrients, temperature, and time may all be adjusted based on the desired result. *E. coli* are so easy to grow that they can grow in mineral salts, water, and a sugar source. The cells can be grown in liquid cultures or as solid cultures on agar plates (Fig. 1.7). Liquid cultures can be stored in a refrigerator for weeks, and the bacteria will not be harmed. Additionally, bacteria can be frozen at −70 °C for 20 years or more, so different strains can be maintained without having to constantly culture them. *E. coli* are normally grown in air but can grow anaerobically if an experiment requires that oxygen be eliminated.

Bacteria are single-celled organisms. The cells in a bacterial culture are identical in contrast to mammalian cells where even a single tissue contains many different types of cells. Each *E. coli* has one circular chromosome with one copy each of about 4000 genes. This is significantly fewer than in humans, who have two copies each of about 25,000 genes on 46 chromosomes. This makes genetic analysis much easier in bacteria (Fig. 1.8).

Escherichia coli is the model bacterial organism used in basic molecular biology and biotechnology research. The organism is simple in structure, grows easily in the laboratory, and contains very few genes.

MANY BACTERIA CONTAIN PLASMIDS

Because many different types of bacteria are found in every environment, competition for nutrients and habitat occurs regularly. Many bacteria compete using a form of biological warfare and secrete toxins, called **bacteriocins**, which kill neighboring bacteria. For example, nisin, a bacteriocin from *Lactococcus lactis*, kills other food-borne pathogens such as *Listeria monocytogenes* and *Staphylococcus aureus*. *E. coli* also produce bacteriocins, called **colicins**. *Bacteriocin* is a general term, whereas *colicin* specifically refers to toxins produced by *E. coli*. (Sometimes *colicin* is used as a general term, but this is not strictly correct.) *E. coli* makes different types of colicins, such as colicin E1 or colicin M, to kill neighboring cells. Colicins act by two main mechanisms. Some puncture the cell membrane, allowing vital cellular ions to leak out, and destroying the proton motive force that drives ATP production. Others encode nucleases that degrade DNA and RNA. These toxins do not affect their producer cells because the cell that makes the toxin also makes an **immunity protein** that recognizes the toxin and neutralizes it.

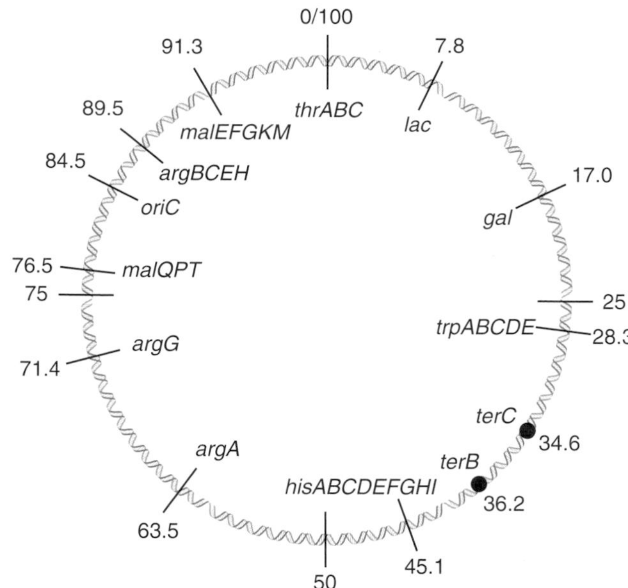

FIGURE 1.8
The *E. coli* Chromosome
The *E. coli* chromosome is divided into 100 map units, arbitrarily starting at the *thrABC* operon. Various genes and their locations are shown. The replication origin (*oriC*) and termination zone (*terB* and *terC*) are indicated.

The ability to make colicin is due to the presence of an extrachromosomal genetic element called a **plasmid**. These are small rings of DNA that exist within the cytoplasm of bacteria and some eukaryotes such as yeast. A colicin-producing plasmid has several genes: the gene for the colicin, the gene for the immunity protein, and genes that control plasmid replication and copy number. In addition, all plasmids contain an **origin** for DNA replication. When the host cell divides, the plasmid divides in step (Fig. 1.9). These colicin plasmids are used extensively for molecular biology. The colicin genes have been removed, and the remaining segments have been greatly modified so that other genes can be expressed efficiently in bacteria. The resulting **recombinant plasmids** are the crux of all molecular biology. All the modern advances in biotechnology started with the ability to express **heterologous** proteins in bacteria (see Chapter 3 for cloning vectors).

> Another useful trait of *E. coli* is the presence of extrachromosomal elements called plasmids. These small rings of DNA are easily removed from the bacteria, modified by adding or modifying genes, and reinserted into a new bacterial cell where new genes can be evaluated.

OTHER BACTERIA IN BIOTECHNOLOGY

Other bacteria besides *E. coli* are used to produce biotechnology products. *Bacillus subtilis* is a Gram-positive bacterium that is used as a research organism to study the biology and genetics of Gram-positive organisms. *Bacillus* can form hard spores that can survive almost indefinitely. It is also used in biotechnology. For industrial production, secreting proteins through the single membrane of Gram-positive bacteria is much easier than secreting them through the double membrane of Gram-negative bacteria; therefore, *Bacillus* strains are used to make extracellular enzymes such as proteases and amylases on a large scale.

Pseudomonas putida is a bacterium that normally lives in water. It is a Gram-negative bacterium like *E. coli* but is commonly used in environmental studies because it is able to degrade many aromatic compounds. *Streptomyces coelicolor* is a soil bacterium that is Gram positive. This organism degrades cellulose and chitin, and also produces a large number of different antibiotics. Another example of a common industrial microorganism is *Corynebacterium glutamicum*, which is used to produce L-glutamic acid and L-lysine for the biotechnology industry.

Many different bacteria are used for biotechnology research because of their unique qualities.

BASIC GENETICS OF EUKARYOTIC CELLS

Most eukaryotes are **diploid**; that is, they have two homologous copies of each chromosome. This is the case for humans, mice, zebrafish, *Drosophila*, *Arabidopsis*, *Caenorhabditis elegans*, and most other eukaryotes. Having more than two copies of the genome is extremely rare in animals, and only one rat from Argentina has been discovered with four copies of its genome. On the other hand, many plants, especially crop plants, are **polyploid** and contain multiple copies of their genomes. For example, ancestral wheat has seven pairs of chromosomes (i.e., its diploid state = $2n$ = 14), whereas the wheat grown for food today has 42 chromosomes. Thus, modern wheat is hexaploid. Domestic oats, peanuts, sugar cane, white potato, tobacco, and cotton also have four to six copies of their genome. This makes genetic analysis very difficult!

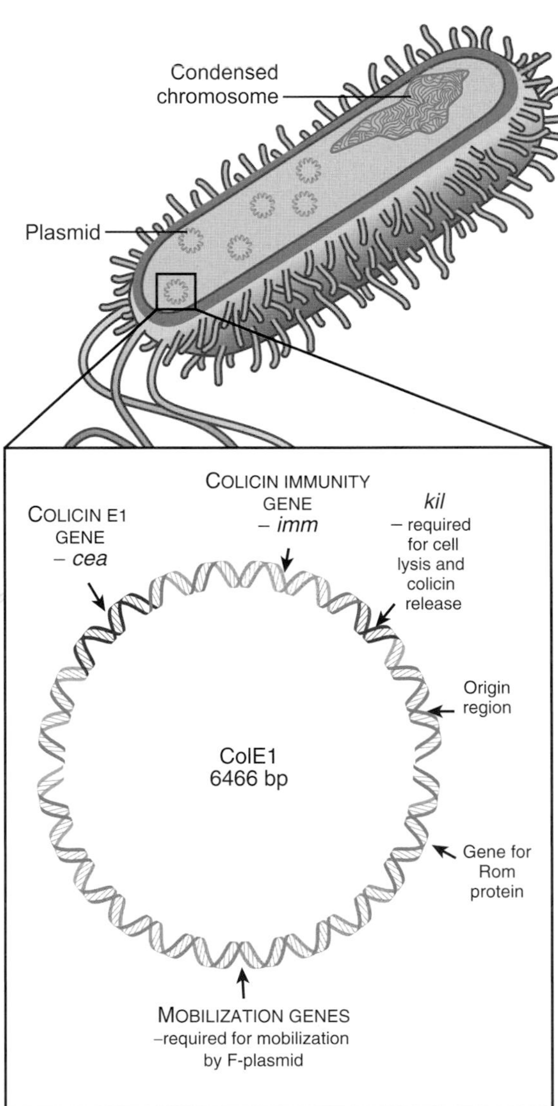

FIGURE 1.9
Plasmids Encode the Genes for Colicin
ColE1 plasmids are extrachromosomal DNA elements that are maintained by bacteria for producing a toxin (*cea* gene). They also carry genes for toxin release and immunity. These plasmids have been modified to carry genes useful in genetic engineering.

In animals, there is a division between **germline** and **somatic** cells. Germline cells are the only ones that divide to give haploid descendents. Diploid germline cells give rise to haploid gametes—the eggs and sperm that propagate the species—by undergoing meiosis. After mating, the two haploid cells fuse to become diploid (forming the zygote). Somatic cells, on the other hand, are normally diploid and make up the individual. Any mutations in a somatic cell disappear when the organism dies, whereas a mutation in a germline cell is passed on to the next generation (Fig. 1.10).

If a somatic cell is mutated early in development, all the somatic cells derived from this ancestral cell will receive the defect. Suppose this ancestral cell is the precursor of the left eye and that this defect prevents the manufacture of the brown pigment responsible for brown eyes. The right eye will be brown, but the mutant left eye will be blue (Fig. 1.11). Blue eyes are not due to blue pigment; they simply lack the brown pigment. People or animals with eyes that don't match are unusual but not incredibly rare. Such events are known as **somatic mutations**. They are not passed on to the offspring. Nonetheless, mutations in somatic cells can cause severe problems, as they are the cause of most cancers (see Chapter 19).

In plants, the division between germline and somatic cells is less distinct because many plant cells are **totipotent**. A single plant cell has the ability to form any part of the plant, reproductive or not. This is not true for the majority of animal cells. Nevertheless, many animal cells do have the potential to form several different types of cells. A cell able to differentiate into multiple cell types is called a **stem cell**. Research on embryonic stem cells has become a hot political topic because of the potential ability to form an embryo. However, researching adult stem cells holds much promise (see Chapter 18). For example, researchers are hoping to identify stem cells that can form new neurons so that patients with spinal cord injuries can be cured.

Eukaryotic cells are more complex than bacteria. Eukaryotic cells are also specialized; that is, some cells are for reproduction, some cells are stem cells that can differentiate into somatic cells, and some cells are specialized in function and shape.

YEAST AND FILAMENTOUS FUNGI IN BIOTECHNOLOGY

Fungi are incredibly useful microorganisms in the world of biotechnology. Anyone who has grown mold on a loaf of bread understands the ease with which these are cultured. Fungi are traditionally used in food applications. Yeasts are used in baking and brewing and other fungi in cheese making, mushroom cultivation, and making foods such as soy sauce. Cheese production uses a variety of fungi. For example, a mold called *Penicillium roqueforti* makes the blue veins in cheeses such as Roquefort, and *Penicillium candidum, Penicillium caseicolum,* and *Penicillium camemberti* make the hard surfaces of Camembert and Brie cheeses. Soy sauce is made from soybeans that are fermented with *Aspergillus oryzae*.

Fungi are responsible for the production of many industrial chemicals and pharmaceuticals. The most famous is penicillin, which is manufactured by *Penicillium notatum*, in large tanks called **bioreactors**. Citric acid is a chemical additive to food that occurs naturally in lemons. It gives the fruit their sour taste. Rather than extracting citric acid from lemons, it has been manufactured since about 1923 by culturing *Aspergillus niger*.

Much like bacteria, yeast has a two-fold purpose in biotechnology. It offers many of the same advantages as bacteria with the additional advantage of being a eukaryote. Yeasts are also important for production of biotechnological products. The most common research strain of yeast is brewer's or baker's yeast, *Saccharomyces cerevisiae*. This is the same little creature that makes the alcohol in beer and makes bread soft and fluffy by releasing carbon dioxide bubbles that get trapped in the dough.

Yeast is a single-celled eukaryote that has its cellular components compartmentalized (Fig. 1.12). Like all eukaryotes, yeasts have their genomes encased in a **nuclear envelope**. The nucleus and cytoplasm are separated, but they communicate with each other through gated channels called **nuclear pores**. *Saccharomyces cerevisiae* has 16 linear chromosomes that have **telomeres** and **centromeres**, two features not found in bacteria. The yeast genome was the first eukaryotic genome sequenced in its entirety. It has 12 Mb of DNA with about 6000 different genes. Unlike higher eukaryotes, yeast genes have very few intervening sequences or **introns** (see Chapter 2). Outside the nucleus, yeast has organelles including the endoplasmic reticulum, Golgi apparatus, and mitochondria to carry out vital cellular functions.

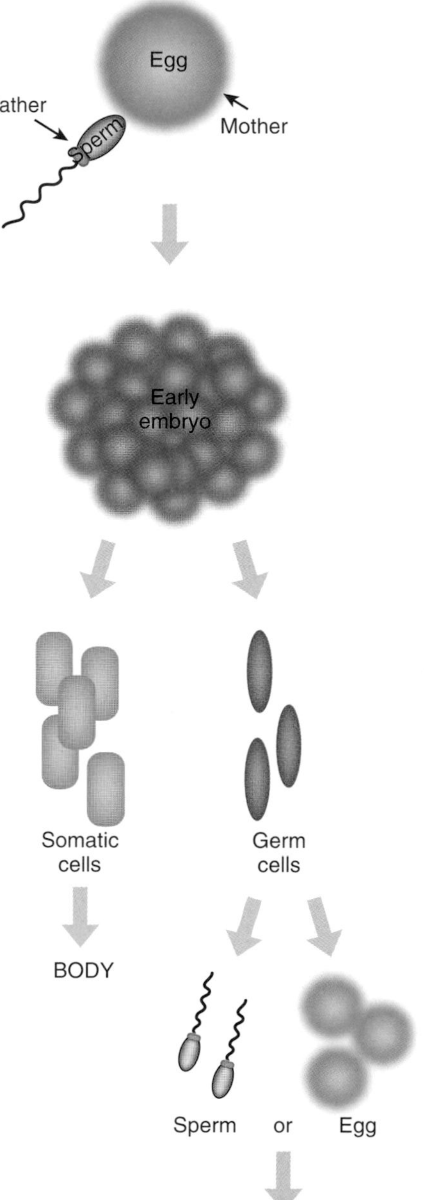

FIGURE 1.10
Somatic versus Germline Cells
During development, cells either become somatic cells, which form the body, or germline cells, which form either eggs or sperm. The germline cells are the only cells whose genes are passed on to future generations.

Like bacteria, yeast grow as single cells. A culture of yeast has identical cells, making genetic and biochemical analysis easier. The culture medium can either be liquid or solid, and the amount and composition of nutrients can be controlled. The temperature and time of growth may also be controlled. Under ideal circumstances, yeast doubles in number in about 90 minutes, as opposed to *E. coli*, which doubles in 20 minutes. Although slower than bacteria, the growth of yeast is fast in comparison to other eukaryotes. Like bacteria, yeast cells can be stored for weeks in the refrigerator and may be frozen for years at −70 °C.

Much like bacteria, some yeast cells also have extrachromosomal elements within their nuclei. The most common element is a plasmid called the **2-micron circle**. Like the

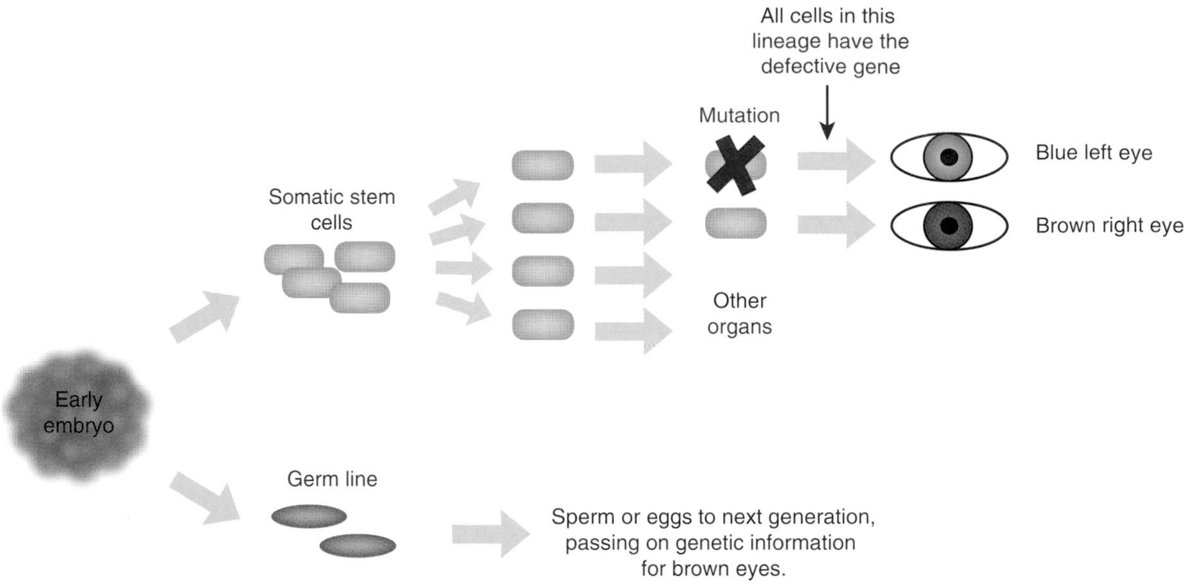

FIGURE 1.11 Somatic Mutations
The early embryo has the same genetic information in every cell. During division of a somatic cell, a mutation may occur that affects the organ or tissue it gives rise to. Because the mutation was isolated in a single precursor cell, other parts of the body and the germline cells will not contain the mutation. Consequently, the mutation will not be passed on to any offspring.

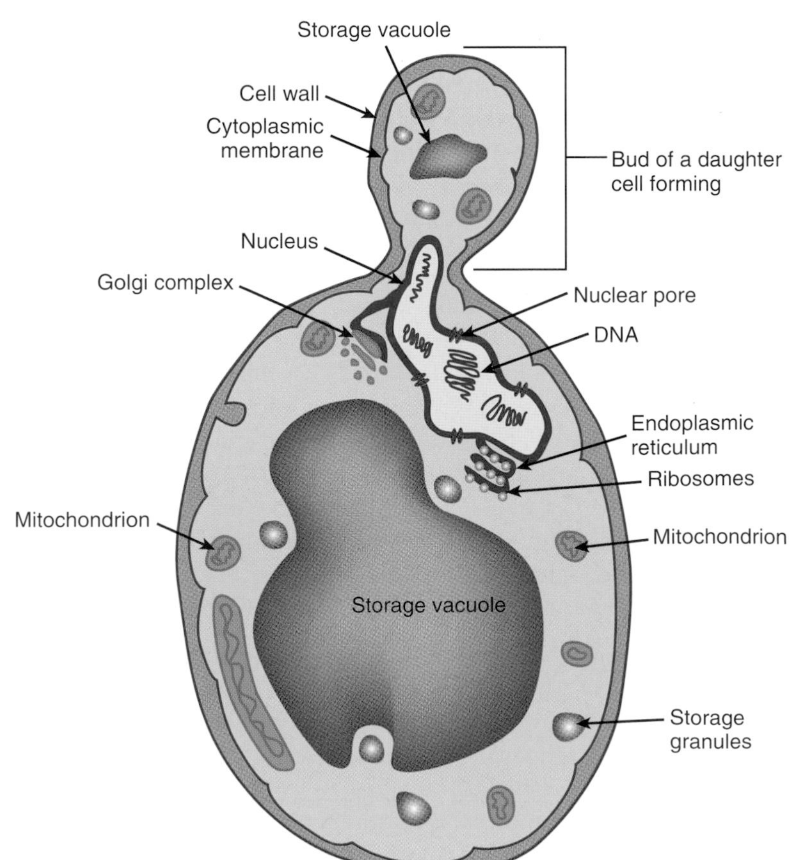

FIGURE 1.12 Structure of Yeast Cell
This yeast cell, undergoing division, is starting to partition components into the bud. Eventually, the bud will grow in size and be released from the mother (lower oval), leaving a scar on the surface of the cell wall.

chromosomes of all eukaryotes, the DNA of this plasmid is also wound around histones. This element has been exploited as a cloning vector (see Chapter 3) to express heterologous genes in yeast. The plasmid has two perfect DNA repeats (*FRT* sites) on opposite sides of the circle. The plasmid also has a gene for **Flp protein**, also called **Flp recombinase** or **flippase**. This enzyme recognizes the *FRT* sites and flips one half of the plasmid relative to the other via DNA recombination (Fig. 1.13). Flippase recombines any DNA segments carrying *FRT* sites, no matter what organism they are in. Consequently, flippase is used in transgenic engineering in higher organisms (see Chapter 16). In plants, a related system, Cre (recombinase) plus *LoxP* sites, is used in a similar way (see Chapter 15).

Yeast offer a variety of advantages to biotechnology. They are single-celled organisms that grow fast. Yeast are eukaryotes with chromosomes that have telomeres and centromeres, like the human genome.

Yeast cells have extrachromosomal elements similar to plasmids that allow researchers to study new genes.

YEAST MATING TYPES AND CELL CYCLE

Yeast cells grow and divide by budding. Cellular organelles such as mitochondria and some cellular proteins are partitioned into the growing **bud**. Finally, mitosis creates another nucleus, and when the bud has reached a sufficient size, the new daughter cell is released, leaving a scar on the surface of the mother cell. Budding creates genetically identical cells because the genome divides by mitosis.

Yeast has diploid and haploid phases in its life cycle, greatly simplifying genetic analysis. Most yeast found in the environment is diploid, having two copies of its genome. Under poor environmental conditions, yeast can undergo meiosis, creating four **haploid spores**, called **ascospores**, contained within an **ascus**. These are released to find a new environment with more nutrients. If the spores find a better environment, they germinate. In the laboratory, the haploid cells can be isolated and grown separately, but in the wild, haploid cells quickly fuse with another, forming diploid cells again (Fig. 1.14). This life cycle allows individual genes to be followed during segregation and inheritance patterns to be analyzed much as with Mendel's peas. However, the shorter life cycle of yeast allows greater numbers to be analyzed.

FIGURE 1.13
The 2-Micron Plasmid of Yeast
Two different forms of the 2-micron plasmid are shown. The enzyme Flp recombinase recognizes the *FRT* sites and recombines them, thus flipping one half of the plasmid relative to the other half.

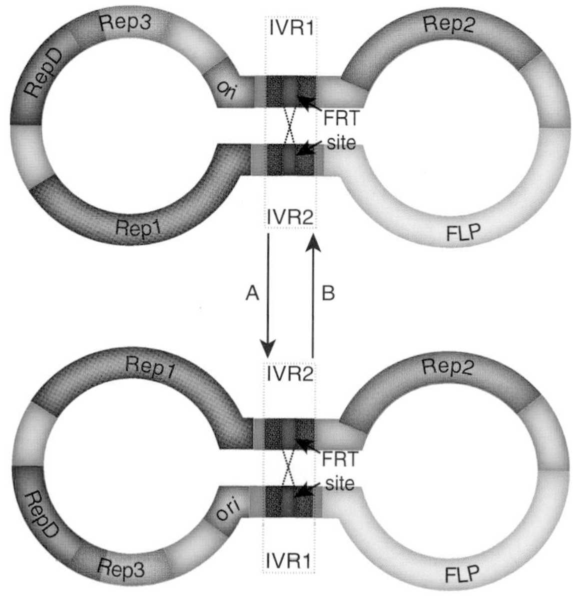

Just as meiosis creates haploid male and female gametes in humans, meiosis in yeast creates haploid cells of two different mating types. Because they are structurally the same, rather than male and female, the yeast mating types are called **a** and α. Fusion may occur only between different mating types; that is, only an **a** plus an α cell can merge forming a diploid. Each mating type expresses a distinct mating pheromone that binds to receptors on the opposite mating type. The pheromones are secreted into the environment. For example, when an **a** cell encounters the α pheromone, a cell surface receptor, the α receptor, binds the α pheromone, readying the yeast for fusion. Conversely, when α cells encounter an **a** pheromone, the cell surface **a** receptor binds the **a** pheromone and readies the cell for mating. The two cells then fuse, combining two different genomes into one. The exchange of genes during sex is important for evolution, as it forms new genetic combinations that may have an advantage in different environments.

Diploid yeast will also form genetic clones by budding when plenty of nutrients are available for growth.

Yeast, like other eukaryotic organisms, can create new genetic combinations with sexual reproduction. The two forms of haploid yeast are **a** and α, which mate to form a new genetically unique diploid cell.

MULTICELLULAR ORGANISMS AS RESEARCH MODELS

Single-celled creatures offer many advantages, but understanding human physiology requires information about cellular interactions. Although single-celled organisms interact with each other, this is not the same as multicellular organisms where one cell is surrounded by other cells on all sides. The location of cells affects both their role and development. The cells in our hair follicles are different from our skin cells. Bone cells differ drastically from the long nerve cells of our spinal cord. Much basic work on cellular interactions, development of multicellular organisms, and understanding cellular physiology in different tissues has been done on the roundworm *Caenorhabditis elegans*. Although this is a multicellular organism, it is still relatively simple compared to mammals or other vertebrates.

Caenorhabditis elegans, a Small Roundworm

C. elegans is a small roundworm that is found in soil, particularly rotting vegetation, where it feeds on bacteria (Fig. 1.15). There are two sexes, a self-fertilizing hermaphrodite and a male, allowing genetic studies on both self- and cross-fertilization. The body is shaped as a simple nonsegmented tube that is encased in a cuticle layer to prevent dehydration. Inside *C. elegans*, there are 959 somatic cells, which include more than 300 neurons. The head has many sense organs that respond to taste, smell, temperature, and touch, but no eyes. There is a nerve ring that serves as the brain and a nerve cord that runs down the back of the body. The digestive system consists of a pharynx followed by intestine and anus. There are 81 muscle cells that control the sinusoidal movement of the worm around its environment. The reproductive system occupies the largest volume within the worm. In the hermaphrodite, the tail is long and tapered, whereas the male has a blunt end. The hermaphrodite has a vulval opening where it lays eggs. The sperm cells come either from itself or from a male *C. elegans* in a sexual encounter.

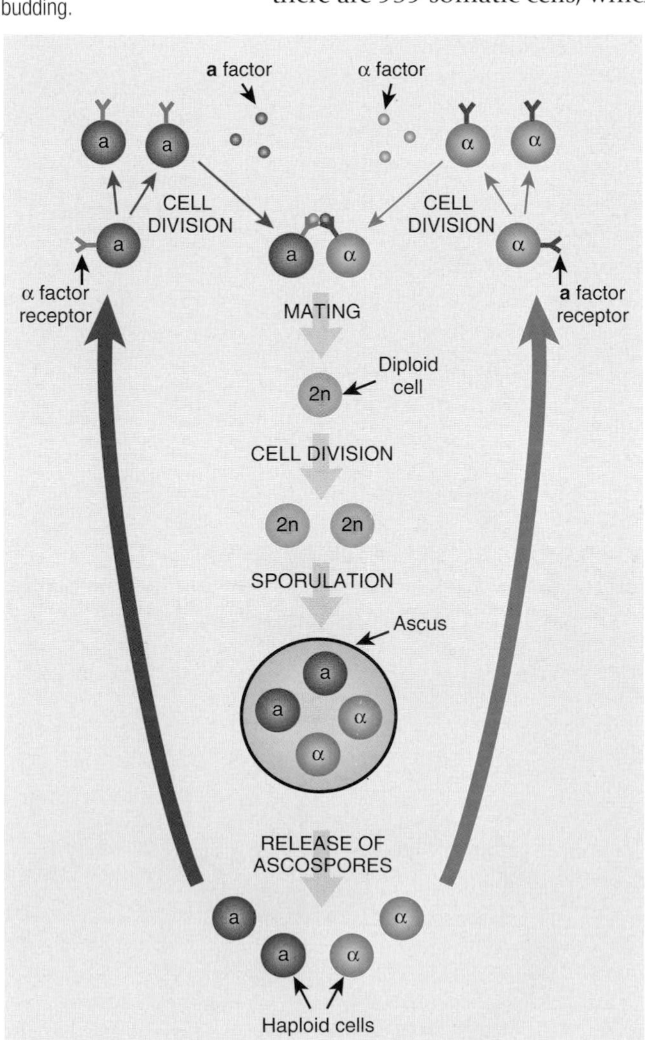

FIGURE 1.14
Alternating Haploid and Diploid Phases of Yeast
Haploid cells come in two different forms: **a** and α. These express mating pheromones, **a factor** and **alpha (α) factor**, which attract the two forms to each other. When the pheromones bind to receptors on the opposite cell type, the two haploid cells become competent to fuse into a diploid cell. Diploid cells sporulate under growth-limiting conditions. Otherwise, the diploid cells form genetic clones by budding.

C. elegans has many advantages for molecular biology and genetics. These creatures are transparent and can be studied in real time using various fluorescent techniques. They have many physiological characteristics similar to higher animals. For example, they undergo programmed cell death, and the genes involved are similar to genes found in humans (see Chapter 20). *C. elegans* is used to study development, aging, sexual dimorphism, alcohol metabolism, cellular differentiation, and many other phenomena that apply to humans.

The life cycle of *C. elegans* is conducive to research. One generation lasts about 3 days. First, a sperm and egg cell fuse, and a single-celled embryo partially develops within the hermaphrodite's body. After the embryo hatches from the chitin shell, the larval stages begin. There are four larval stages that culminate with the adult worm, with the sexual development occurring last (Fig. 1.16). Spermatogenesis is the limiting factor in the number of offspring that a hermaphrodite produces, which is about 300 progeny.

C. elegans is a model multicellular eukaryotic organism. Biotechnology research uses this organism because it is easy to grow, it is transparent, and it is a hermaphrodite, so it can create either genetic clones or novel genetic organisms.

Drosophila melanogaster, the Common Fruit Fly

Another multicellular model organism widely used because of its genetics is *Drosophila melanogaster*, usually referred to simply as *Drosophila*, the common fruit fly. This insect is about 3 mm in length and can often be found around rotting fruit. These flies are easy to grow and maintain in a lab. They need a food source and are kept in a bottle capped with cotton so they cannot escape. Their entire life span is 2 weeks and starts with an egg about 0.5 mm in length (Fig. 1.17). The embryo hatches into a worm-like larva after about 24 hours. There are three larval instars that develop 1 day, 2 days, and 4 days after the first instar larva. Each instar grows and eats continuously and molts to form the next instar. The third larval instar forms a pupa that is immobile. The pupa usually clings to the side of the flask, where it stays for 4–6 days. During this time the larva transforms into the winged adult fly. Wings, legs, antenna, segmented bodies, eyes, and hair are formed.

The main focus of *Drosophila* research is genetics. Many different mutations are available, from simple changes such as longer or shorter body hairs, to dramatic mutations where body segments are duplicated. That is, some

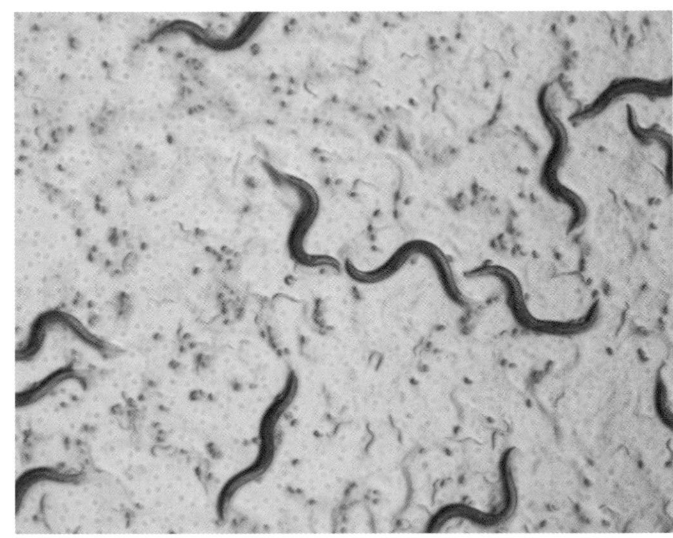

FIGURE 1.15 Caenorhabditis elegans
Plate of *C. elegans*. Small dark spots are embryos that are going to hatch, and the long adult worms are moving in a sinusoidal pattern across the surface. Courtesy of Jill Bettinger, Virginia Commonwealth University, Richmond, VA.

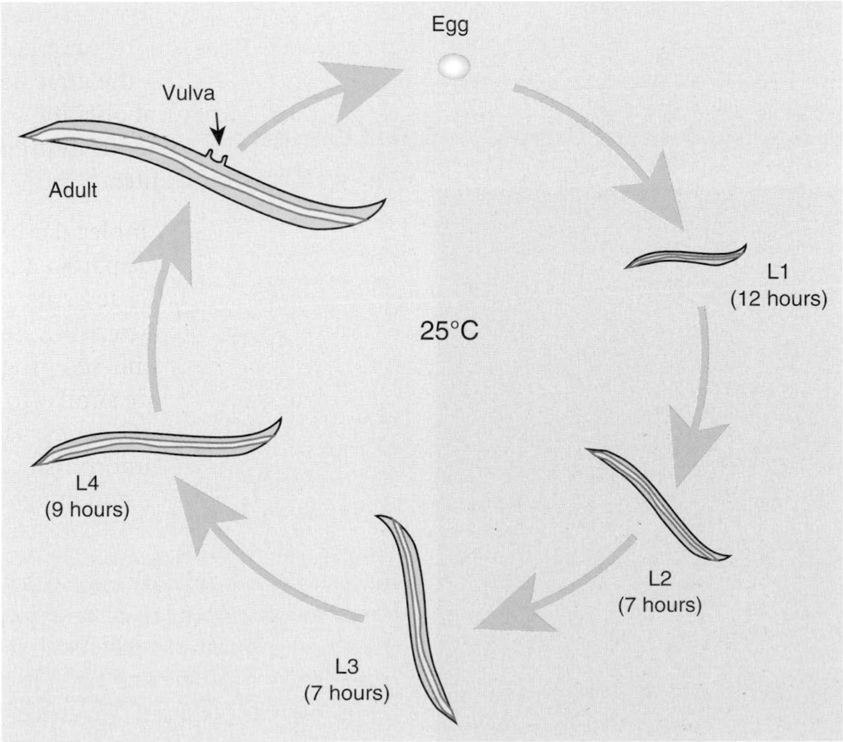

Egg
Vulva
Adult
L1
(12 hours)
25°C
L2
(7 hours)
L3
(7 hours)
L4
(9 hours)

FIGURE 1.16 Life Cycle of *Caenorhabditis elegans*
When the *C. elegans* sperm fuses with an egg, a small worm develops (L1). The larva goes through multiple stages until it reaches the sexually mature adult phase.

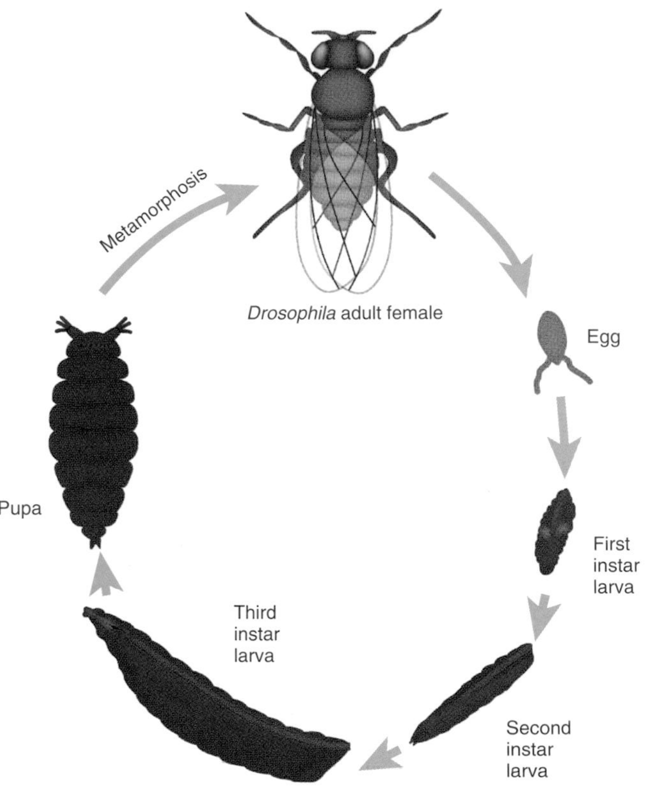

FIGURE 1.17 Life Cycle of *Drosophila melanogaster*
Drosophila fruit flies start as tiny eggs that develop into worms. After a series of larval stages, the worm forms a pupa where the adult form develops.

mutants of *Drosophila* have four wings or extra legs. Studying these mutants has identified many genes that determine basic body patterns in *Drosophila*, and based on homology, humans too. The genome of *Drosophila* has been sequenced and has 165 Mb of DNA, divided among three autosomes and the X/Y sex chromosomes. There are a predicted 12,000 genes in the genome.

During the rapid growth of the larval stages, the number of cells actually remains fairly constant. The size of the cells does increase dramatically, though. In order for these large cells to work, a large amount of extra protein and mRNA needs to be made, and the chromosomes duplicate hundreds of times to provide multiple gene copies. Although they duplicate, they do not divide but stay attached to each other, creating thick **polytene chromosomes** (Fig. 1.18). Because they are so large, they can be visualized under a light microscope. The polytene chromosomes have characteristic banding patterns, with each section of each chromosome being unique. The banding pattern allows some mutations to be localized. For example, a deletion that causes white eyes (in the adult) would alter the banding pattern on the corresponding polytene chromosome. Thus, the mutation can easily be mapped to its chromosome location.

The true sexual reproduction of *Drosophila* allows genetic manipulations, and the complex alterations that occur in the pupal to adult fly metamorphosis are two key characteristics that are studied by researchers.

Zebrafish Are Models for Developmental Genetics

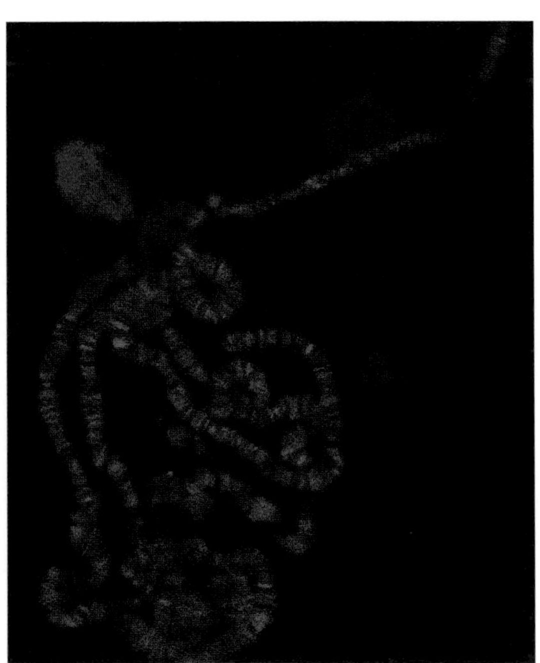

FIGURE 1.18 Polytene Chromosome
Fluorescent staining of polytene chromosome from *Drosophila*. Photo courtesy of LPLT/Wikimedia commons.

The small zebrafish (*Danio rerio*) is a simple vertebrate used in molecular biology research. It is a common fish found in pet stores for keeping in freshwater aquariums. The qualities that have made it so prevalent as a pet also make it attractive for research. It is easy to maintain and breed in an aquarium. A wide variety of mutations exist, which makes the fish handy for genetics research. The adult is about an inch long with black horizontal stripes down its body (Fig. 1.19). The mother lays about 200 eggs at one time, so many offspring can be studied after one mating. Embryonic development occurs outside the mother. The embryos are completely transparent, so the effects of mutations that affect embryo development can be seen with ease. Moreover, different cells can either be destroyed or moved to new locations, and the effect on development can be traced. Such experiments are insightful for deciphering the effect of position on cellular development. The embryos develop from one single cell to a tiny fish in about 24 hours, so studies of development can be done relatively quickly.

FIGURE 1.19
The Zebrafish, *Danio rerio*
This fish is used as a model vertebrate to study genetics, cell biology, and developmental biology. Photo courtesy of Wikipedia commons.

The zebrafish genome has been sequenced. There are 25 pairs of chromosomes with a haploid genome size of 1700 Mb of DNA. About 70% of human genes that code for proteins have orthologs in zebrafish. Thus, when a new gene function is identified in the fish, it suggests possible roles for corresponding human genes, and researchers are turning to mutations in zebrafish to create a disease model organism. For example, in humans, porphyria causes skin sensitivity to light, and porphyrin metabolic precursors to be secreted. Zebrafish with a mutation in uroporphyrinogen decarboxylase (UROD) have the same phenotype, suggesting that mutations in this gene are responsible for this disease. In addition, human studies have identified a potential mutation in a ribosomal protein, RPS19, as the causative agent for Diamond–Blackfan anemia. To confirm that this gene was responsible for the disease, a zebrafish was developed that did not express the RPS19 ortholog. This mutant fish had anemic symptoms, much like the human disease, and recapitulated the disease.

Another key advantage for zebrafish is the large number of offspring and their ability to grow outside their mother. The embryos are easily used in a drug screens to find compounds that treat these diseases. For example, melanocytes in zebrafish arise from the neural crest cells. To treat melanoma cancer in humans, drugs are needed to stop their proliferation. When various zebrafish were grown in different chemical compounds, an inhibitor (leflunomide) was identified to inhibit the developmental migration of neural crest cells (melanocyte precursors). Further studies found the same compound was effective in inhibiting melanoma metastasis, and further studies are underway to study whether this compound will work in humans.

Zebrafish are key organisms to study development of embryos because they have live babies that develop outside the mother. In addition, as many as 70% of our genes have zebrafish orthologs. The combination of their life cycle and the genetic relatedness makes zebrafish a good model organism for drug screens.

Mus musculus, the Mouse, Is Genetically Similar to Humans

The model organism most closely related to humans is the mouse. The mouse genome has about 2500 Mb of DNA on 20 different chromosomes. Less than 1% of the genes have no human gene counterpart, so mouse genetics relates to humans very readily. Mice are easy to manipulate genetically, and animals with one or more genes inactivated (knockout mice) are fairly easy to generate. In addition to genetic deletions, extra genes can be inserted and expressed in the mouse, giving **transgenic** animals (see Chapter 16). The effect of such genetic manipulations on growth, development, or physiology can be determined.

Researchers consider mice very similar to humans because they have so many genes in common.

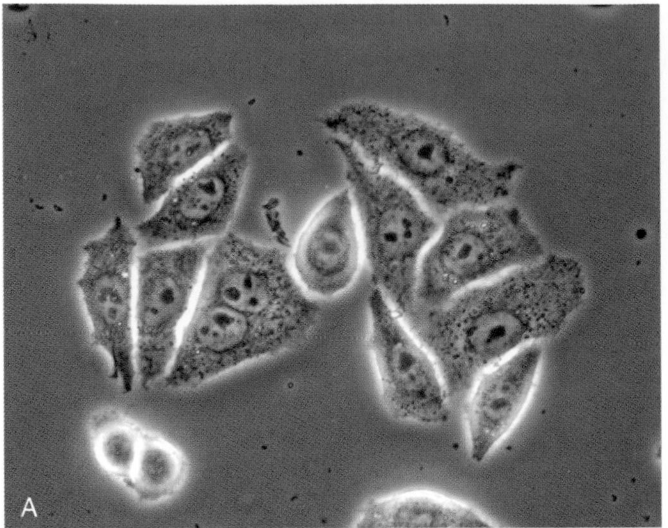

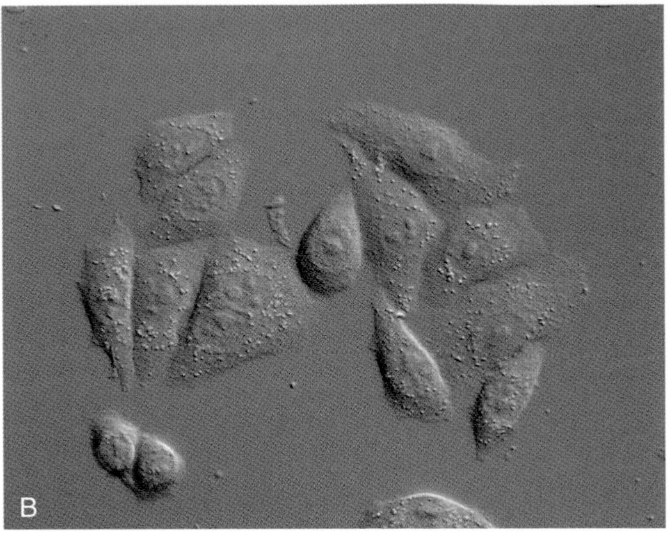

FIGURE 1.20
Human HeLa Cells Grown *In Vitro*
HeLa cells were taken from the tumor of Henrietta Lacks, a woman suffering from cervical cancer in the 1950s and have been cultured continuously ever since. (A) Viewed under phase contrast. (B) Viewed under differential interference contrast. Courtesy of Michael W. Davidson, Optical Microscopy Group, National High Magnetic Field Laboratory, Florida State University, Tallahassee, Florida.

ANIMAL CELL CULTURE *IN VITRO*

Another way to approximate human physiology is by studying mammalian cells cultured *in vitro* (Fig. 1.20). Many different cell lines have been generated from humans and monkeys, and they can be grown in plastic dishes or flasks using culture media containing growth factors and nutrients. Cell lines must be maintained at 37 °C and require an atmosphere rich in carbon dioxide. **Adherent cell lines** stick to and divide on the plastic dishes, whereas **suspension cells** grow and divide in liquid culture. Most cell lines are one particular type of cell from a particular tissue, and many different cell lines have been grown from kidney, liver, heart, and so forth. The original cell lines cannot divide in culture forever (see Chapter 20). Primary cells, as they are called, can be maintained for only a short time. Using cancer cells overcomes this limitation since cancer cells do not stop dividing (see Chapter 19 for discussion). These cell lines are immortal and can, in principle, be grown under the correct circumstances forever.

The best aspect of using cultured human cells is the ability to do genetic studies. Different genes can be expressed in cultured cells, and their effect on cellular physiology can be determined. In addition, gene deletions or mutations can be examined. Cultured mammalian cells are also important for production of recombinant proteins, which are then isolated and purified for medicine, research, and other biotechnology applications.

Cell lines have also been developed from insects (Fig. 1.21). They are primarily used to express heterologous proteins for the biotechnology industry. Insect cells are preferred to mammalian cells because they require fewer nutrients for growth and survive in media free of serum. Mammalian cells require serum from fetal cows, which is very expensive and in limited supply. Insect cells also grow at lower temperatures without carbon dioxide and therefore do not require special incubation chambers.

Insect cells are used in research to study viruses that are transmitted between insects and plants, as well as cell signaling pathways. Insect cell lines are primarily derived from *Spodoptera frugiperda* (fall armyworm), *Trichoplusia ni* (cabbage looper), *Drosophila melanogaster* (fruit fly), *Heliothis virescens* (tobacco budworm), the mosquito, and others. The most common cell lines are those from ovarian tissue of *S. frugiperda*, which include Sf9 and Sf21 cells; those from embryonic cells from *T. ni*, which include the "High Five" cell lines; and those from late-stage *Drosophila* embryos, which include Schneider S2 cells.

Studying cells in a dish rather than an organism provides the researcher with another way to study genes. The cell lines are useful for genetic manipulations such as expressing new genes or deleting existing genes.

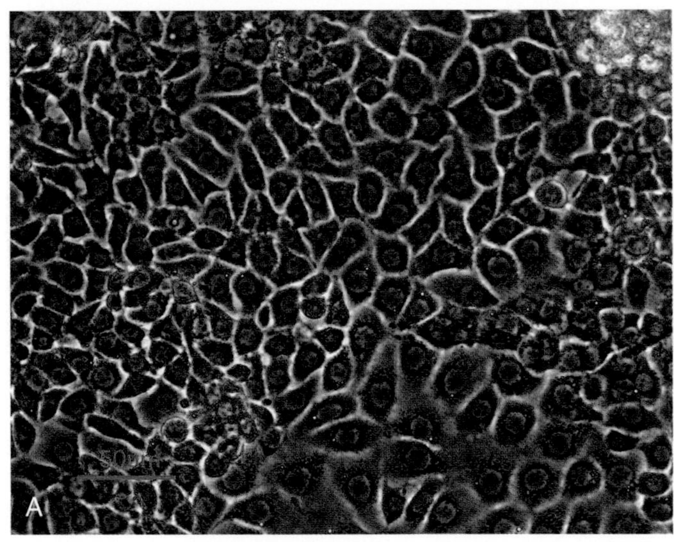

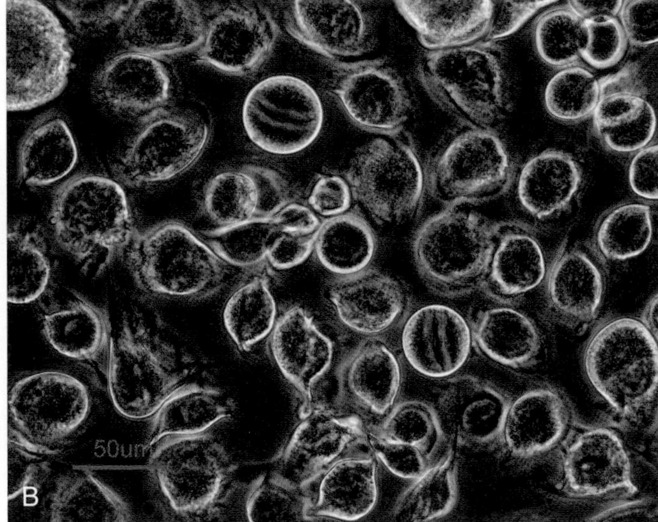

FIGURE 1.21 Insect Cells in Culture
(A) HvT1 cells from tobacco budworm testes are strongly attached to the surface of the dish. (B) TN368 cells from cabbage looper ovary are only loosely attached. Courtesy of Dwight E. Lynn, Insect Biocontrol Lab, USDA, Beltsville, MD.

ARABIDOPSIS THALIANA, A MODEL FLOWERING PLANT

The model organism most widely used in plant genetics and molecular biology is the weed *Arabidopsis thaliana*, wild mustard weed or mouse ear cress (Fig. 1.22). Growing different crops to feed the world population is incredibly important, and much money is invested in research on the crops most used for food, such as rice, soybean, wheat, and corn. These plants have huge genomes, and most are polyploid—even hexaploid (such as wheat). Therefore, a model organism is essential to learn the basic biology of plants. *Arabidopsis* has much the same responses to stress and disease as crop plants. Moreover, many genes involved in reproduction and development are homologous to those in plants with more complex genomes.

Arabidopsis has many convenient features. First, it is easily grown and maintained in a laboratory setting. The plant is small and grows to match its environment. If there is plenty of space and nutrients, the plant can grow to over a foot in height and width. If the environment is a small culture dish in a lab, the plant will grow about 1 cm in height and width. At either size, the plant forms flowers and seeds. An entire generation from seed to adult to seeds is finished in 6–10 weeks, which is relatively quick for a plant. (Note that for corn or soybeans, only one generation can occur in the span of a summer.) In *Arabidopsis*, many seeds are produced on each plant, so aiding genetic analysis. Much like yeast, *Arabidopsis* can be maintained in a haploid state.

Arabidopsis has a small genome for a plant, containing only five chromosomes with a total of 125 Mb of sequence. The genome was completely sequenced in 2000, allowing researchers to identify about 25,000 genes and important sequence features. Rice has also been sequenced and has an estimated 40,000 to 50,000 genes. This tops the number of predicted human genes, and so rice (and doubtless many other plants) may be more "advanced" than us lowly humans.

Plant research also relies on a model organism to study. *Arabidopsis thaliana* is used because of its size, ease of growth, and small genome.

FIGURE 1.22
Arabidopsis thaliana
The plant most used as a model for plant biology research is *A. thaliana*, a member of the mustard family (Brassicaceae). Courtesy of Dr. Jeremy Burgess, Science Photo Library.

VIRUSES USED IN GENETICS RESEARCH

Viruses are entities that border on living. But unlike genuine living organisms, viruses cannot survive outside a host organism. Viruses are pathogens that invade host cells and subvert them to manufacture more viruses. Viruses are simple in principle and yet very powerful. They consist of a protein shell called a **capsid** surrounding a genome made of RNA or DNA. The particle is called a **virion** and, unlike a living cell, has no way to make its own energy or duplicate its own genome. The virus relies on the host to do this work.

Viruses come in many different types and can inhabit every living thing from bacteria to humans to plants. Viral diseases in humans are extremely common, and most cause only minor symptoms. For example, when rhinovirus invades, the victim ends up with a runny nose and other cold symptoms and usually feels miserable for a few days. However, viruses do cause a significant number of serious diseases, such as AIDS, smallpox, hepatitis, and Ebola.

When viruses invade bacteria, the infected bacteria usually die. Bacterial viruses are called **bacteriophage** or **phage**, and they normally destroy the bacterial cell in the process of making new viral particles. When bacteria grow on an agar plate, they form a hazy or cloudy layer (a bacterial lawn) over the top of the agar. If the culture of bacteria is infected with bacteriophage, the viruses eat holes or **plaques** into the bacterial lawn, leaving clear zones where all the bacteria were killed.

Bacteriophages, like other types of viruses, have the following stages of their life cycle (Fig. 1.23):

(a) Attachment of the virion to the correct host cell
(b) Entry of the virus genome
(c) Replication of the virus genome
(d) Manufacture of new virus proteins
(e) Assembly of new virus particles
(f) Release of new virions from the host

Not every virus kills the host cell, and in fact, many viruses have a latent phase in which they lie dormant within the cell, not producing any proteins or new viruses. **Latency**, as it is called in animal cells, is also called **lysogeny** when referring to bacteria. (In contrast, the phase of viral growth in which the host cell is destroyed is called the **lytic phase**.) Sometimes a virus becomes latent by inserting its genome into the genome of the cell. The viral genome integrates into a host chromosome and remains inactive until some stimulus triggers it to reactivate. The integrated virus is called a **provirus** (or a **prophage** if the virus invades bacteria).

The great variety of viruses can be divided into groups based on capsid shape or the type of genome. The three major shapes are spherical, filamentous, and complex. Spherical viruses actually have 20 flat triangular sides and are thus icosahedrons. Complex viruses come in various shapes, but some have legs that attach to the host cell, a linear segment that injects the DNA or RNA genome into the host, and a structure that stores the viral genome. This type of complex virus is common among bacteriophages, several of which are widely used in molecular biology research. Bacteriophage T4, lambda, P1, and Mu all look like the Apollo lunar landers (Fig. 1.24).

Viral genomes are varied in size, but all contain sufficient genetic information to get the host cell to make more copies of the virus genome and make more capsid proteins to package it. At the very least, a virus needs a gene to replicate its genome, a gene for capsid protein, and a gene to release new viruses from the host cell. Bacteriophage Qβ infects bacteria; its entire genome is only 3500 base pairs, and the entire genome consists of only four genes. On the other hand, large complex viruses may have more than 200 genes that are used at different times after infecting the host cell. The genes are then divided into categories based on when they are active. Some genes are considered **early genes** and are active immediately after infecting the host, whereas **late genes** are active only after the virus has been inside the host cell for some time.

Viral genomes are either made from DNA or RNA, can be double-stranded or single-stranded, and can be circular or linear. When viruses use a single strand of RNA as genome, this can either be the **positive** or **plus (+) strand** or the **negative** or **minus (–) strand**. The positive strand corresponds to the coding strand and the negative strand to the template strand (see Chapter 2). When a positive-strand RNA virus injects its genome into the host, the RNA can be used directly as a messenger RNA to make protein. If the RNA virus has a negative-strand genome, the RNA must first be converted into double-stranded form (the **replicative form** or **RF**). Then each strand is used: the negative strand is used as a template to make more positive-stranded genomes, and the positive strand is used to make proteins.

Some viruses actually use both RNA and DNA versions of their genome (Fig. 1.25). **Retroviruses** infect animals and include such members as **human immunodeficiency virus** (HIV). The genome inside a retrovirus particle is a single-stranded RNA that is converted to DNA once it enters the host. **Reverse transcriptase** is the enzyme that manufactures the DNA copy of the RNA genome and is used extensively in molecular biology and genetic engineering (see Chapter 3). Once the DNA copy is made, it is inserted into the host DNA using two repeated DNA sequences at the ends called **long terminal repeats (LTRs)**. Once integrated, the retrovirus becomes part of the host's genome. This is why there is no complete cure for acquired immunodeficiency syndrome (AIDS). The host can never rid itself of the retroviral DNA once it becomes integrated. The viral genes then direct the synthesis of new viral particles that infect neighboring cells. Reverse transcriptase is an example of a viral gene product that is synthesized by the host and packaged inside the virions for use in the next infection cycle.

Retroviral genomes have three major genes, *gag*, *pol*, and *env*, as well as several minor genes. The *tat* and *rev* genes regulate the expression of the other retroviral genes. *Nef*, *vif*, *vpr*, and *vpu* encode four accessory proteins that block the host cell's immune defense and increase the efficiency of virus production. *Gag*, *pol*, and *env* each give single mRNA transcripts that encode multiple proteins. *Gag* encodes three proteins involved with making the capsid. *Pol* gives three proteins: a **protease** that digests other proteins during particle assembly, reverse transcriptase that makes the DNA copy of the genome, and an **integrase** that integrates the viral DNA into the host chromosome. *Env* codes for two structural proteins; one forms the outer spikes, and the other helps the virus enter the host cell.

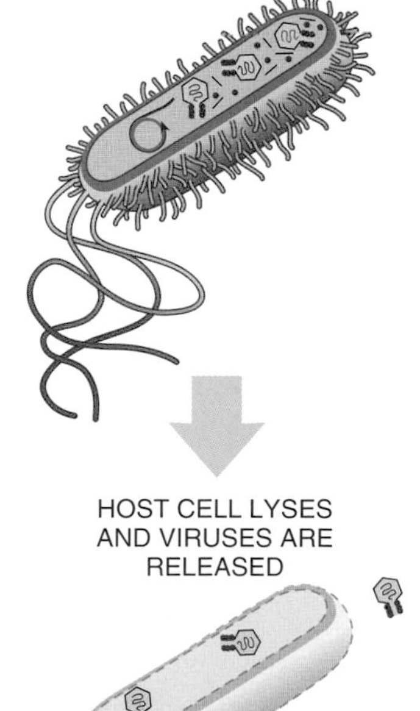

FIGURE 1.23 Virus Life Cycle
The life cycle of a virus starts when the viral DNA or RNA enters the host cell. Once inside, the virus uses the host cell to manufacture more copies of the virus genome and to make the protein coats for assembly of virus particles. Once multiple copies of the virus have been assembled, the host cell bursts open, allowing the progeny to escape and find other hosts to invade.

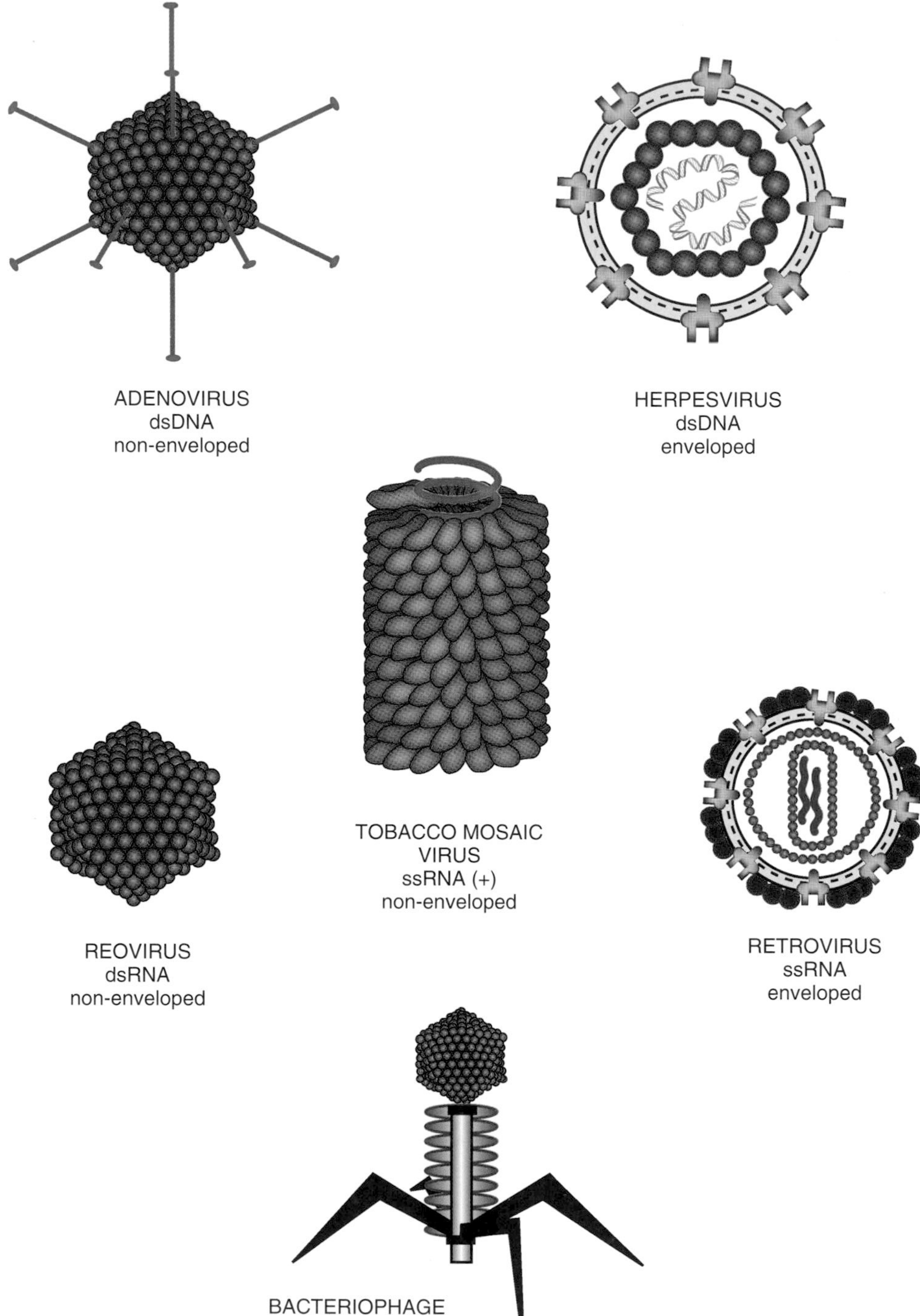

FIGURE 1.24
Examples of Different Viruses
Viruses come in a variety of shapes and sizes that determine whether the entire virus or only its genome enters the host cells.

Viruses are used extensively in biotechnology research because they specialize in inserting their genome into the host genome. They subvert the host into expressing their genes and making more copies of themselves. Researchers exploit these characteristics to study new genes, to alter the genomes of other model organisms, and to do gene therapy on humans.

SUBVIRAL INFECTIOUS AGENTS AND OTHER GENE CREATURES

We have used the term **gene creatures** to refer to various genetic entities that are sometimes called subviral infectious agents. These creatures exist but are not considered living because none of them can produce their own energy, duplicate their own genomes, or live independent of a host. The main advantage a virus has over a gene creature is the ability to survive as an inactive particle outside the host cell. Gene creatures are not normally found outside the host cell.

Satellite viruses are defective viruses. They can either replicate their genome or package their genome into a capsid, but they are unable to do both by themselves. Satellite viruses rely on a **helper virus** to supply the missing components or genes. For example, hepatitis delta virus (HDV) is a small single-stranded RNA satellite virus that infects the liver. Its helper is hepatitis B virus. Bacteriophage P4 is a satellite virus that infects *E. coli*. It is a double-stranded DNA virus that can replicate as a plasmid or integrate into the host chromosome, but it cannot form virus particles by itself. It relies on P2 bacteriophage to supply the structural proteins. P4 sends transcription factors to the P2 genome to control expression of the genes it pirates.

Gene creatures also include genetic elements that may be helpful to the host. For example, the plasmids of *E. coli* and yeast are genetic elements that cannot produce their own energy and rely on the host cell to replicate their genome. They cannot survive outside a host cell. These traits qualify plasmids as gene creatures. Like viruses and satellite viruses, plasmids are **replicons**; that is, they have sufficient information in their genome to direct their own replication. Plasmids may confer positive traits to the host. For example, plasmids can provide antibacterial enzymes, such as bacteriocins, that help their host compete with other bacteria for nutrients (see earlier discussion). Plasmids may carry genes for antibiotic resistance, thus allowing the host bacteria to survive after encountering an antibiotic. Plasmids may confer virulence, making the host bacteria more aggressive and deadly. Finally, some plasmids contain genes that help the host degrade a new carbon source to provide food.

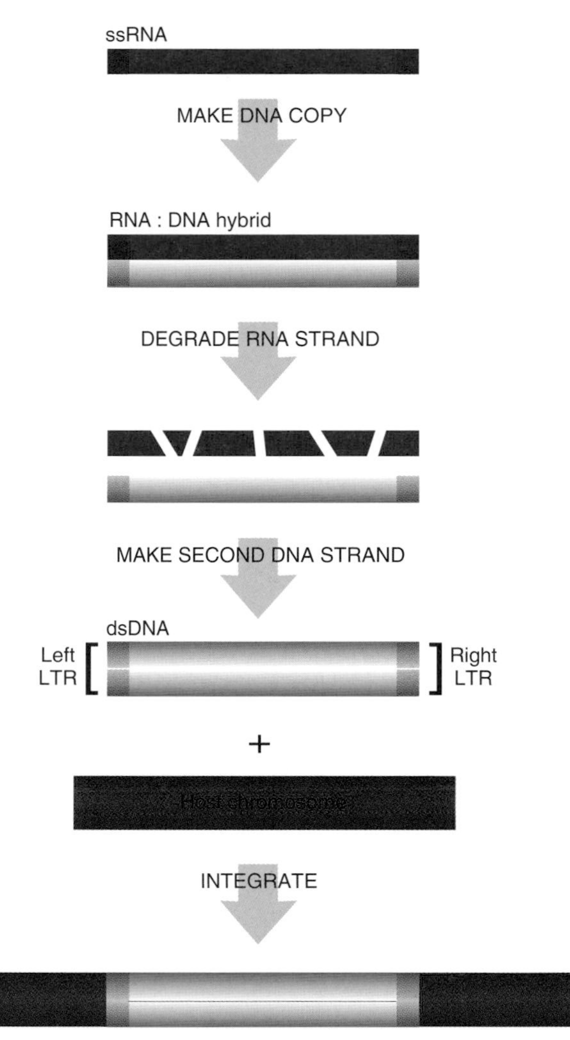

FIGURE 1.25
Retroviral Life Cycle
Retroviral genomes are made of positive RNA. Once the RNA enters the host, a DNA copy of the genome is made using reverse transcriptase. The original RNA strand is then degraded and replaced with DNA. Then the entire double-stranded DNA version of the retrovirus genome can integrate into the host genome.

Plasmids are usually found as circles of DNA, although some linear plasmids have been found. Plasmids come in all sizes, but are usually much smaller than the bacterial chromosome. The genes on plasmids are often beneficial to the host. Because the plasmid coexists within the cytoplasm of the host cell, it does not generally harm its host.

The F plasmid is found in some *E. coli*, and it is about 1% of the size of the chromosome. It was named "F" for fertility because it confers the ability to mate. F plasmids can transfer themselves from one cell to the next in a process called **conjugation** (Fig. 1.26). The plasmid has genes for the formation of a specialized pilus, the sex-pilus, which physically attaches an F^+ *E. coli* to an F^- cell. After contact, a junction—the conjugation bridge—forms between the two cells. During replication of the F plasmid, one strand is cut at the origin, and the free end enters the cytoplasm of the F^- cell via the conjugation bridge. Inside the recipient, a complementary strand of DNA is made and the plasmid is recircularized. The

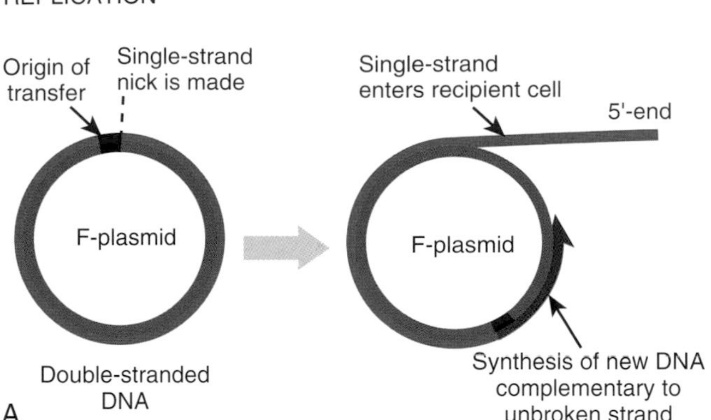

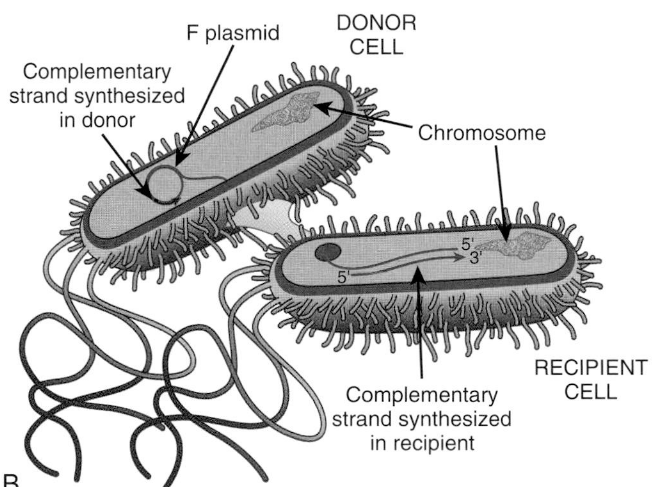

FIGURE 1.26 Conjugation in *E. coli*

(A) During bacterial conjugation, the F plasmid of *E. coli* is transferred to a new cell by rolling circle replication. First, one strand of the F plasma is nicked at the origin of transfer. The two strands start to separate, and synthesis of a new strand starts at the origin (green strand). (B) The single strand of F plasmid DNA that is displaced (pink strand) crosses the conjugation bridge and enters the recipient cell. The second strand of the F plasmid is synthesized inside the recipient cell. Once the complete plasmid has been transferred, it is recircularized.

other strand of the parent plasmid remains in the original F^+ cell and is also duplicated. Thus, after conjugation, both cells become F^+. Occasionally, the F plasmid integrates into the host chromosome. If an integrated F plasmid is transferred to another cell via conjugation, parts of the host chromosome may also get transferred. Therefore, bacteria can exchange chromosomal genetic information through conjugation.

Another gene creature that is very useful in biotechnology is the **transposable element** or **transposon**. This genetic element is merely a length of DNA that cannot exist or replicate as an independent molecule. To survive, it integrates into another DNA molecule. **Mobile DNA** or **jumping genes** are two terms used to describe transposons. When the transposon moves from one location to another, the process is called **transposition**. Unlike plasmids, transposons lack an origin of replication and are not considered replicons. They can only be replicated by integrating themselves into a host DNA molecule, such as a chromosome, plasmid, or viral genome. Transposons can move from site to site within the same host DNA or move from one host molecule of DNA to another. If a transposon loses its ability to move, its DNA remains in place on the chromosome or other DNA molecule.

Transposons come in several varieties and are classified based on the mechanism of movement. Transposons have two inverted DNA repeats at each end and a gene for **transposase**, the enzyme needed for movement. Transposase recognizes the inverted repeats at the ends of the transposon and excises the entire element from the chromosome. Next, transposase recognizes a **target sequence** of 3 to 9 base pairs in length on the host DNA. The transposon is then inserted into the target sequence, which is duplicated in the process. One copy is found on each side of the transposon. When a transposon is completely removed from one site and moved to another, the mechanism is **conservative transposition** or **cut-and-paste transposition** (Fig. 1.27). This leaves behind a double-stranded break that must be repaired by the host cell. Several cellular mechanisms exist to make this type of repair.

An alternative mechanism is **replicative transposition**, where a second copy of the transposon is made. **Complex transposons** use this method. Much as before, transposase recognizes the inverted repeats of the transposon. However, in this case it only makes single-stranded nicks at the ends. Transposase then makes two single-stranded nicks, one at each end of the target site. Each single DNA strand of the transposon is joined to one host strand at the target site. This creates two single-stranded copies of the transposon. The host responds to such single-stranded DNA regions by making the second, complementary strand of the transposon. This gives two copies of the transposon. Notice how the transposon itself does not replicate. It tricks the host into making the replica.

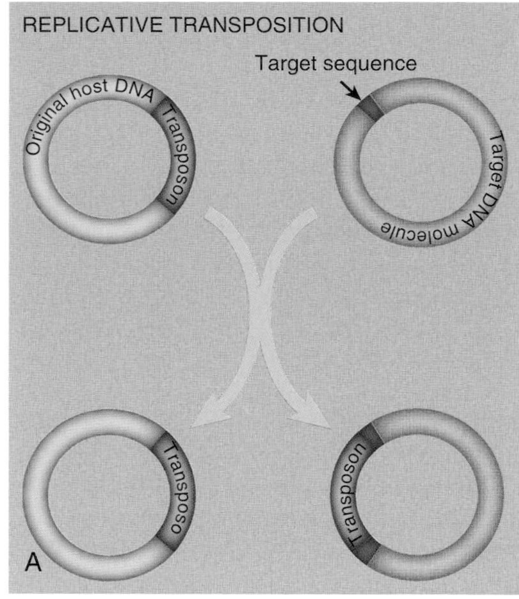

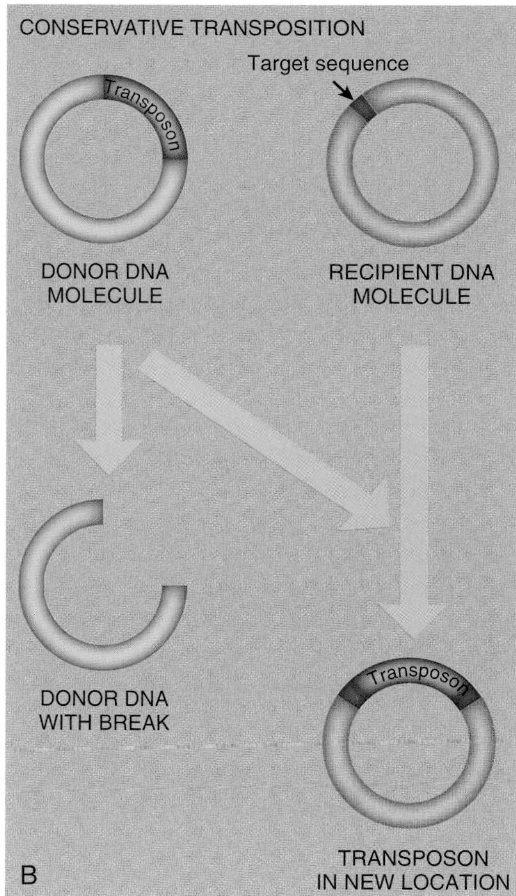

FIGURE 1.27 Transposons Move by Replicative or Conservative Transposition
(A) Replicative transposition leaves the original transposon in its original place, and a copy is inserted at another site within the host genome. (B) During conservative transposition, the original transposon excises from its original site and integrates at a different location.

Transposon movement can cause problems for the host. When the transposon moves, there is a potential for insertions, deletions, and inversions in the host DNA. If two copies of a transposon are found on a plasmid and the target sequence is on the host chromosome, a segment of the plasmid (flanked by the transposons) may be inserted into the host DNA. More generally, when multiple transposons are near each other, the ends of two neighboring but separate transposons may be used for transposition. When the two ends move to a new location, the DNA between them will be carried along. Whole genes or segments of genes may be deleted from the original location in this process. Conversely, regions of chromosome may become duplicated. If transposons are active and move often, the genome will become very damaged, and the host cells often commit suicide (see Chapter 20). Because the transposon will be destroyed along with its host, many transposons move only rarely. Controlling their movement preserves their existence within the genome and keeps the host cell from committing suicide.

Gene creatures is a term to describe genetic elements that exist within the confines of a host cell yet are separate from the original host genome. Some gene creatures include satellite viruses, plasmids, and transposons.

The plasmid is a unique gene creature because it confers positive traits such as resistance to antibiotics, bacteriocins, and the ability to transfer genetic material between two cells.

Transposons do not contain origins for their independent replication as do plasmids. These elements subvert the cell to make their copies by inducing breaks in the genome.

Summary

This chapter introduces the variety of different organisms used to study genes useful for biotechnology. Each organism, even the lowly gene creatures, is based on DNA. DNA and RNA have unique structures that ensure their survival and existence in all facets of life. Each structure has a backbone of alternating phosphate molecules with sugar residues. In DNA, the sugar, deoxyribose, is missing a hydroxyl group on the 2′ carbon. The bases, which attach at the 1′ carbon, form pairs so that adenine joins with thymine and guanine joins with cytosine. These pairs are held together with hydrogen bonds that induce the two backbones to twist into a double-stranded helix. In RNA, the sugar, ribose, has one extra hydroxyl group, and the base thymine is replaced with uracil.

Many different organisms are used in biotechnology research, and they have a particular trait that is useful to study new genes. Bacteria are genetic clones that are easily grown and stored for long periods of time. Two key traits are their simple genomes and availability of plasmids to alter their genetic makeup. Although useful, bacteria are prokaryotes and differ greatly from humans. Therefore, eukaryotic model organisms are also used for research. Yeasts are single-celled eukaryotes that have similar traits to human cells, such as multiple chromosomes, a nucleus, and various organelles. In addition, yeasts also have plasmids in which extra genes can be added to study in a model organism. Finally, the chapter outlines the key traits of multicellular organisms from barely visible roundworms such as *C. elegans* to mice; cultured human, animal, and insect cells; and the model plant organism, *Arabidopsis*.

Besides real organisms, research in biotechnology relies on gene creatures such as viruses, transposons, and plasmids. These genetic vehicles are critical to manipulating the genome of the model organisms. In fact, viruses may be the key to accomplishing gene therapy in humans also.

Viruses are used as vehicles to inject foreign DNA into a host cell. Transposons are also used to deliver new genes into the host DNA. Plasmids are used for the same purpose, but do not work in higher organisms and, therefore, are restricted to cultured cells, yeast, and bacteria. The use of gene creatures and model organisms is key to biotechnology research.

End-of-Chapter Questions

1. Which statement best describes the central dogma of genetics?
   - **a.** Genes are made of DNA, expressed as an RNA intermediary that is decoded to make proteins.
   - **b.** The central dogma only applies to yellow and green peas from Mendel's experiments.
   - **c.** Genes are made of RNA, expressed as a DNA intermediary, which is decoded to make proteins.
   - **d.** Genes made of DNA are directly decoded to make proteins.
   - **e.** The central dogma only applies to animals.
2. What is the difference between DNA and RNA?
   - **a.** DNA contains a phosphate group, but RNA does not.
   - **b.** Both DNA and RNA contain a sugar, but only DNA has a pentose.
   - **c.** The sugar ring in RNA has an extra hydroxyl group that is missing in the pentose of DNA.

d. DNA consists of five different nitrogenous bases, but RNA only contains four different bases.
e. RNA only contains pyrimidines and DNA only contains purines.

3. Which of the following statements about eukaryotic DNA packaging is true?
a. The process involves DNA gyrase and topoisomerase I.
b. All of the DNA in eukaryotes can fit inside of the nucleosome without being packaged.
c. Chromatin is only used by prokaryotes and is not necessary for eukaryotic DNA packaging.
d. Eukaryotic DNA packaging is a complex of DNA wrapped around proteins called histones, and further coiled into a 30-nanometer fiber.
e. Once eukaryotic DNA is packaged, the genes on the DNA can never again be expressed.

4. Which statement about *Thermus aquaticus* is false?
a. *T. aquaticus* was isolated from a hot spring.
b. The DNA polymerase from *T. aquaticus* is used in molecular biology for a procedure called polymerase chain reaction (PCR).
c. The DNA polymerase from *T. aquaticus* is able to withstand very high temperatures.
d. *T. aquaticus* can survive high temperatures and low pH.
e. *T. aquaticus* is found in the frozen lakes of Antarctica.

5. Which statement about *Escherichia coli* is not correct?
a. *E. coli* is called "the workhorse of molecular biology."
b. *E. coli* can grow in a simple solution of water, a carbon source, and mineral salts.
c. All *E. coli* strains are pathogenic, and therefore must be handled accordingly.
d. The chromosome of *E. coli* consists of one circular DNA molecular containing approximately 4000 genes.
e. All of the above answers are correct.

6. Plasmids from bacteria can be described by which of the following statements?
a. Plasmids provide an advantage to the host bacterium to compete against non-plasmid-containing bacteria for nutrients.
b. Plasmids are used as a molecular biology tool to express other genes efficiently in the host bacterium.
c. Plasmids are extrachromosomal segments of DNA that carry several genes beneficial to the host organism.
d. Plasmids have their own origin of replication.
e. All of the above statements describe plasmids.

7. Which of the following statements is not correct about the usefulness of fungi in biotechnology research?
a. Fungi produce the blue veins in some types of cheeses.
b. Yeast is responsible for the alcohol in beer and for bread rising.
c. Fungi are called "the workhorses of molecular biology."
d. The 2-micron circle is a useful extrachromosomal element from yeast that can be utilized in molecular biology research.
e. Fungi produce many industrial chemicals and pharmaceuticals.

(*Continued*)

8. What mechanism does yeast utilize to control mating type in the cells?
   a. Yeast is only able to reproduce through mitosis.
   b. The MAT locus in the yeast genome contains two divergent genes that encode for the pheromones **a** and α, along with the pheromone receptors.
   c. The mating type of yeast is determined by pheromones called **b** and β.
   d. There are no mechanisms to control mating type in yeast because all of the cells are structurally the same.
   e. Yeast mating types are generally referred to as either male or female.

9. Which of the following yeast cellular component is typically not found in bacteria?
   a. centromeres
   b. telomeres
   c. nuclear pores
   d. nuclear envelope
   e. All of the above are found in yeast and not bacteria.

10. Identify the statement about multicellular model organisms that is correct.
   a. *C. elegans* has been used extensively to study multicellular interactions partly because the creature can reproduce by self-fertilization (genetic clones) or sexually (novel genetic organisms).
   b. Based on homology, research on *Drosophila* mutants has identified genes in the human genome responsible for body patterns.
   c. The zebrafish, or *Danio rerio,* are used to study developmental genetics because the embryonic cells are easily destroyed or manipulated and the effects can be observed within 24 hours.
   d. The mouse is a model organism for studying human genetics, physiology, and development because less than 1% of the genes in the mouse genome have no genetic homology in humans.
   e. All of the statements are correct.

11. What is the main advantage for studying cells in culture rather than in a whole organism?
   a. Cell lines in culture are easily manipulated genetically to introduce new genes or delete other genes.
   b. Cell lines are not very stable and therefore, it is more advantageous to study cells within the organism itself.
   c. There is no advantage to studying cells in a cell line rather than in a live organism.
   d. The information obtained from studying cell lines as opposed to live organisms is not relevant to what happens *in vivo.*
   e. None of the above is the main advantage.

12. Why is *Arabidopsis thaliana* used as a model organism for plant genetics and biology?
   a. *Arabidopsis* responds to stress and disease similarly to important crop plants such as rice, wheat, and corn.
   b. *Arabidopsis* is easy to grow and maintain in the laboratory.
   c. The genome of *Arabidopsis* is relatively small compared to other plants.
   d. The generation cycle of *Arabidopsis* is shorter than most other crop plants and produces many seeds for further study.
   e. All of the above statements are reasons for using *Arabidopsis* as a model organism.

13. Why are viruses significant to biotechnology?

- **a.** They are able to insert their genome into the host genome, thus integrating genes in the process.
- **b.** Viruses can be used to alter the genomes of other organisms.
- **c.** Reverse transcriptase, an enzyme used in molecular biology, is encoded in a retroviral genome.
- **d.** Viruses play an important role in delivering gene therapy to humans.
- **e.** All of the above statements are reasons why viruses are significant to biotechnology research.

14. Which statement best describes the F plasmid?

- **a.** F plasmids contain genes for formation of a specialized pilus that initiates the formation of a conjugation bridge between two cells for the purpose of transferring genetic material.
- **b.** The F plasmid does not have an origin of replication and can therefore not replicate itself.
- **c.** The primary host for the F plasmid is *Saccharomyces cerevisiae.*
- **d.** The F plasmid is not important for biotechnology research.
- **e.** All of the above statements describe the F plasmid.

15. Which of the following elements is important in biotechnology research?

- **a.** transposons
- **b.** F plasmid
- **c.** satellite viruses
- **d.** plasmids
- **e.** all of the above

Further Reading

Ablain, J., & Zon, L. I. (2013). Of fish and men: using zebrafish to fight human diseases. *Trends in Cell Biology, 23,* 584–586.

Steele, J. H., & Lutz, R. A. (2001). Hydrothermal vent biota*. In *Encyclopedia of Ocean Sciences,* 2nd ed. Amsterdam: Academic Press (pp. 133–143).

Verma, A. S., Singh, A., Tsuiji, H., & Yamanaka, K. (2014). Animal models for neurodegenerative disorders. In *Animal Biotechnology: Models in Discovery and Translation.* Amsterdam: Academic Press (pp. 39–56).

Verma, A. S., Singh, A., Ram, K. R., & Chowdhuri, D. K. (2014). *Drosophila*: A model for biotechnologists. In *Animal Biotechnology: Models in Discovery and Translation.* Amsterdam: Academic Press (pp. 3–19).

DNA, RNA, and Protein

Biotechnology

http://dx.doi.org/10.1016/B978-0-12-385015-7.00002-8

THE CENTRAL DOGMA OF MOLECULAR BIOLOGY

Two essential features of living creatures are the ability to reproduce their own genome and manufacture their own energy. To accomplish these feats, an organism must be able to make proteins using information encoded in its DNA. Proteins are essential for cellular architecture, giving the cell a particular shape and structure. Proteins include enzymes that catalyze reactions used to make energy. Proteins control cellular processes like replication. Proteins provide channels in the membrane for cells to communicate with each other or share metabolites. Making proteins is a key operation for all living organisms.

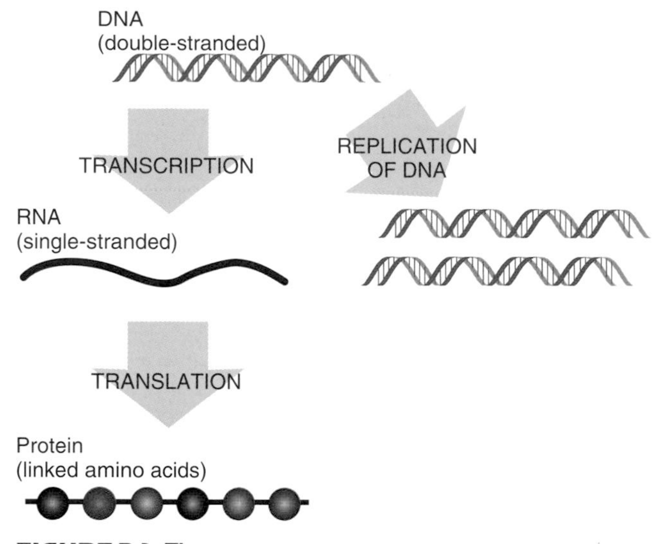

FIGURE 2.1 The Central Dogma
Cells store genetic information as DNA, which is able to replicate so that daughter cells have the same information as the parent. When a protein is needed, DNA is transcribed into RNA, which in turn is translated into a protein. RNA also functions in a cell to regulate gene expression and as ribozymes that carry out catalytic reactions.

The **central dogma** of molecular biology states that information flows from DNA to RNA to protein (Fig. 2.1). First, this chapter focuses on how RNA is made from DNA in a process called **transcription**. Next, the mechanisms used to control transcription are discussed. We then discuss how particular RNA molecules called **mRNA** or **messenger RNA** are used to make protein in a process called **translation**. By examining these processes, the reader will gain an understanding of the complexity involved in engineering cells for the purposes of biotechnology.

The central dogma of molecular biology is that DNA is transcribed into RNA, which in turn is translated into proteins.

TRANSCRIPTION EXPRESSES GENES

Gene expression involves making an RNA copy of the DNA code, a process called transcription. Making RNA involves uncoiling the DNA, melting the strands at the start of the gene and moving any histones out of the way, making an RNA molecule that is complementary in sequence to the **template** strand of the DNA with an enzyme called **RNA polymerase**, and stopping at the end of the gene. The newly made RNA releases from the DNA, which then returns to its supercoiled form.

Two long-standing questions in biology are how a cell turns genes on and off, and what genes are transcribed at what time in development or function. These questions have multiple answers that are based on the different types of genes. **Housekeeping genes** encode proteins that are used continually. Inducible genes are converted to protein only under certain circumstances. For instance, in *Escherichia coli,* genes that encode proteins involved with the utilization of lactose are expressed only when lactose is present (see later discussion). The same principle applies to the genes for using other nutrients. Various inducers and accessory proteins control whether or not these genes are **expressed** or made into RNA; they will be discussed in more detail in upcoming sections.

The final product encoded by a gene is often a protein but may be RNA also. Genes that encode proteins are transcribed to give messenger RNA (mRNA), which is then translated to give the protein. Other RNA molecules, such as tRNA, rRNA, snRNA, and other regulatory noncoding RNAs, are used directly (i.e., they are not translated to make proteins). Some RNA molecules, called **ribozymes**, catalyze enzymatic reactions. One well-researched ribozyme is an rRNA found in the large subunit of the ribosome (see Chapter 5). The genes that ultimately code for a protein via an mRNA intermediate are studied most often since they are historically thought to be the most important to the function of the organism. Coding regions of a gene, sometimes called a **cistron** or a **structural gene**, have the code to make a protein or a nontranslated RNA. (The term *cistron* was originally defined by genetic complementation using the *cis/trans* test.) In contrast, an **open reading frame (ORF)** is a stretch of DNA (or the corresponding RNA) that encodes a protein and therefore is not interrupted by any stop codons for protein translation (see later discussion).

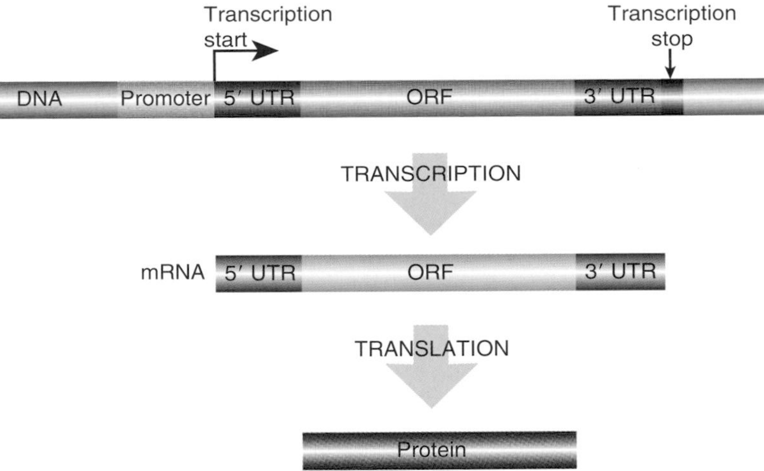

FIGURE 2.2 The Structure of a Typical Gene
Genes are regions of DNA that are transcribed to give RNA. RNA can be translated into protein or used directly. The gene has a promoter region plus transcriptional start and stop points that flank the region that is converted into mRNA. After transcription, the mRNA has a 5′ untranslated region (5′ UTR) and 3′ untranslated region (3′ UTR), which are not translated; only the ORF is translated into protein.

During transcription, the enzymes that make RNA must identify the start site among the DNA code. Every gene has a region upstream of the coding sequence called a **promoter** (Fig. 2.2). RNA polymerase recognizes this region and starts transcription here. Bacterial promoters have two major recognition sites: the −10 and −35 regions. The numbers refer to their approximate location upstream or before the transcriptional start site. (By convention, positive numbers refer to nucleotides downstream, or after the transcription start site; and negative numbers refer to those upstream, or before.) The exact sequences at −10 and −35 vary, but the consensus sequences are TATAA and TTGACA, respectively. When a gene is transcribed continually or **constitutively**, then the promoter sequence closely matches the consensus sequence. If the gene is expressed only under special conditions, **activator proteins** or **transcription factors** are needed to bind to the promoter region before RNA polymerase will recognize it. Such promoters rarely look like the consensus.

The **transcription start site** begins after the promoter and denotes where RNA polymerase starts adding nucleotides complementary to the template strand. Between the transcription start site and the ORF is a region called the **5′ untranslated region (5′ UTR)**. This region is not made into protein but contains translation regulatory elements like the ribosome binding site. Next is the ORF, where no translational stop codons are found. Then there is another untranslated region after the ORF, known as the **3′ untranslated region (3′ UTR)**. This region is not made into protein either, and often contains important regulatory elements that modulate the rate of translation. Finally, transcription stops at the termination sequence.

Bacterial RNA polymerase is made of different protein subunits. The **sigma subunit** recognizes the −10 and −35 regions, and the **core enzyme** catalyzes RNA. RNA polymerase synthesizes nucleotide additions only in a 5′ to 3′ direction. The core enzyme has five protein subunits: a dimer of two α proteins, a β protein, a related β′ subunit, and an ω subunit. The β and β′ subunits form the catalytic site, and the α subunit helps recognize the promoter. The 3D structure of RNA polymerase shows a deep groove that can hold the template DNA and a minor groove to hold the growing RNA.

Genes have a transcriptional promoter, where RNA polymerase attaches to the DNA and begins making an RNA copy of the template strand. The RNA has three regions: the 5′ UTR contains information important for making the protein, the ORF has the actual coding region translated into amino acids during translation, and the 3′ UTR contains other important regulatory elements.

MAKING RNA

In bacteria, once the sigma subunit of RNA polymerase recognizes the −10 and −35 regions, the core enzyme forms a **transcription bubble** where the two DNA strands are separated from each other (Fig. 2.3). The strand used by RNA polymerase is called the template strand (aka **noncoding** or **antisense)** and is complementary to the resulting mRNA. The core

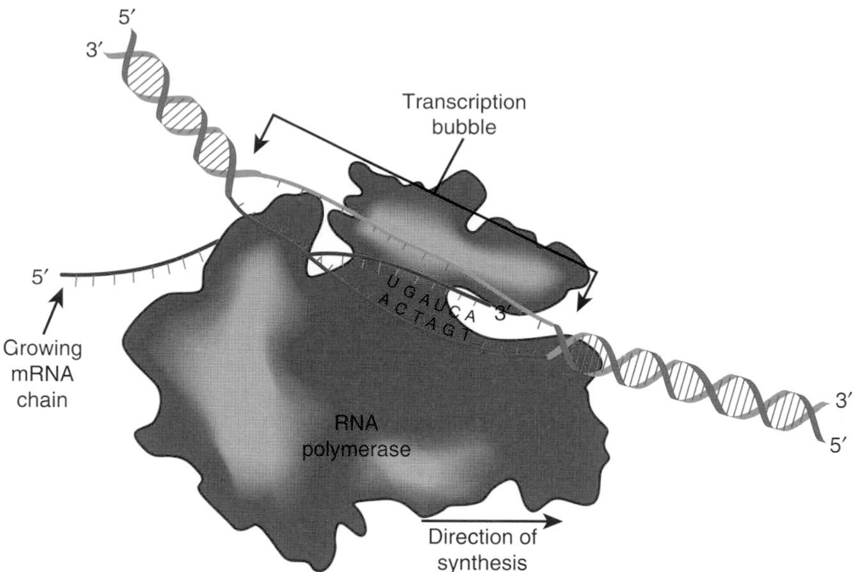

FIGURE 2.3 RNA Polymerase Synthesizes RNA at the Transcription Bubble RNA polymerase is a complex enzyme that can hold a strand of double-stranded DNA open to form a transcription bubble and add ribonucleotides to create RNA complementary to the template strand.

enzyme adds RNA nucleotides in the 5′ to 3′ direction, based on the sequence of the template strand of DNA. The newly made RNA anneals to the template strand of the DNA via hydrogen bonds between base pairs. The opposite strand of DNA is called the **coding strand** (aka **nontemplate** or **sense strand**). Because this is complementary to the template strand, its sequence is identical to the RNA (except for the replacement of thymine with uracil in RNA).

RNA synthesis normally starts at a purine (normally an A) in the DNA that is flanked by two pyrimidines. The most typical start sequence is CAT, but sometimes the A is replaced with a G. The rate of elongation is about 40 nucleotides per second, which is much slower than replication (~1000 bp/sec). RNA polymerase unwinds the DNA and creates positive supercoils as it travels down the DNA strand. Behind RNA polymerase, the DNA is partially unwound and has surplus negative supercoils. DNA gyrase and topoisomerase I either insert or remove negative supercoils, respectively, returning the DNA back to its normal level of supercoiling (see Chapter 4).

RNA polymerase makes a copy of the gene using the noncoding or template strand of DNA. RNA has uracils instead of thymines.

TRANSCRIPTION STOP SIGNALS

RNA polymerase continues transcribing DNA until it reaches a termination signal. In bacteria, the **Rho-independent terminator** is a region of DNA with two inverted repeats separated by about six bases, followed by a stretch of As. As RNA polymerase makes these sequences, the two inverted repeats form a hairpin structure. The secondary structure causes RNA polymerase to pause. As the stretch of As is transcribed into Us, the DNA/RNA hybrid molecule becomes unstable (A/U base pairs have only two hydrogen bonds). RNA polymerase "stutters" and then falls off the template strand of DNA in the middle of the As.

Bacteria also have **Rho-dependent terminators** that have two inverted repeats but lack the string of As. **Rho (ϱ) protein** is a special helicase that unwinds DNA/RNA hybrid double helices. Rho binds upstream of the termination site in a region containing many cytosines. After RNA polymerase passes the Rho binding site, Rho attaches to the RNA and moves along the RNA transcript until it catches RNA polymerase at the hairpin structure. Rho then unwinds the DNA/RNA helix and separates the two strands. The RNA is then released.

Transcription terminates either in a Rho-independent manner or in a Rho-dependent manner.

THE NUMBER OF GENES ON AN mRNA VARIES

Bacterial and eukaryotic chromosomes are organized very differently. In prokaryotes, the distance between genes is much smaller, and genes associated with one metabolic pathway are often found next to each other. For example, the lactose **operon** contains several clustered genes for lactose metabolism. Operons are clusters of genes that share the same promoter and are transcribed as a single large mRNA that contains multiple structural genes or cistrons. Thus, the mRNA transcripts are called **polycistronic mRNA** (Fig. 2.4). The multiple cistrons are translated individually to give separate proteins. In eukaryotes, genes are often separated by large stretches of DNA that do not encode any protein. In eukaryotes, each mRNA has only one cistron and is therefore called **monocistronic mRNA**. If a polycistronic transcript is expressed in eukaryotes, the ribosome translates only the first cistron, and the other encoded proteins are not made.

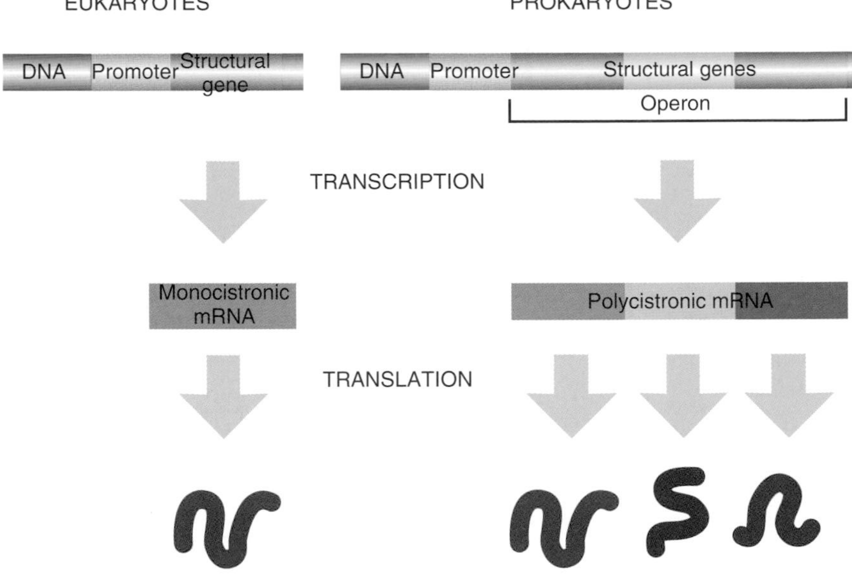

FIGURE 2.4 Monocistronic versus Polycistronic
Eukaryotes transcribe genes in single units, where each mRNA encodes for only one protein. Prokaryotes transcribe genes in operons as one single mRNA and then translate the proteins as separate units.

Bacterial mRNA transcripts have multiple open reading frames for proteins in the same metabolic pathway. Eukaryotes tend to have only one open reading frame in a single mRNA transcript.

EUKARYOTIC TRANSCRIPTION IS MORE COMPLEX

There are several differences between eukaryotic and prokaryotic transcription with more complexity associated with eukaryotic transcription. The simple fact that eukaryotic mRNA is synthesized in a nucleus makes the process more involved than bacterial transcription, but this is only one of the differences.

In contrast to the single RNA polymerase in prokaryotes, eukaryotes have three different RNA polymerases that each transcribe different types of genes. **RNA polymerase I** transcribes the eukaryotic genes for large ribosomal RNA. These two rRNAs are transcribed as one long mRNA that is cleaved into two different transcripts: the 18S rRNA and 28S rRNA. These are used directly and not translated into protein. **RNA polymerase III** transcribes the genes for tRNA, 5S rRNA, and other small RNA molecules. **RNA polymerase II** transcribes the genes that encode proteins and has been studied the most.

Starting transcription of eukaryotic genes is more complex than in bacteria. The layout of the eukaryotic promoter is much different. RNA polymerase II needs three different regions, the **initiator box**, the **TATA box**, and various upstream elements that bind proteins known as transcription factors. The initiator box is the site where transcription starts and is separated by about 25 base pairs from the TATA box. The upstream elements vary from gene to gene and aid in controlling what proteins are expressed at what time.

Many proteins are involved in positioning eukaryotic RNA polymerase II at the transcriptional start site (Fig. 2.5 and Table 2.1). RNA polymerase II requires several **general transcription factors** to initiate transcription at all promoters. In addition, **specific transcription factors** are needed that vary depending on the particular gene (see later discussion). The **TATA binding protein** or **TATA box protein** (TBP) recognizes the TATA box. This factor is used by all three RNA polymerases in eukaryotes. For RNA polymerase II, TBP is found with other proteins in a complex called TFIID. (For the other RNA polymerases, TBP associates with different proteins.) After this complex binds, TFIIB binds to

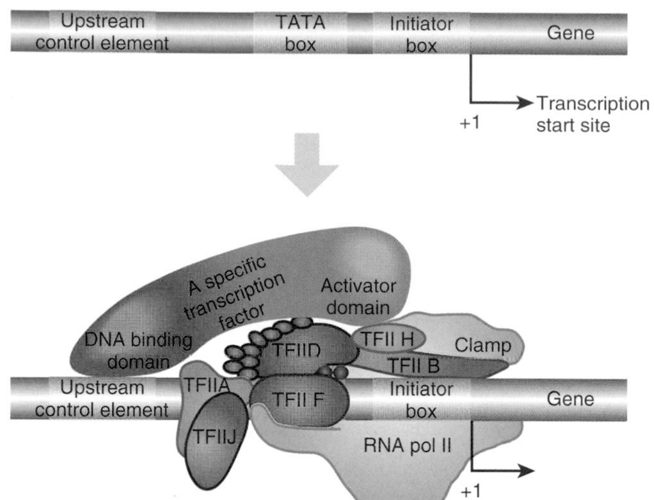

FIGURE 2.5 Eukaryotic Transcription
Many different general transcription factors help RNA polymerase II find the TATA and initiator box region of a eukaryotic promoter. Specific transcription factors bind to upstream control elements and transmit the activation signal to the general transcription factors and RNA polymerase II.

the promoter, which then triggers the binding of RNA polymerase II and TFIIA. RNA polymerase is associated with TFIIF, which probably helps it bind to the promoter. Once RNA polymerase II has bound to the promoter, it still requires TFIIE, TFIIH, and TFIIJ to initiate transcription. In particular, TFIIH phosphorylates the tail of RNA polymerase II, which allows it to move along the DNA. As RNA polymerase II leaves the promoter, it leaves behind all of the general complexes except TFIIH.

Bacterial RNA polymerase can function with a promoter containing no upstream elements. However, in eukaryotes, the upstream elements are essential to RNA polymerase II function, and a promoter with no upstream elements is extremely inefficient at initiating transcription. These elements are from 50 to 200 base pairs in length and vary based on the gene being expressed. They bind regulatory proteins known as specific transcription factors, as opposed to the general transcription factors shared by all promoters that use RNA polymerase II. For example, the specific transcription factors Oct-1 and Oct-2 proteins bind only to the Octamer elements. Oct-1 is found in all tissues, whereas Oct-2 is found only in immune cells. A plethora of specific factors exists, which is beyond the scope of this discussion.

Eukaryotes have three different RNA polymerases that transcribe different genes. RNA polymerase II binds to TATA and initiator boxes of the promoter region of protein-encoding genes. Different general transcription factors facilitate RNA polymerase II binding.

Eukaryotes require specific transcription factors to initiate gene transcription also. There are also a large number of different specific transcription factors.

Table 2.1 General Transcription Factors for RNA Polymerase II

TBP	Binds to TATA box, part of TFIID
TFIID	Includes TBP, recognizes Pol II specific promoter
TFIIA	Binds upstream of TATA box; required for binding of RNA Pol II to promoter
TFIIB	Binds downstream of TATA box; required for binding of RNA Pol II to promoter
TFIIF	Accompanies RNA Pol II as it binds to promoter
TFIIE	Required for promoter clearance and elongation
TFIIH	Phosphorylates the tail of RNA Pol II, retained by polymerase during elongation
TFIIJ	Required for promoter clearance and elongation

REGULATION OF TRANSCRIPTION IN PROKARYOTES

In prokaryotes, various activator and **repressor** proteins control which genes are transcribed into mRNA. The activators and repressors work by binding to DNA in the promoter region and either stimulating or blocking the action of bacterial RNA polymerase. In *E. coli*, about 1000 of the 4000 total genes are expressed at one time. Activator proteins work by **positive regulation**; in other words, genes are expressed only when the activator gives a positive signal. In contrast, repressors work by **negative regulation**. Here, the gene is expressed only when the repressor is removed. Some repressors block RNA polymerase from binding to the DNA; others prevent initiation of transcription even though RNA polymerase has bound.

Regulation of transcription is complex, even in simple prokaryotes. Many genes are controlled by a variety of factors. Some operons in bacteria have multiple repressors and activators. Less often, regulatory proteins may block elongation either by slowing the actual rate of elongation or by signaling premature termination. Conversely, a few **antiterminator proteins** are known that override termination and allow genes downstream of the termination site to be expressed.

Prokaryotes use positive regulation, where activator proteins signal RNA polymerase to transcribe the gene, or negative regulation, where the transcription factor inhibits RNA polymerase.

Prokaryotic Sigma Factors Regulate Gene Expression

Prokaryotic RNA polymerase includes a sigma (σ) subunit, which recognizes the promoter first and binds the catalytic portion of the enzyme (the core enzyme). There are many different sigma subunits, and each one recognizes a different set of genes. The σ70 subunit, or RpoD, is the most commonly used form. It recognizes most of the housekeeping genes in *E. coli*. During the stationary phase, when *E. coli* is not growing rapidly, σ38, or RpoS, activates the necessary genes. (Sigma subunits are named either by σ plus their molecular weight or by Rpo [for RNA polymerase] plus their function: D = default, S = stationary, etc.)

Another sigma factor, RpoH, or σ32, activates genes needed during heat shock. Normally, *E. coli* grows at body temperature, 37 °C, and stops growing at temperatures much above 43 °C. At such higher temperatures, proteins begin to unfold and are degraded. RpoH activates expression of **chaperonins** that help proteins fold correctly and prevent aggregation. RpoH also activates **proteases** that degrade proteins too damaged by the heat to be saved.

The transcription and translation of RpoH depend on temperature. When *E. coli* grows at a normal temperature, very few misfolded proteins are present. DnaK (a chaperonin) and HflB (a protease) are found in the cytoplasm, but since there are few proteins to degrade, they bind to RpoH and degrade it. They even degrade partially translated RpoH protein. When high temperatures promote unfolding and aggregation of proteins, DnaK and HflB bind to the aberrant proteins and no longer destroy RpoH. Now the sigma factor initiates transcription of other genes associated with heat shock.

Sigma (σ) subunits are transcription factors that associate with prokaryotic RNA polymerase and control which genes are transcribed.

Lactose Operon Demonstrates Specific and Global Activation

Many genes require specific regulator proteins to activate RNA polymerase binding and transcription. Some of these proteins exist in two forms: active (binds to DNA in promoter region) and inactive (nonbinding). The forms are interconverted by small **signal molecules** or **inducers** that alter the shape of the protein. For example, the inducer *allo*-lactose controls the activator protein for the lactose operon.

The lactose or *lac* operon is well characterized genetically, and the importance of each of the DNA elements in the promoter region has been studied. The promoter region controls three structural genes: *lacZ*, *lacY*, and *lacA*. The *lacZYA* genes are transcribed as a polycistronic message. Upstream of the promoter, and transcribed in the opposite direction as the *lacZYA*

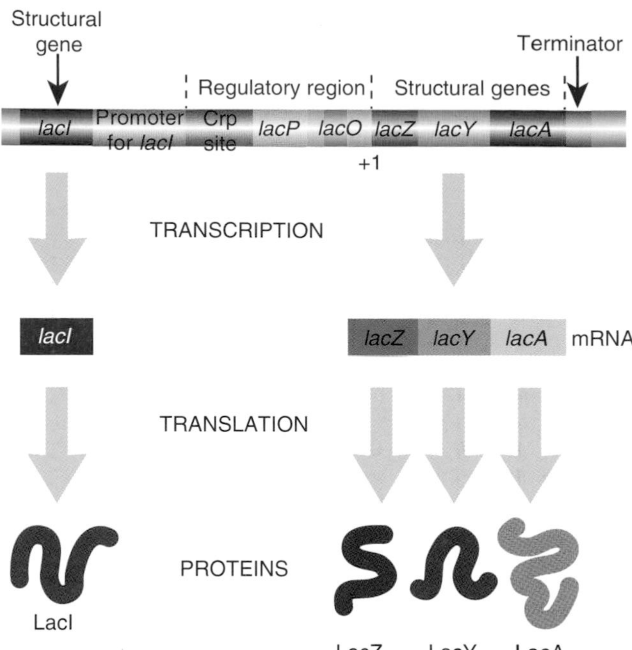

FIGURE 2.6 Components of the *lac* Operon

The *lac* operon consists of three structural genes, *lacZYA*, which are all transcribed from a single promoter, designated *lacP*. The promoter is regulated by binding of the repressor at the operator, *lacO*, and of Crp protein at the Crp site. Note that in reality, the operator partly overlaps both the promoter and the *lacZ* structural gene. The single *lac* mRNA is translated to produce the LacZ, LacY, and LacA proteins. The *lacI* gene that encodes the LacI repressor has its own promoter and is transcribed in the opposite direction from the *lacZYA* operon.

region, is another gene, which encodes the LacI protein, the *lac* operon repressor (Fig. 2.6). The *lacZ* gene encodes **β-galactosidase**, which cleaves the disaccharide lactose into galactose and glucose. The *lacY* gene encodes **lactose permease**, which transports lactose across the cytoplasmic membrane into the bacteria. Finally, the *lacA* gene encodes the protein **lactose acetylase**, with an unknown role. The promoter has a binding site, *lacO*, for the repressor protein, which overlaps the binding site for RNA polymerase. This region is also known as the **operator**, and when the repressor binds, RNA polymerase cannot transcribe the operon.

There is also a binding site for **CRP protein (cyclic AMP receptor protein)**, also known as **CAP (catabolite activator protein)**. This global regulator activates transcription of many different operons for using alternate sugar sources. It is active when *E. coli* does not have glucose to utilize as an energy source.

The environment controls whether or not the lactose operon is expressed (Fig. 2.7). When *E. coli* has plenty of glucose, then the lactose operon is turned off (as well as other operons for other sugars such as maltose or fructose). When glucose is present, levels of a small inducer, **cyclic AMP (cAMP)**, are low. If *E. coli* consumes all the available glucose, the levels of cAMP increase. cAMP binds to Crp, the global regulator, which then dimerizes so that it can bind to the Crp sites in various promoters, such as the lactose operon. Crp binding will not activate the lactose operon alone, and lactose must also be present to activate transcription. If lactose is available, β-galactosidase converts some lactose into *allo*-lactose. This acts as an inducer and binds to the tetrameric LacI repressor protein. This releases the repressor from the promoter. The lactose operon is expressed only when both glucose levels are low and lactose is present. The control relies on two inducer molecules: cAMP binds to the global activator, Crp, and *allo*-lactose binds to the specific repressor, LacI. One control is global (Crp) because it controls many different operons, and one control is specific (LacI) because it regulates only the lactose operon.

Many researchers use the lactose promoter to control expression of other genes. In the lab, a **gratuitous inducer, IPTG (isopropyl-thiogalactoside)**, replaces *allo*-lactose (Fig. 2.8). IPTG is not cleaved by β-galactosidase because its two halves are linked through a sulfur rather than oxygen. Since it is not metabolized, IPTG does not have to be added continually throughout the experiment (as would be the case for *allo*-lactose).

The *lac* operon is important to understand because its inducers and regulators are used to control new genes that are engineered into model organisms.

Control of Activators and Repressors

Various mechanisms control gene activators and repressors. In some cases, the repressor or activator binds to the promoter of its own gene and controls its own transcription; this is called **autogenous regulation**.

Many activators and repressors rely on activation by small molecules, as for Crp and LacI. In some cases, a repressor needs a **co-repressor** in order to be active. For example, ArgR represses the arginine biosynthetic operon when arginine is present. Arginine is a co-repressor and ensures that the bacteria do not make the amino acid when it is not needed.

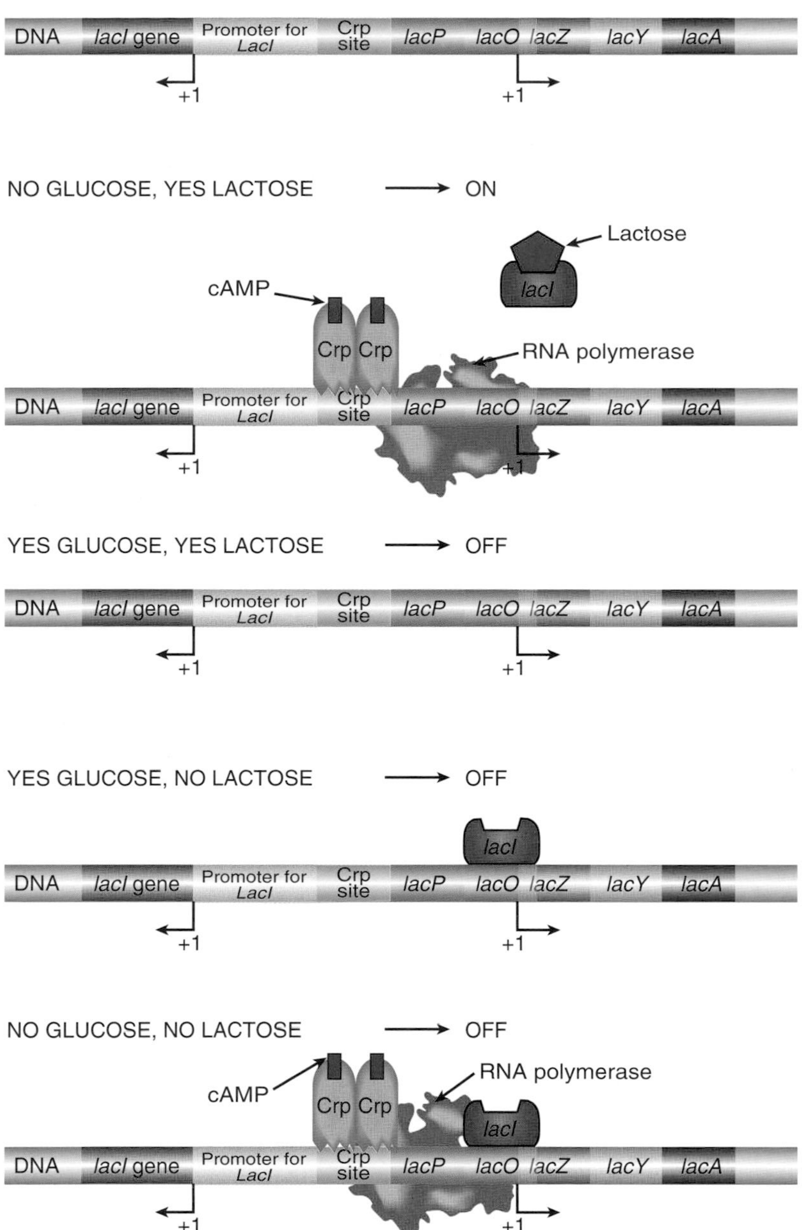

FIGURE 2.7 Control of Lactose Operon
The lactose operon is converted into a polycistronic mRNA only when glucose is absent and lactose is present. When glucose is available, the global activator protein, Crp, does not activate binding of RNA polymerase. When there is no glucose, Crp binds to the promoter and stimulates RNA polymerase to bind. The lack of lactose keeps LacI protein bound to the operator site and prevents RNA polymerase from transcribing the operon. Only when lactose is present is LacI released from the DNA.

In many cases, adding different groups, such as phosphate, methyl, acetyl, AMP-, and ADP-ribose, covalently modifies activators or repressors. The **two-component regulatory systems** of bacteria transfer phosphate groups from a sensor protein to a regulator protein (Fig. 2.9). The first protein, the **sensor kinase**, senses a change in the environment and changes shape. This causes the kinase to phosphorylate itself using ATP. The phosphate group is then transferred to the regulator protein (an activator or repressor), which changes shape to its DNA binding form. The phosphorylated regulator then binds to its recognition site in the target promoter. This either stimulates or represses transcription of the operon.

LACTOSE
(galactose & glucose)

ALLO-LACTOSE

ISOPROPYL-β-D-THIOGALACTOSIDE
(IPTG)

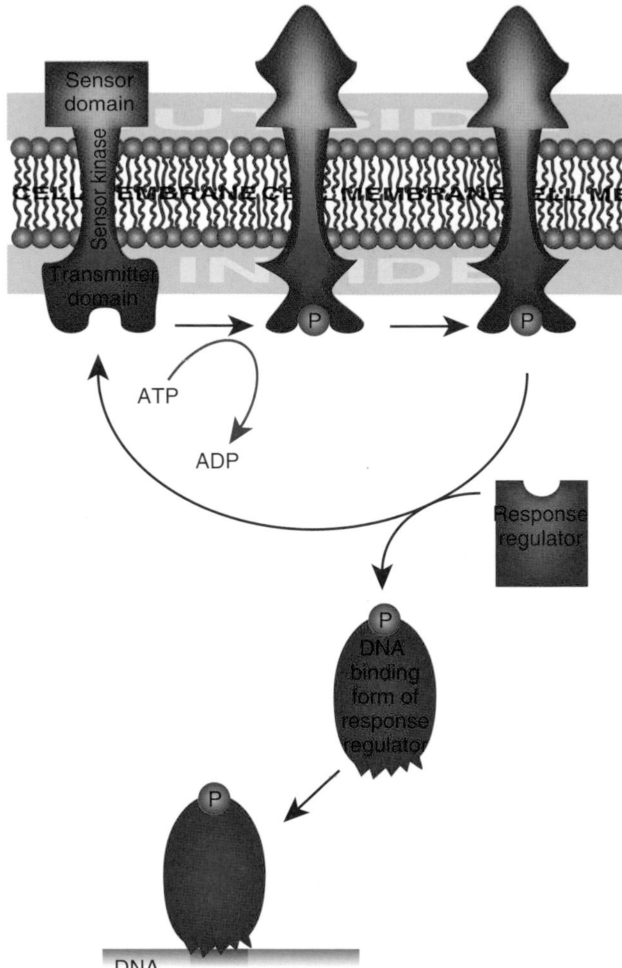

FIGURE 2.8 Structures of Lactose, *allo*-Lactose, and IPTG IPTG is a nonmetabolizable analog of the lactose operon inducer, *allo-lactose.* β-galactosidase cannot break the sulfur linkage and therefore does not cleave IPTG in two.

FIGURE 2.9 Model of Two-Component Regulatory System The two-component regulatory system includes a membrane component (sensor kinase) and a cytoplasmic component (response regulator). Outside the cell, the sensor domain of the kinase detects an environmental change, which leads to phosphorylation of the transmitter domain. The response regulator protein receives the phosphate group and consequently changes configuration to bind the DNA.

There are many different two-component systems in bacteria that respond to a variety of environmental conditions. For example, when there is low oxygen, the ArcAB system modifies gene expression to compensate. The ArcB protein is the sensor kinase, and it has three phosphorylation sites. The ArcA regulator has only one site for phosphorylation. The phosphate group transfers from one site to the next in a **phosphorelay** system, ultimately regulating transcription of the genes. These types of phosphorelays are very common, particularly in eukaryotes where there are often more than two components.

Phosphorylation, co-repressors, and co-activators modulate prokaryotic gene activators and repressors of transcription to express genes only in appropriate conditions.

REGULATION OF TRANSCRIPTION IN EUKARYOTES

Just as the initiation of transcription is more complex in eukaryotes, so is its control. The mechanisms to regulate which gene is expressed at what time are very complicated. The fact that eukaryotic DNA wraps around histones hinders many proteins from binding to the DNA, delaying access by activators and repressors. In addition, the nuclear membrane prevents the access of most proteins to the nucleus. Consider the complexity of the human body, with its multiple tissues. Each gene in each cell needs to be expressed only when needed and only in the amount needed. In addition to normal organ functions and to changes during development, the environment has a huge impact on our bodies, and changes in gene expression help us adapt. Overall, the numbers and types of controls for

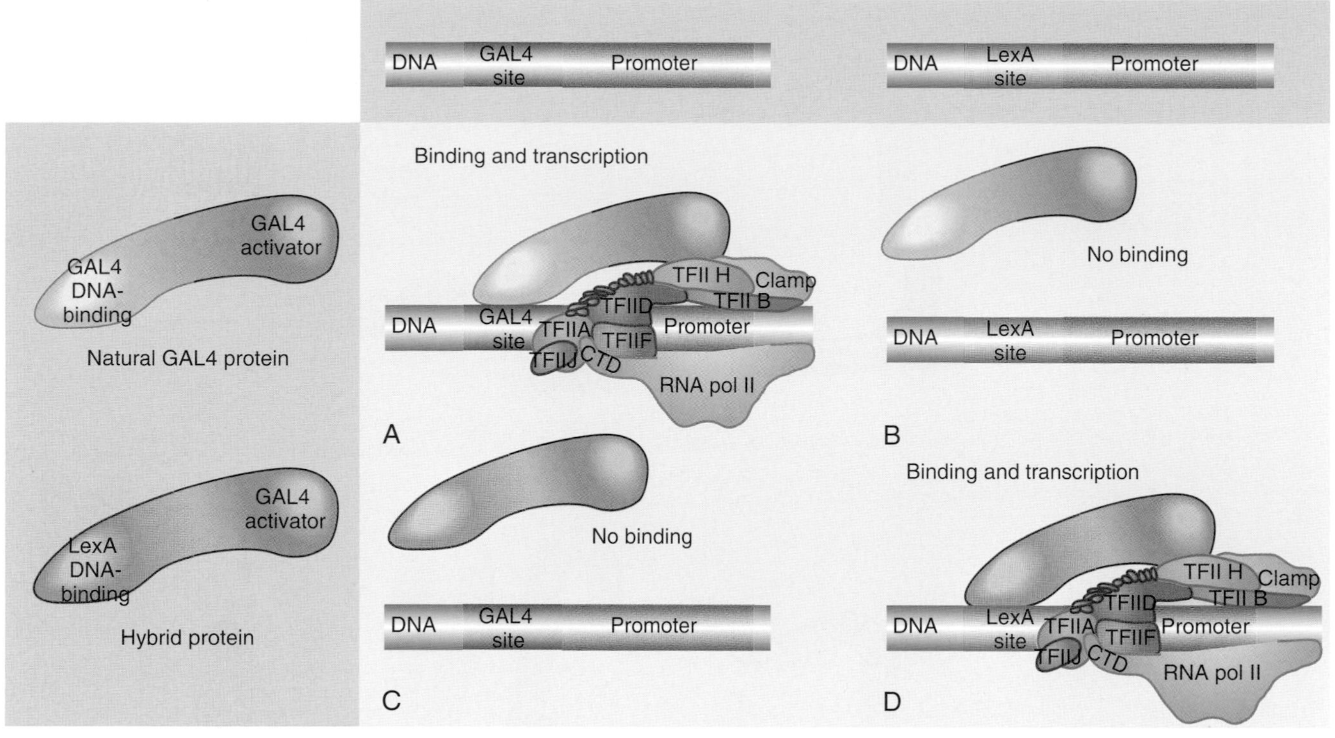

FIGURE 2.10 Transcription Factors Have Two Independent Domains
(A) One domain of the GAL4 transcription factor normally binds to the GAL4 DNA recognition sequence, and the other binds the transcription apparatus. (B) If the LexA site on the DNA is substituted for GAL4 site, the transcription factor does not recognize or bind the DNA. (C) An artificial protein made by combining a LexA binding domain with a GAL4 activator domain will not recognize the GAL4 site on the DNA, but (D) will bind to the LexA recognition sequence and activate transcription. Thus, the GAL4 activator domain acts independently of any particular recognition sequence.

gene expression are staggering. Other eukaryotes such as mice, rats, *Arabidopsis*, *C. elegans*, and even the relatively simple yeast have similarly complex control systems.

There are many different transcription factors for eukaryotic genes, yet they all have at least two domains: one binds to DNA, and the other binds to some part of the transcription apparatus. The two domains are connected yet may function when separated from each other (Fig. 2.10). If the DNA binding domain of one transcriptional regulator is connected to the activation domain of another, the hybrid protein will work, each part retaining its original characteristics. That is, the DNA binding domain will bind the same sequence as it did before, and the activation domain will activate transcription as before. This property can be exploited when trying to identify protein-to-protein interactions with newly characterized proteins in the yeast two-hybrid screen (see Chapter 9).

Transcription factors work via an assembly of many proteins called the **mediator complex** (Fig. 2.11A). These proteins receive all the signals from each of the activator proteins, compile the message, and transmit this to RNA polymerase II. The mediator contains 26 different subunits, most of which make up the core. The presence of other proteins may vary depending on the cell or organism. These accessory proteins were originally thought to be co-activators or co-repressors since their presence varies based on tissue.

The mediator complex sits directly on RNA polymerase II waiting for information from activators or repressors. These may bind to regions just upstream of RNA polymerase II and the mediator complex. However, eukaryotic transcription factors may also bind to DNA sequences known as **enhancers** that may be thousands of base pairs away from the promoter. Even so, the regulatory proteins bind directly to the mediator complex. The enhancer elements work in either orientation but affect only genes that are in the general vicinity. The prevailing theory is that the DNA loops around so that the enhancer is brought near the promoter.

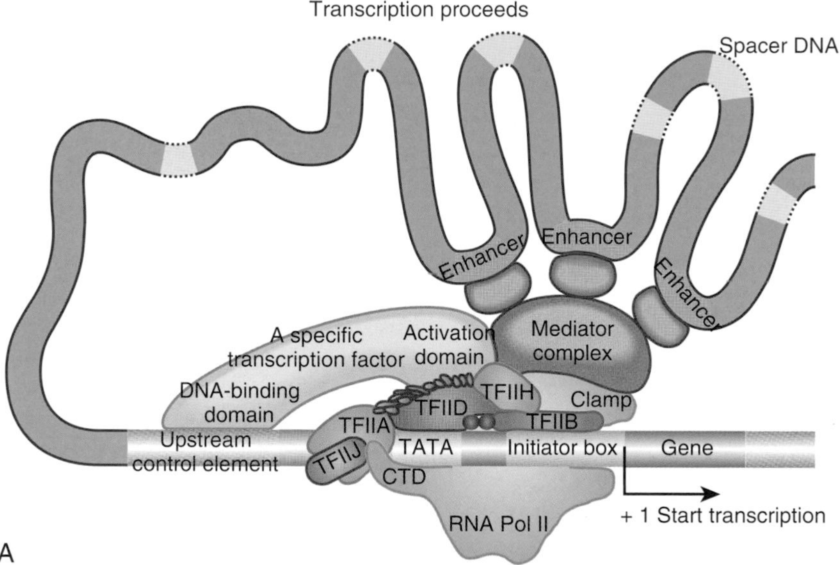

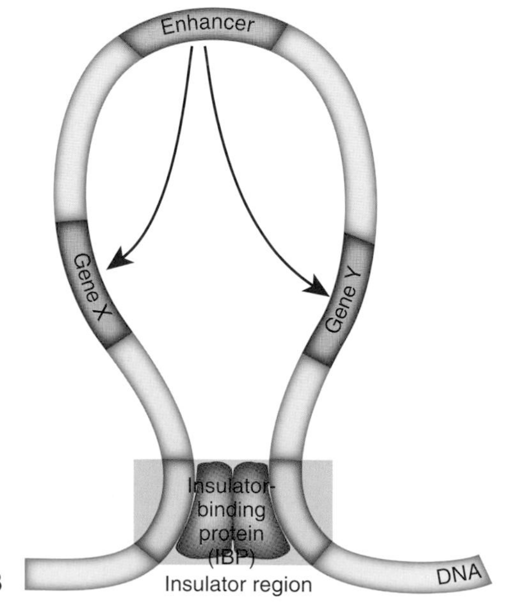

FIGURE 2.11 Enhancer and Insulator Sequences
(A) Enhancer elements are found many hundreds of base pairs from the gene they control. They bind specific proteins that interact with the mediator complex by looping the DNA around. (B) Insulator binding protein (IBP) connects DNA at the insulator binding sequence to form large loops of DNA. This arrangement keeps the correct enhancers associated with the correct genes.

Insulators are DNA sequences that prevent enhancers from activating the wrong genes. Insulators are placed between enhancers and those genes they must not regulate. The **insulator binding protein (IBP)** recognizes the insulator sequences and blocks the action of enhancers that are not within the looped region (see Fig. 2.11B). Insulator sequences may be controlled by methylation. When the DNA sequence is methylated, IBP cannot bind and the enhancer is allowed to access promoters beyond the insulator.

The eukaryotic transcription factor has two domains. The DNA binding domain binds to the DNA at the promoter, and the activator domain has sites for initiating RNA polymerase action.

Eukaryotic transcription factors control gene expression by binding to the mediator complex. Insulator sequences prevent transcription factors from binding to the wrong promoter. Enhancer sequences are far from the promoter but may loop around to directly bind to the mediator complex.

Eukaryotic Transcription Enhancer Proteins

AP-1 (activator protein-1) activates a wide variety of genes and provides an example of the complexity involved in eukaryotic gene expression. This protein affects a variety of genes and responds to a wide range of different stimuli. The most potent stimulators of AP-1 include growth factors and UV irradiation, which are two disparate processes with different end points. The former stimulates cell growth, whereas the latter induces cell death; yet they both work through the same transcription factor, AP-1. The complex effects of this single transcription factor are still being investigated.

AP-1 is actually a dimer of two proteins from the Fos and Jun family of transcription factors. In addition, members of the ATF/CREB family can replace one of the Fos or Jun proteins in the dimer. The dimer recognizes a palindromic sequence 5′-TGA(C/G)TCA-3′ (Fig. 2.12). AP-1 belongs to a family of DNA binding proteins called **bZIP proteins**. The proteins each have a dimerization domain and a DNA binding domain. Jun family members dimerize with themselves or form heterodimers with Fos family members. In contrast, Fos members can bind to Jun but cannot dimerize by themselves. Fos and Jun also have activation domains that receive cellular signals that increase or attenuate their activity.

When AP-1 is stimulated, two different effects are involved. First, the cell makes more Fos and Jun proteins through increased expression of their genes. In addition, the proteins themselves become more stable and are not degraded as quickly. Second, the activity of Fos and Jun are stimulated by phosphorylation of their activation domain by **JNK (Jun amino-terminal kinase)**. Many other cellular signaling proteins can alter Jun and Fos activity, but JNK is the most potent. Phosphorylation of Jun and Fos triggers their interaction with the protein mediator complex and RNA polymerase II. It also affects other signal proteins and triggers other genes.

The eukaryotic transcription factor, AP-1, consists of two different proteins that work as a dimer. These transcription factors control gene expression, but what type of gene activated by the protein depends on the constituents of the dimer, the type and amount of post-translational modifications, and the interaction with other modifier proteins.

Epigenetics and DNA Regulation

Epigenetics is any heritable change in DNA other than changes in nucleotide sequence (Fig. 2.13). The different types of epigenetic changes that affect gene expression include

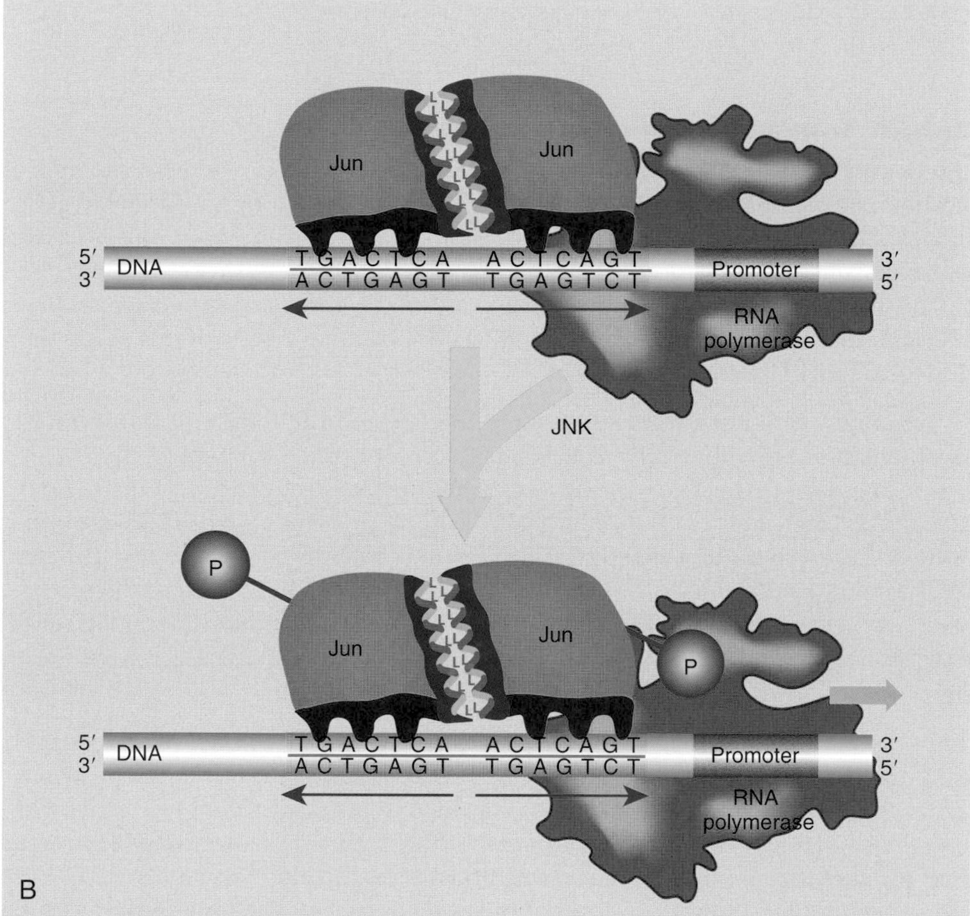

FIGURE 2.12 Eukaryotic Regulation of Transcription
(A) The eukaryotic transcription factor, AP-1, is a dimer of two proteins from the Jun family, Fos family, or ATF/CREB family. These two proteins interact through their leucine zippers. (B) To activate transcription, AP-1 must itself first be activated by phosphorylation by the kinase, JNK. Only then does Jun stimulate RNA polymerase II to transcribe the appropriate genes.

histone post-translational modifications, DNA methylation, nucleosome remodeling, and RNA-associated silencing. These four mechanisms work in conjunction with a variety of transcription factors, enhancers, repressors, and other proteins that modify gene expression to take the DNA that has so little variability and to make the entire biological diversity seen on our planet.

A POST-TRANSLATIONAL MODIFICATIONS OF HISTONES

AGGREGATED

DISAGGREGATED

H3 H3 H3 H3 H2A H2A H4 H4 H2B H4 H4 H2B

Acetyl group

H3 H3 H2A H4 H4 H2B

H3 H3 H2A H4 H4 H2B

Acetyl group

B DNA METHYLATION

C NUCLEOSOME REMODELING

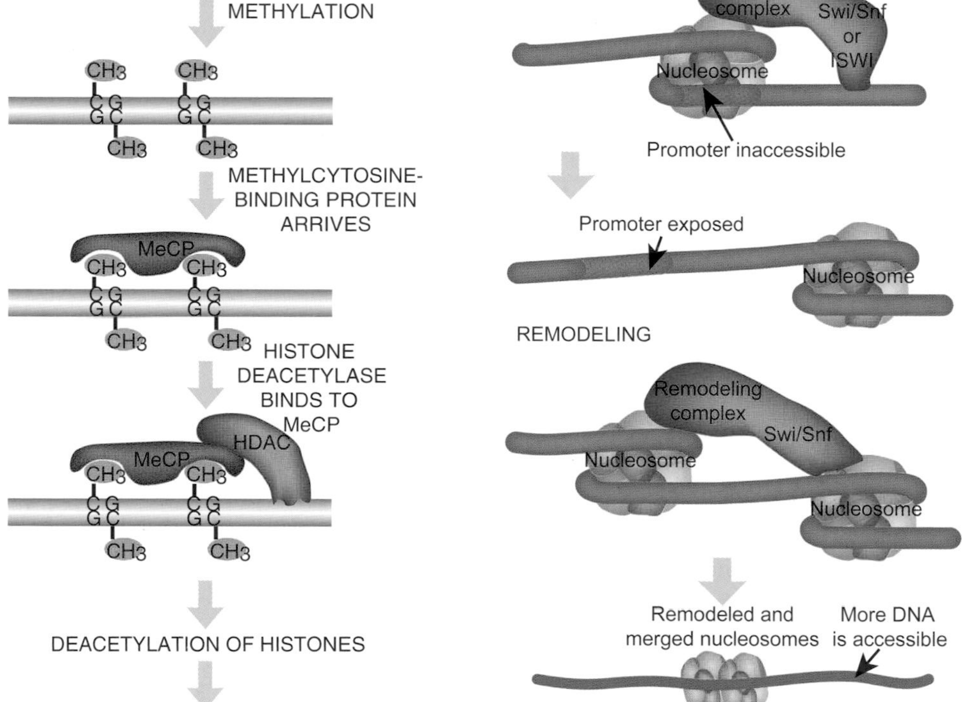

D X-INACTIVATION BY NONCODING RNA

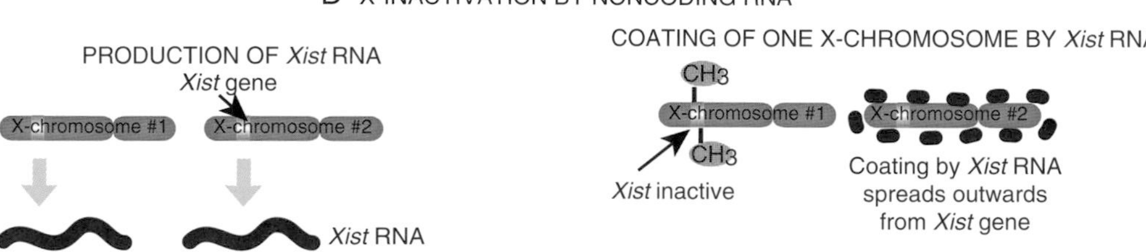

INACTIVATION OF ONE X-CHROMOSOME BY METHYLATION

Xist stays active

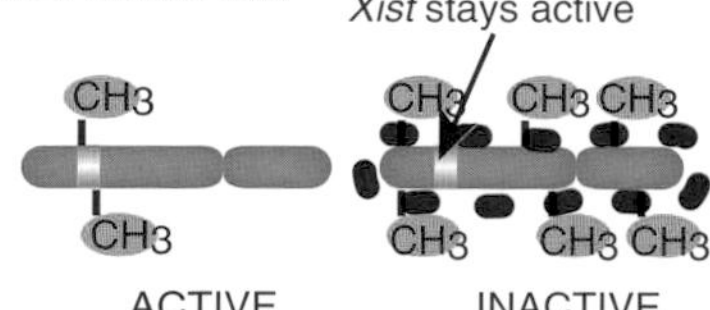

FIGURE 2.13
Epigenetic Changes Control Gene Expression
(A) Histone post-translation modifications such as acetylation can loosen nucleosomes, making them more accessible to various regulatory proteins. (B) DNA methylation can induce heterochromatin formation. First, the area to be silenced is methylated. The methyl groups attract methylcytosine binding protein (MeCP), which in turn attracts histone deacetylases. Once HDAC removes the acetyl groups from the histone tails, the histones aggregate tightly. The closeness of histones excludes any DNA binding proteins and hence turns off gene expression in the area. (C) Nucleosome remodeling moves nucleosomes by sliding them from promoter areas or can remodel histones to be further or closer together. (D) A noncoding RNA called Xist, which is produced from both copies of the X chromosome, covers the surface of one of the X chromosomes, which in turn induces methylation and heterochromatin formation of the X chromosome. The inactivation of one X chromosome ensures equal amount of protein expression with males that have only one X chromosome.

Most epigenetic changes in DNA-associated proteins such as histones modulate the access that various transcription factors and regulatory proteins have to genes. Eukaryotic DNA wraps around histones to form nucleosomes (the "beads on a string" structure of chromatin; see Chapter 1). Loosely packed nucleosomes provide access for transcription factors, whereas tightly packed nucleosomes exclude regulatory proteins. Therefore, controlling the density of nucleosomes regulates transcription initiation.

The histone proteins have protein extensions or "tails" that can be modulated by enzymes called **histone acetyl transferases (HATs)**, which transfer acetyl groups to lysines within the tail. These histone tails normally stabilize DNA by binding neighboring nucleosomes, so aggregating the nucleosomes. When HAT transfers acetyl groups to the tails, they can no longer bind to neighboring nucleosomes, and the structure loosens. To tighten the nucleosomes, **histone deacetylases (HDACs)** remove the acetyl groups, and the histones reaggregate (Fig. 2.13A). Although this is generally true, there are some histone post-translational modifications that decrease the accessibility of the DNA; therefore, post-translational modifications are considered context specific for gene expression.

Another epigenetic modulation of eukaryotic DNA important to gene expression is methylation (see Fig. 2.13B). In prokaryotes, differences in methylation patterns distinguish the newly synthesized strand of DNA from the template during replication. In eukaryotes, methylation is used to **silence** various regions of DNA and prevent their expression. Methylation of cytosine by two different enzymes occurs in the sequence CpG or CpNpG, where p represents the phosphate group that links the cytosine and guanosine nucleotides on the strand of DNA. **Maintenance methylases** add methyl groups to newly synthesized DNA to give the same pattern as in the template strand. ***De novo* methylases** add new methyl groups, and of course, **demethylases** remove unwanted methyl groups. Many genes are located near stretches of DNA containing many CpG sequences, called **CpG islands**. If these are methylated, the nearby genes are not expressed, whereas if they are not methylated, the genes are expressed. Methylation patterns depend on the tissue. For example, muscle cells will not methylate CpG islands in front of genes necessary for muscle function. The muscle-specific genes will have methylated CpG islands in other tissues, though.

Silencing can occur for one gene or for areas as large as an entire chromosome. In order for large regions to become "silenced," CpG or CpNpG sequences are methylated, which then attracts **methylcytosine binding proteins** that block binding of other DNA binding proteins. These proteins also recruit HDACs that deacetylate the histone tails, thus condensing the chromatin. The areas turn into heterochromatin, which prevents any further gene expression.

The third epigenetic modification also affects nucleosome spacing. Chromatin remodeling complexes slide histones, remove histones, remodel histones, alter nucleosome spacing, and affect nucleosome assembly. These complexes move histones to expose promoter regions or can completely remodel histones so that more DNA is accessible (Fig. 2.13C). They affect the overall spacing of nucleosomes around different genes. These complexes also are able to replace histone proteins with variant histones. Histone variants replace normal histones in different regions of the genome and mark regions of active transcription. For example, the SWR1 complex in yeast (a chromatin remodeling complex) replaces histone 2A with histone 2A.Z (H2A.Z), which protects regions from turning into heterochromatin, and therefore marks areas of gene activation.

Methylation patterns have a major impact during development because gene expression is tightly controlled for proper development. Some genes remain methylated in the gamete, whereas other genes must be demethylated so they can become active. **Imprinting** occurs when a gene from the parent is methylated in the gamete and remains methylated in the new organism. This affects relatively few genes. In contrast, genes that are not imprinted change their methylation patterns during development. Imprinting can be different for male gametes and female gametes, setting the stage for sexual differences in gene expression. One special type of imprinting is **X-inactivation,** where the entire second X chromosome in females is silenced by methylation and heterochromatin formation (except for a few loci). This inactivation is triggered by the final epigenetic modification, noncoding RNA regulation (Fig. 2.13D). The Xist gene is expressed in both X chromosomes initially, but one copy is methylated and silenced. The chromosome containing the silenced Xist gene remains active. The other chromosome continues to make Xist mRNA, which then coats the surface of the chromosome and triggers methylation and heterochromatin formation. Therefore, the Xist mRNA is one of the few loci of this X chromosome that is expressed. This ensures that females with two X chromosomes have the same level of gene expression as males who have an X/Y chromosome pair. In addition, other RNAs that affect gene expression include antisense RNA and small interfering RNA (siRNAs).

Epigenetic regulation of DNA includes DNA methylation, histone modification by acetylation, nucleosome remodeling, and silencing of DNA by various types of RNA including antisense RNA, noncoding RNA, and small interfering RNA. The density of nucleosome packing can exclude transcription factors from binding to the enhancer regions.

Adding methyl ($-CH_3$) groups to the cytosine of CpG areas controls expression of nearby genes. These groups can prevent binding of various transcription factors and have implications in setting up male/female differences during development.

EUKARYOTIC mRNA IS PROCESSED BEFORE MAKING PROTEIN

Bacterial mRNA is translated without any processing. Indeed, bacteria often start translating their mRNA while it is still being transcribed (known as *coupled transcription/translation).* However, eukaryotic mRNA is processed in a variety of ways before it leaves the nucleus. First, eukaryotic mRNA must have a **cap** added to the 5′ end of the message (Fig. 2.14). The cap is a GTP that is added in reverse orientation and is methylated on position 7 of the guanine base. Methyl groups may also be added to the first one or two nucleotides of the mRNA.

The second modification of eukaryotic mRNA is adding a long stretch of adenines to the 3′ end—the **poly(A) tail**. Three sequences at the end of an mRNA mediate the addition of the tail: the recognition sequence for the **polyadenylation complex** (AAUAAA); the cut site for cleavage binding factor; and the recognition sequence for polyadenylation binding protein (a length of GU repeats). First, the polyadenylation complex binds to the AAUAAA, and an endonuclease in the complex cuts the mRNA after a CA dinucleotide downstream from the AAUAAA recognition sequence. Next, poly(A) polymerase adds 100 to 200 adenine nucleotides. Finally, the poly(A) binding protein binds to both the poly(A) tail and the cap structure. This circularizes the mRNA.

A third modification made to eukaryotic mRNA is the removal of **introns**. Eukaryotic DNA contains many stretches of intervening sequence (introns) between regions that will ultimately code for a protein (**exons**). First, the entire region is transcribed into an RNA molecule called the **primary transcript**. After capping and tailing, this is processed to remove the introns. The exons are spliced together to form the mRNA. Proteins called **splicing factors** recognize the exon/intron borders, cut the DNA, and join the neighboring exons. These three modifications probably occur with the capping first since this occurs when transcription starts, but the poly(A) tail addition and splicing occur simultaneously. The poly(A) tail addition releases the mRNA from RNA polymerase and demarks the end of transcription. Whether or not splicing has been completed by this endpoint is unknown as of now.

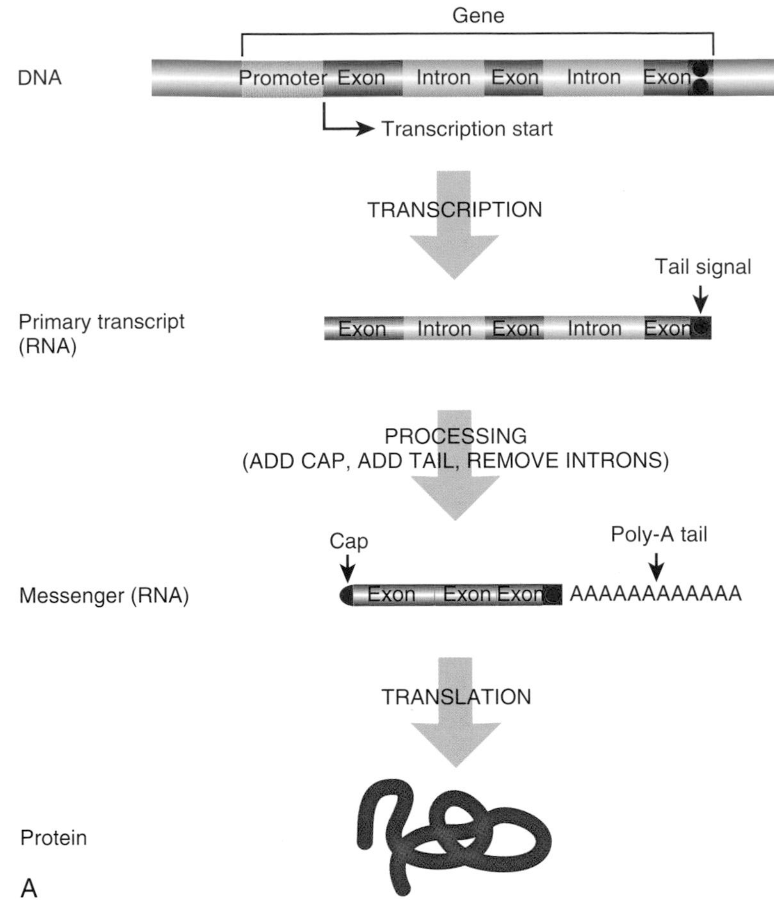

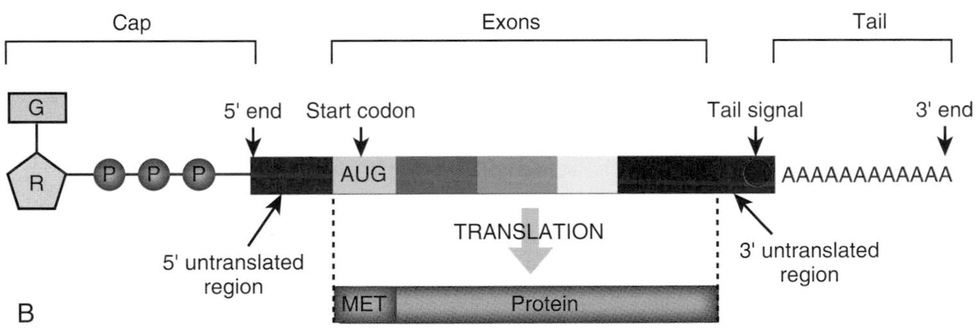

FIGURE 2.14 Processing Eukaryotic mRNA
(A) Eukaryotic RNA is processed before exiting the nucleus for translation into protein. A cap (reversed GTP with a methyl group on position 7 of the guanine base) is added to the 5′ end of the message, a poly(A) tail is added to the 3′ end, and the introns are spliced out. These modifications stabilize the message for protein translation in cytoplasm. (B) Detailed structure of a processed eukaryotic mRNA. The cap structure is followed by the 5′ UTR, the protein coding exons, 3′ UTR, and a poly(A) tail.

Eukaryotic RNA is transcribed as a primary transcript, a cap is added at the 5′ end, a poly(A) tail is added, and the introns are removed. The mRNA is then transported from the nucleus to the endoplasmic reticulum for translation by ribosomes.

TRANSLATING THE GENETIC CODE INTO PROTEINS

The Genetic Code Is Read as Triplets or Codons

Messenger RNA provides the information a ribosome needs to make proteins. This process is known as translation because it involves translating information carried by nucleic acids to give the sequences of amino acids that make up proteins. Before the mechanism is discussed, the code used to assemble proteins must first be understood. Nucleic acids each have four different bases (the T in DNA is equivalent to U in RNA). However, proteins consist of 20 different amino acids. If each nucleotide corresponded to an amino acid, this would encode only four different amino acids. Two nucleotides give only 16 combinations, still not enough. Only groups of three nucleotides provide enough combinations to create all 20 amino acids (Fig. 2.15). Messenger RNA reads groups of three bases, known as **triplets** or **codons**. Each triplet of bases codes for one amino acid. Because there are more than 20 triplets, many are redundant, so multiple codons will be translated into the same amino acid. For example, valine is encoded by GUU, GUC, GUA, or GUG.

	2nd (middle) base				
1st base	U	C	A	G	3rd base
U	UUU Phe UUC Phe UUA Leu UUG Leu	UCU Ser UCC Ser UCA Ser UCG Ser	UAU Tyr UAC Tyr UAA stop UAG stop	UGU Cys UGC Cys UGA stop UGG Trp	U C A G
C	CUU Leu CUC Leu CUA Leu CUG Leu	CCU Pro CCC Pro CCA Pro CCG Pro	CAU His CAC His CAA Gln CAG Gln	CGU Arg CGC Arg CGA Arg CGG Arg	U C A G
A	AUU Ile AUC Ile AUA Ile AUG Met	ACU Thr ACC Thr ACA Thr ACG Thr	AAU Asn AAC Asn AAA Lys AAG Lys	AGU Ser AGC Ser AGA Arg AGG Arg	U C A G
G	GUU Val GUC Val GUA Val GUG Val	GCU Ala GCC Ala GCA Ala GCG Ala	GAU Asp GAC Asp GAA Glu GAG Glu	GGU Gly GGC Gly GGA Gly GGG Gly	U C A G

FIGURE 2.15 The Genetic Code
The 64 codons found in mRNA are shown with their corresponding amino acids. As usual, bases are read from 5′ to 3′ so that the first base is at the 5′ end of the codon. Three codons (UAA, UAG, UGA) have no cognate amino acid but signal stop. AUG (encoding methionine) and, much less often, GUG (encoding valine) act as start codons. To locate a codon, find the first base in the vertical column on the left, the second base in the horizontal row at the top, and the third base in the vertical column on the right.

The genetic code listed in Fig. 2.15 is considered the **universal genetic code**. Not all organisms use precisely this code, although exceptions are rare. For example, UGA normally signals stop, but in *Mycoplasma,* UGA encodes tryptophan and, in the protozoan *Euplotes,* UGA encodes cysteine.

Small RNA molecules known as **transfer RNA (tRNA)** recognize the individual codons on mRNA and carry the corresponding amino acids. Although tRNA is synthesized as a single strand, it folds back on itself to form regions of double-stranded RNA. The final shape of tRNA is a folded "L" shape with the **anticodon** at one end and the **acceptor stem** at the other. The anticodon consists of three bases complementary to those of the corresponding codon, and it therefore recognizes the codon by base pairing. The acceptor stem is the place where the amino acid is added to the free 3′ end of the tRNA (Fig. 2.16).

How does each specific tRNA carry the correct amino acid? A group of enzymes called **aminoacyl tRNA synthetases** attaches the correct amino acid to the corresponding tRNA. These enzymes are very specific and recognize the correct tRNA by its sequence at the anticodon or elsewhere along the RNA structure. There is a specific aminoacyl tRNA synthetase for each amino acid. The enzymes catalyze the addition of the correct amino acid onto the end of the correct tRNA. In fact, some aminoacyl tRNA synthetases also have domains that edit their work, ensuring that the correct amino acid connects to the correct tRNA.

The first base of the anticodon binds the third base of the codon in the mRNA. Because this nucleotide in tRNA is not constrained by neighboring nucleotides, it can **wobble** instead of forming a perfect double helix. This allows nonstandard base pairs to be created. For example, if the first anticodon base were G, it would normally pair with C in the third position of the codon. Because of wobble, G can also pair with U. Thus, the tRNA for

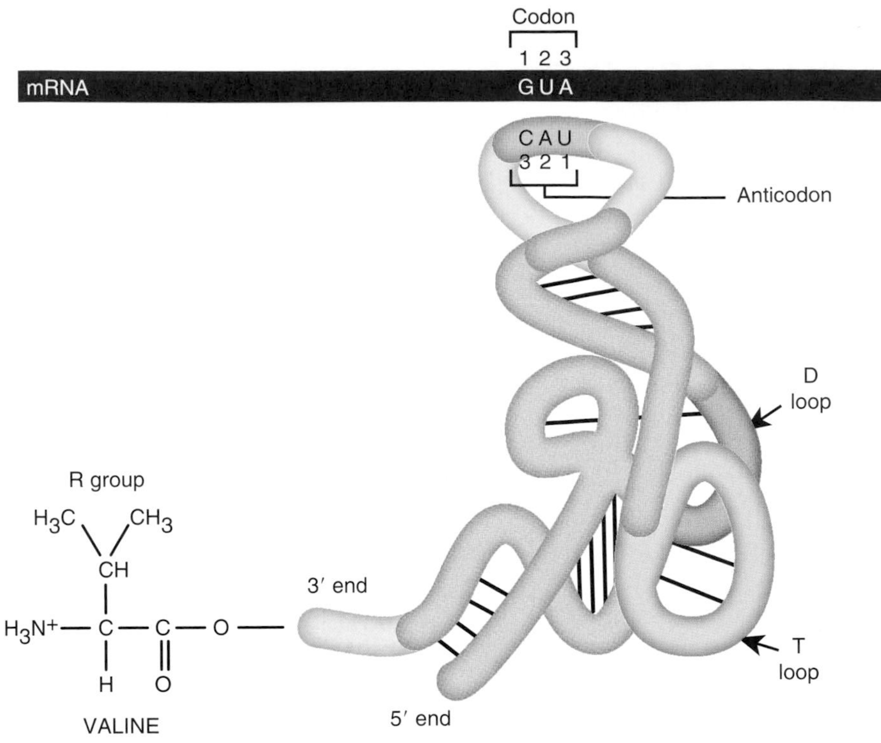

FIGURE 2.16 Structure of tRNA Allows Wobble in the Third Position
Transfer RNA recognizes the codons along mRNA and presents the correct amino acid for each codon. The first position of the anticodon on tRNA matches the third position of the codon.

histidine has the anticodon GUG and recognizes both CAC and CAU in the mRNA. Similarly, U in the first place in the anticodon can base pair with A or G in the third position of the codon. Wobble explains how the same tRNA can read multiple codons all encoding the same amino acid. Each organism has a preference as to which triplet codon is used most often for a particular amino acid. This is called codon bias (see Box 2.1).

During protein translation, each tRNA recognizes a specific three-nucleotide sequence and has the correct amino acid attached to the opposite end. A family of specific enzymes, aminoacyl tRNA synthetases, ensure that each tRNA has the correct amino acid.

Box 2.1 Codon Bias

Several amino acids are encoded by multiple codons and have more than one corresponding tRNA. Thus, valine is encoded by GUU, GUC, GUA, and GUG. One tRNA for valine recognizes GUU and GUC by wobble, but another tRNA is necessary for the other two codons. However, many organisms tend to use only one or two of the codons for amino acids with multiple codons—a phenomenon known as **codon bias**. Consequently, they make low amounts of tRNA for the rarely used codons. Furthermore, different organisms show different codon preferences. This becomes an issue when genes from one organism are expressed in another. Plants and animals often prefer different codons than bacteria for the same amino acids. When bacteria express plant or animal proteins, not enough tRNA is available for the nonpreferred codons, and the ribosomes stall and fall off, making protein yield very low. To remedy this problem, researchers may genetically engineer the genes so that abundant tRNAs recognize their codons (see Chapter 14). Alternatively, bacterial host strains may be engineered to express higher levels of the necessary tRNAs.

Protein Synthesis Occurs at the Ribosome

The molecular machine called a **ribosome** unites mRNA with the appropriate tRNAs and then catalyzes the linkage of amino acids together into a chain. Prokaryotic ribosomes consist of two subunits called the 30S and 50S, which combine to form a functional 70S ribosome. A ribosome consists of several RNA molecules **(ribosomal RNA** or **rRNA)** and many proteins. The 30S subunit has a 16S rRNA plus 21 proteins; the 50S subunit has two rRNAs, the 5S and 23S, plus 34 proteins. The larger subunit has three binding sites for tRNA, called A for acceptor, P for peptide, and E for exit, referring to the action occurring at each site. The 23S rRNA actually catalyzes the addition of amino acids to the growing polypeptide chain and is therefore a ribozyme. (Ribozymes are discussed in Chapter 5.)

In prokaryotes, various factors besides the ribosome are involved in protein synthesis (Fig. 2.17). First, a ribosome must assemble at the start site and begin protein synthesis at the correct start codon. The 5′ untranslated region of the mRNA (see above) has the signal for ribosome binding in front of the start codon. In prokaryotes, translation begins at the first AUG codon after the **Shine–Dalgarno sequence,** or ribosome binding site, which has the consensus sequence UAAGGAGG. The **anti-Shine–Dalgarno sequence** is found in the 16S rRNA of the smaller 30S subunit. So first, the small ribosomal subunit binds the Shine–Dalgarno sequence. A derivative of methionine, **N-formyl-methionine (fMet)**, and a special initiator tRNA **(tRNA$_i$)** are used to initiate translation in prokaryotes. Only initiator tRNA charged with fMet (referred to as **tRNA$_i^{fMet}$)** can bind the small subunit of the ribosome.

Translation factors are proteins needed to recruit and assemble the components of the ribosome and translational complex. **Initiation factors** (IF1, IF2, and IF3) assemble the **30S initiation complex**, which is the 30S ribosomal subunit plus tRNA$_i^{fMet}$. The IF3 factor then leaves the complex, and the 50S ribosomal subunit binds, forming the **70S initiation complex** (see Fig. 2.17A).

Finally, polypeptide assembly can begin (see Fig. 2.17B). The tRNA$_i^{fmet}$ occupies the P-site on the ribosome. Another tRNA recognizes the next codon and enters the A-site; the **peptidyl transferase** activity of 23S rRNA then catalyzes the peptide bond between the first and second amino acids. fMet releases its tRNA, which moves into the E-site. This allows the second tRNA to move into the P-site, and the cycle begins again. A third tRNA, complementary to the next codon, enters the A-site; a peptide bond forms between amino acids 2 and 3; and then the second tRNA moves into the E-site of the ribosome and exits.

Adding successive amino acids is called **elongation** and requires **elongation factors**. EF-T, which is a pair of proteins (EF-Tu and EF-Ts), uses a phosphate group from GTP to catalyze the addition of a new tRNA into the A-site (EF-Tu), thus converting GTP to GDP. After the reaction, the GDP is exchanged for a fresh GTP for the next cycle (EF-Ts). The movement of tRNA from the P-site to the E-site is called **translocation**, and the mRNA simultaneously moves one codon sideways relative to the ribosome. The E-site and A-site cannot be occupied at the same time, and the used tRNA must exit before the next tRNA enters. EF-G oversees the translocation step.

Amino acids are added to the growing chain and the process continues until the ribosome encounters a stop codon (UAA, UAG, or UAA). None of the tRNAs in a cell recognize the stop codon. Instead, proteins known as **release factors** bind the stop codons (see Fig. 2.17C). RF1 and RF2 recognize the different stop codons and stimulate the 23S rRNA to split the bond between the last amino acid and its tRNA. The whole ribosome assembly falls off the mRNA and dissociates. Its components are recycled for translation of another mRNA. The new polypeptide chain folds to form its final structure. In prokaryotes, multiple ribosomes

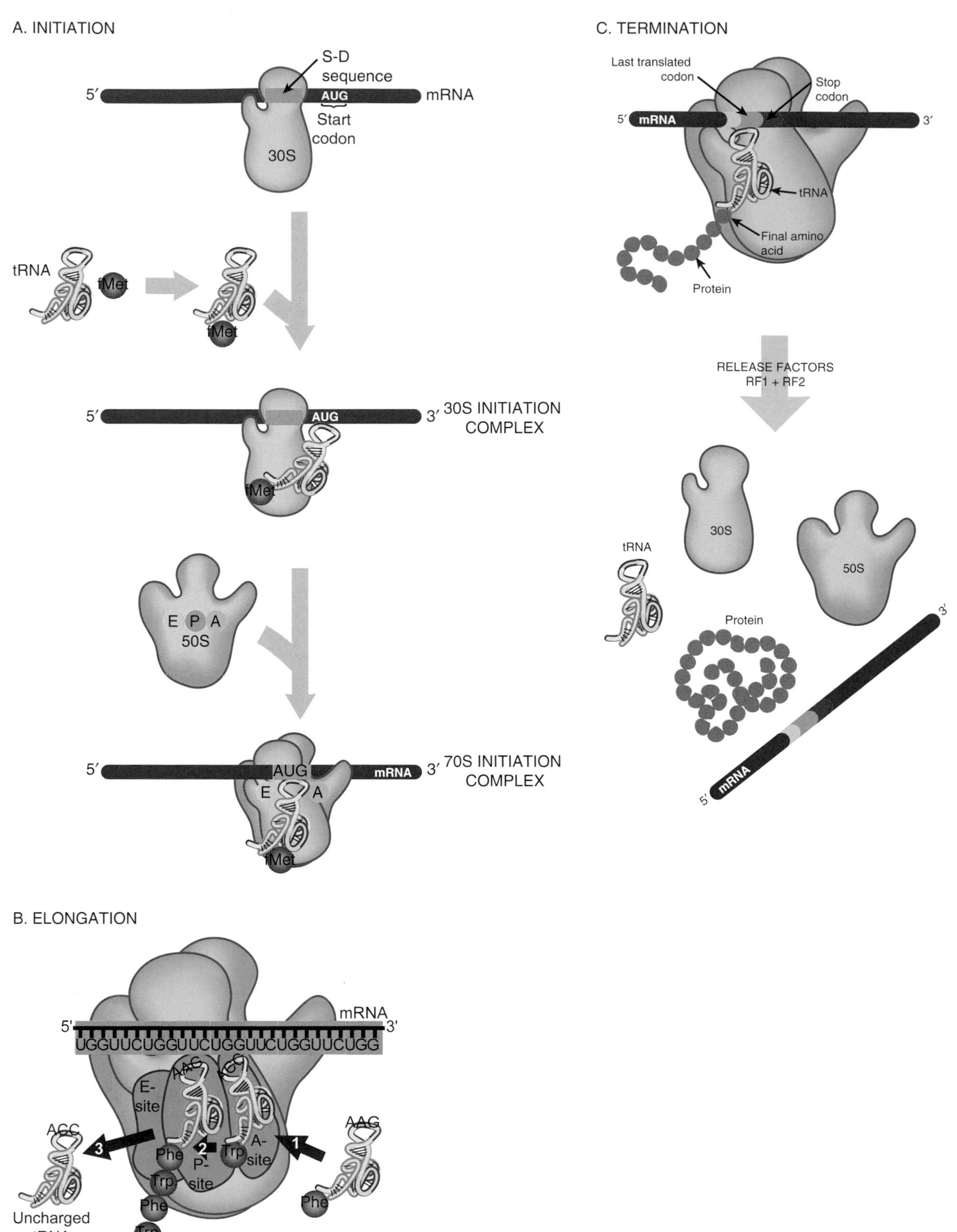

FIGURE 2.17 Translation in Prokaryotes

(A) Initiation. Initiation of translation begins with the association of the small ribosome subunit with the Shine–Dalgarno sequence (S-D sequence) on the mRNA. Next, initiation factors IF1, IF2, and IF3 (not shown) charge or connect the initiator tRNA with fMet. The charged initiator tRNA ($tRNA_i^{fMet}$) associates with the small ribosome subunit and finds the start codon. Finally, the large ribosomal subunit joins the small subunit and situates the initiator tRNA at the P site. (B) Elongation. During elongation, peptide bonds are formed between the amino acids at the A-site and the P-site. The movement of the ribosome along the mRNA and addition of a new tRNA to the A-site are controlled by elongation factors (also not shown). (C) Termination. Termination requires release factors. The various components dissociate. The completed protein folds into its proper three-dimensional shape.

bind to the same mRNA to form a **polysome**. Because there is no nucleus, transcription and translation are often simultaneous. As partially made mRNA comes off the DNA, ribosomes bind and start synthesizing protein.

Translation in prokaryotes starts in the 5′ UTR of the mRNA message, where the ribosome scans for the first start codon. After an initiator methionine is added to the AUG, the ribosome catalyzes the addition of more amino acids. Ribosomes work with elongation factors and release factors to control the movement down the mRNA until the stop codon.

Ribosomes have three different sites of action. The A-site accepts the next tRNA with the correct anticodon and amino acid. The P-site holds the previous tRNA with amino acid. The E-site is occupied briefly after the amino acids are linked as the empty tRNA exits the ribosome.

DIFFERENCES BETWEEN PROKARYOTIC AND EUKARYOTIC TRANSLATION

Translation in eukaryotes differs from prokaryotes in many ways (Fig. 2.18). First of all, mRNA is made in the nucleus, but translation occurs on the ribosomes in the cytoplasm. Therefore, there is no coupled transcription and translation in eukaryotes. Eukaryotic ribosomes have 60S and 40S subunits that combine to form an 80S ribosome, which is a little larger than bacterial ribosomes. Additionally, eukaryotes have more initiation factors than prokaryotes, and they assemble the initiation complex in a different order. Overall, more proteins are involved in eukaryotic translation to deal with the greater complexity of regulation (see Table 2.2).

Despite this, the binding of the mRNA is simpler in eukaryotes. Eukaryotic mRNA does not have a Shine–Dalgarno sequence. Instead, eukaryotic ribosomes recognize the 5′ cap structure, and the Kozak sequence, which is a loosely conserved sequence found around the first AUG. Only one gene per mRNA is found (unlike bacteria, which often have polycistronic messages and whose ribosomes recognize separate Shine–Dalgarno sequences for each coding sequence). The first amino acid in each new polypeptide is methionine, as in bacteria. However, unlike in bacteria, this methionine is not modified with a formyl group. Finally, many eukaryotic proteins are modified after translation by addition of chemical groups. (Although bacteria do modify some proteins, this is much rarer, and the variety of additions is much more limited.)

Eukaryotic mRNA has information for one protein. The ribosome recognizes the cap structure, scans until it finds the first AUG, and starts translating the message into protein. Many different initiation factors, elongation factors, and termination factors are important for eukaryotic translation.

MITOCHONDRIA AND CHLOROPLASTS SYNTHESIZE THEIR OWN PROTEINS

The mitochondria and chloroplasts found in eukaryotes have their own genome and make some of their own proteins. The **symbiotic theory** of organelle origin argues that these organelles were once free-living bacteria or blue-green algae (cyanobacteria) that formed a symbiotic relationship with a single-celled ancestral eukaryote. The bacteria supplied energy to the early eukaryote. Over time, the bacteria gave up many duplicate functions and came to rely on the host for precursor molecules. Eventually, the symbiotic mitochondria and chloroplasts lost the majority of their genes, yet today they still maintain a small version of their genome. These genomes have many genes associated with protein synthesis (Fig. 2.19).

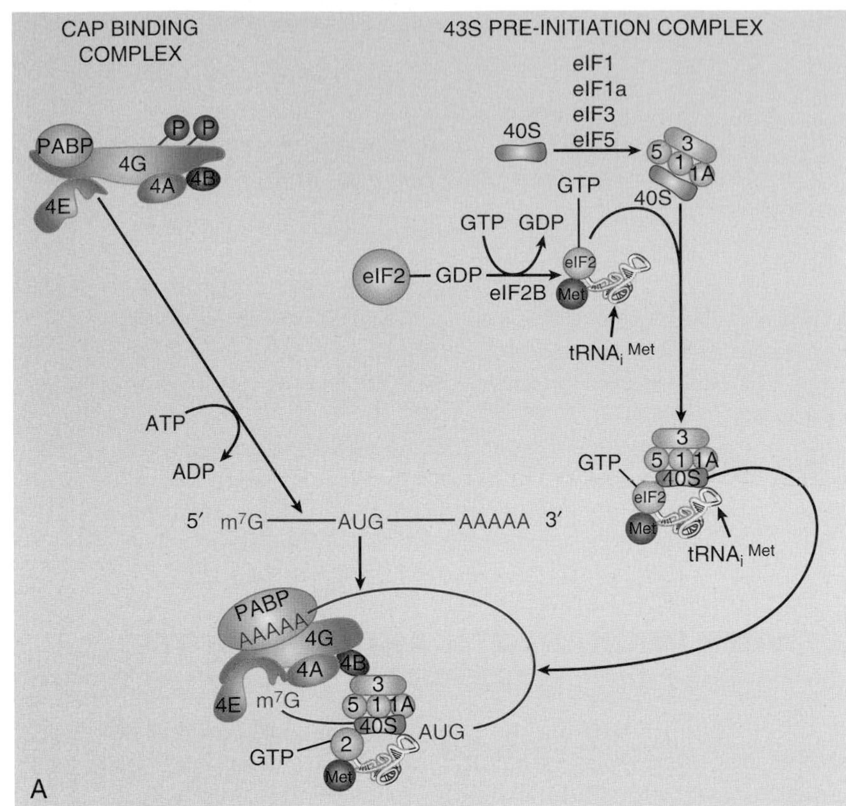

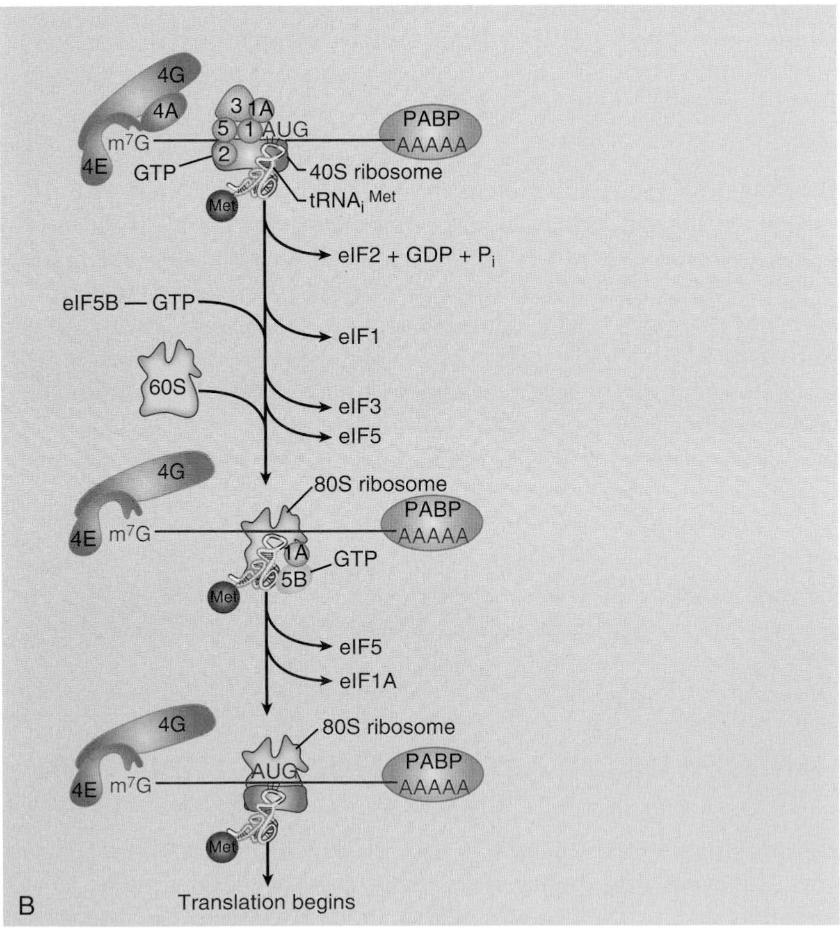

FIGURE 2.18 Translation Initiation in Eukaryotes

(A) The cap-binding complex includes poly(A)-binding protein (PABP), eIF4A, eIF4B, eIF4E, and eIF4G, which is in an unphosphorylated state when unbound to mRNA. ATP transfers a phosphate to the complex to make it competent for binding the mRNA (top left of part A). The 43S initiation complex forms, bringing the small ribosomal subunit together with the tRNA$_i^{met}$. This complex uses GTP to attach the tRNA to the 40S subunit via eIF2. In addition, initiation factors eIF1, eIF1A, eIF3, eIF5, and eIF2B guide and make the complex competent to bind to the 5′ UTR of mRNA (top right of panel A). Finally, the activated cap-binding complex recognizes the cap and poly(A) tail of the mRNA, causing the mRNA to loop around into a circular shape. Then 43S pre-initiation complex can attach and start scanning for the first AUG. (B) After the complex stops at the first AUG, the remaining 60S subunit and associated factors combine to form the final competent 80S ribosome.

Table 2.2 Translation Factors: Prokaryotes versus Eukaryotes

	Prokaryotes	Eukaryotes
Initiation	IF1	eIF1A
	IF2	eIF5B (GTPase)
	IF3	eIF1
		eIF2 (α, β, γ) (GTPase)
		eIF2B (α,β,γ, δ, ε)
		eIF3 (13 subunits)
		eIF4A (RNA helicase)
		eIF4B (activates eIF4A)
		eIF4E (cap binding protein)
		eIF4G (eIF4 complex scaffold)
		eIF4H
		eIF5
		eIF6
		PABP [Poly(A)-binding protein]
Elongation	EF-Tu	eEF1A
	EF-Ts	eEF1B (2–3 subunits)
		SBP2
	EF-G	eEF2
Termination	RF1	eRF1
	RF2	
	RF3	eRF3
Recycling	RRF	
	EF-G	
		eIF3
		eIF3j
		eIF1A
		eIF1

Functionally homologous factors are in the same row.
Adapted from Table 1 of Rodnina MV, Wintermeyer W (2009). Recent mechanistic insights into eukaryotic ribosomes. *Curr Op Cell Biol* **21**, 435–443.

Organelle genes are often more closely related to bacterial genes than to eukaryotic (nuclear) genes. Moreover, the ribosomes in animal mitochondria are 28S and 39S in size, closer to the 30S and 50S subunits of bacteria. The ribosomal RNA of mitochondria and bacteria are also much more similar in sequence than either is to the rRNA encoded by the eukaryotic nucleus.

Mitochondria and chloroplasts have their own genome that includes many genes for transcription and translation. These may have been free-living bacteria that formed a symbiotic relationship with a unicellular eukaryote.

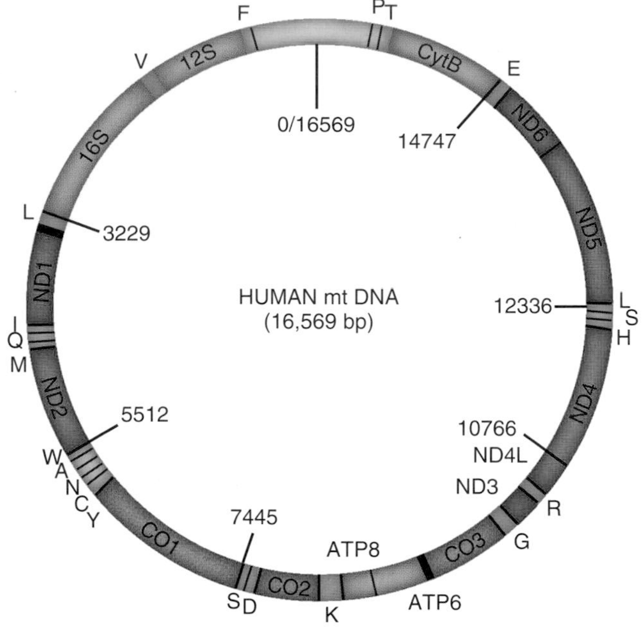

FIGURE 2.19 Human Mitochondrial DNA
The mitochondrial DNA of humans contains the genes for ribosomal RNA (16S and 12S), some transfer RNAs (single-letter amino acid codes mark these on the genome), and some proteins of the electron transport chain.

Summary

This chapter briefly explains the process of transcription and translation, highlighting the differences between eukaryotes and prokaryotes. Transcription occurs when RNA polymerase makes a complementary copy of the gene using ribose, phosphate, uracil, guanine, cytosine, and adenine. The complementary copy is called mRNA, and this form is used to translate into protein. The ribosome holds the mRNA so that two triplet codons starting at AUG are stable. Then a complementary tRNA that is holding the correct amino acid is held close to the mRNA by the ribosome. A second tRNA-amino acid complex moves next to the first, and the ribosome connects the two amino acids using its peptidyl transferase activity. The ribosome translocates to the next triplet codon on the mRNA and continues to link the amino acids to form a polypeptide. These basic mechanisms of transcription and translation are very similar in prokaryotes and eukaryotes.

The regulation of transcription and translation varies significantly between prokaryotes and eukaryotes. First, proteins called transcription factors control the expression of the correct gene at the correct time in the correct amount. In prokaryotes, the lactose operon demonstrates how activator proteins and repressor proteins work together so that lactose utilization genes are only expressed when lactose is the only sugar source for the bacteria. Prokaryotes have different sigma factors, an integral part of RNA polymerase, which specify the correct gene expression. In eukaryotes, many different transcription factors control gene expression by binding to the mediator complex. In addition, eukaryotes use epigenetic changes to control gene expression. These include methylation of DNA; post-translational modification of histones, histone remodeling complexes, and noncoding RNAs; antisense RNAs, and other forms of RNA to control expression of genes. During translation, eukaryotes are actually less complex and express the mRNA transcript as a single message. In prokaryotes, the mRNA may contain multiple coding regions that are translated into proteins simultaneously as the transcript is made from the DNA.

End-of-Chapter Questions

1. Which of the following are important features for transcription?
   - **a.** promoter
   - **b.** RNA polymerase
   - **c.** 5′ and 3′ UTRs
   - **d.** ORF
   - **e.** all of the above
2. Adenine in DNA is complementary to
   - **a.** uracil
   - **b.** adenine
   - **c.** guanine
   - **d.** cytosine
   - **e.** inosine
3. Which of the following is not necessary during Rho-independent termination of transcription?
   - **a.** RNA polymerase
   - **b.** Rho protein
   - **c.** hairpin structure
   - **d.** repeating As in the DNA sequence
   - **e.** All of the above are necessary.
4. Which of the following statements is not true about mRNA?
   - **a.** Prokaryotic mRNA may contain multiple structural genes on the same transcript, known as polycistronic mRNA.
   - **b.** Eukaryotes only transcribe one gene at a time on mRNA, called monocistronic mRNA.
   - **c.** Some eukaryotes are capable of having polycistronic mRNA.
   - **d.** Eukaryotes almost always produce polycistronic mRNA.
   - **e.** The genes for metabolic pathways in bacteria are typically located close together and transcribed on one mRNA.
5. In what way is eukaryotic transcription more complex than prokaryotic transcription?
   - **a.** Eukaryotes have three different RNA polymerases, whereas prokaryotes only have one RNA polymerase.
   - **b.** Eukaryotic transcription initiation is much more complex than prokaryotic initiation because of the various transcription factors involved.
   - **c.** Upstream elements are required for efficient transcription in eukaryotic cells, but these elements are not usually necessary in prokaryotes.
   - **d.** Eukaryotic mRNA is made in the nucleus.
   - **e.** All of the above statements outline ways that eukaryotic transcription is more complex.
6. Why is the *lac* operon of *E. coli* important to biotechnology research?
   - **a.** IPTG is a cheaper additive than lactose to growing cultures.
   - **b.** The *lac* operon is not used in biotechnology research.
   - **c.** The inducers and regulators of the *lac* operon are used to control the expression of genes in model organisms.
   - **d.** The *lac* operon controls the amount of lactose that *E. coli* metabolizes.
   - **e.** All of the above.

(*Continued*)

7. What feature about eukaryotic transcription factors is useful to biotechnology research?
   a. They have two domains, both of which bind to DNA.
   b. They have two domains, both of which bind to separate proteins.
   c. They have two domains: one domain binds DNA and the other binds to some part of the transcription apparatus.
   d. They have only one domain that binds to RNA polymerase.
   e. They have two domains but neither domain can be engineered and are therefore not useful to biotechnology research.

8. Which of the following DNA structure modifications are used to regulate transcription?
   a. acetylation/deacetylation of the histone tails
   b. methylation of specific bases in the DNA sequence
   c. use of non-coding regulatory RNA to alter DNA accessibility
   d. nucleosome remodeling
   e. All of the above are important modifications for transcription regulation

9. Which of the following statements about eukaryotic mRNA processing is not correct?
   a. The mRNA transcript must be exported from the nucleus.
   b. A 5′ cap and a 3′ poly(A) tail must be added.
   c. The introns are removed.
   d. A 3′ cap and a 5′ poly(A) tail must be added.
   e. Exons are spliced together to form the mRNA transcript.

10. Which of the following statements about protein translation is not correct?
    a. The genetic code is read in triplets, also called codons.
    b. The enzyme, aminoacyl tRNA synthetase, is responsible for adding the amino acid to the tRNA.
    c. The anticodon of the tRNA must recognize the codon on the mRNA exactly.
    d. Because of the wobble effect, a tRNA for one amino acid often recognizes multiple codons in the mRNA.
    e. The genetic code is universal.

11. Codon bias can be overcome by which scenario?
    a. Genetically engineering host organisms to express rarer tRNAs
    b. Nothing can be done to overcome codon bias when expressing proteins.
    c. Genetically engineering the gene so that the codons are recognized by more abundant tRNAs
    d. Genetically engineer the gene to remove the codons for rare tRNAs.
    e. Both A and C are suitable scenarios.

12. Choose the statement about translation that is not correct.
    a. The ribosome is comprised of multiple subunits containing both ribosomal RNA and proteins.
    b. The consensus sequence UAAGGAGG is called the Shine–Dalgarno sequence and is recognized by the ribosome.
    c. Translation requires three initiation factors, two elongation factors, and two release factors.
    d. Transcription and translation are coupled in eukaryotes.
    e. There are three sites (E, P, and A) on the ribosome that can be occupied by a tRNA.

13. Which of the following statements does not highlight a difference in eukaryotic and prokaryotic translation?
   a. The first methionine in eukaryotic translation contains a formyl group.
   b. In eukaryotes, mRNA is made in the nucleus but translated in the cytoplasm.
   c. Prokaryotes often couple transcription and translation, forming a polysome.
   d. Eukaryotic mRNA does not have a Shine–Dalgarno sequence, but prokaryotic mRNA does.
   e. Many eukaryotic proteins are chemically modified after translation, which is a much rarer phenomenon in prokaryotes.

14. Why do mitochondria and chloroplasts contain their own genes?
   a. They are free-living prokaryotes, able to survive outside of the host cell.
   b. They are thought to have once been free-living organisms similar to bacteria that formed a symbiotic relationship with a unicellular eukaryote.
   c. They do not contain their own genetic material.
   d. They contain genetic material but do not make their own proteins.
   e. None of the above is correct.

15. Methylation of DNA ______________.
   a. silences gene expression in eukaryotes
   b. enhances the binding of RNA polymerase to the promoters in the methylated region
   c. results in the removal of histones from the methylated regions
   d. causes histone tail fibers to become acetylated
   e. remodels nucleosomes to allow entry of transcription machinery

Further Reading

Bustamante, C., Cheng, W., & Mejia, Y. X. (2011). Revisiting the central dogma one molecule at a time. *Cell, 144*, 480–497.

Clark, D., & Pazdernik, N. (2013). *Molecular Biology*. Waltham, MA: Elsevier Inc.

Kato, S., Yokoyama, A., & Fujiki, R. (2011). Nuclear receptor coregulators merge transcriptional coregulation with epigenetic regulation. *Trends in Biochemical Sciences, 36*, 272–281.

Recombinant DNA Technology

Biotechnology

http://dx.doi.org/10.1016/B978-0-12-385015-7.00003-X

DNA ISOLATION AND PURIFICATION

Basic to all biotechnology research is the ability to manipulate DNA. First and foremost for recombinant DNA work, researchers need a method to isolate DNA from different organisms. Isolating DNA from bacteria is the easiest procedure because bacterial cells have little structure beyond the cell wall and cell membrane. Bacteria such as *E. coli* are the preferred organisms for manipulating any type of gene because of the ease at which DNA can be isolated. *E. coli* maintain both genomic and plasmid DNA within the cell. Genomic DNA is much larger than plasmid DNA, allowing the two different forms to be separated.

To release the DNA from a cell, the cell membrane must be destroyed. For bacteria, an enzyme called **lysozyme** digests the peptidoglycan, which is the main component of the cell wall. Next, a detergent such as sodium dodecyl sulfate (SDS) bursts the cell membranes by disrupting the lipid bilayer. For other organisms, disrupting the cell depends on their architecture. Tissue samples from animals and plants have to be ground up to release the intracellular components. Plant cells are mechanically sheared in a blender to break up the tough cell walls, and then the wall tissue is digested with enzymes that break the long polymers of lignin and cellulose into monomers. DNA from the tail tip of a mouse is isolated after proteinase K degrades the tissue and detergent dissolves the cell membranes. Cells cultured in dishes are probably the easiest since they do not have cell walls or other structures outside their cell membrane. Detergent alone disrupts the cell membrane to release the intracellular components. Every organism or tissue needs slight variations in the procedure for releasing intracellular components including DNA.

Once released, the intracellular components are separated from the insoluble remains such as the cellular membranes, bones, cartilage, and/or cell wall by either centrifugation or chemical extraction. Centrifugation separates components according to size, because heavier or larger molecules sediment at a faster rate than smaller molecules. In addition, materials that are insoluble in the liquid phase form aggregates that sediment to the bottom of a centrifuge tube faster. For example, after the cell wall has been digested, its fragments are smaller than the large DNA molecules. Centrifugation causes the DNA to form a pellet, but the soluble cell wall fragments stay in solution. Another method of separating cellular components, chemical extraction uses the properties of **phenol** to remove unwanted proteins from the DNA. Phenol is an acid that dissolves 60% to 70% of all living matter, especially proteins. Phenol is not very soluble in water, and when it is mixed with an aqueous sample of DNA and protein, the two phases separate, much like oil and water. The protein dissolves in the phenol layer and the nucleic acids in the aqueous layer. The two phases are separated by centrifugation, and the aqueous DNA layer is removed from the phenol.

Once the proteins are removed, the sample still contains RNA along with the DNA. Because RNA is also a nucleic acid, it is not soluble in phenol. Luckily, the enzyme **ribonuclease (RNase)** digests RNA into ribonucleotides. Ribonuclease treatment leaves a sample of DNA in a solution containing short pieces of RNA and ribonucleotides. When an equal volume of alcohol is added, the extremely large DNA falls out of the aqueous phase and is isolated by centrifugation. The smaller ribonucleotides stay soluble. The DNA is then ready for use in various experiments.

DNA can be isolated by first removing the cell wall and cell membrane components. Next, the proteins are removed by phenol, and finally, the RNA is removed by ribonuclease.

ELECTROPHORESIS SEPARATES DNA FRAGMENTS BY SIZE

Gel electrophoresis is used to separate DNA fragments by size (Fig. 3.1). The gel consists of **agarose**, a polysaccharide extracted from seaweed that behaves like gelatin. Agarose is a powder that dissolves in water only when heated. After the solution cools, the agarose hardens.

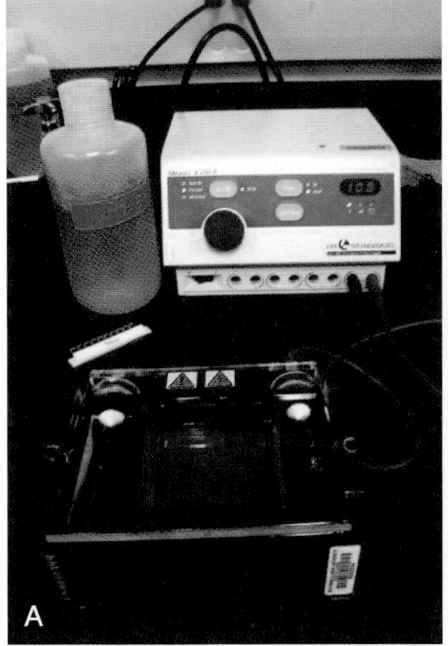

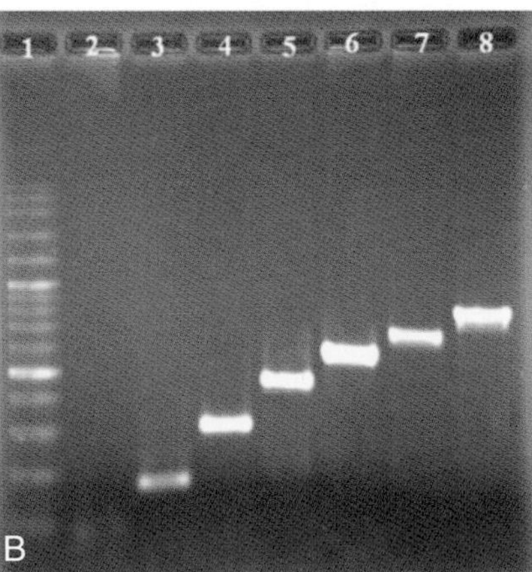

FIGURE 3.1 Electrophoresis of DNA

(A) Photo of electrophoresis supplies. Electrophoresis chamber holds an agarose gel in the center portion, and the rest of the tank is filled with buffer solution. The red and black leads are then attached to an electrical source. FisherBiotech Horizontal Electrophoresis Systems, Midigel System; Standard; 13 × 16-cm gel size; 800 mL buffer volume; Model No. FB-SB-1316. (B) Agarose gel separation of DNA. The size of the fragments can be calculated by comparing them to the standard DNA marker in lane 1. The brighter bands in the marker are 1000 base pairs and 500 base pairs, with the 1000 base-pair marker closer to the wells (marked with numbers 1–8). Used with permission from Ghadaksaz, et al. (2014). The prevalence of some *Pseudomonas* virulence genes related to biofilm formation and alginate production among clinical isolates. *J Appl Biomed* **13(1)**, 61–68. DOI: 10.1016/j.jab.2014.05.002.

For visualizing DNA, agarose solidifies into a rectangular slab about 1/4 inch thick by casting the molten liquid into a special tray. Inserting a comb at one end of the tray before it hardens makes small wells or holes. After the gel solidifies, the comb is removed, leaving small wells at one end.

Gel electrophoresis uses electric current to separate DNA molecules by size. The agarose slab is immersed in a buffer-filled tank that has a positive electrode at one end and a negative electrode at the other. DNA samples are loaded into the wells, and when an electrical field is applied, the DNA migrates through the gel. The phosphate backbone of DNA is negatively charged, so it moves away from the negative electrode and toward the positive electrode. Polymerized agarose acts as a sieve with small holes between the tangled chains of agarose. The DNA must migrate through these gaps. Agarose separates the DNA by size because larger pieces of DNA are slowed down more by the agarose.

To visualize the DNA, the agarose gel is removed from the tank and immersed into a solution of **ethidium bromide**. This dye intercalates between the bases of DNA or RNA, although less dye binds to RNA because it is single-stranded. When the gel is exposed to ultraviolet light, it fluoresces bright orange. Since ethidium bromide is a mutagen and carcinogen, less dangerous DNA dyes such as SYBR Safe® are used in most laboratories now. This DNA dye is also excited by ultraviolet light, emitting a bright orange fluorescence. In Figure 3.1, the DNA fragments are visualized by a positively charged dye from the thiazin family. The dye interacts with the negatively charged backbone of the DNA and is a nontoxic alternative that does not require ultraviolet light sources.

The size of DNA being examined affects what type of gel is used. DNA molecules of the same size usually form a tight band, and the size can be determined by comparing to a set of **molecular weight standards** run in a different well. Because the standards are of known size, the experimental DNA fragment can be compared directly. When DNA samples are separated by size through an agarose gel, DNA fragments from about 200 base pairs to 10,000 base pairs can be separated. For DNA fragments from 50 to 1000 base pairs, polyacrylamide gels are used instead. These gels are able to resolve DNA fragments that vary by only one base pair and are essential to sequencing DNA with the Sanger method (see Chapter 4). For very large DNA fragments (10 kilobases to 10 megabases), agarose is used, but the current is alternated at two different angles. **Pulsed field gel electrophoresis (PFGE)**, as this is called, allows very large pieces of DNA to migrate further than if the current flows in only one direction. Each change in direction loosens large pieces of DNA that are stuck inside the gel matrix, letting them migrate further. Finally, gradient gel electrophoresis can be used to resolve fragments that are very close in size. A concentration gradient of acrylamide, buffer, or electrolyte can reduce compression (i.e., crowding of similar sized fragments) due to secondary structure and/or slow the smaller fragments at the lower end of the gel.

Fragments of DNA are separated by size using gel electrophoresis. A current causes the DNA fragments to move away from the negative electrode and toward the positive. As the DNA travels through agarose, the larger fragments get stuck in the gel pores more than the smaller DNA fragments.

Pulsed field gel electrophoresis separates large pieces of DNA by alternating the electric current at right angles.

RESTRICTION ENZYMES CUT DNA; LIGASE JOINS DNA

The ability to isolate, separate, and visualize DNA fragments would be useless unless some method was available to cut the DNA into fragments of different sizes. Luckily, naturally occurring **restriction enzymes** or **restriction endonucleases** are the key to making DNA fragments. These bacterial enzymes bind to specific recognition sites on DNA and cut the backbone of both strands. They evolved to protect bacteria from foreign DNA, such as viruses. The enzymes do not cut their own cell's DNA because they are methylation sensitive; that is, if one of the nucleotide bases in the recognition sequence is methylated, then the restriction enzyme cannot bind and therefore cannot cut the methylated DNA. Bacteria produce **modification enzymes** that recognize the same sequence as the corresponding restriction enzyme. These methylate each recognition site in the bacterial genome. Therefore, the bacteria can make the restriction enzyme without endangering their own DNA.

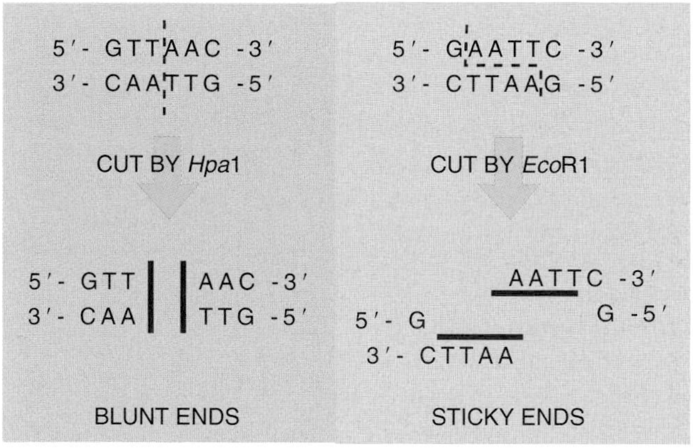

FIGURE 3.2
Type II Restriction Enzymes—Blunt versus Sticky Ends
HpaI is a blunt-end restriction enzyme; that is, it cuts both strands of DNA in exactly the same position. *Eco*RI is a sticky-end restriction enzyme. The enzyme cuts between the G and A on both strands, which generates four base-pair overhangs on the ends of the DNA. Since these ends may base pair with complementary sequences, they are considered "sticky."

Restriction enzymes have been exploited to cut DNA at specific sites, since each restriction enzyme has a particular recognition sequence. Differences in cleavage site determine the type of restriction enzyme. **Type I restriction enzymes** cut the DNA strand 1000 or more base pairs from the recognition sequence. **Type II restriction enzymes** cut in the middle of the recognition sequence and are the most useful for genetic engineering. Type II restriction enzymes can either cut both strands of the double helix at the same point, leaving **blunt ends**, or they can cut at different sites on each strand leaving single-stranded ends, sometimes called **sticky ends** (Fig. 3.2). The recognition sequences of Type II restriction enzymes are usually inverted repeats so that the enzyme cuts between the same bases on both strands. Some commonly used restriction enzymes for biotechnology applications are listed in Table 3.1. Since restriction enzymes recognize a specific nucleic sequence, these can also be used to compare the nucleotide sequence of different organisms or individuals (see Box 3.1).

The number of base pairs in the recognition sequence determines the likelihood of cutting. Finding a particular sequence of four nucleotides is much more likely than finding a six base-pair recognition sequence. So to generate fewer, longer fragments, restriction enzymes with six or more base-pair recognition sequences are used. Conversely, four base-pair enzymes give more, shorter fragments from the same original segment of DNA.

When two different DNA samples are cut with the same sticky-end restriction enzyme, all the fragments will have identical overhangs. This allows DNA fragments from two sources (e.g., two different organisms) to be linked together (Fig. 3.3). Fragments are linked or **ligated** using **DNA ligase**, the same enzyme that ligates the Okazaki fragments during replication (see Chapter 4).

Table 3.1 Table of Common Restriction Enzymes

Enzyme	Source Organism	Recognition Sequence
*Hpa*II	*Haemophilus parainfluenzae*	C/CGG GGC/C
*Mbo*I	*Moraxella bovis*	/GATC GATC/
*Nde*II	*Neisseria denitrificans*	/GATC GATC/
*Eco*RI	*Escherichia coli* RY13	G/AATTC CTTAA/G
*Eco*RII	*Escherichia coli* RY13	/CCWGG GGWCC/
*Eco*RV	*Escherichia coli* J62/pGL74	GAT/ATC CTA/TAG
*Bam*HI	*Bacillus amyloliquefaciens*	G/GATCC CCTAG/G
*Sau*I	*Staphylococcus aureus*	CC/TNAGG GGANT/CC
BglI	*Bacillus globigii*	GCCNNNN/NGGC CGGN/NNNNCCG
NotI	*Nocardia otitidis-caviarum*	GC/GGCCGC CGCCGG/CG
DraII	*Deinococcus radiophilus*	RG/GNCCY YCCNG/GR

/, position where enzyme cuts.
N, any base; R, any purine; Y, any pyrimidine; W, A, or T.

The most common ligase used is actually from T4 bacteriophage. Ligase catalyzes linkage between the 3′-OH of one strand and the 5′-PO_4 of the other DNA strand. Ligase is much more efficient with overhanging sticky ends but can also link blunt ends much more slowly.

Restriction enzymes are naturally occurring enzymes that recognize a particular DNA sequence and cut the phosphate backbone.

When two pieces of DNA are cut by the same restriction enzyme, the two ends have compatible overhangs that can be reconnected by ligase.

Box 3.1 Restriction Fragment Length Polymorphisms Identify Individuals

Restriction enzymes are useful for many different applications. Because the DNA sequence is different in each organism, the pattern of restriction sites will also be different. The source of isolated DNA can be identified by this pattern. If genomic DNA is isolated from one organism and cut with one particular restriction enzyme, a specific set of fragments can be separated and identified by electrophoresis. If DNA from a different organism is cut by the same restriction enzyme, a different set of fragments will be generated. This technique can be applied to DNA from two individuals from the same species. Although the DNA sequence differences will be small, restriction enzymes can be used to identify these differences. If the sequence difference falls in a restriction enzyme recognition site, it gives a **restriction fragment length polymorphism (RFLP)** (Fig. A). When the restriction enzyme patterns are compared, the number and size of one or two fragments will be affected for each base difference that affects a cut site.

(Continued)

Box 3.1 Restriction Fragment Length Polymorphisms Identify Individuals—cont'd

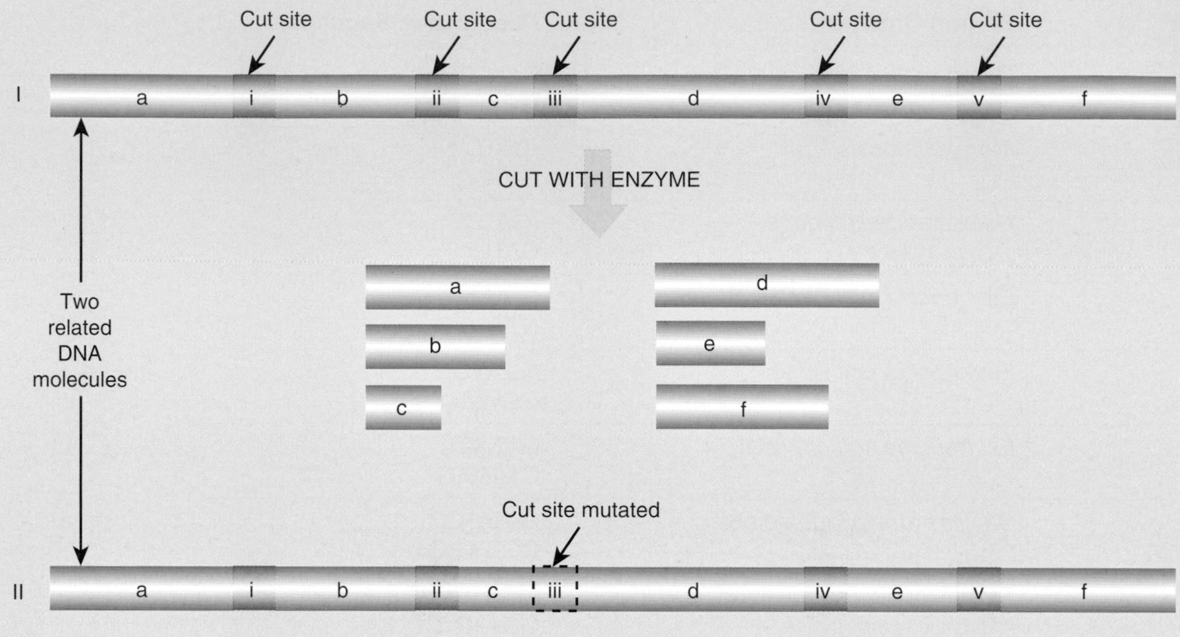

CUT WITH ENZYME

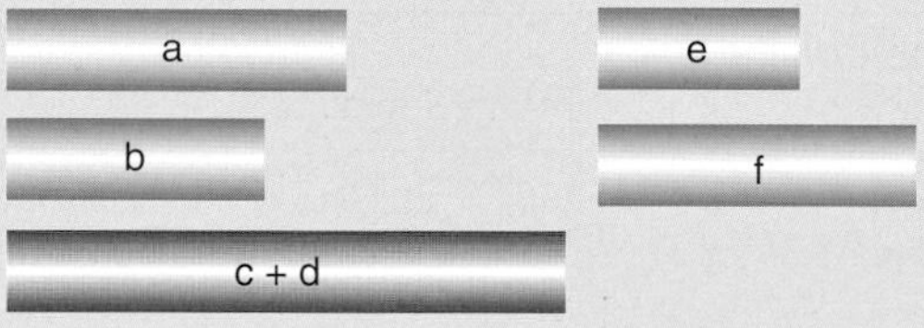

RUN GELS

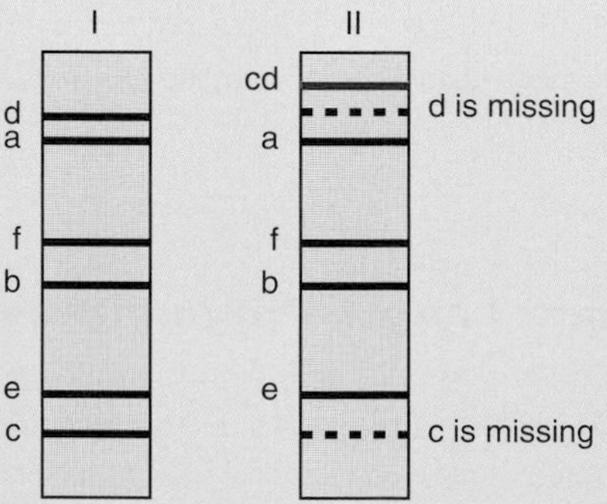

FIGURE A RFLP Analysis

DNA from related organisms shows small differences in sequence that cause changes in restriction sites. In the example shown, cutting a segment of DNA from the first organism yields six fragments of different sizes (labeled *a–f* on the gel). If the equivalent region of DNA from a related organism were digested with the same enzyme, a similar pattern would be expected. Here, a single-nucleotide difference is present, which eliminates one of the restriction sites. Consequently, digesting this DNA produces only five fragments. Fragments *c* and *d* are no longer seen but form a new band labeled *cd*.

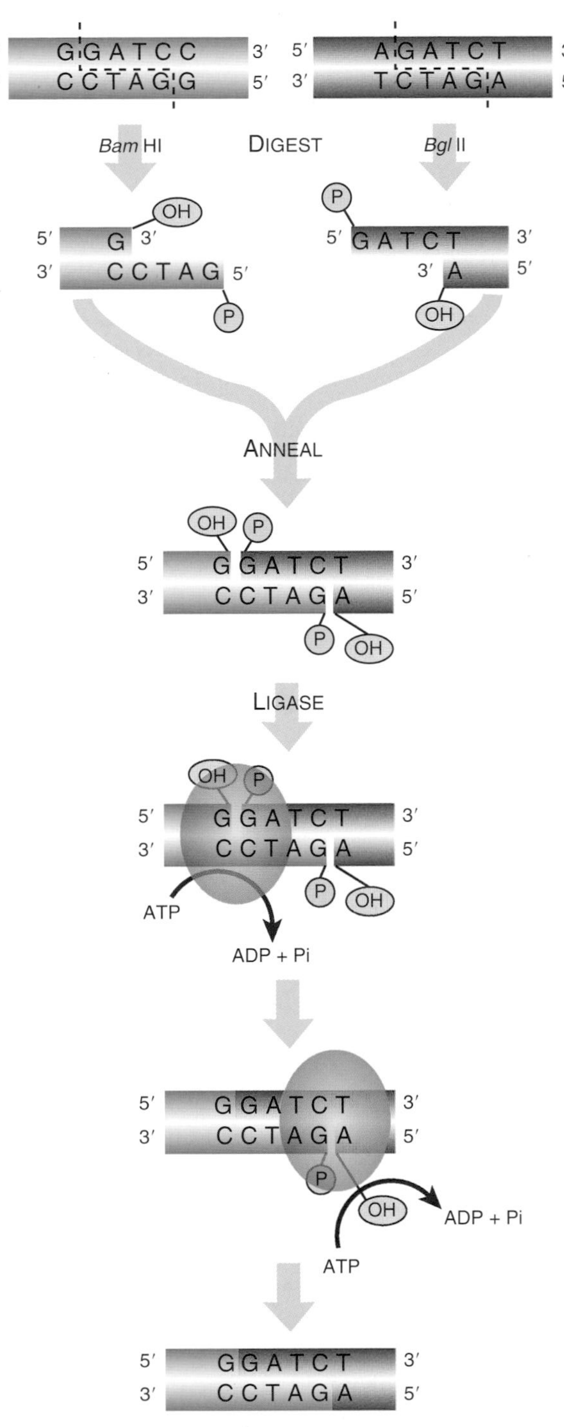

FIGURE 3.3 Compatible Overhangs Are Linked Using DNA Ligase

*Bam*HI and *Bgl* II generate the same overhanging or sticky ends: a 3′-CTAG-5′ overhang plus a 5′-GATC-3′ overhang. These are complementary and base pair by hydrogen bonding. The breaks in the DNA backbones are sealed by T4 DNA ligase, which hydrolyzes ATP to energize the reaction.

METHODS OF DETECTION FOR NUCLEIC ACIDS

Recombinant DNA methodologies require the ability to detect DNA. One of the easiest ways to detect the amount of DNA or RNA in solution is to measure the absorbance of ultraviolet light at 260 nm (Fig. 3.4). DNA absorbs ultraviolet light because of the ring structures in the bases. Single-stranded RNA and free nucleotides also absorb ultraviolet light. In fact, they

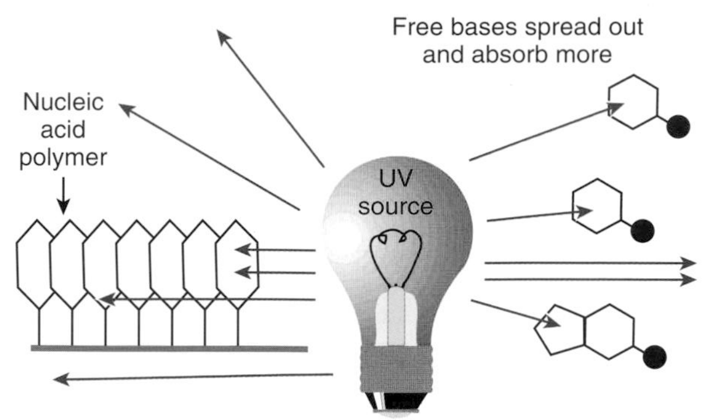

FIGURE 3.4
Determining the Concentration of DNA
All nucleic acids absorb UV light via the aromatic rings of the bases. Stacked nucleotides (on the left) absorb less UV than scattered bases (on the right) because of the ordered structure.

absorb more light because their structures are looser. Since the absorbance of UV light depends on the amount of DNA and the molecular structure, the relationship between UV absorbance and concentration is

Double-stranded DNA concentration (μg/ml) = $(OD_{260}) \times (50\ \mu g\ DNA/ml)/(1\ OD_{260}\ unit)$
RNA concentration (μg/ml) = $(OD_{260}) \times (40\ \mu g\ RNA/ml)/(1\ OD_{260}\ unit)$

In addition to the amount of DNA, a second absorbance reading at 280 nm is commonly used to determine the purity of the sample. The ratio of the 260 nm absorbance value divided by the 280 nm absorbance value will indicate whether the sample is pure. If the sample is pure DNA, then the 260/280 ratio is 1.8; whereas a 260/280 ratio for pure RNA is 2.0. When the ratios deviate from the expected value, there could be residual phenol from the purification or a very low concentration of DNA or RNA.

The concentration of DNA or RNA in a liquid can be determined by measuring the absorbance of UV light at 260 nm.

Radioactive Labeling of Nucleic Acids and Autoradiography

Ultraviolet light absorption is a general method for detecting DNA but does not distinguish between different DNA molecules. DNA can also be detected with radioactive isotopes (Fig. 3.5). During replication, radioactive precursors such as ^{32}P in the form of a **phosphate group** and ^{35}S in the form of **phosphorothioate** can be incorporated. Because native DNA does not contain sulfur atoms, one of the oxygen atoms of a phosphate group is replaced with sulfur to make phosphorothioate. Most radioactive molecules used in laboratories are short lived. ^{32}P has a half-life of 14 days and ^{35}S has a half-life of 68 days, so the isotopes degrade fairly fast. Although radioactive DNA is invisible, photographic film will turn black when exposed to the radioactive DNA. Radioactively labeled DNA is considered **"hot,"** whereas unlabeled DNA is considered **"cold."**

Autoradiography identifies the location of radioactively labeled DNA in the gel (Fig. 3.6). If the gel is thin, like most polyacrylamide gels, it is dried with heat and vacuum. If the gel is thick, like agarose gels, the DNA is transferred to a nylon membrane using capillary action (see Fig. 3.9, later). The dried gel or nylon membrane is placed next to photographic film. As the radioactive phosphate decays, the radiation turns the photographic film black. Only the areas next to radioactive DNA will have black spots or bands. The use of film detects where the hot DNA is on a gel, and the use of ethidium bromide shows where all of the DNA, hot or cold, is. These two methods allow distinguishing one DNA fragment from another.

Radioactive isotopes are incorporated into the DNA backbone during replication. Autoradiography identifies the radioactively labeled "hot" DNA.

Fluorescence Detection of Nucleic Acids

Autoradiography has its merits, but working with and disposing of radioactive waste are costly, both monetarily and environmentally. Using fluorescently tagged nucleotides was

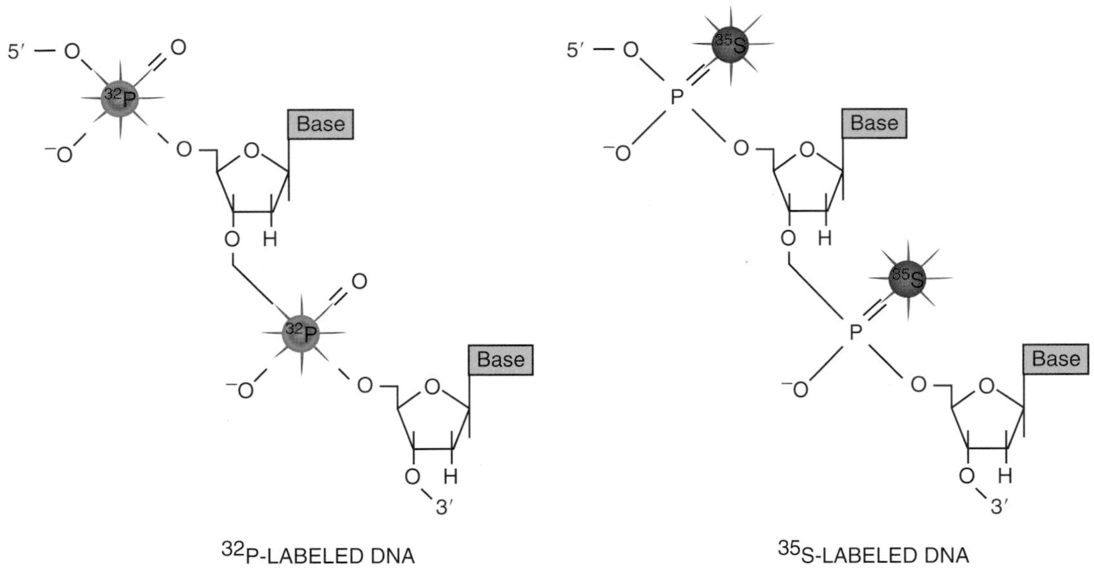

^{32}P-LABELED DNA

^{35}S-LABELED DNA

FIGURE 3.5 Radioactively Labeled DNA
DNA can be synthesized with radioactive precursor nucleotides. These nucleotides have ^{32}P (rather than nonradioactive ^{31}P phosphorus) or ^{35}S (replacing oxygen) in the phosphate backbone.

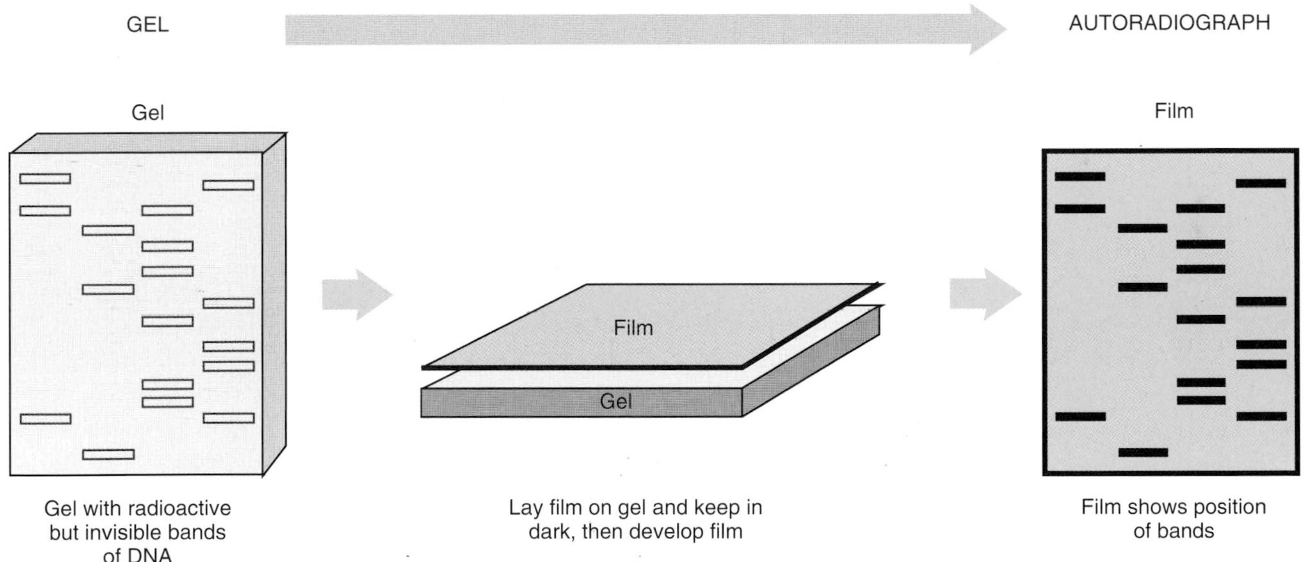

FIGURE 3.6 Autoradiography
A gel containing radioactive DNA (or RNA) is dried and a piece of photographic film is laid over the top. The two are loaded into a cassette case that prevents light from entering. After some time (hours to days), the film is developed and dark lines appear where the radioactive DNA was present.

developed as a better method of DNA detection (Fig. 3.7). Fluorescent tags absorb light of one wavelength, which excites the atoms, increasing the energy state of the tag. This excited state releases a photon of light at a different (longer) wavelength and returns to the ground state. The emitted photon is detected with a **photodetector**. There are many different fluorescent tags, and each emits a different wavelength of light. Some photodetector systems are sensitive enough to distinguish between these different tags; therefore, if different bases have different fluorescent labels, the photodetector can determine which base is present. This is the basis for most modern DNA sequencing machines (see Chapter 4).

Fluorescently labeled nucleotides can be used to incorporate a fluorescent tag on DNA during replication or PCR amplification.

FLUORESCENT TAGGING OF DNA

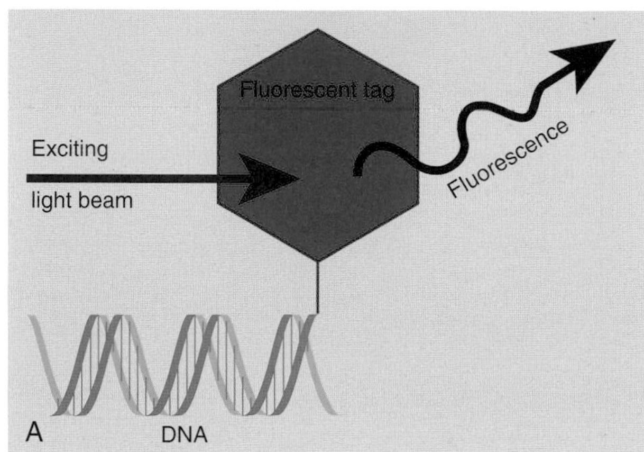

ENERGY LEVELS IN FLUORESCENCE

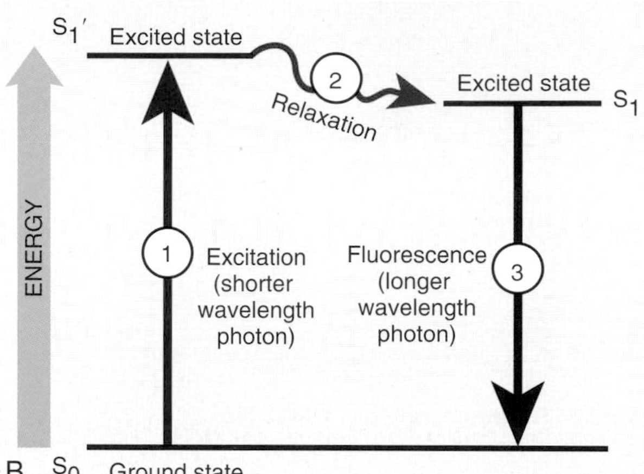

FIGURE 3.7 Fluorescent Labeling of DNA
(A) Fluorescent tagging of DNA. During synthesis, a nucleotide linked to a fluorescent tag is incorporated at the 3′ end of the DNA. A beam of light excites the fluorescent tag, which in turn releases light of a longer wavelength. (B) Energy levels in fluorescence. The fluorescent molecule attached to the DNA has three different energy levels: S_0, S_1', and S_1. The S_0, or ground state, is the state before exposure to light. When the fluorescent molecule is exposed to a light photon, the fluorescent tag absorbs the energy and enters the first excited state, S_1'. Between S_1' and S_1, the fluorescent tag relaxes slightly but doesn't emit any light. Eventually, the high-energy state releases its excess energy by emitting a longer wavelength photon. This release of fluorescence returns the molecule to the ground state.

Chemical Tagging with Biotin or Digoxigenin

Biotin is a vitamin and digoxigenin is a steroid from the foxglove plant. Using these two molecules allows scientists to label DNA without radioactivity or costly photodetectors. To incorporate the label into DNA, a biotin or digoxigenin molecule is chemically linked to uracil; therefore, DNA is synthesized with the labeled uracil replacing thymine using *in vitro* DNA replication as described in Chapter 4. Single-stranded DNA template, DNA polymerase, a short DNA primer, and nucleotides (dATP, dGTP, dCTP, plus dUTP linked to either biotin or digoxigenin) are mixed in a tube and incubated. DNA polymerase synthesizes the complementary strand to the template, incorporating biotin- or digoxigenin-linked uracil opposite all the adenines.

The labeled DNA is visualized in a two-step process (Fig. 3.8). To visualize biotin, a molecule of **avidin** or **streptavidin**, which both have a high affinity for biotin, is added to the DNA sample. Avidin, originally identified in egg whites, has a higher tendency for aggregation and is glycosylated; therefore, streptavidin is used more often. Streptavidin is isolated from *Streptomyces avidinii*, and is not glycosylated. In contrast, a specific antibody is used to visualize digoxigenin. Both avidin and the digoxigenin antibody are conjugated to either a fluorophore or reporter enzyme, which allow for visible detection. One example of a reporter enzyme is **alkaline phosphatase**, an enzyme that removes phosphates from a variety of substrates. Several different chromogenic molecules act as substrates for alkaline phosphatase, but the most widely used one is **X-Phos**. Once alkaline phosphatase removes the phosphate group from X-Phos, the intermediate molecule reacts with oxygen and forms a blue precipitate. This blue color reveals the location of the labeled DNA. Another substrate of alkaline phosphatase is **Lumi-Phos**, which is chemiluminescent and emits visible light when the phosphate is removed. Much like autoradiography, when photographic film is placed over

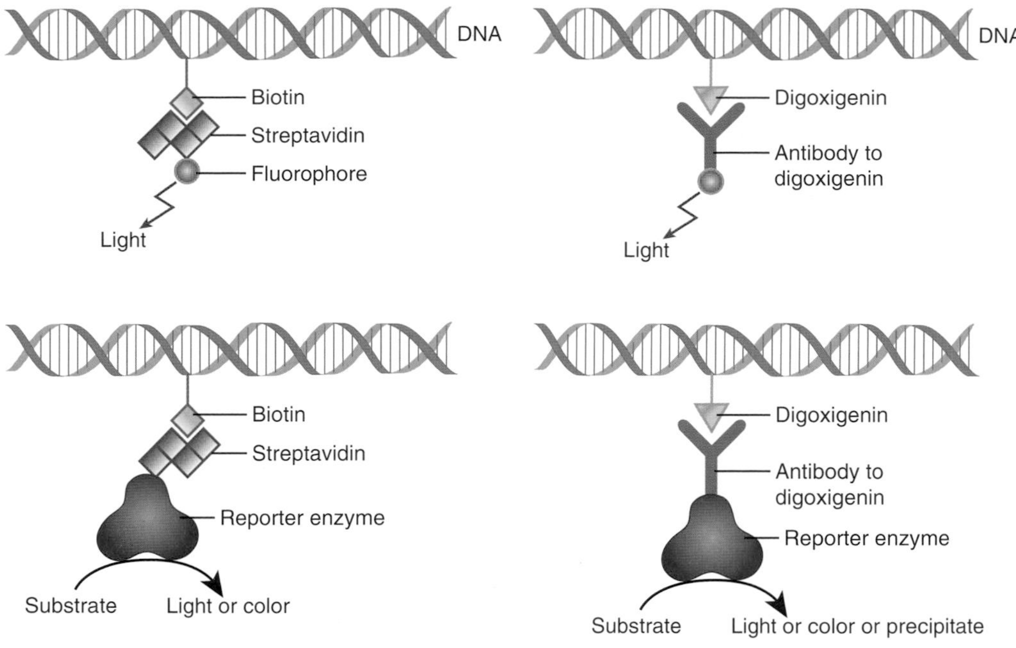

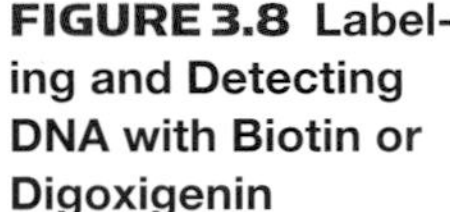

FIGURE 3.8 Labeling and Detecting DNA with Biotin or Digoxigenin
DNA can be synthesized *in vitro* with a uracil nucleotide linked to biotin or digoxigenin. Streptavidin binds tightly to avidin (left panels), and antibody to digoxigenin binds to digoxigenin (right panels). Streptavidin and antibody to digoxigenin can be conjugated to a fluorophore that emits light of specific wavelengths (top panels) or to reporter enzymes such as horseradish peroxidase or alkaline phosphatase (lower panels). Reporter enzymes act on different substrates, some of which release light and others that form a colored precipitate.

labeled DNA treated with Lumi-Phos, emitted light causes the film to turn dark. Another possible reporter enzyme is **horseradish peroxidase** or **HRP**, an enzyme that reacts with luminol to release light. As before, the light can be detected with photographic film.

> Biotin and digoxigenin-labeled DNA are detected using either streptavidin or antibody to digoxigenin. Either label can be conjugated to alkaline phosphatase, which reacts with X-Phos to leave a blue precipitate or Lumi-Phos to emit visible light. Another reporter enzyme commonly used is horseradish peroxidase. These reporter enzymes are used to identify, quantify, or locate the labeled DNA.

COMPLEMENTARY STRANDS MELT APART AND REANNEAL

The complementary antiparallel strands of DNA form an elegant molecule that is able to unzip or **melt** and come back together or **reanneal** (Fig. 3.9). The hydrogen bonds that hold the two halves together are relatively weak. Heating a sample of DNA will dissolve the hydrogen bonds, resulting in two complementary single strands. If the same sample of DNA is slowly cooled, the two strands will reanneal so that G matches with C and A matches with T, as before.

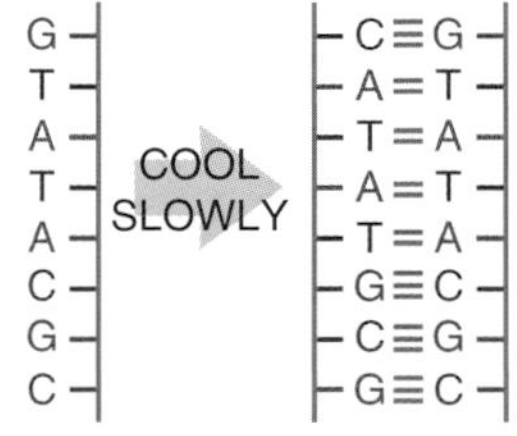

FIGURE 3.9 Heat Melts DNA; Cooling Reanneals DNA
Hydrogen bonds readily dissolve when heated, leaving the two strands intact but separate. When the temperature returns to normal, the hydrogen bonds form again.

The proportion of G/C base pairs affects how much heat is required to melt a double helix of DNA. G/C base pairs have three hydrogen bonds to melt, whereas A/T base pairs have only two. Consequently, DNA with a higher percentage of GC will require more energy to melt than DNA with fewer GC base pairs. The **GC ratio** is defined as follows:

$$\frac{G+C}{A+G+C+T} \times 100\,\%$$

The ability to zip and unzip DNA is crucial to cellular function, and has also been exploited in biotechnology. Replication (see Chapter 4) and transcription (see Chapter 2) rely on strand separation to generate either new DNA or RNA strands, respectively. In molecular biology research, many techniques, from PCR to DNA sequencing, exploit the complementary nature of DNA strands.

The complementary strands of DNA are easily separated by heat and spontaneously reanneal as the DNA mixture cools.

HYBRIDIZATION OF DNA OR RNA IN SOUTHERN AND NORTHERN BLOTS

If two different double helixes of DNA are melted, the single strands can be mixed together before cooling and reannealing. If the two original DNA molecules have similar sequences, a single strand from one may pair with the opposite strand from the other DNA molecule. This is known as **hybridization** and can be used to determine whether sequences in two separate samples of DNA or RNA are related. In hybridization experiments, the term **probe molecule** refers to a known DNA sequence or gene that is used to screen the experimental sample or **target DNA** for similar sequences.

Southern blots are used to determine how closely DNA from one source is related to a DNA sequence from another source. The technique involves forming hybrid DNA molecules by mixing DNA from the two sources. A Southern blot has two components: the probe sequence (e.g., a known gene of interest from one organism) and the target DNA (often from a different organism). A typical Southern blot begins by isolating the target DNA from one organism, digesting it with a restriction enzyme that gives fragments from about 500 to 10,000 base pairs in length, and separating these fragments by electrophoresis. The separated fragments will be double-stranded, but if the gel is incubated in a strong acid, the DNA separates into single strands. Using capillary action, the single strands can be transferred to a membrane as shown in Fig. 3.10. The DNA remains single-stranded once attached to the membrane.

Next, the probe is prepared. First, the known sequence or gene must be isolated and labeled in some way (see earlier discussion). Identifying genes has become easier now that many genomes have been entirely sequenced. For example, a scientist can easily obtain a copy of a human gene for use as a probe to find similar genes in other organisms. Alternatively, using sequence data, a unique oligonucleotide probe can be designed that recognizes only the gene of interest (see Chapter 4). If an oligonucleotide has a common sequence, it will bind to many other sequences. Therefore, oligonucleotide probes must be long enough to have sequences that bind to only one (or very few) specific site(s) in the target genome. To prepare DNA probes for a Southern blot, they are labeled using radioactivity, biotin, or digoxigenin (see earlier discussion). Finally, the labeled DNA is denatured at high temperature to make it single-stranded. (Synthetic oligonucleotides do not require treatment, as they are already single-stranded.)

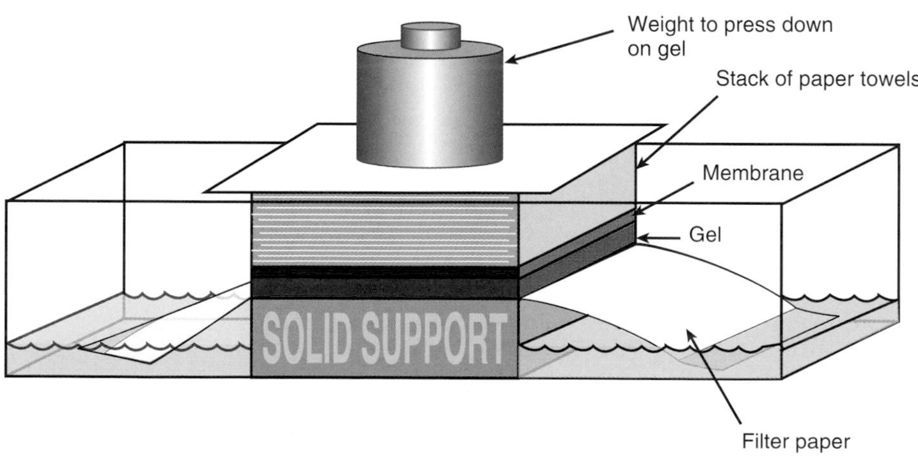

FIGURE 3.10 Capillary Action Transfers DNA from Gel to Membrane
Single-stranded DNA from a gel will transfer to the membrane. The filter paper wicks buffer from the tank, through the gel and membrane, and into the paper towels. As the buffer liquid moves, the single-stranded DNA also travels from the gel and sticks to the membrane. The weight on top of the setup keeps the membrane and gel in contact and helps wick the liquid from the tank.

To perform a Southern blot, the single-stranded probe is incubated with the membrane carrying the single-stranded target DNA (Fig. 3.11). These are incubated at a temperature that allows hybrid DNA strands to form with only a low amount of mismatch. The temperature, and hence the level of mismatching tolerated, can be varied depending on how closely identical the probe and target sequence are expected to be. At a high temperature, the probe will only

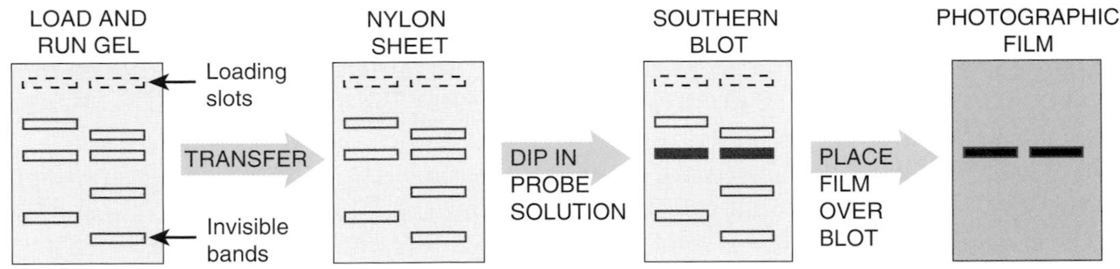

FIGURE 3.11 Hybrid DNA Molecules Can Detect Related Sequences in Southern Blots
Southern blotting requires the target DNA to be cut into smaller fragments and run on an agarose gel. The fragments are denatured chemically to give single strands and then transferred to a nylon membrane. A radioactive probe (also single-stranded) is incubated with the membrane at a temperature that allows hybrids (with some mismatches) to form. When photographic film is placed over the top of the membrane, the location of the radioactive hybrid molecules is revealed.

stay attached at locations with almost identical sequences; whereas at a low temperature, the probe will bind locations with multiple mismatched nucleotides. If the probe is radioactive, then the membrane is exposed to photographic film. If the probe is labeled with biotin or digoxigenin, the membrane may be treated with chemiluminescent substrate to detect the labeled probe and target DNA hybrid, and then exposed to photographic film. Dark bands on the film reveal the positions of fragments with similar sequence to the probe. Alternatively, biotin or digoxigenin labels may be visualized by treatment with a chromogenic substrate. In this case, blue bands will form directly on the membrane at the position of the related sequences.

Northern blots are also based on nucleic acid hybridization. The difference is that RNA is the target in a Northern blot. The probe for a Northern blot is either a fragment of a gene or a unique oligonucleotide just as in a Southern blot. The target RNA is usually messenger RNA. In eukaryotes, screening mRNA is more efficient because genomic DNA has many introns, which may interfere with probes binding to the correct sequence. Besides, mRNA is already single-stranded, so the agarose gel does not have to be treated with strong acid. Much like a Southern blot, Northern blots begin by separating mRNA by size using electrophoresis. The mRNA is transferred to a nylon membrane and incubated with a single-stranded labeled probe. As before, the probe can be labeled with biotin, digoxigenin, or radioactivity. The membrane is processed and exposed to film or chromogenic substrate.

A variation of these hybridization techniques is the **dot blot** (Fig. 3.12). Here, the target sample is not separated by size. The DNA or mRNA target is simply attached to the nylon membrane as a small dot. As in Southern blots, the DNA sample must be made single-stranded before it is attached to the membrane. As before, the dot-blot membrane is allowed to hybridize with a labeled probe. The membranes are processed and exposed to film. If the dot of DNA or mRNA contains a sequence similar to the probe, the film will turn black in that area. Dot blots are a quick and easy way to determine if the target sample has a related sequence, before more detailed analysis by Southern or Northern blotting. Another advantage of dot blots is that multiple samples can be processed in a smaller amount of space.

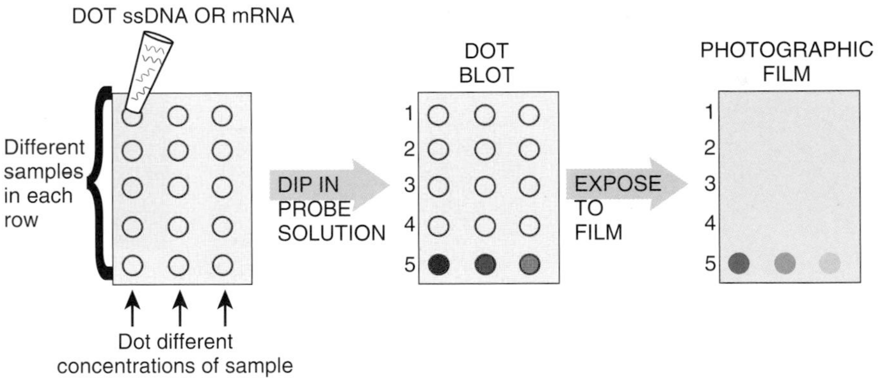

FIGURE 3.12 Dot Blot
Dot blots begin by spotting DNA or RNA samples onto a nylon membrane. Often, different concentrations of the sample are dotted side by side. The membrane is incubated with a radioactive probe and then exposed to photographic film. Samples that contain DNA or RNA complementary to the probe will leave a black spot on the film.

Southern blots form hybrid DNA molecules to determine if a sample of DNA has a homologous sequence to another DNA probe.

Northern blots determine if a sample of mRNA has a homologous sequence to a DNA probe. In large genomes, using mRNA is more efficient because all the introns are removed.

FLUORESCENCE *IN SITU* HYBRIDIZATION (FISH)

The previously discussed hybridization techniques rely on purified DNA or RNA run in an agarose gel. In **fluorescence *in situ* hybridization (FISH)**, the probe is hybridized directly to DNA or RNA within the cell (Fig. 3.13). As described earlier, the probe is a small segment of DNA that has been labeled with fluorescent tags in order to be visualized. The target DNA or RNA is located within the cell and requires some special processing. The target cells may be extremely thin sections of tissue from a particular organism. For example, when a person has a biopsy, a small piece of tissue is removed for analysis. This tissue is preserved and then cut into extremely thin sections to be analyzed under a microscope. These can be used to determine the presence of a gene with FISH. Another source of target cells for FISH is cultured mammalian or insect cells (see Chapter 1). Additionally, blood can be isolated and processed to isolate the white blood cells. (Note: Red blood cells do not contain a nucleus and therefore do not contain DNA.) Chromosomes from white blood cells can be isolated

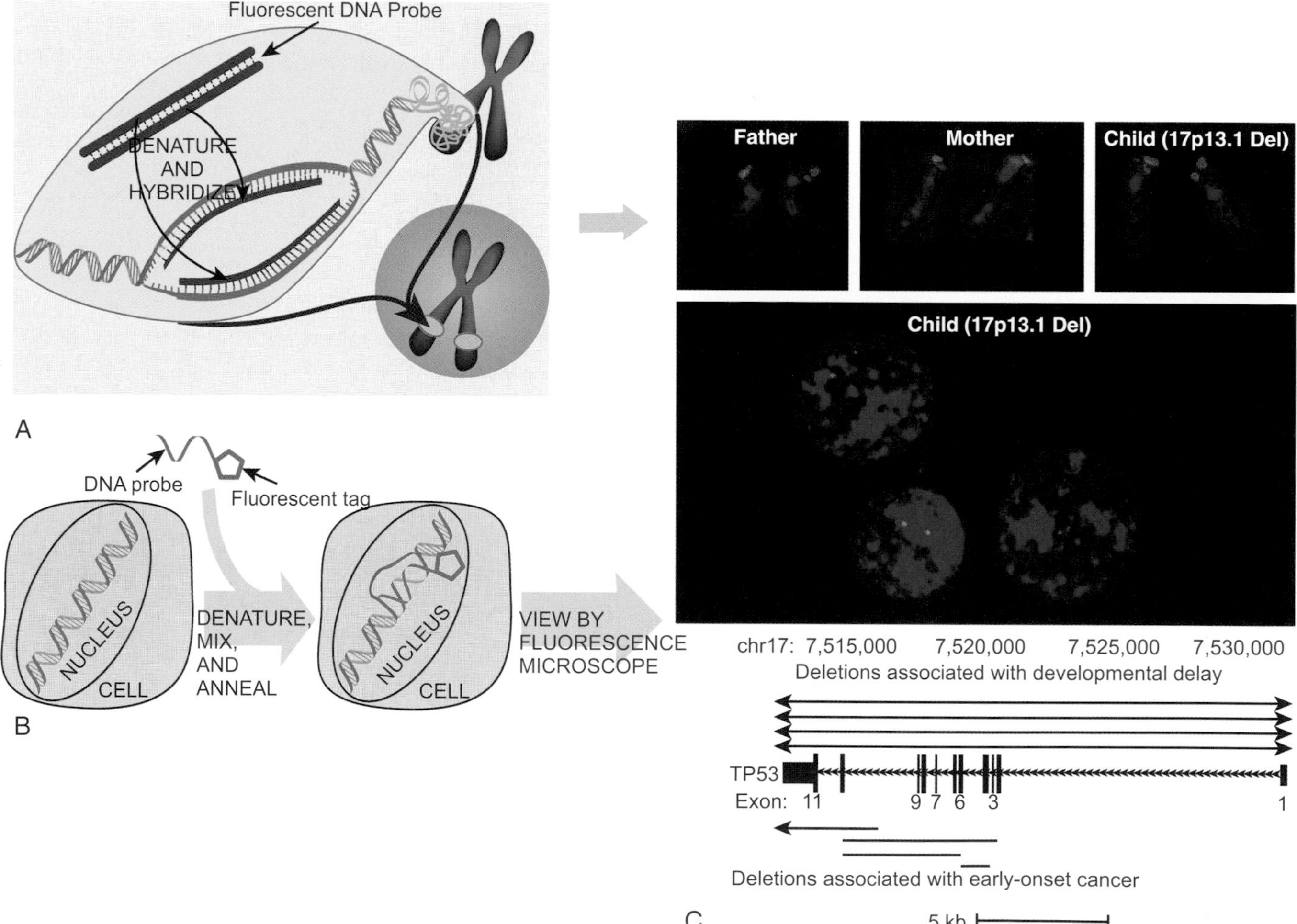

FIGURE 3.13 Gene Location on Chromosomes by FISH

(A) FISH can localize a gene to a specific place on a chromosome. First, metaphase chromosomes are isolated and attached to a microscope slide. The metaphase DNA is denatured into single-stranded pieces that remain attached to the slide. The fluorescently labeled DNA probe hybridizes to the corresponding gene. When the slide is illuminated, the hybrid molecules fluoresce and reveal the location of the gene of interest. (B) A cell with intact DNA in its nucleus is treated to denature the DNA into single strands. The fluorescently labeled DNA probe is added and anneals to the corresponding sequence inside the nucleus. The hybrid molecule will fluoresce when the fluorescent tag is excited by the correct wavelength of light and identifies the location of the gene in the nucleus. (C) An analysis of chromosome structure using TP53 (red) and 17ptel (green) probes. Normal human DNA has two copies of the region of DNA complementary to both the TP53 and 17ptel probes. (Note: Metaphase chromosomes have four copies because the DNA has been replicated.) The parents have a normal DNA structure for these two probe sequences. In comparison, the child has only two red TP53 spots on one chromosome, and the other chromosome has no red spots, suggesting this region of one of the child's chromosomes is deleted. The deletion is visible in both metaphase chromosomes (top) and interphase nuclei (bottom). From Shlien et al. (2010). A common molecular mechanism underlies two phenotypically distinct 17p13.1 microdeletions syndromes. *Am J Hum Gen* **87**, 631–642.

and dropped onto a glass slide. FISH can be done on either interphase or metaphase chromosomes.

Whether the target DNA is from blood, cells cultured in dishes, or actual tissue sections, the cells must be heated to make the DNA single-stranded. Samples where RNA is the target do not need to be heated. The fluorescently labeled probe hybridizes to complementary sequences in the DNA or RNA, and when the cells are illuminated at the appropriate wavelength, the probe location on the chromosome can be identified by fluorescence.

FISH is a technique in which a labeled probe is incubated with cells that have had their DNA denatured by heat. The probe hybridizes to its homologous sequence on the chromosome.

GENERAL PROPERTIES OF CLONING VECTORS

Cloning vectors are specialized plasmids (or other genetic elements) that will hold any piece of foreign DNA for further study or manipulation. The numbers and types of plasmids available for cloning have grown. In addition, other DNA elements are now used, including viruses and artificial chromosomes. Once a fragment of DNA has been cloned and inserted into a suitable vector, large amounts of DNA can be manufactured, the sequence can be determined, and any genes in the fragment can be expressed in other organisms. Studying human genes in humans is virtually impossible because of the ethical ramifications. In contrast, studying a human gene expressed in bacteria provides useful information that can often be applied to humans. Modern biotechnology depends on the ability to express foreign genes in model organisms. Before discussing how a gene is cloned, the properties of vectors are considered.

Useful Traits for Cloning Vectors

Although many specialized vectors now exist, the following properties are convenient and found in most modern generalized cloning plasmids:

- Small size, making them easy to manipulate once they are isolated
- Easy to transfer from cell to cell (usually by transformation)
- Easy to isolate from the host organism
- Easy to detect and select
- Multiple copies, which helps in obtaining large amounts of DNA
- Clustered restriction sites (polylinker) to allow insertion of cloned DNA
- Method to detect presence of inserted DNA (e.g., alpha complementation)

Most bacterial plasmids satisfy the first three requirements. The next key trait of cloning vectors is an easy way to detect their presence in the host organism. Bacterial cloning plasmids often have antibiotic resistance genes that make bacteria resistant to particular antibiotics. When treated with the antibiotic, only bacteria with the plasmid-encoded resistance gene will survive. Other bacteria will die. Other traits have been exploited to detect plasmids. Vectors derived from the yeast 2μ plasmid often carry genes for synthesizing essential amino acids, such as leucine, which allow yeast with mutations in leucine biosynthesis to grow on media lacking leucine.

Plasmids vary in their **copy number**. Some plasmids exist in just one or a few copies in their host cells, whereas others exist in multiple copies. Such multicopy plasmids are in general more useful as the amount of plasmid DNA is higher, making them easier to isolate and purify. The type of origin of replication controls the copy number, since this region on the plasmid determines how often DNA polymerase binds and induces replication.

Most cloning vectors have several unique restriction enzyme sites. Usually, these sites are grouped in one location called the **multiple cloning site (MCS)** or **polylinker** (Fig. 3.14). This allows

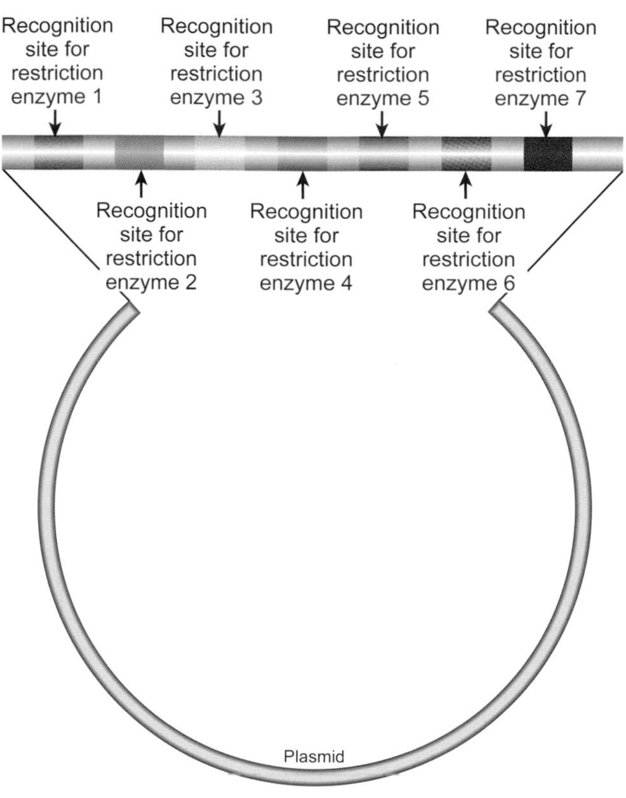

FIGURE 3.14 Typical Polylinker or Multiple Cloning Site
The restriction enzyme sites within the polylinker region are unique. This ensures that the plasmid is cut only once by each restriction enzyme.

researchers to open the cloning vector at one site without disrupting any of the vector's replication genes. Fragments of foreign DNA are digested with enzymes matching those in the polylinker. Ligase connects the vector and insert. Specific restriction enzyme sites can be added using PCR primers or synthetic DNA oligomers (see Chapter 4).

Some vectors have ways to detect whether or not they contain an insert. The simplest way to do this is **insertional inactivation** of an antibiotic gene (Fig. 3.15A). Here, the vector has two different antibiotic resistance genes. The foreign DNA is inserted into one of the antibiotic-resistant genes. Thus, the host bacteria will be resistant to one antibiotic and sensitive to the other.

Alternatively, **alpha complementation** may be used (see Fig. 3.15B). The vector has a short portion of the β-galactosidase gene (the alpha fragment), and the bacterial chromosome has the rest of the gene. If both gene fragments are transcribed and then translated into proteins, the partial proteins combine to form functional β-galactosidase. If DNA is inserted into the plasmid-borne gene segment, the encoded subunit is not made and β-galactosidase is not produced. When β-galactosidase is expressed, the bacteria can degrade X-gal, which turns the bacterial colony blue. If a piece of DNA is inserted into the alpha fragment gene, the bacteria cannot split X-gal and they stay white.

Once an appropriate vector has been chosen for the gene of interest or other insert, the two pieces are ligated into one **construct**. The term *construct* refers to any recombinant DNA molecule that has been assembled by genetic engineering. If both the vector and insert are cut with the same restriction enzyme, the two pieces have complementary ends and require only ligase to link them. Tricks are used to make two pieces of DNA with unrelated ends compatible. Sometimes, short oligonucleotides are synthesized and added onto the ends of the insert to make them compatible with the vector. These short oligonucleotides are called **linkers**, and they add one or a few new restriction enzyme sites to the ends of a segment of DNA. In addition to adding linkers, ends of DNA fragments can be made compatible with a vector multicloning site by PCR amplification. The primers can have extensions of DNA sequence containing the recognition site for a restriction enzyme. After PCR amplification, the DNA can be cut with the restriction enzyme and ligated into the vector (see Chapter 4 for discussion).

Cloning vectors have multiple cloning sites with many unique restriction enzyme sites, they have genes for antibiotic resistance that make the bacterial cell able to grow with the antibiotic present, and they have a way to detect when a foreign piece of DNA is present such as alpha complementation.

SPECIFIC TYPES OF CLONING VECTORS

Because *E. coli* is the main host organism used for manipulating DNA, most vectors are based on plasmids or viruses that can survive in *E. coli* or similar bacteria. Most vectors have

INSERTIONAL INACTIVATION

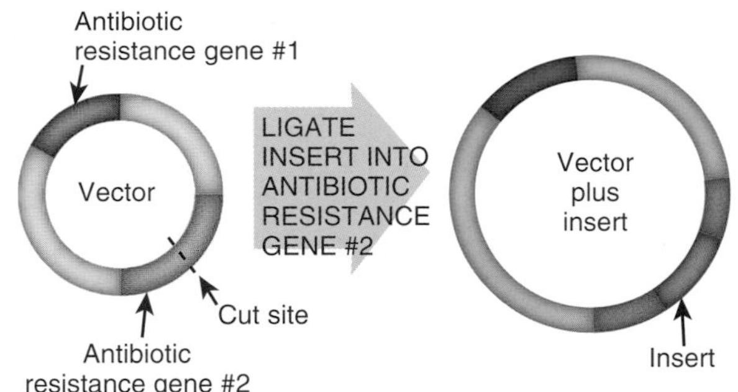

ALPHA COMPLEMENTATION

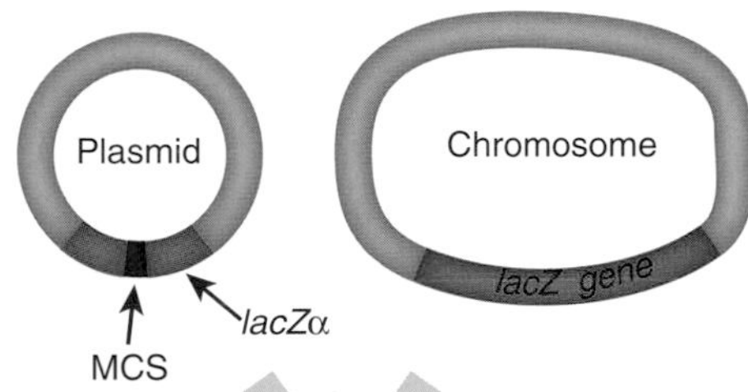

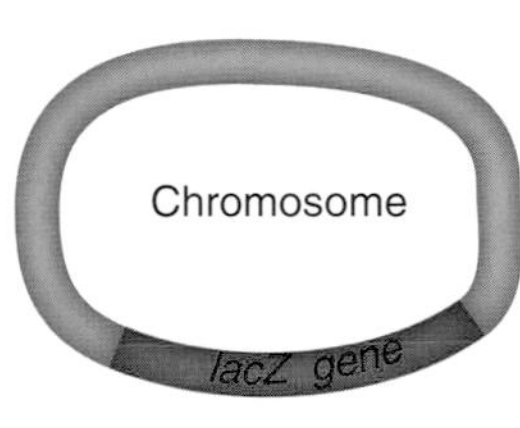

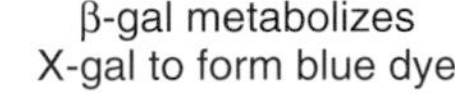

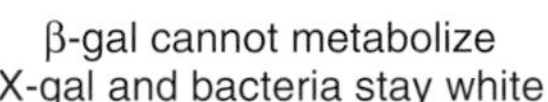

FIGURE 3.15 Detecting Inserts in Plasmids
(A) Insertional inactivation. Cells with an insert become sensitive to the second antibiotic. Cells without an insert remain resistant to the antibiotic. (B) Alpha complementation. Alpha complementation refers to the ability of β-galactosidase to be expressed as two protein fragments, which assemble to form a functional protein. In cells without an insert in the plasmid, β-galactosidase is active and splits X-gal to form a blue dye. In cells with an insert, the alpha fragment is not made and β-galactosidase is inactive. These cells remain white on media with X-gal.

bacterial origins of replication and antibiotic resistance genes. The polylinker or multiple cloning site is usually placed between prokaryotic promoter and terminator sequences (Fig. 3.16A). Some vectors may also supply the ribosome binding site, so any inserted coding sequence will be expressed as a protein. Many other features are present in specialized cloning vectors. The following discussion will introduce some of the different categories of vectors with their essential features.

Many yeast vectors are based on the **2μ circle** of yeast. The native version of the 2μ circle has been modified in a variety of ways for use as a cloning vector. A **shuttle vector** contains origins of replication for two organisms plus any other sequences necessary to survive in either organism (see Fig. 3.16B). Shuttle vectors that are based on the 2μ plasmid have the components needed for survival in yeast and bacteria, plus antibiotic resistance and a polylinker. The *Cen* sequence is a eukaryotic **centromere (Cen) sequence** that keeps the plasmid in the correct location during mitosis and meiosis in yeast. Because yeast cells are eukaryotic and also have such a thick cell

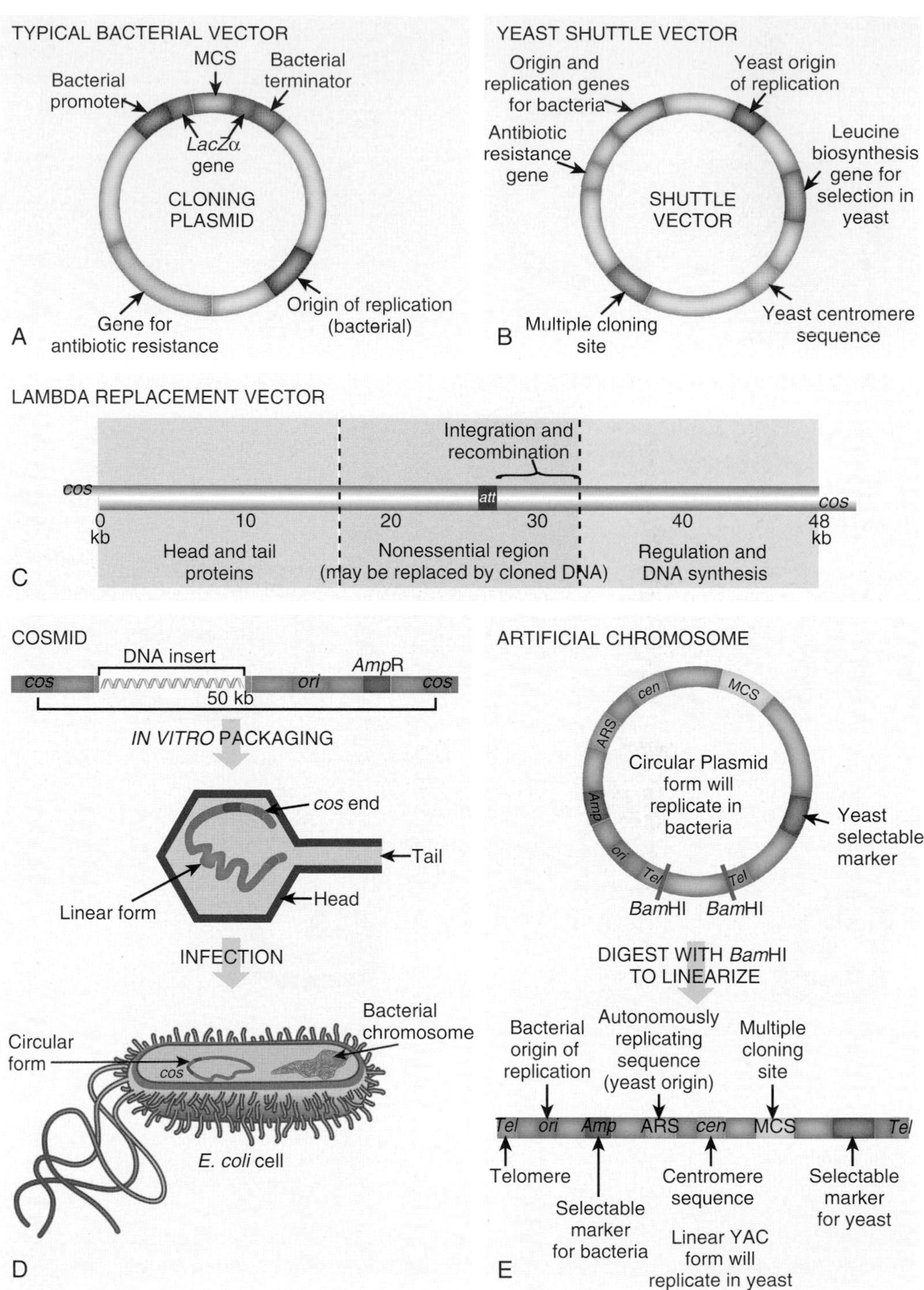

FIGURE 3.16 Various Cloning Vectors
(A) Typical bacterial cloning vector. This vector has bacterial sequences to initiate replication and transcription. In addition, it has a multiple cloning site embedded within the *lacZ* α gene so that the insert can be identified by alpha-complementation. The antibiotic resistance gene allows the researcher to identify any *E. coli* cells that have the plasmid. (B) Yeast shuttle vector. This vector can survive in either bacteria or yeast because it has both yeast and bacterial origin of replication, a yeast centromere, and selectable markers for yeast and bacteria. As with most cloning vectors, there is a polylinker. (C) Lambda replacement vectors. Because lambda phage is easy to grow and manipulate, its genome has been modified to accept foreign DNA inserts. The region of the genome shown in green is nonessential for lambda growth and packaging. This region can be replaced with large inserts of foreign DNA (up to about 23 kb). (D) Cosmids. Cosmids are small multicopy plasmids that carry *cos* sites. They are linearized and cut so that each half has a *cos* site (not shown). Next, foreign DNA is inserted to relink the two halves of the cosmid DNA. This construct is packaged into lambda virus heads and used to infect *E. coli*. (E) Artificial chromosomes. Yeast artificial chromosomes have two forms: a circular form for growing in bacteria and a linear form for growing in yeast. The circular form is maintained like any other plasmid in bacteria, but the linear form must have telomere sequences to be maintained in yeast. The linear form can hold up to 2000 kb of cloned DNA and is very useful for genomics research.

wall, most antibiotics do not kill yeast. Therefore, a different strategy is used to detect the presence of plasmids in yeast. A gene for synthesis of an amino acid, such as leucine, allows strains of yeast that require leucine to grow.

Bacteriophage vectors are viral genomes that have been modified so that large pieces of nonviral DNA can be packaged in the virus particle. Lambda bacteriophages have linear genomes with two cohesive ends—***cos*** **sequences (lambda cohesive ends)**. These are 12-base overlapping sticky ends. When inside the virus coat, the cohesive ends are coated with protein to prevent them from annealing. After lambda attaches to *E. coli*, it inserts just the linear DNA. The proteins that protect the cohesive ends are lost, and the genome circularizes with the help of DNA ligase. The circular form is the **replicative form (RF)**, and it replicates by the rolling circle mechanism (see Chapter 4). Expression of various lambda genes produces the proteins that assemble into protein coats. Each coat is packaged with one genome, and after many of these are assembled, the *E. coli* host explodes, releasing the new bacteriophage to infect other cells.

The lambda bacteriophage is a widely used cloning vector (see Fig. 3.16C). The middle segment of the lambda genome has been deleted and a polylinker has been added. An insert of 37 to 52 kb can be ligated into the polylinker and packaged into viral particles. To work with the bacteriophage DNA without killing the entire *E. coli* culture, the researcher deletes one or more genes necessary for packaging. When the researcher wants to form fully packaged bacteriophages, coat proteins from **helper virus** can be added (Fig. 3.17). The helper viruses do not contain foreign DNA, but supply the missing genes for the coat proteins. Because coat proteins self-assemble *in vitro*, helper lysates are mixed with recombinant lambda DNA, and complete virus particles containing DNA are produced. This is known as *in vitro* packaging.

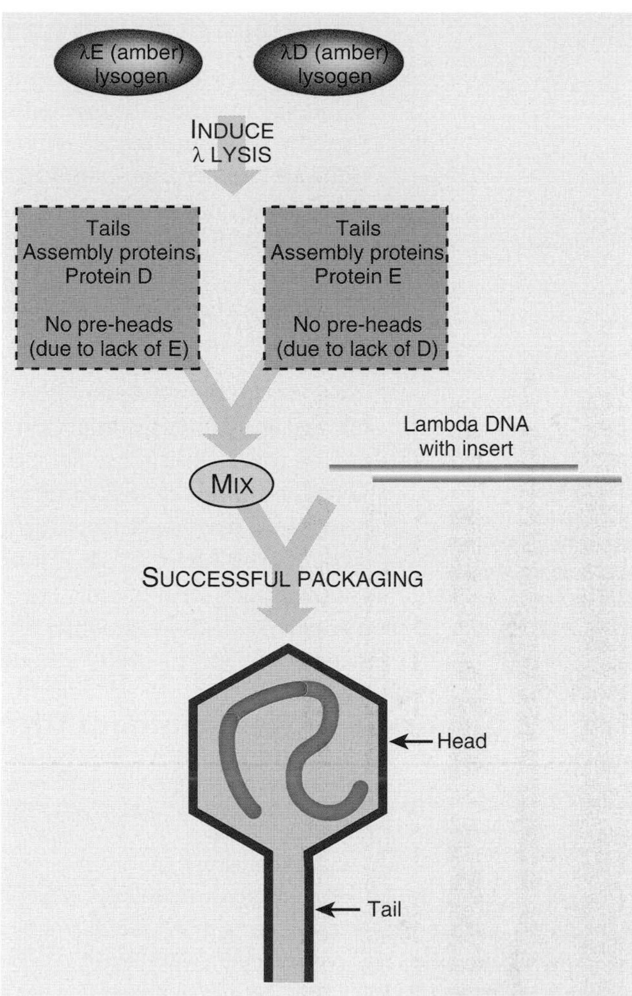

FIGURE 3.17 ***In Vitro* Packaging**
A lambda cloning vector containing cloned DNA must be packaged in a phage head before it can infect *E. coli*. First, one culture of *E. coli* cells is infected with a mutant lambda that lacks the gene for one of the head proteins called E. A different culture of *E. coli* is infected with a different mutant, which lacks the phage head protein D. The two cultures are induced to lyse, which releases the tails, assembly proteins, and head proteins, but no complete heads because of the missing proteins. When these are mixed with a lambda replacement vector, the three spontaneously form complete viral particles containing DNA. These are then used to infect *E. coli*.

Cosmid vectors can hold pieces of DNA up to 45 kb in length (see Fig. 3.16D). These are highly modified lambda vectors with all the sequences between the *cos* sites removed and replaced with the insert. The DNA of interest is ligated between the two *cos* sites using restriction enzymes and ligase. This construct is packaged into a lambda particle produced by helper phage (see Fig. 3.17), and then these are used to infect *E. coli*.

Artificial chromosomes hold the largest pieces of DNA (see Fig. 3.16E). These include yeast artificial chromosomes (YACs), bacterial artificial chromosomes (BACs), and P1 bacteriophage artificial chromosomes (PACs). They are used to contain lengths of DNA from 150 kb to 2000 kb. YACs hold the largest amount of DNA, up to about 2000 kb. YACs have yeast centromeres and yeast telomeres for maintenance in yeast. BACs can be circularized and grown in bacteria; therefore, they have a bacterial origin of replication and antibiotic resistance genes.

Many different cloning vectors are available to biotechnology research. The smaller genes are studied using bacterial plasmids or shuttle vectors, whereas the larger genes are studied in bacteriophage vectors, cosmids, and artificial chromosomes.

Shuttle vectors have sequences that enable them to survive in two different organisms such as yeast and bacteria.

GETTING CLONED GENES INTO BACTERIA BY TRANSFORMATION

Once the gene of interest is cloned into a vector, the construct can be put back into a bacterial cell through a process called **transformation** (Fig. 3.18) (see Box 3.2). Here, the DNA construct is mixed with **competent** *E. coli* cells. To make the cells competent, that is, able to take up DNA, the cell wall must be temporarily opened up. *E. coli* cells are mixed with calcium ions on ice and then shocked at a higher temperature such as 42°C for a few minutes, which destabilizes the membrane and cell wall. Most of the cells die during the treatment, but some survive and take up the DNA. Another method to make *E. coli* cells competent is to expose them to a high-voltage shock. **Electroporation** opens the cell wall, allowing the DNA to enter. This method is much faster and more versatile. Electroporation is used for other types of bacteria as well as yeast.

Bacteria can have different plasmids, but they must have different origin of replications. If there are two different plasmids with the same origin of replication, one of the two will be lost during bacterial replication. For example, if genes A and B are both cloned into the same kind of vector and both cloned genes get into the same bacterial cell, the bacteria will lose one plasmid and keep the other. This is due to **plasmid incompatibility**, which prevents one bacterial cell from harboring two of the same type of plasmid. Incompatibility stems from conflicts between two plasmids with identical or related origins of replication. Only one is allowed to replicate in any given cell. If a researcher wants a cell to have two cloned genes, then two different types of plasmids could be used, or both genes could be put onto the same plasmid.

Box 3.2 Discovery of Recombinant DNA

In 1972, two researchers met at a conference in Hawaii to discuss plasmids, the small rings of extrachromosomal DNA found in bacteria. Herbert W. Boyer, PhD, was a faculty member at the University of California, San Diego, and he was studying restriction and modification enzymes. He had just presented his research on EcoRI. Stanley N. Cohen, MD, was a faculty member at Stanford, and he was interested in how plasmids could confer resistance to different antibiotics. His lab perfected laboratory transformation of *Escherichia coli* using calcium chloride to permeabilize the cells. After the talks ended, the two met over corned beef sandwiches and combined their two ideas.

They isolated different fragments of DNA from animals, other bacteria, and viruses and, using restriction enzymes, ligated the fragments into a small plasmid from *E. coli.* This was the first recombinant DNA made. Finally, they transformed the engineered plasmid back into *E. coli.* The cells expressed the normal plasmid genes as well as those inserted into the plasmid artificially. This sparked the revolution in genetic engineering, and since then every biotechnology lab has used some variation of their technique. Boyer and Cohen applied for a patent on recombinant DNA technology. In fact, Boyer cofounded Genentech with Robert Swanson, a venture capitalist. Genentech is one of the first biotechnology companies in the United States, and under Boyer and Swanson, the company produced human somatostatin in *E. coli.*

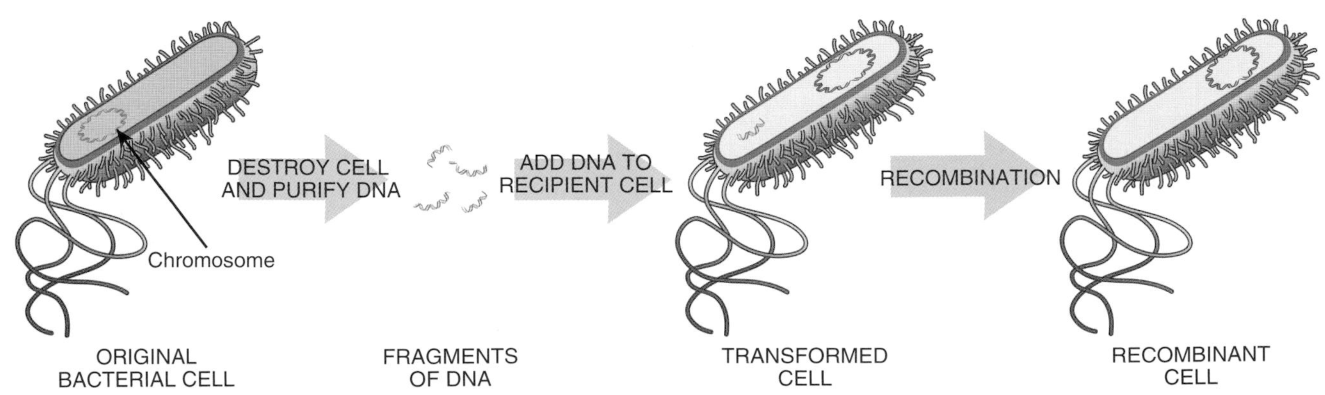

FIGURE 3.18 Transformation

Bacterial cells are able to take up DNA such as recombinant plasmids by incubation with metal ions such as Ca^{++} on ice. This destabilizes the bacterial cell wall, and after a brief heat shock, some of the bacteria take up the DNA or plasmid. If the DNA integrates in the chromosome, the recombinant cell will express any genes found on the DNA. Whole plasmids can also be taken up by a bacterial cell, and these exist as extrachromosomal elements with their own origin of replication.

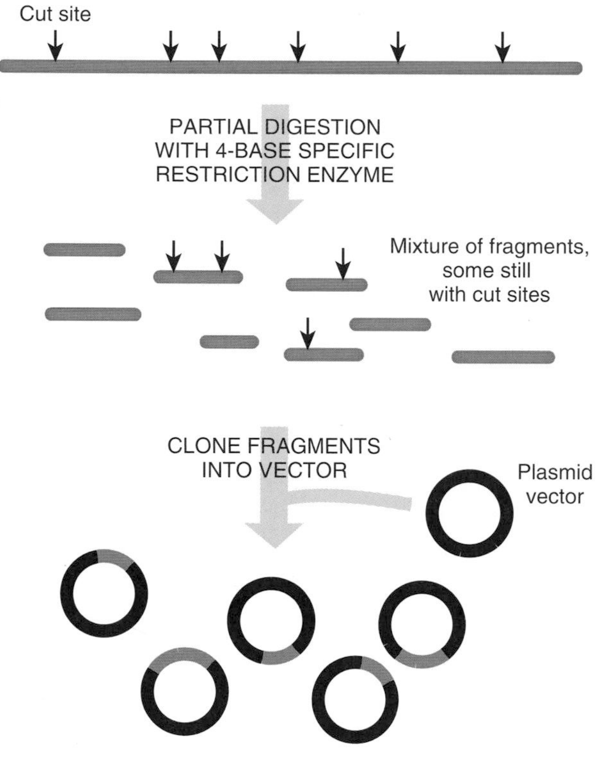

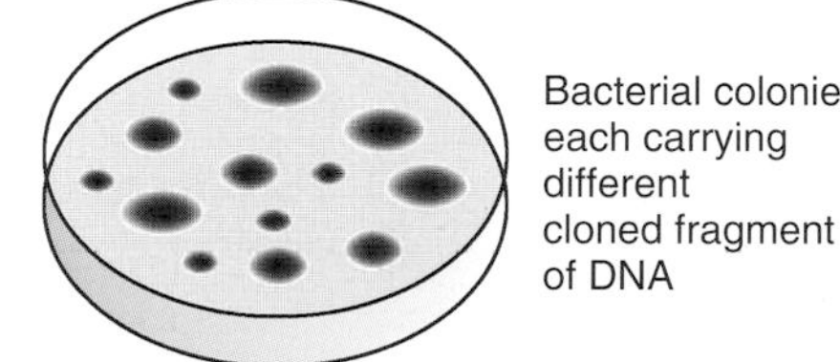

FIGURE 3.19 Creating a DNA Library
Genomic DNA from the chosen organism is first partially digested with a restriction enzyme that recognizes a four base-pair sequence. Partial digestions are preferred because some of the restriction enzyme sites are not cut, and larger fragments are generated. If every recognition site were cut by the restriction enzyme, then the genomic DNA would not contain many whole genes. The genomic fragments are cloned into an appropriate vector, and transformed and maintained in bacteria.

CONSTRUCTING A LIBRARY OF GENES

Gene libraries are used to find new genes, to sequence entire genomes, and to compare genes from different organisms (Fig. 3.19). Gene libraries are made when the entire DNA from one particular organism is digested into fragments using restriction enzymes, and then each of the fragments is cloned into a vector and transformed into an appropriate host.

The basic steps used to construct a library are as follows:

1. Isolate the chromosomal DNA from an organism, such as *E. coli,* yeast, or humans.
2. Digest the DNA with one or two different restriction enzymes.
3. Linearize a suitable cloning vector with compatible restriction enzyme(s) sites.
4. Mix the cut chromosome fragments with the linearized vector and ligate.
5. Transform this mixture into *E. coli.*
6. Isolate large numbers of *E. coli* transformants.

The type of restriction enzyme affects the type of library. Because restriction sites are not evenly spaced in the genome, some inserts will be large and others small. Using a restriction enzyme that recognizes only four base pairs will give a mixture of mostly small fragments, whereas a restriction enzyme that has a six or eight base-pair recognition sequence will generate larger fragments. (Note that finding a particular four base-pair sequence in a genome is more likely than finding a six base-pair sequence.) Even if an enzyme that recognizes a four base-pair recognition sequence is used to digest the entire genome, these sites are not equally spaced throughout the genome of an organism. The digested genome will contain some segments too large to be cloned, and some segments too small. To avoid cutting pieces too small, partial digestion is often used. The enzyme is allowed to cut the DNA for only a short time, and many of the restriction enzyme sites are not cut, leaving larger pieces for the library. In addition, it is customary to construct another library using a different restriction enzyme.

> Gene libraries are used for many purposes because they contain almost the entire genome of a particular organism.

SCREENING THE LIBRARY OF GENES BY HYBRIDIZATION

Once the library is assembled, researchers often want to identify a particular gene or segment of DNA within the library. Sometimes the gene of interest is similar to one from another organism. Sometimes the gene of interest contains a particular sequence. For example, many enzymes use ATP to provide energy. Enzymes that bind ATP share a common signature sequence whether they come from bacteria or humans. This sequence can be used to find other enzymes that bind ATP. Such common sequence motifs may also suggest that a protein will bind various cofactors, other proteins, and DNA, to name a few examples.

Screening DNA libraries by hybridization requires preparing the library DNA and preparing the labeled probe. A gene library is stored as a bacterial culture of *E. coli* cells, each having a plasmid with a different insert. The culture is grown up, diluted, and plated onto many different agar plates so that the colonies are spaced apart from one another. The colonies are transferred to a nylon filter and the DNA from each colony is released from the cells by lysing them with detergent. The cellular components are rinsed from the filters. The DNA sticks to the nylon membrane and is then denatured to form single strands (Fig. 3.20).

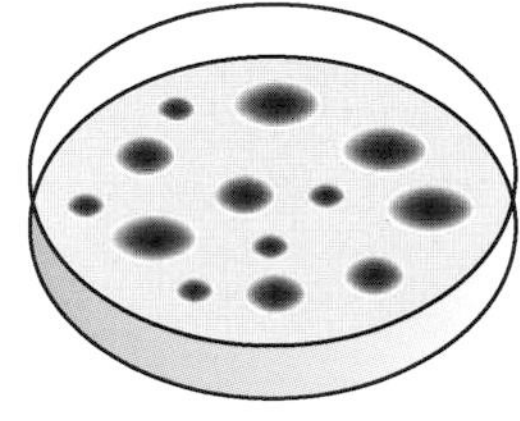

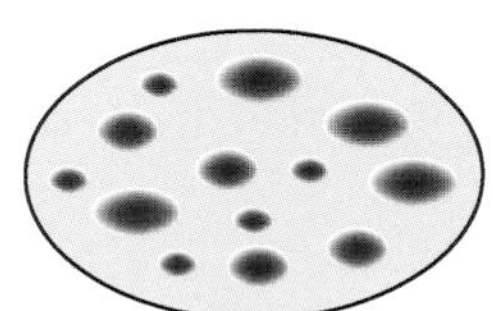

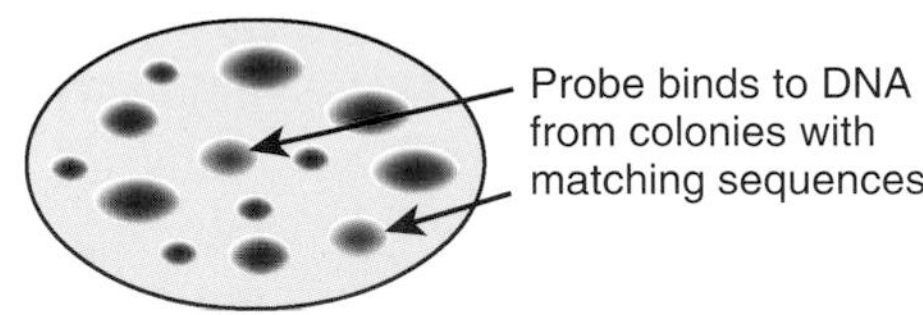

FIGURE 3.20
Screening a Library with DNA Probe
First, bacterial colonies containing the library inserts are grown and plated on large, shallow agar-filled dishes. Many different colonies are plated so that every cloned piece of DNA is present. These colonies are transferred to nylon filters and lysed open. The cell remains are washed away, while the genomic and plasmid DNA sticks to the nylon. The sequences are made single-stranded by incubating the filters in a strong base. When these are incubated with a radioactive single-stranded probe at the appropriate temperature, the probe hybridizes to any matching sequences.

If a scientist is looking for a particular gene in the target organism, the probe for the library may be the corresponding gene from a related organism. The probe is usually just a segment of the gene because a smaller piece is easier to manipulate. The probe DNA is then synthesized and labeled either with radioactivity or with chemiluminescence. Single-stranded probe DNA is mixed with the library DNA on the nylon filters. The probe hybridizes with matching sequences in the library. The level of match needed for binding can be adjusted by incubating at various temperatures. The higher the temperature, the more stringent, that is, the more closely matched the sequences must be. The lower the temperature, the less stringent. If the probe is labeled with radioactivity, photographic film will turn black where the probe and library DNA hybridized. The black spot is aligned with the original bacterial colony. Usually, the most likely colony plus its neighbors are selected, grown, plated, and rescreened with the same probe to ensure that a single transformant is isolated. Then the DNA from this isolate can be analyzed by sequencing (see Chapter 4).

Screening a library has two parts. First, the library clones growing in *E. coli* are attached to nylon filters, and the cellular components washed away and then denatured to form single-stranded DNA pieces. Second, a probe is labeled with radioactivity, is heated to melt the helix into single strands, and finally added to the nylon membranes where it hybridizes to its matching sequence.

EUKARYOTIC EXPRESSION LIBRARIES

In **expression libraries**, the vector has sequences required for transcription and translation of the insert. This means that the insert DNA is expressed as RNA and then translated into a protein. An expression library, in essence, generates a protein from every cloned insert, whether it is a real gene or not. When eukaryotic DNA is studied, expression libraries are constructed using **complementary DNA (cDNA)** to help ensure the insert is truly a gene. Eukaryotic DNA, especially in higher plants and animals, is largely noncoding, with coding regions spaced far apart. Even genes are interrupted with noncoding introns. cDNA is a double-stranded DNA copy of mRNA. cDNA is made by reverse transcriptase, an enzyme first identified in retroviruses (see Chapter 1). It is used in eukaryotic research to eliminate the introns and generate a version of a gene consisting solely of an uninterrupted coding sequence.

In contrast, bacteria have very little noncoding DNA, and their genes are not interrupted by introns; therefore, genomic DNA can be used directly in expression libraries.

Eukaryotic DNA is first made into cDNA in order to construct an expression library (Fig. 3.21). To make cDNA, the messenger RNA is isolated from the organism of interest by binding to a column containing poly(T) (i.e., a DNA strand consisting of repeated thymines). This isolates only mRNA because poly(T) anneals to the poly(A) tail of eukaryotic mRNA.

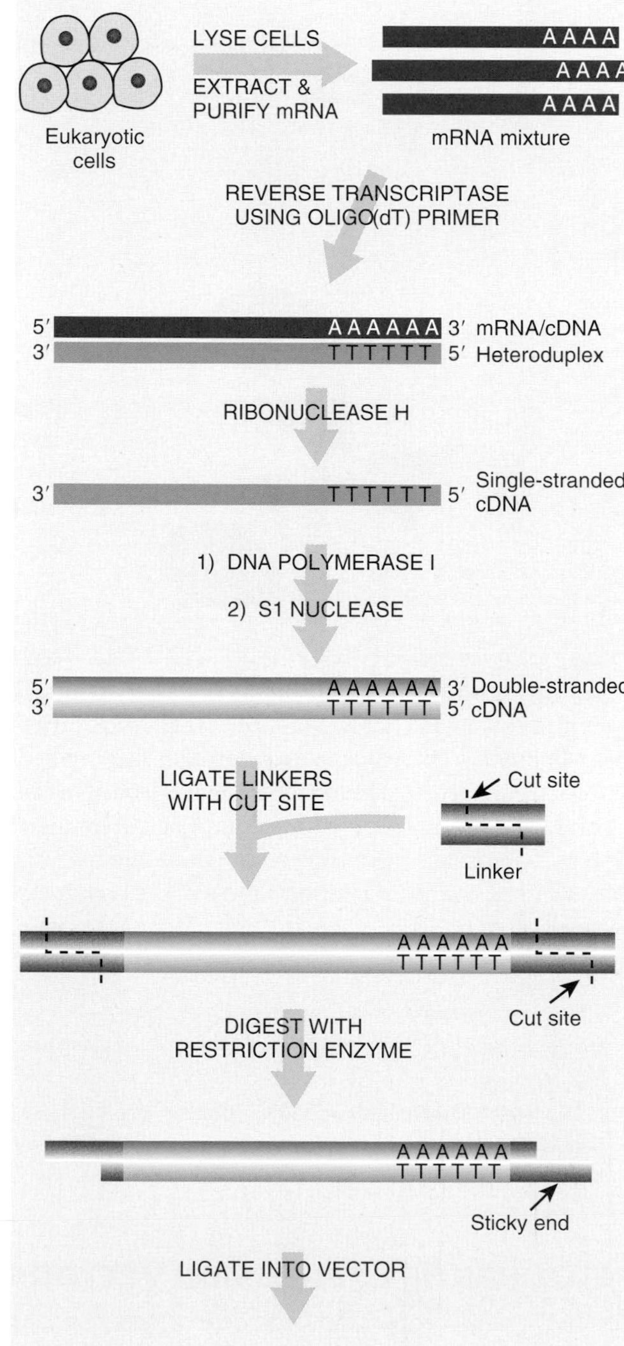

FIGURE 3.21 Making a cDNA Library from Eukaryotic mRNA
First, eukaryotic cells are lysed and the mRNA is purified. Next, reverse transcriptase plus primers containing oligo(dT) stretches are added. The oligo(dT) hybridizes to the adenine in the mRNA poly(A) tail and acts as a primer for reverse transcriptase. This enzyme makes the complementary DNA strand, forming an mRNA/cDNA heteroduplex. The mRNA strand is digested with ribonuclease H, and DNA polymerase I is added to synthesize the opposite DNA strand, thus creating double-stranded cDNA. Next, S1 nuclease is added to trim off any single-stranded ends, and linkers are added to the ends of the dsDNA. The linkers have convenient restriction enzyme sites for cloning into an expression vector.

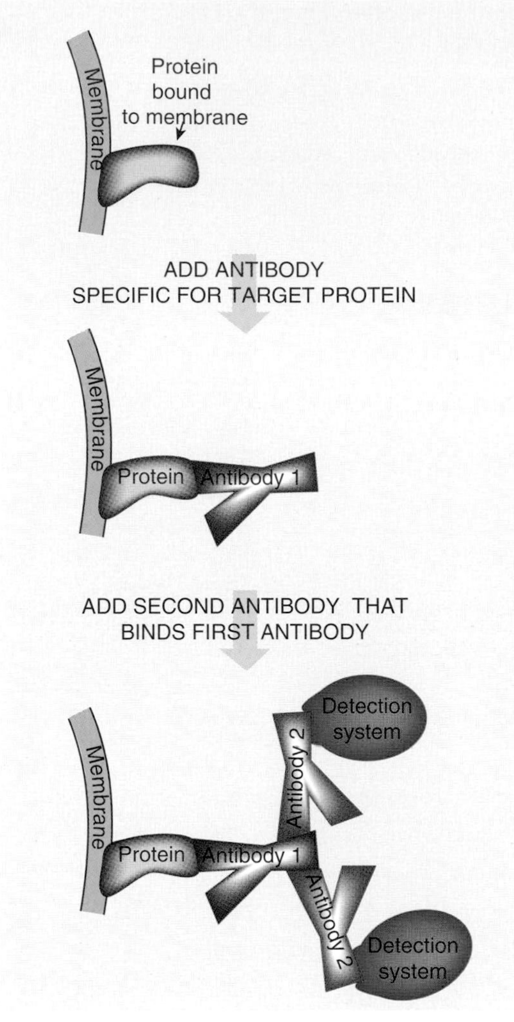

FIGURE 3.22 Immunological Screening of an Expression Library
Bacteria expressing foreign genes are grown on an agar plate, transferred to a membrane, and lysed. Released proteins are bound to the membrane. This figure shows only one attached protein, although in reality many different proteins are present. These include both expressed library clones and bacterial proteins. The membrane is incubated with a primary antibody that binds only the protein of interest. To detect this protein:antibody complex, a second antibody with a detection system such as alkaline phosphatase is added. The bacterial colony expressing the protein of interest will turn blue when X-Phos is added. This allows the vector with the correct insert to be isolated.

The mRNA is converted into cDNA using reverse transcriptase, which synthesizes a DNA complement to mRNA. An enzyme then removes the mRNA part of the mRNA/cDNA heteroduplex, and DNA polymerase makes the second strand of DNA (see Fig. 3.21). The final product is a double-stranded DNA copy of the mRNA sequence.

The cDNA is then ligated into an **expression vector** with sequences that initiate transcription and translation of the insert. In some cases, the insert will have its own translation start site (e.g., a full-length cDNA). If the insert does not contain a translation start, then the reading frame becomes an issue. Because the genetic code is triplet, each insert can be translated in three different reading frames. A protein may be produced for all three reading frames, but only one frame will actually produce the correct protein. To ensure obtaining inserts with the correct reading frame, each cDNA is cloned in all three reading frames by using linkers with several different restriction sites. The number of transformants to screen for a protein of interest is therefore increased.

The cloned genes are transformed into bacteria, which express the foreign DNA. The bacteria are grown on agar, and the colonies are then transferred to a nylon membrane and lysed. The proteins released are attached to the nylon membranes and are screened in various ways. Most often, an antibody to the protein of interest is used (see Chapter 6). This recognizes the protein and can be identified using a secondary antibody that is conjugated to a detection system. Usually, alkaline phosphatase is conjugated to the secondary antibody. The whole complex can be identified because alkaline phosphatase cleaves X-Phos, leaving a blue color where the bacterial colony expressed the right protein (Fig. 3.22). *E. coli* cannot perform most of the post-translational modifications that eukaryotic proteins often undergo. Therefore, the proteins are not always in their native form. Nonetheless, appropriate antibodies can detect most proteins of interest.

Complementary DNA or cDNA is constructed by isolating mRNA and making a DNA copy with reverse transcriptase.

Expression libraries express the foreign DNA insert as a protein because expression vectors contain sequences for both transcription and translation. The protein of interest is identified by incubating the library with an antibody to the protein of interest.

FEATURES OF EXPRESSION VECTORS

Because foreign protein can be toxic to *E. coli*, especially if made in large amounts, the promoter used to express the foreign gene is critical. If too much foreign protein is made, the host cell may die. To control protein production, expression vectors have promoters with on/off switches; therefore, the host cell is allowed to grow and then after sufficient amounts of bacteria are produced, the gene of interested is turned on. One commonly used promoter is a mutant version of the *lac* promoter, *lacUV*, which drives a very high level of transcription, but only under induced conditions (Fig. 3.23). It has the following elements: a binding site for RNA polymerase, a binding site for the LacI repressor protein, and a transcription start site. The vector has strong transcription stop sites downstream of the polylinker region. The vector also has the gene for LacI so that high levels of repressor protein are made, thus keeping the cloned genes repressed. Like all vectors, there is an origin of replication and antibiotic resistance gene for selection in bacteria. When a gene library is cloned behind this promoter, the genes are not expressed due to high levels of LacI repressor. When an inducer, such as IPTG, is added, LacI is released from the DNA, and RNA polymerase transcribes the cloned gene.

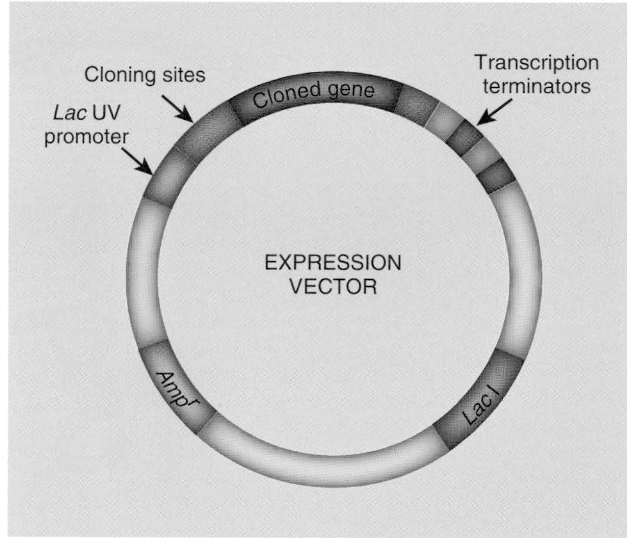

FIGURE 3.23 Expression Vectors Have Tightly Regulated Promoters
An expression vector contains sequences upstream of the cloned gene that control transcription and translation of the cloned gene. The expression vector shown uses the *lacUV* promoter, which is very strong, but inducible. To stimulate transcription, the artificial inducer, IPTG, is added. IPTG binds to the LacI repressor protein, which then detaches from the DNA. This allows RNA polymerase to transcribe the gene. Before IPTG is added, the LacI repressor prevents expression of the cloned gene.

Another common promoter in expression vectors is the lambda left promoter, or P_L. It has a binding site for the lambda repressor. The gene of interest or library fragment is not expressed unless the repressor is removed. Rather than using its natural inducer, a mutant version of the repressor has been isolated that releases its binding site at high temperatures. So when the culture is shifted to 42 °C, the repressor falls off the DNA, and RNA polymerase transcribes the cloned genes.

Another expression system uses a promoter whose RNA polymerase binding site recognizes only RNA polymerase from the bacteriophage T7. Bacterial RNA polymerase will not transcribe the gene of interest. This system is designed to work only in bacteria that have the gene for T7 RNA polymerase integrated into the chromosome and under the control of an inducible promoter.

Some expression vectors contain a small segment of DNA that encodes a protein tag. These are primarily used when the gene of interest is already cloned, rather than for screening libraries. The gene of interest must be cloned in frame with the DNA for the protein tag. The tag can be of many varieties, but **6HIS, Myc,** and **FLAG® tag** are three popular forms (Fig. 3.24). 6HIS is a stretch of six histidine residues put at the beginning or end of the protein of interest. The histidines bind strongly to nickel. This allows the tagged protein to be isolated by binding to a column with nickel attached. Myc and FLAG® are epitopes that allow the expressed protein to be purified by binding to the corresponding antibody. The antibodies may be attached to a column, used for a Western blot, or seen *in vivo* by staining the cells with fluorescently tagged versions of the Myc or FLAG® antibodies. (The histidine tag can also be recognized with a specific antibody, if desired.)

The most important feature of expression vectors is a tightly controlled promoter region. The proteins of the expression library are expressed only under certain conditions, such as presence of an inducer, removal of a repressor, or change in temperature.

Small tags can be fused into the protein of interest using expression vectors. These tags allow the protein of interest to be isolated and purified.

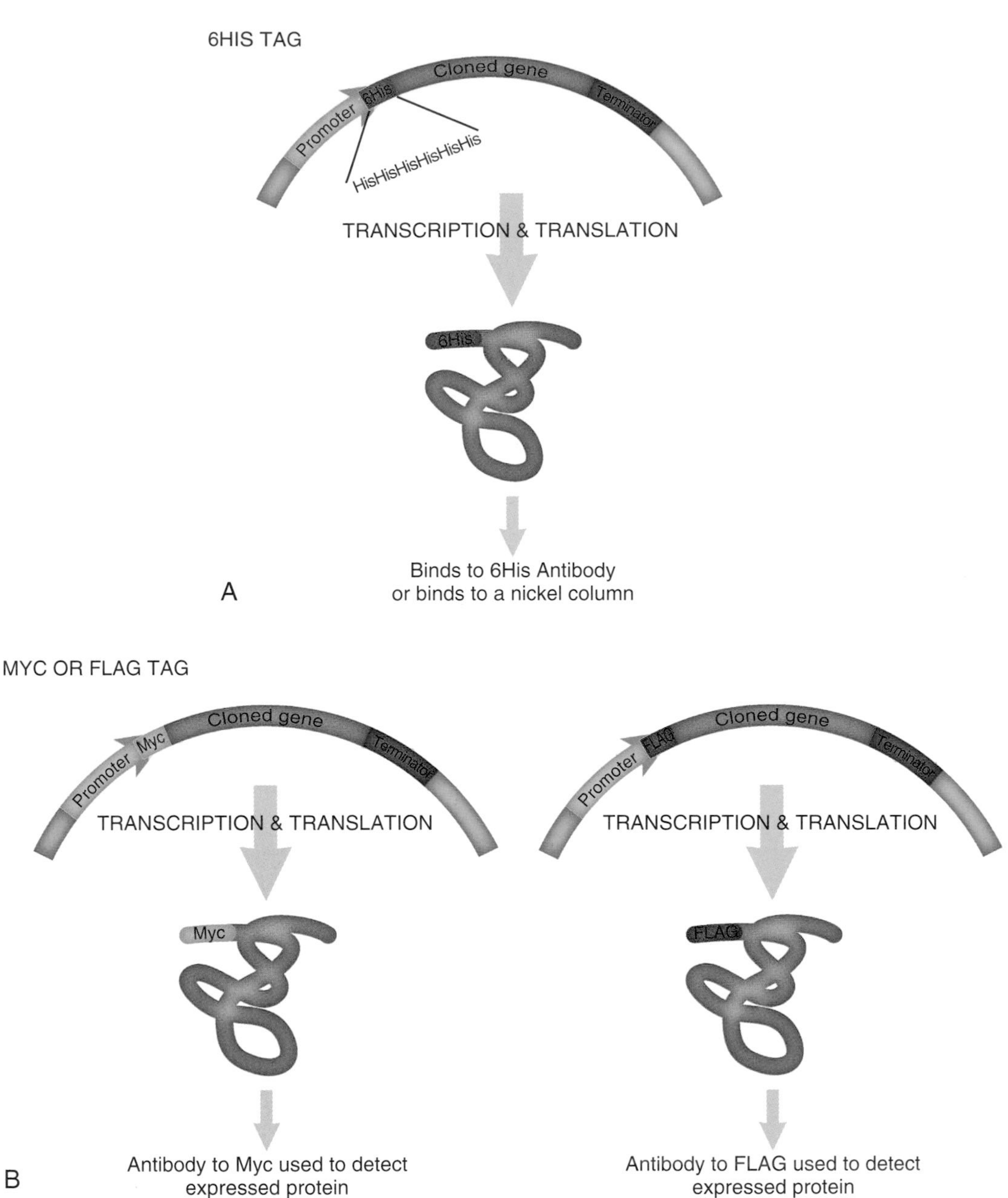

FIGURE 3.24 Using Tags to Isolate Proteins
Some expression vectors have DNA sequences that code for short protein tags. The 6HIS tag (A) codes for six histidine residues. When fused in-frame with the coding sequence for the cloned gene, the tag is fused to the protein. The 6HIS tag specifically binds to nickel ions; therefore, binding to a nickel ion column isolates 6HIS-tagged proteins. Additionally, antibodies to the 6HIS tag can also be used to isolate the tagged proteins. Other tags, such as Myc or FLAG® (B), are specific antibody epitopes that work in a similar manner. Myc-tagged or FLAG®-tagged proteins can be isolated or identified by binding to antibodies to Myc or FLAG®, respectively.

RECOMBINEERING INCREASES THE SPEED OF GENE CLONING

Assembling new DNA vectors with different genes of interest can become difficult when the gene is long, since it can be hard to identify unique restriction enzymes compatible with a polylinker that do not cut within the gene. Large recombinant DNA vectors can be created using homologous recombination, a process called **recombineering** (Fig. 3.25). To facilitate recombination, enzymes from lambda phage called RED are engineered to be expressed by a specific host strain of bacteria. They recognize homologous sequences and recombine them to form a single molecule. These proteins are so efficient that as little as a 45 base-pair region of homology is enough to initiate recombination. In practice, *E. coli* have the genes for the RED proteins under the control of a heat inducible promoter. The gene of interest is electroporated into bacteria that have the lambda RED proteins active in the cytoplasm. They recognize the ends of the gene of interest and find their homologous sequences. In this figure, the homologous sequences are found on the BAC, or bacterial artificial chromosome. The enzymes break the BAC at the appropriate

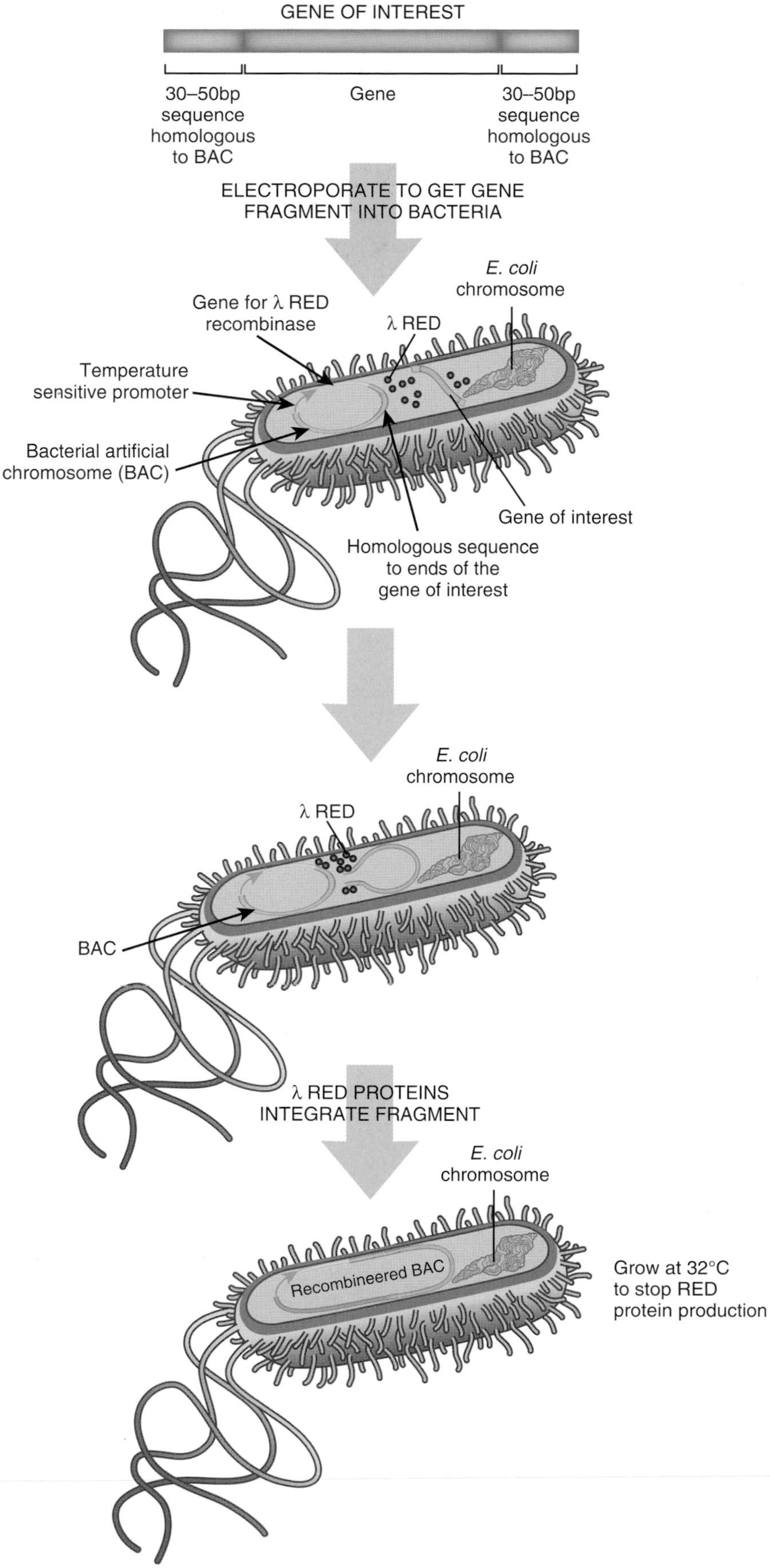

FIGURE 3.25 Recombineering

A gene of Interest is flanked by sequences homologous to the BAC. Once inside the bacteria, the RED proteins recognize the ends of the gene of interest and facilitate homologous recombination between the BAC and gene of interest. The RED proteins are produced only when the bacteria are exposed to high temperatures, since their genes are controlled by a heat-inducible promoter.

location and add the gene of interest. The engineered BAC is removed from this strain of *E. coli* to prevent any residual RED proteins from initiating further recombination.

Identifying which *E. coli* have the gene of interest is different for recombineering because the bacteria have the vector whether or not the insert recombines. Instead of using a positive selection scheme such as antibiotic resistance, a selection/counterselection scheme is used to identify the recombined vector containing the gene of interest (Fig. 3.26). First, the vector contains a gene for *galK*, or galactose kinase, a gene essential for growth on galactose. The bacteria produce galactose kinase and are able to grow on minimal media that contains only galactose as a carbon source. GalK protein also converts 2-deoxygalactose (2-DOG) into a toxic substance, so bacteria expressing GalK die when grown on 2-DOG. After the recombination reaction occurs, the bacteria are plated onto minimal media that have only 2-deoxygalactose. If any bacteria still have *galK*, galactose kinase creates toxin from 2-deoxygalactose, and the bacteria die. When *galK* is replaced with the gene of interest, there is no toxin produced, and the bacteria grow.

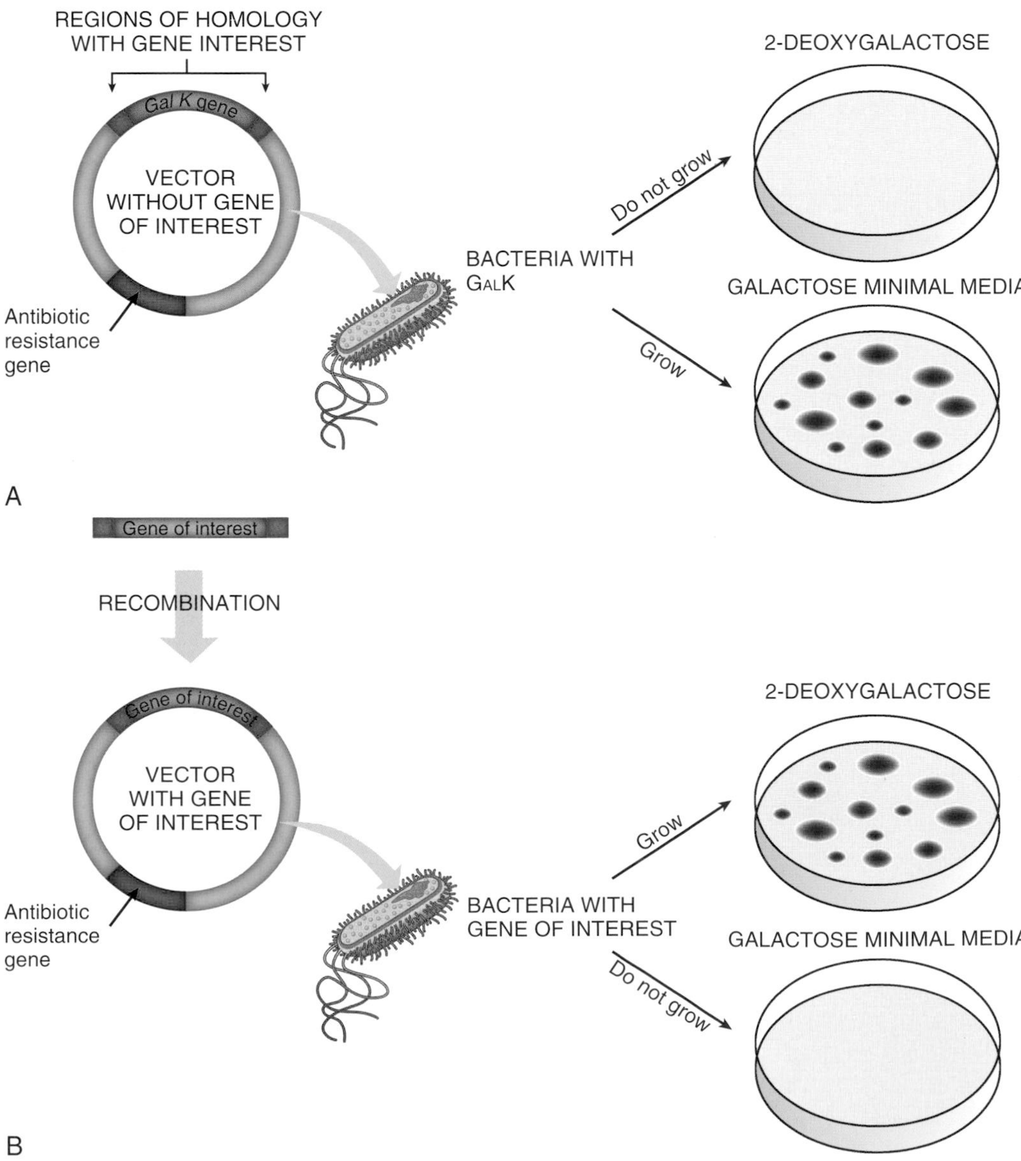

FIGURE 3.26 Selection and Counterselection in Recombineering Recombineering vectors use a selection and counterselection method to identify which bacterium harbors the vector containing the gene of interest. In part A, the gene for *galK* encodes a galactose kinase. When bacteria expressing GalK are grown on 2-deoxygalactose (2-DOG), a toxin is produced, which kills the bacteria (top plate). The GalK also allows the bacteria to grow on galactose minimal media (bottom plate). In part B, recombineering replaces the *galK* gene with the gene of interest, and therefore, the bacteria can no longer grow on galactose minimal media (bottom plate). The lack of GalK allows the bacteria to grow on 2-DOG (top plate).

GATEWAY® CLONING VECTORS

A newer cloning system uses the lambda phage integration and excision sites for cloning genes from one vector to another (Fig. 3.27). Lambda phage exists as a phage but also integrates into the *E. coli* chromosome at the *attB* site to form a prophage. The integration reaction occurs when **integrase** makes staggered cuts in the center of the phage *attP* site and in the center of the bacterial site *attB*. The ends then connect so that the phage DNA is integrated, but notice that the sequences are different than the original. These are called *attL* and *attR* after integration. This reaction can be reversed, but since the two sites are different after integration, another enzyme called Xis, or **excisionase**, removes the inserted DNA and relegates the broken DNA.

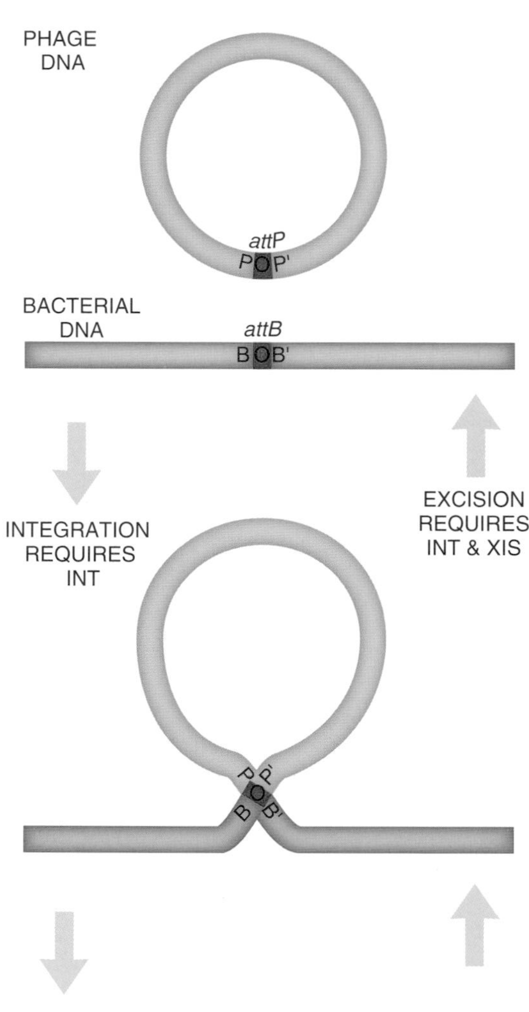

FIGURE 3.27 Integration of Lambda DNA
Phage DNA has an attachment sequence called *attP*. Bacterial DNA has an attachment sequence called *attB*. Bacterial DNA and λ-phage DNA align at the "O" region of the attachment sequences. During integration, int protein induces two double-stranded breaks that are resolved, resulting in the integration of the phage DNA into the bacterial DNA. The process is reversible, and requires int protein and xis protein to excise the phage DNA from the bacterial DNA. Notice that the integrated phage DNA "O" site is flanked with one side from the phage and one side from the bacteria. These are called the *attL* and *attR* sites.

The **Gateway cloning vectors** exploit the lambda phage integration/excision system to study a gene of interest. The vectors include sequences for expressing the gene of interest into protein, adding a protein tag, shuttling the gene between different model organisms, or sequencing the gene. The first step of the system is to get the gene of interest between two *attL* sites (Fig. 3.28). This can be done by cloning the gene into a multicloning site found in the entry clone. The entry clone has a gene called *ccdB* in the middle of the multicloning site, which encodes a toxin that kills the host when expressed. When this gene is expressed in *E. coli*, the bacteria die, which ensures that any bacteria that harbor the original vector die. When the gene of interest replaces the *ccdB* gene, the bacteria are able to grow. (A special strain of *E. coli* with an antitoxin to the *ccdB* gene product allows researchers to maintain the entry clone before the gene of interest is cloned into the vector.)

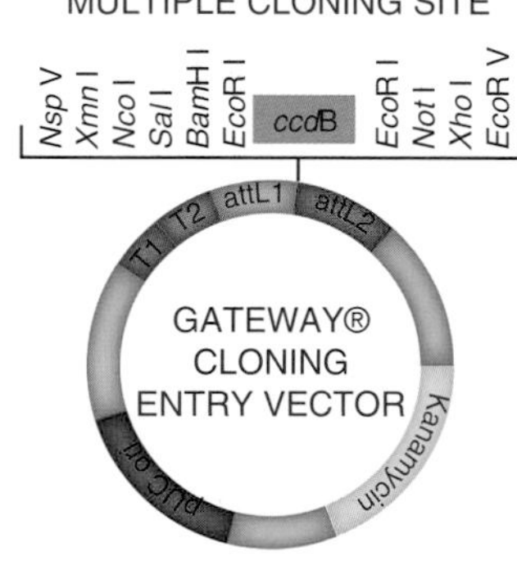

FIGURE 3.28 Gateway® Entry Clone
The entry clone for the Gateway® system has an origin of replication for growing in bacteria, an antibiotic resistance gene for selecting bacteria with the vector, a multicloning site containing the gene *ccdB* in between two *attL* sites (*attL1* and *attL2*). The gene of interest replaces the *ccdB* gene during standard cloning using unique restriction enzyme sites. The *ccdB* gene produces a toxin that kills its host bacteria unless the bacteria has a corresponding gene for an antitoxin.

Once the gene is in the entry vector, two different reactions move it among the different destination clones (Fig. 3.29). The LR reaction removes the gene of interest by cutting at the *attL* sites and moves it into the *attR1* and *attR2* sites in the destination vector. This results in an expression clone that has the gene of interest flanked by *attB1* and *attB2*. The entry vector no longer has the gene of interest, which is replaced with the toxin gene, *ccdB*. Any bacteria receiving this vector die. The only surviving bacteria are those with the gene of interest in the destination vector. Just as with the lambda phage, the reaction is reversible. BP reaction removes the gene of interest from the destination vector and can put it back into any vector with *attL* sites. The ease at which the cloned gene can move between vectors makes this system very adaptable to different research. There are Gateway® cloning vectors for protein expression in bacteria, adding different tags such as HIS6; expressing the gene of interest in insect, human, or mouse cells; or sequencing the gene of interest.

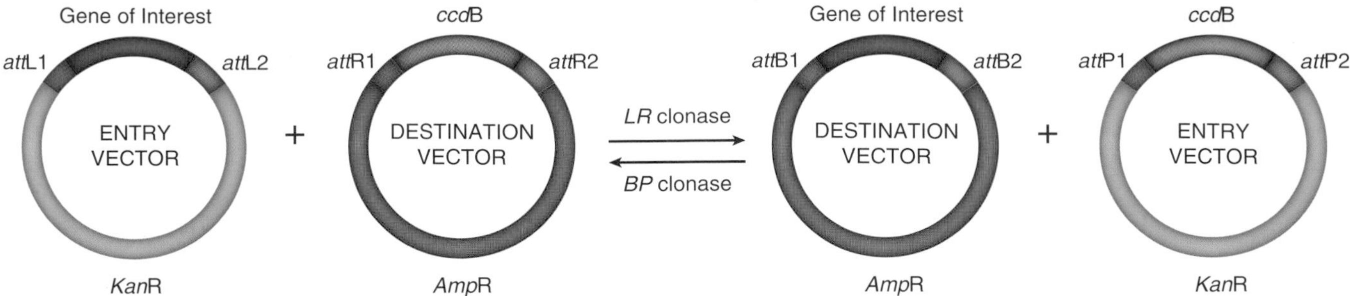

FIGURE 3.29 Gateway® BP and LR Reactions

Moving a gene of interest from the entry clone to the destination vector is done in the LR reaction. The exisionase and integrase enzymes work to remove the gene of interest in the entry clone by cutting at the *attL* and *attR* sites of the entry clone and destination vector. The gene of interest and *ccdB* swap positions, therefore changing the *att* site to become *attB* and *attP*. The BP reaction works in reverse, moving the gene of interest back into the entry clone.

Summary

Recombinant DNA technology is the basis for almost all biotechnology research. Understanding these techniques is tantamount to understanding the rest of the textbook. First, DNA must be isolated from the organism in order to identify novel genes, to recover new recombinant vectors, or to purify a new gene. The DNA is isolated from the cellular components using enzymes followed by centrifugation, RNA digestion with RNase, and precipitation with ethanol. Each organism requires special adaptations of this basic process in order to remove the cellular and extracellular components.

Purified DNA can be manipulated in many different ways. Restriction enzymes cut the phosphate backbone of the DNA into smaller fragments, which can be visualized by gel electrophoresis. Specific DNA pieces are visualized using radioactively labeled nucleotides, followed by autoradiography. To avoid using radioactivity, researchers can synthesize DNA with digoxigenin or biotin-linked nucleotides, which are then linked to an antibody to digoxigenin or streptavidin, respectively. To visualize the antibody or streptavidin, researchers link either one to a fluorophore or reporter enzyme that converts a substrate into light or a colored precipitate. Hybridization of related sequences is a key technique for FISH, Southern blots, Northern blots, and dot blots, as well as the screening of a genomic library for a particular sequence.

The chapter also outlines the key characteristics of vectors, including plasmids, bacteriophage vectors, cosmids, and artificial chromosomes. These extrachromosomal genetic elements vary in their uses but are very important to getting a foreign gene expressed in a host organism. Vectors require a region that is convenient to adding a foreign piece of DNA such as a multicloning site, they need a gene for selection, and they need some easy way to identify whether the vector contains the foreign piece of DNA.

A genomic library simply contains all the DNA of the organism of interest, cut into smaller fragments, and cloned into a vector. The library recombinant vectors are then returned to a host bacterial cell so that only one fragment of the original DNA is inside each bacterium. Expression libraries start with mRNA rather than genomic DNA. The mRNA is converted into cDNA with reverse transcriptase. Libraries are screened for particular DNA sequences of interest by hybridization of related DNA sequences. In the case of expression libraries, each of the DNA pieces is made into protein by the bacteria. These proteins are then screened using antibodies.

Cloning genes using the traditional restriction enzyme digests can be very difficult for large genes, since the gene is likely to have the restriction enzyme sites. To overcome this obstacle, recombineering uses recombination between homologous DNA sequences to insert the gene of interest into a vector. Recombination is also used in the Gateway® cloning system, but instead of using regions of homology, these vectors use the lambda phage *attB*, *attP*, *attR*, and *attL* recognition sequences and the enzymes integrase and exisionase. In both systems, the vector has a gene that produces a toxin (*ccdB*) or converts a specific substrate to a toxin (*galK*). If the gene of interest does not replace *ccdB* or *galK*, then the host bacteria die. If the gene of interest recombines into the vector, then the host bacteria live and propagate the recombinant vector.

End-of-Chapter Questions

1. Which of the following statements about DNA isolation from *E. coli* is not correct?
   - **a.** Chemical extraction using phenol removes proteins from the DNA.
   - **b.** RNA is removed from the sample by RNase treatment.
   - **c.** Detergent is used to break apart plant cells to extract DNA.
   - **d.** Lysozyme digests peptidoglycan in the bacterial cell wall.
   - **e.** Centrifugation separates cellular components based on size.
2. Which of the following is important for gel electrophoresis to work?
   - **a.** Negatively charged nucleic acids to migrate through the gel.
   - **b.** Ethidium bromide to provide a means to visualize the DNA in the gel.
   - **c.** Agarose or polyacrylamide to separate the DNA based on size.
   - **d.** Known molecular weight standards.
   - **e.** All of the above are important for gel electrophoresis.
3. How are restriction enzymes and ligase used in biotechnology?
   - **a.** Restriction enzymes cut DNA at specific locations, producing ends that can be ligated back together with ligase.
   - **b.** Only restriction enzymes that produce blunt ends after cutting DNA can be ligated with ligase.
   - **c.** Only restriction enzymes that produce sticky ends on the DNA can be ligated with ligase.
   - **d.** Restriction enzymes can both cut DNA at specific sites and ligate them back together.
   - **e.** Restriction enzymes randomly cut DNA, and the cut fragments can be ligated back together with ligase.
4. Which of the following is an appropriate method for detecting nucleic acids?
   - **a.** Measuring absorbance at 260 nm.
   - **b.** Autoradiography of radiolabeled nucleic acids.
   - **c.** Chemiluminescence of DNA labeled with biotin or digoxigenin.
   - **d.** Measuring the light emitted after excitation by fluorescent-labeled nucleic acids on a photodetector.
   - **e.** All of the above are appropriate methods for detecting nucleic acids.
5. Why does the GC content of a particular DNA molecule affect the melting of the two strands?
   - **a.** The G and C bond only requires two hydrogen bonds, thus requiring a lower temperature to "melt" the DNA.
   - **b.** Because G and C base-pairing requires three hydrogen bonds and a higher temperature is required to "melt" the DNA.
   - **c.** The percentage of As and Ts in the molecule is more important to melting temperature than the percentage of Gs and Cs.
   - **d.** The nucleotide content of a DNA molecule is not important to know for biotechnology and molecular biology research.
   - **e.** None of the above.

(*Continued*)

6. What is the difference between Southern and Northern hybridizations?
   a. Southern blots hybridize a DNA probe to a digested DNA sample but Northern blots hybridize a DNA probe to, usually, mRNA.
   b. Southern blots use an RNA probe to hybridize to DNA but Northern blots use an RNA probe to hybridize to RNA.
   c. Southern blots determine if a particular gene is being expressed but Northern blots determine the homology between mRNA and a DNA probe.
   d. Southern blots determine the homology between mRNA and a DNA probe but Northern blots determine if a particular gene is being expressed.
   e. Southern and Northern blots are essentially the same technique performed in different hemispheres of the world.

7. What might be a use for fluorescence *in situ* hybridization (FISH)?
   a. For identification of a specific gene in a DNA extraction by hybridization to a DNA probe.
   b. For identification of a specific gene by hybridization to a DNA probe within live cells that have had their DNA denatured by heat.
   c. For identification of an mRNA within an RNA extraction by hybridization to a DNA probe.
   d. For identification of both mRNA and DNA in cellular extracts using an RNA probe.
   e. None of the above.

8. Which of the following are useful traits of cloning vectors?
   a. An antibiotic resistance gene on the plasmid for selection of cells containing the plasmid.
   b. A site that contains unique, clustered restriction enzyme sequences for cloning foreign DNA.
   c. A high copy number plasmid so that large amounts of DNA can be obtained.
   d. Alpha complementation to determine if the foreign DNA was inserted into the cloning site.
   e. All of the above are useful traits.

9. Which of the following vectors holds the largest pieces of DNA?
   a. plasmids
   b. bacteriophage
   c. YACs
   d. PACs
   e. cosmids

10. Besides a high voltage shock, what is another method to make *E. coli* competent to take up "naked" DNA?
    a. high concentrations of calcium ions followed by high temperature
    b. high concentrations of calcium ions and several hours on ice
    c. large amounts of DNA added directly to a bacterial culture growing at 37 °C
    d. high concentrations of minerals followed by high temperature
    e. A high voltage shock is the only way to make *E. coli* competent.

11. Why are gene libraries constructed?
    a. To find new genes.
    b. To sequence whole genomes.
    c. To compare genes to other organisms.
    d. To create a "bank" of all the genes in an organism.
    e. All of the above.

12. Which of the following statements about gene libraries is correct?
- **a.** Genes in a library can be compared to genes from other organisms by hybridization with a probe.
- **b.** A gene library is only necessary to maintain known genes.
- **c.** Every gene in the library must be sequenced first in order to compare genes in the library to genes from other organisms.
- **d.** Gene libraries are only created for eukaryotic organisms.
- **e.** Gene libraries can only be created in prokaryotes.

13. Why must reverse transcriptase be used to create a eukaryotic expression library?
- **a.** Reverse transcriptase is only used to create prokaryotic expression libraries.
- **b.** Reverse transcriptase creates cDNA from mRNA in prokaryotes.
- **c.** Reverse transcriptase ensures the gene is in the correct orientation within the expression vector to create protein.
- **d.** Reverse transcriptase creates cDNA from mRNA because genes in eukaryotes have large numbers of non-coding regions.
- **e.** No other enzymes are used to create expression libraries except restriction enzymes.

14. Which of the following are common features of expression vectors?
- **a.** Small segments of DNA that encode tags for protein purification.
- **b.** Transcriptional start and stop sites.
- **c.** A tightly controlled promoter than can only be induced under certain circumstances.
- **d.** Antibiotic resistance gene.
- **e.** All of the above are common features of expression vectors.

15. Which method is used to construct large, recombinant vectors when polylinker restriction enzymes are not useful?
- **a.** recombineering
- **b.** FISH
- **c.** gene libraries
- **d.** YACs
- **e.** hybridization

16. Which statement about Gateway® cloning is not true?
- **a.** Gateway® cloning vectors exploit a bacteriophage recombination system.
- **b.** Integrase cuts within *attB* and *attP* sites.
- **c.** The *ccdB* gene produces a toxin within host cells carrying the gene of interest.
- **d.** Excisionase regonizes *attL* and *attR* sites to remove the recombined DNAfragment.
- **e.** The cloning into the entry vector is necessary to generate attL sites on the ends of the gene of interest.

Further Reading

Clark, D. P. (2013). *Molecular Biology* (2nd ed.). San Diego, CA: Elsevier Academic Press.

Shlien, et al. (2010). A common molecular mechanism underlies two phenotypically distinct 17p13.1 microdeletions syndromes. *American Journal of Human Genetics*, *87*, 631–642.

DNA Synthesis *In Vivo* and *In Vitro*

Biotechnology

http://dx.doi.org/10.1016/B978-0-12-385015-7.00004-1

INTRODUCTION

Replication copies the entire set of genomic DNA so that the cell can divide in two. During replication, the entire genome must be uncoiled and copied exactly. This elegant process occurs extremely fast in *E. coli,* where DNA polymerase copies about 1000 nucleotides per second. Although the process is slower in eukaryotes, DNA polymerase still copies 50 nucleotides per second. Many biotechnology applications use the principles and ideas behind replication; therefore, this chapter first introduces the basics of DNA replication as it occurs in the cell. We then review some of the most widely used techniques in genetic engineering and biotechnology, including chemical synthesis of DNA, polymerase chain reaction, and DNA sequencing.

REPLICATION OF DNA

To maintain the integrity of an organism, the entire genome must be replicated identically. Even for plasmids, viruses, or transposons, replication is critical for their survival. The complementary two-stranded structure of DNA is the key to understanding its duplication during cell division. The double-stranded helix unwinds, and the hydrogen bonds holding the bases together melt apart to form two single strands. This Y-shaped region of DNA is the **replication fork** (Fig. 4.1). Replication starts at a specific site called an **origin of replication (ori)** on the chromosome. The origin is called *oriC* on the *E. coli* chromosome and covers about 245 base pairs of DNA. The origin has mostly AT base pairs, which require less energy to break than GC base pairs.

Once the replication fork is established, a large assembly of enzymes and factors called a **replisome** assembles to synthesize the complementary strands of DNA (Fig. 4.2). The replisome starts synthesizing the complementary strand on one side of the fork by adding complementary bases in a 5′ to 3′ direction. The **leading strand** is synthesized continuously because there is always a free 3′-OH group. Because DNA polymerase synthesizes only in a 5′ to 3′ direction, the other strand, called the **lagging strand**, is synthesized as small fragments called **Okazaki fragments**. As DNA polymerase makes this strand, the clamp loader must continually release and reattach at a new location. This results in the single-stranded region bubbling out from the replisome. The lagging strand fragments are ligated together by an enzyme called **DNA ligase**. Ligase links the 3′-OH and the 5′-PO_4 of neighboring nucleotides, forming a phosphodiester bond. The final step is to add methyl (-CH_3) groups along the new strand (Fig. 4.3). The original double-stranded helix is now two identical double-stranded helices, each containing one strand from the original molecule and one new strand. This is why the process is called **semiconservative replication**.

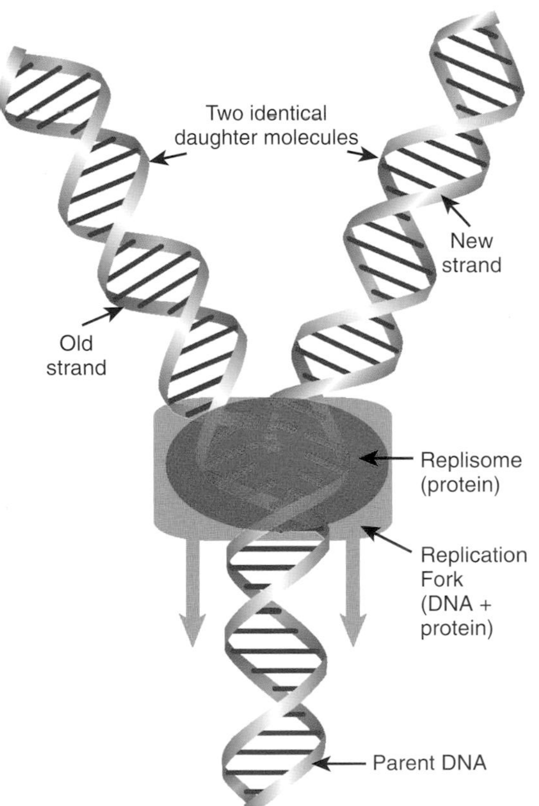

FIGURE 4.1
Replication
Replication enzymes open the double-helix around the origin to make it single-stranded. DNA polymerase adds complementary nucleotides.

In replication, DNA polymerase synthesizes the leading strand as one continuous piece and the lagging strand as Okazaki fragments. Each copy has one strand from the original helix and one new strand.

Uncoiling the DNA

Because DNA is condensed into supercoils in order to fit inside the cell, several different enzymes are needed to open and relax the DNA before replication can start (Fig. 4.2). **DNA helicase** and **DNA gyrase** attach near the replication fork and untwist the strands

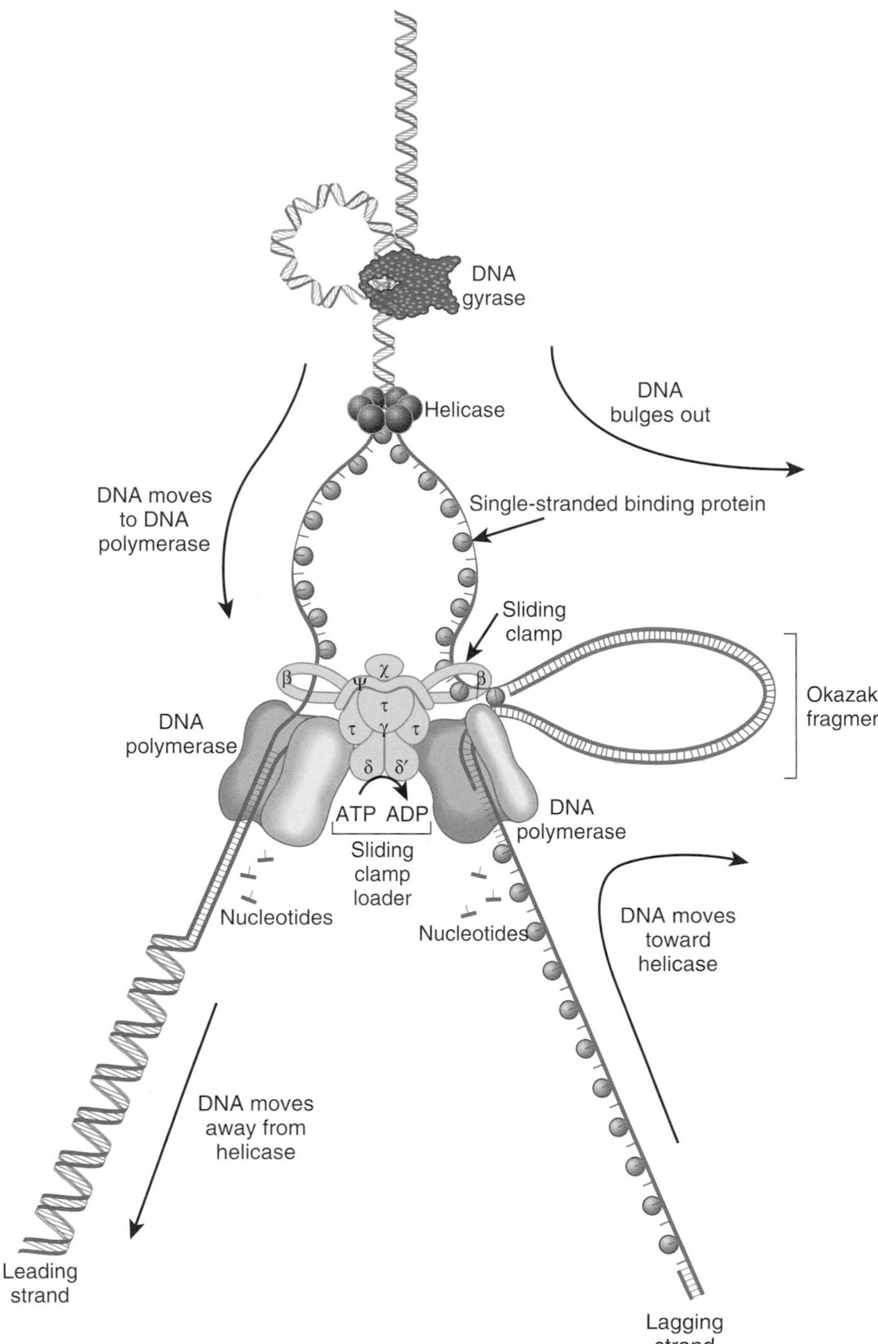

FIGURE 4.2 DNA Polymerase III Replication Assembly During replication, the sliding clamp loader complex makes contacts with single-stranded binding protein and the sliding clamps. This complex stabilizes the two single-stranded DNA strands and provides a stable binding site for two DNA polymerase III molecules. The unwound single-stranded DNA templates move toward the clamp loader complex. On the leading strand (left), the strand is unwinding in a 3′ to 5′ direction, so DNA polymerase can add complementary nucleotides in the 5′ to 3′ direction. On the lagging strand (right), the template strand is antiparallel, and therefore, the strand is unwinding In a 5′ to 3′ direction. Since DNA polymerase III must synthesize the new strand in a 5′ to 3′ direction also, the template strand must move toward the helicase. This causes the lagging strand to bubble out from the complex. Once DNA polymerase III reaches the end of the previous Okazaki fragment, the replicated DNA is released by the clamp loader and a new section of single-stranded DNA is reloaded.

of DNA. DNA gyrase removes the supercoiling, and helicase unwinds the double helix by dissolving the hydrogen bonds between the paired bases. The two strands are kept apart by **single-stranded binding protein**, which coats the single-stranded regions. This prevents the two strands from reannealing so that other enzymes can gain access to the origin and begin replication.

As DNA polymerase travels along the DNA, more positive supercoils are added ahead of the replication fork. Because the bacterial chromosome is negatively supercoiled, initially the new positive supercoils relax the DNA. After about 5% of the genome has been replicated, though, the positive supercoils begin to accumulate and need to be removed. DNA gyrase cancels the positive supercoils by adding negative supercoils. When circular chromosomes are replicated, the two daughter copies may become **catenated**, or connected like two links of a chain (Fig. 4.4). Topoisomerase IV releases catenated daughter strands by introducing double-stranded nicks into one chromosome. The second copy can then pass through the first, giving two separated molecules.

DNA helicase, DNA gyrase, and topoisomerase IV untwist and untangle the supercoiled DNA during replication.

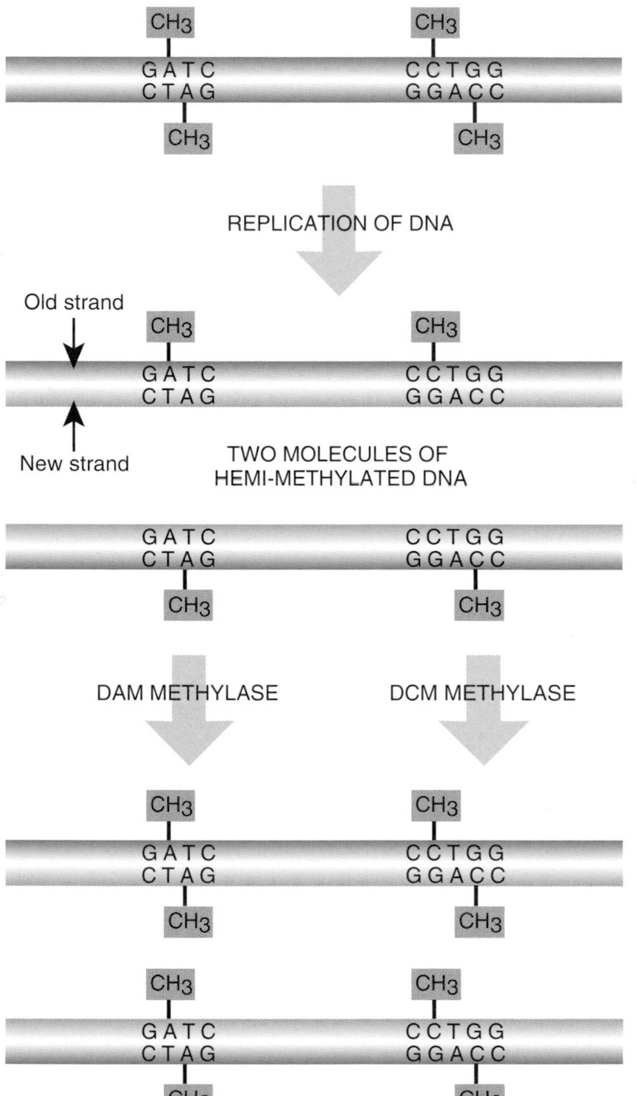

FIGURE 4.3 Hemimethylated DNA: Old Strands versus New
When DNA is replicated, the old strand is methylated, but there is a delay in methylating the new strand, and thus, the DNA double helix is hemimethylated. Dam methylase and dcm methylase add the methyl groups onto the newly synthesized DNA.

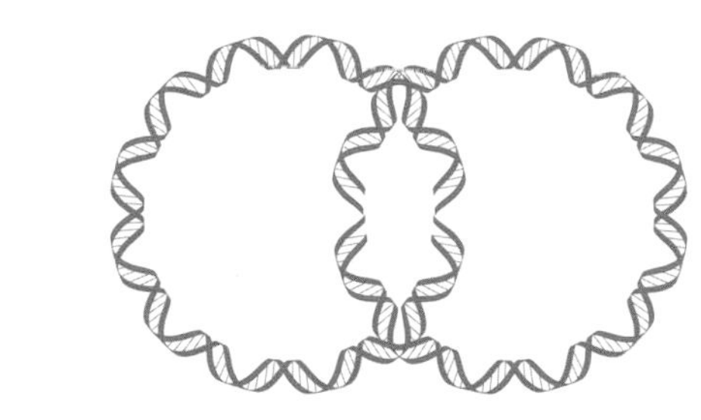

FIGURE 4.4 Untangling Circular Chromosomes
Sometimes after the replication of circular genomes is complete, the two rings are catenated, or linked together like links in a chain. Topoisomerase IV untangles the two chromosomes so they can partition into the daughter cells.

Priming DNA Synthesis

DNA polymerase cannot initiate new strands of nucleic acid synthesis because it can only add a nucleotide onto a pre-existing 3′-OH. Therefore, an 11 to 12 base-pair length of RNA (an RNA primer) is made at the beginning of each new strand of DNA. Since the leading strand is synthesized as a single piece, there is only one RNA primer at the origin. On the lagging strand, each Okazaki fragment begins with a single RNA primer. DNA polymerase then makes DNA starting from each RNA primer. At the origin, a protein called **PriA** displaces the SSB proteins so a special RNA polymerase, called **primase** (DnaG), can enter and synthesize short RNA primers using ribonucleotides. Two molecules of DNA polymerase III bind to the primers on the leading and lagging strands and synthesize new DNA from the 3′ hydroxyls (Fig. 4.5).

Primase, a special RNA polymerase, works with PriA to displace the SSB proteins and synthesize a short RNA primer at the origin. DNA polymerase then starts synthesis of the new DNA strand using the 3′-OH of the RNA primer. This synthesis occurs at multiple locations on the lagging strand.

Structure and Function of DNA Polymerase

DNA polymerase III (PolIII) is the major form of DNA polymerase used to replicate bacterial chromosomes and consists of multiple protein subunits (see Fig. 4.2). The **sliding clamp** is a donut-shaped protein consisting of a dimer of DnaN proteins, also called the β-subunits. Two clamps encircle the two single strands of DNA at the replication fork. A cluster of accessory proteins, called the **clamp loader complex**, loads the clamps onto DNA strands. The two sliding clamps bind two **core enzymes**, one for each strand of DNA. The core enzyme consists of three subunits: DnaE (α subunit), which links the nucleotides together; DnaQ (ε subunit), which proofreads the new strand; and HolE (θ subunit), which stabilizes the two other subunits (not shown in Fig. 4.2). As the α subunit adds new nucleotides, the ε subunit recognizes any distortions and removes any mismatched bases. A correct nucleotide is then added. Bacterial DNA polymerase III can add up to 1000 bases per second, which is an extraordinarily fast rate of enzyme activity.

> The multiple subunits of DNA polymerase III work together to synthesize a new strand of DNA. The core has two essential subunits: the α subunit links the nucleotides, and the ε subunit ensures that they are accurate.

Synthesizing the Lagging Strand

After the new lagging strand of DNA has been made, it has many segments of RNA derived from multiple RNA primers, as well as multiple breaks, or **nicks**, along the backbone that need to be sealed (Fig. 4.6). DNA polymerase I removes the RNA primers from the lagging strand. DNA polymerase I has exonuclease activity that removes the RNA bases, and then its polymerase activity fills in the regions with DNA bases. The RNA bases may also be removed by RNaseH, an enzyme that specifically identifies RNA:DNA heteroduplexes and removes the RNA bases. Finally, the DNA fragments of the lagging strand are linked together with a ligation reaction by DNA ligase. DNA polymerase I and DNA ligase are both very important enzymes in molecular biology and are used extensively in biotechnology.

> Because the lagging strand is synthesized in small pieces, either DNA polymerase I or RNaseH excise the RNA bases and replace them with DNA. DNA ligase closes the nicks in the sugar/phosphate backbone of the new DNA strand.

Repairing Mistakes after Replication

After replication is complete, the **mismatch repair system** corrects mistakes made by DNA polymerase. If the wrong base is inserted and DNA polymerase does not correct the error itself, there will be a small bulge in the helix at that location. Identifying which of the two bases is correct is critical. The cell assumes that the base on the new strand is wrong and the original parental base is correct. The mismatch repair system of *E. coli* (MutSHL) deciphers which strand is the original by monitoring methylation. Immediately after replication, the DNA is **hemimethylated**; that is, the old strand still has methyl groups attached to various bases, but the new strand has not been methylated yet (see Fig. 4.3). Two different *E. coli* enzymes add methyl groups: **DNA adenine methylase (Dam)** adds a methyl group to the adenine in GATC, and **DNA cytosine methylase (Dcm)** adds a methyl group to the cytosine in CCAGG or CCTGG. These enzymes methylate the new strand after replication, but they are slow. This allows mismatch repair to find and fix any mistakes first.

Three genes of *E. coli* are responsible for mismatch repair: *mutS, mutL,* and *mutH* (Fig. 4.7). MutS protein recognizes the bulge or distortion in the sequence. MutH finds the nearest GATC site and nicks the nonmethylated strand—that is, the newly made strand. MutL holds the MutS plus mismatch and the MutH plus GATC site together (these may be far apart on the DNA helix). Finally, the DNA on the new strand is degraded and replaced with the correct sequence by DNA polymerase III.

> In *E. coli*, mismatch repair proteins (MutSHL) identify a mistake in replication, excise the new nucleotides around the mistake, and recruit DNA polymerase III to the single-stranded region to make the new strand without a mistake.

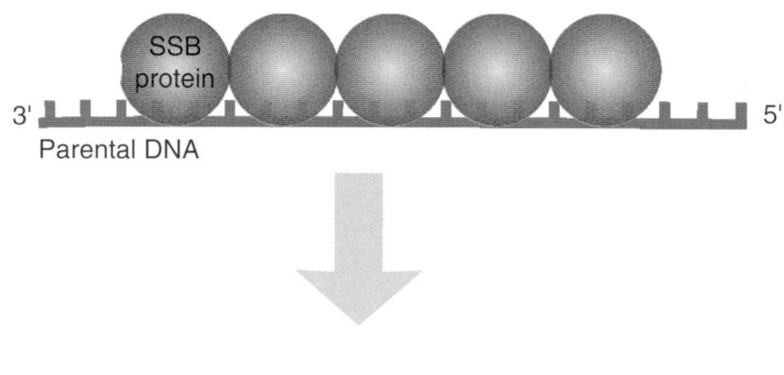

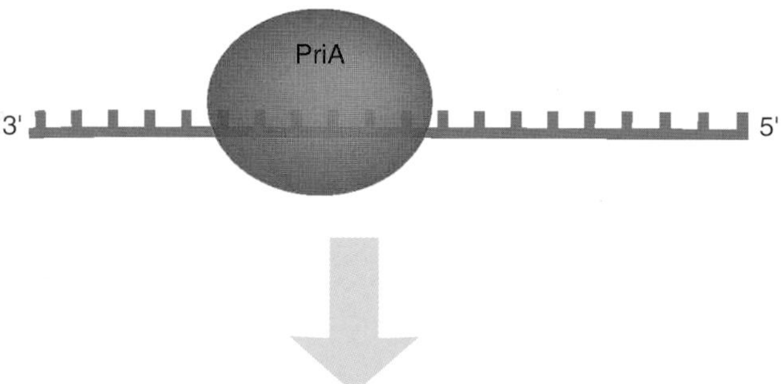

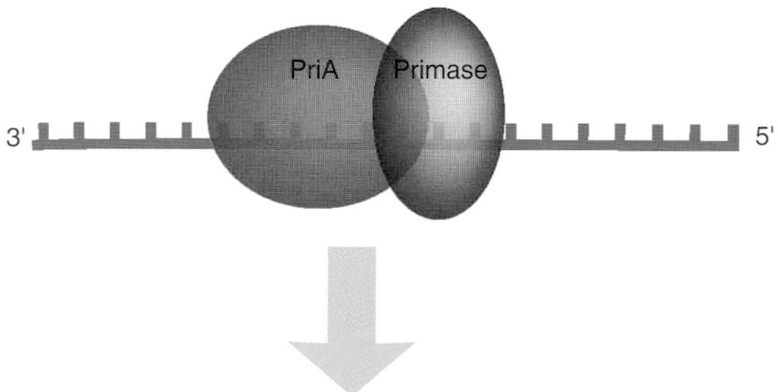

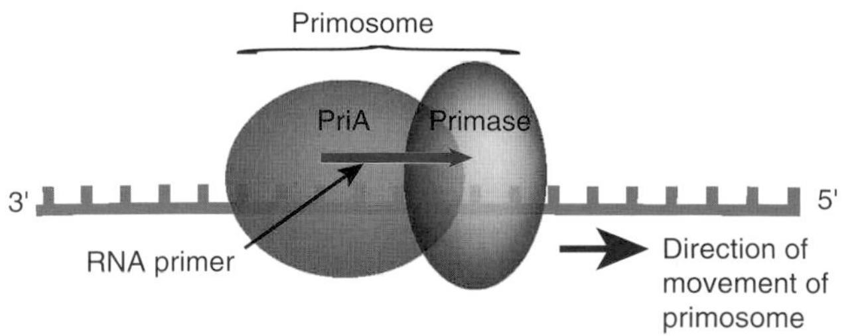

FIGURE 4.5 Strand Initiation Requires an RNA Primer
DNA polymerase cannot synthesize new DNA without a pre-existing 3′-OH. Thus, DNA replication requires an RNA primer to initiate strand formation. (A) First, the PriA protein displaces the SSB proteins. (B) Second, primase associates with thc PriA protein. (C) Last, the primase makes the short RNA primer.

COMPARING REPLICATION IN GENE CREATURES, PROKARYOTES, AND EUKARYOTES

Although the basic mechanism for replication is the same for most organisms, the timing, direction, and sites for initiation and termination are variable. The major differences in replication occur mainly because of the special challenges posed by circular and linear genomes. Normal DNA replication occurs bidirectionally in prokaryotes and eukaryotes, whether the genome is linear or circular. Two replication forks travel in opposite directions, unwinding the DNA helix as they go. In bacteria such as *E. coli*, there is only one origin, *oriC*, and replication occurs in both directions around the circular chromosome until it meets at the other side, the terminus, *terC*. Halfway through this process, the chromosome looks like the Greek letter θ; therefore, this process is often called **theta replication** (Fig. 4.8). The single circular chromosome then becomes two. Theta replication is also used by many plasmids, such as the F plasmid of *E. coli*, when growing and dividing asexually (as opposed to transferring its genome to another cell via conjugation).

Some plasmids and many viruses replicate their genomes by a process called **rolling circle replication** (Fig. 4.9). At the origin of replication, one strand of the DNA is nicked and unrolled. The intact strand thus rolls relative to its partner (hence "rolling circle"). DNA is synthesized from the origin using the circular strand as a template. As DNA polymerase circles the template strand, the new strand of DNA is base-paired to the circular template. Meanwhile the other parental strand is dangling free. This dangling strand is removed, ligated to form another circle, and finally a second strand is synthesized. This process results in two rings of plasmid or viral DNA, each with one strand from the original molecule and one newly synthesized strand.

Some viral genomes use rolling circle replication but continue to make more and more copies of the original circular template. They continue rolling around the circle, synthesizing more and more copies that are all dangling as a long single strand. The long strand of new DNA may be made double-stranded or left single-stranded (depending on the type of virus). Finally, the dangling strand is chopped into genome-sized units and packaged into viral particles. Some viruses circularize these copies before packaging; others simply leave the genomes linear.

Long linear DNA molecules such as human chromosomes pose several problems for replication. The ends pose a particularly difficult problem because the RNA primer is synthesized at the very end of the lagging strand. When the RNA primer is removed by an exonuclease, there is no upstream 3′-OH for addition of new nucleotides to fill the gap. (In eukaryotes, there is no equivalent to the dual-function DNA polymerase I. A separate exonuclease, MF1, removes the RNA primers, and DNA polymerase δ fills in the gaps for the lagging strand.) Over successive rounds of replication, the ends of linear chromosomes get shorter and shorter. Special structures called **telomeres** are found at the tips of each linear chromosome and prevent chromosome shortening from affecting important genes. Telomeres have multiple tandem repeats of a short sequence (TTAGGG in humans). The enzyme **telomerase** can regenerate the telomere by using an RNA template to synthesize new repeats. This happens only in some cells; in others, the telomeres shorten every time the cell replicates its DNA. One theory regards telomere shortening as a molecular clock, aging the cell, and eventually triggering suicide (see Chapter 20).

The length of linear chromosomes also poses a problem. The time it takes to synthesize an entire human chromosome would be too long if replication began at only one origin. To solve this issue, multiple origins exist, each initiating new strands in both directions. These are elongated until they meet the new strands from the other direction.

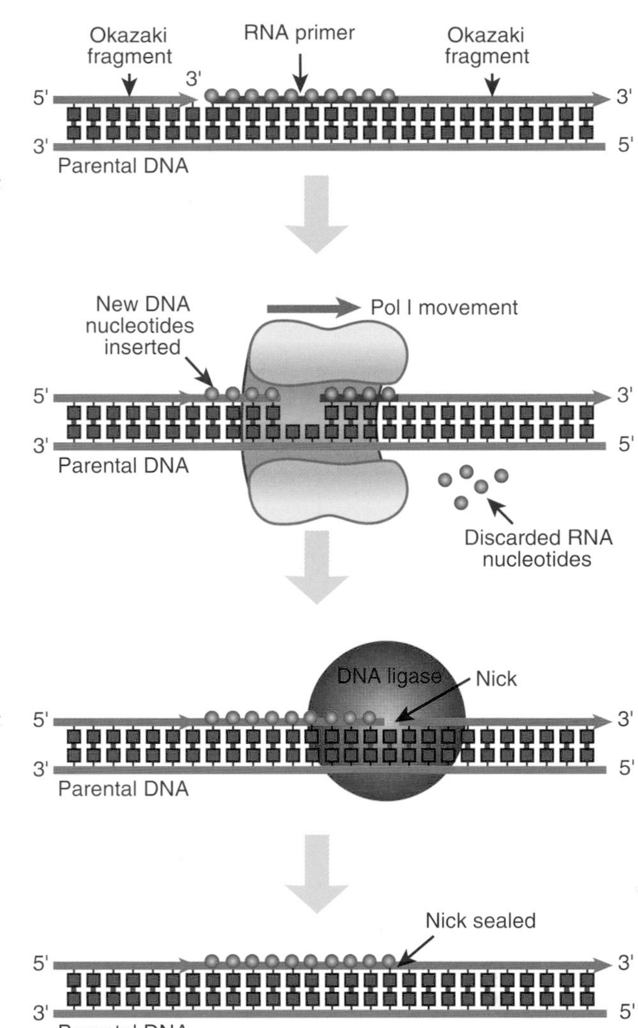

FIGURE 4.6 Joining the Okazaki Fragments
When first made, the lagging strand is composed of alternating Okazaki fragments and RNA primers. Next, DNA polymerase I binds to the primer region, and as it moves forward, it degrades the RNA and replaces it with DNA. Finally, DNA ligase seals the nick in the phosphate backbone.

The cellular structure of eukaryotes also poses some problems for replication. (In bacteria, the chromosome simply replicates, the two copies move to each end of the cell, and a new wall forms in the middle. There are no nuclear membranes or organelles to divide; there is just one chromosome plus, perhaps, some plasmids.) In eukaryotes, the cell has a specific **cell cycle**, with four different phases, and replication occurs at specific points (Fig. 4.10). During G_1, the cell rests for a period before DNA synthesis begins. This period varies, lasting about 25 minutes for yeast. The next phase is S, or synthesis, during which the entire genome is replicated. This is usually the longest phase, lasting about 40 minutes in yeast. The third phase, G_2, is another resting phase before the cell undergoes mitosis, in the M phase. During mitosis, cells divide their walls and membranes into two separate cells, partitioning the new chromosomes and other cellular components into each half. The signal that triggers cell division depends on many factors, including environment, size, and age.

Eukaryotic mitosis is a dynamic process with much movement and repositioning of cellular components. First, the nuclear membrane must be dissolved before the chromosomes can separate. After replication, the two sets of chromosomes are partitioned to separate sides of the cell. The chromosomes attach to long fibers making up the **spindle** via special sequences called **centromeres**. They slide along the spindle fibers until they reach separate ends of the cell. A new cell membrane separating the two halves is then synthesized. Other cellular components including mitochondria, endoplasmic reticulum, lysosomes, and so forth are split between the two daughter cells. Finally, a new nuclear membrane must be assembled around the chromosomes of each new daughter cell. The dynamics of this process are still being investigated, and new proteins and molecules are still being discovered that mediate different parts of mitosis in eukaryotic cells.

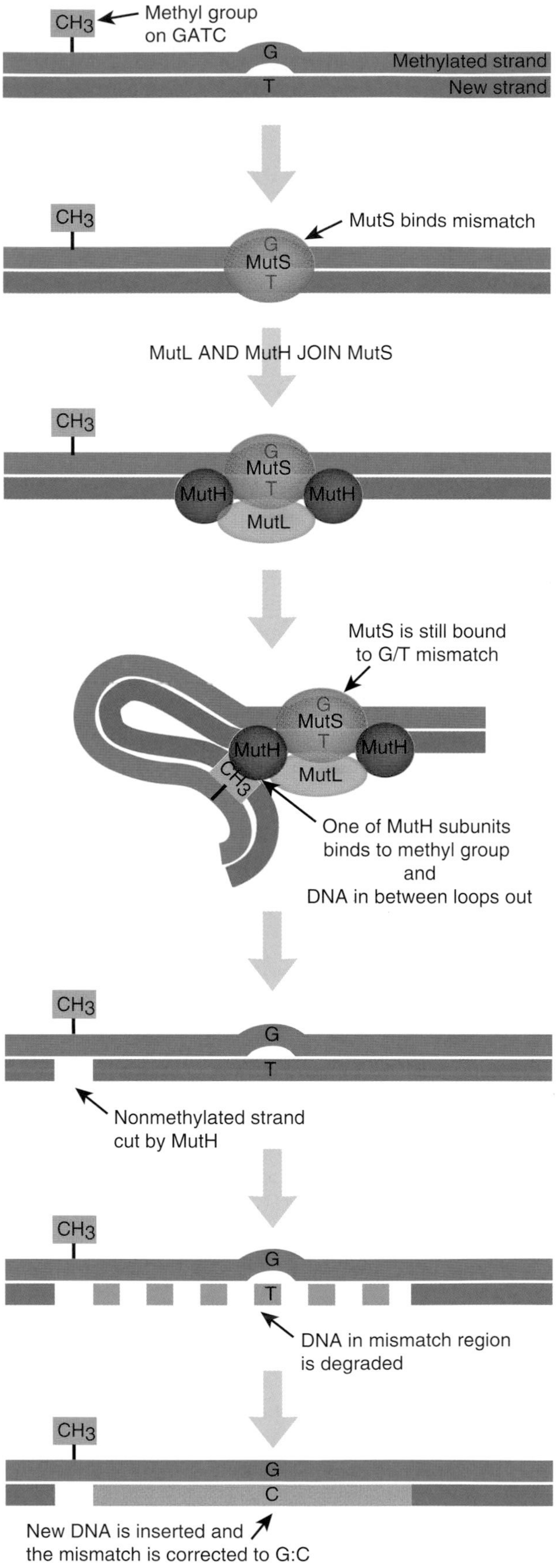

FIGURE 4.7 Mismatch Repair Occurs after Replication

MutS recognizes a mismatch shortly after DNA replication. MutS recruits MutL and two MutH proteins to the mismatch. MutH locates the nearest GATC of the new strand by identifying the methyl group attached to the "parent" strand. MutH cleaves the nonmethylated strand and the DNA between the cut and the mismatch is degraded. The region is replaced, and the mismatch is corrected.

1

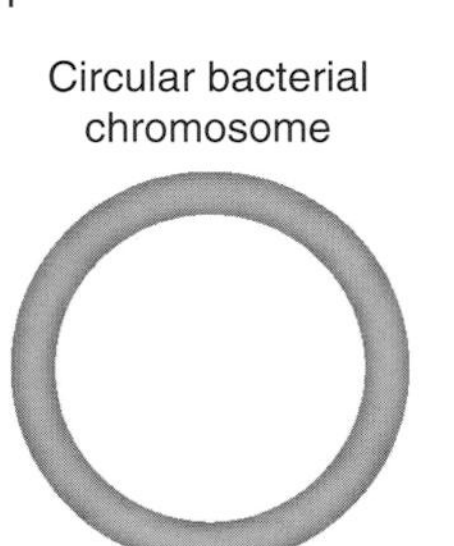

2

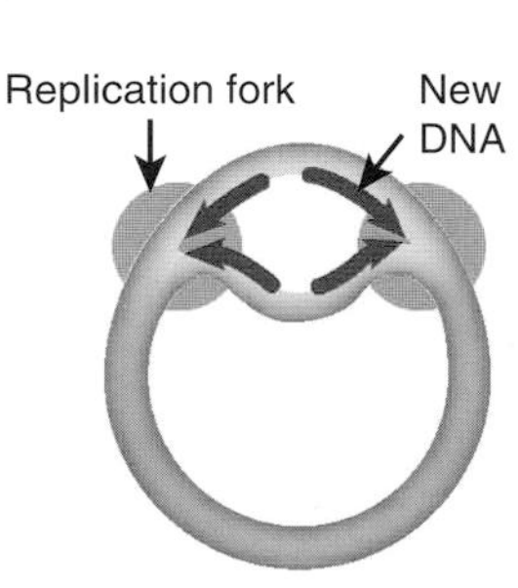

3

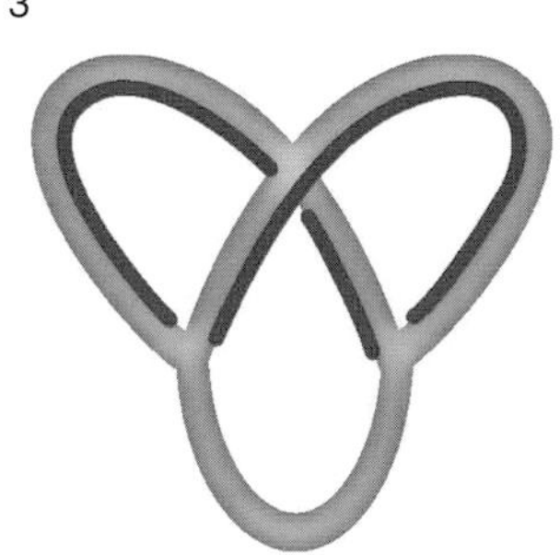

4

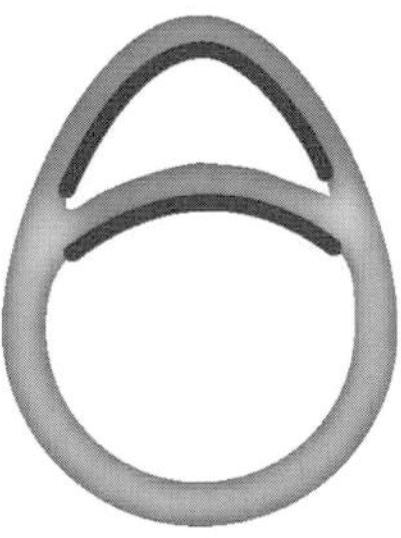

FIGURE 4.8 Theta Replication
In circular genomes or plasmids, replication enzymes recognize the origin of replication, unwind the DNA, and start synthesis of two new strands of DNA, one in each direction. The net result is a replication bubble that makes the chromosome or plasmid look similar to the Greek letter theta (θ). The two replication forks keep moving around the circle until they meet on the opposite side.

Bacteria and viruses use either theta replication or rolling circle replication to create new genomes.

Eukaryotic cells have chromosomes with multiple origins of replication. Telomeres protect the ends of the chromosomes because each round of replication shortens the DNA. Replication in eukaryotes occurs only at a specific point during the cell cycle.

IN VITRO DNA SYNTHESIS

Making DNA in the laboratory relies on the same basic principles outlined for replication (Fig. 4.11). DNA replication needs the following "reagents": enzymes to melt the two template DNA strands apart, an RNA primer with a 3′-hydroxyl for DNA polymerase to synthesize a new DNA strand, a pool of nucleotide precursors, plus DNA polymerase to catalyze the addition of new nucleotides.

To perform DNA replication in the laboratory, the researcher makes a few modifications. First, the enzymes that open and unwind the template DNA are not used. Instead, double-stranded DNA is converted to single-stranded DNA using heat or a strong base to disrupt the hydrogen bonds that hold the two strands together. Alternatively, template DNA can be made by using a virus that packages its DNA in single-stranded form. For example, M13 is a bacteriophage that infects *E. coli*, amplifies its genome using rolling circle replication, and packages the single-stranded DNA in viral particles that are released without lysing open the *E. coli* cell. If template DNA is cloned into the M13 genome, then the template will also be manufactured as in a single-stranded form. This DNA can be isolated directly from the viral particles.

During *in vitro* synthesis of DNA, an RNA primer is not used because RNA is very unstable and degrades easily. Instead, a short single-stranded oligonucleotide of DNA is used as a

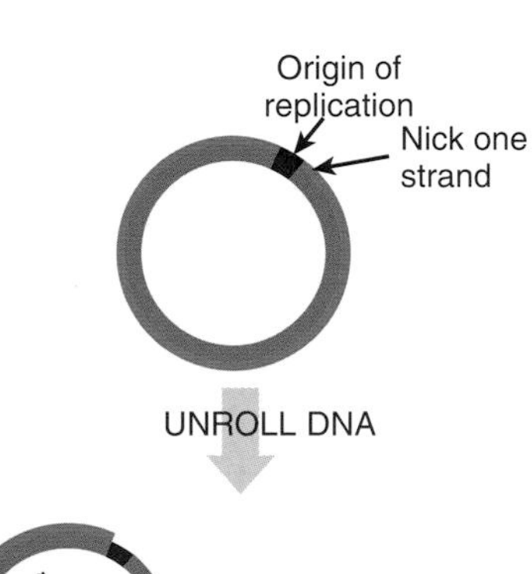

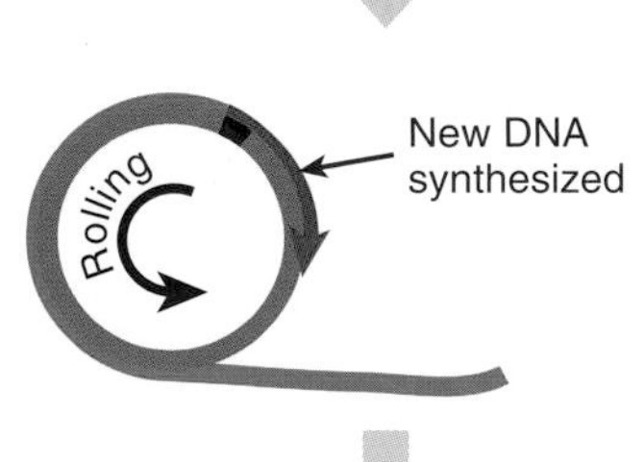

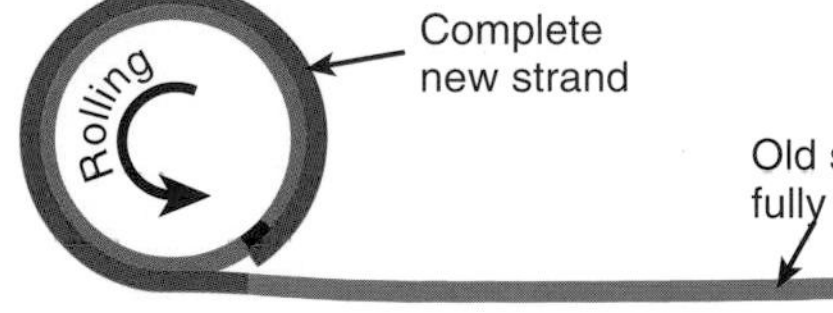

FIGURE 4.9 Rolling Circle Replication
During rolling circle replication, one strand of the plasmid or viral DNA is nicked, and the broken strand (pink) separates from the circular strand (purple). The gap left by the separation is filled in with new DNA starting at the origin of replication (green strand). The newly synthesized DNA keeps displacing the linear strand until the circular strand is completely replicated. The linear single-stranded piece is fully "unrolled" in the process.

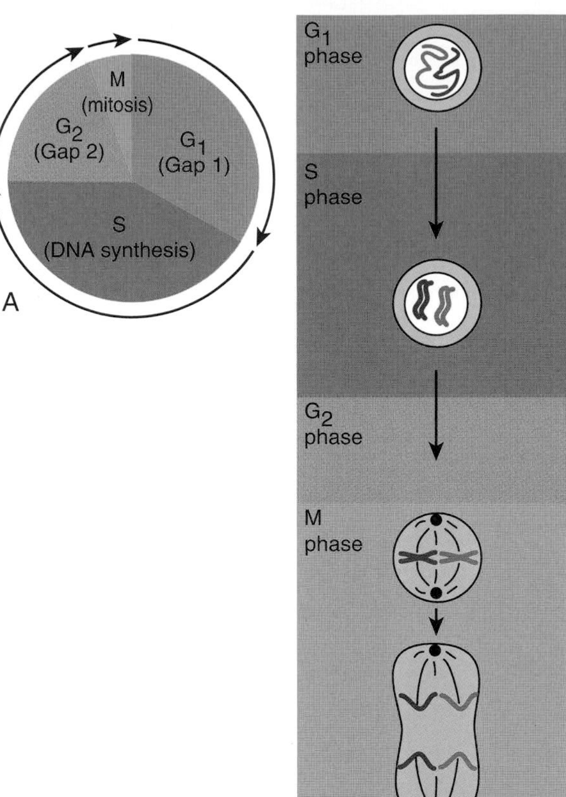

FIGURE 4.10
Eukaryotic Cell Cycle
DNA replication occurs during the S phase of the cell cycle but the chromosomes are actually separated later, during mitosis, or the M phase. The S and M phases are separated by G_1 and G_2.

primer. (As long as the primer has a free 3′-hydroxyl, DNA polymerase will add nucleotides onto the end.) The primers are synthesized chemically (see later discussion) and mixed with the single-stranded template DNA. The oligonucleotide primer has a sequence complementary to a short region on the DNA template. Therefore, at least some sequence information must be known about the template. If the sequence of the template DNA is unknown, it may be cloned into a vector, and the primer is then designed to match sequences of the vector (such as the polylinker region) that are close to the inserted DNA.

Finally, purified DNA polymerase plus a pool of nucleotides (dATP, dCTP, dGTP, and dTTP) is added to the primer and template. The primer anneals to its complementary sequence, and DNA polymerase elongates the primer, creating a new strand of DNA complementary to the template DNA.

In vitro replication requires a single-stranded piece of template DNA, a primer, nucleotide precursors, and DNA polymerase.

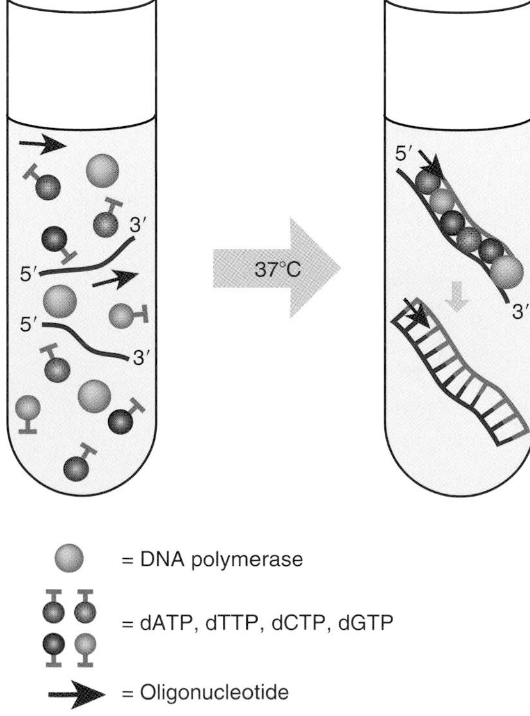

FIGURE 4.11 ***In Vitro* DNA Synthesis**
DNA synthesis in the laboratory uses single-stranded template DNA, plus DNA polymerase, an oligonucleotide primer, and nucleotide precursors. After all the components are incubated at the appropriate temperature, double-stranded DNA is made.

CHEMICAL SYNTHESIS OF DNA

Making DNA chemically rather than biologically was one of the first new technologies to be applied by the biotechnology industry. The ability to make short synthetic stretches of DNA is crucial to using DNA replication in laboratory techniques. DNA polymerase cannot synthesize DNA without a free 3′-OH end to elongate. Therefore, to use DNA polymerase *in vitro*, the researcher must supply a short primer. Such primers are used to sequence DNA (see later discussion), to amplify DNA with PCR (see later discussion), and even to find genes in library screening (see Chapter 3). So a short review of how primers are synthesized is included here.

Research into chemical synthesis of DNA began shortly after Watson and Crick published their research on the crystal structure of DNA. H. Gobind Khorana at the University of Chicago was an early pioneer in the study of **oligonucleotide** synthesis (see Box 4.1). Technically, oligonucleotides are

Box 4.1 Khorana, Nirenberg, and Holley

Har Gobind Khorana, Marshall W. Nirenberg, and Robert W. Holley are pioneers in the field of molecular biology. The three scientists received the Nobel Prize in Physiology or Medicine in 1968 for their combined efforts in identifying which triplet codons coded for which amino acid. Khorana originally began chemical synthesis of DNA in order to help elucidate the role of different enzymes. He wanted to understand the mode of action for nucleases and phosphodiesterases, but without being able to chemically synthesize a defined nucleic acid, the work on enzymes would be very difficult.

Khorana's lab determined ways to synthesize dinucleotide, trinucleotide, and tetranucleotide sequences using chemical synthesis. Rather than using single nucleotide additions, his lab focused on synthesizing nucleotides in blocks. His ability to chemically synthesize blocks of DNA was the backbone experiment, but many other discoveries were instrumental in determining the amino acid codes.

Matthaei and Nirenberg (1961) experimentally determined that polyuridylate [poly(U)] mixed with a bacterial cell-free amino acid incorporating system created polyphenylalanine. This experiment determined that the codon UUU encoded for the amino acid phenylalanine. During this time, Robert Holley was working on tRNA. He specifically identified the structure of the tRNA for alanine by purifying tRNA-alanine from yeast, fragmenting the tRNA into pieces with nucleases, and logically piecing together the size of the fragments and the sites at which the enzymes were recognized. Other important discoveries included the purification of DNA polymerase and RNA polymerase.

These experiments were woven into an elegant method of determining which triplet nucleotide sequence encoded which amino acid. First, Khorana's groups began synthesizing dinucleotide, trinucleotide, and tetranucleotide double-stranded DNA fragments. For example, one of these fragments had the following structure:

5′ TCTCTC 3′
3′ AGAGAG 5′

Arthur Kornberg had previously won the Nobel Prize for his discovery and purification of DNA polymerase I. Khorana's group mixed their short synthesized DNA with pure DNA polymerase to create long polydeoxynucleotides with a known sequence. Next, the DNA pieces were mixed with RNA polymerase to create long polyribonucleotides of known sequence. These were mixed with the cell-free system devised by Matthaei and Nirenberg, which made polypeptides. The preceding dinucleotide example resulted in a polypeptide of repeating serine and leucine. The experiment demonstrated that TCT or CTC encoded serine or leucine, respectively. There was no way to determine definitively which codon matched which amino acid, so more experiments were needed.

The final important contributions to make the final assignments were using purified tRNAs labeled with ^{14}C. Nirenberg and Leder (1964) mixed Khorana's synthetic trinucleotides and mixed them with the labeled tRNAs and ribosomes. (Note: The isolation of pure tRNA was not possible without Robert W. Holley's work.) They looked for binding of the labeled tRNA to the trinucleotide sequence. These experiments provided clear answers to many of the trinucleotide sequences, but many times the results were not very clear. It was the combination of these experiments with Khorana's work that determined the direct genetic code.

any piece of DNA less than 20 nucleotides in length, but today, *oligonucleotide* denotes a short piece of DNA that is chemically synthesized. In 1970, Khorana's lab synthesized an active tRNA molecule of 72 nucleotides (Agarwal et al., 1970). The chemistry he used was inefficient and cumbersome, but some of his ideas are still used in current oligonucleotide synthesis. Today chemical synthesis is done with an automated **DNA synthesizer** that creates DNA by sequentially adding one nucleotide after another in the correct sequence order.

Unlike *in vivo* DNA synthesis, artificial synthesis is done in the 3′ to 5′ direction. The first step is attaching the first nucleotide to a porous material made of **controlled pore glass (CPG)**. The first nucleotide is not attached directly but is linked to the surface via a spacer molecule that binds to the 3′-OH of the nucleotide (Fig. 4.12). The column pores allow reagents to be washed through and removed easily. (Using CPG is one improvement over Khorana's technology. He used polymer beads to couple the reaction but found that the polymer swelled as the reagents passed through the column, which inhibited synthesis. CPG is superior because it does not swell.)

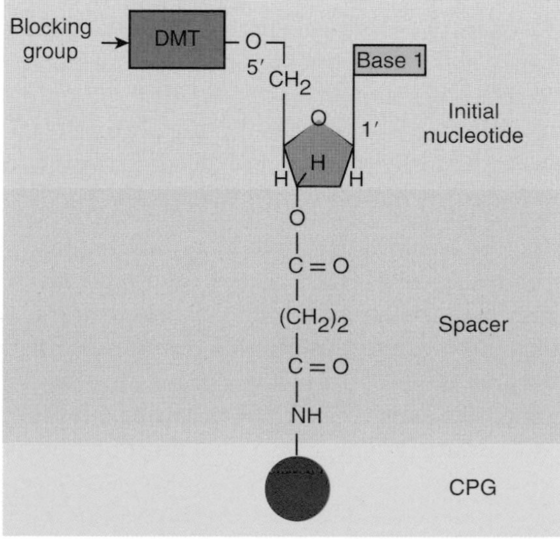

FIGURE 4.12 Addition of a Spacer Molecule and First Base to the CPG
The first nucleotide is linked to a glass bead via a spacer molecule attached to its 3′-OH group. The structure of the spacer varies, but it is important to keep the synthesis away from the glass surface and to allow efficient removal of the completed oligonucleotide.

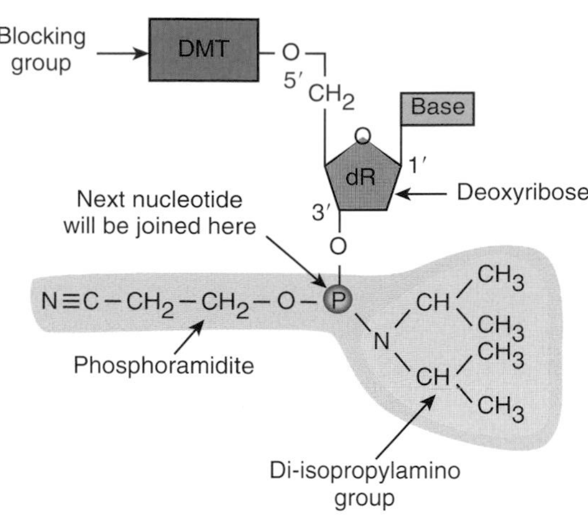

FIGURE 4.13 Nucleoside Phosphoramidites Are Used for Chemical Synthesis of DNA
Nucleotides are modified to ensure that the correct group reacts with the growing oligonucleotide. Each nucleotide has a DMT group blocking its 5′-OH. The 3′-OH is activated by a phosphoramidite group, which is originally also protected by di-isopropylamine.

When the spacer is linked to the nucleotide 3′-OH, a chemical blocking group is attached to the 5′-OH. Thus, the 3′-OH is the only available reactive group. Khorana's early synthesis was revolutionary in this respect because he chose the **dimethyloxytrityl (DMT) group**, which is still used as a blocking group in today's synthesizers. DMT has a strong orange color and is easily removed from the 5′-OH so that another nucleotide can be linked to the first. In practice, the CPG–spacer–first nucleotide is washed, and then the DMT group is removed by mild acid such as trichloroacetic acid (TCA). The 5′-OH is now ready to accept the next nucleotide. The efficiency of removing DMT is critical. If DMT is not removed completely, many of the potential oligonucleotides will fail to elongate. The orange color reveals the efficiency of removal and is easily measured optically.

Each nucleotide is added as a **phosphoramidite**, which is a nucleotide that has a blocking group protecting a 3′-phosphite group (Fig. 4.13). (One problem with early oligonucleotide synthesis technology was branching. Rather than the incoming nucleotide adding to the 5′ end, it sometimes attached to the phosphate linking two nucleotides.) To prevent branching, every added nucleotide has a di-isopropylamine group attached to the 3′ phosphite group, which also stabilizes the nucleotides allowing long-term storage. Before another nucleoside is added, the 3′ phosphite group is activated by tetrazole. The next nucleotide is then added, and it reacts with the phosphite to form a dinucleotide (Fig. 4.14).

If the terminal nucleotide of a growing chain fails to react with an incoming nucleotide, the chain must be capped off to prevent generation of an incorrect sequence by later reactions. The 5′-OH of all unreacted nucleotides is acetylated with acetic anhydride. This terminates the chain so that no other nucleoside phosphoramidites can be added. (Fig. 4.15)

At this stage of the synthesis, the column has CPG–spacer–first nucleoside–phosphite–second nucleoside–DMT. Phosphites are used because they react much faster, but they are unstable. Adding iodine oxidizes the phosphite triester into the normal phosphodiester, which is more stable under acidic conditions (Fig. 4.16).

The column can now be prepared to add the third nucleotide. The DMT is removed with TCA, and the third phosphoramidite nucleotide is added. The chains are capped so that any dinucleotides that failed to react with the third phosphoramidite are prevented from adding any more nucleosides. Finally, the phosphite triester is oxidized to phosphodiester. This process continually repeats until all the desired nucleotides are added and the final oligonucleotide has the correct sequence (Fig. 4.17).

After the final phosphoramidite nucleoside is added, the oligonucleotide still has DMT protecting the 5′-OH, cyanoethyl groups attached to the phosphates, and amino-protecting groups on the bases. (Amino groups would react with the reagents during synthesis; therefore, chemical groups are added to protect the bases before they are added to the column.) All three types of protective groups must be removed. The organic salts of the protecting groups are then removed by desalting, and the final oligonucleotide is cleaved from the CPG surface. Finally, the 5′-OH must be phosphorylated to make the oligonucleotides biologically active. A kinase from bacteriophage T4 is used to transfer a phosphate group from ATP to the 5′ end of the oligonucleotides. The newly synthesized oligonucleotide is now ready for use.

Chemical synthesis of DNA occurs by successively adding phosphoramidite nucleotides to the previous base attached to controlled pore glass (CPG) columns. Synthesis occurs in a 3′ to 5′ direction by removing the 5′-blocking group from the existing nucleotide and adding the new activated phosphoramidite nucleotide. After coupling, the unreacted nucleotides are capped, and the phosphate triester is oxidized to a phosphodiester group. Synthesis ends by removal of all blocking groups from the bases, removing the cyanoethyl groups, and cleavage from the CPG.

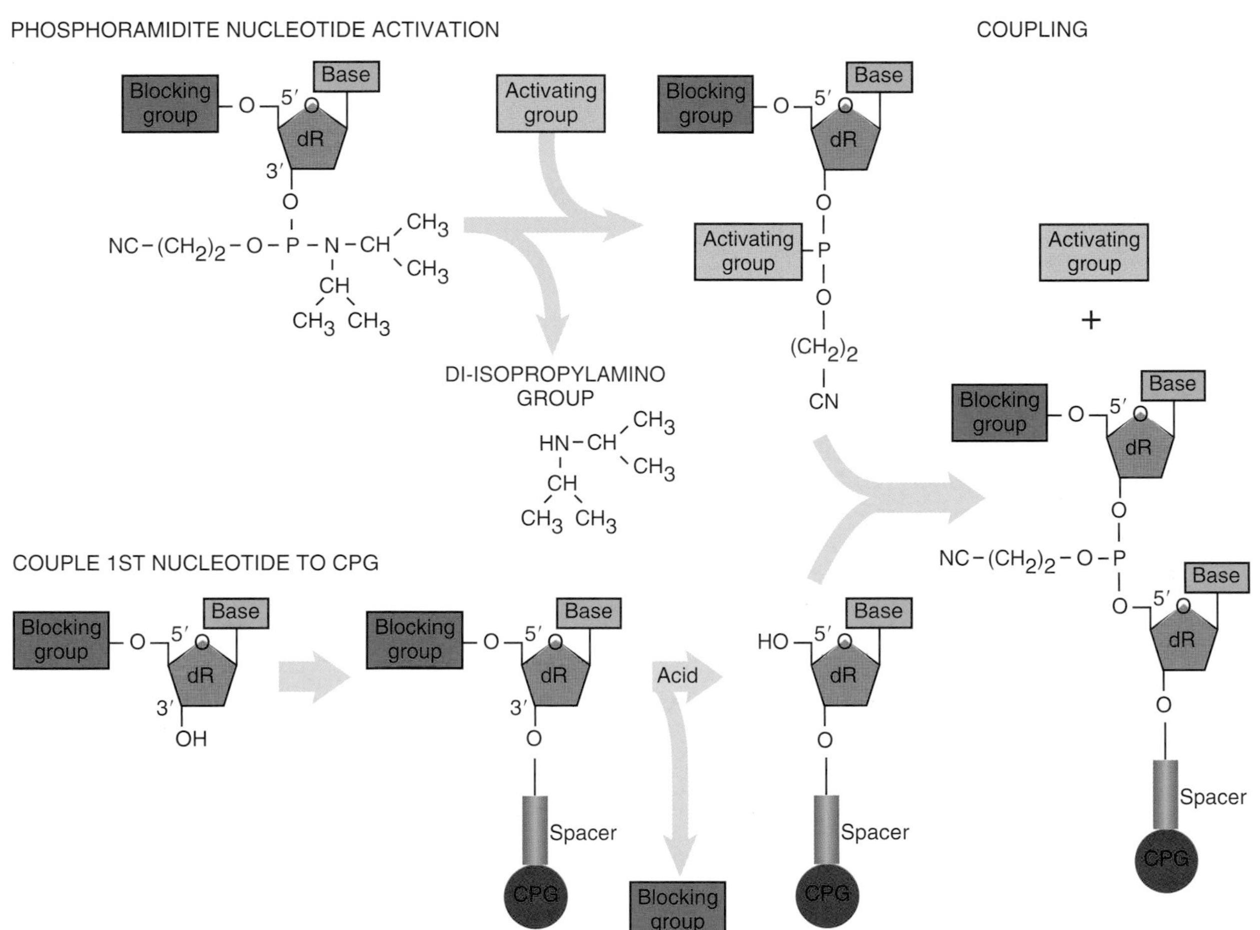

FIGURE 4.14 Adding the Second Nucleotide

During chemical synthesis of DNA, nucleotides are added in a 3′ to 5′ direction (the opposite of *in vivo* DNA synthesis). Therefore, the 3′-OH of an incoming nucleotide must be activated, but the 5′-OH must be blocked (see top nucleotide). For nucleotides already attached to the bead, the opposite must be done. Here, the blocking group on the 5′-OH of nucleotide 1 is removed by treatment with a mild acid. When the second nucleotide is added, it reacts to form a dinucleotide.

CHEMICAL SYNTHESIS OF COMPLETE GENES

As mentioned earlier, at each nucleoside addition in chemical synthesis, a proportion of oligonucleotides do not react with the next base, and these are capped with an acetyl group. The efficiency for nucleoside addition is critical, because if each step has low efficiency, the number of full-length oligonucleotides will decrease exponentially. For example, if the efficiency is 50% at each round, only half of the oligonucleotides add the second base, one-fourth would add the

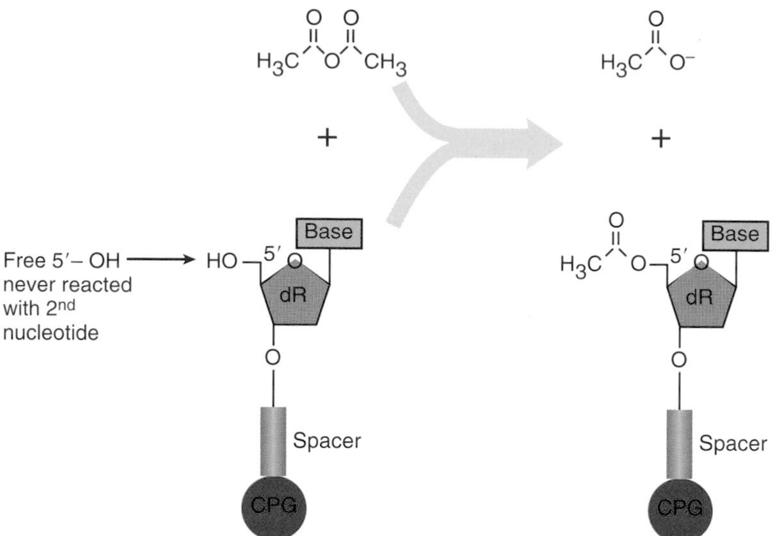

FIGURE 4.15
Capping of Unreacted Nucleotides
If any of the first nucleotides are not coupled to a second nucleotide, these could react with a subsequent nucleotide creating an internal deletion of the oligonucleotide. To prevent this error, the unreacted 5′-OH is capped with an acetyl group from acetic anhydride.

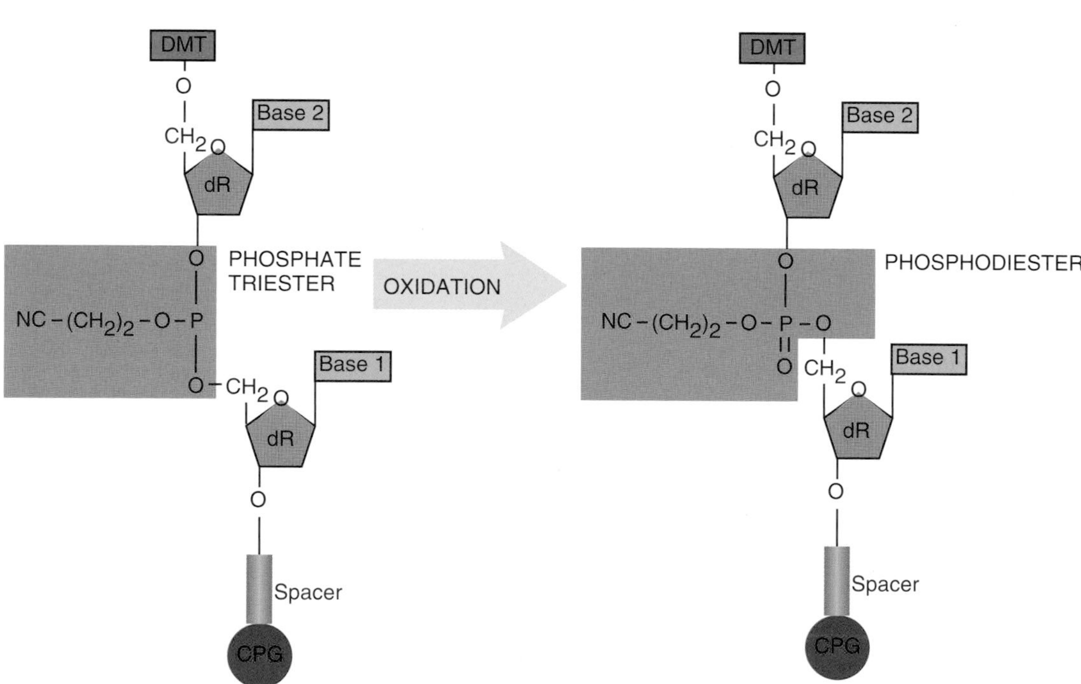

FIGURE 4.16
Oxidation Converts Phosphite Triester into a Phosphodiester
The phosphite triester is oxidized to a phosphodiester by adding iodine. This stabilizes the dinucleotide for further additions.

third base, one-eighth would get four bases; one-sixteenth would get the fifth base, and so on. Even if the final product were merely 10 bases in length, poor coupling would yield minuscule amounts of full-length product. It is critical for DNA synthesizers to have about 99% efficiency in each round, and then truncated products are the minority of the final sample. With high efficiencies, it is possible to synthesize longer segments of DNA. At 99% efficiency, an oligonucleotide that is 100 nucleotides long would give about 30%–40% final yield. If the desired oligonucleotide is separated from the truncated products by electrophoresis (see Chapter 3), it is possible to get plenty of full-length products.

Complete genes can be synthesized by linking smaller oligonucleotides together (Fig. 4.18). If the complete sequence of a gene is known, then long oligonucleotides can be synthesized identical to that sequence. The efficiency of the DNA synthesizer usually limits the length of each segment to about 100 bases; therefore, the gene segments are made with overlapping ends. Because oligonucleotides are single-stranded, both strands of the gene must be synthesized and annealed to each other, and then the segments are linked using ligase. Another strategy for assembly is to create strands that overlap only partially and then use DNA polymerase I to fill in the large single-stranded gaps.

DNA can be synthesized in long segments provided each base is added efficiently. These long segments can be linked into one complete gene.

POLYMERASE CHAIN REACTION USES *IN VITRO* SYNTHESIS TO AMPLIFY SMALL AMOUNTS OF DNA

The **polymerase chain reaction (PCR)** amplifies small samples of DNA into large amounts, much as a photocopier makes many copies of a sheet of paper. The DNA is amplified using the principles of replication; that is, the DNA is replicated over and over by DNA polymerase until a large amount is manufactured. Kary Mullis invented this technique while working at Cetus in 1983. He later won the Nobel Prize in Chemistry for PCR because of its huge impact on biology and science. PCR is used in forensic medicine to identify victims or criminals by amplifying the minuscule amounts of DNA left at a crime scene (see Chapter 23); PCR can identify infectious diseases such as HIV before symptoms emerge (see Chapter 21); PCR can amplify specific segments of genes without the need for cloning the segment first; in fact, PCR is now used in all aspects of the biological sciences.

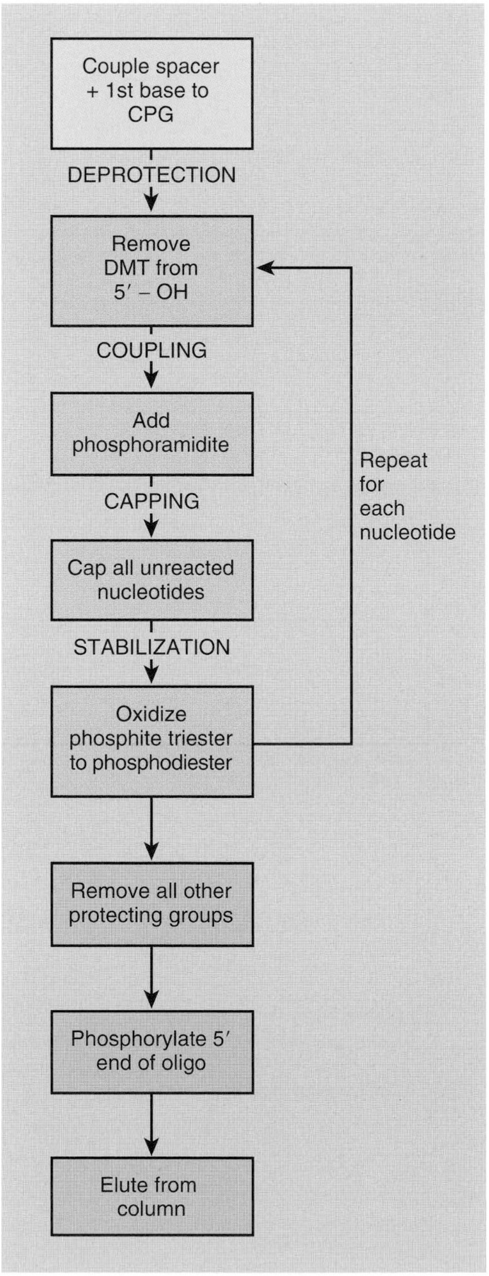

Couple spacer + 1st base to CPG

DEPROTECTION

Remove DMT from 5′ – OH

COUPLING

Add phosphoramidite

CAPPING

Cap all unreacted nucleotides

STABILIZATION

Oxidize phosphite triester to phosphodiester

Repeat for each nucleotide

Remove all other protecting groups

Phosphorylate 5′ end of oligo

Elute from column

FIGURE 4.17 Flow Chart of Oligonucleotide Synthesis
Oligonucleotide synthesis has many steps that are repeated. The first nucleotide is coupled to a bead with a spacer molecule. Next, the 5′-DMT is removed, and activated phosphoramidite nucleotide is added to the 5′ end of the first nucleotide. All the first nucleotides that were not linked to a second nucleotide are capped to prevent any further extension. Next, the phosphite triester is converted to a phosphodiester. These steps (in green) are repeated for the entire length of the oligonucleotide. Once the oligonucleotide has the appropriate length, the steps in tan are performed on the entire molecule.

Just as the photocopier needs more paper, ink, and a machine to make the copies, PCR requires specific reagents. The sample to be copied is called the template DNA, and this is often a known sequence or gene. The template DNA is typically double-stranded, and extremely small quantities are sufficient. The template DNA can be found within a complex mixture such as whole genomic DNA samples or within a fairly simple sample of bacterial plasmid DNA. The second reagent needed for PCR is a pair of oligonucleotide primers, which have sequences complementary to the ends of the template DNA. The DNA primers are oligonucleotides about 8 to 20 nucleotides long. One primer anneals to the 5′ end of the sense strand, and the other anneals to the 3′ end of the antisense strand of the target sequence. The primer sequences specify the exact target region of the DNA sample, thus focusing the reaction on the template DNA even if it is found within a complex mixture of genomic DNA. The third reagent is a supply of

COMPLETE SYNTHESIS OF BOTH STRANDS

SYNTHESIS OF OLIGONUCLEOTIDES
(i.e., single-stranded segments of DNA)

ANNEAL

SEAL NICKS WITH DNA LIGASE

COMPLETE dsDNA

A

PARTIAL SYNTHESIS FOLLOWED
BY POLYMERASE

SYNTHESIS OF OLIGONUCLEOTIDES

ANNEAL

FILL GAPS USING DNA POLYMERASE I

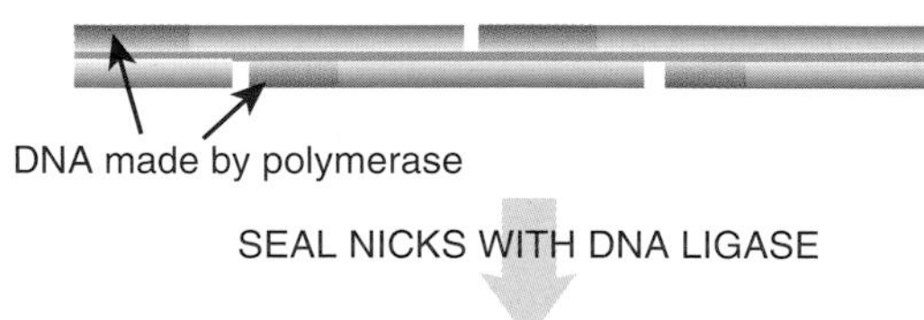

SEAL NICKS WITH DNA LIGASE

B

FIGURE 4.18
Synthesis and Assembly of a Gene
(A) Complete synthesis of both strands. Small genes can be chemically synthesized by making overlapping oligonucleotides. The complete sequence of the gene, both coding and noncoding strands, is made from small oligonucleotides that anneal to each other, forming a double-stranded piece of DNA with nicks along the phosphate backbone. The nicks are then sealed by DNA ligase. (B) Partial synthesis followed by polymerase. To manufacture longer pieces of DNA, oligonucleotides are synthesized so that a small portion of each oligonucleotide overlaps with the next. The entire sequence is manufactured, but gaps exist in both the coding and noncoding strands. These gaps are filled using DNA polymerase I, and the remaining nicks are sealed with DNA ligase.

nucleoside triphosphates, and the final reagent is ***Taq* DNA polymerase** from *Thermus aquaticus,* which actually makes the copies.

The basic mechanism of PCR includes heat denaturation of the template, annealing of the primers, and making a complementary copy using DNA polymerase, each step found in DNA replication. The three steps are repeated over and over until one template strand generates millions of identical copies. An amount of DNA too small to be seen can be copied so that it can be cloned into a vector, or visualized on an agarose gel (see Chapter 3). The process requires changing the temperature in a cyclic manner. Changing temperatures is accomplished by a **thermocycler**, a machine designed to change the temperature of its heat block rapidly so that each cycle can be completed in minutes. The temperature cycles between 94 °C to denature the template; 50 °C–60 °C to anneal the primer (depending on the length and sequence of the primer); and 72 °C for *Taq* polymerase to make new DNA. Before thermocyclers were developed, PCR was accomplished by moving the mixture among three different water baths at different temperatures every few minutes, which was very tedious.

In principle, the PCR cycle resembles DNA replication with a few modifications (Fig. 4.19). Like other *in vitro* DNA synthesis reactions, the double-stranded template is denatured with high heat rather than enzymes. Then the temperature is lowered so that the primers anneal to their binding sites. The primers are made so that each binds to opposite strands of the template, one at the beginning and one at the end of the gene. Then DNA polymerase elongates both primers and converts both single template strands to double-stranded DNA. (Note: During sequencing, only one primer is used and only one strand of the template is replicated, but during PCR both strands are copied.) *Taq* polymerase is the most widely used polymerase for PCR because it is very stable at high temperatures and does not denature at the high temperatures needed to separate the strands of the template DNA. *Taq* polymerase comes from *Thermus aquaticus,* a bacterium that grows in the hot springs of Yellowstone Park, USA. After the first replication cycle, the whole process is repeated. The two DNA strands are denatured at high heat, and then the temperature drops to allow the primers

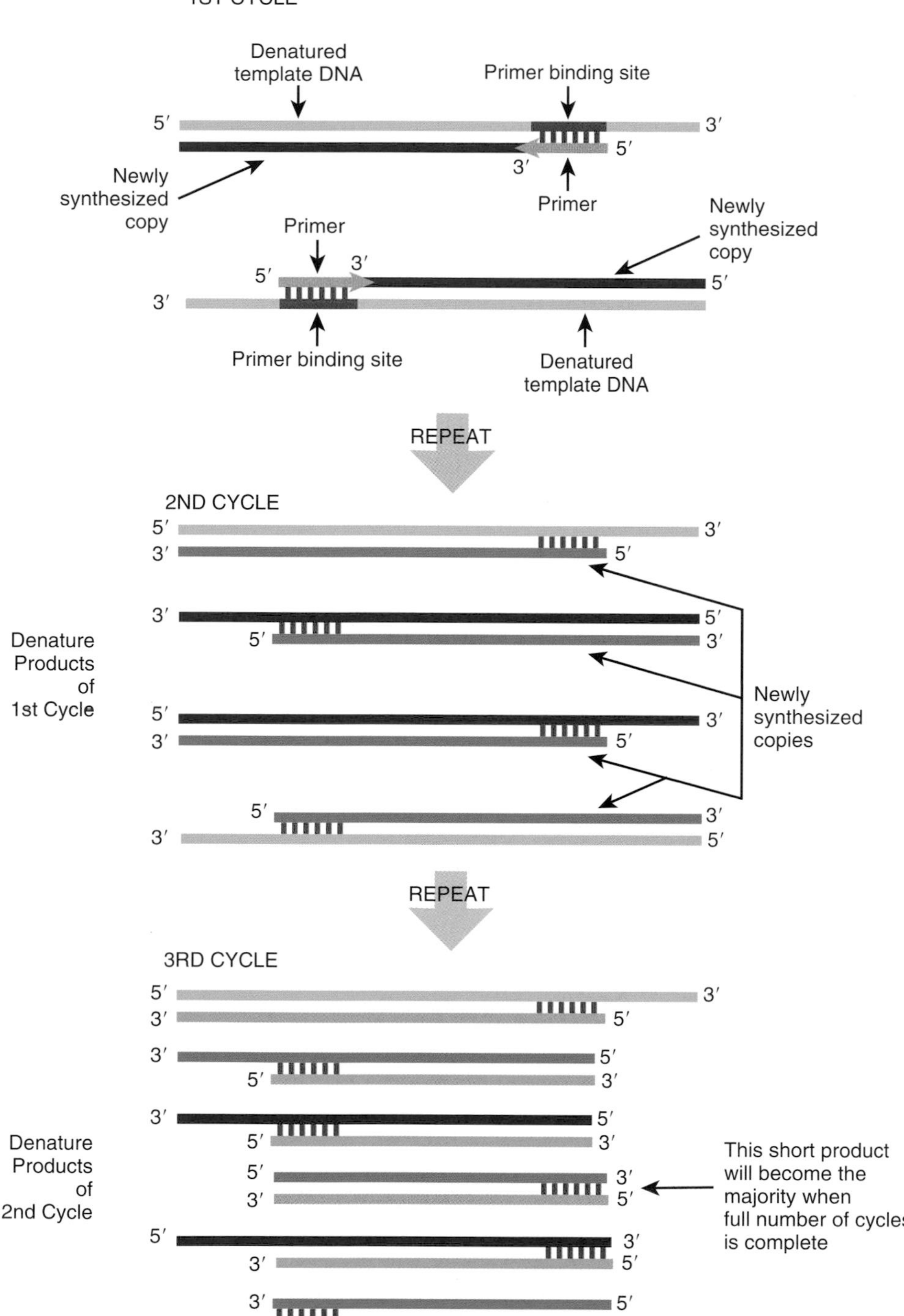

FIGURE 4.19 PCR, the First Three Cycles

In the first cycle, double-stranded template DNA (light purple) is denatured, complementary primers are annealed to the primer binding sites, and a new copy of the template is generated by *Taq* polymerase (red). In the second cycle, the two double-stranded products from the first cycle are denatured to form four single-stranded templates. The same set of primers anneals to the four template strands, and *Taq* polymerase makes each of the four double-stranded (dark purple). In the third cycle, the four double-stranded products from the second cycle are denatured, the primers anneal, and the four products from the second cycle become eight (light blue). Each subsequent round of denaturation, primer annealing, and extension doubles the number of copies, turning a small amount of template into a large amount of PCR product.

to anneal to their target sequences. *Taq* polymerase synthesizes the next four strands, and now there are four double-stranded copies of the target sequence. Early in the process, some longer strands are generated; however, eventually only the segment flanked by the two primers is amplified. Ultimately, the template strands and early PCR products become the minority. The shorter products become the majority.

The primers are key to the process of PCR. If the primers do not anneal in the correct location, if the span between the primers is too large, or if the primers form hairpin regions rather than annealing to the target, then *Taq* polymerase will not be able to amplify the segment. Also, if both primers anneal to the same strand, the reaction will not work. If the template has a known sequence, primers are synthesized based on the sequences upstream and downstream of the region to be amplified. Modifications exist that allow researchers to analyze unknown sequences by PCR (see later discussion).

> PCR is a process that uses DNA polymerase in an *in vitro* sequencing reaction. Here, a double-stranded template is replicated to make two copies. Each of these products is replicated to make four, and the process continues exponentially.

MODIFICATIONS OF BASIC PCR

Many different permutations of PCR have been devised since Kary Mullis developed the basic procedure. All rely on the same basic PCR reaction, which takes a small amount of DNA and amplifies it by *in vitro* replication. Many of these variant protocols are essential tools for recombinant DNA research.

Several strategies allow amplifying a DNA segment by PCR even if its sequence is unknown. For example, the unknown sequence may be cloned into a vector (whose sequence is known). The primers are then designed to anneal to the regions of the vector just outside the insert.

In another scenario, the sequence of an encoded protein is used to generate PCR primers. Remember that most amino acids are encoded by more than one codon. Thus, during translation of a gene, one or more codons are used for the same amino acid. Therefore, if a protein sequence is converted backwards into nucleotide sequence, the sequence is not unique. For example, two different codons exist for histidine and glutamine, and four codons exist for serine. Consequently, the nucleotide sequence encoding the amino acid sequence histidine–glutamine–valine can be one of 16 different combinations.

If primers are made that depend on protein sequence, they will be **degenerate primers** and they will have a mixture of two or three different bases at the wobble positions in the triplet codon. During oligonucleotide synthesis, more than one phosphoramidite nucleotide can be added to the column at a particular step. Some of the primers will have one of the nucleotides, whereas other primers will have the other nucleotide. If many different wobble bases are added, a population of primers is created, each with a slightly different sequence. Within this population, some will bind to the target DNA perfectly, some will bind with only a few mismatches, and some won't bind at all. Of course, the annealing temperature for degenerate primers is adjusted to allow for some mismatches.

Inverse PCR is a trick used when sequence information is known only on one side of the target region (Fig. 4.20). First, a restriction enzyme is chosen that does not cut within the stretch of known DNA. The length of the recognition sequence should be six or more base pairs in order to generate reasonably long DNA segments for amplification by PCR. The target DNA is then cut with this restriction enzyme to yield a piece of DNA that has compatible sticky ends, one upstream of the known sequence and one downstream. The two ends are ligated to form a circle. The PCR primers are designed to recognize the end regions of the known sequence. Each primer binds to a different strand of the circular DNA, and they both point "outward" into the unknown DNA. PCR then amplifies the unknown DNA to give linear molecules with short stretches of known DNA at the ends, and the restriction enzyme site in the middle.

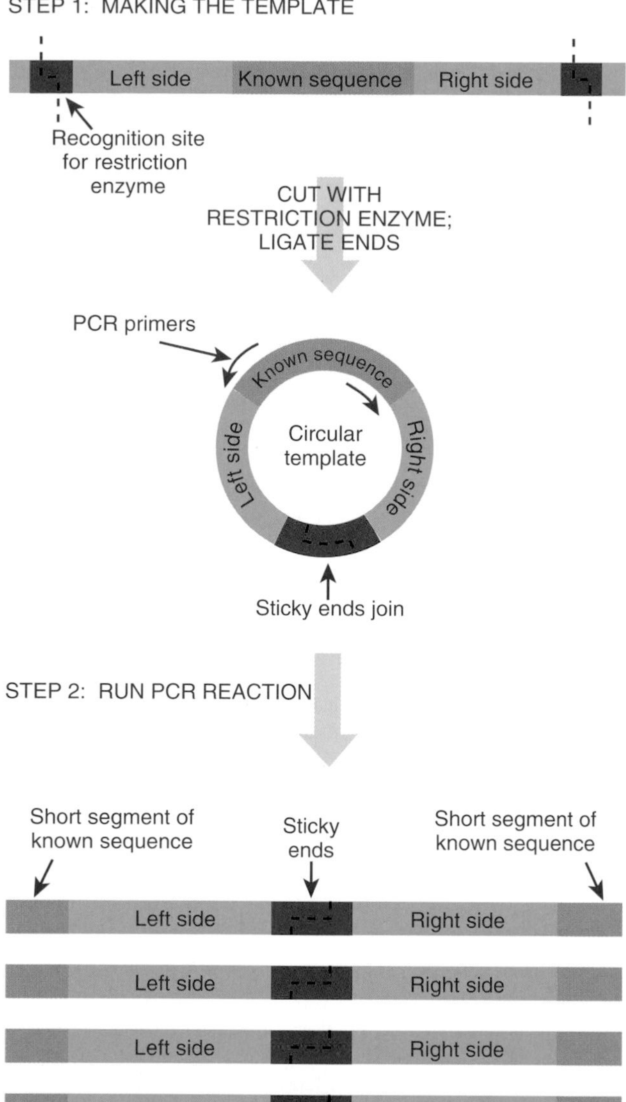

FIGURE 4.20 Inverse PCR

Inverse PCR allows unknown sequences to be amplified by PCR provided that they are located near a known sequence. The DNA is cut with a restriction enzyme that cuts upstream and downstream of the known region but not within it. The linear piece of DNA is circularized and then amplified with primers that anneal in the known region. The PCR products have the unknown DNA from the left and right of the known sequence. These can be cloned and sequenced.

Degenerate primers are designed based on amino acid sequences and contain different nucleotides at the wobble position.

Inverse PCR sequences DNA near a known sequence by finding a restriction enzyme recognition sequence away in the unknown region, cutting out this template, and amplifying the entire piece with *Taq* polymerase.

REVERSE TRANSCRIPTASE PCR

Reverse transcriptase PCR (RT-PCR) uses the enzyme reverse transcriptase to make a cDNA copy of mRNA from an organism and then uses PCR to amplify the cDNA (Fig. 4.21). The advantage of this technique is evident when trying to use PCR to amplify a gene from eukaryotic DNA. Eukaryotes have introns, some extremely long, which interrupt the coding segments. After transcription, the primary RNA transcript is processed to remove all the introns, hence becoming mRNA. Using mRNA as the source of the target

DNA relies on the cell removing the introns. In practice, RT-PCR has two steps. First, reverse transcriptase recognizes the 3′ end of primers containing repeated thymines and synthesizes a DNA strand that is complementary to the mRNA. (The thymines base-pair with the poly(A) tail of mRNA.) Then the RNA strand is replaced with another DNA strand, leaving a double-stranded DNA (i.e., the cDNA). Next, the cDNA is amplified using a normal PCR reaction containing appropriate primers (one usually recognizes the poly(A) tail), *Taq* polymerase, and nucleotides.

RT-PCR uses reverse transcriptase to convert mRNA into double-stranded DNA, and then the gene without any introns can be amplified by regular PCR.

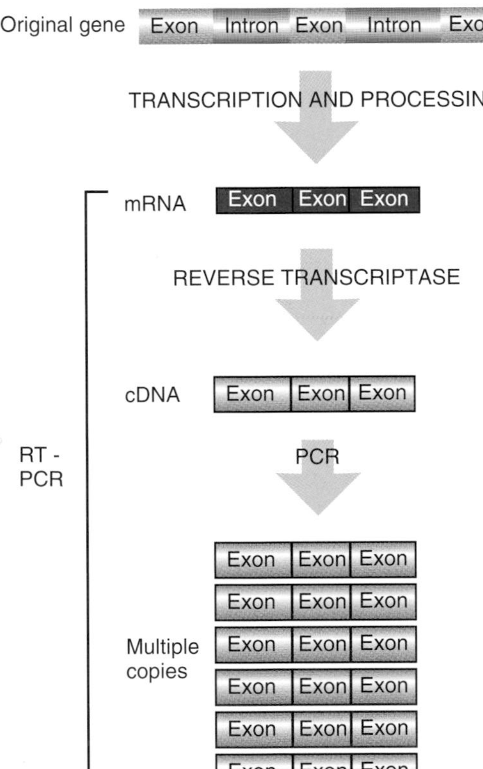

FIGURE 4.21 Reverse Transcriptase PCR
RT-PCR is a two-step procedure that involves making a cDNA copy of the mRNA and then using PCR to amplify the cDNA.

PCR IN GENETIC ENGINEERING

PCR allows scientists to clone genes or segments of genes for identification and analysis. PCR also allows scientists to manipulate a gene that has already been identified. Various modified PCR techniques allow scientists to hybridize two separate genes or genes segments into one, delete or invert regions of DNA, and alter single nucleotides to change the gene and its encoded protein in a more subtle way.

PCR can make cloning a foreign piece of DNA easier. Special PCR primers can generate new restriction enzyme sites at the ends of the target sequence (Fig. 4.22). The primer is synthesized so that its 5′ end has the desired restriction enzyme site, and the 3′ end has sequence complementary to the target. Obviously, the 5′ end of the primer does not bind to the target DNA, but as long as the 3′ end has enough matches to the target, then the primer will still anneal. *Taq* polymerase primes synthesis from the 3′ end; therefore, the enzyme is not bothered by mismatched 5′ sequences. The resulting PCR product can easily be digested with the corresponding restriction enzyme and ligated into the appropriate vector.

Rather than incorporating restriction enzyme sites into the ends of the PCR product, **TA cloning** will clone any PCR product directly (Fig. 4.23). *Taq* polymerase has terminal transferase activity that generates a single adenine overhang on the ends of the PCR products it makes. Special vectors containing a single thymine overhang have been developed, and simply mixing the PCR product with the TA cloning vector plus DNA ligase clones the PCR product into the vector without any special modifications.

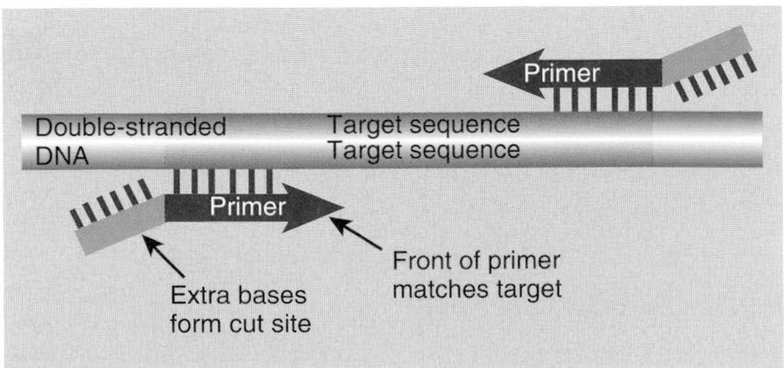

FIGURE 4.22 Incorporation of Artificial Restriction Enzyme Sites
Primers for PCR can be designed to have nonhomologous regions at the 5′ end that contain the recognition sequence for a particular restriction enzyme. After PCR, the amplified product has the restriction enzyme sites at both ends. If the PCR product is digested with the restriction enzyme, this generates sticky ends that are compatible with a chosen vector.

PCR can be used to manipulate cloned genes also. Two different gene segments can be hybridized into one using **overlap PCR** (Fig. 4.24). Here, PCR amplification occurs with three primers: one is complementary to the beginning of the first gene segment, one is complementary to the end of the second gene segment, and a third is half complementary to the end of gene segment 1 and half complementary to the beginning of gene segment 2. During PCR, the two gene segments become fused into one by a mechanism that is hard to visualize, but probably involves looping of some of the early PCR products.

PCR can be used to create large deletions or insertions into a gene (Fig. 4.25). Once again the design of the PCR primers is key to the construction. For example, primers to generate insertions have two regions: the first half is homologous to the sequence around the insertion point; the second half has sequences complementary to the insert sequence. For example, suppose an antibiotic resistance gene such as *npt* (confers resistance to neomycin) is to be inserted into a cloning vector. The primers would have their 5′ ends complementary to the sequence flanking the insertion point on the vector and their 3′ ends complementary to the ends of the *npt* gene. First, the primers are used to amplify the *npt* gene and give a product with sequences homologous to the vector flanking both ends. Next, the PCR product is transformed into bacteria harboring the vector. The *npt* gene recombines with the insertion point by homologous recombination, resulting in insertion of the *npt* gene into the vector.

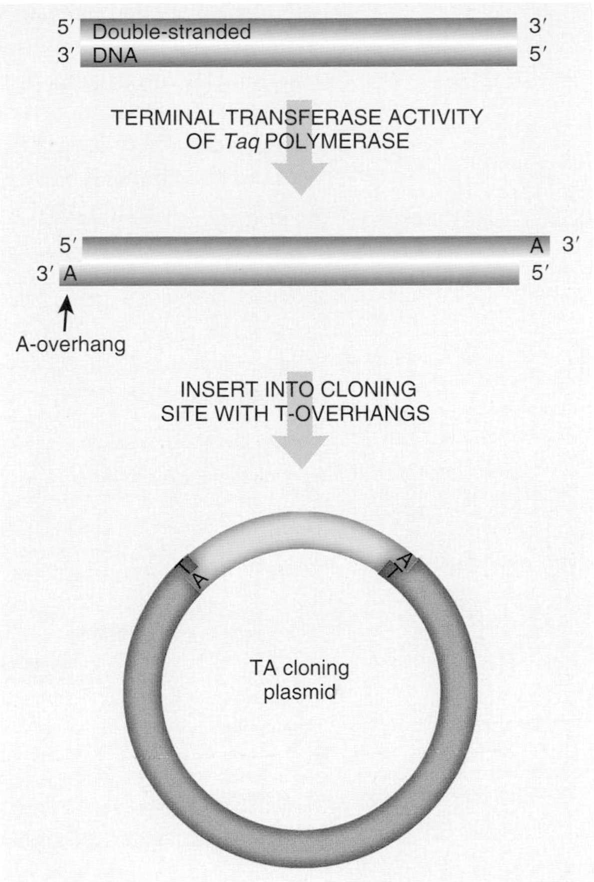

FIGURE 4.23 TA Cloning of PCR Products
When *Taq* polymerase amplifies a piece of DNA during PCR, the terminal transferase activity adds an extra adenine at the 3′ ends. The TA cloning vector was designed so that when linearized, it has a single 5′-thymine overhang. The PCR product can be ligated into this vector without the need for special restriction enzyme sites.

The insertion point(s) will determine whether the antibiotic cassette causes just an insertion or both an insertion plus a deletion. If the two PCR primers recognize separate homologous recombination sites, then the incoming PCR segment will recombine at these two sites. Homologous recombination then results in the *npt* gene replacing a piece of the vector rather than merely inserting at one particular location.

PCR can also generate nucleotide changes in a gene by **directed mutagenesis** (Fig. 4.26). Usually, only one or a few adjacent nucleotides are changed. First, a mutagenic PCR primer is synthesized that has nucleotide mismatches in the middle region of the primer. The primer

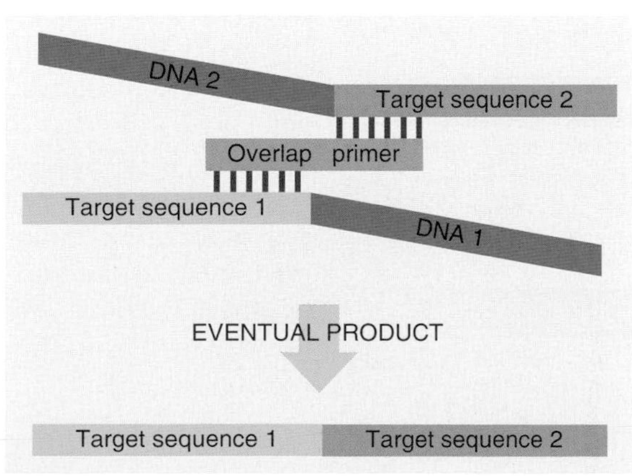

FIGURE 4.24 Overlap PCR
Overlapping primers can be used to link two different gene segments. In this scheme, the overlapping primer has one end with sequences complementary to target sequence 1, and the other half similar to target sequence 2. The PCR reaction will create a product with these two regions linked together.

will anneal to the target site with the mismatch in the center. The primer needs to have enough matching nucleotides on both sides of the mismatch so that binding is stable during the PCR reaction. The mutagenic primer is paired with a normal primer. The PCR reaction then amplifies the target DNA incorporating the changes at the end with the mutagenic primer. These changes may be relatively subtle, but if the right nucleotides are changed, then a critical amino acid may be changed. One amino acid change can alter the entire function of a protein. Such an approach is often used to assess the importance of particular amino acids within a protein.

The 5′ end of PCR primers does not need to be complementary to the template DNA, and can be designed to add restriction enzyme sites to the PCR product. The terminal transferase activity by *Taq* polymerase adds a single adenine onto the 3′ end of the PCR product. These traits allow the PCR product to be cloned into a vector.

PCR can be used to delete, insert, and even fuse different gene segments.

PCR can be used to make small changes in nucleotide sequences by directed mutagenesis.

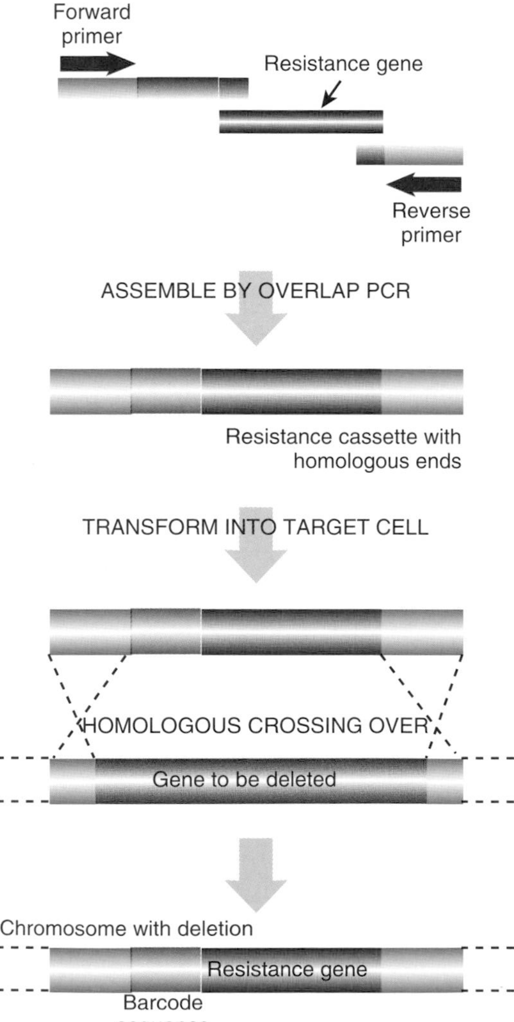

FIGURE 4.25 Generation of Insertions or Deletions by PCR
In the first step, a specifically targeted cassette is constructed by PCR. This contains both a suitable marker gene and upstream and downstream sequences homologous to the target site. The engineered cassette is transformed into the host cell, and homologous crossing over occurs. Recombinants are selected by the antibiotic resistance carried on the cassette. The barcode sequence is a unique DNA sequence only found in the cassette used to identify the location of the cassette.

PCR OF DNA CAN DETERMINE THE SEQUENCE OF BASES

Being able to quickly and easily determine the sequence of any gene has been the driving force for the recent advances made in biotechnology. Frederick Sanger developed a method for sequencing a gene *in vitro* in 1974. He was interested in the amino acid sequence of insulin and decided to deduce the sequence of the protein from the nucleotide sequence. He invented the **chain termination sequencing** method, which is still used today (Fig. 4.27). Much like DNA replication, chain termination sequencing requires a primer, DNA polymerase, a single-stranded DNA template, and deoxynucleotides. During *in vitro* sequencing reactions, these components are mixed and DNA polymerase makes many copies of the original template. The first trick needed to deduce the sequence is to stop synthesis of the newly synthesized DNA chains at each base pair. Consequently, the fragments generated will differ in size by one base pair and, when separated by gel electrophoresis, create a ladder of fragments. The next step is to figure out the identity of the last nucleotide. If the final base pair for each fragment is known, the sequence may be directly read from the gel (reading from bottom to top).

But how do we know what the final base is for each fragment on the sequencing ladder? DNA polymerase synthesizes a new strand of DNA based on the template sequence. The chain consists of deoxynucleotides, each with a hydroxyl group at the 3′ position on the deoxyribose ring. DNA polymerase adds the next nucleotide by linking the phosphate of the incoming nucleotide to the 3′-hydroxyl of the previous nucleotide. If a nucleotide lacks

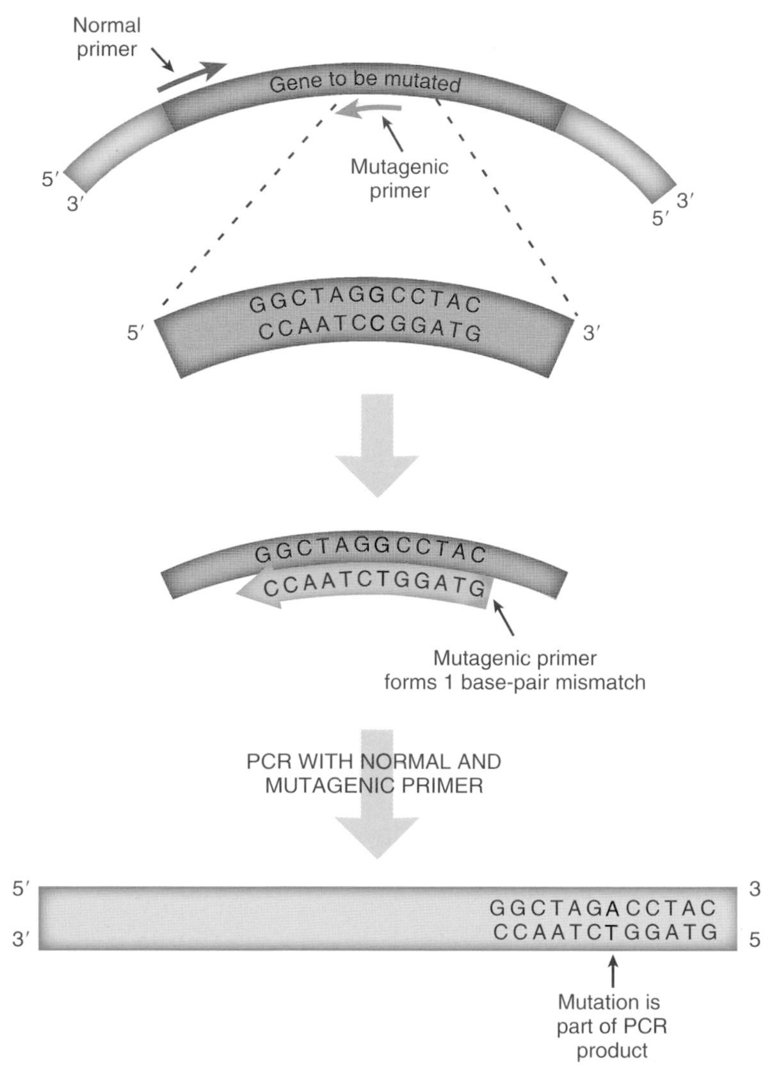

FIGURE 4.26 Direct Mutagenesis Using PCR
The gene to be mutated is cloned, and the entire sequence is known. To alter one specific nucleotide, normal and mutagenic primers are combined in a PCR reaction. The mutagenic primer will have a mismatch in the middle, but the remaining sequences will be complementary. The PCR product will incorporate the sequence of the mutagenic primer.

the 3′-hydroxyl, no further nucleotides can be added and the chain is terminated (Fig. 4.28). During a sequencing reaction, a certain percentage of nucleotides with no 3′-hydroxyl, called **dideoxynucleotides**, are mixed with the normal deoxynucleotides. Such reactions typically have a maximum length of about 800 nucleotides.

The fragments are relatively small for DNA and vary in length by only one nucleotide; therefore, they must be separated by size using polyacrylamide gel electrophoresis (see Chapter 3). The principle is the same as for agarose gel electrophoresis, but polyacrylamide has smaller pores, and so smaller fragments can be separated with higher resolution. The sequence is actually read from the bottom of the gel to the top, because the fragments terminated closest to the primer are smaller (hence run faster) than the ones further from the primer. The bands appear as a ladder, each separated by one nucleotide; therefore, each band represents the fragments ending with the dideoxynucleotide complementary to the template strand.

Automated DNA sequencing uses a PCR-type reaction to sequence DNA. In PCR sequencing, or **cycle sequencing**, the template DNA with unknown sequence is amplified by *Taq* polymerase as any normal PCR reaction. *Taq* DNA polymerase was modified to remove its proofreading ability and increase the speed at which it incorporates nucleotides. Cycle sequencing

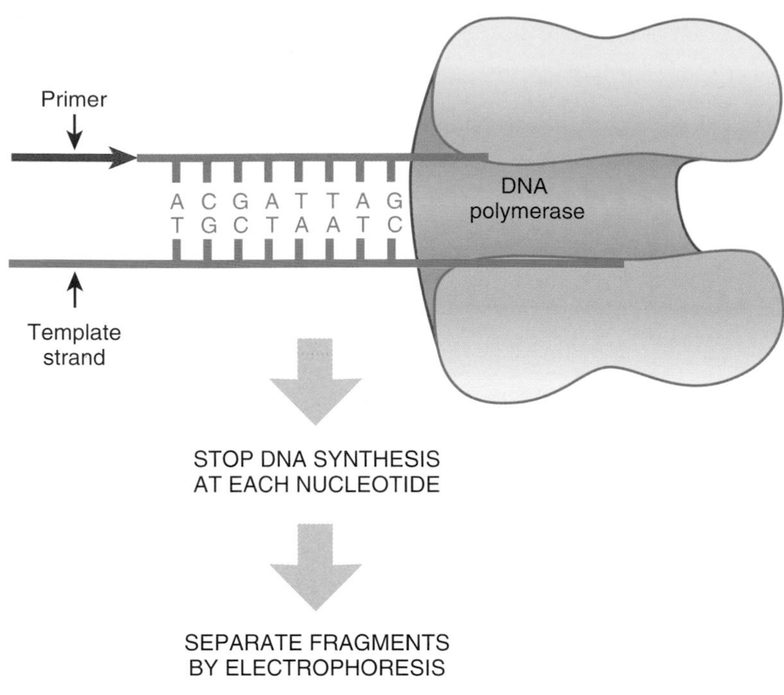

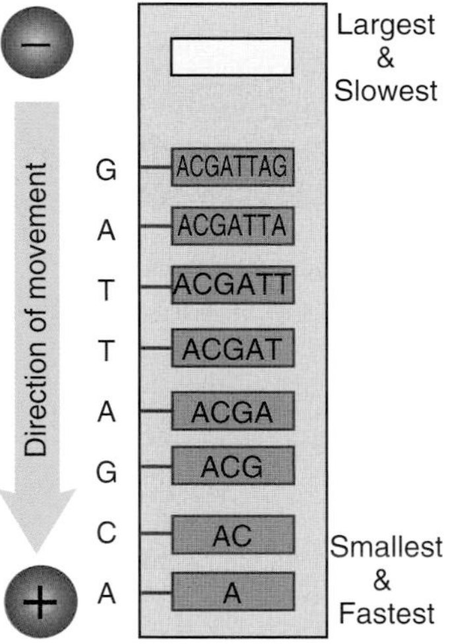

FIGURE 4.27 Chain Termination Method of Sequencing
During chain termination, DNA polymerase synthesizes many different strands of DNA from the single-stranded template. DNA polymerase will stop at each nucleotide, such that strands of all possible lengths are made. They are separated by size using electrophoresis. The smallest fragments are at the bottom and represent the primer plus only the first one or two nucleotides of the template DNA. Longer fragments contain the primer plus longer stretches of synthesized DNA complementary to to the template DNA.

reaction mixtures include all four deoxynucleotides, all four dideoxynucleotides, a single primer, template DNA, and *Taq* polymerase. To discern the identity of the dideoxynucleotide, they are linked to a unique fluorophore for each of the four nucleotides.

The samples are amplified in a thermocycler. First, the template DNA is denatured at a high temperature; then the temperature is lowered to anneal the primer; and finally, the temperature is raised to 72°C, the optimal temperature for *Taq* polymerase to make DNA copies of the template. During polymerization, dideoxynucleotides are incorporated and cause chain termination. The ratio of dideoxynucleotides to deoxynucleotides is adjusted to ensure that some fragments stop at each G, A, T, or C of the template strand. After *Taq* polymerase makes thousands of copies of the template, each stopping at a different nucleotide, the entire mixture is separated in one lane of a sequencing gel (Fig. 4.29). Bands of four different colors are seen, corresponding to the four fluorescently labeled dideoxynucleotides and hence the four bases.

Cycle sequencing has many advantages. During cycle sequencing, each round brings the temperature to 95°C, which destroys any secondary structures or double-stranded regions. Another advantage of cycle sequencing is to control primer hybridization. Some primers do not work well with regular sequencing reactions because they bind to closely related sequences. During cycle sequencing, the primer annealing temperature is controlled and can be set quite high in order to combat nonspecific binding. Finally, cycle sequencing requires very little template DNA; therefore, sequencing can be done from smaller samples.

Another advance in sequencing has been the detection system. **Automatic DNA sequencers** detect each of the fluorescent tags and record the sequence of bases (Fig. 4.30). Some automatic DNA sequencers can read up to 384 different DNA samples using capillary tubes filled with gel matrix to separate the DNA fragments. At the bottom of each capillary tube is a fluorescent activator, which emits light to excite the fluorescent dyes. On the other side is the detector, which reads the wavelength of light that the fluorescent dye emits. As each fragment passes the detector, it measures the wavelength and records the data as a peak on a graph. For each fluorescent dye, a peak is recorded and assigned to the appropriate base. An attached computer records and compiles the data into the DNA sequence.

Automated sequencing has a large startup cost because the sequence analyzer is quite expensive, but they run multiple samples at one time, and thus the cost per sample is quite low. Many universities and companies have a centralized facility that does the sequencing for all the researchers. In fact, sequencing has become so automated that many researchers just send their template DNA and primers to a company that specializes in sequencing.

RANDOM TERMINATION AT "G" POSITIONS

Original sequence:
TCGGACCGCTGGTAGCA

Mixture of dCTP, dATP, dTTP, dGTP (G) and ddGTP (G).

1. TCG
2. TCGG
3. TCGGACCG
4. TCGGACCGCTG
5. TCGGACCGCTGG
6. TCGGACCGCTGGTAG

RUN ON SEQUENCING GEL

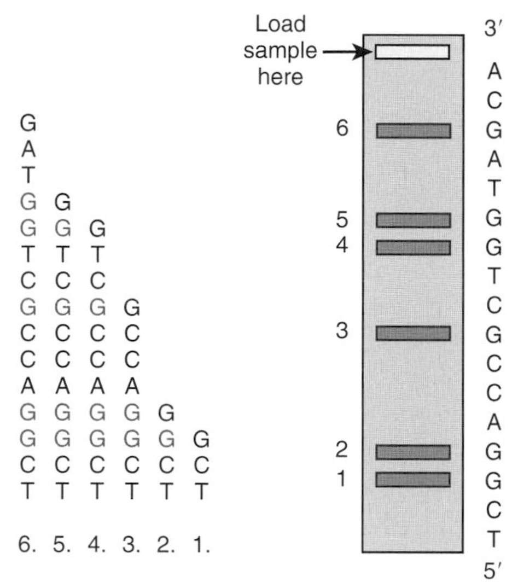

Sequences ending in "G"

FIGURE 4.28 Chain Termination by Dideoxynucleotides
During the sequencing reaction, DNA polymerase makes multiple copies of the original sequence. Sequencing reaction mixtures contain dideoxynucleotides that terminate growing DNA chains. The example here shows a sample reaction, which includes triphosphates of both deoxyguanosine (dG) and dideoxyguanosine (ddG). Whenever ddG is incorporated (shown in red), it causes termination of the growing chain. If dG (blue) is incorporated, the chain will continue to grow. When the sequencing reaction containing the ddG is separated on a polyacrylamide gel, the fragments are separated by size. Each band directly represents the fragment ending in G from the original sequence.

Chain-terminating dideoxynucleotides are the key to determining DNA sequence. When these are incorporated into an *in vitro* replication reaction, DNA polymerase cannot add any more nucleotides and the synthesis reaction ends. In cycle sequencing, a PCR reaction includes a controlled amount of fluorescently labeled dideoxynucleotides. *Taq* polymerase stops adding nucleotides when a dideoxynucleotide is incorporated. The fluorescent tag is used to identify the ending base of each fragment using an automated sequencer.

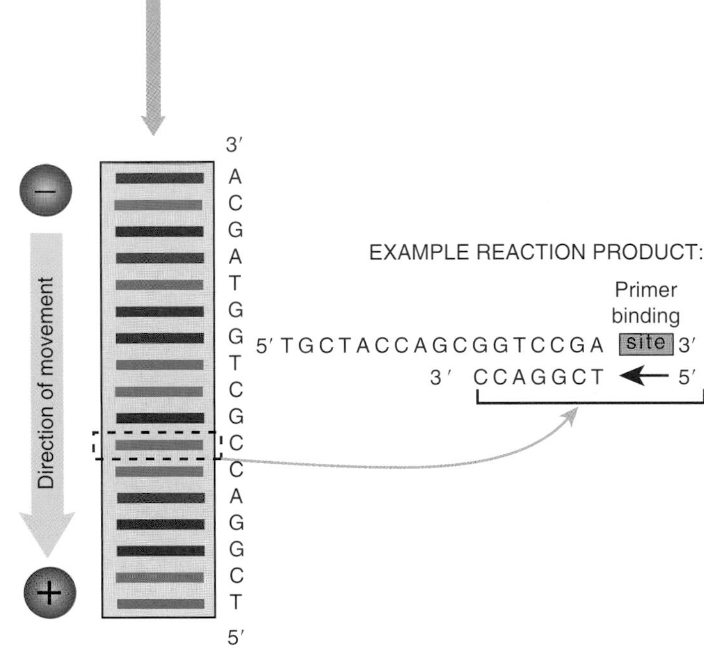

FIGURE 4.29 Cycle Sequencing
During cycle sequencing, the reaction contains template DNA, primer, *Taq* polymerase, deoxynucleotides, and dideoxynucleotides. Each of the different dideoxynucleotides has a different fluorescent label attached. The automated sequencer detects the color and compiles the sequence data.

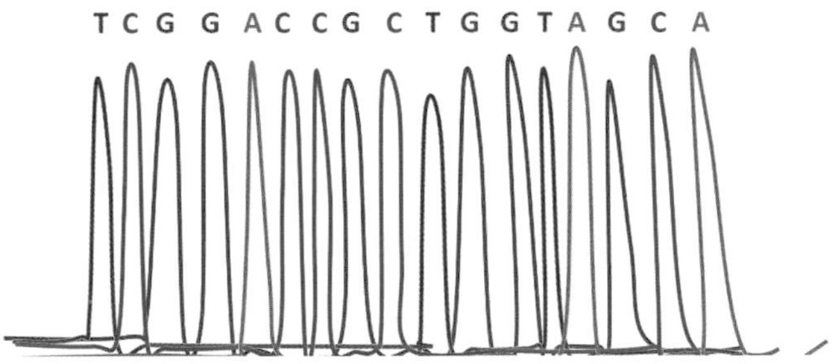

FIGURE 4.30 Data from an Automated Sequencer
A representative set of data from an automated sequencer. The fluorescent peaks for the individual bases are shown. The computer compiles the information into a sequence file for the researcher.

NEXT-GENERATION SEQUENCING TECHNOLOGIES

Sequencing DNA using chain termination was the workhorse for the initial sequencing of the first human genome. Throughout the human genome project, the cost for each base of DNA dropped by making advances in the capillary electrophoresis chain-termination method. The cost for sequencing one million bases of DNA in September 2001, the end of the initial sequence, was $5292, and so for the whole human genome, over $95 million. Because of the advances in chain termination sequencing, the human genome project was done early and under budget. As of October 2013, the cost to sequence one million base pairs of DNA dropped less than 6 cents. The cost to sequence an entire human genome, therefore, is a mere $5096. The incredible decrease in cost stems from the advent of **massively parallel sequencing**, which is a descriptive name for **next-generation sequencing**. These technologies use a type of platform that can hold millions of DNA fragments in separate locations. There are many different chemistries used in next-generation sequencing, and they are rapidly changing. Two sequencing platforms, 454 sequencing and Illumina, are outlined here.

The first step of any next-generation technology is to prepare the DNA for amplification by PCR (Fig. 4.31). Genomic DNA is isolated from the organism of interest according to a standard DNA isolation protocol. The pure DNA is then sheared into small fragments using sonication. To amplify each of the fragments, the end of each piece of DNA must have known sequence. This is impossible, especially for genomes that have never been sequenced. And even if the genome sequence is known, sonication creates random breaks in the DNA, so there is no way to truly know the sequence at each end. The trick to circumvent this problem is to add **linkers** or **adaptors**, which are short DNA pieces with a known sequence. They are added to the ends using the TA cloning technology. The linker or adaptor sequence depends on which of the next-generation sequencing technologies are employed. A **barcode sequence** or an **index sequence** is a key feature of the adaptor. The barcode or index sequence is much like a zip code in your address: the sequence is unique to the sample of DNA, and it allows multiple samples of DNA to be analyzed at the same time, a procedure called **multiplexing**.

Once the DNA sample is fragmented and adaptors are added onto each end, the DNA is attached to a solid surface so that individual DNA fragments are separated from each other. In 454 sequencing, the DNA fragments are attached to beads via the adaptors. The set of beads with small DNA oligonucleotides complementary to the adaptor is mixed with the DNA at a ratio such that one DNA fragment will attach to a single bead. Ensuring that a single DNA from the genome attaches to a single bead is a critical step for sequencing. In the Illumina sequencing methodology, the same principle applies, but the DNA fragments are added to the surface of a flow cell. The surface has DNA primers complementary to the adaptor scattered on the surface. These must also be of sufficient distance from each other to appear as a separate location by sensors at the bottom of the flow cell.

The next step for next-generation sequencing is to create multiple copies of the single piece of DNA using PCR. For 454 sequencing, **emulsion PCR** creates multiple copies of the single piece of DNA that attaches to the bead. The process begins by creating an emulsion of oil and water such that only one bead is found in each of the water droplets. In addition, the water droplets contain free deoxynucleotides, primers complementary to the adaptors, and *Taq* DNA polymerase. Within the droplets, the DNA fragment is amplified using the traditional denaturation, annealing, and elongations steps. The final result is a bead coated with identical copies of the DNA fragment. The emulsion prevents the DNA from one bead diffusing to a different bead.

In a similar fashion, DNA fragments attached to the surface of a flow cell for Illumina sequencing are amplified by incubating the flow cell with deoxynucleotides and DNA polymerase in a process called **bridge amplification**. The primers used for amplifying the DNA fragment are attached to the flow cell, so the DNA anneals to another primer on the surface, forming a bridge. These are amplified and released to form a cluster of identical DNA fragments.

Once a cluster of identical DNA pieces is produced on the bead or flow cell, these pieces are denatured into single-stranded DNAs competent for sequencing. Sequencing for 454 and Illumina occurs as the single-stranded DNA is replicated. Each technology uses a different detection method for identifying the sequence, but in both 454 and Illumina sequencing, each nucleotide is identified one by one; that is, as a nucleotide is added to the complementary strand, the identity is recorded by a sensor and stored by an attached computer. This method of sequencing is called **sequencing by synthesis**. In 454 sequencing, the beads

FRAGMENT DNA AND ADD ADAPTORS TO ENDS

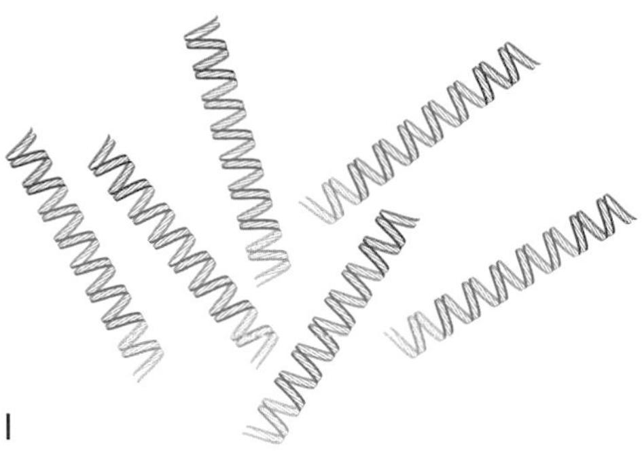

SEPARATE EACH TEMPLATE TO INDIVIDUAL DROPLETS OR SITES ON CHIP

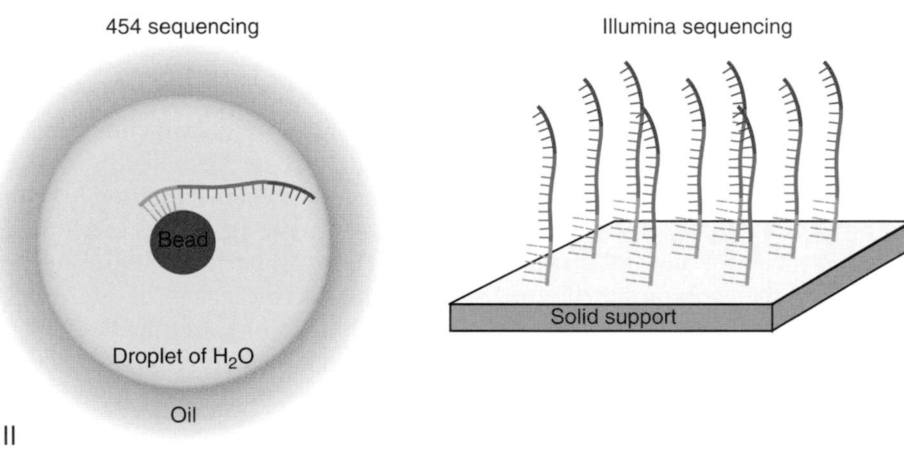

AMPLIFY EACH SINGLE TEMPLATE

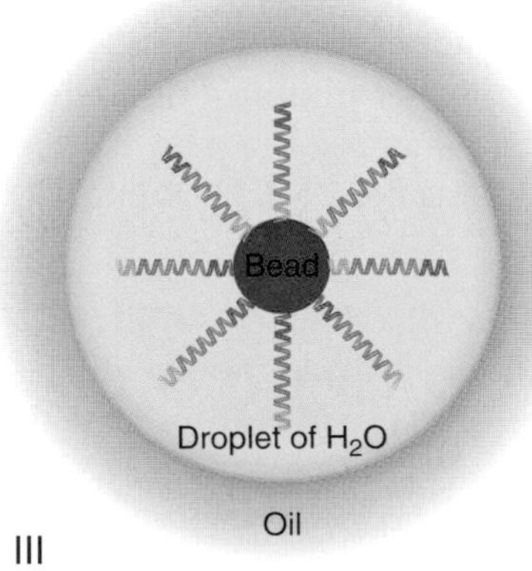

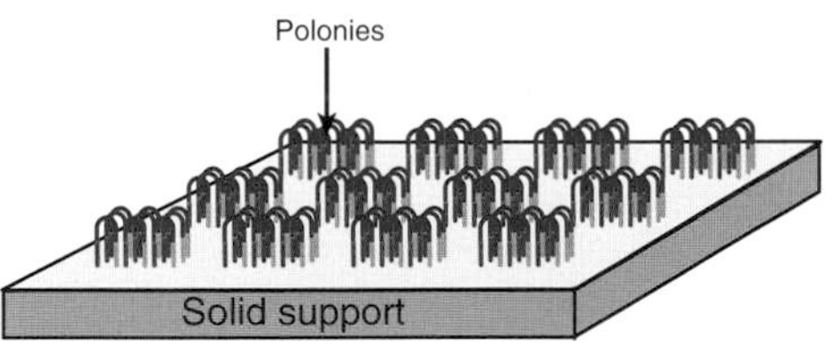

FIGURE 4.31 Next-Generation Sequencing
During next-generation sequencing, the DNA is prepared for amplification by isolating and fragmenting the sample. Adaptors are added to the ends of the fragments and then annealed to complementary oligonucleotides on the surface of the bead for 454 sequencing (left) or flow cell for Illumina sequencing (right). These are added such that one unique fragment attaches to one bead (left), or the spacing of the attached DNA is sufficient for recognition by the detector below the flow cell. Each single DNA template is amplified and denatured to be single-stranded. Sequence is determined after annealing primer to one end of the template and determining the identity of the each individually added nucleotide (see text for details). The sequence is detected by sensors below the well of the picotiter plate or below the flow cell and recorded by an attached computer.

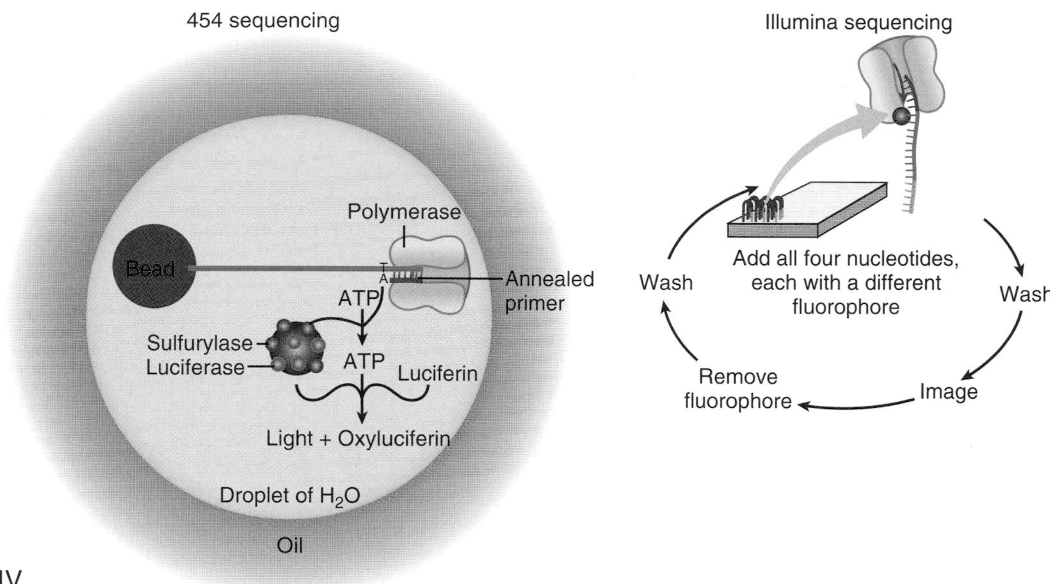

FIGURE 4.31 Cont'd

coated with copies of DNA are separated into a picotiter plate such that only one bead is within each well. A picotiter plate has over a million individual wells or holes in the surface that hold 75 picoliters. The lower surface of the well is optically clear to allow the light to be visualized by the detector. After a single bead enters the individual well or hole, a primer is annealed to the adaptor sequence on the fragment. Just like pyrosequencing, one of the four deoxynucleotides and DNA polymerase are floated across the picotiter plates, and if the template in the well has the complementary base pair, DNA polymerase adds the nucleotide to the primer, releasing a molecule of pyrophosphate. The well of the picotiter plate also contains luciferase and sulfurylase, which react with the released pyrophosphate and release a flash of light. At the bottom of the well, the attached sensor records the flash of light and sends the information to the computer. The flash of light appears only in the cells where the added nucleotide was incorporated, and the other cells remain dark. Each of the four nucleotides is added separately and washed away before adding another.

Illumina sequencing also determines each nucleotide as it is incorporated but uses a unique reversible fluorescent dye for guanine, cytosine, adenine, and thymine. After a primer is annealed to the adaptor sequence on the DNA template, all four nucleotides are added to the flow cell simultaneously. The wavelength of light released at each DNA cluster is recorded by the computer. The fluorescent dye terminator is removed, and another batch of four fluorescently labeled nucleotides is added. Again, the signal for each spot is recorded and stored. As each nucleotide is added, the computer compiles a sequence for each spot on the flow cell.

For both methods, the recording of data from each well of the picotiter plate or spot on the flow cell is compiled as a sequence for each separate DNA. Each of these pieces of sequence information is called a **read**. The technology for each method limits the number of nucleotides that can be determined with certainty, ranging from 50 to 400 base pairs depending on the sequencing machine and sequencing technology. To enhance the quality sequencing data results, **paired end reads** confirm the sequence information by repeating the entire sequence reaction, but using a primer to the opposite side of the fragment, thus essentially sequencing from both ends of the DNA fragment.

The amount of sequence data compiled by next-generation sequencing is tremendous, and without the increase in computer power and storage, there would be no way possible to compile the

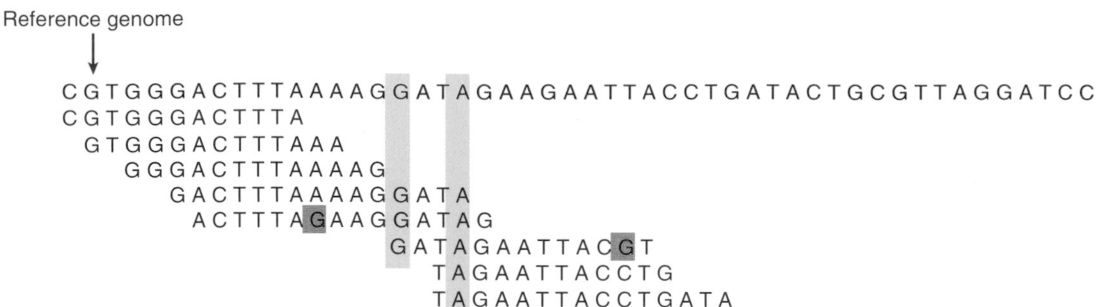

FIGURE 4.32 Data from Illumina Sequencing
The reference sequence is listed across the top. The sequence for the individual reads are shown and aligned under the identical sequence in the reference genome. The yellow boxes represent the location where the read and the reference genome differ. If these differences are found in every read, then they are most likely a true difference. If the difference is seen in only one read, then the change is most likely a sequencing error. Read depth varies from one nucleotide to the next (see blue rectangles). The higher the read depth, the more confidence the researcher has in the data.

data into a linear DNA sequence. Each read— that is, each set of sequence information from each well of the 1.7 million wells of the picotiter plate in 454 sequencing or each of the approximately 2 million DNA clusters on the flow cell in Illumina sequencing—represents a small, unique piece of an entire genome. Computers compare each piece of sequence back to a reference genome, if available, and the final output has the reference sequence across the top with each of the reads aligned below (Fig. 4.32). The number of sequences that map to a region in the genome is called **read depth**, and is used to ensure that a change is not an error in the sequencing process. If, for example, a single nucleotide substitution was found in a single read, but the other 29 reads for that region were identical to the reference genome, this substitution would most likely be considered an error. If, on the other hand, there was a single nucleotide substitution for 29 of 30 reads for that region, then this is most likely a true difference between the genome sequenced and the reference genome. If, for some reason, a region of the genome had a read depth of only 2, and a mutation was identified in this region, the validity would be highly suspect. On the other hand, if there was a mutation in 98% of the reads in a region where the read depth was 200, then the mutation is most likely a real substitution.

The ability to sequence an entire human genome has gone from a multiyear, multibillion dollar project to a simple procedure done in a few hours to few days, depending on the machine and technology used. In fact, the goal is to be able to reduce the cost of a human genome sequence to $1000. The ability to quickly assess the genomic sequence is going to change many disciplines. Understanding one genome can be misleading, but now we have the ability to compare genomes among different organisms and even among different people. In fact, a consortium of different universities and companies has compiled an integrated map of genetic variations by sequencing over 1000 different human genomes. Another major goal for this new technology is to compare cancer genomes to the normal genome. Early studies are identifying what mutations are common to cancers, and identifying mutations in genes that determine whether or not the patient will respond to a particular therapy. The ability to ascertain so much information so fast is bound to have applications that have yet to be discovered.

Summary

This chapter outlines the process of DNA replication. First, to replicate the DNA, DNA gyrase and DNA helicase relax the coiling in the DNA. The relaxed DNA is open and ready for the replisome to assemble at the origin. Single-stranded binding protein coats or binds to the open DNA, which keeps the DNA stable. Then PriA prepares an RNA primer at the origin to provide a 3′-OH group for DNA polymerase to attach the complementary bases during replication. DNA polymerase makes new DNA only in a 5′ to 3′ direction, so on the leading strand, the whole strand is made in one piece. Because the lagging strand is antiparallel, DNA polymerase has to make the strand in smaller segments called Okazaki fragments.

In vitro DNA synthesis can be made by purified DNA polymerase or by chemically linking nucleotides. In reactions done with chemical reagents, the DNA is single-stranded and is short because the process is not very efficient. Chemical synthesis of DNA is primarily used for making short primers or oligonucleotides.

In vitro DNA synthesis by DNA polymerase is very versatile and can be used to amplify a piece of DNA from a few copies to millions using PCR. Modifications of PCR include inverse PCR to amplify unknown regions of DNA, and RT-PCR of mRNA rather than DNA creates copies of genes without any introns. Additionally, PCR can be used to clone copies of genomic DNA into a vector using TA cloning or by adding novel restriction enzyme sites at the end of the PCR product. Finally, PCR can mutate template DNA by inserting or deleting regions, linking two separate regions together, or by mutating single nucleotides.

In vitro DNA synthesis is the basis for determining the sequence of DNA. In cycle sequencing, a single reaction contains four different fluorescently labeled dideoxynucleotides and unlabeled deoxynucleotides at a ratio that ensures one dideoxynucleotide incorporates at each nucleotide position of the template. The final reaction creates a tube filled with DNA fragments that end at each possible position, and that end nucleotide is fluorescently labeled. As the fragments are separated by size in a capillary tube-filled gel matrix, the smallest fragments exit the bottom of the tube first. As each subsequent fragment passes a detector, the identity of the fluorescent tag is determined and recorded. In contrast, next-generation sequencing reads nucleotides one by one as they are added to a primer; 454 sequencing employs pyrosequencing in order to determine what nucleotide is added. Since the release of pyrophosphate is the same for each of the four nucleotides, only one nucleotide is added at a time. The flash of light is recorded for each DNA fragment template where the nucleotide was incorporated. Illumina next-generation sequencing uses reversible 3′-fluorescent dye-linked nucleotides, which are added to the DNA template. Thus, all four nucleotides are added simultaneously to a flow cell containing the DNA templates. After the identity of the nucleotide that is added at each cluster is recorded, the fluorescent dye is removed from the nucleotide and washed away. The main difference between typical chain-termination sequencing and next-generation sequencing is the scale. Chain-termination sequencing occurs on one single template DNA. In contrast, 454 sequencing uses a picotiter plate with 1.7 million wells and Illumina's flow cell with several million DNA clusters, each well or cluster representing a single unique piece of DNA from the genome.

End-of-Chapter Questions

1. Which of the following enzymes aid in uncoiling DNA?
   a. DNA gyrase
   b. DNA helicase
   c. topoisomerase IV
   d. single-stranded binding protein
   e. all of the above
2. Why is an RNA primer necessary during replication?
   a. DNA polymerase III requires a 3′-OH to elongate DNA.
   b. An RNA primer is not needed for elongation.
   c. DNA polymerase requires a 5′-phosphate before it can elongate the DNA.
   d. A DNA primer is needed for replication instead of an RNA primer.
   e. An RNA primer is only needed once the DNA has been elongated and DNA polymerase is trying to fill in the gaps.

(*Continued*)

3. What are the functions of the two essential subunits of DNA polymerase III?
   a. Both subunits synthesize the lagging strand only.
   b. One subunit links nucleotides and the other ensures accuracy.
   c. They both function as a clamp to hold the complex to the DNA.
   d. The subunits function to break apart the bonds in the DNA strand.
   e. One subunit removes the RNA primer and the other synthesizes DNA.

4. Which of the following statements about mismatch repair is incorrect?
   a. MutSHL excise the mismatched nucleotides from the DNA.
   b. Mismatch repair proteins identify a mistake in DNA replication.
   c. The mismatch proteins recruit DNA polymerase III to synthesize new DNA after the proteins have excised the mismatched nucleotides.
   d. MutSHL can synthesize new DNA after a mismatch has been excised.
   e. MutSHL monitors the methylation state of the DNA to determine which strand contains the correct base when there is a mismatch.

5. Which of the following statements is incorrect regarding DNA replication?
   a. Rolling circle and theta replication are common for prokaryotes and viruses.
   b. Each round of replication for linear chromosomes, such as in eukaryotes, shortens the length of the chromosome.
   c. Prokaryotic chromosomes have multiple origins of replication.
   d. Eukaryotic replication only occurs during the S-phase of the cell cycle.
   e. Eukaryotic chromosomes have multiple origins of replication.

6. During *in vitro* DNA replication, which of the following components is not required?
   a. single-stranded DNA
   b. a primer containing a 3′-OH
   c. DNA helicase to separate the strands
   d. DNA polymerase to catalyze the reaction
   e. nucleotide precursors

7. Which of the following is not a step in the chemical synthesis of DNA?
   a. The 3′ phosphate group is added using phosphorylase.
   b. The addition of a blocking compound to protect the 3′ phosphite from reacting improperly.
   c. The 5′-OH is phosphorylated by bacteriophage T4 kinase.
   d. The addition of acetic anhydride and dimethylaminopyridine to cap the 5′-OH group of unreacted nucleotides.
   e. The amino groups on the bases are modified by other chemical groups to prevent the bases from reacting during the elongation process.

8. During chemical synthesis of DNA, a portion of the nucleotides does not react. How can the efficiency of such reactions be increased?
   a. The unreacted nucleosides are not acetylated so that more can be added in subsequent reactions.
   b. The efficiency of the reaction is not critical. Instead, the quality of the final product is more important than the quantity.
   c. The desired oligonucleotide can be separated from the truncated oligos by electrophoresis.
   d. Oligonucleotides should be made using DNA polymerase III instead of *in vitro* chemical synthesis.

e. The reaction times can be increased to allow the reaction to be more efficient.

9. Which of the following components terminates the chain in a sequencing reaction?
- **a.** dideoxynucleotides
- **b.** Klenow polymerase
- **c.** DNA polymerase III
- **d.** deoxynucleotides
- **e.** DNA primers

10. Which of the following statements about PCR is incorrect?
- **a.** The DNA template is denatured using helicase.
- **b.** PCR is used to obtain millions of copies of a specific region of DNA.
- **c.** A thermostable DNA polymerase is used because of the high temperatures required in PCR.
- **d.** Template DNA, a set of primers, deoxynucleotides, a thermostable DNA polymerase, and a thermocycler are the important components in PCR.
- **e.** Primers are needed because DNA polymerase cannot initiate synthesis, but can only elongate from an existing 3′-OH.

11. Which of the following is not an advantage of automated cycle sequencing over the chain termination method of sequencing?
- **a.** The reactions in an automated sequencer can be performed faster.
- **b.** The reactions performed in an automated sequencer can be read by a computer rather than a human.
- **c.** Higher temperatures are used during cycle sequencing, which prevent secondary structures from forming in the DNA and early termination of the reaction.
- **d.** In cycle sequencing, nonspecific interactions by the primer can be controlled by raising the annealing temperature.
- **e.** All of the above are advantages of cycle sequencing.

12. Which of the following statements about degenerate primers is not correct?
- **a.** Degenerate primers have a mixture of two or three bases at the wobble position in the codon.
- **b.** Because of the nature of degenerate primers, the annealing temperature during PCR using these primers must be lowered to account for the mismatches.
- **c.** Degenerate primers are often designed by working backwards from a known amino acid sequence.
- **d.** Degenerate primers are used even when the sequence of DNA is known.
- **e.** Within a population of degenerate primers, some will bind perfectly, some will bind with mismatches, and others will not bind.

13. Which of the following techniques would allow a researcher to determine the genetic relatedness between two samples of DNA?
- **a.** inverse PCR
- **b.** reverse transcriptase PCR
- **c.** TA cloning
- **d.** overlap PCR
- **e.** randomly amplified polymorphic DNA

(Continued)

14. Why would a researcher want to use RT-PCR?
- **a.** RT-PCR is used to compare two different samples of DNA for relatedness.
- **b.** RT-PCR creates an mRNA molecule from a known DNA sequence.
- **c.** RT-PCR generates a protein sequence from mRNA.
- **d.** RT-PCR generates a DNA molecule without the noncoding introns from eukaryotic mRNA.
- **e.** All of the above are applications for RT-PCR.

15. Which of the following is an application for PCR?
- **a.** site-directed mutagenesis
- **b.** creation of insertions, deletions, and fusions of different gene segments
- **c.** amplification of specific segments of DNA
- **d.** for cloning into vectors
- **e.** all of the above

16. In ______________ sequencing, the DNA fragments are bound to a solid surface via a flow cell.
- **a.** Illumina
- **b.** 454
- **c.** chain termination
- **d.** Sanger
- **e.** cycle

17. Flashes of light are emitted whenever a base is added in ______________ sequencing.
- **a.** Illumina
- **b.** 454
- **c.** chain termination
- **d.** Sanger
- **e.** cycle

Further Reading

Agarwal, K. L., Büchi, H., Caruthers, M. H., Gupta, N., Khorana, H. G., Kleppe, K., et al. (1970). Total synthesis of the gene for an alanine transfer ribonucleic acid from yeast. *Nature, 227*, 27–34.

Hillier, L. W., Marth, G. T., Quinlan, A. R., Dooling, D., Fewell, G., Barnett, D., et al. (2008). Whole-genome sequencing and variant discovery in C. elegans. *Nature Methods, 5*, 183–188.

Mardis, E. R. (2008). The impact of next-generation sequencing technology on genetics. *Trends in Genetics: TIG, 24*, 133–141.

Mardis, E. R. (2011). A decade's perspective on DNA sequencing technology. *Nature, 470*, 198–203.

Metzker, M. L. (2010). Sequencing technologies—the next generation. *Nature Reviews. Genetics, 11*, 31–46.

Taft-Benz, S. A., & Schaaper, R. M. (2004). The theta subunit of Escherichia coli DNA polymerase III: a role in stabilizing the epsilon proofreading subunit. *Journal of Bacteriology, 186*, 2774–2780.

Yoo, B., Kavishwar, A., Ghosh, S. K., Barteneva, N., Yigit, M. V., Moore, A., et al. (2014). Detection of miRNA expression in intact cells using activatable sensor oligonucleotides. *Chemical Biology, 21*, 199–204.

RNA-Based Technologies

Biotechnology

http://dx.doi.org/10.1016/B978-0-12-385015-7.00005-3

NONCODING RNA PLAYS MANY ROLES

RNA plays a multifaceted role in biology that is adaptable for many different applications in biotechnology. The most widely understood role of RNA is in protein synthesis, which includes messenger RNA (mRNA), transfer RNA (tRNA) and ribosomal RNA (rRNA) (see Chapter 2). However, RNA plays many other roles. Several small RNAs, such as snRNA, snoRNA, and gRNA, take part in RNA processing by removing introns. Some RNA sequences can catalyze enzyme reactions. **Ribozymes**, as they are called, are found in many organisms, catalyzing cleavage and ligation of various substrates. Between the increased speed and accuracy of sequencing and a heightened awareness of RNA in the cell, an ever-increasing number of roles has been found for RNA in the regulation of gene expression and in cell defense. Entirely new classes of **noncoding RNAs (ncRNAs)** have been discovered and characterized. Table 5.1 summarizes the major RNA classes and their functions.

Indeed, several classes of regulatory RNA modulate gene expression at the stage of translating mRNA into protein rather than transcribing DNA to give mRNA. For example, in some organisms, **antisense RNA** controls protein translation. Antisense RNA binds to the complementary mRNA and blocks translation. From this discovery came the potential use of antisense RNA to block or attenuate synthesis of proteins that cause various diseases. Several of the RNAs in Table 5.1 are subclasses of antisense RNA. For example, microRNA is found in eukaryotes, where it often regulates development and cellular differentiation and many small bacterial regulatory RNAs act via an antisense mechanism.

RNA also takes part in defending the cell against foreign genetic elements, including viruses, plasmids, and transposable elements. In eukaryotes, **RNA interference (RNAi)** plays a major role in protecting against RNA viruses. Here, noncoding small-interfering RNAs (siRNA) identify specific mRNAs and trigger their degradation. This fortuitous finding opened the door to a specific technique for controlling protein translation. Since RNA interference was discovered in 1993, its application has become widespread. Bacteria lack RNA interference but instead possess the CRISPR system that uses small RNAs (crRNA) to identify and combat both DNA and RNA viruses. CRISPR acts by a mechanism quite distinct from RNA interference but still very useful in biotechnology. CRISPR RNAs can be introduced into a eukaryotic cell in order to make small deletions in endogenous genes or to insert different tags or markers (e.g., GFP, FLAG, HA) in specific genes. The use of CRISPR in genome editing is described in Chapter 17, and the basic process is explained later in this chapter.

This chapter presents examples of how RNA affects genome defense, transcription, RNA processing, protein translation, and enzyme function, and it focuses on applications of these different categories in biotechnology.

> In addition to taking part in translation, noncoding RNA plays many roles in molecular biology. Several classes of RNA have found major application in biotechnology. Antisense RNA and RNA interference regulate gene expression, and the CRISPR system is used in genetic engineering.

RNA COORDINATES GENOMIC INTEGRITY IN EUKARYOTES

RNA plays several roles in maintaining genome stability in eukaryotes. It is required for the proper synthesis of chromosome ends (telomeres) and for dosage compensation in diploid animals. Suppressing the replication and movement of transposable elements in the germline also depends on RNA.

Table 5.1 Major Classes of RNA

	Class	Abbreviation	Size in Nucleotides	Role	Distribution
Genomic Integrity and Protection	Piwi interacting RNA	piRNA	25–32	Transposon silencing in germline cells	Eukaryotes
	Small-interfering RNA	siRNA	22	Defense against foreign RNA	Eukaryotes
	Telomerase RNA	TERC	451	Synthesis of telomeres	Eukaryotes
	CRISPR RNA	crRNA	24–48	Defense against foreign RNA and DNA	Bacteria plus Archaea
	Xist RNA	—	17,000	X chromosome inactivation	Eukaryotes
Transcription	Antisense RNA	aRNA	19–25	Genetic regulation	All organisms
	Enhancer RNAs	eRNAs	200–500	Genetic regulation	Eukaryotes
	6S RNA	6S RNA	184 (*E. coli*)	Regulating transcription	Bacteria
	Micro RNA	miRNA	22	Regulating mRNA degradation and translation	Eukaryotes
	Circular RNA	circRNA	1000 or more	Regulation of miRNA abundance	Eukaryotes
	Long noncoding RNA	lncRNA	Wide range	Various regulatory roles	Eukaryotes
	Small RNA regulators	sRNA	<300	Gene regulators (various mechanisms)	Bacteria
RNA Processing	Guide RNA	gRNA		Editing of mRNA	Protozoa
	Small nuclear RNA	snRNA	100–300	Splicing of RNA	Eukaryotes plus Archaea
	Small nucleolar RNA	snoRNA	60–300	RNA nucleotide modification	Eukaryotes plus Archaea

(*Continued*)

Table 5.1 Major Classes of RNA—cont'd

	Class	Abbreviation	Size in Nucleotides	Role	Distribution
Protein Translation	Messenger RNA	mRNA	Wide range	Protein synthesis	All organisms
	Transfer RNA	tRNA	70–90	Protein synthesis	All organisms
	Ribosomal RNA	rRNA	120, 160, 1868, 5025	Protein synthesis	All organisms (sizes shown are for higher animals)
	Transfer-messenger RNA	tmRNA		Rescues stalled ribosomes	Bacteria
	Riboswitch	—	40–140	Controls translation, transcription, or splicing of attached mRNA	All organisms (very rare in eukaryotes)
	Dual-function RNA	—		Protein coding plus various regulatory roles	All organisms
Enzymatic Function	Ribozymes	—	>250	Function as enzymes	All organisms
	Signal recognition particle RNA	7SL RNA or SRP RNA	300	Membrane insertion of proteins	All organisms

Eukaryotic chromosomes consist of a linear DNA molecule with special sequences called **telomeres** at each end. During DNA replication, the ends of chromosomes cannot be replicated since DNA polymerase cannot synthesize DNA without a pre-existing 3′OH. During a typical replication round, DNA synthesis begins with an RNA primer created by RNA polymerase that supplies the 3′OH group. At the ends, the scenario is different. **Telomerase** is an enzyme that uses an RNA component (TERC) to regenerate the ends that are not created during replication, thus maintaining the chromosome structure. Rather than a primer, the RNA component acts as a template to actually increase the length of the ends. Without the RNA component of telomerase, the ends of chromosomes shorten and eventually lead to chromosomal fusions and deletions. In addition, mutations in either the protein or RNA portion of telomerase are associated with cancers and diseases such as dyskeratosis congenita. The biology of telomere maintenance in relationship to aging is discussed in Chapter 20.

Telomerase consists of the RNA template (TERC) plus the telomerase reverse transcriptase subunit (TERT protein) (Fig. 5.1). When the TERC RNA folds into its proper secondary structure, the RNA template sequence is near the reverse transcriptase binding region. From this core unit, three different arms jut out and interact with other accessory proteins that stabilize the structure. The RNA component provides the scaffold for proper telomerase assembly as well as the template sequence.

Gene dosage compensation occurs to equalize the amount of proteins produced from genes on the sex chromosomes in diploid organisms. In both insects and humans, females have two copies of the X chromosome, whereas males have only one copy. Insects and mammals both compensate for this, but by completely different mechanisms. One common factor is that both mechanisms rely on special RNA molecules. In the fruit fly *Drosophila*, the male (XY) equalizes expression in comparison to the female by doubling gene expression from the single X chromosome in males. In *Drosophila*, two noncoding RNAs called *roX1* and *roX2* complex with five different proteins to form the MSL complex. The complex then binds to the genes on the male X chromosome and increases transcription.

In humans and other mammals, the second X chromosome in females is inactivated. Thus, males and females both essentially function with only one active X chromosome. The inactivation is due to a long noncoding RNA called *Xist*, which coats the inactive X chromosome. The *Xist* gene of the active X chromosome is inactivated by methylation, and the *Xist* gene on the inactivated X chromosome is transcribed. Expressing the *Xist* gene thus inactivates the X chromosome that carries it. Furthermore, an antisense RNA, *Tsix*, which is transcribed from the *Xist* locus but in the reverse direction, regulates the expression of the *Xist* gene on the active X chromosome. Using genome editing to move the *Xist* gene to another chromosome, which is then shut down, is being considered as a possible approach to curing Down syndrome (see Chapter 17).

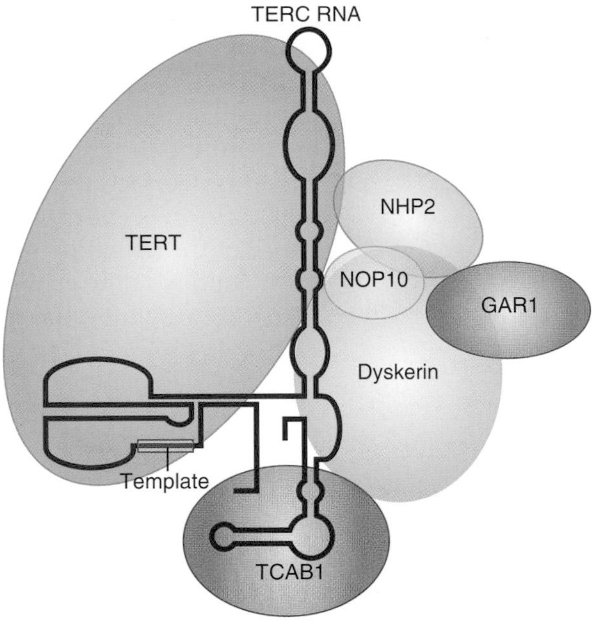

FIGURE 5.1 Arrangement of Proteins around the RNA Core of Telomerase
Telomerase reverse transcriptase, or TERT, is the major telomerase protein and contains the active site for DNA synthesis that uses the TERC RNA as a template. Several other proteins are needed for stability. The protein names shown here are those for human telomerase.

Piwi-interacting RNAs (piRNA) are another a class of small RNAs essential to maintaining the genome in eukaryotes. piRNA are 24–30 nucleotides in length, have a monophosphate group preferably attached to a uridine at the 5′ end, and have a 2′*O*-methyl group at the 3′ end (Fig. 5.2). The piRNAs are encoded in the genome and are found in large clusters or within the introns of other genes. They are complementary in sequence to endogenous transposons, which are clustered in the centromere area or the telomere regions. When piRNA gene clusters are expressed into RNA, members of the Argonaut protein family recognize the piRNA, cleave it into small pieces, and then use these pieces as single-stranded templates to bind and cleave any complementary RNA produced by the transposon. This action prevents endogenous transposons from moving to new locations.

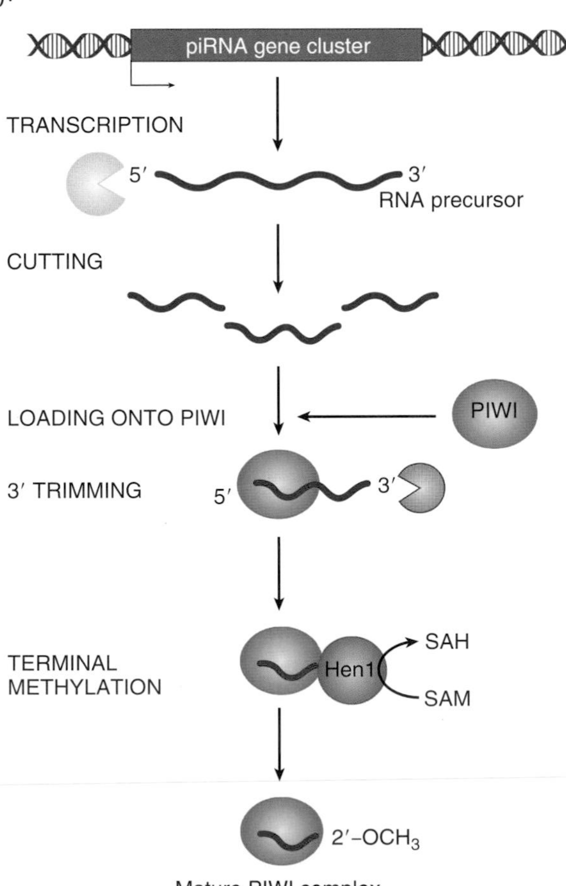

FIGURE 5.2 Role of Piwi-Interacting RNAs
Genomic clusters of piRNA are transcribed into long RNA precursors. They are cleaved into shorter, piRNA precursor molecules. After the PIWI complex binds these molecules, they are trimmed to generate the final piRNA. The PIWI complex then uses piRNA as a template to locate and silence sequences derived from transposable elements. Two variants of the PIWI complex exist: one specialized for nuclear silencing and the other for cytoplasmic silencing.

The arrangement of the eukaryotic genome within the nucleus was originally thought to be an amorphous soup of chromatin fibrils. However, the interior of the nucleus is highly organized. Ribosome assembly is localized in the nucleolus, a spherical structure within the nucleus. Furthermore, the use of chromosome specific markers has revealed that each chromosome is found in a specific domain. Recent studies have shown that expressed genes (in the form of euchromatin) are localized to the central core of the nucleus while nontranscribed DNA (as heterochromatin) occupies the region closest to the nuclear envelope. This organization is critical to function; and the role of RNA, especially lncRNA, in maintenance of the structure is only starting to be understood. These transcripts are produced from regions of the genome, then stay within the nucleus, and partition to the chromatin, suggesting they play a role in chromatin structure or regulation.

Several classes of RNA promote genome integrity in eukaryotes. Maintenance of telomeres, control of gene dosage, and nuclear organization all involve noncoding RNA. In addition, piRNA plays a major role in protecting the genome during reproduction.

RNA PROTECTS GENOMES FROM INVADING VIRUSES

In addition to promoting internal genome stability, RNA is involved in protecting against external genetic elements, especially viruses. In eukaryotes, RNA interference is the major mechanism of RNA-mediated virus protection, whereas in prokaryotes, the CRISPR system operates instead. RNA interference is discussed later since it shows many features in common with the microRNA system that regulates cellular genes. Indeed, the two systems probably share a common evolutionary origin. RNA interference protects only against viruses with RNA genomes but not DNA viruses.

The CRISPR system is found in both bacteria and Archaea, but not in eukaryotes. It varies considerably in its components among different bacteria and is not found in all species. CRISPR differs in its components and mechanism from RNA interference. Moreover, CRISPR can protect against viruses with RNA or DNA genomes, as well as hostile plasmids and transposons. In consequence, CRISPR has been applied to genome editing (see Chapter 17). Here, we outline the basic mechanism of the CRISPR system.

CRISPR is based on memory. The CRISPR system stores an array of short sequence fragments derived from foreign genetic elements. CRISPR, which stands for clustered regularly interspaced short palindromic repeats, refers to the way foreign genetic sequences are stored on the bacterial chromosome. When nucleic acids appear whose sequences contain matches to those stored, they are destroyed (Fig. 5.3). Both DNases and RNases are present among the CAS proteins, and thus, the CRISPR system can defend bacteria against both RNA viruses and DNA viruses. It also prevents the entry into bacteria of foreign plasmids and transposons.

There is considerable variation in the enzyme components of the CRISPR system between different bacteria. Some bacteria lack the CRISPR system entirely, some have very simple systems, and others have multiple CRISPR arrays with many different degradative enzymes. Bacteria that occupy environments where there is a major threat from viruses tend to have the more complex CRISPR systems.

The antiviral defense systems, RNA interference in eukaryotes and CRISPR in prokaryotes, operate using distinct mechanisms but both rely on noncoding RNA.

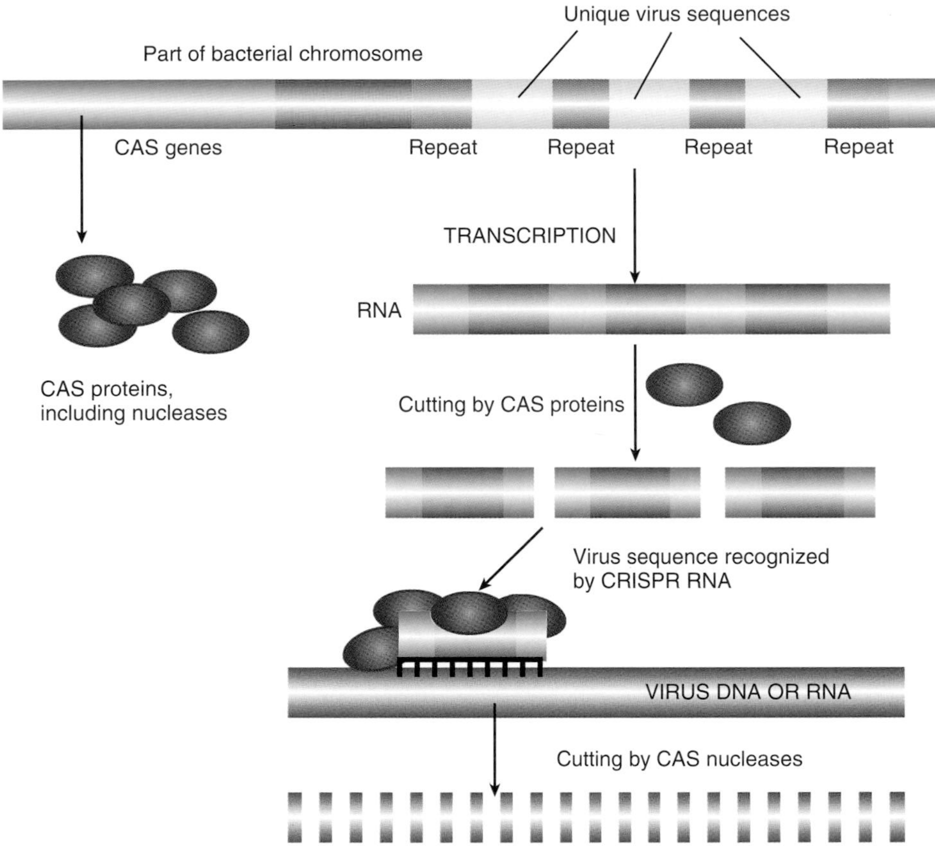

FIGURE 5.3 Overview of the CRISPR System
Foreign DNA sequences are stored as an array on the bacterial chromosome, separated by identical repeats. This region is transcribed into a long RNA and then processed into smaller individual RNA guides (crRNA). The CAS nucleases use these guides to find and destroy intruding foreign nucleic acids, both RNA and DNA.

RNA MODULATES TRANSCRIPTION

All organisms regulate gene transcription during development. In addition, transcription controls homeostasis of organisms, coordinating the proper protein complement for each environment or condition the organism experiences. Proteins known as transcription factors bind immediately before the gene, a region termed the promoter, to activate or repress RNA polymerase. In addition to short-range control of the gene via the promoter, chromatin can loop around so that enhancers that are thousands of base pairs away connect to the transcriptional machinery and activate transcription. RNA controls and modulates gene expression too, which adds a whole extra layer of complexity to gene expression.

In bacteria, a variety of small RNA (sRNA) molecules take part in genetic regulation. Most of them act by using an antisense mechanism, and they bind to mRNA to prevent its translation. However, various other mechanisms are also found. Some sRNA molecules bind to mRNA but activate translation by altering the secondary structure. Other sRNA molecules act via binding to proteins.

In eukaryotes, there is a much greater number and variety of regulatory RNAs. MicroRNAs are short RNAs that act via an antisense mechanism to prevent translation or promote degradation of mRNA. A variety of longer RNAs (e.g., enhancer RNA, circular RNA, lncRNA) are also involved in regulation. The role of many of these is as yet poorly characterized.

> Noncoding RNA takes part in regulating transcription in both bacteria and eukaryotes, although by different mechanisms.

Antisense RNA Modulates mRNA Expression

Antisense refers to the orientation of complementary strands during transcription. The two complementary strands of DNA are referred to as *sense* (=coding or plus) and *antisense* (=noncoding or minus; see Chapter 2). Transcription uses the antisense strand as template, resulting in an mRNA that is identical in sequence to the sense strand (except for the replacement of uracil for thymine). Antisense RNA is synthesized using the sense strand as template; therefore, it has a sequence complementary to mRNA (Fig. 5.4).

Antisense RNA is made in normal cells of many different organisms, including humans. Artificial antisense RNA is also made for manipulating gene expression in laboratory settings. When a cell has both the mRNA (i.e., the sense strand of RNA) plus a

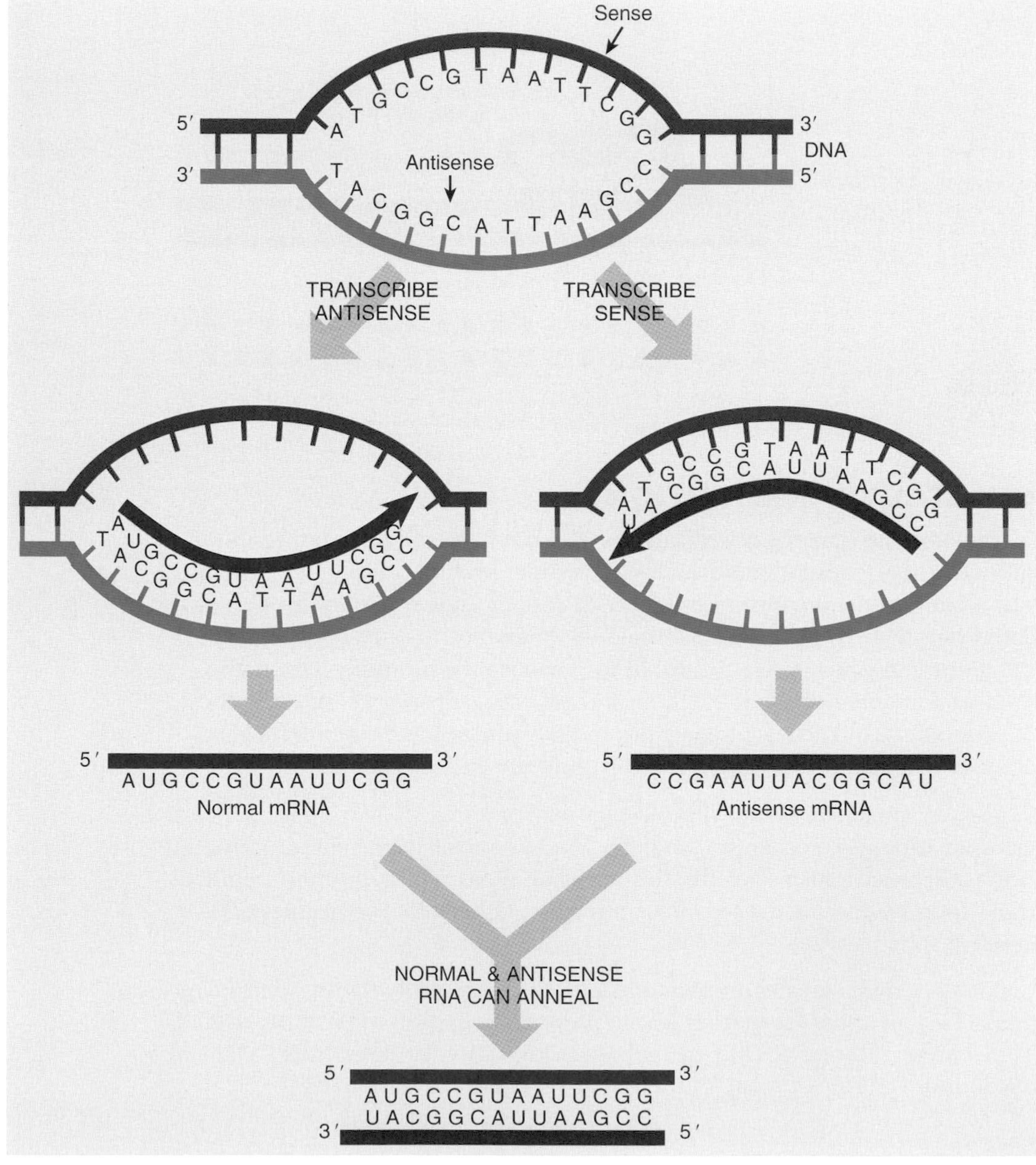

FIGURE 5.4 Antisense RNA Is Complementary to Messenger RNA
Transcription from both strands of DNA creates two different RNA molecules—on the left, the messenger RNA, and on the right, antisense RNA. These two have complementary sequences and can form double-stranded RNA.

complementary antisense copy, the two single strands anneal to form double-stranded RNA. The duplex can either inhibit protein translation by blocking the ribosome binding site, or inhibit mRNA splicing by blocking a splice site (Fig. 5.5A). When antisense sequences are made in the laboratory, they are usually synthesized as DNA because this is more stable than RNA (see Chapter 4). In this case, the DNA:RNA duplex is digested by RNase H (see Fig. 5.5B). RNase H is a cellular enzyme that normally functions during replication. It recognizes and cleaves the RNA backbone of a DNA:RNA duplex, targeting the antisense DNA:mRNA duplex for further degradation. RNase H recognizes a 7-base-pair heteroduplex, so the region of homology between the antisense DNA and target mRNA need not be very long.

Antisense RNA sequences are complementary to a target mRNA. Antisense RNA forms double-stranded regions that block either protein translation or splicing of introns.

Antisense RNA Controls a Variety of Biological Phenomena

Naturally occurring antisense genes have been found that control a variety of different processes. When antisense genes are transcribed, they produce an RNA molecule that is complementary to the mRNA of their target gene(s). One example of natural antisense control is found in *Neurospora*. This fungus follows a strict schedule, based on circadian rhythms, and forms hyphae only at specific times during the day. Many mutants have been identified that do not follow this timetable. The genes affected by these mutations are regulators in the circadian rhythm of *Neurospora*. One of the first to be identified was *frequency (frq)*.

Mutations in this gene change how often the fungus forms hyphae. The amount of normal *frq* mRNA fluctuates, with highest levels during the day and lowest at night. Conversely, antisense *frq* RNA also cycles, but in reverse, with the lowest levels during the day and the highest during the night. Although the exact mechanism is uncertain, *Neurospora* that do not produce antisense *frq* RNA have disrupted circadian rhythms. In addition, both the antisense and sense mRNAs are induced by light and therefore respond directly to the environment to maintain the correct circadian rhythm.

Using antisense to regulate gene expression is so widespread in nature that scientists became curious how many potential antisense/sense partners exist in various genomes. In the human genome, 20%–40% of all protein coding genes also have antisense partners. They can be complementary to promoter, introns, exons, and even the 3′ UTR region of the gene. The position of the antisense gene within the genome categorizes the antisense transcript as ***cis*** or ***trans***, where *cis* antisense partners are found adjacent to or within the complementary gene, and *trans* refers to antisense partners that are found at different locations in the genome. Related to natural antisense transcripts are small noncoding regulatory RNAs called **microRNAs (miRNAs)**, which inhibit gene expression through an antisense mechanism (see later discussion). Using computer searches, around 1000 potential microRNAs have been identified in humans, but because these are only about 20 nucleotides long, identifying them conclusively by computer is very difficult.

EXAMPLES OF NATURAL ANTISENSE CONTROL

- Control of Circadian Rhythm in *Neurospora:* The time of day controls when the fungus forms hyphae by regulating the antisense and sense mRNA for the *frq* gene.

3′
5′
Antisense mRNA
3′
5′
+
Splice sites
5′
3′ Sense mRNA
Ribosome
binding
site
3′
5′
5′
3′
Ribosomes
cannot bind
3′
5′
5′
3′
Splicing factors
cannot bind
A

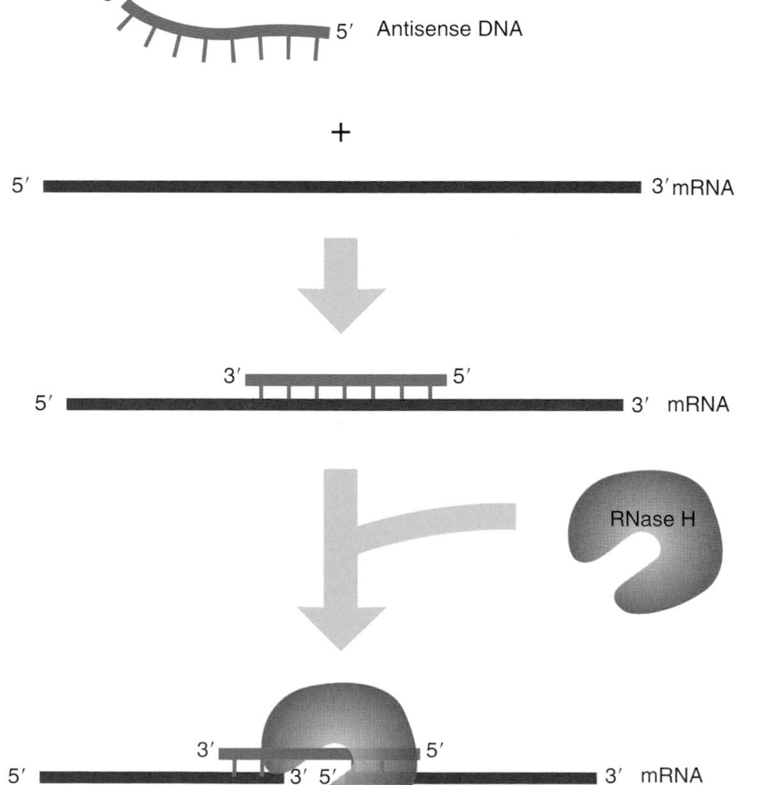

FIGURE 5.5 Antisense RNA Blocks Protein Expression

(A) The complementary sequence of antisense RNA binds to specific regions on mRNA. This can block the ribosome binding sites or splice junctions. (B) Antisense DNA targets mRNA for degradation. When antisense DNA binds to mRNA, the heteroduplex of RNA and DNA triggers RNase H to degrade the mRNA.

- Iron Metabolism in Bacteria: FatB/RNAα are sense/antisense partners that control regulation of iron uptake in the fish pathogen *Vibrio anguillarum*. When iron is plentiful, higher amounts of RNAα prevent *fatA* and *fatB* expression. When iron is scarce, the bacteria need to get iron from the environment. RNAα is degraded, and *fatA* and *fatB* are expressed so *Vibrio* can ingest iron.
- Control of HIV-1 Gene Expression: Antisense *env* mRNA binds to the Rev Response Element (RRE) on *env* mRNA. When antisense blocks the RRE, Env protein is not produced. When antisense *env* mRNA is absent, Env protein is produced.
- Control of Eukaryotic Transcription Factors: The transcription factor hypoxia-induced factor (HIF-1) is a basic helix–loop–helix dimeric protein that turns on genes associated with oxygen and glucose metabolism, including glucose transporters 1 and 3 and enzymes of the glycolytic pathway. Antisense mRNA to the α subunit mRNA controls the expression of the transcription factor. The level of antisense RNA is modulated by the amount of oxygen in the environment.
- Control of RNA Editing: Antisense/sense loops are formed between complementary exon and intron sequences of the gene for the glutamate-gated ion channel in human brain. These loops are recognized by dsRNA-specific adenosine deaminase (DRADA), which converts adenosine to inosine by deamination. This alters the sequence of the final mRNA and hence of the protein, thus reducing the permeability of the ion channel.
- Alternate Splicing of Thyroid Hormone Receptor mRNA: Antisense RNA transcribed from the thyroid hormone locus inhibits splicing of the *c-erbAα* gene. Two alternately spliced transcripts give the authentic thyroid hormone receptor and a decoy receptor that does not bind thyroid hormone. These two forms modulate cellular responses to thyroid hormone.
- Control of ColE1 Plasmid Replication: RNAI and RNAII mRNA are sense/antisense partners that prevent DNA polymerase from initiating plasmid replication. The amount of antisense RNAI controls how often replication is initiated.

Organisms have antisense genes and microRNAs that bind to a target mRNA and prevent its translation into protein. These modulate a large number of systems, including hyphae formation in *Neurospora*, development, replication, and many more.

ANTISENSE TRANSCRIPTS CAN INDUCE FORMATION OF HETEROCHROMATIN

Trans acting antisense transcripts are often transcribed from pseudogenes, and may suppress or activate the regular gene. One example is the gene for phosphatase and tensin homolog (PTEN), a tumor suppressor gene, whose level of expression correlates to the severity of cancer. There is a pseudogene for PTEN (PTENpg) that produces three noncoding RNAs: PTENpg1 sense (green), PTENpg1 antisense α (longer red line), and β (shorter red line) (Fig. 5.6). The PTENpg1 sense sequence is 95% identical to the PTEN gene even though it is made from a different gene. All three RNAs regulates PTEN but by two different mechanisms. The α antisense RNA converts the PTEN genomic region into heterochromatin, repressing further transcription. Curiously, this antisense RNA does not bind to the sense strand even though it is complementary in sequence. Instead, the α antisense transcript attracts two chromatin-modifying proteins, DNMT3A and EZH2, to compact the histones so that RNA polymerase cannot access the promoter. The β antisense transcript is shorter, as it begins at an internal transcription start site. The β antisense RNA forms a complex with the third transcript from the pseudogene, PTENpg1 sense. This resulting double-stranded RNA attracts miRNAs that are targeted against PTEN and therefore blocks gene expression (see later discussion of miRNA function).

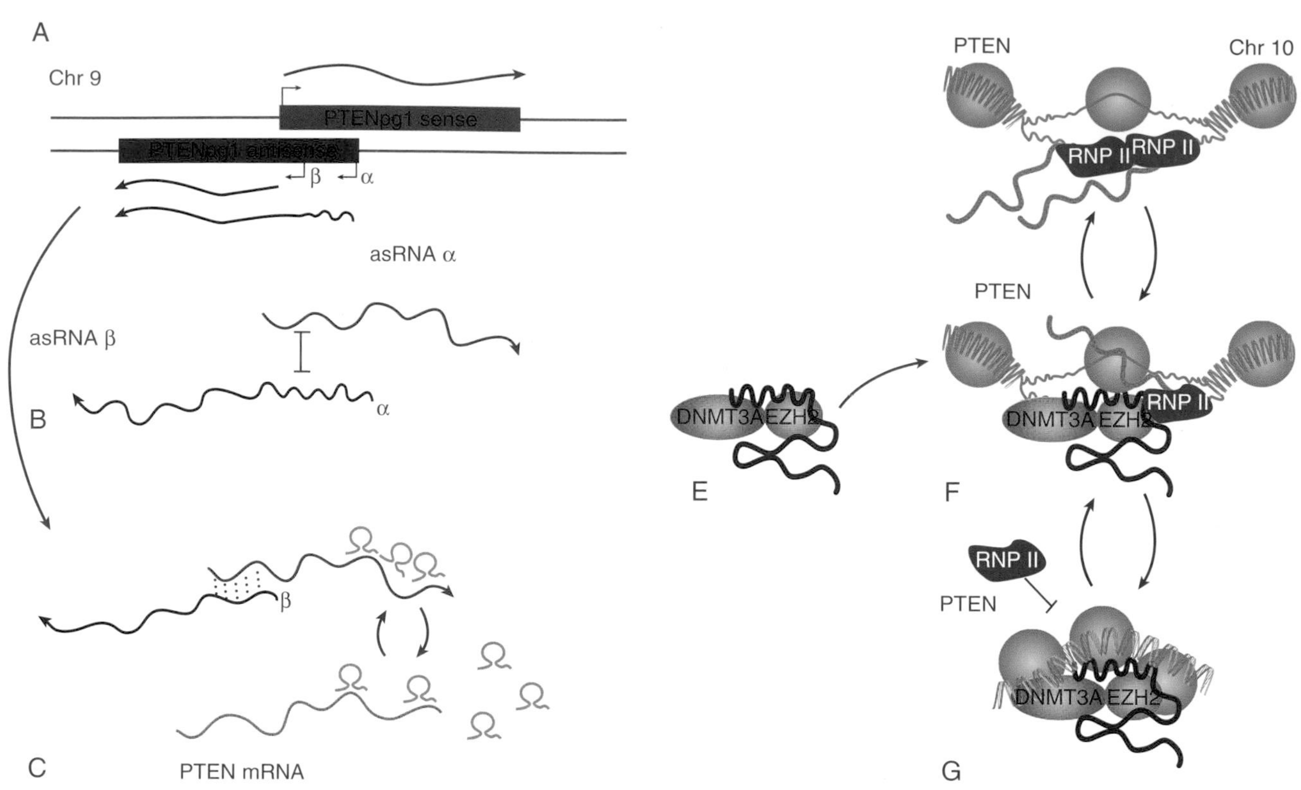

FIGURE 5.6 PTEN Pseudogene Encodes Three Noncoding RNAs That Regulate PTEN Expression
(A) The pseudogene for PTEN has three transcription start sites. Transcription of the top strand (green) produces PTENpg1 sense RNA. Transcription on the lower strand (red) produces two different forms, a longer α antisense RNA and a shorter β antisense RNA. (B–C) PTENpg1 sense RNA and β antisense RNA (asRNA β anneal over complementary areas). This duplex attracts miRNAs (light blue) and prevents the miRNAs from promoting the degradation of PTEN mRNA (from the normal PTEN gene). (E–G) The α form recruits two chromatin modification enzymes, DNMT3A and EZH2, to condense the histones (purple spheres) around the PTEN gene, which excludes RNA polymerase (RNP II) and prevents transcription.

Using Antisense RNA

In the laboratory, antisense RNA can be made by using two different methods (Fig. 5.7). The easiest method is to chemically synthesize oligonucleotides that are complementary to the target gene. The oligonucleotides are then injected or transformed into the target cell (see later discussion). Alternatively, the gene of interest can be cloned in the opposite orientation so that transcription gives antisense RNA. The vector carrying the anti-gene is then transformed into the target organism.

Full-length antisense RNA can be transcribed from a vector that has been inserted into the cell. (See Chapters 15 and 16 for more details on inserting foreign DNA into plant and animal cells.) First, the target gene is cloned in reverse orientation so that antisense RNA is produced instead of sense mRNA (see Fig. 5.7B). This method is believed to inactivate the cellular target mRNA by forming a heteroduplex of sense/antisense RNA. Heteroduplex formation relies on both RNAs to first unfold. If either RNA has a very stable secondary or tertiary structure, then the construct may not work inside the cell.

The advantage of internal synthesis of antisense RNA is that the antisense expression can be controlled. If the antisense gene is cloned behind an inducible promoter, then the antisense RNA is not made until the gene is induced by specific signals or conditions. This capability may be useful to allow organ-specific expression of an antisense gene. Another advantage is that the antisense RNA may be continuously expressed internally over a long-term period. This avoids the inconvenience and expense of constant administration of external antisense oligonucleotides.

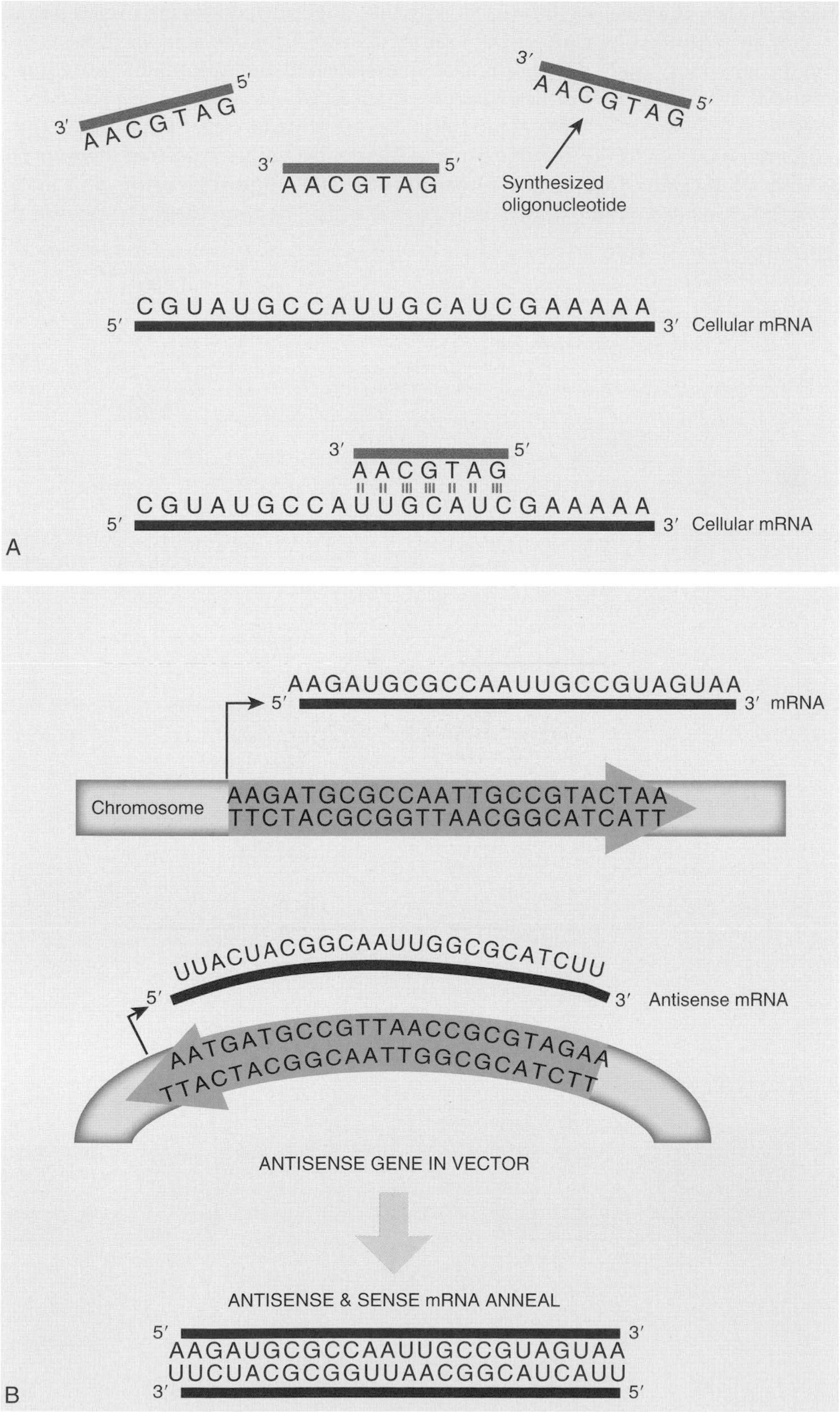

FIGURE 5.7 Making Antisense RNA in the Laboratory

(A) Antisense oligonucleotides. Small oligonucleotides are synthesized chemically and injected into a cell to block mRNA translation. (B) Antisense genes. Genes are cloned in inverted orientation so that the sense strand is transcribed. This yields antisense RNA that anneals to the normal mRNA, preventing its expression.

In practice, shorter, chemically synthesized, antisense oligonucleotides are more often used. In fact, they are traditionally made of DNA rather than RNA for two reasons: DNA is more stable in the laboratory, and DNA synthesis is an established and automated procedure. Inside the cell, DNA oligonucleotides are still very susceptible to degradation by endonucleases; therefore, various chemical modifications are added to increase stability. The most common modification is to replace one of the nonbridging oxygens in the phosphate groups with sulfur (Fig. 5.8) to make a **phosphorothioate oligonucleotide**. This makes the phosphorus a chiral center; one diastereomer is resistant to nuclease degradation, but the

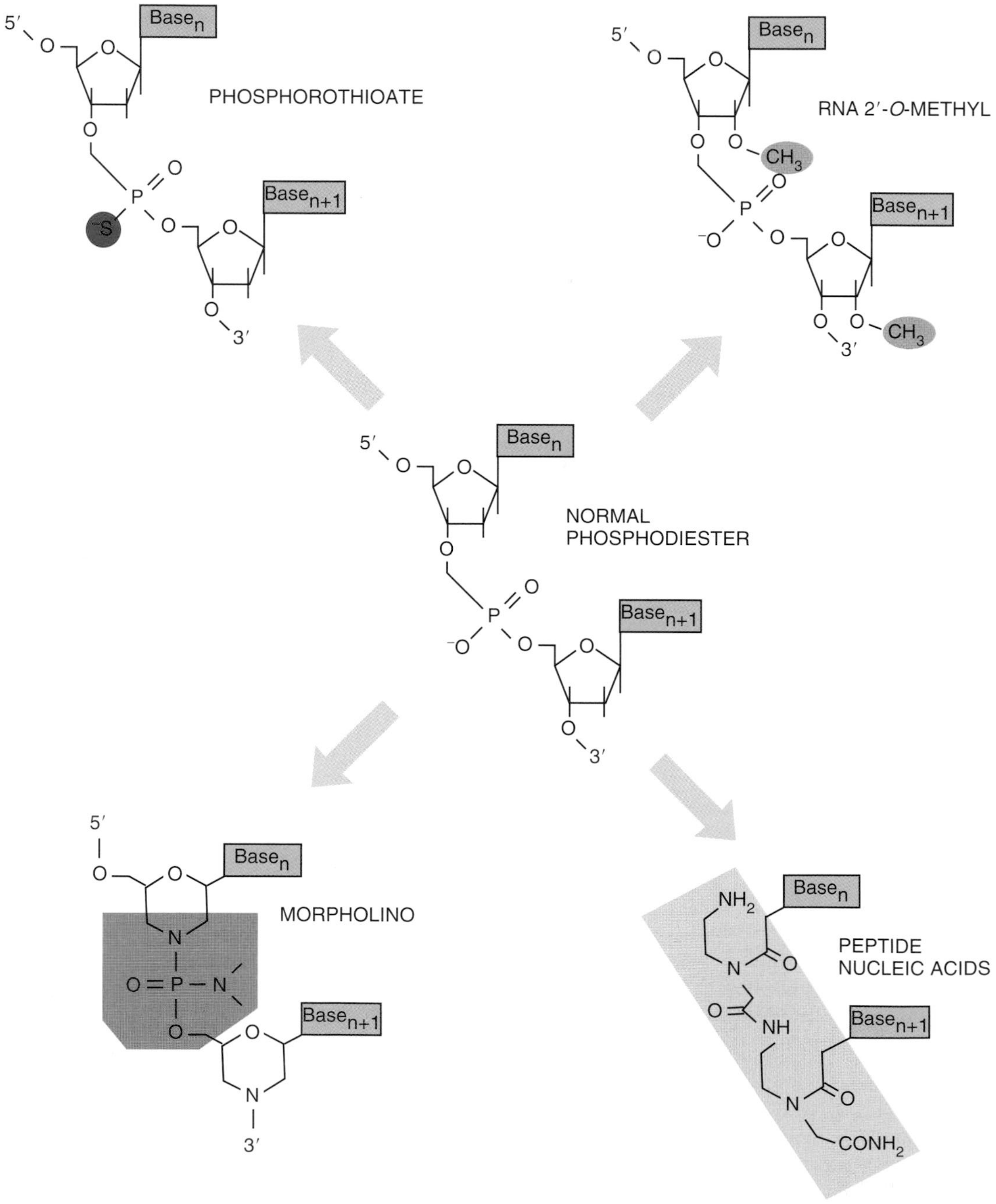

PHOSPHOROTHIOATE

RNA 2′-*O*-METHYL

NORMAL PHOSPHODIESTER

MORPHOLINO

PEPTIDE NUCLEIC ACIDS

FIGURE 5.8 Modifications to Oligonucleotides

Replacing the nonbridging oxygen with sulfur *(upper left)* increases oligonucleotide resistance to nuclease degradation. Adding an *O*-alkyl group to the 2′-OH on the ribose (*upper right)* makes the oligonucleotide resistant to nuclease degradation and also to RNase H. Morpholino-antisense oligonucleotides and peptide nucleic acids are two more substantial changes in the standard oligonucleotide structure *(lower left and lower right)*. Both are resistant to RNase H degradation. The RNase H-resistant oligonucleotides are used to target splice junctions or ribosome binding sites in order to prevent translation of their target mRNA.

other is still sensitive, leaving about half of the antisense molecules functional inside the cell. This modification does not affect the solubility of the oligonucleotides or their susceptibility to RNase H degradation. These types of antisense oligonucleotides have been developed to inhibit cancers such as melanoma and some lung cancers. The most common side effect with phosphorothioate oligonucleotides is nonspecific interactions, especially with proteins that interact with sulfur-containing molecules.

Two other modifications have fewer nonspecific interactions than phosphorothioate oligonucleotides. Adding an *O*-alkyl group to the 2′-OH of the ribose makes the oligonucleotide resistant to DNase and RNase H degradation (see Fig. 5.8). Inserting an amine into the ribose ring, thus changing the five-carbon ribose into a morpholino ring, creates **morpholino-antisense oligonucleotides** (see Fig. 5.8). In addition to the morpholino ring, a second amine replaces the nonbridging oxygen to create a **phosphorodiamidate** linkage. This amine neutralizes the charged phosphodiester of typical oligonucleotides. The loss of charge affects their uptake into cells, but alternative methods have been developed to get these antisense molecules into the cells (see later discussion). Both types of modified oligonucleotides are resistant to RNase H. Therefore, they do not promote degradation of an mRNA:DNA hybrid target. Consequently, their use is restricted to blocking splicing sites in the pre-mRNA transcript or to block ribosome binding sites.

The most different modified oligonucleotides are **peptide nucleic acids (PNAs)**, which have the standard nucleic acid bases attached to a polypeptide backbone (normally found in proteins) rather than a sugar-phosphate backbone (see Fig. 5.8). The polypeptide backbone has been modified so that the RNA bases are spaced at the same distance as the typical oligonucleotide. The spacing is critical to function because the bases of a PNA must match the bases in the target RNA. This molecule is also uncharged and works through non-RNase-H-dependent mechanisms as with morpholino antisense oligonucleotides. Antisense PNA has been developed to inhibit translation of the HIV viral transcript, *gag-pol*, and to block translation of two cancer genes: *Ha-ras* and *bcl-2*.

As noted earlier, antisense oligonucleotides that are resistant to RNase H must be made to target splice sites and/or ribosome binding sites in order to block the target mRNA. In some cases, these sequences are not well characterized in the target mRNA, so modified antisense oligonucleotides become useless. Making mixed or chimeric antisense oligonucleotides can restore the targeting of RNase H to the mRNA while allowing the use of modified structures to prevent degradation (see later discussion). In these chimeric antisense oligonucleotides, the core has a short (~7) base-pair span of phosphorothioate linkages, which are RNase H sensitive, flanked on each side by sequences consisting of one of the RNase H-resistant modifications (Fig. 5.9). The flanking regions contain 2′-*O*-methyl groups, morpholino structures, or even PNA. These chimeric molecules can target any accessible regions of the mRNA, not just splice sites or ribosome binding sites.

A target gene can be cloned in the inverse direction to create an antisense gene that can be carried on a vector. Alternatively, shorter antisense oligonucleotides can be made artificially. The antisense RNA and endogenous mRNA will bind, thus preventing translation of target mRNA into protein. Various structural modifications are incorporated to stabilize artificially made antisense RNA.

Delivery of Antisense Therapies

Getting antisense oligonucleotides into cells requires special techniques because they do not cross cell membranes easily enough on their own to be effective. Moreover, targeting the antisense oligonucleotide to the correct intracellular location poses a further obstacle. Although the natural uptake of oligonucleotides occurs by an unknown mechanism, the process is active and depends on temperature, oligonucleotide concentration, and cell type.

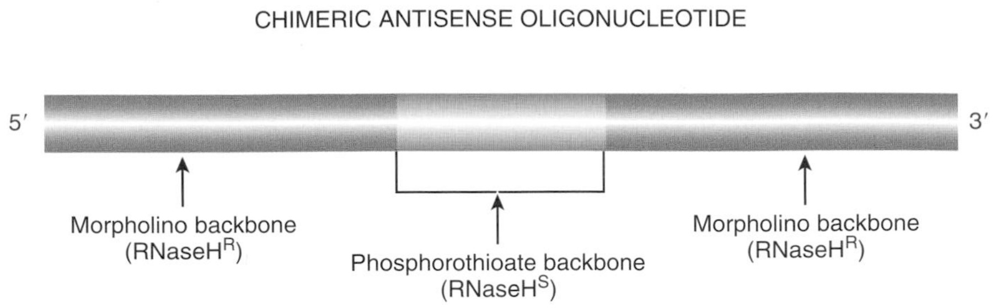

CHIMERIC ANTISENSE OLIGONUCLEOTIDE

5′ 3′

Morpholino backbone ($RNaseH^R$)

Phosphorothioate backbone ($RNaseH^S$)

Morpholino backbone ($RNaseH^R$)

FIGURE 5.9 Chimeric Oligonucleotides with RNase H-Sensitive Cores
Chimeric oligonucleotides are made using different chemistries. The core region maintains RNase H sensitivity, whereas the outer regions are RNase H resistant. When the oligonucleotide hybridizes to target molecules in the cell, RNase H will digest the hybrid of oligonucleotide plus mRNA only where the central domain forms a heteroduplex. RNase H will not digest any nonspecific complexes between the chimeric oligonucleotide ends and the wrong mRNA.

Because oligonucleotides are highly charged, they cannot cross lipid membranes and are probably taken up by endocytosis. This, however, results in the oligonucleotide trapped inside the uptake vesicle rather than free in the cytoplasm. Escape from these vesicles is slow and poorly understood. It has been suggested that oligonucleotides may also enter via membrane-bound receptors, but this suggestion is controversial.

A common method to deliver oligonucleotides to cells is to use **liposomes** (Fig. 5.10A). Liposomes are small vesicles made of bilayers of phospholipids and cholesterol. Whether the liposome is neutral or positively charged depends on the type of phospholipid used to manufacture it. The oligonucleotides ride on the exterior if the liposome is positively charged or reside in the aqueous interior if neutral. Positively charged liposomes are drawn to the cell surface because it is negatively charged, and the entire liposome, oligonucleotides and all, is engulfed by endocytosis. Some liposomes contain "helper" molecules that make the endosomal membrane unstable and release the liposome directly into the cytoplasm. Other delivery "vehicles" are cationic polymers, which include poly-L-lysine and polyethylenimine. They operate via electrostatic interactions as discussed earlier, but they are toxic when taken into the cell; therefore, they are not used very often.

When the endosomal pathway is used for uptake, as with liposomes, there is a good chance that the antisense oligonucleotide will be degraded or not released to the cytoplasm. To alleviate this problem, antisense oligonucleotides may be attached to **basic peptides** (see Fig. 5.10B). They include the Tat protein of HIV-1, the N-terminal segment of HA2 subunit of influenza virus agglutinin protein, and Antennapedia peptide from *Drosophila* (which normally acts as a transcription factor). These peptides are able to enter the cell nucleus. When they are attached, the antisense oligonucleotides are taken directly into the nucleus.

Other methods to get oligonucleotides into the cells require chemically or manually disrupting the membrane. Membrane pores can be generated by **streptolysin O** permeabilization (see Fig. 5.10C) or electroporation (see Chapter 3). Streptolysin O is a toxin from *Streptococci* bacteria that aggregates after binding to cholesterol in the membrane, forming a pore. The oligonucleotide passes through the pore and enters the cytoplasm directly. Antisense oligonucleotides can also be microinjected directly into each cell, but this method cannot be used for treating patients and is useful only for small-scale experiments on cultured cells (Fig. 5.11A). Another mechanical method is called **scrape-loading** (see Fig. 5.11B). Here, adherent cultured cells are gently scraped off the dish while the oligonucleotide bathes the cells. Removal of the cells probably creates small openings that allow the oligonucleotides to enter the cytoplasm.

Antisense oligonucleotides enter the target cell by endocytosis of oligonucleotide-filled liposomes, by riding on basic peptides that normally enter the nucleus, by passing through pores created by streptolysin O, by microinjection, or by mechanical shearing.

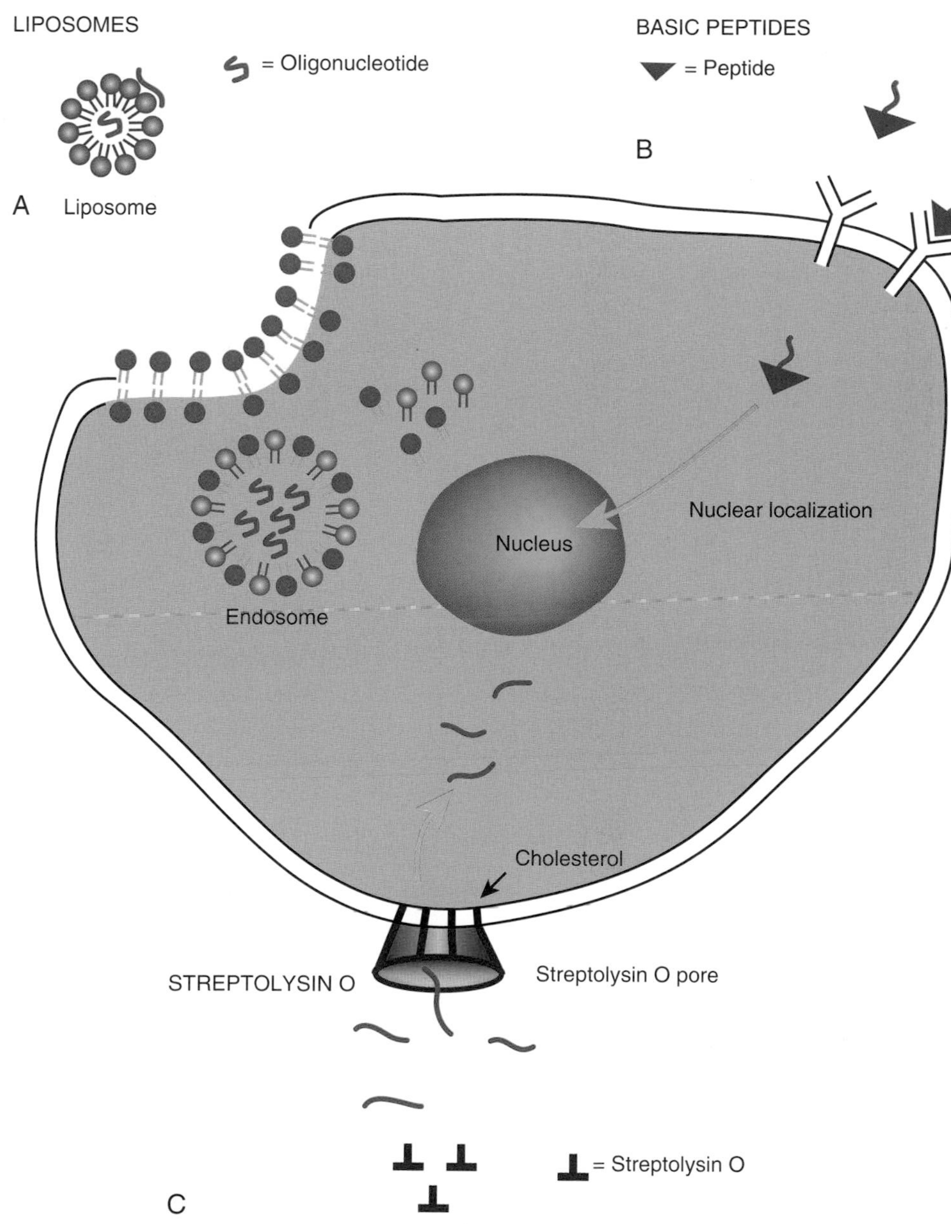

FIGURE 5.10 Methods of Antisense Oligonucleotide Uptake by Cells
(A) Liposomes are spherical structures made of lipids and cholesterol. Oligonucleotides are either encapsulated in the central core or ride on the exterior surface of the liposome. The complexes enter the cell via endocytosis and are released into the cytoplasm. (B) Basic peptides are naturally occurring proteins that normally enter the nucleus of target cells. The oligonucleotide can be fused to these peptides and ride Into the nucleus with the basic peptide. (C) Streptolysin O, a toxin from *Streptococci* bacteria, aggregates at the membrane to form a pore-like structure. The oligonucleotides can pass into the cell through the pore.

RNA Interference Uses Antisense RNA to Silence Gene Expression

RNA interference (RNAi) is a pathway for gene regulation where short double-stranded RNA (dsRNA) segments trigger an enzyme complex to degrade a target mRNA. In essence, the short dsRNA pieces decrease target protein expression by degrading its corresponding mRNA. RNAi was discovered in a variety of different organisms, including plants, fungi, mammals, flies, and worms. Different organisms have variations of the same basic response. Mutations in the enzymes responsible for RNAi affect a wide range of cellular processes. Some affect development of the organism; others affect the ability to fend off viruses, particularly RNA viruses. In still other cases, mutations affecting RNAi

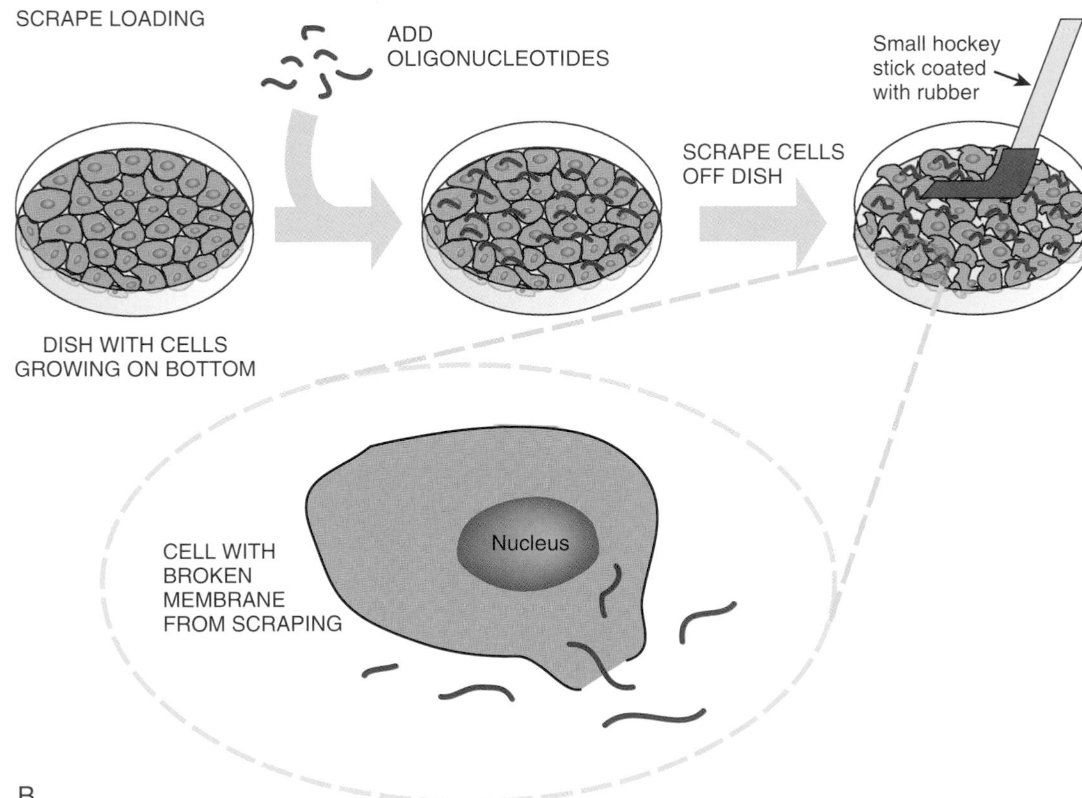

FIGURE 5.11 Microinjection and Scrape-Loading

(A) Oligonucleotides can be injected directly into a cell using a very fine needle. Microinjection can be done on individual cells grown in culture. (B) Scrape-loading is a mechanical method of getting oligonucleotides into cultured cells. As the cells are scraped off the bottom of the culture dish, the membranes break open, allowing the oligonucleotides to enter. When the cell membrane reseals, the oligonucleotides are trapped within the cells.

increase transposon movement, suggesting that RNAi may also prevent transposon jumping. All these processes rely on regulating mRNA translation or mRNA degradation.

RNAi occurs in two different stages: the initiation phase and the effector phase. Initiation begins with the formation of the shortened dsRNA. The full-length dsRNA can arise from three main sources. First, externally infecting RNA viruses replicate through a dsRNA intermediate, which can trigger RNAi. One theory suggests that the RNAi mechanism may have evolved to combat these infecting viruses (see Box 5.1). Another source is the organism's own genomic DNA, which contains sequences that code for microRNAs, specific mediators of RNAi (see later discussion). Finally, dsRNA can be produced from aberrant transcription of a genetically engineered gene (Fig. 5.12A). After the cellular enzymes recognize the dsRNA, an endonuclease called **Dicer** cuts the dsRNA into small fragments about 21 to 23 nucleotides in length called **small-interfering RNAs (siRNAs**; see Fig. 5.12B). Dicer is

Box 5.1 Does RNAi Protect Mammalian Cells from Viral Infections?

It is well established that plants use RNAi to protect themselves from viruses. Virus-derived siRNAs are produced when the plant is infected with either DNA or RNA viruses. Plants with mutations in various RNAi components are more susceptible to viral diseases. In plants, the RNAi signal spreads to uninfected regions, thus protecting the neighboring tissues. Finally, some plant viruses have genes/proteins that suppress the RNAi pathway.

In mammals, RNAi plays a lesser role in protection from viruses. When a mammal is infected with a virus, potent immune responses protect the organism from many different infections. Cellular proteins such as toll-like receptors, protein kinase R, and retinoic acid-inducible gene I are activated by virus entry. These proteins activate many different genes, most notably, type I interferons and nonspecific RNases. These genes work in unison to fight the infection.

Mammals cannot spread the RNAi signal to uninfected tissues, as do plants and invertebrates. Thus, RNAi is not the major mechanism for antiviral defense in mammals. Nonetheless, recent evidence does suggest that RNAi helps to limit virus invasion of mammalian cells. First, some proteins from mammalian-specific viruses target RNAi proteins. For example, NS1 from influenza virus binds to siRNA *in vitro* and suppresses RNA silencing when expressed in plants. Another viral protein, Tat from HIV, has been shown to inhibit purified Dicer *in vitro.*

The experimental work that uses siRNA and plasmid-encoded shRNA to block viral infections in mammalian cells is the most convincing. Many studies have found that administering siRNA or shRNA to animal models reduces virus replication and protects the organism from lethal infections. So although mammalian cells have other defense mechanisms, activating the RNAi system does protect against viral assaults.

a dsRNA-dependent RNA endonuclease that belongs to the RNase III family. The siRNAs have a two-nucleotide overhang on the 3′ ends (characteristic of RNase III–type enzymes). The 5′ ends are phosphorylated by a kinase associated with Dicer, making the siRNAs competent for the next phase.

In the effector phase, Dicer transfers the siRNA to a ribonucleoprotein complex called the **RNA-induced silencing complex (RISC)**. RISC is activated by the siRNAs and uses an RNA helicase to unwind the double-stranded fragments, making single strands. The antisense single-stranded siRNA is then kept as a guide to find complementary sequences in the cytoplasm. When RISC binds complementary sequences, the **Argonaut (AGO) family** member associated with the RISC complex cleaves the target mRNA, which is then degraded by exonucleases in the cytosol. This destroys all of the mRNA that is complementary to the siRNA. Both Dicer and the RISC complex are dependent on ATP for energy. The antisense siRNA specifies which mRNA is targeted, ensuring that no nonspecific mRNAs are degraded.

RNAi does not require many molecules of siRNA. In fact, as few as 50 copies of siRNA may destroy the entire cellular content of target mRNA. The ability to target so many mRNA molecules with so few siRNA copies relies on amplification by the enzyme **RNA-dependent RNA polymerase (RdRP)**, which creates dsRNA. RdRP uses the cleaved target mRNA as template to synthesize more dsRNA. Dicer recognizes the new dsRNA and cleaves it into more siRNA, thus amplifying the number of siRNA molecules (Fig. 5.13).

The final aspect of RNAi is its ability to modulate DNA expression by converting copies of the target gene into heterochromatin (Fig. 5.14). The siRNA can direct the heterochromatin-forming enzymes and proteins to the target gene location. Once the open, expressed DNA conformation is converted into heterochromatin, no more mRNA is produced. Therefore, RNAi can repress gene expression permanently.

RNAi has two phases: The initiation phase forms double-stranded RNA approximately 21 to 23 nucleotides long called siRNA, and the effector phase makes the double-stranded siRNA into a single-stranded template that searches out complementary mRNA and destroys them.

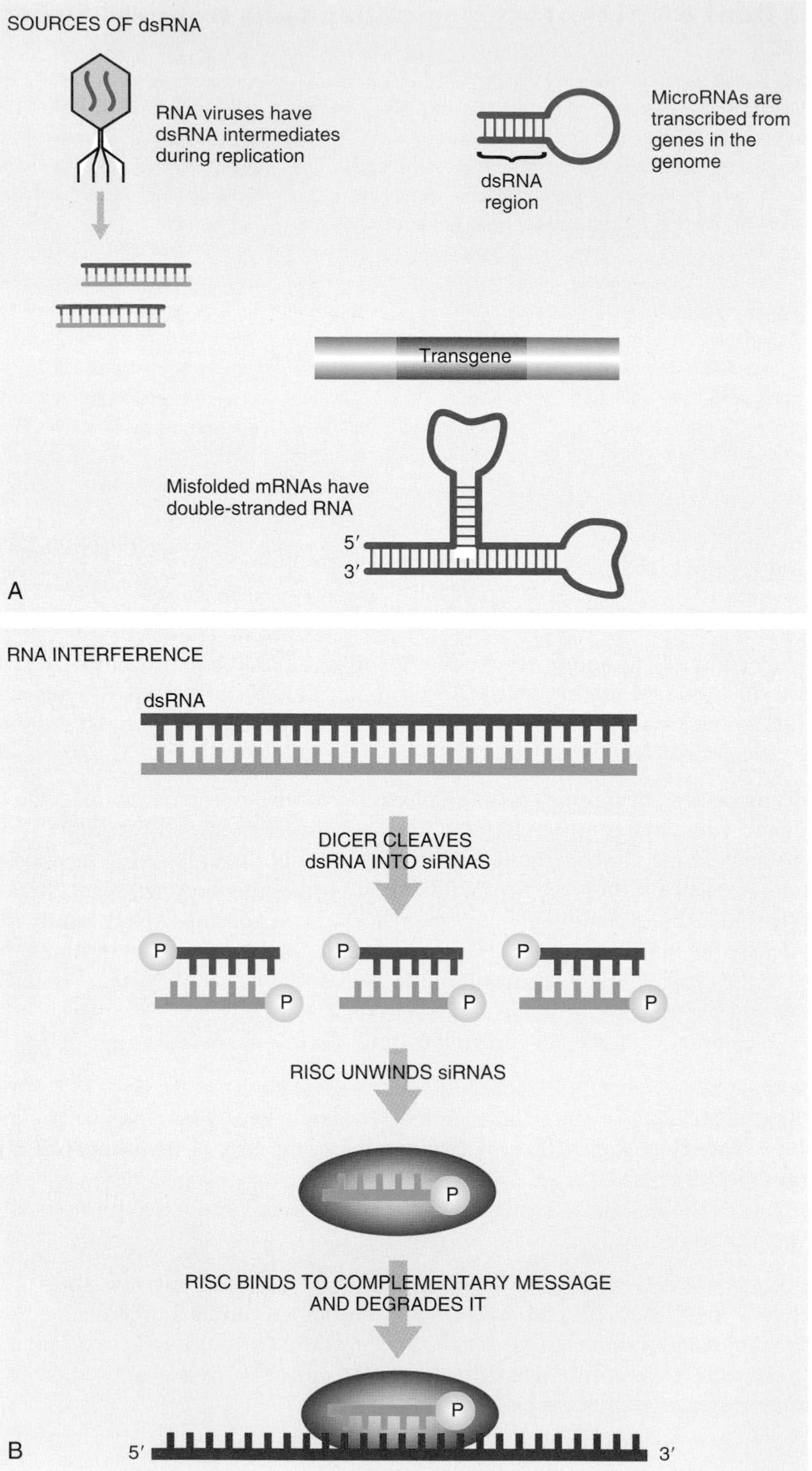

FIGURE 5.12 Cellular Mechanism of RNAi
(A) Double-stranded RNA triggers RNA interference. dsRNA is produced by RNA viruses during infections, microRNA encoded by the genome, or overexpression of transgenes. (B) RNA interference degrades all the RNA that is complementary to segments of double-stranded RNA. First, Dicer recognizes dsRNA and cuts it into pieces of 21–23 nucleotides. A kinase phosphorylates the 5′ end of each piece. Next, RISC unwinds the siRNAs and uses one strand to search out complementary mRNA, which is degraded by associated enzymes.

RNAi IN PLANTS AND FUNGI

RNAi was actually first observed in plants, where it was named **post-transcriptional gene silencing (PTGS)**. This phenomenon was first noted when some early transgenic experiments in plants gave strange results. When an extra copy of a gene was inserted to increase protein production, both the inserted gene (i.e., the transgene) and the resident gene were silenced. The result was a plant that made less of the target protein rather than more. For example, in 1990, researchers inserted a gene to make petunia flowers a darker purple. Instead, the plant made white flowers. Both the transgene and the endogenous gene were suppressed, leaving the flower without any pigment. A similar phenomenon was seen in the fungus *Neurospora*, where it was called **quelling**. After the discovery of RNAi in *Caenorhabditis elegans*, it was recognized that RNAi, PTGS, and quelling all operate via the same mechanism. These processes all act after the stage of transcription. Thus, plenty of mRNA was produced from the silenced transgenes, at least initially. After some time, the mRNA for the transgene was found in two fragments, suggesting an endonuclease cleaved it in two. Later, the target mRNA was found in smaller and smaller fragments, suggesting that exonucleases were digesting the large mRNA segments. Finally, the genes were converted into heterochromatin, and transcription was shut down.

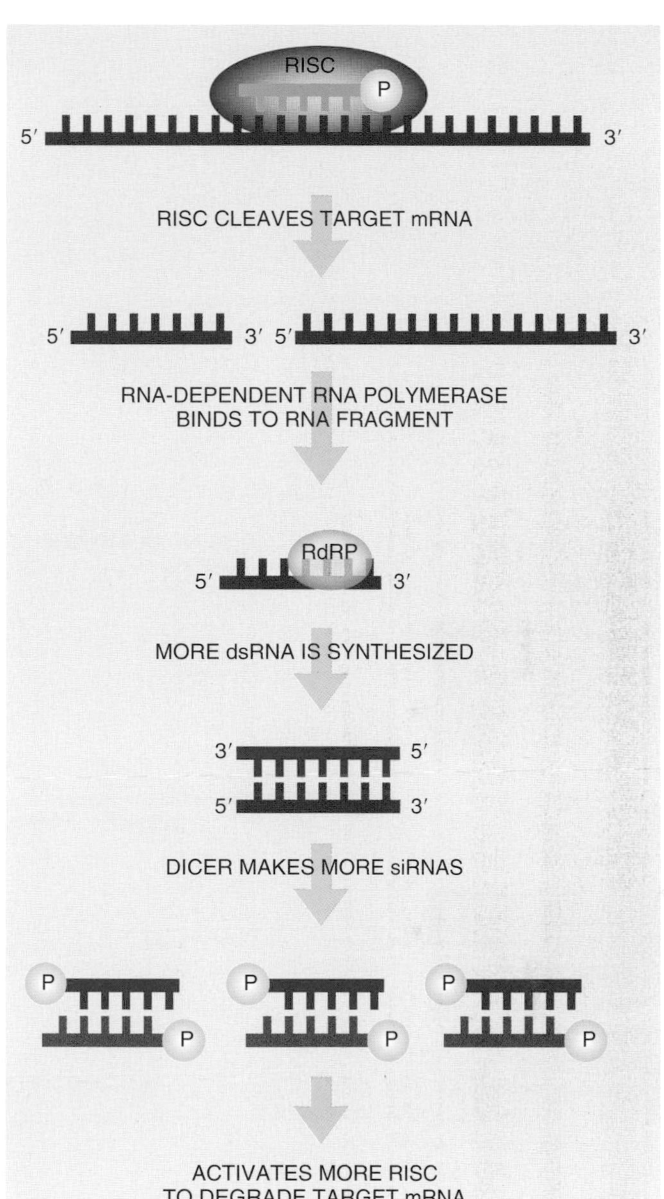

FIGURE 5.13
Amplification of RNAi
After RISC-associated enzymes cleave the complementary mRNA in a cell, another enzyme, RNA-dependent RNA polymerase, binds to some of these fragments. RdRP synthesizes complementary strands, making more double-stranded RNA. Dicer recognizes these fragments and creates more siRNA.

How does an extra copy of a gene induce a system that is triggered by dsRNA? One theory is that overproduction of certain mRNAs triggers RdRP to make dsRNA from the excess. This dsRNA activates Dicer to create siRNAs that quench mRNA, both from the transgene and from any closely related endogenous gene. Alternatively, when certain transgenes are expressed, some regions of the mRNA may fold back on themselves to form **hairpins**. These double-stranded segments may also activate Dicer. Genetic analysis of the model plant, *Arabidopsis*, has shown that the RdRP encoded by the SDE1 gene is necessary for transgene silencing but is not needed for antiviral RNAi. (In the latter case, the virus RNA polymerase would make dsRNA, and the plant RdRP enzyme would therefore not be necessary.) This favors the first model for transgene-triggered silencing.

The most interesting aspect of PTGS is the ability of silencing to propagate from one part of the plant to the next. Plants can be grafted; that is, a leaf or stem can be attached to a different plant. If the grafted piece has a transgene silenced by PTGS, this will also silence the corresponding endogenous gene. The effect of RNAi then travels through the vascular system of the plant and affects regions without the transgene. RNAi in *C. elegans* also has the ability to spread, not only from tissue to tissue, but also from parent to progeny. In contrast, mammals lack the ability to spread the RNAi signal.

The ability of RNAi to spread may not rely solely on siRNA movement. In plants, the potyviruses produce an inhibitor of RNAi called **helper component proteinase (Hc-Pro)**. This protein blocks the accumulation of siRNA. Despite this, the RNAi signal still spreads to other parts of the plant and triggers methylation of DNA, thus turning it into heterochromatin. Other viral genes that inhibit different steps of the RNAi process will, it is hoped, illuminate the mechanism of spreading.

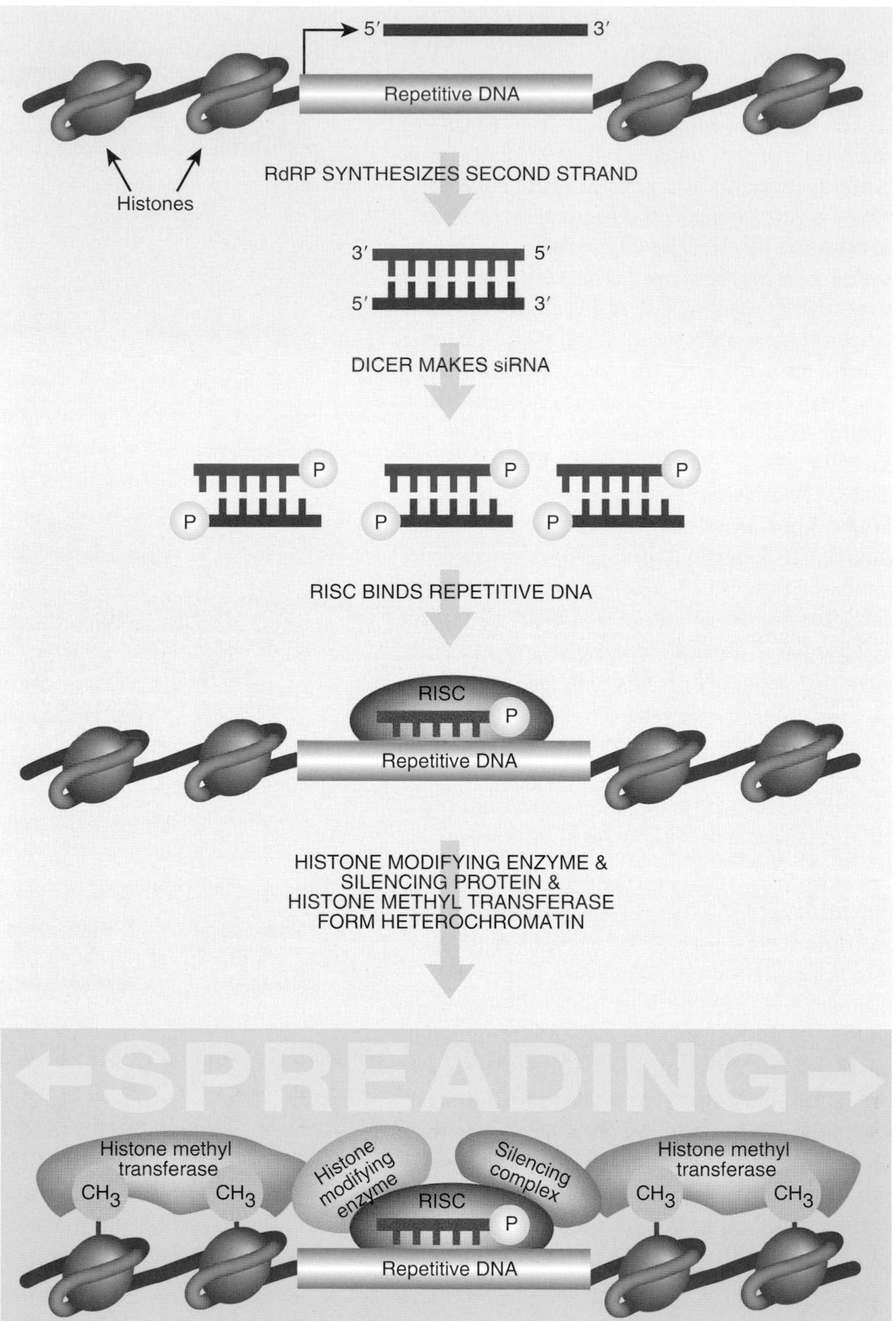

FIGURE 5.14
Heterochromatin Formation by RNAi
The RISC complex containing single-stranded siRNA can also recognize and bind to complementary DNA sequences. When RISC associates with a repetitive DNA element, various histone-modifying enzymes and silencing complexes are activated to turn that region of DNA into heterochromatin, thus silencing the region from any further expression.

Other terms used to describe variants of RNAi are **transcriptional gene silencing**, **cosuppression**, and **virus-induced silencing**. Transcriptional gene silencing refers to the silencing of gene expression by converting the gene into heterochromatin. Cosuppression is an early name for PTGS. Virus-induced silencing occurs when the viral genome has a double-stranded RNA intermediate, which triggers Dicer and RISC. A comprehensive term, **GENE impedance (GENEi)**, has been proposed to encompass all these phenomena but is rarely used.

Early experiments in making transgenic plants identified a phenomenon now called RNAi. RNAi is also known as post-transcriptional gene silencing (PTGS), quelling, transcriptional gene silencing, cosuppression, and virus-induced silencing.

MicroRNAs Modulate Gene Expression

The development from embryo to adult of the worm *C. elegans* requires RNAi to turn off genes at appropriate times. In this case, RNAi is not triggered by intrusion of external sequences such as transgenes or viruses. During development, small noncoding RNA molecules known as **microRNAs (miRNA)** are transcribed from the worm's own genome. These miRNAs regulate gene expression by blocking the translation of target mRNA. MicroRNAs, first identified in *C. elegans*, are now known to be present in plants and animals, including humans. RNAi induced by miRNAs is similar to the mechanism described previously. The mRNA targets are identified by antisense; that is, the miRNA has sequences that are complementary to part of the target mRNA. Some miRNAs bind to the target mRNA and block the initiation of translation. In other cases, the miRNA binds to the 3′ UTR region of the mRNA.

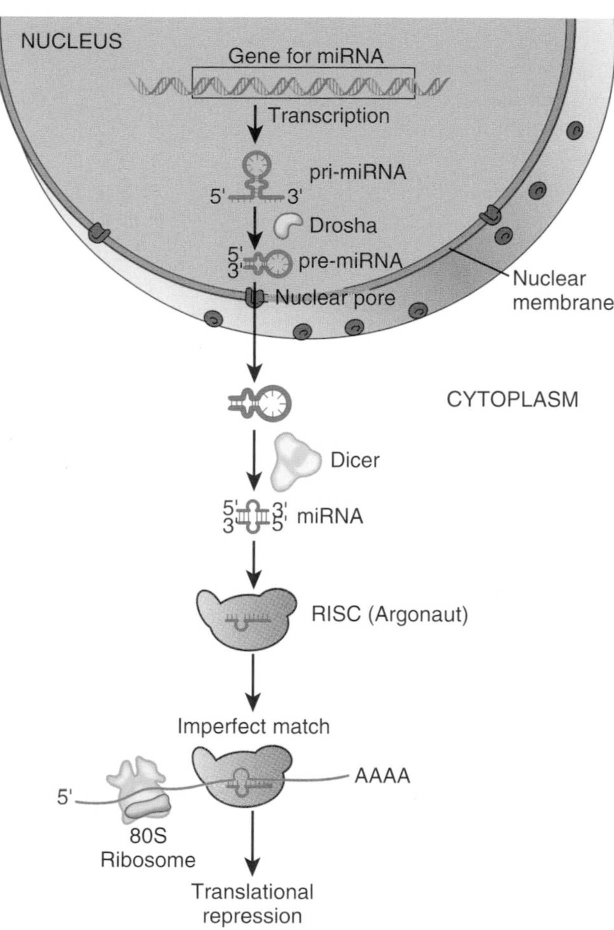

FIGURE 5.15 Pathway of miRNA Inactivation of Gene Translation in *Drosophila*

Two cleavage events create miRNAs. First, the gene for the miRNA is transcribed into an RNA that folds into a stem loop called a pri-miRNA. Drosha cleaves pri-mRNA in the nucleus to create a pre-miRNA, which is then competent to exit the nucleus. In the cytoplasm, Dicer cuts the ends of the pre-miRNA to form a mature miRNA. In some organisms, the miRNA does not have perfectly matched sequences and therefore has regions that bulge due to the mismatches. RISC complex creates the single-stranded template and searches the cytoplasm for any matching sequences. When a match is found, the Argonaut component of RISC cuts the target mRNA, marking it for degradation.

MicroRNAs are transcribed as longer precursor molecules, **pri-microRNAs**, of approximately 70 nucleotides in length. In *Drosophila*, pri-microRNAs are transcribed as polycistronic messages, which are then cleaved by an endonuclease called Drosha. The cleaved products are called **pre-miRNAs**. Pre-miRNAs exit the nucleus. In the cytoplasm, Dicer recognizes the stem-loop and cleaves the loop structure. The RISC complex then separates the two strands. The miRNAs found in animals such as *C. elegans* can tolerate a few mismatched base pairs within the binding domain. In animals, the antisense miRNA strand blocks translation of the target mRNA (Fig. 5.15), which is not degraded. In contrast, in plants, microRNA must have perfect matches and relies on RISC-mediated recognition and cleavage to degrade the target mRNA.

Most microRNA molecules target multiple mRNAs, and they rarely inactivate any mRNA totally. They serve to coordinate and modulate regulation of multiple genes and consequently often play a role in development. A newly discovered class of RNA, known as circular RNA (circRNA), may counteract the effect of miRNA. The circRNA molecules are large (around 1500 nucleotides) and have multiple binding sites (over 70 in some cases) for the corresponding miRNA. The circRNA acts as a molecular sponge that absorbs the miRNA and prevents it from acting in its target mRNAs.

MicroRNAs (miRNAs) modulate expression of various genes during development in many organisms. They are first translated as pri-miRNAs from the organism's own genome, processed into 21 to 23 nucleotide pieces, and then RISC separates the strands and searches the cytoplasm for mRNAs with complementary sequences.

Applications of RNAi for Studying Gene Expression

By inhibiting translation, RNAi allows the elimination of a particular protein from an organism, without the need for genetic modification. RNAi thus provides a powerful tool to study the roles that particular proteins play. The application of RNAi to studying *C. elegans* is especially well understood. Three different methods are used to get dsRNA into *C. elegans* and stimulate RNAi to block expression of a target gene (Fig. 5.16). The little

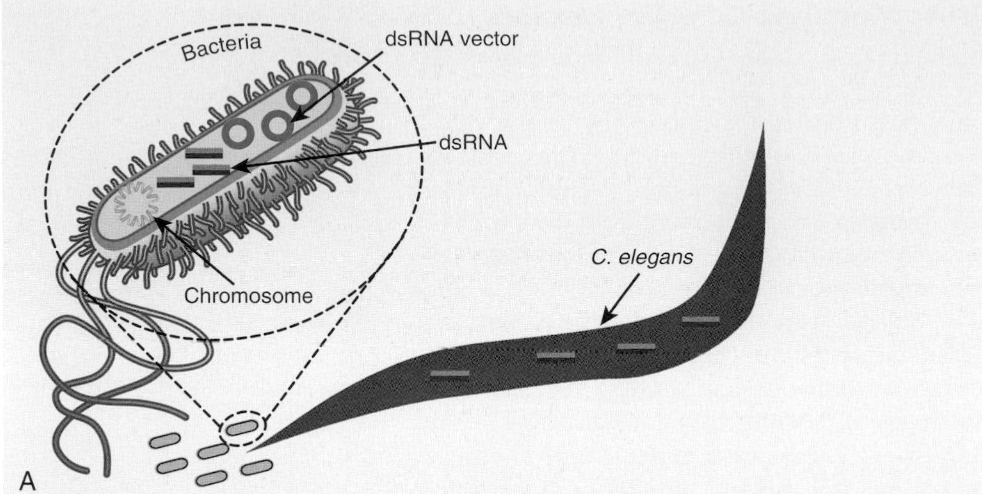

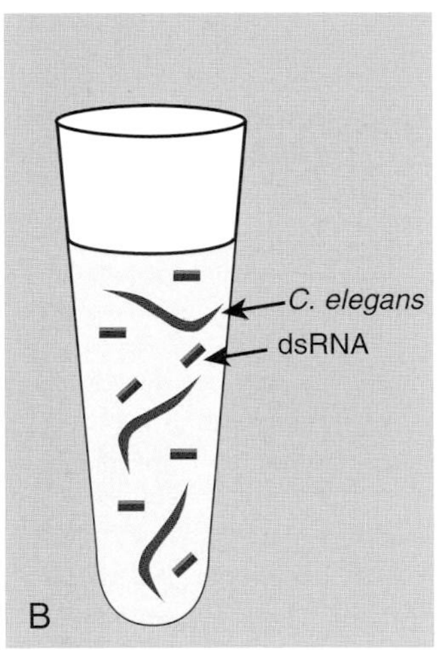

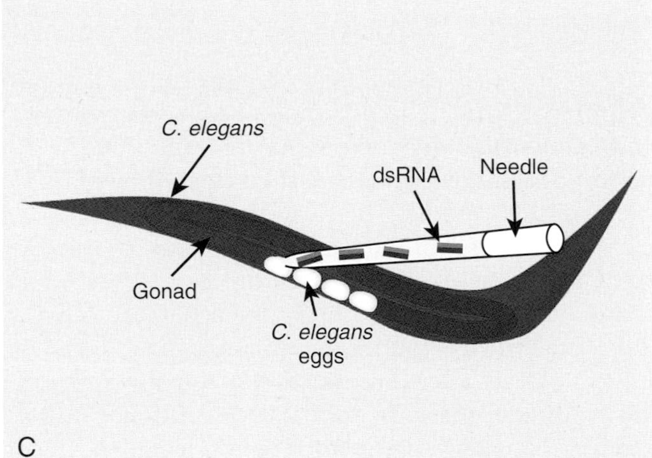

FIGURE 5.16
Delivering dsRNA to *C. elegans*
(A) *C. elegans* can absorb dsRNA expressed in bacteria that they eat. (B) *C. elegans* can absorb dsRNA by swimming in a solution containing dsRNA. (C) Injecting dsRNA into an egg will trigger gene silencing in the developing worm.

worm has an uncanny ability to take up exogenous DNA or RNA. Worms can be fed *Escherichia coli* bacteria expressing the dsRNA of interest. When the worm digests the bacteria, the dsRNA is taken up into the intestinal cells and triggers RNAi. Dicer then cleaves the dsRNA into small siRNAs that activate RISC to block the target mRNA. Another method to deliver dsRNA is simply bathing the worms in a solution of dsRNA. The exogenous dsRNA is absorbed into the worm, where it activates RNAi. A third method to induce RNAi is to inject worm eggs with dsRNA. The worm develops with the dsRNA inside, and RNAi is activated in all the cells. Bathing worms in dsRNA or feeding them dsRNA-expressing bacteria is incomplete—some cells are not penetrated—yet because the signal can spread from cell to cell, the method works. Surprisingly, the RNAi effect can pass from parent to offspring. The progeny of a worm with a silenced gene will silence the same gene. Current research suggests that epigenetic influences such as histone modification play a major role in passing the RNAi gene silencing from parent to offspring.

Delivering dsRNA to *Drosophila* is not quite as easy. The dsRNA must be microinjected directly into a developing *Drosophila* egg. The dsRNA enters the cells as the embryo develops. This method knocks out the protein of interest through all stages of development. Obviously, if the protein is essential for development, activating RNAi too early can stop development and kill the fly. Luckily, *Drosophila* has a feature not found in *C. elegans*. Fly cells can be cultured *in vitro* in nutrient medium, and dsRNA can be transfected into the cultured cells. Therefore, if RNAi kills the embryo, the corresponding protein can still be examined in cultured cells.

C. elegans can take up dsRNA pieces that activate RNAi by ingesting transgenic bacteria that are expressing dsRNA, by bathing in a solution of the pure dsRNA, or by having the dsRNA injected into the eggs. *Drosophila* is also used as a model organism to study protein function with RNAi by microinjecting dsRNA into *Drosophila* eggs or cultured cell lines.

RNAi FOR STUDYING MAMMALIAN GENES

An important application for RNAi is testing the individual roles of human proteins, which can reveal new targets for curing diseases or can identify the causative agent for that disease. Until recently, using RNAi in mammalian systems was not possible. Applying dsRNA to cultured mammalian cells or whole mice induces a potent antiviral response. Interferon is produced, which triggers the cells to degrade all RNA transcripts and shut down protein synthesis. Thus, the methods used in *C. elegans* and *Drosophila* kill mammalian cells.

Recognizing that small-interfering RNA (siRNA) triggers RNAi was the key to its application in mammals. Instead of using long dsRNA as in *C. elegans* and *Drosophila,* exposing mammalian cells to dsRNA shorter than 30 nucleotides activated the mammalian counterparts of Dicer and RISC. This in turn abolished expression of the target mRNA. Such short dsRNA segments thus act as endogenously produced siRNA. As described earlier, siRNAs made by Dicer are short double-stranded RNA about 21 to 23 base pairs in length. In addition, the siRNAs have a two-base 3′ overhang that is more stable when it consists of two uracils. To study a particular target mRNA in mammalian cells, chemically synthesized siRNAs with these characteristics are designed to have the complementary sequence to the target. Like antisense oligonucleotides, these may have modifications to make them more stable, such as methyl groups added to the 2′-OH of the ribose. The most difficult aspect of *in vitro* siRNA construction is determining an effective sequence. The sequence on the target mRNA must be accessible to the siRNA, which may be challenging because of RNA secondary structure. Many suitable siRNA are designed, and each is tested for activating RNAi. These siRNA are delivered to mammalian cells much like antisense oligonucleotides, including transfection, liposomes, and microinjection.

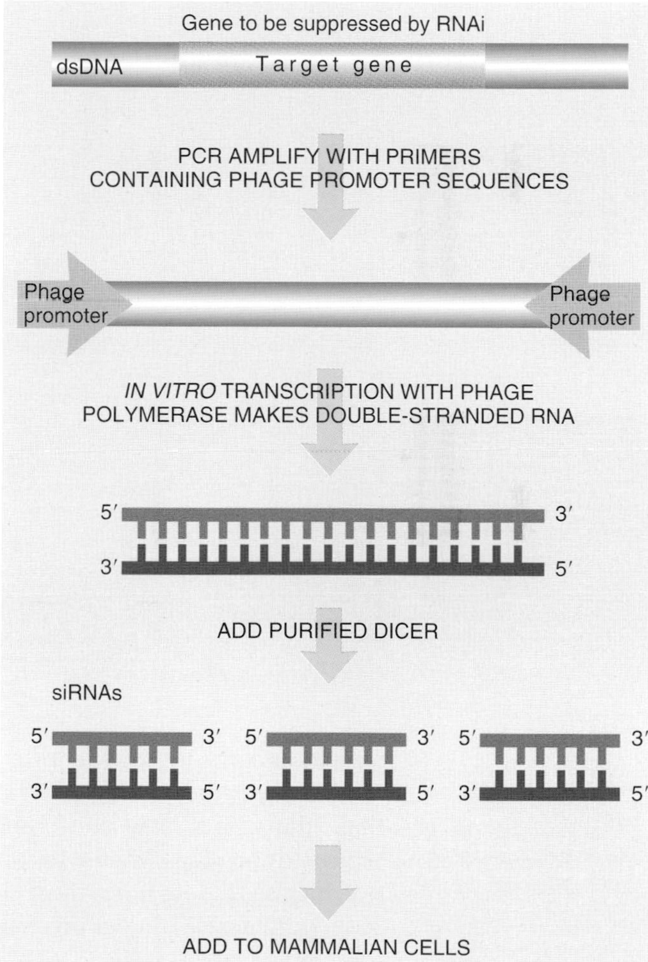

FIGURE 5.17 *In Vitro* Treatment with Dicer Generates siRNAs
The key to making siRNAs *in vitro* is cloning the target gene so that both the sense and antisense strands are expressed into mRNA. The two strands anneal spontaneously, and when purified Dicer is added, small siRNAs are produced.

Rather than chemically synthesizing siRNA, the target mRNA can be mixed with purified Dicer to cleave the mRNA into siRNA pieces (Fig. 5.17). Purified Dicer generates multiple siRNAs as it would *in vivo*. To supply the target mRNA, scientists amplify the chosen target gene with PCR using PCR primers that also contain a promoter sequence for T7 RNA polymerase. The dsDNA is then converted into dsRNA by T7 RNA polymerase. The RNA strands are allowed to anneal spontaneously, and the dsRNA is then mixed with purified Dicer, which digests it into multiple different siRNAs. These are then transfected into mammalian cells.

Mammalian cells can also be induced to activate RNAi by expressing **short hairpin RNAs (shRNAs)** that mimic the structure of microRNAs. A gene for the shRNA is constructed in a vector. The shRNA can be transcribed as two complementary strands by two different promoters facing in opposite directions, or simply made as one transcript with complementary ends interrupted by a sequence that forms a loop (Fig. 5.18). In both constructs, the complementary

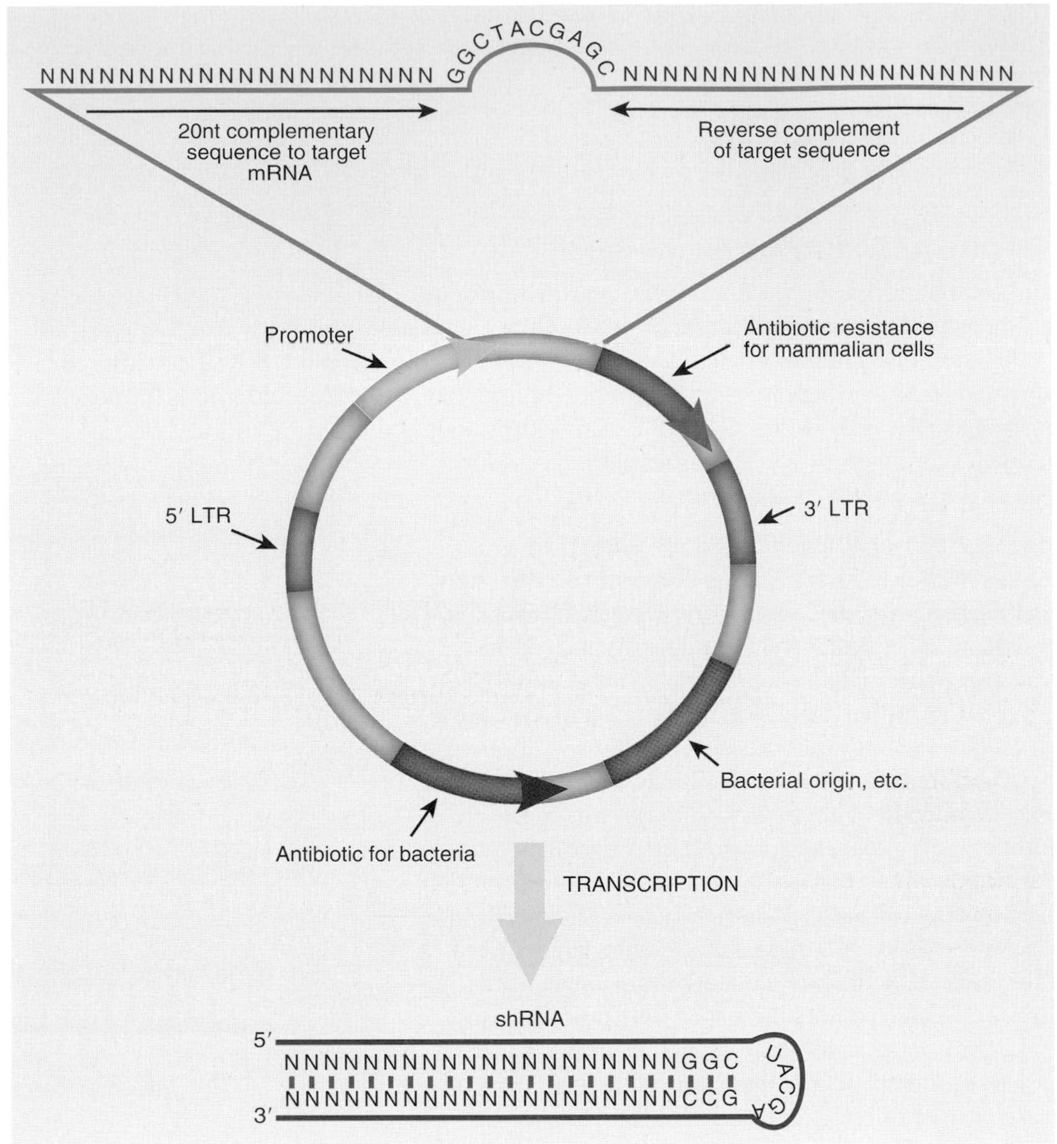

FIGURE 5.18 Design of shRNA Expression Vectors to Activate RNAi
Vectors can be designed to express different shRNA molecules. This retrovirus-based vector has two complementary sequences about 20 nucleotides in length that form a stem, separated by a loop region. When the vector is transformed into a cell, the shRNA is transcribed and activates gene silencing.

sequences come together to form a double-stranded RNA region. The promoter for mammalian RNA polymerase III is most commonly used to express shRNA *in vivo*. This enzyme normally transcribes small noncoding RNAs. RNA polymerase III starts transcription at a specific sequence and stops when it encounters four to five consecutive thymines. In addition, RNA polymerase III does not activate the enzymes that add a cap and a poly(A) tail to the transcript. Thus, polymerase III does precisely what is necessary to create a shRNA that mimics those found in eukaryotic cells.

Rather than designing the shRNA constructs from scratch, another strategy uses pre-existing microRNA. First, the stem portion is replaced with sequences that match the target mRNA. The newly designed microRNA will trigger RNAi for the target mRNA rather than its endogenous target due to the change in sequence.

Vectors that express shRNA have some advantages over adding siRNA. When siRNA is added to mammalian cultures, the effect is temporary. When the siRNA is gone, the effect ends. A shRNA vector has a more sustained effect as it continues to produce shRNA for a considerable time. In addition, expression of the shRNA may be controlled using promoters that are inducible or tissue-specific. In a clinical setting, the ability to deliver the siRNA only to those tissues that need it is crucial, and using a vector system can accomplish this task.

RNAi can be triggered in mammalian cells using chemically synthesized siRNA, creating a shRNA that degrades the target mRNA, or by modifying existing miRNA to recognize a different target mRNA.

FUNCTIONAL SCREENING WITH RNAi LIBRARIES

RNAi can be used to screen the entire genome by constructing an **RNAi library**. In these vector-borne gene libraries, each gene sequence is expressed as dsRNA (rather than DNA). Thus, each library clone targets the corresponding gene for suppression by RNA interference. In theory, such libraries could be used to screen each protein in an organism for its functional role. At present, there are still technical problems with reliability and reproducibility. Nonetheless, some gene products have been characterized—for example, those that affect resistance to several viruses.

An RNAi library containing most of the predicted genes in the genome of *C. elegans* has been constructed in *E. coli*. These dsRNA expressing *E. coli* are then fed to *C. elegans*, thus triggering RNAi and removing one protein from the organism. This allows the effect of suppressing expression of each single protein from the worm to be assayed. Using this library, more than 900 genes have been identified whose suppression kills the embryo or causes gross developmental defects. Most of these genes had no known function before. Similarly, an RNAi library of about 7200 genes (about 91% of the predicted genes in the genome) has been constructed in *Drosophila*. With *Drosophila*, multiple investigators have examined the same regulatory pathways by RNAi screening. Unfortunately, agreement is very low between different studies; often merely 20%–30% of overlap is seen.

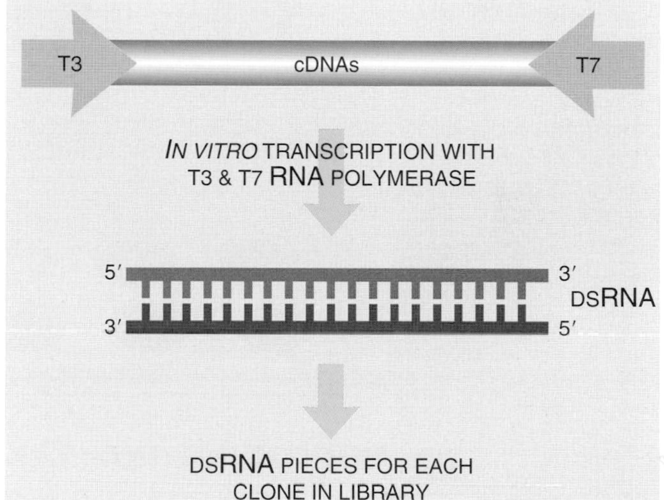

FIGURE 5.19 Constructing an RNAi Library
Each clone in the library must have two different promoters flanking the coding region. When the clone is transcribed into mRNA, both an antisense and a sense transcript will be produced, and the two strands will come together to form double-stranded RNA.

To construct an RNAi library, scientists isolate the genes as cDNA clones (Fig. 5.19) and then amplify them using PCR (see Chapter 4). The PCR primers are specific for each gene and are designed to add two different promoters at the ends of each gene. For example, the 3′ end would have a T7 polymerase promoter, and the 5′ end would have a T3 polymerase promoter. The PCR products are then cloned into a suitable vector. When the vector is present in a cell with both T7 and T3 RNA polymerase, both an antisense and a sense transcript are transcribed for each of the clones. The two strands spontaneously anneal to form dsRNA, which then activates RNAi.

For mammals, RNAi libraries must be constructed with siRNAs or shRNAs rather than full-length dsRNA, because full-length dsRNA is toxic. This kind of RNAi library is analyzed either with multiwell plates or **live cell microarrays** (Fig. 5.20). For multiwell plates, the library is transformed into a large number of cells that are then inoculated into the wells of the plate. The number of cells in each well is adjusted so that only one siRNA is found in each well. Another method for screening a siRNA or shRNA library is to spot each clone onto slides in a microarray. Live cells are then added and take up the siRNA or shRNA at these locations. In either case, the cells are analyzed for any noticeable symptoms.

RNAi libraries are designed to express dsRNA for each gene in the genome. Each library clone targets one protein by promoting degradation of the corresponding mRNA. RNAi libraries are used to identify the role of unknown proteins. Mammalian cells can be screened for defects induced by an RNAi library clone using a live cell microarray or by using a multiwell-plate assay.

FIGURE 5.20
Multiwell Plate Assays and Live Cell Microarrays
(A) Small-interfering RNA and short-hairpin RNA libraries can be transfected into mammalian cells. Each cell can then be assessed for altered phenotypes, such as loss of adherence, mitotic arrest, or changed cell shape. Clones that cause interesting phenotypes are isolated and sequenced to identify the protein that was suppressed. (B) Rather than transforming cells, the siRNA or shRNA can be spotted onto microscope slides. As cells grow and divide on the slide, they take up RNAs. This initiates RNAi. The cells are then screened for phenotype changes over the spot.

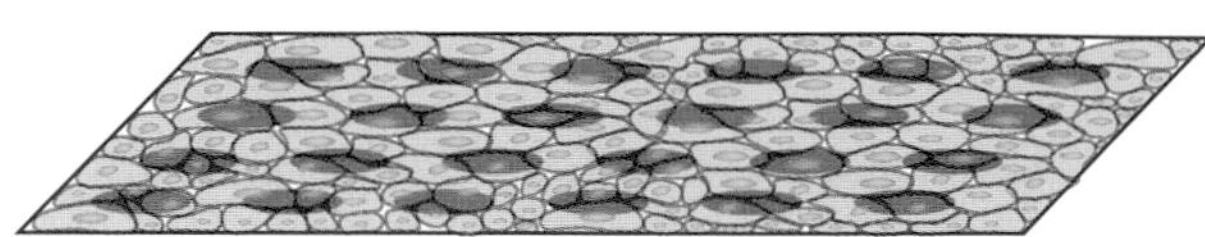

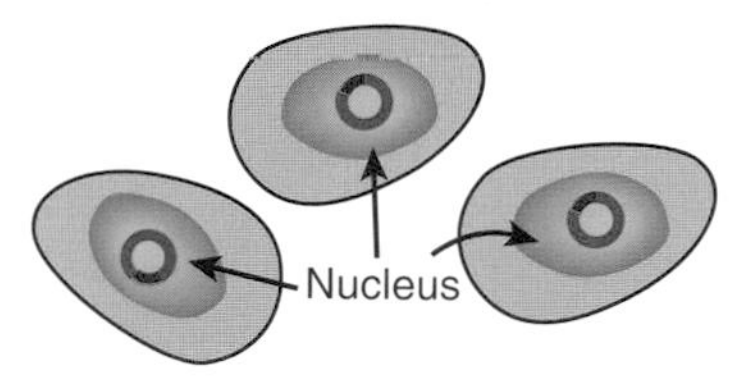

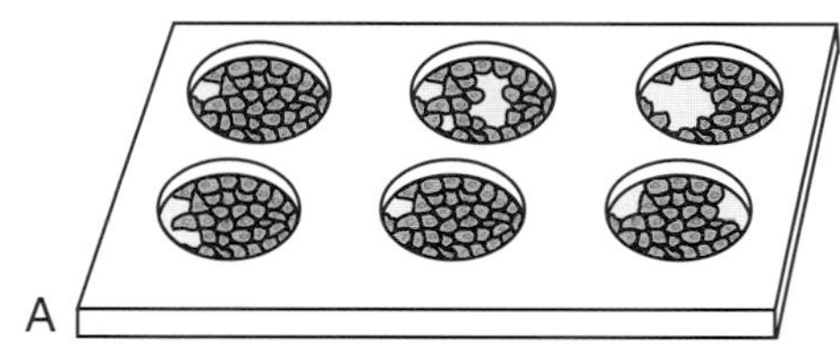

NONCODING RNAs TAKE PART IN RNA PROCESSING

The role of RNA in RNA processing and editing has been known for some time. Noncoding RNAs take part in processing other RNA molecules, such as those involved in protein synthesis: tRNA, rRNA, and mRNA. In all three of these RNA molecules, the structures of the mature RNA is different from the original RNA transcript, and other noncoding RNAs are essential for these modifications. For example, the enzyme ribonuclease P (RNase P) trims both the 5′ and 3′ ends of tRNA precursors. RNase P is unusual in being a complex of RNA and protein, but it is the RNA component that cleaves tRNA. Because of its catalytic activity, RNase P is considered a ribozyme.

Another example of a small noncoding RNA working as a key component of RNA processing occurs when the primary transcript is processed into mRNA in the nucleus of eukaryotes. A noncoding RNA called **snRNA** (small nuclear RNA) removes introns from the primary transcript. snRNA and protein work together in a complex called the **spliceosome** that identifies the intron/exon borders and removes the nonprotein coding introns. The key proteins and snRNA assemble at the splice sites that border the exon and intron sequences. During the splicing reaction, three essential sequences are required. The 5′ splice site is recognized by U1 snRNA, the 3′ splice site is recognized by U2AF protein, and the branch site within the intron is recognized by U2 snRNA. U1 and U2 RNA pull the two exons close together, cleave the RNA at the two splice sites, and ligate the exons into one piece. The intron is released as a lariat structure (Fig. 5.21). Researchers have found that defects in these components may cause diseases like lupus and spinal muscular atrophy.

In addition, noncoding RNAs are also involved in **alternate splicing**. Not all exons within a gene are always used, and therefore the final mRNA may not include all of the available exons. Alternate splicing occurs in almost 95% of human genes, a fact only recently discovered by next-generation sequencing technology. The role of snRNA and other noncoding RNAs in these processes is still being investigated. Alternative splicing of mRNA is widely used to control gene expression in eukaryotes. Mutations affecting this process may cause some forms of inherited disease. Recently, antisense oligonucleotides have been used to correct the genetic errors in alternative splicing that cause beta-thalassemia and some cancers (see Box 5.2).

Another example of noncoding RNA function occurs in the nucleolus. Here, rRNAs are transcribed by RNA polymerase I, but the transcript cannot be used as is. Instead, rRNA ribonucleotides are modified. Noncoding RNAs, known as snoRNAs (small nucleolar RNA), are responsible for RNA modification. The human genome contains 400 snoRNA species, but only half of these have a known target. Most snoRNAs are encoded within introns and are transcribed with their parent gene. When the splicing machinery removes the intron, the snoRNA sequence is spliced from the intron, debranched, and then processed to the correct size by an endonuclease. These snoRNAs guide key proteins to modify rRNA and snRNA. The H/ACA-box (SNORA) family of snoRNAs directs pseudouridylation (the replacement of uridine with pseudouridine), whereas SNORD family members methylate 2′ *O*-ribose. In addition to the role in rRNA maturation, snoRNAs are processed into smaller pieces called snoRNA-derived RNAs. These RNAs have an unknown function, although some studies suggest they play a role in the regulation of cellular growth and therefore affect some cancers.

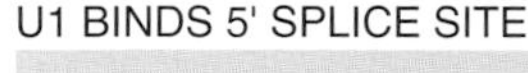

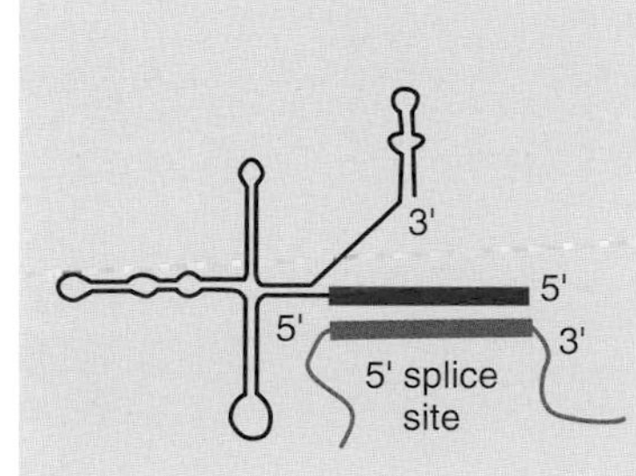

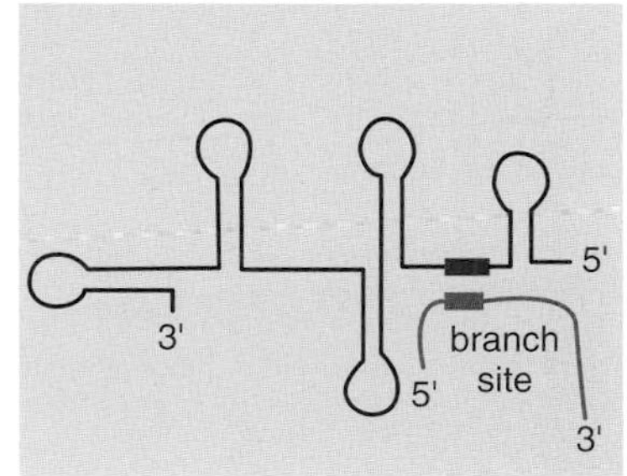

FIGURE 5.21 Operation of the Spliceosome
(A) The spliceosome consists of several ribonucleoproteins (U1 to U6), also known as "snurps," which take part in splicing. These assemble at the splice sites at the intron/exon boundaries. (B) The binding of U1 at the 5′ splice site and of U2 at the branch site is shown in greater detail.

The three major RNAs involved in translation are all processed. In all cases, processing requires the participation of other noncoding RNA species.

Box 5.2 Antisense Oligonucleotides Cure Splicing Defects

Some diseases are caused by aberrant patterns of mRNA splicing. For example, beta-thalassemia is a blood disorder in which red blood cells cannot carry enough oxygen because of defective hemoglobin. Some cases of the disorder arise from aberrant splicing of the beta-globin pre-mRNA. These cases are due to mutations that generate extra splice sites between exons two and three in the beta-globin gene. This results in the inclusion of part of the intron sequence in the mRNA and final protein. Antisense morpholino-oligonucleotides have been designed to target and suppress the extra mutant splice sites. The antisense oligonucleotide corrects the splicing pattern and restores the correct protein in red blood cells taken from patients with this form of beta-thalassemia (Fig. A, part A).

The Bcl-x gene in humans produces two different proteins by alternate splicing. Bcl-xL, the longer protein, includes a segment of coding material between exons 1 and 2. Bcl-xS, the shorter protein, lacks this segment. Bcl-xL protein promotes cell growth by blocking apoptosis, whereas Bcl-xS opposes this and causes cells to die via apoptosis (see Chapter 20). In some cancers, the long form is overexpressed. Thus, devising a method to inhibit Bcl-xL expression and enhance Bcl-xS expression could induce these cancer cells to

(Continued)

Box 5.2 Antisense Oligonucleotides Cure Splicing Defects—cont'd

undergo apoptosis and die. Antisense 2′-O-methyl-oligonucleotides complementary to the splice site in the pre-mRNA prevent the longer form from being made. The Bcl-xS protein then counteracts the cancerous growth by promoting apoptosis (Fig. A, part B). Those cancer cells that were "fixed" by antisense therapy also became more sensitive to chemotherapeutic agents because of the restoration of apoptosis.

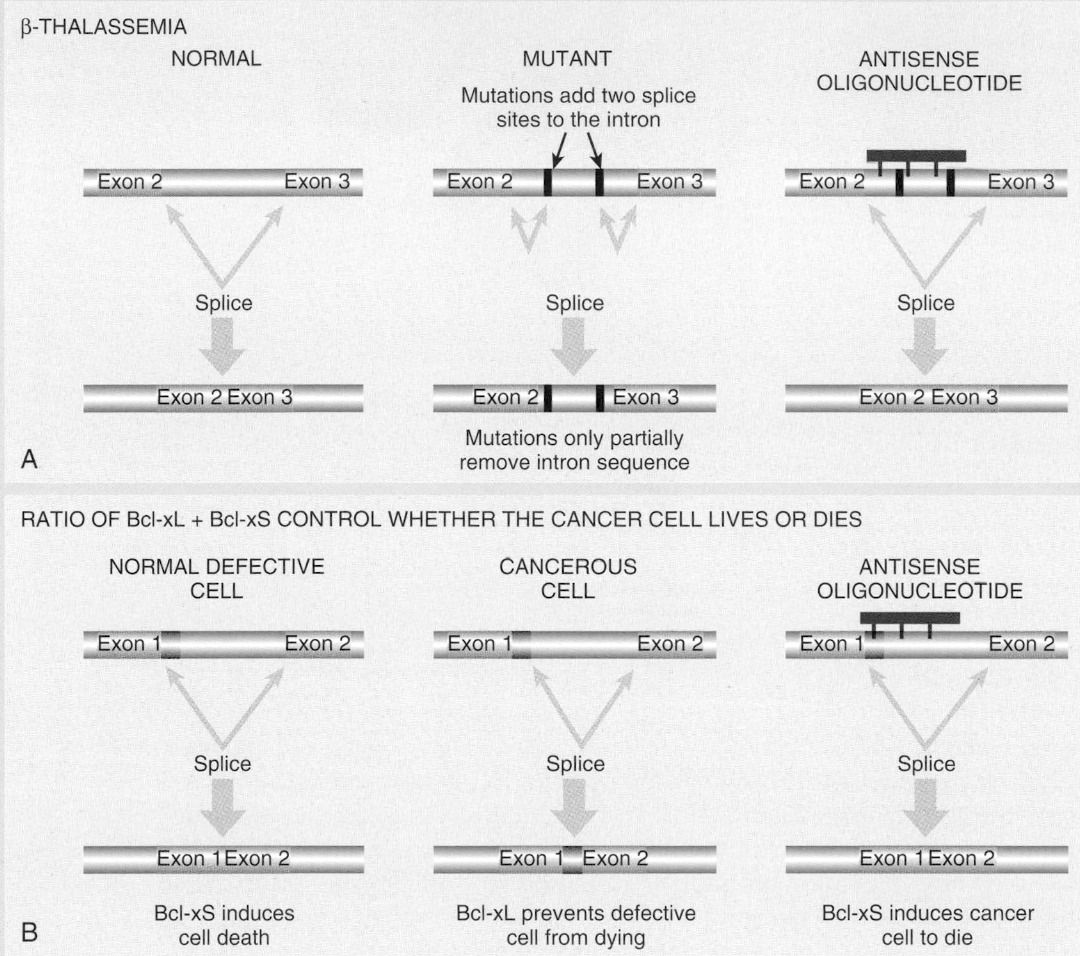

FIGURE A Antisense Oligonucleotides Correct Splicing Errors

(A) Beta-thalassemia is a blood disorder in which extra splice sites are found between exon 2 and exon 3 *(middle)*. Antisense oligonucleotides that block the extra splicing junctions restore the original structure of the gene during splicing *(right)*. (B) The *Bcl-x* gene makes two different proteins through alternate splicing. Bcl-xS is made when a normal cell becomes defective and promotes cell death via apoptosis (*left*). Some cancerous cells do not produce Bcl-xS. Instead, the longer form, Bcl-xL, is produced via alternate splicing and protects the cancerous cell from apoptosis *(middle)*. To resensitize the cancerous cell to apoptosis, antisense oligonucleotides that block the splicing junction for Bcl-xL restore Bcl-xS protein production and, ultimately, sensitivity to apoptosis *(right)*.

RIBOSWITCHES ARE CONTROLLED BY EFFECTOR MOLECULES

Examples of RNA that regulate gene expression have proliferated over the last decade. Such regulation may occur at a variety of levels, including both transcription and translation. **Riboswitches** provide a fascinating example in which the regulatory RNA is actually part of the RNA molecule whose expression is being regulated. Riboswitches are segments of RNA located on mRNA molecules close to the 5′ end. The riboswitch domain alternates between two different RNA secondary structures that determine whether or not the mRNA is expressed. Unlike most regulatory RNA, riboswitches bind small effector molecules, such as amino acids or other nutrients. Binding of the effector molecule triggers a conformation change in the riboswitch. In most cases, effector binding terminates mRNA transcription prematurely or prevents mRNA translation.

The vast majority of riboswitches are found in bacteria, mostly in genes for biosynthetic enzymes. For example, in *E. coli,* the thiamine riboswitch is controlled by thiamine pyrophosphate, a vitamin. When the vitamin is abundant, it binds to the TH1 box (i.e., a riboswitch) close to the 5′ end of the mRNA, and transcription of the mRNA is aborted. When the vitamin is absent, the mRNA is transcribed and translated to give enzymes that make more thiamine. Similar control occurs for riboflavin biosynthesis in *Bacillus subtilis.* The vitamin itself binds to the riboswitch domain of the mRNA and controls whether or not the mRNA is expressed.

Riboswitches normally work by changing the stem and loop structure of the mRNA transcript. In **attenuation riboswitches**, the effector molecule binds to the mRNA as it is being transcribed. If the effector binds, changes in structure create a terminator loop, which causes the transcriptional machinery to fall off prematurely. The incomplete mRNA is degraded. When the effector is in short supply, then the mRNA is transcribed to completion (Fig. 5.22A). Alternatively, some riboswitches work through translational inhibition. Here, the riboswitch controls whether or not protein translation occurs by sequestering the Shine–Dalgarno sequence. When the effector molecule is abundant, its binding changes the stem-loop structure so that the Shine–Dalgarno sequence is not accessible to the ribosomes (see Fig. 5.22B).

A novel riboswitch was identified in *Bacillus subtilis* that controls the expression of a biosynthetic gene *(glmS)* for a cell wall component (Fig. 5.23). As for other riboswitches, a product of the biosynthetic pathway controls whether or not the mRNA is expressed. However, instead of hiding the Shine–Dalgarno sequence or creating a terminator loop, the change in RNA secondary structure creates a self-cleaving ribozyme. The *glmS* gene of *B. subtilis* encodes the enzyme glutamine fructose 6-phosphate amidotransferase, which converts fructose 6-phosphate plus glutamine into glucosamine 6-phosphate (GlcN6P). This is further converted into a component of the cell wall, UDP-GlcNAc. When this is abundant, it binds to *glmS* mRNA, altering the secondary structure. The new structure functions as a ribozyme that cuts the mRNA, preventing any further translation.

Although the vast majority of riboswitches bind small molecules, a few are exceptional. The T box riboswitch binds tRNA, a rather large molecule. This controls expression of amino acid metabolism genes depending on whether the tRNA is charged or uncharged. Other riboswitches do not use an effector molecule at all. Instead, they respond directly to thermal stress. For example, the *rpoH* gene of *E. coli* is involved in the heat shock response. In addition to other forms of

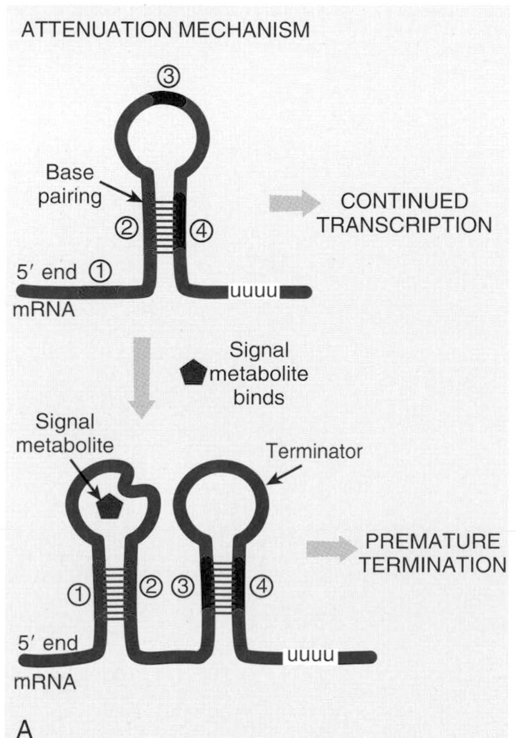

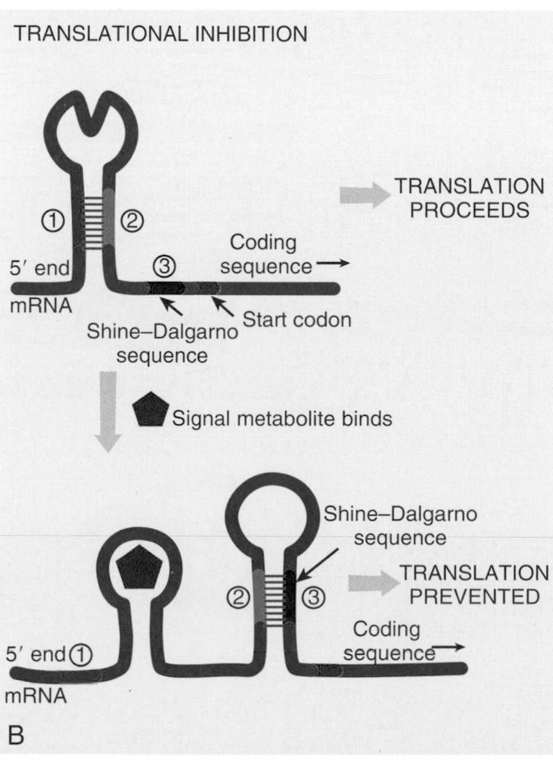

FIGURE 5.22 Riboswitches Control mRNA Expression
Riboswitches alternate between two stem and loop structures depending on the presence or absence of the signal metabolite. (A) In the attenuation mechanism, the presence of the signal metabolite results in formation of the terminator structure, and transcription is aborted. (B) In the translational inhibition mechanism, the presence of the metabolite results in sequestration of the Shine–Dalgarno sequence, which prevents translation of the mRNA.

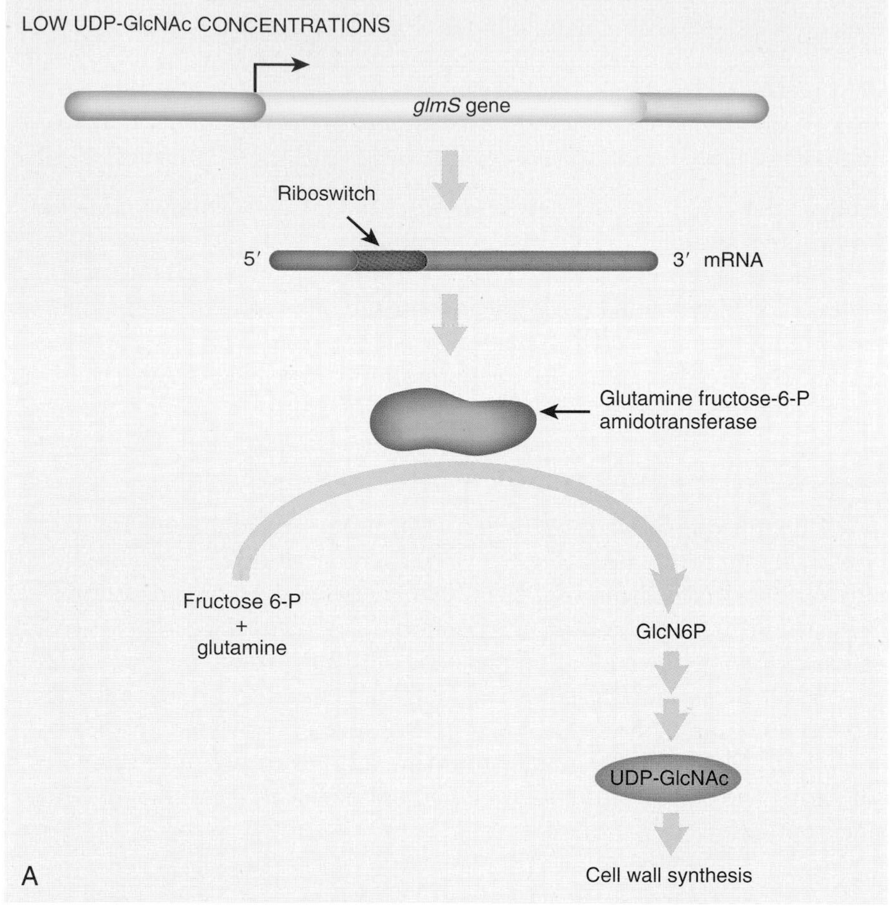

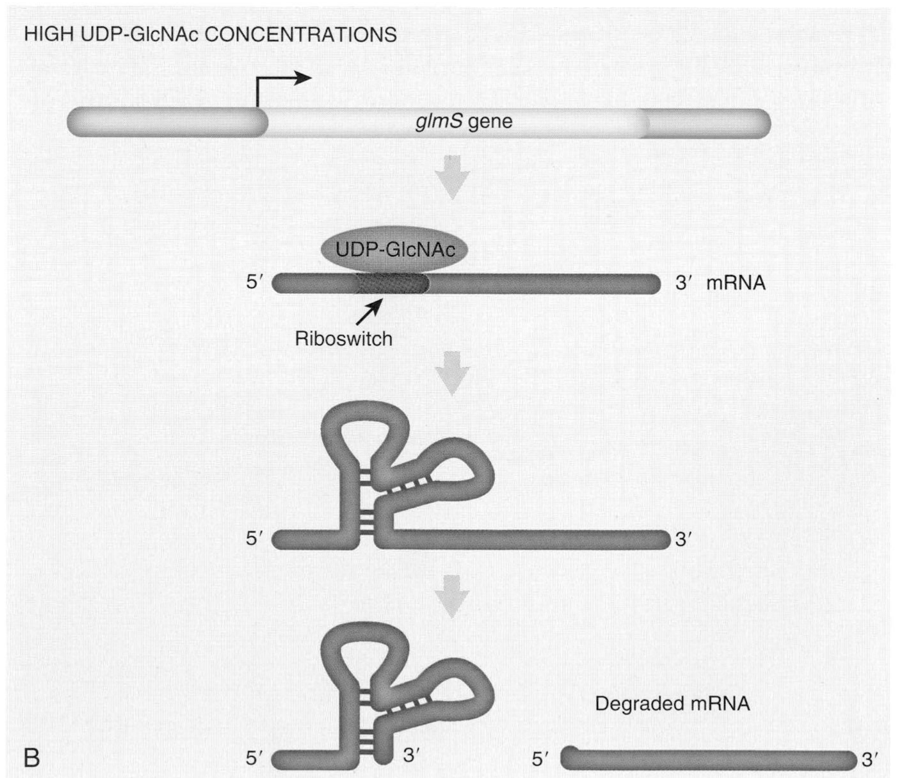

FIGURE 5.23 Ribozyme Riboswitch of *B. subtilis GlmS* Gene

(A) Cell wall synthesis occurs during growth conditions. When the cell is growing, levels of UDP-GlcNAc are low and are quickly converted into cell wall components. (B) If the cell is not growing, UDP-GlcNAc is not incorporated into the cell wall and accumulates. The excess UDP-GlcNAc binds to a riboswitch on the *glmS* gene. Once bound, it activates the self-cleaving ribozyme to degrade the mRNA and halts the production of glutamine fructose 6-phosphate amidotransferase.

regulation, the mRNA contains a thermosensor domain, which controls the amount of translation. At normal temperatures, the thermosensor has a stem-loop structure that prevents translation. When the heat increases, the stem-loop structure falls apart and translation can occur.

The only type of riboswitch found so far in eukaryotes is the thiamine riboswitch. Despite being homologous in sequence to the thiamine riboswitches of bacteria, it operates by a different mechanism and controls alternative splicing of the mRNA precursor. Only fungi and plants possess this thiamine riboswitch, but it is absent in animals.

Riboswitches are mRNA sequences that bind directly to effector molecules to control the expression of the mRNA into protein.

RNA CATALYZES ENZYME REACTIONS

Ribozymes are RNA molecules that bind to specific targets and catalyze enzymatic reactions. Some ribozymes consist of RNA associated with proteins, but the RNA catalyzes the actual reaction. Some ribozymes work like allosteric enzymes; that is, binding an effector molecule alters the ribozyme structure so that the ribozyme becomes competent to cleave its substrate. Ribozymes are naturally occurring, but biotechnology research has started to exploit their unique characteristics for medical and industrial applications.

There are eight known classes of ribozymes at present, with the distinct possibility that more will be identified. Ribozymes are classified as large or small. The large ones range from several hundred nucleotides to 3000 nucleotides in length. Large ribozymes were the first identified, and the first of these were the group I introns of *Tetrahymena.* These intron sequences are found in pre-mRNA that are able to self-splice. They do not use splicing factors such as snRNA (aka snurps). Group I introns are common in fungal and plant mitochondria, in nuclear rRNA genes, in chloroplast DNA, in viruses, and in the tRNA genes of chloroplasts and eubacteria. The important aspect of intron self-cleavage is the RNA structure. RNA is a linear polymer, but because of base pairing between different regions, RNA also has a secondary structure. Multiple stem-loop structures fold into different configurations leading to a three-dimensional structure much like a protein. The example shown is the second group I intron within the *orf142* gene of bacteriophage Twort, which infects *Staphyloccoccus aureus* (Fig. 5.24). The three-dimensional structure of group I introns brings the two exons close together, facilitating removal of the intron between them (Fig. 5.25).

Group II introns are also self-splicing sequences found within genes. They are less common than group I introns, being found only in fungal and plant mitochondria, in chloroplasts of plants and *Euglena,* in algae, and in eubacteria. These introns do not self-splice *in vitro* and require far from physiological conditions to work. The three-dimensional structure of the intron creates these abnormal conditions *in vivo,* affecting the microenvironment to create the correct ionic concentrations. The 3D structure of group II introns brings the two exons together, facilitating intron removal and exon ligation (Fig. 5.26). Interestingly, the structure of these introns is similar to the structure of snRNA, suggesting that group II introns may be evolutionary precursors to the snRNAs and the spliceosome.

Another naturally occurring large ribozyme is RNase P from bacteria. This is an RNA-protein complex, but the RNA component is the catalytic entity. RNase P cleaves the 5′ end of pre-tRNA molecules to remove the leader sequence. RNase P can act on multiple substrates, unlike the group I and group II introns that naturally act only on themselves.

Ribozymes are naturally occurring RNAs that can facilitate an enzymatic reaction. Group I and group II introns are two types of ribozymes that can cleave the phosphate backbone release themselves from the mRNA molecule and rejoin the ends without using any protein enzymes.

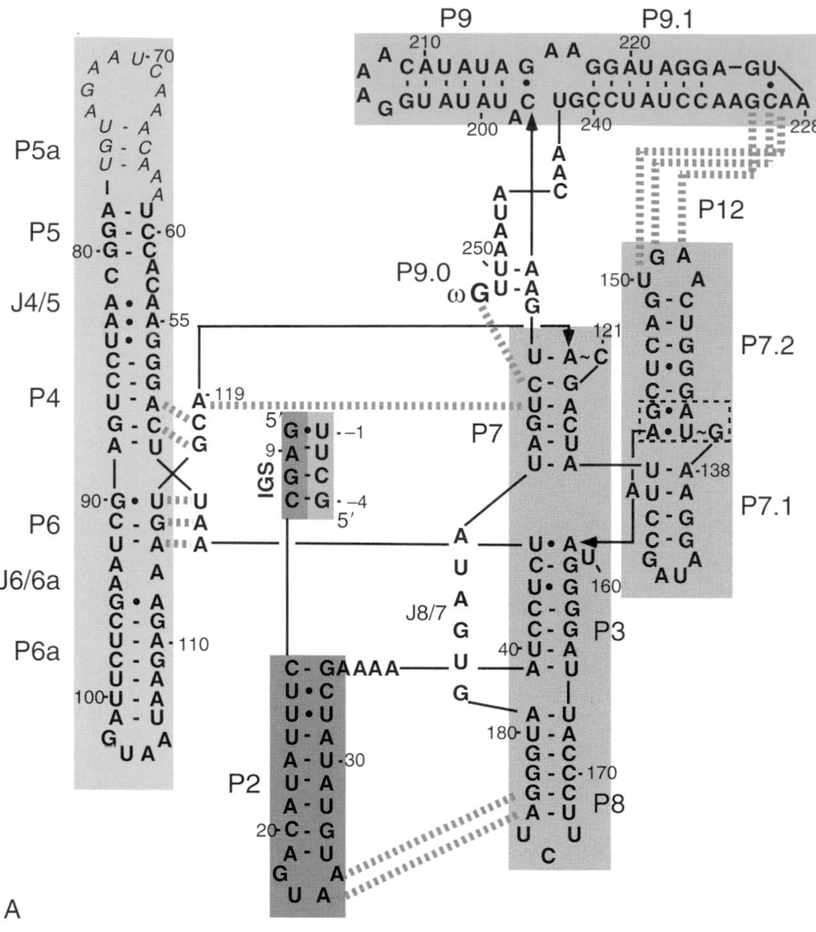

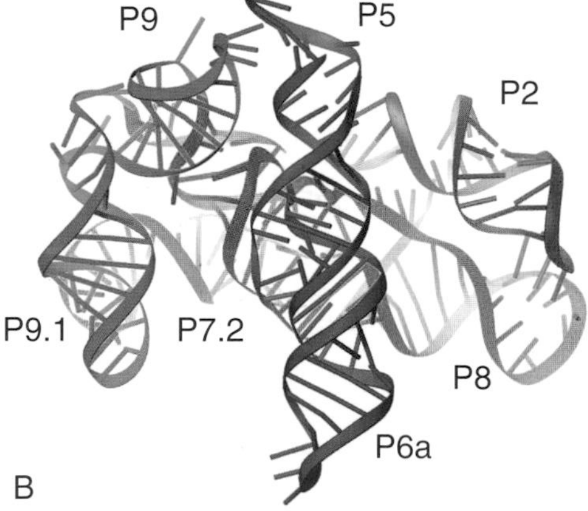

FIGURE 5.24
Structure of the Twort Ribozyme
(A) Primary and secondary structure of the wild-type intron. The P1-P2 domain is highlighted in red, the P3-P7 region is green, the P4-P6 domain is blue, the P9-P9.1 region is purple, the P7.1-P7.2 subdomain is yellow, and the product oligonucleotide is cyan. Dashed lines indicate key tertiary structure contacts. Nucleotides in italics (P5a region) are disordered in the crystal. IGS, internal guide sequence. (B) Ribbon diagram colored as in (A). The backbone ribbon is drawn through the phosphate positions in the backbone. From Golden BL, Kim H, Chase E (2004). Crystal structure of a phage Twort group I ribozyme-product complex. *Nat Struct Mol Biol* **12,** 82–89, Reprinted by permission from: Macmillan Publishers Ltd., copyright 2005.

SMALL NATURALLY OCCURRING RIBOZYMES

In contrast to the large ribozymes, small ribozymes are only about 30 to 80 nucleotides long. Small ribozymes include hammerhead and hairpin ribozymes, hepatitis delta virus ribozyme, Varkud satellite, and twister ribozyme. They are often found in viroids, virusoids, and satellite viruses, which are **subviral agents**. Viroids are self-replicating pathogens of plants that are merely naked single strands of RNA with no protein coat. Satellite viruses are small RNA molecules that require a helper virus for either replication or capsid formation. Their genomes may encode proteins. Virusoids are even less functional and are often considered a subtype of satellite virus. Virusoids are single strands of circular RNA that encode no proteins. They rely on helper viruses for both replication and a protein coat.

The **hammerhead ribozyme** is a small catalytic RNA that can catalyze a self-cleavage reaction. Hammerhead ribozymes take part in the replication of some viroids and satellite RNAs (Fig. 5.27). These both exist as single-stranded RNA genomes that form rod-like structures that are resistant to cellular ribonucleases. During viroid replication, the positive RNA strand is replicated by the host cell RNA polymerase, resulting in a long concatemer of negative-strand genomes. RNA polymerase then uses this as a template to make a positive strand. The long RNA is cut into individual unit genomes by the hammerhead motif. Hammerhead ribozymes first cleave the ribose phosphate backbone of RNA and then ligate the linear unit genomes into circular genomes.

Another small ribozyme is the **hairpin ribozyme** (Fig. 5.28). It is found in pathogenic plant satellite viruses such as tobacco ring spot virus. The hairpin ribozyme from tobacco ring spot virus was originally called the "paperclip" ribozyme, a rather better description of the structure. *In vivo,* hairpin ribozymes cleave the linear concatemers of ssRNA genomes, much like hammerhead ribozymes, and then ligate the linear segments into circular genomes.

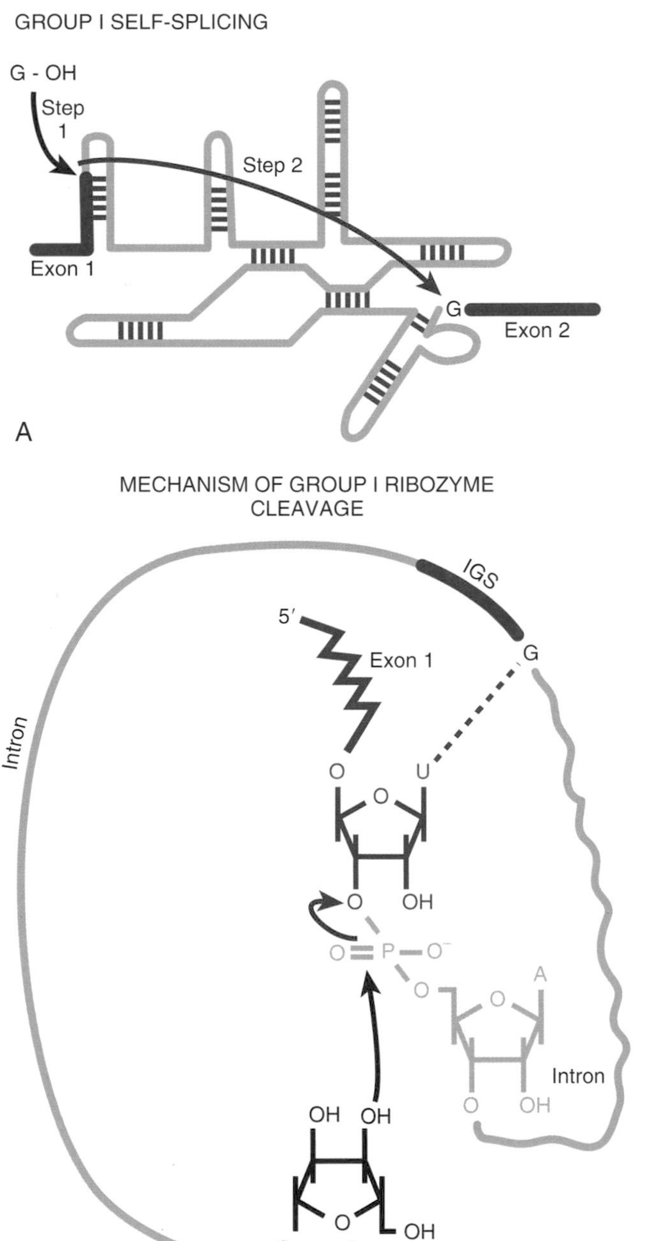

FIGURE 5.25 Mechanism of Group I Self-Splicing Reaction
(A) The secondary structure of group I introns shows multiple hairpins that mediate the cleavage reaction. In step 1, a free guanosine (Red, G-OH) mediates the cleavage of the exon 1–intron boundary. In step 2, the free end of exon 1 cleaves and ligates to exon 2. (B) Mechanism of group I ribozyme cleavage. First, the exon sequences are brought near the catalytic core via the internal guide sequence (IGS). Exon 1 has an important uridine (U) that forms a U=G base pair with the IGS (dotted line). The other end of the ribozyme has a binding site for the nucleophile, a free guanosine (red), which initiates intron removal by attacking the end of exon 1 with the 3′-OH of its ribose. The free 3′-OH on the exon than reacts with the splice site on exon 2. The intron is spliced out, and the two exons are united (not shown). Although it appears that this reaction requires energy, the actual number of bonds stays the same, and no net energy is needed.

Two other small ribozymes are the **Varkud satellite (VS) ribozyme** from *Neurospora* and the **hepatitis delta virus (HDV) ribozyme** of humans. Both use similar reaction mechanisms for self-cleavage and ligation. The VS ribozyme helps replicate the small Varkud plasmid found within the mitochondria of *Neurospora.* HDV is a viroid-like satellite virus of hepatitis B virus. Hepatitis B infects the liver and can cause liver scarring and liver failure. In patients with hepatitis B, the presence of HDV amplifies the symptoms, causing a very severe and often fatal form of the disease. HDV has a single-stranded RNA genome that occurs in both positive (genomic) and negative (antigenomic) forms in liver cells. Both forms have regions that fold into an active ribozyme, which catalyzes RNA cleavage and ligation. Unlike plant viroids, HDV also has an open reading frame encoding the *delta antigen* protein. Delta antigen plus coat proteins from hepatitis B virus are needed to package HDV into small spherical particles. These particles can be spread from cell to cell, and from person to person via bodily fluids such as saliva and semen.

Small naturally occurring ribozymes are found in small subviral agents such as viroids and satellite viruses. They have common motifs that catalyze RNA cleavage.

ENGINEERING RIBOZYMES FOR PRACTICAL APPLICATIONS

Ribozymes can be engineered to suppress the expression of genes, such as those that promote cancer or those from pathogenic viruses. The ribozyme catalytic core is linked to a sequence that recognizes the target gene mRNA, usually an antisense probe, thus combining the two strategies (Fig. 5.29). The target region must be free of secondary structure and have no protein-binding sites. The antisense sequence is split so that the 5′ half is in front of the ribozyme catalytic core and the 3′ half is behind. When this chimeric ribozyme is mixed with target mRNA, the antisense regions base-pair with the target and the ribozyme cleaves the target mRNA. The two halves of the target mRNA are further degraded by other enzymes.

The engineered ribozyme can attack many target mRNA molecules because it is an enzyme, not merely an inhibitor. Using a ribozyme is much better than using antisense

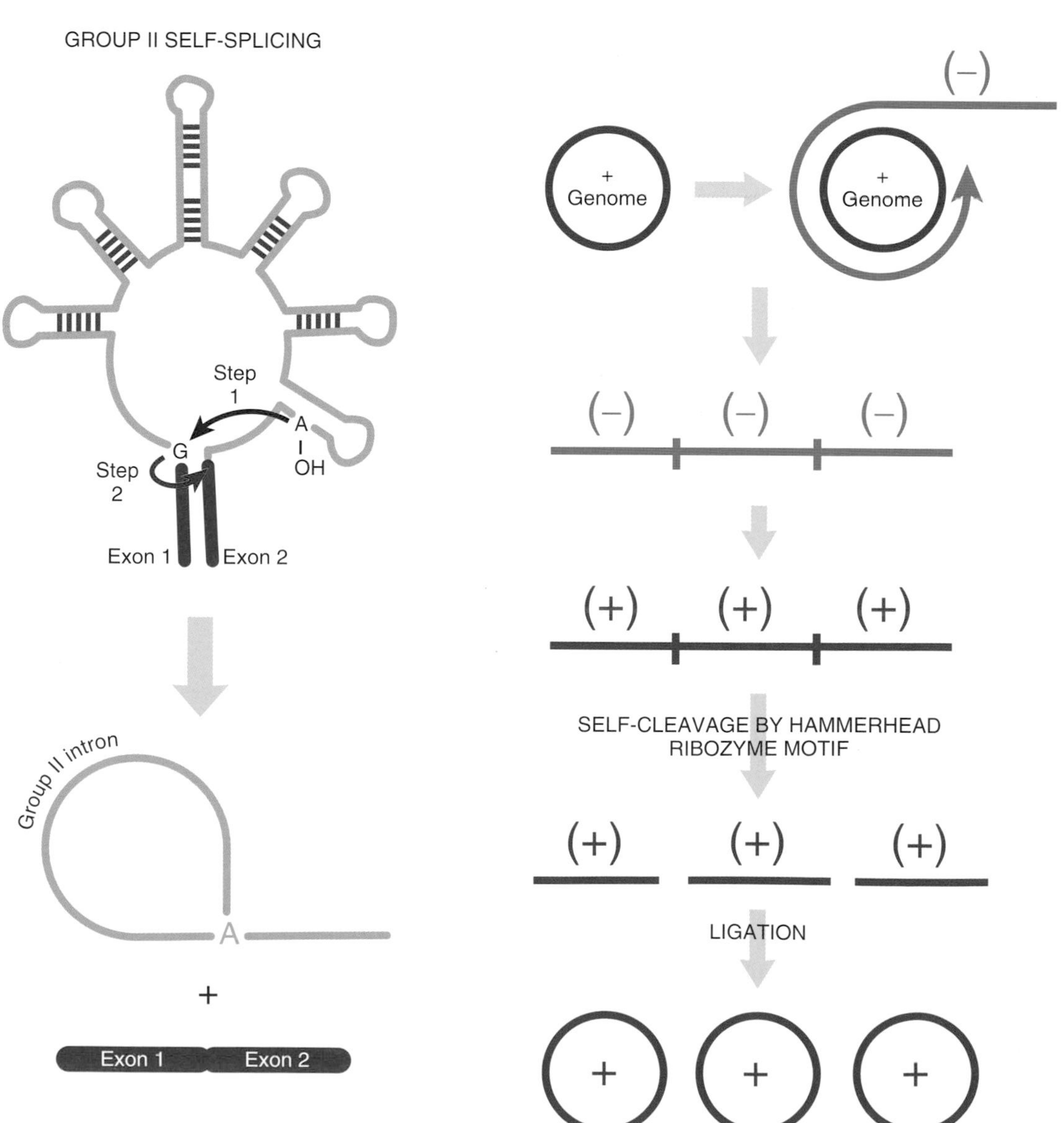

FIGURE 5.26 Group II Intron Splicing Reactions
The secondary and tertiary structure of group II introns brings the two exons together, but the reaction mechanism does not require an external nucleophile. Instead, the 2′-OH of an internal conserved adenine acts as a nucleophile, attacking the 5′ splice site and cleaving the phosphate backbone. The 3′-OH of the 5′ splice site attacks the 3′ splice site, resulting in two ligated exons and a free intron. The intron forms a lariat structure.

FIGURE 5.27 Life Cycle of Viroids
Viroids are single-stranded circular RNA genomes with no protein coat, but they have the ability to self-replicate. First, the plus-stranded genome is converted into a concatemer of negative-stranded genomes with rolling-circle replication. RNA polymerase converts the negative-stranded genomes into plus-stranded genomes, which are separated and ligated into circular genomes. The hammerhead ribozyme, embedded within the viroid genome, catalyzes the separation and ligation into a circle.

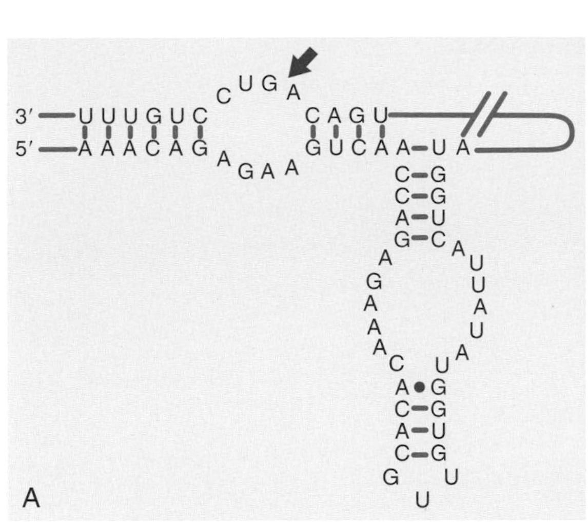

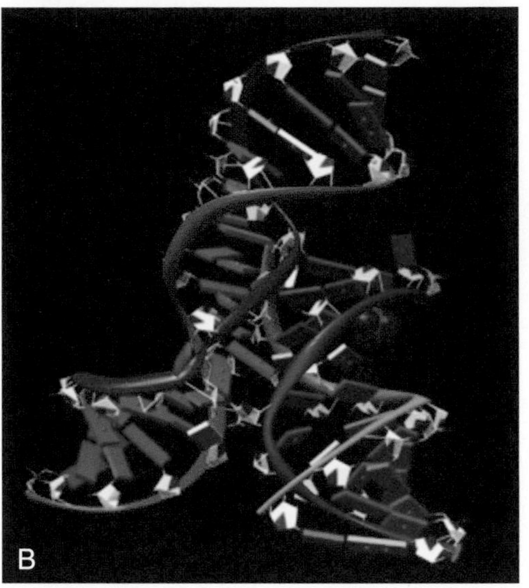

FIGURE 5.28 Secondary Structure of Hairpin Ribozyme
(A) The minus strand of the tobacco ring spot virus genome is shown with the cleavage site indicated by the red arrow. (B) The three-dimensional representation of the hairpin ribozyme. Reproduced from Salter et al., 2006.

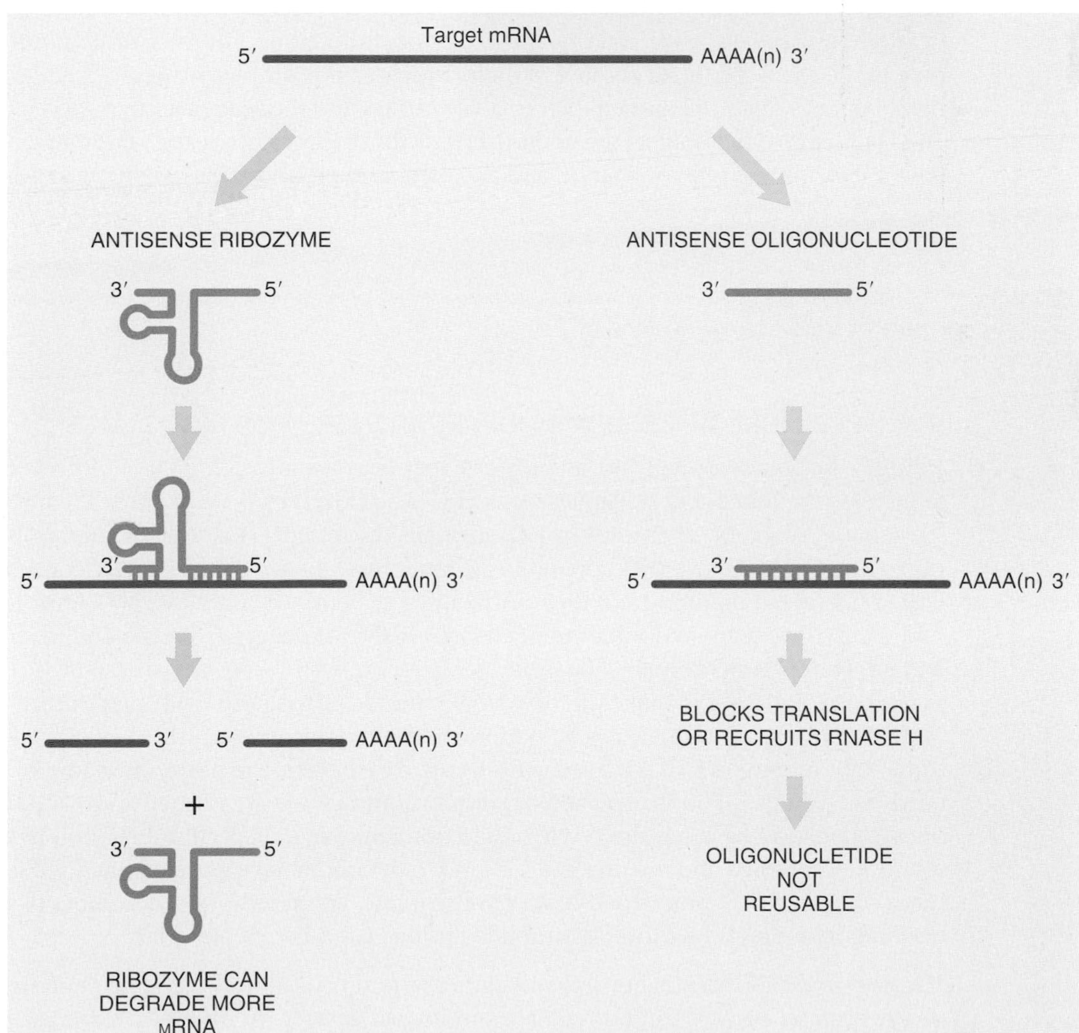

FIGURE 5.29 Antisense Construct with Ribozyme Inactivates Target mRNA
The chimeric antisense ribozyme not only has the ability to bind to a specific target mRNA, but also cleaves the target mRNA. A traditional antisense oligonucleotide must rely on recruiting RNase H to digest the target mRNA. However, RNase H also degrades the antisense oligonucleotide, which cannot therefore be reused. In contrast, antisense ribozymes are not cleaved or degraded; therefore, they can continue to catalyze degradation of target sequences.

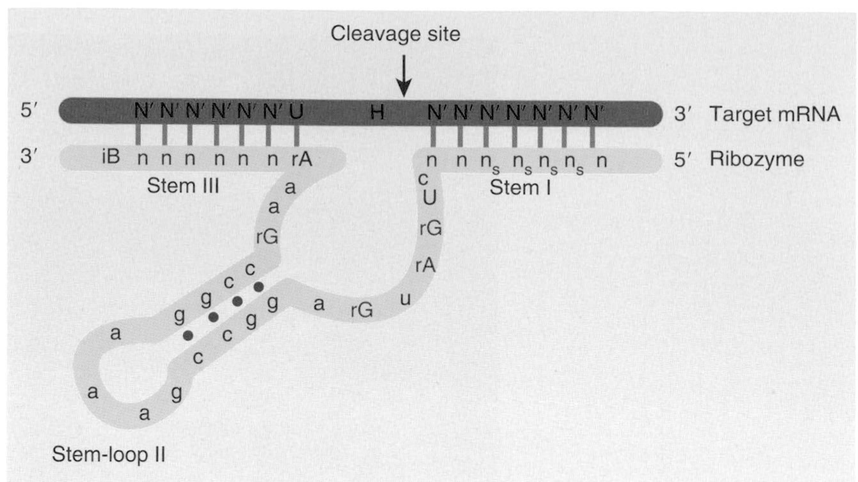

FIGURE 5.30 Nuclease-Resistant Ribozyme Bound to a Target mRNA
The ribozyme consists of 2′-*O*-methyl nucleotides and phosphorothioate linkages. At the 3′ end a 3′-3′ deoxyabasic sugar (iB) is added. All three modifications prevent nuclease degradation of the ribozyme. The five green nucleotides (rA or rG) form the catalytic core and cut the target mRNA at the cleavage site. The H at the cleavage site represents an A, C, or U.

inhibition alone because antisense constructs are degraded along with the target mRNA (see Fig. 5.29).

The catalytic core used for the constructs just described is usually from either hairpin or hammerhead ribozymes (Fig. 5.30). Altering the ribozymes from group I introns, from group II introns, or from RNase P is difficult because of their large size and complex structure. Small ribozymes have a natural division between their catalytic centers and the sequences that specify their target. Thus, it is easy to manipulate the intended target for ribozyme to cleave. Hammerhead ribozymes have a lower propensity for ligation and are often used preferentially over hairpin ribozymes.

Adding a ribozyme motif such as the hammerhead or hairpin region from the small ribozymes can make antisense oligonucleotides more stable because they are not degraded. These constructs cut the target mRNA without the use of RNase H.

RNA SELEX IDENTIFIES NEW BINDING PARTNERS FOR RIBOZYMES

Natural ribozymes normally act on only one specific substrate. One goal of biotechnology is to increase the number of substrates for the known ribozymes. A procedure called **RNA SELEX** (Systematic Evolution of Ligands by EXponential enrichment) isolates new substrates for existing ribozymes from a large (10^{15}) population of random-sequence RNA oligonucleotides (Fig. 5.31). First, a mixture of random DNA oligonucleotides is chemically synthesized. These oligonucleotides are converted into double-stranded DNA (dsDNA) using a 5′ primer and Klenow polymerase. The 5′ primer contains the promoter sequence for T7 RNA polymerase, which is added to the pool of dsDNA to make multiple single-stranded RNA (ssRNA) copies of each oligonucleotide. The ribozyme of interest is then mixed with this large pool of ssRNA oligonucleotides, and those RNA molecules that bind to the ribozyme are isolated. The ribozyme is immobilized on beads to facilitate isolation. Any nonspecifically bound RNAs are washed away, and the specific ones are isolated. Each repeated cycle of selection removes nonspecifically bound RNAs. After the selection is repeated and the final RNA bound to the ribozyme is purified, the RNA is converted into cDNA using a 3′ primer and reverse transcriptase. Because the actual number of specific binding molecules is low, they are amplified using PCR before sequencing.

The use of SELEX extends beyond ribozymes, and it is used in drug design and delivery. The process can be applied to finding DNA binding substrates for different enzymes. In **DNA SELEX**, the initial pool of random-sequence oligonucleotides is not converted to RNA. Instead, the oligonucleotides are used directly in substrate binding and selection.

New substrates for a known ribozyme are found by incubating the pure ribozyme with a large pool of random RNA sequences. Any RNA sequence that binds to the ribozyme is a potential substrate.

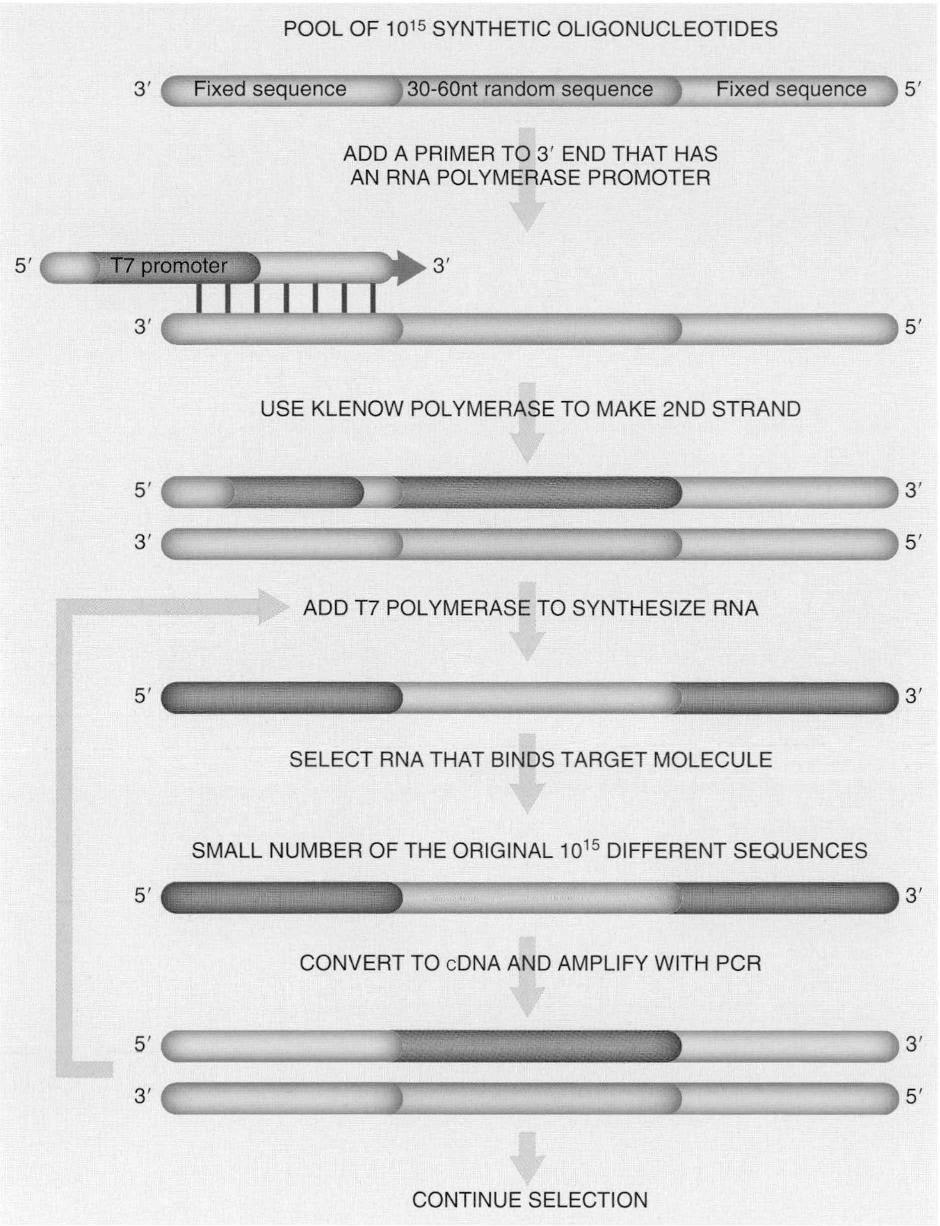

POOL OF 10^{15} SYNTHETIC OLIGONUCLEOTIDES
3′ Fixed sequence 30-60nt random sequence Fixed sequence 5′
ADD A PRIMER TO 3′ END THAT HAS AN RNA POLYMERASE PROMOTER
5′ T7 promoter 3′
3′ 5′
USE KLENOW POLYMERASE TO MAKE 2ND STRAND
5′ 3′
3′ 5′
ADD T7 POLYMERASE TO SYNTHESIZE RNA
5′ 3′
SELECT RNA THAT BINDS TARGET MOLECULE
SMALL NUMBER OF THE ORIGINAL 10^{15} DIFFERENT SEQUENCES
5′ 3′
CONVERT TO cDNA AND AMPLIFY WITH PCR
5′ 3′
3′ 5′
CONTINUE SELECTION

FIGURE 5.31 RNA SELEX Identifies New Ribozyme Substrates
The key to RNA SELEX is using a very large pool of random RNA sequences. First, DNA oligonucleotides are chemically synthesized to create a large pool of random sequences. These are converted into dsDNA with a primer and Klenow polymerase. The primer adds an RNA polymerase binding site. The dsDNA oligonucleotides are transcribed into RNA with RNA polymerase. This large pool of random RNA is then screened for binding to the ribozyme. Any RNA molecules that bind are kept, and the rest are discarded. Those that bind are converted to cDNA and amplified with PCR. The selection process is repeated numerous times to enrich for RNA sequences that bind more tightly to the ribozyme.

IN VITRO EVOLUTION AND *IN VITRO* SELECTION OF RIBOZYMES

It is also possible to generate new ribozymes with novel enzymatic capabilities from large pools of random RNA sequences. Using *in vitro* selection allows new ribozyme reactions to be identified from random nucleotide sequences (Fig. 5.32).

For example, a ribozyme that catalyzes the ligation of a particular sequence can be identified. This approach begins by synthesizing a set of random oligonucleotide sequences. However, these represent the pool of potential ribozymes rather than substrates as seen in RNA SELEX. Each random sequence is flanked by two known sequences. The 5′ end sequence is one substrate for the desired ligation reaction. The 5′ end also has a terminal triphosphate to energize ligation. The 3′ end has a sequence domain that binds a chosen effector molecule, which allows the ligation reaction to be regulated.

The second substrate for ligation is mixed with the pool of potential ribozymes and incubated in conditions that favor ligation. If one of the random RNA sequences ligates the second substrate to its 5′ end, the resulting RNA molecule (i.e., ribozyme plus

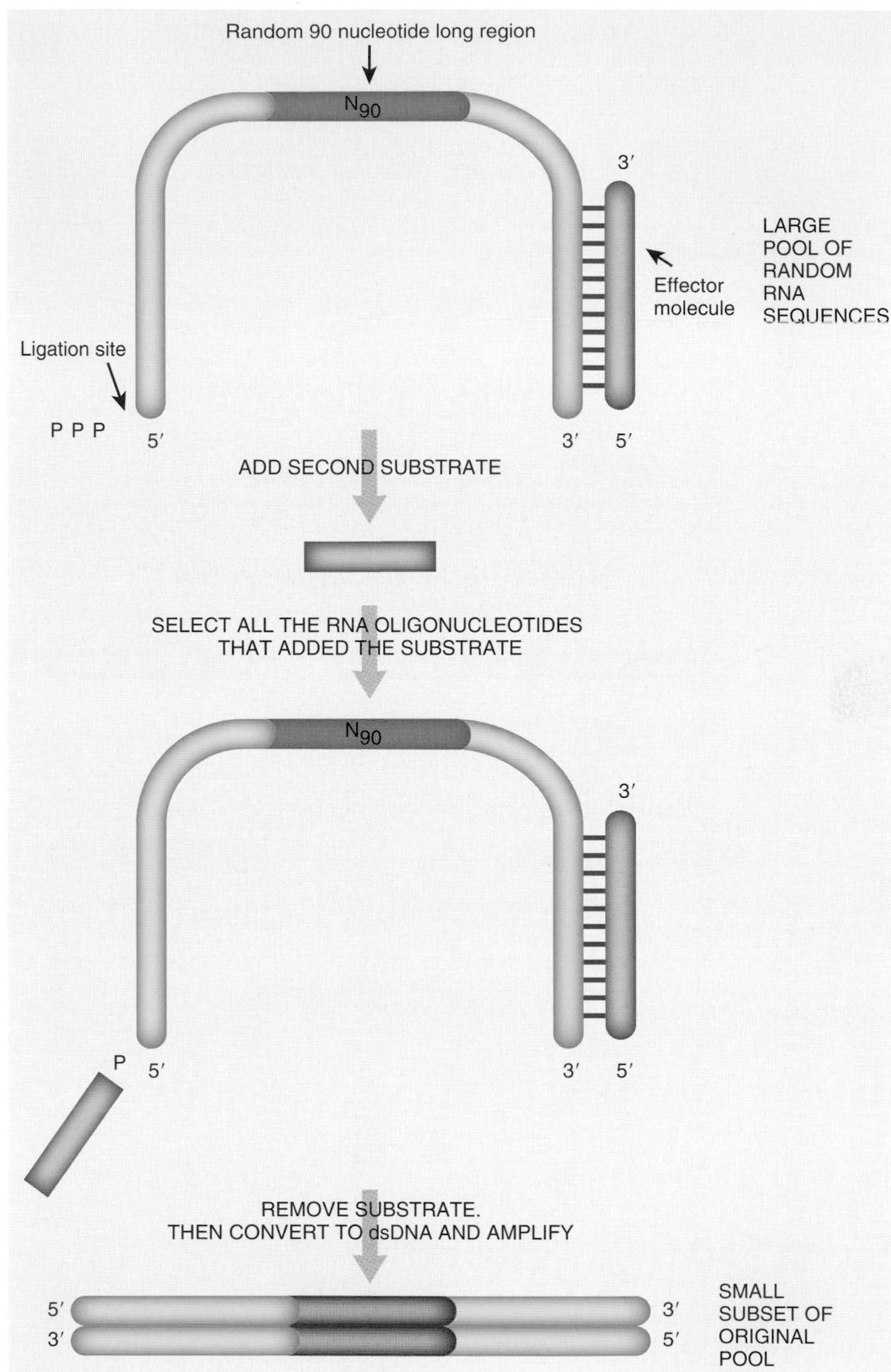

FIGURE 5.32 *In Vitro* Selection of Ribozyme Ligation
The pale pink molecule represents the large pool of random RNA sequences. At the 5′ end, there is a substrate sequence with a terminal triphosphate. At the 3′ end, a blue effector molecule is bound to facilitate ligation. The second substrate (purple) is then incubated with the random pool of oligonucleotides. If any of the random sequences catalyze the ligation of the substrates, the resulting species will be larger and may be separated out by gel electrophoresis. The ligated oligonucleotide is isolated from the gel, amplified by PCR, and finally sequenced.

ligation product) will run more slowly on an agarose gel. The slower molecules are isolated from the gel. The ribozyme suspect is then converted to DNA with reverse transcriptase. Finally, the DNA is amplified with PCR using primers that match the 5′ and 3′ ends of the original RNA constructs.

In vitro evolution enhances *in vitro* selection by adding a mutagenesis step after each cycle of selection (Fig. 5.33). This method begins with a pool of random oligonucleotides, as before. The pool of random sequences is then mutagenized. The most efficient method is to use error-prone PCR (see Chapter 4) to amplify the initial pool of sequences. The pool both becomes larger and the sequences diversify even more. This

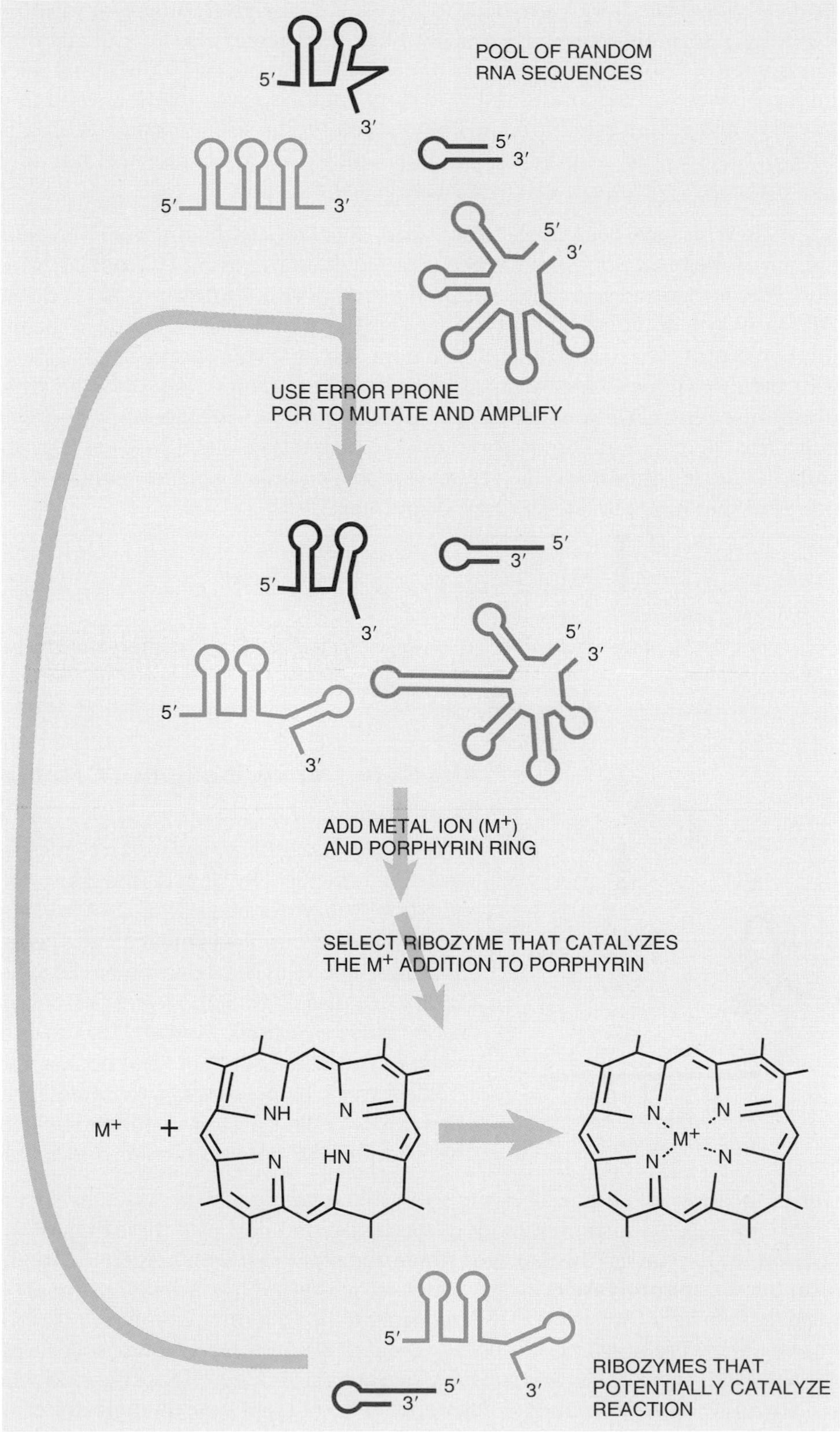

FIGURE 5.33 *In Vitro* Evolution of Ribozymes
In vitro evolution tries to find an RNA sequence that works as a ribozyme. In this example, the researcher is looking for a ribozyme that catalyzes the addition of a metal ion (M^+) to a porphyrin ring. The pool of random RNA sequences is created and amplified with error-prone PCR to increase the odds of finding one or two sequences that catalyze the reaction. Each successive round of selection and mutation improves any ribozymes that are found.

pool is then selected for the sequence that carries out the desired reaction. For example, artificial ribozymes have been evolved to add metal ions to mesoporphyrin IX (see Fig. 5.33). The mutagenesis and selection steps can be repeated over and over to improve the ribozyme. Once an efficient ribozyme is obtained, the sequence is determined after converting the RNA into cDNA.

Artificial ribozymes have been made to carry out nucleophilic attacks at various centers, including phosphoryl, carbonyl, and alkyl halides. There is also an artificial ribozyme that can isomerize a 10-member ring structure. In each of these cases, the initial pool of RNA molecules was selected for the ability to carry out the specific reaction. In both *in vitro* selection and *in vitro* evolution, the key to success is the selection step. It must be stringent enough that most of the nonfunctional RNA molecules are eliminated, but not so stringent that ribozymes with weak activity are eliminated too early.

A few attempts have been made to construct ribozymes with clinical applications. For example, a hammerhead ribozyme has been created that can inhibit HIV uptake into cells. This ribozyme recognized seven sites within the human CCR5 mRNA. CCR5 is the coreceptor for HIV entry into immune cells but is not essential for humans, and, indeed, people without this coreceptor are resistant to HIV infection. The ribozyme was targeted against a human gene because HIV is so highly mutable that it might become resistant to any ribozyme. The ribozyme effectively eliminates CCR5 expression *in vitro*. Another ribozyme, which does target the HIV genome, has also been used to treat HIV-positive patients. The ribozyme is added to CD34+ immune cells (HIV target cells) and then given to patients. However, so far no ribozymes have provided a survival advantage.

In vitro selection can also generate new ribozymes by mixing random sequences that represent potential ribozymes with a specific substrate.

Adding a mutagenesis step to the *in vitro* selection procedure allows the ribozyme to "evolve" into a better enzyme.

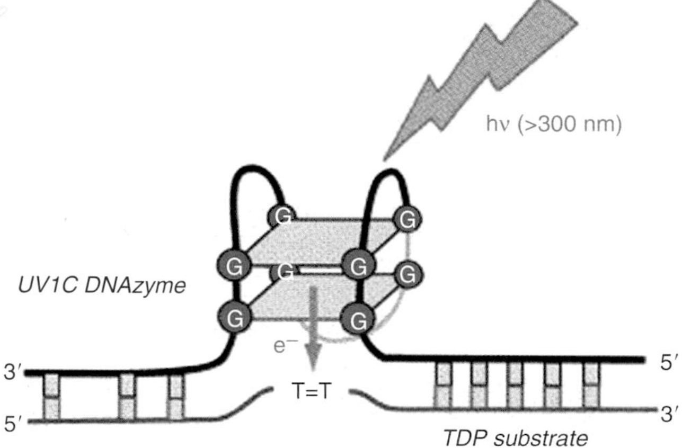

FIGURE 5.34
Deoxyribozyme That Repairs Thymine Dimers
A model for the deoxyribozyme UV1C–substrate complex. Light energy is absorbed by the guanine quadruplex. The thymine dimer is thought to lie close to the guanine cluster within the folded deoxyribozyme. This allows electron flow from the excited guanines to the thymine dimer. From Chinnapen DJ, Sen D (2007). Towards elucidation of the mechanism of UV1C, a deoxyribozyme with photolyase activity. *J Mol Biol* **365,** 1326–1336. Reprinted with permission.

Allosteric Deoxyribozymes Catalyze Specific Reactions

Because RNA may display catalytic properties, researchers investigated whether DNA can do the same. Although no natural DNA enzymes are known, DNA nonetheless can catalyze various reactions in a manner analogous to RNA-based ribozymes. Indeed, *in vitro* selection has been used to create a variety of artificial deoxyribozymes or DNAzymes that catalyze various reactions. Most DNAzymes catalyze reactions involved in processing RNA or DNA because they are easiest for SELEX type schemes to select. Examples include RNA cleavage, DNA cleavage, DNA depurination, RNA ligation, DNA phosphorylation, and thymine dimer cleavage.

One of the most interesting DNAzymes can split thymine dimers caused by UV radiation of DNA. Different organisms have various mechanisms to deal with these dimers. For example, excision repair removes the damaged strand and replaces it with new DNA. Another mechanism involves **photolyase** enzymes, which recognize and repair thymine dimers when activated by blue light. To isolate a DNA sequence to perform the photolyase reaction, scientists carried out *in vitro* selection on a pool of random DNA oligonucleotide sequences. The random sequences were first linked to a substrate that consisted of two DNA oligonucleotides joined via a thymine dimer. If a random DNA oligonucleotide split the thymine dimer after exposure to blue light, then the overall length of the DNA construct would be smaller. The smaller species were isolated by gel electrophoresis. This experiment was successful, and a specific DNAzyme (UV1C) that could catalyze a photolyase reaction was identified (Fig. 5.34).

Deoxyribozymes are DNA sequences that catalyze an enzymatic reaction. All are artificial.

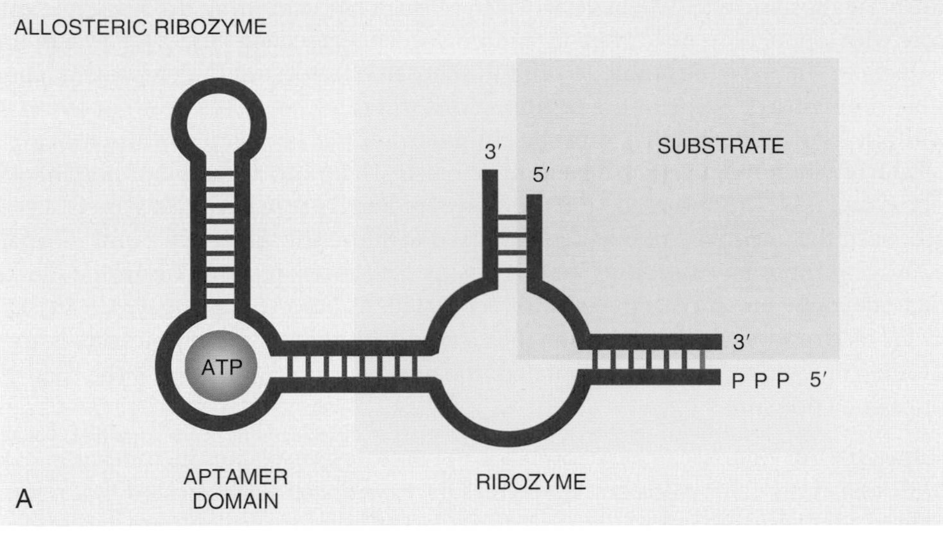

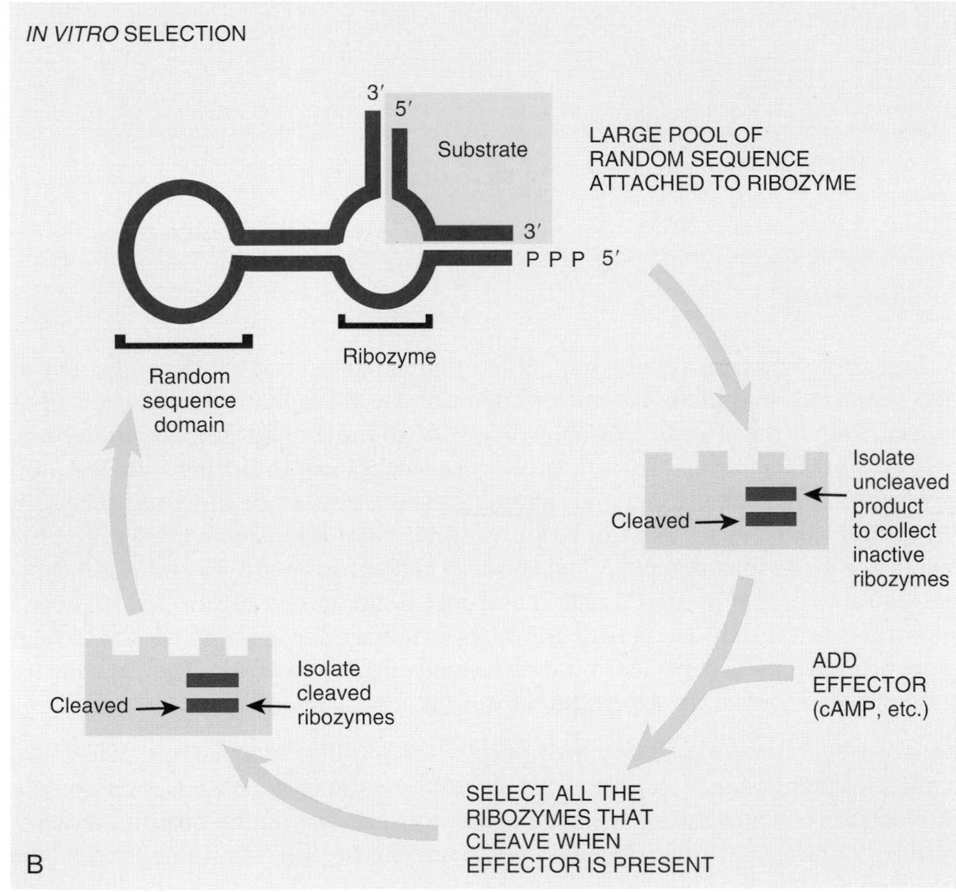

FIGURE 5.35 Designing Allosteric Ribozymes

(A) Modular design of a ribozyme. The ribozyme has three different domains joined together. The substrate domain (light green background) base-pairs with the ribozyme domain (light purple background), and the aptamer domain binds the allosteric effector (ATP in this example). (B) *In vitro* selection scheme to identify ribozymes that are active only when bound to an effector (i.e., are allosteric). First, all ribozymes that cleave substrate in the absence of an effector are removed. If the substrate is cleaved without the effector, the ribozyme will move faster during electrophoresis. Only the uncleaved ribozyme/substrate band is isolated from the gel. Next, the ribozymes are mixed with an effector molecule. This time, the ribozymes that cleave the substrate are isolated. Repeated cycles of isolation will identify a ribozyme that works only with the effector.

Engineering Allosteric Riboswitches and Ribozymes

Artificial or modified ribozymes have enormous potential in medicine and biotechnology. Consequently, the ability to control the activity of a ribozyme would be very advantageous. Ribozymes can be combined with riboswitches to achieve control by using the small effector molecule that triggers the riboswitch. To engineer a ribozyme to cleave only in the presence of a certain effector molecule, scientists use a combination of **modular design** and *in vitro* selection. Modular design takes various domains from different ribozymes and merges them to create a new molecule. For example, the catalytic core of a particular hammerhead ribozyme can be genetically linked to the binding domain of another, changing the binding specificity of the original ribozyme (Fig. 5.35A).

Artificial allosteric riboswitches have been selected by combining the ribozyme catalytic core with a pool of many different random sequences (see Fig. 5.35B). Some of the random sequences will have the ability to bind the chosen effector and thus represent a pool of possible riboswitches. Some of the combinations will catalyze self-cleavage or substrate cleavage without regulation, and they must be eliminated. If the ribozyme construct cleaves itself, the products will move faster during electrophoresis. Therefore, the pool of possible riboswitch/ribozymes is electrophoresed, and the slower moving, uncleaved RNAs are isolated from the gel. Next, the uncleaved ribozymes are mixed with the chosen effector and incubated under cleavage-promoting conditions. In this positive selection step, any ribozyme that undergoes cleavage in the presence of the effector is isolated. As before, the ribozymes are separated by gel electrophoresis, but this time the cleaved (shorter and faster) molecules are isolated. Cloning and sequencing of the isolated ribozyme constructs determine the sequence of the riboswitch domain.

Some effectors that researchers have used to control riboswitches include cyclic GMP, cyclic AMP, and cyclic CMP. Allosteric ribozymes have been artificially created that respond not only to small organic molecules such as cyclic AMP, but also to oligonucleotides, proteins, and even metal ions.

> Ribozymes can be created with riboswitches. The riboswitch controls the ribozyme so that it is active only when the effector molecule is present.

Summary

Although RNA was once thought of as an intermediary in the transfer of genetic information from DNA to protein, it is now recognized that RNA plays a wide variety of other roles. Indeed, RNA is the most functionally diverse of all the biological macromolecules. RNA helps maintain genomic structure, protects genomes from invading viral DNA, modulates transcription and translation, and can even perform enzymatic functions. The roles of tRNA, rRNA, and mRNA in protein transcription and translation are well known. The roles of small regulatory RNAs such as snRNA and snoRNA in processing mRNA and rRNA, respectively, are critical to proper cellular function and are still under investigation as to precisely how, when, and where they act. Translation is also controlled by riboswitches, which are RNA sequences that alter shape after binding a small effector molecule. They may be used in biotechnology to control the expression of various genetic constructs.

Antisense RNAs modulate gene transcription in many different organisms. In fact, a large fraction of protein coding genes have antisense genes found either in *cis* or *trans*. These complementary RNA sequences match the mRNA, bind to the sequence, and inhibit protein translation. The double-stranded RNA cannot be converted to proteins because ribosomes cannot bind. In addition, double-stranded RNA triggers enzymes to degrade the duplex, which obliterates the target mRNA. Other antisense RNAs suppress transcription by converting the gene to heterochromatin. The application of antisense in the laboratory is a natural evolution of this knowledge. Altering the phosphate-sugar backbone is essential to stabilize artificial antisense oligonucleotides. Efficient uptake of oligonucleotides is a major problem, and several approaches have been used.

In RNA interference (RNAi), double-stranded RNA (dsRNA) activates Dicer to cleave the RNA into segments of 21 to 23 nucleotides known small-interfering RNA, or siRNA. The RISC enzyme complex unwinds the double-stranded siRNAs and uses the single-stranded RNA as a template to find similar sequences. When RISC finds complementary sequences, it cleaves them. This destroys both foreign mRNA and the dsRNA characteristic of RNA virus replication. A similar process is used for developmental gene regulation. In this case, the original dsRNA is transcribed from the genome as pri-miRNA that is then processed into microRNAs. RNAi is now widely used in biotechnology and has largely replaced the use of other RNA-based approaches. RNAi may be used to eliminate one protein from the cell at a time to investigate its role. This approach provides insight into proteins whose function is still unknown.

Ribozymes are RNA molecules with enzymatic activity. Naturally occurring ribozymes are classified as large or small. The small ribozymes have compact motifs and therefore are useful for designing ribozymes in the laboratory. For example, the hammerhead motif can be linked to an antisense RNA sequence that recognizes mRNA from a disease-causing virus. The antisense segment will bind the target mRNA, and the hammerhead motif will then cut the target. Large pools of random RNA sequences can be created to either find new substrates for an existing ribozyme or find new sequences that are ribozymes. These can be "evolved" into better ribozymes by adding a mutagenesis step before the selection. In addition to RNA ribozymes, DNA has also been shown to have catalytic power.

End-of-Chapter Questions

1. Which of the following statements about antisense RNA is true?
   - **a.** Antisense RNA binds to form double-stranded regions on RNA to either block translation or intron splicing.
   - **b.** Antisense RNA is transcribed using the sense strand of DNA as a template.
   - **c.** The sequence of antisense RNA is complementary to mRNA.
   - **d.** Antisense RNA is made naturally in cells and also artificially in the laboratory.
   - **e.** All of the above statements about antisense RNA are true.
2. Which biological function is not controlled by antisense RNA?
   - **a.** iron metabolism in bacteria
   - **b.** the circadian rhythm of *Neurospora*
   - **c.** replication of prokaryotic genomic DNA
   - **d.** replication of ColE1 plasmid
   - **e.** developmental control of basic fibroblast growth factor
3. Which of the following is a modification of antisense oligonucleotide structure to increase intracellular stability?
   - **a.** insertion of an amine into the ribose ring to create a morpholino structure
   - **b.** attachment of nucleic acid bases to a peptide backbone instead of a sugar-phosphate backbone
   - **c.** replacement of one of the oxygen atoms in the phosphate group with a sulfur atom to inhibit nuclease degradation in some molecules
   - **d.** addition of an *O*-alkyl group to the 2′-OH of the ribose group to make the molecule resistant to nuclease degradation
   - **e.** all of the above
4. How can antisense RNA be expressed within a cell?
   - **a.** The target gene can be cloned inversely into a vector and under the control of an inducible promoter.
   - **b.** The antisense RNA cannot be expressed within a cell and instead must be delivered via liposomes.
   - **c.** Antisense RNA can be expressed within cells, but this is unfavorable because of the high degree of non-specific interactions.
   - **d.** No system has been designed to express antisense RNA within a cell.
   - **e.** None of the above is correct.

(Continued)

5. Which of the following terms describes when gene regulation occurs by short is dsRNA molecules triggering an enzymatic reaction that degrades the mRNA of a target gene?
   a. post-transcriptional gene silencing
   b. quelling
   c. co-suppression
   d. RNA interference
   e. all of the above

6. Which statement about RNAi is not correct?
   a. RNAi was first discovered in plants.
   b. RNAi has two phases: initiation and effector.
   c. During the initiation phase of RNAi, a protein called Dicer cuts dsRNA into small fragments called siRNAs.
   d. Non-specific interactions between the antisense siRNA and mRNA often cause mRNAs to be degraded that should not have been.
   e. The RNA-induced silencing complex has both helicase and endonuclease activities.

7. Which of the following is not a method for delivering dsRNA for RNAi into *Drosophila* or *C. elegans*?
   a. ingestion of transgenic bacteria that express dsRNA
   b. injection of dsRNA into eggs
   c. bathing in a solution of pure dsRNA
   d. injection of dsRNA into cell culture lines

8. How can RNAi be triggered in mammalian cells?
   a. transfection of siRNA
   b. chemically synthesized siRNA
   c. degradation of target mRNA through shRNA creation
   d. modification of an existing shRNA to recognize a different mRNA
   e. all of the above

9. What information has been obtained through the creation of RNAi libraries?
   a. the function of unknown proteins by degrading all of the mRNA for that protein
   b. the mechanism by which *E. coli* delivers dsRNA to *C. elegans*
   c. the mechanism by which heterochromatic formation occurs after some RNAi
   d. all of the above
   e. none of the above

10. What is a ribozyme?
   a. an enzyme that cuts ribosomes
   b. an RNA molecule that binds to specific targets and catalyzes reactions
   c. an enzyme that catalyzes the degradation of dsRNA
   d. an RNA molecule that catalyzes the degradation of ribonucleases
   e. none of the above

11. Which of the following is a large ribozyme?
   a. hairpin ribozyme
   b. hammerhead ribozyme
   c. Twort ribozyme
   d. hepatitis delta virus
   e. Varkud satellite ribozyme

12. What process is used to identify possible ribozyme substrates?
- **a.** DNA SELEX
- **b.** DNA BLAST
- **c.** RISC
- **d.** RNA SELEX
- **e.** GENEi

13. What property must a ribozyme possess in order to be used in clinical medicine?
- **a.** stability and resistance to degradation
- **b.** no deleterious side effects to the host
- **c.** expression within a diseased cell only
- **d.** be able to be delivered to the correct location
- **e.** all of the above

14. What is a riboswitch?
- **a.** an mRNA sequence that binds directly to an effector molecule to control the translation of the mRNA into protein
- **b.** an enzyme that converts ribozymes into deoxyribozymes
- **c.** the effector molecule responsible for translational control of a particular mRNA
- **d.** an RNA molecule that switches between being translated into protein or being a ribozyme
- **e.** none of the above

15. Which of the following is not an example of an effector molecule for riboswitches?
- **a.** some cyclic mononucleotides
- **b.** oligonucleotides
- **c.** metal ions
- **d.** some proteins
- **e.** antisense RNAs

16. Which RNA is incorrectly paired with its function?
- **a.** snRNA – RNA processing
- **b.** circRNA – transcriptional regulation
- **c.** piRNA – RNA processing
- **d.** Xist – chromosomal structure
- **e.** lncRNA – transcriptional regulation

17. Which of the following helps replicate telomeres?
- **a.** snoRNA
- **b.** Xist
- **c.** TERC
- **d.** gRNA
- **e.** circRNA

18. In *Drosophila*, two non-coding RNAs called *roX1* and *roX2* are used to ________________.
- **a.** double the expression of X chromosomal genes in males
- **b.** inhibit the expression of one X chromosome in females
- **c.** inactivate X chromosomal gene expression in males
- **d.** regulate the replication of sex chromosomes in males and female.

19. What is the role of the alpha (α) antisense form in PTEN expression?
- **a.** Anneals over complementary areas to prevent degradation.
- **b.** Attraction of miRNAs to promote PTEN mRNA degradation.

(Continued)

c. Recruits two chromatin modification enzymes to condense histones surrounding PTEN genes.
d. Activate transcription of PTEN gene by modifying chromatin structure.
e. Activates PTEN mRNA degradation.

20. What role do piRNAs play?
   a. Serve as a template for transposon silencing.
   b. Serve as a guide to mRNA degradation enzymes.
   c. Structural component of some ribozymes.
   d. Antisense RNA involved in RNA processing.
   e. Silences the second X chromosome in human females.

Further Reading

Aravin, A. A., Hannon, G. J., & Brennecke, J. (2007). The Piwi-piRNA pathway provides an adaptive defense in the transposon arms race. *Science, 318*, 761–764.

Arthanari, Y., Heintzen, C., Griffiths-Jones, S., & Crosthwaite, S. K. (2014). Natural antisense transcripts and long non-coding RNA in neurospora crassa. *PLoS One* 9, e91353.

Ashe, A., et al. (2012). piRNAs can trigger a multigenerational epigenetic memory in the germline of C. elegans. *Cell, 150*, 88–99.

Batista, P. J., & Chang, H. Y. (2013). Long noncoding RNAs: cellular address codes in development and disease. *Cell, 152*, 1298–1307.

Benenson, Y. (2012). Synthetic biology with RNA: progress report. *Current Opinion in Chemical Biology, 16*, 278–284.

Cech, T. R., & Steitz, J. A. (2014). The noncoding RNA revolution—trashing old rules to forge new ones. *Cell, 157*, 77–94.

Clark, D. P., & Pazdernik, N. J. (2012). *Molecular Biology* (2nd ed.). Waltham, MA: Elsevier Academic Press/Cell Press.

Derrien, T., Johnson, R., Bussotti, G., Tanzer, A., Djebali, S., Tilgner, H., et al. (2012). The GENCODE v7 catalog of human long noncoding RNAs: analysis of their gene structure, evolution, and expression. *Genome Research, 22*, 1775–1789.

Estrozi, L. F., Boehringer, D., Shan, S.-O., Ban, N., & Schaffitzel, C. (2011). Cryo-EM structure of the E. coli translating ribosome in complex with SRP and its receptor. *Nature Structural & Molecular Biology, 18*, 88–90.

Gavrilov, K., & Saltzman, W. M. (2012). Therapeutic siRNA: principles, challenges, and strategies. *The Yale Journal of Biology and Medicine, 85*, 187–200.

Henkin, T. M. (2014). The T box riboswitch: A novel regulatory RNA that utilizes tRNA as its ligand. *Biochimica et Biophysica Acta, 1839*(10), 959–963.

Hirsch, A. J. (2010). The use of RNAi-based screens to identify host proteins involved in viral replication. *Future Microbiology, 5*, 303–311.

Johnsson, P., Lipovich, L., Grandér, D., & Morris, K. V. (2014). Evolutionary conservation of long non-coding RNAs; sequence, structure, function. *Biochimica et Biophysica Acta, 1840*, 1063–1071.

Juliano, R. L., Ming, X., & Nakagawa, O. (2012). Cellular uptake and intracellular trafficking of antisense and siRNA oligonucleotides. *Bioconjugate Chemistry, 23*, 147–157.

Kassube, S. A., Fang, J., Grob, P., Yakovchuk, P., Goodrich, J. A., & Nogales, E. (2013). Structural insights into transcriptional repression by noncoding RNAs that bind to human Pol II. *Journal of Molecular Biology, 425*, 3639–3648.

Kirwan, M., & Dokal, I. (2009). Dyskeratosis congenita, stem cells and telomeres. *Biochimica et Biophysica Acta, 1792*, 371–379.

Kornfeld, J. W., & Brüning, J. C. (2014). Regulation of metabolism by long, non-coding RNAs. *Frontiers in Genetics, 5*, 57 eCollection.

Lam, M. T. Y., Li, W., Rosenfeld, M. G., & Glass, C. K. (2014). Enhancer RNAs and regulated transcriptional programs. *Trends in Biochemical Sciences, 39*(4), 170–182.

Lee, H. C., Gu, W., Shirayama, M., Youngman, E., Conte, D., Jr., & Mello, C. C. (2012). C. elegans piRNAs mediate the genome-wide surveillance of germline transcripts. *Cell, 150*, 78–87.

Li S1, & Breaker, R. R. (2013). Eukaryotic TPP riboswitch regulation of alternative splicing involving long-distance base pairing. *Nucleic Acids Research, 41*, 3022–3031.

Li, Y., Lu, J., Han, Y., Fan, X., & Ding, S. W. (2013). RNA interference functions as an antiviral immunity mechanism in mammals. *Science, 342*, 231–234.

Maillard, P. V., Ciaudo, C., Marchais, A., Li, Y., Jay, F., Ding, S. W., & Voinnet, O. (2013). Antiviral RNA interference in mammalian cells. *Science, 342*, 235–238.

Martens-Uzunova, E. S., Olvedy, M., & Jenster, G. (2013). Beyond microRNA—novel RNAs derived from small non-coding RNA and their implication in cancer. *Cancer Letters, 340*, 201–211.

Mulhbacher, J., St-Pierre, P., & Lafontaine, D. A. (2010). Therapeutic applications of ribozymes and riboswitches. *Current Opinion in Pharmacology, 10*, 551–556.

Nadal-Ribelles, M., Solé, C., Xu, Z., Steinmetz, L. M., de Nadal, E., & Posas, F. (2014). Control of Cdc28 CDK1 by a stress-induced lncRNA. *Molecular Cell, 53*, 549–561.

Oustric, V., Manceau, H., Ducamp, S., Soaid, R., Karim, Z., Schmitt, C., et al. (2014). Antisense oligonucleotide-based therapy in human erythropoietic protoporphyria. *American Journal of Human Genetics, 94*, 611–617.

Pan, Q., van der Laan, L. J., Janssen, H. L., & Peppelenbosch, M. P. (2012). A dynamic perspective of RNAi library development. *Trends in Biotechnology, 30*, 206–215.

Peng, J. C., & Lin, H. (2013). Beyond transposons: the epigenetic and somatic functions of the Piwi-piRNA mechanism. *Current Opinion in Cell Biology, 25*, 190–194.

Phillips, C. M., Montgomery, B. E., Breen, P. C., Roovers, E. F., Rim, Y.-S., Ohsumi, T. K., et al. (2014). MUT-14 and SMUT-1 DEAD box RNA helicases have overlapping roles in germline RNAi and endogenous siRNA formation. *Current Biology: CB, 24*, 839–844.

Pircher, A., Bakowska-Zywicka, K., Schneider, L., Zywicki, M., & Polacek, N. (2014). An mRNA-derived noncoding RNA targets and regulates the ribosome. *Molecular Cell, 54*, 147–155.

Salter, J., Krucinska, J., Alam, S., Grum-Tokars, V., & Wedekind, J. E. (2006). Water in the active site of an all-RNA hairpin ribozyme and effects of Gua8 base variants on the geometry of phosphoryl transfer. *Biochemistry, 45*, 686–700.

Scheer, U., & Hock, R. (1999). Structure and function of the nucleolus. *Current Opinion in Cell Biology, 11*, 385–390.

Scheer, U., & Weisenberger, D. (1994). The nucleolus. *Current Opinion in Cell Biology, 6*, 354–359.

Serganov, A., & Patel, D. J. (2007). Ribozymes, riboswitches and beyond: regulation of gene expression without proteins. *Nature Reviews. Genetics, 8*, 776–790.

Shirayama, M., Stanney, W., Gu, W., Seth, M., & Mello, C. C. (2014). The vasa homolog RDE-12 engages target mRNA and multiple argonaute proteins to promote RNAi in C. elegans. *Current Biology: CB, 24*, 845–851.

Stahel, R. A., & Zangemeister-Wittke, U. (2003). Antisense oligonucleotides for cancer therapy—an overview. *Lung Cancer, 41*, 81–88.

Tang, X., Lim, S. C., & Song, H. (2014). RNase AS versus RNase T: similar yet different. *Structure, 22*, 663–664.

Towbin, B. D., Gonzalez-Sandoval, A., & Gasser, S. M. (2013). Mechanisms of heterochromatin subnuclear localization. *Trends in Biochemical Sciences, 38*, 356–363.

Ulitsky, I., & Bartel, D. P. (2013). lincRNAs: genomics, evolution, and mechanisms. *Cell, 154*, 26–46.

Ulveling, D., Francastel, C., & Hubé, F. (2011). When one is better than two: RNA with dual functions. *Biochimie, 93*, 633–644.

Voorhees, R. M., & Ramakrishnan, V. (2013). Structural basis of the translational elongation cycle. *Annual Review of Biochemistry, 82*, 203–236.

Wachter, A. (2010). Riboswitch-mediated control of gene expression in eukaryotes. *RNA Biology, 7*, 67–76.

Wachter, A. (2014). Gene regulation by structured mRNA elements. *Trends in Genetics: TIG, 30*(5), 172–181.

Wittmann, A., & Suess, B. (2012). Engineered riboswitches: expanding researchers' toolbox with synthetic RNA regulators. *FEBS Letters, 586*, 2076–2083.

Wood, A. M., Garza-Gongora, A. G., & Kosak, S. T. (2014). A crowdsourced nucleus: understanding nuclear organization in terms of dynamically networked protein function. *Biochimica et Biophysica Acta - Gene Regulation Mechanism, 1839*, 178–190.

Yang, H., Vallandingham, J., Shiu, P., Li, H., Hunter, C. P., & Mak, H. Y. (2014). The DEAD box helicase RDE-12 promotes amplification of RNAi in cytoplasmic foci in C. elegans. *Current Biology: CB, 24*, 832–838.

Immune Technology

Biotechnology

http://dx.doi.org/10.1016/B978-0-12-385015-7.00006-5

INTRODUCTION

The world is full of infectious microorganisms, all looking for a suitable host to infect. Bacteria, viruses, and protozoans are constantly attempting to gain entry into our tissues. If nothing prevented these attempts at invasion, no human could survive. Fortunately, cells of the immune system patrol the organism, protecting the entire body from attack. Any foreign macromolecules that are not recognized as being "self" are regarded as signs of an intrusion and trigger an immune response. In particular, proteins that are exposed on the surfaces of invading microorganisms attract the attention of the immune system. These molecules are called **antigens**. Some of the immune system molecules that recognize and bind to them are called **antibodies** (Fig. 6.1).

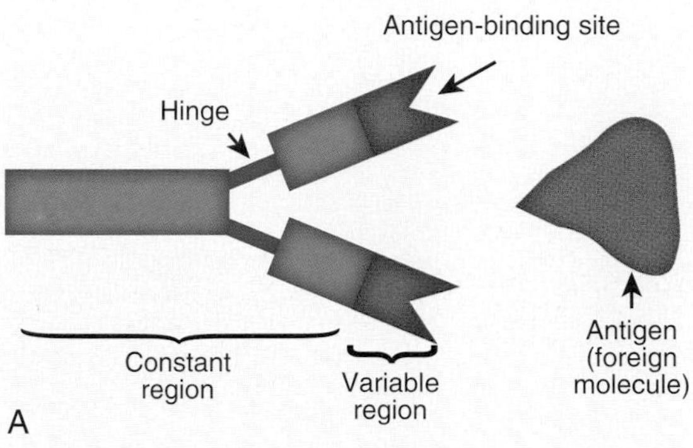

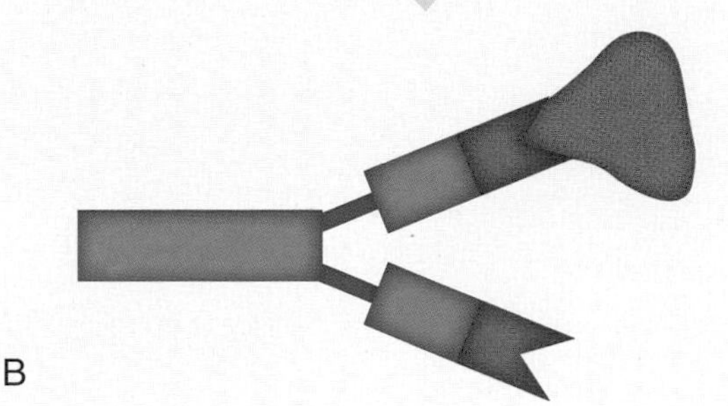

FIGURE 6.1 Foreign Antigens Are Recognized by Antibodies
(A) Antibodies are Y-shaped molecules produced by the immune system in vertebrates. They bind to specific portions of proteins or antigens of any invading pathogen. (B) The variable region mediates the binding of an antigen to the antibody.

To be prepared for any possible invasion, the **B cells** of the **adaptive immune system** generate billions of different antibodies. Most antibodies are secreted into the lymph, but some remain bound to the cell surface and are called **B-cell receptors (BCR)**. Eventually, when a foreign antigen appears, a few of the billions of predesigned antibodies will fit the antigen reasonably well (Fig. 6.2). Those B cells that make antibodies that recognize the antigen now divide rapidly and go into mass production. Thus, the antigen determines which antibody is amplified and produced. Once a matching antibody has bound invading antigens, the immune system brings other mechanisms into play to destroy the invaders.

Although the antibody that originally recognized the invading pathogen was a good fit for the antigen, there is a stage of refinement during which those antibodies that bound to the invading antigen are modified by mutation to fit the antigen better. In addition, the immune system keeps a record of antibodies that are actually used. If the same invader ever returns, the corresponding antibodies can be rushed into action, faster and in greater numbers than before. Vaccines exploit this capacity by stimulating the immune system to store the antibodies that recognize and destroy a pathogenic virus such as smallpox. Yet the vaccines cause no disease symptoms themselves (see later discussion).

The immune system keeps a repertoire of B cells that are poised to make antibodies to invading pathogens. When one of these B-cell antibodies is needed, the B cell starts dividing so that many antibodies can be produced and they are available to attack the pathogen. Some of these clones are refined by mutation to make a more specific antibody.

ANTIBODIES, ANTIGENS, AND EPITOPES

The term *antigen* refers to any foreign molecule that provokes a response by the immune system. In practice, most antigens are proteins made by invading bacteria or viruses. In particular, glycoproteins, which carry carbohydrate residues, and lipoproteins, which carry lipid residues, generate strong immune responses; that is, they are highly antigenic. Other macromolecules can also work as antigens. Polysaccharides are often found as surface components of infiltrating germs and may act as antigens. Even DNA can be antigenic under certain

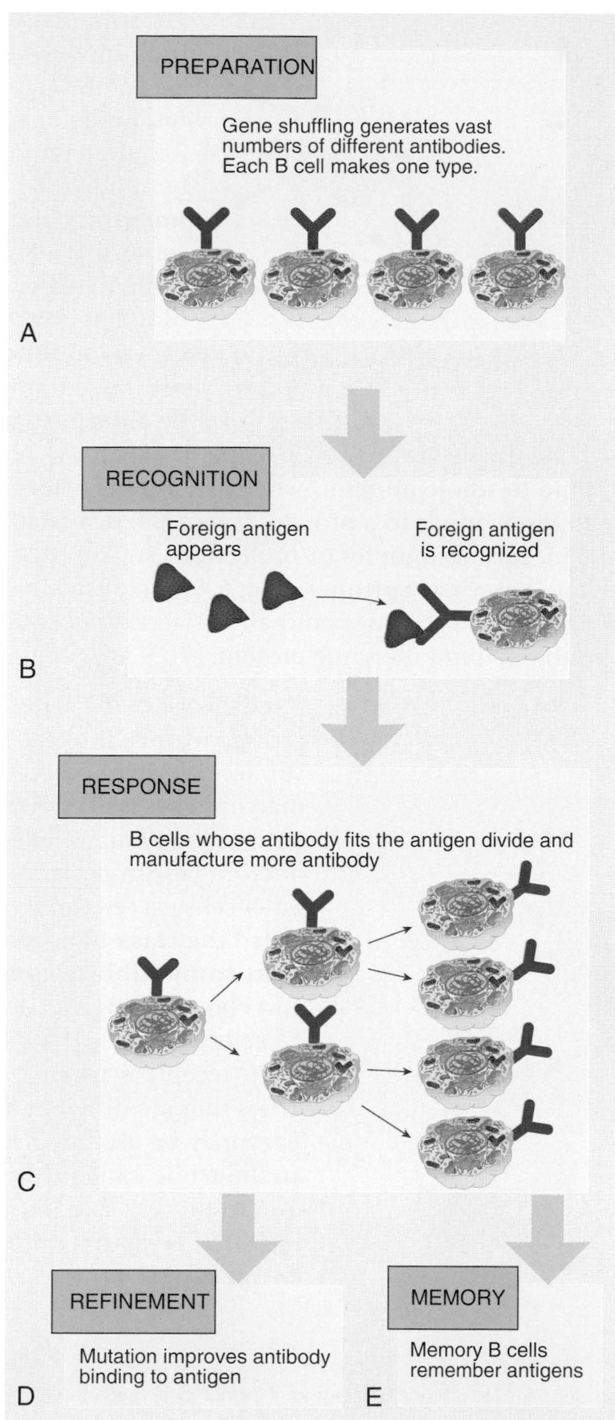

FIGURE 6.2 Predesigned Antibodies Are Ready for Foreign Antigens
Long before an attack by a pathogen, an army of B cells produces a large repertoire of antibodies (A). When one of the antibodies binds to an antigen (B), that particular B cell starts dividing and expanding (C). The majority of the B cells refine the antibody so that the antigen/antibody complex binds more tightly, and they fight the pathogens (D). A small subset of B cells become memory cells that never die, waiting for another attack by the same pathogen (E).

circumstances. Not surprisingly, the antigens exposed on the surface of an alien microorganism will usually be detected first (Fig. 6.3). Later in infection, especially after the cells of some invaders have been disrupted by the immune system, molecules from the interior of the infectious agent may be liberated and also act as antigens.

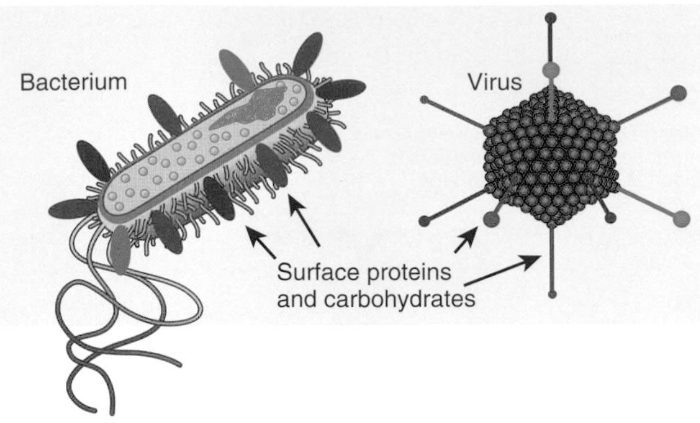

FIGURE 6.3
Surface Antigens of Microorganisms
The surfaces of bacteria and viruses are coated with glycoprotein and lipoproteins that are recognized by antibodies in the host organism.

The immune system mediates immunity to various infectious agents through **specific immunity** or **acquired immunity**. Acquired immunity can be subdivided into **humoral immunity** and **cell-mediated immunity**. Humoral immunity is mediated by antibodies in the blood plasma, which are also called **immunoglobulins**. Cell-mediated immunity is mediated by antigen-specific cells called **T lymphocytes**, which are divided into T_H or T helper cells, and T_C or T cytotoxic cells. Antibodies generally bind to whole proteins, whereas T-cell receptors bind to fragments of protein. When an antibody binds to a protein, it recognizes a relatively small area on the surface of the protein, such as dimples or projections sticking out from the surface. Such recognition sites are known as **epitopes** (Fig. 6.4). Because intact proteins are large molecules, they may have several epitopes on their surfaces. Consequently, several different antibodies may be able to bind the same protein.

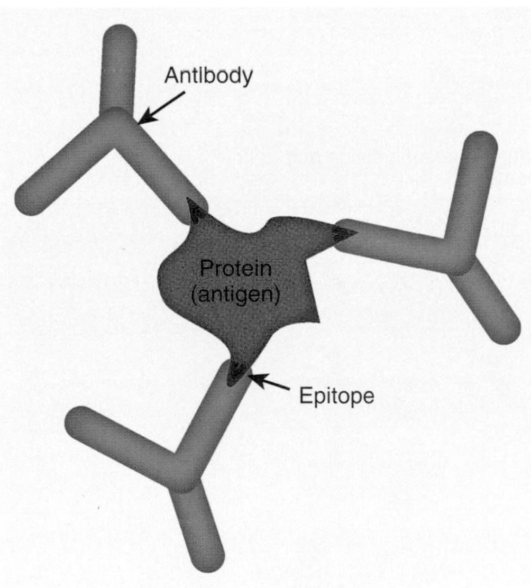

FIGURE 6.4
Antibodies Bind to Epitopes on an Antigen
Antibodies recognize only a small ridge on the surface of a protein. The region of the antigen that binds to the antibody is called an epitope.

T cells work in the same manner but recognize only antigens expressed on the surface of other body cells, particularly macrophages, cells infected with a virus, or antibody-making B cells, as opposed to the microorganism itself. T cells recognize these other cells via cell surface receptor proteins called the **class I** and **class II major histocompatibility complexes (class I** and **class II MHCs)**. Class I MHCs activate T_H cells, and class II MHCs activate T_C cells. MHC receptors are encoded by a family of genes that are different for every person. They may be used to distinguish people and must be matched in organ transplantation to prevent rejection. Another name for the MHC receptors is **human leukocyte antigens (HLAs)**.

Acquired immunity is divided into two branches. Humoral immunity is mediated by antibodies in blood plasma, which are produced by B cells. The second part is cellular immunity, which is mediated by T cells.

Antibodies recognize epitopes or specific regions of the invading pathogen.

T cells recognize cell surface receptors called class I and class II major histocompatibility complexes that are expressed on the surface of body cells that become infected with an invading pathogen.

THE GREAT DIVERSITY OF ANTIBODIES

Since there is an almost infinite variety of possible antigens, a correspondingly vast number of different antibody molecules are needed. The amino acids making up protein molecules can certainly be arranged to give an almost infinite number of different sequences and, therefore, of different shapes. However, this leads to a major genetic problem. If a separate gene encoded each antibody, this would require a gigantic number of genes and a vast amount of DNA. Even if the entire mammalian genome was coding DNA, it could encode only a few million antibodies, which is far too few.

Rather than a unique gene for each antibody, the immune system generates a vast array of different protein sequences from a relatively small number of genes by shuffling gene segments in a process called **V(D)J recombination**. Instead of storing complete genes for each antibody, the immune system assembles antibody genes from a collection of shorter DNA segments. Shuffling and joining these partial genes allows the generation of an immense variety of antibodies. In Figure 6.5, this idea is illustrated using three alternative front ends and three rear ends. Combining them in all possible ways gives nine different genes. The immune system is a fascinating example of how massive genetic diversity can be generated by shuffling relatively few segments of genetic information. Animals can make billions of possible antibodies from only a few thousand gene segments.

The detailed genetics of antibody diversity is a complex issue and is described in textbooks on immunology. The rest of this chapter discusses those aspects of immunology of importance to biotechnology. They include antibody structure, the bioengineering of antibodies, biotechnological techniques that use antibodies, and finally vaccines. The chapter ends with techniques used to identify and produce new vaccines.

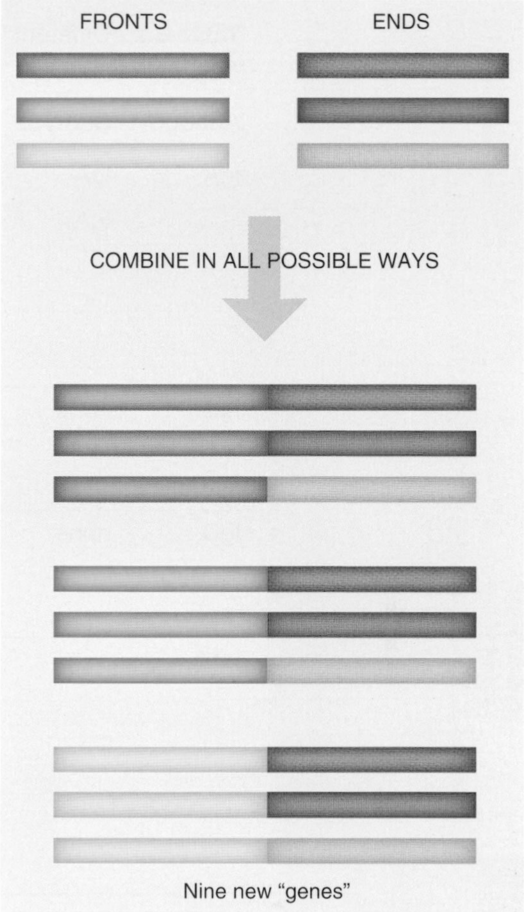

FRONTS

ENDS

COMBINE IN ALL POSSIBLE WAYS

Nine new "genes"

FIGURE 6.5
Modular Gene Assembly
Linking different segments of genes creates exponential numbers of unique combinations.

Antibodies are very diverse in structure so that all the pathogens can be recognized.

Antibodies are produced by shuffling gene segments rather than having one gene code for each different antibody.

STRUCTURE AND FUNCTION OF IMMUNOGLOBULINS

Depending on the type of heavy chain, antibodies are categorized into different classes, and they assume different roles in the immune system (see Table 6.1). The most abundant and typical antibody has a gamma heavy chain and is called **immunoglobulin G (IgG)**. IgG has four different subclasses, but as a whole, IgG is found mainly in blood serum. About 75% of the serum antibodies are IgG, and they are critical to stimulate immune cells to engulf invading pathogens. IgG is the only antibody able to transfer across the placenta during pregnancy. The second most common antibody in serum is secretory IgA. This antibody is also found in mucosal secretions as well as colostrum and breast milk. It is extremely important in fighting respiratory and gastrointestinal infections, especially in infants, in whom gastrointestinal illnesses are particularly deadly. The third most common is IgM, which is usually found as a pentamer. The unusual structure of IgM provides multiple binding sites for antigens (10 in IgM versus 2 in IgG). This structure makes IgM good for clumping microorganisms and then stimulating immune cells to digest the entire complex. IgD is found at low levels, and its role is still uncertain. IgE is the least common antibody in serum and is primarily found attached to mast cells. IgE is the antibody that stimulates allergic responses by releasing the histamines that cause all the common symptoms of allergies, including runny noses, sneezing, and coughing.

Table 6.1 Different Types and Functions of Human Antibodies

Antibody	Subtype	Light Chain	Heavy Chain	Function	Structure
IgA	IgA_1 IgA_2	κ or λ	α_1 α_2	Prevents pathogen attachment	J chain; Secretory piece; SECRETORY; MONOMER
IgE	none	κ or λ	ε	Allergic reaction; inhibits parasites	Extra domain
IgD	none	κ or λ	δ	Activates lymphocytes	Tail piece
IgM	none	κ or λ	μ	Clumps microbes and activates complement	Tail piece; MONOMER; Extra domain; J chain; PENTAMER
IgG	IgG_1 IgG_2 IgG_3 IgG_4	κ or λ κ or λ κ or λ κ or λ	γ_1 γ_2 γ_3 γ_4	Activates complement, activates other immune cells	IgG_1,IgG_2,IgG_4; IgG_3

Note: Light chains are depicted in light blue and heavy chains are purple.

Each IgG antibody consists of four protein subunits, two **light chains** and two **heavy chains**, arranged in a Y-shape (Fig. 6.6). The light chains are encoded by one of two gene loci, κ and λ. Disulfide bonds between cysteine amino acid residues hold the chains together. Each of the light and heavy chains consists of one to four **constant regions** and a single **variable region**. The constant region is the same for all chains of the same class. The variable regions work together to form the **paratope**, which is the region or surface of the antibody that binds to the target molecule, the antigen. There are millions of different variable regions, which are generated by genetic shuffling called V(D)J recombination. In Figure 6.7, the possible segments encoded in the germline are shown at the top; they can be inverted or deleted due to rearrangements. These recombination events occur in the bone marrow during early B-cell development and are initiated by RAG1 and RAG2, which nick the DNA backbones. Nonhomologous end joining (NHEJ) enzymes reconnect the ends to form inversions or deletions. Interestingly, NHEJ enzymes imprecisely reconnect these ends, thus inducing insertions or deletions. During transcription of the recombined segment, various segments are skipped due to alternate splicing, producing a transcript with a single V, single J, and single C. Further processing during translation produces the complete unique κ light chain. The heavy-chain locus is encoded on chromosome 14 and has about

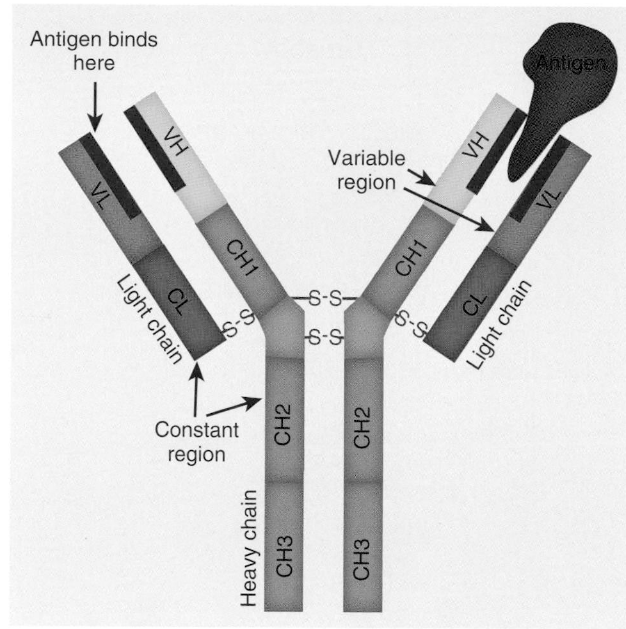

FIGURE 6.6 Structure of an Antibody
Y-shaped antibodies consist of two light chains and two heavy chains. Each consists of segments: CH1, CH2, and CH3 are heavy-chain constant regions; CL is the light-chain constant region; VH is the heavy-chain variable region; and VL is the light-chain variable region. Antigens bind to the variable regions.

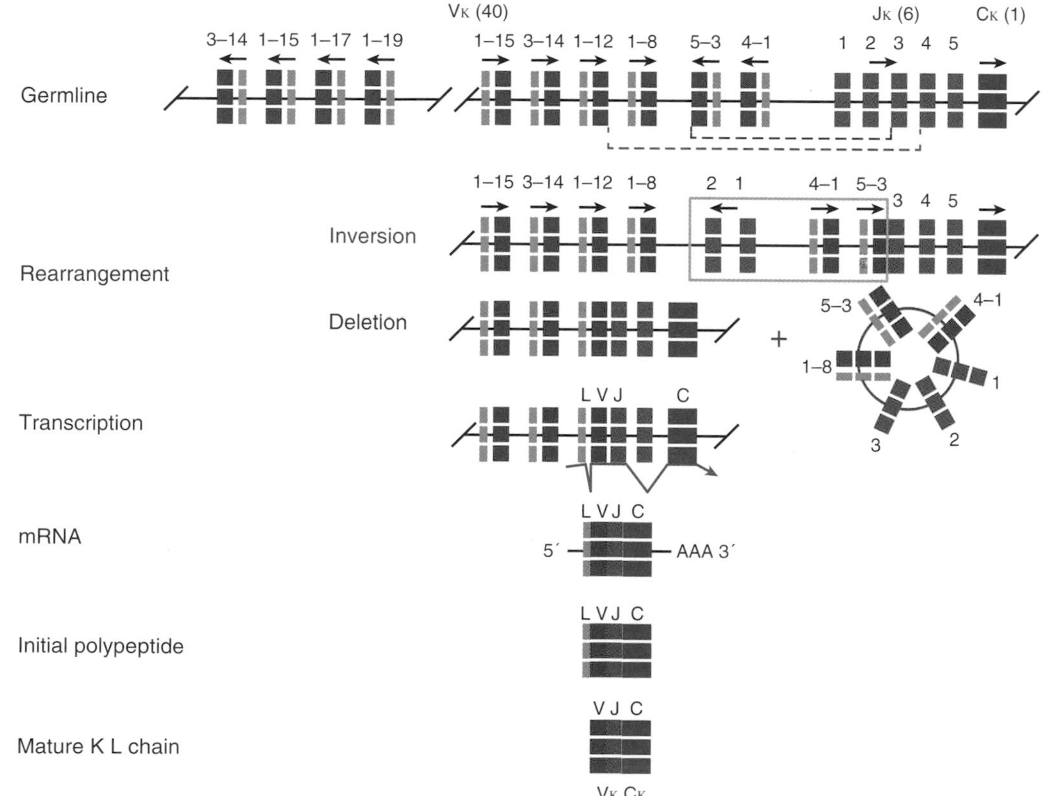

FIGURE 6.7 V(D)J Recombination
The κ locus found on chromosome 2 has the segments for the light chain. In the germline, there are 75 V segments (only around 30 are actually functional), 5 J segments, and 1 C (or constant) region. Each segment has a recombination signal sequence at the end that elicits the recombination between 1 V and 1 J segment. Every light chain has a combination of 1 V segment, 1 J segment, and the constant region (C).

FIGURE 6.8 Fab Fragments and Fc Fragment of an Antibody
Antibodies can be split into two Fab fragments and one Fc fragment by breaking the molecule at the hinge region.

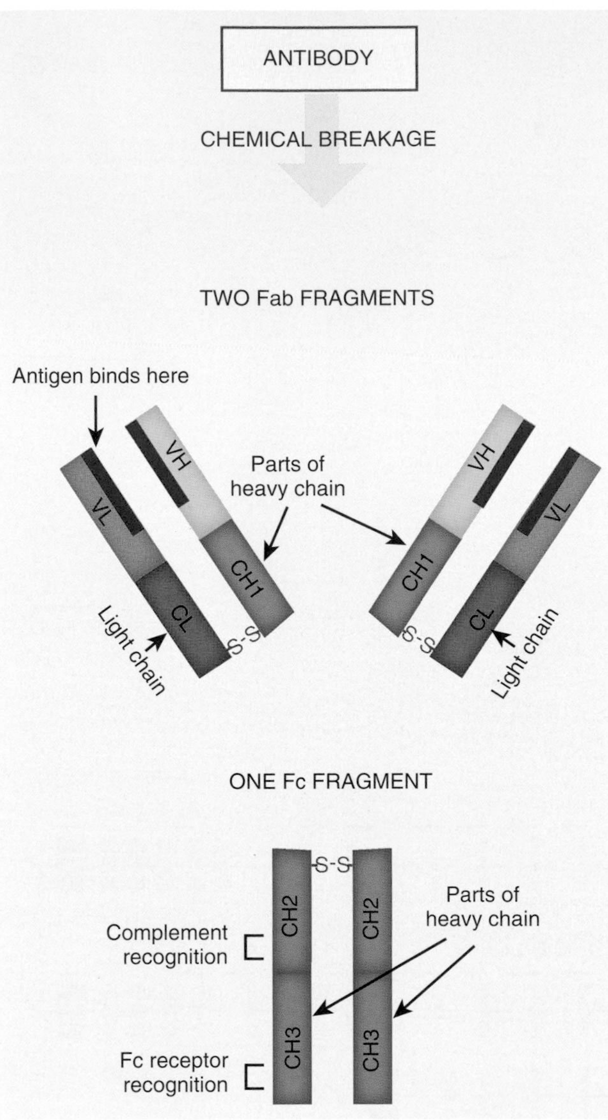

39 functional V segments, 27 D segments, and 6 J segments. When only one of each segment is used, there can be greater than 10^4 possible combinations alone. Further diversity arises from rearrangements of the D segments by alternate splicing, nucleotide insertion or deletion during recombination, and the addition of nucleotides at random by terminal deoxytransferase (TdT) during recombination. The possible combinations increase to greater than 10^7 from a single heavy-chain locus when these other events are included.

Breaking an antibody at the "hinge" where the heavy chains bend yields three chunks: two identical **Fab fragments** and one **Fc fragment** (Fig. 6.8). Fab, meaning "fragment, antigen binding," consists of one light chain plus half of a heavy chain. Fc, meaning "fragment, crystallizable," contains the lower halves of both heavy chains. Other components of the immune system often recognize and bind to the Fc region of an antibody (see later discussion).

IgG antibodies have a Y-shaped structure. The hinge or bend region divides the two Fab fragments from the Fc fragment. There are two light chains and two heavy chains.

MONOCLONAL ANTIBODIES FOR CLINICAL USE

There are many clinical uses for antibodies. They are used in diagnostic procedures (including the ELISA—see later discussion), for pregnancy testing, and to detect the presence of proteins characteristic of particular disease-causing agents. In the future, they may be used to specifically kill cancer cells or destroy viruses. Such uses need relatively large amounts of a pure antibody that specifically recognizes a single antigen. Even if an experimental animal is inoculated with a purified single antigen, its blood serum will contain a mixture of antibodies to that antigen. Remember that a single antigen has multiple epitopes, and thus antibodies will vary in both specificity and affinity. Nowadays, such a mixture is referred to as **polyclonal antibody** because it results from antibody production by many different clones of B cells, which all recognized the same antigen. Such a mixture is of little use either for a specific, accurate assay or for other techniques in biotechnology.

To create large amounts, scientists must isolate and grow a single line of B cells making one particular antibody in culture. Such a pure antibody made by a single line of cells is known as a **monoclonal antibody**. Unfortunately, B cells live for only a few days and survive poorly outside the body. The solution to this problem is to use cancer cells. **Myelomas** are naturally occurring cancers derived from B cells; they therefore express immunoglobulin genes. Like many tumor cells, myeloma cells will continue to grow and divide in culture forever if given proper nutrients. To make monoclonal antibodies, scientists fuse the relatively delicate B cell, which is making the required antibody, to a myeloma cell (Fig. 6.9). (To avoid confusion, scientist use a myeloma that has lost the ability to make its own antibody.) The resulting hybrid is called a **hybridoma**. In principle, the fused cells can live forever in culture and will make the desired antibody.

In practice, an animal, such as a mouse, is injected with the antigen against which antibodies are needed. When antibody production has reached its peak, a sample of antibody-secreting B cells is removed from the animal. These cells are fused to immortal myeloma cells to give a mixture of many different hybridoma cells. The tedious part comes next. Many individual hybridoma cell lines must be screened to find one that recognizes the target antigen. Once it is found, the hybridoma is grown in culture to give large amounts of the monoclonal antibody.

Monoclonal antibodies recognize only one epitope on the antigen and derive from one single B cell.

Fusing antigen-stimulated B cells from a mouse spleen with a myeloma cell line produces an immortalized hybridoma. Each of the cells can be grown *in vitro* and evaluated for its affinity to the original antigen to make a monoclonal antibody.

HUMANIZATION OF MONOCLONAL ANTIBODIES

Monoclonal antibodies could target human cancer cells by recognizing specific molecules appearing only on the surface of cancer cells. Ironically, the main problem with their use as a therapy is that the human immune system regards antibodies from mice or other animals as foreign molecules themselves, and so attempts to destroy them!

One approach that may partly solve this problem is using genetic engineering to make **humanized** monoclonals (Fig. 6.10A). Since the variable, or V-region, of the antibody recognizes the antigen, the constant, or C-region, may therefore be replaced with a humanized version. To accomplish this, scientists isolate and culture the first-generation hybridoma, generally using mouse B cells. Then the DNA encoding the mouse monoclonal antibody is isolated and cloned. The DNA for the constant region of the mouse antibody is then replaced with the corresponding human DNA sequence. The V-region is left alone. The human/mouse hybrid gene is then put back into a second mouse myeloma cell for production of antibody in culture. Although not fully human, the hybrid is less mouse-like and provokes much less reaction from the human immune system.

Further humanization can be accomplished by altering those parts of the V-region that are not directly involved in binding the antigen. A closer look at the V-region of each chain shows that most of the variation is restricted to three short segments that form loops on the surface of the antibody, thus forming the antigen-binding site (see Fig. 6.10B). These are known as hypervariable regions or as **complementarity determining regions (CDRs)**. Overall, each antigen-binding site consists of six CDRs—three from the light chain and three from the heavy chain. Full humanization of an antibody involves cutting out the coding regions for these six CDRs from the original antibody and splicing them into the genes for human light and heavy chains.

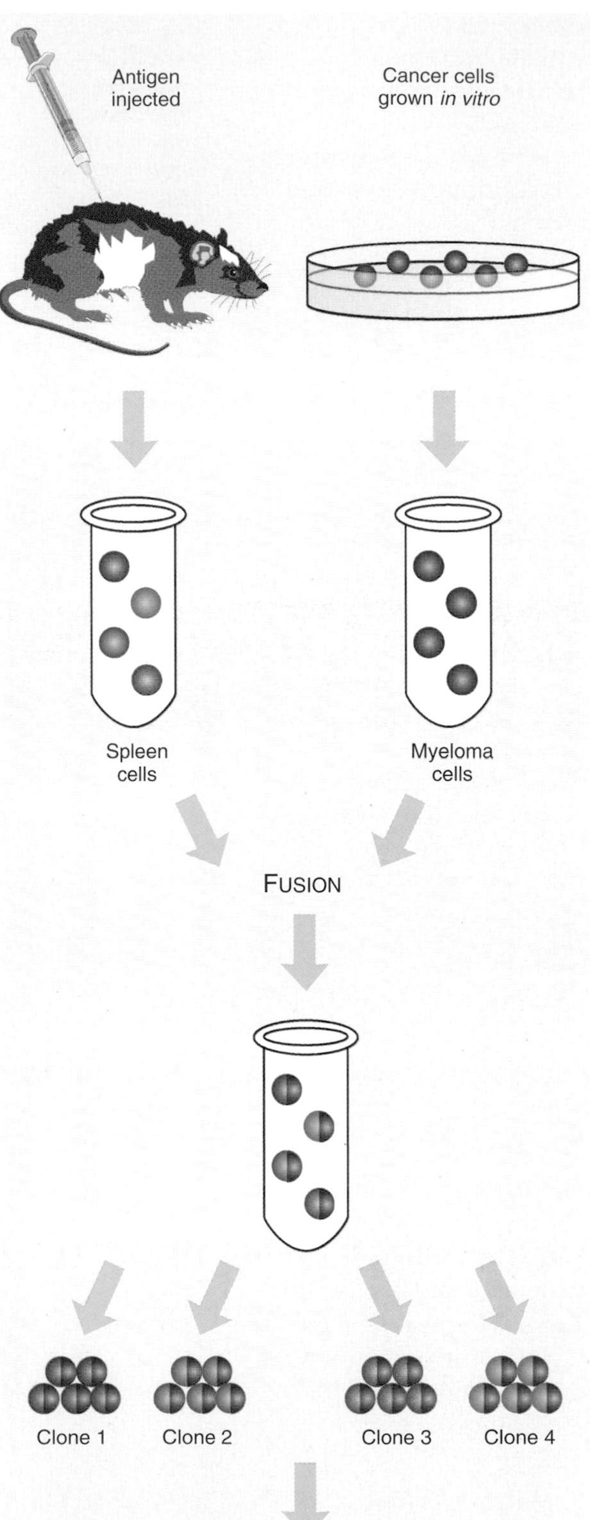

FIGURE 6.9 Principle of the Hybridoma

Monoclonal antibodies derive from a single antibody-producing B cell. The antigen is first injected into a mouse to provoke an immune response. The spleen is harvested because it harbors many activated B cells. The spleen cells are short-lived in culture, so they are fused to immortal myeloma cells. The hybridoma cells are cultured and isolated so each hybrid is separate from the other. Each hybrid clone can then be screened for the best antibody to the target protein.

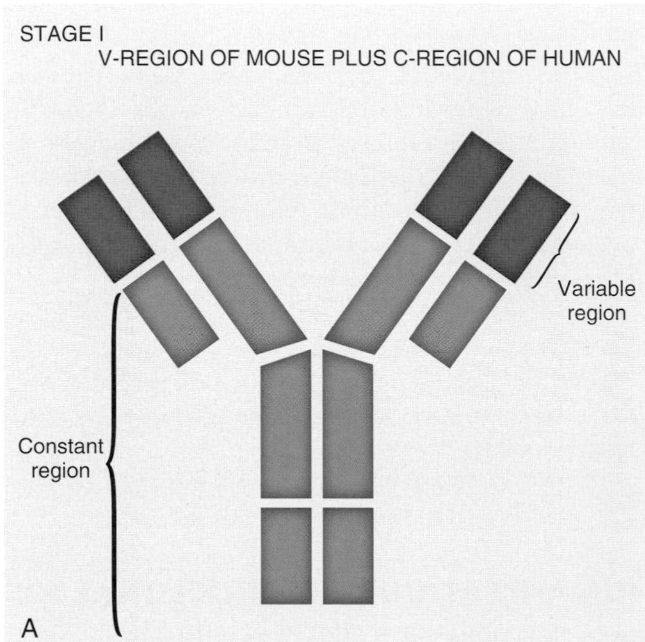

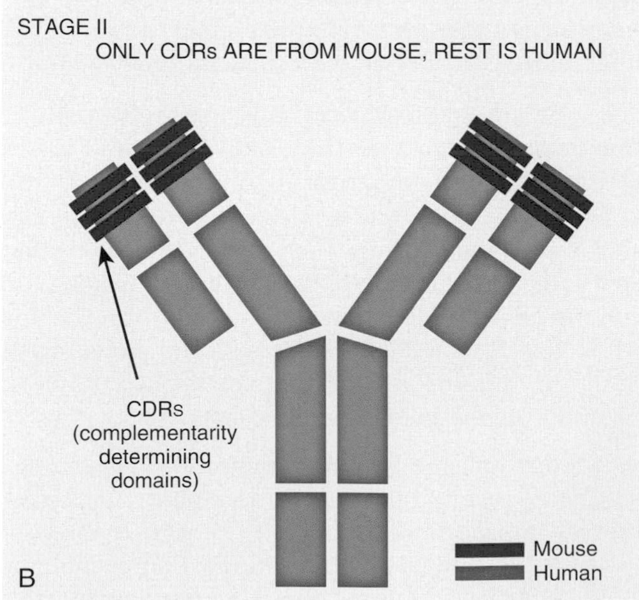

FIGURE 6.10 Humanization of Monoclonal Antibodies

Antibodies from a mouse can be altered to become more like a human antibody. (A) The entire constant region of the heavy and light chain can be replaced with constant regions from a human. (B) Antibodies have six CDRs that determine the actual antigen-binding site. The entire antibody except the CDR region can be replaced with human sequence.

Removing the constant regions of a mouse antibody and replacing them with human constant regions makes humanized antibodies. Human cells do not reject these antibodies.

HUMANIZED ANTIBODIES IN CLINICAL APPLICATIONS

There are currently many different humanized monoclonal antibodies in development to treat a variety of conditions. Many different antibodies have been approved by the Federal Drug Administration (FDA) for many different conditions. Table 6.2 presents a partial list of different FDA-approved antibodies. The first humanized monoclonal antibody approved for clinical use, trastuzumab **(Herceptin)**, is for the treatment of breast cancer. The FDA approved this therapeutic agent in 1998. Herceptin recognizes a cell surface receptor called human epidermal growth factor receptor type 2 (HER2). This receptor is part of a larger family, including HER3, HER4, and the founding member, the epidermal growth factor receptor (EGFR). These receptors control whether a cell proliferates, differentiates, or undergoes programmed suicide by signaling a variety of intracellular proteins that modulate gene expression. In breast cancer patients, when the HER2 receptor is overproduced, the breast cancer is much more resistant to chemotherapy. Excess receptor is thus a good indicator that the patient will not survive as long. Herceptin binds to the extracellular domain of HER2, preventing the receptor from being internalized. This prevents the cancer cell from dividing and induces the immune system to attack the cell (Fig. 6.11). When Herceptin is used in combination with chemotherapy to treat breast cancer, patients survive much longer. The main point to keep in mind is that Herceptin binds one specific protein; therefore, the particular breast cancer must have excess amounts of HER2 in order for the treatment to be effective.

Table 6.2 FDA-Approved Antibodies

Product	Antigen	Target	Trade Name
Murine Monoclonals			
Arcitumomab	Carcinoembryonic antigen	Metastatic colorectal cancer detection	CEA Scan
Capromab pentetate	Tumor surface antigen PSMA	Prostate adenocarcinoma detection	ProstaScint
Chimeric			
Infliximab	TNFα	Crohn's disease	Remicade
Antibody Fragments			
Nofetumomab (Murine Fab)	Antigen associated with cancer	Detection of small cell lung cancer	Verluma
Trastuzumab	Her-2	Metastatic breast cancer	Herceptin
Palivizumab	Respiratory syncytial virus (RSV) F protein	Respiratory tract disease	Synagis
Human Phage Display/Synthetic Antibody			
Adalimumab	TNFα	Immune disorders, Crohn's disease	Humira

Table is a subset of information from Khan FH (2014). Antibodies and their applications. In *Animal Biotechnology* (Oxford, UK and Waltham, MA, USA: Academic Press), p. 482.

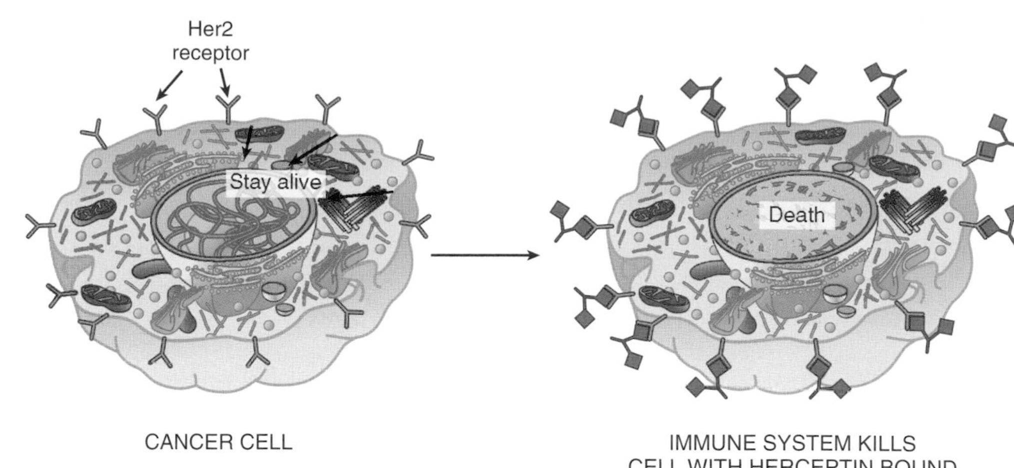

FIGURE 6.11
Herceptin Helps Kill Cancer Cells with HER2
Herceptin is a humanized monoclonal antibody that recognizes the HER2 receptor on breast cancer cells. When the antibody binds to the receptor, the immune system helps destroy the cancer cell, and the cancer cell becomes more sensitive to chemotherapeutic treatments.

Another chimeric antibody approved by the FDA is Remicade (infliximab), which is used to treat rheumatoid arthritis (RA). The antibody targets tumor necrosis factor alpha (TNFα), which is present in joints of people with arthritis. TNFα regulates inflammation and immune system function. Antibodies to TNFα inhibit inflammation in RA by blocking the release of IL-1, a pro-inflammatory cytokine. The researchers first created a hybridoma that expressed antibodies that recognized TNFα and then cloned the variable segments within the heavy-chain gene that were important for binding to the antigen. They also isolated the variable segments of the light-chain gene. These segments were then joined to the human κ light-chain gene, and the heavy-chain variable region was joined to the human constant region. These fusions were then transfected into a new myeloma cell and induced to produce the chimeric antibody. After extensive research on efficacy and safety, the antibody was released for treatment of inflammatory diseases such as RA. The overall cost for treatment is high, but the drug was one of the top ten selling drugs on the market (in sales) for the year 2012.

Monoclonal antibodies to HER2 inhibit breast cancer cells from growing and are used as a treatment for breast cancer patients. Chimeric antibodies to TNFα are used to treat rheumatoid arthritis and are one of the top-selling drugs.

ANTIBODY ENGINEERING

Natural antibodies consist of an antigen-binding site, called the paratope, joined to an effector region that is responsible for activating complement and/or binding to immune cells. From a biotechnological viewpoint, the incredibly high specificity with which antibodies bind to a target protein is useful for a variety of purposes. Consequently, antibody engineering uses the antigen-binding region of the antibody. These antibodies are manipulated and are attached to other molecular fragments.

To separate an antigen-binding site from the rest of the antibody, scientists subclone gene segments encoding portions of the variable antibody chains and express them in bacterial cells. Bacterial signal sequences are added to the amino-terminus of the partial antibody chains, which results in export of the chains into the periplasmic space. Here, the VH and VL domains fold up correctly and form their disulfide bonds. The antibody fragments used include Fab, Fv, and **single-chain Fv (scFv)** (Fig. 6.12). In a Fab fragment, an interchain disulfide bond holds the two chains together. However, the Fv fragment lacks this region of the antibody chains and thus is less stable. This led to development of the single-chain Fv fragment in which the VH and VL domains are linked together by a short peptide chain,

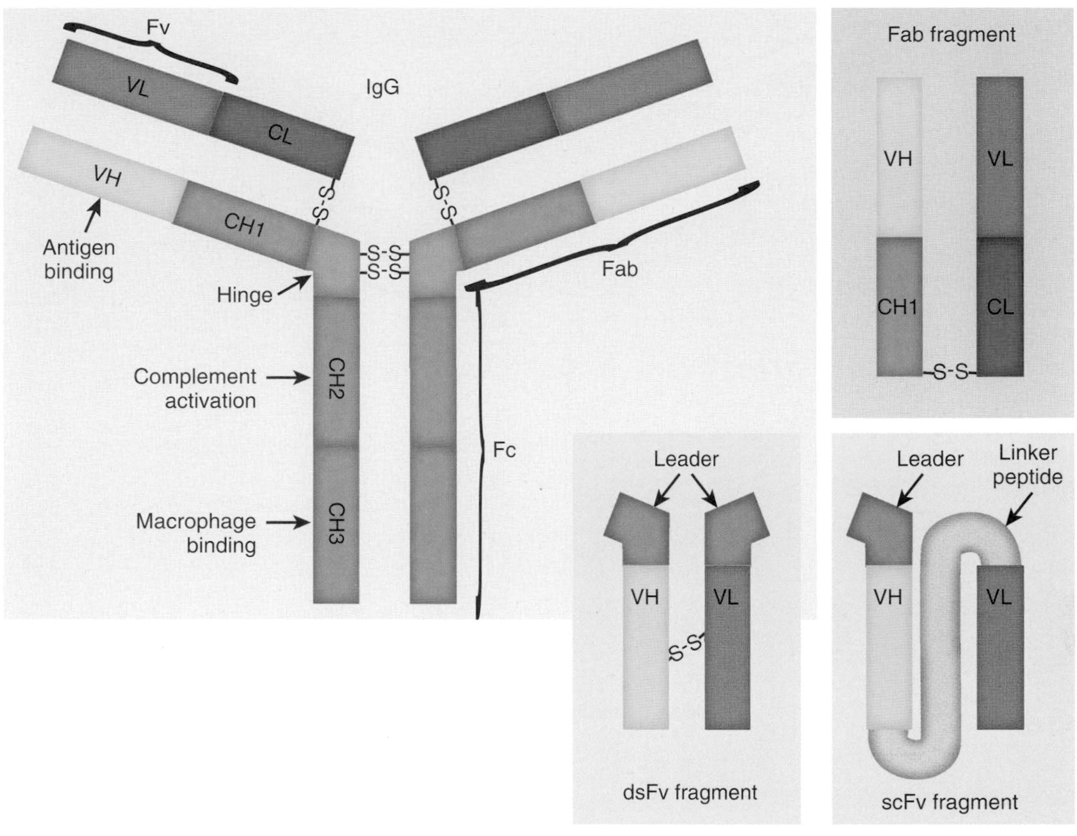

FIGURE 6.12 Fab and Fv Antibody Fragments

Fab fragments are produced by protease digestion of the hinge region. A disulfide bond holds the heavy and light chains together. To make an antibody fragment without any constant region, the genes for the VH domain and the VL domain are expressed in bacteria from a plasmid vector. This structure is unstable because of a lack of disulfide bonds. Therefore, disulfide bonds are engineered into the two halves (dsFv fragment), or a linker is added to hold the VH and VL domains together (scFv fragment).

usually 15 to 20 amino acids long. This chain is introduced at the genetic level so that a single artificial gene expresses the whole structure (VH-linker-VL or VL-linker-VH). A tag sequence (such as a His6-tag or FLAG-tag—see Chapter 9) is often added to the end to allow detection and purification. Such a scFv fragment is quite small, about 25 kDa in molecular weight.

Such scFv fragments are attached to various other molecules by genetic engineering. The role of the scFv fragment is to recognize some target molecule, perhaps a protein expressed only on the surface of a virus-infected cell or a cancer cell. A variety of toxins, cytokines, or enzymes may be attached to the other end of the scFv fragment, to provide the active portion of the final recombinant antibody. In principle, this approach provides a way of delivering a therapeutic agent in an extremely specific manner. At present, the clinical applications of engineered antibodies are under experimental investigation.

Recent work studying camel antibody structure has elucidated a new structure of an antibody not seen in any other model organisms studied to date. Antibodies in camels and their relatives (llamas and alpacas) have only the heavy chain and no light chains, and are called **heavy-chain antibodies** (**hcAb**) (Fig. 6.13). The ends of the heavy chain have the binding sites for the foreign antigens or paratopes. The streamlined structure has major implications for creating antibodies for therapeutic purposes.

The ability to create a small molecule from only the heavy-chain antigen-binding region offers many advantages over other antibody therapeutics. The variable domain of the single heavy-chain antibody called **VHH** is 12–15 kDa in size, which is much smaller than even scFv, and

FIGURE 6.13 Heavy-Chain Antibodies and Nanobodies
A conventional antibody has two heavy chains (purple) and two light chains (orange) held together with disulfide bonds. A heavy-chain antibody derives from a single protein and therefore does not have disulfide bonds. The nanobody is the isolated variable region from the heavy-chain antibody that is very small but has high affinity for its target antigens.

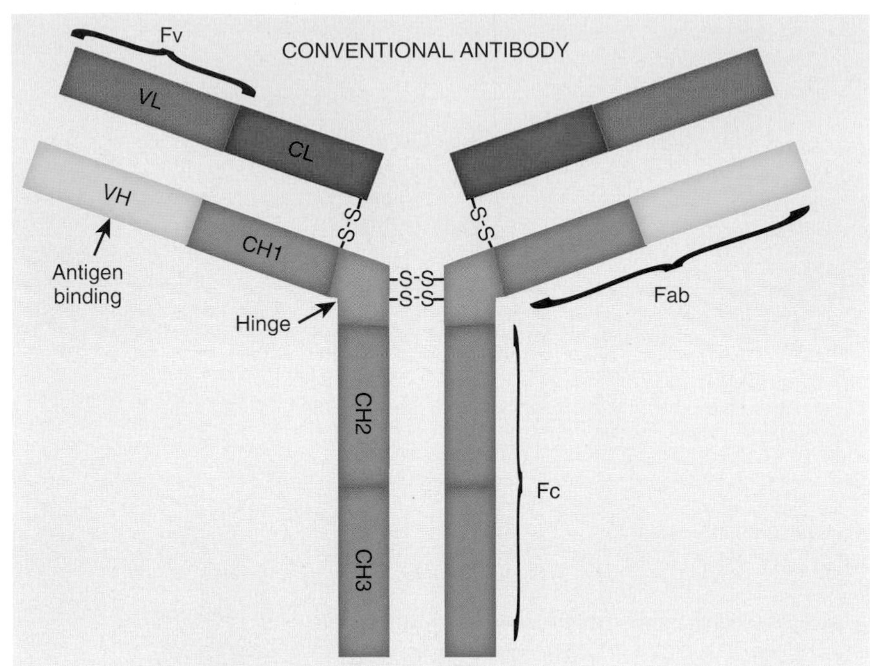

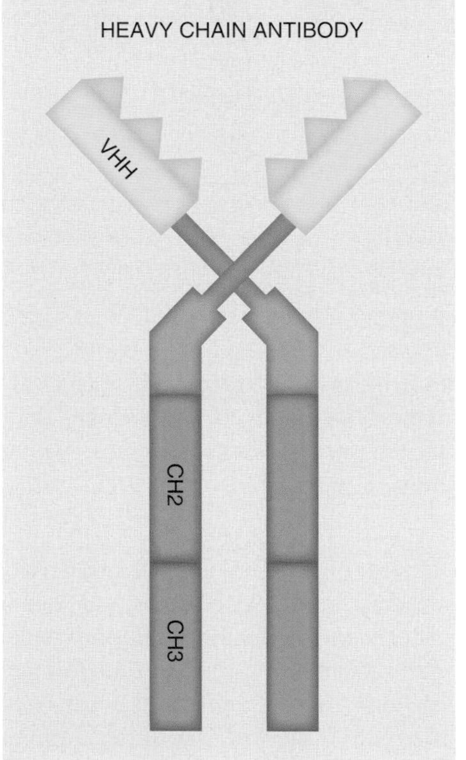

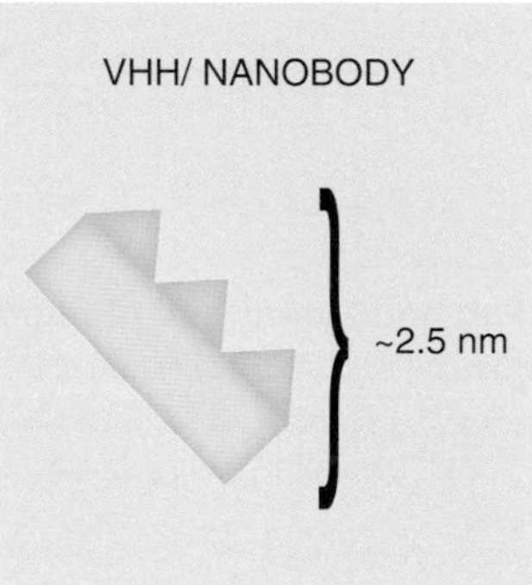

therefore a recombinant protein containing only this domain is called a **nanobody (Nb)** (Fig. 6.13). The structure makes these more amenable to protein engineering since Nbs are small, work as monomers, have no disulfide bonds, and are very stable, even maintaining their structure in high heat or denaturing conditions. They have a very high **affinity** for the antigen, but what is most interesting is the structure. As depicted in Figure 6.4, typical antibodies recognize protruding regions of the antigen, but the paratope of the VHH region actually is flexible. They can recognize epitopes that protrude as regular antibodies, and they can recognize epitopes that are dimples or concave in shape. That means VHH domains can bind directly to enzyme active sites buried within a protein. Another key to their potential function is size; the engineered form of

VHH, without the constant regions, can easily pass through the kidney, so they are rapidly cleared from the body. They can pass through the blood–brain barrier to target regions of the brain. For these reasons, one of the potential uses is for *in vivo* imaging or potentially as a biosensor. Nbs can also be humanized and conjugated to different small molecule therapeutics just as scFvs. As of writing this text, Nbs to TNFα are in clinical trials for treatment of rheumatoid arthritis.

The antigen-binding regions used in antibody engineering may be derived from pre-existing monoclonal antibodies, such as the TNFα antibody that was humanized. Alternatively, a library of DNA segments encoding V-regions may be obtained from a pool of B cells obtained from an animal or human blood sample. Such a library should, in theory, contain V-regions capable of recognizing any target molecule. Using a human source avoids the necessity for the complex humanization procedures described earlier. However, in this case, it is necessary to screen the V-region library for an antibody fragment that binds to the desired target molecule. This may be done by the **phage display** procedure outlined in Chapter 9. The library of V-region constructs is expressed on the surface of the phage, and the target molecule is attached to some solid support and used to screen out those phages carrying the required antibody V-region.

Heavy-chain antibodies from camels can be engineered to create small nanobodies. Nanobodies and single-chain Fvs are linked to various toxins, cytokines, or enzymes to create recombinant antibodies. These antibodies can be used to precisely deliver the toxin, cytokine, or enzyme to the antigen that the scFv or nanobody recognizes *in vivo*.

DIABODIES AND BISPECIFIC ANTIBODY CONSTRUCTS

Various engineered antibody constructs are presently being investigated. A **diabody** consists of two single-chain Fv (scFv) fragments assembled together. Shortening the linker from 15 to 5 amino acids drives dimerization of two scFv chains. This no longer allows intrachain assembly of the linked VH and VL regions. The dimer consists of two scFv fragments arranged in a crisscross manner (Fig. 6.14). The resulting diabody has two antigen-binding sites pointing in opposite directions. If two different scFv fragments are used, the result is a bispecific diabody that will bind to two different target proteins simultaneously. Note that formation of such a bispecific diabody requires that VH-A be linked to VL-B and VH-B to VL-A. It is, of course, possible to engineer both sets of VH and VL regions onto a single polypeptide chain encoded by a single recombinant gene, as shown in Figure 6.14. In the same manner, nanobodies can also be engineered to a bivalent or bispecific arrangement. Doing so increases their potency to their target antigens. Bispecific diabodies have a variety of potential uses in therapy because they may be used to bring together any two other molecules; for example, they might be used to target toxins to cancer cells.

Another way to construct an engineered bispecific antibody is to connect the two different scFv fragments to other proteins that bind together (Fig. 6.15). Two popular choices are **streptavidin** and leucine zippers. Streptavidin is a small biotin-binding protein from the bacterium *Streptococcus*. It forms tetramers, so it allows up to four antibody fragments to be assembled together. Furthermore, binding to a biotin column can purify the final constructs. Leucine zipper regions are used by many transcription factors that form dimers (see Chapter 2). Often, such proteins form mixed dimers when their leucine zippers recognize each other and bind together. Leucine zipper regions from two different transcription factors that associate (e.g., the Fos and Jun proteins) may therefore be used to assemble two different scFv fragments.

Linking two scFv fragments together with either polypeptide linker regions or proteins (e.g., streptavidin or leucine zipper proteins) creates divalent antibodies; that is, each side of the antibody will recognize a different antigen. These constructs are useful to bring two different proteins in close proximity in the cell.

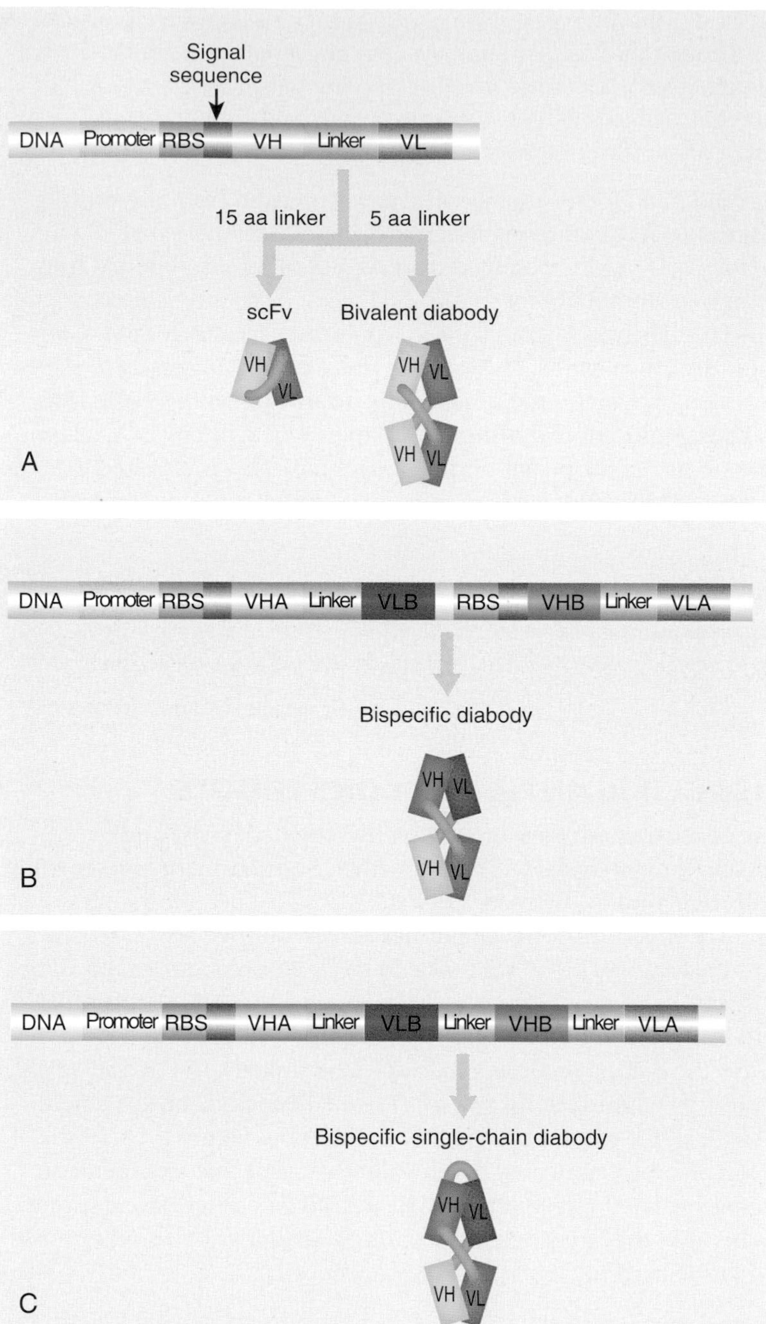

FIGURE 6.14
Engineered Diabody Constructs
(A) Engineering a diabody construct begins by genetically fusing the variable domains of the heavy and light chain (VH and VL) with a linker. The long linker allows a single polypeptide to form into a single antibody-binding domain. The short linker allows two polypeptides to complex into a diabody with two antibody-binding domains. (B) Instead of identical Fv units, two different Fv chains can be coexpressed in the bacterial cell. The two different Fv chains will unite into a diabody with two different antibody-binding domains, a different one on each side. (C) Bispecific antibodies can be made as one single transcript with a linker between VHA and VLB, a linker between the two halves, and finally a linker between VHB and VLA.

ELISA ASSAY

The **enzyme-linked immunosorbent assay (ELISA)** is widely used to detect and estimate the concentration of a protein in a sample. The protein to be detected is regarded as the antigen. Therefore, the first step is to make an antibody specific for the target protein. A detection system is then attached to the rear of the antibody. Usually, this system consists of an enzyme that generates a colored product from a colorless substrate. Alkaline phosphatase, which converts X-Phos to a blue dye (see Chapter 8), is a common choice. The samples to be assayed are immobilized on the surface of a membrane or in the wells of a microtiter dish (Fig. 6.16). The antibody plus detection system is added and allowed to bind. The membrane or microtiter

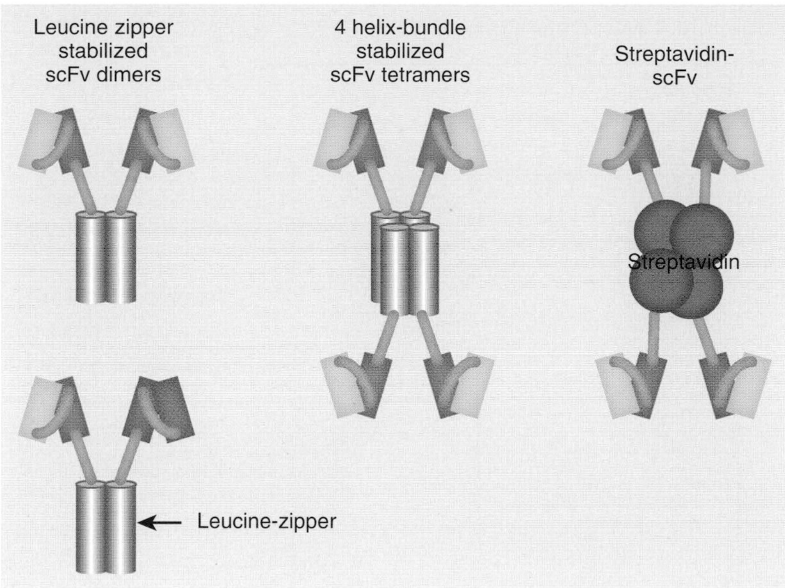

FIGURE 6.15
Engineered Bispecific Antibody Constructs
Instead of genetic linkers to hold diabodies, various proteins can also hold scFv fragments together. Proteins with a leucine zipper domain dimerize; therefore, when scFv genes are genetically fused to these proteins, the scFv domains come together as dimers. Proteins such as streptavidin or those with four helix bundle domains can be genetically fused to scFv domains. When expressed, there are four scFv domains on the outside, providing four different antibody-binding sites.

dish is then rinsed to remove any unbound antibody. The substrate is added, and the intensity of color produced indicates the amount of target protein in the original sample.

A variety of modifications of the ELISA exist. Often, binding and detection are done in two stages, using two different antibodies. The first antibody is specific for the target protein. The second antibody recognizes the first antibody and carries the detection system. For example, antibodies could be raised in rabbits to a series of target proteins. The second antibody, which recognizes rabbit antibodies, could be produced in sheep. These are called *secondary antibodies* and are often described as, for example, sheep anti-rabbit. The secondary antibody has the detection system, and because it will recognize any antibody made in a rabbit, it does not have to be re-engineered for each different target protein. This allows the use of the same final antibody detection system in each assay even if different primary antibodies are used to identify different proteins.

Antibodies are used in ELISA assays to determine the relative concentration of the target protein or antigen in a sample.

Primary antibodies recognize the target protein or antigen. Secondary antibodies recognize the primary antibody and often carry a detection system. Secondary antibodies are made to recognize any antibody that is made in sheep, cow, rabbit, goat, or mouse.

THE ELISA AS A DIAGNOSTIC TOOL

The ELISA is used in many different fields. Diagnostic kits that rely on the ELISA are produced for clinical diagnosis of human disease, dairy and poultry diseases, and even for plant diseases. The diagnostic kits are so simple that most require no laboratory equipment and using them takes as little as 5 minutes. ELISA kits can be used to detect a particular plant disease by crushing a leaf and smearing the leaf tissue on the antibody. When the disease-specific antigen reacts with the antibody, the antibody spot turns blue. In clinical applications, ELISA kits can detect the presence of minute amounts of pathogenic viruses or bacteria, even before the pathogen has a chance to cause major damage. Clinical ELISA kits detect various disease markers. In certain diseases, characteristic

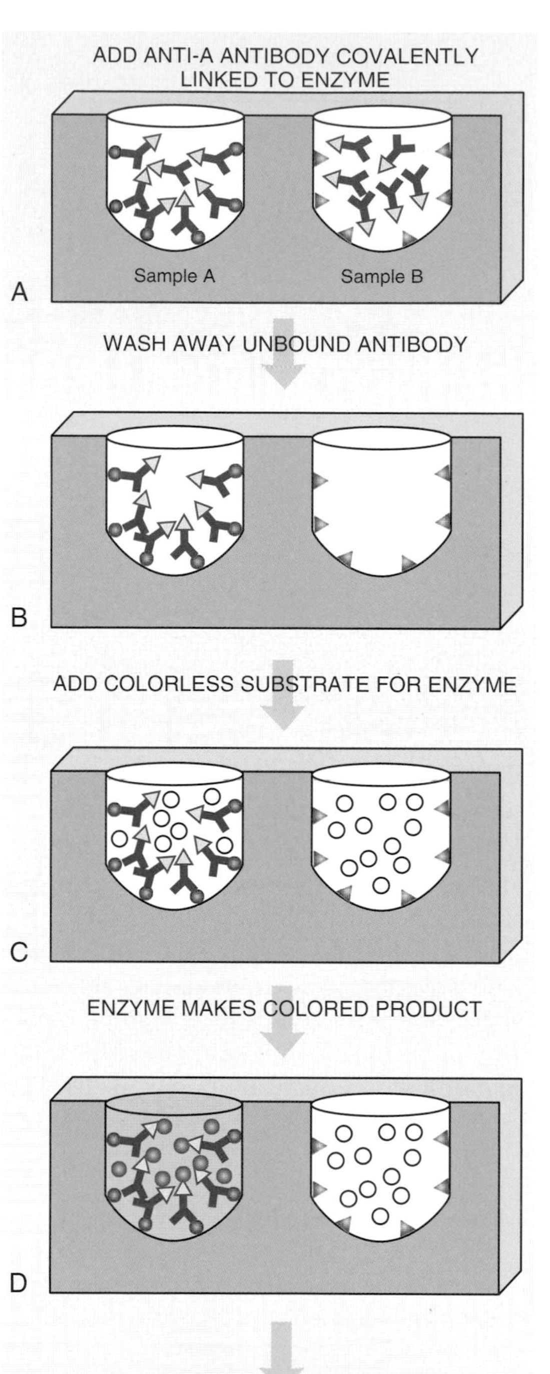

FIGURE 6.16
Principle of the ELISA
ELISA detects and quantifies the amount of a particular protein bound to the well of a microtiter dish. Anti-A antibody is linked to an enzyme such as alkaline phosphatase. The antibody recognizes only the circular protein, and not the triangular protein (A). After the antibody binds to its target, the unbound antibody is washed from the dish (B). A colorimetric substrate of alkaline phosphatase is added to each well (C), and wherever there is antibody, the substrate is cleaved to form its colorful product (D). The amount of color is proportional to the amount of protein.

proteins mark the start of disease progression long before the patient exhibits any symptoms. Detecting such markers can help diagnose and treat a problem before the disease causes serious damage.

ELISA diagnostic testing is even available for you to try at home. Home pregnancy kits are a simple, over-the-counter ELISA assay for **human chorionic gonadotropin (hGC)**. This protein is produced by the placenta and secreted into the bloodstream and urine of pregnant women. The actual pregnancy test has four important features (Fig. 6.17). First, the entire test is on a piece of paper that wicks the urine from one end to the other. This paper has three regions: first, a region where anti-hCG antibody is loosely attached to the paper strip; second, a region called the pregnancy window; and finally, a control window. As the urine wicks up the paper strip, any hCG present is bound by the anti-hCG antibody. If the woman is pregnant, the anti-hCG/hCG complex moves up the paper strip. If the woman is not pregnant, the anti-hCG antibody moves up the paper strip alone. (Even if the woman is pregnant, there is excess anti-hCG, and so unbound anti-hCG antibody is always found.) If the woman is pregnant, the anti-hCG/hCG complex reaches the pregnancy window, where it binds to secondary antibody #1. This is attached to the paper in the shape of a plus sign and cannot move. The secondary antibody has a color detection system attached to it. When the anti-hCG/hCG complex binds to the secondary antibody, it triggers color release and a plus sign forms. The control window contains secondary antibody #2. This control window recognizes only anti-hCG antibody that is not bound to hCG, so its color is activated whether or not the woman is pregnant.

ELISA is a powerful diagnostic tool because antibodies can be made to almost any protein. For pregnancy tests, any hCG in the urine binds to antibodies to hCG, which in turn bind to immobilized secondary antibody to form the plus sign.

VISUALIZING CELL COMPONENTS USING ANTIBODIES

Antibodies can be used to visualize the location of specific proteins within the cell. **Immunocytochemistry** refers to the visualization of specific antigens in cultured cells, whereas **immunohistochemistry** refers to their visualization in prepared tissue sections. In either technique, the first step is to prepare the cells. They must be treated to maintain

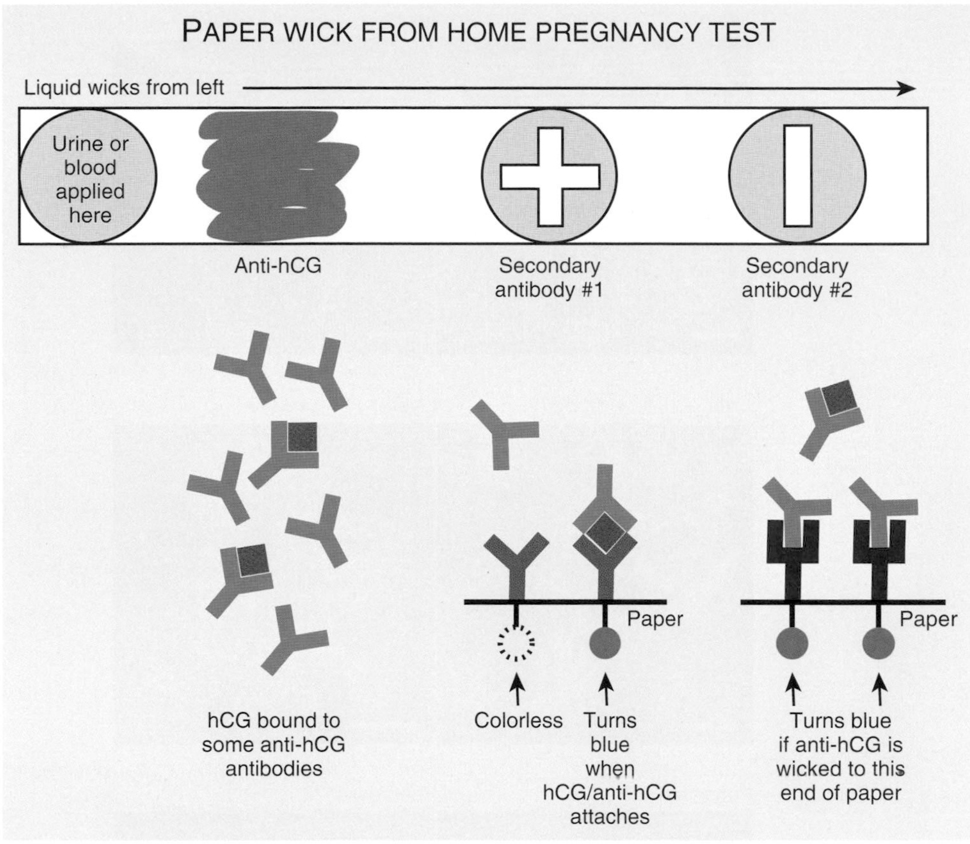

FIGURE 6.17
Home Pregnancy Tests Are an ELISA Diagnostic Tool
The pregnancy test shown here has four important areas along the paper wick. The urine or blood is applied on the far left and wicks to the right. The anti-hCG antibodies loosely attached to the paper are next. If the urine has hGC, this binds to its antibody and travels along the paper as a complex. In the next area, a secondary antibody that recognizes only the hCG–primary antibody complex is firmly attached in a plus pattern. When the hCG complex binds to the secondary antibody, the detection system turns blue. The final spot is a different secondary antibody that recognizes the primary antibody without any hCG. This is a positive control, to ensure that the antibody was released and wicked up the paper with the urine.

their cellular architecture so that the cells appear much as they would if still alive. Usually, the cells are treated with cross-linking agents such as formaldehyde or with denaturants like acetone or methanol. In immunohistochemistry, tissue samples can be frozen and then sliced into small, thin sections (about 4 mm), providing a two-dimensional view of the tissue. Another option is to embed the tissue sample in paraffin wax. Here, the cells are first dehydrated in a series of alcohol solutions and then treated with the wax. The tissue is then sectioned into thin two-dimensional slices as for frozen tissues.

Once a single, thin layer of prepared cells or tissue sections is readied, preserved cells are then permeabilized to make the antigen more accessible to the antibody. If in wax, the tissue sections are dewaxed and rehydrated. Fixed tissue sections can be irradiated with microwaves, which break the cross-links induced by the fixative, or the samples can be heated under pressure. Both methods allow the primary antibody to find its antigen within the sample. A secondary antibody contains the detection system to visualize the location of the antigen. The secondary antibody binds to the primary antibody/antigen complex, and then the appropriate reagents are added to visualize the location of the complex. (In some cases, a single antibody, with an attached detection system, is used.)

Antibody detection systems include enzymes or fluorescent labels. A common enzyme-mediated detection system is alkaline phosphatase, as with the ELISA (see Fig. 6.16). Fluorescently labeled antibodies must be excited with UV light, upon which the fluorescent label emits light at a longer wavelength. Samples are directly visualized with a microscope attached to a UV light source (Fig. 6.18). Fluorescent antibodies tend to bleach out when exposed to excess UV; therefore, the microscope is attached to a camera to record the data as a digital image.

Immunocytochemistry and immunohistochemistry use a primary antibody to a specific cellular target protein to visualize its location within the cell. The primary antibody is visualized by adding a secondary antibody with a detection system.

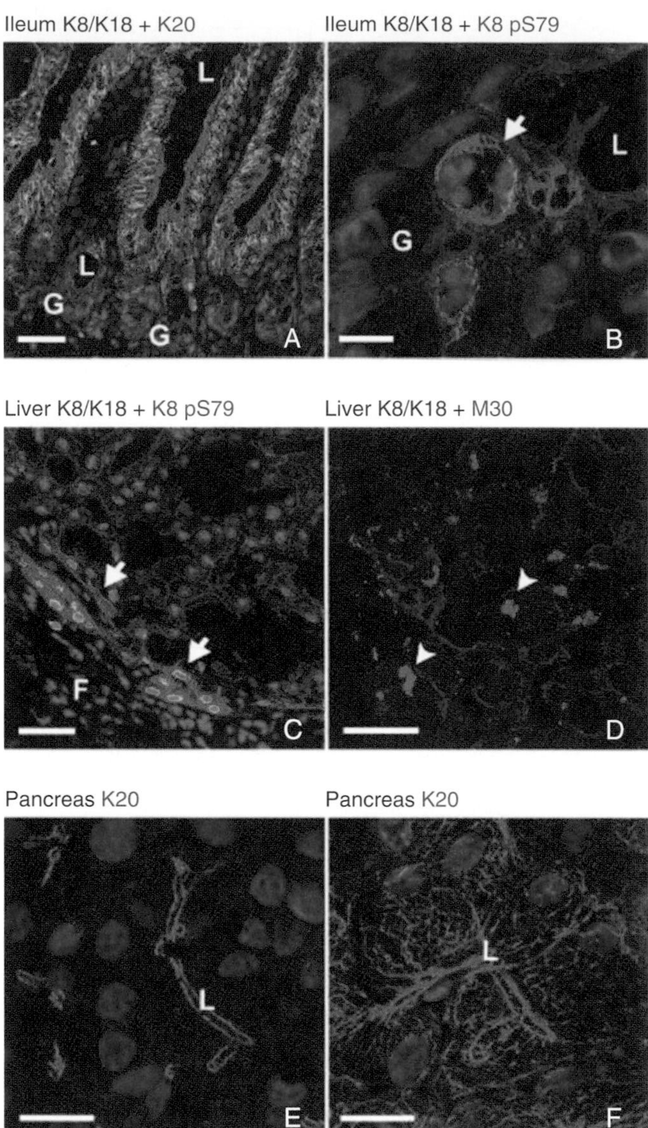

FIGURE 6.18 Fluorescent Antibody Staining

Keratin fibers provide structural integrity to all types of cells, and also comprise skin, hair, and nails. The different keratin forms that are created by post-translational modification can be visualized by using isoform-specific antibodies that recognize the different keratins. (A) Antibodies to keratin K8/K18 isoform were added to a section of mouse intestine (ileum). The secondary antibody for K8/K18 was labeled with a red fluorescent tag. In addition, antibodies from keratin K20 were added to the ileum section, and the secondary to this antibody was labeled with a green fluorescent tag. The areas that had both types of keratins fluoresced yellow. DNA was labeled with propidium iodide, which fluoresced blue. (B) Mouse ileum cells were labeled with two antibodies: one specific to keratin K8/K18 isoforms and one specific to the phosphorylated form of K8. The location where both proteins were found is yellow. DNA fluoresces blue. (C and D) Human cirrhotic liver cells were stained with K8/K18 (red) and antibodies to the phosphorylated form of K8 (green). Arrowheads highlight the phosphorylated isoform. (E and F) Mouse pancreas cell sections were incubated with antibodies to keratin K20 (green). The DNA in the nuclei is labeled blue. From Ku NO, et al. (2004). Studying simple epithelial keratins in cells and tissues. *Methods in Cell Biol* **78**, 489–517. Used with permission.

FLUORESCENCE-ACTIVATED CELL SORTING

As explained in the preceding section, fluorescent antibodies are used to find the location of intracellular proteins. Fluorescent antibodies are also able to bind to surface antigens. Many cells of the immune system have specific antigens on their surface that distinguish them from others. Each immune cell can have over 100,000 antigen molecules on their surface. These surface antigens characterize the different types of immune cells and are systematically named by assigning them a **cluster of differentiation (CD) antigen** number. The antigens were mostly identified before their physiological function was known. For example, CD4 antigens are associated with T-helper cells; and CD8, with killer T cells. Monoclonal antibodies are available to label many CD antigens, especially the most common.

Fluorescence-activated cell sorting (FACS) involves the mechanical separation of a mixture of cells into different tubes based on their surface antigens (Fig. 6.19). Because the antibody attaches to the outside of the cell, the cell does not have to be prepared as described previously. In this example, helper T cells and killer T cells can be separated from other white blood cells based on the presence of CD4 or CD8 surface antigens. First, the cell suspension is labeled with monoclonal antibodies to the surface antigens of interest. In this example, antibodies to CD4 and CD8 are used. Both antibodies have fluorescent labels that are different for the two antibodies. The labeled cell suspension is loaded into a charging electrode. Drops of liquid containing only one cell are released to the bottom, and the fluorescence detector notes whether or not the drop of liquid is labeled for CD4 or CD8 by the color of its fluorescence. If the drop has an antigen, an electrical charger pulls or pushes the droplet to the right or left, separating the two antigens into separate tubes. If the drop has no antigen in it, it gets no electrical charge and goes into a third tube. Usually, two different antibodies are used, but some of the newer FACS machines can sort up to 12 different fluorescently labeled antibodies and can sort up to 300,000 cells per minute.

Flow cytometry is a related technique to analyze fluorescently labeled cells. As with FACS, cells are labeled with monoclonal antibodies to cell-surface antigens. The antibodies are conjugated to a variety of different fluorescent labels, and each antibody is detected based on its fluorescence. The cells are loaded into a charging electrode and released in small droplets. During flow cytometry, the cells are not sorted and saved;

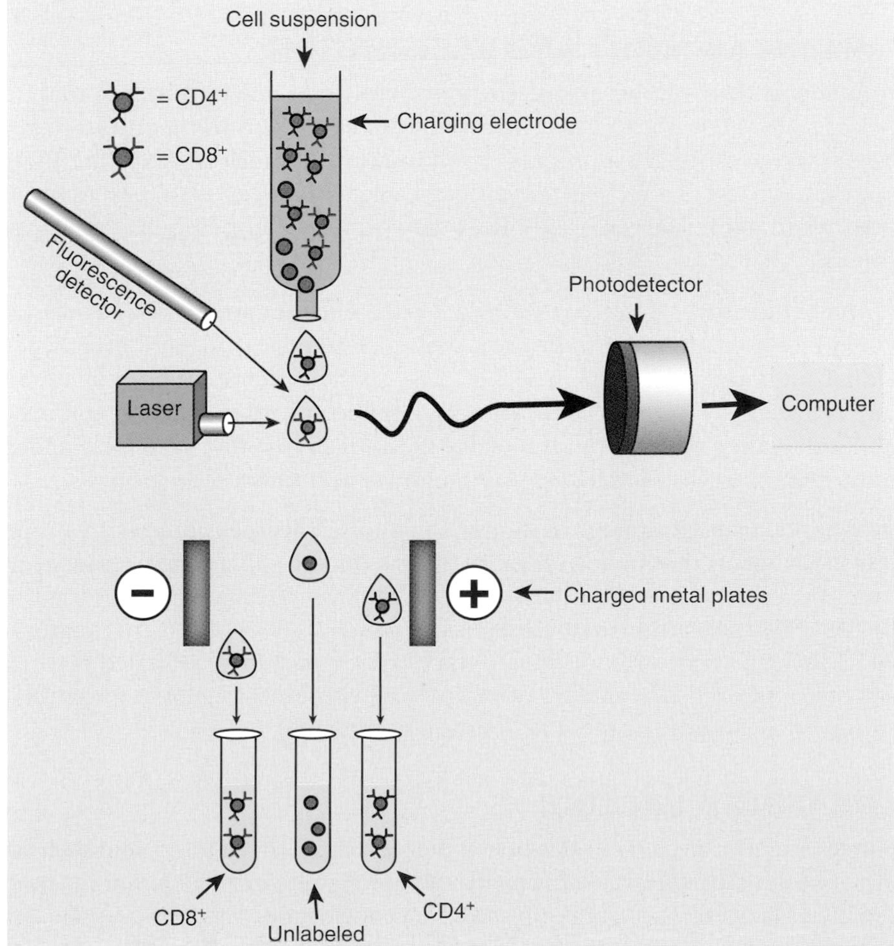

FIGURE 6.19
FACS Separates CD4+ and CD8+ Cells
FACS machines can separate fluorescently labeled cells into different compartments. A mixture of CD4+, CD8+, and unlabeled cells is separated based on their fluorescence. When the fluorescence detector notes green, the charged metal plates pull that drop to the left or minus plate, allowing those cells to collect into the left tube. If no fluorescence is detected, the drop stays neutral and is collected in the middle tube. If the drop fluoresces red, the charged plates pull the drop to the plus side, and it collects in the right tube.

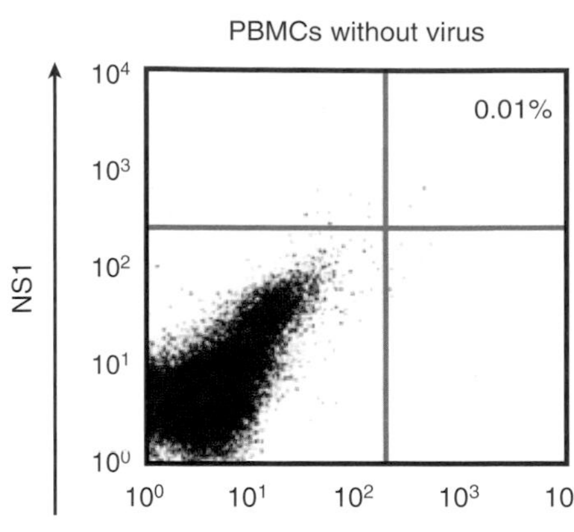

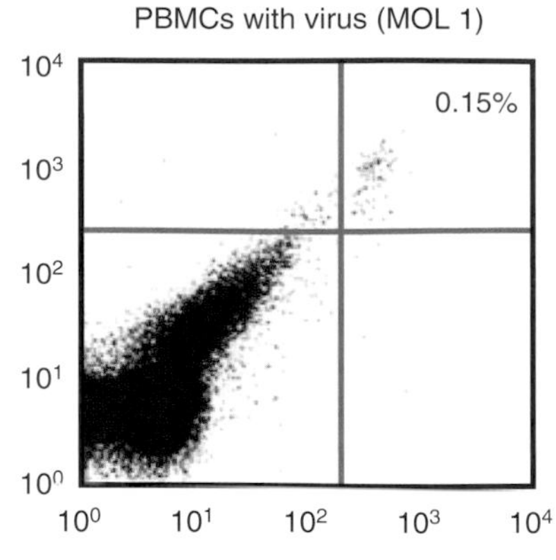

FIGURE 6.20
Example of Flow Cytometry Data
Peripheral blood mononuclear cells (PBMCs) were infected with dengue virus without (*left*) and with an antibody that stimulates viral infection (*right*). The cells were then fixed, permeabilized, and labeled with anti-E protein or anti NS1 antibodies, two viral proteins. The cells that express both viral proteins represent infected PBMCs, which are increased in number in the presence of the stimulatory antibody. From Fu Y, et al. (2014). Development of a FACS-based assay for evaluating antiviral potency of compound in dengue infected peripheral blood mononuclear cells. *J Virological Methods* **196**, 18–24. Reprinted with permission.

instead, the sample of cells is measured and discarded. As the cells pass the detector, the computer records the fluorescence and plots the number of cells with each of the fluorescent labels. These cells are plotted, with a small dot representing each of the cells (Fig. 6.20).

FACS and flow cytometry use monoclonal antibodies to surface antigens. The FACS machine can sort the cells into individual samples, and flow cytometry simply records the fluorescent label and plots the data on a graph.

IMMUNE MEMORY AND VACCINATION

Individuals who survive an infection normally become immune to that particular disease, although not to other diseases. The reason is that the immune system "remembers" foreign antigens, a process called **immune memory**. Next time the same antigen appears, it triggers a far swifter and more aggressive response than before. Consequently, the invading microorganisms will usually be overwhelmed before they cause noticeable illness.

Immune memory is due to specialized B cells called **memory cells**. As discussed earlier, virgin B cells are triggered to divide if they encounter an antigen that matches their own individual antibody. Most of the new B cells are specialized for antibody synthesis, and they live only a few days. However, a few active B cells become memory cells, and instead of making antibodies, they simply wait. If one day the antigen that they recognize appears again, most of the memory cells switch over very rapidly to antibody production.

Vaccination takes advantage of immune memory. **Vaccines** consist of various derivatives of infectious agents that no longer cause disease but are still antigenic; that is, they induce an immune response. For example, bacteria killed by heat are sometimes used. The antigens on the dead bacteria stimulate B-cell division. Some of the B cells form memory cells so, later, when living germs corresponding to the vaccine attack the vaccinated person, the immune system is prepared. The makers of vaccines are constantly trying to find different ways to stimulate the immune system without causing disease.

CREATING A VACCINE

Because vaccines are such a huge part of the biotechnology industry, and such an important part of our health-care system, much research and money are invested in finding new and improved vaccines. Many vaccines are administered to young babies; thus, ensuring the safety and effectiveness of vaccines is critical (see Box 6.1). Many different methods of developing a vaccine exist.

Most vaccines are simply the disease agent, killed with high heat or denatured chemically. Heat or chemical treatment inactivates the virus or bacterium so it cannot cause disease. Yet enough of the original structure exists to stimulate immunity. When the live agent infects the vaccinated person, memory B cells are activated and the disease is suppressed. Such whole vaccines elicit the best immune response, but many diseases cannot be isolated or cultured to

Box 6.1 Vaccine Safety

In the United States, infants receive vaccines for many different illnesses, including diphtheria, tetanus, pertussis (whooping cough), measles, mumps, rubella, chickenpox, polio, and hepatitis A and B. All these vaccines are given to children before they enter school. The list is long, but many of the vaccines are combined into one shot. Paradoxically, the effectiveness of vaccines has made many question their use. Many argue that vaccines are not needed because so few people actually get these diseases. It is easy to forget that the reason why very few people get diphtheria or measles is that so many are vaccinated. In 1980, about 4 million people contracted measles, but only about 10% of the world population had received measles vaccine. In 2012, about 122,000 cases of measles were recorded in the world, but about 84% of the world population had received the vaccine for measles. The percentage of children receiving the vaccine has increased from 72% in 2000 to 84%, which has resulted in a 78% decrease in the number of children diagnosed with measles worldwide. In the United States, the number of cases of measles varies from year to year, with only 37 in 2004 and 220 in 2011. Overall, these numbers are low, so if you are not vaccinated, the likelihood of contracting measles is very slim. However, the more people who opt not to vaccinate their children, the more cases of the disease there will be, and those who remain unvaccinated will gradually be at increased risk. In 2014, a record number of measles cases were being reported in the United States with a total of 644 cases from 27 states reported to the CDC. In early 2015, a multi-state outbreak for measles occurred and was linked to an amusement park in California. The outbreaks are probably due to an unvaccinated child from another country, and demonstrate that the greater number of people who do not receive a vaccine, the more likely diseases can spread. These types of outbreaks raise awareness that a vaccine is not just to protect the individual, but it is to protect the community also. Especially those persons in a community that cannot receive vaccines because of other health issues or allergies to vaccine components. If these individuals are surrounded by vaccinated people, they are less likely to contract the disease.

Other vaccines have been eliminated from the childhood immunization schedule because the diseases have been eradicated. For example, so many people across the world were vaccinated against smallpox that the disease was not seen at all for years. Now, smallpox vaccine is no longer given to the entire population. The only smallpox that exists is kept in two different labs, one in the United States and one in Russia. The fear of smallpox re-emerging as a disease is always present, but massive immunizations are not needed when there are no people with smallpox as of now.

Other vaccines have the opposite issue: Even with widespread vaccination for pertussis, the number of cases of whooping cough is on the rise. In 2012, there were 48,277 cases and 20 deaths from infection, whereas in 2002, the Centers for Disease Control reported 9771 cases in the entire United States. Unfortunately, the deaths were mainly in infants who were too young to be immunized for pertussis. To prevent babies from contracting pertussis, doctors now urge pregnant mothers to get re-immunized for the disease. In addition, another booster shot is now recommended for teens and for any adults who did not receive a booster shot of pertussis vaccine in their teens. Many different theories exist that try to explain the increase in whooping cough. Some attribute the use of a more sensitive test to diagnose whooping cough, and others suggest this may be a natural cycle of *B. pertussis* pathogenicity. Others attribute the increase to waning immunity. Once a child receives the last booster shot at age 5, the immunity to whooping cough wanes after about 10 years.

Vaccines cause some adverse side effects. In most cases, vaccines cause a local reaction, pain, and swelling at the injection site. Other possible side effects are systemic, perhaps a fever or a mild form of the disease, as is the case with the flu shot. Some vaccines can cause allergic reactions because of impurities in the vaccines. Some vaccines are made in eggs, and traces of egg proteins may remain in the vaccines. Often people with allergies to eggs still tolerate the vaccine, but some may have an allergic reaction. Another potential allergenic component is gelatin. Of course, anyone allergic to a vaccine component cannot be vaccinated, and therefore rely upon those people that surround them to be protected from the disease.

Other safety concerns about vaccines are based on the preservatives. Until 1999, the most common preservative was thimerosal, a mercury-containing compound. Thimerosal can cause allergic reactions in some children and has also been thought to cause autism. Unfortunately, the timing for diagnosis of autism and receiving the vaccines coincide, and therefore, many people believe that the shot was the cause for the onset of autism. Although the timing is coincident, there are no studies that show that the vaccine causes autism. The number of children that develop autism is identical for the ones that are vaccinated in comparison to those that are not vaccinated. The true cause of autism is an extremely active area of research, and hopefully some answers will identify the true cause for this devastating diagnosis.

make whole vaccines. Other times, the cost of culturing the pathogen is prohibitive. Moreover, growing live viruses is a dangerous job, with potential exposure of lab workers. With these limitations in mind, many different strategies have been developed to make improved vaccines.

Attenuated vaccines are still-living pathogens that no longer express the toxin or proteins that cause the disease symptoms (Fig. 6.21). Sometimes, viruses or bacteria are genetically engineered to remove the genes that cause disease. Other attenuated vaccines are related but nonpathogenic strains of the infectious agent (see Box 6.2). Making attenuated virus does not pose the same risks as for live virus. However, much research is needed to identify those genes that cause disease. Another disadvantage is that an attenuated virus might revert to the pathogenic version, especially if the attenuated virus has only one of the disease-causing antigens destroyed or mutated.

Subunit vaccines are effective against one component or protein of the disease agent, rather than the whole disease (Fig. 6.22). Subunit vaccines are available only because of recombinant DNA technology. The first step in creating a subunit vaccine is identifying a potential protein

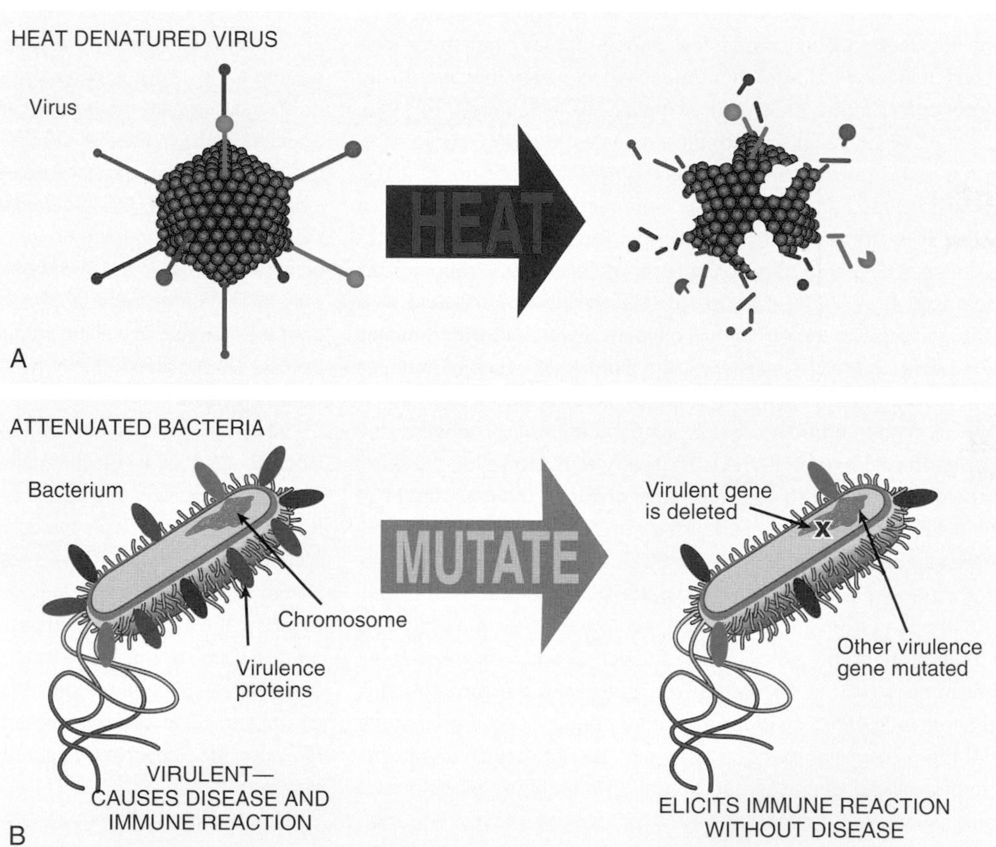

FIGURE 6.21 Whole Vaccines Include Killed or Attenuated Pathogens
(A) High heat or chemical treatment kills pathogens but leaves enough antigens intact to elicit an immune response. Once exposed to a dead virus or bacterium, memory B cells are established and prevent the live pathogen from making the person sick.
(B) Attenuated viruses or bacteria have been mutated or genetically engineered to remove the genes that cause illness. The immune system generates antibodies to kill the attenuated pathogen and establishes memory B cells that prevent future attack.

Box 6.2 Cowpox and Smallpox

Infection with cowpox produces only mild disease but gives immunity to the frequently fatal smallpox. In medieval times, a substantial proportion of the population caught smallpox. About 20% to 30% of those infected died, and the survivors ended up with ugly pockmarks on their faces—hence, the name *smallpox*. Milkmaids rarely suffered from smallpox because most had already caught cowpox from their cows. Consequently, milkmaids were seldom pockmarked and gained a reputation for beauty due to their unblemished skin. This observation led to Edward Jenner's classic experiments in which he Inoculated children with cowpox and demonstrated that inoculation protected against infection with smallpox. The term *vaccination* is derived from *vacca*, the Latin for "cow."

or part of a protein that elicits a good immune response. Most subunit vaccines are made from proteins found on the outer surface of the virus or bacterium because they elicit the strongest immune response. Experiments must be done to evaluate the protein chosen for the subunit vaccine. Once a suitable protein is identified, its gene is isolated and then expressed in cultured mammalian cells, eggs, or some other easily maintained system. The target protein is isolated from other proteins and used to immunize mice to test its effectiveness. After extensive testing in animals, the purified protein can be used as a vaccine.

Sometimes subunit vaccines fail, perhaps because the protein does not form the correct structure when expressed in mammalian cells or eggs. In these cases, **peptide vaccines** are created. These vaccines use just a small region of the protein. Since such peptides are small, they are conjugated to a **carrier** or **adjuvant** to stimulate a stronger immune response (Fig. 6.23).

Other vaccines target multiple proteins from a virus or multiple related viruses in one dose to decrease the number of immunizations administered. These **multivalent** vaccines are common and include the flu vaccine and MMR vaccine (measles, mumps, and rubella). These vaccines have antigens to a number of different related viruses. In the case of the flu vaccine, heptavalent forms include antigens to the seven most commonly found strains of influenza circulating in the population. Injection of the different antigens elicits an immune response to each of the different types. Unfortunately, influenza viruses evolve and change rapidly, so although the vaccine will protect the person from the known seven strains, a newly developed influenza type could still cause an infection in an immunized patient.

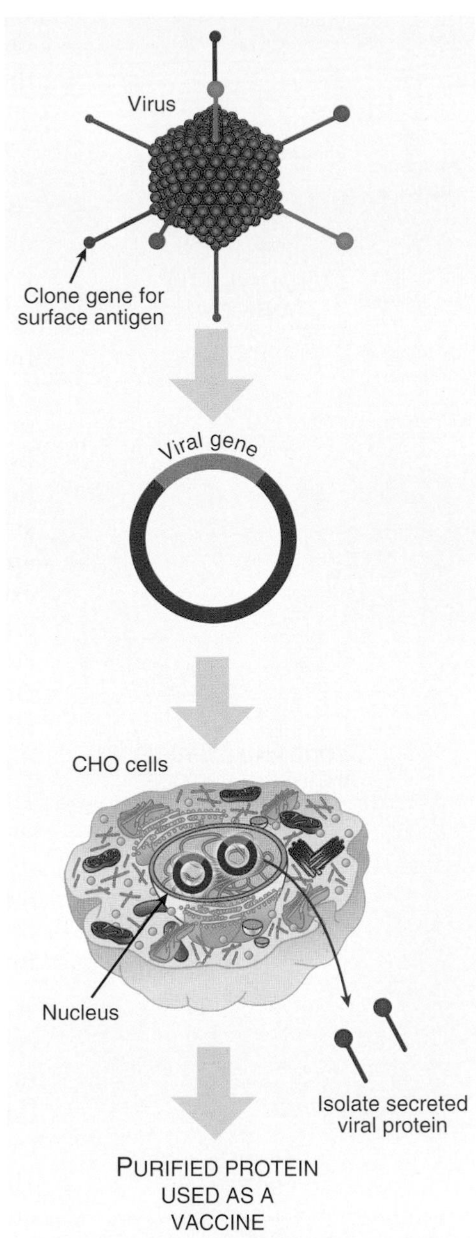

FIGURE 6.22 Subunit Vaccines Rely on a Single Antigen
A single antigenic protein from a pathogen is isolated, and its gene is cloned into an expression vector. The gene is expressed in cultured mammalian cells (such as Chinese hamster ovary [CHO] cells), isolated, purified, and used as a vaccine.

Killed pathogens, attenuated pathogens, single proteins, or epitopes from a disease-causing pathogen are used as vaccines. They are isolated and injected into people to elicit their immune response without causing the disease. Multivalent vaccines contain antigens to different proteins from a pathogen or family of pathogens.

MAKING VECTOR VACCINES USING HOMOLOGOUS RECOMBINATION

Another method of displaying a foreign antigen for use as a vaccine is the **vector vaccine**. Here, genetic engineering creates a nonpathogenic virus or bacterium that expressed an antigen from the disease-causing virus. When this virus or bacterium infects a person, it induces immunity both to the nonpathogenic microorganism and to the attached antigen.

FIGURE 6.23 Peptide Vaccines Are Conjugated to Carrier Proteins
Peptide vaccines are small regions of an antigenic protein from a pathogen. The peptide is often an epitope that elicits a strong immune response. Because the peptide is small, multiple peptides are conjugated to a carrier protein to prevent degradation and to stimulate the immune system.

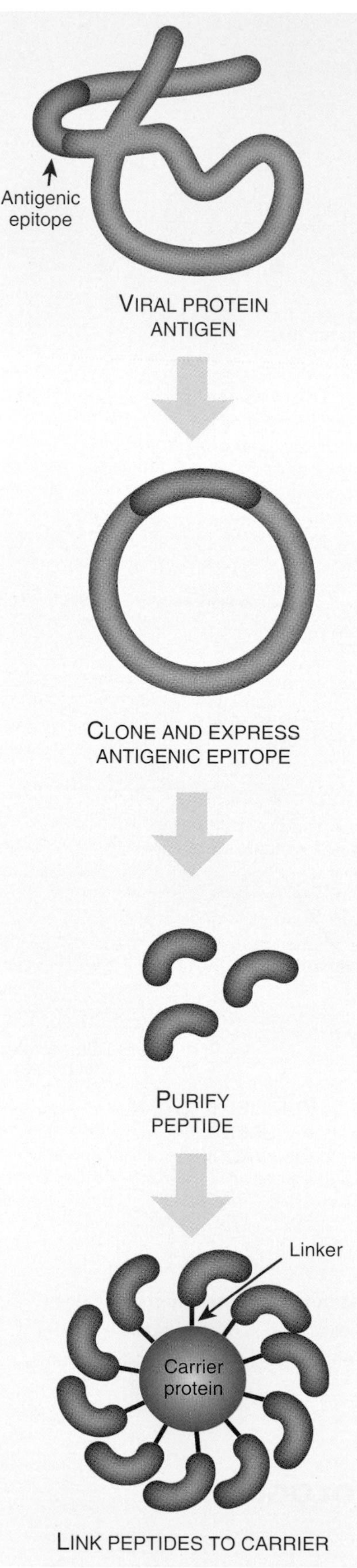

For example, vaccinia virus is a nonpathogenic relative of the smallpox virus. Using vaccinia virus is so effective that smallpox was eradicated. If vaccinia virus expresses an antigen from another deadly virus, the person vaccinated would gain immunity to smallpox and the other virus at the same time. Indeed, multiple genes could be inserted, conferring resistance to multiple diseases. The benefit of using vaccinia virus is that it is very potent and stimulates development of both B cells and T cells.

Inserting genes into the vaccinia genome is awkward because the genome has very few restriction enzyme sites, but genes can be added using **homologous recombination** (Fig. 6.24). In homologous recombination, two segments of similar or homologous DNA sequences align, and one strand of each DNA helix is broken and exchanged to form a **crossover**. A single crossover creates a hybrid molecule; if two crossovers occur close together, entire regions of DNA are exchanged. During homologous recombination in vaccinia, a region of single-stranded DNA is generated from a double-stranded break in the incoming new gene. The single-stranded region invades the double helix of the vaccinia genome to form a triple helix. One of the strands from vaccinia then is free to hybridize to the single-stranded homologous region on the incoming gene. If this occurs on both sides, the foreign gene is inserted into the vaccinia genome.

There are many examples of vaccines that use vaccinia virus as a way to stimulate an immune reaction. For example, a pentavalent vaccinia virus that expresses five different antigens to protect patients from H5N1 influenza virus is under development. The antigens include H5 hemagglutinin, N1 neuraminidase protein, a nucleoprotein (NP), and two matrix proteins (M1 and M2). In addition, the adjuvant contains IL-15, a cytokine produced by the immune system, which functions to stimulate natural killer cells and the innate immune response. Mice that were injected with the IL-15 mixed with adjuvant had a higher serum concentration of antibodies to the five influenza antigens. These antibodies were produced faster and in greater numbers than the control mice that only received the vaccinia vector vaccine without IL-15, suggesting this strategy could provide quicker and stronger immunity to the flu (Fig. 6.25).

Changing a harmless virus or bacteria so that it expresses a protein from a disease-causing pathogen on its surface can trick the immune system into making antibodies to the disease-causing pathogen.

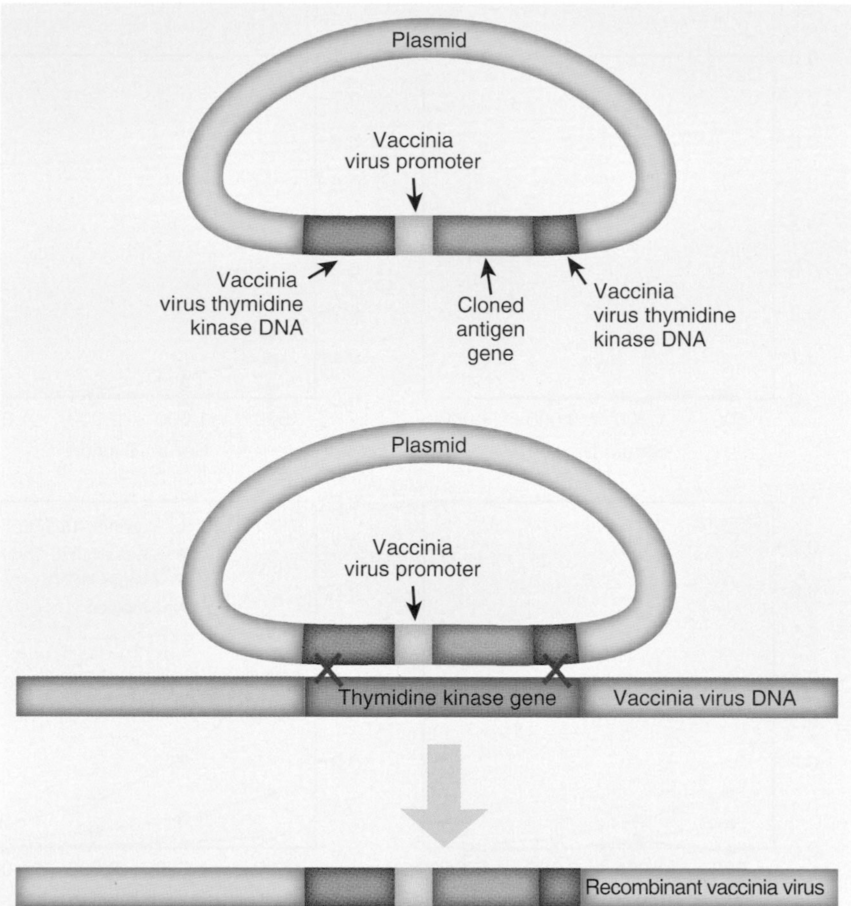

FIGURE 6.24 Homologous Recombination Adds New Genes to the Vaccinia Genome
The plasmid contains two regions homologous to the virus thymidine kinase gene on each side of the cloned antigen gene. When the plasmid aligns with the vaccinia genome, the regions of homology elicit a recombination event. The recombinant vaccinia will acquire the cloned antigen gene and lose the gene for thymidine kinase.

REVERSE VACCINOLOGY

Many genomes from infectious agents have now been sequenced. **Reverse vaccinology** takes advantage of this information to find new antigens for use in immunization (Fig. 6.26). The primary research begins with cloning each of the genes from the infectious organism into an expression library. Each of the proteins in the library is expressed and isolated. Complex mixtures of these different proteins are screened in mice for immune response, and when a pool induces a response, the proteins are subdivided, until each protein is tested for stimulating the immune system and for its ability to protect the mice from the actual infectious agent. The proteins that elicit the best response can either be combined into a subunit vaccine or used as separate vaccines.

Reverse vaccinology has been used to create a vaccine for *Neisseria meningitidis* serogroup B, which is a major cause of meningitis in children. Attenuated bacteria were not effective as vaccines, and until the sequencing of the *N. meningitidis* genome, no vaccine was available. A library of 350 different *N. meningitidis* proteins was expressed in *E. coli,* and purified. Each was individually assessed computationally for its ability to induce an immune response. Surface proteins were then screened to see if they elicited an immune response. Of the 350 tested proteins, only 29 became potential candidates. There are three most promising isolates: Factor H-binding protein (fHbp), Neisserial heparin binding antigen (NHBA), and Neisserial adhesion A (NadA). As of writing this chapter, a fusion protein of NHBA, fHbp, and NadA has been evaluated and is so far successful, but it still is awaiting full approval. Without the ability to sequence genomes, vaccine development was often impossible, but now new and emerging diseases can be studied to find potential vaccines.

Reverse vaccinology uses the expressed genomic sequences to find new potential vaccines. Normal vaccines are created using the pathogenic organism. The term *reverse* refers to the use of expressed DNA over the purified proteins from the organism itself.

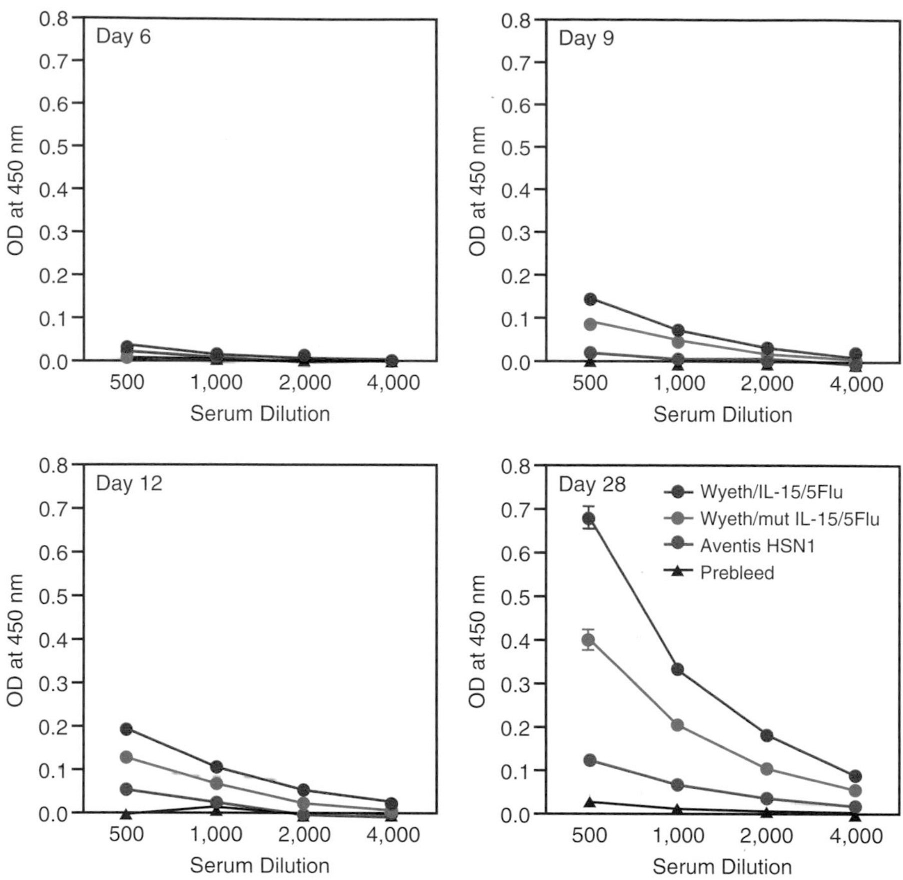

FIGURE 6.25
Amount of Antibodies to H5 Hemagglutinin after Vaccination with Flu Vaccine with or without 1L-15
The amount of antibodies to H5 hemagglutinin was measured by ELISA and recorded as OD at 450 nm. The serum from mice before vaccination (pre-bleed, black line) was compared to mice vaccinated with commercially available H5N1 vaccine (Aventis H5N, green line) and mice vaccinated with the pentavalent vaccine to H5N1 (Wyeth/IL-15/5Flu, red line; and Wyeth/mut IL-15/5Flu, blue line). The addition of IL-15 (red) stimulated a greater amount of antibodies to H5 hemagglutinin in comparison to mice stimulated with a mutated version of IL-15 (blue) that does not stimulate the immune system. Serum titers were compared on four different days after vaccination: 6, 9, 12, and 28. The amount of H5 hemagglutinin antibodies appeared at day 9 to 12 after vaccination, whereas the commercially available vaccine took 28 days for the same response. From Poon LL, et al. (2009). Vaccinia virus-based multivalent H5N1 avian influenza vaccines adjuvanted with IL-15 confer sterile cross-clade protection in mice. *J Immunol* **182**, 3063–3071. Reprinted with permission.

IDENTIFYING NEW ANTIGENS FOR VACCINES

Another approach to creating vaccines is to identify bacterial pathogen genes that are expressed when the pathogen enters the host. These genes usually encode proteins that are different from surface antigens. They encompass a variety of adaptations the pathogen makes in order to live within the host organism. Typically, bacteria that enter animals are engulfed by phagocytes and digested by the enzymes within the lysosome. Some pathogens are engulfed as usual but avoid digestion. Modifications required to live intracellularly include changes in nutrition and metabolism and mechanisms to protect against host attacks. Many different types of genes are needed for this switch, and the products of these genes are potential antigens for vaccine development.

Traditionally, identifying genes that are expressed only in host cells relies on gene fusions. Suspected genes or their promoters are genetically fused to a reporter such as β-galactosidase, luciferase, or green fluorescent protein (GFP), or to epitope tags such as FLAG or myc (see Chapter 9). The fusion gene is introduced into the pathogenic organism, which is then allowed to infect the host. The amount of reporter gene expression correlates with the expression level of that particular genomic region, whether it is a promoter or actual gene with its promoter. For example, if a gene linked to GFP increases in fluorescence after host cell invasion, then the target gene is a potential vaccine candidate because it may be important for bacterial pathogenesis.

Individual gene fusions are fine for suspected genes, but screening for novel genes with this method would be tedious. Instead, **differential fluorescence induction (DFI)** uses a combination of GFP fusions and FACS sorting (see earlier discussion) to identify novel genes involved with host invasion (Fig. 6.27). First, a library of genes or genomic fragments from the pathogenic organism is genetically linked to GFP. The library is transformed into bacterial cells where the gene fusions are expressed to give GFP. The bacteria are then given a specific stimulus related to host invasion. For example, when phagocytes engulf them, bacteria leave a neutral environment (pH 7) and enter a compartment that is acidic (pH 4). To determine if pH change induces gene expression, the bacteria with the fusion library are shifted to an acidic environment. They are then sorted using FACS to collect those with high GFP expression. If the novel gene fused to GFP is truly induced by low pH, its GFP levels should drop when it is shifted back to neutral pH. Therefore, the cells with high GFP expression are shifted to pH 7 and resorted, but this time bacteria with low levels of GFP are collected. The smaller pool of bacteria are again stimulated with low pH and sorted, collecting those with high GFP expression. This sorting scheme eliminates genes that are constitutively expressed plus those that are not induced by low pH. The remaining genes are acid-induced genes that adapt the organism to living within the host. They may then be evaluated as antigens for vaccine development.

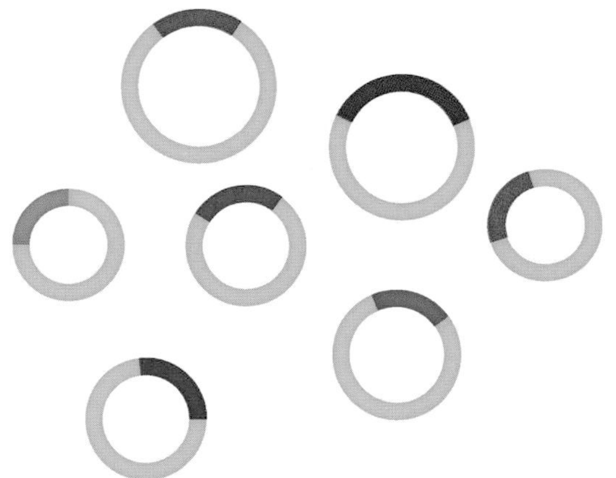

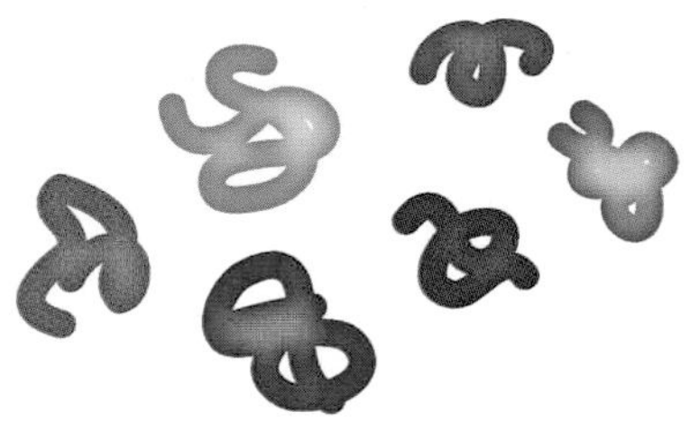

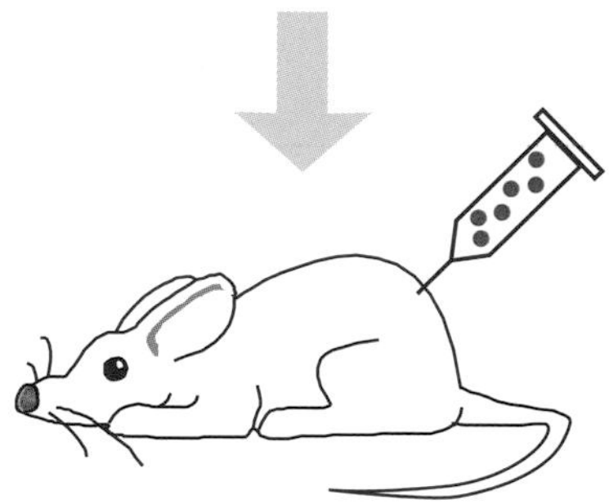

FIGURE 6.26
Reverse Vaccinology
Reverse vaccinology uses the genes identified in the genome of pathogenic agents. First, the genes are cloned into expression vectors and expressed to give proteins. Each potential antigen is screened for an immune response.

Another method to identify new antigens for vaccine development is ***in vivo* induced antigen technology (IVIAT)** (Fig. 6.28). This method takes serum from patients who have been infected with a particular disease to which a vaccine is needed. The serum is a rich source of antibodies against the chosen disease agent. The serum is then mixed with a sample of the disease-causing microorganism. Doing so removes antibodies that bind to cell-surface proteins expressed by the microorganism. This process leaves a pool of antibodies against proteins that are expressed only during infection. To identify the proteins corresponding to these antibodies, scientists construct a genomic expression library containing all the genes from the microorganism. The library is expressed in *E. coli* and is probed by the remaining antibodies. When an antibody matches a library clone, the gene insert is sequenced to identify the protein antigen. This method directly identifies protein antigens that stimulated antibody production during a genuine infection; therefore, antigens identified by this method are likely vaccine candidates.

Pathogens must change their metabolism when changing from a free-living organism to the environment within their host. The proteins that help the pathogen adapt to this switch are potential proteins to which a vaccine could be made.

DFI and IVIAT are two techniques to identify proteins that allow pathogens to live within an organism. DFI fuses the potential proteins to fluorescent tags and selects the clones that are only expressed inside the organism. IVIAT uses serum from patients who have been infected with the pathogen to find the antibodies that bind intracellular pathogenic proteins.

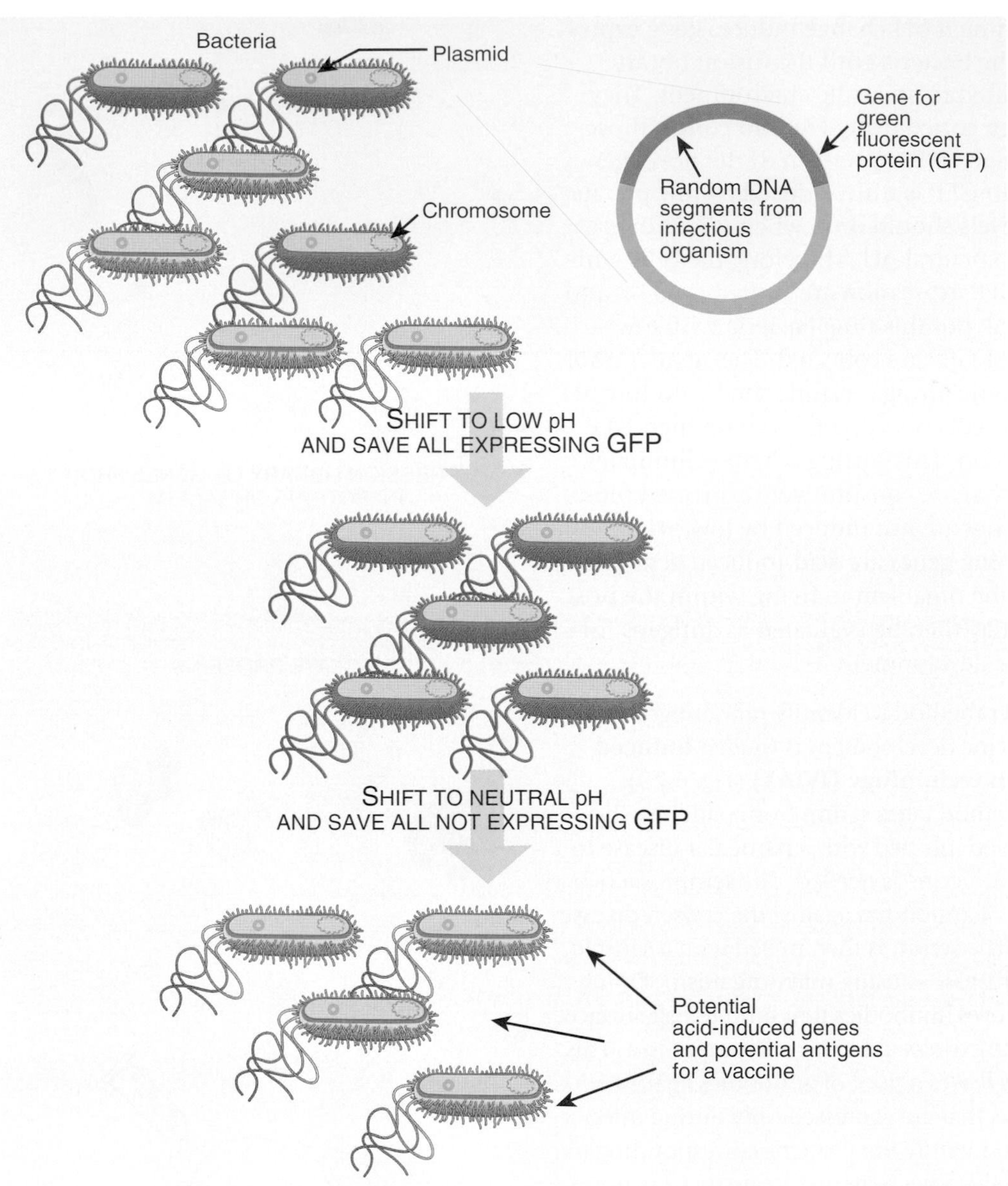

FIGURE 6.27 Differential Fluorescence Induction (DFI)
First, genes from the pathogen of interest are cloned in frame with the GFP gene. The fusion proteins are then expressed in bacteria. The entire recombinant bacterial population is exposed to low pH. The bacteria expressing GFP are isolated. These clones either express the GFP protein constitutively or were induced by the low pH. To isolate the clones that are expressed only at low pH, scientists shift the green cells to neutral pH, and this time, they keep only the colorless cells. Repeating this procedure will ensure a pure set of genes that are induced only under low pH.

DNA VACCINES BYPASS THE NEED TO PURIFY ANTIGENS

The principle of the **DNA vaccine** is to administer just DNA that encodes appropriate antigens instead of providing whole microorganisms or even purified proteins. Naked DNA vaccines consist of plasmids carrying the gene for the antigen under control of a strong promoter. The intermediate early promoter from cytomegalovirus is often used because of its strong expression. The DNA is then injected directly into muscle tissue. The foreign genes are expressed for a few weeks, and the encoded protein is made in amounts sufficient to trigger an immune response. The immune response is localized to the chosen muscle, which helps avoid side effects. In addition, purified DNA is much cheaper to prepare than purified protein and can be stored dry at room temperature, avoiding the need for refrigeration. The best method of delivering DNA is attaching it to a microparticle with a cationic surface (Fig. 6.29) because the surface binds to the negatively charged phosphate backbone. After the DNA-coated microparticle enters the cells, the DNA is slowly released from the bead and is then converted into protein. The slow release of DNA elicits a better immune response than a large direct dose of DNA. The immune system has to create more and more antibodies to the proteins.

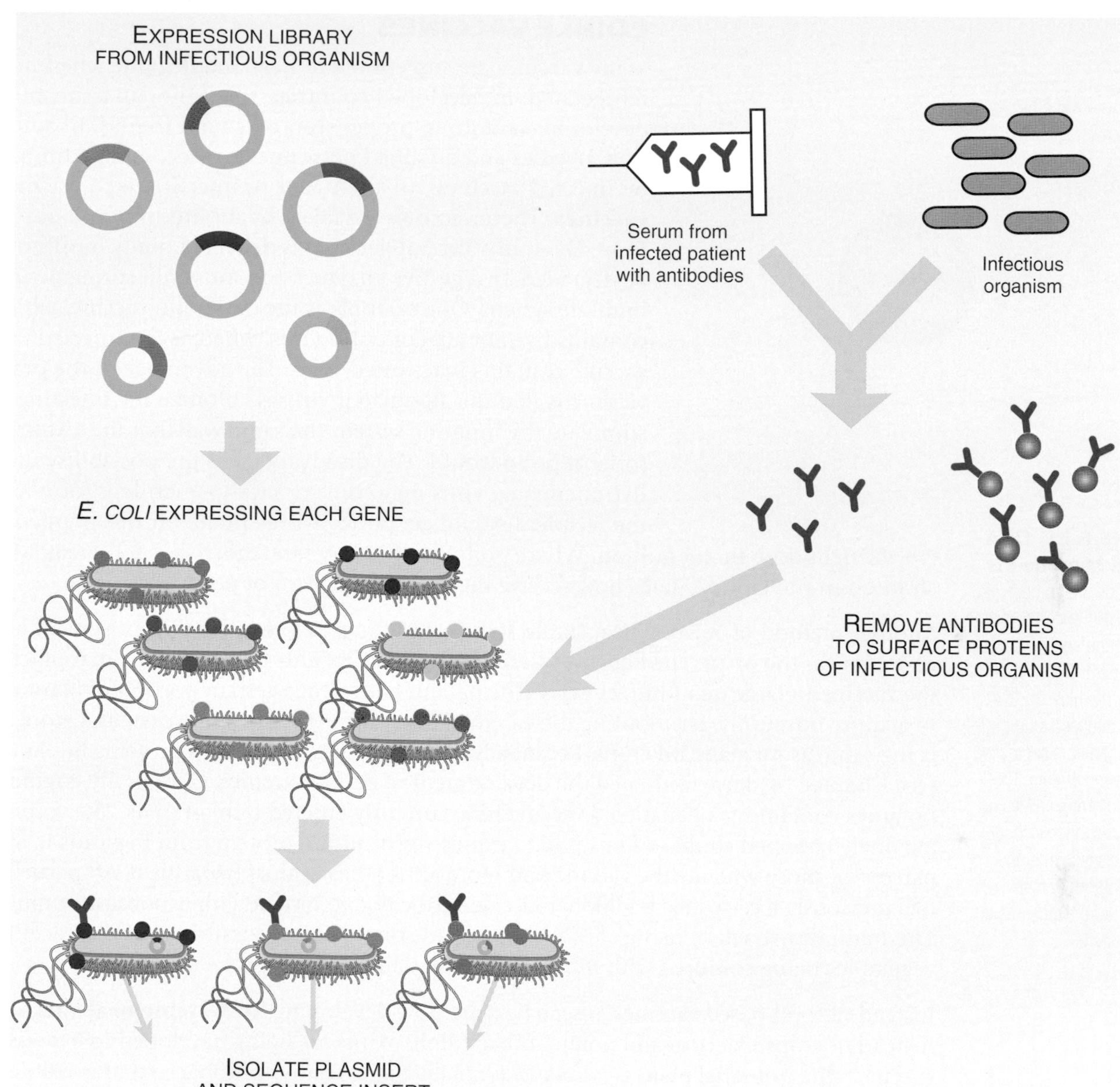

FIGURE 6.28 ***In Vivo* Induced Antigen Technology (IVIAT)**
Finding novel antigens to make a new vaccine relies on identifying proteins that elicit an immune response. IVIAT identifies antigens directly from patients who have been exposed to the pathogenic organism. First, an expression library is established that includes each of the genes from the pathogen of interest. Next, serum from infected patients is collected and preabsorbed to the infectious organism (grown in culture) to remove the antibodies that recognize surface proteins. The remaining antibodies are used to screen the expression library. When an antibody recognizes a cloned protein, the specific DNA clone is sequenced to identify the gene product.

One problem with DNA vaccines is that certain DNA sequences may induce an immune response directly. In particular, some DNA sequence motifs found in bacterial DNA may elicit strong immune responses, which in turn may cause the body to target its own DNA, thus generating an autoimmune response.

Rather than injecting a protein, some vaccines are simply DNA of a gene that will elicit an immune response. After the DNA enters the cell, it is converted into protein, which elicits the immune response to create memory B cells to that protein.

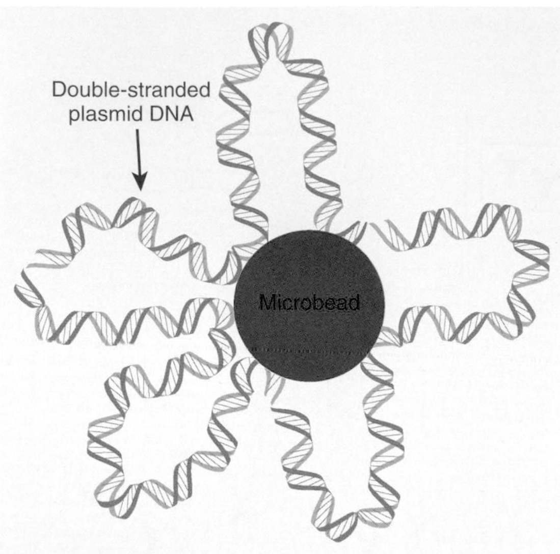

FIGURE 6.29 DNA-Coated Microbeads Microbeads are coated with plasmid DNA encoding an antigen gene and injected into a patient. Once inside the cells, the plasmid DNA is slowly released, and the protein antigen is expressed over a period of time. The expressed protein elicits an immune response without causing disease, thereby vaccinating the person against future exposures to the pathogen.

EDIBLE VACCINES

Many vaccines are susceptible to heat and degrade when not refrigerated. In developed countries, this is not an issue, but in developing countries, proper storage is hard to find. In addition, needles and qualified personnel are needed to administer injected vaccines. An alternative to injection is to use **oral vaccines**. These vaccines are taken by mouth in liquid or pill form. Of course, the antigen that is delivered orally must not be degraded by digestive enzymes and must still stimulate the immune system. One example is the oral polio vaccine, which contains live attenuated polio virus, whereas the injected polio vaccine contains inactivated virus. The advantage of the oral vaccine is that the attenuated viruses colonize the intestine and stimulate the immune system the same way that the virulent form of polio would. The disadvantage is the possibility that the live attenuated virus may convert back to a virulent form and the recipient would get polio. The estimate for this happening is 1 virulent dose in 2.5 million. Where polio itself is very rare, this risk is too great. Most children in the United States now receive the inactivated form of polio vaccine.

Another method of creating heat-stable, low-cost vaccines is to express the antigens in plants and then eat the plant. The benefits of **edible vaccines** include being able to "manufacture" the vaccine in large quantities cheaply. The patient has to eat a certain portion of plant tissue to acquire immunity. Distributing the vaccine in developing countries is easy, and storage is the same as for standard crops. Recent advances in expressing foreign proteins in plants (see Chapter 14) have facilitated the development of edible vaccines. Genetically engineered potatoes containing a hepatitis B vaccine have currently entered human trials. The volunteers ate finely chopped chunks of raw potato expressing a surface protein from hepatitis B. Sixty percent of those who ate the vaccine had more antibodies against hepatitis B. All participants had previously received the traditional vaccine, so the potato vaccine simply boosted immunity. The main drawback of using vaccines in a food source is the possibility of the vaccine vegetables being confused with normal vegetables and used as food.

Instead of food-based vaccines, researchers are now developing **heat-stable oral vaccines**. Instead of crops like corn and potato, other edible plants are being developed to express the vaccine. One potential plant is *Nicotiana benthamiana,* a relative of tobacco that is edible but is not used for food. Another potential plant is the single-celled algae called *Chlamydomonas reinhardtii* that is a great model organism for studying how cilia form and function. In addition, *Chlamy* are useful to study chloroplast function.

Edible vaccines are either live attenuated virus like the oral polio vaccine, or an antigenic protein that is expressed in a food.

Summary

The immune system has two different components: humoral immunity and cell-mediated immunity. Humoral immunity includes the production of antibodies by B cells that are found in the serum and other bodily fluids. The antibodies have a general Y shape that consists of two heavy chains and two light chains. The hinge region of the Y divides the molecule into the Fc (constant) and Fab (variable) regions. Cell-mediated immunity involves the activation of T cells, a subset of white blood cells. The T cells become active when a pathogen invades a cell, and the cell starts presenting fragments of the pathogen on the cell surface major histocompatibility complexes. In both arms of the immune reaction, the antibody or T cell recognizes only small epitopes or distinct regions of the pathogenic proteins. The immune system can make many different antibodies to one protein because only these small areas are recognized.

In the laboratory, antibodies can be made to specific proteins by injecting an animal such as a mouse or rabbit with a pure sample of the protein. To make monoclonal antibodies, scientists fuse mouse B cells to immortal myeloma cells to make hybridomas. Each B-cell fusion makes an antibody to one specific epitope of the protein. Polyclonal antibodies, on the other hand, include all the antibodies to the protein, that is, the antibodies recognize multiple epitopes. Antibodies are used in ELISA, where the amount of the target protein in a mixture can be estimated by the amount of antibody that binds. In immunohistochemistry and immunocytochemistry, an antibody to the target protein is used to localize its position within the cell. Antibodies are also used to sort samples of cells by FACS and are used to count a specific type of cell in flow cytometry.

Vaccines stimulate our immune systems to form antibodies and memory B cells without causing the disease for which the vaccine is providing protection. Vaccines are live attenuated viruses, inactivated or dead viruses, subunits of a virus, or simply peptides from a viral protein. The vaccines could also be made from a related but harmless virus or bacteria that express a protein from the pathogenic virus or bacteria. Reverse vaccines and DNA vaccines are created from genomic DNA sequences that are expressed into protein. Reverse vaccines are made in a laboratory, whereas DNA vaccines are injected directly into the muscular tissue as DNA. Also, some vaccines are made by expressing pathogenic proteins in edible crops. A person can receive resistance to the pathogen by simply ingesting these plants. New antigenic proteins are the key to making a good vaccine. DFI and IVIAT are two methods to identify potential antigenic proteins from the pathogenic organism.

End-of-Chapter Questions

1. What are antigens and antibodies?
    - **a.** Antigens are foreign bodies and antibodies are immune system components that recognize antigens.
    - **b.** Antibodies are foreign bodies and antigens recognize them and work to destroy them.
    - **c.** Antigens are produced by B cells in response to antibody accumulation.
    - **d.** Antigens are foreign bodies and antibodies are a specific cell type from the immune system.
    - **e.** none of the above
2. Which of the following is an accurate description of B and T cells?
    - **a.** B cells recognize antigens expressed on the surface of other cells and T cells produce antibodies.
    - **b.** B cells are components of the cell-mediated immunity and T cells comprise the humoral immunity.
    - **c.** Major histocompatibility complexes are associated with B cells whereas T cells produce antibodies.
    - **d.** B cells produce antibodies and T cells recognize antigens expressed on the surface of other cells.
    - **e.** none of the above
3. How are the variants of antibodies produced?
    - **a.** Each variant is encoded on one gene.
    - **b.** by post-translational modification of the antibodies

(Continued)

  c. by shuffling a small number of gene segments around
  d. by splicing the transcript into various configurations
  e. all of the above

4. Which of the following statements about antibodies is not correct?
  a. Antibodies consist of two light chains and two heavy chains.
  b. Polyclonal antibodies are derived from hybridomas.
  c. Antibodies are classified into classes and have distinct roles in the immune system.
  d. One particular antibody made from a clonal B cell is called a monoclonal antibody.
  e. Monoclonal antibodies are made by fusing B cells to myelomas, culturing the hybridomas, and screening for appropriate antigen recognition.

5. Which of the following statements about humanized antibodies is correct?
  a. Humanized antibodies to the ClfA protein of *S. aureus* may provide a way to eliminate the antibiotic-resistant pathogen in patients with nosocomial infections.
  b. Herceptin has been effective in treating some patients with breast cancer.
  c. Humanized monoclonal antibodies are created by removing the constant regions of mouse antibodies and replacing them with human constant regions.
  d. Full humanization of an antibody involves removing the hypervariable regions and splicing them into the heavy and light chains of human antibodies.
  e. All of the above are correct.

6. How is the creation of recombinant antibodies useful to researchers?
  a. Recombinant antibodies can be used to precisely deliver toxins, cytokines, and enzymes directly to the antigen.
  b. The production of recombinant antibodies is strictly theoretical and probably will serve no purpose to biotechnology research.
  c. Recombinant antibodies allow for more efficient production and isolation of the scFv.
  d. Recombinant antibodies can deliver toxins, cytokines, and enzymes but are disseminated throughout the organism.
  e. none of the above

7. Why is an ELISA used?
  a. to quantify the amount of a specific protein or antigen in a sample
  b. to quantify the amount of DNA in a sample
  c. to determine the amount of antibody within a sample
  d. to dilute out antibody from serum in a microtiter plate
  e. none of the above

8. Which of the following is an example of how ELISA is used?
  a. home pregnancy test
  b. detection of pathogenic organisms
  c. detection of plant diseases
  d. detection of dairy and poultry diseases
  e. all of the above

9. In which application are fluorescent antibodies used?
  a. immunocytochemistry
  b. flow cytometry

c. immunohistochemistry
d. fluorescence activated cell sorting
e. all of the above

10. Which of the following statements about immunity is not true?
   a. Vaccines use a live infectious agent that is still capable of producing disease in order to elicit an immune response.
   b. The immune system remembers foreign antigens through memory B cells.
   c. Vaccines consist of an antigen from an infectious agent that induces an immune response.
   d. Immunity to a fatal disease can often be triggered by infection with a closely related infectious agent, as in the cases of cowpox and smallpox.
   e. Antibody-producing B cells normally live only a few days but memory cells survive for a long time.

11. How are vaccines made so that they do not cause disease?
   a. killing the infectious agent with heat or denaturing the infectious agent with chemicals
   b. using a component or protein of the infectious agent instead of the organism itself
   c. genetically engineering the infectious agent to remove the genes that cause disease
   d. using a related but non-pathogenic strain of an infectious agent
   e. all of the above

12. What is reverse vaccinology?
   a. the removing of B cells from a person's body, exposing them to an infectious agent *in vitro,* and then returning them to the body
   b. the use of expressed genes from an expression library to find proteins that elicit an immune response in mice to create new vaccine candidates
   c. the vaccination of a person with a related, but non-pathogenic, strain to elicit an immune response
   d. the vaccination of a person after he or she has already been exposed to the pathogen
   e. none of the above

13. What is critical to finding novel antigens for vaccine development?
   a. the growth of live infectious agents to create whole vaccines
   b. the engineering of genes to attenuate infectious agents
   c. the identification of proteins that elicit an immune response
   d. the identification of the immune system components unique to specific infectious agents
   e. none of the above

14. Which of the following statements about edible vaccines is not true?
   a. In developing countries, proper storage and availability of needles and personnel to administer the vaccine are limiting factors in vaccinating the population.
   b. Edible vaccines must not be destroyed by the digestive system and must still elicit an immune response.

(*Continued*)

c. A problem with using edible vaccines is the possibility that vaccine vegetables could be mistaken for normal vegetables used as food.
d. Edible vaccines are usually too expensive to be manufactured in large quantities.
e. All of the above are true.

15. Which of the following is not a risk associated with vaccines?
a. adverse side effects
b. allergic reactions
c. preservatives containing mercury
d. induction of autoimmunity in some individuals
e. all of the above are potential risks associated with vaccination

16. Which of the following is the mechanism of action for Herceptin?
a. Herceptin binds to the HER2 promoter to prevent transcription, thus lowering amounts of receptor.
b. Herceptin binds to intracellular HER2 proteins to prevent cancer cells from dividing.
c. Herceptin binds to HER3 and HER4 cell surface receptors to activate the immune system.
d. Herceptin binds to extracellular domain of HER2 and prevents internalization and subsequent cancer cell division.
e. Herceptin targets TNFα.

17. All of the following statements are Remicade are true except_______________.
a. Remicade targets TNFα in the joints of people with rheumatoid arthritis.
b. Remicade is a chimeric antibody.
c. Remicade's mechanism of action includes enhancing the release of IL-1.
d. Remicade is a fusion protein produced in a myeloma cell line.
e. All of the above are true.

18. Heavy chain antibodies have major implications for therapeutic purposes because_______________.
a. they have smaller variable domains and can be purified more easily
b. they only have heavy chains and no light chains and are more easily engineered
c. nanobodies are easily constructed from the constant regions of these antibodies
d. they are derived from camels and related animals, and there would not be recognized by the human immune system
e. they recognize much smaller antigens

19. All of the following are features or functions of nanobodies except_______________.
a. small, monomeric, lack disulfide bonds, and resistance to denaturation
b. high affinity to the antigen
c. recognize protruding and recessed paratopes
d. cannot be humanized
e. consist of only the VHH domain of heavy chain antibodies

Further Reading

Betts, M. R., Brenchley, J. M., Price, D. A., De Rosa, S. C., Douek, D. C., Roederer, M., et al. (2003). Sensitive and viable identification of antigen-specific $CD8^+$ T cells by a flow cytometric assay for degranulation. *Journal of Immunology Methods, 281*, 65–78.

Clark, D. P. (2005). *Molecular Biology: Understanding the Genetic Revolution*. San Diego, CA: Elsevier Academic Press.

Clark, M. (2000). Antibody humanization: a case of the "emperor's new clothes"? *Immunology Today, 8*, 397–402.

Elgert, K. D. (1996). *Immunology: Understanding the Immune System*. New York: Wiley-Liss.

Fischer, O. M., Streit, S., Hart, S., & Ullrich, A. (2003). Beyond Herceptin and Gleevec. *Current Opinion in Chemical Biology, 7*, 490–495.

Glick, B. R., & Pasternak, J. J. (2003). *Molecular Biotechnology: Principles and Applications of Recombinant DNA* (3rd ed.). Washington, DC: ASM Press.

Handfield, M., Brady, L. J., Progulske-Fox, A., & Hillman, J. D. (2000). IVIAT: A novel method to identify microbial genes expressed specifically during human infections. *Trends in Microbiology, 8*, 336–339.

Patti, J. M. (2004). A humanized monoclonal antibody targeting *Staphylococcus aureus*. *Vaccine, 228*, S39–S43.

Scarselli, M., Giuliana, M. M., Adu-Bobie, J., Pizza, M., & Rappuoli, R. (2005). The impact of genomics on vaccine design. *Trends in Biotechnology, 23*, 84–91.

Valdivia, R. H., & Falkow, S. (1997). Probing bacterial gene expression within host cells. *Trends in Microbiology, 5*, 360–363.

Nanobiotechnology

Introduction
Visualization at the Nanoscale
Scanning Tunneling Microscopy
Atomic Force Microscopy
Weighing Single Bacteria and Virus Particles
Nanoparticles and Their Uses
Nanoparticles for Labeling
Quantum Size Effect and Nanocrystal Colors
Nanoparticles for Delivery of Drugs, DNA, or RNA
Nanoparticles in Cancer Therapy
Assembly of Nanocrystals by Microorganisms
Nanotubes
Antibacterial Nanocarpets
Detection of Viruses by Nanowires
Ion Channel Nanosensors
Nanoengineering of DNA
DNA Origami
DNA Mechanical Nanodevices
Controlled Denaturation of DNA by Gold Nanoparticles
Controlled Change of Protein Shape by DNA
Biomolecular Motors

Biotechnology

http://dx.doi.org/10.1016/B978-0-12-385015-7.00007-7

INTRODUCTION

In 1959, Richard Feynman was the first scientist to suggest that devices and materials could someday be fabricated to atomic specifications: "The principles of physics, as far as I can see, do not speak against the possibility of maneuvering things atom by atom."

Molecular biology originated largely from the study of microorganisms. One micrometer is one millionth of a meter, and cells of *Escherichia coli,* the geneticist's favorite bacterium, are roughly 1 micrometer (= "micron") in length. A nanometer is one thousandth of a micrometer (= 10^{-9} meters; Fig. 7.1). The terms *micro-* and **nano-** are both from Greek. *Mikros* means "small." More imaginative is *nanos,* a little old man or dwarf. *Pico-* comes from Spanish, where it means a small quantity, or *beak* (from Latin *beccus,* "beak," ultimately of Celtic origin). Prefixes for even smaller quantities are shown in Table 7.1. As far as length is concerned, these are applicable only to subatomic dimensions. Nevertheless, when dealing with masses and volumes on the nanoscale, we may find femtograms and zeptoliters.

Recently, science has advanced into the area of **nanotechnology**. As the name indicates, the impetus has come from pursuing practical applications, especially in the fields of electronics and materials science, rather than a quest for theoretical knowledge. Nanotechnology involves the individual manipulation of single molecules or even atoms. Building components atom by atom or molecule by molecule in order to create materials with novel or vastly improved properties was perhaps the original goal of nanotechnologists. However, the field has expanded in a rather ill-defined way and tends to include any structures so tiny that their study or manipulation was impossible or impractical until recently. At the nanoscale, quantum effects emerge and materials often behave strangely, compared to their bulk properties.

The internal components of biological cells are on the same scale as those studied by nanotechnology. As a consequence, nanotechnologists have looked to cell biology for useful structures, processes, and information. Cellular organelles such as ribosomes may be regarded as programmable "nanomachines" or "nano-assemblers." Thus, nanotechnology is spilling over into molecular biology. Much of "nanobiotechnology" is in fact molecular biology viewed from the perspective of materials science and described in novel terminology.

All chemical reactions operate at a molecular level. What distinguishes true nanotechnology is that single molecules or nanostructures are assembled following specific instructions. A ribosome does not merely polymerize amino acids into a chain. It takes specified amino acids, one at a time, according to information provided, and links them in a specific order. Thus, the critical properties of a nano-assembler include the ability not merely to assemble structures at the molecular level, but to do so in a specific and controlled manner.

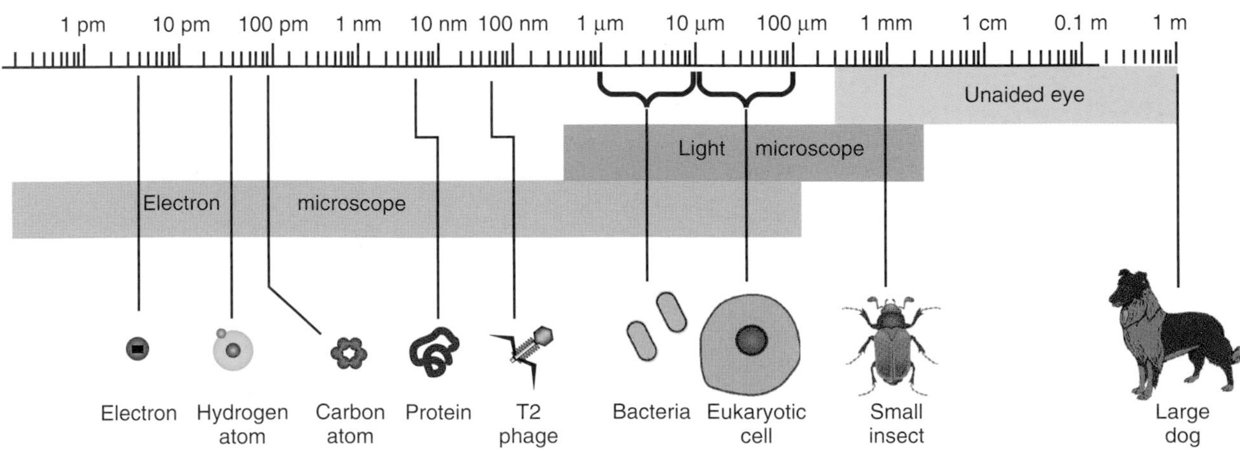

FIGURE 7.1 Size Comparisons
The objects range in size from 1 meter to 1 picometer.

Table 7.1 Prefixes and Sizes

Length Unit	Meters	Examples
5.9 Terameters		Mean distance from Sun to Pluto
Terameter	10^{12}	
150 Gigameters		Distance to the Sun
Gigameter	10^{9}	
380 Megameters		Distance to the Moon
6.3 Megameters		Radius of the Earth
3.2 Megameters		Length of Great Wall of China
Megameter	10^{6}	
Kilometer	10^{3}	
30 Meters		Blue whale
Meter	1	Large dog
Millimeter	10^{-3}	Small insect
Micrometer	10^{-6}	Bacterial cell
500 Nanometers		Wavelength of visible light
100 Nanometers		Size of typical virus
3.4 Nanometers		One turn of DNA double helix
Nanometer	10^{-9}	Molecules
350 Picometers		Molecular diameter of water
260 Picometers		Atomic spacing in solid copper
77 Picometers		Atomic radius of carbon (= resolution limit of atomic force microscope as of 2004)
32 Picometers		Atomic radius of hydrogen
Ångstrom		= 100 picometers = 10^{-10} meter
2.4 Picometers		Wavelength of electron
Picometer	10^{-12}	
Femtometer	10^{-15}	Radius of atomic nucleus
Attometer	10^{-18}	Radius of proton
Zeptometer	10^{-21}	
Yoctometer	10^{-24}	Radius of neutrino

The main practical objectives of nanobiotechnology are using biological components to achieve nanoscale tasks. Some of these tasks are nonbiological and have applications in such areas as electronics and computing, whereas others are applicable to biology or medicine. The purpose of this chapter is to show, by selected examples, how biological approaches can contribute to nanoscience.

Many internal components of biological cells are in the nanoscale range. As nanotechnology advances, it is developing many links with biotechnology and genetic engineering.

VISUALIZATION AT THE NANOSCALE

To manipulate matter on an atomic scale, we need to see individual atoms and molecules. Although individual molecules have been visualized with the electron microscope, it was the development of scanning probe microscopes that opened up the field of nanotechnology. These instruments all rely on a miniature probe that scans across the surface under investigation.

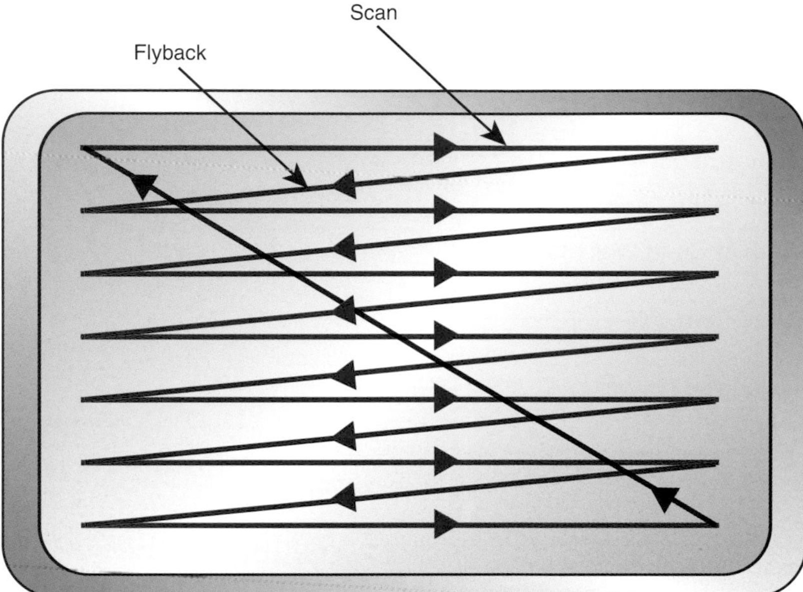

FIGURE 7.2
Principle of Raster Scanning
In raster scanning, the probe moves to and fro across the target region. The probe scans only while moving in one direction ("scan"). When the probe travels in the reverse direction, it moves more rapidly without making contact ("flyback").

All scanning probe microscopes work by measuring some property, such as electrical resistance, magnetism, temperature, or light absorption, with a tip positioned extremely close to the sample. The microscope **raster-scans** the probe over the sample (Fig. 7.2) while measuring the property of interest. The data are displayed as a raster image similar to that on a television screen. Unlike traditional microscopes, scanned-probe systems do not use lenses, so the size of the probe rather than diffraction limits their resolution. Some of these instruments can be used to alter samples as well as visualize them.

The first of these instruments was the **scanning tunneling microscope (STM)**, which was developed by Gerd Binnig and Heinrich Rohrer at IBM (see following section). They received the Nobel Prize in 1986. The STM sends electrons, that is, an electric current, through the sample and so measures electrical resistance. The **atomic force microscope (AFM)** is especially useful in biology and measures the force between the probe tip and the sample.

Visualization of individual molecules or even atoms is possible using scanning probe microscopes.

SCANNING TUNNELING MICROSCOPY

When a metal tip comes close to a conducting surface, electrons can tunnel from one to the other, in either direction. The probability of tunneling depends exponentially on the distance apart. Surface contours can be mapped by keeping the current constant and measuring the height of the tip above the surface. This allows resolution of individual atoms on the surface being studied. This is the principle of the scanning tunneling microscope (Fig. 7.3).

Atoms may also be moved using the STM. In 1989, in perhaps the most famous experiment in nanotechnology, D. M. Eigler and E. K. Schweizer fabricated the IBM logo by arranging 35 xenon atoms on a nickel surface. They chose nickel because the valleys between rows of nickel atoms are deep enough to hold xenon atoms in place, yet small enough to allow the xenon atoms to be pulled over the surface. To move xenon atoms, they placed the STM tip above a xenon atom, using imaging mode. Next, scanning mode was turned off and the tip lowered until the tunneling current increased several-fold ("fabrication mode"). The xenon atom was attracted to the STM tip and was dragged by moving the tip horizontally. The atom was deposited at its new location by reducing the tunneling current. Since then, several diagrams have been made in the same way. Carbon monoxide man is shown in Figure 7.4.

From a biological perspective, the weakness of STM is that it requires a conducting surface, in practice generally a metal layer of some sort. The atomic force microscope (see the following

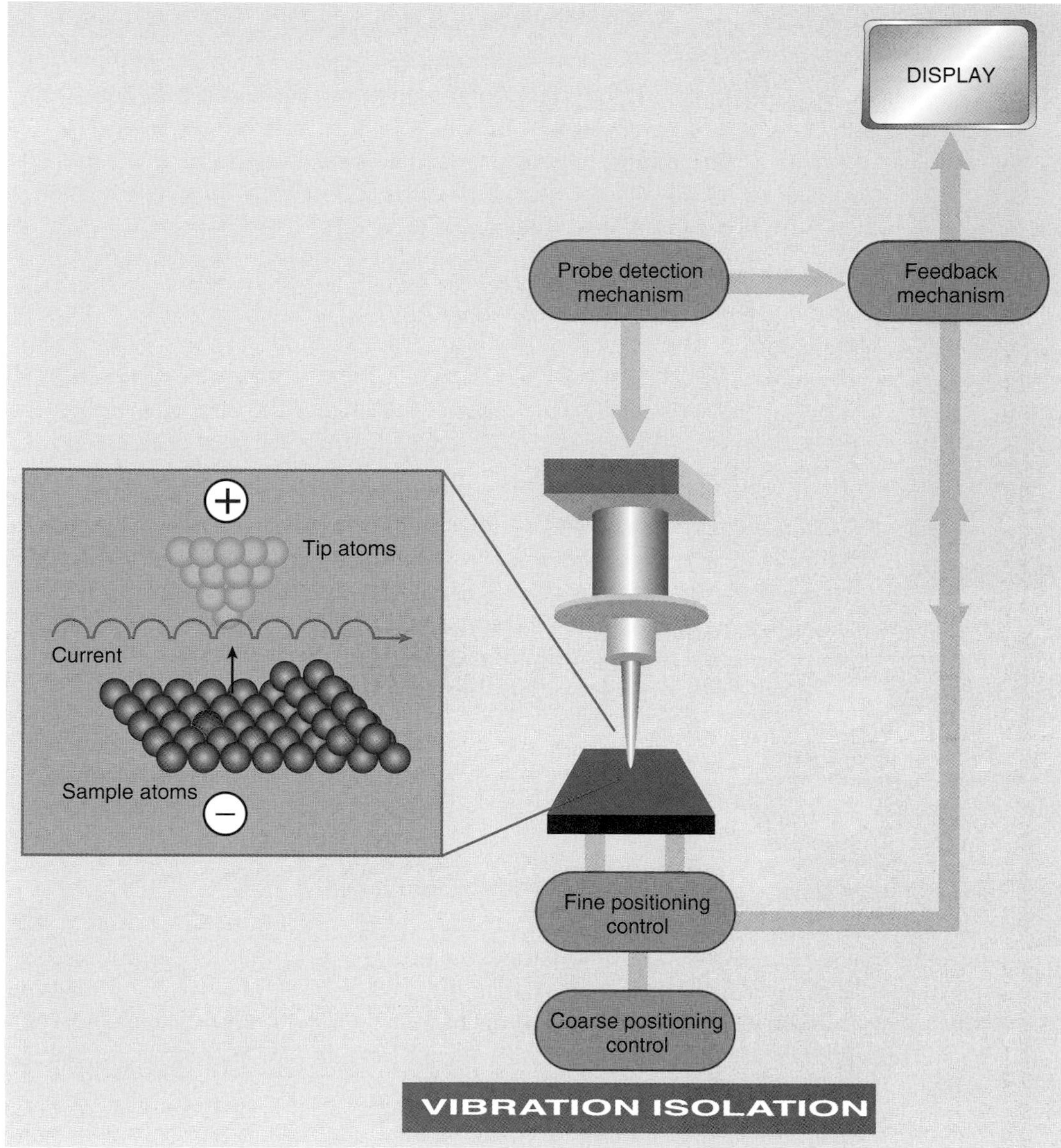

FIGURE 7.3 Principle of Scanning Tunneling Microscope
The probe tip and surface atoms of the sample are shown in the inset.

section) has the advantage of not needing conductive material and has therefore been more widely applied in biology.

The scanning tunneling microscope can be used to detect or move individual atoms on a conducting surface.

ATOMIC FORCE MICROSCOPY

Visualization at the nanoscale is often performed using atomic force microscopy. As the name indicates, it operates by measuring force, not by using a stream of particles such as photons (as in light microscopy) or electrons (as in electron microscopy). Physicists sometimes compare the operation of an AFM to an old-fashioned record player, which uses a needle to scrape the surface of a record. Perhaps to a biologist, the difference between a light microscope and AFM is like the difference between reading text with the eyes and feeling Braille.

The atomic force microscope was invented in 1985 by Gerd Binnig, Calvin Quate, and Christof Gerber. The AFM uses a sharp probe that moves over the surface of the sample and that bends in response to the force between the tip and the sample. The movement of the

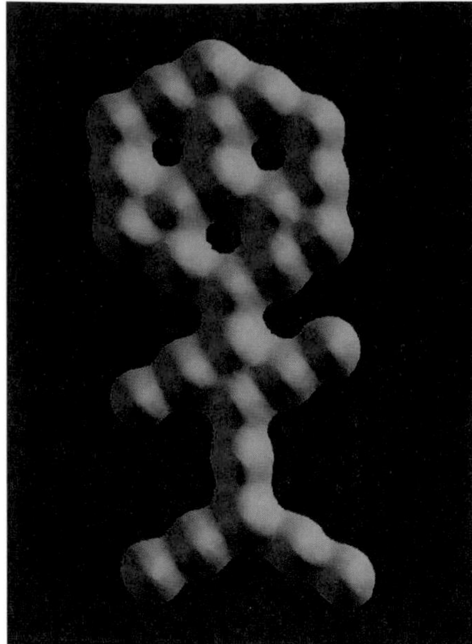

FIGURE 7.4 Carbon Monoxide Man by Zeppenfeld
The atoms were arranged by STM. The medium is carbon monoxide on platinum. Courtesy of International Business Machines Corporation. © 1995, IBM.

probe performs a raster scan, and the resulting topographical image is displayed on-screen.

During scanning, the movement of the tip or sample is performed by an extremely precise positioning device that is made from **piezoelectric ceramics**. (These are materials that change shape in response to an applied voltage.) It usually takes the form of a tube scanner that is capable of sub-Ångstrom resolution in all three directions.

The AFM probe is a tip on the end of a cantilever. As the cantilever bends because of the force on the tip, a laser monitors its displacement, as shown in Figure 7.5. The beam from the laser is reflected onto a split photodiode. The difference between the A and B signals measures the changes in the bending of the cantilever. For small displacements, the displacement is proportional to the force applied. Hence, the force between the tip and the sample can be derived.

The distance between tip and sample is adjusted so that it lies in the repulsive region of the intermolecular force curve; that is, the AFM probe is repelled by its molecular interaction with the surface. The repulsion gives a measure of surface topography, and this is what is generally displayed, with color-coding indicating relative height. It is possible to scan a surface for topography and then raise the AFM probe and rescan to detect electrostatic or magnetic forces. These can then be plotted for comparison with the topography.

As with the STM, it is possible to use the AFM to move single atoms, although this was only achieved in 2003. Researchers at Osaka University in Japan removed a single silicon atom from a surface and then replaced it.

Through use of the AFM, it is possible to visualize polymeric biological molecules such as DNA or cellulose and even to see the individual monomers and, at high resolution, even the atoms of which they are composed. Perhaps more interesting is the use of AFM to monitor the formation of biological complexes, such as the binding of proteins to single molecules of DNA. An example is the formation of the complex between DNA and the bacterial protein RecA, which is involved in repair of damaged DNA (Fig. 7.6). As shown, RecA assembles onto the DNA starting at one end. Eventually, the whole length of the DNA molecule will be coated.

The atomic force microscope can detect atoms or molecules by scanning a surface for shape or electromagnetic properties.

FIGURE 7.5 The Atomic Force Microscope (AFM)
The deflection of the tip of the probe by the surface is monitored by a laser.

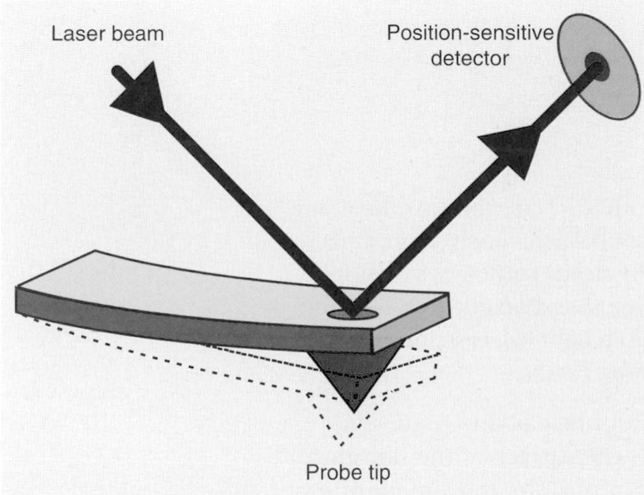

WEIGHING SINGLE BACTERIA AND VIRUS PARTICLES

It has been known for many years that bacteria are on the order of 1000 nanometers in size and 1 picogram in weight. However, in addition to detecting microorganisms via nanotechnology, it is now possible to weigh them individually.

The oscillation frequency of a diving board depends on the mass applied. Scaling down, it is possible to construct a cantilever of micrometer dimensions (approximately 6 microns long by 0.5 micron wide with an end platform about 1 micron square). The oscillation frequency can be measured by using a laser and observing the altered light reflection. Addition of single bacterial cells or even virus particles changes the oscillation frequency of the cantilever. The mass of single cells or virus particles has been measured this way in the laboratory of Harold Craighead at Cornell University (Fig. 7.7).

To hold the bacteria or viruses in place, the cantilever is coated with an antibody that recognizes the microorganism to be weighed. A single cell of *E. coli* was 1430 × 730 nanometers in size and weighed 665 femtograms (665 × 10^{-15} grams). Viruses (weighing around 1 femtogram) can be detected by reducing the size of the cantilever and enclosing it in a vacuum. By mid-2005, this technique had been refined to weigh a single macromolecule—a double-stranded DNA of approximately 1500 base pairs (roughly the size of a typical coding sequence). Today this approach allows measurements of small proteins and other molecules in the zeptogram range (10^{-21} grams).

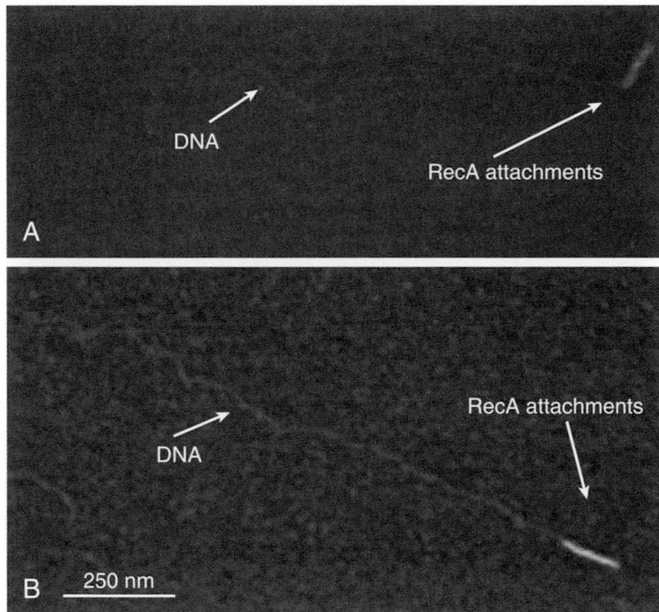

FIGURE 7.6 AFM Shows RecA Attached to DNA at One End
When RecA protein binds to DNA, the complexes form first at one end of the DNA, as shown here by atomic force microscopy. From Li BS, Wei B, Goh MC (2012). Direct visualization of the formation of RecA/dsDNA complexes at the single-molecule level. *Micron* **43**, 1073–1075.

A possible future use for weighing individual particles is in identifying and quantitating infectious agents. Although bacteria change weight as they grow and divide, individual viruses have characteristic weights because they are assembled. Using antibodies attached to the cantilevers to capture the target particle (bacterium, virus, or even proteins) provides specificity. This approach can even be used to detect and count the number of particles of the prion protein responsible for mad cow disease (see Chapter 21 for more information about prions). An array of cantilevers (rather than a single cantilever) would be used when counting particles for quantification.

Laser monitoring of the oscillation of a nanoscale cantilever allows single bacteria or viruses to be individually weighed.

NANOPARTICLES AND THEIR USES

Nanotechnology began with advances in viewing and measuring the incredibly small. It then moved on to building structures at the nanoscale. Simple nanostructures are now being used for a variety of analytical purposes, and a second generation is being developed for clinical use.

As their name indicates, **nanoparticles** are particles of submicron scale—in practice, from 100 nm down to 5 nm in size. They are usually spherical, but rods, plates, and other shapes are sometimes used. They may be solid or hollow and are composed of a variety of materials, often in several discrete layers with separate functions. Typically, there is a central functional layer, a protective layer, and an outer layer allowing interaction with the biological world.

The central functional layer usually displays some useful optical or magnetic behavior. Most popular is fluorescence. The protective layer shields the functional layer from chemical damage by air, water, or cell components and conversely shields the cell from any toxic properties of the chemicals composing the functional layer. The outer layer or layers allow nanoparticles to be "biocompatible." This generally involves two aspects: water solubility and specific recognition. For biological use, nanoparticles are often made water soluble by adding a hydrophilic outer layer. In addition, chemical groups must be present on the exterior to allow specific attachment to other molecules or structures (Fig. 7.8).

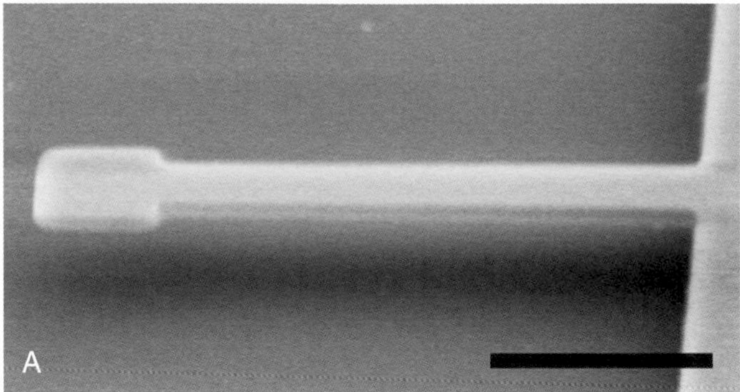

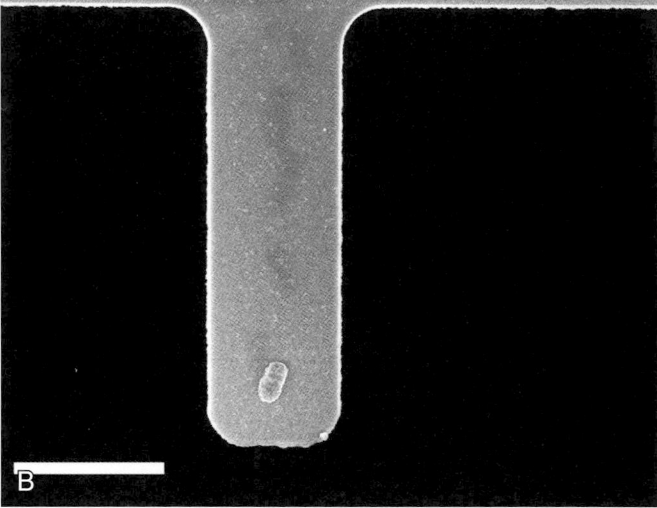

FIGURE 7.7 Weighing a Single Bacterium
(A) Scanning electron micrograph of cantilever oscillator with length 6 microns, width 0.5 microns, and a 1 micron by 1 micron paddle. Scale bar corresponds to 2 microns.
From Ilic B, et al. (2004). Virus detection using nanoelectromechanical devices. *Appl Phys Lett* **95**, 2604–2607. Copyright 2004. Reprinted with permission from the American Institute of Physics.
(B) Scanning EM of a single *E. coli* cell attached to the cantilever by antibody. Courtesy of Craighead group, Cornell University. From Ilic B, et al. (2001). Single cell detection with micromechanical oscillators. *J Vac Sci Technol B* **19**, 2825–2828. Copyright 2001. Reprinted with per mission from the American Institute of Physics.

Nanoparticles (see Box 7.1) have a variety of uses in the biological arena:

(a) Fluorescent labeling and optical coding
(b) Detection of pathogenic microorganisms and/or specific proteins
(c) Purification and manipulation of biological components
(d) Delivery of pharmaceuticals and/or genes
(e) Tumor destruction by chemical or thermal means
(f) Contrast enhancement in magnetic resonance imaging (MRI)

Nanoparticles are now widely used in a range of biological procedures. They include both analytical and clinical applications.

Box 7.1 Trendy Terminology

Nanoparticles are referred to by a variety of nanoterms, depending on their shape and structure. The meanings of nanorod, nanocrystal, nanoshell, nanotube, nanowire, and so forth, should be obvious enough. And, despite what you might think, quantum dots are not a new brand of frozen snack but an alternative name for fluorescent nanocrystals, small enough to show quantum confinement and used in biological labeling.

NANOPARTICLES FOR LABELING

Consider luminescent CdSe nanorods as an example of nanoparticles used for labeling (Fig. 7.9). These nanorods can be used as fluorescent labels for molecular biology because they absorb light from the UV to around 550 nm and emit strongly at 590 nm. They were made—appropriately enough—in the lab of Thomas Nann, in Freiberg, Germany.

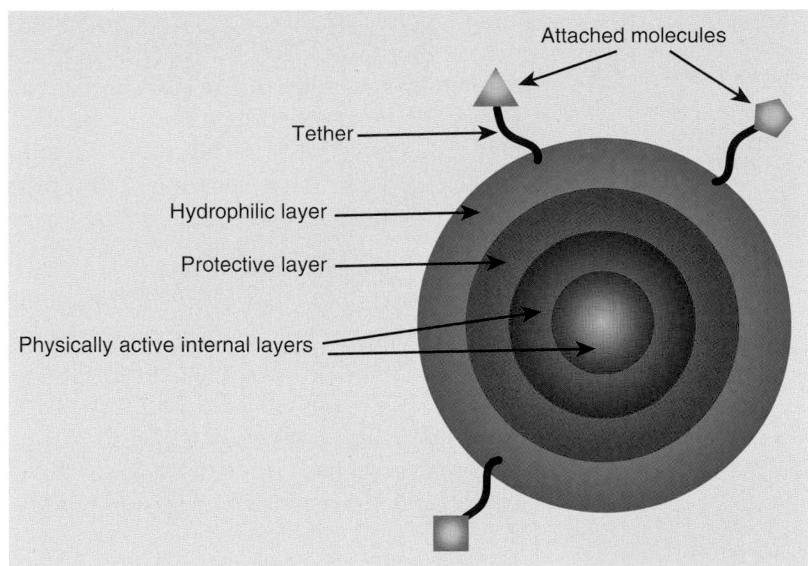

FIGURE 7.8 Typical Layered Structure of Nanoparticles
Several layers surround the physically active core. Chemical groups are often added to the exterior to allow attachment of biological molecules.

These nanorods measure approximately 3 nm in width by 10 to 20 nm in length. A core of luminous cadmium selenide (CdSe) is surrounded by a shell of ZnS (zinc sulfide, wurtzite) that protects the core against oxidation. Outside this is a layer of silica, which allows coupling of phosphonates or amines to the exterior of the nanorod. These hydrophilic groups make the nanorods water soluble. These outer chemical groups also allow attachment of the nanorods to proteins.

The scaffold inside eukaryotic cells is built from cylindrical protein structures known as microtubules. These protein structures are often disassembled into monomers (known as tubulin) and reassembled in different locations. **Nanorods** can be used to follow this remodeling by attaching them to the tubulin monomers. The addition of guanosine triphosphate (GTP) stimulates the assembly of microtubules, and the fluorescent nanorods can be seen aggregating into linear structures.

Why use a complex multilayered nanostructure instead of a simple fluorescent dye?

(a) Although nanocrystals have narrow emission peaks, they have broad absorption peaks (rather than narrow ones like typical dyes). Consequently, they do not bleach during excitation and can therefore be used for continuous long-term irradiation and monitoring.
(b) Nanocrystals have high brightness—the product of molar absorptivity and quantum yield. (**Molar absorptivity** is the absorbance of a one molar solution of pure solute at a given wavelength; the higher it is, the more light is absorbed. The **quantum yield** is the ratio of photons absorbed to photons emitted during fluorescence.)
(c) The emission maximum of a nanocrystal depends on the size and so can be set to any desired wavelength by making crystals of the appropriate size (see later discussion).

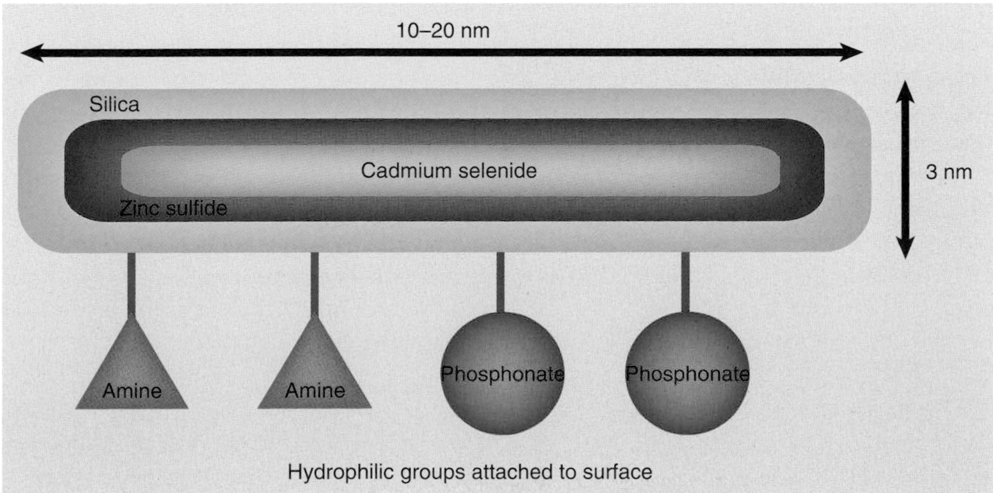

FIGURE 7.9 CdSe Nanorods
Luminescent CdSe nanorods are encased in protective layers of zinc sulfide and of silica. Hydrophilic chemical groups on the outside allow proteins or other biological molecules to be attached.

Nanoparticles can also be targeted to specific tissues, such as cancer cells, by adding appropriate antibodies or receptor proteins to the nanoparticle surface. Fluorescent nanoparticles are often known as **quantum dots** and are now commercially available for a wide range of biological labeling. Although fluorescent dyes can be attached to other molecules, nanoparticles are more versatile in this regard. Quantum dots can be used to label DNA molecules as well as proteins. Thus, labeling of PCR primers with quantum dots results in fluorescently labeled PCR products—a variant referred to as *quantum dot PCR*.

A variety of materials have been used to give better contrast enhancement in MRI (magnetic resonance imaging). Nanoparticles containing assorted compounds are seeing increasing use in this area. For example, super-paramagnetic iron oxide nanoparticles (SPIONs) act as good MRI contrast agents. Their magnetic properties vary with particle size. Larger particles, of greater than 300 nm, are used for bowel, liver, and spleen. Smaller particles, of 20 to 40 nm, show higher diagnostic accuracy for detecting early tumors in lymph nodes than conventional materials. Because they are relatively safe, SPIONs have also been suggested as possible carriers for delivery of both drugs and DNA.

Fluorescent nanoparticles are widely used in biological labeling. They last longer than traditional fluorescent dyes and are often brighter.

QUANTUM SIZE EFFECT AND NANOCRYSTAL COLORS

When materials are subdivided into sufficiently small fragments, quantum effects begin to influence their physical properties. The fluorescent nanoparticles discussed earlier are in fact **semiconductors** that are small enough to show such quantum effects.

Semiconductors are substances that conduct electricity under some conditions but not others. In *N-type* semiconductors (as in normal electric wires), the current consists of negatively charged electrons. In *P-type* semiconductors, the current consists of **holes**. A hole is the absence of an electron from an atom. Although not physical particles, holes can move from atom to atom. Electrons and holes may combine and cancel out, a process that releases energy. Conversely, energy absorbed by certain semiconductors may generate an electron-hole pair whose two components may then move off in different directions.

Nanoparticle labels can be made with different emission wavelengths, covering the UV, visible spectrum, and near infrared. Emission wavelengths obviously vary depending on the semiconductor material. However, in addition, the quantum size effect (Fig. 7.10) allows the same semiconductor to emit at different wavelengths, depending on the size of the nanoparticle. The smaller the nanoparticle, the shorter the wavelength (i.e., the higher the energy) it emits.

Fluorescent nanoparticles may be regarded as miniaturized light-emitting diodes (LEDs). These are semiconductors that work by absorbing energy (either electrical or light) and creating electron-hole pairs. When the electrons and holes recombine, light is emitted. For bulk material, the energy, and hence the wavelength, of the emitted light depends on the chemical composition of the semiconductor. However, at nanoscale dimensions, quantum effects become significant. If the physical size of the semiconductor is smaller than the natural radius (the **Bohr radius)** of the electron-hole pair, extra energy is needed to confine the electron–hole pair. This is referred to as **quantum confinement** and occurs with nanocrystals of around 20 nm or less. The smaller the semiconductor crystal, the more energy is needed and the more energetic (shorter in wavelength) is the light released.

The emission wavelength of a fluorescent nanoparticle depends on its size; therefore, an experimenter may modify it easily.

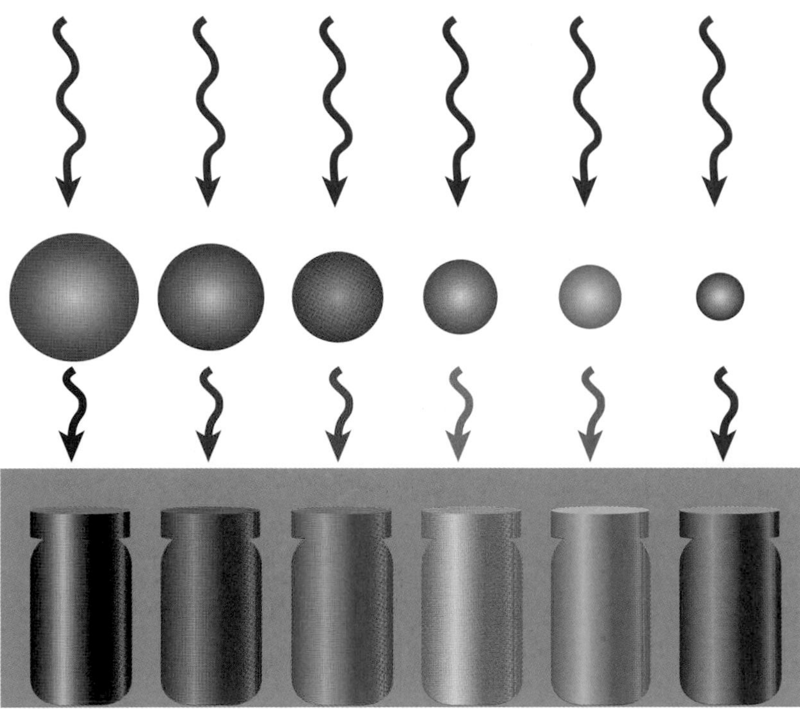

FIGURE 7.10 Quantum Size Effect
Nanocrystals of different sizes absorb UV light and re-emit the energy. The wavelength of emission depends on the size of the nanocrystal. The smaller the crystal, the more energetic the emission. From Riegler J, Nann T (2004). Application of luminescent nanocrystals as labels for biological molecules. *Anal Bioanal Chem* **379**(7–8), 913–919. With kind permission from Springer Science and Business Media.

NANOPARTICLES FOR DELIVERY OF DRUGS, DNA, OR RNA

Because nanoparticles can be targeted to specific tissues, they can be used to deliver a variety of biologically active molecules, including both pharmaceuticals and genetic engineering constructs.

Large polymeric molecules such as DNA may themselves be compacted to form nanoparticles of around 50 to 200 nm in size. This involves the addition of positively charged molecules (e.g., cationic lipids, polylysine) to neutralize the negative charge of the phosphate groups of the nucleic acid backbone. Other molecules may be added to promote selectivity for certain cells or tissues.

Alternatively, hollow nanoparticles (**nanoshells**) may be used to carry other, smaller molecules. Such nanoshells must be made from biocompatible materials such as polyethyleneimine (PEI) or chitosan (Fig. 7.11). The latter alternative is popular because it is both naturally derived and biodegradable. Chitin is a beta-1,4-linked polymer of N-acetyl-D-glucosamine. It is found in the cell walls of insects and fungi and, among biopolymers, is second only in natural abundance to cellulose. Chitosan is derived from chitin by removing most of the acetyl groups through alkali treatment and has been shown as safe for administration to humans.

An interesting combination of two modern technologies is using nanoshells to carry short-interfering RNA (siRNA). Delivery of siRNA triggers RNA interference, which results in the destruction of target mRNA (see Chapter 5). The siRNA may be targeted against mRNA from genes expressed preferentially in cancer cells (see below) or genes characteristic of certain viruses (see Chapter 21).

Hollow nanoparticles may be used to deliver DNA, RNA, or proteins.

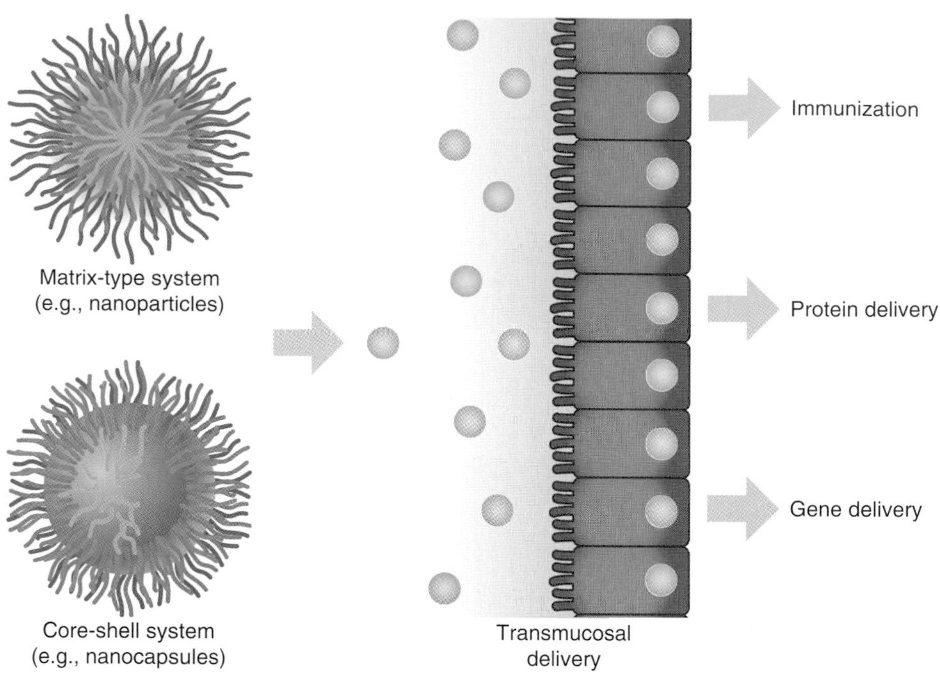

FIGURE 7.11 Chitosan Nanoparticle
The two main types of chitosan nanocarriers: matrix systems (e.g., nanoparticles) and core-shell systems (e.g., nanocapsules). The chitosan interacts strongly with mucosal surfaces, resulting in three potential applications: vaccination, protein delivery and gene therapy. Modified after Garcia-Fuentes M, Alonso MJ (2012). Chitosan-based drug nanocarriers: where do we stand? *J Control Release* **161**, 496 –504.

NANOPARTICLES IN CANCER THERAPY

It is possible to destroy tumor cells through use of a variety of toxic chemicals or localized heating. In both cases, a major issue is delivering the fatal reagent to the cancer cells and avoiding nearby healthy tissue. When toxic chemical reagents are used, the reagent not only must be delivered specifically to the target cells but also prevented from diffusing out of the cancer cells. Both related objectives may be achieved by using hollow nanoparticles to carry the reagent. Nanoparticles may be targeted to tumors by adding specific receptors or reactive groups to the outside of the nanoparticles. These are chosen to recognize proteins that are solely or predominantly displayed on the surface of cancer cells. It is hoped that such nanoparticles will be safe to give by mouth.

Diffusion is more difficult to deal with but may be limited to some extent by designing nanoparticles for slow release of the reagent. A clever alternative is to produce the toxic agent inside the nanoparticle after it has entered the cancer cell. Photodynamic cancer therapy involves generating singlet oxygen by using a laser to irradiate a photosensitive dye. The singlet oxygen is highly reactive and in particular destroys biological membranes via oxidation of lipids. After diffusing out of the nanoparticle, the toxic oxygen reacts so fast that it never leaves the cancer cell (Fig. 7.12).

Nanoparticles may also be used to kill cancer cells through localized heating. In one approach, nanoparticles with a magnetic core are used. An alternating magnetic field is used to supply energy, and it heats the nanoparticle to a temperature lethal to mammalian cells. Another approach uses metal nanoshells. They consist of a core, often silica, surrounded by a thin metal layer, such as gold. Varying the size of the core and thickness of the metal layer allows such nanoparticles to be tuned to absorb from any region of the spectrum, from UV through the visible to the IR. Because living tissue absorbs least in the near infrared, the nanoparticles are designed to absorb radiant energy in this region of the spectrum. This results in external near infrared being specifically absorbed and heating the surrounding tissue.

Another possible way to use nanoparticles against cancer is to control blood vessel development, or angiogenesis. For cancer cells to develop into a genuine tumor, a blood supply is needed. The balance between pro- and antiangiogenic growth factors that bind to cell surface receptors controls the formation of new blood vessels. Gold nanoparticles that carry relatively short peptides on their surfaces have been constructed to bind to these receptors. Peptides that bind to the neuropilin-1 receptor, such as KATWLPPR, block blood capillary formation.

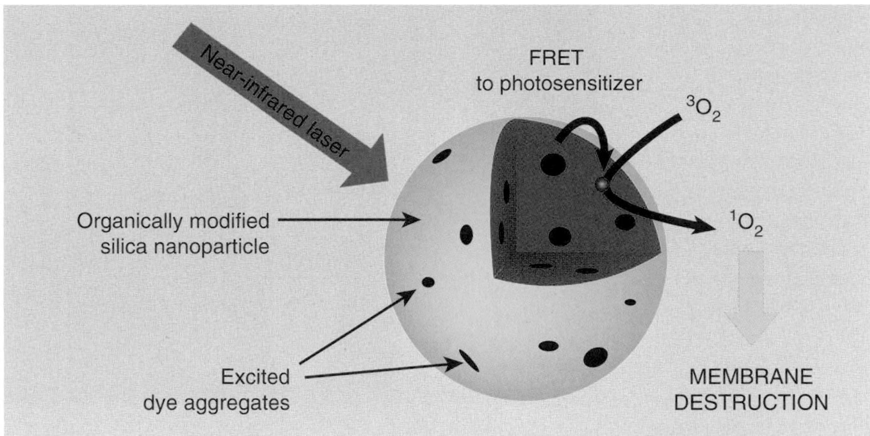

FIGURE 7.12 Nanoparticle for Singlet Oxygen Release
The near-infrared laser excites the dyes attached to the nanoparticle. Energy transfer to photosensitizers by fluorescent resonance energy transfer (FRET) results in conversion of normal (triplet) oxygen to singlet oxygen.

Nanoparticles may be used to kill cancer cells by localized heating or by local generation of a toxic product such as singlet oxygen. Another approach to cancer therapy is using nanoparticles to regulate blood vessel formation.

ASSEMBLY OF NANOCRYSTALS BY MICROORGANISMS

It has been known for many years that bacteria may accumulate a variety of metallic elements and may modify them chemically, usually by oxidation or reduction. For example, many bacteria accumulate anions of selenium or tellurium and reduce them to elemental selenium or tellurium, which is then deposited as a precipitate either on the cell surface or internally. Certain species of the bacterium *Pseudomonas* that live in metal-contaminated areas and the fungus *Verticillium* can both generate silver nanocrystals. However, so-called "nanobacteria" are a mineral artefact (see Box 7.2).

Box 7.2 The Nanobacteria—Nanotechnology Meets Nanomythology

Cryptozoology is the study of "undiscovered" creatures such as the Loch Ness monster or Bigfoot. However, students of microbiology no longer need to feel left out. The new field of nanocryptobiology is here. It is perhaps not surprising that some investigators claimed to have discovered "nanobacteria." They were supposedly 100-fold smaller than typical bacteria, yet capable of growth and replication. They were proposed as causative agents in the formation of kidney stones and then linked to heart disease and cancer. Unfortunately, "nanobacteria" are too small to contain ribosomes or chromosomes, and it has become clear that they are mineral artifacts. Their supposed replication was due to the fact that certain minerals can act as nuclei for further crystallization. It scarcely needs adding that "fossilized nanobacteria" have also been seen in meteorites from Mars and have been claimed as evidence for life on Mars. However, similar mineral structures have been found in both lunar meteorites and terrestrial rocks.

Recently, it has been found that when *E. coli* is exposed to cadmium chloride ($CdCl_2$) and sodium sulfide, it precipitates cadmium sulfide (CdS) as particles in the 2- to 5-nm size range. In other words, bacteria can "biosynthesize" semiconductor nanocrystals. Increasing levels of the sulfhydryl compound, glutathione, improves the yield. When cadmium chloride and potassium tellurite (K_2TeO_3) were provided, a mixture of fluorescent nanocrystals of different sizes and colors made of CdTe was generated (Fig. 7.13). Bizarrely enough, even earthworms can make nanocrystals. If cadmium chloride and potassium tellurite were added to soil, wild-type earthworms made green fluorescent CdTe nanocrystals!

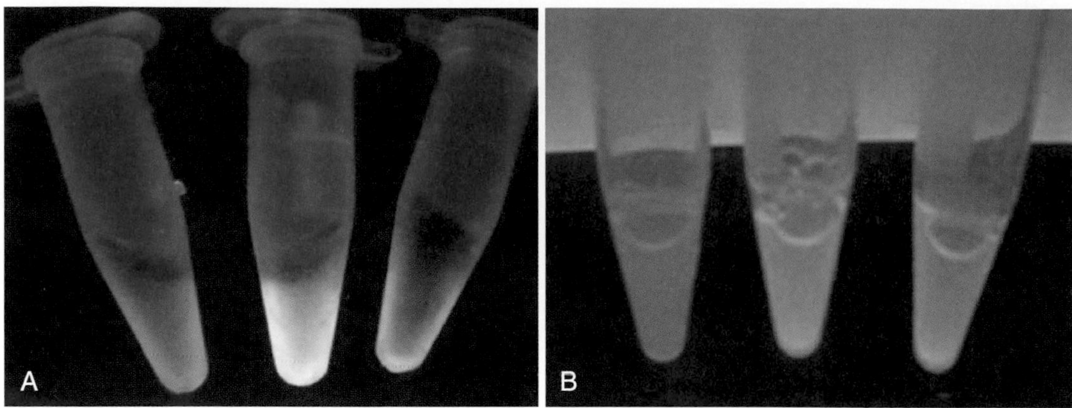

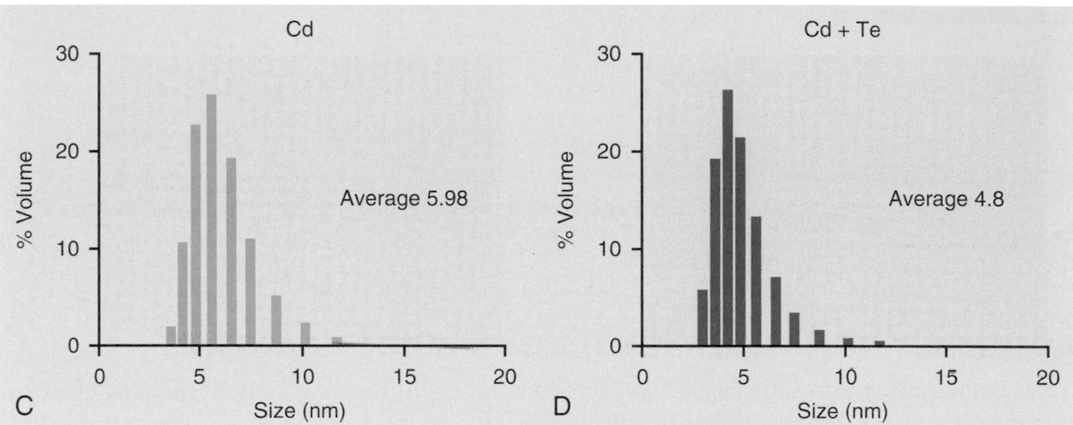

FIGURE 7.13 CdTe Nanoparticles Made by Bacteria
Purification and size determination of CdTe nanoparticles produced by *E. coli.* (A) UV-exposed cell suspensions of *E. coli* AG1/pCA24NgshA, untreated or exposed to $CdCl_2$ or $CdCl_2/K_2TeO_3$ (from left to right). (B) Purified fractions exposed to UV light. (C) and (D) Particle sizes of samples from cells exposed to $CdCl_2$ or $CdCl_2/K_2TeO_3$, respectively. From Monrás JP et al. (2012). Enhanced glutathione content allows the in vivo synthesis of fluorescent CdTe nanoparticles by *E. coli. PLoS One* **7**(11), e48657.

Rather more sophisticated is the use of phage display to select peptides capable of organizing semiconductor nanowires. As described in Chapter 9, phage display is a technique that allows the selection of peptides that bind any chosen target molecule. In brief, stretches of DNA encoding a library of peptide sequences are engineered into the gene for a bacteriophage coat protein. The extra sequences are attached at either the C terminus or N terminus, where they do not disrupt normal functioning of the coat protein. When the hybrid protein is assembled into the phage capsid, the inserted peptides are displayed on the outside of the phage particle. The library of phages is then screened against a target molecule. Those phages that bind the target are kept.

Phage display libraries have been screened to find peptides capable of binding ZnS or CdS nanocrystals. Protein VIII of bacteriophage M13 was used for peptide insertion. For example, ZnS was bound by the peptide VISNHAGSSRRL and CdS by the peptide SLTPLTTSHLRS. Because the bacteriophage capsid contains many copies of the coat protein, the displayed peptide is also present in many copies. Consequently, an array of nanocrystals forms on the phage surface. Because M13 is a filamentous phage, the result is a semiconductor nanowire (Fig. 7.14).

> Nanocrystals and nanowires may be assembled using unmodified bacteria, genetically engineered bacteria, or sophisticated phage display techniques.

NANOTUBES

Carbon **nanotubes** are cylinders made of pure carbon with diameters of 1 to 50 nanometers. However, they may be up to approximately 10 micrometers long. Pure elemental carbon exists as diamond or graphite. In diamond, each carbon is covalently linked to four others forming a 3D tetrahedral lattice that is extremely strong. In contrast, graphite consists of flat sheets of carbon atoms that form a hexagonal pattern. In the sheets of graphite, each carbon atom is covalently bonded to three neighbors, and the sheets can slide sideways over each other because there are no covalent linkages between atoms in different sheets.

To form a nanotube, a single sheet of graphite is rolled into a cylinder. The sheets may be rolled up straight or at an angle to the carbon lattice and may be of various diameters. Depending on the diameter and the torsion, the nanotube may act as a metallic conductor or a semiconductor. Not surprisingly, carbon nanotubes are now finding many uses in electronics, a topic beyond the scope of this book. Single-walled carbon nanotubes, i.e., those consisting of a single layer of graphite, are especially useful in biology as they enter cells very readily.

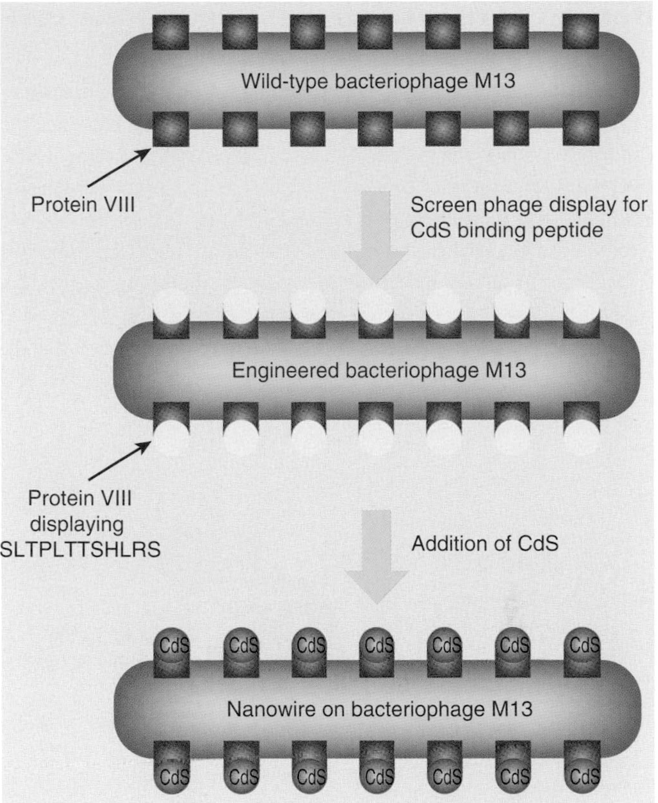

FIGURE 7.14 Nanowire Assembly on Bacteriophage
Phage display yielded engineered versions of the M13 coat protein (protein VIII) with inserted peptides. Some of these are capable of binding CdS. In the presence of CdS crystals, a nanowire forms along the surface of the bacteriophage.

In biotechnology, nanotubes are beginning to find applications. The critical issue is attaching useful biomolecules, such as enzymes, hormone receptors, or antibodies, to the nanotube surface. A major problem in attaching proteins is that the surface of carbon nanotubes is hydrophobic. One approach is to first modify the surface by adding nonionic detergents, such as Triton X100. The hydrophobic portion of the detergent binds to the nanotube surface, and the hydrophilic region can be used to bind proteins. Alternatively, chemical reagents that react with the carbon surface of the nanotube are used to generate side chains carrying reactive functional groups. Proteins can then be linked covalently by reaction with these (Fig. 7.15). Proteins can also be attached to natural, magnetic, nanoparticles (see Box 7.3).

Possible applications of carbon nanotubes in biotechnology and medicine include

(a) Imaging. Even without attached dyes, carbon nanotubes show luminescence in the near infrared. This can be directly used for imaging by near infra-red microscopy (NIRM).

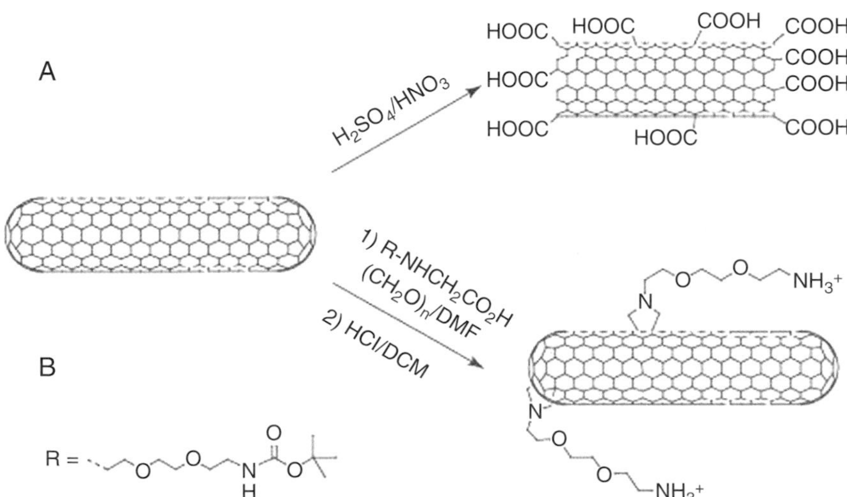

FIGURE 7.15 Attaching Organic Functional Groups to Nanotubes
(A) Carbon nanotubes can be treated with acids to purify them and generate carboxylic groups. (B) Alternatively, they may react with amino acid derivatives and aldehydes to add more complex hydrophilic groups to the external surface. From Bianco A, Kostarelos K, Prato M (2005). Applications of carbon nanotubes in drug delivery. *Curr Opin Chem Biol* **9,** 674–679. Reprinted with permission.

Box 7.3 Magnetosomes: Natural Bacterial Magnetic Nanoparticles

Naturally occurring magnetic nanoparticles are made by magnetotactic bacteria, such as *Magnetospirillum*. These microorganisms can detect magnetic fields and orient themselves in response. They contain **magnetosomes**, consisting of nanosized crystals of magnetic iron oxide (magnetite, Fe_3O_4) or, less often, iron sulfide (greigite, Fe_3S_4) inside an envelope of protein. The magnetosomes are aligned in chains along the cell axis. Synthesis of the protein shell and mineralization of the magnetic core are under genetic control. At least in some cases, the genes responsible for the magnetosome are clustered on the bacterial chromosome.

It is possible to attach other molecules to the outside of magnetosomes by genetically modifying proteins of the magnetosome envelope (Fig. A). The gene for the Mms16 protein of *Magnetospirillum magneticum* has been fused to the genes for luciferase and the dopamine receptor in the lab of Dr. Tadashi Matsunaga of the Tokyo University of Agriculture and Technology. The fused proteins are displayed on the surface of the magnetosomes. After the bacterial cells are disrupted, the magnetosomes carrying the attached proteins can be purified by magnetic separation. This should allow easier analysis of membrane-bound receptors, such as those from the human nervous system, in a simplified system.

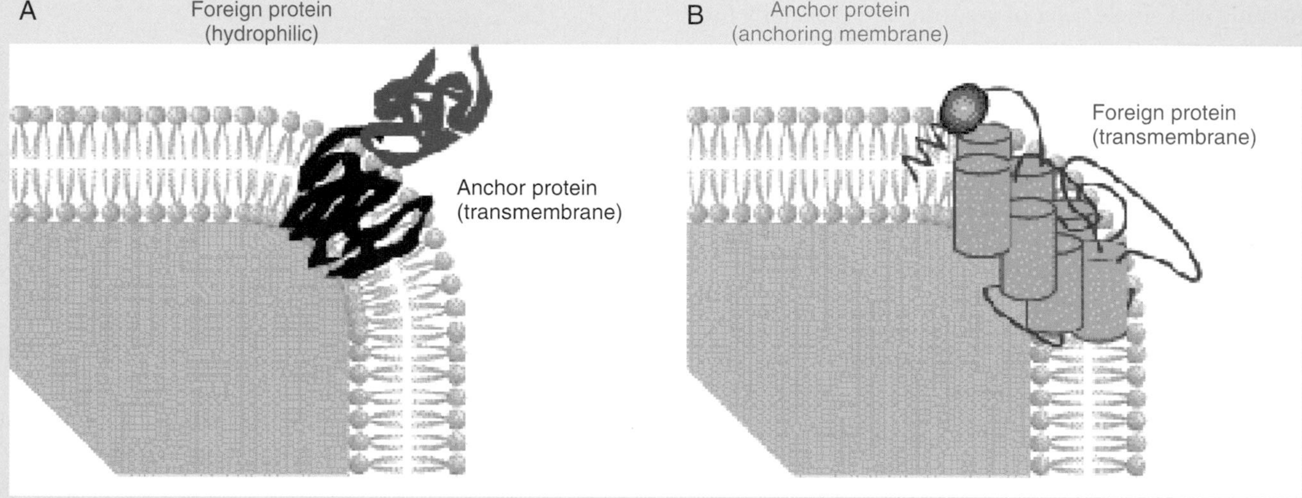

FIGURE A Protein Display on Magnetosome Membrane
(A) Display of hydrophilic protein using MagA as an anchor. (B) Display of transmembrane protein using Mms16 as an anchor. From Matsunaga T, Okamura Y (2003). Genes and proteins involved in bacterial magnetic particle formation. *Trends Microbiol* **11**, 536–542. Reprinted with permission.

(b) Electrochemical sensors. The electrical properties of carbon nanotubes change when attached chemical groups bind or react with other molecules. Thus, electrical signals can be generated upon detection of target molecules.

(c) Photothermal killing of cancer cells. As noted previously, nanoparticles can be used to kill cancer cells. Carbon nanotubes absorb near infrared radiation and can generate local heating that causes cell death. As above, the nanotubes must carry molecules that target them specifically to the cancer cells.

(d) Drug delivery. When using carbon nanotubes, drugs are attached to the outside (rather than being encapsulated). This normally requires an intervening linker molecule, as shown in Figure 7.16.

(e) Tissue regeneration. Nanotubes that carry appropriate chemical groups may be used as scaffolds for regenerating tissues such as bone and nerves. This approach is still experimental.

A range of uses has been found for hollow carbon nanotubes that are fabricated to carry a variety of biologically useful side chains.

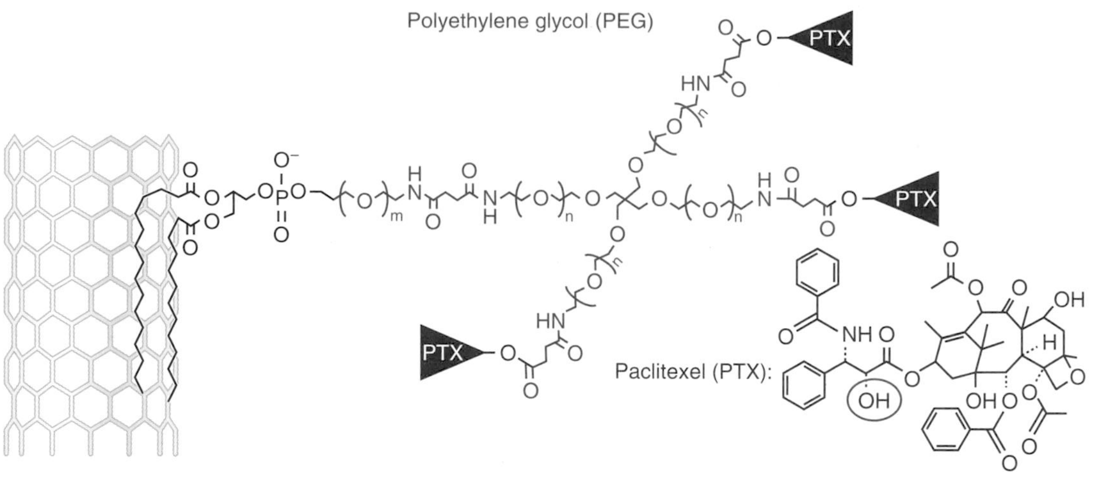

FIGURE 7.16 Drug Delivery by Carbon Nanotubes
A carbon nanotube has been modified for paclitaxel (PTX) delivery. Single-walled carbon nanotubes with bound phospholipids were attached to branched polyethylene glycol (PEG) chains. These in turn were linked to PTX. The OH group inside the blue ring indicates where the PTX was linked to the PEG. From Gomez-Gualdrón DA et al. (2011). Carbon nanotubes: engineering biomedical applications. *Prog Mol Biol Transl Sci* **104**, 175–245.

ANTIBACTERIAL NANOCARPETS

Nanocarpets are structures formed by stacking a large number of nanotubes together vertically, with their cylindrical axes aligned. Nanocarpets capable of changing color and of killing bacteria have been assembled from specially designed lipids that spontaneously assemble into a variety of nanostructures depending on the conditions. In water, nanotubes are formed. Partial rehydration of dried nanotubes generates a side-by-side array—the nanocarpet.

The lipid consists of a long hydrocarbon chain (25 carbons) with a diacetylenic group in the middle of the chain. The individual nanotubes are about 100 nm in diameter by 1000 nm in length. The walls of the nanotubes consist of five bilayers of the lipid. Both the separate lipid molecules and the assembled nanocarpet kill bacteria. Like other long-chain amino compounds, they act as detergent molecules and disrupt the cell membrane. Consequently, the nanocarpet provides a surface lethal to bacteria. This property could be very useful if nanocarpets are used in biomedical applications.

Diacetylenic compounds have the interesting ability to change color. The nanocarpet starts out white, but if exposed to ultraviolet light, it turns deep blue. UV irradiation causes cross-links to form by reaction between acetylenic groups on neighboring molecules. This polymerization stabilizes the nanocarpet. Blue nanocarpets change color on exposure to a variety of reagents. Detergents and acids change them from blue to red or yellow, and the presence of bacteria, such as *E. coli*, gives red and pink shades. Eventually, such materials may be used both as biosensors and for protection against bacterial contamination.

Nanotubes may be assembled to create surfaces (nanocarpets) that are antibacterial or act as biosensors.

DETECTION OF VIRUSES BY NANOWIRES

Nanowires are what their name suggests. They have nanoscale diameters but may be several microns long. They may be metallic and act as electrical conductors, or they may be made from semiconductor materials.

Biosensors can be made using silicon semiconductor nanowires. They may be coated with antibodies that bind to a specific virus. Binding of the virus to the antibody triggers a change in conductance of the nanowire. For a p-type silicon nanowire, the conductance decreases when the surface charge on the virus particle is positive and, conversely, increases if the virus surface is negative. Single viruses may be detected by using this approach (Fig. 7.17). It is also possible to attach single-stranded DNA to the nanowire. In this case, binding of the complementary single strand triggers changes in conductance. Possible future applications include both clinical testing and sensors for monitoring food, water, and air for public health and/or biodefense.

Nanowire sensors are capable of detecting specific individual viruses. Binding of a virus particle changes the conductance of the nanowire.

ION CHANNEL NANOSENSORS

Somewhat more complex than nanotubes and nanowires are **nanoscale ion channels** that are assembled into membranes. These channels are designed so that they can be controlled to permit the movement of ions under only certain conditions. The ion flow generates an electrical current that is detected, amplified, and displayed by appropriate electronic apparatus.

Ion channels can be used as biosensors by attaching a binding site for the target molecule at the entry to the channel. Attached antibodies are often used for the binding sites. The simplest arrangement results in the channel being open in the absence of the target molecule and shut when it is detected. A drop in ion flow therefore signals detection of the target molecule.

At present, such ion channels are being developed using modified biological components. The ion channel itself can be made using the peptide antibiotic gramicidin A (made by the bacterium *Bacillus brevis)*. This transports monovalent cations, especially protons and sodium ions. Natural gramicidin spans half of a standard biological membrane. A short-lived channel is formed when two gramicidin molecules line up, as shown in Figure 7.18. Permanent channels may be made by covalently linking two gramicidin molecules together. Up to 10^7 ions/second flow through a single gramicidin channel. This gives a picoampere current that is easily measured. An alternative is to monitor a change in the pH due to movement of H^+ ions. An optical sensor and a fluorescent pH indicator may be used to do this.

The channels are made responsive by attaching an appropriate ligand molecule to the front end of the gramicidin so that it projects outward from the membrane surface. This ligand is chosen to bind the target molecule and may be an antigen or other small molecule that is recognized by an antibody or protein receptor. It is also possible to attach a single-stranded segment of DNA that will recognize and bind the complementary sequence. Thus, biosensors may be designed to respond to the presence of a variety of biological molecules.

The membrane itself may be a lipid bilayer made using natural membrane lipids. Typical phospholipids span half a membrane (i.e., one monolayer), and the two monolayers can therefore slide relative to one another. Including lipids that span its whole width will stabilize the membrane. Such lipids may be found naturally in certain Archaea or may be synthesized artificially. Lipid bilayers are relatively fragile and in practice must be assembled on some solid support. Building a long-lasting and stable membrane structure has so far proven difficult, and practical ion channel sensors are still under development.

DNA is a long thin molecule that can move through nanoscale channels in response to a voltage. Since DNA is negatively charged, the DNA is attracted through the nanopore to the positive side. A novel DNA sequencing method based on this approach is now being marketed by Oxford Nanopore Technologies. When DNA occupies the channel, the normal ion current is reduced. The amount of reduction depends on the base sequence ($G > C > T > A$). Thus, a computer can measure the current and decipher the sequence based on the differences.

Ion channel sensors operate by opening or closing the channel in response to binding a specific molecule. They may be used to detect a variety of target molecules.

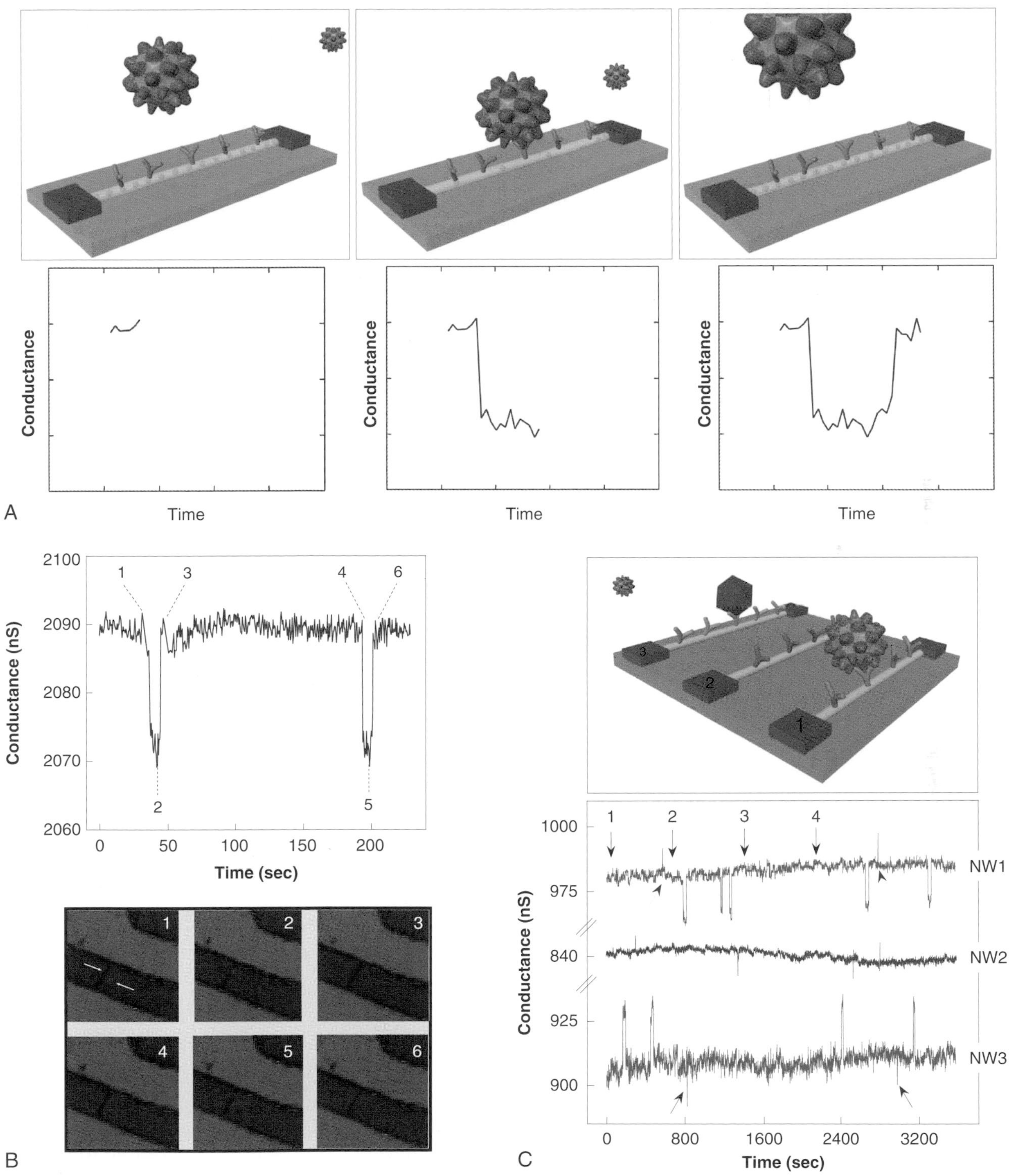

FIGURE 7.17 Nanowire Biosensors

(A) A single virus particle binding to and detaching from the surface of a SiNW nanowire coated with antibody receptors. The corresponding changes in conductance are shown for each step. (B) Conductance and optical data on addition of influenza A virus. (C) Schematic of multiplexed single virus detection. Conductance versus time is recorded simultaneously for three channels specific for different viruses. NW1 responds to influenza A and NW3 to adenovirus. The NW2 channel was not used here. Black arrows 1–4 correspond to addition of adenovirus, influenza A, pure buffer, and a 1:1 mixture of adenovirus plus influenza A. Red and blue arrows indicate conductance changes due to the diffusion of viral particles past the nanowire without specific binding for influenza and adenovirus, respectively. Courtesy of Charles M. Lieber, Harvard University, Cambridge, MA.

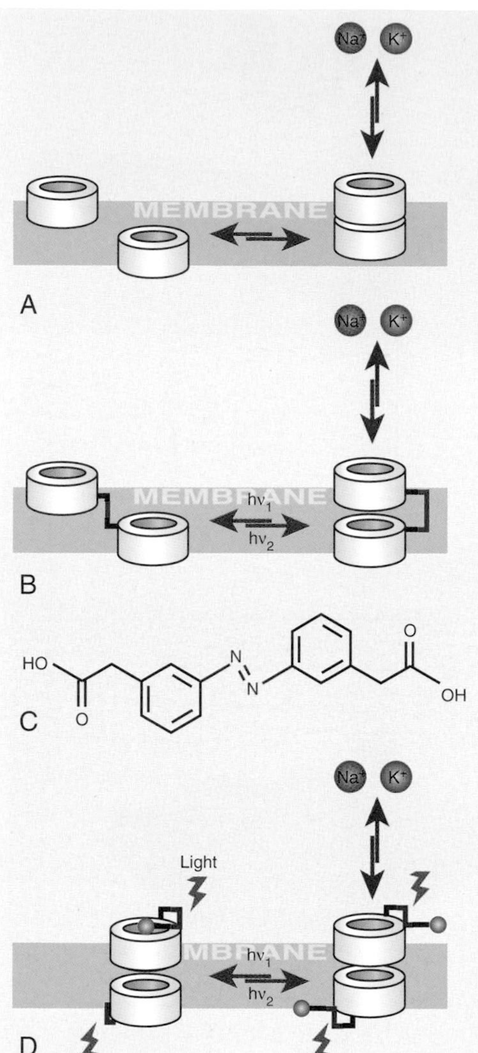

FIGURE 7.18 Natural and Modified Gramicidin Ion Channels
Gramicidin forms a channel for Na^+ and K^+ ions. (A) Natural gramicidin channels are formed when two gramicidin molecules align within a membrane. (B) Two gramicidin monomers joined by a photosensitive linker. (C) Absorption of light changes the conformation of the N = N bond (red) of the linker from *cis* to *trans* and opens or closes the channel. (D) Channels are opened or closed by blocking groups (blue circles) attached by photosensitive linkers to each gramicidin monomer.

NANOENGINEERING OF DNA

In "classical" genetic engineering, the sequence of DNA is deliberately altered in order to generate new combinations of genetic information. Even when major rearrangements are made, in order to function as genetic information, the DNA must remain as a base-paired double-stranded helix with an overall linear structure.

In nanoengineering, the objective is to build structures using DNA merely as structural material, rather than to manipulate genetic information. DNA is attractive because the double helix is a convenient structural module. Moreover, its natural base-pairing properties can be used to link separate DNA molecules together. However, a critical requirement for assembling 3D structures is branched DNA. Although branched structures do form in biological situations (especially the Holliday junction involved in crossing over during recombination), they are not permanent or stable.

Mixing four carefully designed single strands with different sequences can generate cross-shaped DNA. Each strand base-pairs with two of the other strands over half its length (Fig. 7.19). If sticky ends are included in the initial strands, it is possible to link the crosses together into a two-dimensional matrix. The nicks can be sealed by DNA ligase if desired. The principles used in branching can be extended to three dimensions, and it is possible to build cubical DNA lattices.

The DNA double helix is about 2 nm wide with a helical pitch of about 3.5 nm. Hence, it can be used to build nanoscale frameworks. These frameworks can be used for the assembly of other components, such as metallic nanowires or nanocircuits. Note that while DNA is flexible over longer distances, it is relatively rigid over nanoscale lengths, up to about 50 nm.

The cross-shaped DNA molecules (and their 3D counterparts) have the drawback that the junctions are flexible and do not maintain rigid 90-degree angles. Rigid DNA components have been made by using double-crossover (DX) DNA molecules. Two isomers of antiparallel DX DNA exist, with an odd (DAO) or even (DAE) number of half-turns between the crossover points (Fig. 7.20). (Double-crossover molecules with parallel strands also exist, but they behave poorly from a structural viewpoint.) DAE or DAO units can be assembled into a rigid array by providing appropriate sticky ends. It is possible to replace the central, short DNA strand of the DAE structure with a longer protruding strand of DNA (DAE+J). This allows assembly of branched structures.

What is the purpose of building arrays and 3D structures from DNA? Perhaps the most plausible purpose suggested so far is to use the DNA as the framework for assembling nanoscale electronic circuits. So far, normal, unbranched DNA has been used as a scaffold to create linear metallic nanowires. Various metals (gold, silver, copper, palladium, platinum) have been used to coat the DNA, and diameters range from 100 nm down to 3 nm. Eventually, it should be possible to put together these two approaches and build circuits from metal-covered 3D DNA structures.

DNA may be viewed solely as a structural molecule. Three-dimensional frameworks may be built from DNA whose sequence is designed to generate branched structures. Such DNA structures may be used as nanoscale scaffolds for metallic nanowires and circuits.

DNA ORIGAMI

Building nanostructures by assembling multiple different DNA molecules becomes extremely difficult beyond a certain level of complexity. The DNA origami approach greatly simplifies building DNA nanostructures by using one very long DNA strand and folding it up to form a scaffold. A number of much shorter "staple strands" are added in excess to help folding. The "staple strands" bind at specific sites along the longer scaffold strand to drive folding (Fig. 7.21). This approach means that it is no longer necessary to strictly control the ratio of different DNA strands as for the "traditional" DNA folding described in the previous section (see especially Fig. 7.19). Assembly is much faster and yields are much higher with the origami approach.

Except in very simple cases, DNA origami relies on computer-aided design, in particular for specifying the DNA sequences required for building the chosen shapes. Figure 7.22 illustrates the procedure for building a complex structure by using this approach. The traditional approach for such structures would take weeks and involve synthesizing and purifying multiple long strands. These must then be assembled in the correct proportions and in the correct order. DNA origami, in contrast, requires one scaffold strand plus a roughly 10-fold excess of the staple strands. These strands are all mixed together, heated, and then cooled slowly to anneal. This process takes only a few hours.

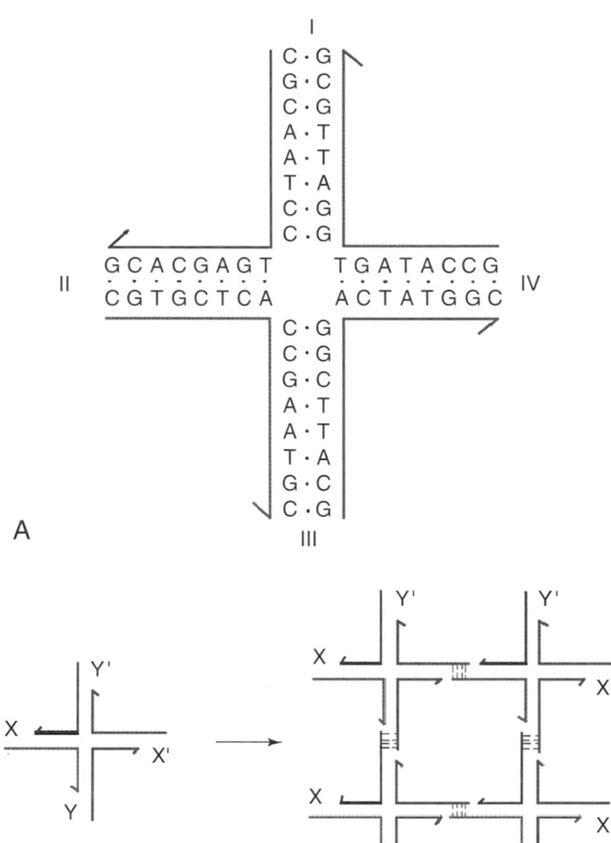

FIGURE 7.19 Branched DNA from Four Single Strands
(A) A branched DNA molecule with four arms. Four different color-coded strands combine to produce four arms (I, II, III, and IV). The branch point of this molecule is fixed. (B) Formation of a two-dimensional lattice from a four-arm junction with sticky ends. X and Y are sticky ends, and X′ and Y′ are their complements. Four monomers are complexed in parallel orientation to yield the lattice structure. DNA ligase can seal the nicks left in the lattice. From Seeman NC (1999). DNA engineering and its application to nanotechnology. *Trends Biotechnol* **17**, 437. Reprinted with permission.

DAE DAO DAE+J

A

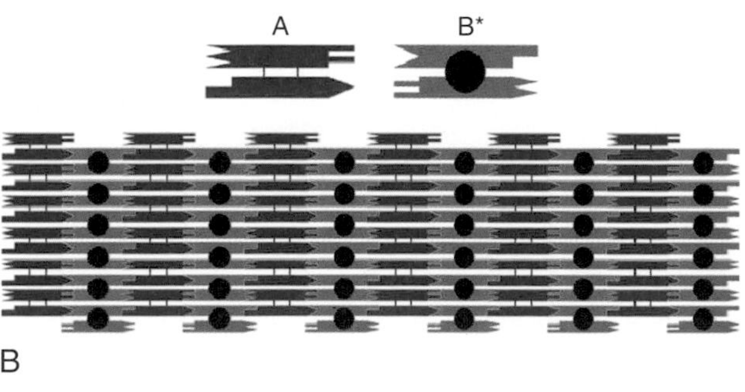

FIGURE 7.20 Rigid DNA Nanomodules
Arrays may be assembled from double-crossover (DX) molecules. (A) DAE and DAO are two antiparallel DX isomers. DAE+J is a DAE molecule in which an extra junction replaces the nick in the green strand of DAE. (B) Two-dimensional array derived from DX molecules. Complementary sticky ends are depicted by complementary geometrical shapes. A is a conventional DX molecule, but B* is a DX+J molecule with a vertically protruding DNA hairpin (black circle). From Seeman NC (1999). DNA engineering and its application to nanotechnology. *Trends Biotechnol* **17**, 437. Reprinted with permission.

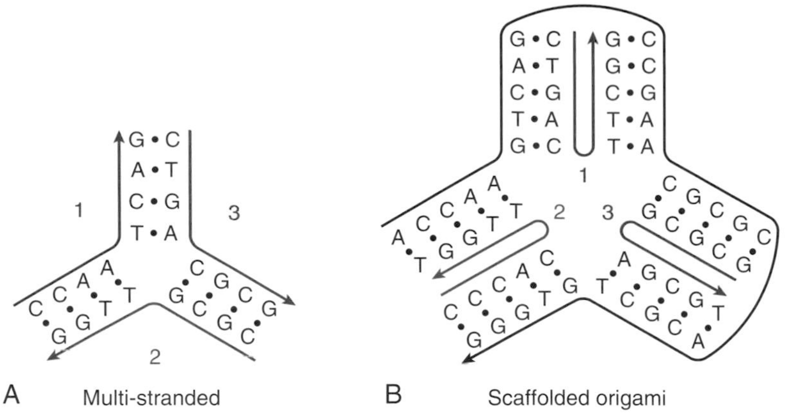

A Multi-stranded B Scaffolded origami

FIGURE 7.21 Principle of DNA Origami
(A) The traditional approach uses multiple strands to build DNA nanostructures. (B) DNA origami uses one long scaffold strand plus several short staple strands that guide folding. From Rothemund PWK (2005). Design of DNA origami. *Proc Int Conf Computer-Aided Design (ICCAD)* 471–479; figure provided by Paul Rothemund, Computation and Neural Systems, Caltech.

In DNA origami, nanostructures are built by folding up a single very long strand of DNA. Many much smaller staple strands assist in the folding.

DNA MECHANICAL NANODEVICES

A rather more futuristic use for 3D DNA structures is as frameworks for mechanical nanodevices. The essential components are moving parts of some kind. Several prototype "DNA machines" have been designed or constructed that illustrate the concept. They all use reversible changes in conformation of a DNA structure driven by changes in base pairing. Such changes may be caused either by changing the physical conditions (heat, salt, etc.) or by adding segments of single-stranded DNA (ssDNA) that base-pair to some region of the DNA machine, as illustrated in Figure 7.23. If ssDNA is used, then another single strand, complementary to the first, is added to convert the machine back to its original conformation. The result is a mechanical cycle that could in principle be used to perform some sort of task. The ssDNA molecules may be regarded as "fuel," and the final waste product is a double-stranded DNA consisting of the two paired ssDNA fuel elements. (Note that this scheme does not involve breaking covalent chemical bonds. It is thus not an enzymatic reaction and is distinct from using DNA as a deoxyribozyme, as described in Chapter 5.)

DNA has been proposed as a framework for nanomachines. Proof-of-concept prototypes have been constructed.

CONTROLLED DENATURATION OF DNA BY GOLD NANOPARTICLES

DNA hybridization is widely used to detect target sequences, both in the laboratory and in clinical diagnosis. Before hybridization can occur, the DNA double helix must be denatured into single strands. This is accomplished by heating bulk DNA. However, newly emerging nanotechnology may allow specific individual DNA molecules to be dissociated when required.

Nanoparticles of about 1.4 nm and containing fewer than 100 atoms of gold are attached to double-stranded DNA. When the structure is exposed to radio waves (generated by an alternating magnetic field), the gold acts as an antenna. It absorbs energy and heats the DNA molecule to which it is attached. This melts the DNA double helix and converts it to

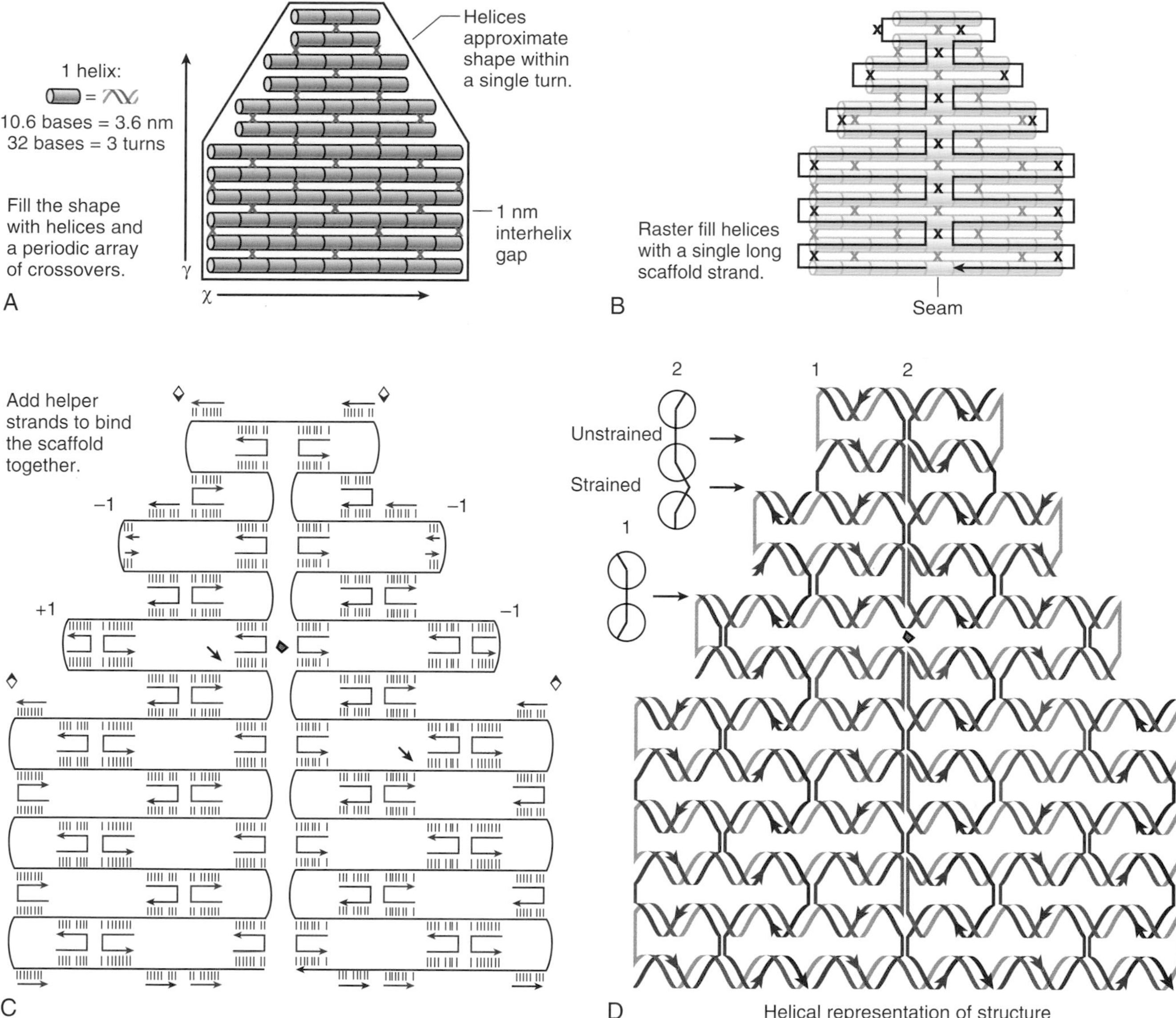

FIGURE 7.22 Steps in Designing DNA Origami

(A) Fill the chosen shape with helixes plus crossovers (needed for stability). (B) Convert design to a single long folded scaffold. (C) Insert staple strands to bind scaffold into shape. (D) Helical representation. From Rothemund PWK (2005). Design of DNA origami. *Proc Int Conf Computer-Aided Design (ICCAD)* 471–479; figure provided by Paul Rothemund, Computation and Neural Systems, Caltech.

single strands. Heating extends over a zone of about 10 nm, so surrounding molecules are unaffected. The heat is dissipated in less than 50 picoseconds, so the DNA may be rapidly switched between the double- and single-stranded states by turning the magnetic field on and off. The procedure may be applied to dsDNA with two separate single strands (Fig. 7.24) or to stem-and-loop structures formed by folding from a single strand of DNA.

Practical applications are several years away. However, because radio waves penetrate living tissue very effectively, it may eventually be possible to control the behavior of individual DNA molecules from outside an organism. Metal antennas of different materials or sizes could be used to tune different DNA molecules to radio waves of different frequencies.

> Attachment of a metallic antenna allows radio waves to melt DNA into single strands. It might eventually be possible to control the behavior of DNA from outside an organism.

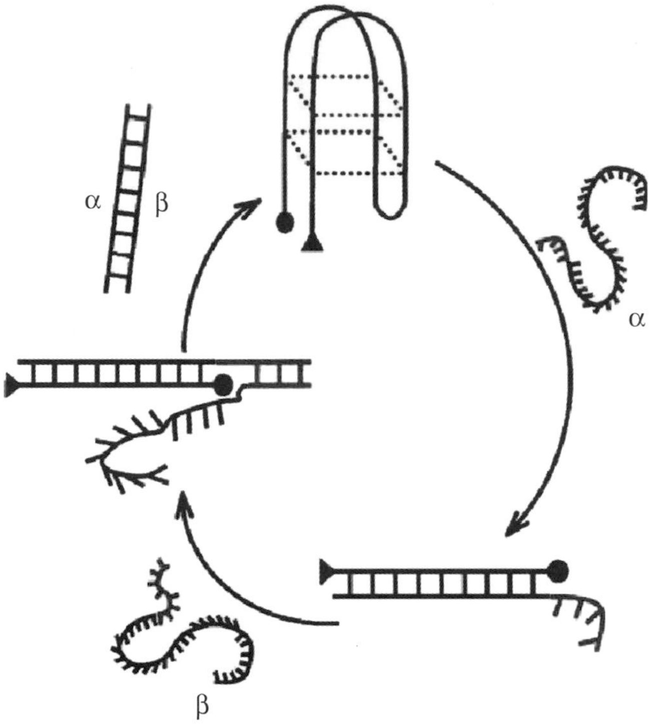

FIGURE 7.23 Prototype DNA Machine
A DNA nanomotor designed by J. J. Li and W. Tan. Successive addition of the complementary DNA strands labeled alpha and beta causes a change in conformation. The DNA nanomotor alternates between a folded quadruplex structure and a double-stranded structure. The nanomotor expands and contracts in a wormlike motion. From Ito Y, Fukusaki E (2004). DNA as a nanomaterial. *J Mol Catalysis B: Enzymatic* **28,** 155–166. Reprinted with permission.

CONTROLLED CHANGE OF PROTEIN SHAPE BY DNA

Allosteric proteins change shape in response to the binding of signal molecules (allosteric effectors) at a specific site. The essence of allosteric control is that the shape change is transmitted throughout the protein and affects the conformation of distant regions of the protein. In allosteric enzymes, binding of an allosteric effector at a distant site alters the conformation of the active site and may change its affinity for the substrate. In this way, some enzymes are switched on and off in response to signal molecules. For example, phosphofructokinase is switched on by the buildup of AMP, which signals that energy is in short supply. The response increases flow into the glycolytic pathway, which increases energy generation. Similarly, many DNA-binding proteins, such as repressors and activators, also change shape on binding small signal molecules.

It is possible to change the shape of a protein artificially by mechanical force. This has been

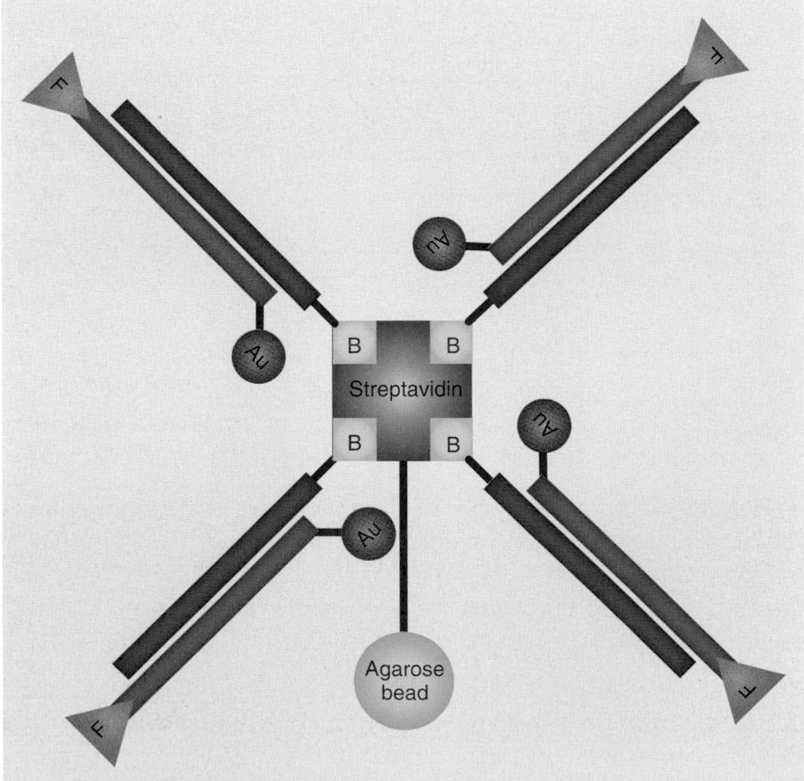

FIGURE 7.24 Controlling DNA Denaturation by Gold Nanoparticles
One strand of DNA has a gold nanoparticle (Au) attached to one end and a fluorescent dye (F) at the other end. The complementary strand has a biotin tag (B) at one end. The biotin is bound by streptavidin and therefore binds the DNA strand to an agarose bead. When the gold absorbs energy, it melts the two strands of DNA. The DNA strand with the fluorescent dye is released into the supernatant, and its fluorescence is monitored.

demonstrated by attaching a single-stranded 60-base segment of DNA between the two poles of a protein. Attaching the DNA requires chemical "handles." These handles are engineered into the target protein by replacing amino acids at appropriate positions with cysteine. The reactive SH group is then used to chemically attach the DNA.

Double-helical DNA is much more rigid than ssDNA. Consequently, the addition of the complementary strand generates tension as it binds and creates a double helix. This approach has been demonstrated in the laboratory of Giovanni Zocchi at UCLA with maltose-binding protein and an enzyme, guanylate kinase. When maltose-binding protein is stretched, its binding site for maltose opens wider than optimum, and the affinity for the sugar decreases. For guanylate kinase (Fig. 7.25), applying tension decreases enzyme activity by lowering affinity for substrate binding. In this case, releasing the tension by adding DNase to digest the DNA switches the enzyme on again.

Potential applications are far in the future. However, it is possible to imagine biosensors that detect DNA sequences based on this mechanism. In addition, it might be possible to externally control enzymes or other proteins by adding appropriate ssDNA (or, of course, RNA).

The shape of a protein may be changed artificially by applying force. This may be demonstrated by attaching DNA strands to the protein. Pairing single-stranded DNA with its complementary strand generates tension and stretches the protein.

BIOMOLECULAR MOTORS

A major aim of nanotechnology is to develop molecular-scale machinery that can carry out the programmed synthesis (or rearrangement) of single molecules (or even atoms) or perform other similar nanoscale tasks. The term (nano)**assembler** refers to a nanomachine that can build nanoscale structures, molecule by molecule or atom by atom. And the term (nano)**replicator** refers to a nanomachine able to build copies of itself when provided with raw materials and energy. This, of course, sounds remarkably like a living cell. Indeed, the organelles of living cells may be regarded as nanomachines and have provided both inspiration and components for nanotechnologists.

To operate, nanomachines will need energy, which will be provided by "molecular motors." At present, such devices are still in development. It has been suggested that biological structures might be used for this purpose. Examples include the ATP synthase, the flagellar motor of bacterial cells, various enzymes that move along DNA or RNA, and assorted motor proteins

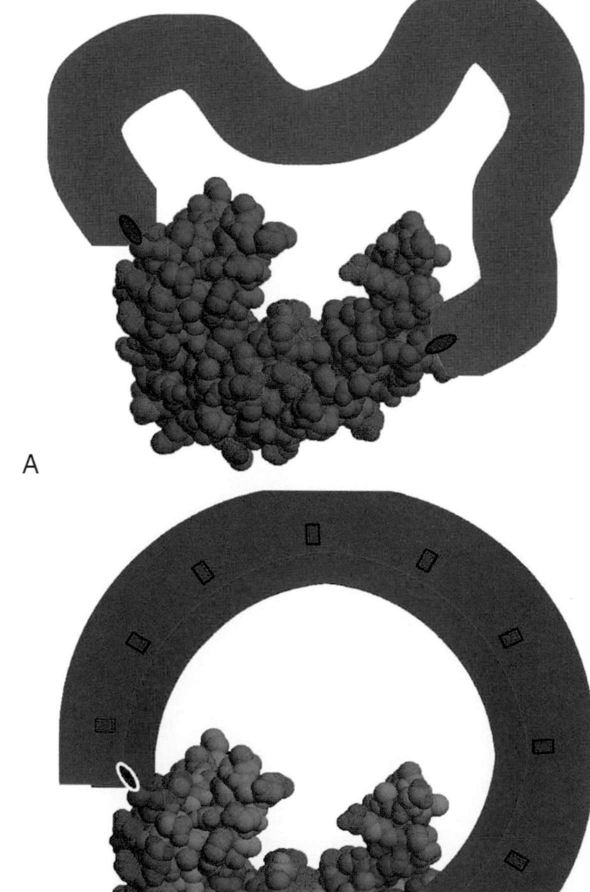

FIGURE 7.25 Controlling Protein Shape by DNA
Protein-ssDNA chimera for guanylate kinase from *Mycobacterium tuberculosis* (PDB structure 1S4Q). The purple attachment points for the molecular spring correspond to mutations Thr75 → Cys and Arg171 → Cys. (A) Unstretched—a single strand of DNA is attached to the protein. (B) The protein is stretched by addition of the complementary DNA strand (purple). Courtesy of Giovanni Zocchi.

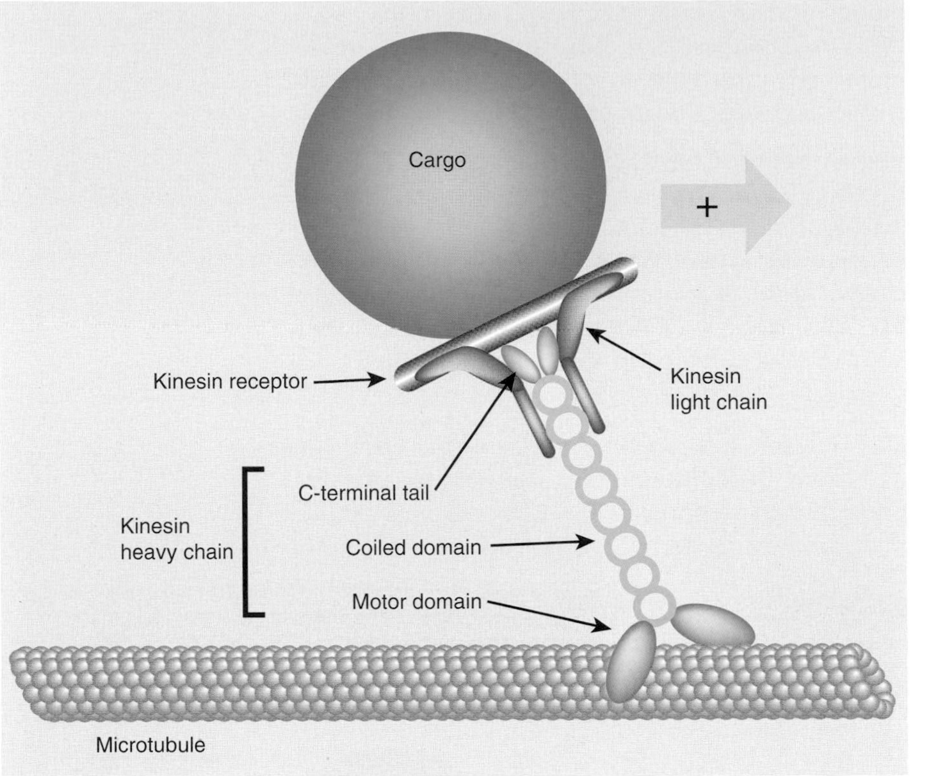

Kinesin receptor
Kinesin light chain
Cargo
+
C-terminal tail
Kinesin heavy chain
Coiled domain
Motor domain
Microtubule

FIGURE 7.26 Kinesin Linear Motor on Microtubules
Kinesin consists of light and heavy chains. The light chains bind to kinesin receptors on vesicles that are to be transported. The heavy chains each include motor domains that use ATP as energy to move kinesin plus the attached cargo along the surface of a microtubule.

of eukaryotic cells. Several of these systems are presently being investigated in the hope of making usable nanodevices that can be coupled to nanomachines to provide energy and/or moving parts.

The ATP synthase is a rotary motor whose natural role is to generate ATP. It is embedded in the mitochondrial membrane and uses energy from the proton motive force. The ATP synthase takes three steps to complete each rotation, and at each step, it makes an ATP. For use in nanotechnology, the F1 subunit would be detached from the membrane and run in reverse (i.e., it would be given ATP as fuel and, from a biological perspective, rotate backward).

Kinesin and dynein are motor proteins that use ATP as energy to move along the microtubules of eukaryotic cells. They therefore act as linear step motors (Fig. 7.26). Their natural role is to transport material. Kinesin moves cargo from the center to the periphery of the cell, whereas dynein carries cargo from the periphery to the center. Kinesin takes steps of 8 nanometers and can move at 100 steps per second (approximately 3 mm/hour!). Each step consumes one ATP for energy. The microtubules they use as tracks are protein cylinders with an outside diameter of 30 nm. In the not so distant future, complex chemical analyses might be carried out on the nanoscale (see Box 7.4).

Proteins that interconvert chemical and mechanical energy have been suggested as possible molecular motors to power future nanomachines.

Box 7.4 From Merely Micro to Truly Nano: Lab-in-a-Cell

Microfluidics (sometimes known as "lab-on-a-chip") refers to the manipulation of liquid samples at the scale of micrometers. Microfluidic devices are available today and are used to process large numbers of small samples. Applications include DNA or protein analysis of blood samples. The volumes involved are usually in the microliter range, although some microfluidics devices can use volumes less than 1 microliter, that is, nanoliter volumes. You might think that this entitles them to be regarded as nanotechnology, but remember that the dimensions of volume are the cubes of linear measure. Thus, a cube with sides of 1 micrometer (10^{-6} m) has a volume of 10^{-18} cubic meters, or 10^{-15} liters (one femtoliter). A nanoliter (10^{-9} liters) is the volume of a cube with sides of 100 micrometers. So handling nanoliters is *not* nanotechnology!

Future prospects are scaling down liquid sample processing to true nanoscale—"lab-in-a-cell." This would involve a microchip platform that uses modified single cells as analytical devices. This idea is still in the conceptual stage, but given the rapid progress in nanotechnology, it may not be so far in the future.

Summary

Many techniques from nanotechnology are now being applied to biological systems. Conversely, biological macromolecules and structures are being used in nanotechnology. Scanning probe microscopes have been used to visualize single molecules of biological importance. Individual bacteria and viruses can be detected and weighed by nanoscale devices. An ever-increasing variety of nanoparticles is already in use for biological labeling and various other analytical purposes. Nanoparticles are also being developed for clinical use and can be used to deliver drugs or to kill cancer cells by localized heating. On the other hand, microorganisms are capable of assembling nanocrystals from inorganic compounds. More complex nanodevices made from protein or DNA components are being assembled and their properties investigated. In particular, DNA is being used to build frameworks for the assembly of nanoscale structures. Some nanoscale sensors and motors are based on biological models and are being built, at least partly, from biological components.

End-of-Chapter Questions

1. What is nanotechnology?
   - **a.** the individual manipulation of molecules and atoms to create materials with novel or improved properties
   - **b.** the creation of new terms to describe very small, almost unimaginable, particles in physics
   - **c.** the term used to describe the size of cellular components
   - **d.** the transition of molecular biology into the physical sciences
   - **e.** none of the above
2. Which property is measured with a scanning probe microscope?
   - **a.** magnetism
   - **b.** electric resistance
   - **c.** light absorption
   - **d.** temperature
   - **e.** all of the above
3. What is considered a weakness of scanning tunneling microscopy (STM)?
   - **a.** the inability to move and arrange atoms to create a design
   - **b.** the possibility of destroying the surface with the metal tip on the microscope
   - **c.** the requirement for a conducting surface to work properly
   - **d.** the inability to apply this technology to biology
   - **e.** all of the above
4. What is an atomic force microscope?
   - **a.** The AFM detects the force between molecular bonds in an object.
   - **b.** The AFM detects atoms or molecules by scanning the surface.
   - **c.** The AFM uses photons to predict the structures present on any surface.
   - **d.** The AFM detects atoms or molecules on a conducting surface.
   - **e.** none of the above
5. Which principle is utilized to weigh a single bacterial cell or virus particle?
   - **a.** Oscillation frequency is dependent upon the mass applied.
   - **b.** It is impossible to weigh a single cell or particle.

(Continued)

c. Oscillation frequency affects the amount of light reflection.
d. Scanning electron microscopy can identify the length and width of a cell, which can further be converted to mass.
e. none of the above

6. What is a potential use of nanoparticles in the field of biology?
   a. delivery of pharmaceuticals or genetic material
   b. tumor destruction
   c. fluorescent labeling
   d. detection of microorganisms or proteins
   e. all of the above

7. What is an advantage to using complex, multilayered nanocrystals over fluorescent dyes?
   a. They do not bleach during excitation because they have broad absorption peaks.
   b. Nanocystals are often brighter than fluorescent dyes.
   c. The emission maximum of nanocrystals can be controlled by adjusting crystal size.
   d. Nanostructures are longer-lived than fluorescent dyes.
   e. All of the above are advantages.

8. Why is chitin the most popular material to construct nanoshells?
   a. Chitin is easy to synthesize.
   b. Chitin has properties that enable it to bind strongly to DNA, RNA, and other small molecules.
   c. Chitin is stable and easy to store at room temperature.
   d. Chitin is naturally derived and biodegradable.
   e. Chitin is easier to manipulate than the alternative for the creation of nanoshells.

9. How can nanoparticles be used to treat cancer?
   a. Nanotubes can create pores in the cancer cells, thus leaking out the cellular components and killing the cell.
   b. Some nanoparticles can bind to specific enzymes in cancer cell metabolism to block reactions.
   c. Nanoparticles can be designed to absorb radiant energy in the IR spectrum, which produces heat that destroys only the cancer cells because living tissue does not absorb IR energy.
   d. Nanoparticles can recruit immune system components directly to the cancer cells.
   e. All of the above are uses.

10. What characteristic of bacteriophage M13 makes it ideal for synthesizing nanowires?
   a. M13 accumulates certain nanoparticle building blocks in high concentrations.
   b. M13 phage is easily manipulated in the laboratory to secrete peptides that nanowires can be assembled upon.
   c. Nanowires can be constructed directly on M13 capsid proteins without any further modifications.
   d. M13 is filamentous.
   e. Nanowires are usually created in bacterial systems, not viral systems.

11. What purpose could nanotubes serve in biotechnology?
- **a.** as a metallic conductor or semiconductor
- **b.** for the creation of components of electronic equipment
- **c.** for attachment of biomolecules, including enzymes, hormone receptors, and antibodies
- **d.** for the detection of a specific molecule in a sample, such as blood
- **e.** all of the above

12. Why do nanocarpets have antibacterial activity?
- **a.** The long-chain amino compounds in the nanocarpet act like a detergent and disrupt the cell membrane.
- **b.** The nanocarpet tubes act as spears and shear the bacterial cells.
- **c.** The nanocarpet binds to bacterial cells and blocks the uptake of nutrients.
- **d.** The nanocarpet immobilizes the bacterial cell so that the cells can be targeted by treatment with UV light.
- **e.** Nanocarpets can act as biosensors so that people can treat the area with antibacterial agents.

13. Which of the following is a structure that can be created by nanoengineering of DNA?
- **a.** cubical structures
- **b.** nanoscale scaffolds for circuits and nanowires
- **c.** frameworks for mechanical nanodevices
- **d.** cross-shaped DNA to create 2D matrices
- **e.** all of the above

14. How might the behavior of individual DNA molecules be controlled from outside the body?
- **a.** exposure to UV light
- **b.** attachment of a metallic antenna, allowing DNA to be melted with radio waves
- **c.** addition of fluorescent tags
- **d.** using an electrical current to align the DNA molecules
- **e.** none of the above

15. Which cellular component is considered to be a nano(assembler)?
- **a.** chromatin
- **b.** lipids
- **c.** ribosomes
- **d.** DNA
- **e.** mRNA

16. Biosynthesis of fluorescent nanocrystals by *Escherichia coli* occurs when the bacteria are exposed __________ to and __________.
- **a.** green fluorescent protein; cadmium chloride
- **b.** cadmium chloride; sodium sulfide
- **c.** sodium sulfide; potassium tellurite
- **d.** cadmium sulfide; potassium tellurite
- **e.** cadmium chloride; potassium tellurite

17. DNA origami __________.
- **a.** involves one long DNA strand and several "staple strands"
- **b.** produces antibacterial nanocarpets

(*Continued*)

c. produces gold antenna to aid in the denaturation of DNA
d. uses a bacteriophage to produce nanotubes
e. produces a nanoscale for the mass measurements of single atoms

18. Addition of _________ improves the yield of biosynthesized nanocrystals.
a. cadmium sulfide
b. fluorophores
c. potassium tellurite
d. glutathione
e. cadmium sulfide

Further Reading

Bartczak, D., Muskens, O. L., Sanchez-Elsner, T., Kanaras, A. G., & Millar, T. M. (2013). Manipulation of *in vitro* angiogenesis using peptide-coated gold nanoparticles. *ACS Nano, 7*(6), 5628–5636.

Billingsley, D. J., Bonass, W. A., Crampton, N., Kirkham, J., & Thomson, N. H. (2012). Single-molecule studies of DNA transcription using atomic force microscopy. *Physical Biology, 9*(2), 021001.

Bronstein, L. M. (2011). Virus-based nanoparticles with inorganic cargo: what does the future hold? *Small, 7*(12), 1609–1618.

Choi, B., Zocchi, G., Wu, Y., Chan, S., & Jeanne Perry, L. (2005). Allosteric control through mechanical tension. *Physical Review Letters, 95*, 78–102.

Garcia-Fuentes, M., & Alonso, M. J. (2012). Chitosan-based drug nanocarriers: where do we stand? *Journal of Controlled Release: Official Journal of the Controlled Release Society, 161*(2), 496–504.

Gu, L. Q., & Shim, J. W. (2010). Single molecule sensing by nanopores and nanopore devices. *Analyst, 135*(3), 441–451.

Hess, H., Bachand, G. D., & Vogel, V. (2004). Powering nanodevices with biomolecular motors. *Chemical European Journal, 10*, 2110–2116.

Ilic, B., Yang, Y., & Craighead, H. G. (2004). Virus detection using nanoelectromechanical devices. *Applied Physiology Letters, 85*, 27.

Kalle, W., & Strappe, P. (2012). Atomic force microscopy on chromosomes, chromatin and DNA: a review. *Micron, 43*(12), 1224–1231.

Li, B. S., Wei, B., & Goh, M. C. (2012). Direct visualization of the formation of RecA/dsDNA complexes at the single-molecule level. *Micron, 43*(10), 1073–1075.

Monrás, J. P., Díaz, V., Bravo, D., Montes, R. A., Chasteen, T. G., Osorio-Román, I. O., Vásquez, C. C., & Pérez-Donoso, J. M. (2012). Enhanced glutathione content allows the *in vivo* synthesis of fluorescent CdTe nanoparticles by *Escherichia coli*. *PLoS One, 7*(11), e48657.

Papazoglou, E. S., & Parthasarathy, A. (2007). *BioNanotechnology (Synthesis Lectures on Biomedical Engineering)*. San Rafael, CA: Morgan and Claypool Publishers.

Saccà, B., & Niemeyer, C. M. (2012). DNA origami: the art of folding DNA. *Angewandte Chemie (International ed. in English), 51*(1), 58–66.

Simmel, F. C. (2012). DNA-based assembly lines and nanofactories. *Current Opinion in Biotechnology, 23*(4), 516–521.

Wahajuddin, A. S. (2012). Superparamagnetic iron oxide nanoparticles: magnetic nanoplatforms as drug carriers. *International Journal of Nanomedicine, 7*, 3445–3471.

CHAPTER 8

Genomics and Gene Expression

Biotechnology

http://dx.doi.org/10.1016/B978-0-12-385015-7.00008-9

INTRODUCTION

The first working draft of the human genome sequence was announced in June 2000. The sequence was refined, and a final high-quality sequence was finished in April 2003 (and published in 2004). In the decade since the completion of the human genome, many advances in technology have increased the speed and accuracy with which an organism's entire DNA sequence is determined. Rather than a multiyear multibillion dollar investment, today's technology allows an entire human genome to be determined for a few thousand dollars and finished in a few days' time.

Scientists have now turned to the interpretation of genomic sequences. As of writing, just shy of 13,000 different genomes from all three domains of life have been sequenced completely, and another 27,000 are partially sequenced. Although much sequence information is known, there is still a disconnection between the sequence data and its functional interpretation. For the human genome, the exact number of genes is still enigmatic, with the current count at 20,805 protein coding genes, but there are also short noncoding genes, long noncoding genes, pseudogenes, and various alternately spliced genes that are not included in this tally. Although the general public expected that knowing the entire genome sequence would clarify genome function, initial analyses have only added to the complexity. For example, some areas of the genome previously thought to serve no function and dubbed "junk DNA" are now known to provide regulatory functions for gene expression, a complexity that was not anticipated at the onset of the Human Genome Project.

Some of the offshoots from the Human Genome Project are listed in Table 8.1. These projects fall into two categories: understanding variation within and among different genomes and association of key sequences with a particular function or dysfunction. For the first category, although a reference genome has been decoded, each person has 4 to 5 million

Table 8.1 Human Genome Project and Related Initiatives

Name of Project	Objective	Website
Human Genome Project	Map and sequence the entire human genome	http://genome.ucsc.edu/
Genome Reference Consortium	Close remaining gaps in the human genome assembly	http://www.ncbi.nlm.nih.gov/projects/genome/assembly/grc/
1000 Genomes Project	Sequence 1000 entire human genomes for comparisons	http://www.1000genomes.org/
International HapMap Project	Identify haplotype maps of human chromosomes to identify sequence variation	http://hapmap.ncbi.nlm.nih.gov
ENCODE project (Encyclopedia of DNA Elements)	Identify all functional elements such as transcription factor binding sites in the human genome	http://www.genome.gov/Encode/
OMIM (Online Mendelian Inheritance in Man)	Compile genes and their phenotypic effects	http://www.ncbi.nlm.nih.gov/omim
Human Epigenome Project	Analyze DNA methylation and other epigenetic modifications	http://www.epigenome.org/
Human Microbiome Project	Identify the microbial inhabitants of the human body	http://commonfund.nih.gov/hmp/index

differences within his or her genome in comparison with other individuals and/or the reference genome. The 1000 Genomes Project aims to sequence over 1000 different human genomes and compile the different variations. The HapMap Project is also investigating variation among individual human genomes. For the second category, the Online Mendelian Inheritance in Man is a list of known phenotypes and the genes responsible. The Human Epigenome Project is focused on identifying DNA methylation patterns in different tissues and different people. Finally, the Human Microbiome Project is working to identify all the different microbial inhabitants of the human digestive tract, skin, mouth, nose, and female urogenital tract. The idea for this project is to determine whether or not these microorganisms are contributing to or protecting from disease, and also to see if differences in the microbiota influence nutrition. This chapter aims to provide a historical perspective on how the human genome was first deciphered, and then discuss how this information is being applied to study genomic variation and gene expression.

Before the advent of next-generation sequencing technology, the assembly of the first human genome sequence was a monumental undertaking. The amount of sequence data to be generated was much larger than any other dataset in the biological sciences. The original plan for sequencing the genome was to create random segments of the genome, sequence each piece, and then use the sequence overlap to assemble the pieces into **contig maps**, which order the clones into longer linear segments without any gaps (Fig. 8.1). The project first mapped large fragments of human DNA (carried by YACs and BACs) to their respective chromosomal locations. The mapping was time consuming but necessary to order the sequence data. At the time of startup, computers were unable to order more sequence data than found in large chromosomal fragments.

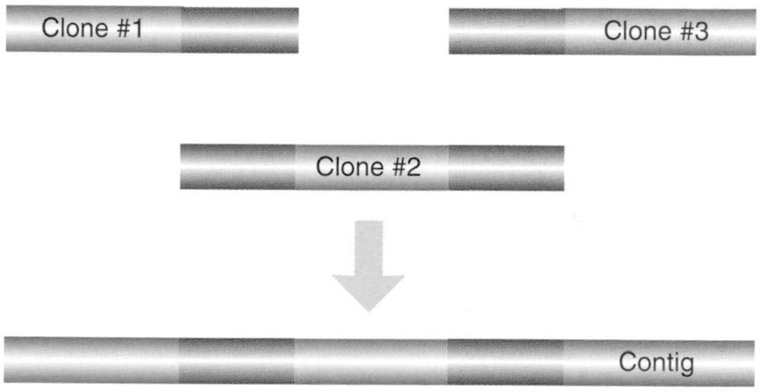

FIGURE 8.1 Contig Mapping
Small clones have regions that overlap with each other. Ordering the small clones into one sequence forms a contig map.

During the 1990s, computing power increased so rapidly that the mapping became less necessary. In 1998, Celera Genomics, led by Craig Venter, decided to sequence the entire human genome faster and cheaper. Celera Genomics proved its point by sequencing the entire 180 Mb genome of the fruit fly, *Drosophila*, between May and December of 1999. Celera used the **shotgun sequencing** (Fig. 8.2) approach, which most researchers thought would not work for such large genomes. Venter sequenced many small fragments of DNA and entered the data into the computer, which then assembled the information into a working draft. Venter was able to sequence the human genome fast largely because of the increase in computer power. It was possible to use computers for most of the genome, although certain problematical regions needed some genetic mapping for correct ordering.

With the emergence of next-generation sequencing, whole genomes may be sequenced in less than a week, but the data generally consist of significantly shorter reads than with Sanger sequencing. To make up for the short read length, the number of reactions in a single sequencing run is huge, numbering in the millions. **Depth of coverage** is the number of individual reads for each section of the genome. Shorter reads are much harder to align by computer, so instead of *de novo* genome assembly, next-gen sequencing data are often compared to a reference genome that was partially determined by Sanger sequencing, or entirely determined by Sanger sequencing such as the Human Genome Project. Even with next-gen sequencing, most genome assembly would be impossible without Sanger sequencing and the genetic and physical mapping techniques developed by the Human Genome Project.

Shotgun sequencing was the technique used to sequence the whole human genome. Each library clone is randomly sequenced, and computer analysis orders the clones by identifying overlapping regions.

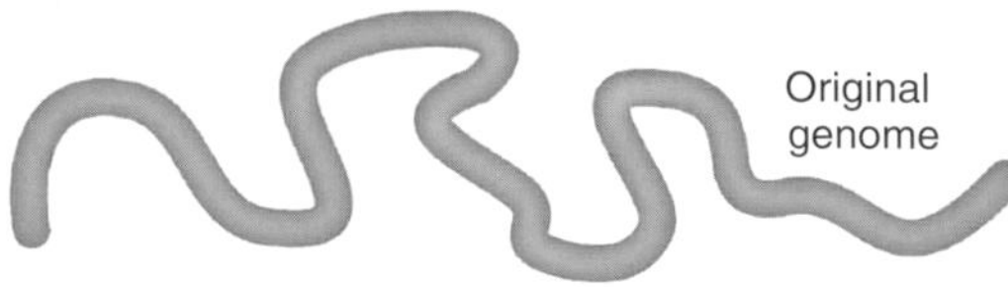

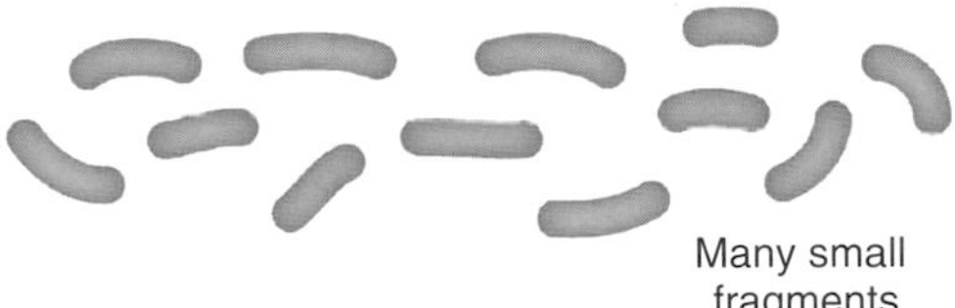

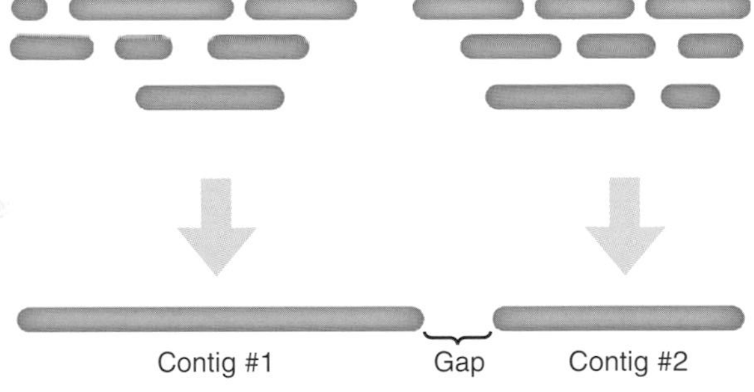

FIGURE 8.2 Shotgun Sequencing
The first step in shotgun sequencing an entire genome is to digest the genome into many small fragments that are cloned and sequenced individually. Computers analyze the sequence data for overlapping regions and assemble the sequences into several large contigs. Because some regions of the genome are unstable when cloned, some gaps may remain even after this procedure is repeated several times.

GENETIC MAPPING TECHNIQUES

Genome maps provide a linear series of landmarks for use when putting together sequence data. There are two different methods for creating genome maps: **genetic mapping** and **physical mapping**. Genetic mapping provides information through mating, pedigree analyses, or gene transfer experiments. Genetic maps, also called **linkage maps**, are based on the relative order of genetic markers revealed by mating experiments of various kinds. Consequently, the actual distance in base pairs between the markers is hard to determine. In contrast, physical maps give the distance between markers in base pairs because the markers are physically associated with a location on a chromosome. Physical location can be determined by radiation hybrid experiments or fluorescence *in situ* hybridization (FISH). The location along the chromosome is useful to determine the correct order of contigs without having to compare sequence. This is especially useful for regions with lots of repetitive sequences. The various approaches to mapping the human genome are summarized in Table 8.2.

Traditional genetic maps are based on **linkage**, that is, the likelihood that two mapping features segregate together after mating. Linkage is a direct result of location; that is, if two markers are far apart or on separate chromosomes, then the likelihood that they segregate together is low. Since eukaryotic cells undergo recombination, even two markers that are on the same chromosome can segregate among progeny after mating, but the closer the two markers are, the less likely they will be separated by recombination. The percentage in which two markers are found together is therefore used to ascertain relative distances along a chromosome.

Genetic maps use landmarks called **markers**. Many different types of markers can be used, and some markers can be used for both physical mapping and genetic mapping. Genes with a specific phenotype are very useful markers for mapping. Unfortunately, not enough genes with visible phenotypes exist to provide a detailed map for the human genome. Genes are usually found in two or more different forms. For example, a gene for plant height may yield a tall plant if one form of the gene is present, or a short plant if a different form of the same gene is present. The different forms, called **alleles**, have various sequence alterations that give different phenotypes to the plant. When multiple alleles are present, a population

Table 8.2 Genetic versus Physical Mapping Techniques

Type of Mapping	Markers Used during Mapping	Methods to Locate Markers in Genome
Genetic	Gene, biochemical trait, DNA markers (RFLP, VNTRs, microsatellites, SNPs)	• Linkage analysis using crosses or matings • Analysis of human family pedigrees
Physical	Sequence tagged sites (expressed sequence tags, VNTRs, microsatellites)	• Restriction enzyme mapping • Radiation hybrid mapping • FISH • Cytogenetic mapping

is considered as **polymorphic**. More than just genes can be polymorphic, and many DNA sequence features show polymorphism (see later discussion). For the Human Genome Project, determining the gene order along the chromosomes was not accomplished by experimental mating (considered unethical in humans). Instead, pedigree analysis determines if different genes segregated together in a family.

Other genetic markers used in mapping a genome show DNA sequence differences like the different alleles of a gene, but these markers do not necessarily associate with a gene or affect the phenotype of an organism. One of the most widely used is **restriction fragment length polymorphisms (RFLP)**. RFLPs are differences in the restriction enzyme recognition sequence, so some members of the population have the restriction enzyme site, and other members do not. The different versions of the RFLP are identified using agarose gel electrophoresis. RFLPs are used because they are easy to identify. For small genomes such as yeast, monitoring the frequency of recombination between two RFLP markers is easy. Diploid yeast cells undergo meiosis and form four haploid cells called a tetrad. Each of these haploid cells can be isolated, grown into many identical clones, and examined. Thus, each RFLP marker can be followed easily from one generation to the next. In humans, following such markers is more challenging, but studies on groups of closely related people, such as large families or small cultures like the Amish, have allowed some RFLPs to be followed in this manner (Fig. 8.3).

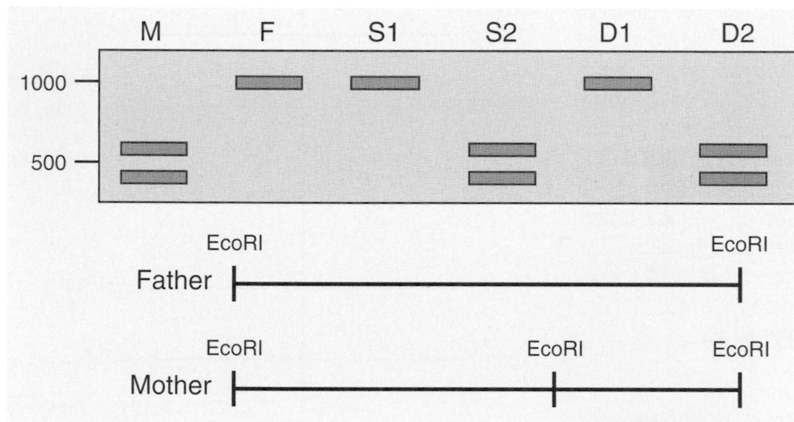

FIGURE 8.3 RFLPs of Family Members
The mother (M) and father (F) of this family have a difference in the sequence of their DNA. In the mother, the difference adds a restriction enzyme site for *Eco*RI. The children (S1, S2, D1, and D2) have inherited one or the other fragment from their parents.

Another marker used in genetic mapping is the **variable number tandem repeat**, or **VNTR** (Fig. 8.4). Sometimes these are called **minisatellites**. These sequence motifs occur naturally in the genome and consist of tandem repeats of 9 to 80 base pairs in length. The number of repeats differs from one person to the next; therefore, these can be used as specific markers on a genetic map. They are also used to identify individuals in forensic medicine or paternity testing. Some repeats are found in multiple locations throughout the genome and cannot therefore be used for making genetic maps, but other repeat sequences are found only in one unique location, making them very useful for mapping experiments.

A fourth type of marker is the **microsatellite polymorphism**, which is also a tandem repeat. However, unlike VNTRs, microsatellites are repeats of 2 to 5 base pairs in length. In animals, they often consist largely of cytosine and adenosine on one strand (hence, mostly G and T on the complementary strand). Plants seem to have less microsatellite base composition bias.

Another type of marker is the **single nucleotide polymorphism**, or **SNP** (pronounced "snip") (Fig. 8.5). SNPs are individual substitutions of a single nucleotide that do not affect the length of the DNA sequence. These changes can be found within genes, in regulatory regions, or in noncoding DNA. When found within the coding regions of genes, SNPs may alter the amino acid sequence of the protein. This in turn may affect protein function;

5'G T A C T A G A C T T A G T A C T A G A C T T A
G T A C T A G A C T T A G T A C T A G A C T T A 3'

FIGURE 8.4 Tandem Repeat of 12 Base Pairs
This individual has only four repeats of this 12 base-pair sequence (the sequence for only one strand of the DNA is shown). Other people may have more or fewer repeats.

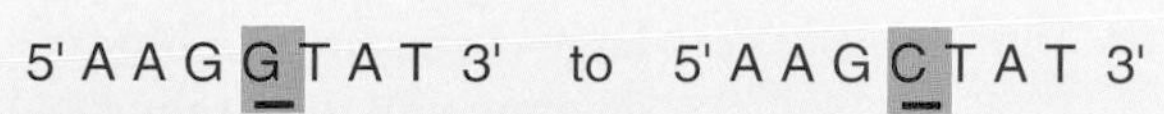

FIGURE 8.5 SNP
The same DNA segment from two different individuals has a single nucleotide difference, that is, an SNP. Such changes are common when comparing DNA sequences between individual people.

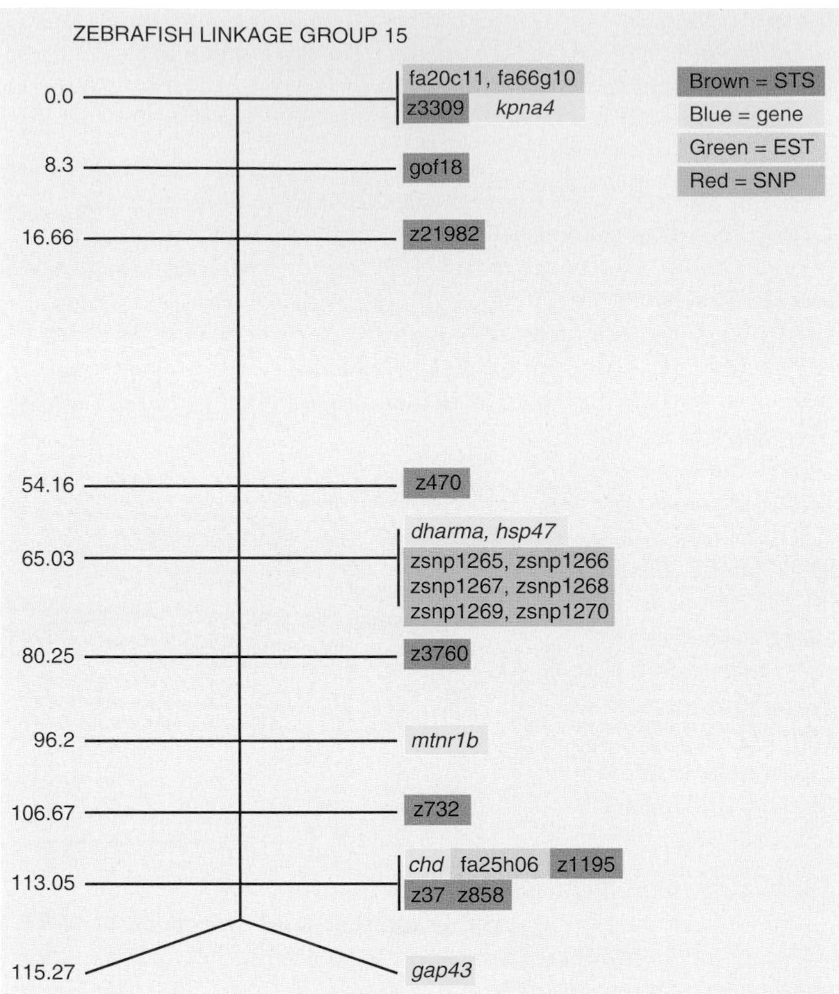

FIGURE 8.6 STS and EST Markers on Zebrafish Linkage Map
The relative positions of various markers are shown on the zebrafish map. The markers include STSs and ESTs that were identified and mapped relative to one another. In addition, the positions of real genes and SNPs are shown relative to the others. The linkages were established using meiotic recombination frequencies and are presented in centimorgans (cM). The GAT linkage group 15 map information for this figure was retrieved from the Zebrafish Information Network (ZFIN), the Zebrafish International Resource Center, University of Oregon, Eugene, OR 97403–5274; http://zfin.org/.

i.e., the SNP corresponds to a difference in an actual gene. If a SNP correlates with a genetic disease, identifying that SNP may diagnose the disease before symptoms appear. When a SNP falls within a restriction enzyme site, it coincides with an RFLP.

Markers such as SNPs, VNTRs, RFLPs, and microsatellites are also used in physical mapping techniques such as restriction enzyme mapping, FISH, or radiation hybrid mapping. These are useful, but for large genomes like the human genome, they still do not provide enough markers. The map builders needed other types of markers, such as **sequence tagged sites (STSs)** (Fig. 8.6). These sites are simply short sequences of 100 to 500 base pairs that are unique and can be detected by PCR. A specialized type of STS is the **expressed sequence tag (EST)**, so called because it was identified in a cDNA library. This means that the EST is expressed as mRNA. These small pieces of sequence data are just portions of larger genes; therefore, many different ESTs may be found within one single gene.

Using physical mapping for markers resembles linkage analysis for genes in the sense that the closer they are, the more likely they will remain together. However, one method of physical mapping is to use restriction enzyme digestion (Fig. 8.7). Either entire genomes or single large clones from a library are digested with a variety of different restriction enzymes. Each enzyme will digest the DNA into different sized fragments, which are then probed for several different STS or EST markers. If two markers are close together, they will often be found on the same restriction fragment, but if they are far apart, they will be on different fragments. The fragment sizes are determined and used to deduce the approximate distance between two markers. The difference between this information and RFLP analysis is that restriction

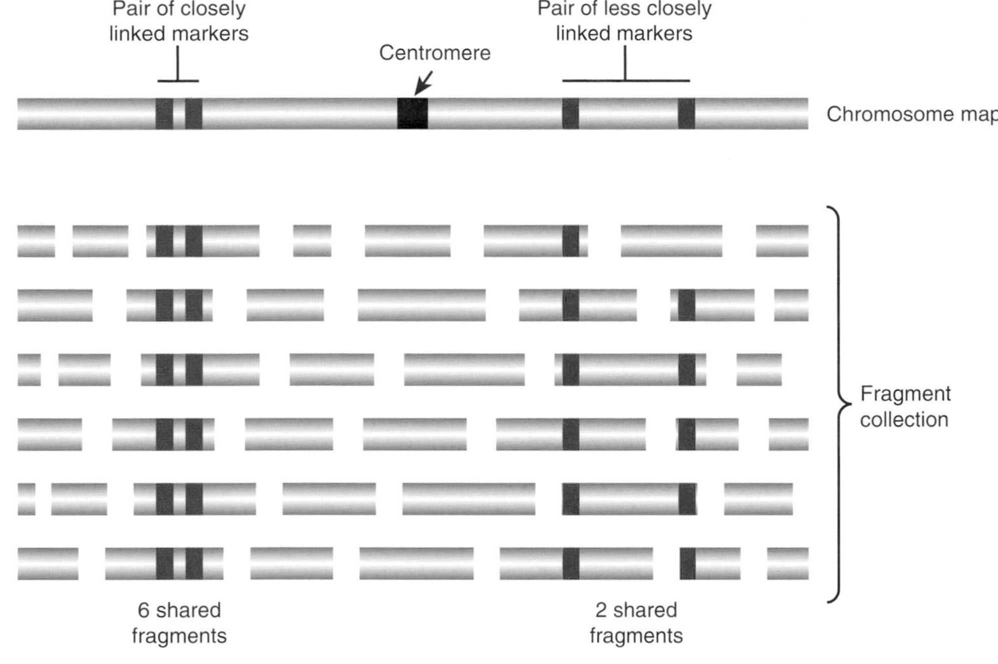

FIGURE 8.7 STS Mapping Using Restriction Enzyme Digests
STS mapping is shown for four STS sites on a single chromosome. Various restriction enzyme digests are performed to cut the chromosome into many different-sized fragments. The number of times two STS sequences are found on the same fragment reveals the proximity of the two markers. In this example, the two purple STSs are found on the same fragment six times and must be close to each other on the chromosome. The two green STSs are found on the same fragment only two times and are therefore farther apart. The purple and green STSs are never found on the same fragment; therefore, they must be far apart.

mapping can determine the distance between two markers as the number of base pairs, whereas RFLP analysis determines how often two different markers are found together but the actual distance is based on cosegregation frequency.

Three other physical mapping techniques are radiation hybrid mapping, cytogenetic mapping, and FISH. FISH analysis is described in detail in Chapter 3. The physical location of a particular DNA probe is determined in metaphase chromosomes by its location relative to the banding pattern. This method provides clues for ordering sequence data into one large contig.

DNA probes can sometimes be unreliable because large cloned segments may actually consist of two fragments of DNA, from different parts of the genome, inserted into the same vector. **Radiation hybrid mapping** overcomes these limitations by examining STSs or ESTs on original chromosomal fragments (Fig. 8.8). To generate these, scientists treat cultured human cells with X-rays or γ-rays to fragment the chromosomes. The radiation dosage controls how often the chromosome breaks, and thus the average length of the fragments. The human cells possess a marker enzyme that allows them to grow on selective media. After irradiation, the human cells are fused to cultured hamster cells using polyethylene glycol or Sendai virus. The hamster cells do not have the selective marker. Consequently, only those hamster cells that fuse with human cells survive. The fragments of human chromosomes become part of the hamster nucleus, and the individual hybrid cell lines can be examined by STS or EST mapping. Because the average fragment length is known, these maps reveal relative distance between two markers.

Cytogenetic mapping is another physical technique that uses original chromosomes. When chromosomes are placed on microscope slides and stained, they form banding patterns that are visible under a light microscope. This **cytogenetic map** shows where a gene or marker lies relative to the stained bands (Fig. 8.9). Cytogenetic maps are very low resolution compared with the other mapping techniques, yet they are useful to compare gene locations on a large scale.

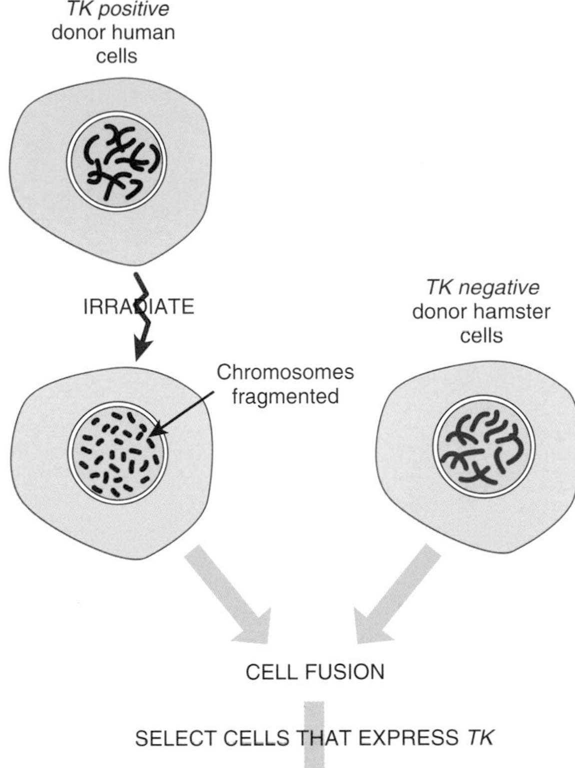

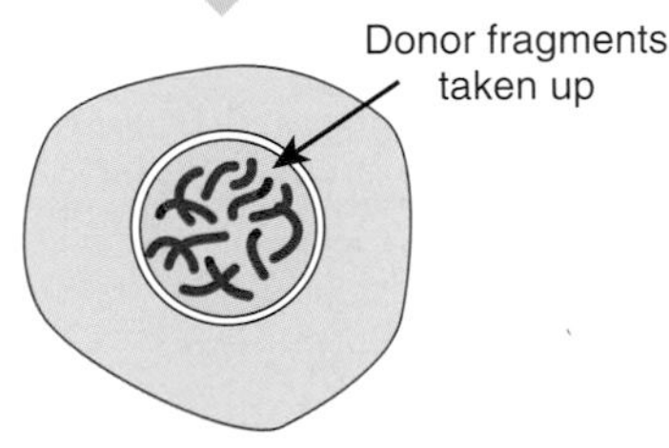

FIGURE 8.8 Radiation Hybrid Mapping
To determine how close STSs and ESTs are to each other, scientists must analyze many large chromosome fragments. Radiation hybrid mapping allows large human chromosome fragments to be inserted into hamster cells. First, the human chromosomes, which carry the thymidine kinase gene (TK^+), are fragmented by irradiation. The human cells are then fused with hamster cells, which are TK^-. Such hybrid cells should express thymidine kinase and will grow on selective medium. Random loss of human chromosome fragments occurs during this process; therefore, each radiation hybrid cell line has a different set of human chromosome fragments, which can be screened for the STSs and ESTs.

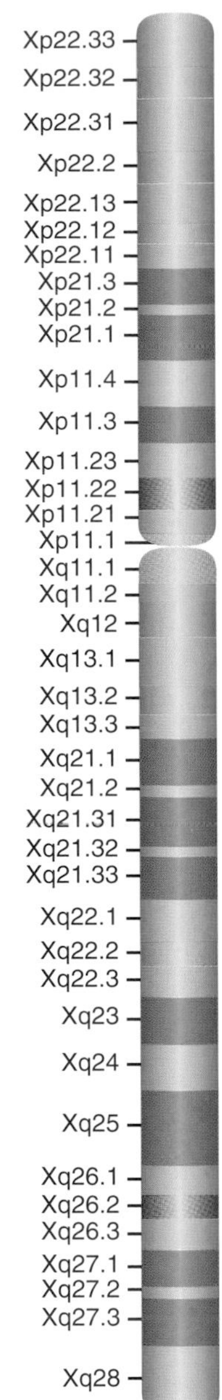

FIGURE 8.9 Banding Pattern of X Chromosome
Staining condensed mitotic chromosomes with various DNA-binding dyes forms a distinct banding pattern. The location of a gene or marker can be determined relative to the bands. For example, a gene located at Xp21.1 is on chromosome X, on the p arm, and on band number 21.1.

Determining the order of various markers allows generation of genetic maps. Markers used include RFLPs, SNPs, VNTRs, and microsatellite polymorphisms. Genetic mapping uses linkage analysis of two markers by mating experiments or pedigree analysis. Physical maps link a marker or DNA sequence to a physical location along a chromosome or large contig. Markers such as ESTs and STSs are used to determine these distances. Techniques include restriction enzyme digestion mapping, FISH, radiation hybrid analysis, and cytogenetic mapping.

GAPS REMAIN IN THE HUMAN GENOME

Although the sequence of the genome is considered complete, there are still gaps. One method of finding the sequence of any gaps is called **chromosome walking**. In this method, a particular clone is sequenced to start the process. Then the new sequence data is used to find overlapping clones (Fig. 8.10). After those are identified and sequenced, more overlapping clones are identified. The process goes in order either up or down the chromosome, compiling the sequence piece by piece. Usually, the first clone is located relative to a particular marker, such as an STS or RFLP.

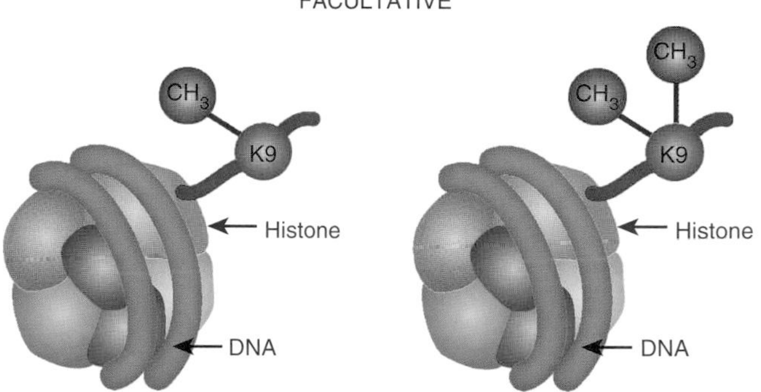

FIGURE 8.10
Chromosome Walking
Researchers identify the downstream and upstream regions of a gene using chromosome walking. In this example, the end of library clone 1 is converted into a probe. The probe is used to screen a library, and a second clone is identified. The two clones overlap and are linked to form a complete gene.

Most of the gaps fall in highly condensed regions of repetitive DNA, known as **heterochromatin**, which is difficult to sequence. Three features characterize heterochromatin: hypoacetylation (i.e., lack of acetyl groups on the histones); methylation of histone H3 on a specific lysine; and methylation on CpG or CpNpG sequence motifs. Heterochromatin is not transcribed and comes in two forms: **facultative heterochromatin** and **constitutive heterochromatin** (Fig. 8.11). The amount of methylation on lysine-9 in histone H3 determines whether or not heterochromatin is considered facultative or constitutive. The constitutive form is found around the centromeres and telomeres of the chromosome and does not change from one generation to the next.

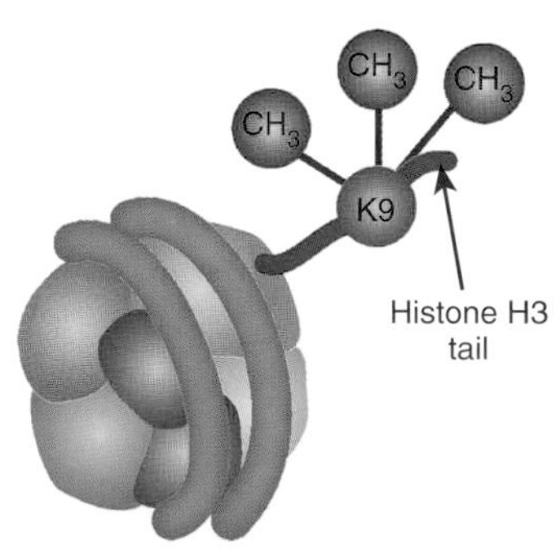

FIGURE 8.11
Facultative versus Constitutive Heterochromatin
The amount of methylation on lysine-9 in histone H3 determines whether or not heterochromatin is considered facultative or constitutive.

Facultative heterochromatin is found in other regions of the chromosomes, and its presence is cell-specific. Once a specific region of a chromosome becomes heterochromatin, all of the cells' descendants will maintain this pattern. The border for facultative heterochromatin is not static, and each cell in a tissue might have a little more DNA condensed than other cells. This is exemplified by the classic inactivation of the *white* gene in *Drosophila*, where the fly will have a mottled red and white eye color because the gene is silenced into facultative heterochromatin in some cells and not others. This genetic variation is called **position effect variegation (PEV)**.

Gaps in genomes can often be sequenced by chromosome walking, where one end of a library clone is used to find other overlapping clones. Most gaps result from heterochromatin, highly condensed repetitive DNA found in specific sites throughout the genome. The physical nature of heterochromatin makes it difficult to sequence.

SURVEY OF THE HUMAN GENOME

The sequence of the human genome is 3.2×10^9 base pairs (3.2 Gbp or gigabase pairs) in length. If the sequence were typed onto paper, at about 3000 letters per page, it would fill 1 million pages of text. This extraordinary amount of information is encoded by the sequence of just four bases: cytosine, adenosine, guanine, and thymine. Most people expected the human genome sequence to reveal the exact number of genes that humans possess. In reality, sophisticated interpretation is needed to identify many of the genes. The tally for the number of protein coding genes is 20,805, but with each refinement of the human genome, the number varies. In addition, many of these genes are alternately spliced, and the actual number of proteins is much higher than the number of genes. Of these genes, we know the function of only around 50%. More than two-thirds of the predicted human proteins are similar in structure to proteins in other organisms.

The genome sizes and estimated gene numbers are given for several organisms in Table 8.3. The animal with the highest number of genes is *Daphnia pulex,* or the water flea, with about 31,000 different genes, of which about one-third are unrelated to any other organism that has been studied thus far (Fig. 8.12). This tiny crustacean is a key organism to understand how the environment affects gene expression. A rare Japanese flower is the current record holder for genome size, with the marbled lungfish the current second place for genome size. The Japanese flower *Paris japonica* has almost 50 times more DNA than humans, approximately 130,000 Mb. In contrast, the mammalian parasite *Encephalozoon intestinalis* has only 2.25 million bases, which is the smallest genome of an organism with a nucleus (Fig. 8.13).

Although wheat has nearly 6 times as much DNA as humans and 95,000 genes, it is hexaploid rather than diploid like most eukaryotes. Thus, the wheat genome may be regarded as combining three sets of around 32,000 genes. Several other plants in Table 8.3 have less DNA but more genes than any multicellular animal sequenced so far. The largest bacterial genomes have more genes than the smaller eukaryotic genomes. An example is *Streptomyces,* famous as the source of many antibiotics. The smallest bacterial genomes, such as *Mycoplasma,* have fewer than 500 genes although—because these bacteria are parasitic, they rely on the eukaryotes—they infect for many metabolites. The protozoan parasite *Trichomonas vaginalis* was proposed to have as many as 60,000 genes when its genome was first sequenced. However, further analysis has reduced this to 46,000, and the total number of genuine genes may be significantly lower. Major problems in annotation were due to many duplicated sequences and the presence of partial gene sequences in them.

In eukaryotic DNA, some genes encompass thousands or even millions of base pairs, most of which comprise introns that are spliced out of the mRNA transcript. For example, the gene for dystrophin (defective in Duchenne's muscular dystrophy) is 2.4 million base pairs long, and some of its introns are 100,000 base pairs or more in length. In contrast, the coding sequence, consisting of multiple exons, is only about 3000 base pairs. In such situations, it is not easy to find coding sequences among the noncoding DNA. On the one hand, this may result in genes (or individual exons) being completely missed. On the other hand, widely separated exons that are, in reality, parts of a single coding sequence may be interpreted as separate genes.

Table 8.3 Estimated Number of Genes for Various Genomes

Organism	Genome Size (Megabase Pairs)	Estimated Genes (Protein Encoding)
Plants		
Wheat (*Triticum aestivum*)	17,000	95,000
Black poplar (*Populus trichocarpa*)	520	45,000
Rice (*Oryza sativa*)	390	38,000
Mustard weed (*Arabidopsis thaliana*)	125	26,000
Japanese flower (*Paris japonica*)	149,000	unknown
Protists		
Paramecium tetraaurelia	72	40,000
Trichomonas vaginalis	160	46,000?
Encephalozoon intestinalis	2.25	1,833
Animals		
Marbled lungfish	130,000	unknown
Human (*Homo sapiens*)	3,200	21,805
Mouse (*Mus musculus*)	2,800	25,000
Roundworm (*Caenorhabditis elegans*)	97	20,493
Fruit fly (*Drosophila melanogaster*)	180	13,600
Daphnia pulex	200	31,000
Fungi		
Aspergillus nidulans	30	9,500
Yeast (*Saccharomyces cerevisiae*)	13	5,800
Bacteria		
Streptomyces coelicolor	8.7	7,800
Escherichia coli	4.6	4,300
Mycoplasma genitalium	0.58	470

Another confounding factor in determining the number of genes is the presence of **pseudogenes**. These duplicated copies of real genes are defective and no longer expressed as proteins. They may be found next to the original, or they may be far away, on different chromosomes. Determining whether or not a "gene" is a pseudogene or genuine gene may be difficult using sequence data alone. Often the expression of a particular region of DNA must be confirmed by finding corresponding mRNA transcripts. DNA microarrays are a popular approach to confirming whether or not a gene is expressed (see later discussion).

The number of genes also hinges on how we define a gene. In addition to the approximately 22,000 protein-encoding genes, there are a thousand or more genes that encode nontranslated RNA. The ribosomal RNA and transfer RNA genes are the most familiar of these. However, a variety of other small RNA molecules are involved in splicing of mRNA and in the regulation of gene expression. Other sequences of DNA may not even be transcribed, yet are nonetheless important. Should these also be regarded as genes?

The number of genes in an organism depends on the definition of *gene* and the distinction between real genes and pseudogenes. The absolute number of genes in any sequence is approximate.

NONCODING COMPONENTS OF THE HUMAN GENOME

The number of protein-encoding genes is only a small fraction of what is important in the human genome. When one is looking at homology with the mouse genome, many regions that do not encode protein are highly conserved. These areas are also conserved among the rat and dog genomes and are called **conserved noncoding elements (CNE)**. There are 500 regions of 200 or more nucleotides that are perfectly conserved and over 10,000 elements that are highly conserved. One estimate suggests that there are millions of regions that are conserved to some degree. Some of these regions are conserved even in fish species, suggesting that these elements are functional, and some theorize that they may act as enhancer elements for gene expression. Another proposed function includes insulator sequences that prevent the wrong enhancer elements from activating the wrong gene, but the actual functions for most of the CNE are unknown. Table 8.4 lists the major components of the human genome.

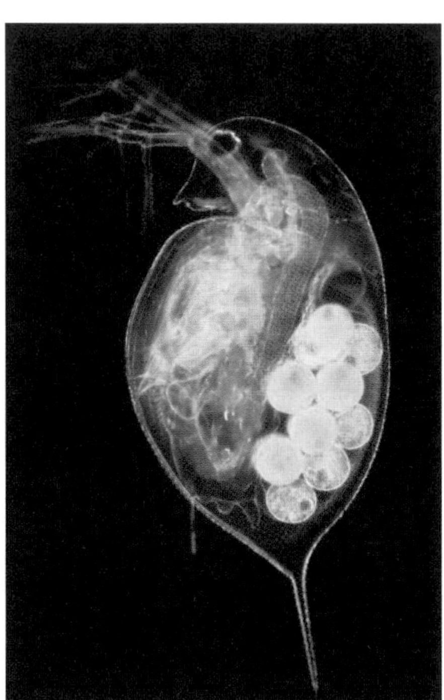

FIGURE 8.12 ***Daphnia pulex*** **(water flea)**The crustacean *Daphnia pulex* has the highest number of genes, 31,000, and it is shown in the photo containing a brood of offspring. Daphnia are able to create clones or mate depending on the environmental conditions. These organisms are used to assess environmental quality. Photo by Paul D.N. Herbert, University of Guelph. Courtesy of Indiana University.

One common type of noncoding DNA is the introns between the coding segments of genes (the exons). About 25% of the human genome consists of genes for proteins. However, of this amount, only 1% is actual coding sequence and the other 24% comprises the introns. Most introns have no function, but occasional examples of whole genes have been found within introns of a different gene. Introns may also contain binding sites for transcription factors and, in that sense, play a role in gene regulation.

FIGURE 8.13 Genome Size Varies Among Organisms
(A) Photo of *Paris japonicum* that has the largest genome size in base pairs. (B) The marbled lungfish has the second largest genome. (C) Encephalozoon intestinalis are the black ovoid structures found in the cell. These spores are at different stages of development, and when all mature, they cause the infested cell to erupt and spread the infection.

Table 8.4 Components of Mammalian Genomes

Unique Sequences	
Protein-encoding genes—comprising upstream regulatory region, exons, and introns	
Genes encoding noncoding RNA (snRNA, snoRNA, 7SL RNA, telomerase RNA, Xist RNA, a variety of small regulatory RNAs)	
Nonrepetitive intragenic noncoding DNA	
Interspersed repetitive DNA	
Pseudogenes	
Short Interspersed Elements (SINEs)	
Alu element (300 bp)	~1,000,000 copies
MIR families (average ~130 bp) (mammalian-wide interspersed repeat)	~400,000 copies
Long Interspersed Elements (LINEs)	
LINE-1 family (average ~800 bp)	~200,000–500,000 copies
LINE-2 family (average ~250 bp)	~270,000 copies
Retrovirus-like elements (500–1300 bp)	~250,000 copies
DNA transposons (variable; average ~250 bp)	~200,000 copies
Tandem Repetitive DNA	
Ribosomal RNA genes	5 clusters of about 50 tandem repeats on 5 different chromosomes
Transfer RNA genes	Multiple copies plus several pseudogenes
Telomere sequences	Several kb of a 6-bp tandem repeat
Mini-satellites (= VNTRs)	Blocks of 0.1 to 20 kbp of short tandem repeats (5–50 bp), most close to telomeres
Centromere sequence (alpha-satellite DNA)	171-bp repeat, binds centromere proteins
Satellite DNA	Blocks of 100 kbp or longer of tandem repeats of 20 to 200 bp, most close to centromeres
Mega-satellite DNA	Blocks of 100 kbp or longer of tandem repeats of 1 to 5 kbp, various locations

Numbers of copies given is for the human genome.

Another feature of the genome is moderately repetitive sequences. Ribosomal RNA genes are found in great numbers because many ribosomes are needed. These genes are considered moderately repetitive elements, but of the coding variety. Noncoding repetitive elements include the **long interspersed element**, or **LINE**, which is found in 200,000 to 500,000 copies and accounts for up to 20% of the human genome (Fig. 8.14). These retrovirus-like elements contain genes inside long terminal repeats (LTRs) similar to retroviruses. They are autonomous; that is, the LINE is able to copy itself and insert new copies into other sites in the genome. However, most copies of LINEs are defective; only a few are still mobile and functional. There are many different types of LINEs, the most common by far in mammals being L1.

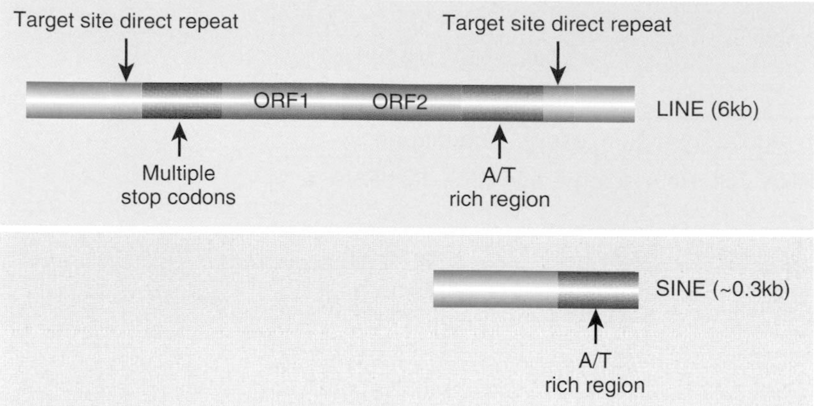

FIGURE 8.14
General Structure of LINE and SINE
LINE elements contain an internal promoter for RNA polymerase II and two open reading frames that encode proteins. The first protein has an unknown function. The second is a bifunctional protein with reverse transcriptase and DNA endonuclease domains. SINE elements usually contain only an internal promoter for RNA polymerase III and some sort of tRNA stem-loop structure, followed by a poly(A) tail.

When a LINE retro-element is active, it moves to a new location via an mRNA intermediate. First, the active LINE is transcribed into mRNA using an internal promoter. The mRNA goes to the cytoplasm, where ribosomes translate it, creating two proteins. One of these, the combined reverse transcriptase/endonuclease protein, binds to the mRNA forming a **ribonucleoprotein** (RNP). This is transported back into the nucleus where the endonuclease domain nicks the DNA at the new location in the genome generating a free 3′-OH end. Next, the reverse transcriptase domain makes a DNA copy of the LINE mRNA using the 3′-OH end as primer. This results in a copy of LINE1 inserted into a new location in the genome. Cellular repair enzymes fill in the gaps to create duplicated sequences flanking both sides of the new LINE.

When a LINE moves to a new location, it may disrupt an essential gene, which would prove fatal to the cell. Control of LINE movement is critical. Too much movement is disruptive and might destroy both the host cell and the LINEs it contains. Conversely, too little movement, and the LINE will fail to reproduce effectively. In humans, many LINEs are found in gene-poor, A/T-rich regions of the genome, suggesting that some mechanism exists to keep these elements away from vital genetic information. Most functional LINE elements are silenced by methylation of their promoters and by methylation of the histones bound to them. In addition, RNA surveillance pathways (piRNA; see Chapter 5) monitor and degrade the LINE RNA transcripts.

In addition to moderately repetitive sequences, the human genome is filled with highly repetitive DNA. The **short interspersed elements**, or **SINEs** (see Fig. 8.14), are retroelements like the LINEs and account for around 13% of the human genome. The most common type of SINE was named the **Alu element** because an *Alu* restriction enzyme site falls within it. The human genome contains about 300,000 to 500,000 Alu elements. SINE elements cannot move to new locations in the genome without help from the LINE reverse transcriptase/endonuclease protein. Unlike LINEs, SINEs are found in gene-rich regions of the human genome, but they are shorter and often inert, so their presence does not usually interfere with gene function.

Another type of highly repetitive element found in the human genome is the minisatellite or VNTR. These were used in mapping the human genome and are scattered around the entire genome (see earlier discussion).

Noncoding genomic DNA includes many different types of elements—for example, LINE, SINE, and satellite DNA.

LINE sequences are transcribed into RNA and translated into proteins. One of these proteins reassociates with the mRNA. The RNP complex re-enters the nucleus, where it integrates a LINE copy into a new genome location.

BIOINFORMATICS AND COMPUTER ANALYSIS

As noted previously, the use of computers has revolutionized the way in which genetic information is gathered and analyzed. The term **bioinformatics** has been coined to describe the scientific discipline of using computers to handle biological information. It encompasses a large number of subfields (Table 8.5). Bioinformatics includes the storage, retrieval, and analysis of data about biomolecules. By far the greatest achievement of the bioinformatics

Table 8.5 Fields of Study Related to Bioinformatics

Field	Description
Computational biology	Evolutionary, population, and theoretical biology; statistical models for biological phenomena
Medical informatics	The use of computers to improve communication, understanding, and management of medical information
Cheminformatics	The combination of chemical synthesis, biological screening, and data mining to guide drug discovery and development
Genomics	The analysis and comparison of the entire genetic complement of one or more species
Proteomics	The global study of proteins
Pharmacogenetics	Using genomic and bioinformatic methods to identify individual differences in response to drugs
Pharmacogenomics	Applying genomics to the identification of drug targets

revolution has been the sequencing of the human genome. The term *bioinformatics* is now used to include analyses of data from DNA microarrays (see later discussion), assessment of the function of genomes, and the comparison of different genomes.

Because bioinformatics is so widely used, it is important to make genomic data available to researchers. The data from the Human Genome Project is available through the National Center for Biotechnology Information website (http://www.ncbi.nlm.nih.gov/). Some other websites that present sequence data are listed below. On the NCBI home page, you can explore the human genome in many different ways. Using "Gene," you can identify a specific gene by name. The record for each gene contains the gene name and description, its location, a graphical representation of the introns and exons for all the protein isoforms that are known, and a summary of all the information known about the gene. Additionally, the various domains within the protein, such as actin-binding sites, are listed with links to explain the domain and its function. Finally, genes and/or regions of DNA from other organisms that are homologous to the gene are shown. The page also contains links to research papers on the gene's function.

Some Bioinformatics Websites:

- GenBank and linked databases
  - http://www.ncbi.nlm.nih.gov/Gene/
  - http://www.ncbi.nlm.nih.gov/mapview/
  - http://www.ncbi.nlm.nih.gov/genome/guide/human/
- Genome Database (GDB) (human genome)
  - http://genome.ucsc.edu/index.html
- European Bioinformatics Institute (including EMBL and Swissprot)
  - http://www.ebi.ac.uk/
- Flybase (*Drosophila* genome)/Wormbase (*C. elegans* genome)/Yeast genome (*Saccharomyces* genome)
  - http://flybase.org/
  - http://wormbase.org
  - http://www.yeastgenome.org/
- RCSB Protein Data Bank
  - http://www.rcsb.org/pdb/
- PIR Protein Information Resource (PIR)
  - http://pir.georgetown.edu/

The program Map Viewer (http://www.ncbi.nlm.nih.gov/mapview/) is used to browse the human genome without any particular gene in mind. For example, individual chromosomes can be explored via a graphical interface that allows you to zoom in and out of various regions. Another **genome browser** can be found at http://www.ensembl.org.

The amount of information generated by the Human Genome Project is tremendous, and understanding this information without the use of computers is too difficult. **Data mining** refers to the use of computer programs to search and interpret the data. Many bioinformatics researchers develop programs that search the genomic data banks and sift, sort, and filter the raw sequence data. Data mining programs often process information using the following steps:

1. Selection of the data of interest.
2. Preprocessing or "data cleansing." Unnecessary information is removed to avoid slowing or clogging the analysis.
3. Transformation of the data into a format convenient for analysis.
4. Extraction of patterns and relationships from the data.
5. Interpretation and evaluation.

These programs can be designed to search for related sequences, determine areas of coding and noncoding DNA by looking at codon bias, or search for known consensus sequences, just to name a few applications. Searching for related sequences or **similarity searches** allows researchers to identify a potential function for a gene. If a gene of unknown function from humans is very similar to a characterized gene from flies, the two encoded proteins may have similar functions. This type of research is called **comparative genomics**. More than one gene can be compared. For example, entire pathways are often similar in different species. Thus, human insulin attaches to a receptor on the cell surface and controls gene transcription via several intracellular proteins. Remarkably, very similar insulin signaling proteins are found in the roundworm *Caenorhabditis elegans.*

Difficulties may arise if scientists use only comparative genomics to study a new protein. Sometimes similar sequences play radically different roles, as similar proteins may evolve new functions. Thus, sequence similarity does not always imply functional similarity. Finally, the databases themselves are not perfect and may contain mistakes that are misleading. Comparative genomics must be complemented with other studies to reliably assign a role to a novel protein.

Other programs determine coding and noncoding areas of the genome by looking at **codon bias**. Identifying coding regions is critical for finding genes and can be accomplished by looking at the wobble position (third base) of the codon. Although a particular amino acid is often encoded by multiple codons, some codons are used preferentially. This codon bias varies from one organism to another. Most of the tRNAs for a particular amino acid will recognize the favored codon(s). For example, *Escherichia coli* genes preferentially use CGA, CGU, CGC, and CGG for the amino acid arginine, but rarely use AGA or AGG. Consequently, very few tRNAs for arginine are produced that recognize the AGA and AGG codons. Codon bias is seen in regions that encode proteins, but in noncoding regions, the wobble position will not maintain this bias. Thus, a potential gene in *E. coli* would contain relatively few AGA and AGG codons.

Finally, programs that identify consensus sequences allow researchers to find various signatures or motifs associated with particular functions. For example, a site that binds to ATP has specific amino acids in specific locations. These sequences in an unknown protein may help identify one of its functions. Other motifs include actin-binding domains, which indicate that the protein binds to the cytoskeleton, or protease cleavage sites that suggest the protein is subject to intracellular modification by proteases. Any potential motif in the sequence must be confirmed experimentally. For example, a protein with an ATP binding site signature must be shown to bind ATP experimentally. Thus, sequence analysis provides a basis for further experiments.

Bioinformatics is the study of biological information using computers.

Data mining uses computer algorithms to study, sort, and compile information from genomic databases. Information about genes can be obtained by comparing sequences from different organisms.

MEDICINE AND GENOMICS

One of the greatest applications for human genomics data is in disease and its diagnosis. Medical applications of genomics are abundant, and the later chapters of this book cover some of them. **Gene testing** is the most common present application. Once genes have been associated with particular diseases, people can be screened for genetic mutations within the gene. Such tests can diagnose diseases such as muscular dystrophy, cystic fibrosis, sickle cell anemia, and Huntington's disease because these are strictly inherited disorders. In diseases with an environmental component, genetic testing offers information that may change how a person lives his or her life. Perhaps those with a genetic predisposition to colon cancer will have more screenings, earlier than usual, and perhaps alter their diet to minimize the chance of cancer developing. Other applications include gene therapy (see Chapter 17).

The use of genomics in medicine has expanded greatly in the last decade and will change rapidly over the coming years. The cost of sequencing an entire human genome is now low enough for this to be used as a diagnostic test. As of writing, next-generation sequencing techniques have been streamlined to identify the causative mutation in diseases. Instead of sequencing the entire genome, many scientists prefer **whole exome sequencing**, where only the exons from the human genome are amplified and sequenced. This smaller dataset allows the identification of mutations in the sequences that specifically affect proteins, and can suggest whether or not that variation is the causative agent of the disease. Since exons comprise less than 2% of the genome, the cost for this analysis is much cheaper.

As of this writing, around 3000 different diseases have been identified using genomics and pedigree analysis. These are termed **Mendelian diseases** because they are inherited as would be expected if the disease resulted from a single mutation in a single gene. Once a causative gene is identified, then a strategy for treatment can be developed. For example, Marfan syndrome is a disease whose features include long arms, legs, and fingers, tall/thin body type, curved spine, sunken chest, flat feet, and flexible joints. The most serious of the problems is the threat of aortic aneurisms due to weak collagen in the aorta. The disease is caused by defects in connective tissue and in some cases may result in hypermobile joints. The disease is typically inherited from parents, but some cases occur spontaneously. Next-generation sequencing of families affected by Marfan syndrome has identified that a mutation in the fibrillin-1 gene increases the amount of transforming growth factor beta (TGFβ) in those affected by the disease. The increase in TGFβ in turn prevents the tissues of the body from developing properly. The breakthrough discovery that too much TGFβ was the cause allowed doctors to try treating patients with TGFβ inhibitors, which allows the tissues to develop properly and prevents the aorta from enlarging from increased pressure.

Besides Mendelian diseases, common diseases can also be studied with genomics. Many common diseases are **polygenic**; that is, they are caused by a variety of genes, not one specific mutation. Since sequencing of the human genome was completed, scientists have associated different genetic variations with common disorders such as Crohn's disease, type 2 diabetes, autoimmune disease, kidney disease, and psychiatric disorders. The association of a genetic variation and a disease is done by **genome-wide association study (GWAS)**. In these studies, whole genome analyses of two distinct populations are compared. For example, all the variations from a population of people without diabetes are compared to a population of people with type 2 diabetes. Any variant that has a frequency greater than 1% in the affected population is a potential variation that can cause this phenotype. This type of study is analogous to a mutagenesis screen in a model organism. The scientist looks for a subgroup with a similar phenotype from a population of individuals. Each of the individuals has mutations that cause the phenotype, but these may or may not be in the same gene. When the scientist looks at enough individuals, the key genes can be identified. In this example, each individual of a population with type 2 diabetes has some genetic change that makes him or her ill. So far, around 40 different loci (or potential genes) are associated with the disease. Some of these loci are associated with insulin secretion, suggesting that problems with insulin release may be a causative factor for the disease. As with all complicated diseases, the actual changes identified so far can account for only 20% of the heritability in type 2 diabetes.

Genomics has a wide-reaching effect on biotechnology and medicine.

DNA ACCUMULATES MUTATIONS OVER TIME

The Human Genome Project has opened the doors to improved analyses for many areas, including evolutionary biology. The sequence features of the human genome arose over millions of years, as **mutations**, or alterations of the genetic material, occurred, and were passed on to successive generations. During the course of human history, many different events have sculpted our genetic history and resulted in our current genetic state. Each individual has undergone some sort of genetic recombination and/or mutation to become unique in physical and emotional constitution. Genetic mutations constantly occur throughout all the cells of our bodies. Most of the defective cells die and undergo apoptosis (see Chapter 20). When a mutation occurs in the **somatic cells**, the children or offspring do not inherit the mutation; only when a mutation occurs in the **germline**, or sex cells, are the mutations passed on to the next generation.

Many different types of mutations cause genetic diversity (Table 8.6). The most common are **base substitutions**, in which one nucleotide is exchanged for another. When a purine base is replaced by another purine, or a pyrimidine is replaced by another pyrimidine, this is called a **transition**. If a pyrimidine is exchanged for a purine, or vice versa, this is a

Table 8.6 Types of Mutations

Base Changes	Normal	Mutant
Transitions	GAACGT	GAGCGT
Transversions	GAACGT	GATCGT
Missense mutation	GAA CGT	GAT CGT
	Glu Arg	Asp Arg
Conservative substitution	ACT CGT	TCT CGT
	Thr Arg	Ser Arg
Radical replacement	GAT CGT	GCTCGT
	Asp Arg	Ala Arg
Nonsense mutation	GAA CGT	TAA CGT
	Glu Arg	Stop
Insertions	GAACGT	GAAACGT
Deletions	GAACGT	GACGT

transversion. These mutations create the SNPs used in genomic maps. Because different human individuals vary by about 1 in 1000 to 2000 bases, there are on average 2.5 million SNPs over the whole genome.

SNPs, or single base substitutions, can fall anywhere in the genome, in either coding or noncoding DNA. When the SNP falls within a gene, it may alter protein sequence and function. When the base substitution alters one amino acid in a protein, the mutation is called a **missense mutation**. Some missense substitutions have little effect on protein structure or function because one amino acid replaces another with similar properties. This is known as a **conservative substitution**. An example would be replacing threonine by serine, as these vary only slightly in size but not in chemistry (both have an –OH group). **Radical replacements**, on the other hand, can alter the protein function or structure because they involve replacing amino acids with others that have a different chemistry. For example, aspartic acid or serine are often involved in hydrogen bonding, and when either is replaced by a neutral amino acid like valine, the protein structure may become unstable. Sometimes missense substitutions create **conditional mutations**, in which the protein will work under certain conditions but not others. A common conditional mutation is a **temperature-sensitive mutation**, in which the mutation does not alter the protein function at the permissive temperature, but the protein is defective at the restrictive temperature. When base substitutions change a codon for an amino acid into a stop codon, this results in a truncated version of the original protein. These are **nonsense mutations**.

Theoretically, mutations may result in the **insertion** or **deletion** of one or more bases. As for single base substitutions, location is the key to what effect the mutation will have. If the deletion or insertion of a few bases falls within a gene, it may alter the reading frame of the protein, which will create random polypeptide after the mutation. Often, the altered reading frame creates a stop codon, which truncates the protein. Large deletions may, of course, completely remove all or part of a gene. Larger segments of DNA can undergo alterations due to **inversions**, **translocations**, and **duplications**. Inversions occur when DNA segments become inverted relative to the original sequence. Translocations occur when DNA segments are moved to new locations. Duplications occur when the DNA segment is copied and then moved, resulting in two identical regions.

Theoretically, mutations occur randomly throughout the genome. However, **mutation hot spots** are regions where mutations occur at much higher frequencies. Mutations often occur at methylated sites because methylated cytosine often loses an amino group, turning into thymine. DNA polymerase can also induce mutations during DNA replication. Occasionally, the proofreading ability of the polymerase fails and single wrong bases are incorporated. More often, DNA polymerase undergoes **strand slippage**, when a segment of DNA is highly repetitive. The result is either a duplication or deletion, depending on the orientation of the slippage. Genetic variation in the human genome reflects recombination hot spots. In fact, most regions of the human genome are passed from one generation to the next in segments called **haplotype blocks**, or **hapblocks**. Because recombination usually occurs only in certain defined spots, the regions in between the two hotspots will be inherited together as a group or block. Each of the blocks has a variety of different variations, and they always segregate as one; thus, the region is called a hapblock.

The rates at which mutations occur help in understanding how mutations have affected the course of evolution. The rate of mutation is low and depends on the organism and even the particular gene being considered. Nonetheless, over long periods of time, many mutations will occur. As Table 8.7 suggests, the rate of mutation is much lower in genomes that are larger. In *E. coli*, mutations occur at 5.4×10^{-7} per 1000 base pairs per generation, but in humans, mutations occur over 10 times more slowly, at only 5.0×10^{-8} per 1000 base pairs. However, when the mutation rate is corrected for effective genome size (i.e., coding capacity rather than total DNA), it is approximately the same for most organisms. This suggests that some mechanism must actively control the mutation rate.

Table 8.7 Mutation Rates in DNA Genomes

Organism	Genome Size (Kilobases)	Mutation Rate per Generation		
		Per kb	Per Genome (Uncorrected)	Per Effective Genome
Bacteriophage M13	6.4	7.2×10^{-4}	0.005	0.005
Bacteriophage Lambda	49	7.7×10^{-15}	0.004	0.004
Escherichia coli	4,600	5.4×10^{-7}	0.003	0.003
Saccharomyces cerevisiae	12,000	2.2×10^{-7}	0.003	0.003
Caenorhabditis elegans	80,000	2.3×10^{-7}	0.018	0.004
Drosophila	170,000	3.4×10^{-7}	0.058	0.005
Human	3,200,000	5.0×10^{-8}	0.16	0.004

Mutations occur in all organisms at random places in the genome with an approximately similar rate. Mutations can be simple base substitutions, inversions, deletions, or insertions. The length of insertions and deletions is variable.

GENETIC EVOLUTION

Molecular phylogenetics is the study of evolutionary relatedness using DNA and protein sequences. Comparing sequences from different organisms shows the number of changes that have occurred over millions of years. All cellular organisms, including bacteria, plants, and animals, have ribosomal RNA. These sequences can be compared, and the differences can be used to determine relatedness. This approach is less subjective than using physical characters for **taxonomy**. The **cladistic** approach assumes that any two organisms ultimately derive from the same common ancestor (if we go far enough back) and that at some point **bifurcation**, or separation into two **clades**, occurred in their line of descent. The difference between the two organisms indicates how long ago the split occurred. Taxonomy may be based on visible characteristics—that is, the phenotype. This works well, to a first approximation, in organisms with plenty of obvious features, such as mammals and plants. But in organisms such as bacteria, the method falls apart. However, molecular phylogenetics allows making family trees for every organism.

When molecular data are used to study relatedness, it is essential that the sequences are correct and have truly come from the organisms under study. This can be complicated in higher organisms because some sequences have been derived from other organisms, such as viruses or bacteria. This problem applies to all organisms, to some extent. For example, many bacterial genomes contain inserted bacteriophage genomes. Another important point is to ensure that sequences being compared are truly homologous; that is, they have all descended from one shared ancestral sequence. When gene sequences are compared, they are **aligned**, so that the regions of highest similarity correspond (Fig. 8.15).

This type of alignment can determine the relatedness of two or more proteins or genes. The relatedness can be represented graphically by drawing **phylogenetic trees**. The tree has various features: a root, nodes, and branches (Fig. 8.16). The root represents the common ancestor, and the branching indicates the separations that occurred during evolution. Individual nodes represent common ancestors between two subgroups of organisms. Branches represent clades, that is, groups of organisms with a common ancestor. The length of the branches indicates the number of sequence changes, so if the branches are short, the two groups of organisms bifurcated relatively recently, and if the branches are long, the split occurred long ago.

```
Human alpha 1      MVLSPADKTNVKAAWGKVGAHAGEYGAEALERMFLSFPTTKTYFPHFDLS   50
Human alpha 2      MVLSPADKTNVKAAWGKVGAHAGEYGAEALERMFLSFPTTKTYFPHFDLS   50
Rat alpha 1        MVLSADDKTNIKNCWGKIGGHGGEYGEEALQRMFAAFPTTKTYFSHIDVS   50
Mouse alpha 1      MVLSGEDKSNIKAAWGKIGGHGAEYGAEALERMFASFPTTKTYFPHFDVS   50
Chicken alpha-A    MVLSAADKNNVKGIFTKIAGHAEEYGAETLERMFTTYPPTKTYFPHFDLS   50
                   ****  **.*:*  : *:..*. *** *:*:*** ::*.*****.*:*:*

Human alpha 1      HGSAQVKGHGKKVADALTNAVAHVDDMPNALSALSDLHAHKLRVDPVNFK   100
Human alpha 2      HGSAQVKGHGKKVADALTNAVAHVDDMPNALSALSDLHAHKLRVDPVNFK   100
Rat alpha 1        PGSAQVKAHGKKVADALAKAADHVEDLPGALSTLSDLHAHKLRVDPVNFK   100
Mouse alpha 1      HGSAQVKGHGKKVADALASAAGHLDDLPGALSALSDLHAHKLRVDPVNFK   100
Chicken alpha-A    HGSAQIKGHGKKVVAALIEAANHIDDIAGTLSKLSDLHAHKLRVDPVNFK   100
                    ****:*.***** .** .*. *::*:..:** *****************

Human alpha 1      LLSHCLLVTLAAHLPAEFTPAVHASLDKFLASVSTVLTSKYR   142
Human alpha 2      LLSHCLLVTLAAHLPAEFTPAVHASLDKFLASVSTVLTSKYR   142
Rat alpha 1        FLSHCLLVTLACHHPGDFTPAMHASLDKFLASVSTVLTSKYR   142
Mouse alpha 1      LLSHCLLVTLASHHPADFTPAVHASLDKFLASVSTVLTSKYR   142
Chicken alpha-A    LLGQCFLVVVAIHHPAALTPEVHASLDKFLCAVGTVLTAKYR   142
                   :*.:*:**.:* * * .:** :********.:*.****:***
```

FIGURE 8.15 Alignment of Related Hemoglobin Sequences

The hemoglobin sequences were aligned using ClustalW (http://www.ebi.ac.uk/clustalw/). Amino acids marked with * are identical in all sequences; those marked with : and . are not identical but are conserved in the type of amino acid.

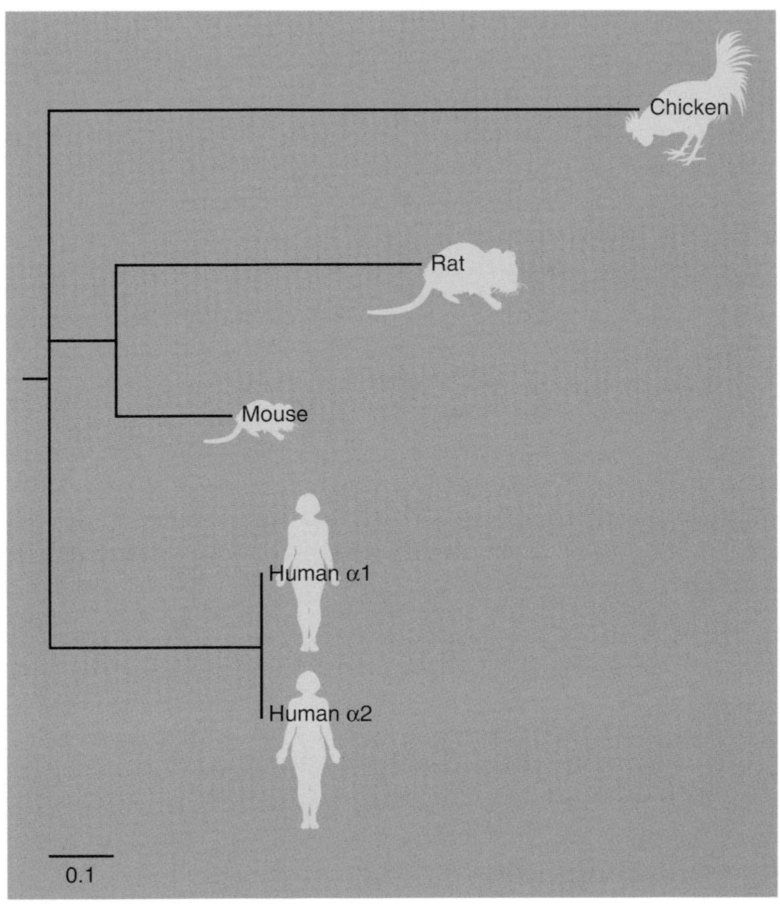

FIGURE 8.16 Phylogenetic Tree of Hemoglobin

The amino acid sequences of chicken globin-A, rat hemoglobin alpha 1 chain, mouse hemoglobin alpha 1 chain, and the alpha 1 and alpha 2 chains from humans were compared. The length of lines represents the number of sequence differences; the longer the line, the more changes in sequence. The differences were analyzed with ClustalW, and the tree was drawn using Phylodendron (http://iubio.bio.indiana.edu/treeapp/).

Based on alignments, genes have been grouped into **families**, groups of closely related genes that arose by successive duplication and divergence. **Gene superfamilies** occur when the functions of the various genes have steadily diverged until some are hard to recognize. For example, the transporter superfamily encompasses many proteins that transport molecules across biological membranes. This superfamily has members that transport

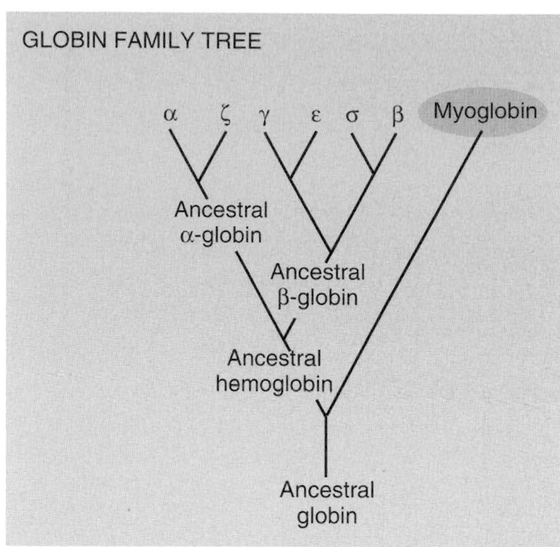

FIGURE 8.17 Globin Family of Genes
Over the course of evolution, a variety of gene duplication and divergence events gave rise to a family of closely related genes. The first ancestral globin gene was duplicated, giving hemoglobin and myoglobin. After another duplication, the hemoglobin gene diverged into the ancestral alpha-globin and ancestral beta-globin genes. Continued duplication and divergence created the entire family of globin genes.

sugars into bacteria, transport water into human cells, and even export antibiotics out of bacteria. They are found in almost all organisms. Another gene superfamily is the globin family (Fig. 8.17). The family includes myoglobin and hemoglobin from different organisms. These proteins all carry oxygen bound to iron; but myoglobin is specific to muscle cells, whereas hemoglobin is specific to blood. The theory is that early in evolution one gene for an ancestral globin existed. At some point, this gene was duplicated and the copies diverged so that one was specialized for blood and the other for muscle. Hemoglobin itself also diverged later into different forms, each used at various stages of development.

New genes may be generated one at a time, but in addition, whole chromosomes or genomes may be duplicated. In some organisms, particularly plants, genome duplications are relatively stable and have occurred quite often. An example is the modern wheat plant. Its ancestors were typical diploids, but modern bread wheat is hexaploid, being derived from three different ancestral plants. The hexaploid varieties arose through hybridization and natural mutation and were exploited because of the higher protein content and better yield. The wheat used to make pasta, durum wheat, is tetraploid, and represents an intermediate step.

Although genomes have characteristic average rates of mutation, individual genes may mutate at very different rates. Essential proteins evolve or mutate more slowly than average. Conversely, the less critical a gene is for survival, the more mutations can be tolerated and the protein evolves more rapidly. Thus, the gene for cytochrome *c*, an essential component in the electron transport chain, has incorporated only 6.7 changes per 100 amino acids in 100 million years. In contrast, fibrinopeptides, which are involved in blood clotting, have had 91 mutations per 100 amino acids in 100 million years. As noted earlier, ribosomal RNA is useful to establish family trees for distantly related organisms. It is found in every organism and is essential to survival; therefore, it is slow to evolve.

What happens if a scientist wants to classify organisms that are closely related? Essential gene sequences do not provide enough genetic variation to differentiate such organisms. Nonessential genes may help, but sometimes even they are too close. In such cases, the wobble position of coding regions or even noncoding regions may be used. As noted in Chapter 2, the wobble position is the third nucleotide of a codon. The same amino acid is often encoded by several codons, which vary only in this third base. Alterations at this position usually have no net effect on protein function or structure and may occur between very closely related species or between individuals of the same species.

Mitochondrial or chloroplast genomes are also compared in order to determine the relatedness of organisms. These genomes accumulate mutations at a higher rate than the nuclear genomes in the same organisms. The organelle genomes vary particularly in the noncoding regions. One drawback to using organelle genomes is that mitochondria and chloroplasts are inherited maternally and thus trace the evolutionary lineage only on the maternal side.

Molecular phylogenetics uses genomic sequences of different organisms to determine their evolutionary relatedness. Essential proteins have fewer mutations over time. Less essential proteins have more mutations over time.

FROM PHARMACOLOGY TO PHARMACOGENETICS

Another field that has undergone many changes due to the Human Genome Project is pharmacology, the study of drugs. Drug development has traditionally been a hit-or-miss matter, with drug discovery often a by-product of other research. Penicillin is one of the 20th century's greatest discoveries, but was found by accident. Alexander Fleming was growing *Staphylococcus* bacteria and left his plates while on vacation. When he came back, mold had contaminated the plates. Miraculously, the staphylococci were not growing close to the mold, which was evidently secreting something that stopped bacterial growth.

Even Viagra was discovered by accident. Scientists were trying to develop heart medications when they noticed the "side effect." One factor that makes drug development costs so high is that many of the paths chosen lead to dead ends.

Another problem of drug development is **adverse drug reaction (ADR)**. Adverse reactions may happen in some patients while others respond well to the same drug. Most drugs are developed with the average patient in mind, yet there is often a subset of people who react badly to the drug. In the United States, the frequency of serious ADRs is approximately 7%. Such ADRs are a significant cause of hospitalizations and death. Differences in drug response often depend on a person's genetic makeup.

Pharmacogenetics is the study of inherited differences in drug metabolism and response, and **pharmacogenomics** is thus the study of all the genes that determine drug response. One major goal of these fields is to reduce the number of ADRs by determining the genetic makeup of the patient before offering a specific drug. The key to "genetic" diagnosis is the use of SNPs (see earlier discussion). Single changes in coding regions can often be correlated with adverse drug reactions. For example, if a certain subpopulation of people does not respond to a drug, then their DNA can be examined for a specific SNP that is absent in patients who do respond. Before the drug is given to new patients, DNA from a blood sample can be tested for the diagnostic SNP. Testing for SNPs can be done by microarray analysis (see later discussion) in the doctor's office, thus reducing the number of office or hospital visits.

SNP analysis is also used to screen for hereditary defects. Specifically, SNPs can be identified using a technique called Zipcode analysis (Fig. 8.18). Here, many different SNPs can be examined simultaneously. First, PCR is used to amplify the regions containing each SNP being investigated. The PCR fragments could be sequenced fully, but because SNPs differ by only one base, single base extension analysis is done instead. For this, a primer is designed to anneal just one base pair away from the SNP location. This primer also carries a "zipcode" region that is used to identify this specific SNP, and each SNP has a different zipcode. After the Zipcode primer anneals to the PCR fragments, DNA polymerase plus fluorescently labeled dideoxynucleotides are added. This results in a single base being added to the primer. (Note that dideoxynucleotides block chain elongation, and so only one base can be added.) Each base is labeled with a different fluorescent dye, allowing it to be identified. Next, beads linked to complementary zipcode (cZipcode) sequences are added to grab the zipcoded primers. The trapped Zipcode primer with the labeled nucleotide has a different color depending on which base was incorporated, and elucidate the identity of the nucleotide in the patient. The different color beads are sorted and counted by fluorescent-activated cell sorting (FACS; see Chapter 6).

One of the spin-offs from the Human Genome Project is the Pharmacogenetics and Pharmacogenomics Knowledge Base (PharmGKB; http://www.pharmgkb.org/). It records genes and mutations that affect drug response. Consider asthma, a condition in which people overreact to inhaled irritants by cutting airflow in and out of the lungs. The muscle cells around the bronchial tubes constrict, decreasing airflow. Albuterol is a drug used to open the bronchial tubes by relaxing the muscle cells. Albuterol affects the beta2-adrenergic receptor, and mutations in this receptor alter the efficacy of albuterol. A single nucleotide change that replaces glycine at position 16 with arginine gives a receptor protein with a better response to albuterol.

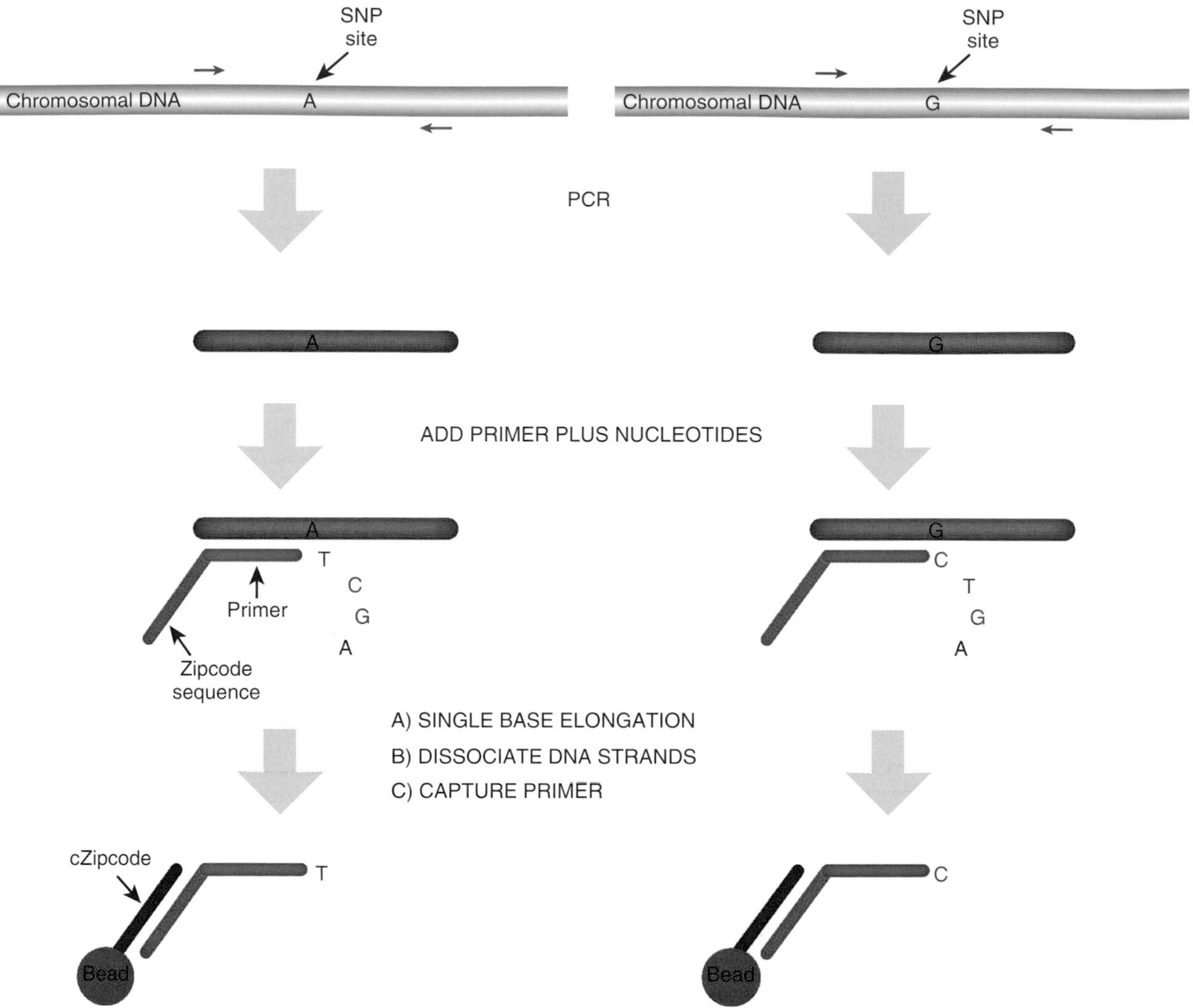

FIGURE 8.18
Zipcode Analysis and Single Base Extension of SNPs
A segment of DNA that includes an SNP site is generated by PCR (only a single strand of DNA is shown here for simplicity). Single base extension is performed with a primer that binds one base in front of the SNP. Person I has an A at the SNP site, and therefore, ddT is incorporated. In person II, a G at the same position results in incorporation of ddC. The bases are labeled with different fluorescent dyes. Use of dideoxynucleotides prevents addition of further bases. The elongated primer is then trapped by binding its Zipcode sequence to the complementary cZipcode, which is attached to a bead or other solid support for easy isolation.

Another key finding concerns the cytochrome P450 family of enzymes. These enzymes play a role in the oxidative degradation of many foreign molecules, including many pharmaceuticals. The CYP2D6 isoenzyme oxidizes drugs of the tricyclic antidepressant class, and different alleles of this enzyme affect how well a person metabolizes these drugs. Much as for albuterol, identifying which allele a patient has will prevent overdosages or adverse reactions. As time goes on, more medical treatments will be designed for the individual rather than the average person.

Pharmacogenetics is the study of inherited differences in drug metabolism and response.

Some SNPs affect how a person metabolizes a certain drug. When scientists determine what SNP correlates with what drug sensitivity, new patients can be screened and possibly avoid adverse drug reactions (ADRs).

GENE EXPRESSION AND MICROARRAYS

As noted earlier, a major issue in determining the correct number of genes is deciding whether or not a sequence is really a gene. Measuring whether a presumed gene is transcribed into mRNA is the first step to deciding if it is genuine. Gene expression was once done on a single gene basis, but now **functional genomics** studies gene expression over the entire genome. Functional genomics encompasses the global study of all the RNA transcribed from the genome—the **transcriptome**; all the proteins encoded by the genome—the **proteome**; and all the metabolic pathways in the organism—the **metabolome**.

Because the entire human genome contains only around 21,000 different genes, using microarrays to study gene expression is feasible. **DNA microarrays**, or **DNA chips**, contain thousands of different unique DNA sequences bound to a solid support, such as a glass slide. Microarrays are based on hybridization between a "probe" and target molecules in the experimental sample. However, in a microarray the probes are attached to the solid support, and the experimental sample is in solution. The microarray often represents the genome of the organism being tested and includes sequences corresponding to each gene in the organism.

To monitor gene expression, scientists test RNA extracted from a cell sample against a microarray. The experimental RNA sample is usually fluorescently tagged. Hybridization of the mRNA to the DNA probes on the solid support indicates whether or not a gene is expressed and to what degree. The level of fluorescence at each point on the array correlates with the level of the corresponding mRNA in the sample.

Microarrays can be used to analyze RNA isolated from cells grown under a variety of conditions—for example, heat shock, acid exposure, cancer, or other disease states. The same array can be hybridized to two or more samples of RNA (control versus experimental) to compare gene expression. Each RNA sample is labeled with a different fluorescent dye, for example, red and green. If a particular RNA is present in only one sample, the corresponding spot on the microarray will be red or green (Fig. 8.19), whereas if the RNA is present in both experimental and control samples, the spot will be yellow (i.e., red plus green). Modern arrays can accommodate thousands or millions of different probes, allowing the entire genome for most organisms to be examined at once. Some arrays are clustered so that all the genes involved in, say, protein synthesis or heat shock are together. The computer reads the color and fluorescence intensity for each of the spots and carries out an analysis.

The results can provide a global view of gene expression in different conditions. For example, slides can be made with every named gene from the yeast genome. These genes may be analyzed for expression at different stages of the cell cycle. For this, a culture of yeast cells is synchronized and arrested at different stages of the cell cycle by adding α factor or by using mutant yeast that freeze at particular stages of mitosis. The gene expression patterns for each stage are compared and compiled (Fig. 8.20).

Array treated with RNA from cells grown under condition 1 and labeled with red fluorescent dye

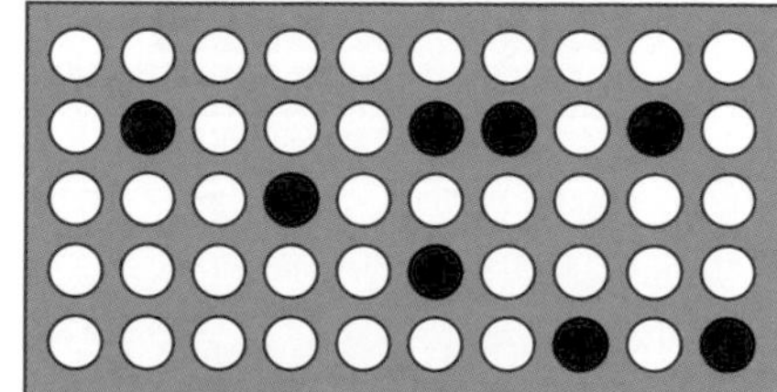

Array treated with RNA from cells grown under condition 2 and labeled with green dye

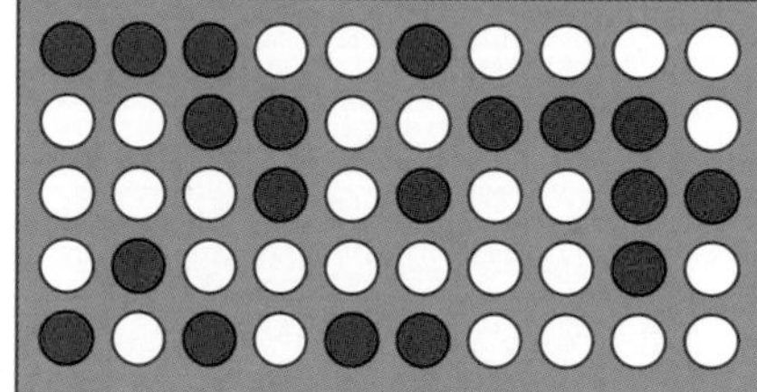

Array treated with both samples of RNA; yellow spots reveal genes expressed under both conditions

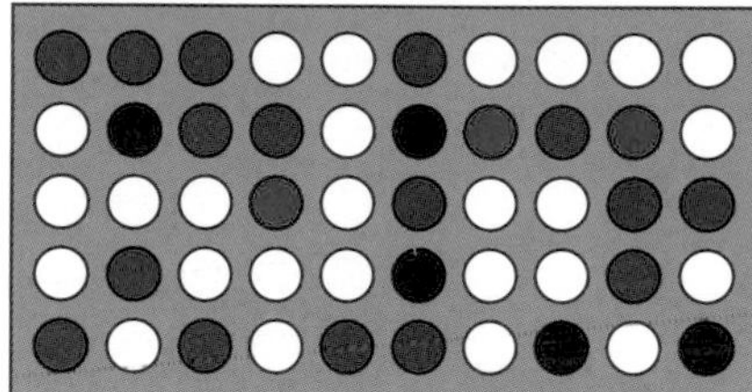

FIGURE 8.19 DNA Chip Showing Detection of mRNA by Fluorescent Dyes DNA chips can monitor many different mRNAs at one time. Each spot on the grid has a different DNA sequence attached. To determine which genes are expressed under which conditions, scientists isolate mRNA and label each sample with a different fluorescent dye. If two different dyes are used (as shown here), the same chip can be used for both. It is then visualized in three different ways: one shows only the red dye, another only the green, and the third merges the two images so overlapping spots look yellow.

Functional genomics includes the global study of all the RNA transcribed from the genome (the transcriptome).

DNA microarrays, or DNA chips, contain thousands of different unique sequences bound to a solid support, such as a glass slide. When fluorescently labeled RNA is incubated with the microarray, complementary sequences hybridize. The level of fluorescence corresponds to the amount of RNA that is bound to the DNA microarray.

MAKING DNA MICROARRAYS

There are two major types of DNA microarrays: one contains cDNA fragments 600 to 2400 nucleotides in length, and the other uses oligonucleotides of 20 to 50 nucleotides in length. Each type of microarray is manufactured differently. When a cDNA microarray is made, each of the different probes must be chosen independently and made by PCR or traditional

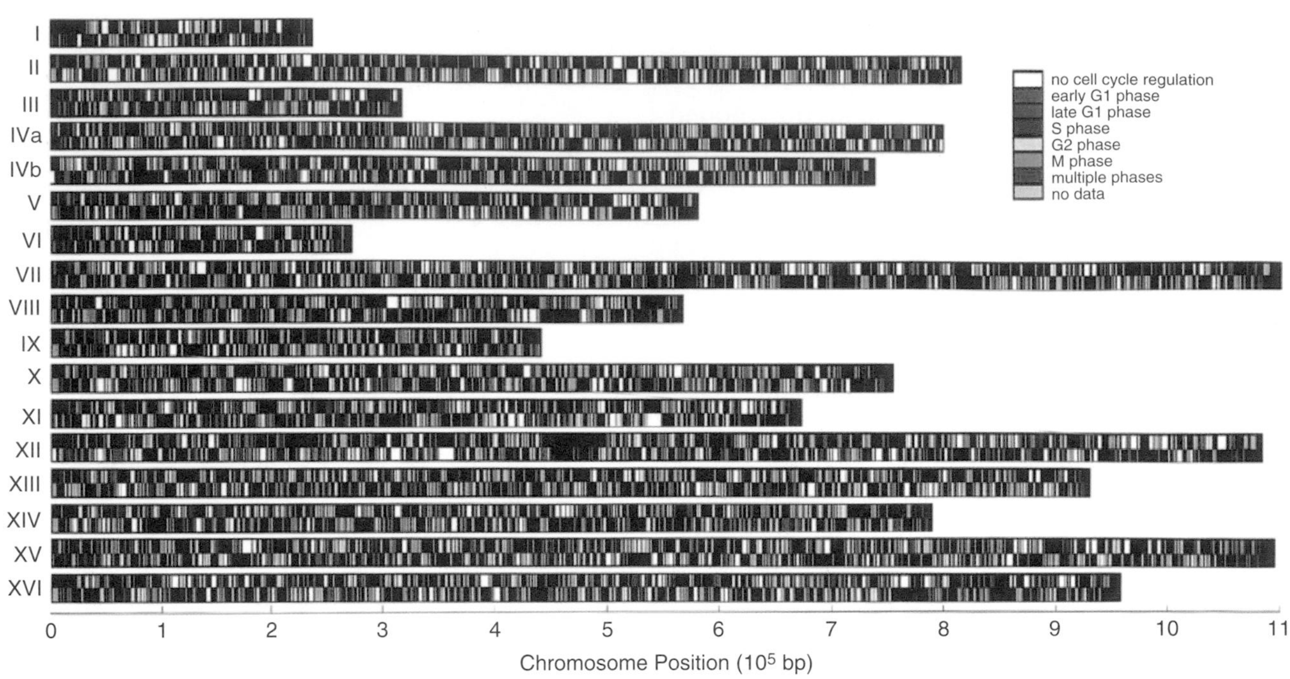

FIGURE 8.20 Gene Expression during Yeast Cell Cycle
Color coding indicates the time of the cell cycle for maximum gene expression. More than 800 different genes that respond to changes in the cell cycle were monitored on the 16 yeast chromosomes. From Cho RJ, et al. (1998). A genome-wide transcriptional analysis of the mitotic cell cycle. *Mol Cell* **2**, 65–73. Reprinted with permission.

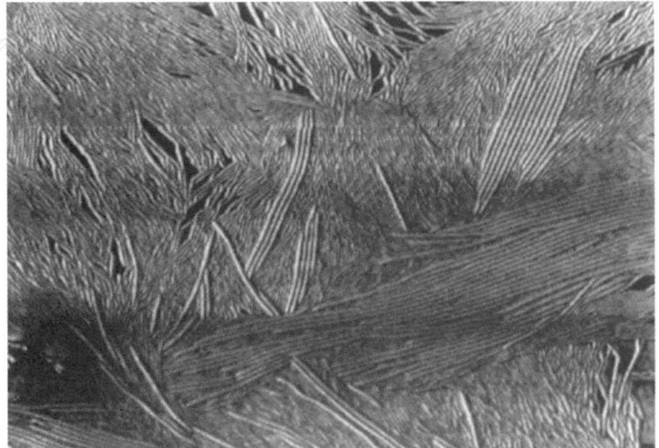

FIGURE 8.21 AFM of DNA on Microarray
A region of a yeast microarray after hybridization. The DNA is clearly deposited in sufficient density to permit many strand-to-strand interactions. The width of the figure represents a scanned distance of 2 micrometers. Reprinted by permission from Macmillan Publishers Ltd.: Duggan DJ, Bittner M, Chen Y, Meltzer P, Trent JM (1999). Expression profiling using cDNA microarrays. *Nature Genetics* **12**, 82–89, copyright 1999.

cloning. Then all the DNA probes are spotted onto the slide. When an oligonucleotide array is made, the oligonucleotides are synthesized directly on the slide.

cDNA Microarrays

The first step in making a cDNA microarray is to determine the numbers and types of probes to attach to the solid support. Since entire genomes have been sequenced for a variety of organisms, identifying potential genes is relatively easy. During the sequencing of these genomes, many cloned segments of DNA containing all or part of various genes were generated. Researchers can either obtain these clones or amplify genes from a sample of DNA using PCR. Each PCR product must be purified before attachment to the glass slide, so that all the extra nucleotides, *Taq* polymerase, and salts are removed and only pure DNA attaches to the slide. Pure cDNA samples can be used directly.

The next step is to create the chip using a microarray robot. Purified samples of each DNA are put into small wells arranged in a grid in microtiter plates. The size of the grid depends on the number of probes. If every predicted human gene is present once, approximately 25,000 different wells are needed. In practice, probes for each gene are attached more than once, in different areas of the chip, to provide several readings for each gene. A grid of pens or quills is dipped into the wells, one pen for each well, using a robotic arm. The pen tips are then touched to a glass slide, where a tiny drop of DNA solution is left behind. The robotic arm continues to manufacture spotted slides until the DNA in the well is used up. Using a robot makes chips cheap and easy to produce. Finally, the DNA is cross-linked to the glass slide with ultraviolet light, which causes thymine in the DNA to cross-link to the glass. Figure 8.21 shows DNA on a microarray grid visualized by an atomic force microscope.

In newer cDNA microarrays, the samples are spotted onto a glass slide using inkjet printer technology. The cDNA samples are sucked into separate chambers of the inkjet printer head

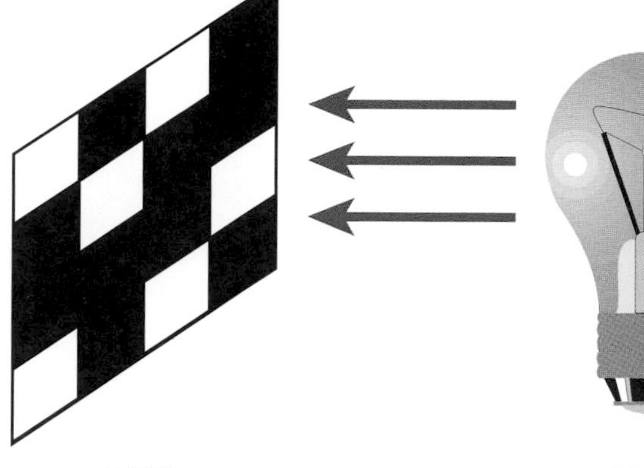

FIGURE 8.22
Photolithography
Light passing through the mask makes a particular pattern on the glass slide. If the slide is coated with a light-activated substance, only the regions that are illuminated will be activated for the addition of another nucleotide.

and then spotted onto the glass slide much as ink is spotted onto paper in a printer. Inkjet technology prevents variations in size and quantity of cDNA in the sample spots. Special adaptors prevent the inkjet sample channels from mixing, thus preventing cross-contamination.

Oligonucleotide Microarrays

Oligonucleotides are traditionally synthesized chemically on beads of controlled pore glass (CPG; see Chapter 4). Therefore, it is a small logical leap to synthesize many different oligonucleotides side by side on a glass slide. The main difference between synthesizing single nucleotides on beads versus making arrays on glass slides is that the array has thousands of different oligonucleotides, and each must be synthesized in its proper location with a unique sequence. To accomplish this, **photolithography** and solid-phase DNA synthesis are combined (Fig. 8.22). Photolithography is a process used in making integrated circuits, where a mask allows a specific pattern of light to reach a solid surface. The light activates the surface it reaches, while the remaining surface remains inactive.

A glass slide is first covered with a spacer that ends in a reactive group. It is then covered with a photosensitive blocking group that can be removed by light. In each synthetic cycle, those sites where a particular nucleotide will be attached are illuminated to remove the blocking group. Each of the four nucleotides is added in turn. At each addition, a mask is aligned with the glass slide. Light passes through holes in the mask and activates the ends of those growing oligonucleotide chains that it illuminates. Much as in traditional chemical synthesis, each nucleotide has its 5′-OH protected. Thus, after each addition, the end of the growing chain is blocked again. These protective groups are light activated, so at each step, a new mask is aligned with the slide, and light deprotects the appropriate nucleotides. The entire process continues for each nucleotide at each position on the glass slide. Making the masks is the key to this technology (Fig. 8.23).

DNA microarrays are made with cDNA or oligonucleotides. cDNA arrays are created by spotting pure samples of cDNA clones onto the glass in a small spot. The DNA is cross-linked to the glass with UV light. For oligonucleotide arrays, DNA is synthesized directly onto the glass slide.

HYBRIDIZATION ON DNA MICROARRAYS

Hybridization on a microarray is similar to what occurs during other hybridization procedures, such as Southern blots or Northern blots. All these techniques rely on the complementary nature of nucleic acid bases. When two complementary strands of DNA or RNA

1

Blocking group

Coupling group

Glass chip

LIGHT TREATMENT

2

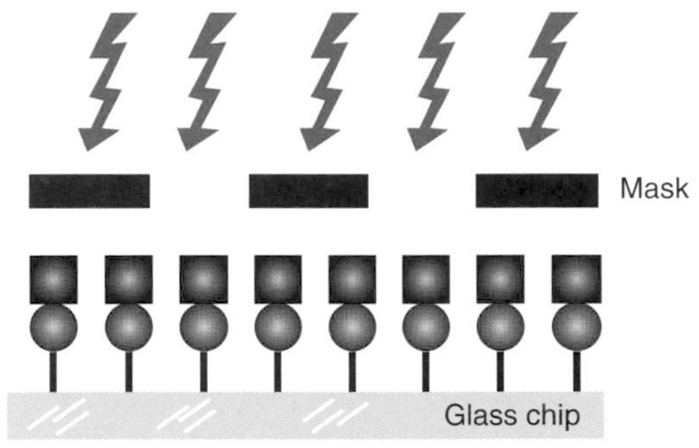

COUPLING GROUPS REVEALED

3

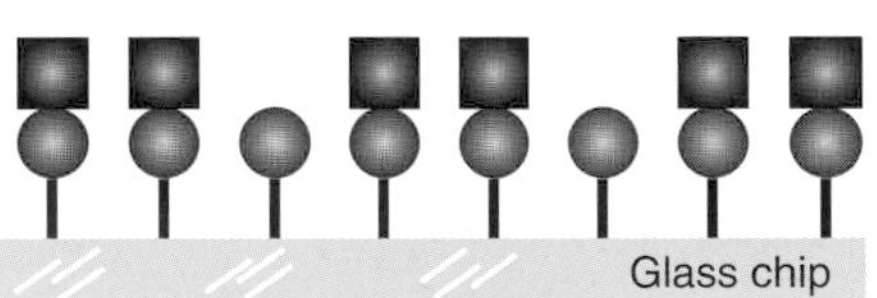

COUPLE WITH ACTIVATED AND BLOCKED G

4

G G

Glass chip

LIGHT TREATMENT

5

G G

Glass chip

COUPLING GROUPS REVEALED

6

G G

Glass chip

COUPLE WITH ACTIVATED AND BLOCKED T

7

T

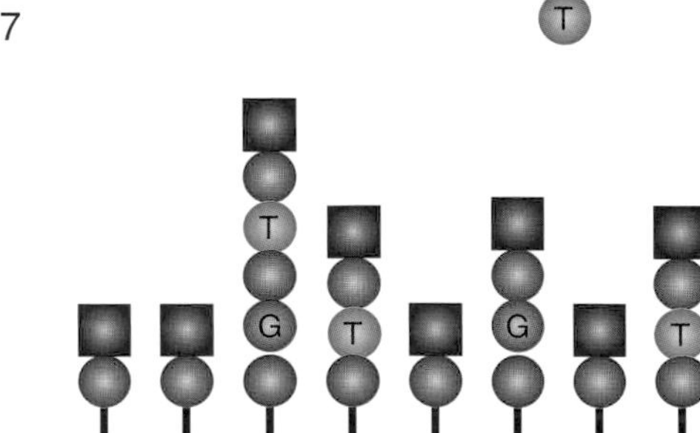

CONTINUE

FIGURE 8.23 On-Chip Synthesis of Oligonucleotides

Arrays may be created by chemically synthesizing oligonucleotides directly on the chip. First, spacers with reactive groups are linked to the glass chip and blocked. Then each of the four nucleotides is added in turn (in this example, G is added first, then T). A mask covers the areas that should not be activated during any particular reaction. Light activates all the groups not covered with the mask, and a nucleotide is added to these. The cycle is repeated with the next nucleotide.

are alongside each other, the bases match up with their complement, that is, thymine (or uracil) with adenine, and guanine with cytosine. On a DNA microarray, hybridization is affected by the same parameters as in these other techniques.

How the DNA is attached to the slide can affect how well the probe DNA and target DNA hybridize, especially for oligonucleotide microarrays (Fig. 8.24). The short length of oligonucleotides requires that the entire piece be accessible to hybridize. The length of the spacer between the oligonucleotides and the glass slide optimizes hybridization. An oligonucleotide attached with a short spacer has many of its initial nucleotides too close to the glass and inaccessible to incoming RNA or DNA. Oligonucleotides with longer spacers may fold back and tangle up. Oligonucleotides with medium-sized spacers are far enough from the glass, but not so far as to get tangled. Thus, medium-sized spacers give the best access for hybridization.

Hybridization of two lengths of DNA (or RNA with DNA) depends on certain sequence features. One important property is the ratio of A:T base pairs to G:C base pairs. G:C base pairs have three hydrogen bonds holding them together, whereas A:T base pairs have only two hydrogen bonds. Thus, more GC base pairs give stronger hybridization. If the sequence has too many A:T base pairs, the duplex may form slowly and be less stable. Another important consideration is secondary structure. If the probe sequence can form a hairpin structure, it will hybridize poorly with the target. If the probe has several mismatches relative to the target, the duplex may not form efficiently. All these issues must be addressed when making an oligonucleotide microarray. Computer programs are available to identify suitable regions of genes with sequences that will produce effective probes.

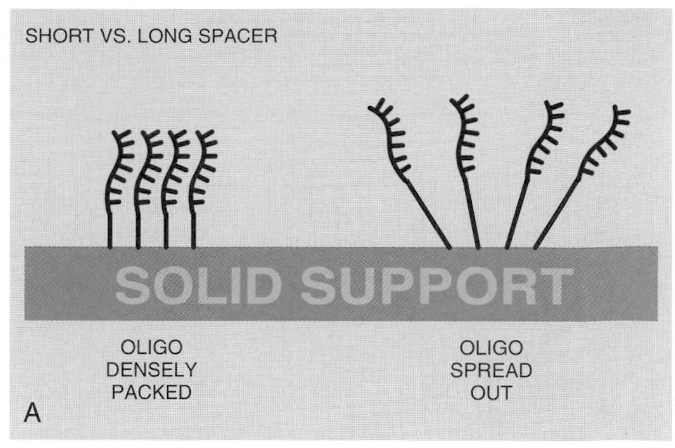

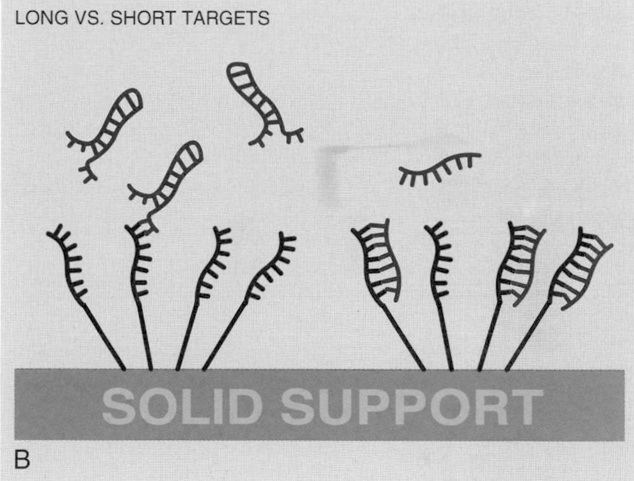

FIGURE 8.24 Length of Spacers and Target Molecules Affect Hybridization on Microarrays

(A) When the spacer between the glass slide and oligonucleotide is too short, the oligonucleotides are condensed and not accessible to hybridize. If the spacer region is too long, the oligonucleotides and spacers tangle and fold, preventing optimal hybridization. (B) When the target for hybridization is too long, the target sequences may form hairpins with themselves rather than bind to the array oligonucleotides.

cDNA arrays are less prone to the problems seen in oligonucleotide arrays. cDNAs are double-stranded, so secondary structures such as hairpins are less likely to be a problem. During a hybridization reaction, cDNA arrays must be denatured either with heat or chemicals, making the probes single-stranded. Then the single-stranded RNA samples are allowed to hybridize on the slide under conditions that promote duplex RNA:cDNA without any mismatches.

Oligonucleotide microarrays must have a sufficient spacer and little secondary structure in order to hybridize with the samples.

MONITORING GENE EXPRESSION USING WHOLE-GENOME TILING ARRAYS

Whole-genome tiling arrays (WGAs) are oligonucleotide microarrays that cover the entire genome. The first entire genome to be represented by a whole-genome array was from *Arabidopsis*. A gene chip was designed with 25-mer oligonucleotides that overlapped each other and covered the entire sequence of the genome. Complementary oligonucleotide sequences were tiled back to back along each entire chromosome and ordered so that the array could be conveniently analyzed for gene expression (Fig. 8.25).

FIGURE 8.25 Whole-Genome Array Designs
Whole-genome arrays (WGAs) contain oligonucleotide probes that cover the entire genome in an overlapping set. In quasi-whole-genome arrays, the probes are spaced equal distances apart through the genome. The probes thus cover the entire genome, except for the gaps between the probes. Splice junction arrays have only probes that span the upstream and downstream regions of known splice junctions in mRNA. Exon-scanning arrays have probes derived from exon sequences only.

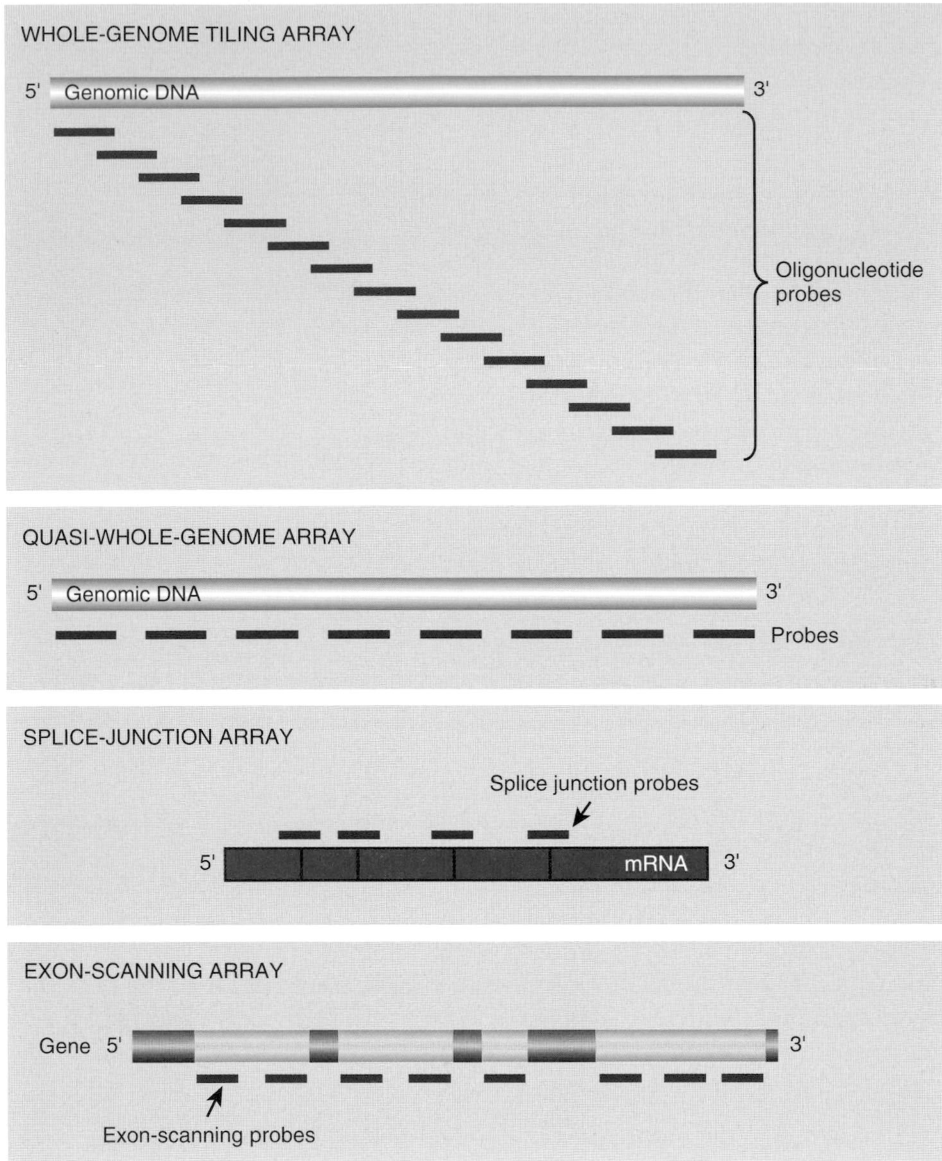

For the human genome, tiling arrays have been made to cover the entire sequences of chromosomes 21 and 22. They also use 25-mer oligonucleotides, but rather than overlapping, the oligonucleotides were spaced 35 base pairs apart along the sequence. These are therefore strictly only "quasi-whole-genome arrays." Compared to arrays that include only known genes, tiling arrays have the potential to identify novel regions that are transcribed, whether these encode unknown protein-encoding genes or nontranslated RNA. The RNA extracted from many different cell lines and tissues has been used to monitor gene expression, assess differences in splicing patterns, find new genes, and find RNA-binding protein target sequences.

The most interesting finding from studying human chromosomes 21 and 22 is that much larger portions of these chromosomes are transcribed into RNA than previously predicted from computer analyses of exon regions. About 90% of the transcribed regions occurred outside the known exons. The majority of the transcribed regions generated noncoding RNA, mostly of less than 75 base pairs in length. This suggests that noncoding RNA may have a much greater role than previously thought. These arrays also identified new exons that were previously unknown. In addition, these arrays can identify novel alternatively spliced mRNAs (and hence protein variants). The WGAs for chromosomes 21 and 22 have also been used to compare the level of expression of exons within the same gene. About 80% of the genes had exons with varied levels of expression, implying most genes have some sort of alternate splicing.

Another use for whole genome arrays is to analyze the results of **chromatin immunoprecipitation (ChIP)**. ChIP begins by cross-linking all the various transcription factors to chromatin, essentially freezing them in place. Next, the chromatin is sheared into smaller fragments, and the DNA/transcription factor complexes are isolated. Affinity purification isolates each particular transcription factor from the others (e.g., using antibodies to the transcription factor, Jun isolates all the Jun/DNA complexes from this mixture). Finally, the DNA sequences that are bound to the chosen transcription factor are identified using WGA. The entire procedure, including the analysis on a gene chip, is called **ChIP-chip**. This analysis can precisely locate transcription factor binding sites. Curiously, binding locations for NF-κB, for example, have been found within both coding and noncoding regions, such as introns or the 3′ ends of genes. These surprising findings suggest that transcription factors may also function outside the traditional upstream promoter region.

Another use for WGA is to identify regions of the genome that are methylated. Methylation prevents the inappropriate expression of various genes, especially those used only during development of young organisms, or those genes from transposons or viruses that could be detrimental. Cancerous cells have methylation patterns much different from those of normal cells, suggesting that this type of regulation is critical to proper growth control. To identify the methylated regions, scientists first treat genomic DNA with **sodium bisulfite**, which deaminates nonmethylated cytosine to uracil yet does not affect methylated cytosine. The treated DNA is then hybridized to a WGA. Those regions with nonmethylated cytosine no longer hybridize to the array because the cytosines have been converted to uracil (which pairs with A, not G). Those regions of the genome that are methylated still hybridize well because methylated cytosine and guanine form a stable base pair.

Finding genetic variations and polymorphisms is critical to genome analysis, and whole-genome arrays offer a nonbiased method to analyze samples. A WGA that has the reference sequence for the human genome can be used to identify and catalog all different types of polymorphisms, including SNPs, VNTRs, and repetitive elements. Indeed, an overlapping WGA made to the entire reference sequence of the human genome and spaced by a single base pair could be used to effectively resequence the entire genome with ease and speed.

Whole-genome arrays are oligonucleotide arrays that have sequences that cover the entire genome. They can be used to identify transcription factor binding sites, regions of methylation, SNPs, VNTRs, repetitive elements, and so forth.

MONITORING GENE EXPRESSION BY RNA-Seq

A completely different approach to analyzing the transcriptome is to isolate total cellular RNA and then sequence all of it. RNA molecules present in multiple copies will be sequenced multiple times, and thus, the number of copies will be revealed. In practice, sequencing RNA is technically unfeasible. Therefore, the RNA is converted to complementary DNA (cDNA) by reverse transcriptase and the DNA is sequenced. Until recently, this approach was impractical due to the colossal amount of sequencing needed. However, major advances in DNA sequencing technology have transformed the field of transcriptomics, since an entire cDNA library can now be sequenced quickly and cheaply.

RNA-Seq, or whole transcriptome shotgun sequencing, creates a cDNA library from fragmented total mRNA, and then every cDNA is sequenced (Fig. 8.26). These sequences are then aligned with the genome of the organism. The relative copy number of each cDNA sequence is an indication of gene expression levels.

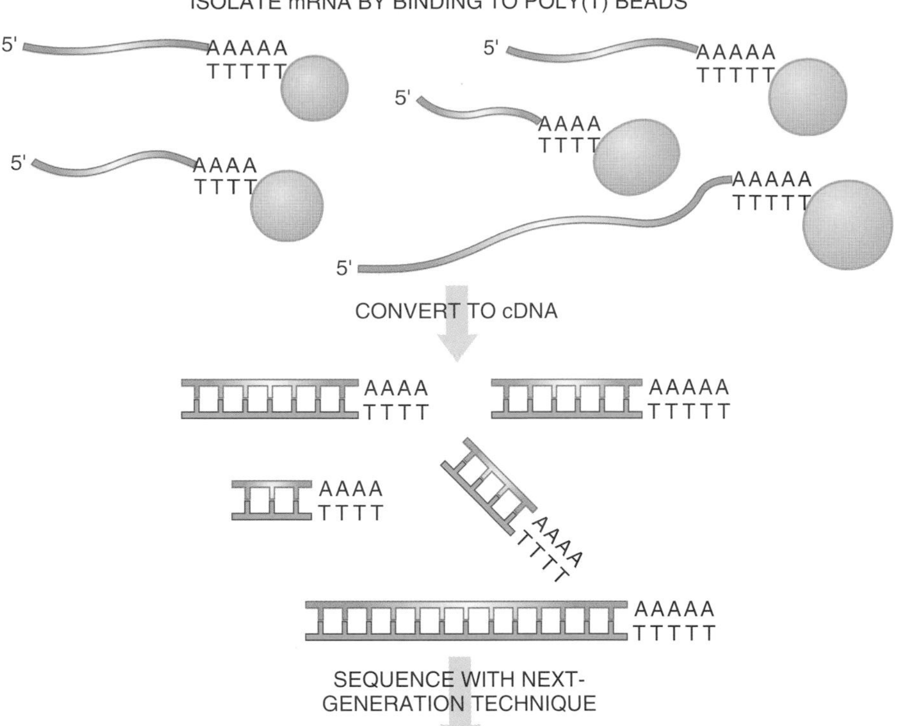

FIGURE 8.26
RNA-Seq Protocol
The entire transcriptome can be identified by sequencing a cDNA library in its entirety. Next-generation sequencing makes this process possible, resulting in the identification of each copy of every RNA that was expressed.

Relative to typical microarrays, the major advantages of RNA-Seq are as follows (Fig. 8.27):

(a) RNA-Seq is probe independent. Consequently it gives more accurate measurements of the relative expression level of different RNA molecules by counting them directly.
(b) RNA-Seq has a greater dynamic range. The fluorescent signals used for microarrays have both a minimum sensitivity and become saturated at high levels. Direct counting is not subject to saturation.
(c) RNA-Seq monitors both coding and noncoding RNA.
(d) RNA-Seq detects alternatively spliced transcripts and yields their relative numbers.
(e) When a diploid (or polyploid) organism contains multiple different alleles of the same gene, RNA-Seq will monitor allele specific expression.
(f) Even when the genome sequence for an organism is not available for reference, RNA-Seq can still be performed.
(g) The small amount of RNA required for RNA-Seq means that it is possible to carry out analyses of single cells.
(h) "Dual RNA-Seq" allows simultaneous analysis of RNA from host cells infected with a pathogen. Computer analysis allocates the different sequences to the two organisms.

As noted, whole genome tiling arrays can also perform some of these measurements (such as detecting noncoding RNA and alternatively spliced transcripts). Nonetheless, RNA-Seq is more sensitive and gives more quantitative data in these cases. Conversely, RNA-Seq does have the disadvantages of being relatively expensive and requiring sophisticated data analysis.

RNA-Seq can be used to compare gene expression of the same cells under different conditions. Alternatively, gene expression of cells from different tissues of higher organisms may be compared. The example shown in Figure 8.28 shows the comparison of placenta versus several other body tissues. Nearly 300 genes showed significantly greater expression in placental tissue, and several transcriptional regulators were identified that were probably involved in this elevated expression.

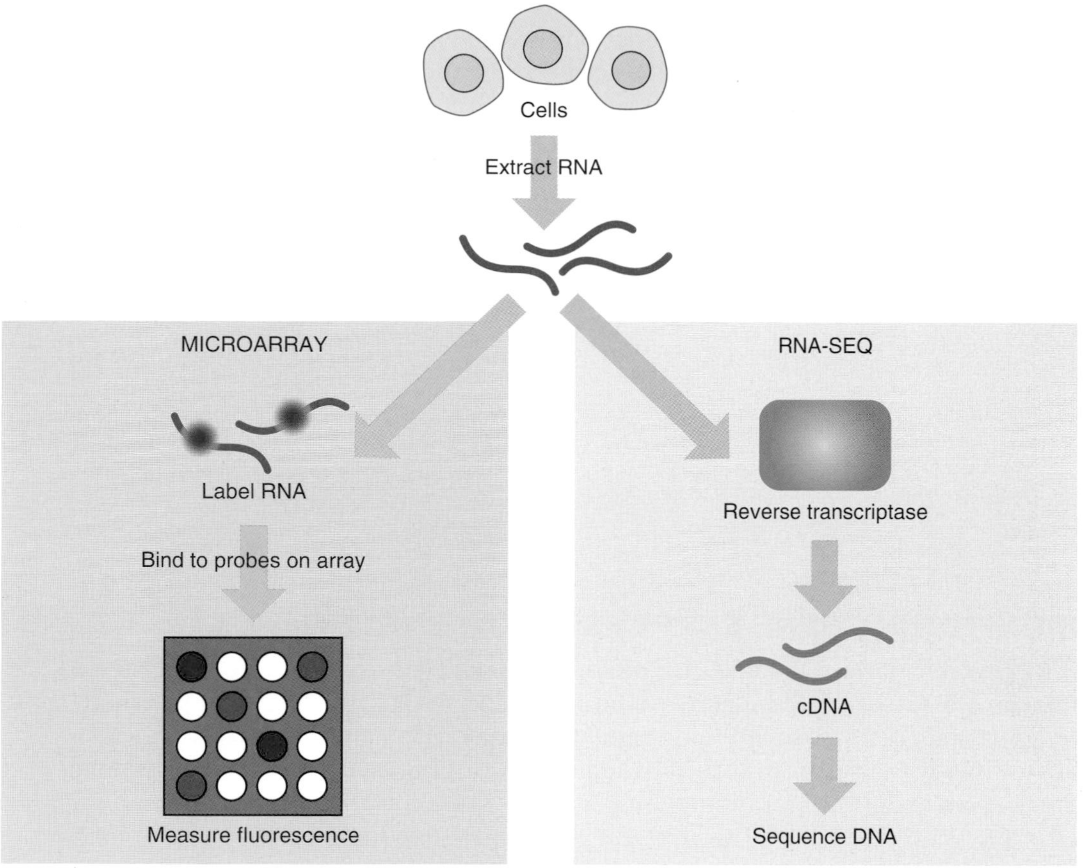

FIGURE 8.27 Microarray versus RNA-Seq
The microarray approach relies on using probes to bind labeled RNA. The level of expression is deduced from the fluorescent signal due to the hybridization of RNA with probe. In contrast, RNA-Seq is independent of probes and relies on directly counting the number of copies of sequenced cDNA derived from the RNA.

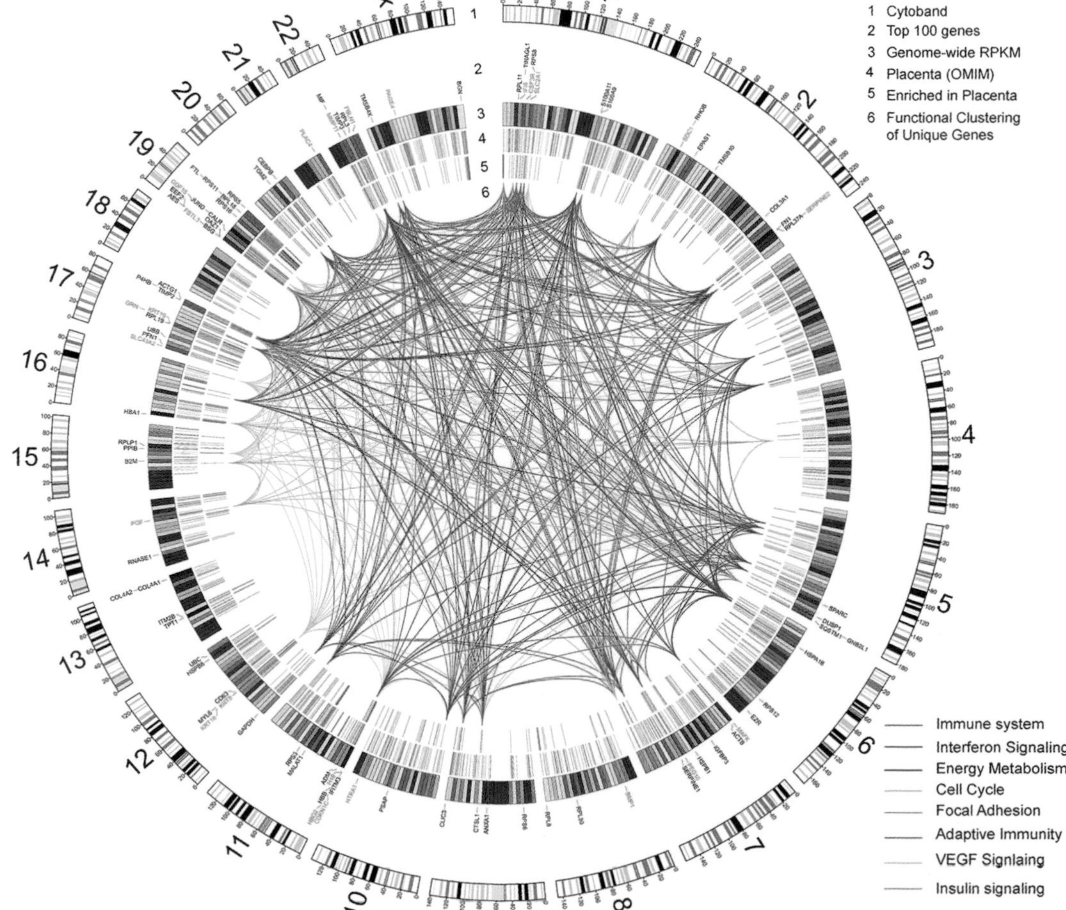

FIGURE 8.28 Placental Transcriptome by RNA-Seq
The transcriptome of the human placenta was compared to that of several other tissues. This Circos diagram shows RNA-seq data for the 23 chromosome pairs. Track 1: Chromosomes with bands. Track 2: Location of top 100 highly expressed placental genes in placenta. Track 3: Average RPKM values (reads per Kb per million) summarized over 6 MB regions. Track 4: Locations of placenta-related genes in OMIM database. Track 5: Genes specifically expressed in placenta (3-fold over 7 other tissues). Track 6: Functional clustering of placental genes. Used with permission from Saben J, Zhong Y, McKelvey S, Dajani NK, Andres A, Badger TM, et al. (2013). A comprehensive analysis of the human placenta transcriptome. *Placenta*, **35(2),** 125–131.

RNA-Seq has many clinical applications. In cancer research, gene fusions occur due to chromosomal rearrangements. RNA-Seq can reveal if the fused genes are expressed into mRNA and estimate the relative abundance of the fusion product. The technique can also identify expressed single nucleotide polymorphisms (SNPs). This type of information can identify the genes responsible for a particular disease by comparing the expression of SNPs from affected individuals and their healthy family members. Restricting comparison to expressed sequences eliminates any irrelevant SNPs found in the DNA, since many SNPs are not in expressed areas of the genome. RNA-Seq can also identify post-translational editing of mRNA that is not evident from looking at just the DNA sequence. This can suggest a new function for a gene.

New generation sequencing technologies have allowed the analysis of RNA amounts by the RNA-Seq approach. This technique can be used to measure global gene expression. In addition, RNA-Seq can be used to monitor noncoding RNA and novel splice variants.

MONITORING GENE EXPRESSION OF SINGLE GENES

Although microarrays and RNA-Seq provide a global view of gene expression, they do not provide extensive information on individual genes. Further details for specific genes of interest are often obtained using **reporter genes**. These are genes whose products are unusually easy to assay, thus allowing the investigator to carry out a large number of analyses on organisms grown under many conditions. In this approach, the gene of interest is physically linked to a reporter gene, creating a **gene fusion**. The regulatory region of the gene of interest is isolated first. This segment is normally found upstream of the gene of interest and includes sites for transcription factors to bind, plus various enhancer elements. The coding sequence of the gene of interest is replaced with the reporter gene so that the regulatory elements now control the reporter gene rather than the original gene of interest.

Reporter genes often encode enzymes whose activity is easy to assay. One of the most widely used reporters is the ***lacZ* gene** from *E. coli*, which encodes the enzyme **β-galactosidase** (Fig. 8.29). This enzyme splits disaccharide sugar molecules into their monomers but also cleaves various artificial substrates. When the substrate ONPG is cleaved, one of the cleavage products forms a visible yellow dye. When X-Gal is cleaved by β-galactosidase, one of the products reacts with oxygen to form a blue dye.

The ***phoA* gene** is another reporter gene that encodes **alkaline phosphatase**, which removes phosphate groups from many different substrates (Fig. 8.30). Artificial substrates are designed so that when the phosphate is removed, they either change color or fluoresce.

Another popular reporter gene is **luciferase**, which emits a pulse of visible light when the correct substrate, **luciferin**, is supplied (Fig. 8.31). Luciferase is an enzyme encoded by the ***lux* gene** in bacteria or the ***luc* gene** in fireflies. The two luciferases are not related and have different enzyme mechanisms. Both genes work well as reporter genes and have been cloned onto vectors that work in a variety of different organisms. Detecting the light emitted by luciferase is difficult because of its low levels and requires special equipment such as a luminometer or scintillation counter.

Another extremely popular reporter protein is **green fluorescent protein**, or **GFP**, which is not an enzyme (Fig. 8.32). This protein has natural fluorescence that does not require any cofactors or substrates. Better still, the fluorescence is active in living tissues, so that when the protein is expressed, the organism gains a green fluorescence. This is

CH2OH CH2OH HO O H O H O H H OH H H OH H OH H H OH H OH

GALACTOSE β(1,4) GLUCOSE
= LACTOSE

β-galactosidase

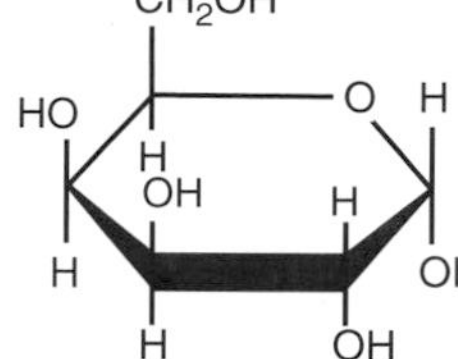

D-GALACTOSE

CH2OH H O H H OH H HO OH H OH

D-GLUCOSE

I

CH2OH NO2 HO O O H OH H H H H OH

o-NITROPHENYL GALACTOSIDE
= ONPG

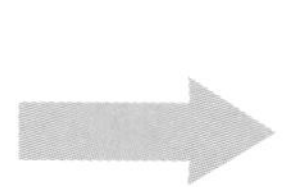

β-galactosidase

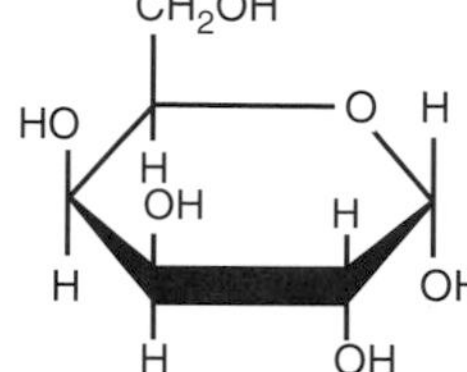

D-GALACTOSE

OH NO2

o-NITROPHENOL
bright yellow

II

CH2OH Cl Br HO O O H OH H N H H H H OH

5-BROMO-4-CHLORO-3-
INDOLYL GALACTOSIDE
= X-GAL

β-galactosidase

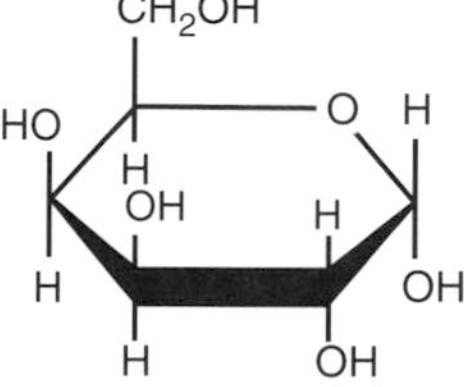

D-GALACTOSE

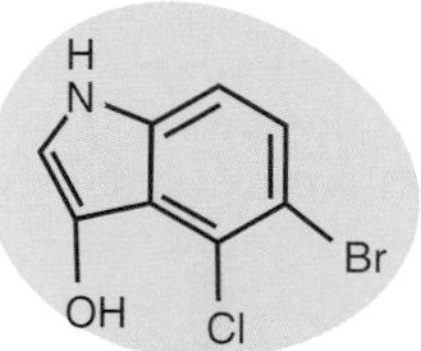

5-BROMO-4-CHLORO-
3-INDOXYL
unstable

SPONTANEOUSLY
REACTS WITH
OXYGEN IN AIR

INDIGO-TYPE DYE
dark blue and insoluble

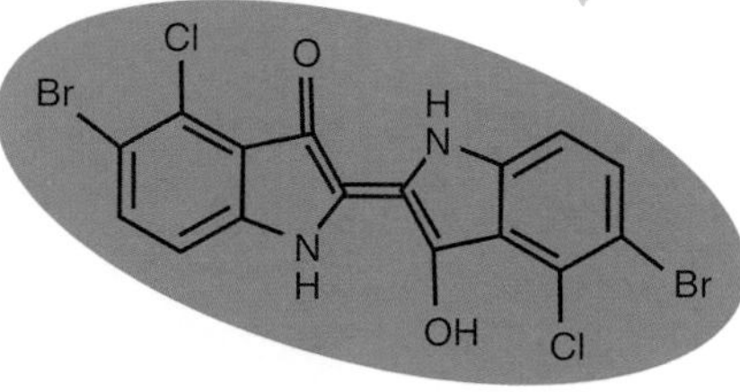

III

FIGURE 8.29 β-Galactosidase Has Multiple Substrates

The enzyme β-galactosidase normally cleaves lactose into two monosaccharides: glucose and galactose. β-Galactosidase also cleaves artificial substrates, such as ONPG and X-Gal, releasing groups that form visible dyes. ONPG releases the bright yellow substance *o*-nitrophenol, whereas X-Gal releases an unstable group that reacts with oxygen to form a blue indigo dye.

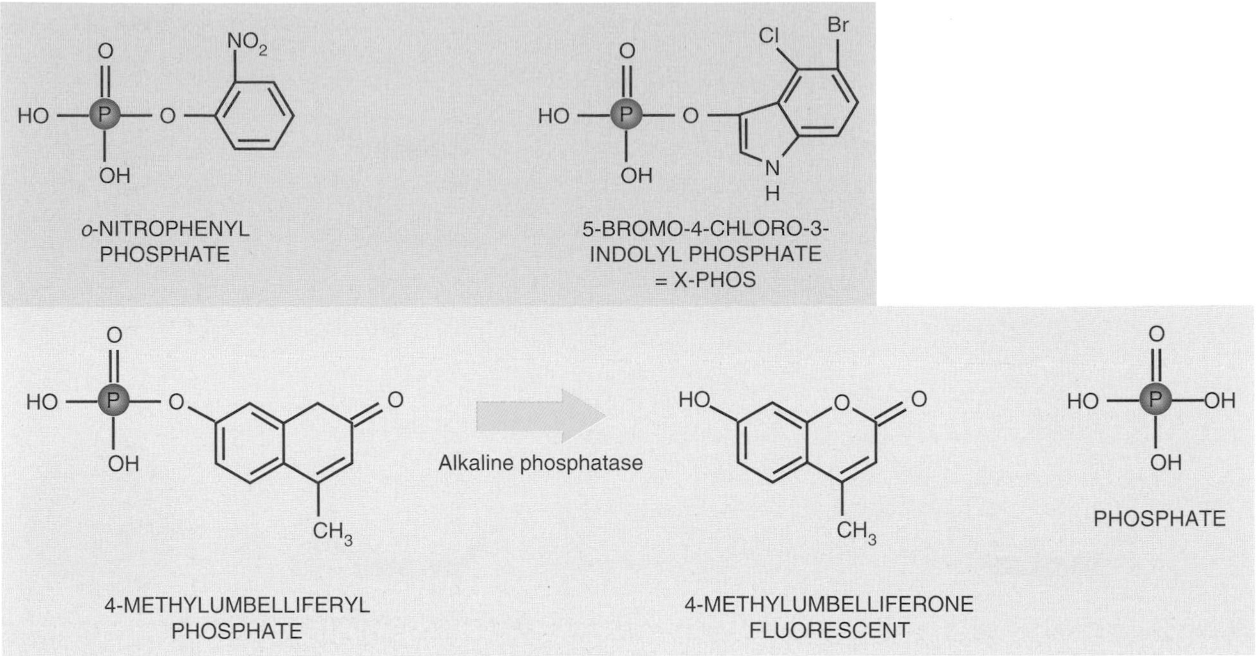

o-NITROPHENYL PHOSPHATE

5-BROMO-4-CHLORO-3-INDOLYL PHOSPHATE = X-PHOS

Alkaline phosphatase

4-METHYLUMBELLIFERYL PHOSPHATE

4-METHYLUMBELLIFERONE FLUORESCENT

PHOSPHATE

FIGURE 8.30 Substrates Used by Alkaline Phosphatase
Alkaline phosphatase removes phosphate groups from various substrates. When the phosphate group is removed from *o*-nitrophenyl phosphate, a yellow dye is released. When the phosphate is removed from X-Phos, further reaction with oxygen produces an insoluble blue dye. When the phosphate is removed from 4-methylumbelliferyl phosphate, this releases a fluorescent molecule.

FIGURE 8.31 The Luciferase Reaction Emits Light from Luciferin
Luciferase from bacteria uses a long-chain aldehyde, oxygen, and the reduced form of the cofactor FMN (flavin mononucleotide) as its luciferin. Firefly luciferase uses ATP, oxygen, and firefly luciferin to produce light.

Bacterial Luciferase:

$FMNH_2 + O_2 + R\text{-}CHO \rightarrow R\text{-}COOH + FMN + H_2O + \text{light}$

Firefly Luciferase:

$\text{Luciferin} + O_2 + ATP \rightarrow \text{oxidized luciferin} + CO_2 + H_2O + AMP + PPi + \text{light}$

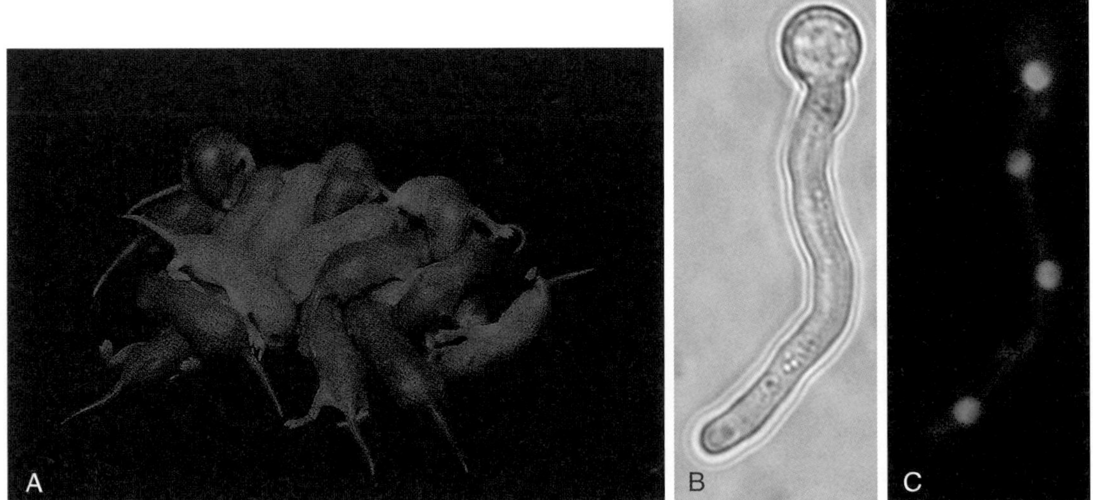

FIGURE 8.32 Transgenic Organisms with Green Fluorescent Protein
The gene for GFP has been integrated into the genome of animals, plants, and fungi. After exposure to long-wavelength UV, the organisms emit green light. (A) Transgenic mice with GFP among normal mice from the same litter. The *gfp* gene was injected into fertilized egg cells to create these mice. GFP is produced in all cells and tissues except the hair. Credit: Eye of Science, Photo Researchers, Inc.(B) Phase contrast and (C) fluorescent emission of germlings of the fungus *Aspergillus nidulans*. Original GFP was used to label the mitochondria and a red GFP variant (DsRed) for the nucleus. From Toews MW, et al. (2004). Establishment of mRFP1 as a fluorescent marker in Aspergillus nidulans and construction of expression vectors for high-throughput protein tagging using recombination in vitro (GATEWAY). *Curr Genet* **45,** 383–389.

especially noticeable when the organism is transparent, like zebrafish or the worm *Caenorhabditis elegans*. GFP is excited by long-wavelength UV light of 395 nm and then emits light at the green wavelength of 510 nm. The original protein is from the jellyfish *Aequorea victoria* and is encoded by the gene *gfp*. Many new variants of GFP have been developed that emit light at different wavelengths, including red, blue, and yellow. The main advantage of using GFP as a reporter is the ability to see expression in living tissues.

Other techniques are useful to confirm or extend gene expression data. Differential display PCR (see Chapter 4) is useful to compare mRNA expression patterns from different tissue samples or experimental conditions. Finally, Northern blot (see Chapter 3) analysis can monitor expression levels of mRNA that vary in different experimental conditions. Gene expression data from various sources can be compiled into regulatory networks where the different gene products (RNA and protein) work together to create an intact organism during development (see Box 8.1).

Fusing regulatory sequences from an individual gene of interest to a reporter gene allows detailed monitoring of the expression pattern of the gene.

Reporter genes encode enzymes such as β-galactosidase, alkaline phosphatase, and luciferase that cleave their substrates to form a visible dye or light. Green fluorescent protein has luminescent properties that allow it to absorb one wavelength of light and emit a longer wavelength.

EPIGENETICS AND EPIGENOMICS

Some phenotypic changes can be inherited despite no accompanying alterations in the DNA base sequence. In the early days of genetics, such events were regarded as exceptions to the laws of Mendelian genetics and often ignored as awkward. Today, we know that inherited changes in gene expression are responsible for such effects, and the phenomenon is referred to as **epigenetic inheritance**. Epigenetics is perhaps best viewed as an "extra" level of inheritance, superimposed on top of the DNA sequence. The **epigenome** is the total number of possible epigenetic changes that can be imposed on any particular genome.

Most epigenetic events are indeed due to changes to the DNA, but to alterations that do not change the base sequence. The simplest and most common examples result from methylation of the DNA. Clearly, this is a chemical alteration to DNA that does not change the base sequence. In higher organisms, another common epigenetic mechanism is the chemical modification of the histone proteins around which eukaryotic DNA is wound. Both of these alterations can greatly affect gene expression.

Epigenetics is not always due to DNA modifications. Both RNA and protein-based mechanisms may occur. RNA interference (RNAi) is discussed in Chapter 5. In some organisms, such as nematode worms and plants, the RNAi response can be inherited as a result of the amplification and persistence of the short-interfering RNA (siRNA) molecules that trigger the response. This therefore counts as RNA-based epigenetics. Protein-based epigenetics is seen in the prions of yeast, discussed in Chapter 21. In this case, regulatory proteins with a changed conformation are inherited and are ultimately responsible for environmental adaptation via altered gene expression.

Such alterations are not true epigenetic effects unless the altered gene expression state is inherited by another generation of cells. For single-celled organisms, this is unambiguous, but in multicellular organisms, epigenetic inheritance may occur at two levels: between cells within the same organism or across generations via the gametes and sexual reproduction. Unfortunately, now that epigenetics and epigenomics have become fashionable, these terms are often wrongly used to include cases in which DNA methylation, or modification of the histone proteins, causes changes in gene expression, even when they are not passed on between generations.

Examples of epigenetics in bacteria due to methylation of DNA that persist between generations have been known for some time, but have rarely been discussed from the viewpoint of epigenetics. They include the methylation of the genome DNA by the modifying enzyme

Box 8.1 Endomesoderm Specification in Sea Urchin Embryos

Using a variety of techniques together can elucidate the network of gene regulation controlling the formation of an entire embryo. In the sea urchin *Strongylocentrotus purpuratus,* the embryo undergoes a specific set of spatial and temporal events that control development of the endomesoderm. The precursor cells of the blastula start the process and continue to develop and divide into the adult sea urchin. Control of development is due both to altering gene expression and varying protein–protein interactions. Perturbing the function of these genes with various techniques, such as morpholino antisense mRNA, mRNA overexpression, and two-hybrid analysis in the sea urchin, together with methods to confirm location such as whole-mount *in situ* hybridization, has allowed a network of gene functions to be proposed (Fig. A). Arrows show each gene that exerts its influence on other genes or proteins. Such networks can be constructed and further tested to refine the model. The latest version of the endomesodermal network is at http://sugp.caltech.edu/endomes/.

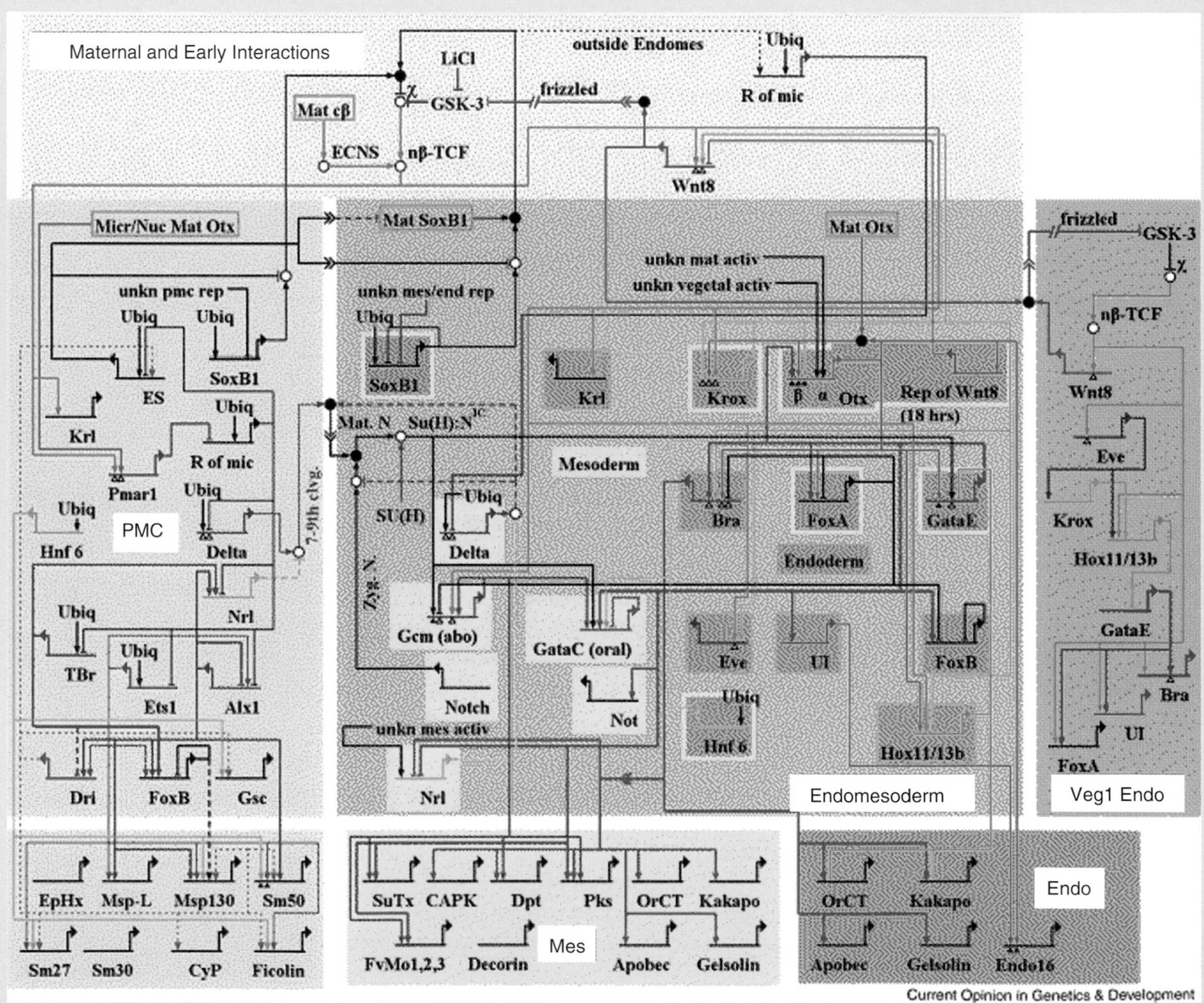

FIGURE A Genome View of Endomesodermal Gene Regulatory Network

The gene regulatory network is divided into spatial domains. Each gene is depicted as a short horizontal line from which extends a bent arrow indicating transcription. Genes are indicated by the names of the proteins they encode. From Oliveri P, Davidson EH (2004). Gene regulatory network controlling embryonic specification in the sea urchin. *Curr Opin Genet Dev* **14,** 351–360. Reprinted with permission.

of restriction/modification systems (see Chapter 3) and regulation of phase variation in those cases in which methylation of DNA is responsible. Phase variation is the random and reversible switching of phenotypes between alternative states: "on" and "off." It is typically seen with surface components of bacteria such as pili, flagella, and outer membrane proteins and can have major effects on bacterial lifestyle and infectivity. Clearly, changing proteins that are accessible on the cell surface can protect against detection by the immune system. Moreover, attachment to or invasion of host cells depends on which bacterial appendages are present.

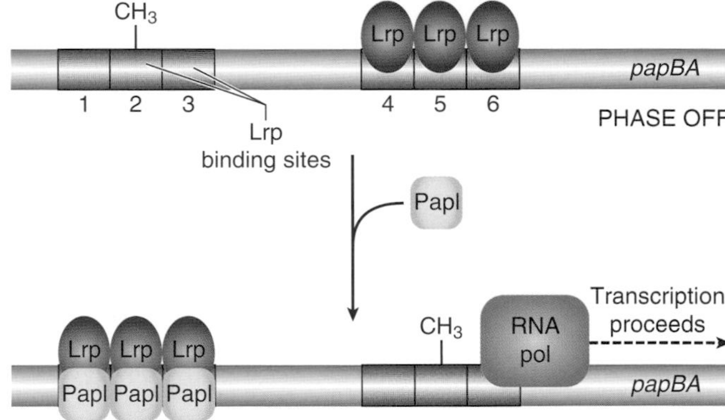

FIGURE 8.33 Phase Variation of the PapAB Pilus

The promoter region of the *papAB* operon found in some pathogenic *E. coli* has two clusters of binding sites for the Lrp DNA-binding protein. Each of these clusters may be methylated or nonmethylated. The Lrp protein will only bind to nonmethylated DNA. When Lrp binds close to the promoter (sites 4, 5, and 6), transcription is blocked. Conversion from one form to the other depends on the PapI protein, which helps Lrp bind to the more distal sites (numbers 1, 2, and 3).

For example, the *papBA* operon of uropathogenic *Escherichia coli* encodes the pyelonephritis-associated pilus that allows binding to membranes of host cells in the kidneys. Synthesis of the pilus is thus required for urinary tract infections to occur. Synthesis of the pilus flip-flops between on and off, depending on the methylation state of two sequences in the promoter of the *papBA* operon (Fig. 8.33). This in turn determines whether the two DNA-binding regulatory proteins, Lrp and PapI, bind.

Epigenetic effects are due to inherited changes in gene expression but are not due to changes in the DNA base sequence. Alterations in DNA methylation or histone modification are often responsible for epigenetic effects.

EPIGENOMICS IN HIGHER ORGANISMS

Regulation of gene expression by methylation of DNA is of much greater significance in eukaryotes and is especially important in the development of multicellular organisms. In humans, methylation of the genome typically occurs on the cytosine of CpG motifs. Many such motifs are present in the upstream regulatory regions in front of genes, and in these cases, methylation usually turns off the genes. Methylation differences accumulate between identical twins during their lives and may account for some of their divergence in metabolism and/or behavior. Overall human DNA methylation decreases with age, and this has been speculatively linked to a variety of diseases including cancer.

The term *methylome* denotes the total number of methylated sites on the DNA, whether the methyl groups are inherited (and hence truly epigenetic) or not. Most methyl groups on eukaryotic DNA are on cytosines, especially on CpG sequences. The methylome is therefore analyzed by bisulfite sequencing. Treatment with sodium bisulfite converts nonmethylated cytosine into uracil, but methylated cytosine is not affected. Consequently, sequencing with and without bisulfite treatment followed by comparison of the two sequences allows methylated sites to be identified (Fig. 8.34).

Some major roles of epigenetic regulation in mammals are as follows:

(a) Genome integrity: Nearly half the human genome consists of assorted mobile elements. In higher plants, the proportion is even higher. Although most of these elements are defective, uncontrolled replication or movement of those still active could cause major damage to the genome. In practice, these elements are mostly covered with DNA methylation and histone modifications that prevent expression. These often vary depending on the nature of the element. Thus, DNA transposons typically have high levels of methylation of Lys9 on histone 3, whereas retrotransposons tend to be methylated on Lys20 of histone 4.

(b) X chromosome inactivation: In female mammals, one of the pair of X chromosomes in each cell is inactivated in order to keep gene dosage the same as in males, where there is only

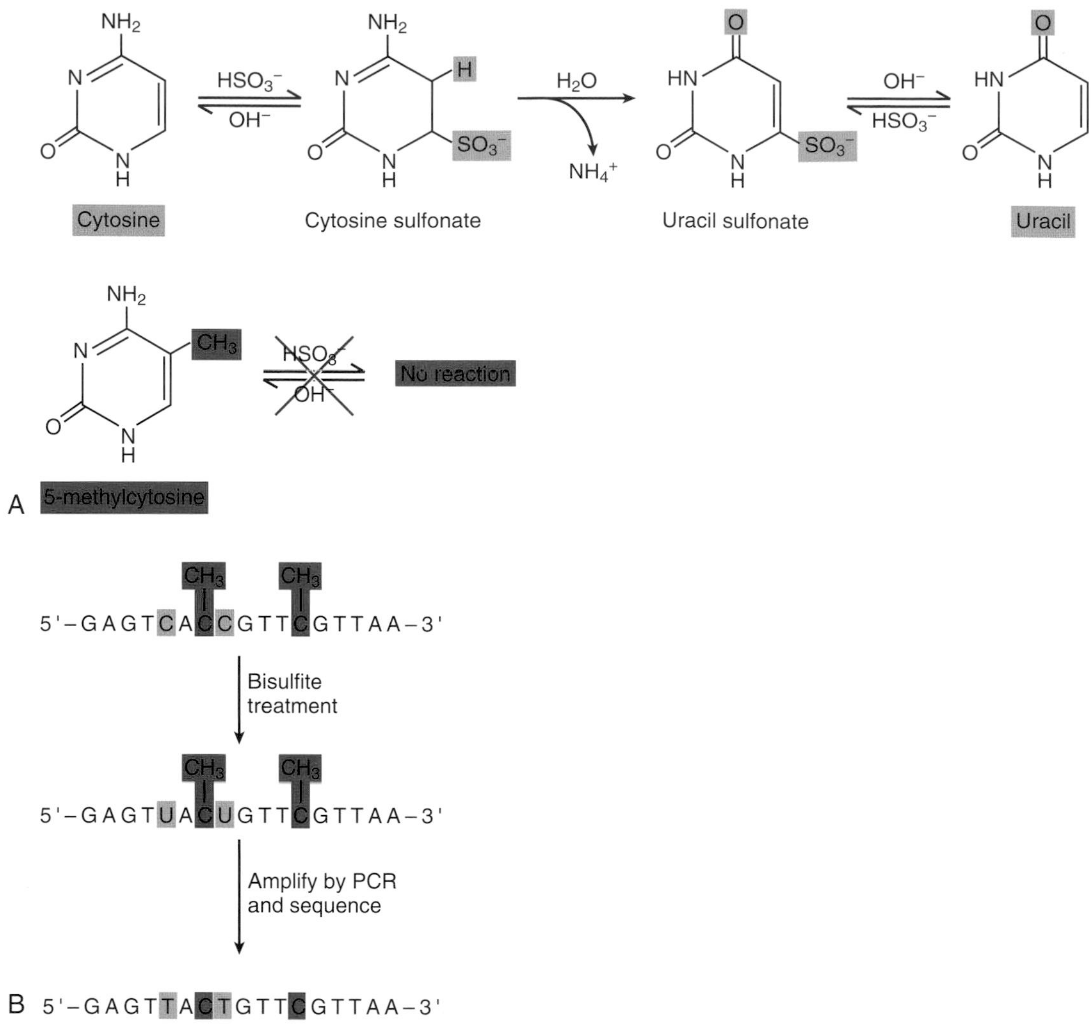

FIGURE 8.34
Bisulfite Sequencing of the Methylome
To discover which cytosines in the genome are methylated, scientists carry out sequencing on both untreated DNA and DNA that has been treated with bisulfite. The bisulfite converts nonmethylated cytosines (green) to uracils (A). During PCR, the uracil is replaced by thymine (B). However, methylated cytosine (red) is protected and remains as cytosine. The two sequences are then compared.

a single X chromosome. This is done by DNA methylation and histone modifications and requires the *Xist* noncoding RNA.

(c) Parental imprinting: As discussed in Chapter 17, in a few genes only one of the pair of alleles is expressed. Which is chosen depends on whether the allele comes from the mother or the father. The allele from the other parent is silenced by methylation of the DNA and histones.

(d) Development and differentiation: The cells of multicellular organisms all share the same genome but perform very different biological roles. A variety of regulatory mechanisms, including epigenetic changes, play a part differentiation. In particular, once cells have reached a final specialized form, they often rely on epigenetic modifications to ensure that all their descendants are of the same type.

Although instances are rare and difficult to analyze, environmental effects can trigger epigenetic changes that affect how genes are expressed in future generations of organisms (not just of cells). In humans, it is hardly surprising that epigenetic changes that occur in the mother may affect the offspring. However, a few surveys have suggested that the grandparents' diet may have effects on gene expression, and hence the health, of children two generations later. Proving this conclusively with humans is not feasible. However, experiments with mice have shown that the fathers' diet can influence the expression of genes involved in metabolism in the offspring, even when all the mothers had the same diet and the new generation never met their fathers.

This makes good sense from an evolutionary perspective. Epigenetics can pre-adapt the new generation to the environment they will probably encounter without the need for permanent alterations in the genome. Untangling the effects of genetics, epigenetics, and environment, especially for complex conditions such as obesity or diabetes, is very difficult and still in its infancy.

In higher animals, epigenetics may involve cell generations within a single organism or distinct generations of individual organisms. A variety of environmental factors can cause epigenetic changes that persist between generations.

Summary

Genomics is the study of the total nucleotide sequence for an organism of interest, including genes, pseudogenes, noncoding regions, and regulatory regions. In human genomics, identifying the sequence of the entire set of chromosomes was a major achievement. The sequence of the human genome was determined by making DNA libraries, sequencing each of the clones, and then compiling the sequences. Physical maps, genetic maps, and computer algorithms were used to arrange the sequences in the correct order. Without the great advances in computing, the Human Genome Project would have taken much longer and cost more money. Data mining of the information has identified many potential protein coding regions, regulatory elements, and different types of repetitive elements. The human genome is predicted to contain about 21,000 genes, plus SINES, LINEs, and tandem repeats such as telomeres, centromeres, and satellite DNA. Computer analysis of such sequence information is called *bioinformatics.*

Genomics has changed many fields of study, including medicine, pharmacology, and evolutionary biology. Understanding and identifying new genes related to diseases is changing the way diseases are treated and diagnosed. Much of this textbook is devoted to these advances. The study of genomics focuses on mutations in the genome, by identifying single nucleotide polymorphisms, methylation patterns, or differences in tandem repeats. Mutations include single nucleotide changes, inversions, deletions, and insertions. Pharmacologists hope to correlate these differences with drug sensitivity, thus preventing adverse drug reactions. In evolutionary biology, physical features have previously been used to determine the relatedness of two organisms. Since the genomes of many organisms have now been sequenced, their DNA sequences can be used to determine relatedness. Over time, mutations accumulate within every genome. The more essential genes change slowly over time, whereas less essential genes incorporate more changes.

Genomics also encompasses gene expression, which was first done on a global scale using genomic DNA microarrays that have either cDNA or synthetic oligonucleotides, linked to a glass slide. Fluorescently labeled mRNA is then added. Where an mRNA hybridizes to the immobilized genomic DNA, that region will fluoresce. The amount of fluorescence correlates to the amount of mRNA and hence gene expression. More recently, arrays have been partly superseded for transcriptome analysis by the use of RNA-Seq. This technique depends on sequencing all the RNA extracted from a cell (after conversion to cDNA) and counting how many copies of each RNA molecule are present. In single gene analysis, specific regulatory regions defined by genomics are linked to a variety of different reporter genes, including β-galactosidase, alkaline phosphatase, luciferase, and green fluorescent protein. These studies replace the gene of interest with the reporter gene, which is much easier to assay and thus allow analysis of the regulatory region under study.

End-of-Chapter Questions

1. Which of the following is utilized in genomic research?
   - **a.** microsatellite polymorphism
   - **b.** restriction fragment length polymorphism
   - **c.** single nucleotide polymorphism
   - **d.** variable number tandem repeat
   - **e.** all of the above
2. What is contig mapping?
   - **a.** Determination of regions that overlap from one clone to the next in a library
   - **b.** The distance in base pairs between two markers
   - **c.** The use of landmarks in the genes to put together sequencing data
   - **d.** The relative order of specific markers in a genome
   - **e.** The mapping that determines if a library sequence is from one continuous gene or two gene segments cloned into one vector
3. Which method was used to sequence the human genome?
   - **a.** cytogenetic mapping
   - **b.** shotgun sequencing
   - **c.** chromosome walking
   - **d.** radiation hybrid mapping
   - **e.** All of the above were used in combination to complete the project.
4. Which organism has the most genes?
   - **a.** *H. sapiens*
   - **b.** *D. melanogaster*
   - **c.** *O. sativa*
   - **d.** *P. trichocarpa*
   - **e.** *A. thaliana*
5. What is a gene?
   - **a.** a segment of DNA that encodes a protein
   - **b.** a segment of DNA that encodes non-translated RNA
   - **c.** sequences of DNA that are not transcribed
   - **d.** a segment of DNA that is transcribed
   - **e.** all of the above
6. Which of the following is considered a field of study related to bioinformatics?
   - **a.** proteomics
   - **b.** computational biology
   - **c.** genomics
   - **d.** cheminformatics
   - **e.** all of the above
7. How is data mining useful to biotechnology research?
   - **a.** It allows researchers to determine sequence similarity, which usually translates into functional similarity.
   - **b.** Data mining allows researchers to use computers to study, sort, and compile the vast amounts of raw data generated through bioinformatics.

c. Data mining is the act of gathering the raw data from research projects such as sequencing into one central location.
d. Data mining usually provides too much information, which only slows down the research project and is therefore not very useful.
e. None of the above

8. Which type of mutation is the most common?
   a. insertion of one or more bases
   b. depletion of one of more bases
   c. base substitutions
   d. inversion of DNA segments
   e. duplications of DNA segments

9. Which of the following statements about mutations is not true?
   a. Mutations occur in all organisms at the same rate.
   b. DNA polymerase never produces mutations during replication because of the proofreading ability of this enzyme.
   c. Mutations often occur at methylated cytosine residues.
   d. Mutations such as duplications or deletions occur due to repetitive sequences causing strand slippage.
   e. When comparing mutation rates to coding capacity, mutation rates are usually the same for most organisms, which suggests a mechanism to control the rate.

10. Which one of the following is often used to establish family trees for organisms because it is present in all organisms and does not accumulate mutations quickly?
   a. rRNA
   b. fibrinopeptides
   c. hemoglobin
   d. chloroplasts
   e. mitochondrial DNA

11. Which of the following statements about DNA microarrays is not correct?
   a. Fluorescently labeled mRNA from the organism hybridizes to the DNA on the glass slide.
   b. DNA microarrays contain thousands of DNA segments on a support, such as a glass slide.
   c. Hybridization to a DNA microarray can only occur once.
   d. The amount of fluorescence correlates with the amount of mRNA in the sample.
   e. The data obtained from DNA microarrays represents a global view of gene expression, even under particular growth conditions.

12. What is the term used to describe the process of synthesizing oligonucleotides directly on the glass slide?
   a. photosynthesis
   b. photolithography
   c. light-activated oligosynthesis
   d. on-chip oligosynthesis
   e. protected oligosynthesis

(*Continued*)

HPLC is very adaptable because of the availability of different types of stationary phase materials. **Size exclusion chromatography** columns contain porous beads that separate mixtures of proteins by size. Large molecules do not enter the pores of the beads and travel through the column quickly, while smaller compounds are delayed. Many different pore sizes are available for mixtures of different size ranges. **Reverse-phase HPLC** uses columns packed with hydrophobic alkyl chains attached to silica-based material. The column binds and delays hydrophobic molecules while hydrophilic molecules elute faster.

Ion-exchange HPLC uses a stationary phase with charged functional groups that bind oppositely charged molecules in the sample. Such molecules remain in the column after the sample has passed through. To elute them, the mobile phase is changed. For example, if the pH of the mobile phase is adjusted, the net charge on many proteins will be altered, and they will be released from the column. Other stationary phases form hydrogen bonds with the analyte and separate based on overall polarity. For **affinity** HPLC, the stationary phase contains a molecule that specifically binds the target protein, for example, an antibody. When a mixture passes over the stationary phase, only the target protein is bound, and other proteins pass through. Changing the mobile phase so as to disrupt the interaction releases the protein of interest.

As molecules exit the column, they must be detected. Many different detectors exist. These usually respond by plotting peaks as molecules pass by. **Refractive index detectors** monitor whether the exiting mobile phase refracts any light by shining a light beam through it. Compounds present in exiting fractions scatter light, and a photo-detector records this as a positive signal. The amount of scatter affects the height of the peak, and the length of time of the scatter determines the width of the peak. **Ultraviolet detectors** have a UV light source and a detector to determine when the passing mobile phase absorbs the UV light. Such detectors may monitor one or more wavelengths depending on the substance being examined. **Fluorescence detectors** detect compounds that fluoresce, that is, absorb and re-emit light at different wavelengths; **radiochemical detectors** detect radioactively labeled compounds; and **electrochemical detectors** measure compounds that undergo oxidation or reduction reactions. An approach that is increasingly used for proteomics is detection by mass spectrometry. This combination allows proteins separated by HPLC to be fed directly into a mass spectrometer for identification (see later discussion for mass spectrometry).

A critical aspect of HPLC is getting a good separation between the different proteins in the sample, that is, good **resolution**. Each peak that comes off the column should be symmetrical and as narrow as possible. For high resolution, many experimental conditions can be adjusted. The most obvious is changing the stationary phase. Sometimes just changing the particle size of the stationary phase improves separation. An alternative is to adjust the composition of the mobile phase. Temperature also affects many separations and may need to be controlled.

An analyte is a sample of molecules in a liquid. This is the mobile phase that moves through a chromatography column.

The stationary phase is the actual column packed with different materials. The properties of these materials determine what proteins are retained in the column and what proteins are expelled from the column.

Different detectors are at the end of the column to determine when the protein exits. These detectors can identify proteins by refractive index, ultraviolet absorption, fluorescence, radioactivity, or electrochemistry.

DIGESTION OF PROTEINS BY PROTEASES

Proteases (also known as **proteinases** or **peptidases**) hydrolyze the peptide bond between amino acid residues in a polypeptide chain. Proteases may be specific and limited to one or more sites within a protein, or they may be nonspecific, digesting proteins into individual amino acids. The ability to digest a protein at specific points is critical to mass spectrometry (see later discussion) and many other protein experiments. For example, proteases are used to cleave fusion proteins or remove single amino acids for protein sequencing.

Proteases are found in all organisms and are involved in all areas of metabolism. During programmed cell death, proteases digest cellular components for recycling (see Chapter 20). Plants deploy proteases to protect themselves from fungal or bacterial invaders. In the biotechnology industry, proteases have many uses. For example, they are included as additives in detergents to digest proteins in ketchup, blood, or grass stains.

Proteases are classified by three criteria: the reaction catalyzed, the chemical nature of the catalytic site, and their evolutionary relationships. **Endopeptidases** cleave the target protein internally. **Exopeptidases** remove single amino acids from either the amino- or carboxy-terminal ends of a protein. Exopeptidases are divided into **carboxypeptidases** or **aminopeptidases** depending on whether they digest proteins from the carboxy- or amino-terminus, respectively. Proteases are also divided based on their catalytic site architecture. **Serine proteases** have a serine in their active site that covalently attaches to one of the protein fragments as an enzymatic intermediate (Fig. 9.7). This class includes the chymotrypsin family (chymotrypsin, trypsin, and elastase) and the subtilisin family. **Cysteine proteases** have a similar mechanism but use cysteine rather than serine. They include the plant proteases (papain, from papaya, and bromelain, from pineapple) as well as mammalian proteases such as calpains. **Aspartic proteases** have two essential aspartic acid residues that are close together in the active site although far apart in the protein sequence. This family includes the digestive enzymes pepsin and chymosin. **Metalloproteases** use metal ion cofactors to facilitate protein digestion and include thermolysin. Finally, the fifth class of proteases, **threonine proteases**, has an active-site threonine.

Researchers who study proteases have not escaped the "-omics" culture, and sometimes the term **degradome** is used for the complete set of proteases expressed at one specific time by a cell, tissue, or organism. Understanding the degradome of an organism relies on much the same techniques as used for proteomics, although modifications are made to look only at proteases rather than at the entire proteome. For example, protease chips contain only antibodies to known proteases rather than to all proteins (see later discussion).

Many different proteases exist in nature, and they are useful for a variety of applications in biotechnology. Specific proteases recognize specific amino acids and cut the peptide bond in a specific location.

MASS SPECTROMETRY FOR PROTEIN IDENTIFICATION

Mass spectrometry is a technique to determine the mass of molecules. In mass spectrometry, a molecule is fragmented into different ions whose masses are accurately measured. The ions generate a spectrum of unique peaks which therefore determines the identity of the original molecule. The molecule 2-pentanol is used as an example in Figure 9.8. Here, an electron beam aimed at the sample fragments the 2-pentanol into different ions: the molecular ion gains an electron, m-1 loses hydrogen from the alcohol group, m-15 loses a methyl group, m-17 loses the alcohol group, and m/e = 45 loses the alkyl chain.

These ions are accelerated into a vacuum tube by an ion-accelerating array. The ions travel through the tube at different speeds due to a magnetic field that causes the ions to follow a curved path within the tube (Fig. 9.9). The curves eliminate ions that are too small or too big. Ions that are too small gain so much momentum from the magnetic field that they collide with the wall. Those that are too big are not deflected by the magnetic field and also collide with the wall. Ions in the right size range are deflected by the magnetic field around

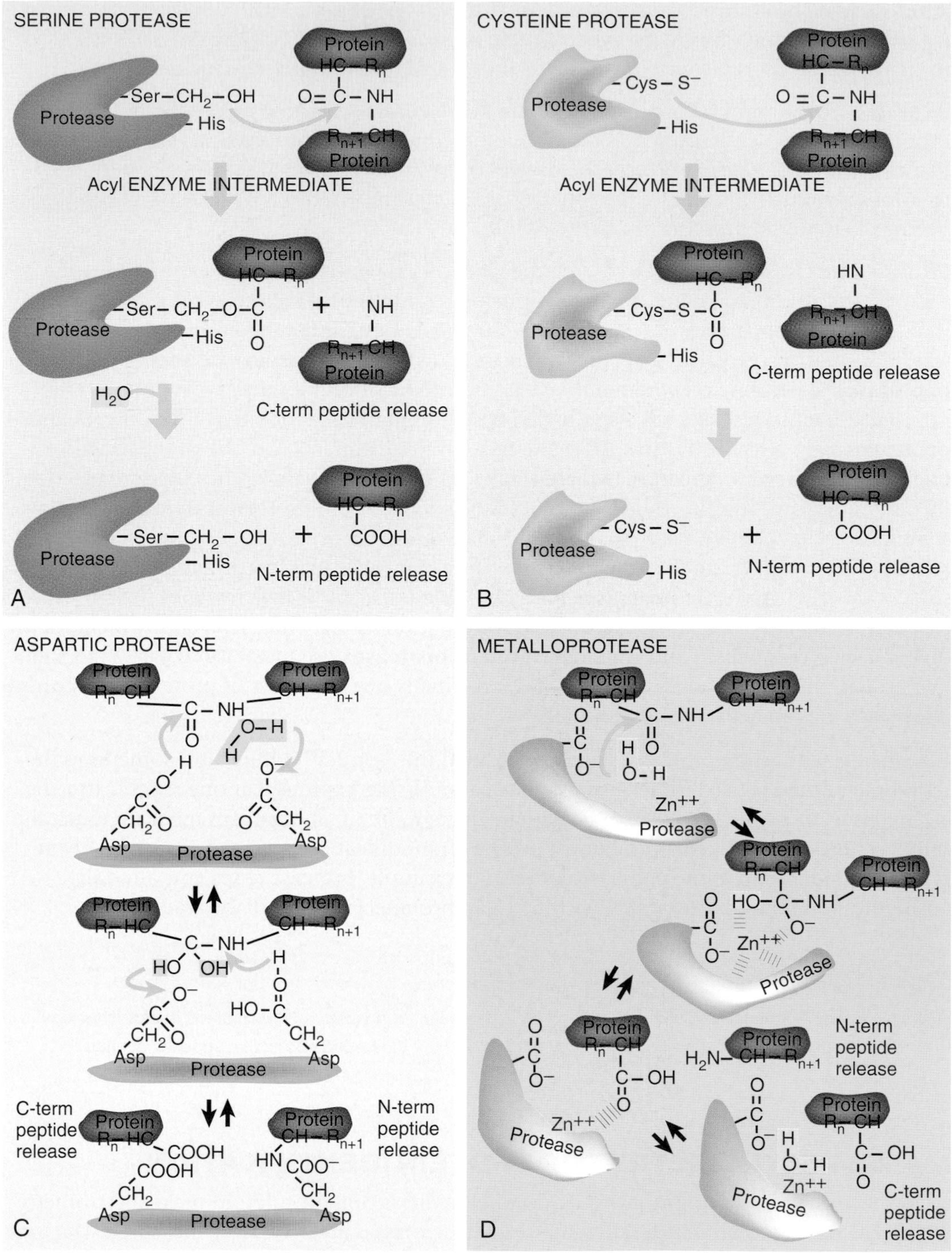

FIGURE 9.7 Mechanism of the Four Classes of Endopeptidase
(A) Serine proteases cleave the peptide bond of a protein by the formation of an acyl enzyme intermediate. The active-site serine forms a temporary bond with the amino-terminal half of the digested protein. (B) Cysteine proteases are similar to serine proteases but use cysteine at the active site. (C) Aspartic proteases have two active-site aspartic acids that coordinate hydrolysis of the peptide bond in the target protein. (D) Metalloproteases hydrolyze a target protein using a metal ion such as Zn^{2+}.

both curves to hit the collector, where they are recorded as peaks in the mass spectrum. The **base peak** is the most intense, and other peaks are measured relative to the base peak. The time ions take from the accelerator to the collector correlates directly to the size of the ion. Each peak is plotted based on the mass to charge ratio (m/z). The losses of mass (such as m-17 or m-15)

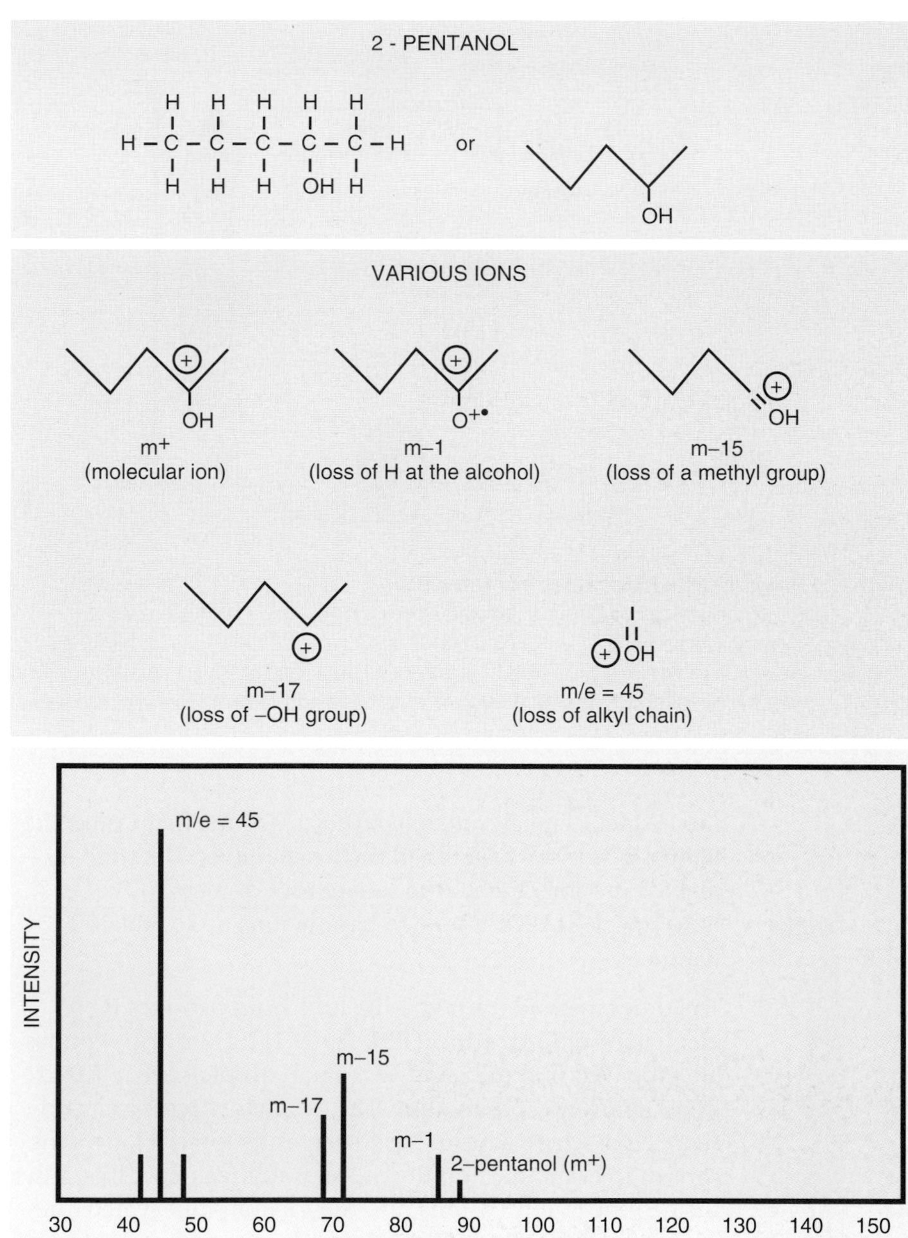

FIGURE 9.8 Basic Mass Spectrometry for 2-Pentanol

Every substance can be fragmented into multiple ions. This example shows the molecular structure of all the ions from 2-pentanol. A mass spectrometer separates these ions by size and graphs the results. The spectrum is always the same for each substance, and so an unknown substance can be identified by comparing its spectrum with a database of known substances.

refer to the loss of specific groups from the parent molecule and are most informative to the structure of the sample molecule because each such group has a characteristic mass.

Until recently, very large molecules such as proteins were beyond the range of mass spectrometry. Two different ionization techniques have been developed that have made proteins manageable. The first technique, which embeds peptides in a solid matrix before ionization, is called **MALDI**, or **matrix-assisted laser desorption-ionization** (Fig. 9.10). Here, the peptides are embedded in a material such as 4-methoxycinnamic acid that absorbs laser light. The matrix absorbs and transfers the laser energy to the peptides, causing them to release different ions. The ions are accelerated through a vacuum tube by a charged grid. At the far end, the **time-of-flight (TOF)** detector records the intensity and calculates the mass. In between is a **flight tube** that is free of electric fields. The ions are accelerated with the

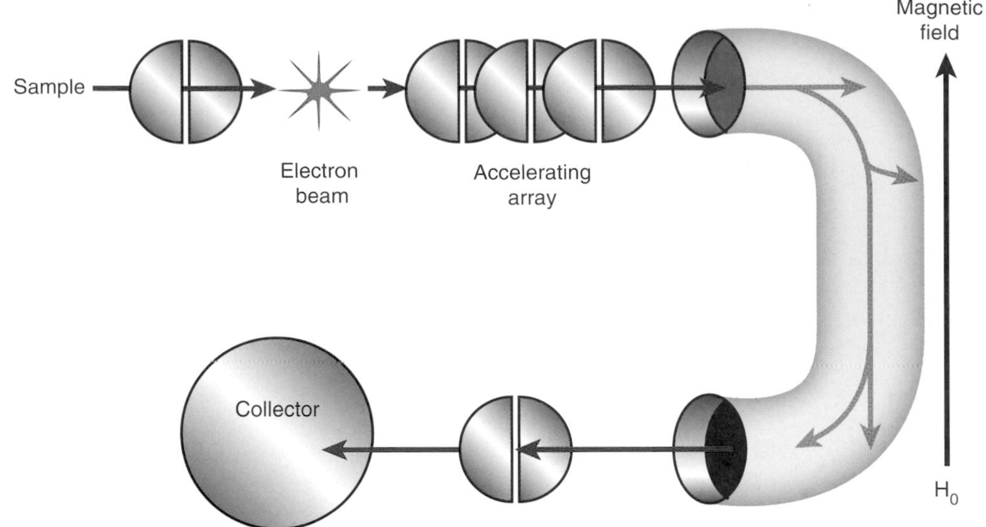

FIGURE 9.9 Schematic Diagram of a Mass Spectrometry Tube
The sample travels through a narrow slit and then passes through a beam of electrons that disrupts it into a mixture of ion fragments. The accelerating array moves the fragments into the C-shaped tube. This is surrounded by a strong magnetic field that prevents ions that are too small or too large from exiting the tube. A collector detects the exiting fragments and measures the time it took for them to travel the tube. The computer then converts the time of travel into size and charge information and plots this as a mass spectrum (not shown).

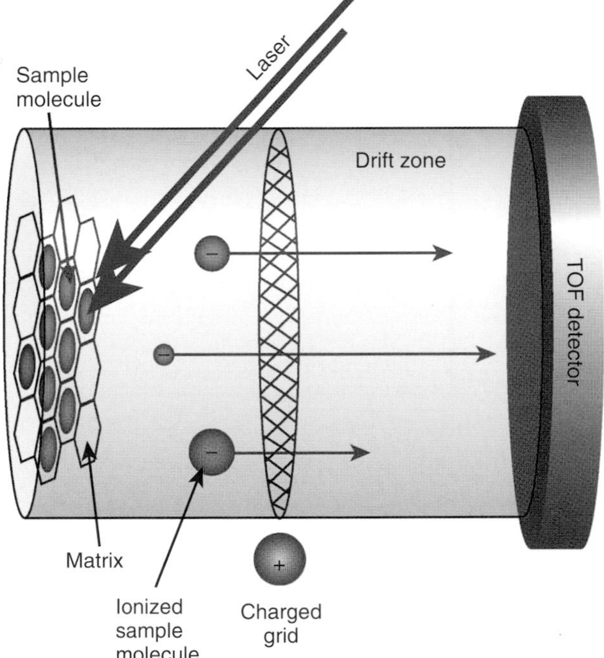

FIGURE 9.10 MALDI/TOF Mass Spectrometer
Mass spectrometry can be used to determine the molecular weight of peptides. The peptides are crystallized in a solid matrix and exposed to a laser, which releases ions from the peptides. These travel along a vacuum tube, passing through a charged grid, which helps separate the ions by size and charge. The time it takes for ions to reach the detector is proportional to the square root of their mass to charge ratio (m/z). The molecular weight of the peptide can be determined from these data.

same kinetic energy, and when they reach the flight tube, the lighter ions move faster than the heavier ions. The time-of-flight is proportional to the square root of mass to charge ratio (m/z). MALDI is able to handle ions up to 100,000 daltons.

The other method for preparing ions from peptides is **electrospray ionization (ESI**; Fig. 9.11). Here, the peptides are dissolved in liquid, and very small droplets are released from a narrow capillary tube. The droplets enter the electrostatic field, where a heated gas, such as hydrogen, causes the solvent to evaporate and the droplets to break up. This causes the peptide to release ions into the vacuum tube, where they are accelerated by the electric field. The detector at the far end varies based on the sample being studied. A TOF detector may be used, as described earlier. Other detectors use quadrupole ion traps or Fourier transform ion cyclotron resonance to determine the mass of the ions. Quadrupole ion traps capture the ions in an electric field. The ions are then ejected into the detector by a second electric field. The electric field controls what size ions can pass to the detector, and varying the field allows different-sized ions to be detected. Combination detectors exist that use both TOF and quadrupole ion traps. The advantage that ESI has over MALDI is that proteins isolated from HPLC (see earlier discussion) require no special preparation and can be used directly. The disadvantage of ESI is that masses of about 5000 are the maximum.

The use of mass spectroscopy will become more prevalent as the methodology improves. For example, **surface-enhanced laser desorption-ionization (SELDI)** mass spectroscopy takes liquid samples and ionizes the peptides that adhere to a treated metal. The technique shows great promise for bodily fluids such as blood and, it is hoped, will help in identifying a particular protein profile for different disease states. Perhaps one day patients will be diagnosed

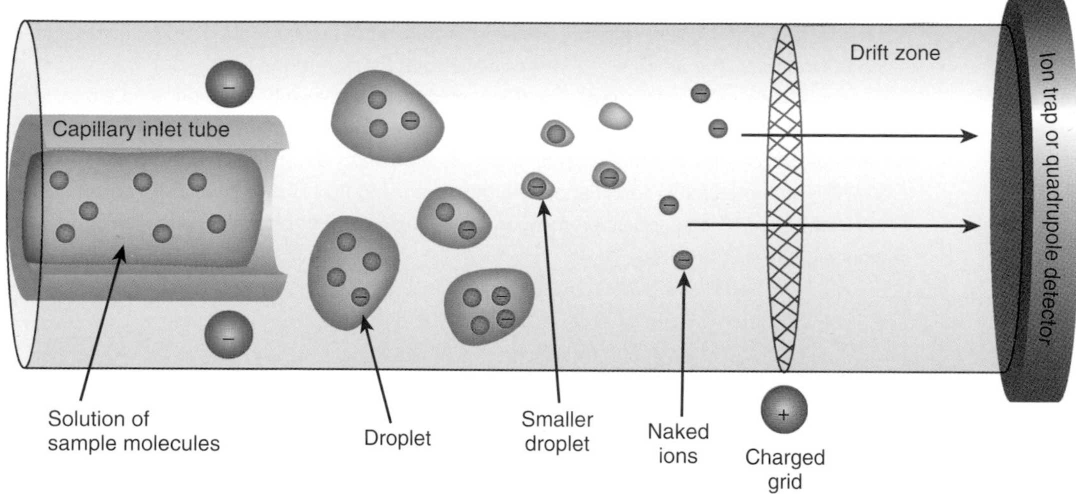

FIGURE 9.11 Electrospray Ionization (ESI) Mass Spectrometer
ESI mass spectrometry uses a liquid sample of the peptide held in a capillary tube. After exposure to a strong electrostatic field, small droplets are released from the end of the capillary tube. A flow of heated gas within the drift zone evaporates the solvent and releases small charged ions. The charged ions vary in size and charge, and the pattern of ions produced is unique to each peptide. The ions are separated by size using a charged grid to either impede or promote the flow toward the detector.

with cancer long before any symptoms are detected. A change in the peptide profile in their blood could denote that cancer cells are forming. MALDI and ESI are sensitive enough to detect changes in proteins due to phosphorylation, glycosylation, and so forth. The technique can identify which amino acid is modified because only one specific ion is altered.

Another improvement for MALDI to characterize protein samples is a technique called **multidimensional protein identification technique (MuDPIT)**. In this method, a more complex mixture of proteins can be fragmented into peptides and then identified by mass spectroscopy. Traditional mass spectroscopy has limited numbers of different proteins that can be identified. Usually, samples with more than 100 proteins are too complex and have to be further fractionated by SDS-PAGE or chromatography. In MuDPIT, a sample with up to 3000 different proteins can be reliably evaluated. The key to the technology is using a **2D LC microcapillary column** to separate the peptide fragments prior to standard mass spectroscopy (Fig. 9.12). This column improves peptide separation over traditional HPLC since there are two different stationary phases packed into a tube with an inner diameter of only 100 microns. The 2D aspect refers to the two stationary phases; the first is a strong cation exchange resin followed by a reverse-phase resin. The peptides are first separated by charge on the cation resin and then by size and hydrophobicity on the reverse-phase column. Next, the proteins are eluted with a combination of two solutions: an ammonium salt gradient that releases the peptides by charge and an organic solvent gradient like acetonitrile that releases the peptides based on their hydrophobicity. As the peptide releases from the column and enters the needle-like exit portal, it is ionized by electrical charge, and the ions are separated using TOF analysis. This technique is powerful because many different proteins can be identified and quantified using one column and one mass spectroscopy experiment.

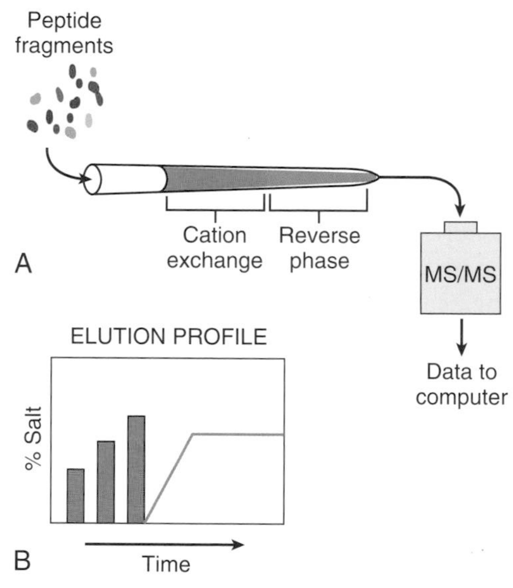

FIGURE 9.12 MuDPIT Uses 2D LC Microcapillary Columns to Separate Peptides Prior to Mass Spectroscopy
(A) Microcapillary column containing two different stationary phases separates proteins for mass spectroscopy. The first part of the column has a strong cation exchange resin, which separates the peptide fragments by charge. The second part of the column has reverse-phase resin, which separates the peptides by size and hydrophobicity.
(B) After the peptides adhere to the column, they are eluted by alternating an increasing ammonium salt concentration with a reverse-phase gradient. The peptides with the lowest charge are eluted first to the reverse-phase portion of the column. Next, an organic solvent gradient releases the lowest charged peptides in order of size and hydrophobicity. The eluted peptides pass through the needle and are ionized for mass spectroscopy.

Mass spectroscopy ionizes a sample into smaller parts and measures the time it takes for these ions to reach the detector. The amount of time correlates with the size of the ion.

Proteins can be ionized for mass spectroscopy after they are embedded either in a matrix for MALDI or in liquid as for ESI. Both techniques can identify the protein by its pattern of fragmentation, and the techniques can identify any modifications such as phosphorylation and glycosylation.

MuDPIT separates more complex protein mixtures using 2D LC microcapillary column separation prior to ionization. The 2D LC column separates by charge, hydrophobicity, and size using two different stationary phases in tandem.

PREPARING PROTEINS FOR MASS SPECTROSCOPY

Determining the sequence of a short peptide is readily achieved using mass spectroscopy techniques (Fig. 9.13). To determine the sequence of a peptide, researchers must obtain a pure sample of the protein either by cutting a spot from a two-dimensional gel or by HPLC purification. First, the proteins are treated with reducing agents to break apart any disulfide bridges. To keep the –SH group from reforming a disulfide bridge, the proteins are also alkylated. The protein is then digested into fragments using a protease such as trypsin, which cuts proteins on the carboxy-terminal side of arginine and lysine. Cutting a protein into peptides helps reduce undesirable characteristics of the entire protein.

For example, membrane proteins are hydrophobic and stick together, and digesting them into peptide fragments destroys this characteristic. Solubility issues can also often be resolved by digesting a protein into peptides. Determining the sequence of these peptides will yield the sequence of the original protein. Usually, the peptide sequence from only one or two fragments is sufficient for identifying the original protein from which they derived.

During mass spectroscopy, the peptide mixture is ionized into multiple fragments (Fig. 9.14). For peptides, common ions include a doubly protonated form $(M + 2H)^{2+}$, where M is the mass of the peptide and H^+ is the mass of a proton. The ion peaks are plotted versus the mass to charge ratio (m/z). For the doubly protonated peptide ion, this would be the mass of the ion divided by 2. For example, if the original peptide was 1232.55 daltons, the double protonated ion would have a mass of 1232.55 daltons + (2 × 1.0073) for each added hydrogen. The peak would appear at 617.2828. (Note: The peak is plotted at the mass to charge ratio. That is, the mass for this ion is 1234.5646, and the charge is +2. The peak appears at m/z, or 1234.5646/2.) When the mass spectrometer separates peptide ions, the first step is to determine the charge state of the ion. Usually, a cluster of peaks occurs for each peptide ion. If peaks are 1 dalton apart, the charge state of the peptide is 1. If the peaks are 0.5 daltons apart, the charge state is 2.

To determine the peptide sequence, researchers use two rounds of mass spectroscopy. This is called **tandem mass spectroscopy** because one ion is produced in the first round of mass spectroscopy; then that ion is fragmented by collision with a gas such as hydrogen, argon, or helium. As before, the ion fragments are separated based on their mass to charge ratio. Each peak usually varies by one amino acid, and the size difference between the peaks determines the amino acid sequence. Sometimes the spectrum obtained for a peptide ion is ambiguous, so databases of peptide ion spectra are used for comparison.

Each amino acid degrades into predicted ions in mass spectroscopy. The amino acid sequence for an unknown peptide can be deduced based on the pattern of ions produced in comparison with the known patterns.

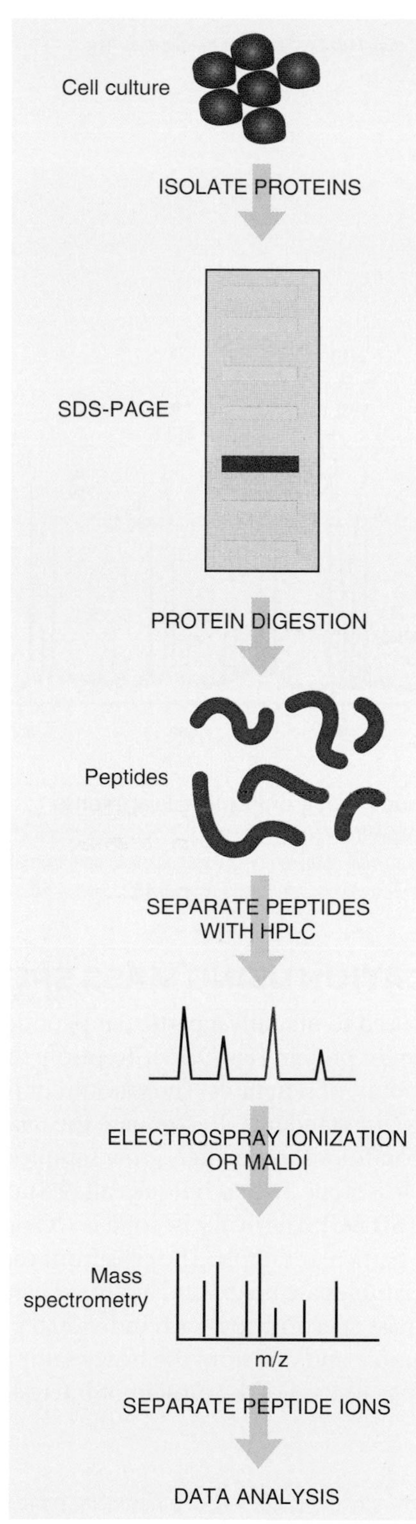

FIGURE 9.13 Preparation of Proteins for Mass Spectrometry
Because mass spectrometry is so sensitive, the use of large whole proteins is limited. Instead, peptide fragments are generated by protease digestion. The peptides are easily separated with HPLC, and then specific peptides are subjected to mass spectrometry.

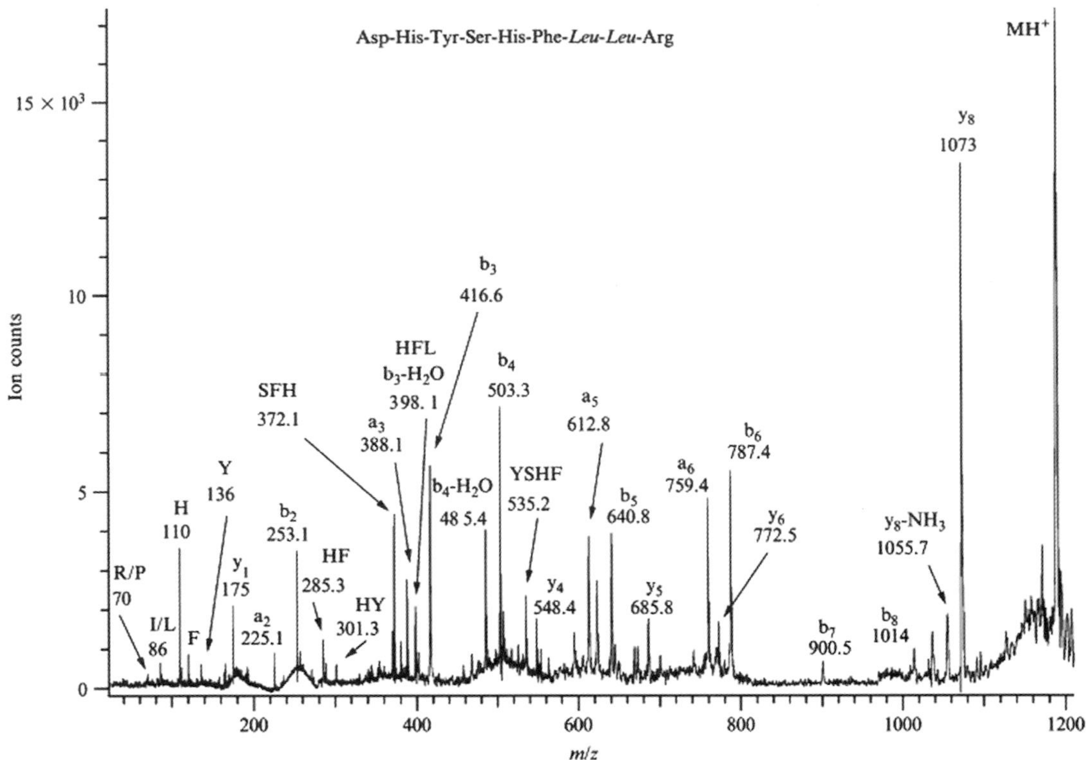

FIGURE 9.14 Mass Spectroscopy Trace of Peptide Fragments

Post-source decay spectrum of a tryptic peptide (m/z 1187.6) from the 50-kDa subunit of DNA polymerase from *Schizosaccharomyces pombe*. The spectrum was acquired on a Voyager mass spectrometer (Applied Biosystems). From Medzihradszky KF (2005). Peptide sequence analysis. *Methods Enzymol* **402,** 209–244. Reprinted with permission.

PROTEIN QUANTIFICATION USING MASS SPECTROMETRY

Mass spectrometry can also be used to quantify a particular peptide from a protein, which directly correlates to the amount of protein (Fig. 9.15). To purify the protein, researchers prepare the sample and add a small amount of standards. The amount of peptide and therefore protein is determined by comparison to the standards. To compare the relative amounts of one protein in two different experimental conditions, researchers grow samples of cells with and without amino acids tagged with a stable isotope, in a technique called **Stable Isotope Labeling by Amino acids in Cell culture (SILAC)**. The heavy isotope is ^{13}C or ^{15}N. These isotopes increase the mass of all proteins in that particular sample. The cells from each condition are lysed, and the proteins isolated. The two samples are mixed and analyzed using one of the forms of chromatography followed by mass spectroscopy. Each individual peptide will now have two peaks: one from the normal sample and one from the heavy sample. The ratio of the two peaks will determine the relative change in level of the protein of interest between the samples.

> Changing experimental conditions affects the amount of a particular protein. This change can be determined using mass spectroscopy by adding heavy isotopes to one of the samples in a process called SILAC.

PROTEIN TAGGING SYSTEMS

Protein tagging systems are tools for the isolation and purification of single target proteins from a mixture. The target protein is genetically fused to a segment of DNA that codes for a "tag," creating a hybrid gene. This hybrid coding system is inserted into a vector with the

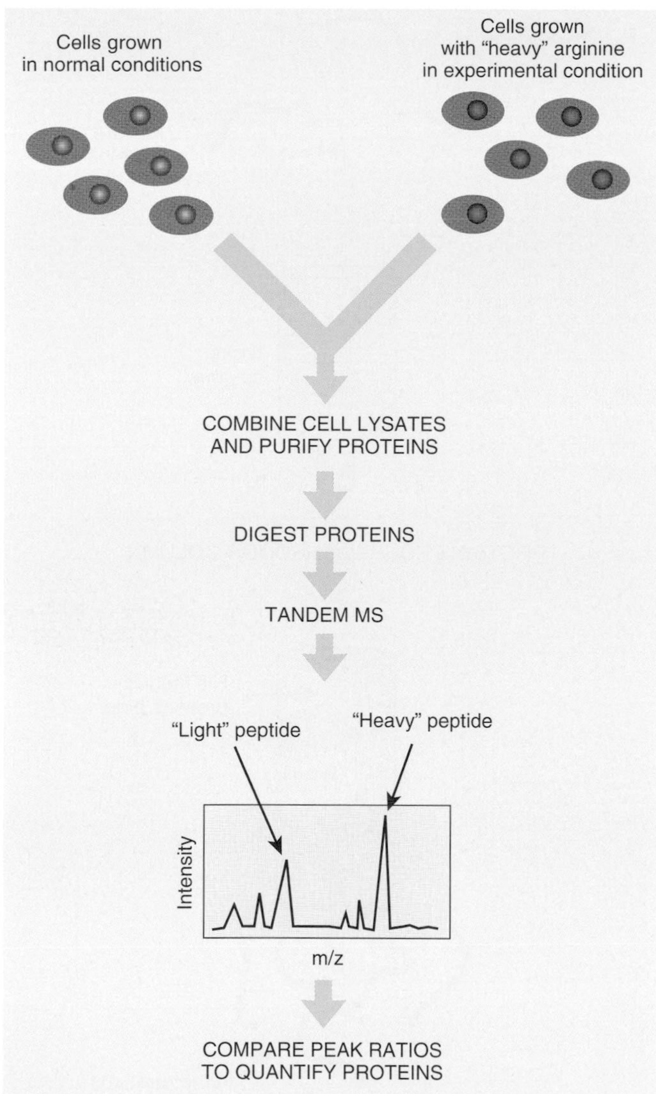

FIGURE 9.15 Quantification of Peptides Using Mass Spectrometry
Mass spectrometry can compare the amount of a particular protein in samples from two different conditions. Cells in one condition are grown with a "heavy" amino acid, such as arginine, which is incorporated during protein synthesis. The proteins are digested with proteases into small peptide fragments. These are ionized by ESI mass spectrometry. The analysis gives pairs of light and heavy peaks for each peptide. Peak sizes correlate to the amount of peptide and hence protein. The "heavy" peak is more abundant in this example; thus, this protein is more abundant under the experimental conditions.

appropriate promoters and terminators to express the tagged protein of interest. The gene construct is transformed into a suitable host organism for expression. When the cells are grown and disrupted to release the proteins, the target protein can be easily isolated because of its tag.

Many different types of tags are used to isolate proteins because the chemistry and size of the tag may affect the protein of interest in a negative way. The first widely used tag, called the **polyhistidine**, or **His6 tag**, consists of six histidine residues in a row (Fig. 9.16). Histidine binds very tightly to nickel ions; therefore, His-tagged proteins are purified on a column to which Ni^{2+} ions are attached. Once attached to the column, the His6-tagged protein is removed by disrupting the Ni^{2+}–His interaction with free histidine or imidazole. The polyhistidine tag may be attached to the carboxy- or amino-terminal end of the protein of interest. Because the His6 tag is very short, the target protein is rarely affected by adding it.

Other short tags for proteins include FLAG, which is recognized by a specific antibody. FLAG has the peptide sequence AspTyrLysAspAspAspAspLys. As before, the gene for the target

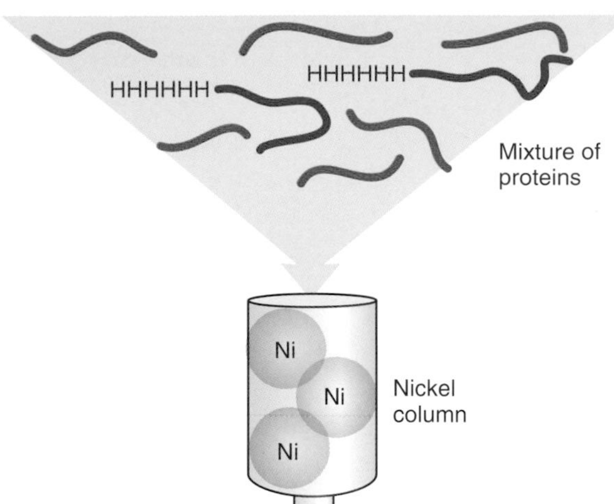

PROTEINS POURED THROUGH COLUMN

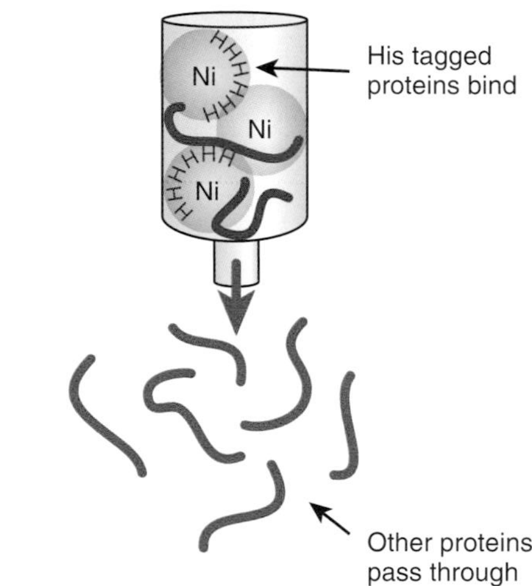

ELUTE WITH HISTIDINE OR IMIDAZOLE

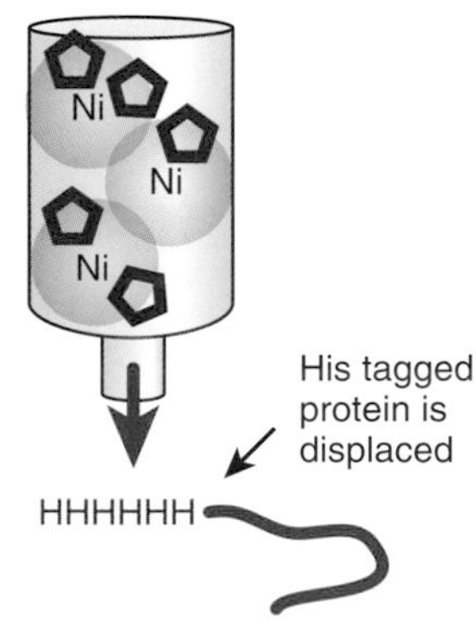

FIGURE 9.16 Nickel Purification of His6 Tagged Protein
To isolate a pure sample of one specific protein, the gene for the protein is genetically linked to a coding region for six histidine residues. The expressed fusion protein can be isolated from a mixture of proteins because of the chemistry of histidines. The histidines bind to nickel-coated beads, and the remaining untagged proteins pass through the column. The histidine-tagged protein is then eluted by passing free histidine or imidazole over the column.

protein plus a short DNA segment encoding FLAG is cloned into a vector, and the hybrid protein is produced in either bacteria or a cell line. FLAG-tagged proteins can be isolated from a cell lysate using the anti-FLAG antibody either bound to beads or attached to a column. Only the tagged protein attaches to the beads/column. Finally, the FLAG-tagged protein is separated from the antibody by adding free FLAG peptide. The short peptide is present in surplus and competes for antibody with the tagged protein, which is therefore eluted from the beads or column.

Another short tag is the "Strep" tag, provided by a short DNA segment that encodes a 10-amino-acid peptide with a similar 3D structure to biotin (see Chapter 3). The biotin-like peptide binds tightly to the proteins avidin or streptavidin, so Strep-tagged proteins are isolated by binding to streptavidin-coated beads or a streptavidin column.

Besides short tags, longer tags that consist of entire proteins are used for some applications. Four popular tags include **green fluorescent protein (GFP)** from jellyfish, **protein A** from *Staphylococcus*, **glutathione-S-transferase (GST)** from *Schistosoma japonicum*, and **maltose-binding protein (MBP)** from *Escherichia coli*. Just like the short tags, the genes for these longer tags are genetically fused to the target gene. The hybrid gene constructs are expressed by using appropriate transcriptional promoters and terminators. Once the host cells express the hybrid gene, the fusion protein is isolated by purifying the protein tag. GFP-tagged proteins offer a specific advantage since the tag also autofluoresces under UV light. The exact location in which a protein is expressed can be determined by microscopy first, and then binding antibodies to GFP can purify the protein. In a similar manner, specific antibodies to protein A can be used to purify protein A-tagged proteins. Purified GFP or protein A-tagged proteins can be released from the antibody by lowering the pH. MBP binds to maltose (attached to beads or a column), and the fusion protein is released by adding free maltose. GST binds to its substrate, glutathione (on beads or a column), and free glutathione is used to release the hybrid protein. Once the fusion protein has been isolated, it must be cleaved to separate the target protein from the tag protein.

FIGURE 9.17 Maltose-Binding Protein Fusion Vector
Vectors such as pMAL have polylinker regions for cloning a target gene in frame with the gene for a tag protein such as MBP (the *malE* gene). The fusion protein is easily isolated because MBP binds to maltose columns. The fusion protein also has a binding site for the protease, factor Xa. When the fusion protein is bound to the column, factor Xa will release the target protein and leave behind the MBP domain.

A useful feature of the longer tags is the presence of a protease cleavage site between the target protein and the tagging protein. The vector for pMAL (New England Biolabs, Inc., Ipswich, MA) has the gene for MBP, followed by a spacer region, a recognition site for factor Xa, then the polylinker region for cloning the target gene (Fig. 9.17). Factor Xa is a specific protease used in the blood clotting system, and inserting its recognition sequence allows the MBP portion of the hybrid protein to be cleaved from the target protein. After the hybrid protein is eluted from the purification column with maltose, the original protein is isolated through protease treatment. This is extremely useful when pure, native protein is needed for analysis.

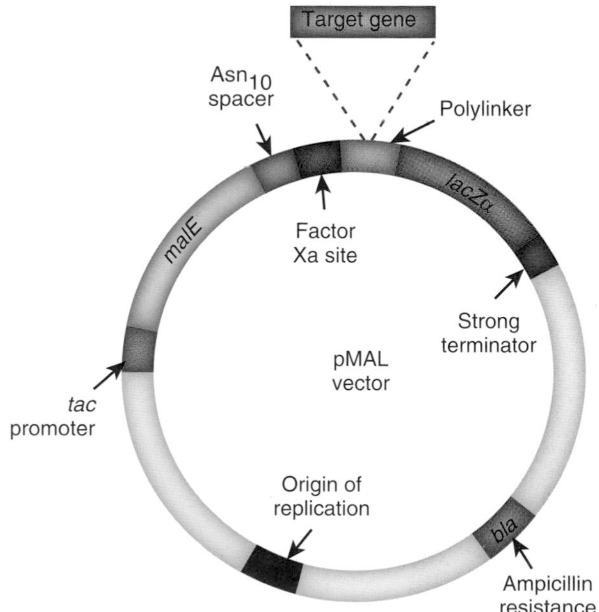

An even easier way to obtain a pure sample of native protein is the self-cleavable **intein** tag. The approach is based on inteins, self-splicing intervening segments found in some proteins. Inteins are the protein equivalent to introns in RNA. The intein removes itself from its host protein via a branched intermediate (Fig. 9.18).

The Intein Mediated Purification with Affinity Chitin-binding Tag (IMPACT) system from New England Biolabs uses a modified intein from the *VMA1* gene of *Saccharomyces cerevisiae*. Intein cleavage is used to release the target protein after purification of the fusion protein (Fig. 9.19). This yeast intein originally cleaved both its N-terminus and its C-terminus, but it has been modified so that it cleaves only its N-terminus. The chitin-binding tag of this system is the small chitin-binding domain (CBD) from the chitinase A1 gene of *Bacillus circulans*. (Chitin is the substance that forms the exoskeleton of insects.) The vector has a

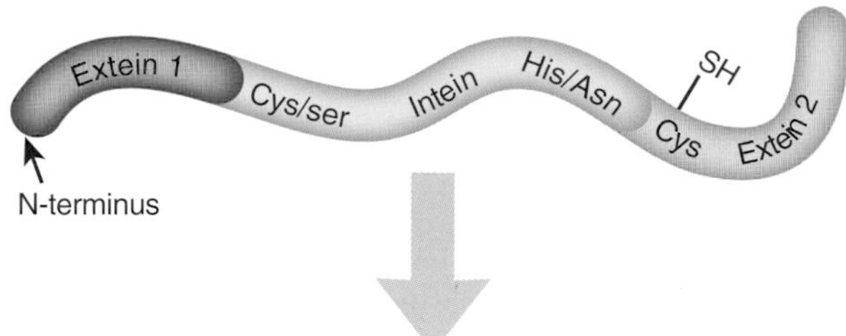

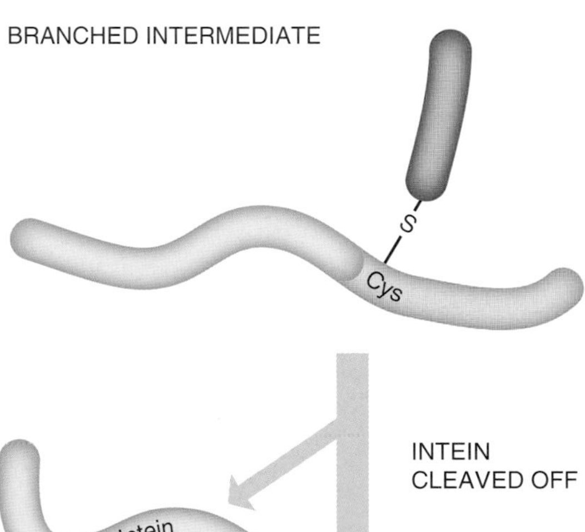

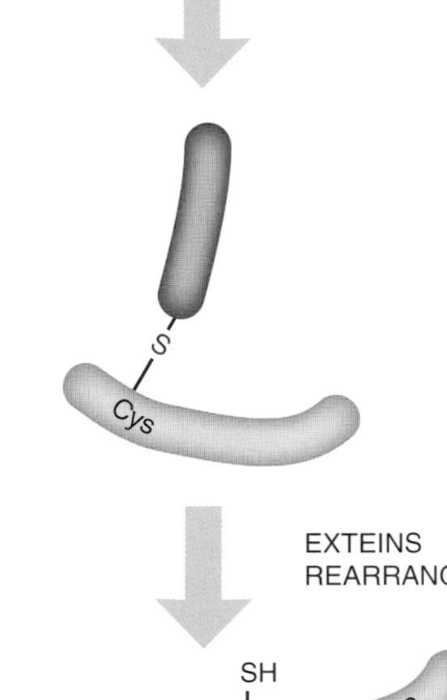

FIGURE 9.18 Mechanism of Intein Removal

The intervening intein segment splices itself out in two stages. The intein has a Cys or Ser at the boundary with extein 1 and a basic amino acid at its boundary with extein 2. The downstream extein (2) has a Cys residue at the splice junction. Extein 1 is cut loose and attached to the sulfur side chain of the cysteine at the splice junction. This forms a temporary branched intermediate. Next, the intein is cut off and discarded, and the two exteins are joined to form the final protein.

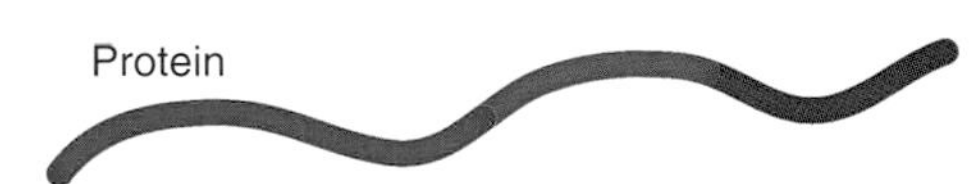

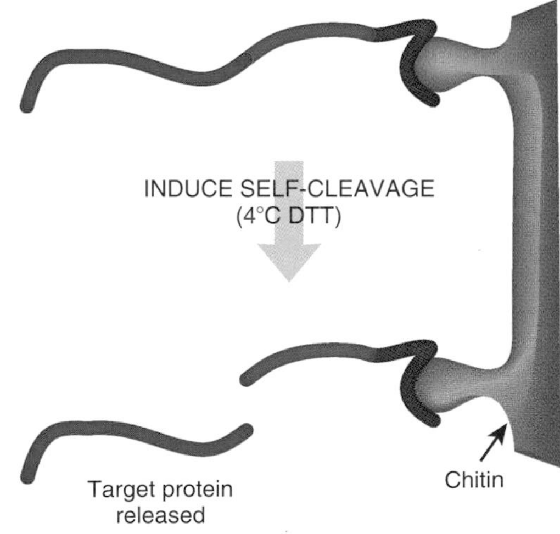

FIGURE 9.19 Intein-Mediated Purification System

Inteins that can self-cleave at their amino-terminus allow specific proteins to be purified and cleaved from a fusion protein in one step. First, the fusion protein is purified by passing over a column made of chitin. (Note: The CBD, or chitin-binding domain, recognizes and binds the chitin molecule.) The column is incubated with DTT in a refrigerator. The intein cleaves itself at its amino terminus and releases the target protein.

polylinker or cloning site for the target gene followed by the DNA segment encoding the intein, followed by the CBD. The fusion protein is expressed, and cell lysates containing the hybrid protein are isolated. When the lysate passes through a chitin column, the hybrid protein binds to the column via the CBD, and the remaining cellular proteins elute. The column is then incubated at 4 °C with dithiothreitol (DTT), a thiol reagent that activates the intein to cleave its N-terminus. Thus, the target protein is released from the column, leaving behind the intein and CBD regions.

Protein tags are either short peptides or entire foreign proteins fused genetically to a protein of interest. Protein tags provide a means to isolate the protein of interest from the rest of the cellular proteins. In the case of GFP-tagged proteins, the cellular location can also be identified using microscopy.

PHAGE DISPLAY LIBRARY SCREENING

A **phage display** library is a collection of bacteriophage particles that have segments of foreign proteins protruding from their surface (Fig. 9.20). Normal bacteriophages have outer coats made of proteins. The outer coat of M13 bacteriophage has about 2500 copies of the major coat protein (gene VIII protein) and about five copies of the minor coat protein (gene III protein). Gene III protein is located at the end of the cylindrical bacteriophage particle with its N-terminus facing outward. One popular phage display system fuses the foreign sequence to the N-terminus of the gene III protein. The result is that M13 now has about five copies of the foreign protein segment on its surface at

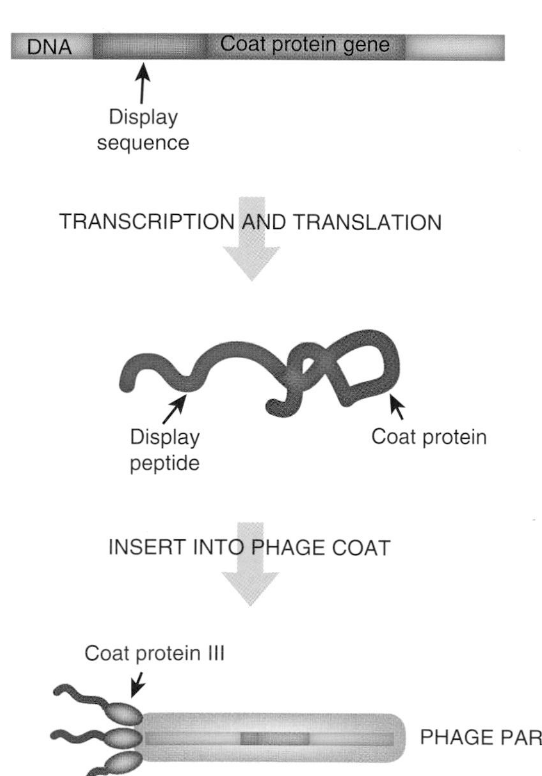

FIGURE 9.20 Principle of Phage Display

To display a peptide on the surface of a bacteriophage, researchers must fuse the DNA sequence encoding the peptide to the gene for a bacteriophage coat protein. In this example, the chosen coat protein is encoded by gene III of phage M13. The N-terminal portion of gene III protein is on the outside of the phage particle, whereas the C-terminus is inside. Therefore, the peptide must be fused in frame at the N-terminus to be displayed on the outside of the phage.

one end of the bacteriophage particle, as shown in Figure 9.20. M13 is especially convenient because it does not lyse the bacteria it infects; the viral particles are simply secreted through the bacterial cell envelope.

For the phage to display a foreign protein, the gene for that protein must be fused to gene III to produce a hybrid protein. The gene of interest must be in frame with gene III for proper expression. The M13 genome can accommodate extra DNA because the filamentous bacteriophage particles are simply made longer if a larger genome needs to be packaged. The M13 genome containing the gene of interest is transformed into *E. coli*, where the bacteriophage DNA directs the synthesis of new particles containing the protein of interest in the coat.

Bacteriophage can display artificial peptides as well as segments of natural proteins. Random oligonucleotides generated by PCR can be cloned and fused to gene III of M13. Each random oligonucleotide will encode a different peptide. These constructs are transformed into *E. coli*, and each transformant produces bacteriophage with different foreign peptides fused to gene III protein.

The collection of displayed protein segments or peptides can be screened by **biopanning** to find those with a particular property, perhaps a specific protein-binding domain, or a specific peptide structure that binds an antibody (Fig. 9.21). In biopanning, the library of phages displaying the foreign peptides is incubated with a target protein, such as an antibody, bound to a bead or membrane. All the recombinant phages that bind to that antibody adhere to the solid support, and the others are washed away. All the bound phages are released and incubated with *E. coli* to replicate the phage. The procedure is usually repeated in order to enrich for peptides that bind specifically because some non-specific binding could occur. Once a phage with a useful peptide is identified, the clone is sequenced to determine the structure of the peptide.

Full-length protein libraries can also be studied using phage display, but they pose some extra problems. When M13 is used, coding sequences for full-length proteins must be cloned

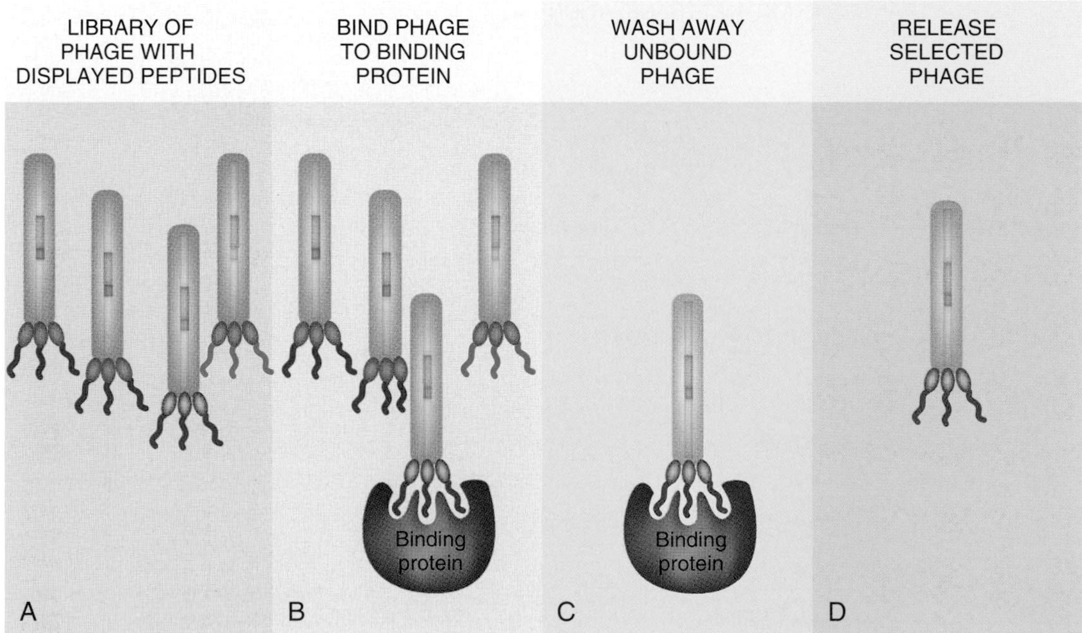

FIGURE 9.21 Biopanning of Phage Display

Biopanning is used to isolate peptides that bind to a specific target protein, which is usually attached to a solid support such as a membrane or bead. The phage display library (A) is attached to the binding protein (B). Those phages that bind to the target protein will be retained (C), but the others are washed away. The phage that does recognize the binding protein can be released, isolated, and purified.

in frame with both a signal sequence at their N-terminal end and gene III at their C-terminal end. (The signal sequence is required to direct the hybrid protein to the viral coat.) Ensuring the correct reading frame is reasonable for one or two genes, but for an entire library, there is too much room for error. Besides, the possible creation of a stop codon at either fusion junction would prevent the hybrid protein from being expressed. The solution is to use T7 bacteriophage for libraries of full-length proteins. T7 has a coat protein whose C-terminal tail is exposed to the outside. To be expressed on the bacteriophage surface, the protein library must therefore be fused to the C-terminus of the coat protein. This requires only one fusion junction. Furthermore, even if library sequences are cloned out of frame or contain stop codons, the coat protein itself is unaffected and still assembled, although the attached library proteins will be defective.

Being able to express full-length proteins for biopanning is very useful to proteomics researchers. To identify a protein that binds to a particular cell surface receptor, a researcher can biopan a phage display library for receptor binding. Another example is finding RNA binding proteins. Here, RNA is anchored to a solid support, and the phage display library is incubated with this RNA "bait." The phages that stick to the RNA bait are isolated and enriched by repeating the procedure. Each isolated clone can then be sequenced to identify which proteins bind RNA.

Phage display is a technique in which foreign proteins or peptides are fused to a coat protein on the surface of the phage. The phage then displays them for analysis.

Biopanning identifies binding partners for a protein of interest. The protein of interest is incubated with a phage display library. When a phage binds to the protein of interest, it is isolated and the sequence for the displayed peptide is determined.

PROTEIN INTERACTIONS: THE YEAST TWO-HYBRID SYSTEM

In addition to protein function and expression, proteomics attempts to find relevant protein interactions. For those who like "-omics" terminology, the total of all protein–protein interactions is called the **protein interactome**. For example, hormones usually bind to receptors that pass on the signal. Often this involves a protein relay in which one protein activates another, which in turn activates yet another. To understand hormone function, researchers must identify all the proteins in the signal cascade. Phage display is one way to identify interactions, but the displayed proteins may not fold correctly, or specific cofactors may be missing when mammalian proteins are expressed in bacteria.

An approach to overcoming these difficulties is to use the yeast **two-hybrid system**, in which the binding of two proteins activates a reporter gene. The binding of a transcriptional activator protein, GAL4, to the promoter region of the reporter gene activates transcription, which results in synthesis of the reporter protein. GAL4 contains two domains needed to turn on the reporter gene. The DNA-binding domain (DBD) recognizes the promoter element and positions the second domain, the activation domain (AD), next to RNA polymerase, where it activates transcription. These two domains can be expressed as separate proteins but cannot activate the reporter gene unless they are brought together (Fig. 9.22).

In the two-hybrid system, the two domains are each fused to different proteins by creating hybrid genes. The **bait** is the DBD genetically fused to the protein of interest, and the **prey** is the AD fused to proteins that are being screened for interaction with the bait. When the bait and prey bind, the DBD and AD activate transcription of the reporter gene.

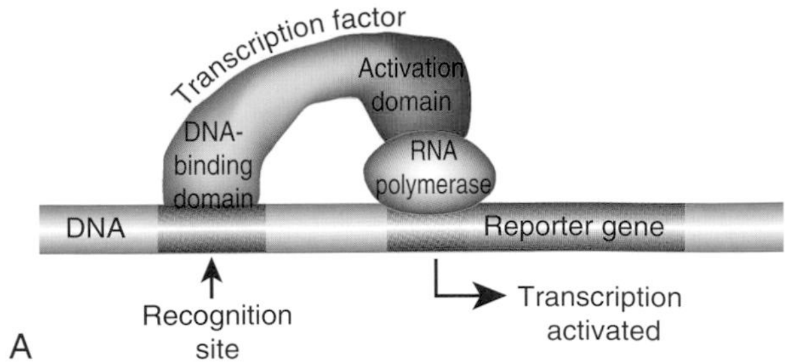

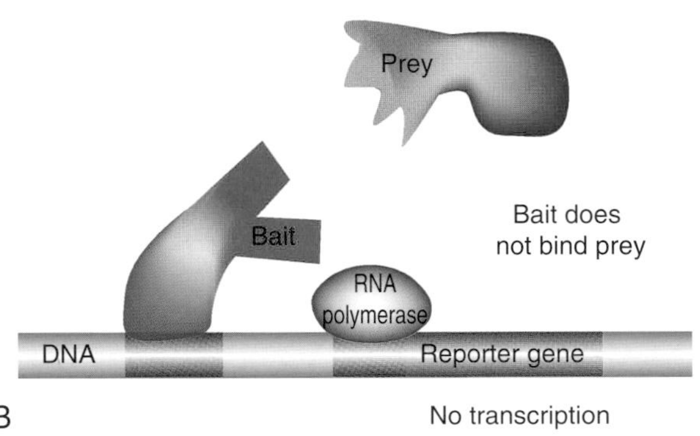

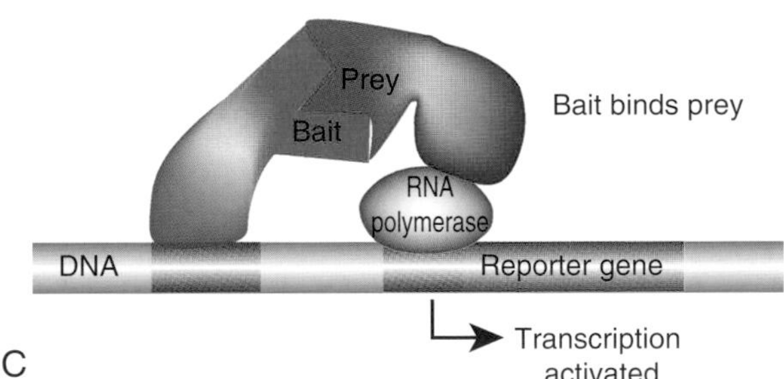

FIGURE 9.22 Principle of Two-Hybrid Analysis

(A) Yeast transcription factors have two domains: the DBD (purple) recognizes regulatory sites on DNA, and the AD (red) activates RNA polymerase to start transcription of the reporter gene. For two-hybrid analysis, two proteins (Bait and Prey) are fused separately to the DBD and AD domains of the transcription factor. The Bait protein is joined to the DBD and the Prey protein to the AD. (B) The Bait protein and Prey protein do not interact, and the reporter gene is not turned on. (C) The Bait binds the Prey, bringing the transcription factor halves together. The complex activates RNA polymerase, and the reporter gene is expressed.

Two vectors are needed to perform two-hybrid analysis (Fig. 9.23). The first vector has a multiple cloning site for the bait protein at the 3′-end of the GAL4-DBD; therefore, the fusion protein has the Bait protein as its C-terminal domain. The second vector has a multiple cloning site for the Prey protein at the 5′-end of the GAL4-AD, and the fusion protein has the Prey protein as its N-terminal domain. Both plasmids must be expressed in the same yeast cell. If the bait and prey proteins interact, the reporter gene is turned on.

The reporter genes must be engineered to be under control of the GAL4 recognition sequence. Common reporter genes include *HIS3*, which encodes an enzyme in the histidine pathway and whose expression allows yeast cells to grow on media lacking histidine, or *URA3*, which allows growth without uracil. These reporter systems require yeast host cells

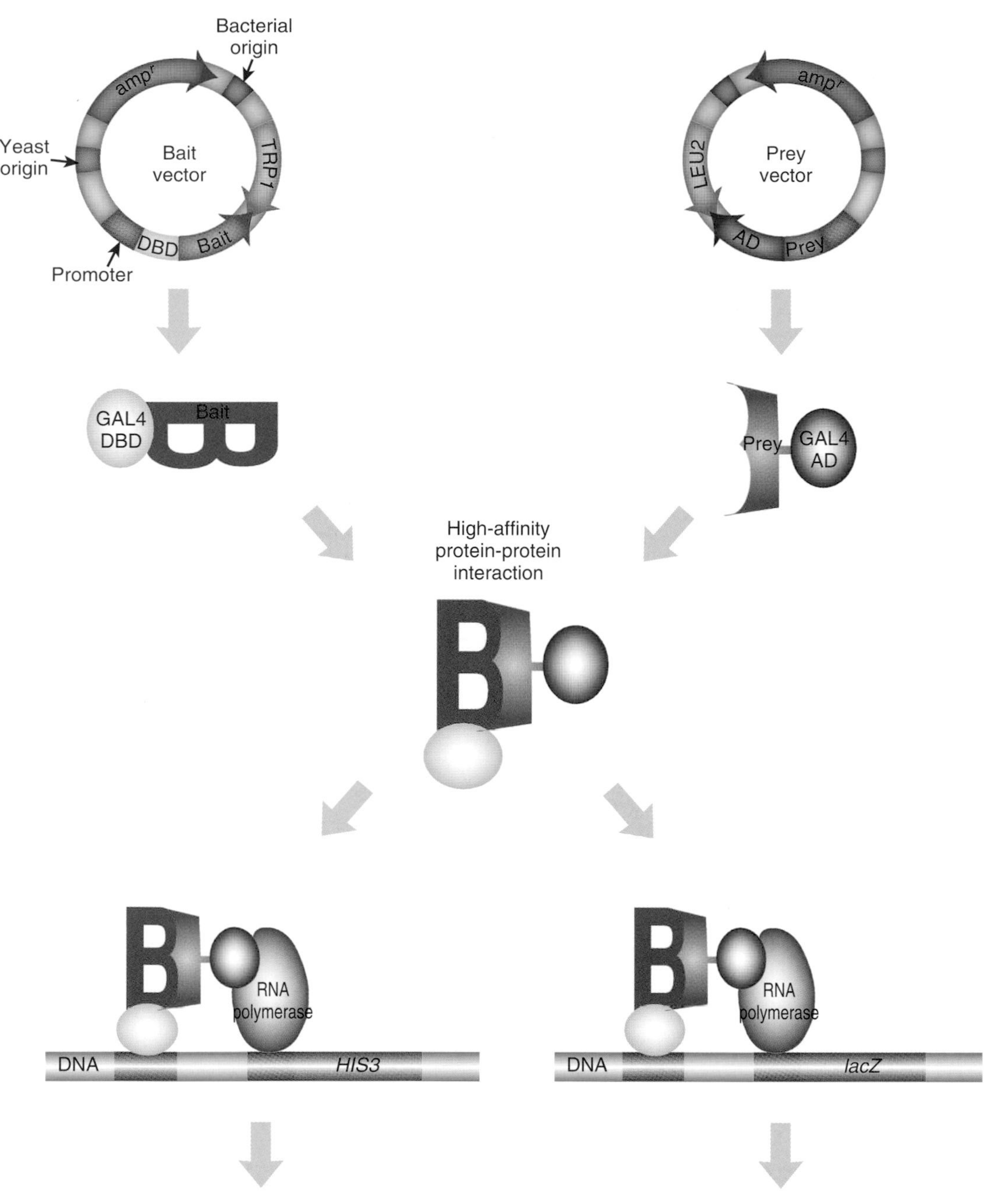

FIGURE 9.23 Vectors for Two-Hybrid Analysis
Two different vectors are necessary for two-hybrid analysis. The bait vector has coding regions for the DBD and for the Bait protein. The Prey vector has coding regions for the AD and for the Prey protein. These two different constructs are expressed in the same yeast cell. If the Bait and Prey interact, the reporter gene is expressed. Two reporter systems are shown here. The *His3* gene allows yeast to grow on histidine-free media. The *lacZ* gene encodes β-galactosidase, which cleaves X-gal, forming a blue color.

that are defective in the corresponding genes. However, they do allow direct selection of positive isolates.

Another reporter used is *lacZ* from *E. coli,* which encodes β-galactosidase. Both bacteria and yeast that express *lacZ* turn blue when grown with X-Gal. β-galactosidase cleaves X-Gal, releasing a blue product. The reporter genes are usually integrated into the yeast genome rather than being carried on a separate vector.

The yeast two-hybrid system has been used to identify all the protein interactions in the yeast proteome by mass screening with mating (Fig. 9.24). Yeast has about 6000 different proteins, and each of them has been cloned into both vectors via PCR. This way, each protein can be used as both bait and prey. All the bait vectors were transformed into haploid yeast of one mating type and the prey vectors into the other mating type. Haploid cells carrying bait are fused to haploid cells with prey, and the resulting diploid cells are screened for reporter gene activity. This analysis thus examined 6000 × 6000 combinations for protein interaction.

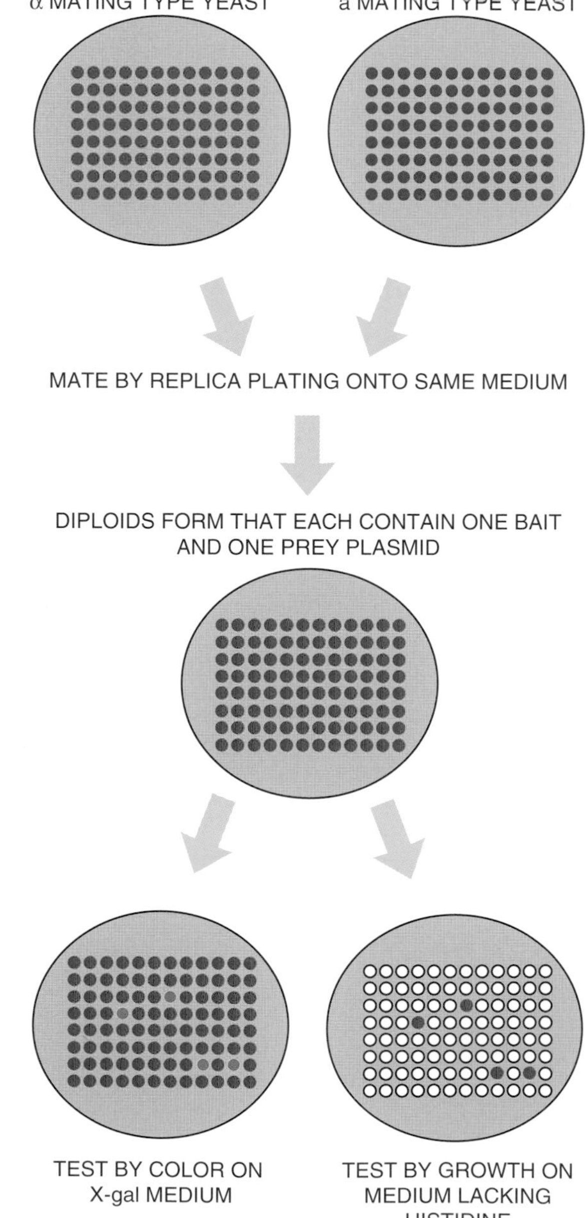

FIGURE 9.24 Two-Hybrid Analysis: Mass Screening by Mating
To identify all possible protein interactions using the two-hybrid system, haploid α yeast are transformed with the Bait library, and haploid **a** yeast are transformed with the Prey library. When the two yeast types are mated with each other, the diploid cells will each contain a single bait fusion protein and a single prey fusion protein. If the two proteins interact, they activate the reporter gene, which allows the yeast to grow on media lacking histidine (yeast *His3* gene) or turn the cells blue when growing on X-gal medium (*lacZ* from *E. coli*). This process can be done for all 6000 predicted yeast proteins using automated techniques.

The yeast two-hybrid system has significant limitations. Because transcription factors must be in the nucleus to work, the target proteins must also function in the nucleus. For some proteins, entering the nucleus may cause the protein to misfold. For other proteins, the nucleus does not contain the proper cofactors and the protein may be unstable. Large proteins may not be expressed well or may be toxic to the yeast, leading to false negative results. When protein interactions are checked by other methods, it is clear that the two-hybrid system misses many interactions and generates a significant number of false positive interactions. Performance can be improved by using multiple sets of two-hybrid vectors that use other transcription factors than GAL4 or systems that fuse the Bait protein to the N-terminus of the DNA-binding domain and the Prey protein to the C-terminus of the activation domain.

Yeast two-hybrid analysis finds proteins that bind together. Cellular proteins are linked to the AD of GAL4 or the DBD of GAL4. When two cellular proteins bring the AD and DBD together, the completed GAL4 binds to the reporter gene promoter. Reporter gene products allow the yeast to grow on histidine-free media or turn blue on media that contain X-Gal.

PROTEIN INTERACTIONS BY CO-IMMUNOPRECIPITATION

Co-immunoprecipitation is a technique to examine protein interactions in the cytoplasm rather than the nucleus (Fig. 9.25). Here, the target protein is expressed in cultured mammalian cells, which are lysed to release the cytoplasmic contents. The target protein is precipitated from the lysate with an antibody. Other proteins that are associated with the target protein remain associated with the antibody-protein complex. (If no antibody exists for the target protein, a small tag [such as FLAG or His6; see earlier discussion] can be engineered onto the protein.) Protein A from *Staphylococcus*, in turn, binds the antibodies. The protein A is attached to beads before it is added to the cell lysate. This generates very large target protein/antibody/protein A/bead complexes, which are gently isolated from the rest of the cellular proteins by centrifugation. The complexes are separated by size with SDS-PAGE. The gel should show the target protein, the antibody, protein A, and other bands that represent interacting proteins. These can be identified with protein sequencing and/or mass spectrometry.

Co-immunoprecipitation is often used to confirm the results from yeast two-hybrid analysis, especially for mammalian proteins. Many two-hybrid experiments reveal novel uncharacterized proteins. To confirm the interaction, researchers tag both proteins for easy isolation. Adding a tag is

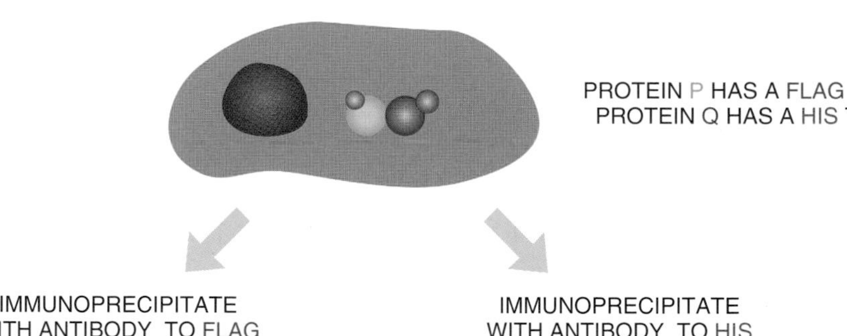

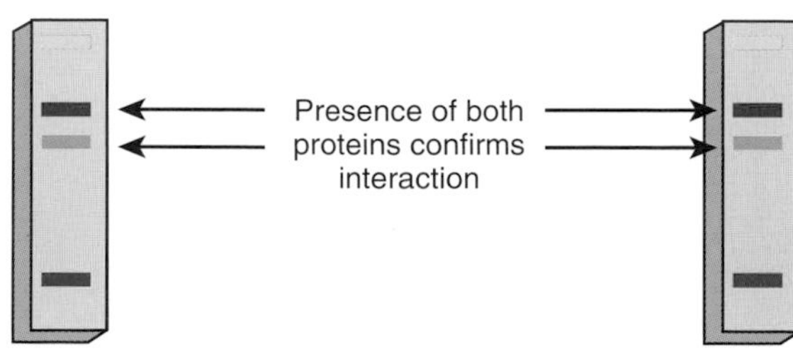

FIGURE 9.25 Co-immunoprecipitation
To determine whether protein P and Q interact within the cytoplasm, researcher fuse each protein to a different tag for easy isolation. Each fusion protein is expressed in mammalian cells, which are then lysed to release the cell proteins. The cells must be lysed gently to avoid disrupting the protein interactions. The fusion proteins are isolated using the tag sequence. Each tagged protein and all its associated proteins are isolated independently. For example, on the left, Flag-tagged protein P is isolated with an antibody to the Flag sequence, and on the right, His6-tagged protein Q is isolated with an antibody to the His6 sequence. The protein complexes are separated by SDS-PAGE. This example shows the two tagged proteins P and Q interacting.

much easier than generating a specific antibody to each new protein. For example, protein P is tagged with FLAG, while protein Q is tagged with His6. Each vector construct is transformed into a mammalian cell line, and each protein is expressed. The cell lysate is harvested and divided into two samples. The protein P complexes are isolated from the first sample, whereas the protein Q complexes are isolated from the second sample. Each of the complexes is isolated with protein A-coated beads. The different proteins from each sample are separated by SDS-PAGE. If the two proteins interact, both proteins will be found in both samples.

Co-immunoprecipitation determines whether two proteins bind together in the cytoplasm.

PROTEIN ARRAYS

Protein-detecting arrays may be divided into those that use antibodies and those based on using tags. In the **ELISA assay** (see Chapter 6), antibodies to specific proteins are attached to a solid support, such as a microtiter plate or glass slide. The protein sample is then added and if the target protein is present, it binds its complementary antibody. Bound proteins are detected by adding a labeled second antibody.

Another antibody-based protein-detecting array is the **antigen capture immunoassay** (Fig. 9.26). Much like the ELISA, this method uses antibodies to various proteins bound to a solid surface. The experimental protein sample is isolated and labeled with a fluorescent dye. If two conditions are being compared, proteins from sample 1 can be labeled with Cy3, which fluoresces green, and proteins from sample 2 can be labeled with Cy5, which fluoresces red. The samples are added to the antibody array, and complementary proteins bind to their cognate antibodies. If both sample 1 and 2 have identical proteins that bind the same antibody, the spot will fluoresce yellow. If sample 1 has a protein that is missing in sample 2, then the spot will be green. Conversely, if sample 2 has a protein missing from sample 1, the spot will be red. This method is good for comparing protein expression profiles for two different conditions.

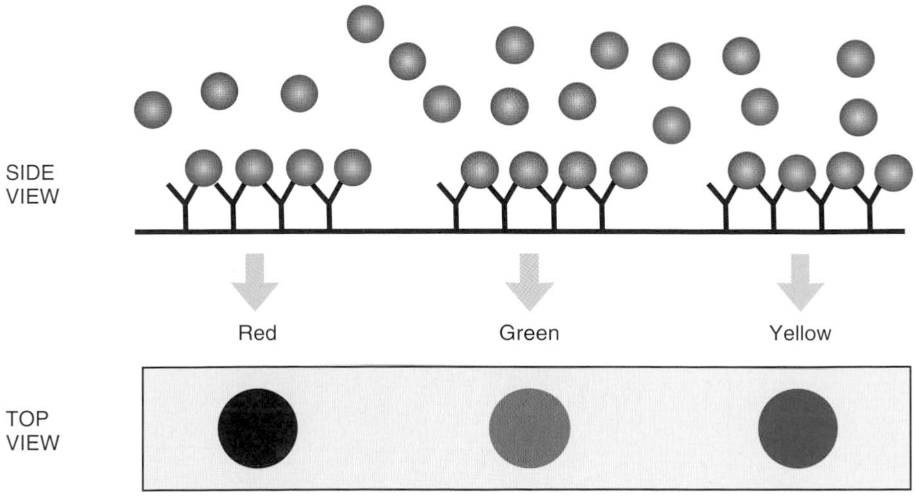

FIGURE 9.26 Ideal Results for Antigen Capture immunoassay
Various different antibodies are fused to different regions of a solid surface. Each spot has a different antibody. If the antibody recognizes only proteins labeled with Cy5, the region will fluoresce red *(left)*. If the antibody recognizes only proteins labeled with Cy3, the region will fluoresce green *(middle)*. If the antibody recognizes proteins in both conditions, the spot will fluoresce yellow *(right)*.

In the third method, the **direct immunoassay** or **reverse-phase array**, the proteins of the experimental sample are bound to the solid support (Fig. 9.27). The proteins are then probed with a specific labeled antibody. Both presence and amount of protein can be monitored. For example, proteins from different patients with prostate cancer can be isolated and spotted onto glass slides. Each sample can be examined for specific protein markers or the presence of different cancer proteins. The levels of certain proteins may be related to the stages of prostate cancer. This immunoassay helps researchers to decipher these correlations.

The main problem with immunology-based arrays is the antibody. Many antibodies cross-react with other cellular proteins, which generates false positives. In addition, binding proteins to solid supports may not be truly representative of intracellular conditions. The proteins are not purified or separated; therefore, samples contain very diverse proteins. Some proteins will bind faster and

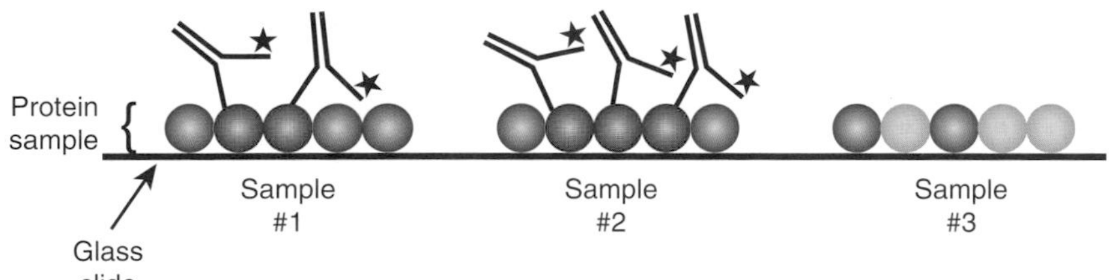

FIGURE 9.27 Direct Immunoassay
The direct immunoassay binds the protein samples to different regions on a solid support. Each spot has a different protein sample. Next, an antibody labeled with a detection system is added. The antibody binds only to its target protein. In this example, the antibody recognizes only a protein in patient samples 1 and 2.

better than others. Also, proteins of low abundance may not compete for binding sites. Another problem is that many proteins are found in complexes, so other proteins in the complex may mask the antibody-binding site.

Rather than using antibodies, protein interaction arrays use a fusion tag to bind the protein to a solid support (Fig. 9.28). The use of protein arrays to determine protein interactions and protein function is a natural extension of yeast two-hybrid assays and co-immunoprecipitation. Protein arrays can assess thousands of proteins at one time, making this a powerful technique for studying the proteome. Protein arrays were first used systematically in yeast because its proteome contains only about 6000 proteins. Libraries have been constructed in which each protein is fused to a His6 or GST tag. The proteins are then attached by the tags to a solid support such as a glass slide coated with nickel or glutathione. To build the array, researchers isolate each protein individually and spot it onto the glass slide. The tagged proteins bind to the slide, and other cellular components are washed away. Each spot has only one unique tagged protein.

Once the array is assembled, the proteins can be assessed for a particular function. In the laboratory of Michael Snyder at Yale University, the yeast proteome has been screened for proteins that bind **calmodulin** (a small Ca^{2+} binding protein) or phospholipids (Fig. 9.29). Both calmodulin and phospholipid were tagged with biotin and incubated with a slide coated with each of the yeast proteins bound to the slide via His6-nickel interactions. The biotin-labeled calmodulin or phospholipid was then visualized by incubating the slide with Cy3-labeled streptavidin. (Streptavidin binds very strongly to biotin.) The results identified 39 different calmodulin-binding proteins (only six had been identified previously), and 150 different phospholipid-binding proteins.

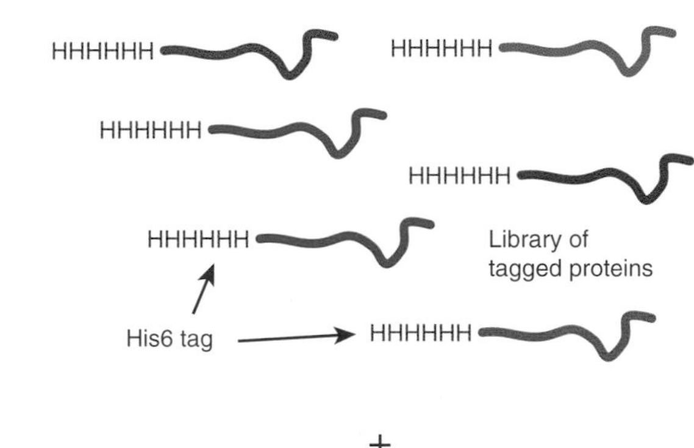

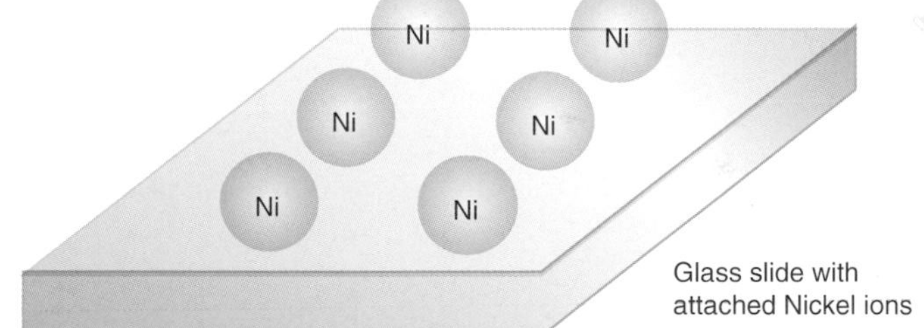

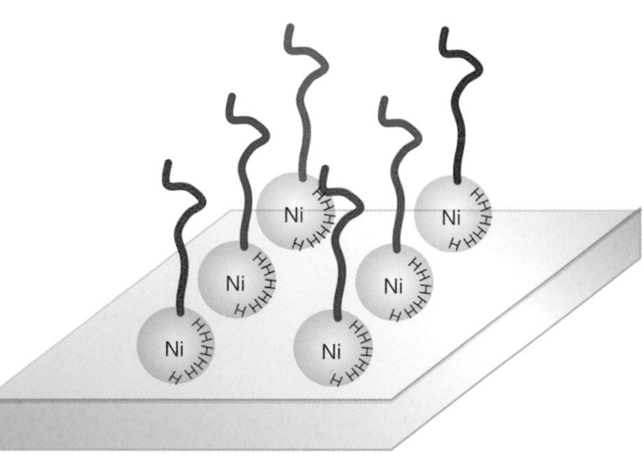

FIGURE 9.28 Protein Interaction Microarray—Principle
To assemble a protein microarray, researchers incubate a library of His6-tagged proteins with a nickel-coated glass slide. The proteins adhere to the slide wherever nickel ions are present.

Protein microarrays ready for screening are now commercially available for yeast and humans. The ProtoArray® Human Protein Microarray, available from Invitrogen, includes about 9000 human proteins (about 40% of the human proteome).

Proteins to be screened

His6 tag

Ni Ni Ni Ni Ni Ni

Slide

PHOSPHOLIPID TAGGED WITH BIOTIN

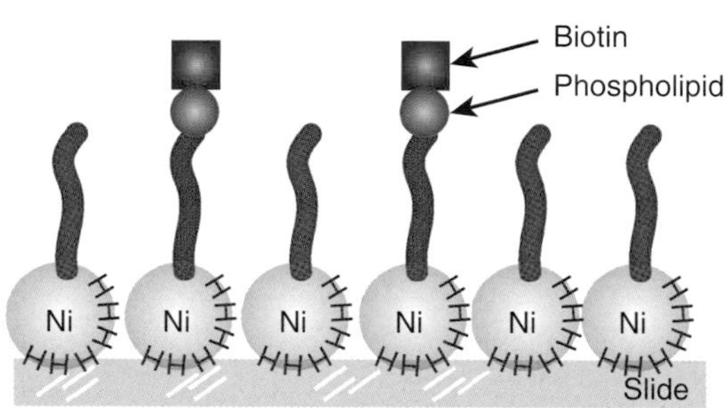

AVIDIN WITH CY3 FLUORESCENT LABEL

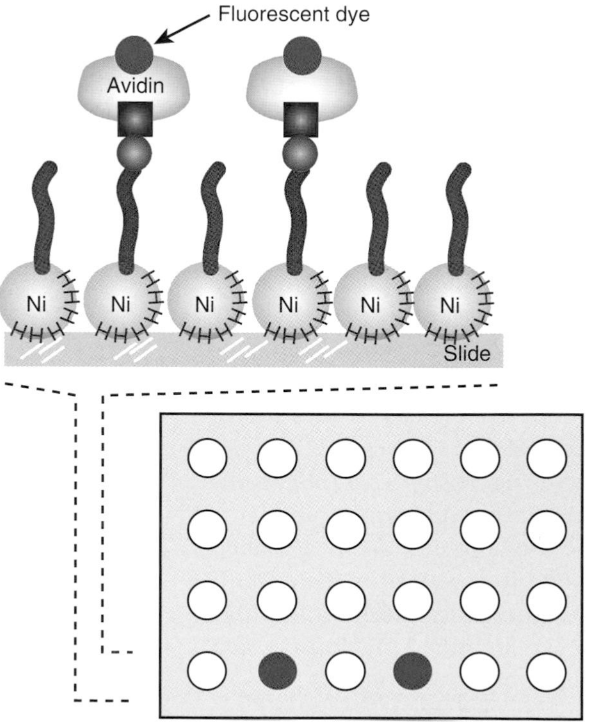

FIGURE 9.29 Screening Protein Arrays Using Biotin/Streptavidin
Protein microarrays can be screened to find proteins that bind to phospholipids. The protein microarray is incubated with phospholipid bound to biotin. Then the bound phospholipid is visualized by adding streptavidin conjugated to a fluorescent dye. Spots that fluoresce represent specific proteins that bind phospholipids.

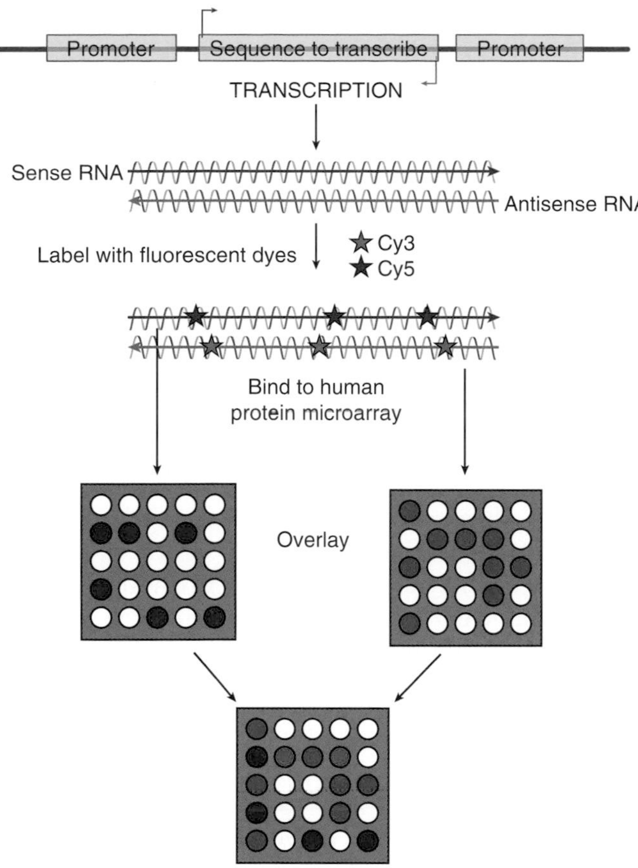

FIGURE 9.30 RNA Binding to Human Protein Microarray

A human microarray was screened for binding of RNA molecules. Both sense and antisense strands were used for each RNA molecule examined. The two strands were labeled with fluorescent dyes of different colors (red for sense and green for antisense). If a protein bound both strands, then a yellow spot was seen. Modified after Siprashvili Z et al. (2012). Identification of proteins binding coding and non-coding human RNAs using protein microarrays. *BMC Genomics* **13**, 633.

Proteins in an array can be screened for binding other proteins or small molecules such as enzyme substrates or signal molecules. It is thus possible to screen a proteome for enzymes that use a particular substrate, provided that the substrate is available and can be fluorescently labeled without preventing activity. Recently, protein arrays have been screened for binding of small hairpin RNA molecules (Fig. 9.30).

Various arrays are used to screen proteins. They may be divided into immunology-based approaches and tag-based approaches. The immunoassays depend on binding of antibodies to their target proteins. Either the antibody or target protein is labeled with a fluorescent dye. In tag-based arrays, proteins are attached to a support via tags such as His6 or GST. They are then screened with a variety of target molecules that carry fluorescent dyes.

METABOLOMICS

As methods for identifying small molecules become more accurate and sensitive, metabolic research has become more global. The **metabolome** consists of all the small molecules and metabolic intermediates within a system, such as a cell or whole organism, at one particular time. Understanding the metabolome is complex because small metabolites affect many other components of a cell. Metabolites flow in a complex network and form many different transient complexes. The network of metabolites may be compared to city streets. At each corner, a decision on which route to take must be made, and such decisions continue until the final destination is reached. Each metabolite molecule follows a specific pathway, often with several potential branches, and at each junction, a decision is made before moving on to the next step. Characterizing the metabolome under particular conditions is known as **metabolic fingerprinting**.

Several techniques that involve separating and/or identifying many small metabolites simultaneously have made metabolomics possible. Nuclear magnetic resonance (NMR) of extracts from cells grown with ^{13}C-glucose has allowed simultaneous measurement of multiple metabolic intermediates. Metabolites have also been identified by thin layer chromatography after growth in ^{14}C-glucose or by HPLC with UV or fluorescence detection. These methods are not very sensitive, and some metabolites may not be separated or identified. Nonetheless, NMR is widely used for the following reasons: (a) it is highly reproducible; (b) it is nondestructive; (c) it is rapid (little sample processing and no chemical derivatization are necessary); and (d) it yields detailed structural data.

Overall, mass spectroscopy offers the best way to analyze whole metabolomes. The technique can identify many different metabolites (even novel ones) and is extremely sensitive. Mass spectroscopy can determine the exact molecular formula for a compound, so every metabolite can be identified. Even if isomers exist, their fragmentation patterns will be different although the molecular formula is the same.

The use of mass spectroscopy is often combined with other separation methods to simplify analysis. Different types of chromatography are used to separate the complex cellular extract into different fractions, which can then be analyzed by mass spectroscopy. These methods include liquid chromatography, gas chromatography, and capillary electrophoresis. The disadvantages of these mass spectroscopy-based methods are that they are slow and separation by the associated chromatography techniques is often difficult to reproduce accurately. In practice, a good approach is an initial survey by NMR followed by more detailed analysis via mass spectroscopy.

Metabolomics is especially valuable in studying plants because metabolites affect the pigments, scents, flavors, and nutrient content. These traits are all commercially important, and using mass spectroscopy to analyze these metabolites will aid in developing better-tasting and fresher produce. For example, in strawberries, 7000 metabolites can be identified by mass spectroscopy (Fig. 9.31). Comparing white and red strawberries has identified which of the metabolite peaks in the mass spectrum corresponds to the intermediates in pigment synthesis.

As with other "-omics," recent technical advances now allow metabolomics to be performed on single cells or even organelles. Both plant cells (Fig. 9.32) and animal cells have been analyzed by using this approach. Different types of cells within higher organisms carry out very diverse roles and also respond very differently to stimuli. Consequently, they show highly varied metabolomes. Single cell analysis is valuable in revealing cellular diversity and allowing comparison of different cell types. For example, when pollen lands on the stigma of a female plant, it germinates to produce a tube. Use of mass spectroscopy coupled with gas chromatography can detect over

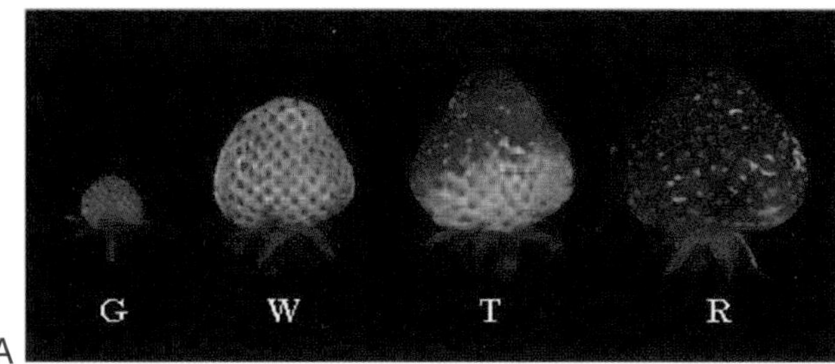

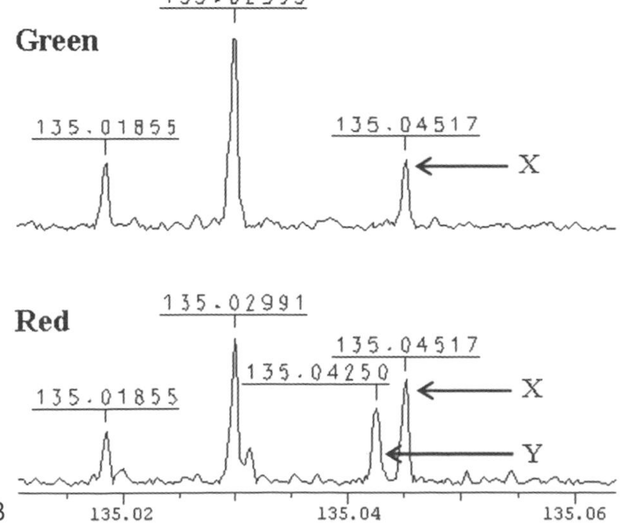

FIGURE 9.31 Metabolome Analysis of Strawberry
Nontargeted metabolic analysis in strawberry. (A) Four consecutive stages of strawberry fruit development (G, green; W, white; T, turning; R, red) were subjected to metabolic analysis using Fourier transform mass spectrometry (FTMS). Similar fruit samples were used earlier to perform gene expression analysis using cDNA microarrays. (B) An example of high-resolution (>100,000) separation of very close mass peaks in data obtained from the analysis of green and red stages of fruit development. Peaks marked with an X have the same mass, whereas peak Y is different by a mere 3 ppm. Courtesy of Phenomenome Discoveries Inc., Saskatoon, Canada.

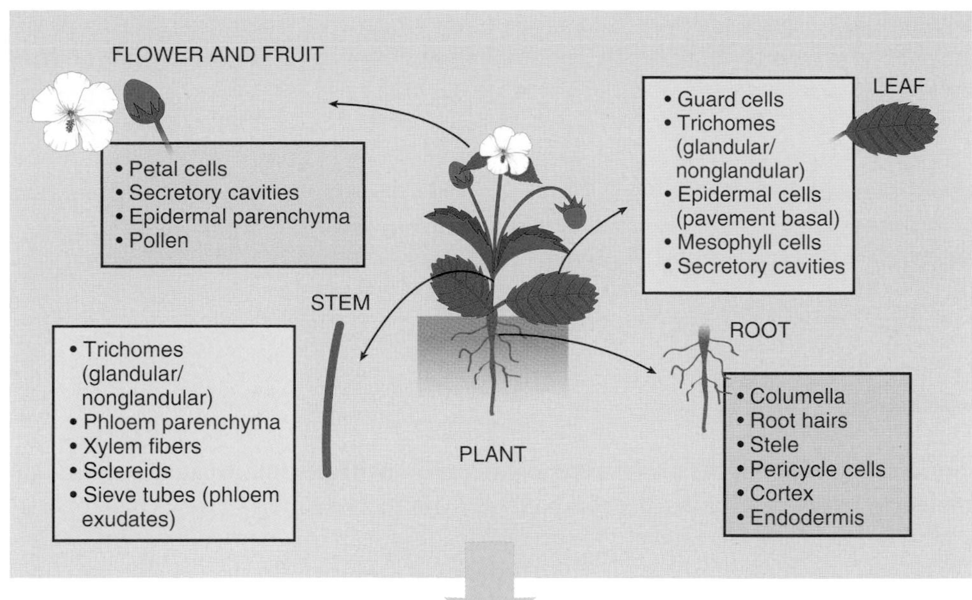

• LASER MICRODISSECTION
• MICROMANIPULATION
• MECHANICAL ISOLATION
• PROTOPLASTING
• CELL SORTING

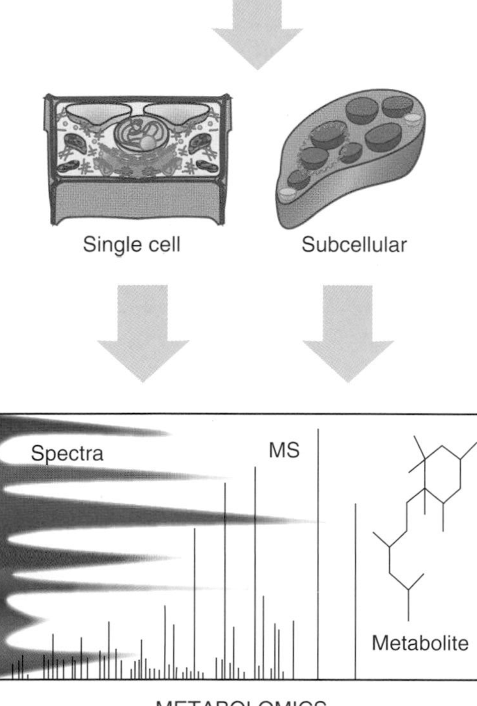

FIGURE 9.32 Plant Single-Cell Metabolomics
Single cells can be isolated by laser microdissection, micromanipulation, mechanical isolation, protoplasting, and cell sorting. It is also possible to isolate some subcellular components, such as chloroplasts or vacuoles. Cells from different tissues, such as stems, roots, leaves, and flowers, will give very different metabolomes. Analysis is carried out by multiple techniques, especially NMR and mass spectroscopy. Modified after Misra BB, Assmann SM, Chen S (2014). Plant single-cell and single-cell-type metabolomics. *Trends Plant Sci* **19**, 637–646.

250 metabolites in single pollen cells. As the pollen tube grows, energy-generating pathways are activated as shown by increases in intermediates belonging to glycolysis and the citric acid cycle.

The metabolome consists of the entire complement of small molecules and metabolites within a system, such as a cell or whole organism. Because the metabolome is dynamic, a metabolic fingerprint is the entire complement of small metabolites at one particular point in time.

Summary

Proteomics is the study of the protein complement for an organism. Because proteins change in response to many conditions, the proteome is dynamic, adapting to new challenges and environments. The term *translatome* refers to the proteome at one particular point in time. SDS-PAGE separates proteins by size. These proteins are trapped inside a gel but can be transferred to nitrocellulose for Western blot analysis. In the Western blot, a protein of interest can be visualized by adding labeled antibodies. The Western blot can be used to determine the relative abundance of the protein of interest. Another useful tool to separate proteins is HPLC. This method keeps the proteins in a liquid, and the column materials separate the proteins of interest from the mixture. These columns vary greatly, making HPLC a key method in separating protein mixtures.

Many proteins are difficult to isolate because of their chemistry or size. Proteases are enzymes that break the peptide bond. There is a huge variety of proteases, some with specific binding/cutting sites and others with nonspecific effects. Using proteases to digest a protein makes them more manageable for research.

Mass spectroscopy breaks molecules into their ions and records their mass to charge ratio, which is calculated by the time it takes for the ions to travel through the flight tube. Until the advent of MALDI and ESI, proteins were too large and complex for mass spectroscopy. MALDI and ESI are two new methods for preparing the protein for ionization. ESI is particularly useful because the proteins are ionized from a liquid solvent. This method can be linked with HPLC, as in MuDPIT. Here, a two-dimensional liquid chromatography column separates complex peptide mixtures by size, charge, and hydrophobicity before analyzing them by mass spectroscopy.

Once the protein of interest has been identified by HPLC, mass spectroscopy, or SDS-PAGE, the gene sequence can be found. Once cloned, the gene then can be expressed into protein under the control of a regulated promoter. To identify the protein of interest from the remaining cellular proteins, researchers use genetic fusions to either short tags or full proteins. A function for the protein of interest is often hard to determine, and being able to express the protein and isolate the protein via these tags is key to studying its function.

Sometimes finding new protein-binding partners can further the understanding of the protein's function. In yeast two-hybrid analysis, the protein of interest is genetically fused to one half of a transcription factor, GAL4. Potential binding partners are genetically fused to the other half of GAL4. When the protein of interest binds to a different protein, then the transcription factor turns on the reporter gene. This changes the yeast physiology, marking the cell. The gene for the binding partner can be isolated from the yeast, sequenced, and possibly identified. In a related experiment, co-immunoprecipitation finds binding partners for a protein of interest.

Just as microarrays are used for mass screening of nucleic acids, protein arrays can be used for global protein analysis. They consist of arrays with many different proteins applied as spots on a solid support. One type of protein array contains antibodies for screening. The other kind represents the proteome and consists of many different cellular proteins attached by tags. Such arrays may be screened for binding of a variety of labeled molecules to the proteins. These include enzyme substrates, other proteins, and RNA.

Metabolomics is a newer field of research that looks at changes in metabolites. This method is useful in understanding how cells change biochemical pathways in response to the environment or during development. Metabolites are analyzed using mass spectroscopy techniques.

End-of-Chapter Questions

1. Why is SDS used in the electrophoresis of proteins?
   - **a.** SDS coats the protein with a negative charge so that the sample can run through the gel.
   - **b.** SDS is a specific protease that digests large proteins in the sample.
   - **c.** SDS allows the coomassie blue stain to bind to the proteins in the gel so that they may be visualized.
   - **d.** SDS adds more molecular weight to each sample so that the proteins do not run off the end of the gel.
   - **e.** none of the above
2. What is an issue with using 2D-PAGE?
   - **a.** Hydrophobic proteins may not run as expected due to the hydrophobic surfaces.
   - **b.** Highly expressed proteins may cover up proteins that are not as abundant but running in the gel nearby.
   - **c.** Some proteins may not migrate through polyacrylamide and therefore not be represented on the gel.
   - **d.** Rare cellular proteins are hard to visualize with Coomassie blue protein stain.
   - **e.** All of the above are issues with 2D-PAGE.
3. Which one of the following is not used during Western blotting?
   - **a.** secondary antibody with a conjugated detection system
   - **b.** agarose gel electrophoresis
   - **c.** non-fat dry milk
   - **d.** primary antibody that recognizes the protein
   - **e.** nitrocellulose membrane
4. Which of the following statements about HPLC is not correct?
   - **a.** There are two phases to HPLC: mobile and stationary.
   - **b.** Separation, identification, and purification of proteins are just a few of the applications for HPLC.
   - **c.** The downside to HPLC is that it is not very adaptable due to the availability of stationary phase material.
   - **d.** Adjusting the experimental conditions, changing the particle size of the stationary phase, and controlling temperature are factors that affect resolution.
   - **e.** All of the above statements are true.
5. Which of the following is not an example of protease activity?
   - **a.** Some proteases cleave the phosphodiester bond between nucleic acid residues.

(*Continued*)

b. Some proteases cleave within a protein sequence and other proteases snip off residues from either end.
c. Some proteases contain serine, cysteine, threonine, or aspartic acid residues within their active sites.
d. Proteases hydrolyze the peptide bond between amino acid residues.
e. Metalloproteases contain metal ion cofactors within their active site.

6. Which of the following statement about mass spectroscopy is incorrect?
a. MS ionizes the sample and then measures the time it takes for the ions to reach the detector.
b. SELDI-MS has great potential for analyzing protein profiles of body fluids and may, in the future, be used to identify diseases before symptoms appear.
c. Glycosylation and phosphorylation of proteins can be identified using ESI or MALDI.
d. ESI is able to handle much larger ions than MALDI.
e. The time-of-flight for ions is directly correlated with the mass of the ion in mass spectroscopy.

7. Which of the following statements is not true about sequencing peptides with mass spectroscopy?
a. The entire protein can be sequenced all at once using mass spectroscopy.
b. Some purified proteins must be digested with proteases to eliminate undesirable characteristics such as hydrophobicity and solubility.
c. Two rounds of mass spectroscopy are used to determine the sequence.
d. In order to determine the sequence, a pure sample of protein is obtained through 2D-PAGE or HPLC.
e. A database of protein ion spectra is used to compare the peaks of the unknown peptide to determine the sequence.

8. Which of the following is used to quantify proteins with mass spectroscopy?
a. ^{2}H
b. ^{33}P
c. ^{35}S
d. ^{32}P
e. ^{125}I

9. Why are protein tags useful?
a. Protein tags are exactly the same thing as reporter fusions and perform similar functions.
b. Tags allow the protein to be isolated and purified from other cellular proteins.
c. Tags allow the protein to be quantitated.
d. Protein tags enable the protein to which they are fused to perform their function more readily.
e. None of the above.

10. How is biopanning useful to proteomics research?
a. To express large amounts of protein on the cell surface of yeast.
b. To screen expression libraries in *E. coli*.
c. To alter the cell membrane structures of cells by expressing foreign proteins on the cell surface.

- **d.** To isolate specific peptides that bind to a specific target protein.
- **e.** all of the above

11. Which of the following is needed to perform yeast two-hybrid assays?
- **a.** Two vectors are needed to express the bait and prey proteins.
- **b.** A reporter gene under the control of the GAL4 recognition sequence.
- **c.** The DBD of a transcription factor genetically fused to the protein of interest, also called the bait.
- **d.** The AD domain of a transcription factor genetically fused to proteins that are being screened for interactions with bait.
- **e.** All of the above are needed.

12. For what is co-immunoprecipitation used?
- **a.** to determine if a protein-of-interest binds to a specific DNA sequence
- **b.** to examine protein–protein interactions in the nucleus instead of in the cytoplasm
- **c.** to examine protein–protein interactions in the cytoplasm instead of the nucleus
- **d.** to allow a protein to be expressed in mammalian cell culture
- **e.** none of the above

13. What is a problem associated with immuno-based arrays?
- **a.** Proteins that are bound to solid supports may not be representative of intracellular conditions.
- **b.** The antibody may cross-react with other cellular proteins, producing a false positive.
- **c.** Low concentrations of some proteins may not be able to compete for active sites compared to those that are in abundance.
- **d.** Proteins that are often found in complexes may have the antibody binding site masked by the other proteins in the complex.
- **e.** All of the above are problems associated with immuno-based arrays.

14. Which of the following has been extensively studied using protein interaction arrays?
- **a.** proteins in yeast that bind calmodulin or phospholipids
- **b.** proteins in yeast that bind to glutathione-S-transferase
- **c.** proteins that are able to bind to biotin and streptavidin
- **d.** proteins that are able to bind to various cofactors present in the sample
- **e.** none of the above

15. Which of the following methods is the best way to analyze a metabolome?
- **a.** high-pressure liquid chromatography
- **b.** mass spectroscopy
- **c.** nuclear magnetic resonance
- **d.** thin layer chromatography
- **e.** ELISA

16. Which of the following would be the best method to identify thousands of different proteins simultaneously?
- **a.** MALDI-TOF
- **b.** traditional MS
- **c.** SILAC
- **d.** MuDPIT
- **e.** SELDI

(Continued)

17. Proteins in a protein arrays can be screened for binding to all of the following except ______________.
    a. small hairprin RNAs
    b. enzyme substrate
    c. proteins
    d. signaling molecules
    e. DNAs

18. All of the following statements concerning metabolomics is true except ______________.
    a. metabolomics can help produce fresher produce
    b. metabolomics is the identification of small molecules and metabolic intermediates within a system
    c. NMR and MS methods in metabolomics are not very sensitive methods for studying the metabolome
    d. prior to MS, cell extract is separated by chromatography
    e. in higher organisms, different cell types have different metabolomes

Further Reading

Bell, M. R., Engleka, M. J., Malik, A., & Strickler, J. E. (2013). To fuse or not to fuse: what is your purpose? *Protein Science, 22*(11), 1466–1477.

Gubbens, J., Zhu, H., Girard, G., Song, L., Florea, B. I., Aston, P., et al. (2014). Natural product proteomining, a quantitative proteomics platform, allows rapid discovery of biosynthetic gene clusters for different classes of natural products. *Chemical Biology, 21*, 707–718.

Hamdi, A., & Colas, P. (2012). Yeast two-hybrid methods and their applications in drug discovery. *Trends in Pharmacological Sciences, 33*(2), 109–118.

Hong, Y. S. (2011). NMR-based metabolomics in wine science. *Magnetic Resonance in Chemistry: MRC, 49*(Suppl 1), S13–21.

Jorge, I., Burillo, E., Mesa, R., Baila-Rueda, L., Moreno, M., Trevisan-Herraz, M., et al. (2014). The human HDL proteome displays high inter-individual variability and is altered dynamically in response to angioplasty-induced atheroma plaque rupture. *Journal of Proteomics, 106C*, 61–73.

Medzihradszky, K. F. (2005). Peptide sequence analysis. *Methods in Enzymology, 402*, 209–244.

Oliveira, B. M., Coorssen, J. R., & Martins-de-Souza, D. (2014). 2DE: the phoenix of proteomics. *Journal of Proteomics, 104*, 140–150.

Misra, B. B., Assmann, S. M., & Chen, S. (2014). Plant single-cell and single-cell-type metabolomics. *Trends in Plant Science, 19*(10), 637–646.

Paget, T., Haroune, N., Bagchi, S., & Jarroll, E. (2013). Metabolomics and protozoan parasites. *Acta Parasitologica, 58*(2), 127–131.

Rakonjac, J., Bennett, N. J., Spagnuolo, J., Gagic, D., & Russel, M. (2011). Filamentous bacteriophage: biology, phage display and nanotechnology applications. *Current Issues in Molecular Biology, 13*(2), 51–76.

Rotilio, D., Della Corte, A., D'Imperio, M., Coletta, W., Marcone, S., Silvestri, C., et al. (2012). Proteomics: bases for protein complexity understanding. *Thrombosis Research, 129*, 257–262.

Salzano, A. M., Novi, G., Arioli, S., Corona, S., Mora, D., & Scaloni, A. (2013). Mono-dimensional blue native-PAGE and bi-dimensional blue native/urea-PAGE or/SDS-PAGE combined with nLC-ESI-LIT-MS/MS unveil membrane protein heteromeric and homomeric complexes in Streptococcus thermophilus. *Journal of Proteomics, 94*, 240–261.

Siprashvili, Z., Webster, D. E., Kretz, M., Johnston, D., Rinn, J. L., Chang, H. Y., & Khavari, P. A. (2012). Identification of proteins binding coding and non-coding human RNAs using protein microarrays. *BMC Genomics, 13*, 633.

Stynen, B., Tournu, H., Tavernier, J., & Van Dijck, P. (2012). Diversity in genetic *in vivo* methods for protein–protein interaction studies: from the yeast two-hybrid system to the mammalian split-luciferase system. *Microbiology and Molecular Biology Reviews: MMBR, 76*(2), 331–382.

Sun, H., Chen, G. Y., & Yao, S. Q. (2013). Recent advances in microarray technologies for proteomics. *Chemical Biology, 20*(5), 685–699.

Toniolo, L., D'Amato, A., Saccenti, R., Gulotta, D., & Righetti, P. G. (2012). The Silk Road, Marco Polo, a bible and its proteome: a detective story. *Journal of Proteomics, 75*(11), 3365–3373.

Valiente, M., Obenauf, A. C., Jin, X., Chen, Q., Zhang, X. H. -F., Lee, D. J., et al. (2014). Serpins promote cancer cell survival and vascular co-option in brain metastasis. *Cell, 156*, 1002–1016.

Washburn, M. P., Wolters, D., & Yates, J. R. (2001). Large-scale analysis of the yeast proteome by multidimensional protein identification technology. *Nature Biotechnology, 19*, 242–247.

Zenobi, R. (2013). Single-cell metabolomics: analytical and biological perspectives. *Science, 342*(6163), 1243259.

Recombinant Proteins

Proteins and Recombinant DNA Technology
Expression of Eukaryotic Proteins in Bacteria
Insulin and Diabetes
Cloning and Genetic Engineering of Insulin
Translation Expression Vectors
Codon Usage Effects
Avoiding Toxic Effects of Protein Overproduction
Inclusion Bodies and Protein Refolding
Increasing Protein Stability
Improving Protein Secretion
Protein Fusion Expression Vectors
Protein Glycosylation
Expression of Proteins by Eukaryotic Cells
Expression of Proteins by Yeast
Expression of Proteins by Insect Cells
Expression of Proteins by Mammalian Cells
Expression of Multiple Subunits in Mammalian Cells
Comparing Expression Systems

Biotechnology

http://dx.doi.org/10.1016/B978-0-12-385015-7.00010-7

PROTEINS AND RECOMBINANT DNA TECHNOLOGY

Proteomics has opened the door to identify more and more clinically relevant proteins. Once identified, these proteins need to be studied in detail, including expression of the protein in model organisms by using recombinant DNA techniques (see Chapter 3). Some proteins will become therapeutic agents, and large amounts of purified protein will be required.

Once a gene has been cloned, the protein it encodes can be produced in large amounts with relative ease. Smaller, nonprotein molecules, which seem simpler to an organic chemist, would need half a dozen proteins (enzymes) working in series to synthesize them. Thus, paradoxically, proteins, despite being macromolecules, have been more susceptible to genetic engineering than simpler products such as antibiotics. Pathway engineering to produce small organic molecules will be discussed in Chapter 13.

Today, over 100 **recombinant proteins** are in use as therapeutic agents. Of these, nearly half are monoclonal antibodies (discussed in Chapter 6). In terms of sales (2010 data), antibodies accounted for about $50 billion; insulin and analogs, for about $16 billion; blood clotting factors and erythropoietin, $16 billion; and the rest, $25 billion. Here, we consider those proteins that are not antibodies and that may be subdivided by function as follows:

(a) Replacements for proteins that are missing or defective
(b) Increasing the amount of proteins already present
(c) Inhibition of infectious agents, such as bacteria or viruses
(d) Carriers for other molecules (mostly still in development)

These categories are not mutually exclusive; for example, use of interferons to combat virus infection falls into both classes (b) and (c). Most of those used therapeutically are human proteins, although those from animals are also found. Some examples of recombinant proteins in clinical use are given in Table 10.1.

Table 10.1 Proteins Produced by Recombinant Technology

Protein	Function
Erythropoietin	Promoting red blood cell formation in the treatment of anemia
Factor VIII	Helping blood clots form in hemophiliacs
Filgrastim and sargramostim (blood cell–stimulating bone marrow factors)	Boosting white blood cell counts after radiation therapy or transplantation
Insulin	Treating diabetes
Insulin-like growth factor 1 (IGF1)	Treating certain growth problems
Interferon (alpha)	Treating hepatitis B and C, genital warts, certain leukemias and other cancers
Interferon (beta)	Treating multiple sclerosis
Interferon (gamma)	Treating chronic granulomatous disease
Interleukin-2	Killing tumor cells
Somatotropin	Treating growth hormone deficiency
Tissue plasminogen activator (t-PA)	Dissolving blood clots to prevent heart attacks and lessen their severity

Expressing a gene for large-scale production brings extra problems compared to a laboratory setting. The more copies of a gene that a cell contains, the higher the level of the gene product. Thus, cloning a gene onto a high-copy-number plasmid will usually give higher yields of a gene product. However, high-copy plasmids are often unstable, especially in the dense cultures used in industrial situations. Although the presence of antibiotic resistance genes on most plasmids provides a method to maintain the plasmid in culture, antibiotics are expensive, especially on an industrial scale. One solution to prevent plasmid loss is to integrate the foreign gene into the chromosome of the host cell. This, however, decreases the copy number of the cloned gene to one. Attempts have been made to insert multiple copies of cloned genes in tandem arrays. However, the presence of multiple copies results in instability due to recombination between homologous sequences of DNA.

Recombinant proteins are clinically relevant proteins produced in large scale. The gene for the protein of interest is cloned into a vector and expressed into protein in a model organism.

EXPRESSION OF EUKARYOTIC PROTEINS IN BACTERIA

In Chapter 3, we discussed the basics of cloning genes onto a variety of vectors. Obviously, bacterial genes will usually be expressed when carried on cloning vectors in bacterial host cells, provided that they are next to a suitable bacterial promoter. Special plasmids known as **expression vectors** are often used to enhance gene expression. As noted in Chapter 3, these vectors provide a strong promoter to drive expression of the cloned gene. Expression vectors also contain genes for antibiotic resistance to allow selection of the vector and therefore the recombinant protein. In addition, they must have an origin of replication appropriate to the host.

The expression of eukaryotic proteins is more problematic. Although eukaryotic cells can be used to express eukaryotic proteins, bacteria are simpler to grow and manipulate genetically. Therefore, it is often desirable to express eukaryotic proteins in bacteria (Fig. 10.1). Because eukaryotic promoters do not work in bacterial cells, it is necessary to provide a bacterial promoter. In addition, bacteria cannot process introns; therefore, it is standard procedure to clone the cDNA version of eukaryotic genes, which lacks the introns and consists solely of uninterrupted coding sequence. In fact, the cDNA version of eukaryotic genes is generally used, even for expression in eukaryotic cells, not only to avoid possible processing problems, but also because the amount of cloned DNA is much smaller and consequently easier to handle.

Even if a cloned eukaryotic gene is transcribed at a high level, production of the encoded protein in bacteria may

FIGURE 10.1 Expression of Eukaryotic Gene in Bacteria—Overview
Eukaryotic genes must be adapted for expression in bacteria. First, the mRNA from the gene of interest is converted to cDNA to provide uninterrupted coding DNA. The cDNA is cloned between a bacterial promoter and a bacterial terminator so the bacterial transcription and translation machinery express the coding sequence.

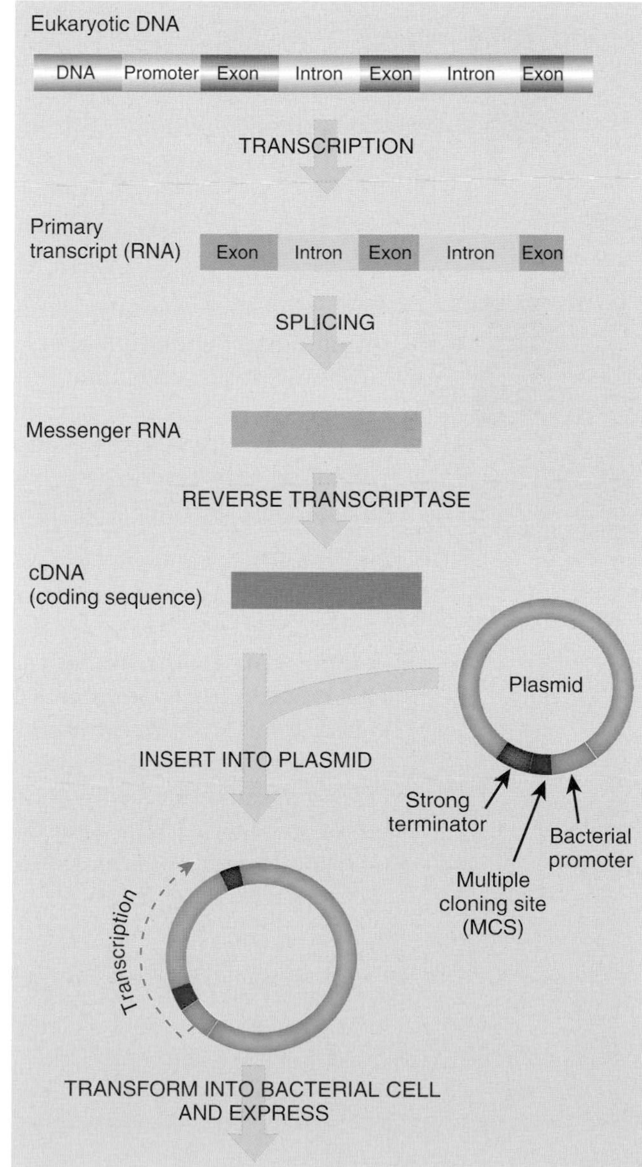

be limited at the stage of protein synthesis. Different mRNA molecules are translated with differing efficiencies. Several factors are involved:

(a) The ribosome-binding site may interact poorly, or not at all, with the ribosome.
(b) mRNA may be unstable or have strong secondary structure.
(c) Codons common in the cloned gene may be rare in the bacterial host and have a limited supply of the corresponding tRNAs.

In addition to standard expression vectors, more sophisticated vectors exist to optimize these other aspects of protein production. In this chapter we discuss the use of translation vectors and fusion vectors to increase the synthesis of a recombinant protein from a cloned gene.

Expression vectors are used to make eukaryotic proteins in bacteria. The vector has the ribosome-binding site, terminator sequences, and a strong regulated promoter. The eukaryotic gene is a cDNA copy of the mRNA.

INSULIN AND DIABETES

Insulin was the first genetically engineered hormone to be made commercially available for human use. Before cloned human insulin was available, people with diabetes gave themselves injections of insulin extracted from the pancreas of animals such as cows or pigs. Although this approach worked well on the whole, occasional allergic reactions occurred, usually to low-level contaminants in the extracts. Today, genuine human insulin (Humulin, marketed by Eli Lilly Inc.) made by recombinant bacteria is available.

Diabetes mellitus is actually a group of related diseases in which the level of glucose in the blood and/or urine is abnormally high. There is considerable variation in the detailed symptoms, and multiple genes are involved. Many cases of diabetes are due to the absence of insulin, a small protein hormone made by the pancreas, which controls the level of sugar in the blood. Lack of insulin results in high blood sugar and causes a variety of complications. In patients with insulin-dependent diabetes mellitus (IDDM), injections of insulin keep blood sugar levels down to near normal. Other defects affect the **insulin receptor**, and so they do not respond to insulin treatment.

Insulin is a protein made of two separate polypeptide chains: the A- and B-chains (Fig. 10.2). Disulfide bonds hold the two chains together. Although the final protein has two polypeptide chains, insulin is actually encoded by a single gene. The original gene product, **preproinsulin**, is a single polypeptide chain, which contains both the A- and B-chains together with the C- (or connecting) peptide and a signal sequence. Preproinsulin itself is not a hormone but must first be processed to give insulin. The signal sequence at the N-terminal end is required for secretion and is then removed by signal peptidase. This leaves **proinsulin**. Removal of the **C-peptide** requires **endopeptidases** that cut within the polypeptide chain. They recognize pairs of basic amino acids at the junctions of the C-peptide with the A- and B-chains. Finally, the terminal Arg and Lys residues are trimmed off by **carboxypeptidase H**.

Insulin is a hormone produced as preproinsulin. The signal sequence is removed by signal peptidase, the C-peptide is removed by endopeptidase, and the final arginine and lysine are trimmed by carboxypeptidase H. After processing, insulin has two chains, A and B, linked by disulfide bonds.

Some diabetics do not produce any insulin and require the insulin as a shot. Other diabetics do not have the insulin receptor, so insulin cannot act on the target cells.

CLONING AND GENETIC ENGINEERING OF INSULIN

As noted above, insulin was the first hormone to be cloned and made by recombinant bacteria for clinical use. Its production is rather unusual compared to many recombinant proteins, because of complications due to processing. If the insulin gene is cloned and directly expressed in bacteria, preproinsulin is made. Because bacteria lack the mammalian processing enzymes, the preproinsulin cannot be converted into insulin (see Fig. 10.2).

Another problem is that disulfide bonds do not readily form in the cytoplasm of *E. coli*. However, if proteins are secreted into the periplasm, some disulfide bond formation occurs. Indeed, proteins such as human growth hormone, with relatively simple arrangements of disulfides, may be correctly folded. This is due to the Dsb proteins of *E. coli*, which form and reshuffle disulfides. Unfortunately, proteins with more complex multiple disulfides, especially those with more than one polypeptide chain, such as insulin, often form incorrect disulfide linkages. Overexpression of DsbC protein, which reshuffles disulfides, often improves the yield of such proteins. This approach is still experimental.

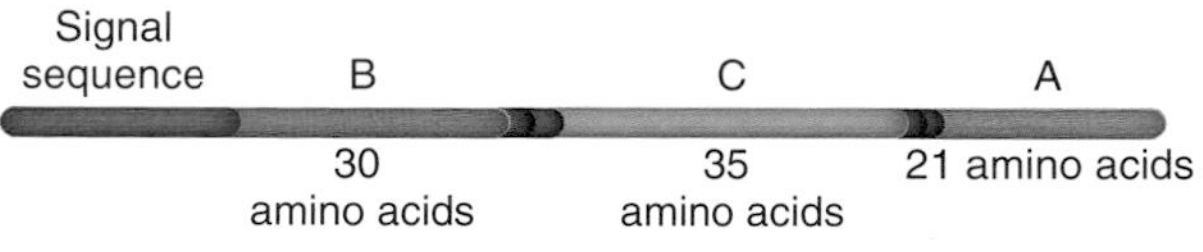

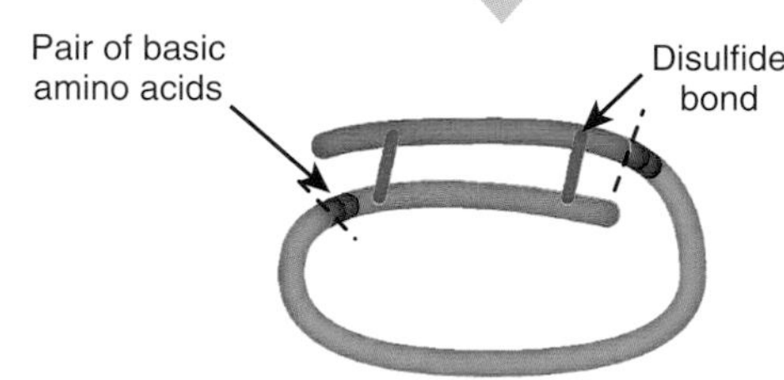

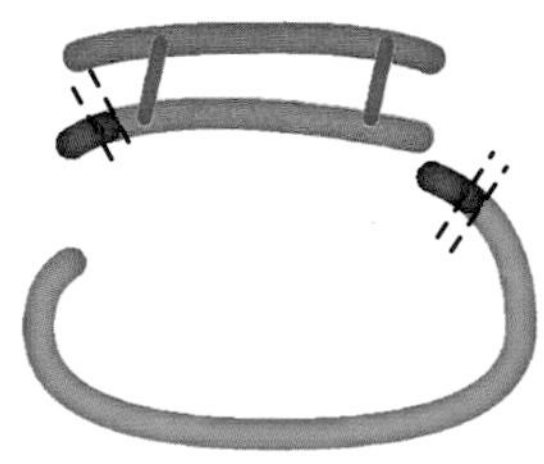

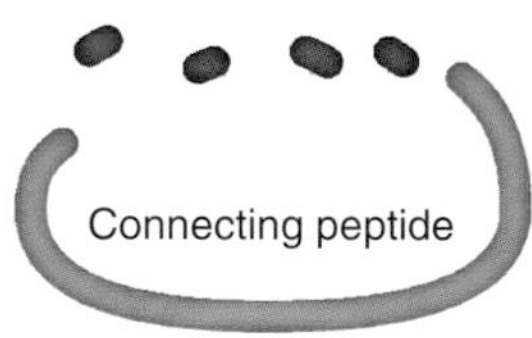

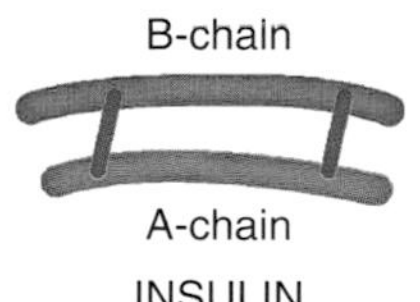

FIGURE 10.2
Processing of Insulin
The gene for insulin produces one transcript that is translated into a single protein called preproinsulin. The signal sequence of preproinsulin is removed after secretion. Next, an endopeptidase removes the C-peptide. This leaves the A- and B-chains held together by disulfide bonds. Last, carboxypeptidase H trims terminal Arg and Lys residues, leaving active insulin.

One approach to expressing cloned insulin would be to purify the preproinsulin and treat it with enzymes that convert it into insulin. This means the processing enzymes would have to be manufactured as well. Clearly, this process is overly complex. The solution chosen was to make two artificial **mini-genes**, one for the insulin A-chain and the other for the insulin B-chain (Fig. 10.3). Two pieces of DNA, encoding the two insulin chains, were synthesized chemically. The two DNA molecules were inserted into plasmids that were put into two separate bacterial hosts. Thus, the two chains of insulin were produced separately by two bacterial cultures. They were then mixed and treated chemically to generate the disulfide bonds linking the chains together.

This approach gives insulin that works well. Nonetheless, natural insulin, even natural human insulin, is not perfect, and tends to form hexamers. This clumping covers up the surfaces by which the insulin molecule binds to the insulin receptor, thus preventing most of the insulin from activating its target cells (Fig. 10.4). *In vivo*, insulin is secreted from the pancreas as a monomer and is distributed rapidly by the bloodstream before it gets a chance to clump. However, when insulin is injected, a high concentration of insulin

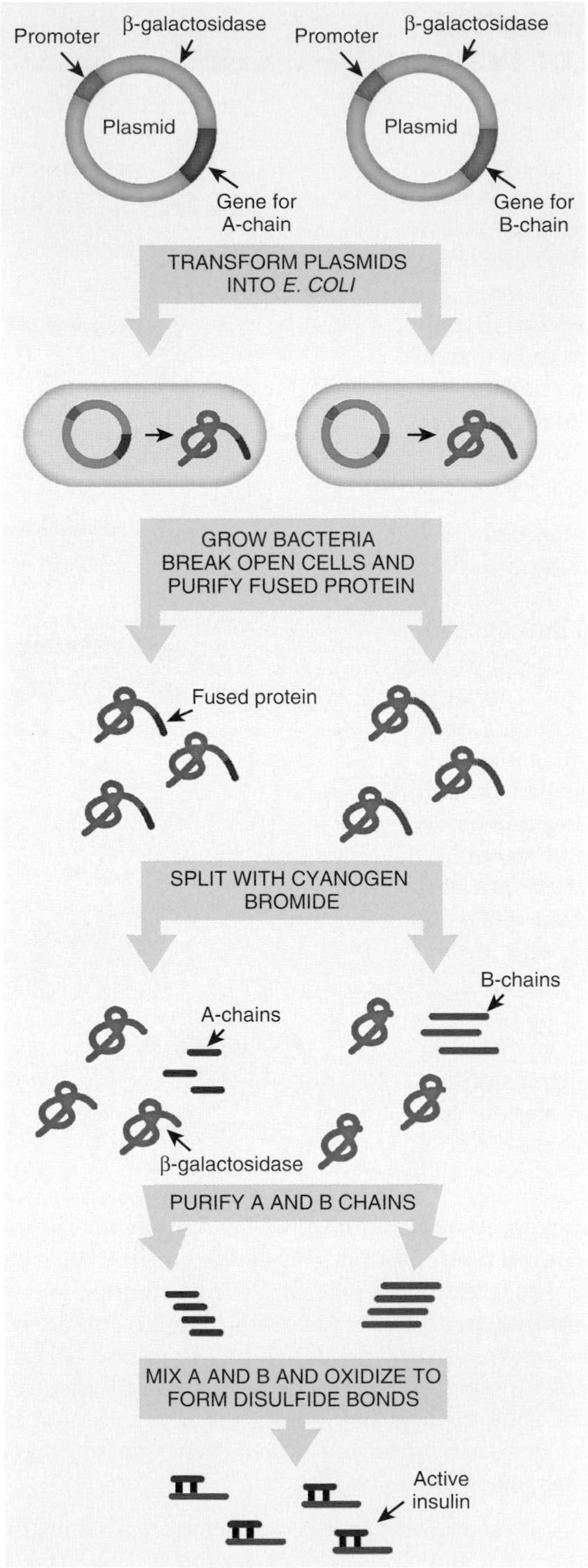

FIGURE 10.3 Cloning of Insulin as Two Mini-Genes

The genes for the A- and B-chains of insulin were cloned on two separate plasmids. Both mini-genes were fused to β-galactosidase because this protein is easy to purify. The plasmids were transformed into bacteria and expressed in separate cultures. The bacteria from each culture were harvested and the fusion proteins purified. The A- and B-chains were cleaved from β-galactosidase by cyanogen bromide and then mixed under oxidizing conditions to form the disulfide bonds, thus making human insulin.

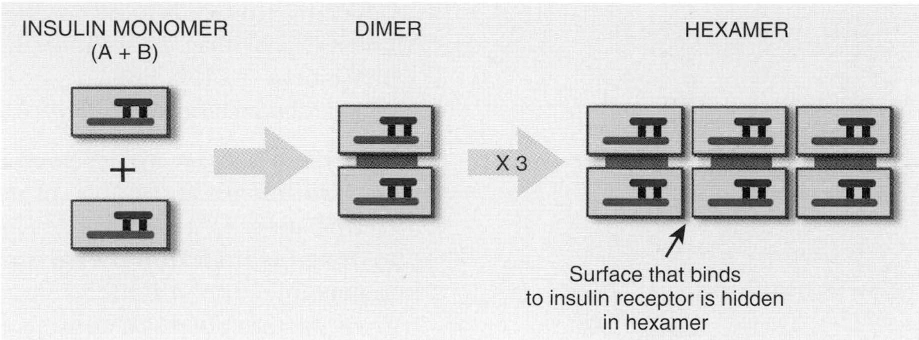

FIGURE 10.4 Insulin Forms Hexamers
High concentrations of insulin cause the monomers to clump into hexamers. The proteins stick to one another by their receptor binding sites.

is present in the syringe, and clumping occurs. After injection, it takes a while for the hexamers to dissociate, and it may take several hours for the patient's blood glucose to drop to normal levels.

Insulin was genetically engineered to prevent clumping. The DNA sequence of the insulin gene was altered to change the amino acid sequence of the resulting protein. A proline in the B-chain that is located at the surface where the insulin molecules touch each other when forming the hexamer was replaced with aspartic acid, whose side chain carries a negative charge. So when two modified insulin molecules approach each other, they are mutually repelled by their negative charges and no longer clump (Fig. 10.5). The altered insulin causes a faster drop in blood sugar than native insulin. The ProB28Asp insulin (NovoLog) was the first fast-acting insulin and is marketed by the Danish pharmaceutical company Novo. Other modifications to generate fast-acting variants of insulin have been introduced by other pharmaceutical companies.

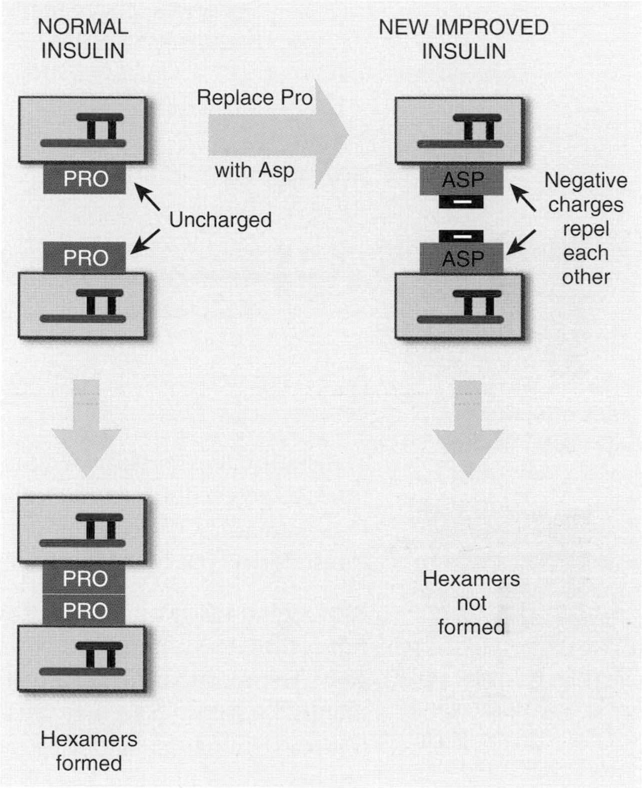

Fast-acting insulin is better for Type 1 diabetes, which results from destruction of the insulin producing cells of the pancreas. However, the converse modification, slow-acting insulin is preferred for Type 2 diabetes, which is caused by obesity. The obesity epidemic has resulted in a relative increase in Type 2 diabetes. Nowadays, about 10% of cases of diabetes are Type 1 and 90% are Type 2. The major slow-acting variant in use today (Lantus; sold by Sanofi-Aventis) has two extra Arg residues added to the end of the B-chain plus an AspA21Gly replacement in the A-chain. The increased positive charges result in precipitation upon injection. The Arg–Arg extension is nibbled off by exopeptidases, and the insulin slowly redissolves over 16–24 hours.

FIGURE 10.5 Engineered Fast-Acting Insulin
Natural insulin has a sticky patch around a proline residue, which causes two insulin molecules to dimerize and eventually form a hexamer. Using genetic engineering, the proline was replaced with a negatively charged aspartic acid residue. The negative charges repel each other and prevent hexamer formation.

> Insulin used for treating diabetes is now produced as a recombinant protein in bacteria. Recombinant insulin is expressed as two mini-genes rather than expressing preproinsulin. Changing the proline to aspartic acid prevents recombinant insulin from clumping.

TRANSLATION EXPRESSION VECTORS

As discussed in Chapter 2, bacterial ribosomes bind mRNA by recognizing the ribosome-binding site (RBS) (also known as the Shine–Dalgarno sequence). The RBS base pairs with the sequence AUUCCUCC on the 16S rRNA of the small subunit of the ribosome. The closer the RBS is to the consensus sequence (i.e., UAAGGAGG), the stronger the association.

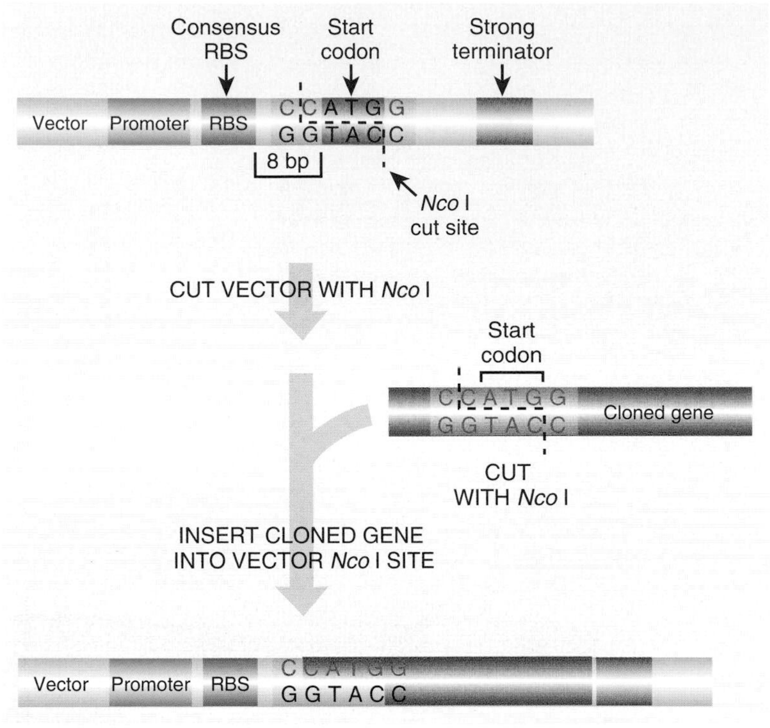

Consensus RBS
Start codon
Strong terminator
Vector Promoter RBS
8 bp
Nco I cut site
CUT VECTOR WITH *Nco* I
Start codon
Cloned gene
CUT WITH *Nco* I
INSERT CLONED GENE INTO VECTOR *Nco* I SITE
Vector Promoter RBS

Generally, this leads to more efficient initiation of translation. In addition, for optimal translation, the RBS must be located at the correct distance from the start codon, AUG.

Expression vectors are designed to optimize gene expression at the level of transcription (see Chapter 3). However, it is also possible to design **translational expression vectors** to maximize the initiation of translation. These vectors possess a consensus RBS plus an ATG start codon located an optimum distance (8 bp) downstream of the RBS. The cloned gene is inserted into a cloning site that overlaps the start codon. The restriction enzyme *Nco*I is very convenient because its recognition site (C/CATGG) includes ATG. Therefore, it is possible to insert a cloned gene so that its ATG coincides exactly with the ATG of the translational expression vector (Fig. 10.6). The gene to be expressed is cut with NcoI at its 5′-end and with another convenient restriction enzyme at its 3′-end. If necessary, an artificial NcoI site may be introduced into the gene by site-directed mutagenesis or by PCR.

FIGURE 10.6 Translational Expression Vector
The recognition site for NcoI is C/CATGG, which has an ATG in the middle. The gene of interest and the translation vector each have an NcoI site, allowing the gene of interest to be cloned exactly into the correct site for highest protein expression. The vector also has a consensus RBS spaced 8 base pairs from the ATG, which provides an optimum binding site for ribosomes. The terminator sequence(s) is also strong to prevent transcriptional run-on.

Translational expression vectors also possess a convenient selective marker, usually resistance to ampicillin or some other antibiotic, and a strong, regulated transcription promoter. Downstream of the cloning site are two or three strong terminator sequences, to prevent transcription continuing into plasmid genes.

Although translational expression vectors can optimize the initiation of translation, they do not control the other factors listed earlier, which depend on the sequence of each individual gene. The secondary structure of the mRNA may greatly influence the level of translation. If the mRNA folds up so that the RBS and/or start codon are blocked, then translation will be hindered. In particular, the sequence of the first few codons of the coding sequence should be checked for possible base pairing with the region around the RBS. If necessary, bases in the third (redundant) position of each codon may be changed to eliminate such base pairing. Active translation plays a major role in preventing mRNA instability. When mRNA is not being actively translated, it becomes subject to degradation, which decreases the protein yield.

Translational expression vectors provide the optimal ribosomal binding site. They also have strong regulated promoters and multiple terminator sequences for controlled transcription.

CODON USAGE EFFECTS

When genes from one organism are expressed in different host cells, the problem of **codon usage** arises. As explained in Chapter 2, the genetic code is redundant in the sense that several different codons may encode the same amino acid. So even though a protein has a fixed amino acid sequence, there is still considerable choice over which codons to use. In practice, different organisms favor different codons for the same amino acid. For example, both AAA and AAG encode the amino acid lysine. In *E. coli*, AAA is used 75% of the time and AAG only 25%. In contrast, *Rhodobacter* does the exact opposite and uses AAG 75% of the time, even though *E. coli* and *Rhodobacter* are both gram-negative bacteria.

Codons are read by transfer RNA (see Chapter 2). When a cell uses a particular codon only rarely, it has lower levels of the tRNA that reads the rare codon. So if a gene with many AAA codons is put into a cell that rarely uses AAA for lysine, the corresponding tRNA may be in such short supply that protein synthesis slows down (Fig. 10.7).

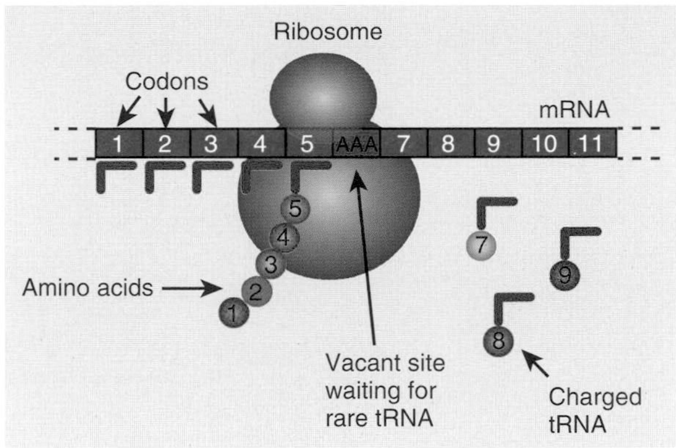

FIGURE 10.7 Codon Usage Affects Rate of Translation
Bacteria prefer one codon for a particular amino acid to other redundant codons. In this example, the ribosome is stalled because it is waiting for lysine tRNA with a UUU anticodon. *Escherichia coli* does not use this codon very often, and there is a limited supply of this tRNA.

Consequently, to optimize protein production, codon usage must be considered. Although it takes a lot of work, genes may be altered in DNA sequence so as to change many of the bases in the third position of redundant codons. This is done by artificially synthesizing the DNA sequence of the whole engineered gene. For a whole gene, individual segments of DNA that overlap are synthesized and then assembled, as discussed in Chapter 4. Genes that have been codon-optimized for new host organisms may show a 10-fold increase in the level of protein produced, due to more rapid elongation of the polypeptide chain by the ribosome. This approach has been used successfully for several of the insect toxins, originally from *Bacillus thuringiensis,* which have been cloned into transgenic plants (see Chapter 15 for details).

Another, less labor-intensive, approach to the codon issue is to supply rare tRNAs. The tRNAs that correspond to rare codons are in short supply. For example, the rarest codons in *E. coli* are AGG and AGA, which both encode arginine and occur only at frequencies of about 0.14% and 0.21%, respectively. The *argU* gene encodes the tRNA that reads both the AGA and AGG codons. Normally, this tRNA is made only in very small amounts. If the *argU* gene is supplied on a plasmid, the host *E. coli* no longer has a tRNA shortage. Consequently, recombinant proteins whose genes have high levels of AGA and AGG codons are made efficiently. Plasmids carrying all the genes for rare codon tRNAs are available commercially. These plasmids have a p15A origin of replication. This makes them compatible with most expression vectors, which typically have ColE1 replication origins (see Chapter 3 for plasmid incompatibility).

A recent survey of 90 human proteins expressed in *E. coli* showed that codon optimization improved yields in 70% of cases. Codon optimization was always superior to supplying rare tRNAs. In a few cases, codon optimization actually decreased protein yields. The reason is probably that some proteins need to be made relatively slowly in order to fold correctly. This has been directly demonstrated in a few cases, such as the circadian clock proteins of the fungus *Neurospora*.

Each organism has certain preferred codons for an amino acid. When the sequence for the recombinant protein encodes a rarely used tRNA, the protein production slows. Adding the rare tRNA or changing the rare codon wobble position solves this problem.

AVOIDING TOXIC EFFECTS OF PROTEIN OVERPRODUCTION

Although higher yields are usually desirable, overproduction of a recombinant protein may harm the host cell. In bacteria, when too much protein is manufactured too fast, the surplus forms **inclusion bodies**. These bodies are dense crystals of misfolded and nonfunctional protein (see following discussion). Thus, expression systems for recombinant proteins have features to control when and how much protein the host cell makes. Two common expression systems are the pET and pBAD systems for *E. coli*. These have control mechanisms to switch recombinant protein production on or off, and the pBAD system can also modulate the amount of protein produced.

The pET vectors have a hybrid *T7/lac* promoter, multiple cloning site, and T7 terminator. Transcription from the hybrid *T7/lac* promoter requires T7 RNA polymerase. The host *E. coli* has the gene for T7 RNA polymerase engineered into the chromosome, but the *lac* operon repressor, LacI, represses the gene. When IPTG is added, this induces release of the LacI protein and expression of the gene. T7 RNA polymerase is then made and binds to the *T7/lac* hybrid promoter, and the cloned protein of interest is manufactured (Fig. 10.8).

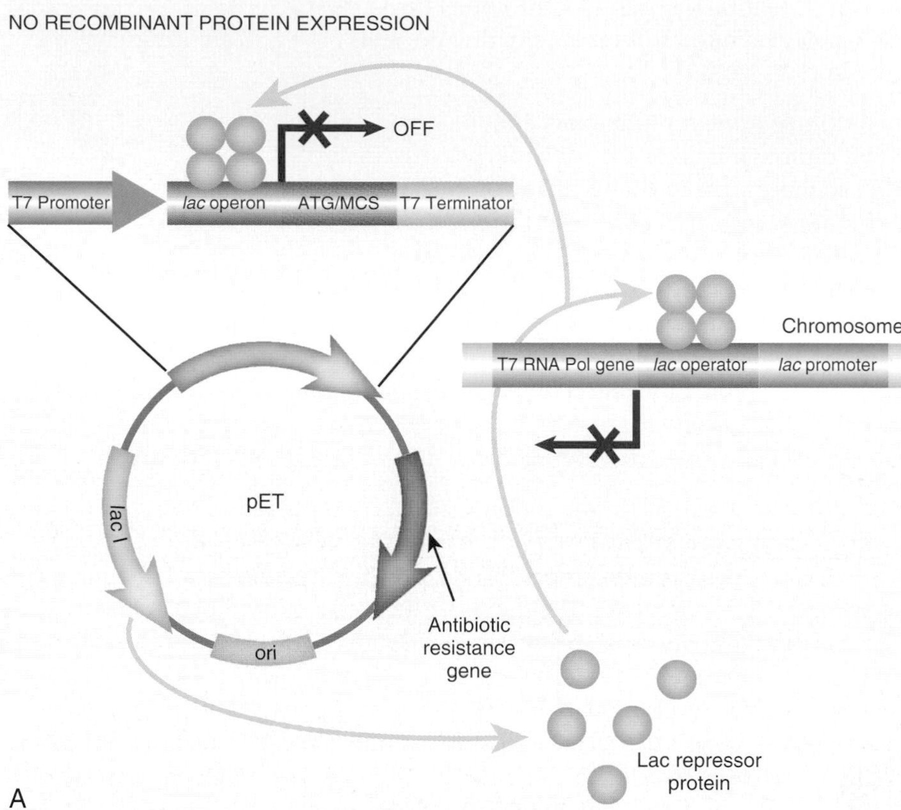

FIGURE 10.8 pET Protein Expression System
(A) Recombinant proteins are inserted into the multiple cloning site (MCS) but are not expressed in the cell until induced. The pET plasmid carries the gene for LacI protein, which represses both the gene for T7 RNA polymerase on the bacterial chromosome and the recombinant protein gene on the pET plasmid. (B) When the inducer IPTG is added, it causes release of LacI from both promoters. The gene for T7 RNA polymerase is then expressed. This polymerase then transcribes the gene for the recombinant protein.

The pBAD system is based on the arabinose operon. This is induced by adding arabinose, which binds to the AraC regulatory protein. Activated AraC exits the O_2 site and binds to the I site (Fig. 10.9). This activates transcription of the cloned gene. The pBAD system is modulated by the amount of arabinose added to the culture. If a lot of recombinant protein is needed, then more arabinose is added. However, if the recombinant protein is toxic to the host cells, then less arabinose is added, and less recombinant protein is made.

The pET and pBAD expression vectors are used to control protein production by controlling the gene expression with inducers and repressors.

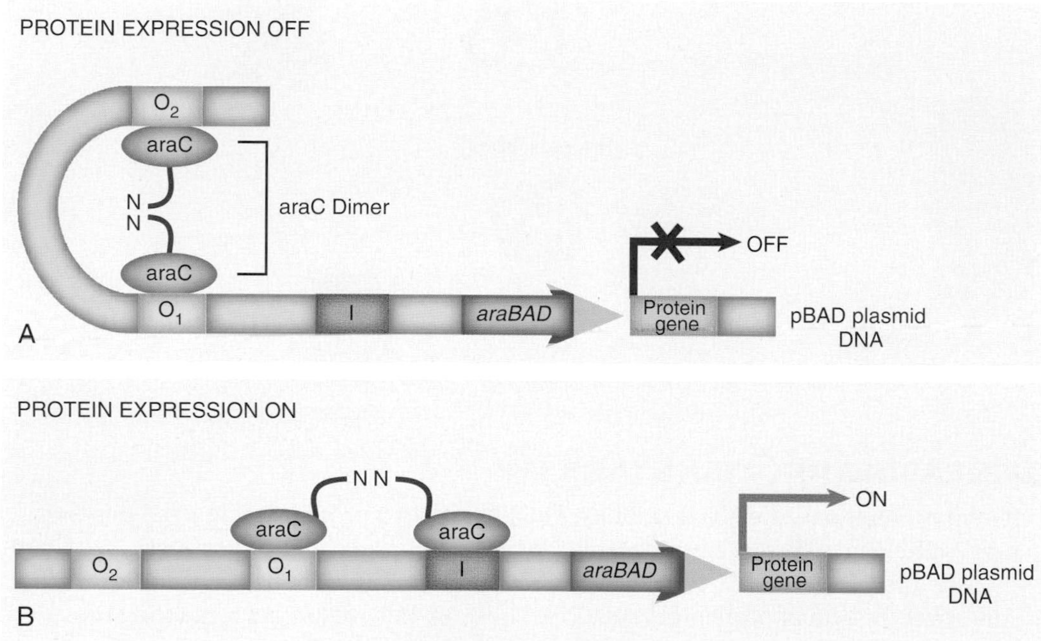

FIGURE 10.9 pBAD Expression System
(A) Recombinant proteins are not expressed when the AraC protein dimer binds to the O_1 and O_2 regulatory regions on the pBAD plasmid. (B) When arabinose is present, the sugar induces AraC to switch conformation, and it now binds to the O_1 and I sites. This conformation stimulates transcription of the recombinant protein.

INCLUSION BODIES AND PROTEIN REFOLDING

In some instances, proteins, which would be active and stable if folded correctly, fold aberrantly during synthesis. The misfolded proteins form inclusion bodies, which are dense particles visible with optical microscopy. Inclusion bodies are usually formed when recombinant proteins are poorly soluble or produced too fast.

One method to alleviate the aggregation is to provide **molecular chaperones** that help fold the recombinant protein correctly. Molecular chaperones attach themselves to polypeptides while they are being translated and help keep the protein unfolded until translation is complete. Then the protein can fold into its correct shape (Fig. 10.10). Another approach is to secrete the cloned protein into either the periplasmic space or the culture medium (see following discussion).

In practice, chaperones are rarely used, and it may be better to accept the formation of inclusion bodies and attempt refolding the protein. Proteins in inclusion bodies are often relatively pure, and sometimes the misfolding is only moderate in extent. Consequently, it may be possible to solubilize and refold the protein, hopefully obtaining biological activity. The principles for protein solubilization are well known, but precise conditions vary greatly for different proteins. Typically, chaotropic agents and urea or guanidine are used to solubilize proteins and then removed later.

To enhance protein stability, the recombinant protein can be coexpressed with a molecular chaperone that helps the protein fold correctly. Alternatively, inclusion bodies can be purified and attempts made to refold the protein.

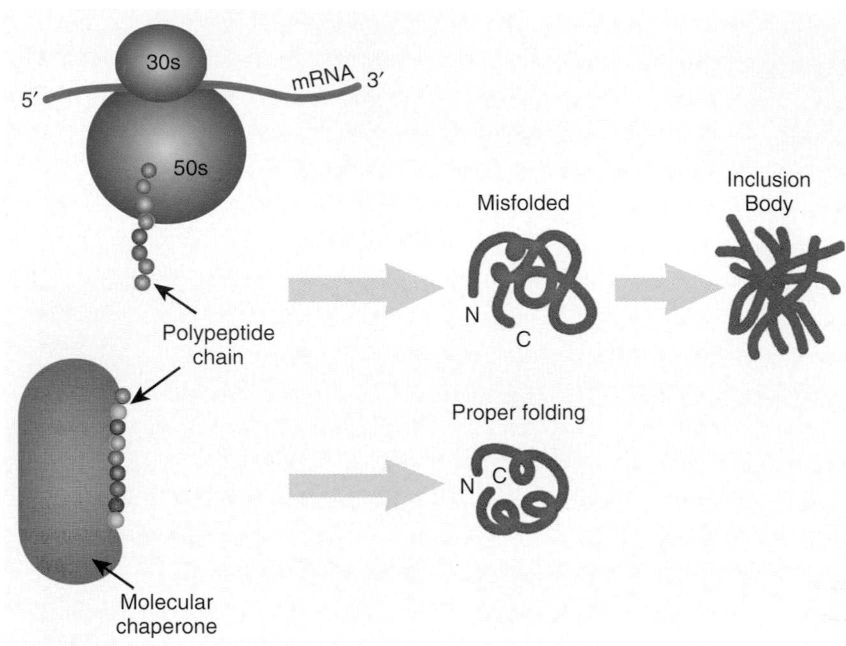

FIGURE 10.10 Molecular Chaperones Help Prevent Inclusion Body Formation
If recombinant proteins fold aberrantly, they aggregate and form inclusion bodies. If, during translation, molecular chaperone proteins hold the recombinant protein, the new polypeptide will fold properly and become fully functional.

INCREASING PROTEIN STABILITY

Different proteins vary greatly in stability. The lifetime of a protein inside a cell depends mainly on how fast the protein is degraded by proteolytic enzymes or proteases (see Chapter 9). Apart from the overall 3D structure of the protein, there are two specific factors that greatly affect the rate of protein degradation. These are believed to act by altering the recognition of proteins by the degradation machinery:

(a) The identity of the N-terminal amino acid greatly affects the half-life of proteins. Although all polypeptide chains are made with methionine as the initial amino acid, many proteins are trimmed later. Therefore, the N-terminal amino acid of mature proteins varies greatly. The N-terminal rule for stability is shown in Table 10.2.

(b) The presence of certain internal sequences greatly destabilizes proteins. **PEST sequences** are regions of 10 to 60 amino acids that are rich in P (proline), E (glutamate), S (serine), and T (threonine). The PEST sequences create domains whose structures are recognized by proteolytic enzymes.

Table 10.2 N-Terminal Rule for Protein Stability

N-Terminal Residue	Approximate Half-Life (Minutes)
Met, Gly, Ala, Ser, Thr, Val	120
Ile, Glu, Tyr	30
Gln, Pro	10
Leu, Phe, Asp, Lys, Arg	2–3

Alteration of the N-terminal sequence of a protein by genetic engineering is relatively simple. A small segment of artificially synthesized DNA encoding a new N-terminal region can be incorporated by standard methods, e.g., PCR (see Chapter 4). The effectiveness of this approach has been demonstrated experimentally for β-galactosidase. Changing an internal PEST sequence without disrupting protein function is a vastly more complex undertaking, even though in theory it could lead to greater stability and therefore greater yields.

To enhance recombinant protein stability, the N-terminal amino acid can be altered, and the PEST sequences can be removed.

IMPROVING PROTEIN SECRETION

When a bacterial cell synthesizes a recombinant protein, it may end up in the cytoplasm, the periplasmic space between the inner and outer membranes, or be exported out of the cell into the culture medium. Secretion across the inner, cytoplasmic, membrane is directed by the presence of a hydrophobic signal sequence at the N-terminal end of the newly synthesized protein. The signal sequence is cut off after export by signal peptidase (also known as leader peptidase). Although bacteria such as *Escherichia coli* export few proteins, special secretion systems do exist that allow export of proteins across both membranes into the culture medium.

It is obviously easier to purify a secreted protein than one that remains in the cytoplasm with the majority of the bacterial cell's own proteins. On the other hand, many proteins are unstable when outside the cellular environment and/or exposed to air. Generally, such proteins are not used in biotechnology since they are awkward to purify or to use. Most proteins and enzymes chosen for practical use are relatively stable outside the cell. A standard goal is to arrange for them to be secreted in order to help isolation and purification. Several approaches exist for this:

(a) A signal sequence is engineered into the cloned gene. Consequently, the recombinant protein will have an N-terminal signal sequence when newly synthesized. This will direct its export to the periplasm by the **general secretory system**. The bacterial cells are then harvested and treated so as to permeabilize or remove the outer membrane, thus releasing the recombinant protein. One major problem is that large amounts of a recombinant protein may overload the secretory machinery and aggregate into inclusion bodies. Substantial amounts of the recombinant protein may accumulate inside the cell with the signal sequence still attached. Secretion may sometimes be increased dramatically by providing elevated levels of the secretory system proteins (Sec proteins). Extra copies of the cloned *sec* genes can be expressed on plasmids to increase the protein amounts. Mutant *E. coli* strains have also been developed with improved secretion.

(b) The recombinant protein may be fused to a bacterial protein that is normally exported (see the following section on fusion vectors). The maltose-binding protein of *E. coli* is efficiently exported to the periplasmic space and is a favorite carrier for recombinant proteins. The recombinant protein is later released from the carrier protein by protease cleavage. This technique is especially useful for relatively short peptides, including many hormones and growth factors, which are often unstable alone.

(c) The gene of interest can be expressed in gram-positive bacteria, such as *Bacillus,* which lack an outer membrane. Consequently, exported proteins go out into the culture medium. Although this is convenient, the genetics of *Bacillus* is still far behind that of *E. coli.* Animal cells are another alternative as they have only a single cytoplasmic membrane, and therefore secrete proteins directly into the medium.

(d) Secretion across both membranes of Gram-negative bacteria such as *E. coli* may be achieved by specialized export systems (Fig. 10.11). Most of these export systems are used

FIGURE 10.11 Secretion across Both Membranes
Protein secretion in *E. coli* involves a general secretory system that recognizes a signal sequence and exports those particular proteins. The protein is only transported to the periplasmic space. Type I secretory systems transport the protein from the cytoplasm, through the periplasmic space, to the outside of the cell. Type II secretory systems take proteins already in the periplasmic space and transport them across the outer membrane.

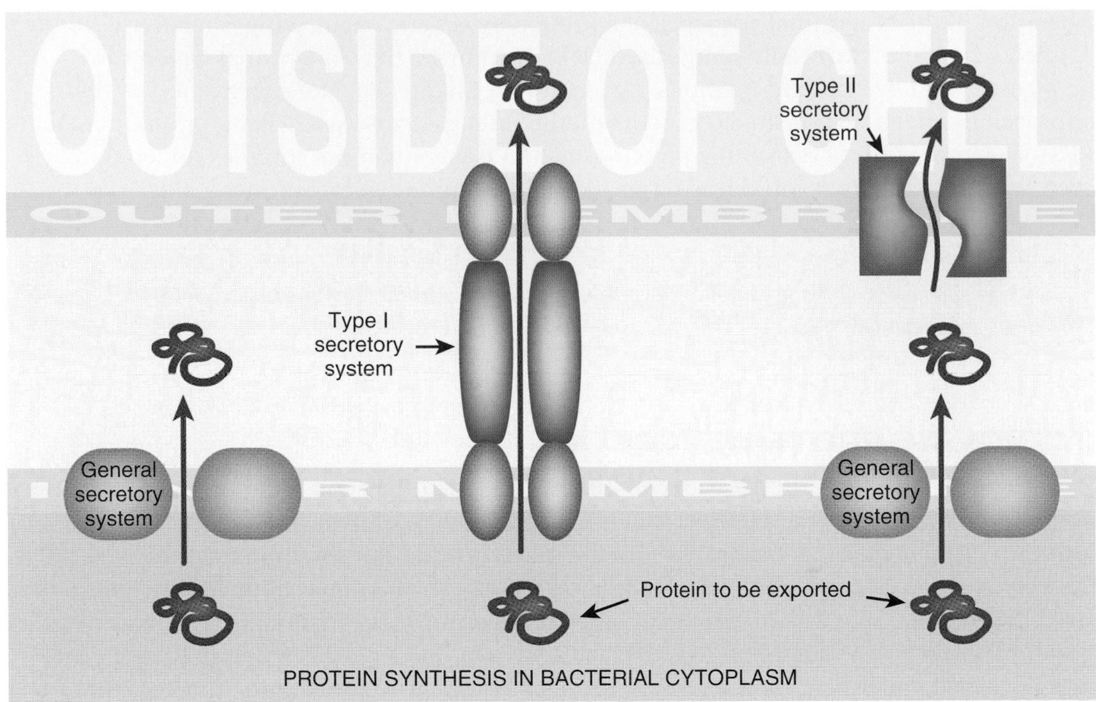

naturally for the secretion of toxins by pathogenic strains of bacteria. Thus, the **type I secretory system** spans both inner and outer membranes and is used to secrete hemolysin by some *E. coli* strains that cause urinary infections. The **type II secretory system** spans the outer membrane only. Export into the medium therefore requires the general secretory system (i.e., the Sec system) to export the protein across the inner membrane first. This two-step process is used by *Pseudomonas* to secrete exotoxin A, for example. Strains of *E. coli* engineered to deploy these alternative export systems are available. The protein to be exported must be modified by adding the appropriate signal sequences so that it is recognized and exported by the chosen export system. The discovery of autotransporter proteins should simplify this approach (see Box 10.1).

Recombinant proteins are usually exported out of the cell into the culture medium. The use of bacterial secretory systems facilitates the export.

Box 10.1 Using Autotransporter Proteins for Secretion

The recently analyzed autotransporter proteins belong to the type V export system. Autotransporter proteins are found naturally as virulence factors of Gram-negative bacteria. They are secreted efficiently despite being large proteins of 100 kd or more. Their export is simpler than with typical type II systems. The general secretory system is used to cross the inner membrane, but the autotransporter beta-domain allows transit of the outer membrane. The name autotransporter refers to the fact that the exported protein carries all information required for export across both membranes, including both a signal sequence and a beta-domain that inserts into the outer membrane (Fig. A). The signal sequence is cleaved after crossing the inner membrane, and the beta-domain is cleaved after crossing the outer membrane. This releases the passenger domain on the outside of the cell. While crossing the periplasmic space, several chaperone proteins assist in folding. Outer membrane transit requires the Bam complex, which catalyzes the insertion of proteins into the outer membrane. When proteins are engineered for export using this system, they are fused to the passenger domain of an autotransporter protein.

The autotransporter proteins include adhesins that enable pathogenic bacteria to bind to host cells as well as enzymes that degrade

Box 10.1 Using Autotransporter Proteins for Secretion—cont'd

host proteins or lipids. The best characterized is the *E. coli* secreted protease, EspP. The plasmid borne *espP* gene is carried by several pathogenic strains of *E. coli*, including the notorious *E. coli* O157:H7. The EspP protein has 1300 amino acid residues (MW = 142 kDa). It is a serine protease that degrades several proteins found in mammalian blood, including coagulation factors and components of the antibacterial complement system.

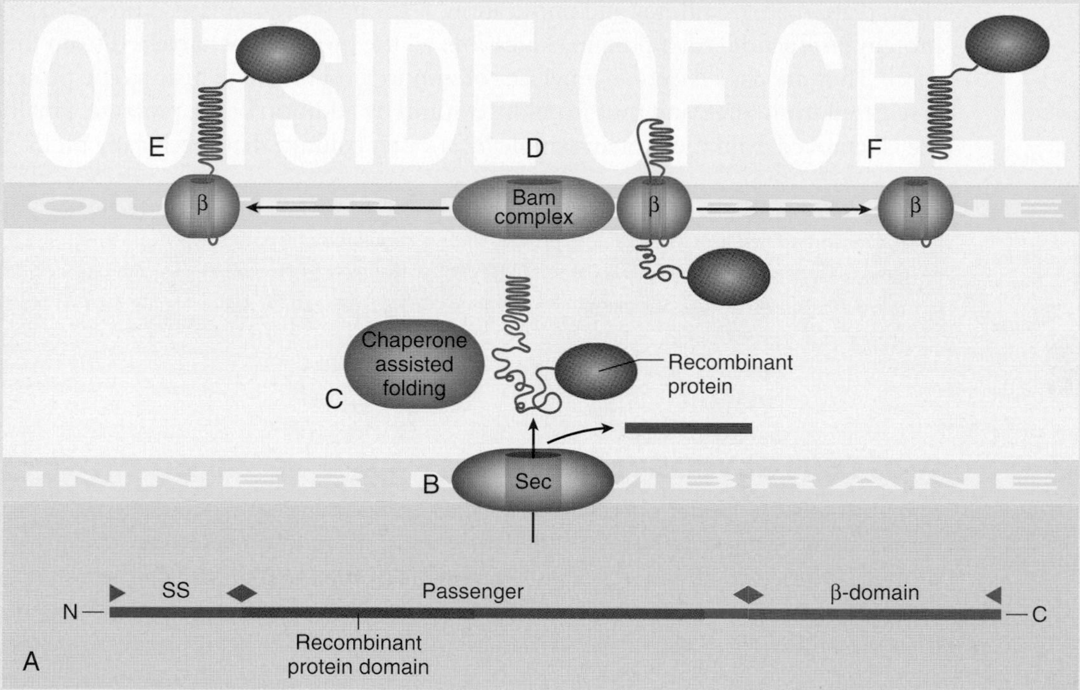

FIGURE A Autotransporter System in *Escherichia coli*
(A) Arrangement of protein domains in autotransporter proteins. The signal sequence (SS; orange) is at the N-terminus. The recombinant protein (olive) is fused between the signal sequence and the original passenger domain (purple). Then come the autochaperone domain (blue) and the beta domain (red). (B) Autotransporter proteins cross the inner membrane via the general secretory system. (C) Chaperone proteins help fold the autotransporter protein in the periplasmic space. (D) The Bam complex is needed for insertion into the outer membrane. After crossing the outer membrane, the passenger domain may remain attached to the beta domain (E) or may be released (F). *(Inset).* Modified from Jong WS, Saurí A, Luirink J (2010). Extracellular production of recombinant proteins using bacterial autotransporters. *Curr Opin Biotechnol* **21**, 646–652.

PROTEIN FUSION EXPRESSION VECTORS

Joining the coding sequences of two proteins together in frame makes a **protein fusion** (see Chapter 9). Consequently, a single, longer polypeptide is made during translation. We have already discussed the use of peptide tags, such as His6, FLAG, or GST to aid in protein purification (see Chapter 9). Here, we are concerned with more sophisticated protein fusions that allow secretion of the fusion protein. If the first (i.e., N-terminal) protein is normally secreted, then the fusion protein will be secreted, too. Thus, it is possible to achieve export of a recombinant protein by joining it to a protein that the cell normally exports.

Protein fusions also help address the issues of stability and purification. Many eukaryotic proteins are unstable inside the bacterial cell. This is especially true of growth factors, hormones, and regulatory peptides, which are often too short to fold into stable 3D configurations. Attaching them to the C terminus of a stable bacterial protein protects them from degradation. If the carrier protein is carefully chosen, purification may be greatly facilitated.

A **protein fusion vector** is a plasmid that allows the gene of interest to be fused to the gene for a suitable carrier protein. Useful carriers are exported proteins that are easy to purify. In addition, the carrier protein gene should have a strong ribosomal binding site and be translated efficiently. One of the best examples is the **MalE protein** of *E. coli*. This protein is normally exported into the periplasmic space, where it transports maltose from the outer membrane to the inner membrane. MalE protein is purified by binding to an amylose resin.

The gene of interest is cloned downstream and in frame with the MalE coding sequence. Between the coding sequences is a protease cleavage site, which allows the fused proteins to be cleaved apart after synthesis and purification. A strong promoter is also provided to maximize protein production. After protein expression, all the proteins are harvested from the periplasmic space. They are passed over an amylose column to bind the MalE fusion. The protein of interest is released from MalE and thus from the column by addition of the protease. Finally, the protease is removed from the protein sample by another column that specifically binds protease.

Sophisticated fusion protein vectors provide strong, regulated promoters, optimal RBSs, and a fusion gene for a secreted protein. The recombinant protein is fused to the secreted protein to facilitate export. There is a protease digestion site at the fusion junction to release the recombinant protein from the fusion protein during purification.

PROTEIN GLYCOSYLATION

Many proteins of higher organisms, especially animals, are glycosylated; that is, they have short chains of sugar derivatives added after translation. **Glycosylation** is required for proper biological function of many proteins. In particular, many cell surface proteins are glycosylated and will not assemble correctly into membranes or function properly if lacking their sugar residues. Around 70% of recombinant human proteins used in therapy have N-glycosyl groups.

Although bacteria glycosylate some proteins, they attach the glycosyl groups to side-chain oxygens (O-glycosylation). Eukaryotes generally have N-linked glycosyl groups. Recently, enzymes performing N-linked glycosylation have been discovered in the bacterium *Campylobacter*, a mild pathogen. The genes for this system have been expressed in *E. coli* and allow the glycosylation of recombinant proteins. In the future, this should allow more human proteins to be produced in *E. coli*.

One solution to expressing glycosylated proteins is to use insect cells. Mammals and insects share the pathway for post-translational glycosylation of asparagine residues up to the addition of mannose (Fig. 10.12). However, the pathways diverge beyond this. Nonetheless, partial glycosylation is better than none. Furthermore, insect cell lines that have been engineered to express the full mammalian glycosylation pathway are now available.

Just as changing the amino acid sequence may generate proteins with novel properties, so altering the glycosylation pattern of a glycoprotein may result in useful alterations in protein behavior. One widely used glycoprotein that is produced by genetic engineering is recombinant human erythropoietin, marketed as Epogen by the Amgen Corporation. Erythropoietin stimulates the production of red blood cells (erythrocytes). It is used to treat anemia and reduce the need for blood transfusions. Natural erythropoietin has four attached oligosaccharides. When erythropoietin was engineered to have extra N-linked glycosylation sites (Asn-Xxx-Ser/Thr), its activity was changed in a paradoxical manner. Although its affinity for its receptor decreased, its *in vivo* activity increased. This was due to a longer half-life after administration. In practice, the longer half-life outweighed the lowered receptor affinity, and the resulting protein had significantly increased clinical activity. This reengineered variant is marketed as Aranesp. It seems likely that similar glycoengineering could be applied to other glycoprotein hormones.

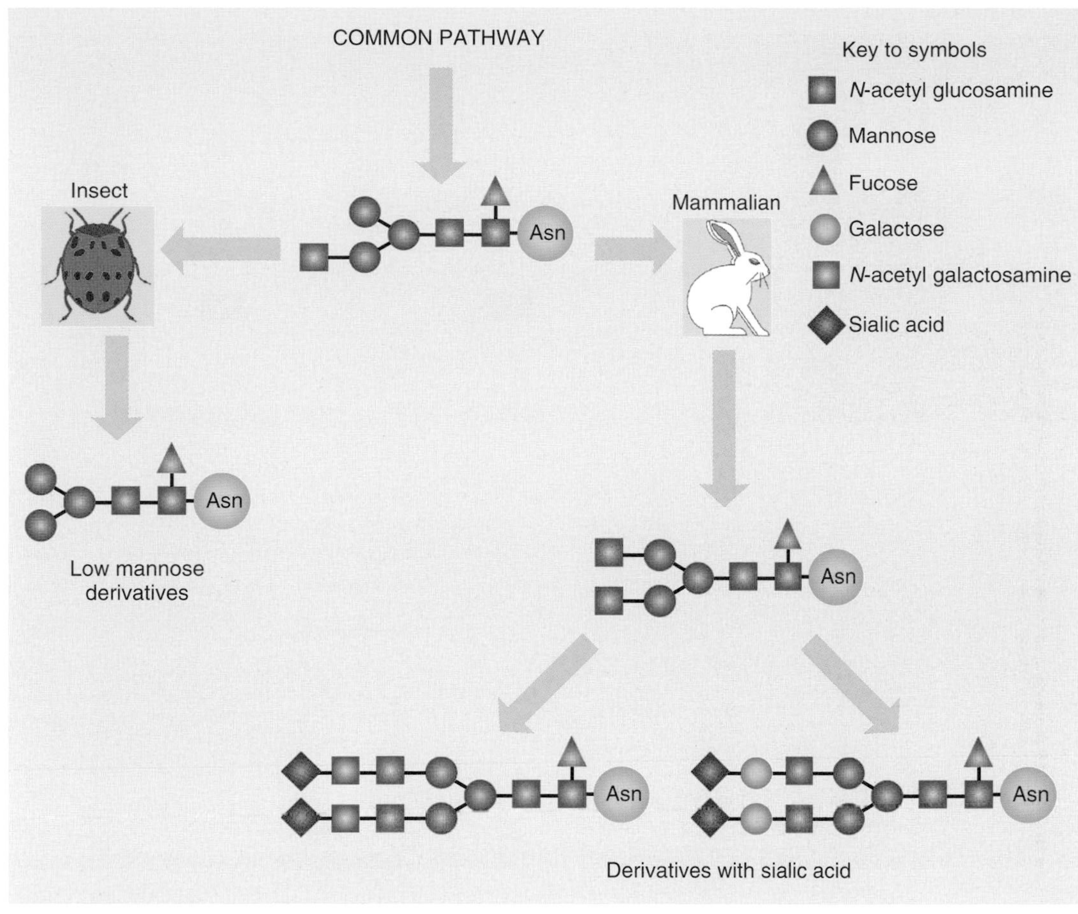

FIGURE 10.12
Protein Glycosylation Pathways
Mammals and insects share several steps in the pathway for post-translational glycosylation of asparagine residues in proteins. However, the final modifications differ as shown. In particular, mammals add sialic acid residues to the ends of the glycosyl chains, whereas insects do not.

Bacteria do not glycosylate eukaryotic proteins correctly. However, insect cells can be used to glycosylate recombinant proteins.

EXPRESSION OF PROTEINS BY EUKARYOTIC CELLS

Although bacterial cells have successfully expressed many eukaryotic proteins, in some cases it is best to express eukaryotic proteins using eukaryotic cells. As noted previously, disulfide bonds are not usually formed inside bacterial cells and are rarely formed properly even when bacterial hosts secrete eukaryotic proteins. Another advantage of expressing eukaryotic proteins in eukaryotic cells is that contamination with bacterial components is avoided. Despite purification, bacterial components that are toxic or promote immune reactions or cause fever may be present in traces if bacteria are used for production.

Furthermore, some eukaryotic proteins are unstable or inactive after being made by bacterial cells. This is especially true of proteins that require post-translational modification. A variety of modifications may occur after the polypeptide chain has been made (Fig. 10.13). They include

- **(a)** Chemical modifications that form novel amino acids in the polypeptide chain.
- **(b)** Formation of disulfide bonds between correct cysteine partners (e.g., for insulin).
- **(c)** Glycosylation (see above).
- **(d)** Addition of a variety of extra groups, such as fatty acid chains, acetyl groups, phosphate groups, and sulfate groups.
- **(e)** Cleavage of precursor proteins may be needed for secretion, correct folding, and/or activation of proteins. This may occur in several stages, as illustrated by insulin.

The enzymes required for these modifications are normally absent from bacterial cells, making it necessary to express certain eukaryotic proteins in eukaryotic cells. Related processing

NOVEL AMINO ACIDS

PYRROLYSINE

LYSINE

CYSTEINE

SELENOCYSTEINE

HISTIDINE

DIPHTHAMIDE

A

DISULFIDE BOND

CYSTEINE

CYSTINE

B

GLYCOSYLATION

Sugar residues

OUTSIDE

CELL MEMBRANE

Protein

INSIDE

C

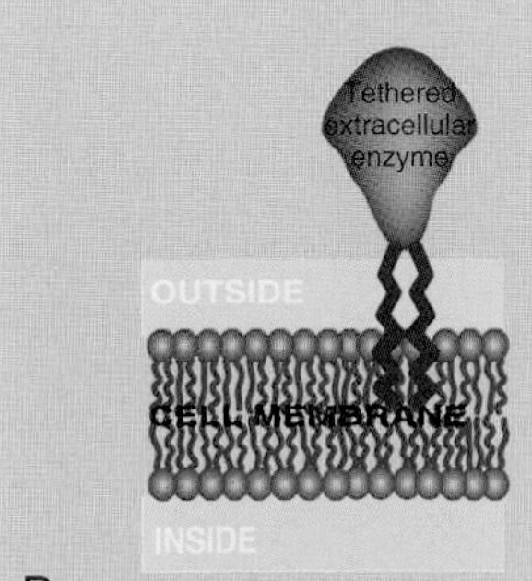

ACETYLATION

PHOSPHORYLATION

D MYRISTOYLATION

FIGURE 10.13 Protein Modifications in Eukaryotes

(A) Novel amino acids may be inserted during translation (e.g., pyrrolysine, selenocysteine) or made after translation by modifying other amino acids (e.g., diphthamide from histidine). (B) Disulfide bonds link two cysteine molecules together by their side chains. (C) Glycosylation is the addition of sugar residues to the surface of proteins. (D) Myristoylation, acetylation, and phosphorylation all add different groups to amino acids within a protein. These added groups alter the function of a protein. Myristoylation adds fatty acid chains that tether a protein to the cell membrane. Acetylation and phosphorylation often activate or deactivate proteins.

enzymes are often present in a range of higher organisms; thus, it is rarely absolutely necessary to express a protein in its original organism. Here, we are concerned with protein production in cultured cells. However, as discussed in Chapters 15 and 16, it is now possible to engineer whole transgenic animals or plants to produce recombinant proteins.

As discussed in Chapter 3, shuttle vectors are designed to move genes between different groups of organisms. Because genetic engineering is more difficult for eukaryotes, most expression vectors for eukaryotic cells are in fact shuttle vectors. Such vectors allow genetic engineering to be carried out in bacteria, usually *E. coli,* and allow transfer to other organisms for gene expression. We will consider the use of yeasts, insect cells, and mammalian cells to express recombinant proteins.

Many proteins require special modifications that are available only in eukaryotic cells.

EXPRESSION OF PROTEINS BY YEAST

As already discussed, brewer's yeast, *Saccharomyces cerevisiae,* has been used as a model single-celled eukaryote in molecular biology (Chapter 1). The yeast genome has been sequenced, and many genes have been characterized. From the viewpoint of biotechnology, several other factors favor the use of yeast for production of recombinant proteins:

(a) Yeast can be grown easily on both a small scale or in large bioreactors.
(b) After many years of use in brewing and baking, yeast is accepted as a safe organism. It can therefore be used to produce pharmaceuticals for use in humans without needing extra government approval.
(c) Yeast normally secretes very few proteins. Consequently, if it is engineered to release a recombinant protein into the culture medium, this can be purified relatively easily.
(d) DNA may be transformed into yeast cells either after degrading the cell wall chemically or enzymatically, or, more usually, by electroporation (see Chapter 3).
(e) A naturally occurring plasmid, the 2-micron circle, is available as a starting point for developing cloning plasmids.
(f) A variety of yeast promoters have been characterized that are suitable for driving expression of cloned genes.
(g) Although only a primitive single-celled organism, yeast nonetheless carries out many of the post-translational modifications typical of eukaryotic cells, such as addition of sugar residues (glycosylation). However, yeast only glycosylates proteins that are secreted.

Recombinant proteins may be engineered for secretion by yeast (Fig. 10.14). Adding a signal peptide upstream of the coding region flags the protein for secretion. The signal sequence used is taken from the gene for mating factor α, a protein that is normally secreted (see Chapter 1). Yeast signal peptidase recognizes the sequence Lys-Arg and cleaves off the signal peptide once the protein has crossed the cell membrane. It is therefore necessary to position the two codons for Lys-Arg immediately upstream of the coding sequence for the recombinant protein to ensure that the engineered protein has a correct N-terminal sequence.

Various vectors exist for use with yeast. They may be classified into three main classes:

(a) Plasmid vectors. These are used most often and are typically shuttle vectors that can replicate in *E. coli* as well as yeast. The yeast replication system is normally derived from the yeast 2-micron circle (see Chapter 1).
(b) Vectors that integrate into the yeast chromosomes. Because plasmids may be lost, especially in large-scale cultures, this approach has the advantage of stability. The disadvantage is that only a single copy of the cloned gene is present, unless tandem repeats are used. These, however, are unstable because of crossing over. In practice, this approach is rarely used.

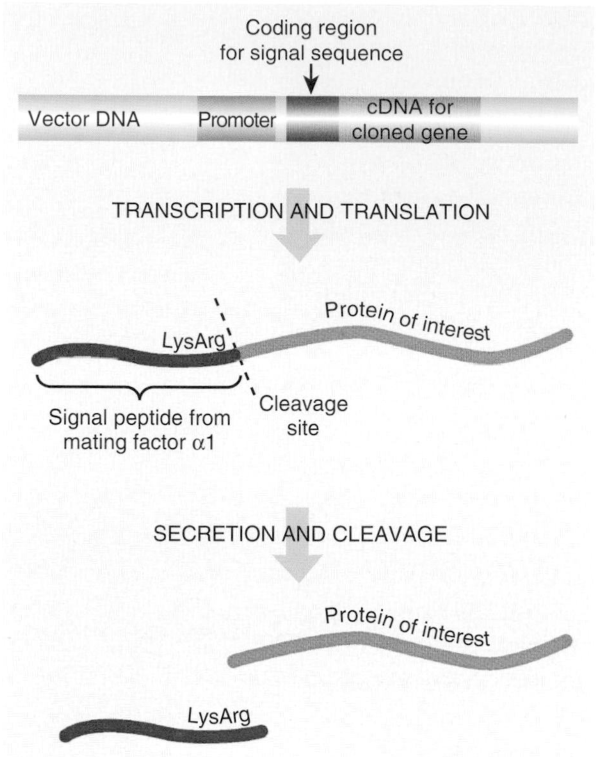

Coding region for signal sequence
Vector DNA Promoter cDNA for cloned gene
TRANSCRIPTION AND TRANSLATION
LysArg Protein of interest
Signal peptide from mating factor α1 Cleavage site
SECRETION AND CLEAVAGE
Protein of interest
LysArg

FIGURE 10.14 Engineering Proteins for Secretion by Yeast
To make yeast cells secrete the protein of interest, a small signal sequence is fused just upstream of the coding region. The signal sequence ends with a Lys-Arg, which is recognized by signal peptidase. Cleavage occurs after the Arg during secretion.

(c) Yeast artificial chromosomes (see Chapter 3). These are used for cloning and analysis of large regions of eukaryotic genomes. However, they are not convenient for use as expression vectors.

Assorted proteins for human use are now produced using *Saccharomyces cerevisiae.* They include insulin, factor XIIIa (for blood coagulation), several growth factors, and several virus proteins (from human immunodeficiency virus, hepatitis B, hepatitis C, etc.) used as vaccines or in diagnostics.

Although many recombinant proteins have been expressed successfully in yeast, yields are often low. The problems include

(a) Expression plasmids are lost during growth in large bioreactors.
(b) Proteins supposed to be secreted are often retained in the space between membrane and wall rather than exiting into the culture medium.
(c) Glycosylation of proteins is often excessive, and the final protein may have too many sugar residues attached to function correctly. Recently, however, yeast has been engineered to produce human-like glycan structures.

Yeasts provide many advantages to producing recombinant proteins because they are true eukaryotes, have plasmid vectors, grow fast and easily, and finally because they have well-known characteristics.

EXPRESSION OF PROTEINS BY INSECT CELLS

Mammalian cells are relatively delicate and have complex nutritional requirements. This makes them both difficult and expensive to grow in culture compared to bacteria or yeasts. In contrast, cultured insect cells are relatively robust and can be grown in simpler media than mammalian cells. Consequently, insect-based expression systems have been developed. These systems have the additional advantage of providing post-translational modifications that are very similar to those found in mammalian cells.

The vectors used in cultured insect cells are almost all derived from a family of viruses, the **baculoviruses**, which infect only insects (and related invertebrates such as arachnids and crustaceans). Baculoviruses are unusual in forming packages of virus particles, known as **polyhedrons**. Some infected cells release single virus particles that can infect neighboring cells within the same insect. But when the host insect is dead or dying, packages of virus particles embedded in a protein matrix are released instead. The matrix protein is known as **polyhedrin**, and the polyhedron structure protects the virus particles while they are outside the host organism in the environment. When swallowed by another insect, the polyhedrin is dissolved by digestion, and the polyhedron falls apart. This releases individual virus particles that can infect the cells of the new host insect.

The polyhedrin gene has an extremely strong promoter, and late during infection, the polyhedrin protein is made in massive amounts. Because polyhedrin is not actually needed for virus infection of cultured insect cells, the polyhedrin promoter may be used to express recombinant proteins. The polyhedrin coding sequence is removed and replaced by cDNA encoding the protein to be expressed.

There are many different baculoviruses, and the one most often used is **multiple nuclear polyhedrosis virus (MNPV)**. It infects many insects and replicates well in many cultured insect cell lines. A popular cell line used to propagate this baculovirus is from the fall armyworm (*Spodoptera frugiperda*). Yields of polyhedrin—and therefore of a recombinant protein using the polyhedrin promoter—are especially high in this cell line.

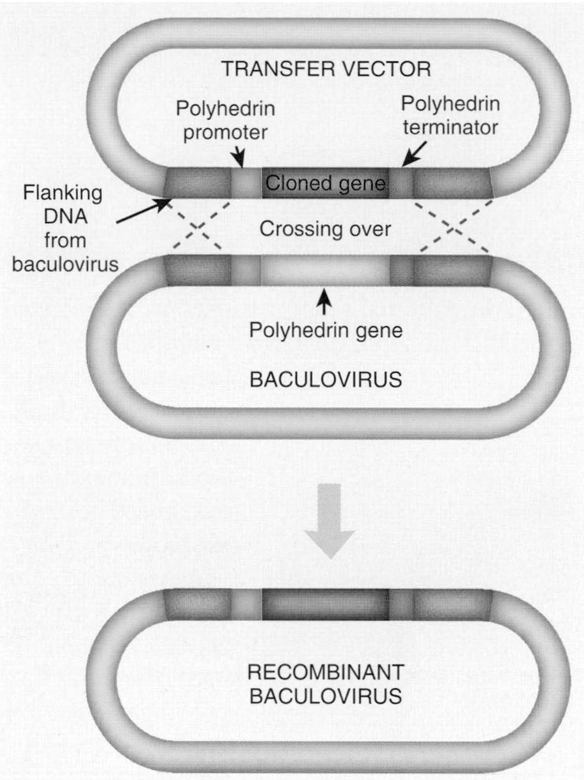

FIGURE 10.15 Baculovirus Expression Vector
To express a gene in insect cells, the gene must be inserted into a baculovirus genome. First, the gene of interest is cloned into a transfer vector containing the baculovirus polyhedrin gene promoter followed by a multiple cloning site and the polyhedrin terminator. This is done in *E. coli*. The construct is transfected into insect cells along with the normal baculovirus genome. A double crossover between the polyhedrin promoter and terminator replaces the polyhedrin gene with the gene of interest.

Because the expression vector is a virus, construction is carried out in two stages (Fig. 10.15). In the first stage, a "transfer" vector is used to carry the cDNA version of the cloned gene. The transfer vector is an *E. coli* plasmid that carries a segment of MNPV DNA, which initially included the polyhedrin gene and flanking sequences. The polyhedrin coding sequence was then replaced with a multiple cloning site. When the cloned gene is inserted, it is under control of the polyhedrin promoter. Construction up to this point is done in *E. coli*. In the second stage, the segment containing the cloned gene is recombined onto the baculovirus, thus replacing the polyhedrin gene. To achieve this, insect cells are transfected with both the transfer vector and with MNPV virus DNA. A double-crossover event generates the required recombinant baculovirus.

Because this procedure sometimes gives undesirable results, other insect vectors have been constructed. One possibility is a shuttle vector that replicates as a plasmid in *E. coli* and as a virus in insect cells. Such baculovirus–plasmid hybrids are referred to as **bacmids** (Fig. 10.16). They consist of an almost entire baculovirus genome into which a segment of DNA from an *E. coli* plasmid has been inserted. This region carries a bacterial origin of replication, a selective marker, and a multiple cloning site. The inserted segment replaces the polyhedrin gene of the baculovirus. The cloned gene is inserted into the MCS, giving a recombinant bacmid. Bacmid DNA is then purified from *E. coli* and transfected into insect cells, where it replicates as a virus.

The recombinant baculovirus may then be used to infect cultured insect cells, and the desired protein is produced over several days.

Several hundred eukaryotic proteins have been successfully made experimentally using the baculovirus/insect cell system, and the vast majority were correctly processed and modified. However, the baculovirus system has rarely been used commercially. Its major use has been in vaccine production, in particular in the manufacture of Cervarix™, a vaccine against human papillomavirus, intended to protect against cervical cancer. Several veterinary vaccines are also made using the baculovirus system.

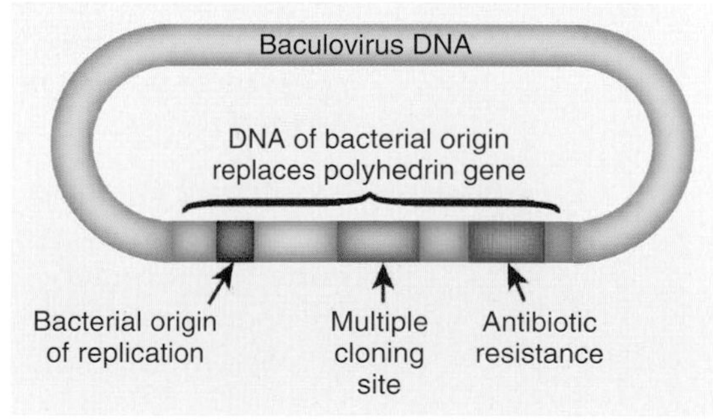

FIGURE 10.16 Bacmid Shuttle Vector
Bacmids are primarily baculovirus genomes with the addition of a bacterial origin of replication, a multiple cloning site, and an antibiotic resistance gene. These sequences allow the bacmid to survive in *E. coli* but still infect insect cells.

Insect cells are easy to grow for making recombinant protein. Insect cells are infected with a virus genome that has the recombinant protein gene. Rather than making viral particles, the infected insect cells make recombinant protein.

EXPRESSION OF PROTEINS BY MAMMALIAN CELLS

Some cloned animal genes may need to be expressed in cultured mammalian cells. Shuttle vectors that can replicate in both mammalian cells and bacteria are used for this (Fig. 10.17). As usual, such vectors contain a bacterial origin of replication and an antibiotic resistance gene allowing selection in bacteria. These vectors must also possess an origin of replication that works in mammalian cells. Usually, this is taken from a virus that infects animal cells, such as SV40 (simian virus 40). Viral promoters are often used because they are strong, producing copious amounts of protein. Alternatively, promoters from mammalian genes that are expressed at high levels (e.g., the genes for metallothionein, somatotropin, or actin) may be used. The multiple cloning sites lie downstream of the strong promoter. Since animal genes are normally cloned as the cDNA, the vector must also provide a polyadenylation signal (i.e., tail signal) at the 3′-end of the inserted gene. Finally, single cells showing high production levels must be isolated (see Box 10.2).

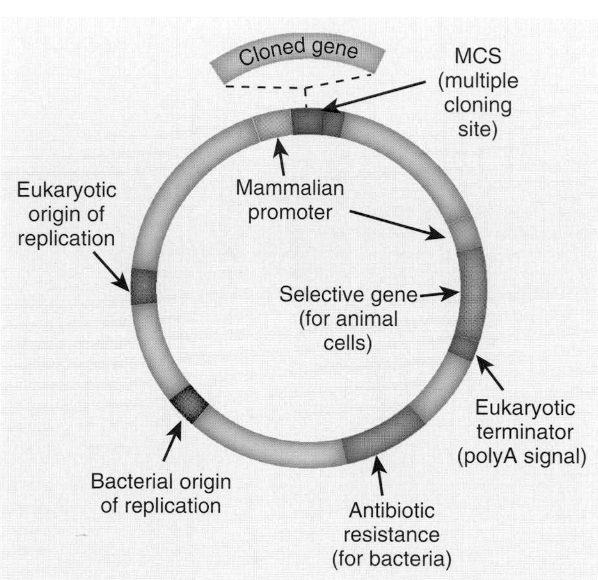

FIGURE 10.17
Mammalian Shuttle Vector
Mammalian shuttle vectors contain an origin of replication and an antibiotic resistance gene for growth in bacteria. The vector has a multiple cloning site between a strong promoter and a polyadenylation signal as well as a eukaryotic origin of replication and a eukaryotic selection gene, such as *npt*.

Several selective markers may be used with animal cells. Very few antibiotics kill animal cells. However, the antibiotic **genetecin**, also known as **G-418**, kills animal cells by blocking protein synthesis. G-418 is related to antibiotics such as neomycin and kanamycin that are used against bacteria. The *npt* (NeoR) gene, encoding neomycin phosphotransferase, adds a phosphate group to neomycin, kanamycin, and G-418 that inactivates these antibiotics. Consequently, *npt* can be used to select animal cells using G-418 or bacterial cells using neomycin or kanamycin.

Because of the lack of antibiotics for mammalian cells, mutant eukaryotic cells lacking a particular enzyme are sometimes used for selection. The plasmid then carries a functional copy of the missing gene. This rather inconvenient approach has been used in both yeast and animal cells. In yeast, genes for amino acid biosynthesis have been used (see Chapter 1). In animal cells, the ***DHFR* gene**, encoding **dihydrofolate reductase**, is sometimes used. This enzyme is required for synthesis of the essential cofactor folic acid and is inhibited by **methotrexate**. Mammalian cells lacking the *DHFR* gene are used, and a functional copy of the *DHFR* gene is provided on the shuttle vector. Treatment with methotrexate inhibits DHFR and hence selects for high-level expression of the DHFR gene on the vector. Methotrexate levels can be gradually increased, which selects for a corresponding increase in copy number of the vector. (The chromosomal *DHFR* genes must be absent to avoid selecting chromosomal duplications rather than the vector.)

An alternative approach is to use a metabolic gene as a dominant selective marker. The enzyme glutamine synthetase protects cells against the toxic analog, **methionine sulfoximine**. The resistance level depends on the copy number of the glutamine synthetase gene. Therefore, multicopy plasmids carrying the glutamine synthetase gene can be selected even in cells with functional chromosomal copies of this gene.

Mammalian shuttle vectors have features for producing recombinant proteins in mammalian cell cultures.
Because most antibiotics do not harm mammalian cells, a different method of selecting the mammalian shuttle vector must be used. Some vectors have the *DHFR* gene, which protects mammalian cells from methotrexate. Other vectors use the glutamine synthetase gene, which protects the cell from methionine sulfoximine.

Box 10.2 Selecting Mammalian Cell Lines That Yield High Levels of Recombinant Protein

A major issue with protein expression in mammalian cells is finding a cell with high levels of the recombinant protein. This single chosen cell is then used to generate large cultures. This ensures that the recombinant protein is identical throughout the culture. If more than one cell gives rise to a culture, individual cells may produce slightly different variants of the protein. Many techniques are available to isolate single cells with high levels of recombinant protein.

- *Limiting Dilution Cloning:* Cells are transfected with the recombinant protein vector and then suspended in medium. The suspension is diluted, and small amounts are added to small wells in a microtiter dish at a density of less than one cell per well. After the cells grow, small amounts of the medium from each well are screened for the amount of recombinant protein (usually by ELISA; see Chapter 6). High producers are isolated and grown up for production.
- *Cell Sorting:* Cells are transfected with a vector carrying genes for the recombinant protein plus a reporter protein such as green fluorescent protein or dihydrofolate reductase. The transfected cells are sorted based on the expression of the reporter gene using flow cytometry, or FACS (see Chapter 6). The level of reporter protein correlates with the level of recombinant protein in most cases. This method is indirect as it monitors the reporter rather than the actual recombinant protein.
- *Gel Microdrop Technology:* Cells are transfected with a vector carrying the recombinant protein. The cells are then suspended in biotinylated agarose, that is, agarose polymers with biotin attached at points along the chain. A biotinylated antibody recognizes and binds the secreted recombinant protein. Avidin simultaneously binds the biotin on the antibody and on the agarose. Overall, this links the target protein to the agarose. The amount of recombinant protein is determined with a second, fluorescently labeled antibody. This method directly monitors the amount of recombinant protein rather than a coexpressed reporter. Immobilizing the cell in a gel makes it easy to locate and isolate.
- *Automated Cell Selection:* Many automated systems have been developed to identify single mammalian cells that produce large amounts of recombinant protein. Automated systems are very efficient and can screen large numbers of transfected cells. Laser-enabled analysis and processing (LEAP) works by killing all the low-producing cells. The transfected cells are first immobilized in a matrix, and then a fluorescently labeled antibody to the recombinant protein is added. The brightest cell is marked, and a laser kills the remaining cells. Automated cell pickers work by using a similar approach. The brightest cells are chosen, but rather than killing the rest of the cells, an automated pipetting system removes the bright cells.

EXPRESSION OF MULTIPLE SUBUNITS IN MAMMALIAN CELLS

To further complicate recombinant protein expression, many mammalian proteins have multiple subunits that must assemble inside a mammalian cell. In such cases, manufacturing the separate subunits in separate cultures and mixing them later fails to yield active protein.

Consequently, it may be necessary to synthesize more than one polypeptide in the same cell (Fig. 10.18). This may be done in three main ways:

(a) Two separate vectors are used, each carrying the gene for one of the subunits.
(b) A single vector is used that carries two separate genes for the two subunits, each under control of its own promoter.
(c) A single vector is used that carries an artificial "operon" in which the two genes are expressed using the same promoter. Transcription gives a single polycistronic mRNA carrying both genes. As discussed in Chapter 2, eukaryotic cells normally translate only the first open reading frame on an mRNA. However, certain animal viruses have sequences known as **internal**

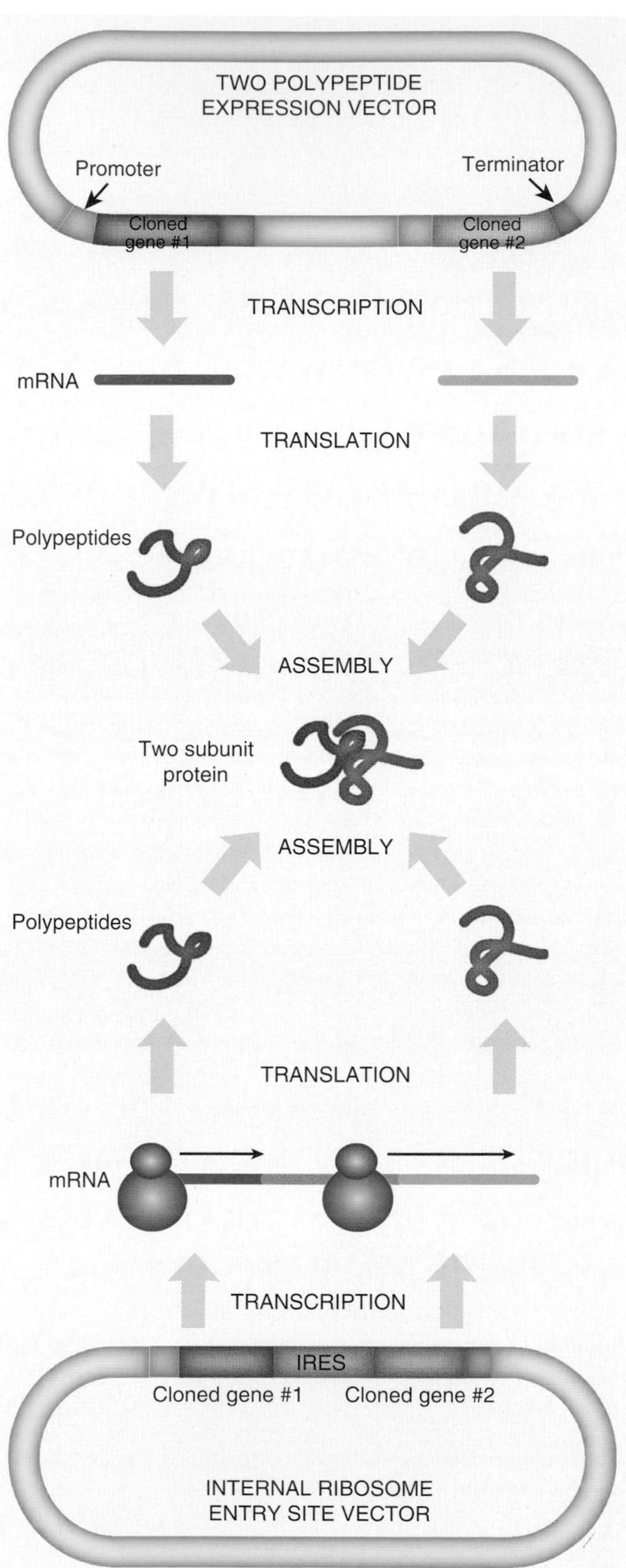

FIGURE 10.18 Expression of Multiple Polypeptides in the Same Cell
Some proteins have two subunits that assemble properly only when inside a mammalian cell. These can be expressed on one shuttle vector with two different promoters (blue), two multiple cloning sites, and two polyadenylation sites (purple). After transfection into a mammalian cell, both proteins are expressed. The two protein subunits assemble using the necessary cellular components. Alternatively, the vector may carry the two genes separated by an internal ribosome entry site (IRES), which allows ribosomes to bind to the mRNA upstream of both genes. As before, the two proteins are made and associate.

ribosomal entry sites (IRESs) that allow the translation of multiple coding sequences on the same message. Note that the mRNA is read twice by two different ribosomes that bind in two different places (the normal 5′-end and the IRES). This is different from the translation of polycistronic mRNA in bacteria where a single ribosome proceeds along the mRNA and translates all the open reading frames in the operon.

If the recombinant protein has multiple subunits, each subunit must be produced separately. Sometimes each subunit is expressed from separate vectors, sometimes one vector contains separate genes for each of the subunits, and finally, some vectors have the subunits expressed as an operon.

COMPARING EXPRESSION SYSTEMS

Each expression system has its own unique strengths and weaknesses when it comes to characteristics such as cost, speed, and ability to glycosylate or fold proteins, as well as government regulations. Bacterial cells are by far the cheapest and fastest method for recombinant protein expression, but they do not glycosylate proteins or fold mammalian proteins correctly. Cultured mammalian cells glycosylate and fold proteins properly, but they are expensive to maintain. Therefore, determining what system to use to express recombinant proteins depends on the protein and its particular idiosyncrasies (Fig. 10.19).

Of the FDA-approved recombinant proteins in therapeutic use, about 60%–70% are made using mammalian cells; about 20% are made using *E. coli*; and the remainder use baculovirus, yeast, and transgenic goats.

No expression system is perfect, and each recombinant protein must be evaluated for which system will work the best.

FIGURE 10.19 Comparison of Recombinant Protein Expression Systems
Each protein expression system falls on a continuum of worst to best for characteristics such as speed, cost, glycosylation, folding, and government regulations. Transgenic animals (rabbit) and plants are discussed in Chapters 15 and 16. The other symbols include cultured mammalian cells, insect cells, yeast, and bacteria.

Summary

To make recombinant proteins, the gene is isolated and cloned into an expression vector. Most recombinant proteins in therapeutic use are from humans but are expressed in other organisms such as bacteria, yeast, or animal cells in culture. Human genes are very complex, often containing large introns. Therefore, an intron-free version of the gene is often made by converting the mRNA into cDNA. Because the cDNA lacks regulatory regions, the expression vectors must provide the promoter, ribosome-binding site, and terminator sequences.

Many issues affect the expression of recombinant proteins. Because organisms differ in which codons are preferred for each amino acid, the level of corresponding tRNAs varies between organisms. If the gene for a recombinant protein uses codons that are rare in the host organism, the low amount of tRNA becomes rate limiting for expression. Adding the rare tRNA or changing the coding sequence for the gene of interest can resolve this issue. The gene for a recombinant protein can also be altered to make the protein more stable. Changing the N-terminal amino acid, removing the PEST sequences, or coexpressing a molecular chaperone all increase protein stability. If too much recombinant protein is produced too fast, it may aggregate into inclusion bodies. Using regulated expression vectors can control the rate at which recombinant protein is produced. Another factor is protein export. It is easier to isolate and purify recombinant proteins if they are exported outside the cell.

Many recombinant proteins require protein modifications, such as glycosylation, that are available only in eukaryotic cells; this sometimes leads to the use of yeast, insect cells, and mammalian cell culture systems.

End-of-Chapter Questions

1. Which of the following is a clinically relevant protein produced by recombinant technology?
   a. insulin
   b. interferon
   c. factor VIII
   d. tissue plasminogen activator
   e. all of the above
2. Which of the following is associated with translation efficiency for the mRNA of genes cloned for protein expression?
   a. mRNA stability
   b. strength of the ribosomal binding site
   c. codon usage
   d. mRNA secondary structure
   e. all of the above
3. Which of the following is not a feature of translational expression vectors?
   a. a selectable marker (i.e., antibiotic resistance)
   b. terminator sequences
   c. the native promoter for the gene that is to be cloned into the vector
   d. consensus RBS plus the ATG initiation codon
   e. a strong, but regulated, transcriptional promoter

4. Which method to overcome codon usage problems is the least labor-intensive?
   a. Supply the genes for the rare tRNAs on a separate plasmid.
   b. Codon-optimize the gene of interest, usually in the third position of the codon.
   c. Express the protein in the organism from which the gene originated.
   d. Grow the host cells that are expressing the protein longer so that they have more time to make the protein containing rare codons.
   e. all of the above
5. In expression systems, which of the following forms when too much protein is manufactured too quickly?
   a. activated AraC
   b. inclusion bodies
   c. T7 RNA polymerase
   d. functional protein
   e. inactivated Lac repressor
6. How can recombinant protein production be enhanced?
   a. alteration of the C-terminal sequence
   b. inclusion of PEST sequences
   c. addition of a detergent to help prevent inclusion body formation
   d. co-expression of a molecular chaperone
   e. all of the above
7. What can be done to ensure a recombinant protein is excreted outside the cell?
   a. addition of a signal sequence
   b. using specialized export systems
   c. fusion to a protein that is normally exported
   d. expressing the protein in an alternative system, such as Gram-positive bacteria
   e. all of the above
8. What could be used to increase the stability and purification of eukaryotic proteins from bacterial cells?
   a. a peptide tag
   b. a protein fusion
   c. a protease site
   d. PEST sequence
   e. a signal sequence
9. Why would it be necessary to express eukaryotic proteins in eukaryotic cells rather than in bacterial cells?
   a. Many eukaryotic proteins are modified and bacteria usually do not have the enzymes needed for these modifications.
   b. Codon usage between eukaryotes and prokaryotes is different.
   c. Molecular chaperones are more readily available in eukaryotes.
   d. Protein production in eukaryotic cells never produces inclusion bodies, which sometimes form in bacterial cells due to overexpression.
   e. none of the above

(Continued)

10. Which of the following is not a problem associated with low yields of proteins expressed in yeast?
- **a.** Proteins are not getting secreted into the medium.
- **b.** Yeast has no known naturally occurring plasmids from which to construct expression vectors.
- **c.** Expression plasmids are lost in large batch cultures of yeast.
- **d.** Too much modification of the protein occurs (i.e., glycosylation).
- **e.** None of the above is a problem associated with low yield.

11. What is a bacmid?
- **a.** a virus used to infect bacterial cells
- **b.** a recombinant protein from bacteria that is expressed in insect cells
- **c.** a shuttle vector that replicates as a plasmid in *E. coli* and a virus in insect cells
- **d.** a general term for bacterial expression vectors
- **e.** none of the above

12. Are insect cells fully able to glycosylate foreign proteins?
- **a.** yes
- **b.** no
- **c.** partially, only up to the addition of mannose
- **d.** uncertain, has never been experimentally tested
- **e.** none of the above

13. Which of the following is not used as a selectable marker to maintain mammalian shuttle vectors within mammalian cells?
- **a.** genetecin
- **b.** fungicides
- **c.** *DHFR* gene
- **d.** glutamine synthetase gene
- **e.** all of the above

14. What function does an internal ribosomal entry site serve?
- **a.** as a way to translate polycistronic mRNA in eukaryotes that carries the genes for multiple subunits of a protein
- **b.** as a protease cleavage site in between two proteins expressed on one transcript
- **c.** as a way to control the expression of two different proteins
- **d.** as a way to differentially express two proteins on the same transcript
- **e.** none of the above

15. Which factor affects which protein expression system to utilize?
- **a.** cost
- **b.** ability to glycosylate or fold proteins
- **c.** speed
- **d.** government regulations
- **e.** all of the above

16. Which of the following is not involved in the manufacturing of insulin?
- **a.** endopeptidases
- **b.** carboxypeptidase H
- **c.** signal peptidase
- **d.** phosphodiesterase 5
- **e.** All of the above are involved in the manufacturing of insulin.

17. Which of the following statements about insulin manufacturing is not true?
- **a.** Diabetics used to have to inject themselves with insulin that had been purified from the pancreas of either cows or pigs.
- **b.** Bacteria are able to express preproinsulin and process this peptide into insulin.
- **c.** Natural insulin forms hexamers, which causes the hormone to clump.
- **d.** A proline converted to an aspartic acid keeps insulin from clumping.
- **e.** Recombinant insulin is expressed as two mini-genes.

18. The recently discovered autotransporter proteins of bacteria:
- **a.** Transport themselves without using any secretory system.
- **b.** Are able to cross membranes because they are very small proteins.
- **c.** Require the Type III secretory system to cross the bacterial outer membrane.
- **d.** Require the BAM complex to cross the bacterial outer membrane.
- **e.** Rely on attached lipid molecules to function.

Further Reading

Dimitrov, D. S. (2012). Therapeutic proteins. *Methods in Molecular Biology (Clifton, N. J.), 899*, 1–26.

Fisher, A. C., Haitjema, C. H., Guarino, C., Çelik, E., Endicott, C. E., Reading, C. A., et al. (2011). Production of secretory and extracellular N-linked glycoproteins in Escherichia coli. *Applied and Environmental Microbiology, 77*, 871–881.

Jong, W. S., Saurí, A., & Luirink, J. (2010). Extracellular production of recombinant proteins using bacterial autotransporters. *Current Opinion in Biotechnology, 21*, 646–652.

Pandyarajan, V., & Weiss, M. A. (2012). Design of non-standard insulin analogs for the treatment of diabetes mellitus. *Current Diabetes Reports, 12*, 697–704.

van Oers, M. M. (2011). Opportunities and challenges for the baculovirus expression system. *Journal of Invertebrate Pathology, 107*(Suppl), S3–S15.

Wang, L. X., & Lomino, J. V. (2012). Emerging technologies for making glycan-defined glycoproteins. *ACS Chemical Biology, 7*, 110–122.

Weiss, A., & Brockmeyer, J. (2012). Prevalence, biogenesis, and functionality of the serine protease autotransporter EspP. *Toxins (Basel), 5*, 25–48.

Young, C. L., Britton, Z. T., & Robinson, A. S. (2012). Recombinant protein expression and purification: a comprehensive review of affinity tags and microbial applications. *Biotechnology Journal, 7*, 620–634.

Zhu, J. (2012). Mammalian cell protein expression for biopharmaceutical production. *Biotechnology Advances, 30*, 1158–1170.

Protein Engineering

Biotechnology

http://dx.doi.org/10.1016/B978-0-12-385015-7.00011-9

INTRODUCTION

A variety of enzymes have been in industrial use since before genetic engineering appeared. However, a mere couple of dozen enzymes account for over 90% of total industrial enzyme use. Some common examples are listed in Table 11.1. These proteins are used under relatively harsh conditions and are exposed to oxidizing conditions not found inside living cells. Consequently, these particular proteins are unusually robust and stable and are not at all representative of typical enzymes in this respect. It is notable that most of them are **hydrolases** that degrade either carbohydrate polymers or proteins.

Increasing the range of industrial enzymes has three facets. First, modern biology has identified many novel enzyme-catalyzed reactions that may be of industrial use. Second, as discussed in the preceding chapter, it is now possible to produce desired proteins in large amounts because of gene cloning and expression systems. Third, the sequence of the protein itself may be altered by genetic engineering to improve its properties. This is known as **protein engineering** and is the subject of the present chapter.

Methods for manipulating DNA sequences and for expressing the encoded proteins have been discussed in previous chapters. Therefore, we omit these details here. Rather, we emphasize the possibilities for altering the biological properties of proteins. In practice, most protein engineering has so far been concerned with making more stable variants of useful enzymes. The objective here is to engineer proteins so that they may be used under industrial conditions without being denatured and losing activity. However, it is also possible to alter proteins to change the specificity of their enzyme activities or even to create totally new enzyme activities. Ultimately, it may be possible to design proteins from basic principles. This will require the ability to predict three-dimensional protein structure from the polypeptide sequence.

Many enzymes are used, often under harsh conditions, by the biotechnology industry. There is a growing need for both tougher and novel enzymes.

Table 11.1 Important Proteins Used Industrially

Protein	Function
Amylases	Hydrolysis of starch for brewing
Lactase	Hydrolysis of lactose in milk processing
Invertase	Hydrolysis of sucrose
Cellulase	Hydrolysis of cellulose from plant materials
Glucose isomerase	Conversion of glucose to fructose for high-fructose syrups
Pectinase	Hydrolysis of pectins to clarify fruit juices, etc.
Proteases (ficin, bromelain, papain)	Hydrolysis of proteins for meat tenderizing and clarification of fruit juices
Rennet	Protease used in cheese making
Glucose oxidase	Antioxidant in processed foods
Catalase	Antioxidant in processed foods
Lipases	Lipid hydrolysis in preparing cheese and other foods

ENGINEERING DISULFIDE BONDS

The formation of **disulfide bonds** is a major factor in maintaining the 3D structure of many proteins. Disulfide bonds are especially important for those proteins found outside the cell in oxidizing environments. In practice, most enzymes used industrially will be exposed to such oxidizing conditions, and therefore, disulfide bonds are particularly relevant.

Introduction of extra disulfide bonds is a relatively straightforward way to increase the stability of proteins. The first step is to introduce two cysteine residues into the polypeptide chain. Then, under oxidizing conditions, these will form a disulfide bond provided that the polypeptide chain folds so as to bring the two cysteines into close contact. Obviously, for this approach to work, the tertiary structure of the protein must be known so that the cysteines can be inserted in appropriate positions (Fig. 11.1). In general, the longer the loop of amino acids between the two cysteines, the greater is the increase in stability. Note, however, that formation of a disulfide linkage can create a strained conformation if the two cysteines are not properly aligned. This may result in a decrease in stability of the protein.

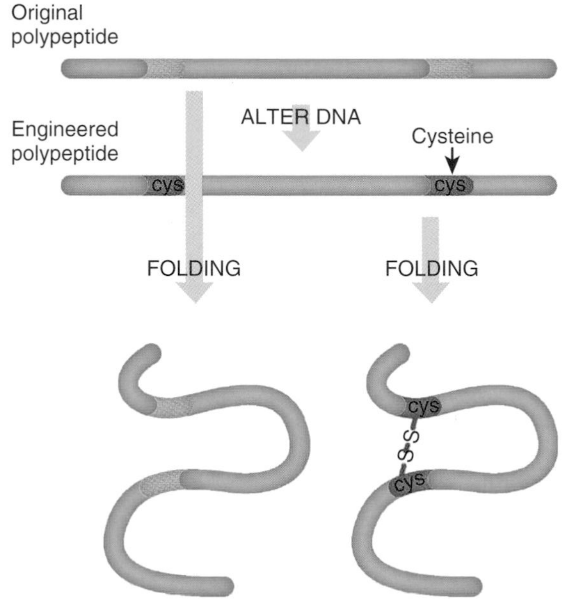

FIGURE 11.1
Introduction of Disulfide Bonds
A disulfide bond can be added to a protein by changing two amino acids into cysteines by site-directed mutagenesis. When the engineered protein is put under oxidizing conditions, the two cysteines form a disulfide bond, holding the protein together at that site.

This approach has been demonstrated using the enzyme lysozyme from bacterial virus T4. The structure of this enzyme has been solved by X-ray crystallography. The polypeptide chain of 164 amino acids folds into two domains and has two cysteines, neither involved in disulfide bond formation in the wild-type protein. One of these, Cys54, was first mutated to Thr to avoid formation of incorrect disulfides. The other, Cys97, was retained for use in disulfide formation. Extensive analysis of possible locations for disulfides was carried out. Those disulfides that might impair other stabilizing interactions in the protein were eliminated. This left three possible disulfide bonds that should theoretically promote stability, located between positions 3 and 97, 9 and 164, and 21 and 142 (Fig. 11.2).

To test these experimentally, five amino acids (Ile3, Ile9, Thr21, Thr142, and Leu164) were converted to Cys in various combinations. Engineered proteins with each individual disulfide as well as proteins with two or three disulfide linkages were tested for stability and for enzyme activity (Table 11.2). Stability was measured by thermal denaturation; the **melting temperature, Tm**, is the temperature at which 50% of the protein is denatured. Although all three disulfides increased stability of the protein, the engineered T4 lysozyme with the 21–142 disulfide had lost its enzyme activity. The precise reason is uncertain, but presumably some structural distortion has altered the active site. Nonetheless, the engineered protein with both of the two other disulfides showed a massive increase in stability and retained almost all of its enzyme activity.

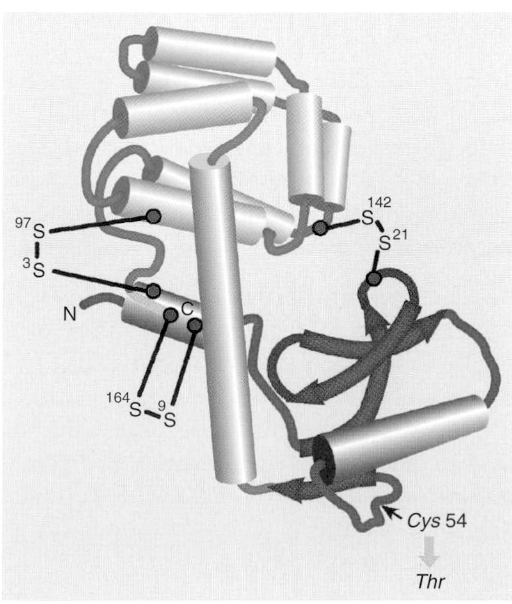

FIGURE 11.2
Disulfide Engineering of T4 Lysozyme
T4 lysozyme has two domains. The N-terminal region is shown in green and red, and the C-terminal region is in blue. These are linked by an alpha helix (purple). Disulfide bonds were added at three locations in T4 lysozyme to increase the stability. The first disulfide was between positions 9 and 164. This links the first alpha helix at the N terminus with the C-terminal tail. The second disulfide is between positions 3 and 97, which links the N- and C-terminal domains. Finally, the third disulfide links position 21 in the N-terminal domain with 142 in the C-terminal domain. The figure depicts alpha helices as cylinders and beta-sheets as green arrows.

Table 11.2 Disulfide Stabilization of T4 Lysozyme

Protein	Disulfide Bonds Present			Stability as Tm	Activity (%)
	3–97	9–164	21–142		
Original	–	–	–	41.9	100
1	+	–	–	46.7	96
2	–	+	–	48.3	106
3	–	–	+	52.9	0
4	+	+	–	57.6	95
5	–	+	+	58.9	0
6	+	+	+	65.5	0

As with all biological samples, the aqueous environment surrounding an enzyme can affect the rate of folding and unfolding, and therefore, the solution contains a complex mixture of folded and unfolded forms of the enzyme. Engineering disulfide bonds tilts the thermodynamic balance to having more fully folded functional proteins in this mixture. For most enzymes, disulfide bonds that prevent the first stages of unfolding enhance the enzymes' stability more than bonds that affect the later stages of unfolding. Computational approaches to identify disulfide bonds that have a positive effect on protein stability and kinetic stability are being used to first assess the changes before making a variety of different mutations, thus saving time and money during the production of enzymes.

Introduction of extra disulfide bonds, by altering the coding sequence for an enzyme, often provides major increases in protein stability.

IMPROVING STABILITY IN OTHER WAYS

Although the introduction of disulfide bonds is the simplest and most effective way to increase protein stability, a variety of other alterations may also help. Because of the effects of entropy, the greater the number of possible unfolded conformations, the more likely a protein is to unfold. Decreasing the number of possible unfolded conformations therefore promotes stability. This may be done in several ways, in addition to introducing extra disulfide linkages as described earlier. Glycine, whose R-group is just a hydrogen atom, has more conformational freedom than any other amino acid residue. In contrast, proline, with its rigid ring, has the least conformational freedom. Therefore, replacing glycine with any other amino acid or increasing the number of proline residues in a polypeptide chain will increase stability. Such replacements must avoid altering the structure of the protein, especially in critical regions. When tested experimentally, such changes do contribute small increases in stability.

Because hydrophobic residues tend to exclude water, these residues tend to cluster in the center of proteins and avoid the outer surface. If cavities exist in the hydrophobic core, filling them should increase protein stability. This may be done by replacing small hydrophobic residues, which are already in or near the core, with larger ones. For example, changing Ala to Val or Leu to Phe will achieve this. However, most proteins have hydrophobic cores that are already fairly stable and have few cavities. Furthermore, inserting larger hydrophobic amino acids to fill these cavities often causes twisting of their side chains into unfavorable conformations. This cancels out any gains of stability from packing the hydrophobic core more completely.

Because of its asymmetrical structure, the alpha helix is actually a dipole with a slight positive charge at its N-terminal end and a slight negative charge at its C-terminal end. The presence of amino acid residues with the corresponding opposite charge close to the ends of an alpha helix promotes stability. In natural proteins, the majority of alpha helixes are stabilized in this manner. However, in cases in which such stabilizing residues are absent, protein engineering may create them.

Asparagine and glutamine residues are relatively unstable. High temperature or extremes of pH convert these amides to their corresponding acids: aspartic acid and glutamic acid. The replacement of the neutral amide by the negatively charged carboxyl may damage the structure or activity of the protein. This may be avoided by engineering proteins to replace Asn or Gln by an uncharged hydrophilic residue of comparable size, such as Thr.

A variety of alterations to the amino acid sequence may yield proteins with moderate increases in stability.

CHANGING BINDING SITE SPECIFICITY

In addition to altering the stability of a protein, it is possible to deliberately change the active site. The most straightforward alterations are those that change the binding specificity for the substrate or a cofactor but do not disrupt the enzyme mechanism. Changing the specificity for a cofactor or substrate may be useful, either to make the product of the enzyme reaction less costly or to change it chemically.

This principle has been demonstrated with several enzymes that use the cofactors **NAD** or **NADP** to carry out dehydrogenation reactions. Both cofactors carry reducing equivalents, and both react by the same mechanism. Although a few enzymes can use either NAD or NADP, most are specific for one or the other. Generally, NAD is used by **dehydrogenases** in degradative pathways, and the respiratory chain oxidizes the resulting NADH. In contrast, biosynthetic enzymes often use NADP. These cofactors differ structurally only in NADP having an extra phosphate group attached to the ribose ring (Fig. 11.3). This gives NADP an extra negative charge and, not surprisingly, enzymes that prefer NADP have somewhat larger binding pockets with positively charged amino acid residues at the bottom. Enzymes that favor NADH often have a negatively charged amino acid residue in the corresponding position.

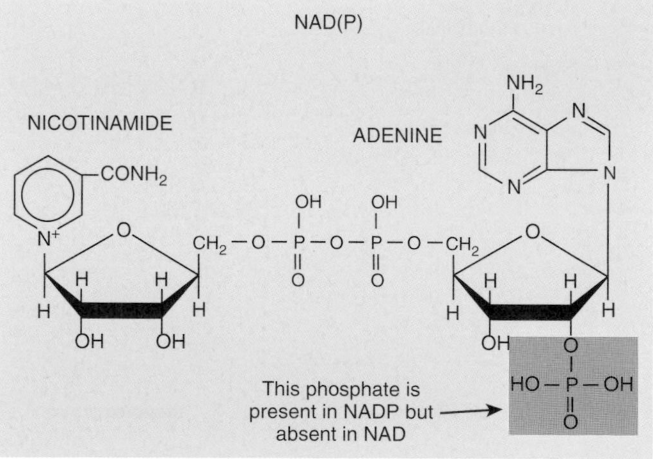

FIGURE 11.3 Difference in Structure between NAD and NADP
NAD (nicotinamide adenine dinucleotide) differs from NADP by one single phosphate (yellow).

Several enzymes that use NAD or NADP have been engineered to change their preference. For example, the **lactate dehydrogenase (LDH)** of most bacteria uses reduced NAD, not NADP, to convert pyruvate to lactate. A conserved aspartate provides the negative charge at the bottom of the cofactor-binding pocket that excludes NADP. If this is changed to a neutral residue, such as serine, the enzyme becomes able to use both NADP and NAD. If, in addition, a nearby hydrophobic residue in the cofactor pocket is replaced by a positively charged amino acid (such as lysine or arginine), the enzyme now prefers NADP to NAD (Fig. 11.4).

The specificity of LDH for its substrate can be altered in a similar way. The natural substrate lactate is a three-carbon hydroxyacid. It is possible to alter several residues surrounding the substrate-binding site without impairing the enzyme reaction mechanism. When a pair of

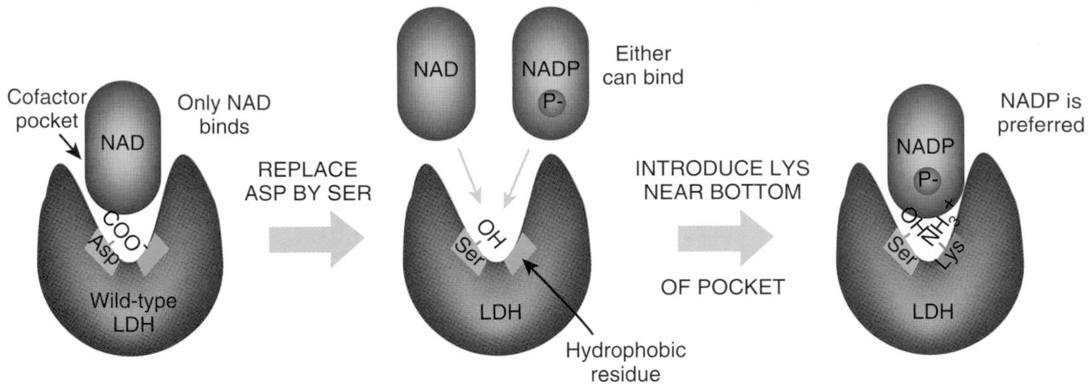

FIGURE 11.4 Changing Cofactor Preference of Lactate Dehydrogenase
Lactate dehydrogenase (LDH) preferentially binds NAD because the binding pocket has an aspartic acid. The negatively charged carboxyl repels the negatively charged phosphate of NADP Changing the aspartic acid to serine allows either NAD or NADP to bind to LDH. Adding a positively charged lysine makes the pocket more attractive to the NADP.

alanines is replaced with glycines, the binding site can be made larger. When hydrophilic residues (Lys, Gln) are replaced with hydrophobic ones (Val, Met), the site becomes more hydrophobic. Alteration of multiple residues gives an engineered LDH that accommodates five or six carbon analogs of lactate and uses them as substrates.

> Altering amino acids in and around the active site may change the specificity of an enzyme toward substrates or cofactors.

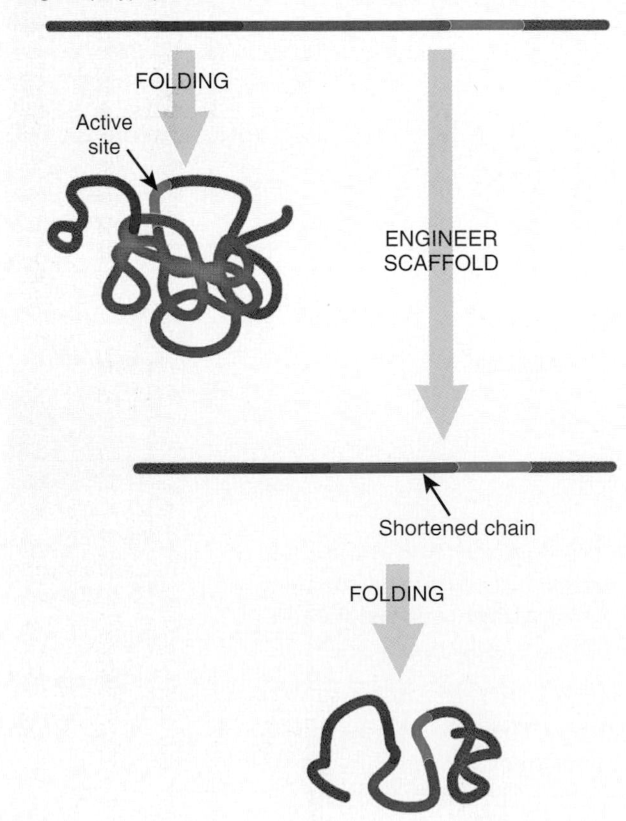

FIGURE 11.5 Scaffold Minimization
Proteins often have large structural domains (orange) outside the active site (blue/red). These domains can be removed, which enhances the proteins' level of expression or makes the isolation and purification easier.

STRUCTURAL SCAFFOLDS

Relatively few of the amino acid residues in a protein are actually involved in the active site. Most of the protein provides the 3D platform or scaffold needed to correctly position the active site residues. Quite often the scaffold is much larger than really necessary. For example, the β-galactosidase of *Escherichia coli* (LacZ protein) has approximately 1000 amino acids, whereas most simple hydrolytic enzymes have only 200 to 300. Presumably it should be possible to redesign a functional β-galactosidase that is only 25% to 30% the size of LacZ protein. From an industrial viewpoint, such a smaller protein would obviously be more efficient.

The concept of using only the active site region of useful proteins has already been applied to antibody engineering, as discussed in Chapter 6. In addition, the technique of phage display (see Chapter 9) has been used to select relatively short peptides for binding to a variety of target molecules. Once such a binding domain has been selected, it can be grafted on to another protein. Such techniques will allow the engineering of smaller proteins whose biosynthesis consumes less energy and material (Fig. 11.5).

Many proteins are larger than necessary from the industrial viewpoint. Removing extra sequences and simplifying the protein may increase industrial efficiency.

DIRECTED EVOLUTION

The protein engineering techniques described to this point require knowledge of protein structure plus detailed knowledge of active-site function. Very few enzymes have been studied this intensively. **Directed evolution** is a powerful technique to alter the function of an enzyme without the need for exhaustive structural and functional data (Fig. 11.6). Directed evolution can be used to change substrate specificity, either changing the enzyme to recognize a totally different substrate, or making subtle changes in cases in which the substrate is slightly different. Regulatory proteins may also be altered (see Box 11.1). The main premise of directed evolution is the random mutagenesis of the gene of interest, followed by a selection scheme for the new desired function. As described in Chapter 5, novel ribozymes may be isolated using a similar principle.

Directed evolution screens for new enzyme activities by constructing a library of different enzymes derived from the same original protein. Random mutagenesis during amplification of the gene creates a library of slightly different genes and, when expressed, slightly different proteins. Random mutants may be generated over the entire length of the gene sequence, or certain target amino acids can be replaced. Recombination (homologous or nonhomologous) can also be used create different permutations of the enzyme. These mutant genes are then screened for production of a protein with the new, desired enzyme activity after insertion into a suitable expression vector and host cell.

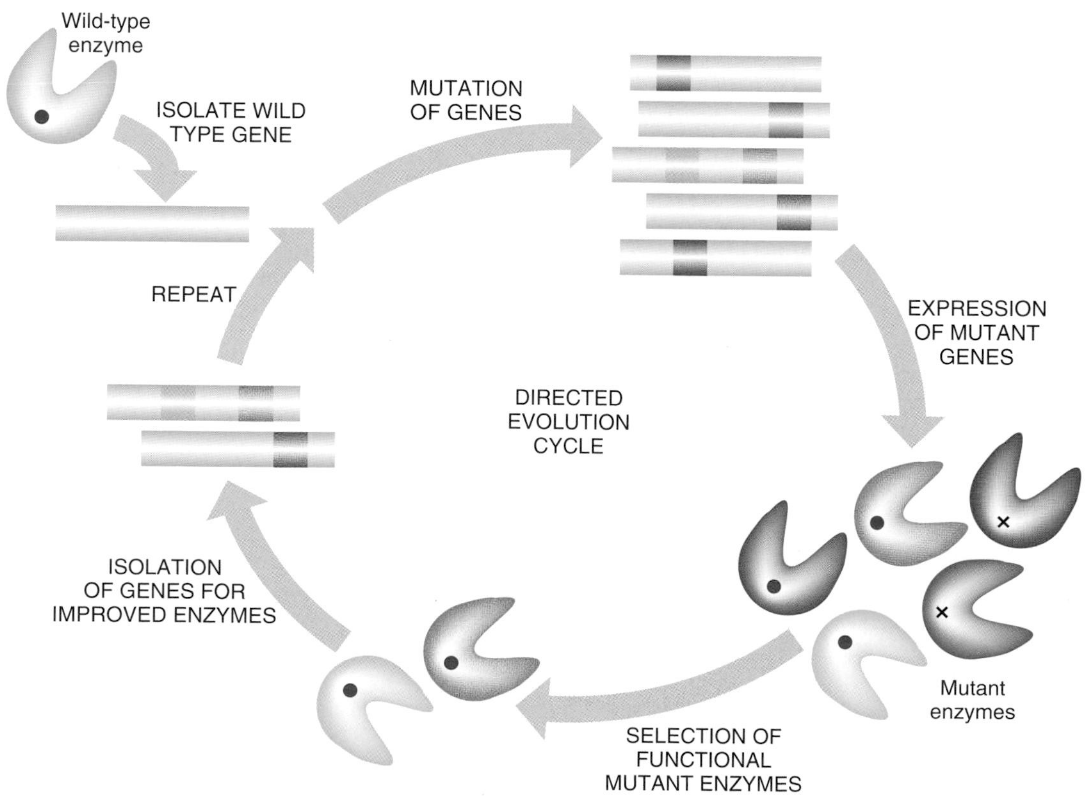

FIGURE 11.6 Directed Evolution Cycle
The gene for the wild-type enzyme (pink bar) is randomly mutated (mutations are represented by small colored blocks). The mutant genes are expressed; and their products, a set of randomly mutated enzymes, are screened for desired properties. Survivors may be subjected to successive cycles of mutation and selection.

Box 11.1 Inverting the Role of the LacI Repressor

The LacI repressor protein is part of the classic bacterial regulatory system, the *lac* operon. Normally, LacI binds to DNA in the absence of inducer and represses transcription. If inducer is present, the LacI repressor binds the inducer, changes conformation, and releases the DNA. This allows transcription of the *lac* operon.

Mutant LacI proteins that do not bind to DNA in either the presence or absence of inducer have long been known. These are known as "constitutive" mutants since the *lac* operon is expressed constitutively in their presence. Recently, directed evolution has been used to select mutants of LacI whose role is inverted. In these anti-LacI mutants, transcription of the *lac* operon is repressed in the presence of inducer and expressed in its absence.

The selection procedure requires an engineered *lac* operon whose expression can be selected both for and against. This is achieved by inserting two extra genes into the operon: *sacB* and *cat*. The *lacZ* gene is retained and is used as a reporter to monitor expression of the operon (Fig. A). The *sacB* gene (from *Bacillus subtilis*) encodes levansucrase. Expression of *sacB* in the presence of sucrose is detrimental because the levansucrase converts sucrose into polymer chains that cripple cell wall formation. However, in the absence of sucrose, expression of *sacB* has no effect. Bacteria cannot grow in the presence of sucrose when this gene is active; therefore, this gene confers negative selection. The *cat* gene encodes chloramphenicol acetyl transferase, which provides resistance to the antibiotic chloramphenicol. Bacteria grow in the presence of chloramphenicol when this gene is expressed; therefore, this confers positive selection.

Although there are several mutations that convert wild-type LacI to constitutive LacI, only the serine-97 to proline (S97P) mutation allows further evolution to anti-LacI (Fig. B). Two more steps are required, as shown, R207L and T258A. Either may be selected first, but whichever route is taken, the final anti-LacI mutant must have the three mutations: S97P, R207L, and T258A. The selection procedure consists of multiple cycles of the following steps:

(a) Mutagenesis by error-prone PCR.
(b) Select in culture medium with sucrose plus the inducer IPTG.
(c) Select in culture medium with chloramphenicol (but no IPTG).

What is most notable is that the final emergence of the anti-LacI mutant depends on the presence of earlier enabling mutations. In other words, the emergence of a particular selected characteristic may depend on the genetic history of the population that is undergoing evolutionary selection.

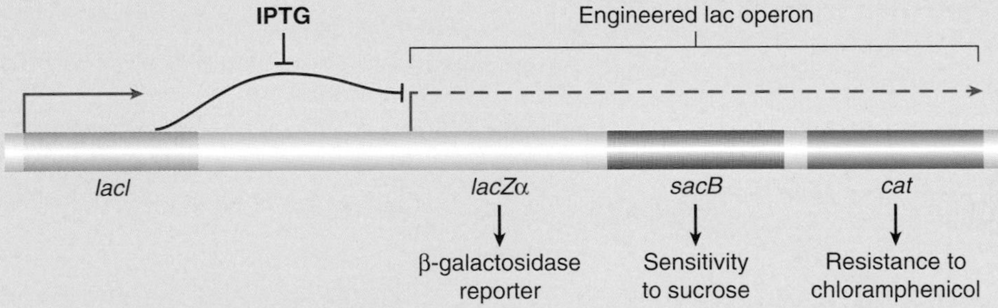

FIGURE A Engineered *lac* Operon for Selecting Anti-Lac
The *lac* operon was modified by inserting the *sacB* and *cat* genes and discarding *lacYA*. The *lacZ* gene for beta-galactosidase was retained. All three genes are expressed simultaneously when the engineered *lac* operon is transcribed.

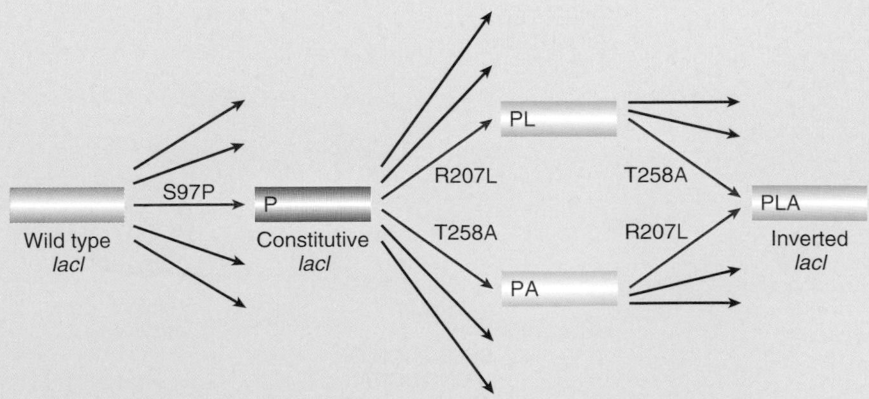

FIGURE B Directed Evolution of Anti-LacI
The evolution of an *anti-lacI* gene requires three steps, as shown.

Random mutagenesis usually starts with a gene whose product functions in a similar manner to the desired catalytic event. The gene is randomly mutated throughout the entire sequence using **error-prone PCR**. Different methods exist to induce errors during PCR amplification. The most straightforward is to add $MnCl_2$ to the PCR reaction. *Taq* polymerase has a fairly high rate of incorporating the wrong nucleotide, and $MnCl_2$ stabilizes the mismatched bases. The error will be copied in subsequent rounds of amplification. Adding nucleotide analogs such as 8-oxo-dGTP and dITP, which form mismatches on the opposite strand, can also enhance the PCR error rate. These analogs in combination with $MnCl_2$ can induce a wide variety of different mutations along the length of a gene. Some random mutations that occur outside the active site may cause subtle changes with profound effects on substrate recognition and enzyme function.

Target mutagenesis is much more focused and requires some knowledge of the enzyme structure, including the active site. For example, **tyrosyl-tRNA synthetase** from *Methanococcus jannaschii* was mutated through site-specific directed evolution. The gene for this enzyme had been sequenced, but no structural data were available. By comparing the sequence with that of another tyrosyl-tRNA synthetase, whose structure had been solved, the researchers identified amino acids potentially involved in tyrosine recognition. These residues were then randomly mutagenized. Altering residues with known, or suspected, functions in substrate recognition is more likely to have potent effects.

An alternative approach to form novel enzymes relies on combining domains from different proteins. This ranges from deliberately joining two or three domains of known function (as discussed next) to methods that can be regarded as directed evolution, such as DNA shuffling and combinatorial protein libraries. These issues are discussed further below.

Proteins may be altered by directed evolution, which consists of random mutagenesis of the coding sequence, followed by biological selection for improved or novel properties.

RECOMBINING DOMAINS

Another approach to generating novel enzymes is to deliberately recombine functional domains from different proteins. An example is the creation of novel restriction enzymes by linking the cleavage domain from the restriction enzyme *Fok*I with different sequence-specific DNA-binding domains. *Fok*I is a **type II restriction enzyme** with distinct N-terminal and C-terminal domains that function in DNA recognition and DNA cutting, respectively. By itself, the endonuclease domain cuts DNA nonspecifically. However, when the nuclease domain is attached to a DNA-binding domain, this domain determines the sequence specificity of the hybrid protein. The two domains may be joined via a sequence encoding a linker peptide such as $(GlyGlyGlyGlySer)_3$ (Fig. 11.7). Cleavage of the DNA occurs several bases to the side of the recognition sequence, as in the native *Fok*I restriction enzyme.

Several different DNA-binding domains have been combined with the nuclease domain of *Fok*I. For example, the **Gal4 protein** of yeast is a transcriptional activator that recognizes a 17 base-pair consensus sequence. Gal4 has two domains: a DNA-binding domain and a transcription-activating domain. The N-terminal 147 amino acids of Gal4 can be fused to the nuclease domain of *Fok*I, giving a hybrid protein that binds to the Gal4 consensus sequence and cleaves the DNA at that location.

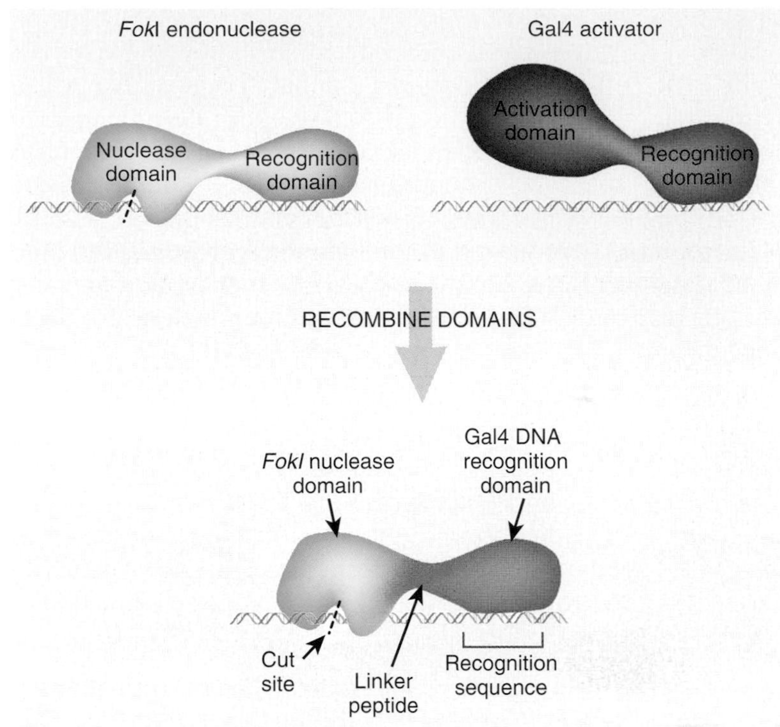

FIGURE 11.7 Recombining Domains to Create a Novel Endonuclease
The *Fok*I endonuclease has separate nuclease and sequence recognition domains. Using genetic engineering, the recognition domain of *Fok*I can be replaced with a Gal4 recognition domain, which binds to a different DNA sequence. The two domains are joined with an artificial linker peptide. The new hybrid enzyme now cuts DNA at different locations from the original *Fok*I protein.

Zinc finger domains have also been joined to the nuclease domain of *Fok*I. The zinc finger is a common DNA-binding motif found in many regulatory proteins. The zinc finger consists of 25 to 30 amino acids arranged around a Zn ion, which is held in place by binding to conserved cysteines and histidines. Each zinc finger motif binds three base pairs, and a zinc finger domain may possess several motifs. In the example shown in Figure 11.8, domains approximately 90 amino acids long and comprising three zinc finger motifs, and which therefore specifically recognize nine base DNA sequences, were connected to the *Fok*I nuclease domain.

Both the amino acid sequence and the corresponding DNA recognition specificity are known for a wide range of zinc fingers. This information allows zinc finger domains to be designed to read any chosen DNA sequence. DNA segments that encode zinc finger domains may be derived either from naturally occurring DNA-binding proteins or by artificial synthesis, because individual zinc finger motifs are about 25 amino acids (75 nucleotides) in length. In principle, engineered proteins can now be provided with a zinc finger domain that will enable them to bind specifically to any chosen DNA sequence. Engineered zinc finger nucleases are now being used in genome editing, as described in Chapter 17.

Many proteins consist of discrete domains with individual functions. Recombining domains from different proteins can generate hybrid proteins with novel combinations of properties.

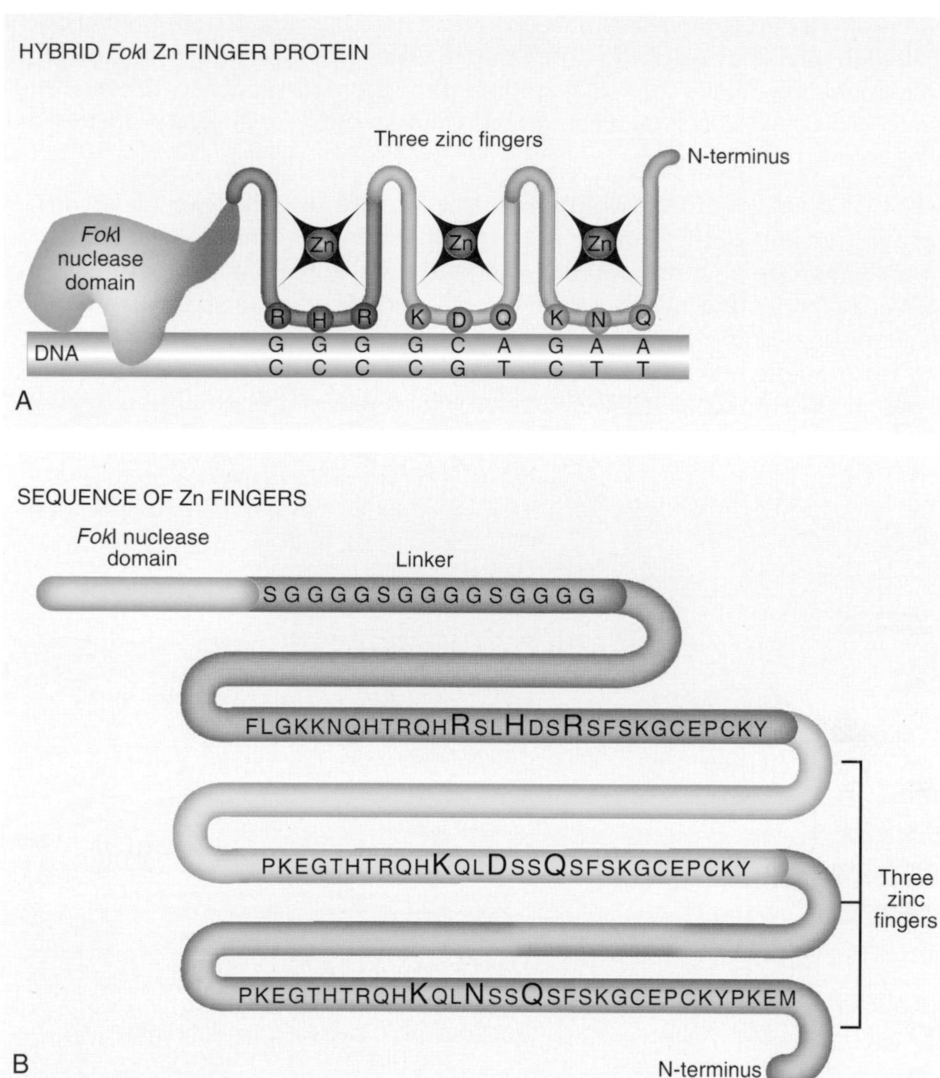

FIGURE 11.8 Assembly of Zinc Finger Domains

(A) The nuclease domain of *FokI* can be linked to a zinc finger domain containing three zinc finger motifs. Zinc fingers recognize three nucleotides each; therefore, any nine-base-pair recognition sequence can, in principle, be linked to the nuclease domain. (B) The sequence of the hybrid between the *FokI* nuclease and the zinc finger domain. The letters represent the amino acid sequence. The amino acids in large letters recognize and bind the DNA sequence.

DNA SHUFFLING

Natural selection works on new sequences generated both by mutation and recombination. **DNA shuffling** is a method of artificial evolution that includes the creation of novel mutations as well as recombination. The gene to be improved is cut into random segments around 100 to 300 base pairs long. The segments are then reassembled by using a suitable DNA polymerase with overlapping segments or by using some version of overlap PCR (see Chapter 4). This recombines segments from different copies of the same gene (Fig. 11.9).

Mutations may be introduced in several ways, including the standard mutagenesis procedures already described. In addition, the DNA segments may be generated using error-prone PCR (see earlier discussion) instead of by using restriction enzymes. Alternatively, mutations may be introduced during the reassembly procedure itself by using a DNA polymerase that has impaired proofreading capability. The result is a large number of copies of the gene, each with several mutations scattered at random throughout its sequence. The final shuffled and mutated gene copies must then be expressed and screened for altered properties of the encoded protein.

A more powerful variant of DNA shuffling is to start with several closely related (i.e., homologous) versions of the same gene from different organisms. The genes are cut at random with appropriate restriction enzymes and the segments mixed before reassembly. The result is a mixture of genes that have recombined different segments from different original genes (Fig. 11.10). Note that the reassembled segments keep their original natural order. For example, several related **β-lactamases** from different enteric bacteria have been shuffled. The shuffled genes were cloned onto a plasmid vector and transformed into host bacteria. The bacteria were then screened for resistance to selected β-lactam

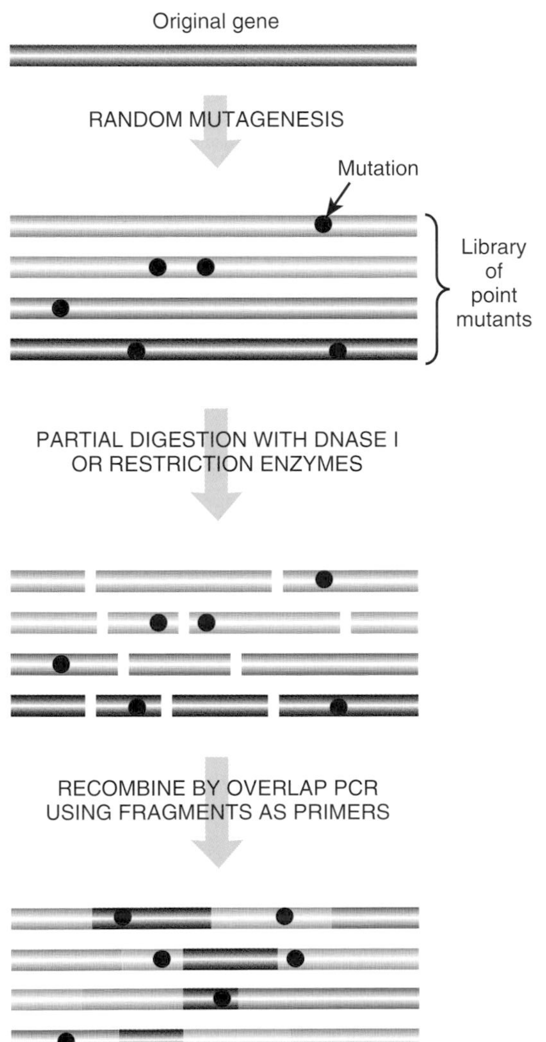

FIGURE 11.9 DNA Shuffling for a Single Gene
Introducing point mutations and shuffling gene segments can generate a better version of a protein. First, many copies of the original gene are generated with random mutations. The genes are then cut into random segments. Last, the fragments are reassembled using overlap PCR. The new constructs are then assessed for enhanced protein function.

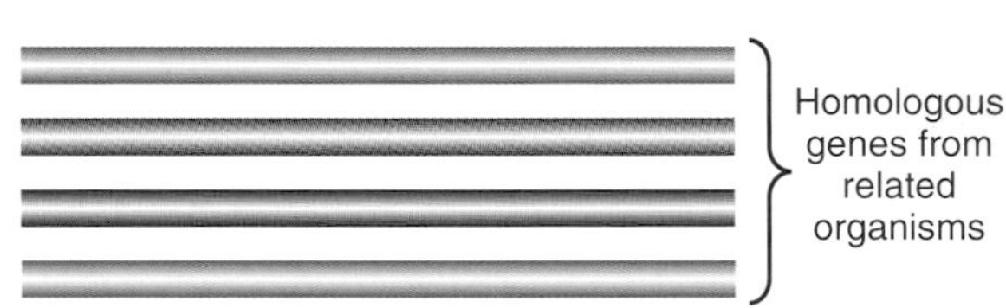

PARTIAL DIGESTION WITH DNASE I OR RESTRICTION ENZYMES

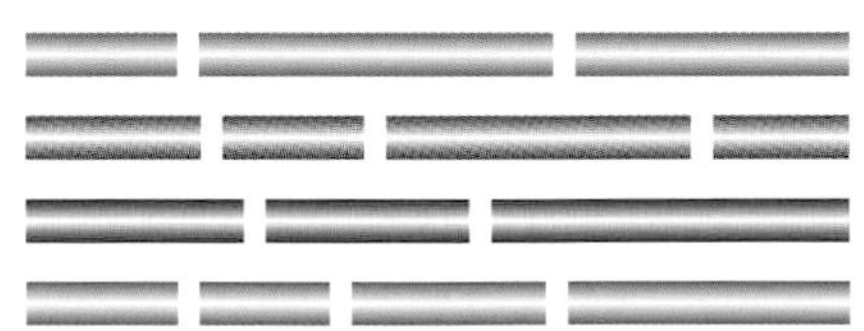

REASSEMBLY BY PCR

FIGURE 11.10 DNA Shuffling for Multiple Related Genes
Shuffling segments from related genes can also enhance the function of a particular protein. The original set of related genes is digested into small fragments and reassembled using PCR. The new combinations are tested for a change in function.

antibiotics. This approach yielded improved β-lactamases that degraded certain penicillins and cephalosporins more rapidly and so made their host cells up to 500-fold more resistant to these β-lactam antibiotics.

In DNA shuffling, the coding sequence for a protein is rearranged in the hope of generating novel or improved activities. Mutations may also be introduced during the procedure to provide more variation.

COMBINATORIAL PROTEIN LIBRARIES

So far we have discussed ways to modify a useful protein that already exists. Another approach to protein engineering is to generate large numbers of different protein sequences and then screen them for some useful enzyme activity or other chemical property. (Screening is often done by phage display or related techniques, as described in Chapter 9.) Rather than merely generating large numbers of random polypeptides, **combinatorial screening** usually uses premade modules of some sort to create a **random shuffling library**. For example, protein motifs known to provide a binding site for metal ions or metabolites might be combined with segments known to form structures such as an alpha helix.

In a common approach, DNA modules of around 75 base pairs (i.e., 25 codons) are made by chemical DNA synthesis. Several modules are then assembled to give a new artificial gene. The modules are usually joined by PCR using overlapping primers (Chapter 4). Modules may be joined in a chosen order or in a randomized manner. For proper expression of the assembled sequence, the front and rear modules are normally specified to provide suitable promoter and terminator sequences. The intervening modules may then be randomly shuffled to generate more possible variation (Fig. 11.11).

The resulting protein library is then screened for some activity associated with the modules used. For example, if modules for binding an organic metabolite and for binding Fe ions were included, then the library of products might be screened for an enzymatic activity that oxidizes the metabolite via iron-mediated catalysis.

A related approach is based on the idea that the exons of eukaryotic genes encode modular segments of proteins, such as binding sites and structural motifs. Although it is by no means

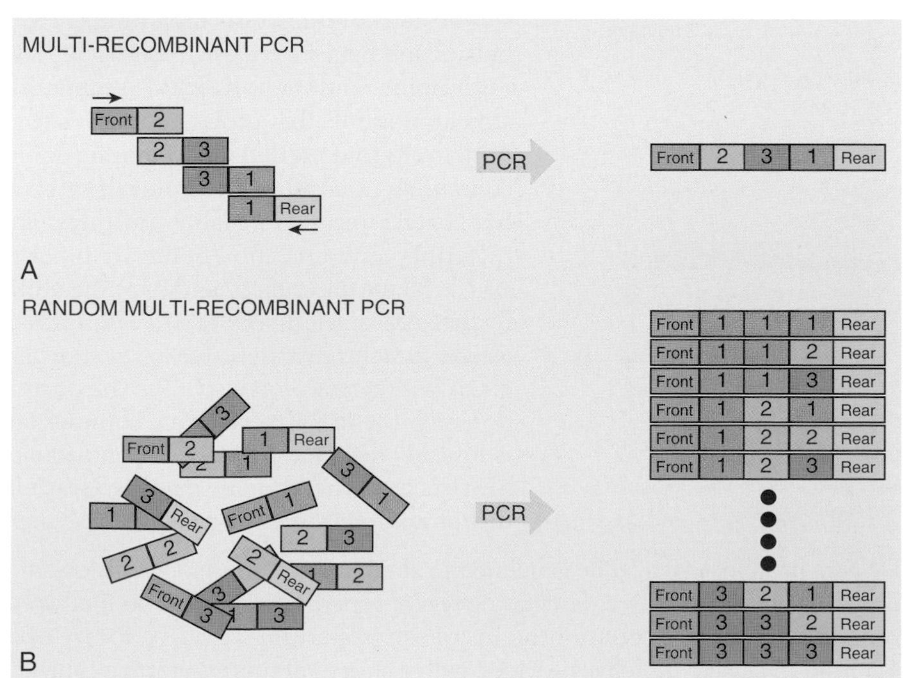

FIGURE 11.11 Generation of Random Shuffling Library

To create a library of new proteins, different modules can be randomly joined together. The first module (yellow) has sequences for the promoter; therefore, this is always added at the front. Similarly, the last module (purple) has the terminator sequences. Using overlapping PCR primers, the modules are joined together in a particular order (part A), or randomly (part B). Because random assembly creates many different combinations, this method creates a library of new proteins that can be screened for a particular function or set of functions.

always true, in many cases exon boundaries do correspond to the ends of structural domains within the encoded proteins. Consequently, it is believed that at least some eukaryotic genes have evolved by the natural shuffling of exons.

In the exon shuffling approach, a **combinatorial library** is generated from an already existing eukaryotic gene. Each of the exons of the eukaryotic gene is generated by a separate PCR reaction. The segments are then mixed and reassembled by overlap PCR. Two variants exist, depending on the design of the overlap primers for the PCR assembly. Reassembling the exons in random order generates a random splicing library, much as described earlier. Less radically, an **alternative splicing library** retains the order of the exons but includes or excludes any particular exon at random (Fig. 11.12). Before final products are checked for the desired activity, they are screened to obtain sequences long enough to encode most of the original exons, as this obviously greatly increases the chances of a functional protein.

Another shuffling technique to generate novel proteins is to use premade protein modules. The coding sequences for the modules are combined in various combinations by a PCR-based approach.

FIGURE 11.12 Generation of Alternative Splicing Library
The exons from an original gene can be recombined such that one exon is missing in each novel construct. The new genes are then screened for new or altered function.

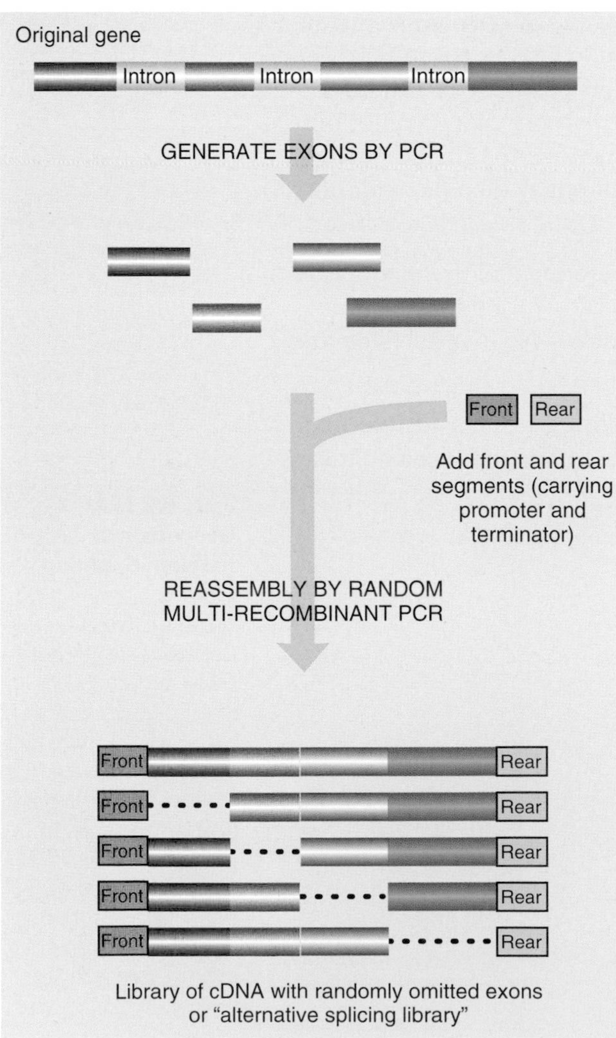

CREATION OF *DE NOVO* PROTEINS

The ultimate in generating novel proteins is to create wholly novel polypeptide sequences—***de novo* proteins**. These proteins are made by first generating random DNA sequences, which are then expressed in a host cell. Obviously, most random polypeptides will not be functional, so it is necessary to screen the libraries of *de novo* proteins for some sort of activity. The vast majority of random polypeptides will not fold into stable or functional structures. Therefore, in practice, the procedures used are modified to increase the chances of generating a functional protein. Thus, while *de novo* proteins are indeed novel, they are not completely random.

The sequence characteristics required to produce certain simple and common structural motifs, such as alpha helices and beta sheets, are understood. Thus, it is possible to combine random polypeptide sequences with artificial structural motifs that provide a nucleus for the *de novo* proteins to fold around. For example, polypeptide libraries containing three alpha helix plus three beta sheet segments have been screened for stable structures. This was followed by further randomization of the amino acids between the structural motifs and screening for the ability to bind easily detected target molecules. The result was the creation of *de novo* proteins that are stable in structure and tightly bind green fluorescent protein (GFP). The purpose of generating *de novo* proteins with ever more complex structural motifs is to study the relationship between protein sequence and 3D structure. The ultimate goal is to learn how to fully predict protein 3D structure given the polypeptide sequence.

De novo proteins can also be generated to shed light on protein evolution, in particular the origin of novel proteins. A fascinating recent experiment has shown that *de novo* proteins can functionally replace certain proteins of the bacterium *E. coli*. A library of approximately 1.5 million *de novo* proteins, provided with a structural motif to form a four-helix bundle, was

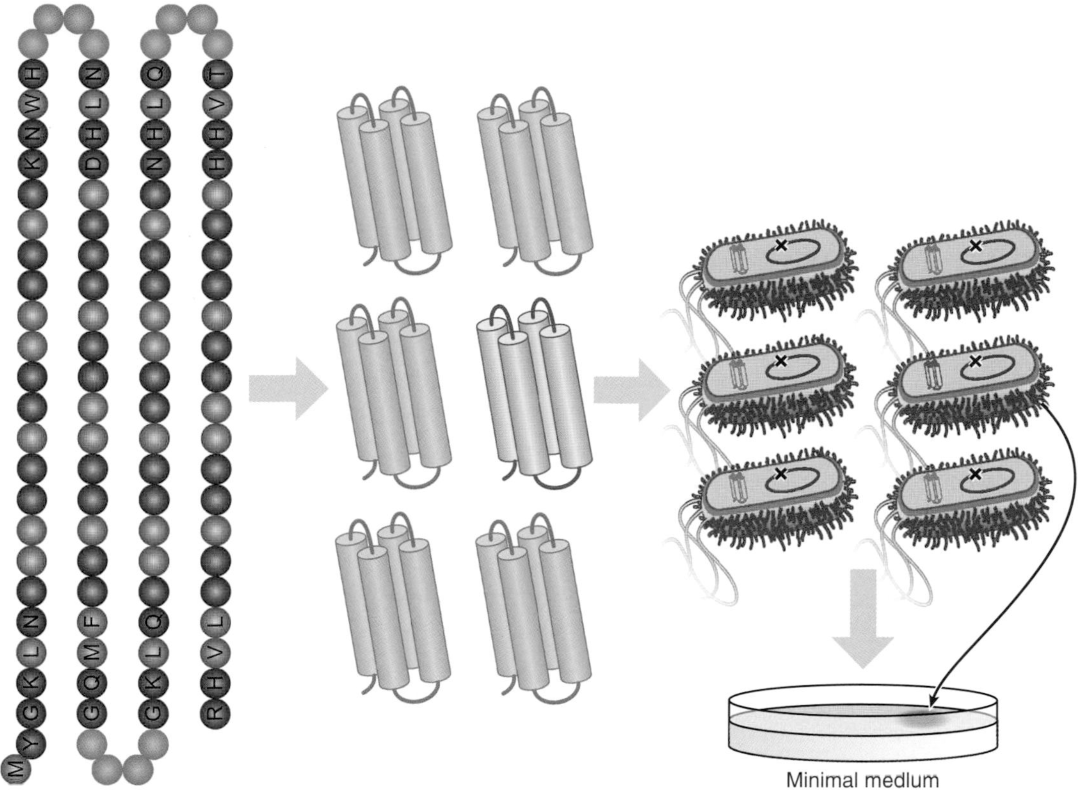

FIGURE 11.13 ***De novo*** **Proteins Rescue** ***E. coli*** **Mutants**
A pattern of polar (red) and nonpolar (yellow) residues was designed to fold into four-helix bundles. Circles with letters show fixed residues, and empty circles show variable positions. The experimental library of 1.5 million synthetic genes cloned on a high-expression plasmid was transformed into strains of *E. coli* from which genes needed for growth on minimal medium had been deleted (black X). Colonies were selected on minimal medium to find novel sequences (purple) that rescue the cells. Modified after Fisher MA et al. (2011). De novo designed proteins from a library of artificial sequences function in *Escherichia coli* and enable cell growth. *PLoS One* 6:e15364.

transformed into a selection of around 30 *E. coli* mutants that each lacked a specific protein, due to a deletion (Fig. 11.13). In four cases (*serB, gltA, ilvA, fes*), *de novo* proteins were found that rescued the mutants. For three of these mutants, several similar *de novo* proteins were isolated. Not surprisingly, high levels of the *de novo* proteins were required, and the rescued mutants grew only slowly. Some of these *de novo* proteins have been evolved for higher levels of activity and show improved rescue (i.e., faster growth) of the *E. coli* mutants. The *de novo* proteins were smaller and of much simpler structure than the natural proteins they replaced. The mechanisms by which the *de novo* proteins rescued the mutant bacteria are mostly unknown, although several indirect effects, such as induction of alternate pathways, were ruled out. However, in one case, it appears that the *de novo* protein did have the enzymatic activity of the natural protein it replaced.

A few proteins from the same *de novo* protein library have been screened by microarrays for possible binding to thousands of small molecules. Several small molecules that bound were found in each case. More limited screening of larger numbers of these *de novo* proteins revealed many that bound the heme cofactor, and several of these showed low levels of peroxidase activity. This result is not surprising given that mixtures of heme with random proteinoids showed similar activity. (These discoveries were made by investigators carrying out experiments to illustrate the chemical origins of life. Mixtures of amino acids were polymerized into random polypeptides by mild heat, and traces of cofactors such as heme were added.) Also, the *de novo* proteins can be used in directed evolution experiments as a single protein, plus its corresponding gene, is available. This allows selection of evolved *de novo* proteins with increased enzyme activity.

Semirandom DNA sequences may be used to generate libraries of *de novo* proteins. These may then be screened for possible binding activities or biological functions. Several *de novo* proteins have been found that can replace natural proteins.

EXPANDING THE GENETIC CODE

Many **non-natural amino acids** have different functional groups that are useful in protein engineering or biochemical analysis. Incorporating a non-natural amino acid into a protein can be done on a small scale by isolating the tRNA for a particular amino acid and attaching the non-natural amino acid. The charged tRNA is then added to an *in vitro* protein translation system, which incorporates the non-natural amino acid into the growing polypeptide chain. This method is too costly and time consuming for large-scale use. For large-scale incorporation, *E. coli* may be genetically modified to insert the non-natural amino acid *in vivo*.

As discussed earlier, the universal genetic code contains codons for 20 different amino acids plus three stop codons (see Fig. 2.15). Two other genetically encoded amino acids are known: **selenocysteine** and **pyrrolysine**. They are inserted during protein synthesis into a few proteins at special stop codons that have been reallocated (UGA for selenocysteine and UAG for pyrrolysine). Both selenocysteine and pyrrolysine have dedicated tRNAs with specific **aminoacyl tRNA synthetases** to charge them. Reallocating stop codons for selenocysteine and pyrrolysine relies on specific recognition sequences nearby. These sequences form stem-loop structures that bind amino acid-specific proteins (SelB protein for selenocysteine).

These exceptional cases suggest that it should be possible to insert totally non-natural amino acids during ribosomal protein synthesis by manipulating pairs of tRNA plus cognate aminoacyl-tRNA synthetase to target chosen stop codons (Fig. 11.14). Such pairs are often referred to as "orthogonal" relative to the normal genetic code of the host organism. Even more radical are attempts to modify the codon size from three bases to two or four (see Box 11.2).

Inserting a non-natural amino acid during *in vivo* protein synthesis requires a mutant aminoacyl-tRNA synthetase that charges a mutant tRNA with the non-natural amino acid. The laboratory of Peter G. Schultz, at the Scripps Research Institute, developed an *E. coli* strain that incorporates *p*-benzoyl-L-phenylalanine (*p*Bpa) at amber codons (UAG). Directed evolution was used to mutate a tyrosyl-tRNA synthetase from *Methanococcus jannaschii* (a thermophilic member of the Archaea). This enzyme was chosen because it does not recognize any tRNA naturally encoded by *E. coli*. The gene for its cognate tRNA (i.e., the specific partner tRNA) must also be provided. The cognate tRNA must also be altered to recognize the amber stop codon (UAG) instead of its original natural codon. Consequently, when the mutant pair of tyrosyl-tRNA synthetase plus cognate tRNA is expressed in *E. coli*, the chosen non-natural amino acid is inserted at amber stop codons.

To alter the tyrosyl-tRNA synthetase, researchers generated a library of mutant enzymes by random mutation of each amino acid residue involved in recognition of the amino acid substrate (originally tyrosine). The library of mutant tRNA synthetase genes was transformed into *E. coli* that possess a gene for the partner tRNA, and a gene for chloramphenicol resistance with an amber codon in the middle (this yields a truncated inactive protein; hence, the bacteria are sensitive to chloramphenicol). The *E. coli* were grown in the presence of *pBpa* and chloramphenicol. If the mutant tRNA synthetase allowed insertion of an amino acid at the amber stop codon, then the chloramphenicol resistance protein was functional, and the cells survived. Otherwise, the cells died. This was the positive selection (Fig. 11.15).

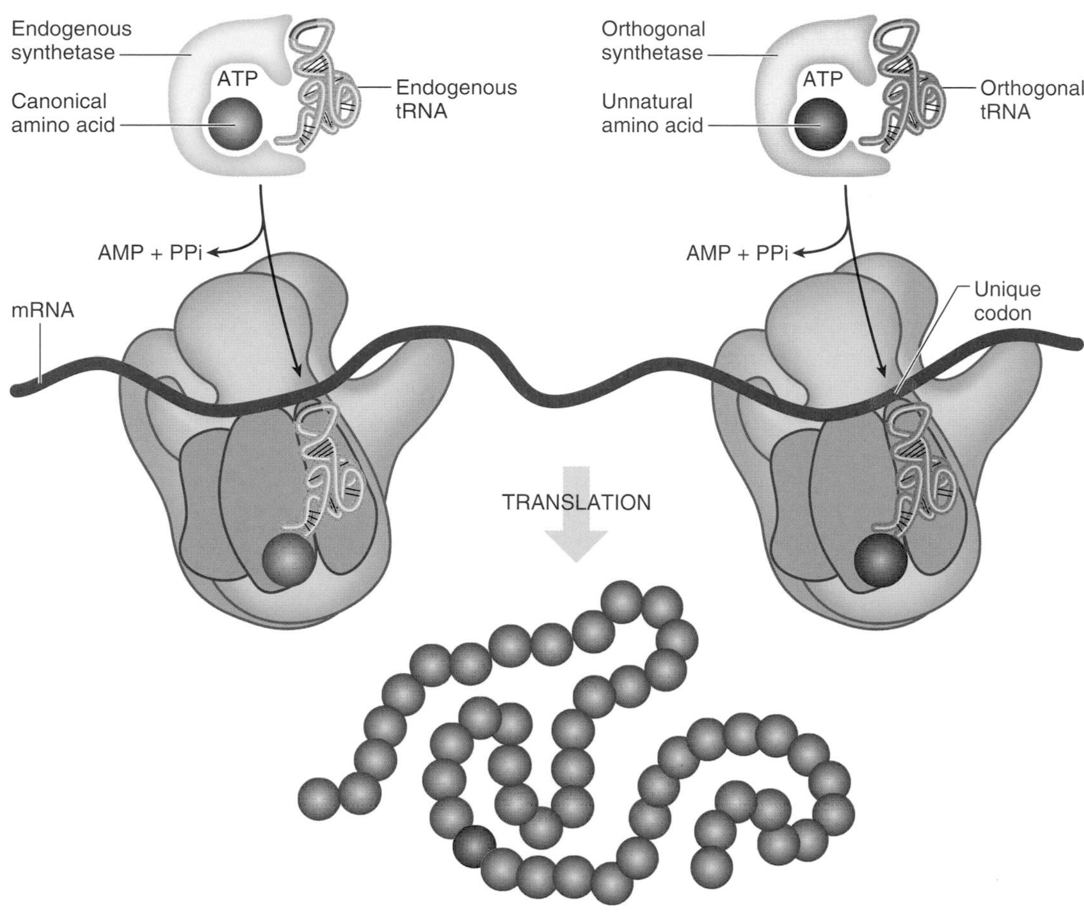

FIGURE 11.14 Method for Genetically Encoding Non-Natural Amino Acids
Both the synthetase that attaches the amino acid to tRNA and its cognate tRNA must be engineered. The synthetase is altered to recognize the chosen non-natural amino acid. The tRNA is altered to recognize a reallocated stop codon. Modified after Wang Q, Parrish AR, Wang L (2009). Expanding the genetic code for biological studies. *Chem Biol* **16**, 323–336.

This positive selection does not exclude mutant tRNA synthetases that charge the amber tRNA with a natural amino acid, so a negative selection scheme was used next. The plasmids carrying the mutant tRNA synthetases were isolated and transformed into a different *E. coli* strain. This *E. coli* had a toxin gene with an amber suppressor mutation plus the amber tRNA. Here, the mutants were grown without any *p*Bpa. If the mutant tRNA synthetase could charge the amber tRNA with a natural amino acid, the toxin would be made and the *E. coli* would die. This eliminated mutant tRNA synthetases that used natural amino acids. The two selections were alternated numerous times, and finally a mutant tRNA synthetase was isolated (Fig. 11.15) that still recognized the amber tRNA and specifically linked *p*Bpa to it.

This altered tRNA synthetase can be used to insert the non-natural amino acid *p*Bpa into various proteins. To achieve this, the researcher must insert an amber codon into the gene that encodes the protein of interest and express the mutant gene in an *E. coli* host cell that expresses the orthogonal *p*Bpa tRNA synthetase plus tRNA pair.

Using an orthogonal synthetase/tRNA pair will result in the chosen non-natural amino acid being inserted at multiple stop codons, most of which are not involved in the experiment. One rather tedious way to avoid this is to eliminate all copies of a particular codon from the genome, thus freeing up a codon for experimental use (i.e., for incorporation of a non-natural amino acid). At present, this approach is feasible only in bacteria such as *E. coli* and, even then, only for the least common of the three alternative stop codons, TAG. Like other organisms, *E. coli* uses TAA, TGA, and TAG as stop codons. In *E. coli*, these are used in the

Box 11.2 Bigger and Smaller Genetic Codes

In addition to adding a single extra amino acid to the genetic code by changing the meaning of a particular stop codon, some more radical alterations have been attempted.

Mutant tRNAs that read a four-base codon, instead of the normal three bases, have been known for some time. These tRNAs were originally isolated as frameshift suppressors and have an expanded anticodon loop. By reading four bases instead of three, they can restore the reading frame of frameshifts that are due to insertion of an extra base. Since the number of combinations of four bases is larger than with three, using a four-base codon to insert non-natural amino acids would allow a wider range of choices. Selection for a four-base suppressor tRNA is similar in principle to that described previously for tRNA/synthetase pairs that insert a non-natural amino acid at a stop codon. However, engineered genes with frameshifts, rather than premature stop codons, are used.

In this case, the starting tRNA was for the rare amino acid, pyrrolysine (see Fig. C). This tRNA and its partner aminoacyl tRNA synthetase are also from a member of the Archaea. As for the archaeal TyrRS enzyme and tRNA pair used previously, they were chosen because the synthetase does not recognize any tRNA naturally encoded by *E. coli*.

Instead of expanding the genetic code, it is possible (in principle) to shrink it. Shrinking would involve generating proteins completely lacking a particular amino acid and then reallocating all codons that encoded the chosen amino acid. So far, this has been demonstrated only in cell extracts using green fluorescent protein and tryptophan (the rarest amino acid).

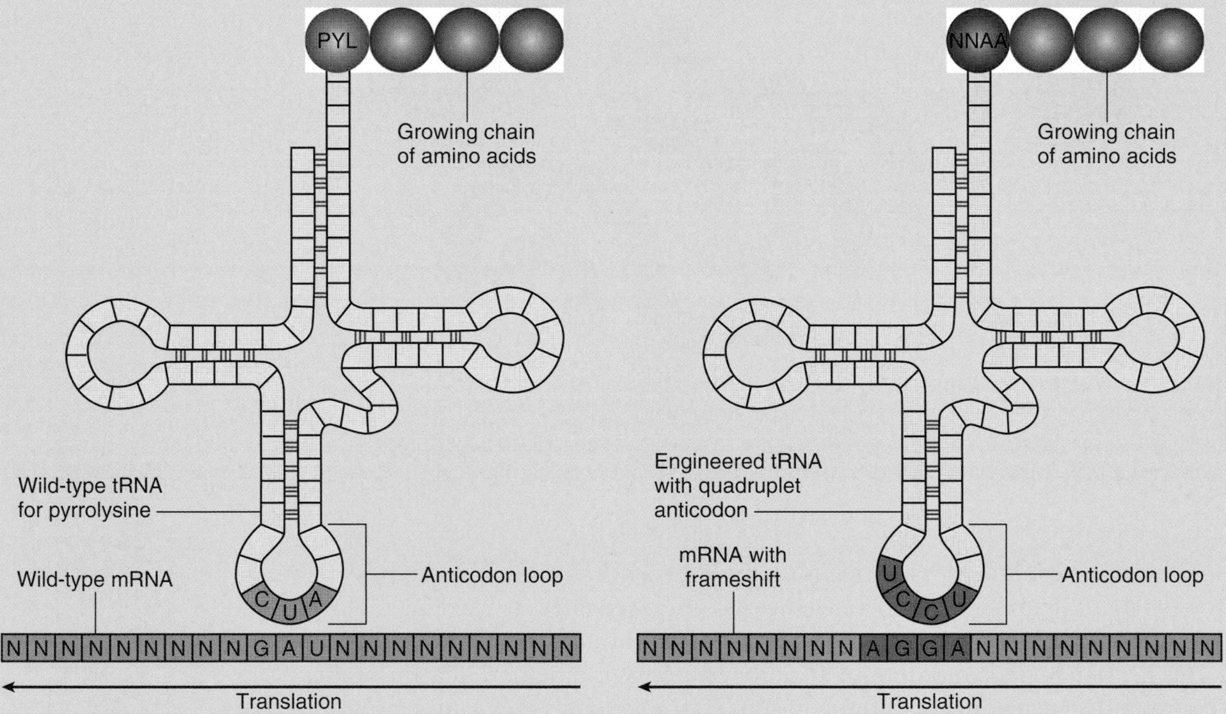

FIGURE C Insertion of Non-Natural Amino Acid at Quadruplet Codon
The wild-type tRNA inserts pyrrolysine (PYL) at an amber stop codon (UAG). The engineered tRNA with a quadruplet anticodon recognizes four bases in the mRNA and can therefore suppress frameshifts. The engineered target gene has the corresponding quadruplet codon at the location of the frameshift. An engineered aminoacyl tRNA synthetase (not shown) charges the quadruplet tRNA with the chosen non-natural amino acid (NNAA).

approximate ratio of 20:9:2, and there are just over 300 TAG stop codons. An *E. coli* has been generated in which all TAG codons have been replaced by TAA. Multiple strains with about 10 such alterations each were combined by conjugation to generate strains with increasing numbers of replacements. This strain allows insertion of a variety of non-natural amino acids.

A variety of schemes have been devised to insert non-naturally occurring amino acids into proteins. The most sophisticated schemes consist of changing the coding properties of selected codons.

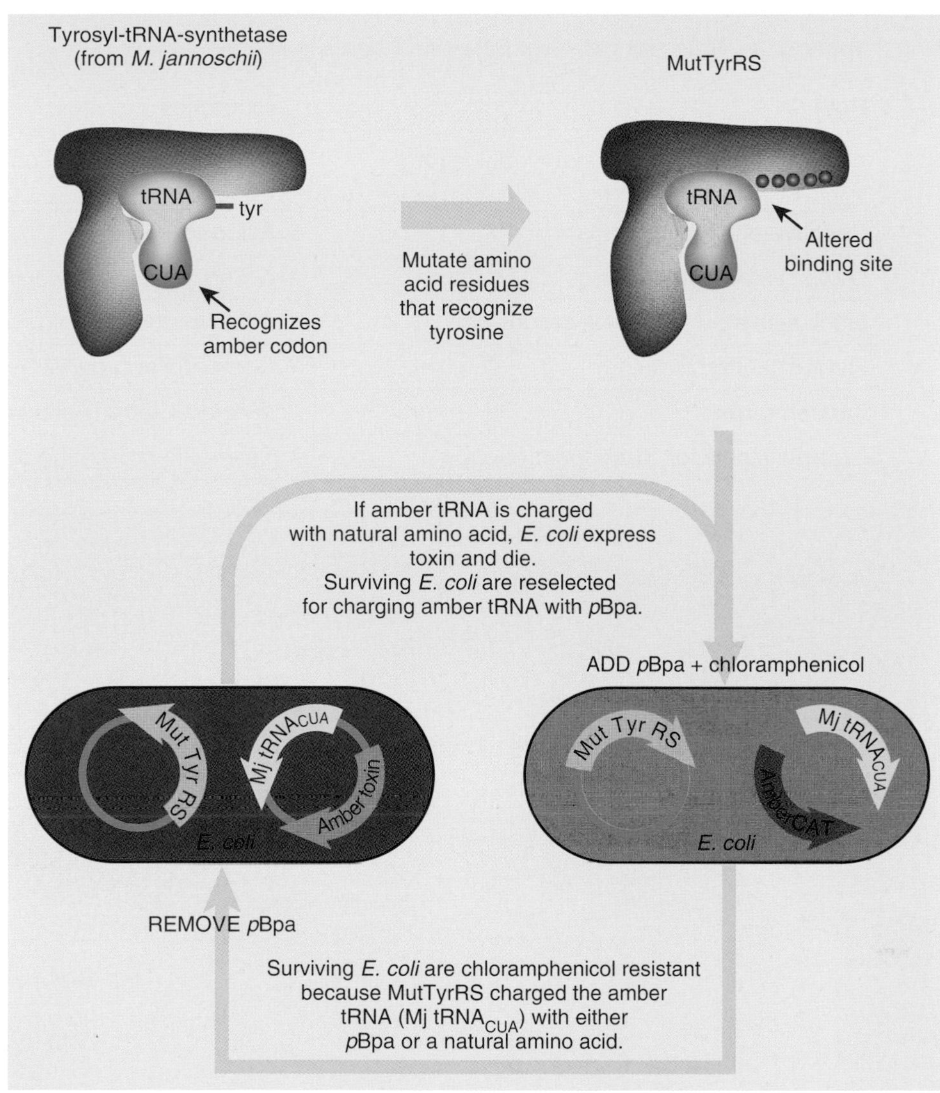

FIGURE 11.15 Positive and Negative Selections for Mutant tRNA Synthetase

Tyrosyl-tRNA synthetase normally attaches tyrosine to its cognate tRNA. This tRNA has been previously engineered so that it now carries the anticodon CUA, which recognizes the amber codon (UAG). The amino acids at the tyrosine recognition site of tyrosyl-tRNA synthetase were randomly mutagenized (TyrRS refers to the two subunits of this enzyme). This gave a library of many different tRNA synthetases (MutTyrRS) that might bind different amino acids (but would still recognize the same tRNA). This library was expressed in cells of *E. coli* containing genes for the engineered tRNA (Mj $tRNA_{CUA}$) plus chloramphenicol acetyl transferase (CAT; confers chloramphenicol resistance). However, the CAT gene has been inactivated by insertion of an amber stop codon (amberCAT). For cells to regain chloramphenicol resistance, an amino acid must be inserted at this amber codon. The cells are grown with *pBpa* and chloramphenicol. MutTyrRS must charge the Mj $tRNA_{CUA}$ with *p*Bpa (or another amino acid) to express active CAT, which protects the *E. coli* from chloramphenicol. The library clones that survive this selection are expressed in a different *E. coli* host *(left)*. This strain has the gene for the amber tRNA (Mj $tRNA_{CUA}$) plus a toxin gene inactivated by an amber codon. No *p*Bpa is present. If MutTyrRS charges the amber tRNA with a natural amino acid, the toxin will be made and kill the cells. This eliminates clones that charge the amber tRNA with amino acids other than *p*Bpa. The positive and negative selections are alternated to find the best mutant tRNA synthetase.

ROLES OF NON-NATURAL AMINO ACIDS

Non-natural amino acids are inserted in the experimental manipulation of a wide range of protein properties. Novel amino acids with different functional groups may be chosen for specific uses. Table 11.3 summarizes the most popular uses for non-natural amino acids, and a few selected examples are illustrated in Figure 11.16.

The first category consists of amino acids that are indeed natural in the sense of being found in proteins but are not genetically encoded. Such amino acids are made by modifying genetically encoded amino acids after translation. A common example is phosphoserine

Table 11.3 Roles for Non-Natural Amino Acids in Proteins

Role	Examples
Analyzing post-translational modification	Serine phosphorylation Lysine acetylation
Reactive chemical groups	Azidohomoalanine
UV activated cross-linking	*p*-Benzoyl-L-phenylalanine
Light-activated functional groups	4,5-Dimethoxy-2-nitrobenzyl-cysteine
Fluorescent probes	Coumarin and dansyl derivatives
Infrared probes	*p*-Azido-L-phenylalanine
Isotopic labels for NMR spectroscopy	Several fluoro-amino acids

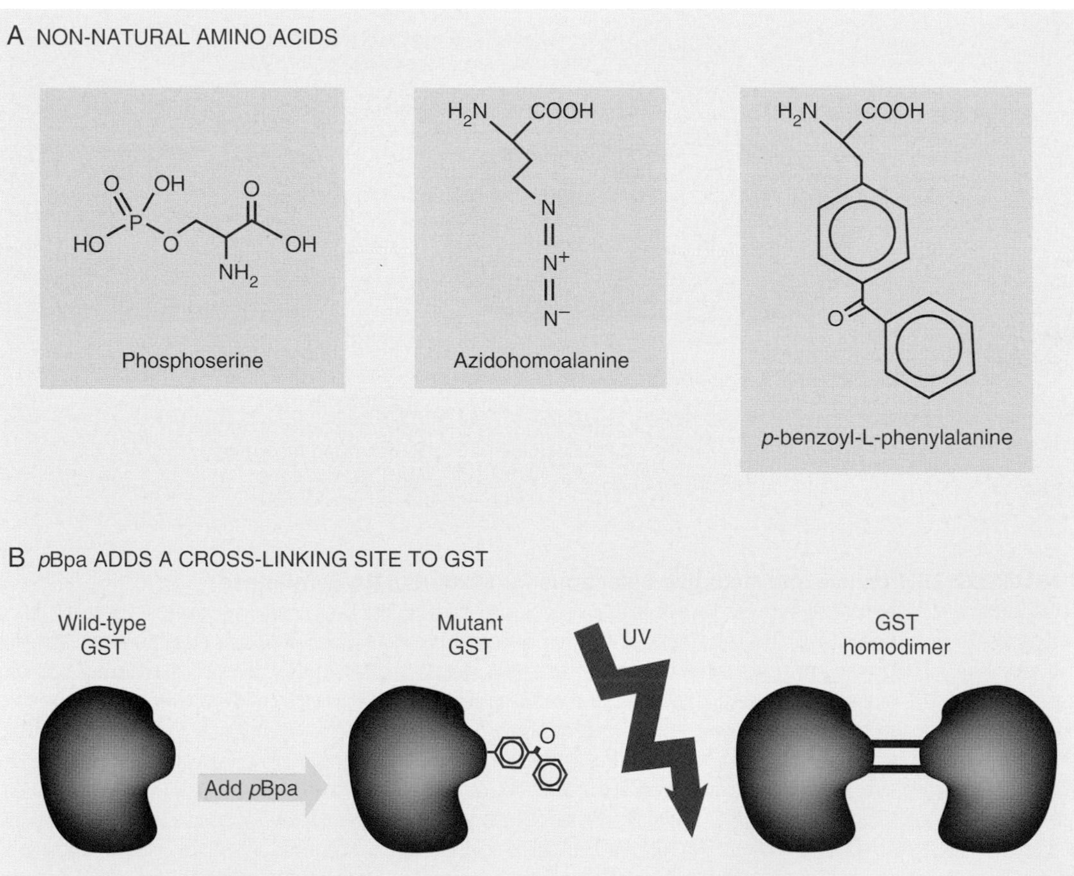

FIGURE 11.16 Adding New Functional Groups to Proteins
(A) Inserting amino acids that are not genetically encoded allows the incorporation of new functional groups. (B) The non-natural amino acid *p*-benzoyl-L-phenylalanine (*p*Bpa) cross-links the GST mutant protein to form a homodimer.

(Fig. 11.16), a common post-translational modification of serine that plays roles in both signal transduction and controlling protein function. Introducing amino acids that are normally made by post-translational modification helps in investigation of their biological roles.

A variety of non-natural amino acids are used to provide reactive chemical groups. These are subdivided in Table 11.3. Some are used for attaching other molecules, including other proteins. Azidohomoalanine (Fig. 11.16) carries a reactive azide group that allows chemical cross-linking to molecules that carry terminal alkyne groups. It is often used for cross-linking two different proteins. The non-natural amino acid *p*-benzoyl-L-phenylalanine (*p*Bpa) that was discussed previously is widely used to insert a cross-linking group that can be activated by UV irradiation. Thus, when the protein glutathione-S-transferase (GST) has *p*Bpa inserted, UV irradiation creates a covalently linked homodimer (Fig. 11.16).

Another use for light-activated chemical groups is to control the activity of the target protein. So-called photo-caged amino acids are converted to natural amino acids upon activation. For example, UV light converts 4,5-dimethoxy-2-nitrobenzyl-cysteine to cysteine, with the loss of a dimethoxynitrobenzyl group. Incorporation of photo-caged amino acids has been used to make ion channel proteins that are inactive but can be activated by UV light in living nerve cells.

Finally, non-natural amino acids are used as probes for several spectroscopic techniques, including fluorescence, infrared, and NMR spectroscopy (Table 11.3). Fluorine derivatives are often used for NMR because the 19F nucleus is naturally abundant and NMR active.

Non-natural amino acids may be used for a variety of experiments. Some provide reactive groups for attachment or cross-linking, whereas others act as spectroscopic probes.

BIOMATERIALS DESIGN RELIES ON PROTEIN ENGINEERING

In the medical field, biomaterials are crucial for reconstructive surgery, tissue engineering, and regenerative medicine. Biomaterials include vascular grafts and cartilaginous tissue scaffolds that facilitate growth of new tissue by providing support and structure. The materials used in these products are based on proteins, and therefore, protein engineering can be used to improve them both mechanically and biochemically.

Many biomaterials are based on extracellular matrix proteins that provide support and structure *in vivo*. For example, collagen and elastin are proteins found in cartilage. *In vivo*, these proteins are secreted from cells known as chondrocytes and form a hard elastic support that cushions our joints. **Elastin-like polypeptides (ELPs)** are engineered proteins similar to native elastin. ELPs possess a repeated peptide sequence such as $(VPGZG)_n$, where Z is any amino acid except proline (Elastin itself contains these repeats, but Z is restricted to Ala, Leu, or Ile). If the amino acid repeat is changed, the physical properties of the final material will also change. If lysine residues are inserted, then two ELP strands can be chemically cross-linked by treatment with hydroxymethyl phosphine derivatives (Fig. 11.17). Varying the location and number of lysines can create various types of films. Alternatively, UV-activated cross-linking groups may be engineered into the ELP peptide. The peptides stay as soluble strands until exposed to UV light. The ability to control gel formation allows a doctor to inject the liquid form at the desired location and then cross-link the ELPs to form a gel.

Besides supplying support, these materials can promote healing and tissue regeneration by attracting cells to the area. Adding different protein-binding domains to the repeated peptide can promote cell migration and adherence. For example, the cell membrane receptor integrin recognizes the extracellular matrix protein fibronectin. If the integrin-binding domain of fibronectin is alternated with the ELP repeat, then integrin-expressing cells will recognize and migrate into the ELP structure. Another example is the peptide Val-Ala-Pro-Gly, which is recognized by a membrane-bound receptor on vascular smooth muscle cells. When this peptide is alternated with the ELP sequence, the resulting material promotes the movement and growth of vascular smooth muscle cells only.

Another method to induce cellular migration and growth is to encapsulate various growth factors in the ELP matrix. For example, if vascular endothelial growth factor or platelet-derived growth factor is mixed with ELPs, then blood vessels are induced to form within the matrix.

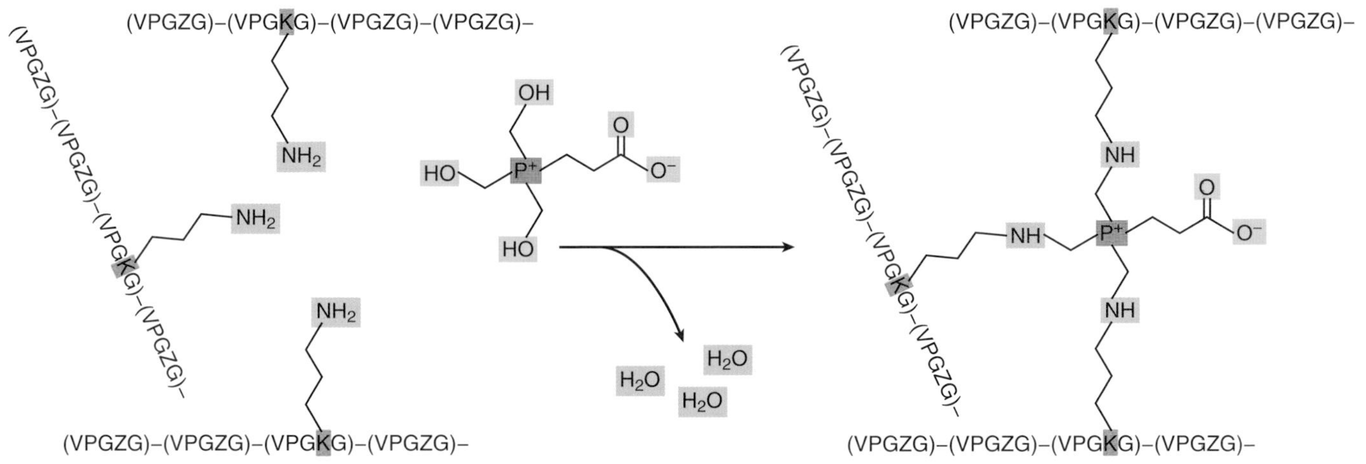

FIGURE 11.17 Chemical Cross-Linking of Elastin-Like Polypeptides

Insertion of lysine (K, green) into the sequence of an elastin-like polypeptide (ELP) made of repeating VPGZG units allows multiple chains to be cross-linked by reaction of the side chain amino groups of lysine with hydroxymethyl phosphine derivatives such as THPP ([Tris(hydroxymethyl)phosphino]propionic acid).

Biomaterials for medical use, especially extracellular proteins used in reconstructive surgery, may be improved by protein engineering.

ENGINEERED BINDING PROTEINS

Making drugs specific for a particular organ can eliminate many unwanted side effects. One way of achieving this is to attach the drug to a reagent that recognizes proteins specific to particular tissues. As noted in Chapter 6, antibodies are the most widely used reagents for binding specific target proteins. However, antibodies require disulfide cross-links to function, and these are often hard to maintain during large-scale manufacture. Some researchers have therefore been seeking alternatives to antibodies.

Most research into finding nonantibody binding partners has focused on certain protein structural domains. Because many different proteins have been crystallized and their structures have been determined, many different binding domains can be compared. The binding domains of each family of proteins share the same general structure, such as a β-barrel or three-helix bundle (Fig. 11.18). To generate novel binding domains, a binding protein with a known structure is chosen and the amino acid residues associated with binding are identified. The binding protein is modified by mutation of these residues and then screened for new binding partners. It is hoped that the targeted directed evolution approach will find new, more easily isolated proteins for targeting drugs to specific target cells within our bodies.

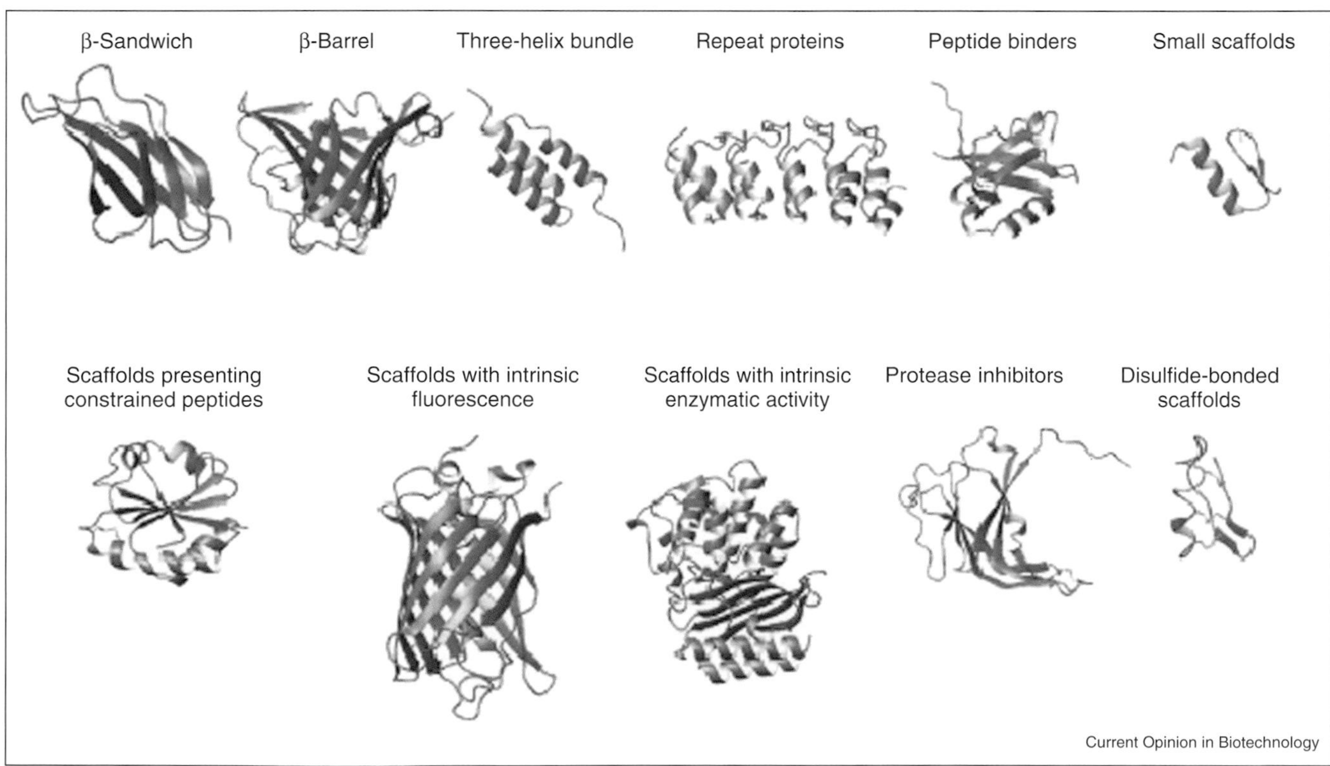

FIGURE 11.18 Structural Domains Involved in Protein–Protein Binding

Some types of protein backbones used as scaffolds for protein-binding agents. The proteins used (with their Protein Data Bank ID numbers) are beta sandwich (1FNA, fibronectin); beta barrel (A chain of 1BBP, lipocalin); thee-helix bundle (1Q2N, SpA domain); repeat proteins (1MJO, AR protein); peptide binders (chain A of 1KWA, PDZ domain); small scaffolds (chain F of 1MEY, zinc finger protein); scaffolds presenting constrained peptides (chain A of 2TrX, thioredoxin A); proteins with intrinsic fluorescence (chain A of 1GFL, GFP) or intrinsic enzyme activity (1M40, beta-lactamase); protease inhibitors (1ECY ecotin); and disulfide-bonded scaffolds (chain A of 1CMR, scorpion toxin). From Binz HK, Plückthun A (2005). Engineered proteins as specific binding reagents. *Curr Opin Biotechnol* **16**, 459–469. Reprinted with permission.

Creating binding proteins for a variety of uses is still in the experimental stages. However, it may eventually prove possible to replace cumbersome antibodies with smaller and more stable proteins.

Summary

Many enzymes are used industrially, especially in food processing. Protein engineering uses genetic technology both to improve enzymes already in use and to create proteins with novel properties. Methods range from introducing single specific changes into protein sequences to screening large numbers of semirandom polypeptides for new activities.

During production or use of enzymes for industry, the enzyme experiences conditions such as high heat, acidity or alkalinity, and other non-native conditions, which can inactivate or destabilize the enzyme. Protein engineering reduces the effects of the harsh conditions by creating stabilized proteins. Adding disulfide bonds between flexible regions of a protein prevents the protein from unfolding in higher temperature, as measured by the Tm. Lysozyme is an example of a protein in which engineered disulfide bridges increased the Tm without affecting the enzymatic activity. Other methods of increasing protein stability include changing glycine to proline, which has less conformational freedom; packing a hydrophobic protein core more completely; stabilizing alpha helices with an opposite-charged amino acid at one end; and changing asparagine and glutamine to other unchanged hydrophilic residues.

Besides stabilizing enzymes for use in industrial applications, protein engineering works to change substrate or cofactor specificity. Some enzymes use expensive substrates or cofactors to catalyze a reaction, and therefore, protein engineering changes the substrate or cofactor to a cheaper, more abundant version. In addition, smaller enzymes are cheaper to manufacture because they require fewer starting materials. Protein engineering can lessen the cost of enzyme production by making the scaffolding regions smaller.

Since all of the preceding techniques require thorough knowledge of the enzyme structure and active site geometry, the techniques cannot be used on newly identified enzymes. In order to work with enzymes of unknown structure, a technique called directed evolution changes substrate specificity or even creates enzymes with novel catalytic functions. In this method, a series of random mutations is created throughout the protein, and then the library of new proteins is screened for the new function. When an enzyme has some structural information, a more focused, directed evolution strategy called target mutagenesis creates random mutations in specific locations of the protein.

Other methods of protein engineering focus on recombining different protein domains. DNA shuffling simply divides a gene or a family of related genes into segments and then reshuffles the pieces into new proteins. A random shuffling library, in contrast, uses known protein motifs rather than gene segments. The protein motifs are combined using PCR, and then the new proteins are screened for a specific function. Combinatorial libraries are another permutation and consist of an assembly of eukaryotic exons from one gene or perhaps a family of genes. These exons can be assembled at random, or they can be assembled in the correct order, just omitting different exons, which is called an alternative-splicing library.

Protein engineering can create *de novo* proteins. Creating random polypeptides establishes a library that can be screened for a specific enzymatic activity. Since the majority of random sequences will be nonfunctional, this approach usually begins with an assembly of simple structural motifs such as alpha helices and beta sheets that are randomly assembled and then screened for activity.

Another method of creating new proteins is to alter the amino acids by using non-natural isoforms. This approach requires a new tRNA that recognizes a specific codon and a non-natural amino acid, and a new aminoacyl tRNA synthetase that connects the non-natural amino acid to the correct altered tRNA. These can then incorporate the non-natural amino acid during protein translation. Non-natural amino acids provide chemical groups that can cross-link different parts of the protein with UV light or chemical groups that mimic post-translational modifications. In addition, some non-native amino acids are "photo-caged" and do not function until exposure to light. In addition, non-natural amino acids are used to "visualize" the protein by spectroscopy.

In summary, protein engineering is essential to many biotechnology applications. It creates more stable enzymes for industrial use, enzymes with new substrate or cofactor specificity, novel enzymes with directed evolution, combinatorial libraries, or the use of non-natural amino acids. In addition, protein engineering creates new biomaterials in the medical field, such as elastin-like peptides, and binding proteins that target a drug or reagent to a specific tissue or organ.

End-of-Chapter Questions

1. What is protein engineering?
- **a.** alteration of a protein through genetic engineering to improve the protein's properties
- **b.** modification of the amino acid residues after a protein has been translated
- **c.** engineering a protein to be expressed in an artificial expression system
- **d.** production of large quantities of proteins for use in industrial settings
- **e.** none of the above

2. Which alteration to the protein sequence provides a major increase in protein stability?
- **a.** glycosylation of amino acid residues
- **b.** acetylation of amino acid residues
- **c.** oxidation of the protein
- **d.** introduction of cysteine residues for disulfide bond formation
- **e.** reduction of the protein

3. Which of the following concepts increases protein stability?
- **a.** decreasing the number of possible unfolded conformations
- **b.** decreasing the number of hydrophobic interactions
- **c.** increasing the number of hydrophilic interactions
- **d.** decreasing the number of disulfide bonds
- **e.** none of the above

4. What is a benefit to changing the active site or cofactor binding site of an enzyme?
- **a.** to prevent a reaction from taking place
- **b.** to produce the enzyme's product more cost-effectively
- **c.** to make the enzyme more stable
- **d.** to ensure that only the native substrate or cofactor are able to bind to the enzyme
- **e.** none of the above

5. Why might it be sensible to simplify an industrially relevant enzyme?
- **a.** to decrease the possibility of misfolding in the protein
- **b.** to increase the amount of protein being made
- **c.** to increase efficiency of the enzyme
- **d.** to decrease the amount of amino acid supplements to the growth medium for the growing cells
- **e.** none of the above

(*Continued*)

6. What term describes the process of creating random mutations in a gene and then selecting for improved function or altered specificity of the resulting protein product?
   a. indirect evolution
   b. protein mutagenesis
   c. translational evolution
   d. directed evolution
   e. instant evolution

7. Which of the following is required to insert a nonnatural amino acid during *in vivo* protein synthesis, but is not required for *in vitro* protein synthesis?
   a. a mutant aminoacyl-tRNA synthetase
   b. a tRNA charged with the nonnatural amino acid
   c. a natural aminoacyl-tRNA synthetase
   d. amino acids
   e. none of the above

8. Which of the following is a way to create novel enzymes?
   a. glycoengineering
   b. using nonnatural amino acids to add new functional groups to amino acids within binding sites
   c. recombining functional domains from different enzymes
   d. DNA shuffling
   e. all of the above

9. How were improved β-lactamases made?
   a. by glycoengineering the N-terminal sequence of β-lactamase
   b. through the addition of nonnatural amino acids within the active site of the enzyme
   c. by DNA shuffling several related β-lactamase genes from different enteric bacteria
   d. by recombining the functional domains of different β-lactamases
   e. none of the above

10. With regards to exons, how is a combinatorial library constructed?
   a. The exons of a eukaryotic gene are created by PCR and then reassembled randomly.
   b. The order of the exons from a eukaryotic gene is maintained but the exclusion or inclusion of a particular exon is random.
   c. The exons are digested with restriction enzymes and then religated.
   d. The exons are assembled and screened by phage display.
   e. none of the above

11. With regards to exons, how is an alternative splicing library created?
   a. The exons of a eukaryotic gene are created by PCR and then reassembled randomly.
   b. The order of the exons from a eukaryotic gene is maintained but the exclusion or inclusion of a particular exon is random.
   c. The exons are digested with restriction enzymes and then religated.
   d. The exons are assembled and screened by phage display.
   e. none of the above

12. What are ELPs?
  a. engineered proteins that are similar to collagen
  b. engineered proteins that contain regenerated tissues
  c. engineered proteins that adhere to cell surface receptors
  d. engineered proteins that are similar to elastin
  e. none of the above

13. Which of the following is a structural domain associated with protein–protein binding?
  a. β-barrel
  b. three-helix bundle
  c. beta sandwich
  d. peptide binders
  e. all of the above

14. Which of the following statements about protein engineering is not correct?
  a. Protein engineering uses genetic technology to improve already useful industrial enzymes.
  b. Protein engineering has the capacity to create enzymes with novel functions.
  c. Protein engineering can improve biomaterials both biochemically and mechanically.
  d. The methods that exist to engineer protein sequences are often not useful.
  e. Enzymes can be made more robust and industrially stable through protein engineering.

15. Which of the following is a not a genetically encoded amino acid?
  a. selenocysteine
  b. pyrrolysine
  c. pBpa
  d. tyrosine
  e. all of the above are genetically encoded

16. Which of the following statements about phosphoserine is not correct?
  a. Phosphoserine is added to a growing polypeptide chain during translation.
  b. Phosphoserine results from the post translational modification of serine.
  c. Phosphoserine is not genetically encoded.
  d. Phosphoserine is involved in signal transduction.
  e. More than one answer above is incorrect.

17. Which of the following is incorrectly matched?
  a. lysine acetylation – post translational modification
  b. coumarin and dansyl derivatives – fluorescent probes
  c. Azidohomoalanine – reactive chemical groups
  d. 4,5-dimethoxy-2-nitrobenzyl-cysteine – reactive chemical groups
  e. fluoro-amino acids – isotopic labeling

Further Reading

Binz, H. K., & Plückthun, A. (2005). Engineered proteins as specific binding reagents. *Current Opinion in Biotechnology, 16*, 459–469.

Böttcher, D., & Bornscheuer, U. T. (2010). Protein engineering of microbial enzymes. *Current Opinion in Microbiology, 13*(3), 274–282.

Branden, C., & Tooze, J. (1998). *Introduction to Protein Structure* (2nd ed.). New York: Garland Publishing.

Fisher, M. A., McKinley, K. L., Bradley, L. H., Viola, S. R., & Hecht, M. H. (2011). *De novo* designed proteins from a library of artificial sequences function in *Escherichia coli* and enable cell growth. *PLoS One, 6*, e15364.

Kawahara-Kobayashi, A., Masuda, A., Araiso, Y., Sakai, Y., Kohda, A., Uchiyama, M., et al. (2012). Simplification of the genetic code: restricted diversity of genetically encoded amino acids. *Nucleic Acids Research, 40*(20), 10576–10584.

Krueger, A. T., Peterson, L. W., Chelliserry, J., Kleinbaum, D. J., & Kool, E. T. (2011). Encoding phenotype in bacteria with an alternative genetic set. *Journal of the American Chemical Society, 133*(45), 18447–18451.

Li, Q., Yi, L., Marek, P., & Iverson, B. L. (2013). Commercial proteases: present and future. *FEBS Letters, 587*(8), 1155–1163.

Nettles, D. L., Chilkoti, A., & Setton, L. A. (2010). Applications of elastin-like polypeptides in tissue engineering. *Advanced Drug Delivery Reviews, 62*(15), 1479–1485.

Neumann, H., Wang, K., Davis, L., Garcia-Alai, M., & Chin, J. W. (2010). Encoding multiple unnatural amino acids via evolution of a quadruplet-decoding ribosome. *Nature, 464*(7287), 441–444.

Neumann, H. (2012). Rewiring translation—genetic code expansion and its applications. *FEBS Letters, 586*(15), 2057–2064.

Qian, Z., Fields, C. J., Yu, Y., & Lutz, S. (2007). Recent progress in engineering alpha/beta hydrolase-fold family members. *Biotechnology Journal, 2*, 192–200.

Smith, B. A., & Hecht, M. H. (2011). Novel proteins: from fold to function. *Current Opinion in Chemical Biology, 15*(3), 421–426.

Socha, R. D., & Tokuriki, N. (2013). Modulating protein stability—directed evolution strategies for improved protein function. *FEBS Journal, 280*(22), 5582–5595.

Wang, Q., Parrish, A. R., & Wang, L. (2009). Expanding the genetic code for biological studies. *Chemical Biology, 16*(3), 323–336.

Environmental Biotechnology

Biotechnology

http://dx.doi.org/10.1016/B978-0-12-385015-7.00012-0

INTRODUCTION

The environment around us has always been a source of new products and stimulated our imagination in developing new technologies. Our species is very successful at harnessing the environment for our benefit. We are also good at destroying or harming the environment for immediate gains. The readily visible world has been charted and mapped, but areas under the ocean and in the deep recesses of jungles are still unknown. In fact, many parts of the visible world still harbor unknown life forms invisible to the naked eye, including bacteria and viruses. These are found in the air, water, and land. Many have unique metabolisms, and some have abilities never seen before. Many can live in extreme environments once thought too hot or too dry for life to exist.

The world has many life forms and creatures that are still undiscovered. Estimates predict that about 10^{31} to 10^{32} viral particles are present in the biosphere, an order of magnitude more than host cells. The **virosphere**, as it is sometimes known, is probably one of the biggest sources of novel genes. In addition to viruses and phage, a large number of microorganisms inhabit the earth, an estimated 5×10^{30} cells. In fact, the human body harbors 10^{14} bacteria, a 10-fold higher number of cells in comparison to the number of our own cells (only 10^{13}). At the time of writing, only about 0.1% to 1% of microorganisms have been cultured. Even the majority of those found growing at moderate temperatures in soil or other normal habitats have not been cultured. In addition to true life forms as such, there is much free DNA in the environment that might also be a source of new genes. The field of environmental biotechnology has revolutionized the study of these previously hidden life forms and DNA. What kinds of secrets do they harbor? What kinds of new enzymes and proteins can be identified?

Molecular biology techniques are now being applied directly to the environment to investigate the uncultured viruses and bacteria. PCR is routinely used to amplify random sequences from many environmental samples in the hope of identifying new genes. After PCR, the DNA is sequenced (see Chapter 4 for PCR and sequencing). Then bioinformatics reveals whether or not the sequence (or a close relative) has already been identified or if it is completely novel (see Chapter 8). Microarrays are also being created to compare the numbers and types of organisms present in different environments (see Chapter 8). Almost every recombinant DNA methodology discussed in the first half of this book can be applied to environmental samples.

Environmental biotechnology is divided into different areas. They include direct studies of the environment, research with a focus on applications to the environment, or research that applies information from the environment to other venues. This chapter focuses on methods used to identify the different processes or genes that are present in the environment and then identifies a few applications for these methods, including a discussion of how algae can be used to create biofuels and other novel ways to create energy from renewable natural resources.

The environment is filled with invisible life forms such as viruses, bacteria, and other gene creatures.

IDENTIFYING NEW GENES WITH METAGENOMICS

Metagenomics is the study of the genomes of whole communities of microscopic life forms without culturing of the organisms. Approaches include next-generation DNA sequencing, PCR, RT-PCR, and microarrays. Metagenomic research allows us to identify microorganisms, viruses, or free DNA that exist in the natural environment by identifying genes or DNA sequences from the organisms. Metagenomics applies the knowledge that all creatures contain nucleic acids that encode various protein products; therefore, organisms do not have to be cultured but can be identified by a particular gene sequence, protein, or metabolite. The term *meta-*, meaning more comprehensive, is also used in *meta-analysis*, which is the process of statistically combining separate analyses. Metagenomics is the same as genomics in its approach. The difference between genomics and metagenomics is the nature of the sample. Genomics focuses on one organism, whereas metagenomics deals with a mixture of DNA from multiple organisms, "gene creatures" (i.e., viruses, viroids, plasmids, etc.), and/or free DNA.

Metagenomics has a variety of different applications, and many different studies use metagenomic analysis. The studies range from identifying and cataloguing large-scale natural environments like oceans or soil to small-scale but equally complex studies like studying the microbiome of the human gut. Metagenomic analysis assays how the environment affects the microbial populations, has been used to identify new beneficial genes from the environment, assays how microorganisms form symbiotic relationships with their hosts, and compares the same gene across a variety of species to define how related the gene family is in nature (see Box 12.1).

Metagenomics has been used to find novel antibiotics, enzymes that biodegrade pollutants, and enzymes that make novel products (Table 12.1). Historically, studying microbes in the environment has identified many useful products. In the early 1900s, Selman Waksman was studying actinomycetes in soil when he discovered the antibiotic streptomycin, which was the first effective treatment for tuberculosis. He was awarded a Nobel Prize in 1952 for this finding. His research focused on culturing soil samples and then screening them for the ability to kill *Mycobacterium tuberculosis*. He cultured more than 10,000 samples to find one single agent that was effective against tuberculosis. In 2002, a more blind approach was taken to identify new antibiotics. Gillespie and colleagues created a library of different DNA fragments isolated from a sample of soil. Each fragment was cloned into an expression vector and then transformed into *E. coli*. Each bacteria expressed the genes found on the random DNA fragment from the environment. They were looking for hemolysin-related genes and discovered two colored bacterial colonies, one orange and one dark brown. Further investigation of these clones found two novel antibiotics, turbomycin A and B (Fig. 12.1). Although this experiment takes random samples, there is still bias since the piece of DNA from the soil must contain the entire gene, and then the gene must be expressed into protein for the

Box 12.1 Sequencing the Sargasso Sea

In 2004, *Science* magazine published one of the largest-scale metagenomic analyses (*Science* 304, 66–74). Craig Venter (who led his company to sequence the human genome) took on a large-scale effort to sequence DNA isolated from the Sargasso Sea. This is part of the North Atlantic Ocean that has a relatively high salt content and is thus thought to have less biodiversity. He took samples of water from a 5-foot depth at various points along a journey on his sailboat. The water was passed through a series of three successively finer filters and then frozen. At intervals, the filters were sent to Maryland for DNA sequencing. Analysis revealed that the Sargasso Sea has a more diverse population than expected, with about 1800 different species of microbes, 150 new bacterial and Archaea species, and more than 1.2 million new genes. Interestingly, more than 700 different bacteriorhodopsin genes were identified. Bacteriorhodopsin is used by bacteria to harvest light for energy and is closely related to rhodopsin, a protein that detects light in mammalian eyes. The sequences from this exploration are found online at the National Center for Biotechnology Information (NCBI; http://www.ncbi.nlm.nih.gov) and are free to the public. The research is part of the Global Ocean Sampling (GOS) Expedition and is published with the accession number PRJNA13694.

The results of this study revealed such a large number of new and interesting bacteria in a nutrient-poor region of the sea, that the Sorcerer II sailboat set sail for different parts of the world to see if there was even more diversity (see Fig. A). The crew took samples in Halifax, Canada, along the east coast of the United States, through the Gulf of Mexico, in the Galapagos Islands, in the central and south Pacific Oceans, Australia, Indian Ocean, South Africa, and back through the Atlantic. The results of this and other surveys of the metagenome are published in a collection of articles called the Ocean Metagenomics Collection by PLOS (see plos.org). The synopsis summarizes the data, which has a whopping 6 million open reading frames that could encode proteins. The protein data set is interesting because it shows that some proteins that were thought to be found only in Eukarya are also found in marine microbes. Examples include indoleamine 2,3-dioxygenase (IDO), which is an enzyme for the immune system. In addition, there were many more members of existing protein families, such as DNA damage repair proteins, phosphatases, nitrogen metabolic enzymes, and protein kinases. These increases in the number and diversity of protein sequences provide a framework for biologists to understand the evolutionary relationships between species, both prokaryotic and eukaryotic.

(Continued)

Box 12.1 Sequencing the Sargasso Sea—cont'd

FIGURE A Sorcerer II
Image showing the sailing yacht Sorcerer II as it journeyed to collect seawater samples. The samples were passed through a progressively finer filter to recover different-sized organisms, from bacteria to small viruses, floating in the ocean.

Table 12.1 Genes and Proteins Identified via Gene Mining

Gene	Environmental Sample
Esterases	Urania deep-sea hypersaline anoxic basins in the Mediterranean Sea
Lipases	Soil outside University of Göttingen, Germany; soil from the Madison, Wisconsin, USA, Agricultural Research Station
Amylases	Soil outside University of Göttingen, Germany; soil from the Madison, Wisconsin, USA, Agricultural Research Station
Chitinases	Estuarine and coastal seawater collected in Delaware Bay, USA
Antibiotics	
Turbomycin A and B	Soil from the Madison, Wisconsin, USA, Agricultural Station
Polyketide synthases	Soil from an arable field in La Côte Saint André, Iserè, France
Vitamin biosynthesis	Avidin-enriched forest soil from outside Göttingen, Germany; sandy soil from a beach near Kavalla, Greece; volcanic soil from Mt. Hood, Oregon, USA
4-Hydroxybutyrate dehydrogenase	Soils from sugar beet field near Göttingen, meadow near Northeim, and the River Nieme valley, Germany

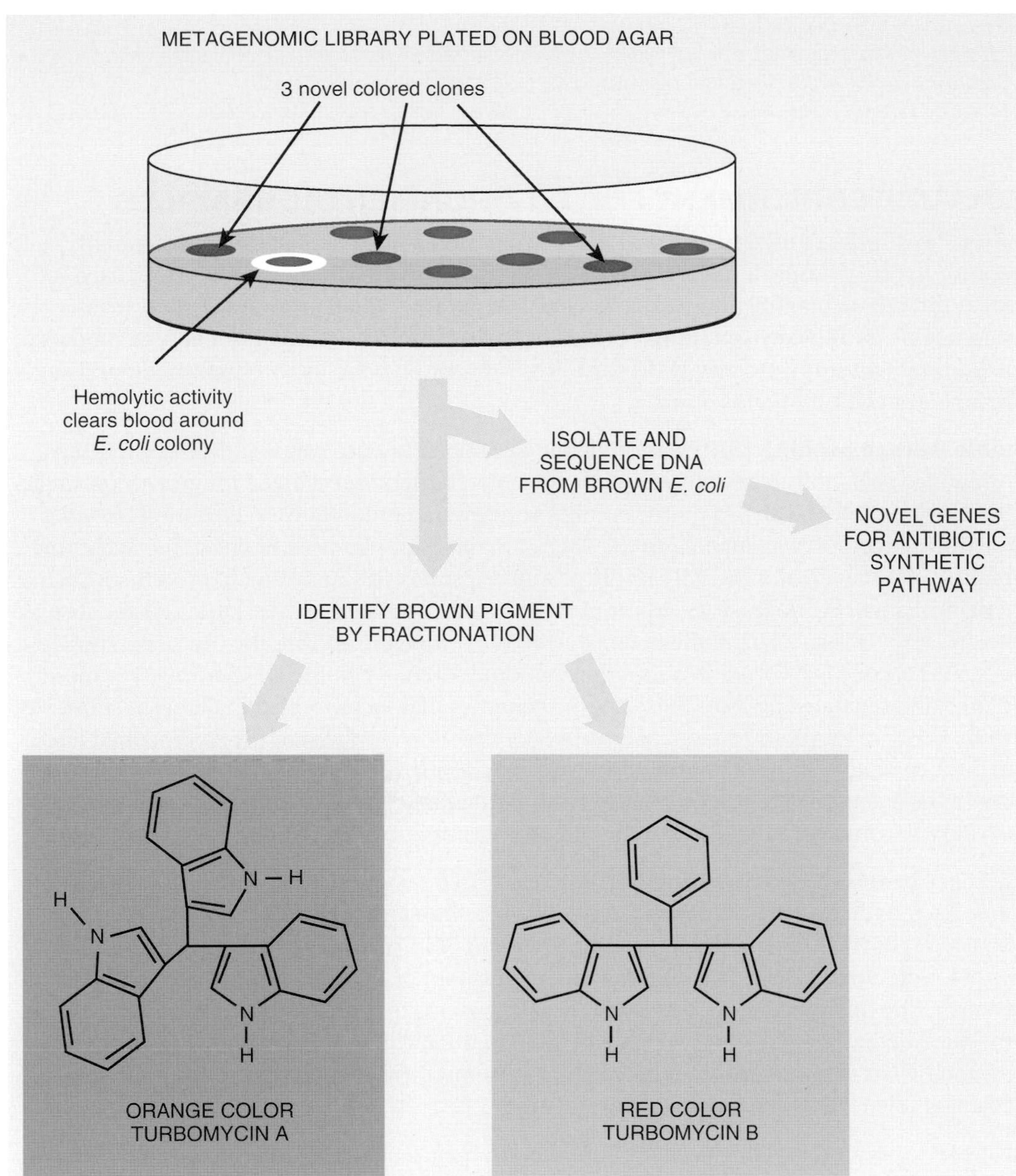

FIGURE 12.1 Novel Antibiotics Discovered in a Metagenomic Library
While screening a metagenomic library for hemolytic activity, three *E. coli* colonies were identified that had a dark brown color. The colored pigment was secreted and found to be a mixture of two different pigments (red and orange). The pure pigments have strong antimicrobial activity against both Gram-positive and Gram-negative bacteria. Sequencing the cloned insert identified the genes responsible for synthesizing the antibiotics.

identification of the antimicrobial agent. The most recent research to identify genes from the environment involves next-generation sequencing (see Chapter 4). Small pieces of random DNA are the perfect sample for these sequencing techniques. Using this approach, researchers at Washington University in St. Louis, USA, identified 2489 contigs that had at least one gene that conferred antibiotic resistance. Although the number of genes was not surprising, the source of microorganisms was surprising. The samples were isolated from healthy infant feces. The authors have identified that microbes in healthy infants contain a large diversity of antibiotic resistance genes and have coined the term *gut-associated resistomes* (Moore et al., 2013). The resulting findings suggest that antibiotic resistance genes are present and ready for any pathogenic strain of bacteria to use. This type of research is a warning that antibiotic resistance could pose a greater threat to future generations.

Metagenomics is the study of the genomes of whole communities of microscopic life forms. Many new proteins have been identified from examining DNA collected from an environment.

CULTURE ENRICHMENT FOR ENVIRONMENTAL SAMPLES

Various methods are used to enhance the starting material for metagenomics research, because a metagenomic library is only as good as its contents. Processing is either used to enrich for a particular fraction from the sample, amplify a small amount of DNA from a large sample, or remove contaminants that degrade the DNA sample. Enrichment strategies include stable isotope probing (SIP) for DNA, RNA, or protein; BrdU enrichment; and suppressive subtraction hybridization.

Stable isotope probing (SIP) was originally developed to trace single carbon compounds during their metabolism by cultured methylotrophs (bacteria specialized for growth on single-carbon compounds). Labeled precursor carbons were traced into fatty acids during bacterial growth. The method was adapted to the environmental samples used to create metagenomic libraries. Here, an environmental sample of water or soil is first mixed with a precursor such as methanol, phenol, carbonate, or ammonia that has been labeled with a stable isotope such as ^{15}N, ^{13}C, or ^{18}O (Fig. 12.2). If the organisms in the sample metabolize the precursor substrate, the stable isotope is incorporated into their genome. Then, when the DNA from the sample is isolated and separated by centrifugation, the genomes that incorporated the labeled substrate will be "heavier" and can be separated from the other DNA in the sample. As described later, the DNA can either be used directly for next-generation sequencing or cloned into vectors to make a metagenomic library. This technique is particularly useful to find new organisms that can degrade contaminants, such as phenol in the example given.

Adding **5-bromo-2-deoxyuridine (BrdU)** is a related technique for enriching the DNA of active bacteria in an environmental sample. Rather than entering only a metabolic subset of bacteria, BrdU is incorporated into the DNA and RNA of any actively growing bacteria or viruses. Note that bacteria and viruses that are dormant or dead, as well as free DNA, will not be labeled by this method. As before, the soil or water is isolated and incubated with BrdU. Any bacteria that are actively growing will take up the nucleotide analog and incorporate it into their DNA. Next, the BrdU-labeled DNA is isolated either with antibodies to BrdU or by density gradient centrifugation (see Fig. 12.2).

RNA-SIP focuses on isolating RNA from the environmental sample (rather than DNA). **Small subunit ribosomal RNA (SSU rRNA)**, that is, the 16S rRNA of bacteria or the 18S rRNA of eukaryotes, is an excellent biomarker because it is essential to all cellular life, it is very abundant within a cell, it is variable among different species, and there is an enormous database of different SSU rRNA sequences making identification relatively easy. In RNA SIP, the SSU rRNA in the environmental sample is labeled. As described earlier, ^{13}C-labeled precursors are supplied to the environmental sample. These are incorporated into SSU rRNA independently of cell division because ribosomal RNA is produced in any cell that is making proteins, not just cells undergoing replication. This technique provides information on bacteria that are not dividing as well as those that are more actively reproducing. Much as before, the RNA is isolated and separated on a gradient by centrifugation. The rRNA bands tend to aggregate together during centrifugation. Therefore, each fraction must be repeatedly separated from the others. The final SSU rRNA fraction may still contain some contaminating nonlabeled rRNAs, so the fraction must be evaluated with care.

RNA-SIP can be used to identify a variety of microorganisms in environmental samples. For example, water from an aerobic industrial wastewater plant was evaluated for phenol-degrading microorganisms. The water was incubated with ^{13}C-labeled phenol, and the SSU rRNA was isolated by centrifugation. The rRNAs were isolated and amplified with RT-PCR followed by denaturing gradient gel electrophoresis. The bands were subjected to mass spectrometry to identify which rRNA sequence was most abundant. Interestingly, an organism belonging to

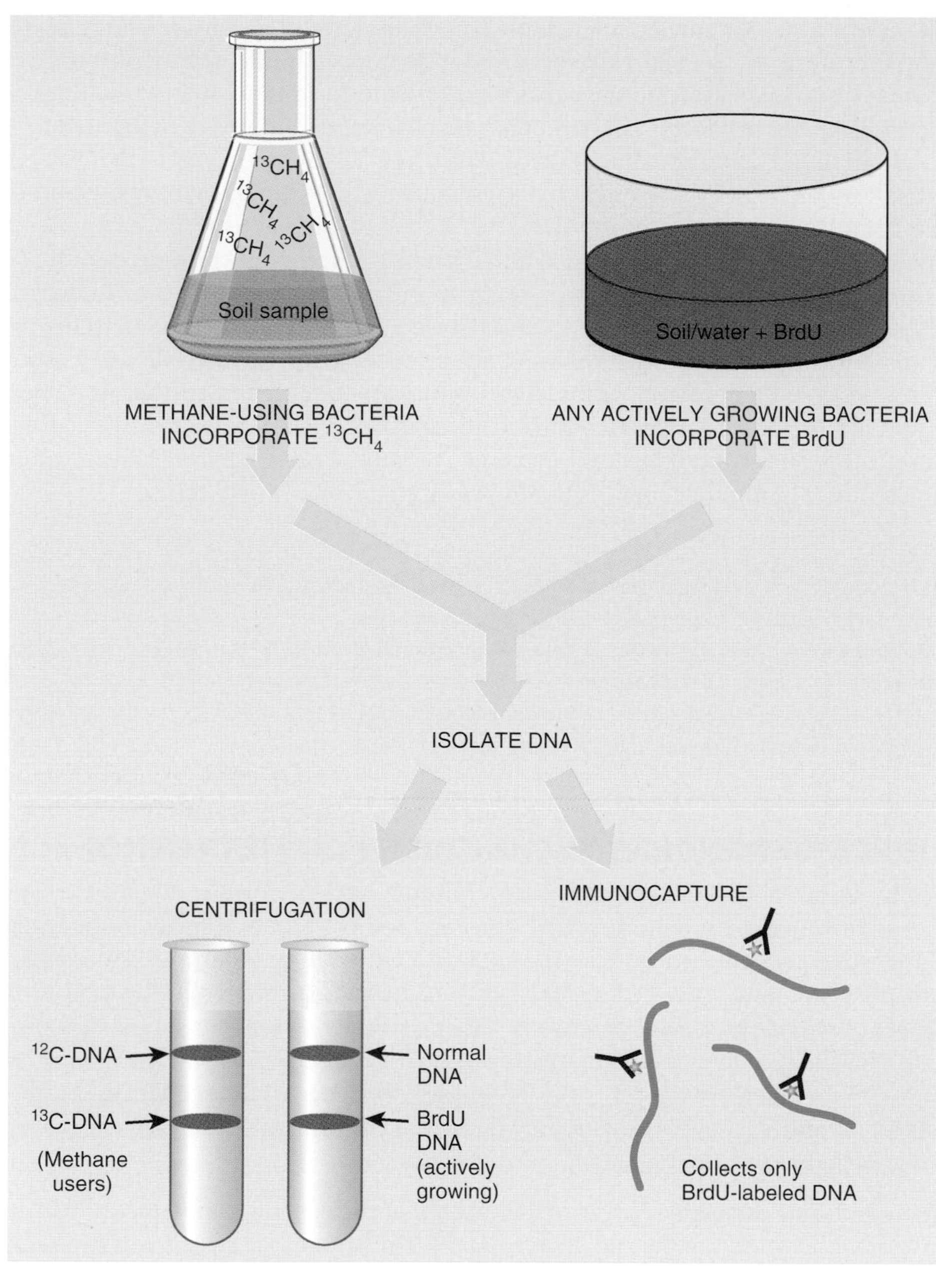

FIGURE 12.2 Stable Isotope Probing and BrdU Enrichment
Samples from the environment can be enriched for particular bacterial populations by incubating them with a labeled precursor such as ^{13}C-methane (*left*) or BrdU (*right*). After the DNA is isolated from the sample, the labeled DNA can be separated from the rest of the soil DNA using centrifugation or immunocapture.

the genus *Thauera* in the β-Proteobacteria was abundant, even though this organism was most usually found in denitrifying conditions. It was previously thought that pseudomonads were degrading the phenol.

Another culture-enrichment technique, **suppressive subtraction hybridization (SSH)**, takes advantage of the genetic differences between samples from two different areas. During standard subtractive hybridization, two different samples are hybridized, and the mRNA that is the same is removed, leaving only mRNA that is different between the two samples. SSH works by the same principle. First, two different conditions must be established. For example, a soil sample from a polluted site could be compared with nearby soil that is not contaminated. The two soil samples will differ in their content of microorganisms, and those microorganisms enriched in the contaminated site could potentially metabolize the contaminant.

When DNA from each sample is isolated, the contaminated soil is considered the **tester** DNA, and the "normal" soil is the **driver** sample. The tester sample is divided into two, and two different linkers are added to the ends of the DNA to form tester A and tester B. Tester A (with linker A), tester B (with linker B), and driver DNA are all mixed, denatured to make them single-stranded, and then rehybridized. The driver DNA is in excess to the testers, which ensures that DNA fragments from bacteria outside the contamination site outnumber those from the contaminated site. The driver DNA will anneal to all the common DNA fragments, making these double-stranded and with only one strand connected to the linker. All the tester DNA that is unique and not found in the uncontaminated soil will be free to hybridize with itself, forming A:A, B:B, or A:B hybrids. PCR primers are added to the hybridization mix; one primer recognizes linker A and the other primer is for linker B. As shown in Figure 12.3, PCR will amplify only those hybrids that are tester:tester. Furthermore, because the A:A and B:B hybrids have inverted linkers, these hybrids will form a "panlike" structure during annealing and will not be amplified by PCR. Thus, only A:B hybrids are amplified, and these represent unique sequences found only in the contaminated site.

Stable isotope probing is adding heavy isotopes to an environmental sample, which distinguishes the actively growing bacteria or viruses from the remaining organisms.

RNA-SIP is similar to SIP, but the heavy isotope is incorporated into the SSU rRNA. These sequences do not require the organism to be undergoing cell division, but they must be metabolically active.

Suppressive subtraction hybridization (SSH) compares the microorganisms present in two different samples by removing the identical DNA and evaluating the unique sequences.

SEQUENCE-DEPENDENT TECHNIQUES FOR METAGENOMICS

Historically, sequence-based metagenomics techniques were performed directly on samples from the environment after culturing. In 1985, Pace and colleagues directly sequenced the 5S and 16S rRNA gene sequences from the environment without culturing. This was technically difficult at the time. After PCR techniques became more prevalent in the early 1990s, the analyses of specific groups of organisms became simpler. Researchers used PCR primers specific to 16S rRNA sequences to identify the different organisms within the environmental sample. Yet these techniques rely on some prior knowledge of the sequences. Moreover, this method identifies a new organism only by its 16S rRNA sequence. It does not reveal the physiology or genetics of the organism.

Using **metagenomic analysis** is a way to identify the entire genetic complement of an environmental sample, without culturing the organisms. The physiology and identification of the organisms from the sample can be determined from the genome sequences. DNA (or RNA) from the environment is isolated (and in some cases enriched as described earlier). The DNA, which is a mixture of fragments from multiple genomes, is then subjected to next-generation sequencing. The first step of this technique is the shearing of genomic DNA into small pieces, which is basically the starting material from the environmental sample. Alternatively, in **metatranscriptomics**, the environmental sample can be used to isolate RNAs, which are then converted to cDNA and analyzed by next-generation sequencing also (Fig. 12.4).

Many environmental biologists want to develop or find new enzymes that degrade pollutants or contaminants. Screening the library with a sequence containing conserved domains of known enzymes can find new enzymes. Along the same lines, PCR primers to conserved domains of known enzymes can amplify never-before-seen genes from the library. For example, aromatic oxygenases facilitate the degradation of aromatic hydrocarbons found in oil and coal. These contaminants can be degraded by a variety of different organisms. The genes and pathways involved are key targets for pathway engineering (see Chapter 13). Using PCR primers to conserved regions of known aromatic oxygenases amplifies the genes for novel (but related) oxygenases found in the environment.

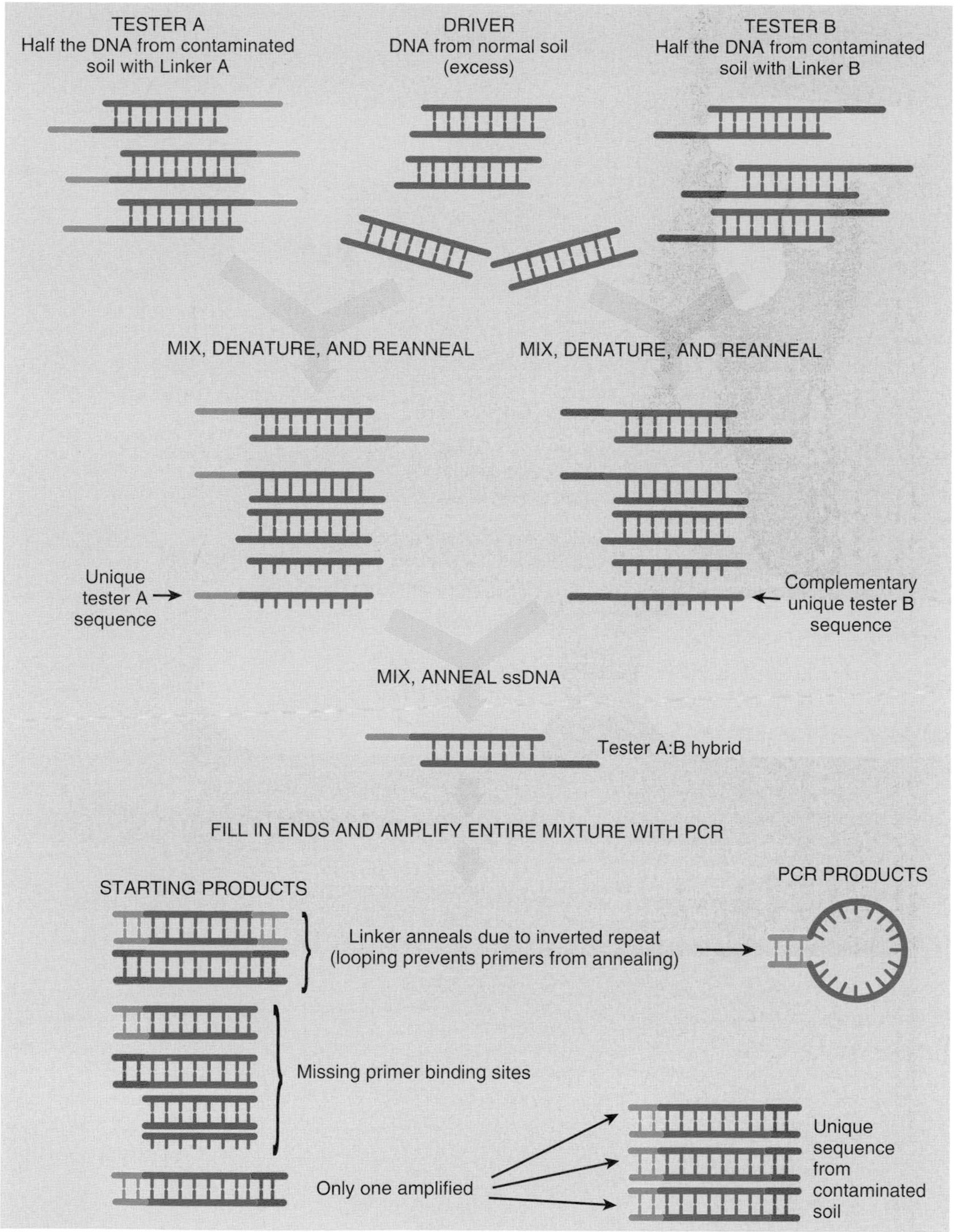

FIGURE 12.3 Suppressive Subtractive Hybridization

SSH begins by dividing the experimental sample in half and adding two different linkers to each half. Each of these is mixed with an excessive amount of normal DNA, which will create an abundance of tester:driver hetero-hybrids (pink/purple strands). Each sample will still have some single-stranded pieces that did not find a complementary strand, including some normal or driver DNA and some tester DNA with a linker. The two pools are then mixed together (they are not denatured this time). The single-stranded pieces will anneal from tester A and tester B to form a hybrid molecule with two different linkers on each side. The entire pool is prepared for PCR by filling in the single-stranded regions, and then using primers to the two different linkers to amplify the unique DNAs. PCR only amplifies the tester A:tester B hetero-hybrids because the other hybrids either are missing a primer binding site or self-anneal via the linker to form a loop structure.

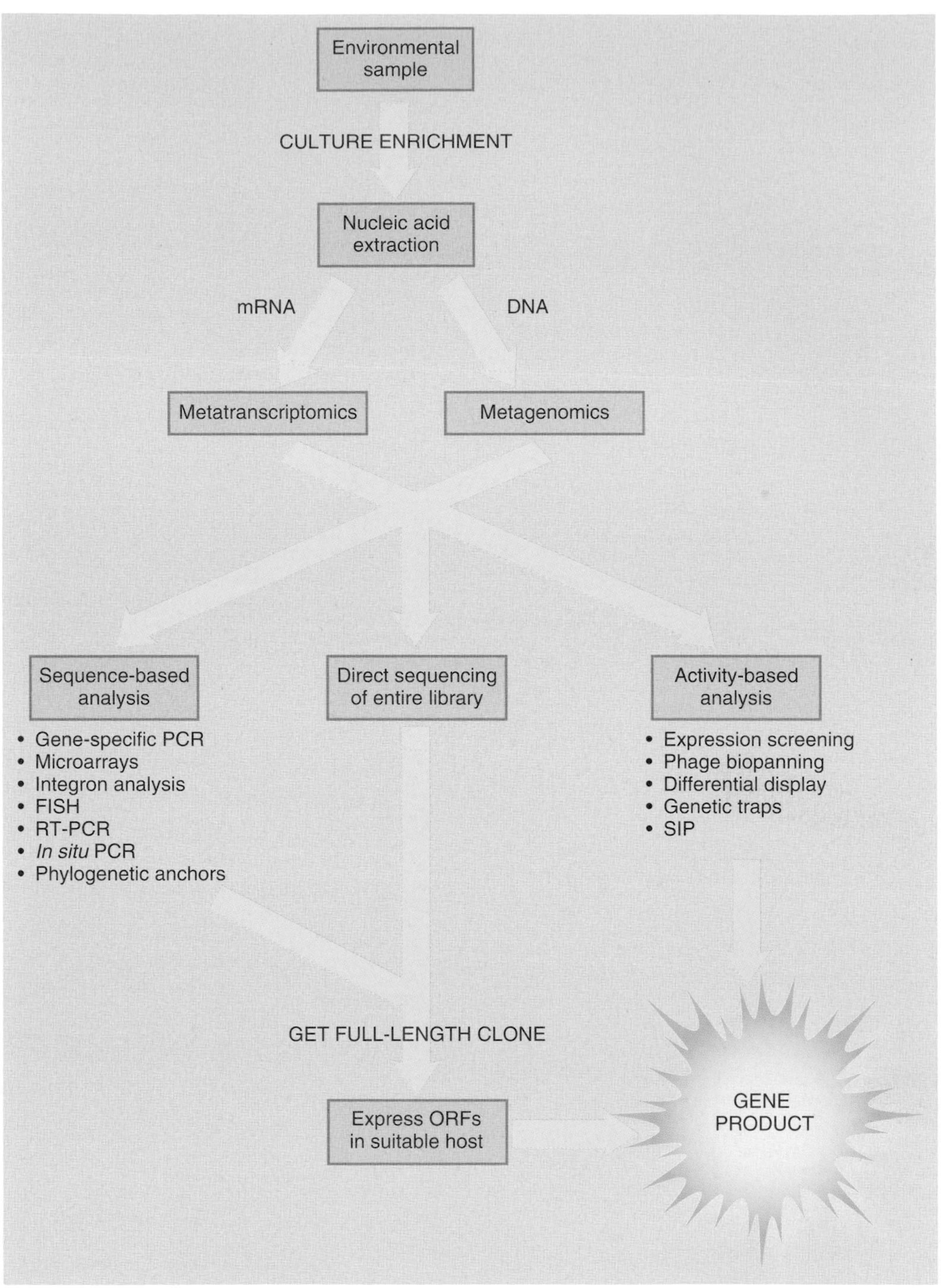

FIGURE 12.4 Techniques to Study Environmental Samples Many different techniques can be applied to DNA or mRNA isolated from the environment. These approaches all converge on discovering the gene product and identifying its function in the environment.

Some sequence-based analyses for metagenomic libraries include microarrays (see Chapter 8), FISH (see Chapter 3), RT-PCR (see Chapter 4), sequencing using phylogenetic anchors, and integron analysis (see later discussion). Sequencing with phylogenetic anchors begins by identifying the sequence of a known gene. Often a marker gene such as the 16S rRNA gene is identified first, and then the regions upstream or downstream of the marker are sequenced. For example, a 16S rRNA sequence from seawater classified a particular genomic fragment to the γ-Proteobacteria. Adjacent to the 16S rRNA gene was a gene similar to bacteriorhodopsin, a transmembrane proton pump that responds to light. The genes for bacteriorhodopsin were originally thought to exist only in the Archaea, but this analysis revealed that other inhabitants of the ocean had similar genes.

Microarrays can also be used to identify the types and numbers of different organisms in an environmental sample. First, a microarray of unique sequences from known organisms is created,

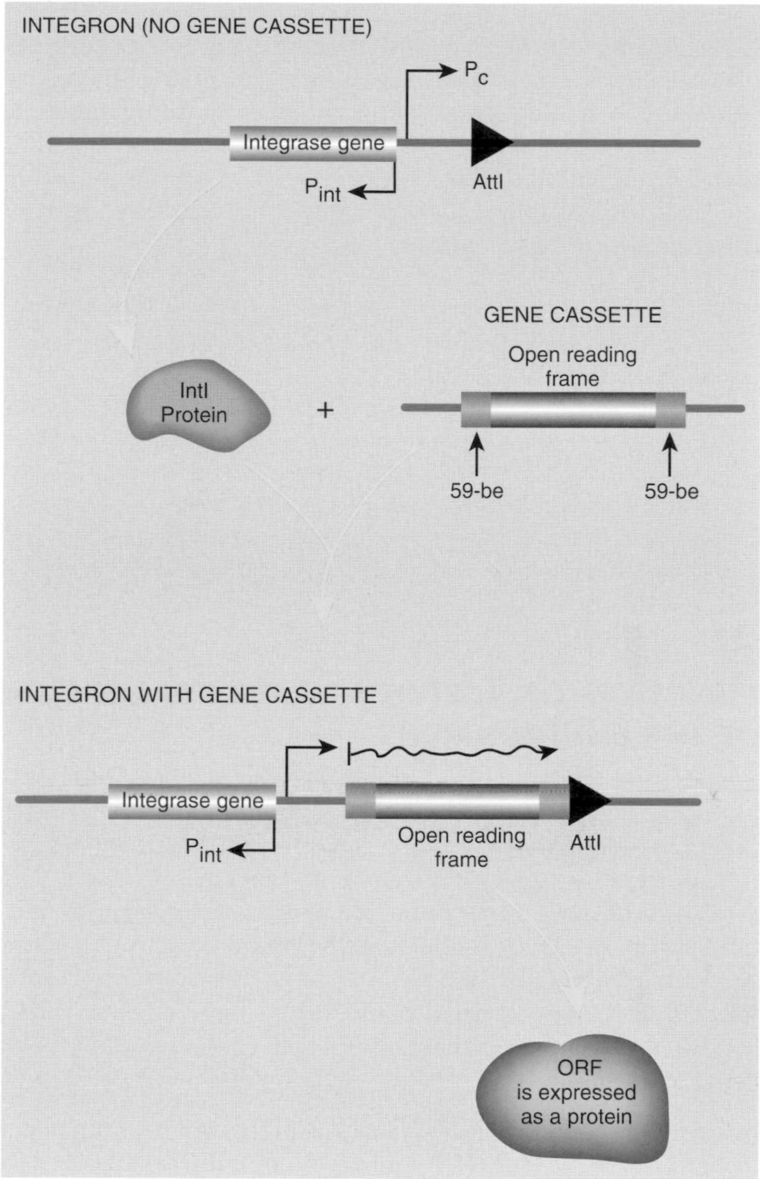

FIGURE 12.5
Integron Analysis
Integrons are genetic elements that capture and express gene cassettes. The integron has three main components: an integrase gene under the control of its own promoter (P_{int}), an *attI* site for integration of the gene cassette, and a promoter to express the gene cassette (P_c). When integrase is expressed, it searches the genome for gene cassettes. Integrase then excises the gene cassette and integrates it into the integron at the *attI* site. Using PCR primers to the 59-be sites allows isolation of any open reading frames in gene cassettes and potentially identifies new genes.

and then fluorescently labeled environmental DNA is hybridized to the array. The results can confirm whether or not a particular bacterium, virus, or gene creature is present, and the relative abundance can be determined by the intensity of fluorescence. FISH and RT-PCR can yield similar information, but the environmental DNA is analyzed with only few probes or primers, respectively. The results are much more direct and focus on identifying known organisms.

Integron analysis identifies open reading frames that are used by integrons and can identify more novel unknown genes than many previous techniques (Fig. 12.5). **Integrons** are related to transposons (see Chapter 1) but are particularly important for the spread of genes for antibiotic resistance and other properties that give the host a growth advantage in a particular environment. Integrons are genetic elements that contain a site *(attI)* to integrate a segment of DNA known as a **gene cassette**, a promoter to express the gene cassette (P_c), and a gene for **integrase *(intI)***, the enzyme that recombines the gene cassette into the integron. Gene cassettes are segments of DNA with one or two open reading frames (ORFs) that lack promoters and are flanked by **59-base elements** (**59-be,** also known as *attC* sites). When the integrase recognizes the 59-be sites, it excises the gene cassette and integrates it downstream of the P_c promoter. This allows the open reading frame to be expressed into protein. The 59-be sequence may vary in length but must contain a conserved, seven-nucleotide sequence.

The gene cassettes are the interesting part of the scenario and the key to integron analysis. Gene cassettes were first identified because many encode antibiotic resistance genes, but they may encode any type of gene. Screening metagenomic libraries using PCR primers that recognize the 59-be elements amplifies these open reading frames. This approach has identified novel genes related to DNA glycosylases, phosphotransferases, and methyl transferases. Additionally, new antibiotic resistance genes may be identified. Integron analysis is also useful to study bacterial evolution and gene transfer because these elements can pass from bacterium to bacterium during conjugation.

Metagenomic libraries identify the entire genetic complement of newly discovered life forms, without the need to culture the organisms.

Environmental DNA can be analyzed using a variety of methods, such as FISH, microarrays, sequencing, and RT-PCR. Searching the regions adjacent to known genes identifies novel genes with many diverse functions. This is called sequencing with phylogenetic anchors.

Integrons are like transposons but have the ability to capture genes from different organisms and move them to others. Integron analysis of a metagenomic library uses PCR to amplify the sequence between the 59-be sites, which identifies the genes captured by the integron.

FUNCTION- OR ACTIVITY-BASED EVALUATION OF THE ENVIRONMENT

Besides these sequence-based approaches, metagenomic libraries can be screened for various functions (Fig. 12.6). Screening a metagenomic library by sequencing has its limits. The function of many genes from exotic organisms cannot be identified by their sequence. In addition, using known genes to screen for new members of a gene family might miss an entire novel class of genes. For example, if a researcher was trying to find enzymes similar to bacteriorhodopsin, sequences similar to bacteriorhodopsin would be found, and some genes would be identified, but others would be missed if their sequences were too divergent. Functional screening, on the other hand, would identify any clone that encodes a bacteriorhodopsin-like protein, that is, a protein that captures light energy to pump the protons across the membrane of the host cell.

Before the function of the various gene pieces can be evaluated, the DNA pieces must be cloned into vectors to create a **metagenomic library**. To create a library of gene sequences from the environment, the DNA ends are treated to be compatible with an expression vector for *E. coli*. The vector contains the necessary sequences for the host organism to express the sequence into a protein. One easy method to make the environmental sequence complementary to the vector is to link a known sequence onto each end. The addition of linkers replaces the ends of the unknown DNA sequence with a known one. This provides a means for the sequence to be amplified by PCR, as in next-generation sequencing, or to be cloned into a vector, in order to make a library. Once the sequences are in the vector, the sequence is transcribed and translated using the transcriptional and translational start and stop sequences of the vector. Then, an easy assay for the target function must be devised. For example, the library clones might be plated on a particular toxic pollutant. If any library inserts encoded an enzyme that metabolizes the pollutant, that particular library clone would grow. Sequencing that particular library clone would provide a brand new gene and its function.

Another functional screen involves fusing the metagenomic DNA in frame with a promoterless gene for green fluorescent protein (GFP). Rather than looking for a particular coding region of a gene, this approach identifies regulatory regions of genes, such as promoter and enhancer regions. Therefore, if this type of DNA is cloned in front of promoterless GFP, the promoter or enhancer region from the environmental gene will regulate the GFP gene. This

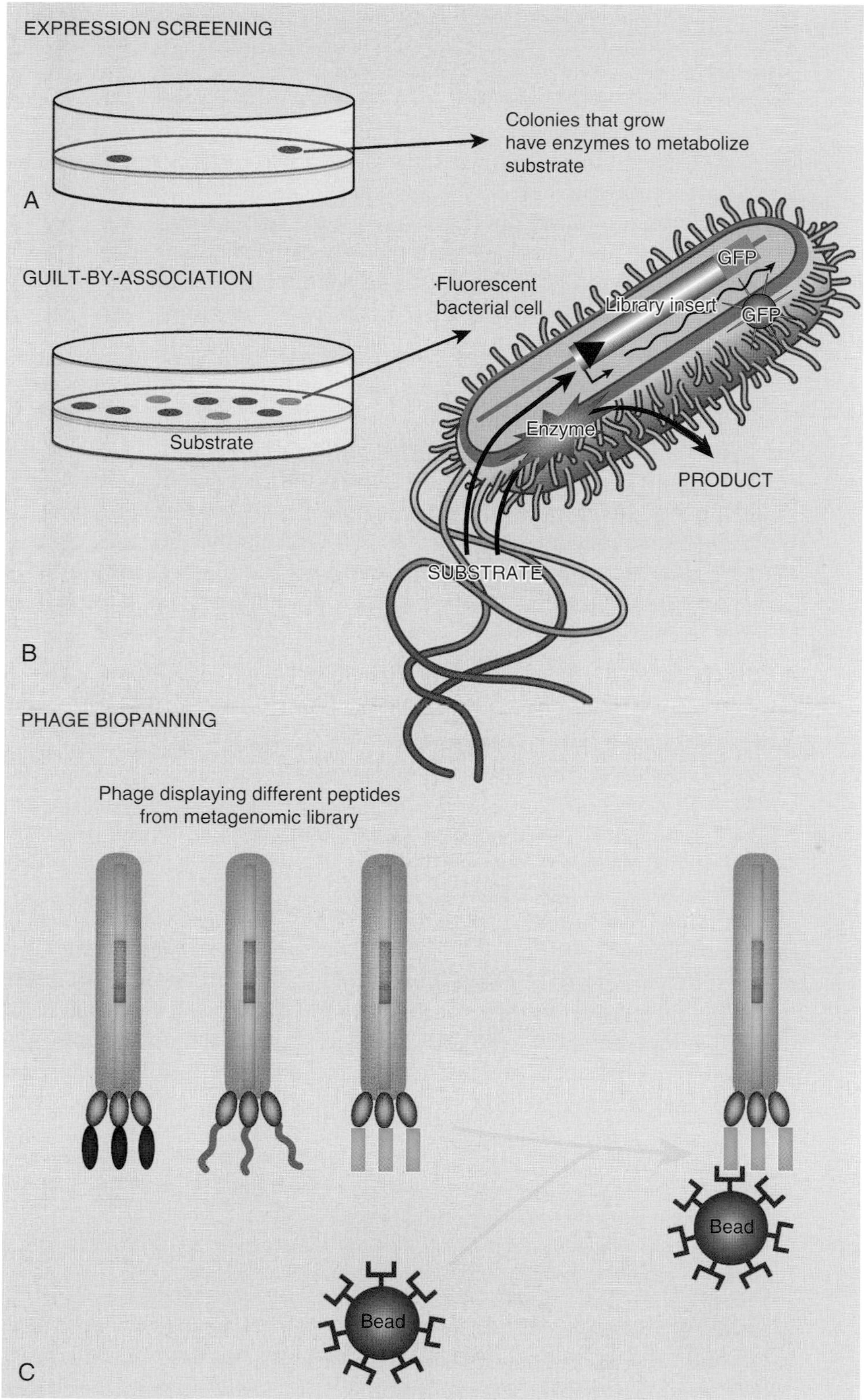

FIGURE 12.6 Function-Based Approaches for Metagenomics

(A) Expression screening identifies those library clones that can grow by metabolizing a particular substrate. To suppress the growth of the other library clones, the substrate must be the only source of carbon, nitrogen, or sulfur in a minimal medium. (B) Guilt-by-association identifies expressed genes that may metabolize a particular substrate. The premise is that if the library insert is responsible for metabolizing the substrate, the natural promoter for the genes responsible is also present. If this promoter is active, it can express a reporter gene such as *gfp*, encoding green fluorescent protein. Positive cells can be isolated from the others by FACS sorting (see Chapter 6). (C) Phage biopanning identifies protein binding partners. Peptides from a metagenomic library are fused to a coat protein from a bacteriophage and expressed on its extracellular surface. When the phage is mixed with the binding partner of interest, immobilized on a bead, those metagenomic peptides that bind to the protein of interest will be attached to the bead. These are easily isolated from the rest of the phage clones.

type of screen can identify potential metabolites that control the promoter and enhancer regions, as well as promoter and enhancer regions that are constitutively active. For example, a promoter induced by glucose would not cause the host bacteria to express GFP and subsequently glow under UV light, unless the host was growing on media containing glucose. In contrast, if the promoter is constitutively active, the GFP would be expressed no matter what nutrients were in the growth media. Besides glucose, other potential metabolites or environmental conditions that could cause a difference in GFP expression include heat or cold, metal ions, or antibiotics. Fluorescence-activated cell sorting (FACS; see Chapter 6) provides a quick and easy way to isolate the fluorescent clones rather than growing them on solid surfaces. The cells can be grown in cultures, and then FACS would sort any bacteria that are fluorescent.

The main barrier to function-based analyses is successful gene expression. Getting a library host such as *E. coli* to express foreign genes is hit or miss because some may be toxic to the host, some may require other factors for expression, and others may have very low activity. Another problem is simply volume. The number of potential clones that have any chosen gene of interest is usually low, and excessive numbers of clones must be screened in order to identify just one or two genes. For example, the lipases identified from German soil (see Table 12.1) were only found in 1 of 730,000 different metagenomic clones. In another example, only two novel Na^+/H^+ antiporters were found after screening 1,480,000 clones. This is why culture enrichment strategies are an important aspect of creating metagenomic libraries (see earlier discussion).

Functional approaches also include expression screening with **phage biopanning**. Even stable isotope probing (see earlier discussion) could be categorized as a function-based approach if the labeled substrate is a specific metabolite that enriches the culture based on metabolic function. Metagenomic phage biopanning uses basically the same method as phage display (see Chapter 9). Cloned DNA inserts are expressed as fusion proteins with a phage coat protein and displayed on the phage surface. Because the displayed proteins carry only segments of foreign protein, the problems associated with heterologous expression are lessened. The cloned DNA is from any organisms found within the environmental sample. Thus, the expressed protein segments could be any part of an enzyme, a membrane protein, etc.; therefore, the success with a phage display library relies on the screening method. For example, phage biopanning can identify binding partners for a particular pollutant, metabolite, or even another protein. If the target molecule is immobilized on a bead, then any phage carrying a protein segment that binds the target will stick to the bead. The phage can then be isolated, and the DNA insert sequenced to identify the sequence responsible.

Screening the metagenomic expression library for a particular function such as growth on a pollutant can identify novel genes that metabolize that pollutant.

In expression screening, the metagenomic library is an expression library; that is, the vector contains the transcription and translation initiation and terminator sequences. When *E. coli* expressing the DNA from the environment is grown with a single substrate, only those clones that can metabolize the substrate actually multiply.

Guilt-by-association relies on the library insert to carry the natural promoters that are regulated by the substrate. When the substrate activates its natural promoter, the library vector sequences contain a reporter gene that is expressed. The *E. coli* expressing this gene is isolated from the rest of the library by FACS or plating on solid media.

Metagenomic phage biopanning can alleviate problems associated with expressing so many diverse genes from a metagenomic library because only small pieces of the protein are expressed. This method identifies new binding domains for a particular protein or substrate.

ECOLOGY AND METAGENOMICS

As noted in the introduction to this chapter, metagenomics research analyzes bacteria, viruses, and even simple gene creatures found within an environmental sample. The results obtained from metagenomic research have major potential for many different applications, including the study of ecology. Metagenomics techniques have been used to identify the entire genome sequence of symbiotic organisms. For example, *Buchnera* are symbiotic bacteria that live within aphids. These bacteria produce amino acids essential to the aphid, and in return, the aphids provide carbon and energy sources to the bacteria. The relationship is so intertwined that neither organism can live without the other. The bacteria have lost so many of their original functions that they are almost organelles. Because there is no possible way to culture the bacteria outside the aphid, a traditional genomic library cannot be established. Instead, a metagenomic library containing both aphid and *Buchnera* DNA was constructed and sequenced. Only when both genomes were examined was the true level of dependence deciphered.

The same scenario was used to sequence the entire genome of the bacteria that coexist within deep-sea tube worms. Tube worms live near thermal vents that are rich in sulfide and reach temperatures of 400°C. The worms lack mouths and digestive tracts and rely completely on symbiotic members of the Proteobacteria to provide nutrition. The bacteria live within a specialized structure called a trophosome where they oxidize hydrogen sulfide to make energy. The energy is used to manufacture amino acids that feed the worm. In return, the worm collects hydrogen sulfide, oxygen, and carbon dioxide and transports these to the bacteria. The metagenomic library contained both worm and bacterial genomes, but yielded information about the bacteria previously unknown. For instance, the bacterial genome had genes for flagella, suggesting that the bacteria may also have a motile phase. Indeed, other observations suggest that the bacteria move through the seawater to colonize juvenile worms.

Metagenomics can also help in understanding microbial competition and communication. This research may have far-reaching applications to all environments, whether they are within the digestive tract of humans or in the deep-sea vents in the oceans. Functional metagenomics can identify small molecules important to microbial survival, such as antibiotics. Metagenomic libraries can be assessed for antimicrobial activity using functional assays to identify new antibiotics. Additionally, sequence-based analysis of metagenomic libraries can identify synthases that make novel polyketides (antibiotics related to erythromycin and rifamycin). Other functional metagenomic screens have been used to identify quorum-sensing molecules. These are indicators of bacterial population density. Because many bacteria infect only eukaryotic cells or make toxins when they are present in sufficient numbers, interference with quorum sensing provides a new approach to antibacterial therapy. Thus, this area is of direct clinical importance. New quorum-sensing molecules have been identified by a detection system that used the reporter GFP. When clones expressed a quorum-sensing molecule, this activated the expression of GFP, making the bacteria fluorescent. The quorum-sensing metagenomic clone can then be isolated with FACS or microscopy, and then sequenced to determine the identity of the genes involved.

Identifying new genes using metagenomics has much promise for understanding the world around us and potentially for solving some of our health problems.

NATURAL ATTENUATION OF POLLUTANTS

Metagenomics can also be applied to biogeochemical research. Perhaps identifying how bacteria affect the environment will help us figure out ways to maintain our species. Understanding how bacteria can live in extreme environments can reveal useful biochemical processes for biotechnology. Most important, finding out how bacteria cope with contaminated sites may provide useful enzymes for cleaning up our pollution.

Bioremediation is one avenue in which biotechnology has made rapid advances. Many different human-made compounds have contaminated the environment around us through everyday use, accidental spillage, or intentional dumping. Many environmental biotechnologists are working on "biological" means of cleaning the environment. In fact, releasing an organism that can degrade a pollutant would provide a very easy, low-cost way of cleaning up a polluted site.

Naturally occurring microorganisms often have the ability to degrade human-made pollutants. For example, *Rhodococcus* has a highly diverse repertoire of pathways to degrade pollutants, such as short- and long-chain alkanes, aromatic molecules (both halogenated and nitro-substituted), and heterocyclic and polycyclic aromatic compounds, including quinolone, pyridine, thiocarbamate, *s*-triazine herbicides, 2-mercaptobenzothiazole (a rubber vulcanization accelerator), benzothiophene, dibenzothiophene, MTBE (see later discussion), and the related ethyl *tert*-butyl ether (ETBE). *Rhodococcus* has several features that contribute to its ability to degrade so many compounds. First, it has a range of different enzymes, including cytochrome P450, that degrade toxic compounds. These are very efficient and versatile in oxidation pathways and catalyze a variety of reactions, including epoxidation (Fig. 12.7). Other enzymes that catalyze key degradation steps include monooxygenases and dioxygenases, which help degrade aromatic compounds (see Chapter 13).

Furthermore, several strains of *Rhodococcus* can survive in solvents such as ethanol, butanol, dodecane, and toluene, which would kill many other bacteria. The oil-degrading strains actually adhere to oil droplets! *Rhodococcus* species are found in all types of environments, including nuclear waste sediments; tropical soil; Arctic soil; and sites in Europe, Japan, and the United States. Genetically speaking, *Rhodococcus* also has unique attributes that are advantageous in biodegradation. The genome of *Rhodococcus* sp. strain RHA1 has 9.7 Mb of DNA, including one chromosome and three large linear plasmids. The plasmids may be critical because they are important for gene transfer and recombination events. The genes for the catabolic enzymes are often found in clusters, flanked by inverted repeats, suggesting that they are acquired and passed from one strain to another by recombination. Such horizontal gene transfer can also transfer these catabolic regions to other bacteria, including *Pseudomonas* and *Mycobacterium*. Chapter 13 describes some plasmids that encode pollutant-degrading enzymes.

Different pollutants are degraded in various ways. Sometimes a single naturally occurring organism can completely degrade a pollutant. Other pollutants require more than one type

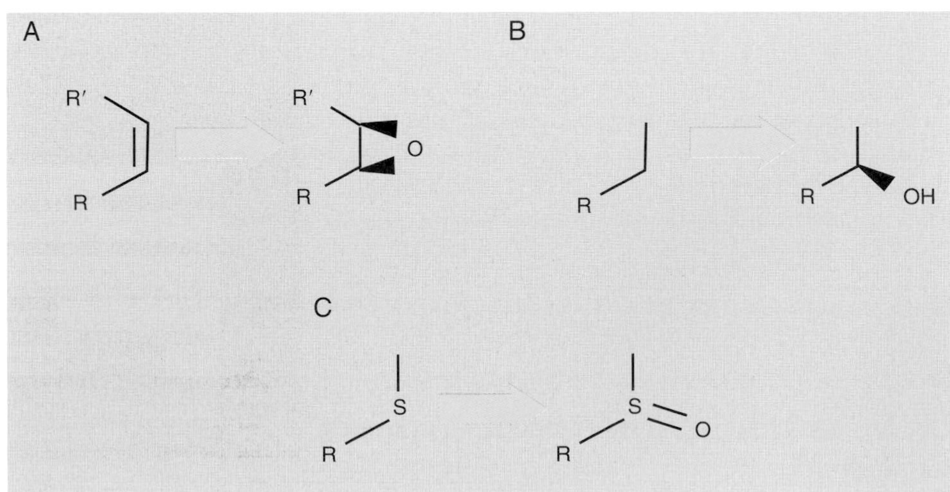

FIGURE 12.7 Oxygenase Reactions Catalyzed by *Rhodococcus*
Rhodococcus can catalyze various reactions, including epoxidation by cytochrome P450 enzymes (A), hydroxylation of alkyl groups by alkane monooxygenases (B), and sulfoxidation of sulfide to sulfoxide (C).

of bacterium to achieve complete degradation. Some pollutants are degraded very slowly. Heavy metals cannot be chemically degraded, so they pose more of a challenge than organic molecules. Therefore, most environmental biotechnologists look for microorganisms that sequester the heavy metal in a solid phase. The conversion of such heavy metals as uranium from an aqueous phase to a solid phase can clean drinking water supplies. Thus, certain anaerobic microorganisms can reduce uranium(VI) to uranium(IV) by utilizing the metal as a terminal electron acceptor. This converts the uranium from a soluble to an insoluble form. In one uranium-contaminated site (Old Rifle, Colorado, USA), one study injected acetate into the groundwater. Acetate is an electron donor that stimulates metal-reducing bacteria to sequester uranium into the solid phase. Within 50 days, some contaminated wells had uranium concentrations lower than the regulated level. Although these results are promising, over time the acetate dissipated and the soluble uranium levels increased again. More research is necessary to find a permanent means to keep uranium out of groundwater.

Another recalcitrant compound that can contaminate groundwater, **methyl tert-butyl ether (MTBE)**, oxygenates gasoline so that it burns more efficiently. Many cases of MTBE contaminating groundwater have been reported. Finding a natural method to clean these sites has great applicability. The United States Environmental Protection Agency (EPA) has classified MTBE as a possible human carcinogen, and drinking water must contain less than 20 to 40 µg/L. One MTBE-contaminated site in South Carolina, USA, had a large plume of MTBE-contaminated gasoline leaking from an underground storage tank at a gas station. The plume ended at a drainage ditch. The concentration of MTBE in the water was low, and in the 2-meter gap between the anaerobic and aerobic zones, the MTBE was metabolized by naturally occurring microorganisms. This led to studies that determined that MTBE could be degraded by bacteria such as *Methylibium petroleiphilum* PM1 in areas that transition from anaerobic (anoxic) to aerobic (oxic). In fact, if anaerobic regions of the MTBE plume in South Carolina were injected with a compound that released oxygen, the concentration of MTBE decreased from 20 mg/L to 2 mg/L, suggesting that **biostimulation** may be a good approach to clean up contamination. Biostimulation is the release of nutrients, oxidants, or electron donors into the environment to stimulate naturally occurring microorganisms to degrade a contaminant. In other areas of MTBE contamination, the site may have to undergo **bioaugmentation**;, that is, specific microorganisms plus their energy sources may need to be added to the site. Such microorganisms may be naturally occurring, a mixture of different organisms, or even genetically modified (see Chapter 13).

Bioremediation uses biologicals, such as bacteria, to clean up a polluted area. Different pollutants are degraded in various ways. Sometimes a single naturally occurring organism can completely degrade a pollutant. Other pollutants require more than one type of bacterium to achieve complete degradation.

Biostimulation is the release of nutrients, oxidants, or electron donors into the environment to stimulate naturally occurring microorganisms to degrade a contaminant. Bioaugmentation is adding specific microorganisms plus their energy sources to decontaminate a polluted area.

BIOFUELS AND BIOENERGY

The consumption of fossil fuels in the world is increasing at a rate that is not sustainable. Approximately 87% of the world energy production is from fossil fuel sources, and the use is only increasing. In the year 2000, a total of approximately 400 quadrillion BTUs were consumed by the world, which increased to over 524 quadrillion BTUs by 2010. The projected growth for the upcoming years is even higher based on the larger number of people across the globe that are no longer living in poverty. The increased use of fossil fuels will eventually deplete our supply of these nonrenewable resources. To be able to supply the projected growth of fuel consumption, researchers are investigating renewable sources for their use to create biofuel and other types of biologically derived energy sources.

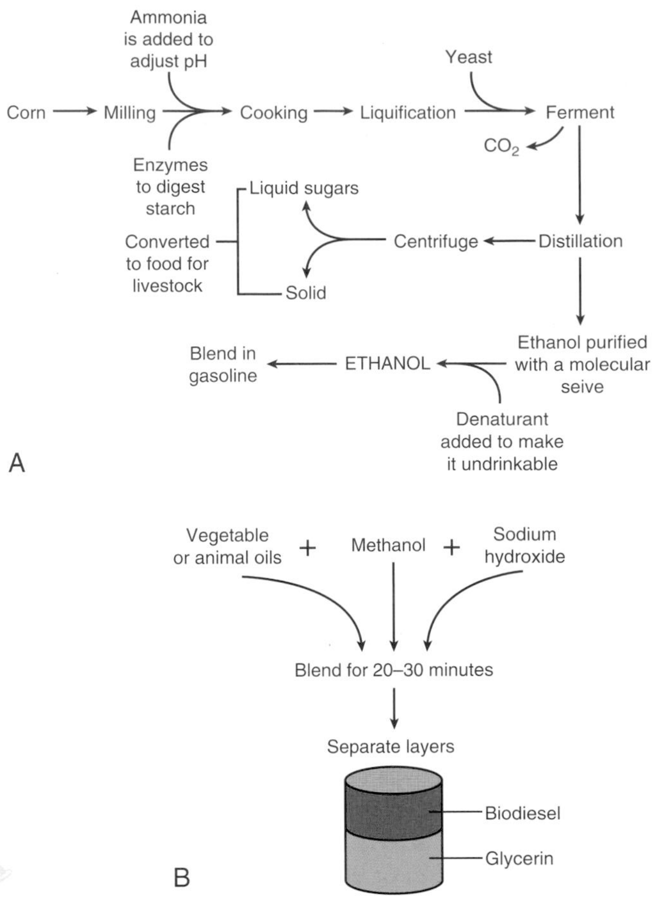

FIGURE 12.8 Production of Biofuels from Corn or Vegetable Oil

(A) Production of ethanol from corn requires the addition of ammonia to adjust the pH and enzymes to help digest starch and yeast to ferment the corn mash. The ethanol is distilled to 95% purity and then purified further by passing through molecular sieves. The purified ethanol is rendered undrinkable by addition of denaturants. The remaining corn products after distillation include liquid sugars and solids that are used to feed livestock. (B) Biodiesel is created by the blending of methanol, sodium hydroxide, and vegetable oil for 20–30 minutes. The process creates biodiesel and glycerin that separate naturally because glycerin has a higher density than biodiesel.

Today there are two commonly produced biofuels: ethanol from corn or sugarcane and biodiesel from oil crops such as soybeans or oil palm (Fig. 12.8). Ethanol is added to gasoline to improve the emission of carbon dioxide and other greenhouse gasses from the combustion of gas. Because ethanol is produced from corn or sugarcane means that the resource is easily replaced year after year. The other major biofuel in use is biodiesel, which is made by taking oil—either vegetable oil from soybeans or animal-based fats— and converting it into fuel by adding alcohol. This fuel is not an additive but replaces regular diesel fuel completely. Although both products have decreased our dependence on fossil fuels, the overall impact is small, mainly due to cost. In addition, the sources for either biological-based fuel are key crops in the food supply, and increasing the use of food to create energy will increase food costs across the world and drive even more people into poverty, starvation, and malnutrition.

The use of algae to create biofuels leads the field of research to find a less costly nonfood alternative to corn, sugarcane, and soybeans. Today's fossil fuels were created by years of heat and pressure on layers and layers of dead creatures from the sea. Since a large portion of the creatures that were fossilized into oil were probably algae, it seems appropriate to look to the original source for a renewable fuel. The use of these is still experimental but offers many advantages to the current sources of biofuel. Algae fall into two different categories, **microalgae** and **macroalgae**, where the first group is represented by microscopic photosynthetic creatures, and the macroalgae are represented by large multicellular plant-like organisms commonly known as seaweeds. There are many different varieties of seaweed, and these have a varied content of oil and sugar. There are macroalgae species that have 50% of their mass as sugar, which could be easily fermented to form bioethanol in a manner similar to corn. In addition, many species have a high lipid content—as much as 50% to 60% of their dry weight—which could be easily converted to biodiesel fuel. At the time of writing, though, algae-derived fuels cost much more than traditional fossil fuels and are not used as alternative biofuels. One idea that would decrease the cost of algal production of bioenergy would be to use both the carbohydrate fraction for ethanol and the lipid portion for biodiesel. The double use of one culture of algae would decrease the overall cost.

In addition to ethanol and biodiesel, algae can also produce biohydrogen and other biogases such as biomethane. Macroalgae species that have high levels of carbohydrates can be converted to biomethane by anaerobic fermentation. In addition, photosynthetic bacteria such as cyanobacteria can produce hydrogen by fermentation of organic acids from sewage waste or agricultural waste. Cyanobacteria use photosynthesis to drive the reaction, which releases hydrogen gas that can be collected. The combination of using this method for treatment of waste products without the input of anything other than sunlight and photosynthetic bacteria to produce a fuel source is very promising but still in the investigation stage. No commercial fuel is produced in this manner as of now.

Although not derived from microorganisms, another potential renewable source of fuel is CO_2. The process of making CO_2 into fuel has been known for some time, but it requires large amounts of heat. Recent research has found new catalysts that make the reaction more feasible. Carbon dioxide is converted to carbon monoxide using carbon nanofibers that contain reduced carbons that pull oxygen from CO_2. The CO molecules can then be converted into fuel products.

Algae can be grown and harvested, and then the carbohydrates can be extracted to create ethanol by fermentation. The lipids can also be used to make biodiesel. Using algae as a renewable source to make biofuels is better than the current method of creating ethanol from corn or biodiesel from soybeans. Algae are easier to grow and cost less to produce, but most importantly, they are not a significant food source for people. Algae can also be used to make biohydrogen or biomethane.

MICROBIAL FUEL CELLS

In microbial fuel cells, microbes such as bacteria catalyze electrochemical oxidations or reductions at an anode or cathode, respectively, to produce an electric current (Fig. 12.9). These fuel cells were originally inefficient and only served the purpose of a battery in very remote areas. Advances in the understanding of the microorganisms have increased the efficiency for the reactions.

Microbial fuel cells can harvest electricity from **electrode-reducing organisms** that donate electrons to the anode. While the microorganism oxidizes organic compounds or substrates into carbon dioxide, the electrons are transferred to the anode. The mechanism of electron transfer can occur by three different pathways (Fig. 12.10). First, electrons can be transferred to the anode through a soluble mediator in the solution bathing the electrode. Second, electrons can be transferred directly to the anode through proteins found on the outer membrane of the bacteria. For example, microorganisms from the *Geobacteraceae* family transfer electrons to electrodes using cytochromes on the outer membrane. *Shewanella oneidensis* also uses cytochrome *c* to transfer electrons but requires an anaerobic environment to convert lactate to acetate. In some instances, bacteria form a thick film on the cathode, so it may be the pili or nanowires that transmit the electrons to the anode. In contrast, **electrode-oxidizing organisms** use electrons from the cathode to reduce substances in the cathode chamber. In aerobic chambers, microorganisms can reduce oxygen to water. In anaerobic environments, nitrate or sulfate can be reduced to nitrite, nitrogen, or sulfur ions. Another potential reduction for these bacteria is the conversion of carbon dioxide to methane or acetate. The process uses acetyl-CoA as an intermediate to build even longer chain fatty acids and alcohols. For example, *G. sulfurreducens* reduces fumarate to succinate with electrons obtained from the cathode. Interestingly, the substrates that these organisms need for the redox reactions can be readily obtained from wastewater or contaminated water, which would both provide energy and clean up the environment.

The use of microbial fuel cells is still not optimized, and the level of electric current generated by such systems is low, but the potential for such systems is great. For example, if a microbial fuel cell were to reduce carbon dioxide to make electricity, not only would there be a renewable source of fuel, but the excess carbon dioxide put into the atmosphere by burning fossil fuels could be used. The best microorganism for producing an electric current is *Sporomusa ovata*, which is an anaerobic, Gram-negative bacterium that converts hydrogen and carbon dioxide to acetate by fermentation. In comparison to a standard hydrogen electrode, this fuel cell produces −400 mV. However, the current generated is small. Current research is now trying to identify what proteins are essential for the various reactions that transfer electrons from the bacteria to the anode or take the electrons from the cathode to reduce substrates. These types of studies should identify ways to optimize the reactions to get the most energy from the bacteria. In addition, researchers are still investigating the best materials

e⁻
e⁻
ANODE
CATHODE
Organic matter
CO_2
H^+
H^+
Aerobic
O_2
H_2O
Anaerobic
CO_2
Nitrate
Sulfate
Acetate
Nitrate
N_2
S^{2-}
Bacteria
Ion exchange membrane
Bacteria
A

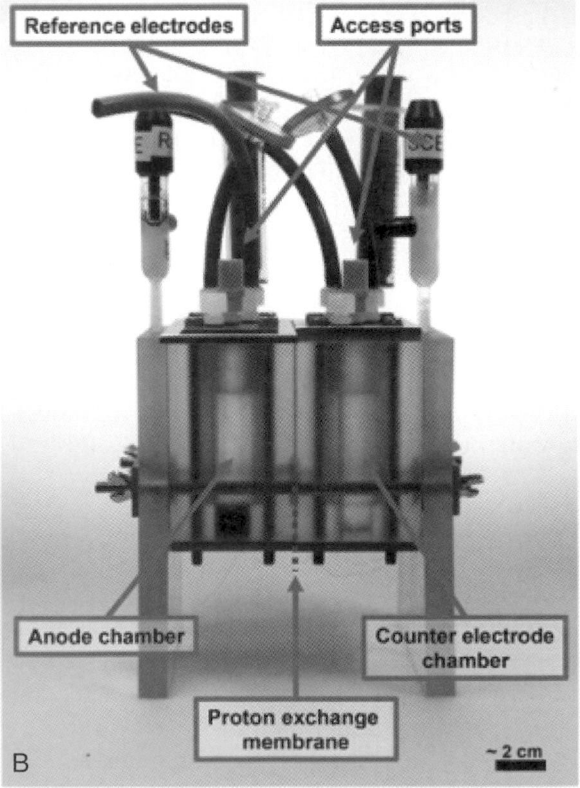

FIGURE 12.9
Microbial Fuel Cell
(A) Schematic showing the cathodic and anodic chambers of a microbial fuel cell. On the anode, microorganisms use organic matter such as wastewater or added nutrients to create electrons, protons, and carbon dioxide. The electrons then flow through the electric meter to the cathode. At the cathode, microorganisms can convert the electrons to reduce oxygen to water under aerobic conditions, or convert nitrate to nitrite or N_2, or convert CO_2 to acetate. These reactions can create fuel precursors. (B) Actual microbial fuel cell showing the anode chamber (left) and cathode chamber (right). These are separated by a membrane that allows protons to freely pass from anode to cathode. From Dolch et al. (2014) Characterization of microbial current production as a function of microbe-electrode interaction. *Bioresource Technol* **157**, 284–292.

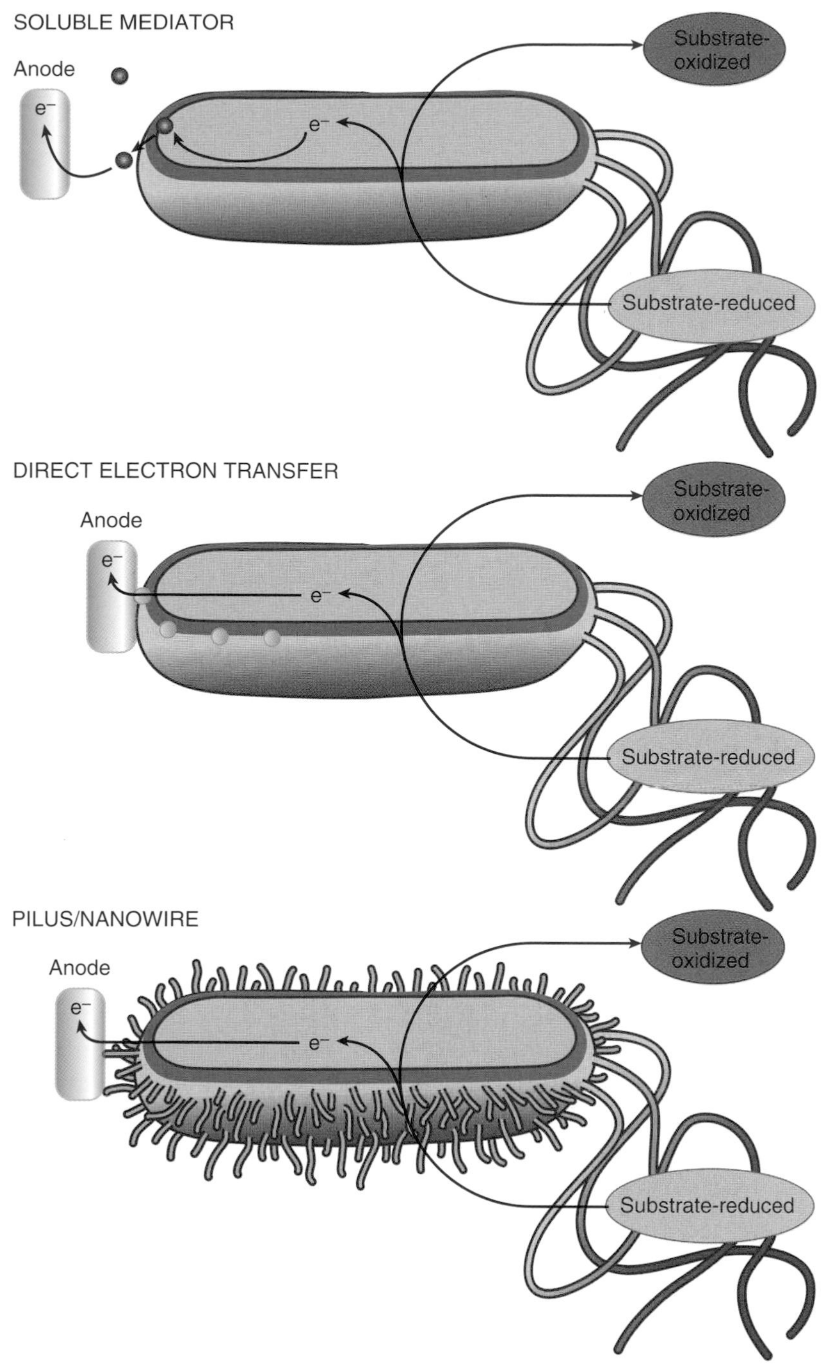

FIGURE 12.10 Transfer of Electrons to the Anode in a Microbial Fuel Cell
Three different methods exist for bacteria to pass electrons from the oxidizing reaction to the anode. There can be an extracellular mediator that absorbs the electrons and passes them onto the anode (top). The bacteria can transfer electrons through outer membrane proteins such as cytochrome c (middle). The electrons can pass from the bacteria to the anode via nanowire structures such as pili (bottom).

for the cathode and anode, as well as the solutions in which to grow the cells. Nevertheless, using microbial fuel cells may help reduce environmental contaminants such as wastewater, reduce atmospheric carbon dioxide by using it to rebuild fuels, and may potentially provide a renewable energy source.

Microbial fuel cells create electricity through the use of microorganisms. Organisms that transfer electrons to the anode are called electrode-reducing organisms. They can pass electrons through a mediator molecule in the solution, directly through proteins in their outer membrane, or through nanowires or pili that coat the outer surface of the bacterium. Electrode-oxidizing organisms take electrons from the cathode to reduce various substances, such as carbon dioxide to acetate.

Summary

The environment has a huge biodiversity of bacteria, viruses, and gene creatures. The majority of these species are unknown and have never been cultured in a lab. The unique niches in which these organisms live make culturing the organisms almost impossible. The use of metagenomics forgoes the culturing of individual organisms and focuses on just sequencing the DNA from the environmental niche.

Since the diversity is so great in a single sample, ways to enrich the sample for certain functions or organisms is critical. Stable isotope probing (SIP) is a method in which precursor molecules with stable isotopes are added to the environmental sample. All the active bacteria and viruses take in the precursors and incorporate the isotope into their cells. The labeled organisms can then be isolated from the unlabeled ones. If the precursors are targeted for incorporation into RNA, the environmental sample can be enriched for RNA rather than DNA or protein. Another method, suppressive subtraction hybridization (SSH), takes advantage of the genetic differences between samples from two different areas. This technique removes all the sequences that are identical between the two areas. The leftover sequences represent potential sequences that are unique to one region. Growing a culture with 5-bromo-2-deoxyuridine (BrdU) enriches the culture for bacteria undergoing DNA synthesis since BrdU is incorporated during replication.

Metagenomic analysis uses next-generation sequencing technology to identify genes or DNA fragments from a mixture of organisms, either from the environment (soil, air, water, etc.); from skin, intestine, or other organs with known microorganisms; or from symbiotic organisms to identify what genes are from the host and what genes are from the symbiont. Metatranscriptomics isolates RNA from these types of tissues and then analyzes them with next-generation sequencing. Integron analysis is a method to identify new genes that are flanked by 59-be sequences. These sequences flank integrons, which are transposon-like elements that capture genes from one organism and move them into different organisms.

Metagenomic libraries can also be screened for functions, such as finding new genes that metabolize a specific substrate, that carry out a specific enzyme reaction, or that simply bind to a known protein. Two techniques that identify proteins which metabolize a particular substrate are expression screening and guilt-by-association. For expression screening, the metagenomic library is made with an expression vector, which contains the transcription and translation initiation and terminator sequences. In contrast, guilt-by-association relies on the library insert to carry the natural promoters that are regulated by the substrate. When the substrate activates its natural promoter, the library vector sequences contain a reporter gene that is expressed. Finally, new proteins from a metagenomic library can be identified by phage biopanning, where a known protein is used to find a novel binding partner.

Ecology examines the relationship between organisms and their environment, and the application of metagenomics to these studies has changed the field of research. Metagenomic sequencing projects such as the one in the Sargasso Sea have identified many more organisms than previously thought existed. Symbiotic relationships can be studied in detail, because the genome of the symbiont can be isolated and identified. Extreme environments are hard to duplicate in the laboratory; therefore, organisms from these areas were impossible to study. Now, the genomic sequences from these organisms can be used to elucidate their metabolisms and special characteristics, which will, it is hoped, find new ways to combat human diseases.

The chapter discusses biological methods of removing pollutants, called *bioremediation*. Many naturally occurring bacteria, such as *Rhodococcus*, are able to metabolize a variety of different pollutants. Biostimulation increases the growth and metabolism of the naturally occurring bacteria to remove the pollutant by giving them the necessary food or precursor substrate. In other cases, bioaugmentation adds both the specific microorganism and its energy source to the contamination site. These techniques have been used to clean MTBE from spill sites.

Microorganisms are also be used to create biofuels and bioenergy. Current biofuels are produced by fermenting corn or sugarcane to make ethanol or converting vegetable or animal oils from soybeans or cows into biodiesel. Ethanol is used as a fuel additive, whereas biodiesel is used directly in the gas tank. These two sources are renewable, but the starting material is potential food. Thus, this approach has a major impact on food availability and price. Instead of using crops, researchers are trying to find microalgae or macroalgae that will provide a large amount of carbohydrate for making ethanol or a large amount of oil for making biodiesel, or potentially both. Right now, the cost is too high, but eventually, the process may come down in price. Algae can also be used to create biogas, such as biohydrogen and biomethane, which are also valuable sources of fuel. Finally, microorganisms have the potential to create electricity by passing electrons onto cathodes in a microbial fuel cell. In contrast, at the anode, microorganisms can use electrons to reduce substances like carbon dioxide into methane or acetate, which can reduce atmospheric carbon dioxide and create precursors for fuel production.

End-of-Chapter Questions

1. According to the book, about what percentage of microorganisms have been cultured?
   a. <0.1%
   b. 0.1%–1%
   c. 1%–10%
   d. 10%–100%
   e. unknown percentage
2. To date, what has metagenomics research been able to identify?
   a. turbomycin A and B
   b. enzymes that can reduce oil contaminants in the soil
   c. bacteria that live in environments contaminated with radioactivity
   d. novel biochemical pathways
   e. all of the above
3. Why must samples be enriched?
   a. The diversity in one sample is too great and enriching selects for a specific organism or function.
   b. Enriching the sample allows all of the organisms to grow in culture.
   c. Enriching the sample prevents organisms from releasing toxins.
   d. The samples do not need to be enriched to identify the organisms growing.
   e. none of the above
4. Which method is better suited for the identification of actively growing bacteria or viruses?
   a. BrdU-enrichment
   b. suppressive subtraction hybridization
   c. stable isotope probing
   d. RT-PCR
   e. DNA sequencing

(*Continued*)

5. Why is small subunit ribosomal RNA (SSU rRNA) an excellent biomarker for RNA-SIP?
   a. abundant within the cell
   b. essential to all life
   c. database exists containing SSU rRNA sequences
   d. variable among different species
   e. all of the above
6. What does suppressive subtractive hybridization do?
   a. hybridizes sample mRNA to known mRNA sequences
   b. identifies the sequences from two areas that are identical, and then uses the leftover sequences to identify unique sequences to one area
   c. suppresses the unknown sequences by binding to all of the known sequences
   d. identifies the sequences of one sample that are already known
   e. none of the above
7. Which of the following is not a potential downfall regarding metagenomic libraries?
   a. contaminated DNA or mRNA samples
   b. DNA that is difficult to isolate and therefore underrepresented in the sample and library
   c. sheared DNA
   d. lack of a cultured organism
   e. none of the above
8. Why are integrons important?
   a. Integrons can spread genes for antibiotic resistance or for a growth advantage under certain environmental conditions.
   b. Integrons encode the industrially relevant enzyme called *integrase.*
   c. Integrons are important for overexpressing certain proteins.
   d. Integrons only contain noncoding DNA.
   e. none of the above
9. What genes have been identified using integron analysis?
   a. phosphotransferases
   b. DNA glycosylases
   c. new antibiotic resistance genes
   d. methyltransferases
   e. all of the above
10. Which of the following was not identified by Venter from the Sargasso Sea?
   a. 1.2 million new genes
   b. 1800 different species of microbes
   c. 700 different bacteriorhodopsin genes
   d. 150 new bacterial/archaea species
   e. all of the above
11. What is a potential problem associated with function-based analyses of genes?
   a. The foreign protein has a toxic effect to the host organism.
   b. Successful expression of the gene of interest in the host organism may fail.
   c. Other cofactors are needed for proper activity that the host organism cannot provide.
   d. The number of clones to screen for the gene of interest is usually too many.
   e. All of the above are problems associated with this type of analysis.

12. What is the general term used to describe the degradation of pollutants using a biological approach?
- **a.** biostimulation
- **b.** bioremediation
- **c.** biodegradation
- **d.** bioprocessing
- **e.** bioaugmentation

13. Which genera of microorganisms have the most diverse pathways for bioremediation?
- **a.** *Mycobacterium*
- **b.** *Pseudomonas*
- **c.** *Rhodococcus*
- **d.** *Escherichia*
- **e.** *Methylobium*

14. What is biostimulation?
- **a.** the release of nutrients, oxidants, or electron donors into the environment to stimulate microorganisms to degrade a pollutant
- **b.** the addition of bacteria to a specific contaminated site for bioremediation
- **c.** the research term used to describe how bacteria can sequester certain heavy metals from a contaminated site
- **d.** the act of decreasing the amount of microorganisms in a contaminated site
- **e.** none of the above

15. Which pollutant has been cleaned from contaminated sites using either biostimulation or bioaugmentation?
- **a.** MTBE
- **b.** acetate
- **c.** aromatics
- **d.** ETBE
- **e.** alkanes

16. How might the cost of algal production of bioenergy be <u>best</u> decreased?
- **a.** Breeding algae to have higher sugar contents.
- **b.** Genetic engineering of microalgae with better fermentation pathways.
- **c.** The consumer switch to using algae-derived biohydrogen and biomethane as the primary energy sources.
- **d.** The use of both the carbohydrate and lipid portions of macroalgae for bioenergy production.
- **e.** The use of microbial fuel cells.

17. Which organism – potential biofuel function is not correctly paired?
- **a.** High-lipid macroalgae – biodiesel
- **b.** Algae – biomethane
- **c.** Cyanobacteria – biohydrogen
- **d.** High-carbohydrate microalgae – ethanol
- **e.** High-carbohydrate macroalgae - biomethane

18. The best microorganism for the production of an electric current in a microbial fuel cell is _______________.
- **a.** yeast
- **b.** *Sporomusa oyata*
- **c.** *Geobacteraceae* family
- **d.** *Shewanella oneidensis*
- **e.** All of the above are appropriate.

Further Reading

Bae, Y. J., Ryu, C., Jeon, J.-K., Park, J., Suh, D. J., Suh, Y.-W., Chang, D., & Park, Y.-K. (2011). The characteristics of bio-oil produced from the pyrolysis of three marine macroalgae. *Bioresource Technology, 102*, 3512–3520.

Debabov, V. G. (2008). Electricity from microorganisms. *Microbiology, 77*, 123–131.

Delgado, A. G., Kang, D.-W., Nelson, K. G., Fajardo-Williams, D., Miceli, J. F., Done, H. Y., Popat, S. C., & Krajmalnik-Brown, R. (2014). Selective enrichment yields robust ethene-producing dechlorinating cultures from microcosms stalled at cis-dichloroethene. *PLoS One, 9*, e100654.

Detter, J. C., Johnson, S. L., Bishop-Lilly, K. A., Chain, P. S., Gibbons, H. S., Minogue, T. D., Sozhamannan, S., Van Gieson, E. J., & Resnick, I. G. (2014). Nucleic acid sequencing for characterizing infectious and/or novel agents in complex samples. In R.P. Schaudies (Ed.), *Biological Identification* (pp. 3–53). Waltham, MA, USA and Kidlington, UK: Woodhead Publishing.

Dolch, K., Danzer, J., Kabbeck, T., Bierer, B., Erben, J., Förster, A. H., Maisch, J., Nick, P., Kerzenmacher, S., & Gescher, J. (2014). Characterization of microbial current production as a function of microbe–electrode interaction. *Bioresource Technology, 157*, 284–292.

Gillespie, D. E., Brady, S. F., Bettermann, A. D., Cianciotto, N. P., Liles, M. R., Rondon, M. R., Clardy, J., Goodman, R. M., & Handelsman, J. (2002). Isolation of antibiotics turbomycin A and B from a metagenomic library of soil microbial DNA. *Applied and Environmental Microbiology, 68*, 4301–4306.

Gross, L. (2007). Untapped bounty: sampling the seas to survey microbial biodiversity. *PLoS Biology, 5*, e85.

Jones, C. S., & Mayfield, S. P. (2012). Algae biofuels: versatility for the future of bioenergy. *Current Opinion in Biotechnology, 23*, 346–351.

Kumar, B., Asadi, M., Pisasale, D., Sinha-Ray, S., Rosen, B. A., Haasch, R., Abiade, J., Yarin, A. L., & Salehi-Khojin, A. (2013). Renewable and metal-free carbon nanofibre catalysts for carbon dioxide reduction. *Natural Communication, 4*, 2819.

Larkin, M. J., Kulakov, L. A., & Allen, C. C. R. (2005). Biodegradation and *Rhodococcus*—masters of catabolic versatility. *Current Opinion in Biotechnology, 16*, 282–290.

Lasken, R. S. (2009). Genomic DNA amplification by the multiple displacement amplification (MDA) method. *Biochemical Society Transactions, 37*, 450–453.

Lovley, D. R. (2008). The microbe electric: conversion of organic matter to electricity. *Current Opinion in Biotechnology, 19*, 564–571.

Mata, T. M., Martins, A. A., & Caetano, N. S. (2010). Microalgae for biodiesel production and other applications: a review. *Renew Sustain Energy Review, 14*, 217–232.

Moore, A. M., Patel, S., Forsberg, K. J., Wang, B., Bentley, G., Razia, Y., Qin, X., Tarr, P. I., & Dantas, G. (2013). Pediatric fecal microbiota harbor diverse and novel antibiotic resistance genes. *PLoS One, 8*, e78822.

Nevin, K. P., Woodard, T. L., Franks, A. E., Summers, Z. M., & Lovley, D. R. (2010). Microbial electrosynthesis: feeding microbes electricity to convert carbon dioxide and water to multicarbon extracellular organic compounds. *MBio, 1*, e00103-10.

Padmanabhan, R., Mishra, A. K., Raoult, D., & Fournier, P.-E. (2013). Genomics and metagenomics in medical microbiology. *Journal of Microbiological Methods, 95*, 415–424.

Rosenbaum, M. A., & Henrich, A. W. (2014). Engineering microbial electrocatalysis for chemical and fuel production. *Current Opinion in Biotechnology, 29C*, 93–98.

Scott, K. (2014). Microbial fuel cells: transformation of wastes into clean energy. In A. Gugliuzza & A. Basile (Eds.), *Membranes for Clean and Renewable Power Applications* (pp. 266–300). Cambridge, UK: Woodhead Publishing.

Steward, G. F., Culley, A. I., Mueller, J. A., Wood-Charlson, E. M., Belcaid, M., & Poisson, G. (2013). Are we missing half of the viruses in the ocean? *The ISME Journal, 7*, 672–679.

Uhlik, O., Leewis, M.-C., Strejcek, M., Musilova, L., Mackova, M., Leigh, M. B., & Macek, T. (2013). Stable isotope probing in the metagenomics era: a bridge towards improved bioremediation. *Biotechnology Advances, 31*, 154–165.

Venter, J. C., Remington, K., Heidelberg, J. F., Halpern, A. L., Rusch, D., Eisen, J. A., et al. (2004). Environmental genome shotgun sequencing of the Sargasso Sea. *Science, 304*, 66–74.

Whiteley, A. S., Manefield, M., & Lueders, T. (2006). Unlocking the "microbial black box" using RNA-based stable isotope probing technologies. *Current Opinion in Biotechnology, 17*, 67–71.

Wooley, J. C., Godzik, A., & Friedberg, I. (2010). A primer on metagenomics. *PLoS Computational Biology, 6*, e1000667.

CHAPTER 13

Synthetic Biology

Biotechnology

http://dx.doi.org/10.1016/B978-0-12-385015-7.00013-2

INTRODUCTION

From the genetic viewpoint, the production of a small molecule such as ethanol may well be more complex than production of a protein such as somatotropin. Although proteins are macromolecules, single genes encode them, whereas small molecules must be made by biochemical pathways that require several steps, each catalyzed by a separate enzyme. Thus, multiple genes are involved, together with their regulatory systems. **Pathway engineering** is the assembly of a new or improved biochemical pathway, using genes from one or more organisms. Most efforts to date have consisted of modification and improvement of existing pathways, rather than the assembly of completely new synthetic schemes. However, totally novel pathways will no doubt begin to appear over the next few years.

This brings us to the concept of **synthetic biology**. This emerging area is constantly changing and is impossible to define precisely. The overall idea is the design of novel biological pathways, devices, or systems. It overlaps with pathway engineering but is more fashionable! Perhaps one key difference is that pathway engineering takes an existing pathway and improves it, whereas in synthetic biology the pathway or device is designed first and then assembled from available components.

Engineered bacteria may be used to degrade agricultural waste, pollutants, including industrial chemicals, as well as excess herbicides, weed killers, and so forth, in a process called **bioremediation**. More recently, bacteria have been designed for use as biosensors or even to kill pathogenic bacteria.

In addition, microorganisms are used to manufacture a variety of products including alcohol, solvents, biofuels, food additives, dyes, and antibiotics. The most efficient pathways are those that convert otherwise useless material into useful products. We start by considering one such scheme: alcohol fermentation. This process was developed long before modern science and is probably mankind's earliest venture into biotechnology.

Genetic engineering can be used to assemble novel or more efficient metabolic pathways. Both degradative and biosynthetic pathways may be engineered.

ETHANOL, ELEPHANTS, AND PATHWAY ENGINEERING

Humans weren't the first to appreciate alcohol. Elephants, monkeys, and other wild animals deliberately consume fruit that has naturally fermented, yielding alcohol. Indeed, elephants in both Africa and Asia may run amok after consuming fermented fruit. Occasionally, elephants will even raid local villages and knock over houses to "recover" fermented liquids from their human competitors! There is even a species of the fruit fly *Drosophila*, from the sherry-producing regions of Spain, that relies on sherry as its sole source of nutrition. These insects spend their lives circling around in the caves where sherry is processed and presumably do not need to fly straight.

The earliest cultural remains from human alcohol consumption date to about 9000 years ago. Analysis of a residue found in Neolithic pottery from Jiahu in northern China showed that it was derived from wine. Similar residues on pottery, dating to about 7000 years ago, were found in Iran.

Alcohol is made from sugar (Fig. 13.1). Sugars are components of the carbohydrates making up much of the bulk of plant matter. So, in principle, alcohol can be made from almost any plant-derived material. Yeast is used to ferment sugars derived from grain or grapes, which produces an alcoholic liquid—the basis of beer or wine, respectively. Distillation is then used to make concentrated liquors such as whiskey or vodka. The lone exception to yeast is the use in Mexico of a bacterium, *Zymomonas*, which ferments sugar from the sap of the agave plant to give a liquid known as pulque. Distillation converts this into tequila.

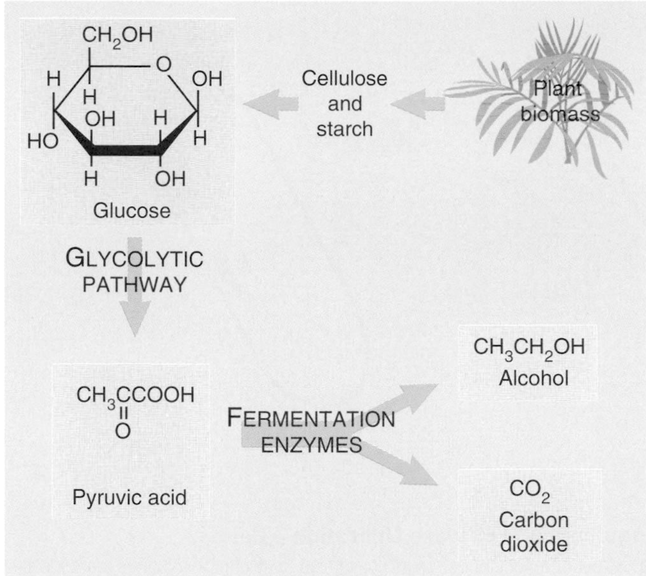

FIGURE 13.1 From Sugars to Ethanol
Plant material contains polysaccharides such as starch and cellulose. Enzymes degrade the starch and release the glucose molecules. The glycolytic pathway then converts glucose to pyruvic acid, which is fermented into alcohol and carbon dioxide.

There is little need for genetic engineering in the area of alcoholic drinks. However, alcohol may be blended with gasoline to give "gasohol," which works well in most internal combustion engines. Conversion of waste biomass to fuel alcohol would not only get rid of large amounts of waste but would also reduce gasoline consumption. If the United States converted the 100 million tons of waste paper it generates each year into fuel-grade alcohol, this could replace 15% of the gasoline used. Ethanol can also be made from corn, which is very economical because many acres of corn are grown each year in the United States, and a large surplus is generated. Unlike wood pulp for paper, corn can be regenerated in 1 year.

The advantage of using *Zymomonas* and yeast is that they make only alcohol during fermentation, whereas most microorganisms generate mixtures of fermentation products. For example, *Escherichia coli* makes a mixture of ethanol, acetate, succinate, lactate, and formate. Although many fermentation products are potentially useful, purification is an expensive drawback. The problem with *Zymomonas* is that it lives entirely on glucose and lacks the enzymes to break down other sugars, let alone those needed to degrade carbohydrate polymers such as starch and cellulose. Yeast is almost as narrow in its growth requirements. *Zymomonas* grows faster than yeast and makes alcohol faster as well. On the other hand, yeasts are more alcohol resistant and are therefore capable of accumulating higher concentrations of ethanol in the medium before growth is halted.

Genetic engineering is being used to make improved strains of both yeast and *Zymomonas* that can use a wider range of sugars. In addition, genes for enzymes capable of breaking down starch, cellulose, or other plant polysaccharides can be inserted (see later discussion). Finally, these organisms can also be engineered for improved resistance to alcohol or for other properties that optimize growth and production under industrial conditions.

Xylose is a five-carbon sugar that is a major component of various polysaccharides (xylans) found in plant cell walls (see later discussion). Vast amounts of waste material from plants are available for possible biodegradation. Breakdown of the polysaccharide polymers would release large amounts of xylose. Consequently, it is worthwhile to develop strains of *Zymomonas* that efficiently ferment xylose to ethanol. This has been done in two stages. First, the genes for metabolism of xylose itself must be introduced, because *Zymomonas* does not naturally use this sugar. The *xylA* and *xylB* genes encode the enzymes xylose isomerase and xylulose kinase, respectively, which convert xylose to xylulose and then to xylulose 5-phosphate. These two genes were placed on a shuttle vector that carries replication origins for both *E. coli,* in which the genetic engineering was done, and *Zymomonas* (Fig. 13.2).

The strain with just the extra *xylAB* genes grew poorly because it accumulated xylulose 5-phosphate as well as the phosphates of other pentose sugars, including ribose 5-phosphate. The genes for transketolase *(tktA)* and transaldolase (*tal*), two enzymes that convert pentose phosphates back into hexose phosphates, were then included on the plasmid, under control

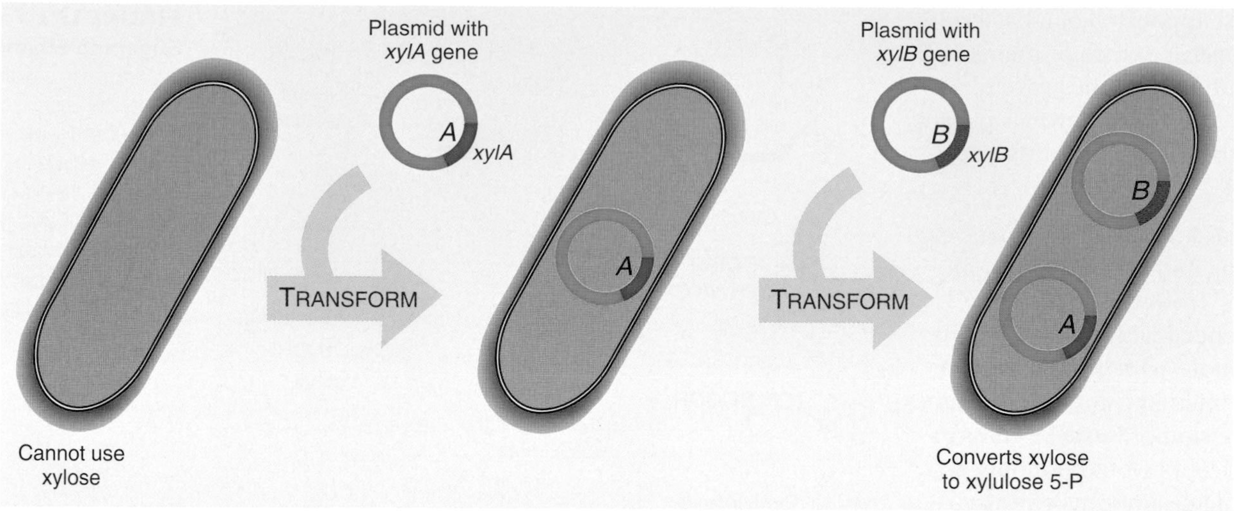

FIGURE 13.2 Pathway Engineering for Xylose Utilization—Genes
Xylose must be degraded by a specific set of reactions before its conversion to alcohol. Two genes are necessary for the initial xylose degradation: *xylA* and *xylB*. The XylA protein converts xylose to xylulose, and XylB phosphorylates this to form xylulose 5-phosphate. The two genes are carried on shuttle plasmids and transformed into bacteria such as *Zymomonas*.

FIGURE 13.3 Pathway Engineering for Xylose Utilization—Reactions
(A) Xylose is converted into xylulose 5-phosphate using the *xylA* and *xylB* genes. (B) Transketolase *(tktA)* converts two five-carbon (C_5) sugar molecules into one three-carbon (C_3) and one seven-carbon sugar (C_7). Next, transaldolase (*tal*) convert these products into a four-carbon sugar and fructose 6-P, a six-carbon sugar. Fructose 6-P is degraded by glycolysis into ethanol. The four-carbon sugar (C_4) and another pentose 5-P (C_5) are converted by transketolase into a second six-carbon sugar (C_6) and a three-carbon sugar (C_3), which both feed into the glycolytic pathway to make ethanol.

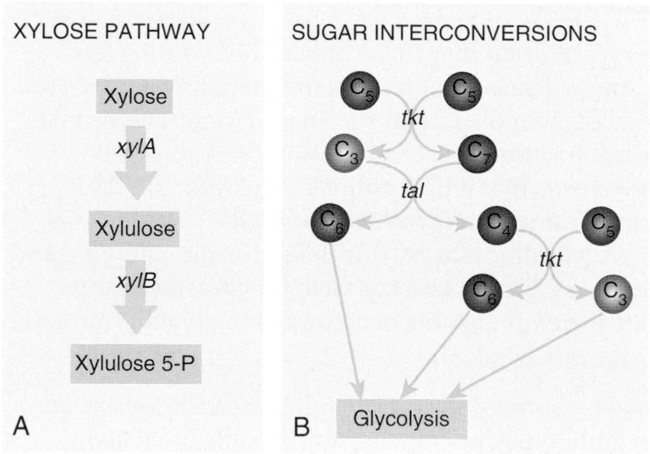

of a separate promoter. The resulting *Zymomonas* was then able to convert xylose first to xylulose 5-phosphate and then to fructose 6-phosphate and glyceraldehyde 3-phosphate. Finally, these central intermediates were fermented efficiently to ethanol (Fig. 13.3).

Not surprisingly, engineered strains of yeast that ferment 5-carbon sugars by the pathway shown in Figure 13.3 are also being developed. In addition, mutations that protect against toxic breakdown products from the lignin component of plant cell walls are being inserted. An alternative to such highly engineered individual strains is to use mixed cultures in which two or more microorganisms carry out separate stages of the fermentation. A major problem with mixed cultures is that less robust microorganisms may decline in relative numbers and even be lost.

Conversion of sugars to alcohol is one of the oldest industrial processes. Pathway engineering may help convert waste plant-derived material, including paper, into fuel alcohol.

DEGRADATION OF STARCH

Starch is a storage polysaccharide found in many plants. It is a polymer of glucose linked by α-1,4 bonds. Starch actually consists of a mixture of linear polymers—**amylose**—and branched polymers—**amylopectin**. The branches of amylopectin are due to α-1,6 bonds, and they occur approximately every 20 glucose residues along the polymer chain. Chain

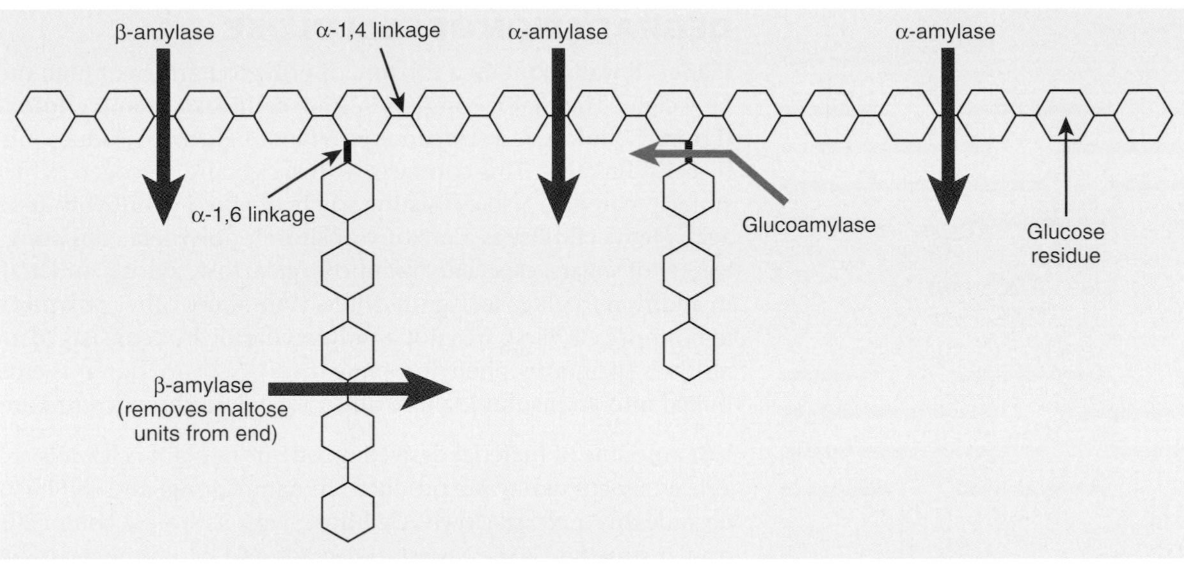

FIGURE 13.4 Enzymatic Breakdown of Starch by Amylases
Starch consists of long chains of glucose residues with other glucose chains branching off the main backbone. The main chain has glucose residues linked by alpha-1,4 linkages, and the side chain starts with an alpha-1,6 linkage. Amylase cuts between the glucose residues. Alpha-amylase cuts the alpha-1,4 linkages in the main chain, whereas beta-amylase cuts units of two glucose residues (maltose) from the ends of the chains. Glucoamylase cuts the alpha-1,6 linkage to remove the side branches from the main glucose chain.

lengths vary from 100 to 500,000 glucose residues for amylose and up to 40 million for amylopectin. The proportions of linear and branched polymers also vary depending on the source of the starch. **Glycogen** is a storage polysaccharide of animals. It is essentially starch with a high proportion of branched polymers.

Starch is used in the food and brewing industries and is mostly converted to glucose by using the purified enzymes α-amylase and glucoamylase, rather than by microorganisms (Fig. 13.4). α-Amylase cleaves the linear regions of the starch chains at random, whereas glucoamylase cuts the branches. The glucose may then be converted to fructose by the enzyme glucose isomerase or to alcohol by microbial fermentation. Because of the large size of the food and brewing industries, α-amylase and glucoamylase plus glucose isomerase account for over 25% of the cost of all enzymes used industrially. Many microorganisms make these enzymes. However, industrially, α-amylase is usually obtained from the bacterium *Bacillus amyloliquefaciens* and glucoamylase from the fungus *Aspergillus niger*.

A variety of improvements to the process of starch degradation might be made by genetic engineering. Recombinant organisms could be made that produce more enzyme. Furthermore, the enzymes themselves could be engineered for better thermal stability or higher rates of reaction as discussed in Chapter 11.

The gene for glucoamylase has been cloned from *Aspergillus niger* and inserted into a suitable yeast strain. The fungal gene was placed under control of a strong yeast promoter and carried on a yeast plasmid. The engineered yeast was able to degrade solubilized starch and ferment the glucose released to alcohol. Ultimately, it may be possible to engineer yeast strains that also express (and secrete) high enough levels of α-amylase to completely convert raw starch to ethanol.

Starch is a widely available raw material. Higher enzyme levels and more stable enzymes would improve the breakdown of starch to sugars.

DEGRADATION OF CELLULOSE

Plant cell walls contain a mixture of polysaccharides of high molecular weight. The major components are **cellulose**, **hemicellulose**, and **lignin**. Cellulose is a structural polymer of glucose residues joined by β-1,4 linkages. This contrasts with starch and glycogen, which are storage materials also consisting solely of glucose, but with α-1,4 linkages. Hemicellulose is a mixture of shorter polymers consisting of a variety of sugars, especially mannose, galactose, xylose, and arabinose, in addition to glucose. Lignin differs from these other polymers in two major respects. First, it is not a polysaccharide but consists of aromatic residues (primarily phenylpropane rings). Second, lignin is cross-linked into an insoluble three-dimensional meshwork structure.

Vast amounts of material derived mostly from plant cell walls are available as agricultural waste products. In nature, fungi and soil bacteria degrade this material slowly. Cellulose, with its simple composition and regular structure, is the easiest to degrade and lignin the most difficult. Paper (plus cardboard and related materials) accounts for the largest fraction of the trash of industrial nations. Because paper consists almost entirely of cellulose, this might potentially be converted to glucose by cellulose-degrading microorganisms.

The polymer chains of cellulose are packed tightly side by side in a crystalline array with loosely packed, noncrystalline zones at intervals (Fig. 13.5). The challenge is to break down cellulose, yielding glucose that can be turned into alcohol or other products. Breakdown of cellulose requires several steps (Fig. 13.5), each catalyzed by a separate enzyme, as follows:

1. Endoglucanase snips open the polymer chains in the middle. This enzyme can attack only the polymer chains in the loosely packed "amorphous" zones.
2. Cellobiohydrolase cuts off molecules with 10 or more glucose units from the free ends created by endoglucanase.
3. Exoglucanase chops off units of two or three glucose units from the exposed ends, which are called cellobiose and cellotriose, respectively.
4. β-Glucosidase (also known as cellobiase) converts cellobiose and cellotriose to glucose.

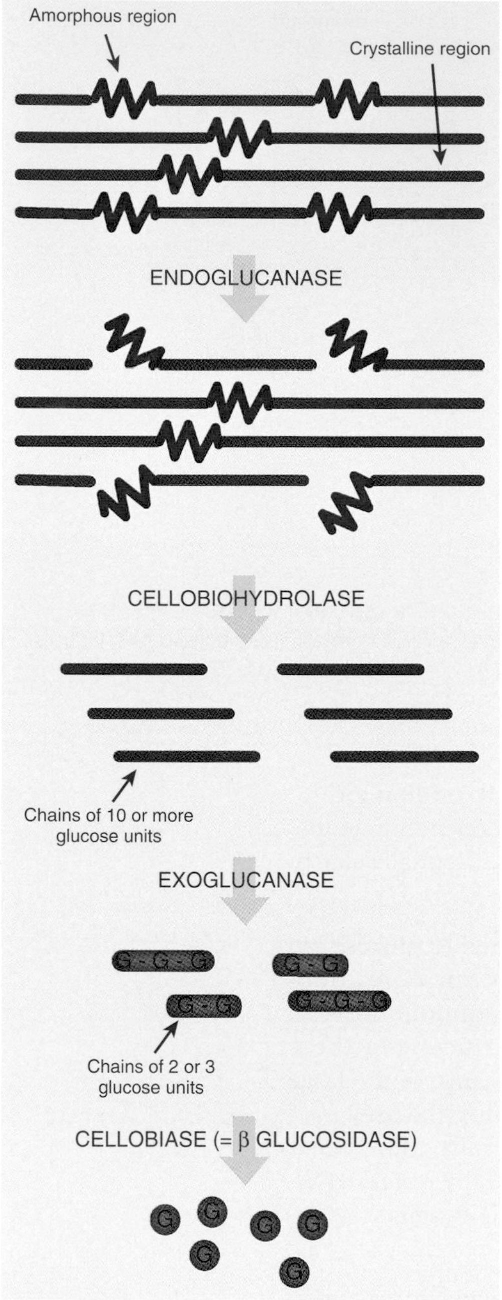

FIGURE 13.5 Cellulose Degradation
Cellulose is broken down into glucose molecules by a series of enzymatic reactions. Cellulose has both amorphous and crystalline regions. Endoglucanase digests within the amorphous areas to begin the degradation. Each of the subsequent enzymes shortens the glucose chains further. See text for details.

The genes for each of these four enzymes have been cloned from various microorganisms. Because cellulose is too big to enter the cell, the first three enzymes must be secreted and work outside. Cellobiose and cellotriose (respectively, the β-1,4–linked dimer and trimer of glucose) are released from cellulose outside the cell and may then be transported inside. They are finally broken into individual glucose molecules, which may be fermented to alcohol. So far, assorted pilot projects have demonstrated the degradation of cellulose from waste paper to glucose by adding separate enzymes purified from different sources. Alternatively, a series of cellulose-degrading microorganisms, each chosen for high levels of one particular enzyme, are used. Finally, yeast or *Zymomonas* is added to convert the glucose to alcohol.

Such multistage procedures are inefficient. Cellobiose acts as a feedback inhibitor of cellulose degradation, and similarly, glucose inhibits hydrolysis of cellobiose. Therefore, the end products of cellulose breakdown must be rapidly removed to allow continuous degradation of the starting materials. Cellobiose degradation is the limiting step in many natural cellulose degraders.

Thus, cloning the gene for cellobiase, placing it under a strong promoter, and putting it back into the same organism could improve degradation.

Overall, what is desirable is a complete recombinant organism, which possesses genes for all four enzymes, expresses them at high levels, and efficiently converts cellulose to alcohol on its own. In practice, genes for cellulose degradation from both bacteria and fungi are first cloned and expressed in *E. coli* for ease of manipulation. At present, genes for the individual stages have been isolated and characterized. Eventually, a cellulose degradation pathway may be assembled in either yeast or *Zymomonas* in order to convert waste cellulose materials to alcohol.

Cellulose is relatively difficult to metabolize, compared to sugars or starch. Assembling an effective pathway in an easily cultivated organism would be of great value in biodegrading waste paper and related material.

SECOND-GENERATION BIOFUELS

At present, biofuels are made from sugar cane, corn, or vegetable oil, all materials that could be used for food production ("first-generation biofuels"). Obviously, it would be a great improvement to use lignocellulose, the mixture of cellulose, hemicellulose, and lignin that makes up plant cell walls and is presently largely wasted. The term *second-generation biofuels* refers to biofuels derived from waste material. Table 13.1 shows the present situation and proposed future biofuels. Figure 13.6 shows the advantages and disadvantages of possible sources for future biofuel development.

Present biofuels use materials that could be used for food production. Proposed future biofuels—second-generation biofuels—would instead use waste materials.

BIODIESEL

Diesel fuel from petroleum (petrodiesel) consists of about 75% saturated hydrocarbons, including branched isomers, plus 25% aromatic hydrocarbons. The majority of molecules range in size from 12 to 15 carbons. Biodiesel is derived from plant oils and animal fats. Treating oils and fats with methanol and sodium hydroxide converts them to fatty acid methyl esters (FAME = biodiesel) plus glycerol and water (waste products). The waste materials must be removed before use. Biodiesel is usually mixed with petrodiesel but occasionally is used alone in engines specially designed for such use.

Adding biodiesel to petrodiesel reduces the levels of carbon monoxide, particulates, hydrocarbons, and sulfur oxides in the exhaust. The European Union is the largest

Table 13.1 First- and Second-Generation Biofuels

Petroleum (2010 Data)		Biofuels	
Application	Percentage	First Generation	Second Generation
Diesel	22%	Biodiesel	Biodiesel
Jet Fuel	9%	—	Biokerosene, sesquiterpenes
Gasoline	42%	Ethanol, biogas (methane)	Ethanol, other alcohols, hydrogen
Other	27%	—	—

FIGURE 13.6 Future of Biofuels
Four approaches to engineering next-generation biofuel producers are shown. An ideal biofuel producer should give high yields of fuel but grow on a cheap renewable source of biomass. (A) Engineer both traits into model organisms. (B) Use native biofuel producers and engineer ability to use biomass for growth. (C) Use organisms that are naturally capable of growth on renewable biomass and engineer biofuel pathways. (D) Use autotrophic organisms that naturally use CO_2 as a carbon source and engineer biofuel pathways. Modified after Gronenberg LS, Marcheschi RJ, Liao JC (2013). Next generation biofuel engineering in prokaryotes. *Curr Opin Chem Biol* **17**, 462–471.

MODEL ORGANISMS

Advantages:
- Well-established genetic tools
- Significant knowledge of metabolic networks
- Higher tolerance to fuels

Challenges
- Feedstock limited to simple sugars
- Two pathways must be introduced

A

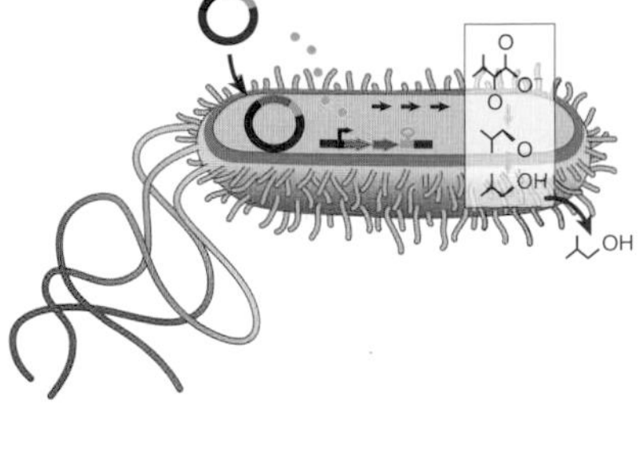

NATIVE BIOFUEL PRODUCERS

Advantages:
- Genes for fuel production are present

Challenges
- Genetic tools not well-established
- Feedstock often limited to simple sugars
- Low tolerance to fuel molecules

B

CELLULOLYTIC ORGANISMS AND THERMOPHILES

Advantages:
- Ability to degrade biomass
- High temperatures advantageous for biomass degradation and isolation of biofuel

Challenges
- Genetic tools not well-established
- Fuel production pathways must be introduced

C

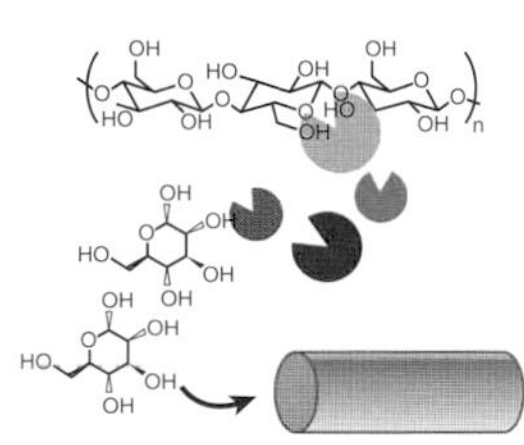

AUTOTROPHIC ORGANISMS

Advantages:
- CO_2 required as sole carbon source

Challenges
- Genetic tools not well-established
- Fuel production pathways must be introduced
- Engineering challenge of scale-up

D

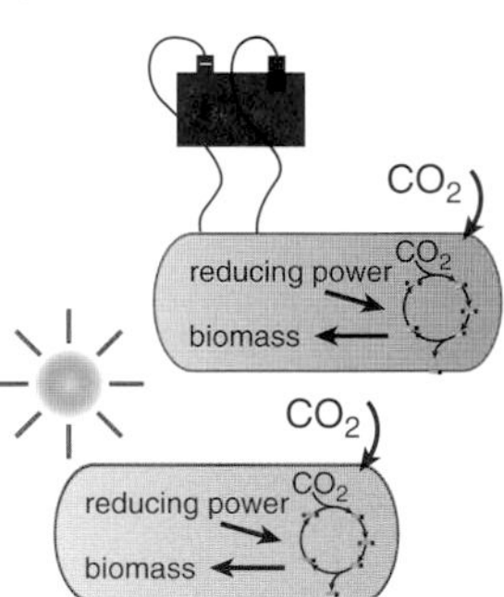

biodiesel producer (53% of global production in 2010). Today, biodiesel is made mostly from plant oils that could also be used as foodstuffs (e.g., soybean, sunflower, palm oils). Clearly, using nonfood oils would be a major advance. Several possibilities exist for second-generation biodiesel, such as using oils from nonedible plants that grow on wasteland or using engineered algae or bacteria to make high levels of fatty acids.

The much-used bacterium *Escherichia coli* can be engineered to greatly increase the production of free fatty acids. Further modification, using genes from other bacteria, can result in engineered *E. coli* that convert the free fatty acids to generate alkanes, which are superior to methyl esters as biodiesel. The first stage consists of modifying the existing genes of *E. coli* to decrease fatty acid degradation and decrease the incorporation of fatty acids into membrane lipids (Fig. 13.7). Increasing the flux of precursors into fatty acid synthesis and improving the release of free fatty acids from carriers also help.

The second stage of engineering consists of introducing two sets of foreign genes that introduce branches into the hydrocarbon chain and that convert the fatty acids into alkanes (Fig. 13.8). Although *E. coli* and related Gram-negative bacteria do not make branched chain

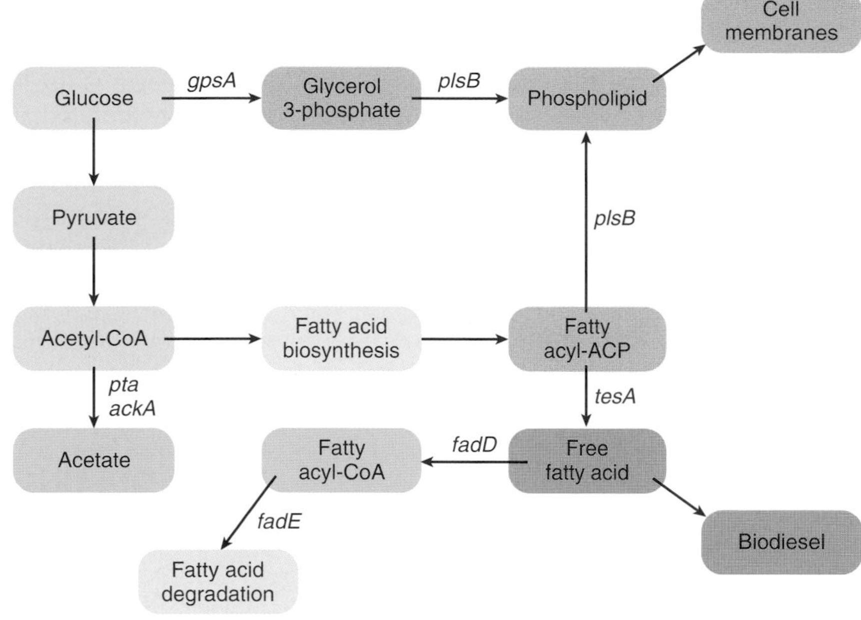

FIGURE 13.7 Increasing Free Fatty Acid Production by *E. coli*

The objective is to generate surplus free fatty acids (brown) for conversion to biodiesel. Biosynthetic pathways are shown in blue (for fatty acids) and green (for phospholipids). Selected genes involved in degradative pathways (*pta, ackA, fadD, fadD*; purple) are inactivated. Genes involved in the synthesis of phospholipids (*gpsA*, *plsB*; green) cannot be wholly inactivated. However, their level of expression is decreased to divert more fatty acids to biodiesel. Conversely, the level of expression of the *tesA* gene that encodes the thioesterase that releases free fatty acids from ACP is increased.

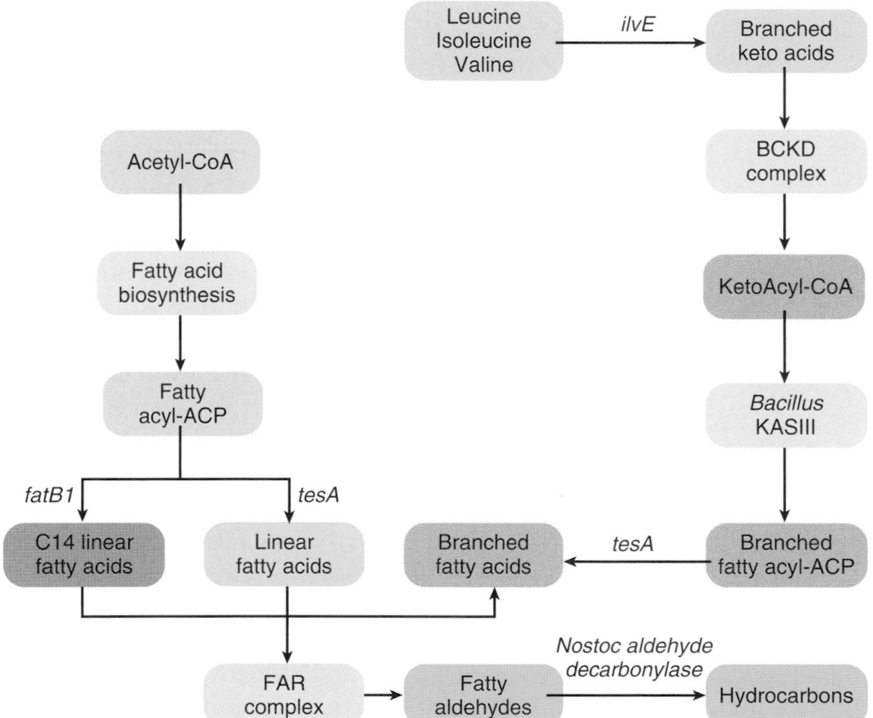

FIGURE 13.8 Converting Fatty Acids to Branched Hydrocarbons

Free fatty acids are converted to a mixture of linear and branched hydrocarbons (a superior biodiesel to methyl esters). Biosynthetic pathways shown in blue are native to *E. coli*. Those in gray are from *Bacillus subtilis* and are needed for forming branched fatty acids. BCKD = branched ketoacid dehydrogenase and KASIII = ketoacyl synthase III. The *fatB1* gene from *Cinnamomum camphora* (camphor) encodes a thioesterase specific for C14 (brown). Fatty acids are converted to aldehydes by the fatty acid reductase (FAR) complex from the luciferase system of *Photorhabdus luminescens* (gray). The final conversion of aldehydes to hydrocarbons utilizes the aldehyde decarbonylase from *Nostoc punctiforme* (purple).

fatty acids, many Gram-positive bacteria, including *Bacillus subtilis*, do so. The branched amino acids leucine, isoleucine, and valine have their amino groups removed by IlvE (of *E. coli*) and are then processed to precursors for fatty acids by the BCKD complex from *Bacillus subtilis*. Elongation of fatty acids requires several enzymes including ketoacyl synthase. Because the *E. coli* enzyme cannot handle branched chains, the *Bacillus subtilis* enzyme must be provided. Inserting the FatB1 thioesterase from a plant (camphor) increases the proportion of shorter fatty acid chains. Finally, the fatty acid reductase (FAR) complex from

Photorhabdus luminescens converts fatty acids to long-chain aldehydes, and the aldehyde decarbonylase from *Nostoc punctiforme* converts the aldehydes to hydrocarbons.

The main difference between gasoline and diesel is the chain length of the hydrocarbon mixture. In gasoline, the range is from 4 to 12 carbons, considerably shorter than in diesel. Alcohols have been used as a gasoline replacement (gasohol; see earlier discussion) but are not as efficient as petroleum-derived gasoline. A more realistic biological replacement for gasoline would require bacteria to generate shorter alkanes. Attempts to do this are in progress and resemble those for biodiesel, except that genes and enzymes specific for shorter chain lengths are used.

Presently, synthesis of biodiesel uses edible plant oils that could be used for food production. Engineering of bacteria may allow the conversion of waste materials to biodiesel.

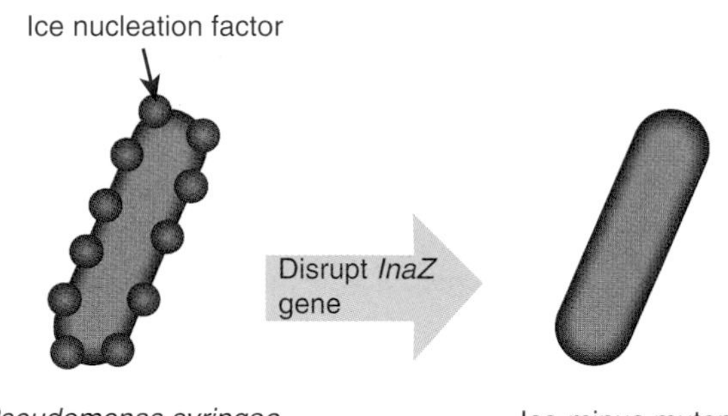

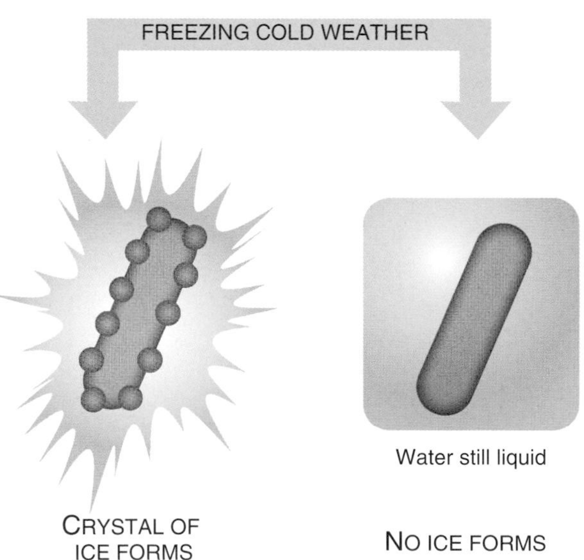

FIGURE 13.9 Disruption of *inaZ* Prevents Ice Nucleation
Cell surface proteins of *P. syringae* provide a nucleation point for ice. The *inaZ* gene encodes an ice-nucleating protein. Under freezing temperatures, wild-type *P. syringae* allow ice crystals to form, disrupting any plant tissues the bacteria are on or within. If the *inaZ* gene is disrupted, the *P. syringae* mutant will not nucleate any ice crystals, allowing the water to supercool.

ICE-FORMING BACTERIA AND FROST

Perhaps the simplest "pathway" of all is the conversion of water to ice. This process may be "catalyzed" by proteins known as **ice nucleation factors**. On a microscopic scale, solidifying water forms crystals of ice. However, ice crystals need a microscopic nucleus, or "seed," to form around. In the absence of structures allowing nucleation, water will supercool down to −8 °C without solidifying. Ice nucleation factors are specialized proteins, mostly found in certain bacteria, which provide nuclei for crystallization.

Each year, frost causes more than a billion dollars in damage to crops in the United States alone. It is not the low temperature itself that does the damage. When water freezes to form ice, it expands, damaging plant tissues. The seeding of ice crystals on and within plants is mostly due to proteins on the surface of bacteria, especially *Pseudomonas syringae* and related species, which live on plants. The ice crystals that form damage the plant tissues and disrupt the vessels (xylem and phloem) that carry water and nutrients throughout the plant. If ice-nucleating bacteria are absent, ice fails to form and instead the water supercools, leaving the plants unharmed.

The ***inaZ* gene** of *Pseudomonas syringae* encodes the best-known ice nucleation protein. Like most bacteria, *E. coli* does not normally promote ice formation, although if it expresses a cloned *inaZ* gene, it will gain ice-nucleating ability. Conversely, when the *inaZ* gene of *Pseudomonas syringae* is disrupted, ice-nucleating ability is lost (Fig. 13.9). The wild, "ice-plus" strains of *Pseudomonas syringae* can be displaced by spraying the "ice-minus" mutants onto crops that are at risk from frost damage. Subsequently, even if the temperature falls below freezing, very few ice crystals form and most of the plants are unharmed.

Bacterial ice nucleation proteins are anchored in the outer membrane by their N-terminal domain and project outward. The central part of the protein consists of repeating domains that act as the templates for the formation of ice crystals. These proteins can be used as platforms to display other proteins on the outside of bacterial cells (Fig. 13.10). The ice nucleation proteins may be engineered by removing some of their domains, depending on the length of tether required.

One application is in the degradation of lignocellulose (see earlier discussion). Enzymes from cellulose-degrading bacteria have been displayed externally on both *E. coli* and *Zymobacter* (an ethanol producer) and shown to function correctly. Another, quite different use is to bind virus particles in water or other fluids. This can be used both for detecting virus contamination and removing the viruses. This was demonstrated by attaching the human poliovirus receptor to a shortened ice nucleation protein that was expressed on the surface of *E. coli*. In both of these cases, truncated ice nucleation proteins, consisting only of the N-terminal domain, were used (Fig. 13.10).

Bacteria that express ice nucleation proteins cause major damage to crop plants. Disruption of the corresponding gene abolishes ice nucleation. The ice nucleation protein has been used to tether other proteins to the outside of bacterial cells.

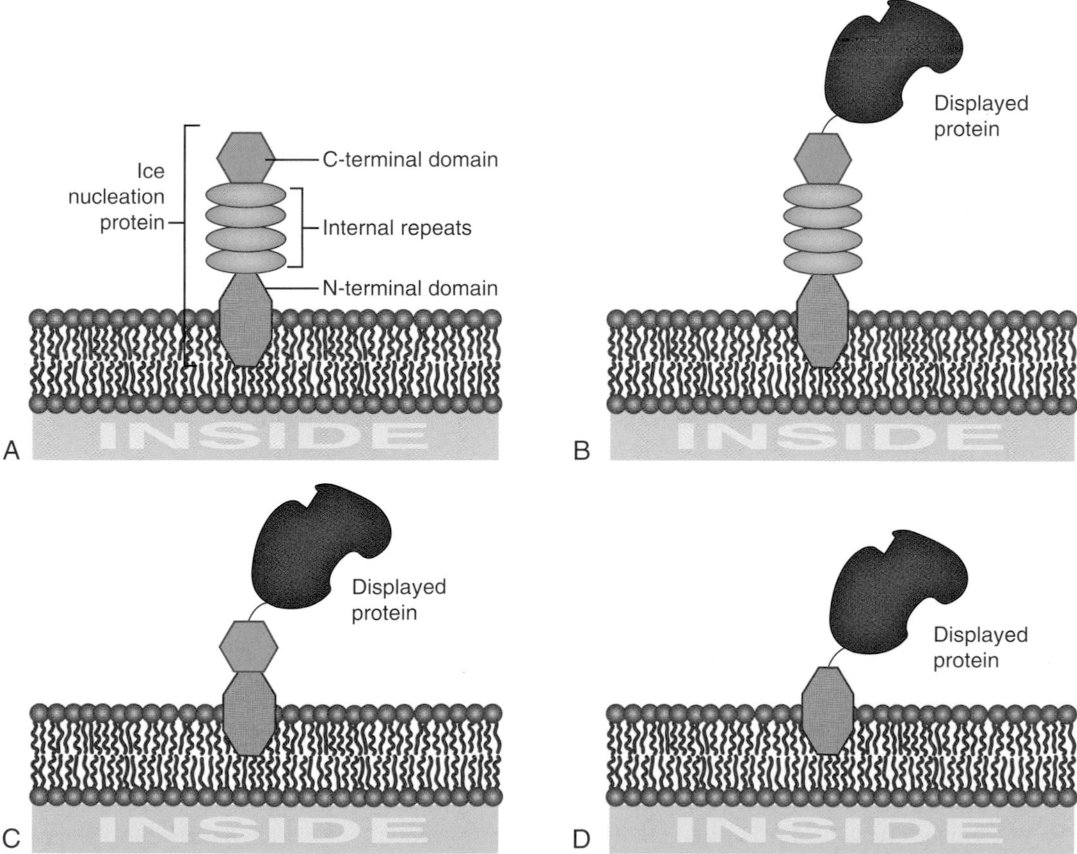

FIGURE 13.10 Cell Surface Display Using InaZ Protein as Anchor
The ice nucleation protein, InaZ, has been used to display other proteins on the surface of bacteria such as *E. coli.* (A) Natural InaZ protein showing arrangement on cell surface. The N-terminal domain is responsible for membrane binding, and the internal repeats promote ice formation. (B) Displayed protein attached to C-terminus of full length InaZ. (C) The InaZ carrier has been shortened by deletion of the internal repeats. (D) Displayed protein attached to N-terminal domain InaZ. Modified after van Bloois E et al. (2011). Decorating microbes: surface display of proteins on *Escherichia coli. Trends Biotechnol* **29**, 79–86.)

BIOREFINING OF FOSSIL FUELS

The growth of industrial civilization, in particular the use of fossil fuels for energy and the development of the organic chemical industry, has led to the pollution of the environment with a wide range of compounds of nonbiological origin. Many fossil fuel deposits, of both coal and oil, contain a high percentage of sulfur—up to 5% sulfur for many coals from eastern Europe or the American Midwest. Burning high-sulfur coal releases large quantities of sulfur dioxide into the atmosphere, which leads to the formation of acid rain. Among the possible solutions to this problem is to develop bacteria capable of removing the offending sulfur compounds from the coal (or oil) before combustion.

Several naturally occurring sulfur bacteria such as *Thiobacillus* and *Sulfolobus* can convert pyrites (FeS_2), the major form of inorganic sulfur found in coal, into soluble sulfate that can be rinsed away. The crucial issue, therefore, is to remove the organic sulfur, especially that found in **thiophene** rings, which typically accounts for 70% or more of the organic sulfur (Fig. 13.11). Although compounds containing thiophene rings are almost never found in modern-day living organisms, they form a substantial part of the organic sulfur fraction of fossil fuels such as coal and oil. The major quinone of Archaebacteria such as *Sulfolobus* is caldariellaquinone, which contains a thiophene ring fused to a benzoquinone. Conceivably, the fused thiophenes of coal are the metabolic fossils of archaebacterial metabolism.

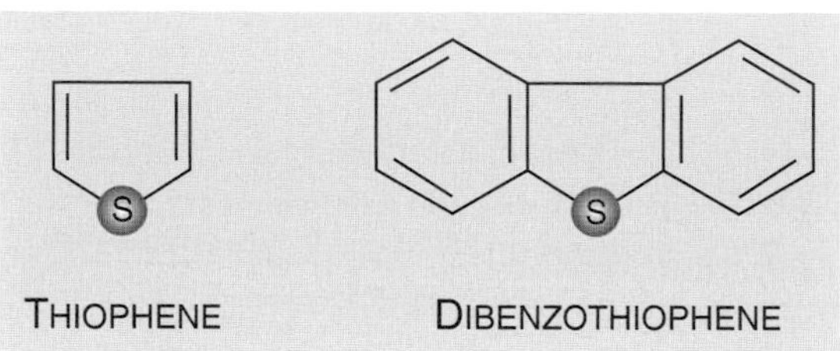

FIGURE 13.11
Thiophene and Dibenzothiophene
Thiophene and dibenzothiophene are the sulfur-containing molecules that account for much of the organic sulfur found in coal and fossil fuels.

Dibenzothiophene (DBT) is widely used as a model compound to represent the organic sulfur in coal and oil (Fig. 13.11). Biodegradation of DBT and removal of sulfur requires several steps, a scheme known as the **4S pathway** (Fig. 13.12). Most bacteria capable of degrading thiophene derivatives show only partial breakdown. Many do not completely remove the sulfur, and others only use DBT that is already oxidized to the sulfone or sulfoxide level. Certain bacteria, especially species of *Rhodococcus,* do indeed completely desulfurize dibenzothiophene, as well as degrade related heterocyclics such as dibenzofuran and xanthones. The ***dszABC*** **operon** of *Rhodococcus* is carried on a linear plasmid and encodes three enzymes responsible for the 4S pathway. In addition a flavin reductase encoded by the *dszD* gene is needed to supply reduced FMN (see Fig. 13.12). The *dszD* gene is not linked to the *dszABC* operon. The enzymes are as follows:

- Third step: DszA = dibenzothiophene sulfone oxygenase
- Fourth step: DszB = benzene sulfinate desulfinase
- First and second steps: DszC = dibenzothiophene oxygenase
- $FMNH_2$ supply: DszD = flavin reductase

However, there are several problems with these natural isolates:

(a) *Rhodococcus* is poorly characterized genetically, so modification is difficult.
(b) Desulfurizing coal or oil requires bacteria that can be grown easily to high density. Although not feeble, *Rhodococcus* is not especially robust.
(c) *Rhodococcus* normally uses its desulfurizing pathway to obtain sulfur (not to degrade benzothiophenes for carbon and energy). Consequently, the *dszABC* genes are expressed only at a low level because only small amounts of sulfur are needed for bacterial growth. Moreover, other sulfur compounds, both inorganic and organic, repress the operon. Thus, the inorganic sulfur present in most high-sulfur coal and oil would repress the *dszABC* genes. Furthermore, pregrowth of the bacteria in organic media with amino acids such as cysteine or methionine also represses these genes.

The *dszABC* gene cluster has therefore been cloned from *Rhodococcus* and placed onto suitable plasmids for expression in *E. coli* and certain robust *Pseudomonas* strains. The *dszABC* genes have been placed under the control of strong promoters that may be induced as required and are not repressed by sulfur compounds (Fig. 13.13). The resulting strains desulfurize DBT better than the original *Rhodococcus*.

High-level operation of the 4S pathway requires a large flow of reducing equivalents. In cells carrying a cloned *dszABC* operon, reduction of FMN by flavin reductase becomes the limiting factor in removal of sulfur from dibenzothiophene. However, flavin reductases from several other bacteria work as well as or better than the *Rhodococcus* DszD enzyme. For example, the HpaC enzyme from *E. coli* has been cloned and expressed at high levels, and it greatly speeds desulfurization. As a further modification, the *hpaC* gene and the *dszABC* genes have been joined together to form a single operon under control of the ***tac* promoter**. (This is a strong promoter that is induced by IPTG. It is a hybrid of the ribosome-binding site from the trp promoter with the operator of the lac promoter.) Thus, the combined desulfurization module can be induced by IPTG when required.

Despite adding oxygen to the sulfur of dibenzothiophene, the DszC enzyme is closely related to the ring dioxygenases that add two hydroxyl groups to aromatic rings. Thus, phenanthrene dioxygenase, which hydroxylates phenanthrene (a three-ringed aromatic hydrocarbon), can also convert DBT to its sulfone. Modification of biphenyl dioxygenase by gene shuffling (see Chapter 11) gives mutant enzymes capable of using dibenzothiophene and related compounds. In these cases, the same enzyme hydroxylates the aromatic rings and also adds oxygen to the thiophene sulfur, giving the sulfoxide and then the sulfone. The DszC enzyme itself has also been mutated to broaden its substrate range. For example, a Val261Phe mutation allows oxidation of methylbenzothiophene and alkyl thiophenes.

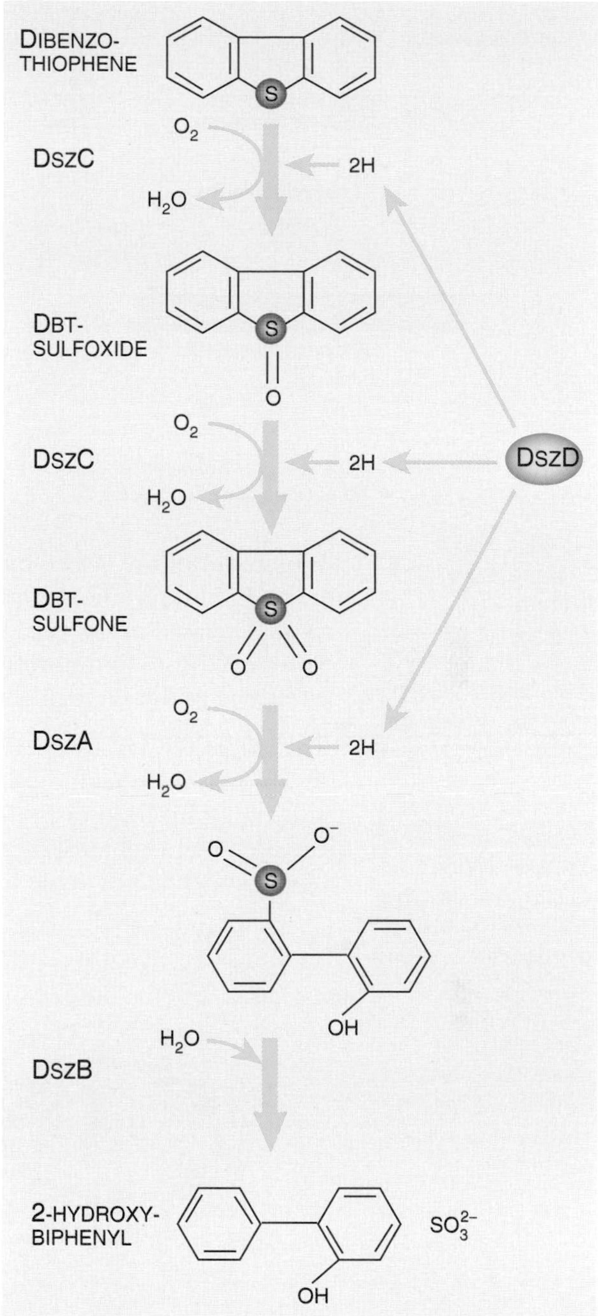

DIBENZO-THIOPHENE
DszC
O_2 H_2O 2H
DBT-SULFOXIDE
DszC
O_2 H_2O 2H
DszD
DBT-SULFONE
DszA
O_2 H_2O 2H
DszB
H_2O
2-HYDROXY-BIPHENYL
SO_3^{2-}

FIGURE 13.12 4S Pathway for Removal of Sulfur from Thiophene Rings
Dibenzothiophene is converted into a sulfoxide by the oxygenase DszC, aided by DszD, a flavin reductase, which supplies the reducing equivalents. The same two enzymes convert the sulfoxide into a sulfone. Next, DszA, another oxygenase, breaks the ring structure. Finally, DszB releases the sulfur as sulfite.

A major achievement of pathway engineering has been the assembly of a pathway to degrade the sulfur-containing thiophene rings often found in fossil fuels.

BIOSYNTHESIS OF β-LACTAM ANTIBIOTICS

Although we have covered the production of alcohol, we have really been considering degradative pathways. The fermentation schemes that produce alcohol are designed to release energy from sugar degradation. However, various natural products are made industrially that rely on genuine biosynthetic pathways. Examples include amino acids and sterols. Paradoxically, molecules of intermediate complexity, such as antibiotics, are the most difficult to tackle by genetic engineering. The reason is that their synthetic pathways may have 20 or more steps. Each step requires a separate enzyme, encoded by its own gene. In addition, many of these pathways are branched and/or interact with other metabolic pathways. Consequently, their regulation is often complex. Analyzing, cloning, and expressing all the genes that encode the enzymes and regulatory proteins for long and complex pathways require a great deal of effort.

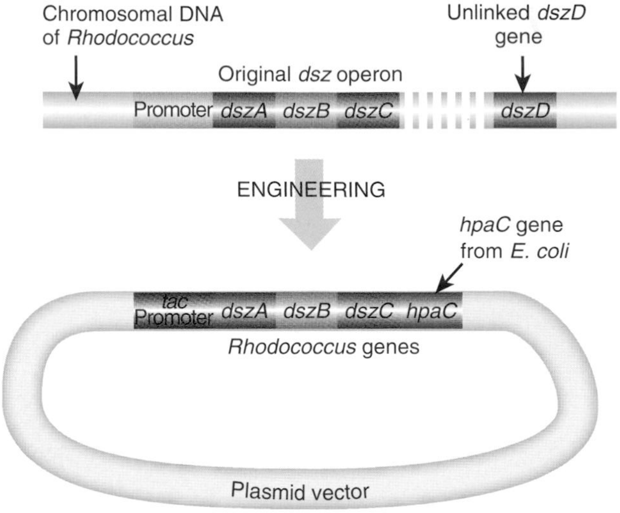

FIGURE 13.13 Engineering of Dsz System
The genes that encode the desulfurization pathway for dibenzothiophene have been cloned from *Rhodococcus* and moved onto a plasmid. The three enzymes DszC, DszA, and DszB need a flavin reductase to supply reducing equivalents. *Rhodococcus* uses the product of its *dszD* gene for this, but the *E. coli* flavin reductase (encoded by *hpaC*) works better. The *E. coli hpaC* gene is therefore overexpressed on the same plasmid. Transcription of the entire cluster of genes is controlled by the inducible *tac* promoter.

Alexander Fleming discovered the mold that makes **penicillin** in the 1920s. The story is a classic case of chance favoring the prepared mind. Fleming left Petri dishes containing bacterial cultures lying around for long enough to get moldy. He then noticed a clear zone, in which the bacteria had been killed and had disintegrated, around a blue mold of the *Penicillium* group. He found that the mold excreted a chemical toxic to bacteria but harmless to animals—penicillin. Fleming called the mold ***Penicillium notatum***. A related mold, *Penicillium chrysogenum*, makes a related antibiotic called **cephalosporin C**. Both antibiotics are members of the **β-lactam** family and are made by separate branches of the same biosynthetic pathway (Fig. 13.14).

The original β-lactams made by molds can be altered chemically to give many different antibiotics. Although cephalosporin C itself has only feeble antibacterial activity, it is the starting point for a vast array of broad-spectrum antibiotics made by chemical modification. First, cephalosporin C must be converted to 7-ACA (7-aminocephalosporanic acid), which is not made by any known organism. Originally, this was done chemically

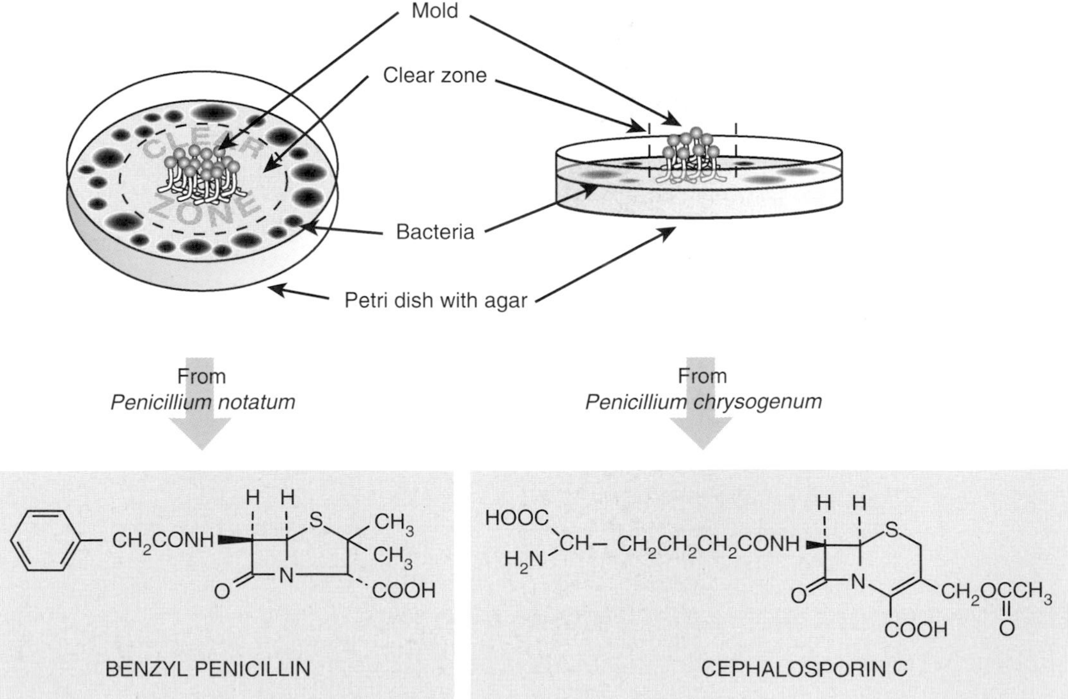

FIGURE 13.14 Biosynthesis of β-Lactam Antibiotics
When certain molds grow on agar originally covered with bacteria, a clear zone appears around the mold where no bacteria are able to grow. The clearing is due to release of antibiotics such as penicillin and cephalosporin C from the mold.

and gave very low yields. However, molds that make cephalosporin C can be engineered to convert this to 7-ACA. Two extra genes are inserted to create the extended pathway (Fig. 13.15). The gene for D-amino-acid oxidase comes from a fungus *(Fusarium solani)* and the cephalosporin acylase gene from a bacterium *(Pseudomonas diminuta)*.

The 7-ACA is used as base compound for a massive range of chemical modifications that provide antibiotics with different properties. Among these are variants that are resistant to bacterial β-lactamases, others that penetrate bacterial cell walls better, as well as antibiotics with better pharmacological properties (e.g., superior absorption from the intestine).

The emergence of antibiotic-resistant bacteria, coupled with the scarcity of newly developed antibiotics, comprises a looming crisis in health care. Only about 20 new antibiotics have been produced since the year 2000, and about 20–30 more are undergoing clinical trials. In particular, there is a shortage of new classes of antibiotics. Of the 20 new antibiotics, about half are based on natural products and half are totally synthetic.

Most of the new natural product derivatives are β-lactams, and of these the majority belong to the relatively novel group of carbapenems (Fig. 13.16). Thienamycin was the first carbapenem to be discovered and is the parent for most of the subsequent members of this class. Carbapenems are distinct in being resistant to cleavage by many β-lactamases and, in some cases, acting as β-lactamase inhibitors, as well as being effective broad-range antibacterial agents.

CEPHALOSPORIN C

D-amino-acid oxidase

H_2O_2

7-β-(5-carboxy-5-oxopentanamido)-cephalosporanic acid

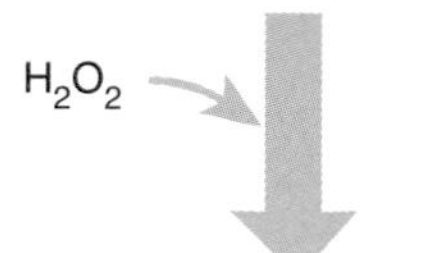

7-β-(4-carboxy-butanamido)-cephalosporanic acid

Cephalosporin acylase

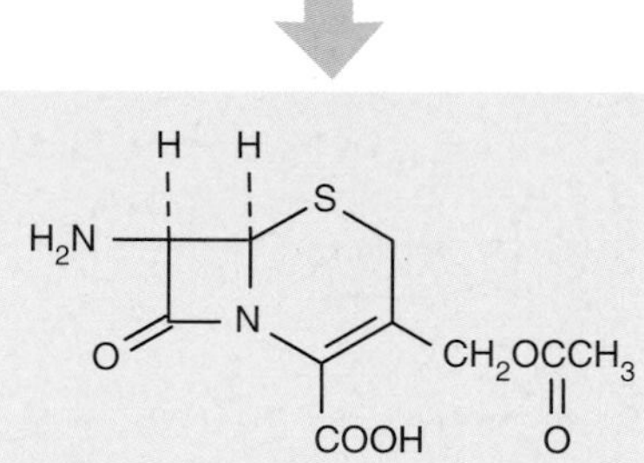

FIGURE 13.15
Engineered Pathway to 7-ACA
To engineer new antibacterial compounds, cephalosporin C must be converted into 7-aminocephalosporanic acid (7-ACA). The enzymes involved in this conversion are D-amino acid oxidase and cephalosporin acylase. The genes for these enzymes have been isolated, cloned, and expressed in different bacteria, as well as in molds producing cephalosporin C itself.

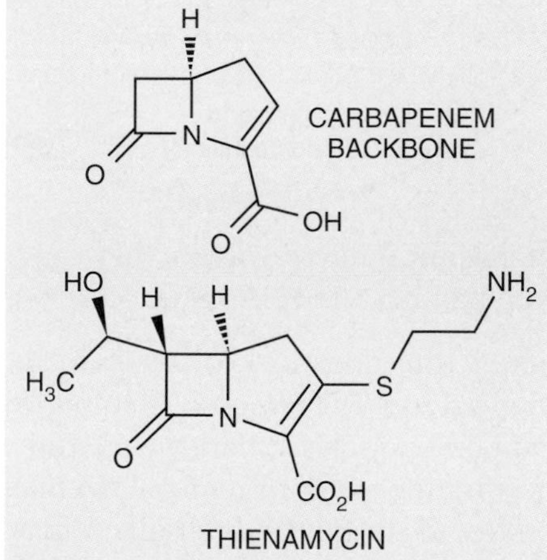

FIGURE 13.16
Carbapenem Backbone and Thienamycin
Note that carbapenems lack the sulfur in the five-membered ring that is characteristic of penicillins. However, carbapenem derivatives have sulfur attached to the ring as shown for thienamycin.

Most β-lactam antibiotics are the result of chemical modification. However, engineering the original β-lactam pathway of the *Penicillium* molds helps to increase the yield of some of these antibiotics.

BIOSYNTHETIC PLASTICS ARE ALSO BIODEGRADABLE

Plastics are polymers built from chains of monomer subunits, like proteins and nucleic acids. However, most plastics consist of the same monomer mindlessly repeated over and over again. Sometimes two or more closely related monomers may be mixed together and follow each other at random.

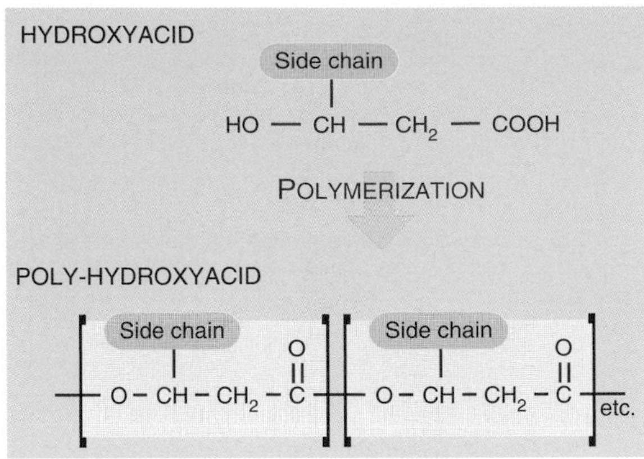

FIGURE 13.17 Structure of Polyhydroxyalkanoate Plastics
Plastics are long chains of repeating subunits. Polyhydroxyalkanoates have repeated hydroxyacid subunits linked through their carboxyl and hydroxyl groups. The various possible side groups alter the final characteristics of the plastic.

Certain bacteria make and store a group of related plastics known as **polyhydroxyalkanoates (PHAs)**. Their composition is shown in Figure 13.17. The bacteria accumulate PHAs when they have surplus carbon and energy but are running low on other essential nutrients, such as nitrogen or phosphorus. When conditions improve, the PHA is broken down and used as a source of energy. The most commonly found PHA has four-carbon (hydroxybutyrate, HB) subunits and is therefore called **polyhydroxybutyrate (PHB)**. However, a plastic made by randomly mixing in 10% to 20% of five-carbon (hydroxyvalerate, HV) subunits has much better physical properties. Still other PHAs containing a proportion of subunits that are eight-carbon or longer give materials that are more rubbery.

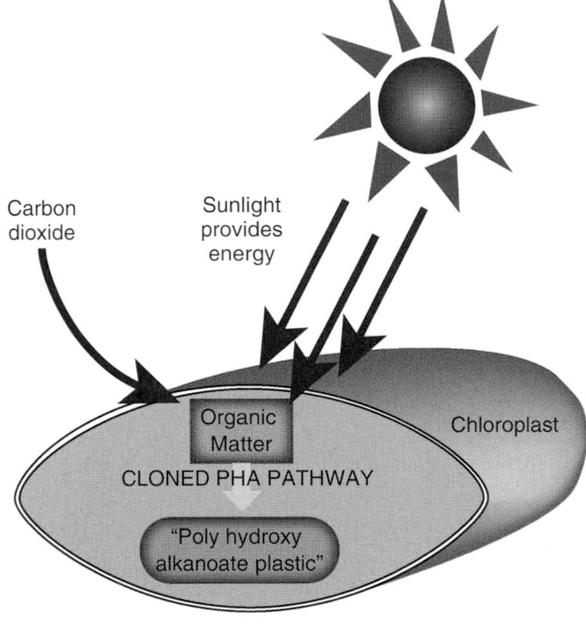

FIGURE 13.18 Expression of PHA Pathway in Plants
The PHA synthesis enzyme pathway has been cloned from *Alcaligenes eutrophus*. The genes were inserted into the chloroplast genome so that the enzymes are expressed only in the chloroplast. The enzymes use the newly synthesized organic matter from photosynthesis to create PHA.

A poly-HB/HV copolymer is manufactured by mutant bacteria of the species *Alcaligenes eutrophus* and is marketed by the Zeneca Corporation (United Kingdom) under the trade name of Biopol. It is more expensive than plastics made from oil, but it is completely biodegradable. Consequently, at present PHAs are restricted to specialized uses. For example, because they break down slowly inside the body to give natural, biochemical intermediates, they can be used for making slow-release capsules. Some bacteria can manufacture PHAs at up to 80% of their dry weight.

Future directions for bioplastics include using Gram-positive bacteria to make PHA for biomedical use. The reason is that the Gram-negative bacteria used at present (e.g., *Alcaligenes*) contain lipopolysaccharide (i.e., endotoxin) in their outer membranes and this requires extra purification steps for removal. To make PHAs economically competitive, they will need to be produced cheaply and in bulk. A major step in this direction would be to use low-cost waste material for bacterial growth.

Another approach is to insert the genes for their synthesis into suitable crop plants. This approach is still in the experimental stage, but the genes for making PHA from *Alcaligenes eutrophus* have been successfully inserted into *Arabidopsis,* a plant widely used for genetic experiments. The engineered PHA genes were designed for expression inside the chloroplasts of the plant (Fig. 13.18). Because chloroplasts are the sites of photosynthesis, newly synthesized organic matter appears here first. Locating the PHA pathway in chloroplasts, rather than in the main compartment of the plant cells, gives a 100-fold increase in PHA yield. The next step will be to move the pathway into a genuine crop plant such as rapeseed or soybeans. Perhaps this will give new meaning to the phrase "plastic flowers!"

Polyhydroxybutyrate (PHB) is a plastic polymer that can be made by bacteria or engineered plants. Related plastics may be engineered for desirable properties and, as a bonus, are biodegradable.

THE INTEGRATED CIRCUITS APPROACH

Engineering new pathways can be approached in a more general way that focuses on the arrangement of genes and promoters rather than specific proteins and metabolites. Scientists who focus on **genetic circuits** often use model genes rather than work with a particular pathway. The reason is partly that novel arrangements may develop new behavior. Thus, different arrangements of the same genes can give highly stable production of the final gene product, or the final product may oscillate at a particular frequency, or the product may be expressed only if a particular environmental signal is present, thus creating a **biosensory system**.

For example, we might desire a particular pollutant (the input signal) to cause a cell to fluoresce green (the output). This can be engineered by making the presence of the pollutant turn on the gene for green fluorescent protein (GFP). This in turn requires arranging a set of control elements such as activators and repressors to control whether or not GFP is made. Changing the arrangement of the control elements can make the cells more or less sensitive to the pollutant.

Some motifs used to regulate gene expression are shown in Figure 13.19. The first is a simple on/off dosage compensation motif. In our example, the pollutant may bind to the repressor protein that controls the GFP gene. If the pollutant is present, the repressor is released from the GFP gene and the cells fluoresce. Another type of circuit is the feed forward motif, where

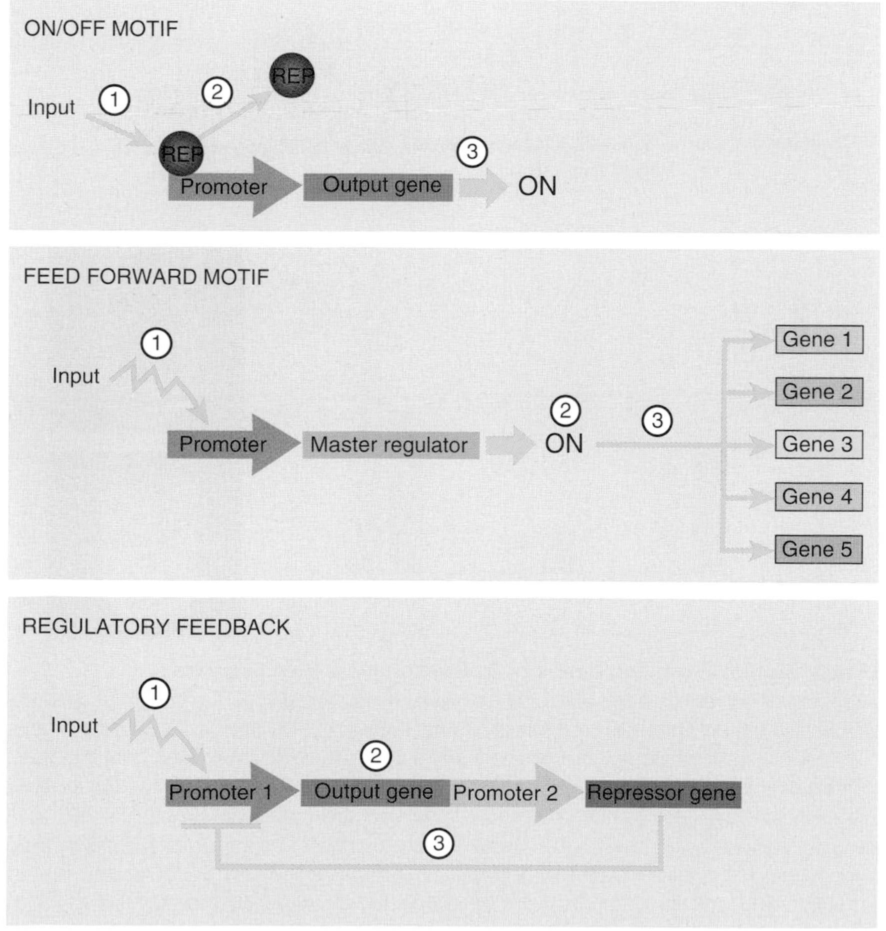

FIGURE 13.19 Schematic Representation of Different Genetic Circuits

A simple on/off motif (*top*) has a repressor protein that prevents the transcription and translation of the output gene. When the input signal is received, the repressor is released from the promoter and the output gene is made. In the feed forward motif (*middle*), a master regulator gene controls a variety of different genes (labeled 1–5). The input signal activates the promoter on the master regulator, which in turn activates genes 1 through 5. In the regulatory feedback motif (*bottom*), two genes are linked in tandem with separate promoters. Promoter 2 is expressed continually to make repressor protein. This prevents the expression of promoter 1 unless an input signal is received. Then the repressor is released, and the output gene is expressed.

one master gene regulates multiple pathways. A third motif involves regulatory feedback, where the final gene product represses the entire pathway. All of these motifs are found in nature and can be used in pathway engineering to control the production of the final protein.

Novel genetic circuits may be engineered into *E. coli*. For example, the system illustrated in Figure 13.20 caused *E. coli* to produce a biofilm in response to DNA damage. The genetic circuit includes the input, either UV light or mitomycin C that cause DNA damage, a regulatory circuit of two repressor genes (*lacI* and lambda *cI*), and an output module containing the biofilm gene *(traA)* controlled by a promoter that responds to the lambda cI repressor. The cells produce a biofilm only when DNA damage activates the RecA protein. RecA protein cleaves the cI repressor. This stops the regulatory circuit from repressing the output module, and therefore output—that is, biofilm—is made.

Integrated circuits have been built using genetic regulatory components, including repressors, activators, and DNA binding sites. Such circuits can respond to signals and act as sensors.

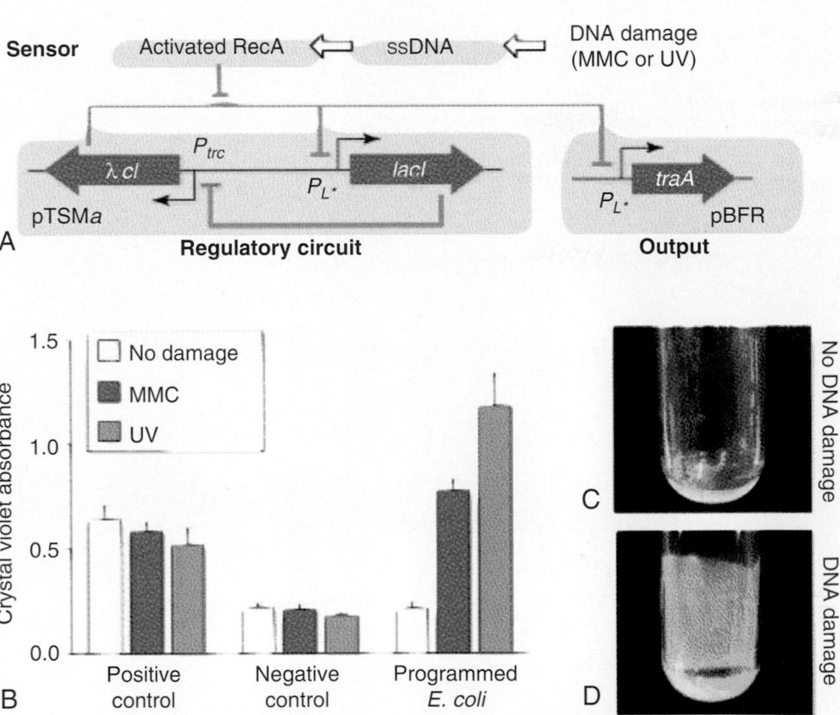

FIGURE 13.20 Biofilm Produced by *E. coli* in Response to DNA Damage
Novel input–output systems engineered to sense DNA damage and induce biofilm formation. (A) Genetic network connecting DNA damage (input) and biofilm formation (output). Repressor cI normally inhibits expression of *lacI* from promoter PL*. DNA damage activates RecA, which then cleaves cI, making it inactive. This allows expression of *lacI*, which inhibits cI expression. Without cI, the cell also produces TraA, which results in biofilm formation. (B) Crystal violet absorbance of control and engineered strains after exposure to the DNA-damaging agents mitomycin C (MMC) and UV radiation. Increased crystal violet absorbance indicates biofilm formation. (C) No biofilm formation is seen in a tube containing engineered cells that were not exposed to DNA-damaging agents. (D) Biofilm formation is seen in a tube containing engineered cells exposed to DNA-damaging agents. From McDaniel R, Weiss R (2005). Advances in synthetic biology: on the path from prototypes to applications. *Curr Opin Biotechnol* **16**, 476–483. Copyright 2005. National Academy of Sciences, USA.

SYNTHETIC GENETIC MATERIALS: xDNA AND XNA

What are the requirements for a macromolecule to carry heritable genetic information? Life on earth uses the two closely related nucleic acid polymers, DNA and RNA. However, many modifications have been made to these molecules both by living cells and in the laboratory.

Although some of these modifications interfere with their biological role, in many cases the ability to act as the genetic material remains intact.

How far can we change the molecular basis yet retain function? Recent work has shown that both the bases and the backbone can be replaced with unnatural components yet allow biological function to continue, albeit under contrived circumstances. In Chapter 5 (refer to Figure 5.8), we described modified oligonucleotides developed for resistance to degradation by nucleases. Some had minor base modifications, such as methylation. Others had rather more radical changes to the backbone: in morpholino nucleic acids, ribose is replaced by a morpholino ring; and in peptide nucleic acids, a polypeptide backbone, characteristic of proteins, is used.

Recently, nucleic acids with normal backbones but with a full set of four unnatural bases have been made. In size-expanded "xDNA," the four bases are homologs of the normal bases, A, T, G, and C, which have extra benzene rings incorporated into their structures. DNA made with xDNA bases in one strand that are paired with normal bases in the complementary strand forms relatively normal double-helix structures *in vitro*. However, the base pairs are wider than normal by 2.4 Å (Fig. 13.21). *In vivo* experiments showed that replacing segments of up to four bases of DNA with the wider xDNA bases allowed replication, but longer tracts failed to function.

Another approach is to alter the nucleic acid backbone. The pentose sugar of the nucleic acid may be replaced with several possible alternative ring structures, giving rise to so-called XNA (where X is the sugar/ring). Typically, different five-membered rings have been incorporated, although occasionally other rings of similar shape and size may be used (Fig. 13.22). Some examples of XNAs under investigation include

Arabino nucleic acid (ANA; R = OH) and its fluoro derivative, 2′-fluoro-arabino nucleic acid (FANA; R = F).
Locked nucleic acid (LNA; 2′-O, 4′-C-methylene-β-D-ribonucleic acid) in which a methylene bridge links the 2′-OH and carbon 4′.

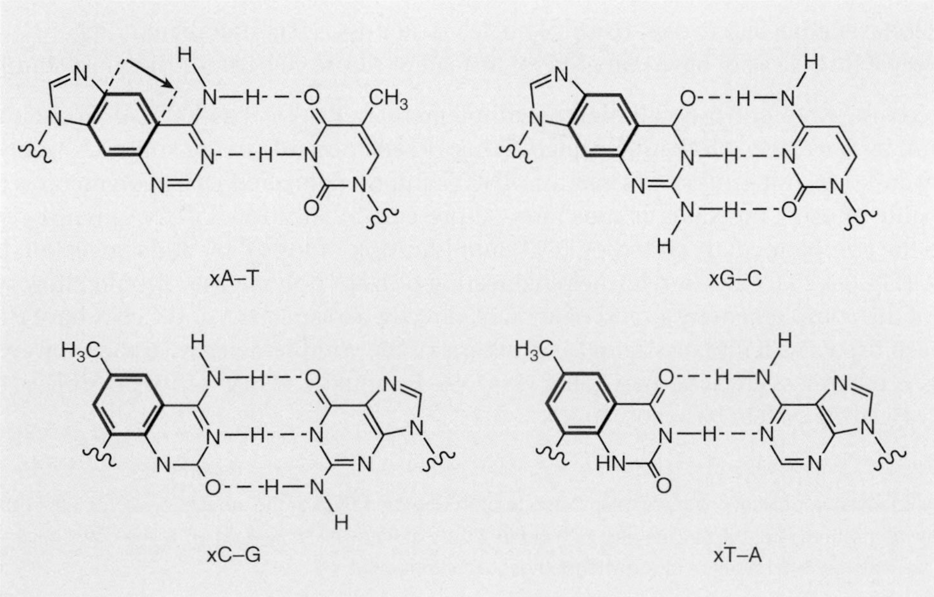

FIGURE 13.21 Size-Expanded DNA (xDNA)
In size-expanded DNA, the expanded bases are paired with natural bases on the opposite strand. The resulting base pairs are wider by 2.4 Å than normal base pairs. Adapted with permission from Krueger AT et al. (2011). Encoding phenotype in bacteria with an alternative genetic set. *J Am Chem Soc* **133**, 18447–18451. Copyright American Chemical Society. Improved figure kindly provided by Eric Kool.

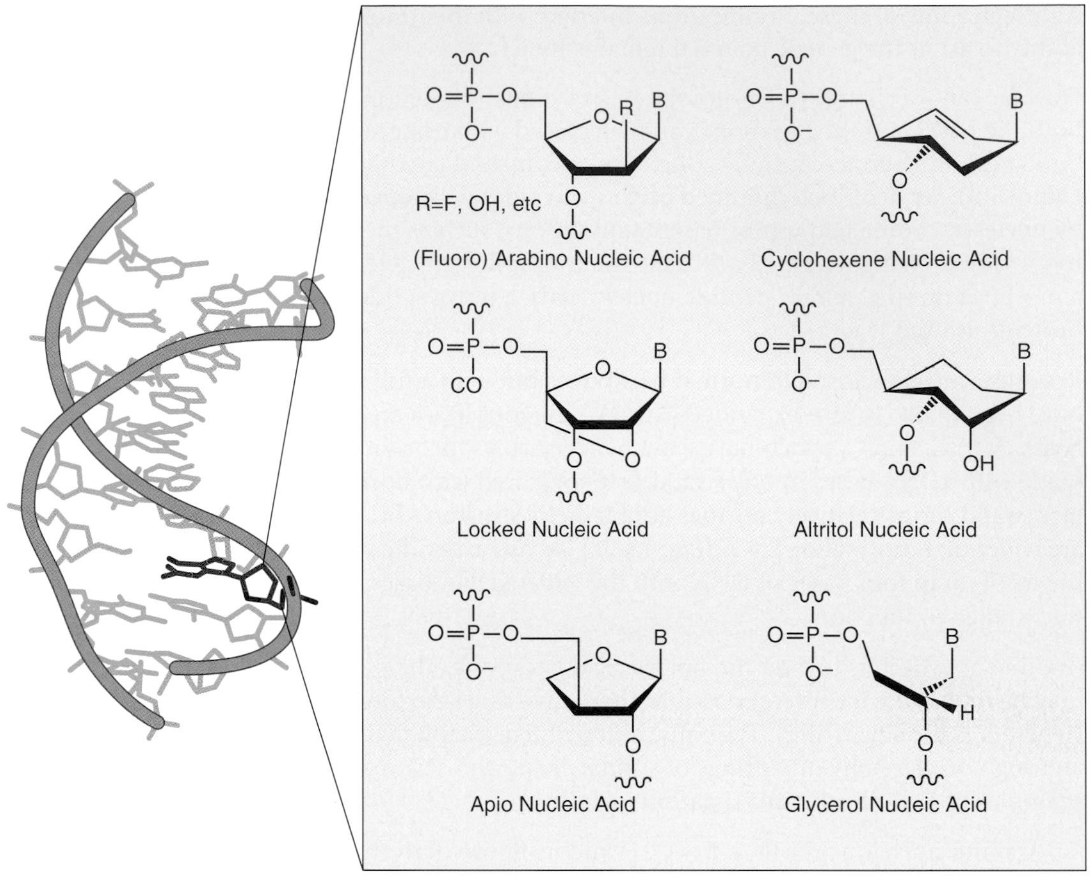

FIGURE 13.22 XNA Analogs of Nucleic Acids
Normal DNA is composed of a deoxyribose–phosphate backbone with bases attached. The structures of several non-natural nucleic acid analogs are shown. B = base. From Chaput JC, Yu H, Zhang S (2012). The emerging world of synthetic genetics. *Chem Biol* **19**, 1360–1371.

Cyclohexene nucleic acid and altritol nucleic acid possess six-membered rings.
Glycerol nucleic acid, based on glycerol instead of ribose, is perhaps the most simple

The preceding XNA polymers all bind to complementary RNA or DNA and are resistant to digestion by nucleases. More interestingly, it has proven possible to use some XNA polymers to carry and transmit genetic information. This requires engineered DNA polymerases that are capable of using the XNAs as substrates. At present, "replication" of XNA involves conversion to the complementary strand of DNA, amplification of the DNA, and conversion back to XNA (Fig. 13.23). However, further engineering of DNA polymerases should allow synthesis of the complementary strand of an XNA directly. So far, FANA and CeNA have successfully been used as a template strand for synthesis of the complementary strand; however, the process is much less efficient than going via DNA. Eventually, it may prove possible to create a self-replicating system based on an XNA.

Many chemical modifications have been made to both DNA and RNA in the laboratory. Some affect the bases, and others alter the nucleic acid backbone. Some of these allow replication of the novel nucleic acid derivatives with retention of the encoded genetic information.

DESIGNER BACTERIA

So-called designer bacteria have been made for a variety of purposes. Bacteria engineered for biofuel production or biowaste degradation could be regarded as designer bacteria. However,

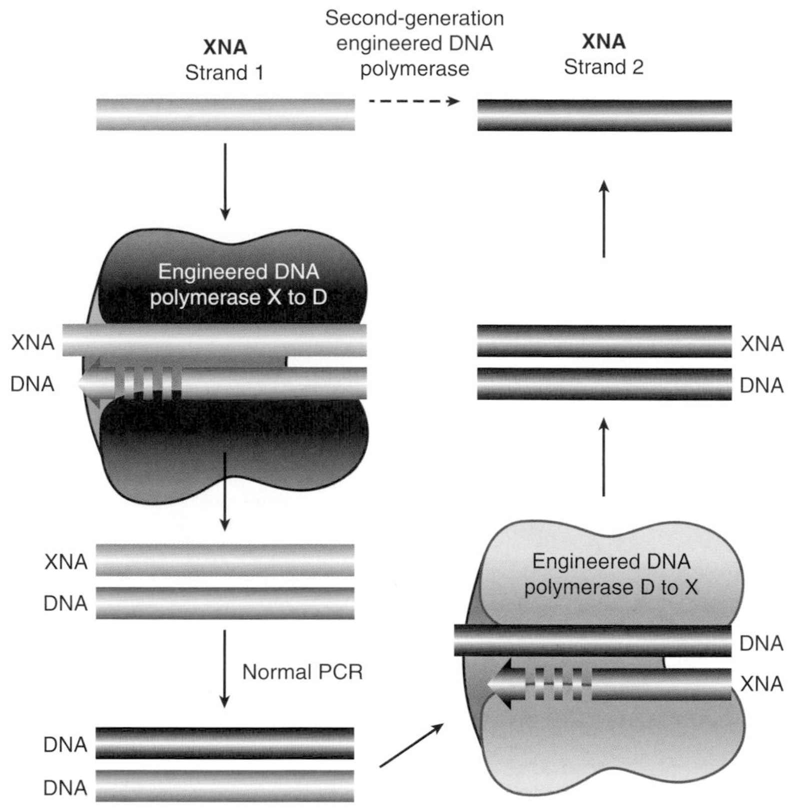

FIGURE 13.23 Replication of XNA

Two separate engineered polymerases are used to convert XNA to complementary DNA and DNA to complementary XNA. The DNA is amplified by a standard PCR reaction in between the two XNA reactions. Further engineering allows direct XNA replication, but at present this is extremely inefficient.

the term is typically limited to bacteria designed to perform a specific "job" of sorts. Examples discussed here include those designed to mend cracks in concrete or those engineered to kill antibiotic-resistant pathogens.

Mending concrete may seem an unlikely career for bacteria. However, many humanmade concrete structures are now old and developing cracks. Mending existing concrete avoids replacement, which is both expensive and causes pollution. "BacillaFilla" consists of common spore-forming soil bacteria, known as *Bacillus subtilis*, which have been modified to detect concrete, fill cracks, and self-destruct if they move away from concrete. The spores of BacillaFilla germinate on contact with concrete, triggered by the highly alkaline pH (Fig. 13.24). The bacteria then enter the cracks. When they reach the ends of cracks, they accumulate and clump. This activates the concrete repair system via quorum sensing. The BacillaFilla then differentiates into three kinds of specialized cells: those that make calcium carbonate crystals, those that make glue, and those that act as fibers for reinforcement.

A more aggressive role is that of the *E. coli* engineered to specifically assassinate *Pseudomonas aeruginosa*. Strains of *P. aeruginosa* are opportunistic pathogens that infect patients with damaged immune systems and quite often grow in the mucus that collects in the lungs of cystic fibrosis patients. They are often antibiotic resistant and form biofilms, which make them hard to treat. When the engineered *E. coli* detects the presence of *P. aeruginosa,* it produces a toxin, pyocin, which is specific for killing *P. aeruginosa*. The *E. coli* assassins killed around 99% of freely swimming *Pseudomonas* and around 90% if they were protected by biofilms.

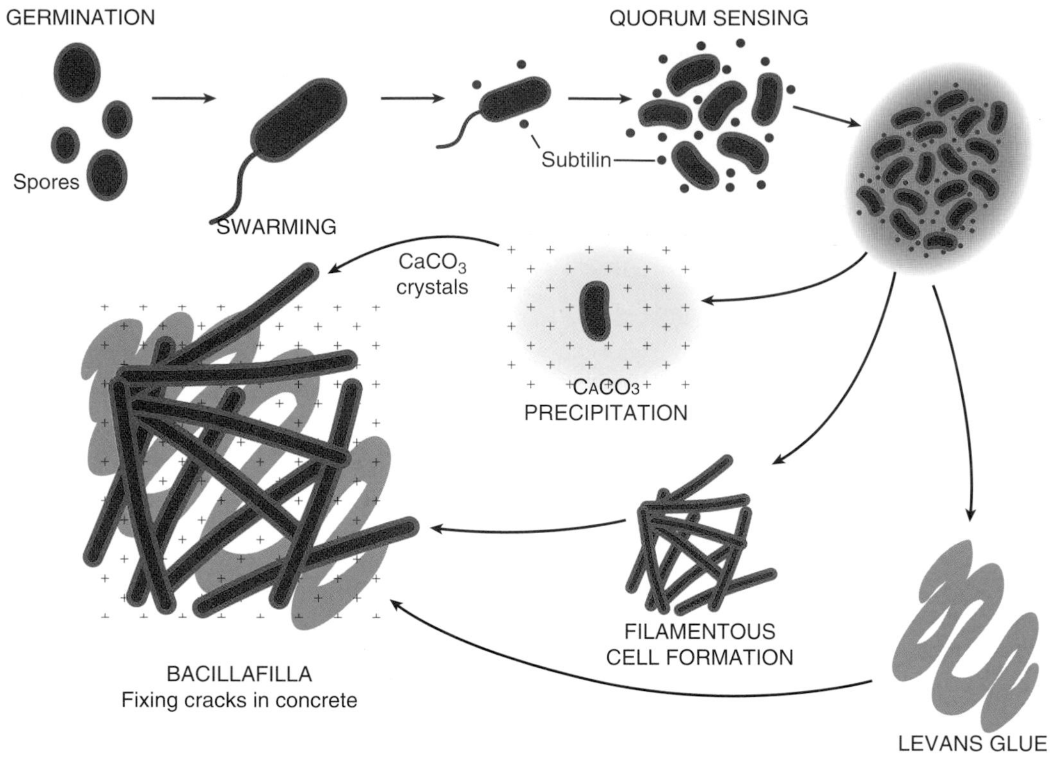

FIGURE 13.24 BacillaFilla Bacteria Mend Concrete
The bacterial spores germinate upon contact with concrete. Three specialized versions of the bacteria are generated by differentiation. Together, they fill cracks with a fibrous mixture of bacterial cells and calcium carbonate. Modified after the University of Newcastle iGEM Team website, http://2010.igem.org/Team:Newcastle, courtesy of Dr. Anil Wipat.

This system takes advantage of two unique characteristics of the target *P. aeruginosa*. Many bacteria communicate by quorum sensing, which allows them to cooperate for group activities such as biofilm formation. One class of widely used signaling molecules is the acyl homoserine lactones (AHLs). These vary among different bacteria in chain length and functional groups. *P. aeruginosa* uses an AHL with a 3-oxo-C12 chain that is specifically recognized by its own LasR protein. The engineered *E. coli* carry the *lasR* gene and make the LasR protein, which allows them to specifically detect *P. aeruginosa* (Fig. 13.25).

Many bacteria kill other, closely related bacteria, with extremely specific protein toxins known as bacteriocins. The purpose of this appears to be competition for the same environmental niche—hence, the high specificity. *P. aeruginosa* uses proteins known as pyocins to kill competing strains of the same species. The engineered *E. coli* produce pyocin S5 upon sensing *P. aeruginosa*. To release the pyocin, the *E. coli* also synthesize the E7 lysis protein, which bursts the cells (Fig. 13.25).

Bacterial cells may be radically redesigned to carry out specific roles. Examples range from mending cracks in concrete to killing specific pathogenic bacteria.

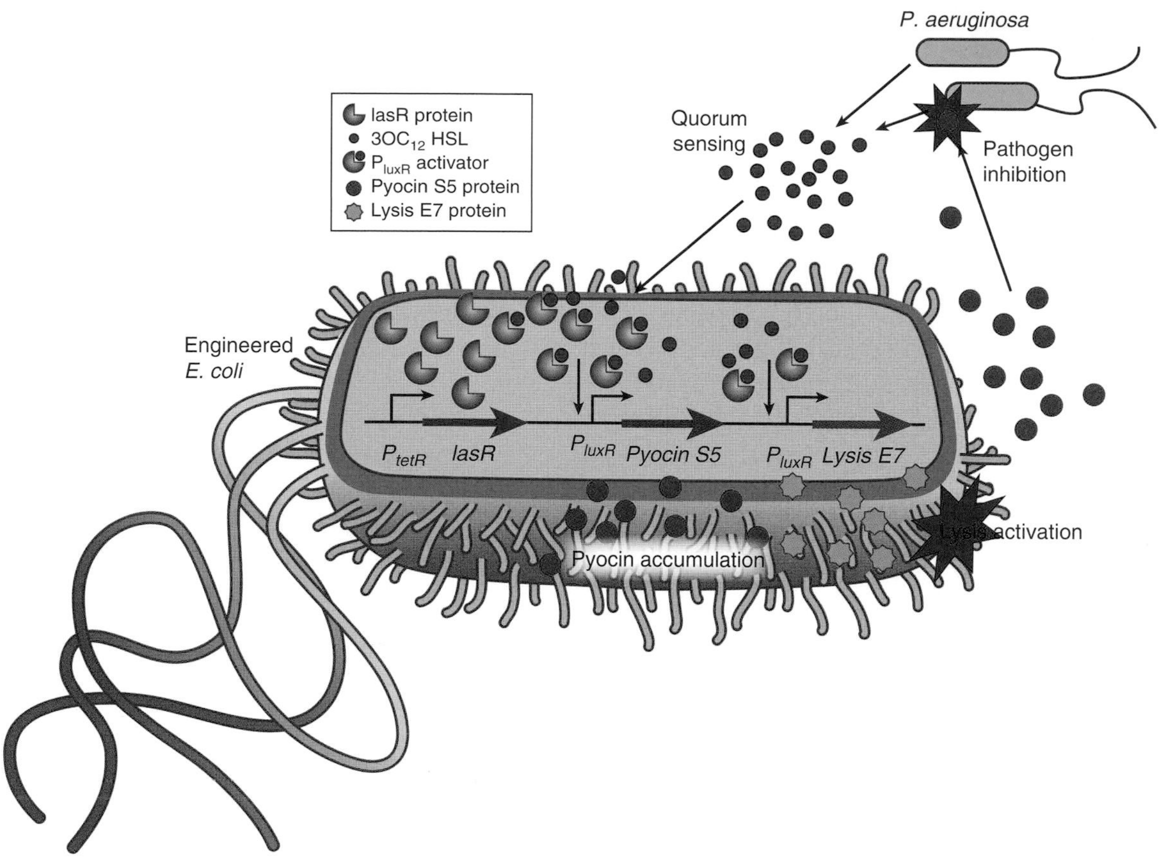

FIGURE 13.25 Pathogen Sensing and Killing System
Engineered "assassin" *E. coli* carries genes allowing both detection and killing of target strains of *Pseudomonas*. The *lasR* gene is controlled by the tetR promoter. After the LasR protein binds the signal molecule (3-oxo-C12 HSL), it activates the *luxR* promoter. This causes the production of both pyocin S5 and lysis protein E7. After the E7 protein attains the threshold concentration that causes the *E. coli* cell to lyse, the accumulated S5 is released into the environment to kill the *P. aeruginosa*. From Saeidi N, et al. (2011). Engineering microbes to sense and eradicate Pseudomonas aeruginosa, a human pathogen. *Mol Syst Biol* **7**, 521.

Summary

Metabolic pathways may be created or modified by assembling genes from one or more organisms. Individual genes may themselves be subjected to engineering to produce regulatory proteins or enzymes with novel activities or binding specificities. Such pathway engineering has been successfully applied to the production of alcohol and other biofuels and to the biodegradation of agricultural waste and industrial pollutants. More complex pathways include the biosynthesis of modified antibiotics. More sophisticated redesign of natural systems is referred to as synthetic biology. This includes construction of generalized genetic control circuits and their application to specific problems. Relatively new areas of research range from radical modification of the information encoding nucleic acids to redesigning bacterial cells with specific technological missions.

End-of-Chapter Questions

1. What is the advantage of using either yeast or *Zymomonas* during fermentation?
   a. *Zymomonas* and yeast generate a mixture of products, whereas other microorganisms only produce alcohol.
   b. *Zymomonas* and yeast make only alcohol, but other microorganisms generate a mixture of products.
   c. *Zymomonas* and yeast have very small growth requirements.
   d. *Zymomonas* and yeast can only break down glucose and no other sugars or carbohydrates to produce alcohol.
   e. none of the above

2. Which of the following best explains how pathway engineering is useful?
   a. the conversion of plant-derived waste from paper manufacturing into alcohol used for fuel
   b. the creation of better-tasting beers
   c. the genetic engineering of *E. coli* strains to grow only on one sugar source, such as glucose
   d. the genetic engineering of plants to decrease the use of cellulose, which is difficult to break down
   e. none of the above

3. How can the process of breaking down starch into sugar be improved?
   a. by beginning with glycogen rather than starch to generate sugar molecules
   b. by chemically treating the starch to break it down into sugars
   c. by identifying other enzymes through proteomics that can perform the same functions as the presently used enzymes
   d. by engineering an organism to produce higher levels and more stable enzymes
   e. all of the above

4. Which of the following enzymes is not involved in the breakdown of cellulose?
   a. cellulase
   b. endoglucanase
   c. cellobiohydrolase
   d. β-glucosidase
   e. exoglucanase

5. What are ice nucleation factors?
   a. Ice nucleation factors are proteins present within *E. coli* cells.
   b. Ice nucleation factors provide a way to prevent ice crystals from forming on plants and destroying plant tissue.
   c. Ice nucleation factors are specialized proteins in bacteria that provide a seed on which ice crystals are formed.
   d. When ice nucleation factors are present, water supercools instead of forming ice crystals.
   e. none of the above

6. Which of the following is an example of a Second Generation Biofuel?
   a. biodiesel
   b. ethanol
   c. bio-kerosene
   d. hydrogen
   e. all of the above

7. Biodiesel is typically produced from _________.
   a. fats and oils
   b. proteins
   c. nucleic acids
   d. starch

8. Ice nucleation factors _________.
   a. are used to prevent ice formation on the surfaces of plants
   b. can be used in biotechnology to display proteins on the surfaces of cells
   c. are involved in the production of carbapenem antibiotics
   d. generate synthetic nucleic acids, such as XNAs and XDNAs
   e. are none of the above

9. The advantage of using carbapenems to generate antibiotics is that carbapenems are _________.
   a. are effective against all microbes, including fungi and protozoans
   b. generated from waste products, thus recycling them
   c. resistant to β-lactamases and even act as inhibitors
   d. not known to increase resistant to antibiotics in the targeted pathogen
   e. typically effective against a narrow range of pathogens

10. Which of the following is not true regarding synthetic nucleic acids?
   a. XNAs containing alterations to the pentose sugars of the nucleic acid backbone.
   b. Synthetic nucleic acid cannot be replicated *in vitro* or *in vivo.*
   c. XDNAs contain benzene rings in the nitrogen-containing bases.
   d. Double-stranded nucleic acid containing XDNA is more narrow then natural dsDNA.
   e. Replication of XNAs requires an engineered DNA polymerase.

11. Which of the following is not associated with BacillaFilla?
   a. mending cracks in concrete
   b. soil bacteria, such as *B. subtilis*
   c. quorum sensing
   d. production of pyocin
   e. designer bacteria

12. In designer bacteria, which of the following is produced by *E. coli* and specifically targets *P. aeruginosa* for destruction?
   a. AHL
   b. pyocin
   c. LasR
   d. E7 lysis protein
   e. 3-oxo-C12

(*Continued*)

13. What is the 4S pathway?
- **a.** a set of reactions for the breakdown of DBT and removal of the sulfur
- **b.** a set of reactions for the synthesis of sulfur-containing compounds
- **c.** a pathway for the incorporation of FeS clusters into enzymes for stability
- **d.** a pathway for the synthesis of the four nucleotides used in the genetic code
- **e.** none of the above

14. What can be produced upon chemical modification of β-lactams?
- **a.** β-lactam antibiotics that are resistant to β-lactamases
- **b.** antibiotics that are more easily absorbed in the intestines
- **c.** antibiotics that can penetrate bacterial cell walls more efficiently
- **d.** increased yields of some β-lactam antibiotics
- **e.** all of the above

15. How is the production of PHA being made more economical?
- **a.** by mass-marketing the product as an environmentally friendly alternative
- **b.** by engineering the PHA genes from *A. eutrophus* into the chloroplasts of *Arabidopsis*, and expression in plants
- **c.** by production in massive fermenters at Zeneca Corporation
- **d.** by expression of the PHA genes in all plant tissues, not just in the chloroplasts
- **e.** none of the above

Further Reading

Abbaszadegan, M., Alum, A., Abbaszadegan, H., & Stout, V. (2011). Cell surface display of poliovirus receptor on Escherichia coli, a novel method for concentrating viral particles in water. *Applied and Environmental Microbiology, 77*(15), 5141–5148.

Butler, M. S., & Cooper, M. A. (2011). Antibiotics in the clinical pipeline in 2011. *The Journal of Antibiotics, 64*(6), 413–425.

Chanprateep, S. (2010). Current trends in biodegradable polyhydroxyalkanoates. *Journal of Bioscience and Bioengineering, 110*(6), 621–632.

Gronenberg, L. S., Marcheschi, R. J., & Liao, J. C. (2013). Next generation biofuel engineering in prokaryotes. *Current Opinion in Chemical Biology, 17*(3), 462–471.

Howard, T. P., Middelhaufe, S., Moore, K., Edner, C., Kolak, D. M., Taylor, G. N., et al. (2013). Synthesis of customized petroleum-replica fuel molecules by targeted modification of free fatty acid pools in *Escherichia coli*. *Proceedings of the National Academy of Sciences of the United States of America, 110*(19), 7636–7641.

Huffer, S., Roche, C. M., Blanch, H. W., & Clark, D. S. (2012). *Escherichia coli* for biofuel production: bridging the gap from promise to practice. *Trends in Biotechnology, 30*(10), 538–545.

Krueger, A. T., Peterson, L. W., Chelliserry, J., Kleinbaum, D. J., & Kool, E. T. (2011). Encoding phenotype in bacteria with an alternative genetic set. *Journal of the American Chemical Society, 133*(45), 18447–18451.

McGovern, P. E., Zhang, J., Tang, J., Zhang, Z., Hall, G. R., Moreau, R. A., et al. (2004). Fermented beverages of pre- and proto-historic China. *Proceedings of the National Academy of Sciences of the United States of America, 101*(51), 17593–17598.

Papp-Wallace, K. M., Endimiani, A., Taracila, M. A., & Bonomo, R. A. (2011). Carbapenems: past, present, and future. *Antimicrobial Agents and Chemotherapy, 55*(11), 4943–4960.

Peccoud, J., & Isalan, M. (2012). The PLoS One synthetic biology collection: six years and counting. *PLoS One, 7*(8) e43231.

Pinheiro, V. B., & Holliger, P. (2012). The XNA world: progress towards replication and evolution of synthetic genetic polymers. *Current Opinion in Chemical Biology, 16*(3–4), 245–252.

Pinheiro, V. B., Taylor, A. I., Cozens, C., Abramov, M., Renders, M., Zhang, S., et al. (2012). Synthetic genetic polymers capable of heredity and evolution. *Science, 336*(6079), 341–344.

Saeidi, N., Wong, C. K., Lo, T. M., Nguyen, H. X., Ling, H., Leong, S. S., Poh, C. L., & Chang, M. W. (2011). Engineering microbes to sense and eradicate *Pseudomonas aeruginosa*, a human pathogen. *Molecular Systems Biology, 7*, 521.

van Bloois, E., Winter, R. T., Kolmar, H., & Fraaije, M. W. (2011). Decorating microbes: surface display of proteins on Escherichia coli. *Trends in Biotechnology, 29*(2), 79–86.

Wang, J., Xiong, Z., Meng, H., Wang, Y., & Wang, Y. (2012). Synthetic biology triggers new era of antibiotics development. *Sub-cellular Biochemistry, 64*, 95–114.

From Cell Phones to Cyborgs

Biotechnology

http://dx.doi.org/10.1016/B978-0-12-385015-7.00014-4

INTRODUCTION

Several areas of biotechnology are advancing more rapidly than others. Perhaps the two most obvious are speedier DNA sequencing technology and emerging novel roles for RNA. This chapter is new to this edition of this book and tackles a third area. Unlike the other two, this area is on the boundary where biotechnology merges with computer science and electronics.

Increased computing power, together with the miniaturization of electronic components, has played a major role not only in the progress of molecular biology but also in the emergence of cell phones and tablets. Recently, we have seen a massive increase in the popularity of such handheld smart devices. Perhaps not surprisingly, these two areas are beginning to interact as molecular biology starts to take advantage of these devices. Although the first scientific uses of cell phones and tablets were in education and analysis, more recently experimental applications have begun to appear. This chapter deals with the use of such mobile smart devices in biotechnology.

A related area is the direct connection of electronic devices to living organisms. Progress here has also been driven by increased computing power plus miniaturization of electronic components. Today, insect cyborgs are paving the way for the bionic man.

Rapid increases in computing power, together with the miniaturization of electronic components, has played a major role in the emergence of handheld electronic devices that are now used in molecular analysis. These advances have also facilitated direct connection of electronic devices to biological systems.

CELL PHONES

A wide range of gadgets is now available that connect to your cell phone. Some are intended for amusement, but others are designed for technical analysis. New attachments are constantly being developed and improved. Table 14.1 lists a selection of these applications. Some are available and others still in development. First-generation devices usually require a cable connection to the cell phone, but second-generation devices often connect via WiFi or Bluetooth.

The first half dozen items in Table 14.1 are modifications of clinical monitors that are already in widespread use but have been remodeled to work in concert with a cell phone. The next few items involve modification or construction of analytical devices designed to take advantage of cell phone capabilities. However, all are based on analytical procedures that are already used routinely. These include ELISA immune assays (see Chapter 6 for details of ELISA) and several approaches to DNA detection for PCR assays (see Chapter 4 for details of PCR). It is not our intention to describe pre-existing technology, so we will cover only the last two items in detail because they involve novel approaches.

A novel detection system for bacteria based on the use of fluorescent quantum dots has been designed for attachment to cell phones (Fig. 14.1). The cell phone detects the fluorescence emission and analyzes the data. The device mounted over the cell phone camera emits fluorescence upon detecting specific kinds of bacteria. The detection system relies on antibodies specific to the target bacteria (Fig. 14.2). One antibody immobilizes the bacteria by binding them to a glass slide. The other antibody links the bacteria to a quantum dot (via biotin and the biotin-binding protein, streptavidin). Light from the light emitting diodes (LEDs) excites the quantum dots, which then emit fluorescence. The sample is held inside an array of capillary tubes containing the antibodies and quantum dots. The prototype device was tested with the bacterium *E. coli* that often contaminates food or water. Consequently, a cheap, mobile testing system would be very useful. This device is presently under development in the laboratory of Aydogan Ozcan at UCLA.

Table 14.1 Applications of Cell Phones in Biotechnology

Application	Description
"Traditional" Clinical Monitors	
Glucose monitor	A glucose monitor reads blood test strips and attaches to a cell phone.
Blood pressure	This standard blood pressure cuff as used by doctors includes a connector to plug into a cell phone.
Heart rate	(a) A cell phone camera monitors how light reflects off the face as blood flows through the vessels in the skin. (b) A more advanced approach is a device that attaches to the phone and performs an electrocardiogram.
Eye exam	A plastic eyepiece snaps onto a cell phone accompanied by an app that carries the user through diagnostic tests on the screen.
Otoscope	This device is used to look inside ears and throat. A cell phone attachment allows the user to take pictures inside the ear for transmission to a doctor.
Endoscope	Endoscopes are devices to look inside the digestive tract. They may be coupled to cell phones instead of regular cameras.
Molecular Biological Monitors	
ELISA assay	A miniature device based on microfluidics performs ELISA assays to detect virus components. The data are transmitted to and analyzed by a cell phone via a wireless connection.
DNA analysis for PCR	A variety of modified and miniaturized devices that run PCR (or related DNA amplification procedures) use the cell phone camera for recording the results and accompanying apps to analyze the data.
Bacterial detection by quantum dot fluorescence	A novel detection system for bacteria relies on antibodies and quantum dots to generate fluorescence upon detecting specific bacteria. The cell phone camera monitors the fluorescence.
On-screen diagnosis	Touch screens are sufficiently sensitive to detect the presence of macromolecules in solution.

Future advances in the cell phone detection of fluorescence emissions allow the detection of nanoscale particles, including some larger viruses, such as herpes viruses, that have been labeled with antibodies carrying fluorescent dyes.

The final item in Table 14.1 does not require an attachment; rather, it involves dripping samples directly onto smart phone screens. It takes advantage of the ability of touch-sensitive screens to detect the extremely small electrical signals due to contact with a fingertip. The sensitivity of touch screens is actually much greater than needed for detecting human fingers. Indeed, it should be possible to detect macromolecules, such as proteins and DNA, by the same approach. Ultimately, tiny droplets of blood, saliva, or other fluids would be applied to a touch screen for diagnosis of infections or clinical problems. This approach is still in the experimental stage. So far, touch screens have been shown to distinguish between different concentrations of bacterial DNA in droplets of only 10 microliters.

Cell phones and other handheld electronic devices are now being adapted and combined for use in clinical and molecular analysis. An ever-increasing number of uses, especially in clinical diagnostics, are appearing. The ultimate approach is to directly apply samples for analysis to the smart phone screen.

FIGURE 14.1 Quantum Dot Bacterial Detection by Cell Phone
The cell phone plus attachment are shown. The attachment weighs ~28 grams (~1 ounce) and has dimensions of ~3.5 × 5.5 × 2.4 cm with a field-of-view of 11 × 11 mm. It can monitor approximately 10 capillary tubes in parallel. Photo kindly provided by Ozcan Research Lab at UCLA.

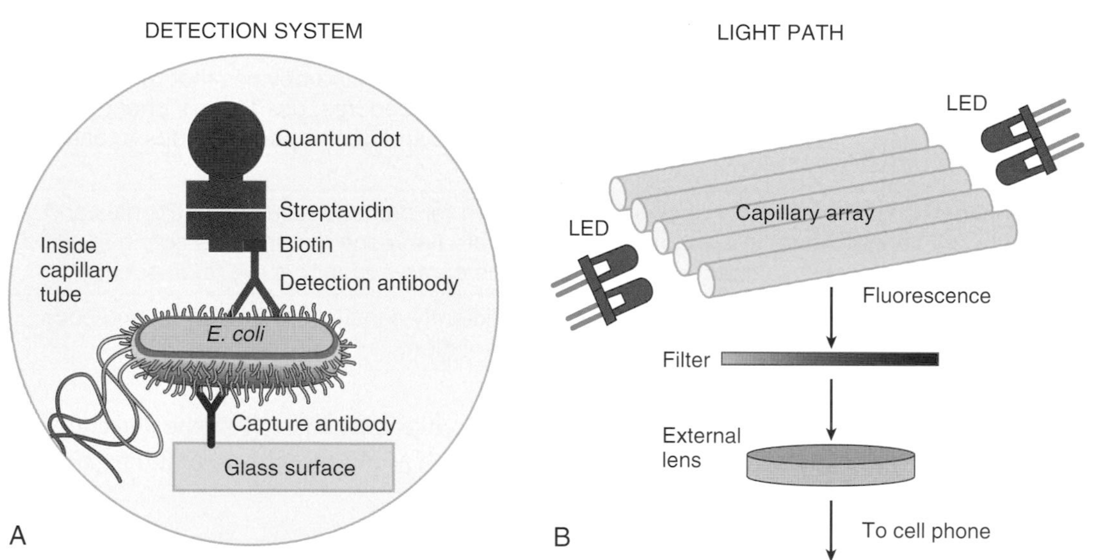

FIGURE 14.2 Mechanism of Quantum Dot Bacterial Detection
The mechanism is shown here. (A) The detection system that captures bacteria and attaches quantum dots to them is located inside capillary tubes. (B) The capillary tubes form an array that is illuminated by LEDs. Fluorescence emitted from the quantum dots is ultimately detected and analyzed by the cell phone. Redrawn from material kindly provided by Ozcan Research Lab at UCLA.

ROBOTICS

Robots in science fiction have traditionally taken human shapes and acted as mechanical replacements for human beings. However, industrial and laboratory robots are quite different, being designed for specific operations. Perhaps the most common are those resembling arms on fixed or mobile bases and often equipped with sensors to monitor their tasks. Robotics and remote-controlled devices are also increasingly used for clinical operations. Although these applications may be regarded as biotechnology in the wider sense, they are beyond the scope of this book.

Robots have recently taken on more specialized tasks in biotechnology. An interesting example is the CoralBot. This is an underwater robot designed to maintain coral reefs. If damaged, coral reefs may take many years to recover. Although this recovery can be speeded by human intervention, doing so is tedious and labor intensive. Enter the CoralBots. These robots, which are being developed in Scotland, work as a group and are programmed to detect damage, reattach broken coral fragments, and carry out other maintenance. Although such reefs are less famous than tropical coral reefs, the west coast of Scotland is bordered by plentiful coral reefs that provide homes for a variety of fish, including sharks.

More practical and closer to deployment is the GasBot (Fig. 14.3). This detects gases released from landfills or other sources. Its intended present target is methane, which is often released from landfills, but other gases such as hydrogen sulfide could be detected in the future. Methane is a major greenhouse gas that contributes to global warming and also damages the ozone layer.

Two relatively new and related areas involve the merger of robotics with biology. These are the remote control of insect behavior by implanted electrodes—insect cyborgs—and the creation of robots modeled on biological organisms—so-called soft robotics. Before discussing these, we will deal with the simpler idea of remotely controlling single genes.

Robotics is widely applied in the biological and clinical arenas. Autonomous robots may also be used in environmental monitoring and repair.

FIGURE 14.3 GasBot Detects Methane
The GasBot is an autonomous robotic vehicle that detects and locates leaks of the gas methane. The GasBot website has more information plus an image gallery: http://aass.oru.se/Research/MRO/gasbot/index.html (Photograph kindly provided by Victor Hernandez and Achim Lilienthal, Örebro University, Sweden.)

RADIO-CONTROLLED GENES

Radio-controlled walking dinosaurs or flying helicopters have been available as children's toys for several years. Recently, the first steps were taken to apply radio controls to living cells, both in culture and in whole living mice.

Unlike visible light, radio waves readily pass through living tissue but are absorbed by many metals. To take advantage of this, iron nanoparticles may be provided to the target cells or tissues. When these cells or tissues absorb radio waves, heat is generated locally and may be used to trigger a variety of responses. In particular, temperature-sensitive proteins may be coupled to the Fe nanoparticles.

A clever illustration of this takes advantage of a temperature-sensitive protein channel (TRPV1) that controls the entry of calcium ions (Ca^{2+}) into nerve cells. The TRPV1 channel had a His-tag attached. This, in turn, was used to bind iron oxide nanoparticles that carried antibodies to the His-tag (Fig. 14.4). When exposed to radio waves, the nanoparticles absorbed the radiation and heated up. This opened the channels, thus allowing in more calcium and triggering action potentials in cultured nerve cells. The influx of Ca^{2+} can also be used to regulate gene activity. In the illustration, an altered version of the insulin gene was used that was driven by a Ca^{2+}-sensitive promoter (Fig. 14.4). The overall result of this setup is that radio waves trigger activation of the insulin gene. Cells containing this system were injected into mice. When the mice were exposed to radio waves, insulin was released and caused a drop in blood glucose. Future refinement of this system may allow the delivery of proteins at specific locations within the body upon exposure to radio waves.

Nanotechnology may be melded with genetic engineering to create hybrid systems that allow genes within living organisms to be remotely controlled by radio waves.

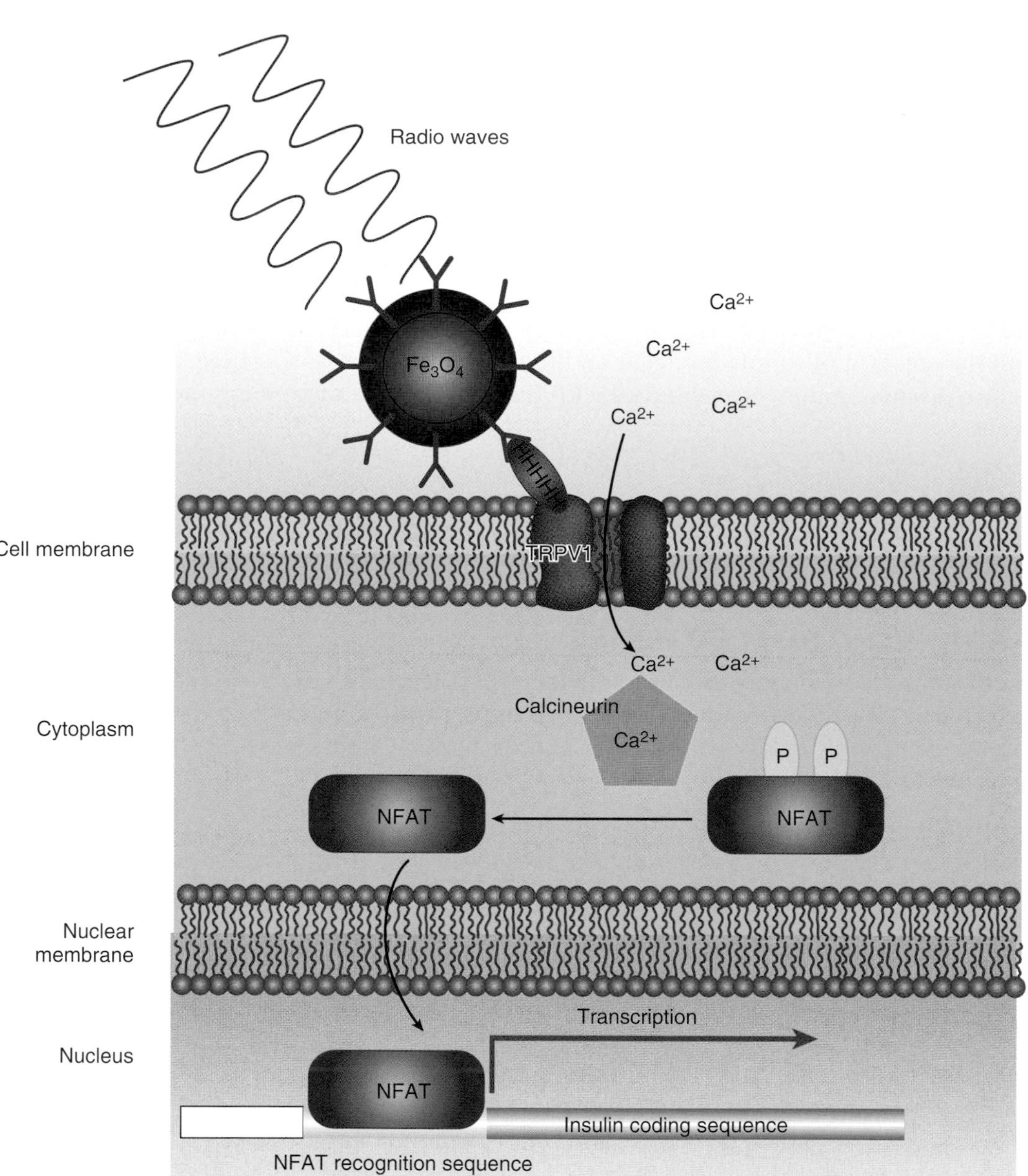

FIGURE 14.4 Radio Waves Used to Control Gene Expression Radio waves are absorbed and converted to heat by an iron oxide nanoparticle that is bound to the TRPV1 channel protein via an antibody that recognizes the His-tag (HHHHH). When TRPV1 is activated by the local heating, it allows calcium ions across the membrane into the cell. In this experiment, the calcium is bound by the protein calcineurin, which activates the NFAT transcription factor by removal of phosphate groups. The activated NFAT enters the nucleus, where it binds to the NFAT recognition sequence in the promoter of the engineered insulin gene. This activates transcription of the insulin gene. The net result is increased production of insulin in response to radio waves.

INSECT CYBORGS

Remote controls can be applied to whole organisms as well as genes. Insects have often been regarded as rather robot-like and dominated by rigid instincts. Some recent research has begun to build on this theme by modifying insects to put them under remote control. This has resulted in the creation of so-called insect cyborgs.

Electrical implants that take control of insect wing or leg movements have been demonstrated with several insects, including moths, beetles, and cockroaches. The implants are activated by radio waves, thus bringing insect behavior under remote control. The insect itself carries inserted electrodes plus a miniature radio antenna, a microcontroller, and a microbattery (Fig. 14.5). Although miniaturized, these components are still significant in size and weight compared to many insects. This is one reason that beetles and cockroaches are often used—they are relatively larger and tougher than many insects.

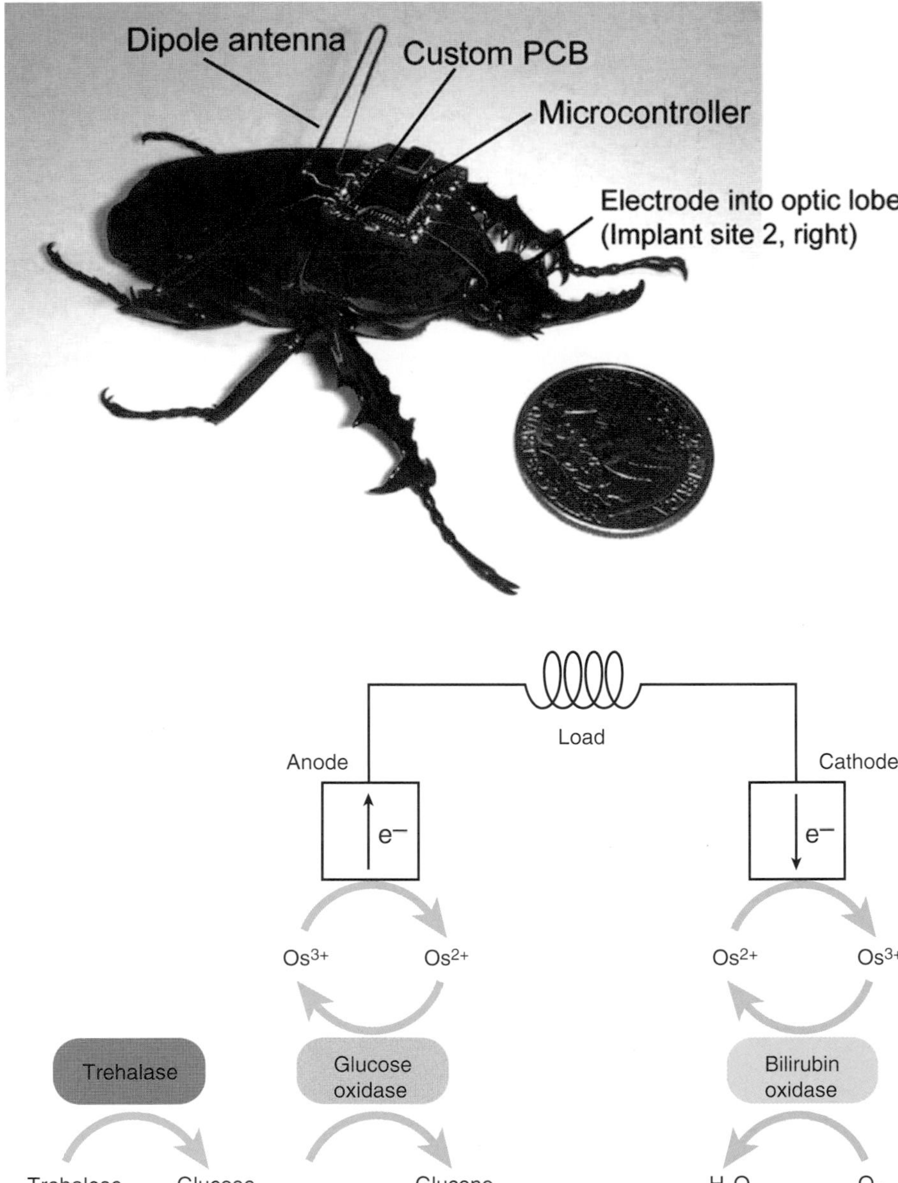

FIGURE 14.5 Radio Waves Used to Control Beetle Flight Radio waves received by the dipole antenna control the flight of this beetle. The custom-printed circuit board (PCB) carries a microbattery and a microcontroller connected to an implanted electrode. From Sato H et al. (2009). Remote radio control of insect flight. *Front Integr Neurosci* **3**, Article 24.

FIGURE 14.6 Biofuel Cell for Cyborg Insects At the anode, trehalase and glucose oxidase act in series to oxidize trehalose to gluconolactone. At the cathode, bilirubin oxidase reduces oxygen to water. The electrodes include osmium complexes that link the biological reactions to the electrical circuit.

Inserting electrodes into insects is not especially difficult, and Backyard Brains (an educational company) is selling a kit that allows customers to construct their own cyborg cockroaches. This has resulted in considerable publicity with accompanying controversy (for a further discussion of RoboRoach, see Chapter 24). Even more controversial is the application of bionics to humans (see Box 14.1).

Our primary focus here is on the biological modifications rather than the electronics. As noted previously, battery weight is a major limiting factor for insect cyborgs. Improvements to batteries are one way to increase cyborg performance. However, it is also possible to avoid using batteries entirely by harnessing insect metabolism to run fuel cells. Implantable fuel cells have been constructed that use oxidation of sugars to generate energy (Fig. 14.6). Unlike mammals whose major blood sugar is glucose, insects use the sugar trehalose, a disaccharide consisting of two glucose molecules linked together.

At the anode of the fuel cell, two enzymes are used (Fig. 14.7). First, the enzyme trehalase splits trehalose into glucose, and then glucose oxidase oxidizes the glucose to gluconolactone. At the cathode, the enzyme bilirubin oxidase converts oxygen to water. The electrons flow from anode to cathode around a circuit coupled by osmium complexes that powers the implanted electronic devices.

Insect cyborgs consist of insects with inserted electrodes that allow their behavior to be remotely controlled. Using miniaturized components and harnessing metabolically generated energy are key factors for improved performance.

Box 14.1 Epidermal Electronics and the Bionic Man

"We have the technology." This catchphrase comes from the 1970s TV series *The Six Million Dollar Man*, which featured a terminally injured human rebuilt as a "bionic man." Today, bionics is no longer science fiction. A variety of bionic components have been developed and are in clinical use. Both bionic limbs and sensory organs have been developed. Body–machine interfaces allow humans to control their bionic devices. With bionics, we are leaving the biological sector and entering the world of electronics, computer technology, and smart materials.

Epidermal electronics refers to the construction of thin, flexible electronic components that can be attached to the surface of the skin (Fig. A). The critical problem here is that electronic components are typically rigid and brittle, whereas skin is flexible and stretches. The solution is to use silicon in the form of ultra-thin layers that can be stretched when attached to the skin. Solar cells, LEDs, transistors, and antennas are among the components that have been made in this manner.

The materials are designed so that the most brittle material overlaps with the layer in a material where there is zero strain. This is referred to as the neutral mechanical plane (NMP; Fig. A, part B). The silicon layer, which also includes chromium and gold segments, is attached to a thin layer of prestrained rubber. The whole assembly is a mere 1.5 microns thick.

Epidermal electronics are being developed for clinical use, especially in electrophysiological measurements (e.g., electroencephalograms and electromyograms).

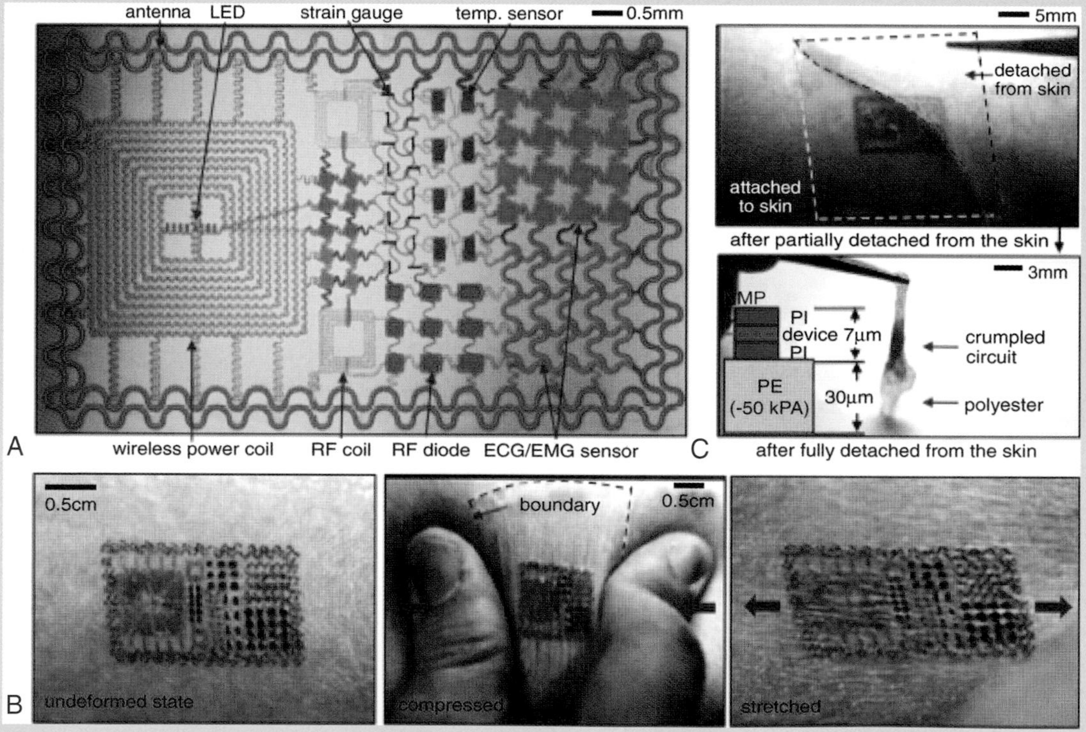

FIGURE A Epidermal Electronics

Flexible electronic components may be mounted directly onto the skin. (A) Typical epidermal electronic circuit. (B) Circuit may be peeled off the skin. (Inset) Schematic of the structure with a red dashed line showing the neutral mechanical plane (NMP). (C) Epidermal electronics shown undeformed (left), compressed (middle), and stretched (right) states. From Martirosyan N, Kalani MY (2011). Epidermal electronics. *World Neurosurg* **76**, 485–486.

SOFT ROBOTICS

Traditional robots are angular and mechanical. While insects fit reasonably well into this paradigm, much of biology may be loosely described as soft and squishy. Many living organisms use soft structures to move: the tentacles of squids and octopuses, the tails of fish, the whole body of snakes and worms—even the flagella of bacteria and sperm cells. **Biomimetics** refers to the imitation of biological structures in engineering. **Soft robotics** is the subdivision of biomimetics that consists of designing robots, or robotic components, inspired by soft biological structures.

In practice, soft robotics is more engineering than biology and relies on the use of modern electronics and, in many cases, novel advanced materials such as metal alloys with shape-changing and/or memory properties. As this is primarily a biology book, we will restrict our discussion to a brief survey that emphasizes the biological inspiration rather than the engineering details.

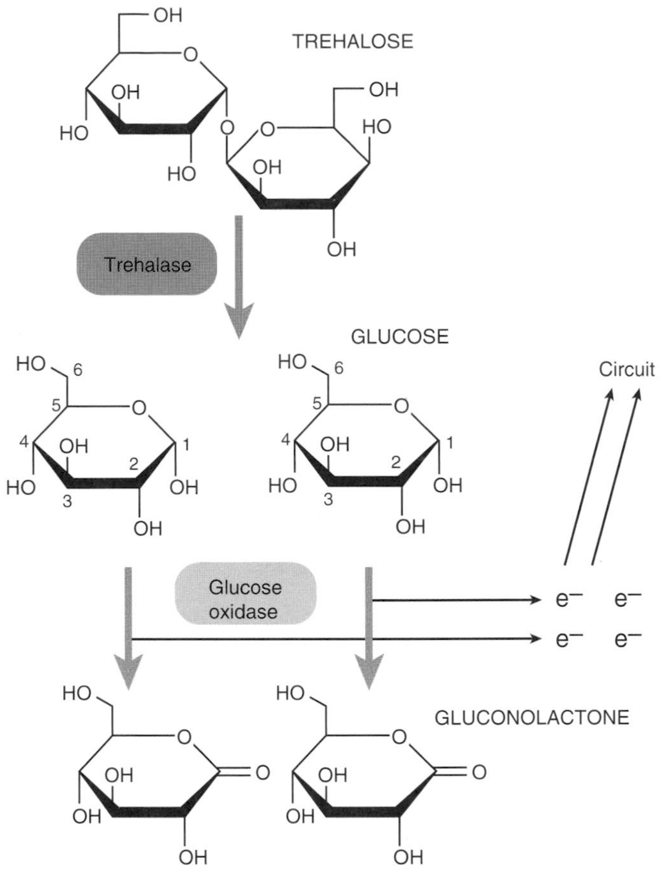

FIGURE 14.7
Enzyme Reactions at Anode of Trehalose Fuel Cell
The two reactions at the anode convert trehalose to gluconolactone. First, trehalose is split into two glucose molecules. Next, these are oxidized to two gluconolactone molecules. The electrons released power the electrical circuit.

Wriggling robots inspired by worms and snakes are a relatively straightforward example of soft robotics. To the engineer, worms are "fixed-volume hydrostats." That is, although they change shape, the volume stays constant. When they contract their lengthwise muscles to shorten the body, they get fatter. Conversely, when they contract circumferential muscles, the body gets thinner but longer. Annelid worms, such as earthworms, are divided into segments, whereas nematode worms are not. Nonetheless, worms of both types have muscle blocks arranged in segments along their bodies (Fig. 14.8). When these muscle blocks alternately contract and relax in turn, this generates a wave of contraction and expansion along the body of the organism. This behavior results in the characteristic wriggling motion.

This type of wriggling motion can be generated by several alternative engineering approaches. Air pressure and metal springs can be used instead of liquid and muscles, for example. More advanced worm robots use shape memory alloys (SMAs). An example of a worm-mimic robot based on SMA springs is shown in Figure 14.9. This robot is divided into rigid segments linked by paired springs. One spring in each pair is made of shape memory alloy, and the other is passive. The actuator and controller units for each SMA spring are installed inside the neighboring rigid segment. The SMA springs are used to generate waves of contraction that run along the body of the robot and so cause movement. Wriggling motion is also used by Micro-Bio-Robots (see Box 14.2).

Box 14.2 Micro-Bio-Robots

Soft robotics has been applied at the microscopic level. Both bacterial and animal flagella have been considered for conferring mobility on microscopic structures.

Artificial bacterial flagella are nonbiological structures whose design is based on that of bacterial flagella. They are actually metallic and consist of two parts: a square, flat head and a helical tail. The head is magnetic and is made from thin metal layers of chromium, nickel, and gold. The tail is a flat ribbon-like (nonmagnetic) metal helix. Applying a rotating magnetic field makes the head rotate, and the attached helical tail follows suit. This generates a spiral motion like that of a bacterial flagellum and generates thrust.

Perhaps more practical, and definitely more attention grabbing, is the "sperm-bot." This is neither a soft robot nor a true cyborg. Rather, a sperm cell is trapped inside a metallic microtube. The sperm cell is used as a motor. The metallic microtubes are magnetic; consequently, external magnets can remotely control their direction (Fig. B). The tubes used are 50 microns long by 5–8 microns wide. The sperm-bots are made by putting microtubes into liquid containing bull sperm. The tubes are wide enough at one end for sperm to enter, but too narrow at the far end for them to escape.

At present, sperm-bots are experimental. Suggested uses include delivering drugs to specific locations or guiding specific sperm to a chosen egg cell. The sperm are inherently harmless and provide their own power.

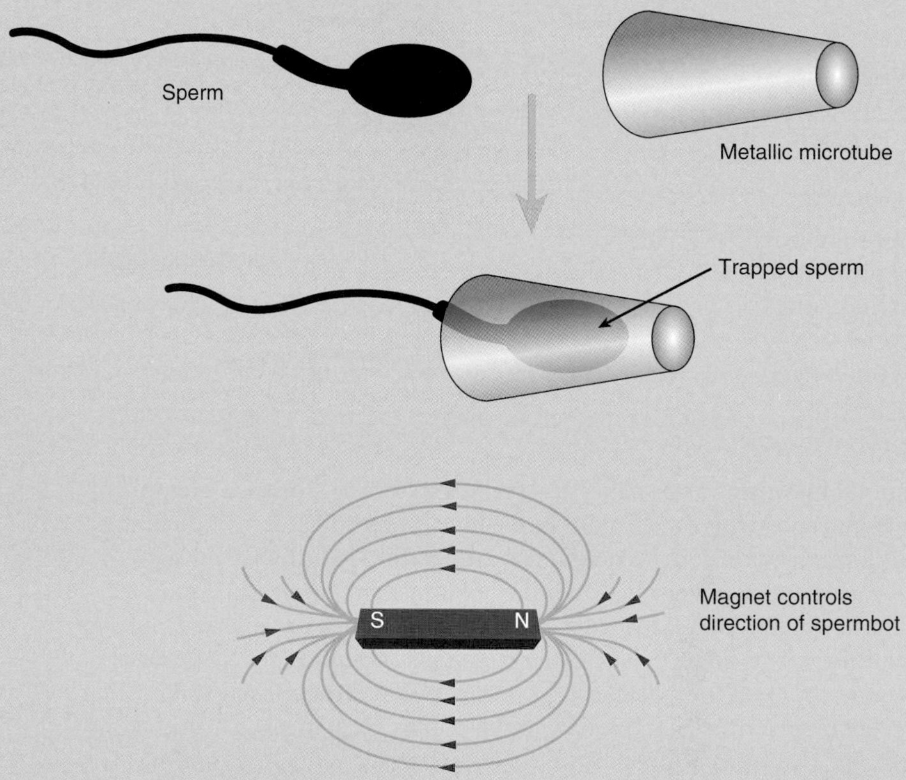

FIGURE B Micro-Bio-Robot Driven by Sperm
An external magnet guides a sperm cell trapped inside a metallic tube.

Soft robotics uses nonrigid biological structures as a model for constructing robots. Both whole organisms, such as worms, and organelles, such as flagella, have been used as models.

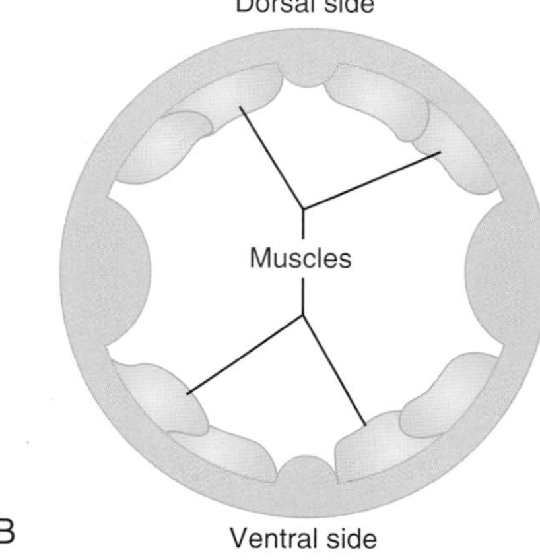

FIGURE 14.8
Muscle Arrangement in Nematode Worm
(A) In cross-section, nematode worms have four muscle blocks. (B) Even though they are not divided into segments externally, their muscle blocks are arranged segmentally along the worm. In *Caenorhabditis elegans,* there are 24 such segmental units that overlap and which are controlled by 12 nerve ganglia. From Yuk H et al. (2011). Shape memory alloy-based small crawling robots inspired by C. elegans. *Bioinspiration Biomimetics* **6**(4) Article 046002 with permission. Figures kindly provided by Jennifer Shin, Department of Mechanical Engineering, KAIST, Daejeon, Republic of Korea.

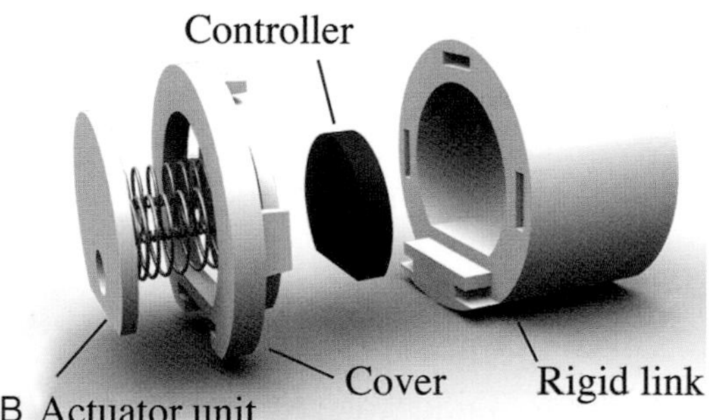

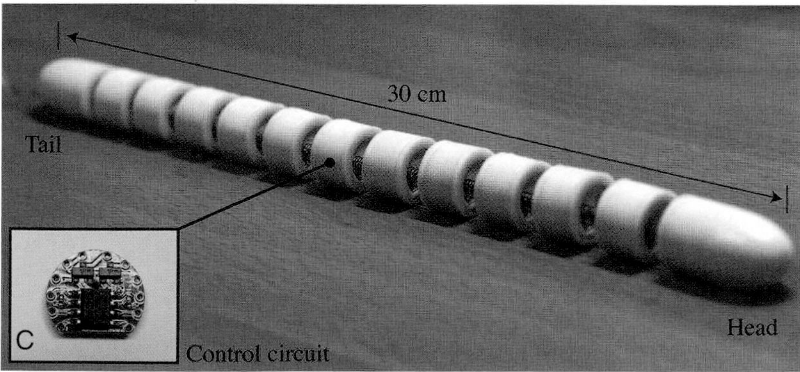

FIGURE 14.9
Assembly of Nematode-Inspired Robot
(A) Each actuator unit consists of a printed circuit board plus a pair of springs: one made of shape memory alloy and a passive spring. (B) Each segment of the worm robot consists of an actuator plus a controller that is inserted inside a rigid link. (C) The complete robot has 12 segments. (D) Close-up of wiring and actuator unit. From Yuk H et al. (2011). Shape memory alloy-based small crawling robots inspired by C. elegans. *Bioinspiration Biomimetics* **6**(4) Article 046002 with permission. Figures kindly provided by Jennifer Shin, Department of Mechanical Engineering, KAIST, Daejeon, Republic of Korea.

Summary

Small, computerized gadgets have massively infiltrated modern-day human culture. Not surprisingly, these gadgets have also impacted the area of biotechnology. The ubiquitous cell phone is now being used for an ever-growing variety of diagnostic and analytical operations. Remote control is another booming area of communication technology that has also benefited from miniaturization. The behavior of genes, cells insects, and higher organisms has been placed, to a varying extent, under remote control. Not only has robotics been applied to living organisms, but the reverse is also happening: living creatures are now being used as inspiration for robot design, an area known as soft robotics.

End-of-Chapter Questions

1. Which of the following is a molecular biology monitor for cell phones that uses quantum dot technology?
- **a.** blood glucose
- **b.** ELISA
- **c.** PCR analysis
- **d.** bacterial detection
- **e.** on screen diagnosis

2. Antibodies are used in which of the cell phone attachment applications?
- **a.** bacterial detection
- **b.** blood glucose
- **c.** endoscopy
- **d.** on screen diagnosis
- **e.** heart rate

3. Which word/phrase does not describe CoralBot?
- **a.** autonomous
- **b.** underwater
- **c.** coral reef maintenance
- **d.** cyborg
- **e.** all of the above describe CoralBot

4. Soft robotics would be used to mimic ______________.
- **a.** bone
- **b.** exoskeletons
- **c.** flagella
- **d.** scalpels
- **e.** armor

5. Which statement is not true?
- **a.** Temperature-sensitive proteins may be coupled to iron nanoparticles, which act as a radio-controlled antenna.
- **b.** A potential model of radio control of genes involved a membrane-bound protein called Calcineurin.
- **c.** TRPV1 is a temperature-sensitive protein channel.
- **d.** Iron oxides absorb radio waves and generate localized heat.
- **e.** TRPV1 controls calcium ion entry.

6. To increase insect cyborg battery life, implantable fuel cells use _______________ to generate energy.

a. glucose
b. trehalose
c. chitin
d. monosaccharides
e. starch

7. The imitation of biological structures in engineering is called _______________.

a. soft robotics
b. hard robotics
c. SMAs
d. Micro-Bio-Robots
e. biomimetics

8. "Sperm bots" _______________.

a. are neither soft robots nor true cyborgs
b. use sperm as motors
c. might provide a novel mechanism for drug delivery
d. are sperm trapped within metallic microtubes
e. are all of the above

Further Reading

Chin, C. D., Cheung, Y. K., Laksanasopin, T., Modena, M. M., Chin, S. Y., Sridhara, A. A., et al. (2013). Mobile device for disease diagnosis and data tracking in resource-limited settings. *Clinical Chemistry, 59*(4), 629–640.

Hopkins, J. K., Spranklin, B. W., & Gupta, S. K. (2009). A survey of snake-inspired robot designs. *Bioinspiration Biomimetics, 4*(2) Article 021001.

Huang, H., Delikanli, S., Zeng, H., Ferkey, D. M., & Pralle, A. (2010). Remote control of ion channels and neurons through magnetic-field heating of nanoparticles. *Nature Nanotechnology, 5*, 602–606.

Kim, S., Laschi, C., & Trimmer, B. (2013). Soft robotics: a bioinspired evolution in robotics. *Trends in Biotechnology, 31*(5), 287–294.

Lee, D., Chou, W. P., Yeh, S. H., Chen, P. J., & Chen, P. H. (2011). DNA detection using commercial mobile phones. *Biosensors and Bioelectronics, 26*(11), 4349–4354.

Magdanz, V., Sanchez, S., & Schmidt, O. G. (2013). Development of a sperm-flagella driven micro-bio-robot. *Advanced Materials, 25*(45), 6581–6588.

Rasmussen, M., Ritzmann, R. E., Lee, I., Pollack, A. J., & Scherson, D. (2012). An implantable biofuel cell for a live insect. *Journal of the American Chemical Society, 134*(3), 1458–1460.

Sato, H., Berry, C. W., Peeri, Y., Baghoomian, E., Casey, B. E., Lavella, G., et al. (2009). Remote radio control of insect flight. *Frontiers in Integrative Neuroscience, 3* Article 24.

Stanley, S. A., Gagner, J. E., Damanpour, S., Yoshida, M., Dordick, J. S., & Friedman, J. M. (2012). Radio-wave heating of iron oxide nanoparticles can regulate plasma glucose in mice. *Science, 336*, 604–608.

Stedtfeld, R. D., Tourlousse, D. M., Seyrig, G., Stedtfeld, T. M., Kronlein, M., Price, S., et al. (2012). Gene-Z: a device for point of care genetic testing using a smartphone. *Lab on a Chip, 12*(8), 1454–1462.

Yeo, W. H., Kim, Y. S., Lee, J., Ameen, A., Shi, L., Li, M., et al. (2013). Multifunctional epidermal electronics printed directly onto the skin. *Advanced Materials, 25*(20), 2773–2778.

Yuk, H., Kim, D., Lee, H., Jo, S., & Shin, J. H. (2011). Shape memory alloy-based small crawling robots inspired by C. elegans. *Bioinspiration Biomimetics, 6*(4) Article 046002.

Zhang, L., Peyer, K. E., & Nelson, B. J. (2010). Artificial bacterial flagella for micromanipulation. *Lab on a Chip, 10*(17), 2203–2215.

CHAPTER 15

Transgenic Plants and Plant Biotechnology

Biotechnology

http://dx.doi.org/10.1016/B978-0-12-385015-7.00015-6

INTRODUCTION

One of the most promising fields within biotechnology is the development of transgenic plants. The ability to transfer genes from any organism into a plant unleashes a nearly boundless capacity to improve crops and other plants of interest. In this chapter, we explore the basis of this powerful and revolutionary technology.

HISTORY OF PLANT BREEDING

For thousands of years, humans have improved crop plants and domestic animals by selective breeding, mostly by trial and error. Over many years, breeders learned that improving their crops and animals has a biological basis. In fact, the father of genetics, Gregor Mendel, experimented with the common pea. He studied easily identified traits such as round versus wrinkled seeds or yellow versus green seeds. Mendel would take the pollen from one plant and put it on the stigma of another plant (Fig. 15.1), a procedure called **cross-pollination**. His experiments showed that plants have some traits that may dominate others. For example, a cross between a yellow-seeded pea plant and a green-seeded pea plant gave only seeds that were yellow. This work was published in 1865, but no one understood the importance of these findings until well after Mendel's death.

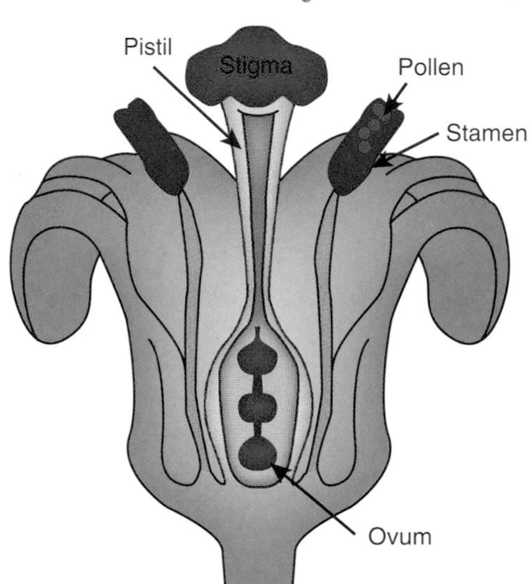

FIGURE 15.1
The Reproductive Organs of a Typical Flowering Plant
Pollen grains are the male reproductive cells of the plant. They are made in the anther (brown), the top portion of the stamen. The female reproductive cells, the ova, are sequestered in the ovary. Pollen reaches the ova via the stigma, which is attached to the ovary by the pistil.

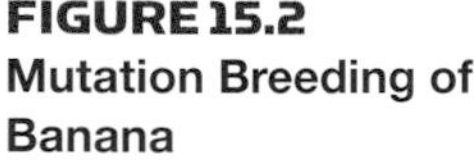

Many scientists around the world still use traditional breeding techniques to enhance crop yields, increase resistance to various pests or diseases, or increase the tolerance of a particular crop to heat, cold, drought, or wet conditions. Although simply crossing a high-yielding plant with another can produce offspring with even higher yields than either parent, the process is long and tedious. Many thousands of plants must be cross-pollinated to find the one offspring with higher yield. The crosses must be done by hand; that is, pollen must be taken from one plant and manually placed on another. In addition, the possibility of finding improved traits is limited by the amount of genetic diversity already present in the plants. Consequently, if the two plants that are crossed share many of the same genes, the amount of possible improvement is limited. If a plant has no genes for disease resistance, there is no way traditional cross-pollination will develop that trait. Therefore, scientists have searched for better ways to improve plants.

In the 1920s, scientists realized that mutations could be induced in seeds by using chemical mutagens or by exposure to X-rays or gamma rays. Although useful, the outcome of such treatments is even less predictable than traditional breeding methods. Nonetheless, **mutation breeding** has been successful, especially in the flower world. For example, new colors and more petals have been expressed in flowers such as tulips, snapdragons, roses, chrysanthemums, and many others. Mutation breeding has also been tried on vegetables, fruits, and crops. Some of the varieties of food we eat today were developed using this method. For instance, peppermint plants that are resistant to fungus were generated this way.

FIGURE 15.2
Mutation Breeding of Banana
Explants of banana are grown from shoots that were treated with chemical mutagens. The resulting plants will be tested for improved disease resistance, higher yield, and heritability. Photo taken at the IAEA's Plant Breeding Unity, Seibersdorf, Austria (Photo D. Calma/IAEA).

To perform mutation breeding, researchers expose a large number of seeds or totipotent tissues to the mutagen to generate various mutations in their DNA. The seeds are then planted and cultivated (Fig. 15.2). However, the majority of seeds or tissues are killed by the treatment. After the viable seeds or tissues are grown, the fruit, flower, or grain of the plant is tested for improvements. If one plant is found with a desirable trait, then its progeny are tested for the trait. Obviously, novel traits are useful only if they are **heritable**, that is, passed from one generation to the next. Because only one original mutant plant would gain any particular desirable trait, this plant would need to be propagated a long time before any of the fruit, grain, or flower would be sold to market. It is important to note that the actual products sold to the consumer were never exposed to the mutagen. Today, chemical mutagens are still used, but molecular biology techniques can identify the actual gene associated with the desired phenotype. (See later discussion.)

Recently, the emergence of molecular biology has opened the door to a much more predictable way to enhance crops. Scientists have discovered ways to move genes from foreign sources into a specific plant, resulting in a **transgenic plant**. The foreign gene, or **transgene**, may confer specific resistance to an insect, protect the plant against a specific herbicide, or enhance the vitamin content of the crop. The major difference between transgenic technology and traditional breeding is that a plant can be transformed with a gene from any source, including animals, bacteria, or viruses as well as other plants, whereas traditional cross-breeding methods move genes only between members of a particular genus of plants. Furthermore, transgenes can be placed in precise locations within the genome and have known functions that have been evaluated extensively before being inserted into the plant. In traditional breeding, on the other hand, the identity of genes responsible for improving the crop is rarely known.

Traditional plant breeding techniques are slow, tedious, and introduce random and uncontrollable changes to the genome. Transgenic plant technology greatly improves this process. It allows for the introduction of completely foreign genes from other organisms, called transgenes, into precise locations within the genome.

PLANT TISSUE CULTURE

One major advantage of plants is that they can often be regenerated from just a single cell. In other words, many plant cells—for example, those in the apical meristem—are **totipotent** and retain the ability to develop into fully mature plants (Fig. 15.3). There is no absolute separation of the germline from the somatic cells in plants, unlike animals. This unique feature of plants allows scientists to grow and manipulate plant cells in culture and then regenerate an entire plant from the cultured cells.

Plant tissue culture can be done on either a solid medium in a Petri dish, called **callus culture**, or in liquid, called **suspension culture** (Fig. 15.4). In both cases, a mass of tissue or cells, known as an **explant**, must be removed from the plant of interest. In callus culture, the tissue can be an immature embryo, a piece of the apical meristem (the region where new plant shoots develop), or a root tip. For suspension culture, cells must be dissociated from one another. Suspension culture usually uses **protoplasts** (plant cells from which the cell wall is removed), microspores (immature pollen cells), or macrospores (immature egg cells). The cells are then cultured with a mixture of nutrients and specific plant hormones that induce the undifferentiated cells to grow.

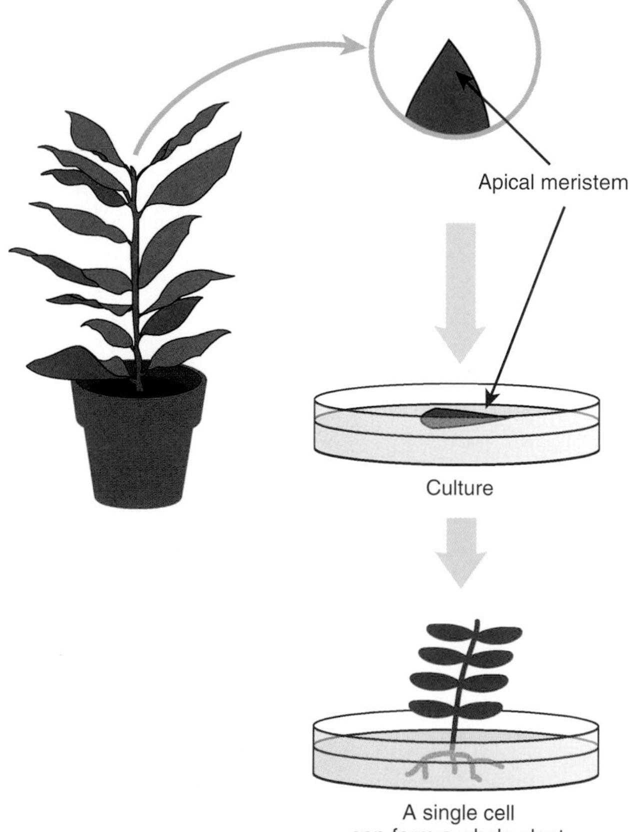

FIGURE 15.3 An Entire Plant Can Be Regenerated from a Single Cell
Small samples of tissue, or even single plant cells, may be cultured *in vitro*. Under appropriate conditions, they may regenerate into complete plants.

CALLUS CULTURE

Nutrients and hormone

Callus

Shoots form

Roots form

A

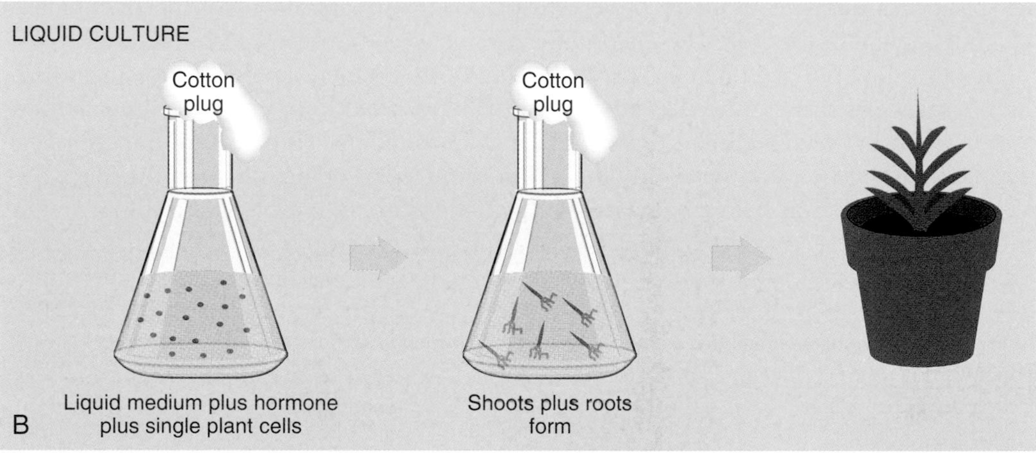

FIGURE 15.4
Callus or Suspension Culture of Cells Can Regenerate Entire Plants

(A) In callus culture, a mass of undifferentiated cells grows on a solid surface. (B) In suspension culture, separated single cells are grown in liquid. Both types of cultures can develop shoots and roots with appropriate manipulation of plant hormone levels. (C) *Oldenlandia affinis* plants growing in callus culture (a), liquid suspension culture (b), shoot tips inoculum (c), and agitated hydroponic culture (d). Used with permission from Dörnenburg H, Frickinger P, Seydel P (2008). Plant cell-based processes for cyclotides production. *J Biotechnol* **135**(1), 123–126.

Box 15.1 2,4-D: Herbicide or Plant Hormone?

The compound 2,4-dichlorophenoxyacetic acid, or 2,4-D, is a herbicide used to kill any dicot plant tissue. The substance is a synthetic auxin, which is a type of plant hormone that is absorbed by the leaves of a plant. When a concentrated amount of the hormone is applied to a dicot, the tissues respond with uncontrolled and unsustained growth, which ultimately kills them. Although 2,4-D is very effective for dicots or broad leaf plants, the substance has no effect on any monocots, such as wheat, corn, rice, and grass. That is why it is marketed as a grass herbicide that can kill weeds such as dandelions but not harm the grass plants.

Different types of plants respond to different hormones. Wheat cells respond to 2,4-dichlorophenoxyacetic acid (2,4-D), an analog of the plant hormone auxin, and tomato plants respond to the hormone cytokinin (see Box 15.1). Both trigger the cells to dedifferentiate and grow. In callus culture, undifferentiated cells form a crystalline white layer on top of the solid medium, called the **callus**. After about a month of growth, the mass of undifferentiated cells can be transferred to a medium with a lower concentration of hormone or with a different hormone. Decreasing the amount of hormone allows some of the undifferentiated callus cells to develop into a plant shoot. In most cases, the small shoots look like new blades of grass growing from the mass of cells. After another 30 days, the hormone is removed completely, which allows root hairs to start growing from some of the shoots. After another 30 days, small plants can be isolated and transferred into soil. In liquid culture, hormones are also used, but the shoot and root tissues grow simultaneously.

Plant shoots can also be cultured using shoot tip inoculum or agitated hydroponic cultures (Fig. 15.4). The two systems rely on growing a tip of the plant that has been cleaned and sterilized by soaking in a dilute solution of bleach. The pieces of the plant are then incubated with nutrients and hormones as before to initiate more shoot growth or root growth. They are then transplanted to make them hardy for growing in dirt.

Because plant tissue culture allows many plants to be produced from one source, the technique is useful for making clones of one particular plant. If a very rare plant is identified, it can be propagated using tissue culture. Only a small cutting is needed to generate many identical progeny. Certain special plant varieties that are hard to maintain by producing seed can be maintained for the long term in culture. Plant cell culture has also been used for mutation breeding. Rather than using seeds, researches expose undifferentiated cells to the mutagen and regenerate plants from the mutagenized cells. Mutagens are more effective on the exposed cells of a callus rather than the protected cells within a seed.

Merely growing plants by tissue culture may induce mutations. The process of regenerating a plant from a single cell may cause three different types of alterations. First, temporary physiological changes can occur in the regenerated plant. For example, when blueberry plants are regenerated via tissue culture, the plants are much shorter. These changes are not permanent, and after a few years growing in the field, the regenerated blueberries are no different than any other blueberries. Second, tissue culture may induce an **epigenetic change**. This is a modification that persists throughout the lifetime of the regenerated plant and may be passed on to a few subsequent generations, but the change does not permanently alter the cultivar. Epigenetic changes are often due to alterations in DNA methylation (see Chapter 8). Third, true **genetic changes** (i.e., mutations) affect the regenerated plant and all its progeny. These changes may be due to point mutations, changes in ploidy level, chromosome rearrangements, activation of transposable elements, or changes to the chloroplast or mitochondrial genomes. These types of mutations are relatively common, but they can be decreased by carefully regulating the available nutrients and hormones during tissue culture.

Because many plant cells are totipotent, plant tissue can be grown in culture. Callus culture and suspension culture cultivate plant cells from an undifferentiated state into small plants by altering hormone concentrations. Plant tissue culture can induce different types of alterations: some that are truly genetic, but others that are temporary, such as epigenetic and physiological changes.

GENETIC ENGINEERING OF PLANTS

In general, plants are genetically engineered to increase yield or to confer resistance to drought or pests. Finding desirable genes can be difficult because multiple interacting genes usually control such traits. In addition, such genes may play other roles in plant physiology or development. So far, most successful genetic engineering of plants has relied on inserting one or a few genes that supply simple, yet useful, properties.

For example, resistance to the herbicide glyphosate is due to a single gene. Making a crop such as soybeans resistant to glyphosate allows the farmer to kill the weeds in the field without harming the soybeans. Another desirable trait often due to a single gene is the production of toxins that kill harmful insects. Also, a two-gene pathway has been engineered into rice to make it more resistant to drought. Before discussing these examples, however, we must first describe how transgenic plants are created.

Ti Plasmid

Plants suffer from tumors although they are quite different from the cancers of animals. The most common cause is the Ti plasmid (tumor-inducing plasmid), which is carried by soil bacteria of the *Agrobacterium* group. The most important aspect of the infection is that a specific segment of the Ti plasmid DNA is transferred from the bacteria to the plant. Because most DNA transfers occur only between closely related organisms, the ability of *Agrobacterium* to transfer DNA from one domain to another makes it an important tool for the genetic engineering of plants.

In nature, *Agrobacterium* is attracted to plants that have minor wounds by phenolic compounds such as acetosyringone, which are released at the wound (Fig. 15.5). These chemicals induce the bacteria to move and attach to the plant via a variety of cell surface receptors. The same inducers activate expression of the virulence genes on the Ti plasmid that are responsible for DNA transfer to the plant. This is under control of a two-component regulatory system (see Chapter 2). At the cell surface, the sensor, VirA, is autophosphorylated when it detects the plant phenolic compounds. Next, VirA transfers the phosphate to the DNA-binding protein, VirG, which activates transcription of the *vir* genes of the Ti plasmid. Two of the gene products (VirD1 and VirD2) clip the T-DNA borders to form a single-stranded immature T-complex. VirD2 then attaches to the 5′ end of the T-DNA, and bacterial helicases unwind the T-DNA from the plasmid. The single-stranded gap on the plasmid is repaired, and the T-DNA is coated with VirE2 protein to give a hollow cylindrical filament with a coiled structure. This is the mature form of T-DNA and traverses into the plant.

T-DNA transfer to the plant is similar to bacterial conjugation. First, *Agrobacterium* forms a pilus. This rod-like structure forms a connection with the plant cell and opens a channel through which the T-DNA is actively transported into the plant cytoplasm. Both pilus and transport complex consist of proteins that are *vir* gene products. Once inside the plant cytoplasm, T-DNA is imported into the nucleus. Both VirE2 and VirD2 have nuclear localization signals that are recognized by plant cytosolic proteins. These proteins take the T-complex to the nucleus, where it is actively transported through a nuclear pore. The single T-DNA strand is integrated directly into the plant genome and converted to a double-stranded form. The integration requires DNA ligase, polymerase, and chromatin remodeling proteins, all of which are supplied by the plant.

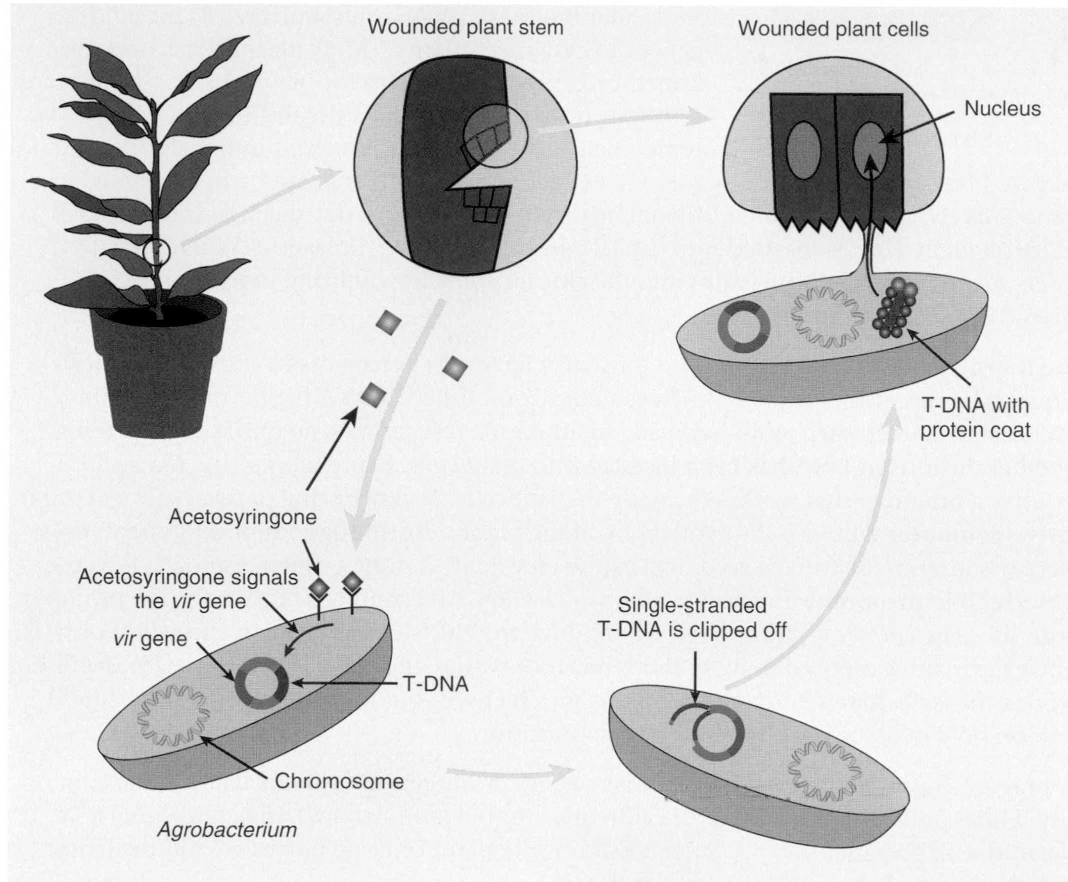

FIGURE 15.5 ***Agrobacterium* Transfers Plasmid DNA into Infected Plants** *Agrobacterium* carrying a Ti plasmid is attracted by acetosyringone from a wounded plant stem. The Ti plasmid is cut by endonucleases to release single-stranded T-DNA, which is covered with protective proteins and transported into the plant cell through a conjugation-like mechanism. The T-DNA enters the plant nucleus, where it integrates into plant chromosomal DNA.

Once integrated, the genes in the T-DNA are expressed. These genes have eukaryote-like promoters, transcriptional enhancers, and poly(A) sites and therefore are expressed in the plant nucleus rather than in the original bacterium. The proteins they encode synthesize two plant hormones: auxin and cytokinin. Auxin makes plant cells grow bigger, and cytokinin makes them divide. The infected plant cells begin to grow rapidly and without control, resulting in a tumor (Fig. 15.6). T-DNA also carries genes for the synthesis of a variety of different amino acid and sugar phosphate derivatives called **opines**. Strains of *Agrobacterium* are differentiated based on the kinds of opines they produce. Opines are made by plant cells that contain T-DNA, but they cannot be utilized by the plant. The Ti plasmid, which is still inside the *Agrobacterium,* carries genes that allow the bacteria to utilize them. Thus, the bacterium has hijacked the plant's metabolic machinery into supplying the bacteria with food. Other bacteria, which might be present by chance, cannot use opines because they do not possess the genes for their uptake and metabolism.

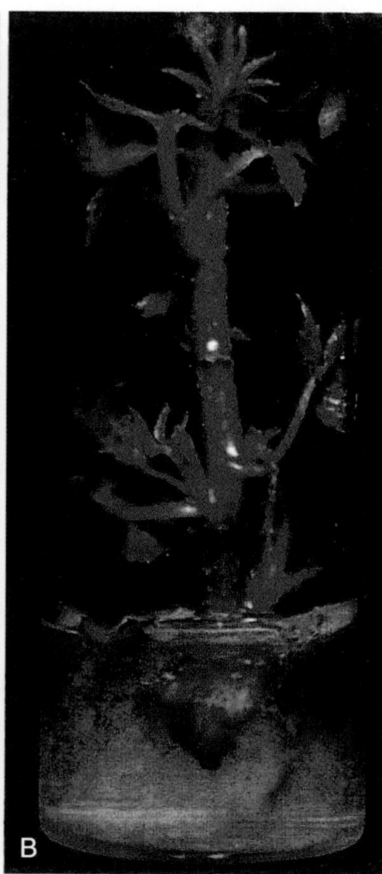

FIGURE 15.6 Plant Tumors Crown gall tumors are caused by *Agrobacterium tumefaciens* (A), but plants expressing inhibitors of key proteins for *Agrobacterium* infection do not develop any tumors (B). Reprinted with permission from Escobar MA, Leslie, CA, McGranahan, GH, Dandekar, AM (2002). Silencing crown gall disease in walnut (*Juglans regia* L.) *Plant Sci* **163**, 591–597.

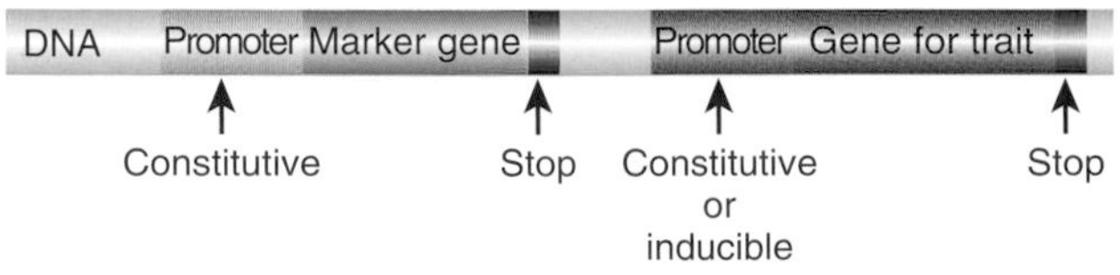

DNA Promoter Marker gene Promoter Gene for trait

Constitutive Stop Constitutive or inducible Stop

FIGURE 15.7 Essential Elements for Carrying a Transgene on Ti Plasmids
The T-DNA segment contains both a transgene and a selective marker or reporter gene. These have separate promoters and termination signals. The marker or reporter gene must be expressed all the time, whereas the transgene is often expressed only in certain tissues or under certain circumstances and usually has a promoter that can be induced by appropriate signals.

Molecular biologists have commandeered Ti plasmids to genetically engineer plants. The Ti plasmid has been disarmed by removing the genes for plant hormone and opine synthesis from the T DNA and streamlined the plasmid by removing genes that are not involved in transferring the T-DNA. (These smaller plasmids are much easier to work with and can be manipulated in *Escherichia coli* rather than their original host, *Agrobacterium*.) A transgene of interest, such as an insect toxin gene, is inserted into the T-DNA region of the Ti plasmid. When the T-DNA enters the plant cell and integrates into the chromosome, it will bring in the transgene instead of causing a tumor.

The transferred region of the plasmid must also have other elements in order for this technique to be successful (Fig. 15.7). The genetically modified T-DNA region must contain a selectable marker, such as an herbicide or antibiotic resistance gene that is used to track whether the foreign DNA has been inserted into plant cells. Expression of the transgene requires a promoter that works efficiently in plant cells. It may be one of two types. A **constitutive promoter** will turn the gene on in all the plant cells throughout development; thus, every tissue, even the fruit or seed, will express the gene. A more refined approach is to use an **inducible promoter** that acts as an on/off switch. An example of this is the *cab* promoter from the gene encoding chlorophyll *a/b* binding protein. This promoter is turned on only when the plant is exposed to light; therefore, root tissues and tubers such as potatoes will not express the gene. Many different promoters may be used, but ideally, the promoter should turn on only in tissues that need transgene function.

In practice, *Agrobacterium* is used to transfer genes of interest into plants using tissue culture. Either protoplasts or a piece of callus are cultured with *Agrobacterium* harboring a Ti plasmid with modified T-DNA. After coculture, the plant cells are harvested and incubated with the herbicide or antibiotic used as the selectable marker. This kills all the cells that were not transformed with T-DNA or failed to express the genes on the T-DNA (Fig. 15.8). The transformed plant cells are then induced to produce shoot and root tissue by altering the hormone conditions in the medium. The small transgenic plants can then be screened for transgene expression levels. (See later discussion.)

Recently, a method for ***in planta Agrobacterium* transformation** was developed and has revolutionized plant transgenics. *In planta* transformation is also known as the **floral dip** method (Fig. 15.9). The method was developed using the model plant *Arabidopsis* but has been extended to other plants, such as wheat and maize. First, *Arabidopsis* plants are grown until flower buds begin to form. These buds are removed and allowed to regenerate for a few days. Once they begin to regenerate, the plants are dipped into a suspension of *Agrobacterium* containing a surfactant, which decreases surface tension and allows the *Agrobacterium* to adhere to the plant and transfer its T-DNA. Because the flower buds are just beginning to form, the T-DNA becomes part of the germline through the ovarian tissue. The plant is allowed to finish growing and produce seed. These seeds are harvested and grown in selective media to find those that have integrated and expressed T-DNA. Although the method gives a low percentage of transformants, so many seeds can be screened that the overall procedure works well.

Particle Bombardment

Another strategy for getting a transgene into plant tissue is particle bombardment (Fig. 15.10). A gun blasts microscopic metal particles carrying DNA through the tough plant cell walls. Unlike Ti plasmid transfer by *Agrobacterium*, this technique works with all types of plants. Though the technique is nonspecific, it has been very successful.

First, either a leaf disk (a round piece of leaf tissue) or a callus is isolated from the plant, placed on a dish, and put in a vacuum chamber. The DNA to be inserted (carrying the

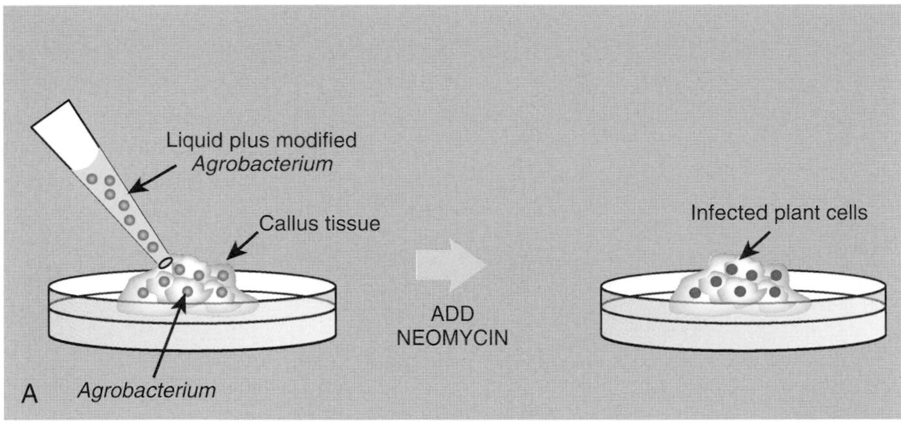

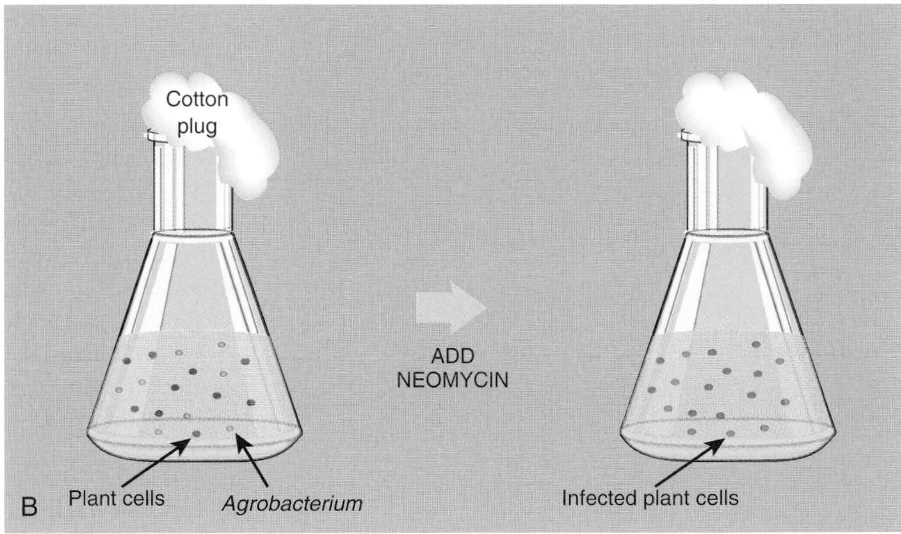

FIGURE 15.8 Transfer of Modified Ti Plasmid into a Plant *Agrobacterium* carrying a Ti plasmid is added to plant tissue growing in culture. The T-DNA carries an antibiotic resistance gene (neomycin in this figure) to allow selection of successfully transformed plant cells. Both callus cultures (A) and liquid cultures (B) may be used in this procedure.

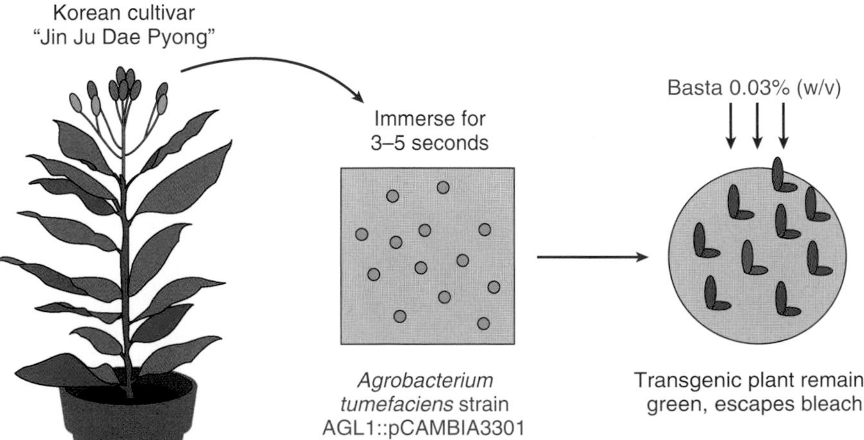

FIGURE 15.9 Floral Dip Method of Plant Transformation Flower buds exposed to *Agrobacterium* containing modified T-DNA can result in the production of transgenic seeds. Adapted from Curtis IS (2003). The noble radish: past, present and future. *Trends Plant Sci* **8**, 305–307.

transgene, proper regulatory elements, and selectable marker) is coated on microscopic gold beads. The beads are placed at the end of chamber. One variant of the method uses a blast of air or helium to drive the filter containing the gold beads toward the stop screen and sample. In the first gene guns, actual firearm blanks were used to accelerate the bullet. Between the bullet and plant tissue is a stop plate. Filter and gold beads hit this stop screen, the DNA-coated beads are thrown forward into the plant tissue. An alternative method is to accelerate the beads by a strong electrical discharge. The high voltage vaporizes a water droplet, and the resulting shock wave propels a thin metal sheet covered with the particles at a mesh screen. The screen blocks the metal sheet but allows the DNA-coated particles to accelerate through into the plant tissue. One advantage of this method is that the strength of the electrical discharge can be controlled; therefore, the amount of penetration into the tissue can be regulated.

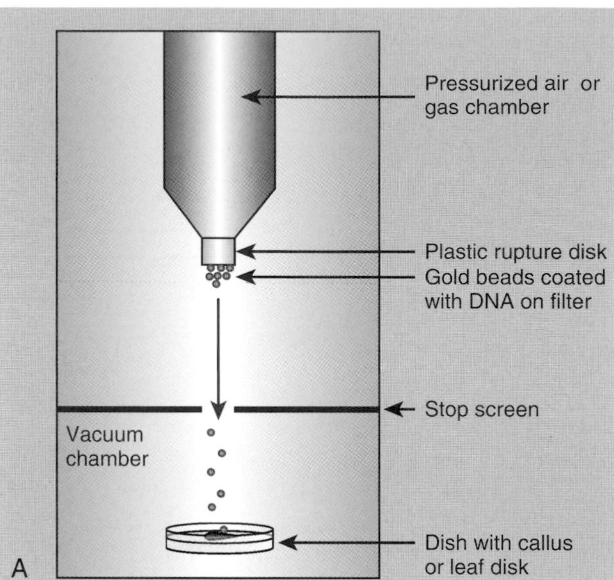

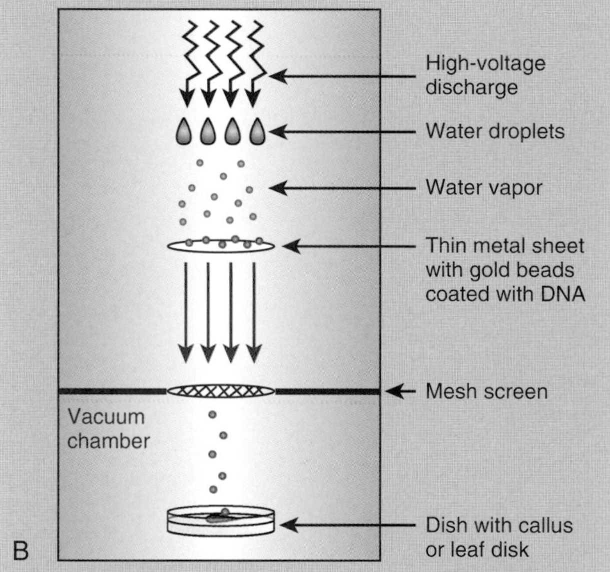

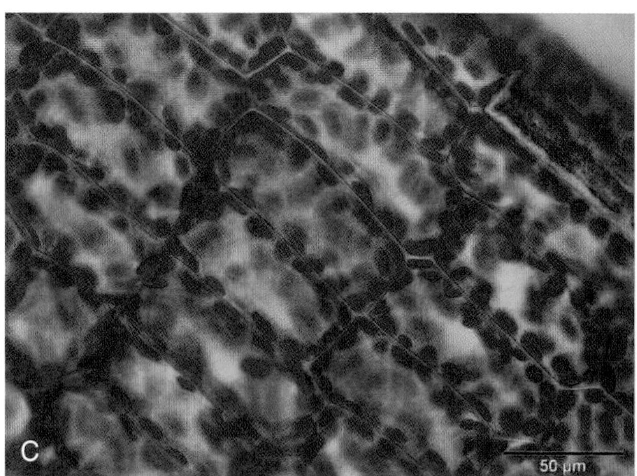

FIGURE 15.10 Two Types of Particle Guns for DNA
A gene gun that operates via (A) pressurized air or (B) high-voltage discharge is depicted. In both cases, the stop plate halts the projectile, and the microscopic metal particles carrying the DNA penetrate the plant tissue. In (C), the image shows a typical plant cell with thick cell walls in which the gold beads must penetrate.

When the beads penetrate the tissue, some will enter the cytoplasm or nucleus of the leaf or callus cells. The DNA dissolves off the beads inside the cells, and it is free to recombine with the chromosomal DNA of the plant (Fig. 15.11). The leaf or callus tissue is then transferred to selection media where the cells that integrated the DNA carrying the selectable marker are able to grow, but other cells die. The transformed plants are regenerated using tissue culture techniques and then screened for the gene of interest.

Particle guns have also been used to transform animal cells, the mitochondria of yeast, the germline of *C. elegans,* and the chloroplasts of *Chlamydomonas,* a small green alga.

Detecting the Inserted DNA

There are different ways to detect the inserted DNA. Perhaps the simplest is to include a selectable marker or reporter gene on the same segment of DNA as the transgene. One widely used reporter gene is ***npt***, which encodes **neomycin phosphotransferase**. This enzyme confers neomycin resistance by attaching a phosphate group to the molecule. Transformed cells are directly selected with the antibiotic neomycin, which kills any cells that did not integrate the DNA.

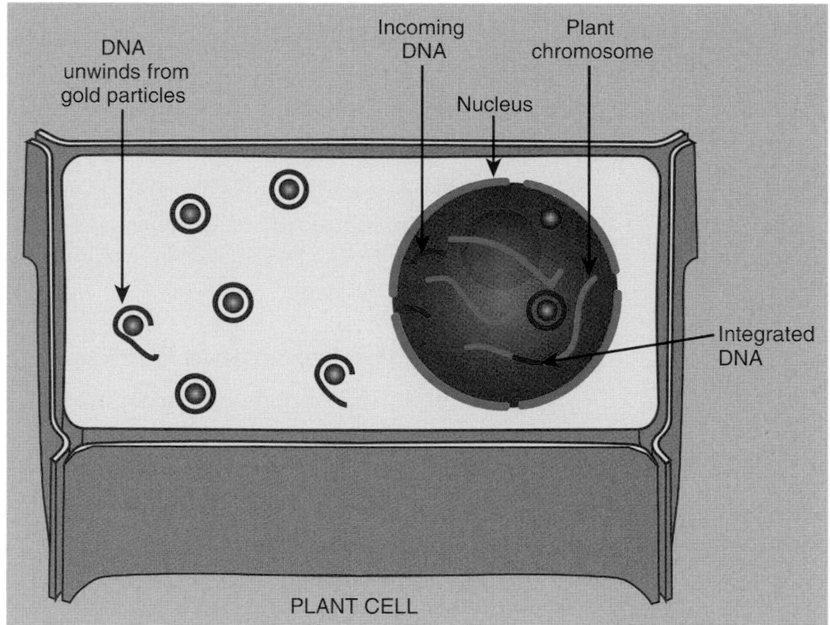

FIGURE 15.11 DNA Carried on Microscopic Gold Particles Can Integrate into Plant Chromosomes After penetrating the cell, the DNA unwinds from around the gold carrier particle. Some of the DNA enters the nucleus and successfully integrates into the plant chromosomes.

An alternative reporter gene encodes for **luciferase**. This enzyme emits light when provided with its substrate, **luciferin** (Fig. 15.12A). Luciferase is found naturally in assorted luminous creatures, from fireflies to luminous squid. If DNA carrying the eukaryotic ***luc*** gene is successfully incorporated into a target plant cell, light will be emitted when luciferin is added. Although high-level expression of luciferase can be seen with the naked eye, usually the amount of light is small and must be detected with a sensitive electronic apparatus such as a scintillation counter, a photocell detector, or a charge-coupled device (CCD) camera.

This reporter gene has a key advantage. The luciferase protein is not stable for long in the plant, so the amount of active protein correlates with the level of gene expression at any

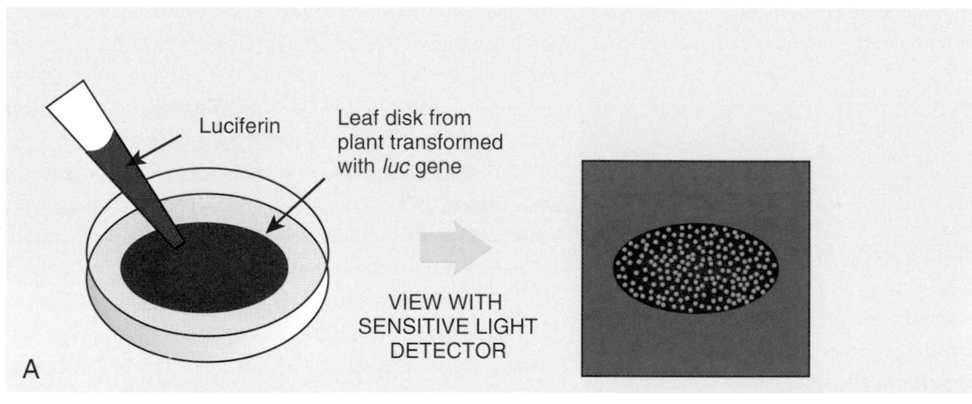

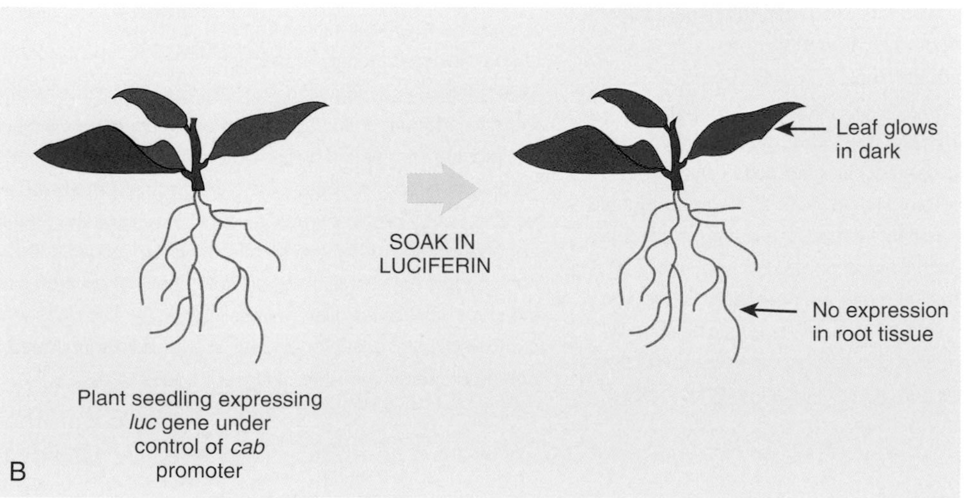

FIGURE 15.12 Luciferase as a Reporter in Plant Tissues Plant tissue carrying the *luc* gene for firefly luciferase emits blue light when provided with the substrate luciferin. (A) A leaf disk is viewed by a photocell detector. (B) The *luc* gene in a seedling is expressed under control of an inducible promoter.

given time. Therefore, *luc* can be used to determine the activity of specific promoters. When a particular condition activates the promoter, luciferase is produced and the plant glows. The *cab* promoter, for instance, is induced in the presence of light. Transgenic *Arabidopsis* plants containing the *luc* construct under control of the *cab* promoter therefore produce luciferase only in photosynthetic tissue that is exposed to light (Fig. 15.12B).

Once cells successfully expressing the reporter gene are identified, they are regenerated into plants using tissue culture techniques. The plants obtained are then screened for the transgene. PCR and DNA sequencing are used to confirm the presence of the transgene and its location in the genome.

Removing the Selectable Marker

One complaint about transgenic technology in agriculture is the inclusion of a selectable marker or reporter gene. In particular, many consumers have unfounded concerns about the presence of an antibiotic resistance gene in the food supply. (See Box 15.2.) However, genetic techniques are available that allow removal of the reporter gene, after integration of the DNA insert has been verified.

Box 15.2 Common Myths and Half-Truths about Transgenic Crops

Genetically modified organisms (colloquially referred to as "GMOs") are extremely controversial, mostly because those who oppose transgenic technology do not understand it. *All* crops have been genetically modified. The oldest variety of edible corn, for example, bears no resemblance to its ancestor, teosinte. These changes have occurred through cross-pollination and selective breeding. Thus, it is more accurate to use the term *transgenic crop* instead of *genetically modified crop*. Here, we dispel many of the myths and half-truths surrounding transgenic crops.

MYTH: Transgenic crops are not regulated.
FACT: Transgenic crops are heavily regulated. The FDA, EPA, and USDA—in addition to foreign governments—all have a say in the process.
MYTH: Transgenic crops are unsafe for human consumption or the environment.
FACT: A review of 1,783 papers and other reports, published in *Critical Reviews in Biotechnology* in 2013, concluded that there was little evidence to suggest that transgenic crops were unsafe for humans or the environment. There was also little evidence to suggest that transgenic crops adversely affected biodiversity.
MYTH: Bt toxin kills Monarch butterflies.
FACT: This myth began with a controversial *Nature* paper, which concluded that Monarch caterpillars were killed when provided milkweed (their sole food source) that was dusted with corn pollen carrying the Bt gene. However, the study had two key flaws. First, it was conducted in a laboratory. In the field, Monarch larvae and egg-laying adults often avoid milkweed contaminated with Bt corn pollen. Second, about 100 Bt corn pollen grains per square centimeter are necessary for Monarch caterpillars to be affected, but this level of pollen occurs only on milkweed directly surrounding or within a cornfield, not in neighboring fields. At a distance of merely 2 meters from a cornfield edge, Bt corn pollen density drops to 14 grains per square centimeter. Additionally, rainfall can remove more than half the pollen from milkweed.
Furthermore, only certain types of Bt toxins are actually harmful to Monarch butterflies. The only pollen that was consistently toxic was Cry1Ab (Event #176), which made up only about 1% of the entire corn crop in 2000. In 2001, this transgene event was no longer approved in the United States. Other Bt toxins that exhibited lower toxicity made use of different promoters or different versions of the toxin, such as Cry1F and Cry9C. It is not a surprise, therefore, that a study published in *PNAS* concluded that Bt corn pollen had a negligible impact on Monarch butterfly populations.
Another study, published in the journal *Insect Conservation and Diversity* in 2013, claims that a decrease in milkweed plants is the reason Monarch butterfly populations are in decline. Milkweed is killed by the herbicide glyphosate, and the authors imply that the use of more glyphosate-resistant ("Roundup Ready") crops, combined with an increase in application of glyphosate, is responsible for the decline. More research will help reveal if this is in fact true. If it is, possible solutions include creating milkweed sanctuaries or creating glyphosate-resistant milkweed.
MYTH: Transgenes cause allergies.
FACT: Molecular biologists do not use genes or proteins that are known allergens. Through the FDA's "voluntary consultation" process, companies provide information on the potential for allergenicity and toxicity. Several years ago, soybean plants were transformed with a gene from the Brazil nut. The gene was intended to increase the methionine content of soybeans, which would improve them as cattle feed. Because many people are allergic to Brazil nuts, the FDA ordered skin prick tests and immunoassays to determine allergenicity. The transgene was found to cause allergic reactions, and the work was discontinued. None of the transgenic plants were ever released to the public.

Box 15.2 Common Myths and Half-Truths about Transgenic Crops—cont'd

MYTH: Selectable markers cause antibiotic resistance.
FACT: Antibiotic resistance is caused by the overuse and misuse of antibiotics, not by molecular biology techniques. Besides, all transgenic crops in use today have their antibiotic resistance genes removed.
MYTH: Food containing GMOs should be labeled.
FACT: While government policy is determined by more than just scientific considerations, it should be noted that there is no scientific justification to support mandatory labeling because transgenic and nontransgenic foods are nutritionally equivalent. Also, requiring labels may have an unintended consequence: Keeping transgenic and nontransgenic foods separate is logistically difficult and would likely cause an increase in food prices.
HALF-TRUTH: Transgenes can escape into the environment.
FACT: The review discussed previously concluded that transgenes have been detected in wild plants, but it does not appear to be a major problem. Remember, genes also can "escape" from nontransgenic crops because they have been genetically modified by farmers for millennia through artificial selection. Yet few would consider this a substantial problem. Besides, research has shown that planting transgenic corn 50 to 100 meters away from nontransgenic corn greatly reduces cross-pollination.
HALF-TRUTH: Transgenic crops lead to an increase in pesticide use.
FACT: Research has shown that herbicide (e.g., glyphosate) use has increased, but insecticide use has decreased. Note that insecticides are more harmful than herbicides for the environment. Also, it is not necessarily a bad thing that herbicide use has increased. Applying herbicides allows farmers to implement no-till farming, which reduces soil erosion and soil runoff into lakes and rivers.

Bacteriophage P1, which naturally infects *E. coli,* has a simple system for genetic recombination, the **Cre/*loxP*** system. Cre stands for "causes recombination" and is a recombinase enzyme that recognizes a specific 34 base-pair DNA sequence, the ***loxP*** site. The Cre protein catalyzes recombination between two *loxP* sites (Fig. 15.13). When one *loxP* site is placed on both sides of a segment of DNA, the enclosed region may be deleted by Cre recombination. To accomplish this, researchers also include the *cre* gene in the transgenic construct and express it when it is time to delete the unwanted DNA segment. This approach allows selectable marker genes to be removed from plant DNA. The segment of DNA that is recombined out of the chromosome has neither an origin of replication nor any telomeres and will be either degraded or lost during mitosis and meiosis.

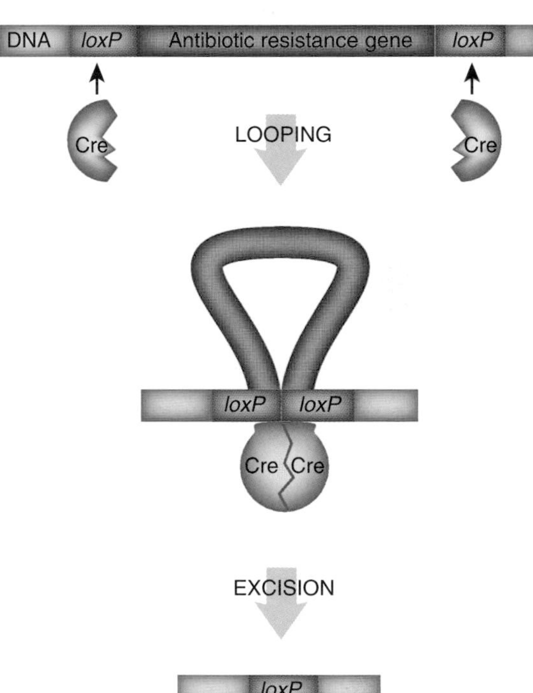

FIGURE 15.13 Cre/*loxP* System of Bacteriophage P1
The Cre protein binds to *loxP* recognition sites in the DNA. Two nearby *loxP* sites are brought together, and recombination between them eliminates the intervening DNA. A single *loxP* "scar" site remains in the target DNA molecule.

The *cre* gene can be added to the system by cross-pollination of two different plants: One plant carrying the transgene plus a selectable marker that is flanked by two *loxP* sites is crossed with another plant carrying the *cre* gene (Fig. 15.14). First, the pollen from the plant with the *cre* gene is added to the stigma of the plant with the transgene. The resulting seeds are grown and checked for sensitivity to the selective agent (e.g., neomycin). If the Cre protein is present in the progeny, the selectable marker gene will be excised and lost during growth. This plant now has the transgene and the *cre* gene, but no longer has the gene for antibiotic resistance. If another cross is made between this transgenic plant and a wild-type plant, some of the progeny will have the transgene but lack the *cre* gene. Using a cross-pollination scheme such as this will ensure that the final transgenic plant has only one extra gene, the transgene. This system has proven so easy and useful that every new variety of transgenic plant released to the public contains only the single transgene of interest.

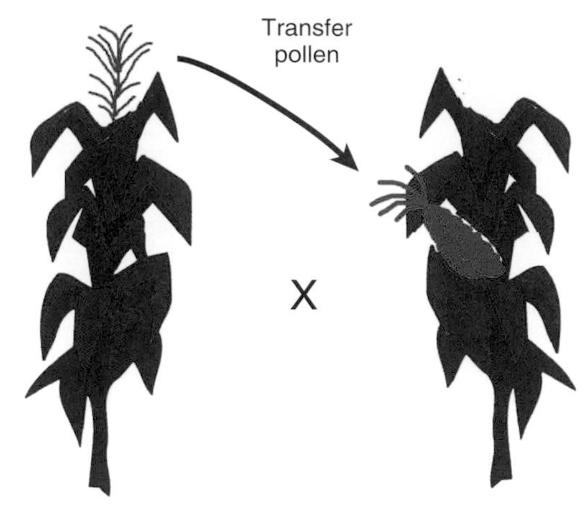

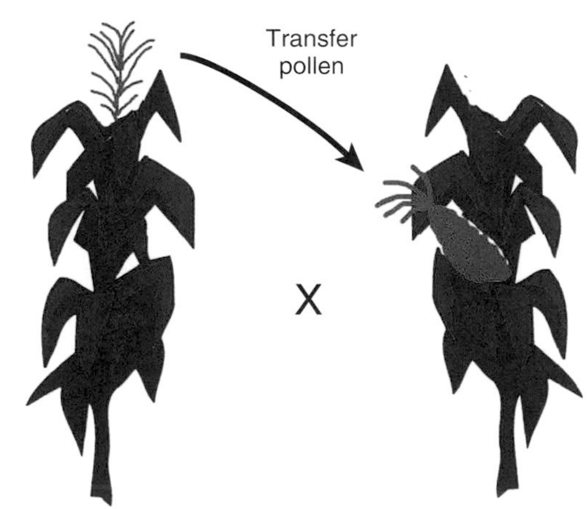

FIGURE 15.14 Cross-Pollination Scheme to Remove Unwanted Marker Genes
Cross-pollination and Mendelian assortment can be used to remove unwanted marker genes. Pollen from a plant carrying the *cre* gene is transferred to a first-generation transgenic plant, still carrying a selectable marker gene (e.g., antibiotic resistance). Expression of Cre protein in the progeny results in excision of the marker gene. Those plants that have kept the transgene but lost the antibiotic resistance marker are then crossed with wild-type plants. Some of the progeny from this cross will have kept the transgene but lost the *cre* gene. The final transgenic plant possesses only the transgene of interest.

TALENs, CRISPR, and Other Genome Editing Techniques

Genome editing techniques such as TALENs, CRISPR/Cas9, and zinc finger nucleases (ZFNs) are beginning to change how genome modifications are made in all organisms, but these techniques are particularly useful for plant engineering (see Chapter 17 for a more in-depth discussion of these methods). Each of these techniques relies on the creation of double-stranded DNA breaks at the precise location of interest in the plant genome. The double-stranded break is repaired using **nonhomologous end joining (NHEJ)** or **homologous recombination (HR)** (Fig. 15.15). When there is a transgene present, these pathways can integrate the new gene at a precise location, or if no transgene is used, the repair process can introduce small insertions or deletions at the recognition site, thus producing plants with inactivating mutations at a particular gene.

There are two main methods to generate the double-stranded DNA breaks for genome editing. The first method relies on endonucleases or restriction enzymes that have very long recognition sequences (up to 40 bases). These are called homing endonucleases, **zinc finger nucleases (ZFNs)**, or **TALE nucleases (TALENs)** (Fig. 15.16). The longer the recognition sequence, the more specific the break, and less likely there will be other regions of the genome that will be affected. These enzymes function just like any other restriction enzyme; that is, the endonuclease recognizes its target sequence and then cuts both strands of the DNA backbone between the phosphate and sugar. The recognition sites for some of these have been engineered to recognize even longer sequences to make them more specific. ZFNs have two domains: a zinc finger domain that has a specific recognition sequence to target the exact location of the intended double-stranded break and the DNA cleavage domain from the *FokI* restriction enzyme. TALENs are also a fusion of two domains. The DNA recognition domain is from TALE (Transcription Activator-Like Effector) proteins and the DNA cleavage domain from *Fok*I.

The second method to induce DNA breaks into a genome at precise locations is done by the CRISPR/Cas9 system. The nuclease is Cas9, which has two DNA cleavage domains. The CRISPR system provides the sequence specificity, which is based on an associated RNA sequence. This system is easier to change to cut a new site within the genome because only the RNA component needs to be altered. In fact, the recognition sequence of the genomic modification site is simply synthesized and cloned into a vector that converts it to RNA. These systems have been used to target various genes from plants such as *Arabidopsis*, tobacco, rice, and wheat. This technique is very recent, so none of these modifications have been done in other crops, and none of the resulting transgenic plants are on the market yet.

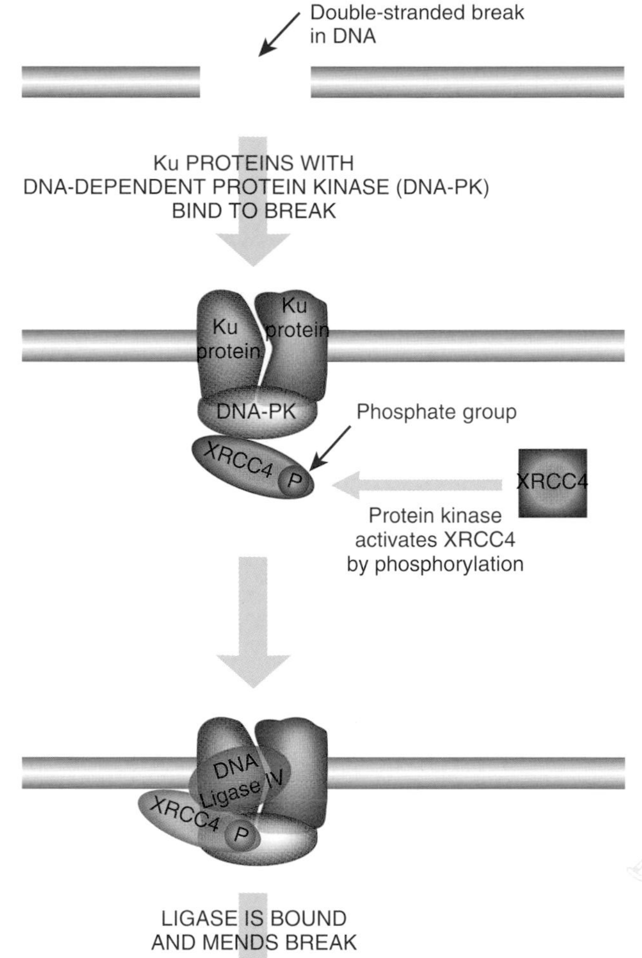

FIGURE 15.15 Nonhomologous End Joining and Homologous Recombination
Double-stranded breaks are recognized by the Ku proteins, which bind one to each end. The two Ku proteins recruit DNA-PK to the complex. DNA-PK phosphorylates XRCC4 protein, which then recruits DNA ligase IV to join the two broken ends.

Product Development and Regulatory Approval

Testing the transformed plants is the most time-consuming part of transgenesis. The expression level of the transgene may vary considerably, depending on the number of integrated copies and their location. Transgenes that land in a region of heterochromatin, for instance, may never be expressed at all. This is particularly problematic for random insertions that are created with microparticle bombardment. On the other hand, the upcoming use of CRISPR/Cas9 for targeting a transgene to a specific location will lessen the likelihood that the transgene will be silenced. Molecular biologists use the term **event** to refer to each independent case of transgene integration.

The first safety issue to address is whether the transgene causes any harmful side effects to the plant or ecosystem. If no harmful effects are found, the transgene is transferred from the experimental plant to one with a much higher yield. Most transgenic plants are made from old varieties that are easy to transform and regenerate new plants, but are not the best variety for use directly in seed production. Furthermore, as discussed earlier, the regeneration of plants through tissue culture may itself cause mutations. To overcome these problems, researchers move the transgene through traditional cross-breeding into high-yielding varieties that farmers are already using.

First, the pollen from the plant with the transgene is harvested and put onto the stigma of the high-yielding variety. The seeds from this cross are harvested and grown. This is the F1 generation, and the plants containing the transgene are selected. For example, if the transgene makes the plant resistant to an herbicide, the F1 generation is sprayed to kill the plants without the transgene. The pollen from the F1 plants that survive is **back-crossed** to the original high-yielding parent. The seeds are grown, plants with the transgene are selected, and the whole process is repeated about five times. This crossing scheme will ensure that roughly 98% of the genes in the final plant are from the high-yielding variety, and the remaining genes are from the original transgenic plant. Because it takes an entire summer for one generation of corn, soybeans, or cotton to grow, this back-crossing scheme can take many years to complete.

Once the transgene is back-crossed into a suitable variety, field tests are performed to determine how the transgene affects growth, yield, disease resistance, and other important traits. Field tests can take years to complete because they are performed at many different locations in order for soil type, terrain, rainfall, and other variables to be considered. Only plants that consistently give the highest yield across a variety of environments with the best disease resistance are selected. The other plants are discarded.

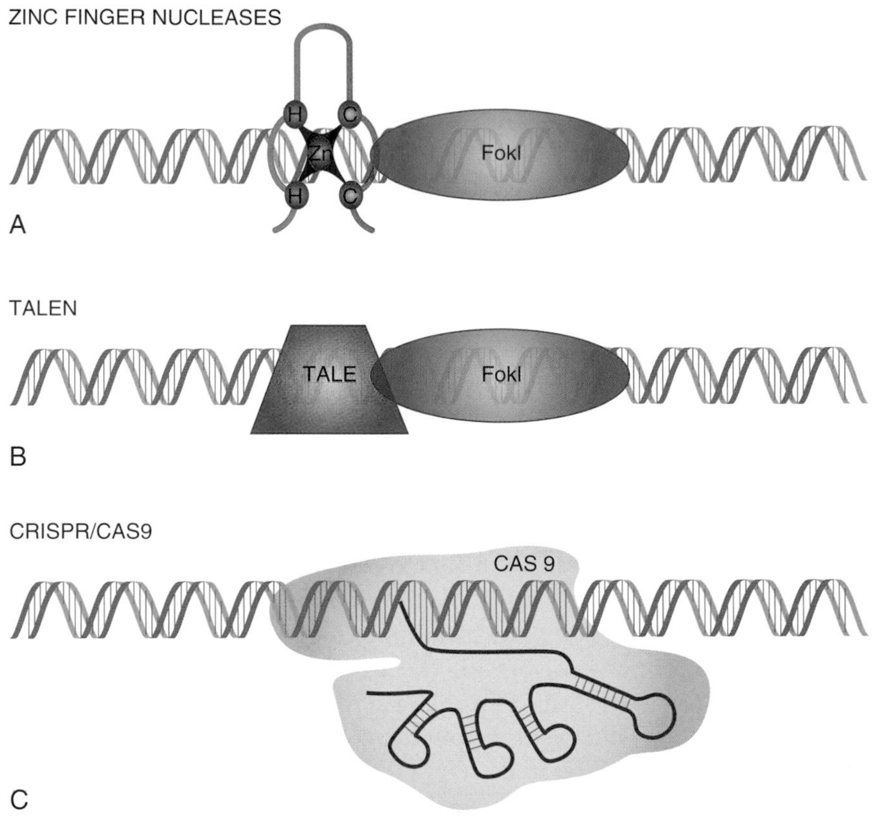

FIGURE 15.16 ZFNs, TALENs, and Cas9 Induce Double-Stranded DNA Breaks
There are two main methods of inducing double-stranded breaks at specific locations in the genome. (A) ZFNs combine the restriction enzyme DNA cleavage domain for *FokI* with the zinc finger transcription factor DNA binding domain. (B) TALENs combine the *FokI* restriction enzyme DNA cleavage domain with a TALE transcription factor DNA recognition domain. (C) The Cas9 protein is unique since it uses a guide RNA to recognize the DNA sequence and then the Cas9 endonuclease to cut the DNA.

Before being released to the public, transgenic plants must also adhere to government regulations. Various agencies regulate all stages of the transgenic construction process. An Institutional Committee for Biosafety regulates how the transgene is handled when making the transgenic plant, whether in *E. coli*, *Agrobacterium*, or the plant itself. These committees are usually associated with the university or company where the work is done, but they all follow guidelines from the National Institutes of Health (NIH). The guidelines regulate the laboratory and greenhouse conditions in which the transgenic plant is grown. To conduct field tests of the transgenic crop, scientists must notify the Animal and Plant Health Inspection Service of the U.S. Department of Agriculture (USDA), which must approve the plan. Scientists must provide extensive data on the transgene and its potential effect on the plant and the ecosystem.

Two other agencies are also involved in regulating transgenic crops. If humans or animals consume the crop as food, it is reviewed by the Food and Drug Administration (FDA). Currently, there is a "voluntary consultation" process, but in practice, most companies submit to the consultation. The company provides information on potential allergenicity and toxicity, as well as on any changes in the nutritional quality of the product. If the transgenic crop satisfies the FDA's guidelines, it is deemed to be "substantially equivalent" to its nontransgenic counterpart, but the FDA does not issue a formal approval. The Environmental Protection Agency (EPA) evaluates the transgenic crop for potential effects on the environment and on animals or insects that inhabit the farmers' fields. Of course, transgenic crops must also satisfy regulations in all other countries in which they are sold.

One method to create transgenic plants involves the Ti plasmid, which was isolated from a pathogen called *Agrobacterium tumefaciens* that infects wounded plants. Normally, the bacterium injects a portion of the Ti plasmid known as T-DNA into the plant, which tricks the plant into making food for the bacterium and forming a tumor in which the bacteria can live. Molecular biologists have re-engineered the Ti plasmid for transferring transgenes into plants. The foreign gene is placed in the T-DNA region. *In planta Agrobacterium* transformation (floral dip) creates transgenic plants by integrating T-DNA into regenerating floral tissues.

Alternatively, gold beads coated with transgene DNA can be blasted into plant cells using a gene gun. The transgene DNA dissolves from the bead and integrates into the plant genome. The latest technology uses TALENs and CRISPR.

Selectable markers or reporter genes are used to help confirm the integration of the transgene. Neomycin phosphotransferase inactivates the antibiotic neomycin, making any cells with this selectable marker resistant to neomycin. The luciferase gene makes the cell glow in the presence of luciferin. The transgene itself is verified using PCR or DNA sequencing. Selectable markers and reporter genes can be removed with the Cre/*loxP* system. Cre is a recombinase that binds to *loxP* sites that flank the reporter gene, removing any DNA between the two sites.

For a transgenic plant to be grown for commercial use, the transgene must be crossed into a high-yielding variety of the plant. It is then scrutinized by a variety of government agencies for possible threats to human health and the environment.

BIOTECHNOLOGY IMPROVES CROPS

Plant pathologist Norman Borlaug played a key role in what became known as the "Green Revolution" by developing high-yielding varieties of wheat that were also resistant to various diseases. As a result, global food production exploded. However, with the human population continuing to grow, innovative techniques may be necessary to help feed future generations.

Saving Crops from Pests

For thousands of years, farmers have struggled against two common enemies: weeds and insects. Much of the effort in producing transgenic crops, therefore, has been aimed at reducing the impact of these two pests.

Herbicide resistance. Herbicides cost the world's farmers more than $14 billion each year. Despite this massive investment, around 10% of crops is lost due to weeds. One problem is that many of the herbicides used do not discriminate between crops and weeds. One solution is to make the crops resistant to the herbicide by genetic engineering. Therefore, when the herbicide is sprayed on the weeds and crop, only the crop will survive.

One of the best herbicides on the market is **glyphosate**. Glyphosate is environmentally friendly because it quickly breaks down to nontoxic compounds in the soil. The glyphosate molecule is a phosphate derivative of the amino acid glycine. Glyphosate kills plants by blocking the synthetic pathway for the aromatic amino acids phenylalanine, tyrosine, and tryptophan by inhibiting one particular enzyme, **EPSPS** (5-enolpyruvoylshikimate-3-phosphate synthase), which is the product of the *aroA* gene and localized to the chloroplast (Fig. 15.17). This target enzyme is found naturally in all plants, fungi, and bacteria, but not in animals. Aromatic amino acids are therefore *essential* to the diets of all animals, including humans, because those organisms cannot produce them. When glyphosate is sprayed onto plants, the herbicide penetrates the chloroplasts and binds to EPSPS, blocking the pathway for aromatic amino acids. The plant essentially starves to death.

Developing herbicide resistance directly in plants is difficult, so scientists isolated a glyphosate-resistant EPSPS enzyme from bacteria. Mutant strains of *Agrobacterium* that are

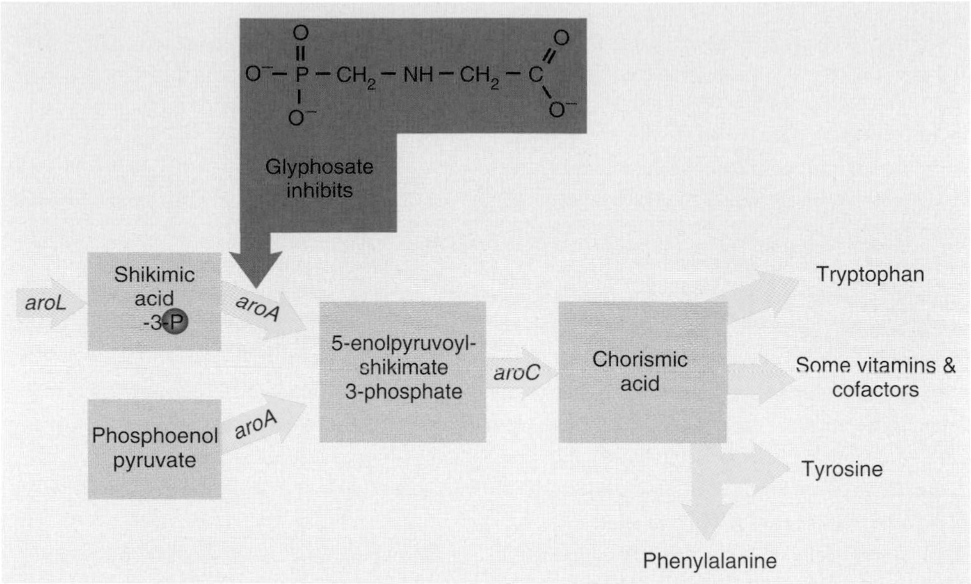

FIGURE 15.17 Glyphosate Inhibits EPSPS in the Aromatic Pathway
The enzyme 5-enolpyruvoylshikimate-3-phosphate synthase (EPSPS) is the product of the *aroA* gene and makes 5-enolpyruvoylshikimate-3-phosphate, a precursor in the pathway to aromatic amino acids and cofactors. Glyphosate, an analog of phosphoenolpyruvate, inhibits EPSPS.

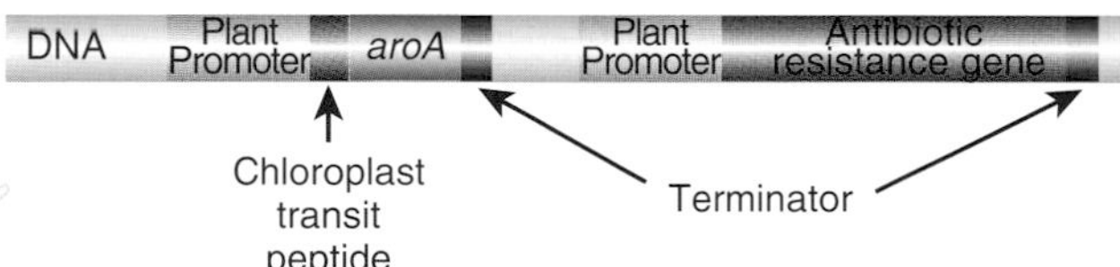

FIGURE 15.18 Expression of the *Agrobacterium aroA* Gene in Plants
The bacterial *aroA* gene must be placed under control of a promoter active in plants. Correct localization of the AroA protein (EPSPS) into the chloroplast requires addition of a chloroplast transit peptide at the N-terminus of the protein.

resistant to glyphosate can be directly selected by plating onto medium containing glyphosate. Such mutants produce an EPSPS enzyme that is resistant to glyphosate but still enzymatically active. A glyphosate-resistant version of the *aroA* gene was then cloned and modified for expression in plants. The bacterial promoter and terminator sequences were replaced with plant promoters and terminators. An antibiotic resistance gene was also added to the construct to allow for selection. Finally, because EPSPS is localized to the chloroplast, DNA encoding a small **chloroplast transit peptide** was added to the front of the gene. The chloroplast transit peptide, present at the N-terminus of the protein, targets EPSPS to the chloroplast but is cleaved off while crossing the chloroplast membrane (Fig. 15.18). Only the functional enzyme enters the chloroplast. The glyphosate-resistant *aroA* gene from *Agrobacterium* has been transformed into several different crops, including soybean, cotton, and canola.

Comparison of the mutant bacterial *aroA* gene with the sensitive wild-type version revealed which amino acid changes were needed for glyphosate resistance. Because bacterial and plant *aroA* genes are homologous, equivalent changes should result in glyphosate resistance in plant *aroA* genes as well. Indeed, this information allowed the *aroA* gene from corn to be engineered by altering its DNA sequence *in vitro*. The altered corn *aroA* gene provided glyphosate resistance after being introduced back into corn plants with the gene gun.

Other herbicide tolerance genes have been used to make transgenic crops, although these are not as widely used as the glyphosate-resistant *aroA* gene. For example, plants can be made resistant to sulfonylureas (Fig. 15.19) and imidazolinones, which inhibit an enzyme in the pathway that synthesizes the branched amino acids leucine, isoleucine, and valine. Plants resistant to these herbicides are quite common because resistance results from a single amino acid substitution in the appropriate enzyme. Another example is resistance to glufosinate, an herbicide that blocks synthesis of glutamine. Glufosinate was originally discovered as an antibiotic produced by *Streptomyces*. Scientists identified the enzyme that prevented *Streptomyces* from being poisoned by its own antibiotic and transformed it into crops.

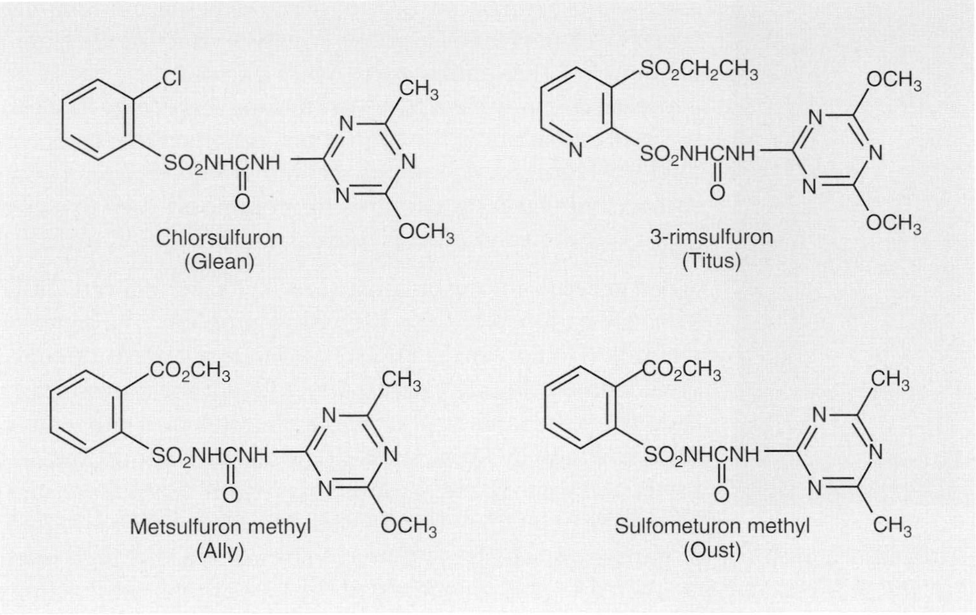

Chlorsulfuron (Glean)

3-rimsulfuron (Titus)

Metsulfuron methyl (Ally)

Sulfometuron methyl (Oust)

FIGURE 15.19 Various Sulfonylureas and Their Trade Names
Sulfonylureas can be used as herbicides. Used with permission from Coly A, Aaron J (1999). Photochemically-induced fluorescence determination of sulfonylurea herbicides using micellar media. *Talanta* **49**, 107–117.

Insect resistance. Spraying crops with insecticides is a very costly and hazardous procedure. Insecticides are often more toxic to humans than are herbicides because insecticides target species closer to our own. Many insect biochemical pathways are found not only in humans, but also in rodents or birds that may inhabit crop fields. Luckily, naturally occurring toxins exist that are lethal to insects but harmless to mammals.

The prime example is the toxin from a soil bacterium called *Bacillus thuringiensis*. **Bt toxin** has been sprayed on crops (including organic crops) to prevent insects such as the cotton bollworm and European corn borer from destroying cotton and corn, respectively. Damage from the European corn borer plus the cost of insecticides to control it cost farmers about $1 billion annually. Damage by the corn borer also makes corn plants susceptible to infection with *Aspergillus*, a fungus that produces aflatoxin that can harm humans if ingested.

Bacteria of the genus *Bacillus* produce spores that contain a crystalline, or **Cry protein**. When insects eat *Bacillus* spores, the Cry protein breaks down and releases the Bt toxin. This toxin binds to the intestinal lining of the insect and generates holes, which cripple the digestive system, and the insect dies (Fig. 15.20). Different species of *Bacillus* produce a family of related Cry proteins that exhibit toxicity toward various groups of insects.

Instead of spraying crops with Bt toxin, scientists have used transgenic technology to insert the *cry* genes directly into plants. When a cloned toxin gene was inserted into tomato plants, for instance, it partially protected against tobacco hornworm. However, the plants made only low

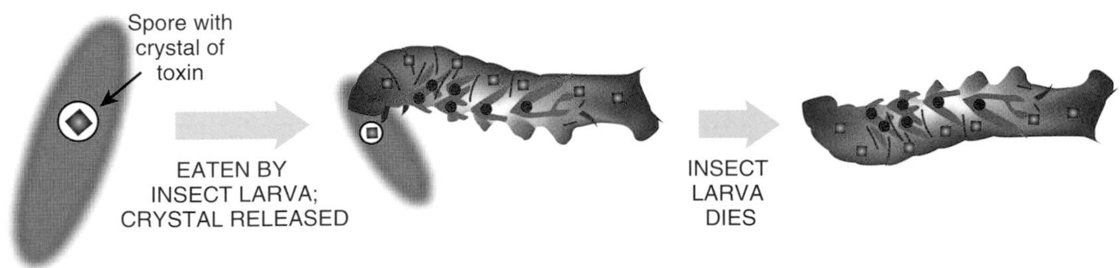

FIGURE 15.20 Insect Larvae Are Killed by Bt Toxin
Bacterial spores of *Bacillus* are found on food eaten by caterpillars. The crystalline protein is released by digestion of the spore, and its breakdown produces a toxin that kills the insect larvae.

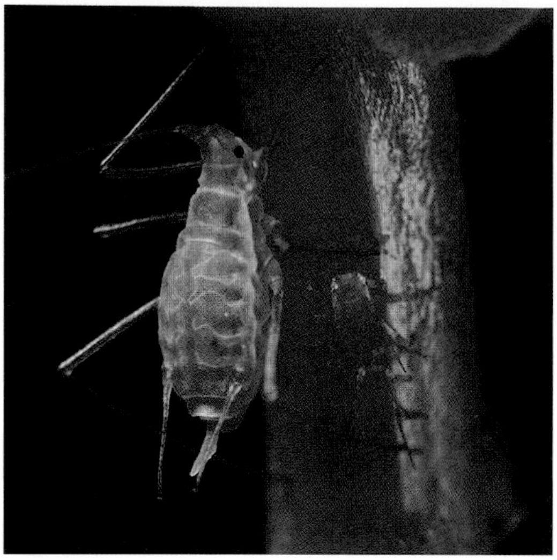

levels of the toxin because the bacterial gene was not optimized to express well in plants. Therefore, scientists removed the latter 506 of the protein's 1156 amino acids, which allowed the plant to produce a truncated (but still effective) toxin using less energy. The toxin gene was also placed under the control of a promoter that gives constant high-level expression in plants. Certain promoters from plant viruses, such as cauliflower mosaic virus, are commonly used to increase transgene expression and often provide a 10-fold boost in toxin production.

When genes from one organism are expressed in a very different host cell, codon usage also becomes a problem. The genetic code is redundant in the sense that several different codons can encode the same amino acid (see Chapter 2). Different organisms favor different codons for the same amino acid, a phenomenon known as **codon bias**, and have different levels of the corresponding tRNAs. If a bacterial transgene uses codons that require tRNA molecules that are rare in the plant, the rate of protein synthesis will be limited. This is particularly a problem for transgenes, which often need to be expressed at high levels. Therefore, the Bt toxin gene was altered by changing many of the bases in the third position of redundant codons. Almost 20% of its bases were altered to make the gene more plant-like in codon usage. Because such tweaks do not change the amino acids encoded, the toxin protein sequence was not affected by the procedure. However, the rate at which plant cells made the protein greatly increased and gave another 10-fold increase in toxin production.

FIGURE 15.21
Aphids
Aphids feed on sap, but plants expressing (E)-β-farnesene can scare them away. Photo courtesy of Shipher Wu and Gee-way Lin, National Taiwan University.

There are two primary advantages to using Bt transgenic crops, such as cotton and corn. First, the toxin does not have to be sprayed, which reduces both the amount of work necessary and the potential for contaminating nearby fields. Second, planting Bt crops has resulted in a dramatic reduction in the amount of insecticides required. According to a recent meta-analysis (Klumper and Qaim, 2014), genetically modified crops have reduced pesticide use by 37%, increased crop yields by 22% and increased farmer profits by 68%. The gain in yield and the decreased use of pesticides are largely due to the crops that are resistant to glycophosate, and other related herbicides. When looking at insect resistance alone, these crops have increased yield by 25% and have decreased pesticide use by 42%. These changes in use are one reason that farmers will continue to use transgenic crops.

An alternative to toxins is to combat insects using pheromones or other compounds that induce changes in behavior. (E)-β-farnesene (EBF) is released by aphids as an alarm pheromone. Aphids avoid plants that express EBF and, simultaneously, aphid predators, such as ladybugs, are attracted to the plants (Fig. 15.21). One group is field-testing a variety of wheat that was transformed with synthetic genes whose encoded enzymes produce EBF. Compared to Bt toxin, the likely benefits of the pheromone approach are that the target insects are unlikely to develop resistance and nontarget insects will not be harmed.

Saving Crops from Disease and Drought

Not only are pests a problem, but farmers also must contend with infectious crop diseases. Bacteria, viruses, and fungi present a major threat to several crops of regional or global importance.

In the 1990s, the papaya industry was severely threatened by the papaya ringspot virus (PRSV), which is spread between plants by aphids. Farms in Brazil, Taiwan, and Hawaii were infected. To counter the infection, scientists utilized a phenomenon known as **parasite-derived resistance**, in which the expression of a gene from a parasite confers resistance to that same parasite. In this case, transgenic papaya was created that expressed the coat protein from PRSV. The scheme worked, and the transgenic papaya helped rescue the industry from decimation.

Today, pathogens are threatening other crops. Since the late 1800s, orange trees in Southeast Asia have been suffering from citrus greening disease, which is spread by psyllids, also known as jumping plant lice (Fig. 15.22). Recently, the infection has spread to orange trees in Brazil and the United States, and scientists are actively developing transgenic orange trees that are resistant to the disease. Bananas are a staple crop in many parts of Africa, playing an important role in the local diet and economy. Wilt, caused by the bacterium *Xanthomonas campestris* or the fungus *Fusarium oxysporum*, potentially poses a serious threat to the banana crop (Fig. 15.23). Transgenic bananas are currently being developed to counter the threat from both pathogens. Similarly, efforts are underway to protect the cassava, another staple crop in the developing world, from viral infection. Transgenic potatoes are being developed that are resistant to fungal blight, which caused the Irish Potato Famine in the mid-1800s. Another type of fungal blight nearly wiped out all 4 billion American chestnut trees in the eastern United States. Using a gene from wheat, one group of researchers intends to repopulate the countryside with a transgenic variety of this once ubiquitous tree. Still, others have suggested creating disease-resistant transgenic grape varieties for more efficient wine production.

FIGURE 15.22 Citrus Greening Disease
Caused by a bacterium, citrus greening disease threatens the world's orange trees. Reused with permission from Vallero DA, Letcher TM (2013). Invasions. In *Unraveling Environmental Disasters* (Waltham, MA, USA, Oxford, UK, and Amsterdam, The Netherlands: Elsevier), pp. 321–351.

With drought a recurring and growing problem in many parts of the world, there is a need for drought-resistant crops. The sugar trehalose can be used to enhance a plant's ability to handle drought because it can absorb and release water molecules. Two different enzymes synthesize trehalose, and a fusion gene encoding a protein with both enzymatic activities was transformed into rice. The transgenic rice became much more tolerant to both drought and high salt concentrations. Many other groups are working on drought-resistant wheat.

One interesting idea for creating drought-resistant crops is to alter how certain plants conduct photosynthesis. There are three major types of photosynthesis: C_3 and C_4 photosynthesis, named after the number of carbon atoms that the molecule into which CO_2 is first incorporated, and CAM photosynthesis. Most plants are C_3, including important crops like wheat and rice. C_4 plants (e.g., grass and corn) and CAM plants (e.g., jade plants and cacti) are better adapted for growth in hotter and dryer environments. By transferring genes from C_4 or CAM plants into C_3 plants, scientists may be able to grow them under less hospitable conditions.

Boosting Nutrients and Reducing Toxins

Researchers are also interested in boosting the nutrient content of crops, especially rice. In many areas of the world, particularly Africa and southeast Asia, rice is a staple of the diet. Unfortunately, rice does not contain vitamin A. People who have few other sources of nutrition are susceptible to vitamin A deficiency, a disease that the World Health Organization estimates to cause blindness in 250,000 to 500,000 children annually, half of whom die within one year. "Golden Rice" expresses the biosynthetic pathway for a vitamin A precursor called β-carotene, and a humanitarian effort to provide the developing world with this transgenic crop is slowly underway (Fig. 15.24). A similar effort to fortify bananas with extra vitamin A is also ongoing. Other researchers have fortified rice with extra zinc and iron. Soybean and canola crops modified to produce omega-3 fatty acids have also been created.

Adding nutritious compounds to plants is not the only way to use transgenic technology to improve the food supply. Another way is to reduce the amount of toxins naturally found in plants or to prevent them from being infected with toxin-producing fungi. Rapeseed,

FIGURE 15.23 ***Fusarium*** **Wilt**
Fusarium oxysporum causes banana plants to wilt rapidly. Reused with permission from Fourie G et al. (2011). Current status of the taxonomic position of *Fusarium oxysporum formae specialis cubense* within the *Fusarium oxysporum* complex. *Infect Genet Evol* **11**, 533–542.

FIGURE 15.24
Golden Rice
"Golden rice" is so named because of its yellowish color, which is due to fortification with β-carotene, a vitamin A precursor. Normal rice is shown in (A). The golden rice shown in (B) has the phytoene synthase gene from daffodil, whereas the rice in (C) has the phytoene synthase from maize or corn. Photograph courtesy of Aron Silverstone and used with permission from Al-Babili S, Beyer P (2005). Golden rice—five years on the road—five years to go? *Trends Plant Sci* **10**, 565–573.

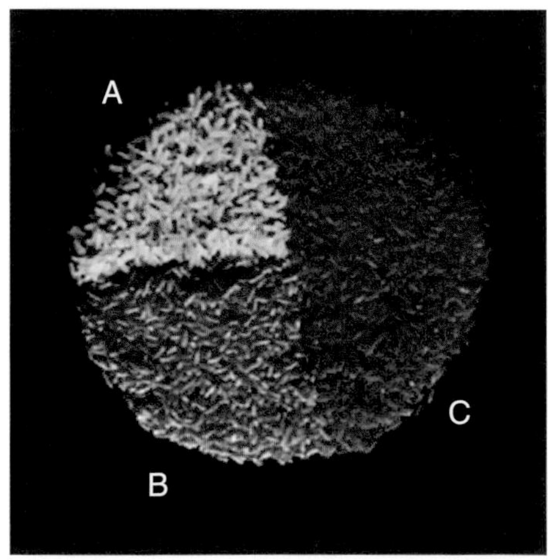

which is grown for oil, can also be fed to livestock, such as pigs and chickens. However, the plant produces a toxin that limits how much feed can be given to animals. Researchers have identified two genes in *Arabidopsis*, a model for plant geneticists and a relative of rapeseed, that control transport of the toxin into seeds. Mutating both genes in *Arabidopsis* prevented the seeds from accumulating the toxin. It is hoped that a similar manipulation in rapeseed will allow for the production of toxin-free animal feed.

The fungus *Aspergillus*, which infects corn, wheat, and several other important crops, produces the carcinogenic **aflatoxin** (Fig. 15.25). The FDA requires the level of aflatoxin in the human food supply to be kept below 20 ppb (ng/g) and below 0.5 ppb (ng/g) for milk. Currently, scientists are identifying genes associated with *Aspergillus* resistance in certain lines of corn. Complicating matters is the fact that resistance appears to be a polygenic trait. Still, the hope is that, once these genes are identified, traditional cross-breeding techniques or transgenic technology could be used to develop safer varieties of corn and other crops. Similarly, the fungus *Fusarium* causes head blight in barley, a crop that is fermented to produce beer. *Fusarium* releases a toxin called deoxynivalenol, more commonly known as "vomitoxin" because its presence in beer can induce vomiting. Developing barley varieties that are resistant to *Fusarium* infection is therefore desirable.

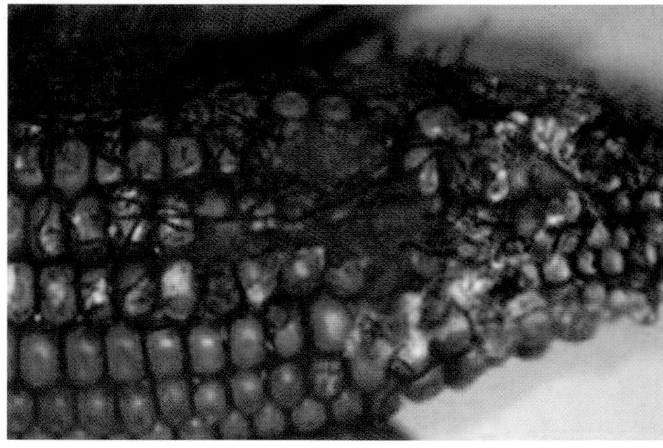

FIGURE 15.25
Aspergillus flavus
Aspergillus flavus growing on an ear of corn. Used with permission from Bhatnagar D, Ehrlich KC, Moore GG, Payne GA (2014). Aspergillus: *Aspergillus flavus*. In *Encyclopedia of Food Microbiology*, 2nd ed. Batt CA, Tortorello ML, eds. (London, UK, Burlington, MA, USA, and San Diego, CA, USA: Academic Press), pp. 83–91.

Other Uses for Transgenic Crops

Many other creative uses for transgenic crops have been proposed or are in development.

Manufacturing pharmaceuticals. One idea is to use plants as miniature manufacturing facilities for pharmaceuticals. Transgenic plants are advantageous because they can help lower production costs in addition to making it easier to isolate and purify medicines in bulk. Furthermore, in plants, post-translational modification of proteins is easier and the risk of contamination with mammalian pathogens is reduced. In 2012, the FDA approved the first-ever plant-made pharmaceutical intended for human use. Patients suffering from Gaucher's, a rare lysosomal storage disease, are deficient in the enzyme glucocerebrosidase. A company called Protalix Biotherapeutics engineered carrot cells to synthesize a replacement enzyme, taliglucerase alfa (trade name Elelyso). The enzyme is targeted to the plant cell vacuole. Inside, the protein undergoes post-translational modification and is protected from degradation. To extract the enzyme, scientists solubilize plant cells in detergent and purify the enzyme using chromatography. This product is cheaper than its rivals, both of which are produced in mammalian cell lines.

Even infectious diseases can be treated using products derived from transgenic plants. Scientists from the U.S. Army Medical Research Institute of Infectious Diseases (USAMRIID) transiently transfected a relative of the tobacco plant (*Nicotiana benthamiana*) to produce antibodies against the Ebola virus, while another group did the same to produce an antitumor IgM monoclonal antibody. It is also possible to use this technique to produce vaccines. A subunit vaccine consisting of a recombinant hemagglutinin protein from the H1N1 influenza virus that caused a pandemic in 2009 is currently undergoing clinical trials.

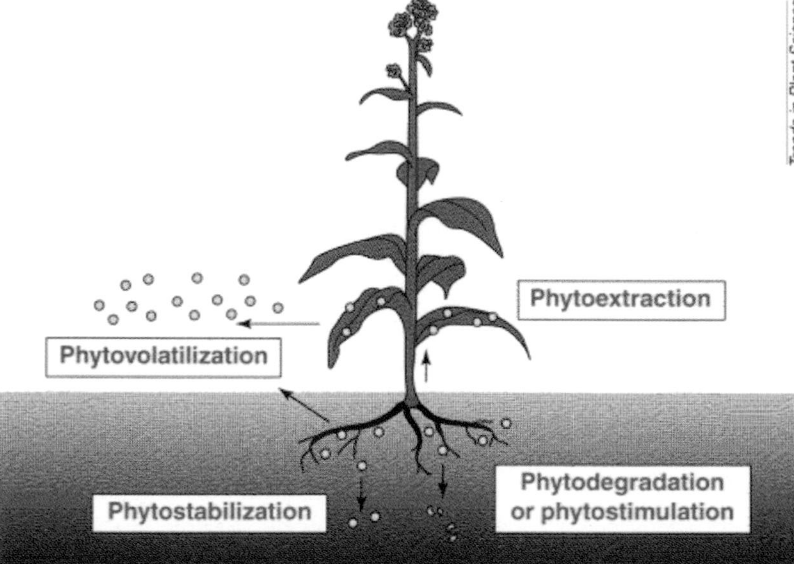

FIGURE 15.26
Phytoremediation
Phytoremediation is an umbrella term for five different strategies: phytostabilization, phytodegradation, phytostimulation, phytoextraction, and phytovolatilization. Used with permission from Pilon-Smits E, Pilon M (2000). Breeding mercury-breathing plants for environmental cleanup. *Trends Plant Sci* **5**, 235–236.

Transgenic plants could also be engineered to encode all of the enzymes necessary to synthesize small molecules, a technique known as pathway engineering (see Chapter 13)

Improving the Environment. Using plants to clean up the environment, for instance, by sequestering soil or water contamination, is known as **phytoremediation**. There are five ways to do this (Fig. 15.26). In the first scenario, *phytostabilization*, plants provide ground cover for a contaminated site by protecting against erosion due to wind and water. Although unmodified plants can perform phytostabilization, the use of transgenics could increase the root system or enhance tolerance to the contaminant. In the second method, *phytodegradation*, the plant breaks down the pollutant, while in the third method, *phytostimulation*, the plant stimulates microbes to degrade it. In the fourth, called *phytoextraction*, a plant assimilates the contaminant into its tissues, after which it is harvested and disposed of properly. In the fifth method, *phytovolatilization*, pollutants are absorbed from the soil and released into the atmosphere, usually after being converted to a less toxic form.

Some plants have natural systems to assimilate heavy metal ions. One natural accumulator is the brake fern, *Pteris vittata*, which can accumulate up to 7500 micrograms per gram of arsenic from a contaminated site. Some plants can concentrate 200 times more arsenic in their leaves than is found in the soil. In other cases, transgenic plants are created to remove the toxic contaminant. *Arabidopsis*, tobacco, and poplar trees that have had the *merB* and *merA* genes from bacteria added to their genome can convert the highly toxic form of methylmercury into Hg(II), which is less bioavailable. MerB converts methylmercury $[CH_3Hg]^+$ and a proton to Hg(II). MerA converts Hg(II) to Hg(0), which is a less reactive elemental form.

Beyond phytoremediation, transgenic technology could also be used to address the problem of climate change. Ongoing research is aimed at turning plants into better sources of biofuel. This could be accomplished by increasing a plant's sugar content or by making its cellulose more amenable to fermentation. (Both sugar and cellulose are converted into ethanol.) Others have proposed more radical solutions, such as engineering plants to absorb more carbon dioxide or replacing streetlights with glowing trees!

Creating Novelty Products. There are less serious, but potentially lucrative, uses for transgenic plants. Okanagan Specialty Fruits is currently awaiting approval—which it is likely to receive—for its product, the "Arctic Apple," which does not turn brown after being sliced. Typically, slicing an apple damages cell walls, triggering the enzyme polyphenol oxidase to react with phenolic compounds that turn the fruit brown. The Arctic Apple, however, is genetically modified to produce less of this enzyme, and the apple stays fresher longer. Another group is investigating the possibility of engineering a tomato that tastes better. Many consumers believe that most tomatoes currently on the market are bland, possibly because the breeding process was focused more on producing a shippable tomato than a tasty one. Molecular techniques could reintroduce flavor-enhancing genes back into the tomato. Additionally, people who suffer from allergies may one day be able to enjoy fresh flowers: One group recently invented pollen-free geraniums.

Glyphosate inhibits EPSPS in the plant aromatic amino acid biosynthetic pathway. EPSPS is a good target because humans and other animals do not have this enzyme. Glyphosate-resistant EPSPS was isolated from bacteria and modified to be expressed in plants.

Transgenic plants have been made to express Bt toxin, the Cry protein from the soil bacterium *Bacillus thuringiensis*. The toxin kills insect pests such as the European corn borer and cotton bollworm. To enhance cry expression in plants, scientists truncated the gene. Furthermore, changing the wobble position in several codons solved the issue of codon bias.

The uses for transgenic plants are nearly limitless. Fighting crop disease and drought, boosting nutrients, reducing toxins, manufacturing pharmaceuticals, and improving the environment are all potential applications of transgenic technology.

RESISTANCE: NATURE RESPONDS TO TRANSGENIC PLANTS

Nature does not sit idly by while humans tamper with it. Evolution is an ongoing process and, just as nature responded to antibiotics, it is now responding to the presence of transgenic plants.

For example, some insects are developing resistance to Bt toxin. The Western corn rootworm is developing resistance to two variants of Bt toxin: Cry3Bb1 and mCry3A. The cotton bollworm is developing resistance to Cry1Ac. Just as with antibiotics, the solution is for biotechnologists to stay ahead of the game. They do this by monitoring resistance trends and studying molecular mechanisms of resistance. With this knowledge, scientists are developing new variants of Bt toxin. Other strategies rely on preventing the development and spread of resistance in the first place. Creating transgenic crops that express very high doses of Bt toxin, which is likelier to kill insects, can help achieve this result. Also, creating transgenic crops expressing two or more toxins can help. Much of the cotton crop planted in the United States now expresses two Bt toxins. Alternatively, refuges consisting of nontransgenic crops can serve as safe havens for nonresistant insects. These nonresistant insects will then mate with resistant insects, diluting the resistance genes in the gene pool.

Herbicide resistance was a problem long before the widespread adoption of transgenic crops. Although it is unlikely that transgenic crops will ever solve this problem, it is hoped that they will not exacerbate it. Unfortunately, some research suggests that as farmers plant more glyphosate-resistant ("Roundup Ready") crops, more weeds are becoming resistant to glyphosate. The reason is that the increased use of glyphosate creates a selective pressure on weeds. One way to tackle the problem is to create transgenic crops resistant to other herbicides, such as 2,4-D, or to make them resistant to more than one herbicide. However, weeds with multiple herbicide resistance have already been detected. Consequently, farmers should implement what is known as *integrated pest management*, a practice that incorporates multiple techniques such as crop rotation, herbicide rotation, and mechanical weed control.

> Nature has responded to transgenic crops in the form of Bt toxin-resistant insects and herbicide-resistant weeds. This is evolution in action. The solution is to create more transgenic crop varieties and to implement farming practices, such as integrated pest management, aimed at reducing the spread of resistance.

FUNCTIONAL GENOMICS IN PLANTS

Because the complete DNA sequences of several plants are known, *functional genomics* allows scientists to study the entire genome rather than one specific gene at a time. Most of this work is still done with *Arabidopsis*, but some has now moved into crop species such as rice, corn, and soybeans. A variety of techniques are being used to study novel genes and metabolic pathways in the hope of improving our current crops.

Insertions are one method to determine the function of unknown genes. For instance, transposons or T-DNA insertions can generate plant mutants. However, instead of including a transgene, the inserted DNA contains only a reporter gene. When the transposon or T-DNA integrates into the plant chromosome, it may disrupt a plant gene. If the insertion knocks out the gene's function, the resulting phenotype may be screened and assessed. Cloning and sequencing the regions upstream and downstream of the insertion will identify the plant gene involved.

Gene silencing is another method to identify the function of unknown plant genes. As described in Chapter 5, gene silencing by RNA interference (RNAi) is a phenomenon that was originally described in plants. RNAi is triggered by double-stranded RNA, which is cut into short segments called short-interfering RNA (siRNA). The RISC enzyme complex uses siRNA to identify homologous RNA (usually mRNA) and digest it, blocking expression of the gene into protein. This phenomenon is exploited in the laboratory by transforming a plant with small oligonucleotides that stimulate RISC to abolish the expression of a chosen gene. The plant can then be studied for any visible phenotype associated with the gene knockdown.

Another method for generating gene knockouts is **fast neutron mutagenesis**. This method uses **fast neutrons**, which are created by nuclear processes such as fission, to induce DNA deletions. With a kinetic energy close to 1 MeV, these free neutrons cause random deletions in exposed DNA. The dose of fast neutrons and, consequently, the number of deletions per genome can be controlled. Seeds treated with fast neutrons, known as M1 seeds, are grown into plants. Each plant has a different deletion or set of deletions and a potentially different phenotype. The seeds from each of these plants, called the M2 seeds, are collected. Most are saved as stock for later use, and others are grown into plants. The DNA from these plants is isolated and collected into successively smaller pools, which are then screened by PCR to find specific genes with deletions. PCR primers are designed to amplify a target gene from the largest pool of DNA. If a deletion was generated within the target gene in one of the plants, the PCR primers will amplify two bands—the wild-type gene plus a shorter segment from the deleted gene. The smaller DNA pools are then screened for the deletion until a specific M2 seed can be associated with the mutation (Fig. 15.27).

Yet another method of creating plant mutations is called **TILLING** (**targeting-induced local lesions in genomes**). First, point mutations are created in a collection of M1 seeds by soaking them in a chemical mutagen, such as ethyl methane sulfonate (EMS), which induces G-C and A-T transitions in DNA. As before, some M2 seeds are saved and the others are grown

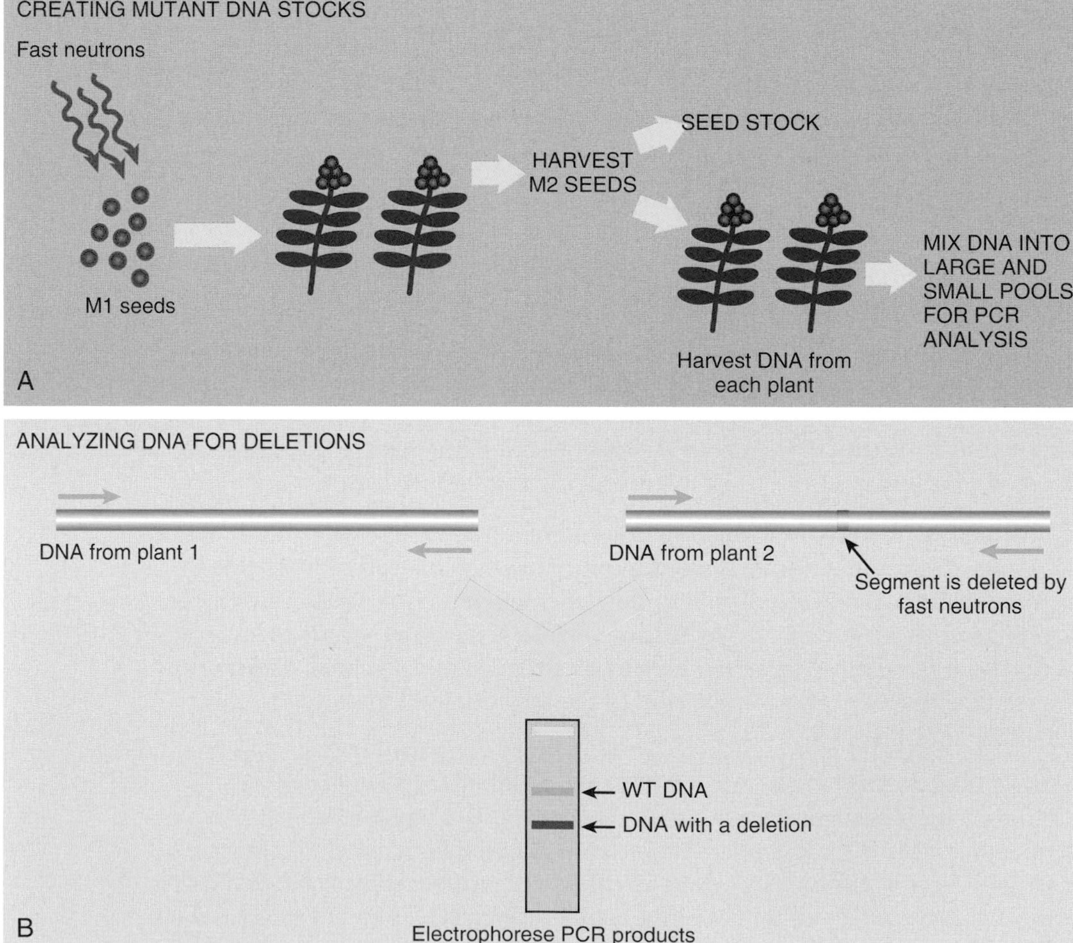

FIGURE 15.27 Identifying Fast Neutron Mutants with PCR

(A) M1 seeds are mutagenized by exposure to fast neutrons. The M2 seeds are grown and DNA harvested from each plant. The DNA is mixed to form large pools from many M2 seeds and successively smaller pools from fewer M2 seeds. (B) The seeds are analyzed for deletions using PCR. The primers recognize specific locations in the plant genome. If the DNA pool contains any deletions, the PCR primer will produce two bands—one from the wild-type (full-length) gene and one from the plant with the deletion.

into plants. DNA is harvested and placed into a large pool and successively smaller pools, and PCR primers amplify selected regions of the DNA. This time, however, each PCR primer carries its own fluorescent label. Consequently, the PCR products have two different labels, one at each end.

The key to identifying point mutations (as opposed to deletions) is to create heteroduplexes of mutant and wild-type DNA. The PCR products are denatured to single strands and then slowly cooled so that the DNA strands reanneal. During reannealing, some mutant strands will anneal with wild-type strands, and the heteroduplex will have a mismatched nucleotide. The enzyme CEL-1 cleaves mismatched DNA. If the PCR product is cleaved by CEL-1, it will have only one fluorescent label, whereas uncleaved DNA (with no mismatches) will have both fluorescent labels. When the PCR products are separated by gel electrophoresis, the digested mutant strands can be identified (Fig. 15.28).

Disrupting random genes can identify plant gene functions. Various techniques can be used, such as insertions, gene silencing, fast neutron mutagenesis, and TILLING.

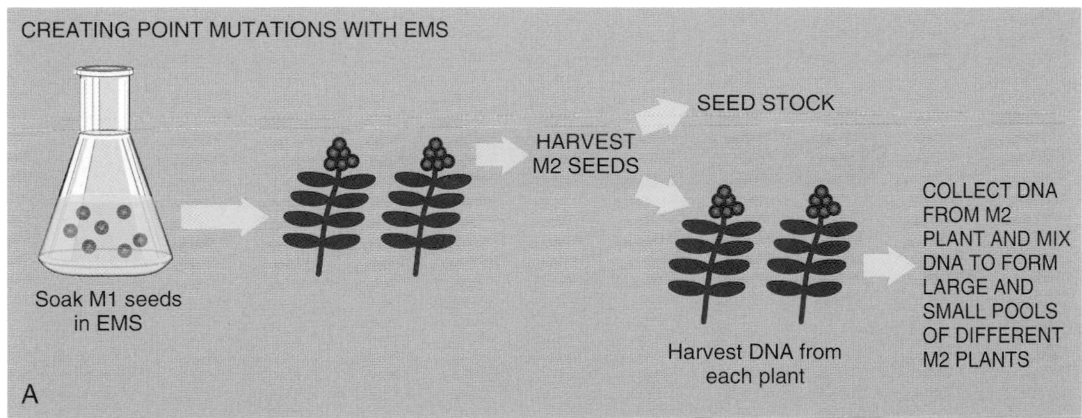

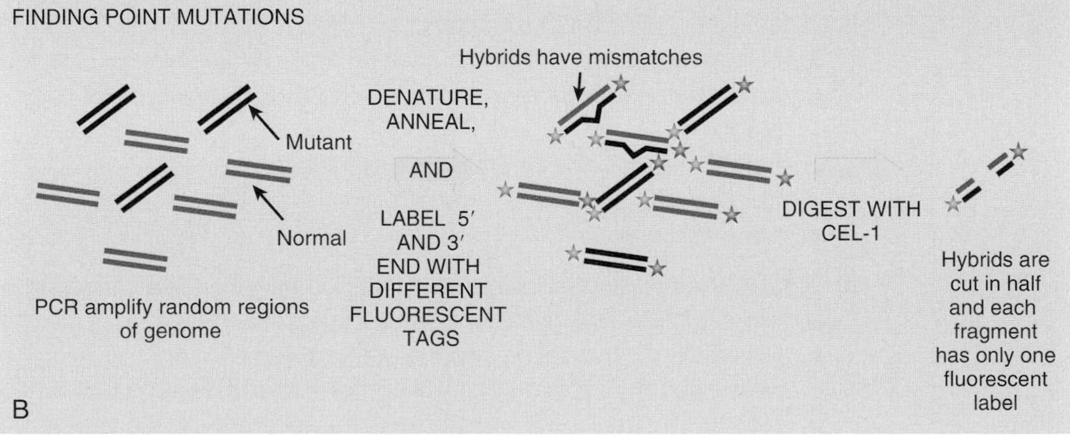

FIGURE 15.28 Identifying Point Mutations with TILLING

TILLING identifies point mutations in a library of plant DNA. (A) EMS, a chemical mutagen, induces point mutations in seeds. The M1 seeds are grown into plants, and the M2 seeds are harvested. Most M2 seeds are stored as a stock, while the remaining M2 seeds are grown into plants. DNA is harvested from each plant and pooled. The larger pools contain DNAs from all the M2 plants, and the smaller pools contain DNA from one or two different M2 plants. (B) Point mutations are identified in the DNA pools using PCR to randomly amplify different areas of the plant genomes. Some PCR products will contain point mutations, and others will be normal. These are denatured and reannealed so that some of the normal and mutant strands form hybrids. The reannealed PCR products are labeled at each end with a different fluorescent tag. The PCR products are then digested with the enzyme CEL-1, which cuts only where the helix has a mismatch. This leaves any mutant:normal hybrids with a single fluorescent tag.

Summary

Most transgenic crops are created to make plants resistant to different herbicides or insects. Others are engineered to become more resistant to disease and drought. Still, other transgenic crops are designed to have more nutrients or fewer toxins. Transgenic crops are highly regulated. Before a crop is allowed to go to the marketplace, the EPA, FDA, and USDA all provide oversight.

To make transgenic plants, scientists must culture plants in liquid suspension culture or in callus culture. Each method uses totipotent cells that retain the ability to develop into an entirely new plant. Transgenes are introduced using a modified Ti plasmid from *Agrobacterium*. Alternatively, particle guns blast DNA constructs into plant cells. Reporter genes, usually conferring antibiotic resistance, are transformed along with the gene of interest, making identification of the transgenic plants easy. Typically, these genes are removed. Newer genome editing techniques, such as CRISPR and TALENs, may bypass the need for reporter genes.

Nature is responding to the presence of transgenic crops in the form of Bt toxin-resistant insects and herbicide-resistant weeds. Newer varieties of transgenic crops and better farming practices can mitigate the spread of resistance.

Transgenic technology also allows for the study of functional genomics in plants. Randomly disrupting genes allows scientists to identify unique properties.

End-of-Chapter Questions

1. Which trait can be conferred onto a plant by a transgene?
   - **a.** resistance to diseases
   - **b.** enhance the nutritional value
   - **c.** resistance to an insect
   - **d.** protection from herbicides
   - **e.** all of the above
2. What is the definition of totipotent?
   - **a.** cells that retain the ability to develop into any cell type of a mature plant
   - **b.** cells that cannot be reprogrammed to produce any type of cell in a mature plant
   - **c.** plant cells that can naturally take up transgenes
   - **d.** transgenic plant cells
   - **e.** none of the above
3. What is the first and most important step in genetically engineering plants?
   - **a.** growth of the plant cells in a culture system
   - **b.** insertion of a transgene into the plant genome
   - **c.** identification of a gene that will confer a specific useful trait onto the plant
   - **d.** cloning the transgene into the appropriate plant shuttle vectors
   - **e.** none of the above
4. Which one of the following statements about the Ti plasmid is true?
   - **a.** The Ti plasmid originates from *Agrobacterium tumefaciens.*
   - **b.** The Ti plasmid has been used by genetic engineers to transfer genes into plants to confer a particular trait.
   - **c.** The Ti plasmid produces tumors on plant roots.

   - **d.** The Ti plasmid carries genes for opine uptake and metabolism, as well as genes for virulence.
   - **e.** all of the above

5. What advantage does an inducible promoter have over a constitutive promoter to express transgenes in plants?
   - **a.** Transgenes under the control of an inducible promoter are only expressed in the tissue that requires that particular trait.
   - **b.** Constitutive promoters ensure that the transgene is turned on even when it is not necessary.
   - **c.** An inducible promoter ensures that the transgene is always turned on, even when it is not necessary.
   - **d.** Transgenes under the control of a constitutive promoter can produce toxins, even in parts of the plant that are consumed by animals.
   - **e.** none of the above

6. What is DNA coated onto when transforming plant cells with a particle gun?
   - **a.** silver
   - **b.** aluminum
   - **c.** helium
   - **d.** gold
   - **e.** calcium

7. Why is either the gene for luciferase or the gene for neomycin phosphotransferase included when transforming DNA into plant cells?
   - **a.** The gene products, particularly luciferase, create interesting petal colors.
   - **b.** These genes encode proteins that can be used to select for or identify cells that have been successfully transformed.
   - **c.** The products of these genes can recruit *A. tumefaciens* to transfer T-DNA.
   - **d.** These genes encode proteins that add nutritional value to the plant tissues.
   - **e.** The products of these genes encode proteins that confer resistance to insects and diseases.

8. What is the significance of using the Cre/*loxP* system in plant biotechnology?
   - **a.** This system prevents the transgene from recombining into the plant genome.
   - **b.** This system creates a more efficient way to integrate useful genes into the plant chromosome.
   - **c.** This system provides a way to remove the selectable marker or reporter gene from the transgenic plants.
   - **d.** This system provides no added benefit to plant genetic engineering.
   - **e.** none of the above

9. Which agency is responsible for regulation of all transgenic technology?
   - **a.** FDA
   - **b.** EPA
   - **c.** USDA APHIS
   - **d.** NIH
   - **e.** all of the above

10. Which of the following statements about glyphosate is not true?
   - **a.** Glyphosate is toxic to the environment according to the EPA.
   - **b.** Glyphosate inhibits the EPSPS enzyme in the synthesis pathway for aromatic amino acids.

(Continued)

- **c.** Glyphosate is a particularly good herbicide because it is environmentally friendly and has no affect on humans since humans do not have the EPSPS enzyme.
- **d.** A mutant strain of *Agrobacterium* provided the *aroA* gene that conferred resistance to glyphosate in plants.
- **e.** All of the above are true.

11. Which of the following statements about Bt toxin is true?
- **a.** Bt toxin is produced by the soil bacterium *Bacillus thuringiensis.*
- **b.** Bt toxin kills insects like cotton bollworms and corn borers.
- **c.** Bt toxin is released by the Cry proteins of *Bacillus* spores that are ingested by insects.
- **d.** Transgenic crop plants expressing the Cry proteins have been very beneficial in decreasing the use of insecticides.
- **e.** All of the above are true.

12. What happened when the genes for drought and stress tolerance were placed under the control of a constitutive promoter and transformed into rice?
- **a.** The plant failed to produce rice.
- **b.** The plant grew faster and taller.
- **c.** The plant produced more rice than usual.
- **d.** The growth of the rice plant was stunted.
- **e.** nothing

13. Which of the following introduce mutations in plant genes?
- **a.** TILLING
- **b.** transposon insertions
- **c.** RNAi
- **d.** fast neutron mutagenesis
- **e.** all of the above

14. Which Cry protein, in high densities, was toxic to Monarch caterpillars?
- **a.** Cry1F Event #176
- **b.** Cry1Ab Event #176
- **c.** Cry9C Event #176
- **d.** Cry2A Event #176
- **e.** none of the above

15. Fast neutron mutagenesis produces ______________.
- **a.** point mutations
- **b.** insertions
- **c.** deletions
- **d.** A/T transitions
- **e.** chromosomal structural changes

Further Reading

Aboul-Ata, A.-A. E., Vitti, A., Nuzzaci, M., El-Attar, A. K., Piazzolla, G., Tortorella, C., et al. (2014). Plant-based vaccines: novel and low-cost possible route for Mediterranean innovative vaccination strategies. *Advances in Virus Research, 89*, 1–37.

Al-Babili, S., & Beyer, P. (2005). Golden rice—five years on the road—five years to go? *Trends in Plant Science, 10*, 565–573.

Bhatnagar, D., Ehrlich, K. C., Moore, G. G., & Payne, G. A. (2014). Aspergillus: *Aspergillus flavus*, In Batt, C. A., Tortorello, M. L. (Eds.) *Encyclopedia of Food Microbiology* (2nd ed.). London, UK, Burlington, MA, USA, and San Diego, CA, USA: Academic Press (pp. 83–91).

Bizily, S. P., Rugh, C. L., Summers, A. O., & Meagher, R. B. (1999). Phytoremediation of methylmercury pollution: *merB* expression in *Arabidopsis thaliana* confers resistance to organomercurials. *Proceedings of the National Academy of Sciences of the United States of America, 96*, 6808–6813.

Borem, A., Diola, V., & Fritsche-Neto, R. (2014). Plant breeding and biotechnological advances. In A. Borem, & R. Fritsche-Neto (Eds.), *Biotechnology and Plant Breeding: Applications and Approaches for Developing Improved Cultivars* (pp. 1–17). London, UK, Waltham, MA, USA, and San Diego, CA, USA: Academic Press.

Borem, A., Diola, V., & Fritsche-Neto, R. (2014). Transgenic plants. In A. Borem, & R. Fritsche-Neto (Eds.), *Biotechnology and Plant Breeding: Applications and Approaches for Developing Improved Culture* (pp. 179–199). London, UK, Waltham, MA, USA, and San Diego, CA, USA: Academic Press.

Borland, A. M., Hartwell, J., Weston, D. J., Schlauch, K. A., Tschaplinski, T. J., Tuskan, G. A., et al. (2014). Engineering crassulacean acid metabolism to improve water-use efficiency. *Trends in Plant Science, 19*, 327–338.

Coly, A. (1999). Photochemically-induced fluorescence determination of sulfonylurea herbicides using micellar media. *Talanta, 49*, 107–117.

Curtis, I. S. (2003). The noble radish: past, present and future. *Trends in Plant Science, 8*, 305–307.

Dörnenburg, H., Frickinger, P., & Seydel, P. (2008). Plant cell-based processes for cyclotides production. *Journal of Biotechnology, 135*, 123–126.

Escobar, M. A., Leslie, C. A., McGranahan, G. H., & Dandekar, A. M. (2002). Silencing crown gall disease in walnut (*Juglans regia* L.). *Plant Science, 163*, 591–597.

Faino, L., & Thomma, B. P. H.J. (2014). Get your high-quality low-cost genome sequence. *Trends in Plant Science, 19*, 288–291.

Fourie, G., Steenkamp, E. T., Ploetz, R. C., Gordon, T. R., & Viljoen, A. (2011). Current status of the taxonomic position of *Fusarium oxysporum formae specialis cubense* within the *Fusarium oxysporum* complex. *Infection, Genetics and Evolution, 11*, 533–542.

Gaj, T., Gersbach, C. A., & Barbas, C. F. (2013). ZFN, TALEN, and CRISPR/Cas-based methods for genome engineering. *Trends in Biotechnology, 31*, 397–405.

Gasiunas, G., & Siksnys, V. (2013). RNA-dependent DNA endonuclease Cas9 of the CRISPR system: Holy Grail of genome editing? *Trends in Biotechnology, 21*, 562–567.

Goff, S. A., Schnable, J. C., & Feldmann, K. A. (2014). Genomes of herbaceous land plants. In Paterson A. H., (Ed.), *Advances in Botanical Research*, Vol. 69, (London, UK, Amsterdam, The Netherlands, Oxford, UK, Waltham, MA, USA, and San Diego, CA, USA: Academic Press), pp. 47–90.

Klumper, W., & Qaim, M. (2014). A meta-analysis of the impacts of genetically modified crops. *PLoS ONE, 9*, e111629. http://dx.doi.org/10.1371/journal.pone.0111629.

Lauersen, K. J., Berger, H., Mussgnug, J. H., & Kruse, O. (2013). Efficient recombinant protein production and secretion from nuclear transgenes in *Chlamydomonas reinhardtii*. *Journal of Biotechnology, 167*, 101–110.

Lozano-Juste, J., & Cutler, S. R. (2014). Plant genome engineering in full bloom. *Trends in Plant Science, 19*, 284–287.

Pilon-Smits, E., & Pilon, M. (2000). Breeding mercury-breathing plants for environmental cleanup. *Trends in Plant Science, 5*, 235–236.

Shukla, V. K., Doyon, Y., Miller, J. C., DeKelver, R. C., Moehle, E. A., Worden, S. E., et al. (2009). Precise genome modification in the crop species *Zea mays* using zinc-finger nucleases. *Nature, 459*, 437–441.

Vallero, D. A., & Letcher, T. M. (2013). Invasions. In *Unraveling Environmental Disasters*. (Waltham, MA, USA, Oxford, UK, and Amsterdam, The Netherlands: Elsevier), pp. 321–351.

Weinthal, D., Tovkach, A., Zeevi, V., & Tzfira, T. (2010). Genome editing in plant cells by zinc finger nucleases. *Trends in Plant Science*, *15*, 308–321.

Zhang, F., Maeder, M. L., Unger-Wallace, E., Hoshaw, J. P., Reyon, D., Christian, M., et al. (2010). High frequency targeted mutagenesis in *Arabidopsis thaliana* using zinc finger nucleases. *Proceedings of the National Academy of Sciences of the United States of America*, *107*, 12028–12033.

Transgenic Animals

Biotechnology

http://dx.doi.org/10.1016/B978-0-12-385015-7.00016-8

NEW AND IMPROVED ANIMALS

For thousands of years people have improved crop plants and domestic animals by selective breeding, mostly at a trial-and-error level. Woollier sheep and smarter sheep dogs have both been improved through many generations of selective breeding. Obviously, the more we know about genetics, the faster and more effectively we can improve our crops and livestock.

Today it is possible to alter plants, animals, and even humans by genetic engineering. Most early experiments in animal transgenics were done with mice, but many larger animals have now been engineered, including livestock such as sheep and goats, pets such as cats and dogs, and even monkeys. In a **transgenic** animal, every cell carries new genetic information. In other words, novel genetic information is introduced into the germline, not merely into some somatic cells as in gene therapy (discussed in Chapter 17). Consequently, the novel genes in a transgenic animal are passed on to its descendants.

This novel genetic information generally consists of genes transferred from other organisms and so referred to as **transgenes**. They may be derived from animals of the same species, from distantly related animals, or even from unrelated organisms such as plants, fungi, or bacteria. The transgenes are themselves often engineered before being inserted into the host animal. The most frequent alteration is to place the transgene under control of a more convenient promoter. This may mean a stronger promoter or a promoter designed to express the transgene under specific conditions. These assorted manipulations have been dealt with in previous chapters. Here, we consider techniques to create transgenic animals and some applications of this technology.

Humans have modified animals by selective breeding for thousands of years. Today, it is possible to insert foreign genes, thus creating transgenic animals with improved qualities.

CREATING TRANSGENIC ANIMALS

Once a suitable transgene is available, the standard scenario for the creation of a transgenic animal by **nuclear microinjection** is as follows:

1. The transgene is injected into fertilized egg cells (Fig. 16.1). Just after fertilization, the egg contains its original female nucleus plus the male nucleus from the successful sperm. These two **pronuclei** will soon fuse together. Before this happens, the DNA is injected into the male pronucleus, which is larger and therefore a better target for microinjection. Nuclear microinjection requires specialized equipment and great skill. The success rate varies from 5% to 40% among various laboratories.
2. The egg is kept in culture during the first few divisions of embryonic development.
3. The engineered embryos are then implanted into the womb of a female animal, the **foster mother.** Here, they develop into embryos and, if all goes well, into newborn animals.
4. Some of the baby animals will have the transgene stably integrated into their chromosomes. In others, the process fails and the transgene is lost. Those that received the transgene and maintain it stably are called **founder animals.** A male and female are mated together to form a new line of animals carrying two copies of the transgene (Fig. 16.1). Note that the founder animals contain only a single copy of the transgene on one chromosome and are heterozygous for the transgene.

When two such founder animals are bred together, 25% of the progeny will get two copies of the transgene and will be homozygous, 25% will get zero copies, and the remaining 50% will get one copy. Homozygous transgenic animals are most useful because if they are further interbred, all of their descendants will get two copies of the transgene.

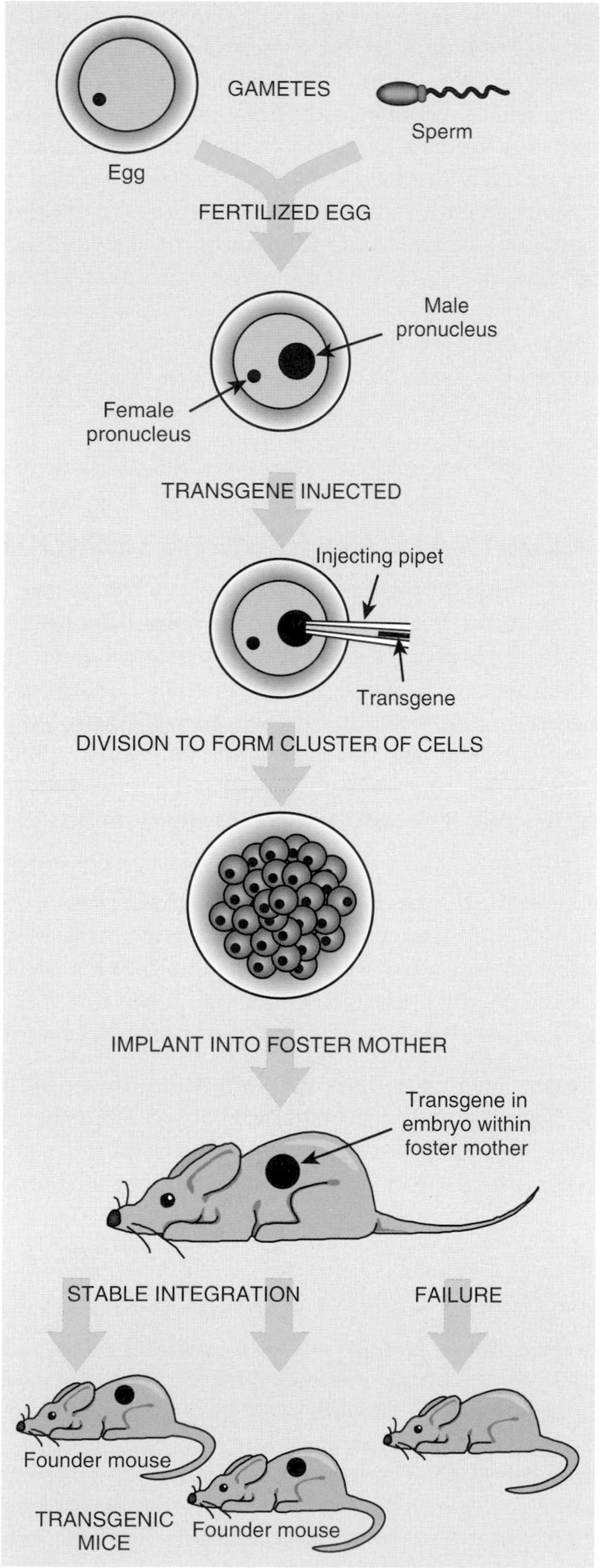

GAMETES
Egg
Sperm
FERTILIZED EGG
Male pronucleus
Female pronucleus
TRANSGENE INJECTED
Injecting pipet
Transgene
DIVISION TO FORM CLUSTER OF CELLS
IMPLANT INTO FOSTER MOTHER
Transgene in embryo within foster mother
STABLE INTEGRATION
FAILURE
Founder mouse
Founder mouse
TRANSGENIC MICE

FIGURE 16.1 Creation of Transgenic Animals by Nuclear Injection
In vitro fertilization is used to start a transgenic animal. Harvested eggs and sperm are fertilized, and before the pronuclei fuse, the transgene is injected into the male pronucleus. The embryo continues to divide in culture and is then implanted into a mouse. The "foster mother" mouse has been treated with hormones so that she accepts the embryo and carries on with the pregnancy. The offspring are screened for stable integration of the transgene. Founder mice have one copy of the transgene.

There is some variability in the timing of integration of DNA injected into the male pronucleus. In some cases, the DNA integrates more or less immediately, so all cells of the resulting animal will contain the transgene. Less often, several cell divisions may happen before DNA integration occurs, and the final result will be a **chimeric animal**, in which some cells contain the transgene and others do not. Sometimes multiple tandem copies of the transgene integrate into the same nucleus; such constructs are often unstable, and the extra copies are often deleted out over successive generations. Integration of the incoming DNA into the host cell chromosomes occurs at random. Often, this is accompanied by rearrangements of the surrounding chromosomal DNA. This suggests that integration often occurs at the sites of spontaneous chromosomal breaks.

Transgenic animals are often generated by microinjection of DNA into the nucleus of a fertilized egg cell. Founder animals arising from such engineered eggs have one copy of the transgene. They must be bred to yield animals that are homozygous for the transgene.

LARGER MICE ILLUSTRATE TRANSGENIC TECHNOLOGY

The classic illustration of transgenic technology was the creation of larger mice by inserting the rat gene for growth hormone. Growth hormone, or **somatotropin**, consists of a single polypeptide encoded by a single gene. In 1982 the somatotropin gene of rats was cloned and inserted into fertilized mouse eggs. The eggs were then placed into foster mother mice that gave birth to the genetically engineered mice. These transgenic mice were larger (about twice normal size), although not as large as rats. This was the first case in which a gene transferred from one animal to another not only was stably inherited, but also functioned more or less normally. Many transgenic mice now exist. A selection appear in Box 16.1.

To express the rat somatotropin gene, it was put under the control of the promoter from an unrelated mouse gene, **metallothionein**, which is normally expressed in the liver (Fig. 16.2). Instead of being made in the pituitary gland, the normal site for growth hormone, the rat somatotropin was mostly manufactured in the liver of the transgenic mice. Even though made in the "wrong" location, the hormone worked and made the mice larger. Human somatotropin has also been expressed in mice and also gives bigger mice.

Size does not depend solely on growth hormone. Defective growth hormone receptors also inhibit growth. African pygmies rarely grow taller than 4 feet, 10 inches, yet they have normal levels of human somatotropin. It appears that pygmies have defective growth hormone receptors. These normally bind the somatotropin circulating in the bloodstream and are

Box 16.1 Trendy Transgenic Mice

Genetically engineered mice have become all the rage and every so often hit the headlines:

- **Marathon Mouse** can run about 1800 meters—more than a mile—before exhaustion. This is twice as far as a normal mouse can last. Marathon mouse has enhanced PPAR-delta—a regulator of several genes involved in burning fat and in muscle development.
- **Mighty Mouse** was engineered to lack myostatin, a protein that slows muscle growth. The result is colossal muscle development. There is one known case of a human with a genetic defect leading to lack of myostatin. A German boy, born in Berlin in 2000, has muscles twice the size of other children his age.
- **Fierce Mouse** has both copies of the *NR2E1* gene deleted and shows abnormal aggression.
- **Smart Mouse** is genetically modified to have improved learning and memory. It has extra copies of the *NR2B* gene encoding the NMDA receptor found in the synapse region of nerve cells in the brain.

And last but not least, the K14-Noggin mouse has an extra Noggin gene from chickens. This affects development of skin and external organs. In particular, the male is extra hairy and has a longer penis. As yet, this mouse, made in 2004, has no official nickname.

The improved athletic abilities of Marathon Mouse and Mighty Mouse have raised the issue of tampering with humans—so-called "gene-doping" in order to promote athletic prowess (see Chapter 24 for further discussion). As for K14-Noggin mouse, one can only imagine the implications!

necessary for the hormone to work on its target tissues. Dwarfism, among nonpygmies, may be due either to defects in production of somatotropin or to a shortage of receptors. **Recombinant human somatotropin (rHST)** is now used to treat the hormone-deficient type of dwarf. However, receptor-deficient dwarfs cannot yet be successfully treated.

In one of the first transgenic experiments, the gene for growth hormone from larger animals was inserted into mice. It was expressed under control of the metallothionein promoter, allowing it to be induced by traces of zinc. The result was larger mice.

FIGURE 16.2 Large Transgenic Mice
A DNA construct containing the rat somatotropin gene under the control of the mouse metallothionein promoter was used to make a transgenic mouse. The transgene causes the mouse to grow to twice its normal body size.

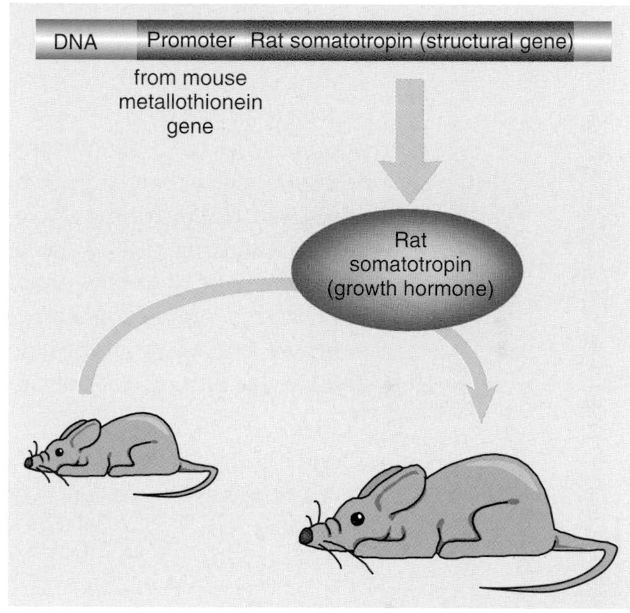

RECOMBINANT PROTEIN PRODUCTION USING TRANSGENIC LIVESTOCK

The somatotropin gene from cows has been cloned and expressed in bacteria, thus allowing the production of large amounts of the hormone, which is known as **recombinant bovine somatotropin (rBST).** The rBST is used in the dairy industry to increase milk production. Unlike in mice, boosting an adult cow's somatotropin levels by injection of extra hormone results in increased milk production, rather than giant cows. Milk from treated cows is now widely marketed.

At present, bacteria, such as *Escherichia coli,* are cultured to make most recombinant proteins, such as human insulin or somatotropin. Such products are expensive and require a highly trained work force. However, using livestock to express these products may be cheaper. Dairy cows produce 10,000 quarts of milk each per year, and an industry to collect and process milk already exists. To take advantage of this industry, several recombinant proteins are now being produced in the milk from transgenic cows or other farm animals. To achieve this result, cloned genes are placed under the control of a regulatory region that will allow gene expression only in the mammary gland. Consequently, the gene product will come out in the milk (Fig. 16.3). For small-scale production of proteins for clinical use, transgenic goats are often used. For example, transgenic goats have been made to produce **recombinant tissue plasminogen activator (rTPA),** which is used for dissolving blood clots.

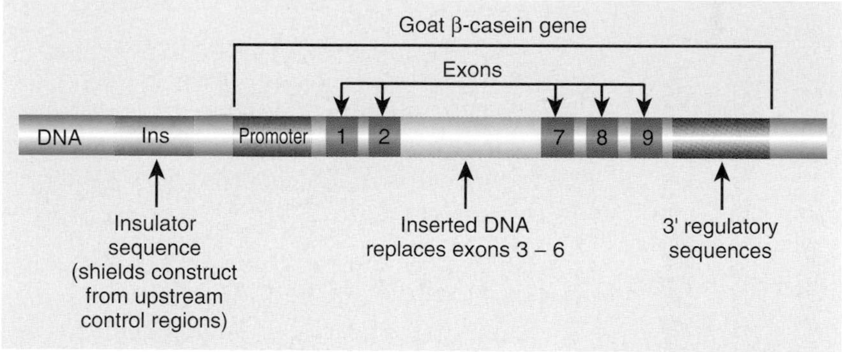

FIGURE 16.3 Milk Expression Construct for Transgenic Goats
To express a recombinant protein in goat milk, the gene of interest is inserted in place of the β-casein gene. The transgene will be expressed using the endogenous promoter and 3′ regulatory elements that restrict β-casein expression to goat milk. The construct also has insulator sequences that block other regulatory elements from affecting expression (see later discussion).

So far, no primate has been successfully cloned, although transgenic rhesus monkeys have been generated. The first successful engineering of a transgenic primate resulted in the birth of ANDi, a rhesus monkey carrying the *gfp* gene, in late 2000. ANDi stands for "inserted DNA" (read backward). A crippled retrovirus vector was used to deliver the gene for GFP to unfertilized eggs that were later fertilized *in vitro*. Treatment of 224 egg cells gave 20 embryos, 5 pregnancies, and eventually, 3 live male monkeys. Only one of these, ANDi, was transgenic and expressed GFP. ANDi does not fluoresce green because GFP levels are too low (also, rhesus monkeys have brown fur over much of their bodies). Transgenic rhesus monkeys are being used as models for some human neurological diseases.

Recombinant proteins may be manufactured by expressing the corresponding genes in transgenic cattle or goats. Transgenic monkeys are used to study disease.

KNOCKOUT MICE FOR MEDICAL RESEARCH

Transgenic animals, mostly mice, are of great value in the genetic analysis of inherited diseases and cancer. Here, we are interested not so much in adding a cloned transgene as in discovering the function of genes already present. The general approach is to inactivate, or "knock out," the gene of interest and then ask what defect this causes.

The target gene is first cloned to achieve this result. Then a **DNA cassette** is inserted into its coding sequence to disrupt the gene. (Most DNA cassettes also include an antibiotic resistance gene for easy detection.) The intruding DNA segment prevents the gene from making the correct protein product and thus abolishes its function. The inactive copy of the gene is then put back into the animal by following the procedure outlined earlier for transgene insertion. The incoming DNA, carrying the disrupted gene, sometimes replaces the original, functional, copy of the gene by homologous recombination. Founder mice are obtained that have one copy of the disrupted gene. When they are bred together, mice with both copies disrupted are obtained. Such mice are known as **knockout mice** and will completely lack gene function (Fig. 16.4). If the gene in question is essential, homozygous knockout mice may not survive or may live only a short time.

Knockout mice have selected genes inactivated. They are widely used in medical research to investigate gene function.

ALTERNATIVE WAYS TO MAKE TRANSGENIC ANIMALS

Although nuclear injection was the first approach used to make transgenic animals, and is still the most widely used, there are several alternative procedures. As discussed in Chapter 17, engineered retroviruses have been used in gene therapy to introduce DNA into the chromosomes of animal cells. Retroviruses can infect the cells of early embryos, including embryonic stem cells (see Chapter 18). Hence, retrovirus vectors may introduce transgenes.

The advantages of using a retrovirus are that only a single copy of the retrovirus plus transgene is integrated into the genome. In addition, use of retroviruses does not require skill in microinjection. The retrovirus carrying the transgene construct is added to the fertilized egg and allowed to infect as normal. The egg is then transplanted into a pseudo-pregnant female mouse. The remainder of the procedure is as before.

The disadvantages are that virus DNA is introduced along with the transgene and that retroviruses can carry only limited amounts of DNA. Furthermore, founder animals made using retroviruses are always chimeras, because insertion of the virus does not occur precisely when the nuclei fuse. Consequently, retroviruses are rarely used in attempts to create fully transgenic animals. On the other hand, partially transgenic animals that have some sectors or tissues altered are useful because the transgenic tissues may be compared with normal tissue within the same animal. This process can eliminate doubts on whether the defects or changes caused by the transgene are merely due to differences among animals or truly due to transgene expression.

Embryonic stem cells may also be used to generate transgenic animals. **Stem cells** are the precursor cells to particular tissues of the body. Embryonic stem cells are derived from the **blastocyst**, a very early stage of the embryo, and retain the ability to develop into any body tissue, including the germline. Embryonic stem cells can be cultured and DNA can be introduced as for any cultured cell line. For successful creation of transgenic animals, embryonic stem cells must be maintained under conditions that avoid differentiation.

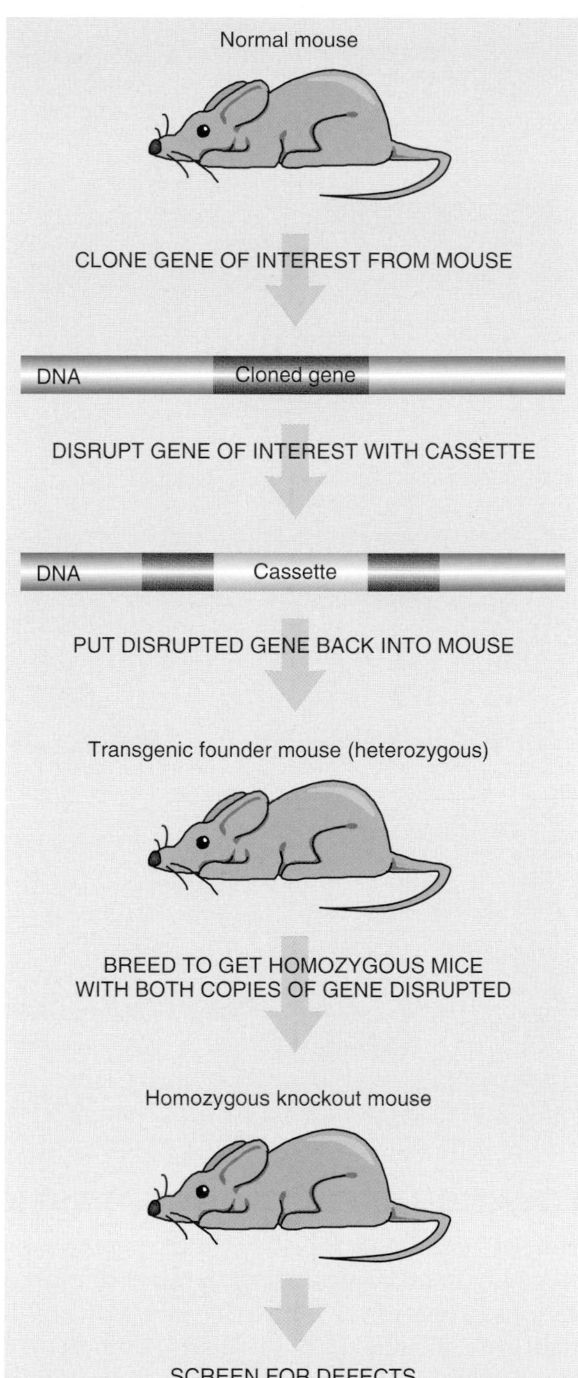

FIGURE 16.4
Knockout Mice
Like traditional transgenic mice, knockout mice are generated *in vitro*. The target gene is cloned and disrupted by inserting a DNA cassette. This work is usually done in bacteria. Once the construct is made, it is put back into a mouse by injection into the male pronucleus during fertilization (or by other methods outlined later). After the transgenic offspring are born, two heterozygotes are crossed to create a homozygous knockout mouse. These are then screened for defects due to inactivation of the target gene.

Engineered embryonic stem cells are then inserted into the central cavity of an early embryo at the blastocyst stage (Fig. 16.5). This process creates a mixed embryo and results in an animal that is a genetic chimera consisting of some transgenic tissues and others that are normal.

If the host embryo and the embryonic stem cells are from different genetic lines with different fur colors, the result is an animal with a patchwork coat. This allows the transgenic sectors of the animal to be identified easily. This chimeric founder animal must then be mated with a wild-type animal. If the embryonic stem cells have contributed to the germline, then the coat color characteristic of this cell line will be transmitted to the offspring. In mice, black (recessive) and **agouti** (dominant) coat colors are often used. The embryonic stem cells are usually taken from an agouti line because the dominant fur color enables the transgenic cell lines to be tracked easily. Both the embryonic stem cells and the host embryo are usually taken from males because the resulting male chimeras can father many children when crossed with wild-type females.

Transgenic animals may be generated by several alternative methods. One approach is to use retrovirus vectors that insert into host chromosomes. Another is to engineer embryonic stem cells in culture before returning them to an early embryo.

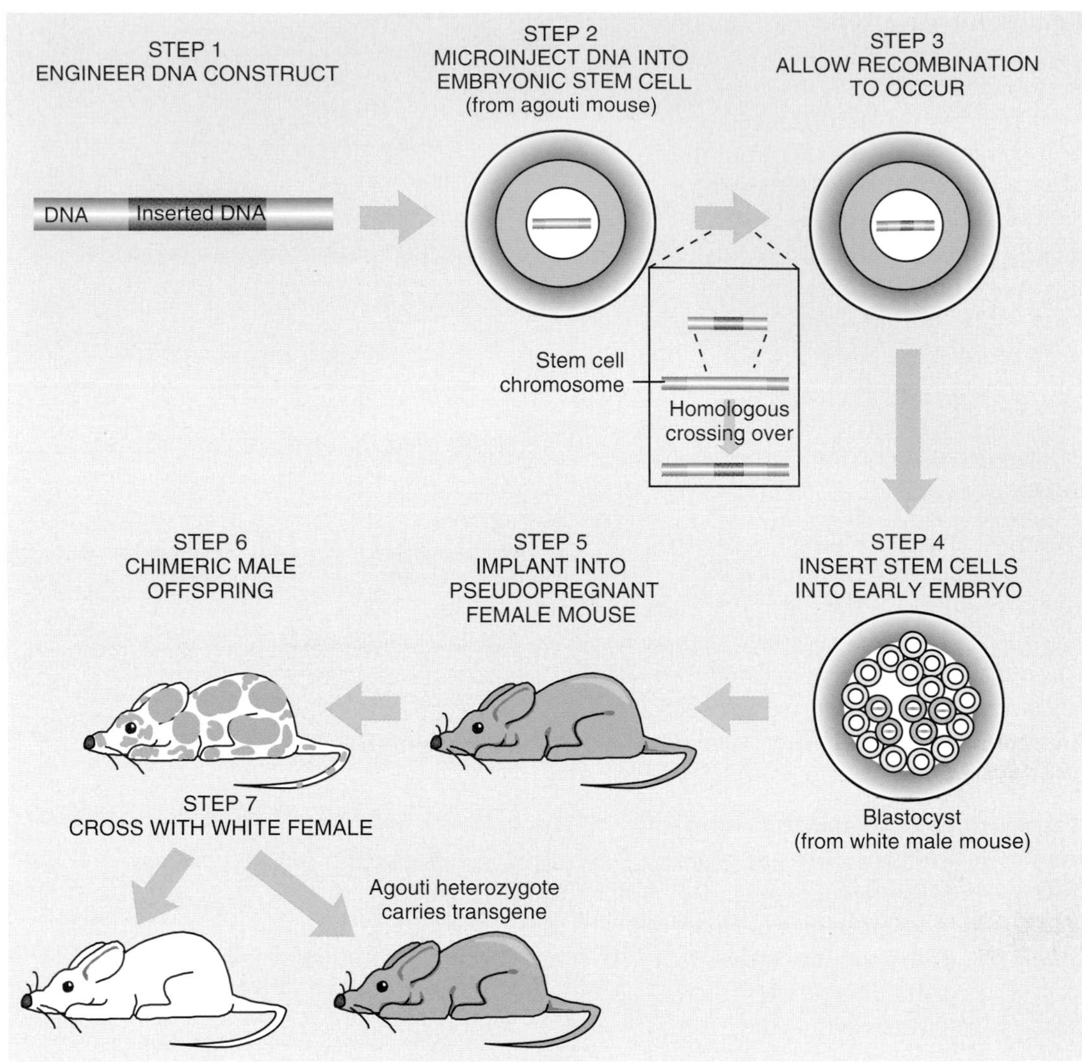

FIGURE 16.5 Use of Embryonic Stem Cells
To create a transgenic animal with embryonic stem cells, a researcher must first insert the transgene into these cells. The stem cells shown here are from an agouti mouse—that is, a mouse with grizzled brown fur. The stem cells are transformed with the transgene, which integrates by homologous recombination. Then the stem cells are injected into an early male embryo (blastocyst) from a white mouse. The embryo is put into a pseudopregnant female mouse. The offspring are chimeras because the majority of cells in the injected blastocyst are normal. The chimera will have a white coat with patches of brown derived from the injected stem cells. The chimera is crossed with a white female, and any fully brown (agouti) offspring will have the transgene incorporated into the germline.

LOCATION EFFECTS ON EXPRESSION OF THE TRANSGENE

Transgenic animals (or plants) carrying the same inserted transgene often differ considerably in expression. Both the level of expression and the pattern of expression in various tissues of the body may vary. Many of these effects are due to the location of the transgene. Expression of the inserted transgene will be affected by any nearby regulatory elements already present in the host animal chromosome. In particular, enhancer sequences work over considerable distances and will affect the expression of any transgenes integrated nearby. In addition, the physical state of the DNA is important. If the transgene is integrated into a region that consists largely of heterochromatin, the transgene will be expressed poorly or not at all. In such regions, the DNA is tightly packed, often methylated, covered with nonacetylated histones, and consequently usually nontranscribed.

Such position effects have been confirmed experimentally by extracting transgene DNA from a transgenic animal in which the transgene was not expressed. This DNA was then used to construct another line of transgenic animals. If some of the new transgenic animals show proper expression of the transgene, this demonstrates that the gene itself is intact and its failure to express in the original host animal was due to its location (Fig. 16.6).

Combating Location Effects on Transgene Expression

Location effects may be avoided by targeting the transgene to a specific site (see later discussion). Alternatively, appropriate regulatory elements can be built in to the transgene construct itself:

(a) **Enhancer sequences.** These sequences control nearby genes or clusters of genes in a dominant manner. For example, the **locus control region (LCR)** in front of the β-globin gene cluster confers high expression (Fig. 16.7). Note that the LCR is distinct from the individual promoters and affects several clustered genes. Such enhancer sequences generally dominate over other nearby regulatory sequences and thus provide position-independent expression. Such sequences may be placed in front of transgenes to confer high-level expression that is independent of chromosomal location.

(b) **Insulator sequences** or boundary elements. These sequences block the activity of other regulatory elements. If a gene is flanked by two insulator sequences, it is protected from the effects of any regulatory elements beyond the insulators (Fig. 16.8). Hence, transgenes can be protected from position effects by including insulator sequences in the transgene construct. Transgenes flanked by insulators probably form independent loops of DNA from which heterochromatin is excluded.

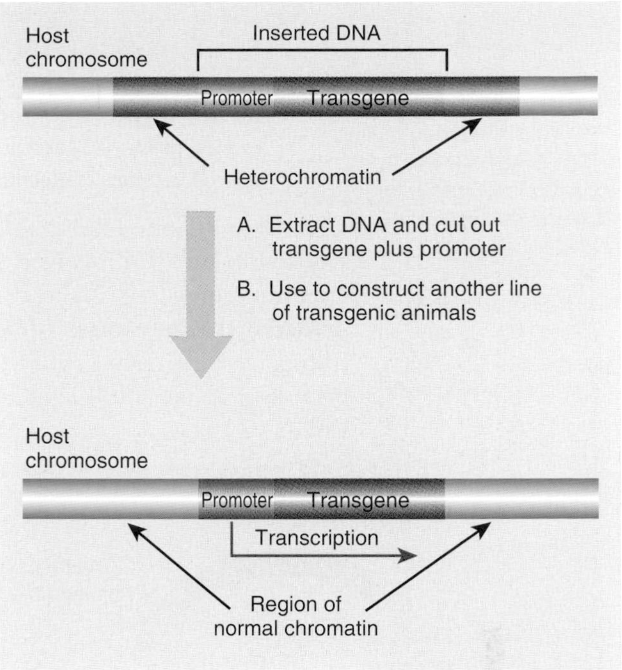

FIGURE 16.6 Failed Expression due to Transgene Location
DNA carrying a transgene was inserted to generate a transgenic animal. In this instance, the DNA was inserted into a region of heterochromatin. Even though transgenic animals were obtained, the transgene was not expressed. The inserted DNA was removed and used to make another transgenic animal. The transgene was expressed in the second animal, showing that it was intact in the first transgenic animal. Because the location of integration was different, the earlier lack of expression must have been due to a position effect.

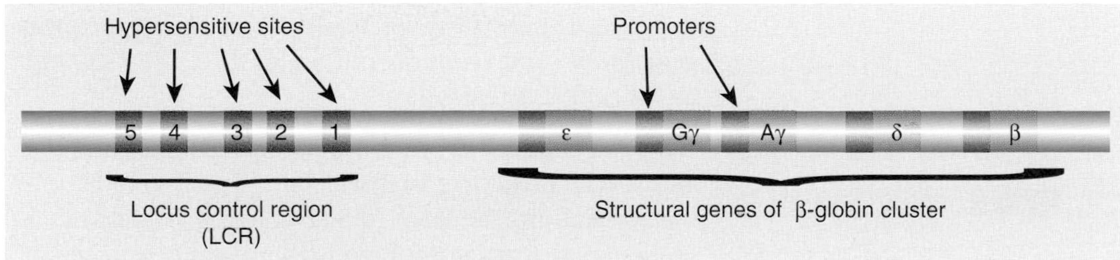

FIGURE 16.7 Locus Control Region (LCR)
The LCR of the β-globin gene cluster enhances expression of all five genes. This control region is outside the individual promoters. The LCR has five DNase I hypersensitive regions, which have multiple consensus sequences for transcription factor binding sites.

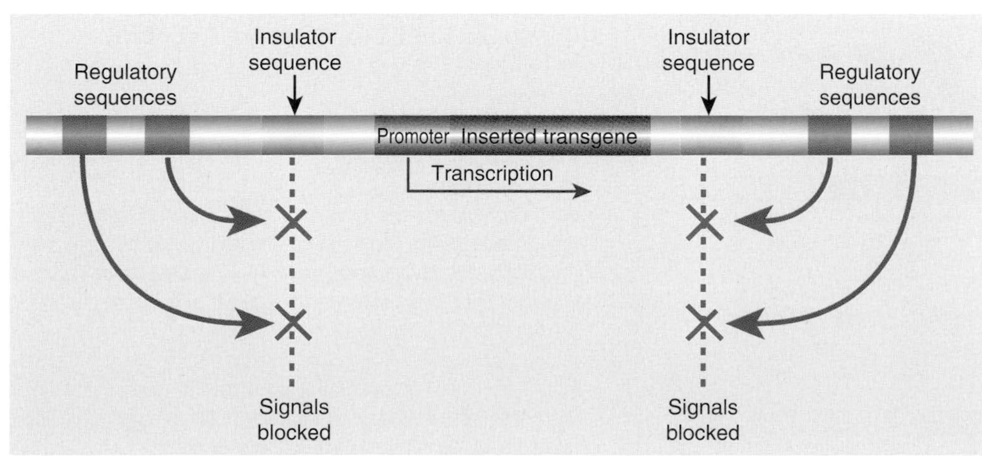

FIGURE 16.8 Protection of a Gene by Insulator Sequences
Insulator sequences are placed flanking a transgene. They protect the transgene from regulatory sequences outside the insulator sequences.

(c) Use of natural transgenes. Most transgenes actually consist of cDNA and therefore differ from the original wild-type version of the gene in lacking introns. Furthermore, most transgenes are under control of viral or artificial promoters, which are shorter and more convenient to engineer than the natural promoters from the original gene. Nonetheless, full-length natural eukaryotic genes are often more resistant to position effects than the shortened engineered versions, especially if both upstream and downstream control elements are included.

Cloning and manipulating full-length animal genes is inconvenient because of the excessive lengths of DNA involved. Nonetheless, it is possible to carry such genes on **artificial chromosomes** (see Chapter 3). In a few cases, natural length transgenes carried on **yeast artificial chromosomes (YACs)** have been used to construct transgenic animals. YAC-based transgenes have been used in the study of long-range regulatory elements. They have also been used to insert full-length genes for humanized monoclonal antibodies into mice (see Chapter 6).

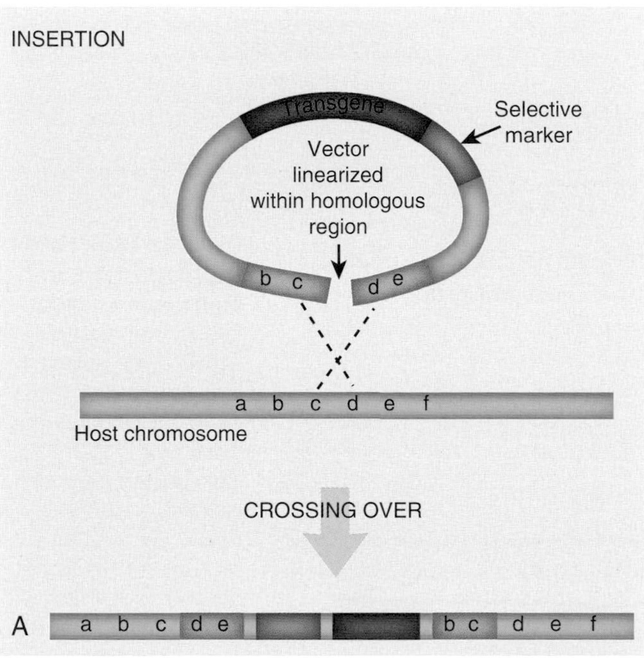

Targeting the Transgene to a Specific Location

Targeting the incoming transgene to a particular location on the host chromosome requires homologous recombination, as opposed to the random integration that usually happens with injected DNA. Inserting a transgene in a specific location may be desirable for several reasons. First, chromosomal location often affects the expression of a transgene (see earlier discussion). Second, the transgene is not necessarily a novel gene. Sometimes the objective of genetic engineering is to replace the original version of a particular gene with an altered version. In this case, it is obviously preferable to insert the incoming gene in the same location and under the same regulation as the gene it is replacing.

Gene targeting relies on homologous recombination, and special **targeting vectors** are designed to direct the integration. The DNA to be inserted is flanked by sequences that are homologous to those at the target location. Targeting vectors may be subdivided into those designed to insert novel DNA and those that replace DNA (Fig. 16.9). Targeting vectors are often linearized just before transforming the DNA into the cell because this promotes more efficient recombination. Integration of the required DNA segment may be selected by antibiotics or by some other positive selection.

FIGURE 16.9 Targeting Vectors Rely on Homologous Recombination

(A) Targeting vectors can insert a transgene into a host chromosome at a specific location. The vector has sequences (blue) homologous to the insertion point on the host chromosome (pink). The linearized vector triggers a single crossover, thus integrating the transgene and selective marker into the host chromosome. (B) Some targeting vectors promote gene replacement. These vectors have two separate regions flanking the transgene that are homologous to the host. When the linearized vector enters the nucleus, homologous regions align, and crossovers occur on each side of the transgene. The host gene is replaced with the transgene and selective marker.

The chromosomal location of a transgene can have major effects on its level of expression. Further engineering may be performed to protect transgenes from positional effects, usually by inserting an appropriate regulatory sequence next to the transgene. Targeting vectors use homologous recombination to insert transgenes at specific chosen locations in the host genome.

DELIBERATE CONTROL OF TRANSGENE EXPRESSION

In many cases, it would be helpful to control the expression of a transgene. For industrial production of a protein product, high-level expression of the transgene is usually preferred, but this is not always true—some proteins are toxic in high amounts, and gene expression needs to be kept low while establishing the line of transgenic animals. However, when transgenes are used for functional analysis, it is clearly convenient to be able to switch gene expression on and off as required. This capability is especially important for those genes that are normally expressed only in particular cell lines or at certain developmental stages. A variety of systems exist that allow the experimenter to control transgenes.

Inducible Endogenous Promoters

Early transgenic constructs often used natural promoters from the host animal (i.e., endogenous promoters) that respond to certain stimuli. For example, rat growth hormone was put under control of the mouse metallothionein promoter when inserted into mice. This promoter is induced by heavy metals such as lead, cadmium, and zinc. The least toxic of these, zinc, is actually used for induction, but even so there will be toxicity problems if continuous long-term induction is needed.

The heat shock promoter from the *Drosophila Hsp70* gene is another example of a natural promoter used to drive expression of transgenes. In this case, the promoter is inactive at room temperature; increasing the temperature to 37 °C is the inducing stimulus.

These simple inducible promoters have several drawbacks. First, there is often significant expression even in the absence of the inducing signal, and the level of induction is often low—perhaps 10-fold or less. Second, there are often toxic side effects. They may be due directly to the inducing signal (e.g., zinc, high temperature), or they may be due to the induction of other, natural, genes that respond to the same signal. Third, the inducing signal may be taken up slowly and/or penetrate only some tissues of the organism. Some of these problems can be avoided by using foreign or artificially constructed promoters.

Recombinant Promoter Systems

Bacterial repressors have been used to control transgenes in animals and plants. Both the **LacI repressor** and the **TetR repressor** have been used in this manner. For use in transgenic animals, the *lac* system must be modified as illustrated in Figure 16.10. The LacI repressor is made by the *lacI* gene, which is modified by adding a Rous sarcoma virus (RSV) promoter to ensure expression in animal cells. Another animal virus promoter, from SV40 (simian virus 40), controls the transgene. In addition, a *lacO* operator site is inserted into this promoter. The LacI repressor binds to the operator site and prevents expression of the transgene via the SV40 promoter (see Fig. 16.10). When IPTG is added, the LacI protein binds the IPTG and is released from the operator site. Thus, transgene expression is induced by IPTG.

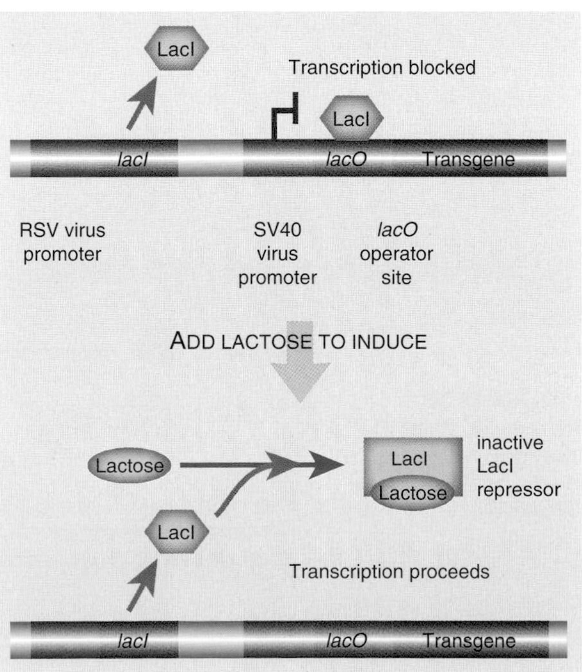

FIGURE 16.10 LacI Control of a Transgene
A Rous sarcoma virus (RSV) promoter drives transcription of the *lacI* gene. When the construct is expressed in a transgenic animal, LacI protein is produced. The LacI repressor binds to the operator site upstream of the transgene and blocks expression. When the inducer IPTG is added, it binds to LacI, which falls off the operator site. The transgene is then transcribed from the SV40 promoter.

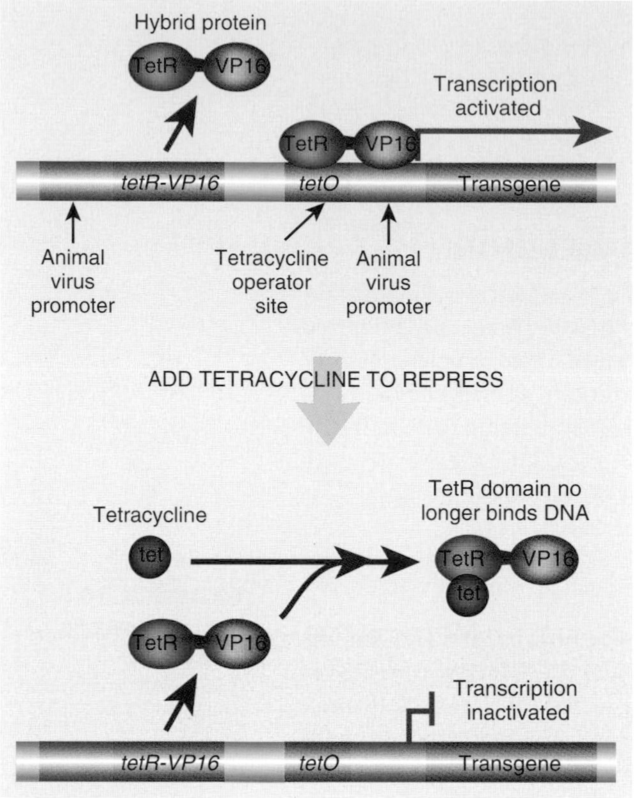

FIGURE 16.11
Hybrid TetR-VP16 Transactivator Systems
A hybrid protein made from TetR and VP16 controls transcription of a transgene. The hybrid protein binds to a TetR operator site, and the VP16 domain activates transcription. When tetracycline is added, the TetR domain binds this and is released from the *tetO* site. Therefore, tetracycline inhibits transgene expression.

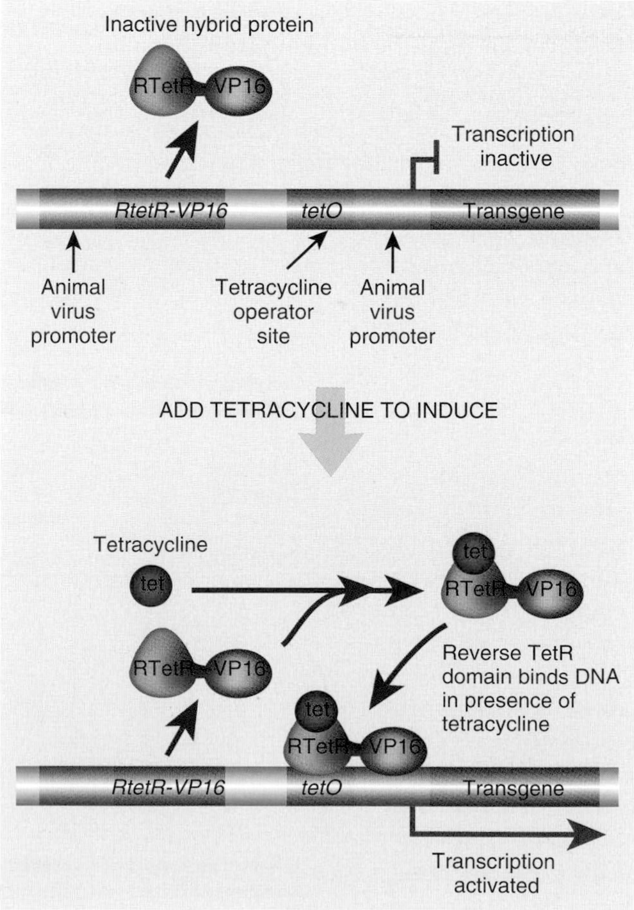

FIGURE 16.12
Reverse TetR-VP16 Transactivator System
The reverse TetR-VP16 hybrid protein binds to *tetO* only when tetracycline is present. So, in contrast to the previous case (see Fig. 16.11), the transgene is expressed only in the presence of tetracycline.

The *tet* system operates in a similar manner to the *lac* system. The ***tet* operon** confers resistance to the antibiotic tetracycline. (It is usually found as part of a transposon, e.g., the Tn10 transposon of *E. coli.)* The TetR repressor protein binds to an operator site and so prevents expression of the tetracycline resistance genes. When tetracycline is present, it binds to the TetR protein, which is therefore released from the DNA. Consequently, the *tet* operon is induced by tetracycline. The *tet* system may be used in higher organisms like the *lac* system. The transgene has a *tetO* operator site inserted into its promoter and will then be induced by tetracycline.

Both the *lac* and *tet* systems work well in eukaryotic cells. They both have low background expression and a high ratio of induction. In particular, the *tet* system may show up to 500-fold induction. One problem is the need for constant high-level expression of LacI or TetR protein. These bacterial DNA-binding proteins may become toxic to eukaryotic cells at high levels.

The *tet* system has been modified to avoid the need for high levels of TetR protein. Fusing TetR protein to eukaryotic transcription factors can give hybrid activator proteins. For example, the **VP16 activator** protein from herpes simplex virus (HSV) has been linked to TetR. When the hybrid TetR-VP16 protein binds to DNA, it recognizes the **tetO operator** sequence (Fig. 16.11). However, when bound, it causes activation (not repression) because the VP16 domain activates transcription. Thus, the TetR recognition sequence now functions as a positive regulatory site. When tetracycline is added, the TetR-VP16 protein is released from the DNA, and the gene is no longer expressed. Here, then, addition of tetracycline prevents gene expression. This system can give as much as 100,000-fold induction, and background expression when the gene is turned off is extremely low.

A further improvement is the use of a mutant TetR protein whose DNA-binding behavior has been inverted. Hence, it binds to DNA in the presence of tetracycline and is released in its absence. Consequently, hybrid activator proteins made using this version of TetR activate the target gene in the presence of tetracycline. This is known as the reverse *tet* transactivator system (Fig. 16.12). Plasmid and virus vectors have been constructed that allow all components of this system, including the cloned transgene, to be inserted into the animal together.

Transgene Regulation via Steroid Receptors

A problem with molecules such as tetracycline is that they often penetrate different tissues unequally. Steroid hormones are lipophilic and consequently penetrate cell membranes rapidly. Once inside the cell, steroids bind to receptor proteins, which in turn bind directly to DNA and regulate gene expression. Steroids are also eliminated within a few hours, and consequently, they are convenient molecules for inducing transgenes.

Here, the problem is that steroids naturally induce many eukaryotic genes. One way to avoid inducing host genes that respond to steroids is to use a steroid not found naturally in the host animal. For example, the steroid **ecdysone** from insects may be used in transgenic mammals and, conversely, the mammalian **glucocorticoid hormones** may be used in transgenic insects or plants. Consequently, the binding protein for the chosen hormone must also be supplied. A recognition sequence for this protein is placed in the upstream region of the target transgene (Fig. 16.13). The result is that the transgene is induced when the steroid is administered.

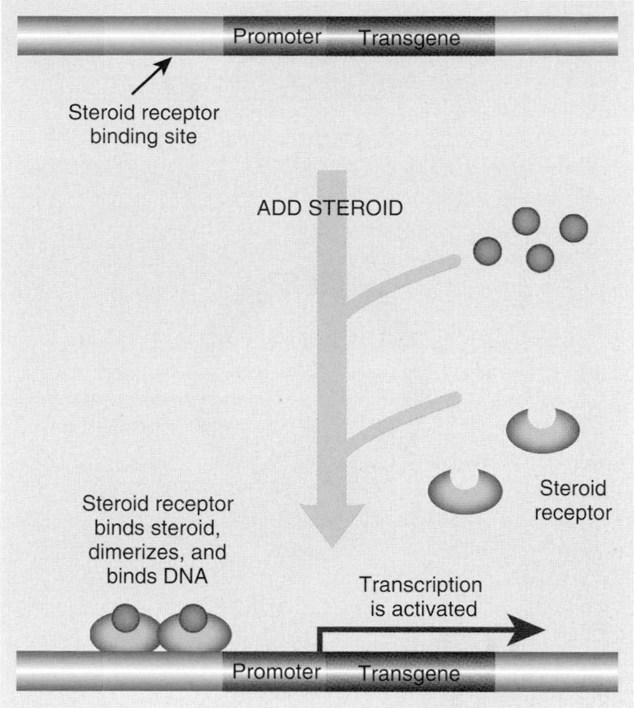

FIGURE 16.13
Steroid Hormone to Activate Transgene
Some transgenic animals have the transgene under the control of a steroid-controlled promoter. When the animal is treated with the steroid, the steroid diffuses across the cell membrane and binds to its receptor in the cytoplasm. The receptor-steroid complex then enters the nucleus, where it binds to the transgene promoter and turns on transcription.

Altered and/or hybrid **steroid receptors** have been used for improved responses. One curious example is the use of a mutant version of the mammalian progesterone receptor. The mutant version can no longer bind progesterone but still binds the antagonist **mifepristone** (RU486, active component of the abortion pill). When mifepristone is used as inducer, the concentration needed is 100-fold less than required for its action in abortion. This receptor can be used without any interference with endogenous steroids, and treatment with RU486 does not induce abortion in the transgenic animal.

Transgenes may be artificially regulated by a variety of control systems. Engineered versions of the bacterial *lac* and *tet* regulators are widely used.

GENE CONTROL BY SITE-SPECIFIC RECOMBINATION

Another way to control the expression of a transgene is via site-specific recombination. In this approach, segments of DNA are physically removed or inverted to achieve activation of the transgene. These DNA manipulations are done after the DNA carrying the transgene and associated sequences has been successfully incorporated into the germline chromosomes of the host animal.

Site-specific recombination involves recognition of short specific sequences by DNA-binding proteins. Recombination then occurs between two of the recognition sequences. Some site-specific recombination systems require several proteins for recognition and crossing over. Others need only a single protein to bind the two recognition sequences and recombine them. These systems are obviously far more useful in genetic engineering. Two such **recombinase** systems have been widely used: the **Cre recombinase** from bacterial virus P1 and the **Flp recombinase (flippase)** from the 2-micron plasmid of yeast. Both Cre and Flp recognize 34 base-pair sites (known as ***loxP*** and **FRT**, respectively) consisting of 13 base-pair inverted repeats flanking a central core of 8 base pairs.

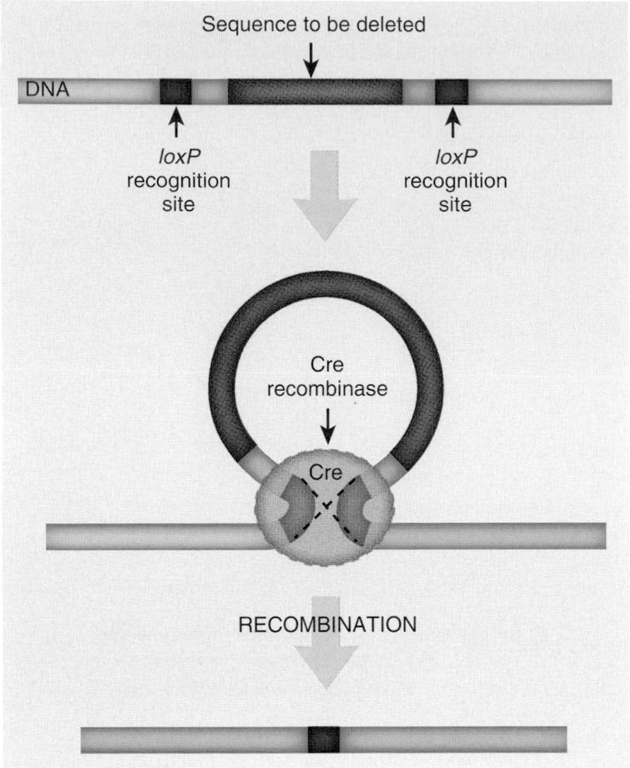

FIGURE 16.14
Site-Specific Deletions in Transgenic Animals
Two *loxP* sites must be inserted into the DNA, flanking the sequence to be deleted. When Cre recombinase is activated, it recombines the two *loxP* sites and deletes the segment of DNA between them.

We have already described the use of the *Cre/loxP* system in plants (Chapter 15) to delete unwanted DNA segments after integration of incoming transgenic DNA. Similar manipulations can be carried out in animals using *Cre/loxP* or Flp/FRT. The specific removal of unwanted segments of DNA requires them to be flanked by *loxP* or FRT sites (Fig. 16.14). (Segments flanked by two *loxP* sites are sometimes referred to as "floxed.") This approach may be used for a variety of purposes, some of which are summarized as follows:

(a) Removal of selective markers. Once transgenic DNA has been successfully integrated, antibiotic resistance genes used for selection and reporter genes used for screening are no longer needed. If they are enclosed between *loxP* or FRT sites, they can be removed, leaving a transgenic organism with only the actual transgene (plus a single copy of the *loxP* or FRT site).

(b) Activation of transgene. Here, the original transgenic construct is made with a blocking sequence between the promoter and the transgene. The blocking sequence is flanked by *loxP* or FRT sites. After integration of the incoming DNA, the blocking sequence is removed by Cre or Flp recombinase, thus activating the transgene.

(c) *In vivo* chromosome engineering. Large-scale deletions or rearrangements of eukaryotic chromosomes may be generated *in vivo* by using the Cre/*loxP* system. Two rounds of DNA insertion introduce *loxP* sites at two separate specific locations. Then the Cre recombinase is activated and deletions are generated.

(d) Creation of conditional knockout mutants. Transgenic constructs may be designed so that a specific gene can be deleted *in vivo*. Generally, two *loxP* sites are inserted into the introns flanking an essential exon of the target gene. On recombination, the exon will be deleted and the target gene will be inactivated. This allows investigation of genes whose knockout mutations are lethal at the embryonic stage. The animal can be allowed to grow into an adult before the recombinase is activated to generate the knockout.

When these recombinase systems are used, the recognition sequences are included in the transgenic constructs. Later, the recombinase itself is provided by one of three methods:

(a) The gene for recombinase may be carried on a plasmid that is transformed into the animal. The recombinase will be expressed transiently, assuming the plasmid does not integrate or survive over the long term.

(b) The recombinase gene may itself be part of the transgenic construct and be induced by some external stimulus.

(c) Two separate lines of transgenic organisms are used. The transgene plus recognition sites are present in one host line, and a second line of transgenic organisms expresses the recombinase (Fig. 16.15). The two lines are then mated together, and the deletions occur in their progeny.

This approach can save a lot of work. Instead of making separate transgenic constructs for each gene under each condition, two sets of transgenic animals, usually mice, are generated and are then crossed to investigate a wide range of genes and environmental conditions. One set of mouse lines has Cre under the control of a variety of promoters that are specific for different tissues or induced by different signals. The second set of mouse lines has a series of different target genes flanked by *loxP* sites. Thus, the role of any particular target gene may be investigated under any of the available conditions by crossing the appropriate pair of strains.

Some bacteriophages possess the ability to integrate themselves into the chromosome of their host bacteria. In some cases, a single protein, an integrase, carries out this process. Several such integrases have been used in genome modification in animals, including human cells. One of the best known is the **ΦC31 (phiC31) integrase** from *Streptomyces* phage ΦC31. The reaction scheme is relatively simple (Fig. 16.16). The genomes of both the bacteriophage and bacterial host possess attachment sites (*att*), which are recognition sites for the integrase. These are similar but not quite identical and are designated *attP* (for phage) and *attB* (for bacterium). The integration reaction normally inserts the circular bacteriophage genome into the bacterial genome. Crossing over between the original *att* sites generates two hybrid *att* sites flanking the insert. The reaction is irreversible because the hybrid sites are not recognized by the ΦC31 integrase.

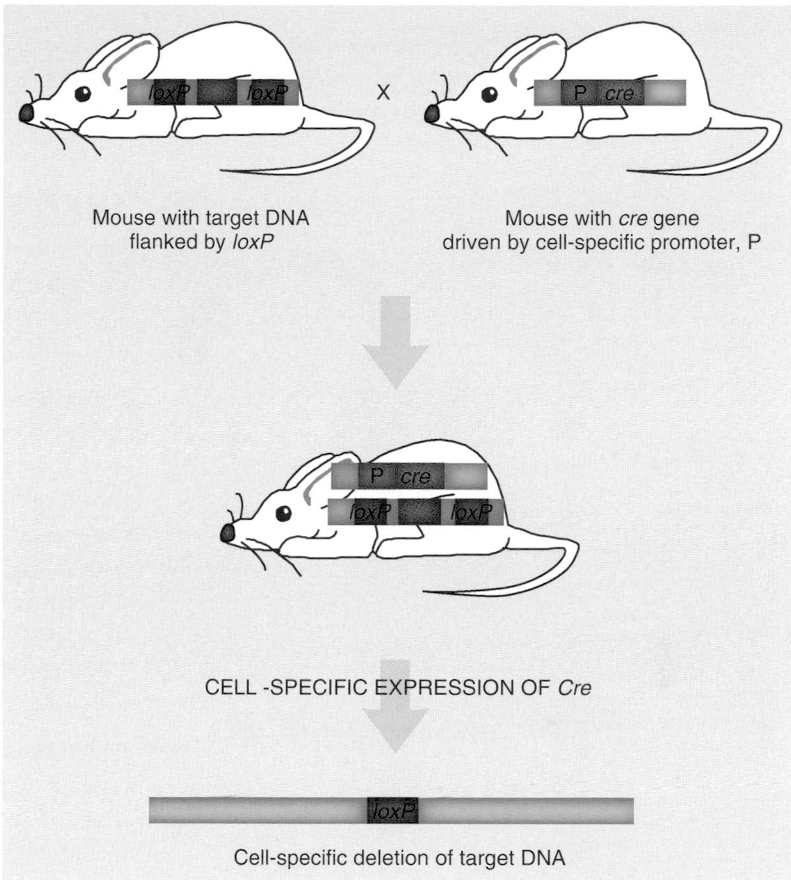

FIGURE 16.15 Two Mouse Cre/Lox System
One mouse has the target DNA that is to be deleted flanked by two *loxP* sites. A second mouse has the gene for Cre recombinase under the control of a tissue-specific or inducible promoter. When these two mice mate, some of the offspring will receive a copy of both genetic constructs. When the Cre protein is induced, it will direct the deletion of the target DNA.

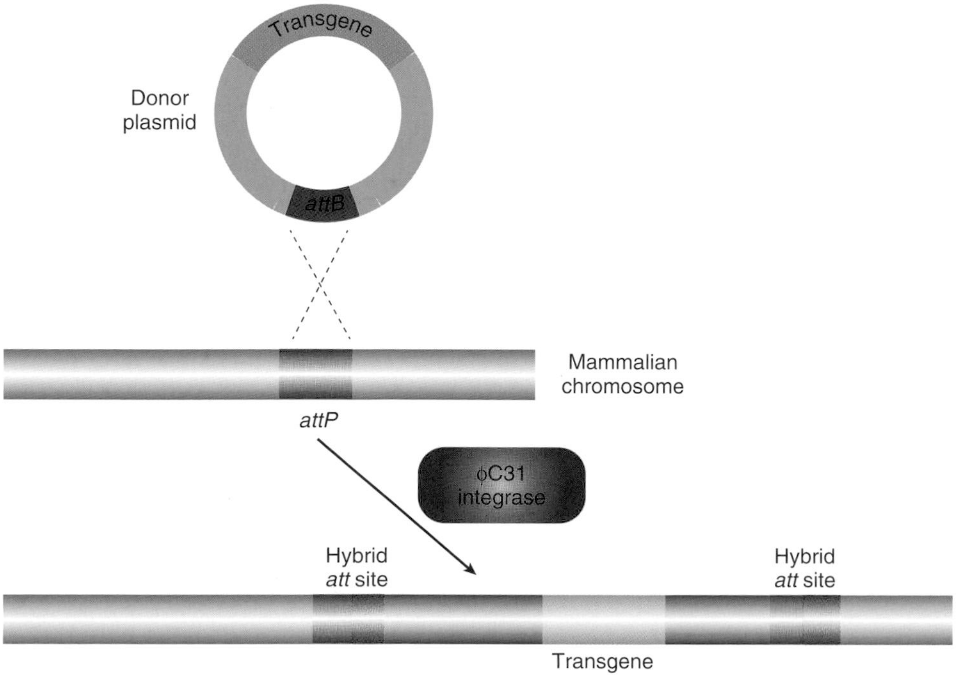

Transgene
Donor plasmid
attB
Mammalian chromosome
attP
ϕC31 integrase
Hybrid *att* site
Hybrid *att* site
Transgene

FIGURE 16.16 The ΦC31 Integrase System
The bacterial ΦC31 integrase inserts segments of foreign DNA by recombining the *attP* and *attB* sites. This generates two hybrid *att* sites flanking the insert. Since the ΦC31 integrase cannot recognize the hybrid sites, the reaction is irreversible.

Transgene expression may be controlled by site-specific recombination. Either the Cre or Flp recombinase can rearrange segments of transgenic DNA to turn transgenes on or off. A similar approach uses bacterial recombinases, such as the ΦC31 integrase.

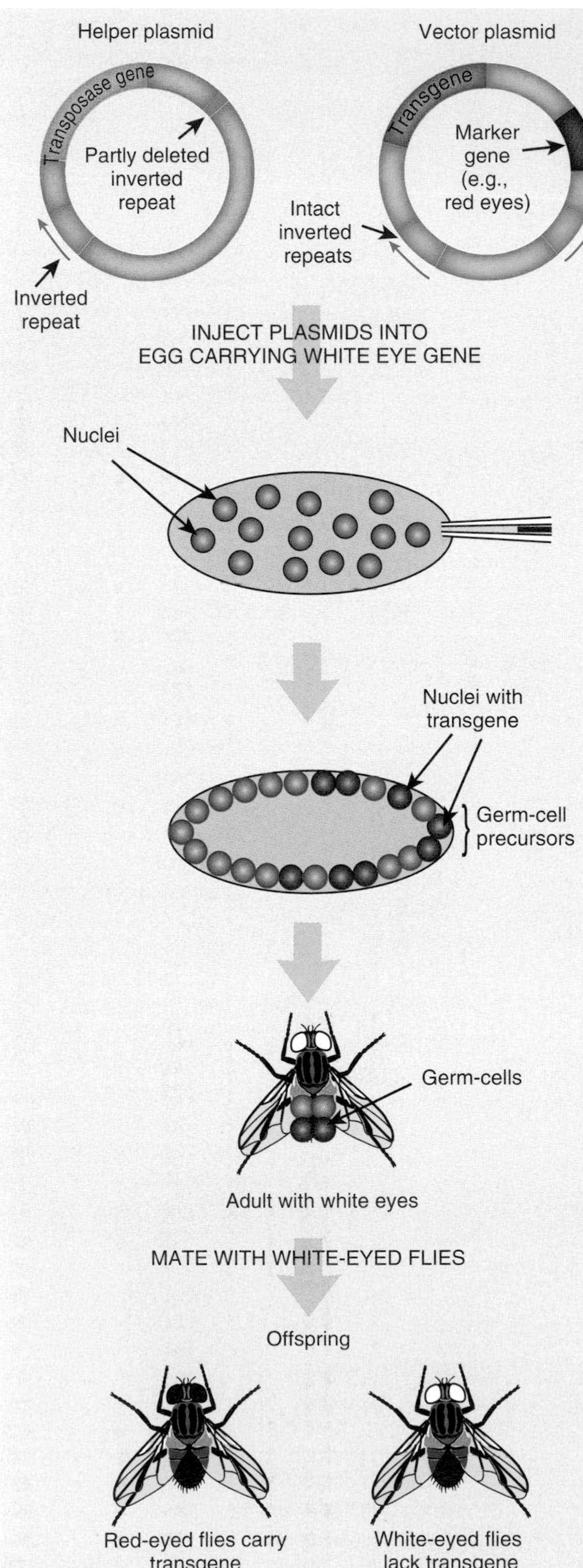

TRANSGENIC INSECTS

Several insects can now be genetically modified. The fruit fly *Drosophila* has been investigated at the molecular level for a long time, and not surprisingly, methods exist for introducing novel genetic material into these flies.

P elements are transposons found in *Drosophila* and other insects, where they cause **hybrid dysgenesis**. In flies carrying a P element, the frequency of transposition is very low because of synthesis of a repressor protein encoded by the resident P element. When P-carrying males are crossed with P-negative females, the transposition frequency in the fertilized egg is very high for a brief period, due to lack of repressor. Random insertion of P elements then causes a high mutation rate and lowers the proportion of viable offspring, that is, hybrid dysgenesis.

P elements are flanked by perfect 31 base-pair inverted repeats. Any DNA sequence that is included between these inverted repeats will be transposed. Therefore, engineered P elements can be used to introduce any sequence of DNA into a strain of fruit flies or other susceptible insects.

DNA may be microinjected into embryos of P-negative strains of *Drosophila.* In fruit flies, the diploid nucleus resulting from fusion of the sperm and egg nuclei divides multiple times without cell division, resulting in a giant cell with many nuclei, known as a **syncytium**. Microinjection is normally done at this stage, and incoming DNA usually integrates into at least some of the nuclei that will give rise to the future germline cells (Fig. 16.17). These nuclei are clustered at one end of the fertilized egg—the *posterior end.* The nuclei then migrate to the outer membrane where a cleavage furrow forms around each nucleus. These furrows expand to form individual cells for each nucleus. The center part remains undivided and acts as a yolk, providing nutrients to the developing larva.

FIGURE 16.17 P Element Engineering in *Drosophila*
Two different plasmids are used to insert transgenes into *Drosophila.* The helper supplies the transposase. It carries an immobile P element with one of the inverted repeats deleted but with functional transposase. The second plasmid has the transgene plus marker (a gene for red eyes) flanked by the P element inverted repeats. Both vectors are injected into the posterior end of an egg, which has 2000 to 4000 nuclei within one membrane. The transposase is expressed, which results in random transposition of the transgene (plus marker) into various chromosomes in different nuclei. Hopefully, some insertions occur in germ cell nuclei. The egg is then allowed to form an adult (with white eyes in this case). This fly is then mated to another white-eyed adult. If insertion into the germline was successful, some offspring will express the marker gene and have red eyes.

The incoming P element is normally carried on a bacterial plasmid that was constructed in a bacterial host. The P element transposes into the *Drosophila* chromosomes, and the plasmid sequences are left behind. In practice, two P elements are often used. One, the helper, provides the transposase but cannot itself move because of defective 31 base-pair inverted repeats (see Fig. 16.17). The other P element, the vector, carries the desired transgene and has intact 31 base-pair inverted repeats, but lacks the transposase gene. Transposition of the vector depends on transposase made by the helper. Once the P element vector has inserted into a particular location on the insect chromosome, it cannot move in future cell generations because it has no transposase of its own. Ideally, it will be inherited stably.

Marker genes are used to monitor the presence of the P element. Selectable markers used in flies include *neo* (neomycin resistance) and *adh* (alcohol dehydrogenase). Alternatively, eye color genes may be used to reveal the presence of a P element vector. Eye color cannot be positively selected; instead, flies are screened for changes in eye color. For example, flies defective in the ***rosy* gene** may be used as host. These flies have brown eyes because of lack of xanthine dehydrogenase, which is involved in synthesis of red eye pigment. If a wild-type copy of the *rosy* genes is included in the P element vector, it will restore the red eye color. If the offspring of a rosy$^{-/-}$ transgenic fly has red eyes, this implies that the transgene was inserted into the germline, and all the cells in the offspring will have the transgene.

Transposons known as P elements are widespread in *Drosophila* and other insects. They have been used to introduce transgenic DNA into insects.

PRACTICAL TRANSGENIC ANIMALS

Transgenic animals are extensively used in the laboratory to model diseases and to study biological mechanisms at the cellular and molecular level. However, there are other, more practical uses for transgenic animals.

Transgenic Animals Help Fight Disease

Perhaps the most practical application of genetic modification in animals will be in the war against infectious and chronic diseases. Many diseases of livestock and other farm animals can spread to humans. The most notorious of these, influenza, can easily spread between humans, birds, and pigs. Preventing pandemics therefore requires stopping the spread of the influenza virus not only among humans, but also among animals.

To that end, researchers have genetically engineered chickens with a greatly reduced capacity to spread H5N1 influenza when housed with uninfected chickens. The chickens were modified to carry a gene encoding a decoy RNA hairpin that interferes with transmission of the virus. The decoy RNA may bind to its RNA-dependent RNA polymerase (i.e., the viral RNA replicase). The genetically modified chickens that were experimentally infected died, but they failed to transmit the virus to other chickens.

Transgenic animals have also been employed in the battle against bacterial infection. This modification is done not for the animals' benefit, but for that of humans. Both cattle and goats have been genetically modified to express human lysozyme in their milk. Lysozyme, an enzyme that protects from infection by destroying the peptidoglycan in bacterial cell walls, is found in high concentration in human milk. It is thought that this helps prevent diarrheal disease, such as from *E. coli*. As children are weaned, they often switch to drinking milk from cows or goats, which contains less lysozyme. This is not a problem in the developed world, but in developing countries, diarrhea is a major and often deadly childhood illness. Transgenic cows and goats that produce milk fortified with extra lysozyme may therefore help avoid such preventable infections. It may also be possible to genetically engineer goats or cows to produce drinkable vaccines (Fig. 16.18).

FIGURE 16.18
Transgenic Goats
Goats can be genetically engineered to produce milk containing antimicrobials, vaccines, or other pharmaceuticals. Source: Gavin W, et al. (2014). Transgenic cloned goats and the production of recombinant therapeutic proteins. pp. 329–342, in *Principles of Cloning*, 2nd ed., Elsevier.

Transgenic goats can also be used to produce pharmaceuticals. Antithrombin is a protein that blocks blood coagulation, and people who have antithrombin deficiency are more susceptible to blood clots. rEVO Biologics manufactures the drug ATryn, which is recombinant antithrombin extracted from the milk of transgenic goats. It is used to prevent blood clots in antithrombin-deficient patients who are pregnant or about to undergo surgery. The benefit of extracting antithrombin from transgenic goat milk is that it circumvents the need to extract it from human blood.

Genetically Modified Mosquitoes

The three diseases that kill most people are acquired immunodeficiency syndrome (AIDS), tuberculosis, and malaria. Each year, malaria infects some 300 to 500 million people and kills around 2 million, mostly African children. The ***Plasmodium*** parasite is transmitted by mosquitoes—as are several other major diseases including yellow fever, dengue fever, and filarial nematodes. At the moment malaria is spreading, and mosquitoes that are resistant to insecticides such as DDT are emerging. The 260-Mb genome of the mosquito ***Anopheles gambiae***, which transmits malaria, has been fully sequenced, and a variety of genetic markers is available. The genome of ***Aedes aegypti***, which transmits yellow fever, is about three times larger (800 Mb), and its sequencing is under way.

DNA can be inserted into germline cells of mosquitoes, as for flies. Several transposons have been used to insert DNA into the genomes of mosquitoes through an approach similar to the use of the P element in *Drosophila* described earlier. The ***piggyBac*** transposon from the cabbage looper (a butterfly) and the ***Minos*** transposon from *Drosophila hydei* are the most widely used. Eye color genes and green fluorescent protein (GFP) have been used as genetic markers. The transgenes are usually expressed from *Drosophila* promoters because they often work well in other insects.

One approach to controlling mosquito-borne diseases is to genetically engineer mosquitoes that are resistant to colonization by the disease agent. The noncarrier mosquitoes would then be released into the wild, where they would displace the population of disease-transmitting mosquitoes. Several attempts have been made to engineer mosquitoes that will no longer carry malaria, or at least carry far fewer malarial parasites. So far, engineering has been done with species of malaria that attack birds or mice. It is hoped that similar approaches will work against human species of malaria and in the *Anopheles* mosquito, which carries them.

After a mosquito takes in a meal of infected blood from a human or animal, its immune system attacks the incoming malarial parasites and does in fact kill a substantial proportion (Fig. 16.19). One approach to engineering mosquitoes therefore attempts to increase expression of proteins such as **defensin A** that belong to the mosquitoes' own immune system. Proteins from other species have also been expressed in the mosquito midgut in attempts to block transmission. For example, transgenic mosquitoes that express bee venom **phospholipase** from a midgut-specific promoter destroy 80% to 90% of the incoming malarial parasites.

Another approach uses genetically engineered human antibodies. For example, artificial genes for **single-chain antibodies** or **scFv fragments** (see Chapter 6) to the **circum-sporozoite protein** of malaria have been constructed. (The sporozoite is the form of the parasite transferred from mosquito to mammal; see Fig. 16.19.) When expressed in the mosquito salivary glands, the antibody greatly reduced the numbers of malaria sporozoites.

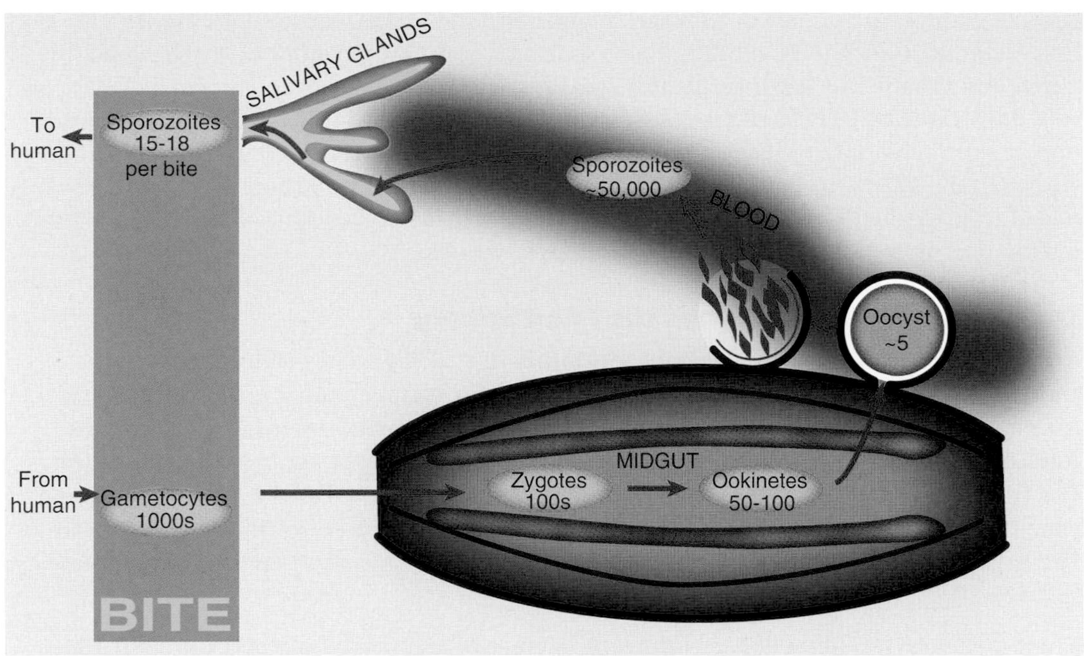

FIGURE 16.19 Development of *Plasmodium* in the Mosquito
When a mosquito bites a human with malaria, the human blood has thousands of *Plasmodium* gametocytes *(bottom of figure)*. They travel into the midgut of the mosquito, where they produce hundreds of zygotes. About 50 to 100 of these develop into mobile ookinetes, which then migrate into the hemolymph or blood and give rise to about 5 oocysts. Each oocyst releases about 50,000 sporozoites, which migrate to the mosquito salivary gland. When the mosquito bites another human, only about 15 to 18 sporozoites enter the bloodstream. *Plasmodium* life stages are shown in green ovals, and mosquito structures are labeled in uppercase.

After engineering a mosquito that no longer transmits malaria, the next problem is replacing the wild mosquito population with the engineered ones. One way to do this is to use a genetic suicide system consisting of two genes, A and B, which must both be inherited together for survival. Such a system is similar to those responsible for programmed cell death in bacteria (see Chapter 20). Engineered males with two copies each of gene A and gene B would be released.

Insects that inherit A and B together will survive, but those that get A without B or vice versa will die. Hybrids between engineered males and wild females will survive because they inherit one A and one B. However, in the next generation, when the hybrids mate with wild mosquitoes, some of the offspring will receive just gene A or gene B but not both and will therefore die. This generates a selective pressure that drives the A and B genes through the population, because the offspring of wild flies die more often than those of engineered or hybrid flies. Genes that kill the malaria parasite or make mosquitoes susceptible to insecticides would be linked to the suicide genes and would therefore spread through the population with them. Computer modeling suggests that modifying a mere 3% of the population is enough to spread the genes.

As mentioned previously, mosquitoes also carry dengue fever. Dengue is rarely lethal but can be debilitating. Globally, it infects around 100 million people per year. Dengue is presently spreading in many warmer regions, including Florida and parts of Australia. In many ways, dengue is simpler to tackle than malaria. Whereas several species of the *Anopheles* mosquito spread malaria, dengue is spread by the single mosquito species *Aedes aegypti*. Malaria is due to a eukaryotic parasite, whereas dengue is a viral infection.

Several attempts are underway to make engineered *Aedes aegypti* mosquitoes that can no longer transmit dengue. As of 2013, the first dengue-resistant mosquitoes are being field-tested in Vietnam. The key ingredient is actually a symbiotic bacterium, known as *Wolbachia*, which lives inside the cells of many insects. Indeed, around two-thirds of all insects carry one or another species of *Wolbachia*. When *Wolbachia* is present, it prevents the replication of viruses of the Flavivirus family, which includes dengue, yellow fever, and West Nile. However, mosquitoes normally lack these bacteria. Once *Wolbachia* was introduced into *Aedes aegypti* mosquitoes, they lost the ability to transmit these viruses.

Transgenic Animals Improve the Food Supply

Another promising use of transgenic animals is in making food production more efficient and sustainable. The first transgenic animal for human consumption to receive FDA approval will likely be an Atlantic salmon genetically modified by the company AquaBounty Technologies; it is expected to grow twice as fast as conventional salmon. The transgenic fish contains the EO-1 alpha transgene consisting of a growth hormone gene from the Chinook salmon under the control of a promoter from the ocean pout. The fish are further manipulated to be triploid (ensuring sterility) and entirely female (Fig. 16.20).

Combined with land-based aquaculture, the chance of an escaped fish spreading the transgene into wild salmon populations is essentially zero. FDA approval was expected in 2014, but is still pending. Meanwhile, several thousand grocery stores have already refused to stock the transgenic fish due to unfounded safety concerns.

Transgenic animals can also be used to alter the nutrient content of the food supply. For instance, pigs and sheep have been genetically modified to produce more omega-3 fatty acids, which are beneficial to human health. Additionally, genetic modification can be used to remove allergens from food. Some people are allergic to the protein β-lactoglobulin found in cow's milk. In 2012, researchers created a transgenic cow that expresses microRNAs that eliminate production of the allergen. However, the transgenic cow, whose name was Daisy, was born without a tail for reasons that are not entirely clear.

Not only can genetics improve animals themselves, it can also improve animal waste! From the University of Guelph in Canada comes Enviropig™ (a trademark of Ontario Pork). These transgenic pigs have received the *appA* gene from *E. coli* under control of the parotid secretory protein (PSP) promoter. As a result, they secrete the enzyme phytase (or acid phosphatase; product of *appA*) in their saliva. This enzyme degrades phytate (inositol hexaphosphate, typically found in the outer layers of cereal grains), which pigs cannot otherwise use. Consequently, these ecosensitive swine no longer need phosphate supplements in their diet. More important, the phosphorus content of the manure is reduced by as much as 60%. This result is significant because high phosphorus runoff into ponds and streams causes algal blooms that taint the water and rob it of oxygen, leading to the death of fish and other aquatic life. On the artistic side, transgenic silkworms have been used to make fluorescent silk (see Box 16.2).

Other Uses for Transgenic Animals

There are other, perhaps less conventional, uses for transgenic animals. Zebrafish genetically modified to express green fluorescent protein in the presence of certain chemicals have been created. This modification allows researchers to examine which tissues are most affected by the chemical of interest. Additionally, it raises the possibility of releasing transgenic animals into the environment to serve as biosensors for all sorts of different pollutants, such as xenoestrogens (Fig. 16.21).

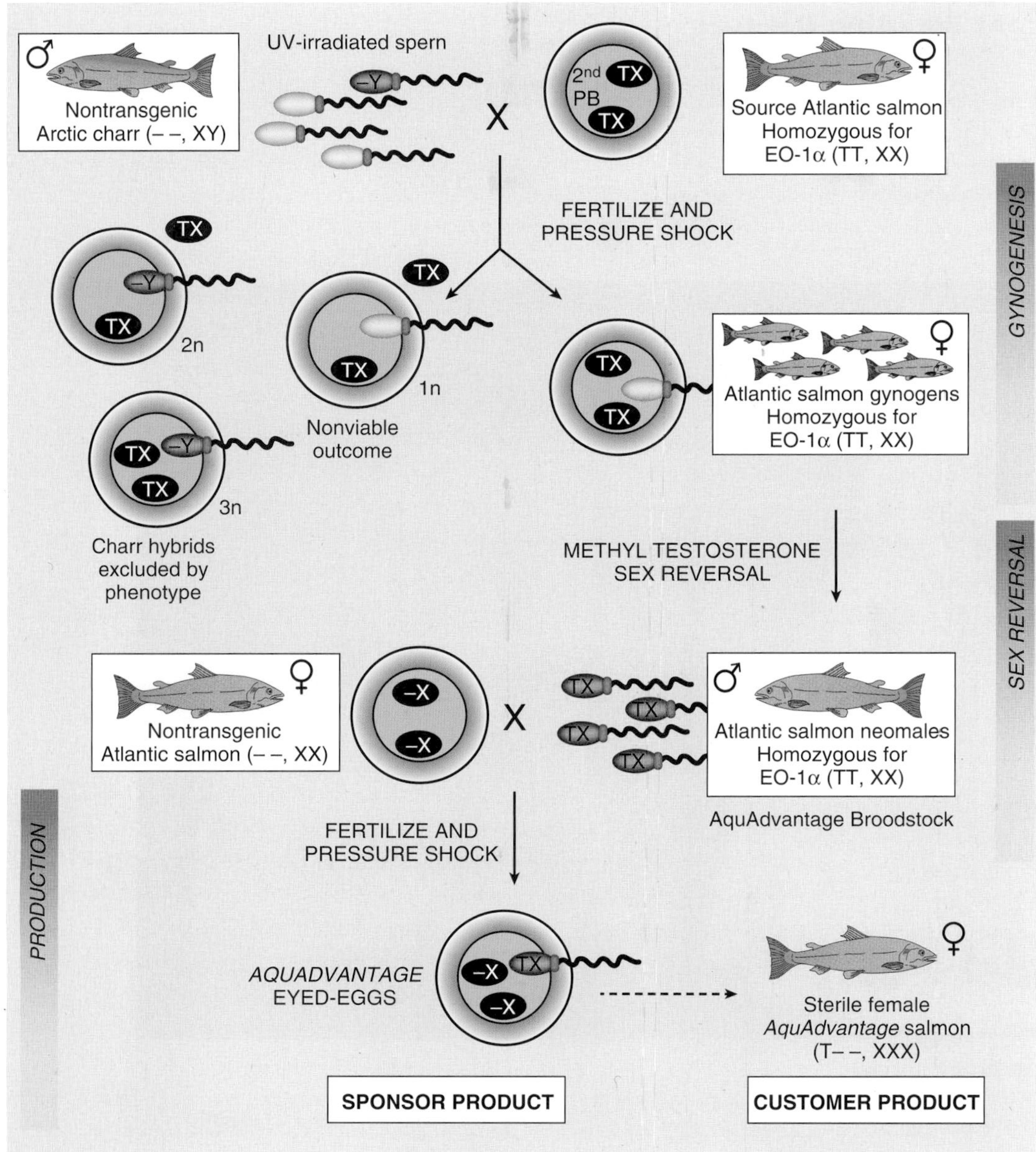

FIGURE 16.20 AquAdvantage Salmon

A complex mating process is used to generate triploid female fish for consumption. Eggs from female fish that are homozygotes for the EO-1 alpha transgene are fertilized with UV-inactivated sperm from an unrelated species. The sperm contributes no genetic material and serves only to activate the egg. Pressure shocking the egg causes it to retain the second polar body, thereby creating a diploid organism. These young fish are then treated with testosterone, converting them into "neomales," which are genotypic females but phenotypic males. Neomale sperm is then crossed with eggs from a conventional female salmon. Once again, the fertilized eggs are pressure shocked, causing retention of the second polar body. This time, the resulting fish are triploid, female, and carry a single copy of the transgene.

Along similar lines, a genetically modified mouse, known as MouSensor, is adept at sniffing out explosives. The neurons in MouSensor's olfactory bulb express many more receptors for an explosive similar to TNT. Its creators believe that upon MouSensor's smelling the explosive, its behavior is likely to change. By monitoring such changes, with a chip implanted under the mouse's skin, researchers could use the mouse to detect landmines in field trials.

Box 16.2 Fluorescent Wedding Dresses

Transgenic animals can also be used to generate innovative materials. Spider silk has superior mechanical properties compared to silkworm silk. Consequently, one group of researchers created a transgenic silkworm that produced a chimeric silkworm/spider silk. Another group created transgenic silkworms that produced fluorescent silk. The researchers placed genes for green fluorescent protein (derived from a jellyfish) or *Discosoma* red fluorescent protein or *Kusabira* Orange (both derived from corals) under the control of a promoter that regulates the gene for fibroin H chain, one of the main protein subunits found in silk. Upon exposure to blue light, the silk fluoresces (Fig. A).

FIGURE A Fluorescent Silk Wedding Dress
A wedding dress made from fluorescent silk photographed under normal light (*left*) and under a blue light using a yellow filter (*right*). Adapted with permission from Tetsuya et al. (2013). Colored fluorescent silk made by transgenic silkworms. *Adv Funct Mater* **23**, 5232–5239. Copyright John Wiley & Sons. Figure kindly provided by Toshiki Tamura.

Transgenic animals have a variety of practical uses. They can be used to improve human nutrition, monitor pollution, and make novel materials, such as fluorescent silk. Livestock have been modified to resist disease, and engineered mosquitoes that no longer transmit malaria or dengue are being constructed.

APPLICATIONS OF RNA TECHNOLOGY IN TRANSGENICS

As discussed earlier, transgenic technology at the DNA level allows us to add genes and to inactivate gene expression by insertion. It is also possible to manipulate genes at the RNA level by the use of **antisense RNA**, **ribozymes**, or **RNA interference**. We have already discussed these topics (see Chapter 5); therefore, this section simply summarizes their application in transgenics. So far, these approaches are still mostly experimental. Using antisense RNA and ribozymes has largely been discontinued in favor of RNA interference, which is much easier to engineer.

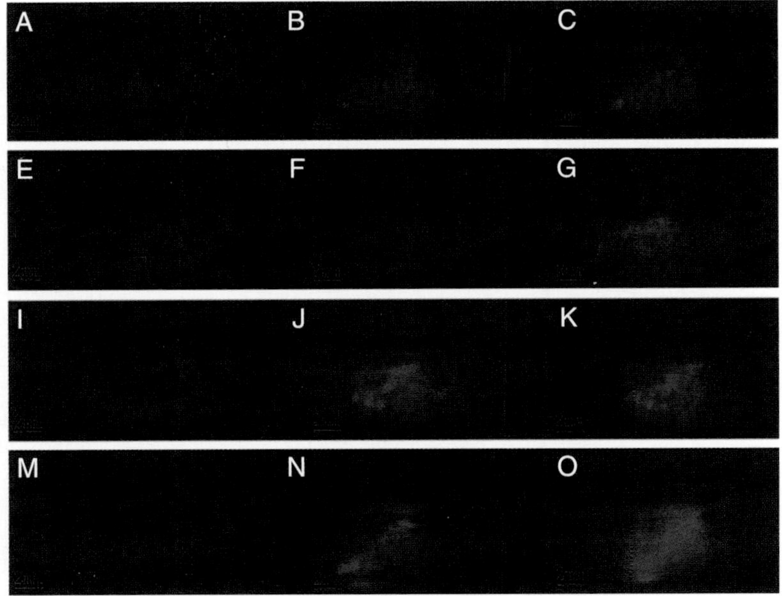

FIGURE 16.21 Induction of GFP by Xenoestrogens in Transgenic Zebrafish
Different concentrations of the xenoestrogen 17alpha-ethynylestradiol were tested against transgenic male zebrafish carrying a fusion of GFP to the *ZVTG* gene. (A–C) 0.1 ng/L; (E–G) 1 ng/L; (I–K) 10 ng/L; (M–O) 100 ng/L. Days after exposure: A, E, I, M, 0 days; B, 7 days; F, 4 days; J, 3 days; N, 2 days; C, 9 days; G, 6 days; K, 5 days; O, 5 days. VTG is a precursor to vitellin, an egg-yolk protein that provides reserves for embryonic development. VTG is normally expressed in females' response to internal estrogens. It may also be activated by xenoestrogens, even in males, as shown here. The zebrafish has 7 genes for VTG, of which the *ZVTG* gene was chosen for its high expression. From: Chen H et al. (2010). Generation of a fluorescent transgenic zebrafish for detection of environmental estrogens. *Aquat Toxicol* **96**, 53–61.

It is possible to insert an antisense transgene when constructing a transgenic organism. The simplest way to construct an antisense gene is to invert the DNA sequence of the original gene. In higher organisms, the cDNA version of the gene is used as the starting point to avoid complications due to intervening sequences. Such an antisense gene will be transcribed to give antisense RNA, which will then bind to the mRNA of the corresponding sense gene (see Chapter 5). This results in inhibition of gene expression at the level of translation. Insertion of antisense genes into mice has resulted in up to 95% inhibition of gene expression, although the effectiveness varies greatly. The antisense RNA may be full length or just a short segment corresponding to part of the sense gene. Short segments complementary to the 5′-untranslated region of the sense mRNA are often very effective at blocking translation. Obviously, antisense constructs can be placed under the control of inducible promoters. This allows antisense silencing of gene expression to be regulated by the experimenter.

It is also possible to insert transgenic constructs that express ribozymes. For example, the ribozyme might be designed to cleave a specific mRNA. When the ribozyme is expressed, it cleaves the mRNA of the target gene and so reduces gene expression. Occasional cases are known in both *Drosophila* and mice where transgenic expression of ribozymes has been successful in decreasing gene expression.

RNA interference covers several related phenomena whose mechanisms involve the formation of double-stranded RNA (dsRNA). Neither prokaryotic nor eukaryotic cells normally make dsRNA. Indeed, it is regarded by cells as evidence of viral activity and therefore is degraded, although by different mechanisms in prokaryotes and eukaryotes. RNA interference is used only by eukaryotes.

When both the sense and corresponding antisense RNA are present in eukaryotic cells, dsRNA forms. A massive drop in expression of the corresponding gene follows this, due to degradation of the mRNA. Only a few molecules of dsRNA are needed to trigger this effect. RNA interference is the favored procedure for gene inactivation in *Caenorhabditis elegans* and is also effective in other animals and plants. Adding RNA directly or expressing the corresponding transgenic DNA constructs can trigger RNA interference. A construct with neighboring sense and antisense sequences will produce hairpin RNA (Fig. 16.22). Alternatively, it is possible to use a single transgene flanked by two promoters, one pointing in each direction.

Gene expression may be decreased by insertion of genes encoding antisense RNA or encoding ribozymes targeted to cleave a specific mRNA. RNA interference may be used to knock down expression of chosen genes. Various RNAi constructs are available, some of which may be incorporated into the host genome and inherited.

NATURAL TRANSGENICS AND DNA INGESTION

Historically speaking, we are all transgenic. The human genome contains a significant number of genes of bacterial origin. These genes have presumably been picked up at various stages in evolutionary history from assorted bacteria. In addition, we carry quite a few genes that came originally from other higher organisms, some of which have been transmitted by retroviruses. Such movement of genes between organisms that are not direct descendants is generally known as lateral or horizontal gene transfer, but might just as well be called "natural transgenesis."

There are two major pathways for naturally acquiring genetic material from other organisms. One is by viral transduction, and the other is by direct intake of DNA. In microorganisms, DNA from the environment may be taken up by transformation. In the case of animals, DNA is constantly ingested along with other components of the diet. Certain protozoa such as *Paramecium* and some amoebas live by ingesting bacteria. It is thought that several genes involved in fermentative metabolism in *Entamoeba* are derived from such ingested bacteria.

Humans and other mammals, whether carnivorous or vegetarian, are constantly eating food that contains substantial amounts of DNA. Although it is generally assumed that ingested DNA is degraded into nucleotides in the intestinal tract, this is not entirely true. Recent findings suggest that a very small proportion of the DNA survives as fragments of moderate size (up to 1000 base pairs) and actually crosses the intestinal wall into the bloodstream of the animal in this form. At least transiently, DNA from food can be traced to several different organs and can also cross the placenta to fetuses and newborns.

DNA from the bacterial virus M13 and the gene for green fluorescent protein have been used as test molecules. More recently, naturally occurring DNA sequences for the plant-specific ribulose-1,5-bisphosphate carboxylase (Rubisco) gene have been tracked by PCR. All of these foreign DNA sequences have been found both in the bloodstream and in the cell nuclei of various tissues in animals that ingested the DNA. Sometimes the foreign DNA appears to be integrated into host chromosomal DNA. However, so far no evidence for expression of genes carried by ingested DNA has been found, nor has incorporation into the germline been seen for DNA eaten by mice. The overall likelihood of food-borne DNA infiltrating the genome and being expressed is unknown, although it seems low. Nonetheless, because vast numbers of animals have eaten DNA every day of their lives over millions of years, it seems likely that this must occur now and then. Furthermore, the fact that our genomes do include a small percentage of foreign genes argues that germline insertion does occur, albeit very rarely.

From a purely theoretical viewpoint, we should remember that living cells contain much more RNA than DNA. Thus, the amount of RNA ingested in the food is at least 10-fold greater than the amount of DNA. Whether any of this RNA survives long enough to cross the

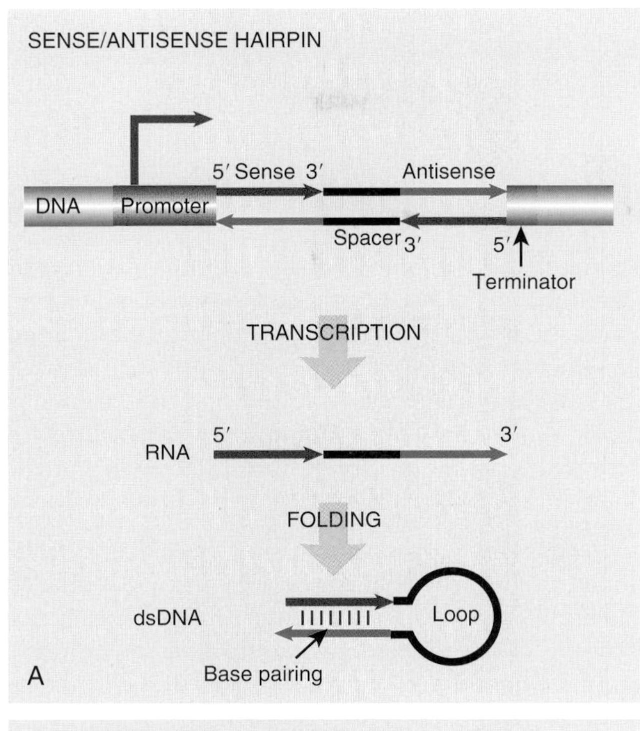

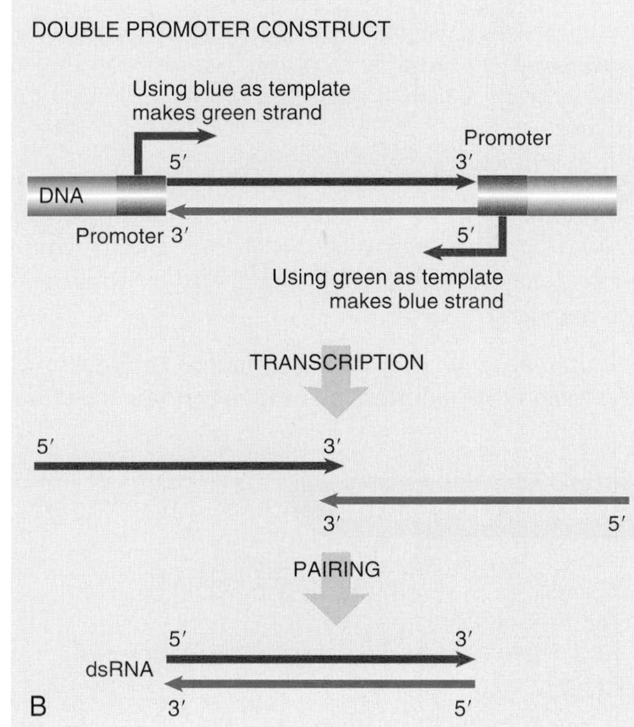

FIGURE 16.22 RNA Interference Constructs

RNA interference occurs when both the sense and antisense RNA of a gene are present and form dsRNA. Two constructs are shown that direct the synthesis of a dsRNA molecule. The first construct (A) has both sense and antisense regions that base-pair. A spacer separates the sense and antisense regions and forms a loop at the end of the hairpin. The double promoter construct (B) has two promoters: one for the sense strand and the other for the antisense strand. The two resulting RNA molecules are complementary and form a dsRNA molecule. The presence of dsRNA triggers degradation of mRNA from the corresponding gene.

intestinal wall and enter animal cells has so far not been investigated. Reverse transcriptase from the assortment of retro-elements present in most animal genomes could, in theory, reverse transcribe such incoming RNA, so generating a DNA copy that might occasionally integrate into the host cell genome.

Short fragments of DNA in the diet can be taken up and incorporated into host chromosomes. Whether such DNA can be incorporated into the mammalian germline is unknown.

Summary

Genetic engineering combined with cloning has allowed the creation of an ever-growing range of genetically modified organisms. Several methods are used to create transgenic animals. In nuclear microinjection, DNA carrying the transgene is injected into the nucleus of a fertilized egg cell. Other approaches are to use retrovirus vectors that insert into the host genome or to engineer embryonic stem cells in culture and then reinsert them into an early embryo. Some transgenic animals have novel or improved capabilities, whereas others are constructed with inactivated genes for medical research. RNA-based methods for transgenic technology are still in their infancy but may provide alternative approaches in the future.

The location of a transgene within its host genome may have major effects on its level of expression. Further modifications may be made to overcome positional effects, usually providing specific regulatory sequences for the transgene. Alternatively, transgenes can be inserted at specific locations in the host genome. Controlling the expression of inserted transgenes is often problematic and may require assembly of highly sophisticated artificial control systems with components derived from several sources. Engineered versions of the bacterial *lac* and *tet* regulators are widely used. Segments of DNA can also be rearranged to control transgene expression. The Cre or Flp recombinases can rearrange DNA segments, thus turning transgenes on or off. Bacterial recombinases, such as the ΦC31 integrase, are also increasingly used.

Transgenic animals have a wide range of practical uses. They can be used to improve human nutrition, monitor pollution, and make novel materials. Domestic animals have been modified to resist disease, both to protect the animals themselves and to avoid transmission of infections to humans. Transgenic insects include engineered mosquitoes that no longer transmit diseases such as malaria or dengue.

Short fragments of DNA can be taken up naturally from food and incorporated into host chromosomes. Whether such DNA modifies the human genome over time is unknown.

End-of-Chapter Questions

1. What process is used to create transgenic animals?
   - **a.** particle bombardment
   - **b.** nuclear microinjection
   - **c.** nuclear fusion
   - **d.** germ line transformation
   - **e.** none of the above
2. In the first transgenic animal experiment, which gene from rats was cloned into mice?
   - **a.** somatotropin
   - **b.** metallothionein
   - **c.** myostatin
   - **d.** plasminogen
   - **e.** PPAR-delta

3. According to the book, which protein is produced by transgenic goats?
  a. recombinant tissue plasminogen activator
  b. recombinant tissue fibrinogen activator
  c. recombinant bovine somatotropin
  d. recombinant human somatotropin
  e. none of the above

4. Why are embryonic stem cells important?
  a. They can be passed from one generation to the next.
  b. These cells carry retroviral genes.
  c. They can develop into any tissue in the body, including the germ line.
  d. The cells are differentiated and can therefore be manipulated.
  e. none of the above

5. How can location effects of transgenes in animals be avoided?
  a. by placing LCR sequences in front of the transgene
  b. by inclusion of insulator sequences on both ends of the transgene
  c. by using natural transgenes instead of the cDNA version
  d. by targeting the transgene to a specific location
  e. all of the above

6. What is used by targeting vectors to insert transgenes at specific locations within the host genome?
  a. homologous recombination
  b. transfection
  c. transduction
  d. conjugation
  e. all of the above

7. Which of the following systems can be used to control the expression of the transgene?
  a. a modified LacI repressor system
  b. the TetR recombinant promoter system
  c. use of a heat shock promoter from *Drosophila*
  d. the use of steroid receptors
  e. all of the above

8. Why are Cre/*loxP* or Flp/FRT used in transgenic animals?
  a. activation of a transgene by removing blocking sequences flanked by the *loxP* or FRT sites
  b. large-scale deletions and rearrangements of the chromosomes
  c. removal of selectable markers that are no longer needed
  d. creation of conditional knockout mutants
  e. all of the above

9. What can be engineered to introduce transgenes into *Drosophila?*
  a. Cre recombinase
  b. S elements
  c. P elements
  d. YACs
  e. none of the above

(*Continued*)

10. How can mosquito-borne diseases be controlled through biotechnology?
    - a. by genetically engineering mosquitoes that are resistant to the disease
    - b. by creating more efficient pesticides
    - c. through the cloning, expression, and purification of novel therapeutic agents
    - d. by engineering non-pathogenic malaria, dengue, and other mosquito-borne diseases
    - e. none of the above
11. Which of the following is derived from bacteria and useful for generating transgenes or modulating gene expression?
    - a. Cre recombinase
    - b. Flp recombinase
    - c. Flippase
    - d. phiC31 integrase
    - e. P elements
12. Transgenic mosquitoes can be produced using a transposon from the cabbage looper called ________.
    - a. *piggyBac*
    - b. *Minos*
    - c. P element
    - d. 2-micron plasmid
    - e. *Wolbachia*
13. Concerning transgenic mosquitoes, which of these is not a strategy for the control of disease spread?
    - a. Overexpression of defensin A from the mosquito's own immune system.
    - b. Expression of phospholipase from bees.
    - c. Expression of a human antibody in mosquito saliva.
    - d. Production of lysozyme in the mosquito gut.
    - e. All of the above are potential strategies.
14. Transgenic ________.
    - a. chickens can help decrease the spread of H5N1 influenza virus
    - b. pigs produce fewer phosphates in their excrement
    - c. mosquitoes can limit the spread of diseases such as malaria and dengue fever
    - d. silkworm can produce fluorescent silk
    - e. animals are all of the above
15. Antisense RNAs can be introduced as transgenes and used in ________.
    - a. gene silencing
    - b. gene activation
    - c. overexpression
    - d. chemical reactions
    - e. none of the above

Further Reading

AquaBounty Technologies, Inc. (2010). *Environmental assessment for AquAdvantage salmon*. Washington, DC: U.S. Food and Drug Administration.

Chen, H., Hu, J., Yang, J., Wang, Y., Xu, H., Jiang, Q., et al. (2010). Generation of a fluorescent transgenic zebrafish for detection of environmental estrogens. *Aquatic toxicology*, *96*(1), 53–61.

Cooper, C. A., et al. (2013). Consuming transgenic goats' milk containing the antimicrobial protein lysozyme helps resolve diarrhea in young pigs. *PLoS One*, *8*(3), e58409.

Iizuka, T., et al. (2013). Colored fluorescent silk made by transgenic silkworms. *Advanced Functional Materials*, *23*, 5232–5239.

Jabed, A., et al. (2012). Targeted microRNA expression in dairy cattle directs production of β-lactoglobulin-free, high-casein milk. *Proceedings of the National Academy of Sciences of the United States of America*, *109*(42), 16811–16816.

Lister, J. A. (2011). Use of phage ΦC31 integrase as a tool for zebrafish genome manipulation. *Methods in Cell Biology*, *104*, 195–208.

Lyall, J., et al. (2011). Suppression of avian influenza transmission in genetically modified chickens. *Science*, *331*, 223–226.

Niu, Y., Yu, Y., Bernat, A., Yang, S., He, X., Guo, X., et al. (2010). Transgenic rhesus monkeys produced by gene transfer into early-cleavage-stage embryos using a simian immunodeficiency virus-based vector. *Proceedings of the National Academy of Sciences of the United States of America*, *107*(41), 17663–17667.

Putkhao, K., Kocerha, J., Cho, I. K., Yang, J., Parnpai, R., & Chan, A. W. (2013). Pathogenic cellular phenotypes are germline transmissible in a transgenic primate model of Huntington's disease. *Stem Cells and Development*, *22*(8), 1198–1205.

Tetsuya, I., et al. (2013). Colored fluorescent silk made by transgenic silkworms. *Advanced Functional Materials*, *23*, 5232–5239.

Teulé, F., et al. (2012). Silkworms transformed with chimeric silkworm/spider silk genes spin composite silk fibers with improved mechanical properties. *Proceedings of the National Academy of Sciences of the United States of America*, *109*, 923–928.

Yang, B., et al. (2011). Characterization of bioactive recombinant human lysozyme expressed in milk of cloned transgenic cattle. *PLoS One*, *6*(3), e17593.

Inherited Defects and Gene Therapy

Biotechnology

http://dx.doi.org/10.1016/B978-0-12-385015-7.00017-X

INTRODUCTION

Genetic defects vary from trivial to life threatening. Although we tend to think of inherited conditions such as cystic fibrosis and muscular dystrophy as diseases, we often refer to cleft palates or color blindness as inherited defects. However, they are all due to mutations in DNA, the genetic material. Not only are some diseases directly caused by mutations, but susceptibility to infectious disease and to factors such as radiation also is influenced by a variety of genes.

Precise numbers are difficult to estimate, but for humans and apes, the mutation rate is around 5.0×10^{-8} per kilobase of DNA per generation, or 0.16 per genome per generation (see Chapter 8 for more details on mutations and Table 8.7 for data on mutation rates). For rodents, the rate is some 10-fold less because fewer cell divisions are needed to form gametes from ancestral germ cells. Thus, mutations constantly accumulate in the germline of humans and other animals. Most of these have little or no effect, but a small percentage give rise to serious hereditary defects. A full listing of human hereditary defects, known as Online Mendelian Inheritance in Man (OMIM), is available on the Internet at http://www.ncbi.nlm.nih.gov/omim.

It is now possible to cure some of these inherited defects. Genetic engineering refers to altering the genome of an organism so that the changes are stably inherited. For multicellular organisms, this implies deliberate alteration of the DNA in the germline cells. In contrast, **gene therapy** (occasionally called **genetic surgery**) is less permanent. The patient is cured, more or less, by altering the genes in only part of the body. For example, cystic fibrosis patients might be partially cured by introducing the wild-type gene into the lungs. However, these changes are not inherited, and the alleles in the germline cells remain defective. Curiously enough, detailed information on gene and protein structure is allowing certain genetic defects to be partly suppressed by custom-designed pharmaceuticals. We have restricted our coverage largely to genetic defects where some form of therapy exists or is in development.

True human genetic engineering is still in the future. At present, genetic engineering is restricted to nonhumans and has resulted in the creation of transgenic plants and animals as described in Chapters 15 and 16. **Eugenics** refers to deliberate improvement of the human race by selective breeding. Early eugenic proposals were based on choosing superior parents by visual inspection or medical screening and breeding them in much the same way as for prize pigs and pedigreed dogs. Today, we have reached the position where direct alterations of the human genome at the DNA level are technically feasible although still clumsy.

Mutations in the human genome are responsible for a wide variety of inherited defects and diseases. Genetic engineering may create organisms with changes that are stably inherited. Gene therapy uses genetics to cure a disease but does not alter the germline cells. Consequently the changes are not heritable.

HEREDITARY DEFECTS IN HIGHER ORGANISMS

If the DNA of a single-celled organism is mutated, the mutation will be passed on to all of its descendants when it divides. The situation in multicellular creatures is more complex. In animals, the germline cells are reserved for reproductive purposes and give rise to the eggs and sperm in mature adults. The somatic cells forming the rest of the body are not passed on to the next generation. However, mutation of somatic cells is involved in cancer, which is dealt with separately in Chapter 19.

Higher organisms such as animals and plants are normally diploid and have two copies (i.e., two alleles) of each gene. Thus, if one copy is damaged by mutation, the other copy can compensate for the loss. Because most mutations are relatively rare, it is unlikely that both alleles of the same gene will carry mutations. Furthermore, most detrimental mutations are recessive to the wild type (Fig. 17.1 and Table 17.1). That is, a single functional allele is sufficient for normal growth, and the defective copy has no noticeable effect on the phenotype.

Nonetheless, we humans all have quite a few mutations randomly scattered among our 25,000 genes. Obviously, close relatives tend to share many defects. Individuals share half of their genetic information with their father and mother although, of course, not the same half with each of them. Brothers and sisters share half of their genetic information on average, although the proportion in any given case varies widely, due to recombination during meiosis when the gametes are formed. Therefore, a child from the mating of close relatives (for example, as in brother/sister or father/daughter) has a much greater chance of getting two copies of the same defect, that is, of being homozygous for the recessive allele (Fig. 17.2).

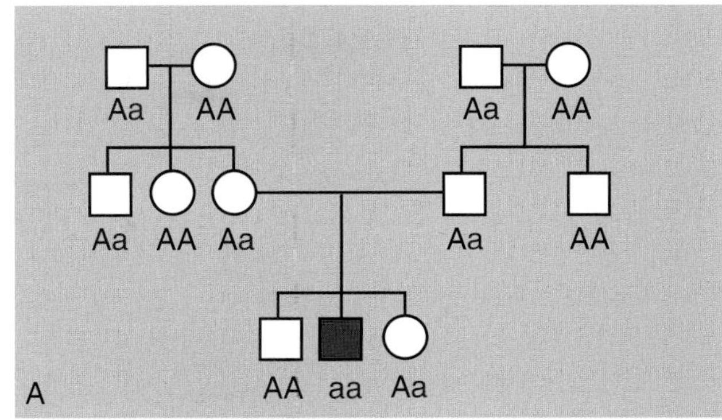

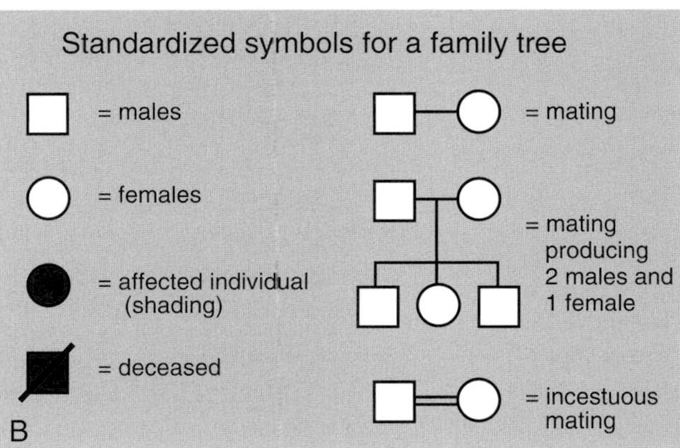

FIGURE 17.1 Inheritance of Recessive Mutations
A defective mutation usually occurs in only one copy of a gene. Therefore, most affected individuals will have one normal copy (A) and one mutated copy (a) of the gene. When two people, both carrying a recessive mutation in the same gene, have children, 25% of the children will inherit both mutant copies and exhibit the disease.

Mating between relatives makes genetic disease more likely. The reason is that a rare recessive allele present in one ancestor may be passed down both sides of the family and two copies may end up in one particular child. In Europe, royal families insisted that their children marry other royalty; therefore, many marriages were between closely related individuals. This led to a high incidence of defects among the offspring, the best known being hemophilia. This is the underlying reason for the taboos most societies have against incest, although the degree of forbidden relationships varies considerably between different cultures. In marriages between unrelated people, recessive diseases will affect the children only if by chance an ancestor from each side of the family carried a copy of the defective allele.

Many genetic defects are fatal to sperm cells, fertilized eggs, or early embryos, so they are never seen in adults. Nevertheless, a wide range of genetic defects is observed in living humans. Some of the most common defects due to a single gene are listed in Table 17.1. Unless otherwise noted, these are recessive and both alleles must be defective for symptoms to appear. Large genes are bigger targets for random mutation. The genes responsible for cystic fibrosis, muscular dystrophy, and phenylketonuria are all abnormally large, and all three genetic defects are relatively common.

Another factor that increases the frequency of a deleterious recessive mutation is a single copy of the defective gene benefitting the heterozygote, even though the homozygote is severely defective. The two best-known cases of this are sickle cell anemia and cystic fibrosis (see Table 17.1). A single copy of the sickle cell mutation provides resistance to malaria, and a single copy of the cystic fibrosis mutation provides resistance to diseases that cause severe diarrhea, including cholera, some *Salmonella* infections, and rotavirus.

Many well-known hereditary defects are due to homozygous recessive mutations in a single gene.

Table 17.1 Some Human Defects due to Single Genes

Disease	Frequency
Adenosine deaminase deficiency Lack of enzyme causes immune deficiency First defect approved for human gene therapy	Rare
Cystic fibrosis Defective ion transport indirectly affects mucous secretion in lungs	1/2000 (whites) Rare in Asians
Duchenne's muscular dystrophy Disintegration of muscle tissue Giant gene is a frequent target for mutations	1/3000 males (sex-linked)
Fragile X syndrome Common form of X-linked mental retardation	1/1500 males 1/3000 females (sex-linked)
Hemophilia Defect in blood coagulation	1/10,000 males (sex-linked)
Myotonic dystrophy Genetically dominant form of dystrophy	1/10,000
Phenylketonuria Mental retardation due to lack of enzyme Detected in newborns by urine analysis Special diet prevents symptoms	1/5000 (Western Europe) Rare elsewhere
Sickle cell anemia Defect in hemoglobin beta chain Heterozygotes are resistant to malaria but homozygotes are sick Common in Africa First molecular disease to be identified	1/400 (U.S. blacks)

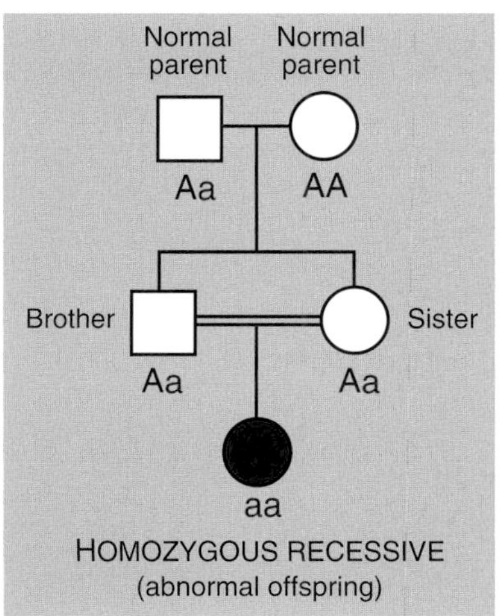

FIGURE 17.2 Homozygous Recessive from Inbreeding
A brother and sister may both inherit a mutant gene (a) from their one parent (their father in this illustration), but will not be affected because they each have a normal gene from their mother. If the brother and sister have offspring, their child may inherit two copies of the defective gene and then exhibit the symptoms of the defect.

HEREDITARY DEFECTS DUE TO MULTIPLE GENES

Several well-known hereditary defects are missing from Table 17.1 because they involve more than one gene. They may be divided into two types. Some multigene defects are due to the interaction of several individual genes. Examples include cleft palate, spina bifida, certain cancers, and diabetes.

Other multigene defects are due to the presence of an extra copy of an entire chromosome. Although most errors involving whole chromosomes are lethal, a few are viable. The best known of these is **Down syndrome**, which causes mental retardation, and is due to an extra copy of chromosome 21. The overall frequency of Down syndrome is about 1 in 800, but like most instances of getting an extra chromosome, it is due to an accident during cell division in the newly fertilized egg and it is not normally inherited. The relative chances of such a mishap increase with maternal age, but even so, most Down syndrome children are born to young women, simply because more younger women have babies. Extra sex chromosomes are found in about 1 in every 1000 people; there are three relatively common possibilities: XXY, XYY, and XXX. The XYY individuals are best known for their supposed tendencies toward violent crime, but the others show various abnormalities also.

Hereditary defects due to multiple interacting genes are difficult to analyze.

DEFECTS DUE TO HAPLOINSUFFICIENCY

Loss-of-function mutations are usually recessive, and a defect in one of the two copies of a diploid gene usually has little effect. Only rarely is one functional copy of a gene insufficient. This situation is known as **haploinsufficiency**. Three main reasons explain most cases where gene dosage is important:

(a) Some structural proteins are needed in very high amounts in certain tissues. Thus, a single functional gene may not allow sufficiently high levels of gene expression.
(b) Some proteins interact with other proteins in strict ratios. Perturbing the amount of one component may have damaging effects.
(c) Some regulatory networks respond in a quantitative manner. Consequently, the absolute level of certain regulatory proteins may be critical for correct operation.

An example of the first case is the protein **elastin**, encoded by the *ELN* gene located on chromosome 7 in band 7q11. Elastin is found in the elastic tissues of skin, lung, and blood vessels. In people with a single defective *ELN* gene, most elastic tissues still function correctly. However, two copies of the *ELN* gene are needed to make a normal aorta, which is extremely elastic. People with one copy of *ELN* cannot make sufficient elastin for this tissue, resulting in a narrowing of the aorta that sometimes requires surgery and is known as congenital supravalvular aortic stenosis (SVAS).

Although no therapy is yet available, two approaches to increase expression level from the single functional copy of the *ELN* gene seem promising. One is to construct an engineered transcriptional activator targeted to the gene of interest. (Construction of hybrid zinc finger proteins is discussed in Chapter 11.) This approach has proven successful using cultured cells.

An alternative is the use of RNA interference (RNAi; discussed in Chapter 5). RNAi is used to antagonize selected RNA molecules based on their sequence. Usually, the targets are messenger RNAs. However, in this case, the target is a regulatory RNA, microRNA-29. This miRNA represses genes for multiple proteins found in connective tissue, including elastin. Antagonizing miR-29 does indeed increase expression of the *ELN* gene in cultured cells from patients suffering from elastin haploinsufficiency and induces higher levels of elastin.

Usually, one functional copy of a gene is sufficient. Occasionally, two functional copies are required for optimal activity, and a defect in one copy consequently results in a deficiency.

DOMINANT MUTATIONS MAY BE POSITIVE OR NEGATIVE

Occasionally, a mutation causes a change of function or even a gain of function in the corresponding gene product. In this case, a single mutant allele may cause significant phenotypic effects—that is, the mutation is dominant. Such gain-of-function mutations are relatively rare in inherited disease but are frequent among the somatic mutations causing cancer (see Chapter 19).

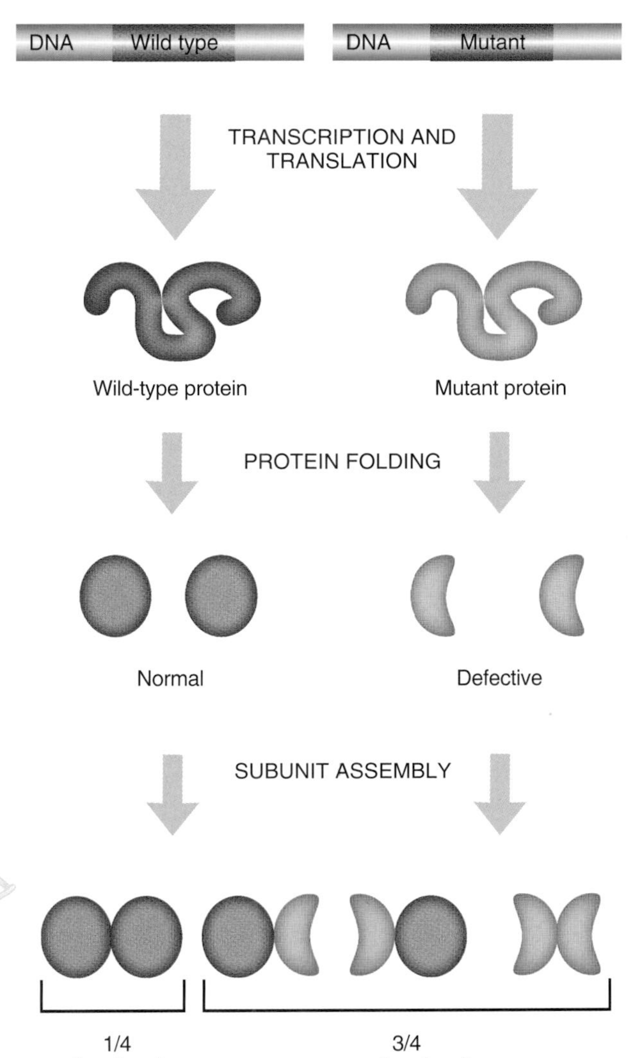

FIGURE 17.3
Dominant Negative Mutations
Dominant negative mutations occur when the defective copy of a gene interferes with the functional copy. For example, the defective protein may bind to and interfere with the normal protein. In this scenario, the proteins function as dimers. If the mutant protein is defective but still forms dimers, then three-fourths of the complexes will be defective.

Gain-of-function mutations are much more specific than those causing loss of function. For example, **achondroplasia**, or short-limbed dwarfism, is due to mutation of a single copy of the *FGFR3* gene that encodes fibroblast growth factor (FGF) receptor #3. FGF receptors receive signals at the cell membrane and transmit the information to the nucleus. Normally, they are activated only when they bind FGF. Although most mutations inactivate the receptor, a few extremely rare mutations yield receptors that are active even in the absence of FGF. Only mutations replacing the amino acid glycine at position 380 with arginine (Gly380Arg) cause achondroplasia. Other mutations in this gene cause other symptoms.

In dominant negative mutations, the mutant protein not only loses its own activity but also interferes with the function of another protein. Thus, an altered, defective protein must be present. Mutations in the same gene that merely abolish protein synthesis are usually recessive. In the simplest scenario for a dominant negative mutation, the affected protein forms oligomers. If a mutation blocks protein synthesis from one allele, there will simply be a 50% drop in the amount of protein. However, if the mutant allele produces an altered protein, this may bind to normal copies of the same protein and give inactive complexes. The overall result may be that almost no active protein is available (Fig. 17.3). Many transcription factors bind to DNA as dimers or tetramers and are susceptible to dominant negative effects. An example is the role of the p53 protein in cancer, discussed in Chapter 19.

No current therapy exists for defects due to dominant mutations. It has been suggested that RNA interference to specifically reduce expression of the mutant allele might be possible.

Hereditary defects due to dominant mutations are rare. They are often due to an altered protein that not only is defective itself but also interferes with the function of other proteins.

DELETERIOUS TANDEM REPEATS AND DYNAMIC MUTATIONS

Occasional disorders are due to tandem repeats of three bases within protein coding regions. The three bases are usually CAG (in the coding strand), which encodes glutamine. The wild-type alleles contain several CAG repeats that give rise to a run of several glutamines within the encoded protein. The mutant alleles have more repeats and give rise to proteins with longer **polyglutamine tracts** (Fig. 17.4). Below a certain number (generally 5–30), these repeats are relatively harmless and stable. Beyond this threshold, the mutations cause disease. In addition, the number of repeats is unstable and tends to expand each successive generation, sometimes to over 100. Hence, these are sometimes known as **dynamic mutations**. Almost all are autosomal dominants.

The known CAG repeat defects all cause neurodegenerative diseases with late onset. Generally, the more repeats, the more severe the pathogenic effects and the earlier the onset.

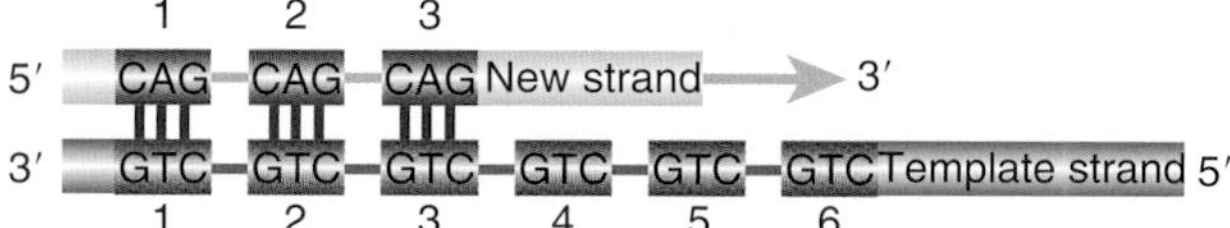

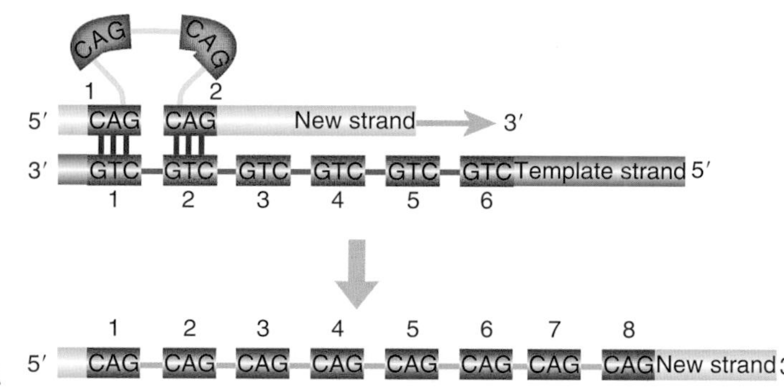

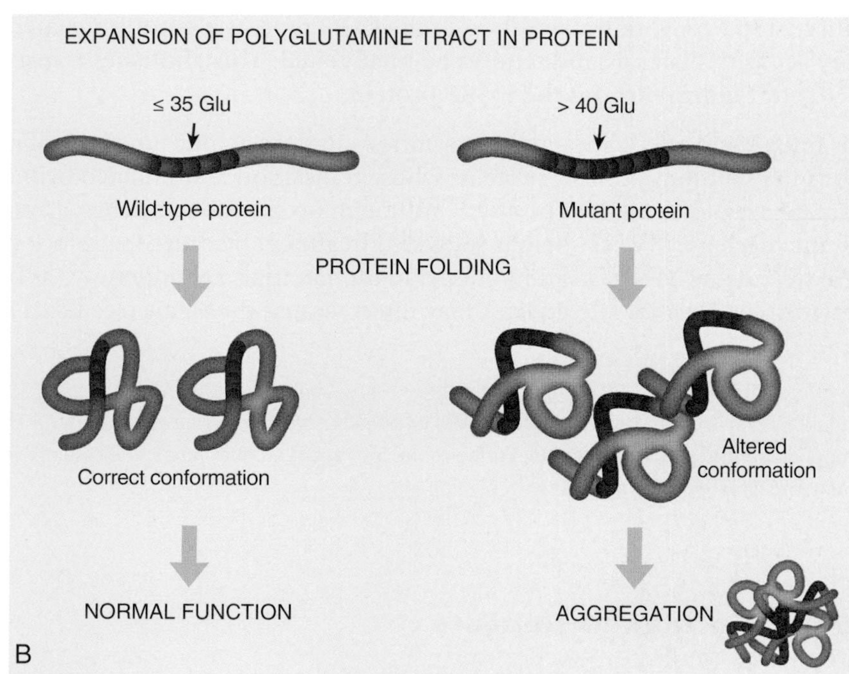

FIGURE 17.4 Tandem CAG Repeats and Polyglutamine Tracts

(A) Long stretches of CAG in DNA can cause errors during replication. If a germline cell has a stretch of six CAG repeats, DNA polymerase may sometimes slip during replication, causing the new strand to loop out. During meiosis, one of the gametes will have eight CAG repeats rather than six. (B) During protein synthesis CAG encodes glutamine and the number of CAG repeats can affect protein structure. Normally, the protein has fewer than 30 glutamines and forms the correct conformation. If there are more than 40 CAG repeats, the extra glutamines cause the protein to misfold, which in turn can cause the protein to aggregate into clumps.

The extended polyglutamine tracts may cause the proteins to clump into aggregates that ultimately kill nerve cells. Different CAG repeat diseases affect different proteins in different nervous tissues. Consequently, symptoms vary. The best known is **Huntington's disease**, which results in loss of control of the limbs, impaired cognition, and dementia. The CAG repeat is in the first exon of the gene encoding huntingtin protein, located on chromosome 4 at 4p16.

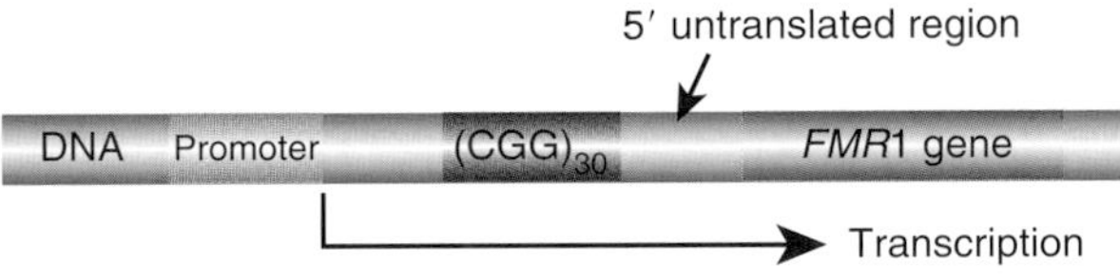

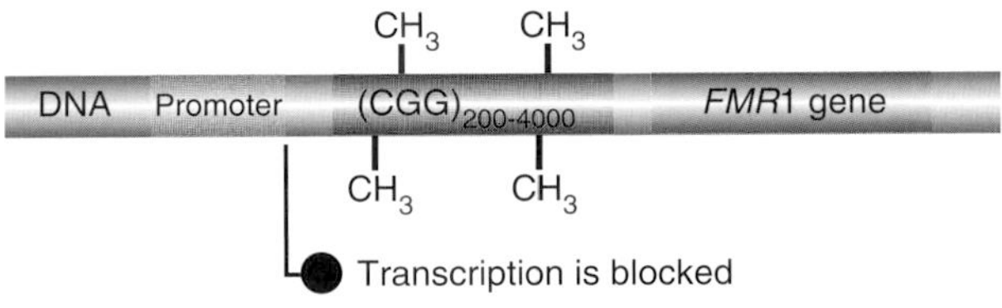

FIGURE 17.5
Fragile X
Between the promoter and gene for *FMR1* is a series of about 30 CGG repeats. They are usually present in the mRNA transcript but are not translated into protein. During replication, DNA polymerase errors cause the number of repeats to expand. When more than 200 repeats occur, the CG sites become highly methylated. RNA polymerase is unable to transcribe the gene, and the protein is not produced.

Unstable expanding tandem repeats may also occur outside coding sequences. Some of these are harmless. Others, such as is **fragile X syndrome**, affect the expression of nearby genes and have devastating results. The name refers to a fragile site within the long arm of the X chromosome. The tandem repeats may cause the chromosome to fragment into two (or just remain held together by a thin thread of material). Fragile X affects about 1 in 4500 males and causes mental retardation. The syndrome is less common and much milder in females. This is largely due to females having two X chromosomes, whereas males have only one.

Fragile X syndrome is due to tandem repeats of CGG within the 5′-untranslated region of the *FMR1* gene in band Xq27 on the affected X chromosome (Fig. 17.5). In the wild type, there are around 30 repeats of CGG. As with CAG repeats, the number of CGG repeats is unstable and tends to expand each generation. No symptoms are observed in those with up to 200 repeats, but such carriers often have children with more than 200 repeats, who do show symptoms. The longest versions may have up to 1300 CGG repeats. Although the CGG repeats are not translated into protein, they act as CG islands and tend to be methylated. This abolishes transcription of the *FMR1* gene, which encodes the FMRP protein.

FMRP binds mRNA and affects local protein synthesis in the synaptic junctions of nerve cells. FMRP inhibits the synthesis of those proteins whose translation is stimulated by metabotropic glutamate receptors (mGluRs) 1 and 5. Although no gene therapy cure is available, discovery of the mechanism of action has suggested treatment by drugs that block mGluR action. These treatments have worked in mice, and human trials are underway (as of 2013). In some other triplet diseases, altered RNA may directly cause problems (see Box 17.1).

Certain mutations consist of multiple tandem repeats of three bases (that is, of a whole codon). The severity depends on the number of repeats. Furthermore, the repeat number tends to increase every generation, giving dynamic mutations.

Box 17.1 RNA Gain of Function in Tandem Repeats

Some triplet-repeat diseases have complex symptoms that cannot fully be explained by alterations in protein structure or in gene expression. Instead, it appears that an altered RNA molecule, carrying the triplet repeat sequence, may directly cause harmful effects. This unusual situation is referred to as RNA gain of function. An example is myotonic dystrophy, which is caused by an unstable CTG repeat. This repeat is located in the 3′-untranslated region of the *DMPK* gene on chromosome 19q13. The repeat sequence is present in the transcribed RNA (as a CUG repeat) but not in the final protein. This repeat hinders processing of the mRNA and its transport into the cytoplasm. Consequently, there is a deficiency in the amount of the dystrophia myotonica protein kinase (DMPK) protein. However, knockout mice with defective *DMPK* genes show only some of the symptoms of myotonic dystrophy.

These further symptoms are due to direct action of the RNA with the CUG repeat (Fig. A). In RNA, a long series of tandem CUG repeats will form a stable hairpin with a single long stem. This RNA is retained in the nucleus, where it binds to several proteins, in particular the double-stranded RNA-dependent protein kinase, PKR. This enzyme is part of the antiviral defense system and is activated by binding to dsRNA. Activated PKR phosphorylates several target proteins, including translation factor eIF2α. The intention is to shut down protein synthesis in virus-infected cells. A variety of other damaging effects occur as a result of sequestration of other nuclear RNA-binding proteins and interference with their normal roles in RNA processing and splicing.

(*Continued*)

Box 17.1 RNA Gain of Function in Tandem Repeats—cont'd

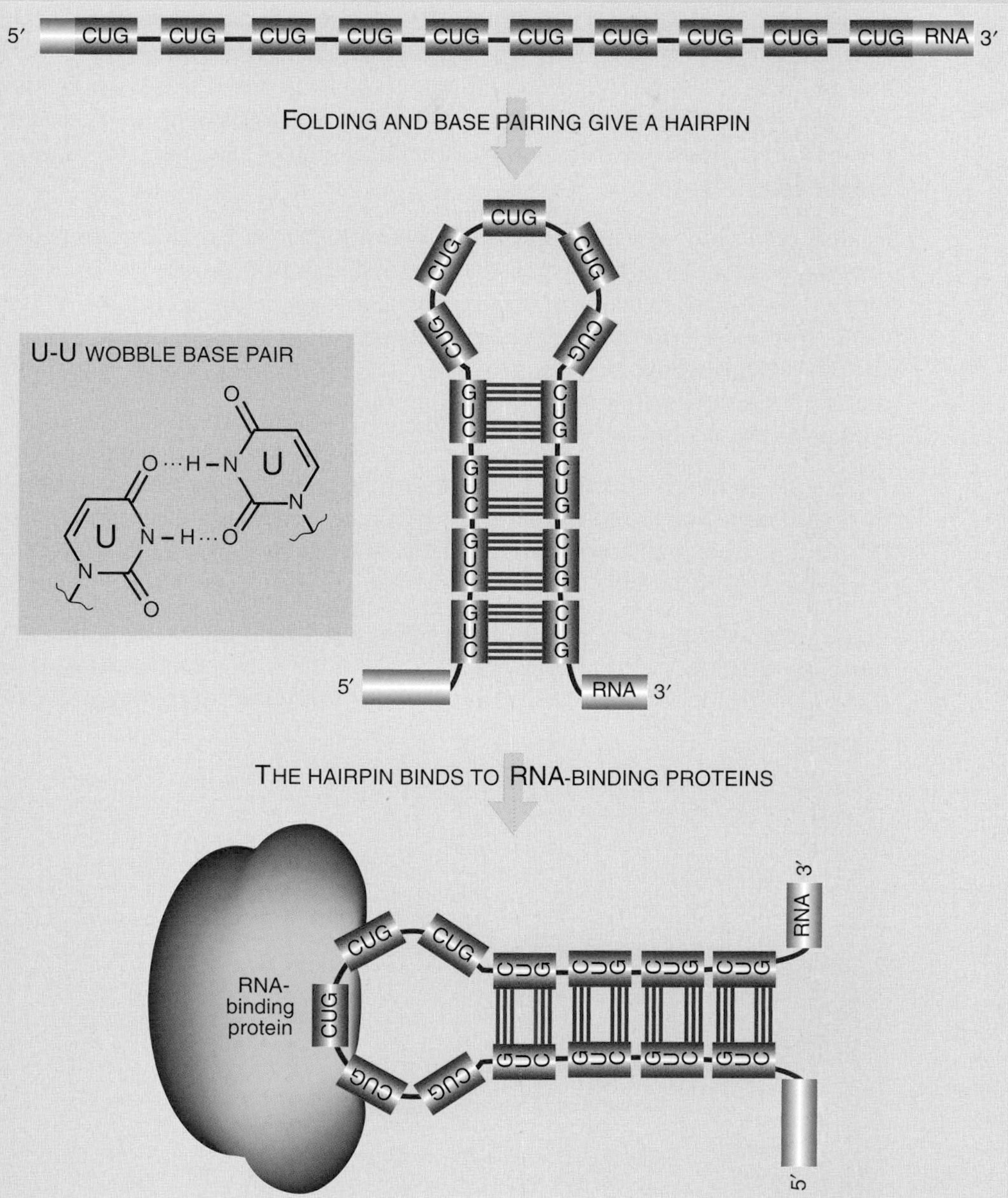

FIGURE A RNA Gain of Function
Tandem CUG repeats form stable hairpins in RNA that may bind tightly to a variety of RNA-binding proteins. The structure includes noncanonical U-U base pairs.

DEFECTS IN IMPRINTING AND METHYLATION

In mammals, somatic cells typically contain two copies of each autosomal gene, one inherited from each parent. When a gene is expressed, both parental alleles are usually transcribed, and there is no difference in expression between maternally and paternally derived alleles. However, there is a small subset of genes whose expression does depend on whether the allele comes from the mother or the father. In such cases, one allele is silenced, a phenomenon known as imprinting.

Genetic imprinting is due to the effects of methylation on gene expression. Typically, one copy of a gene is methylated to prevent its expression. Imprinting occurs when methylation patterns present in the gametes survive to affect gene expression in the new organism. For a few genes, methylation patterns differ in the male and female gametes. How the methylation machinery recognizes and modifies the genes to be imprinted is still uncertain. Whether a particular copy of an imprinted gene is expressed depends on whether it was inherited from the father or the mother. Such inherited changes in gene expression that are not due to alterations in the DNA sequence are known as **epigenetic**.

Prader–Willi syndrome and Angelman's syndrome are due to defects in neighboring genes on chromosome 15 at 15q11-q13 that are subject to imprinting. The imprinting patterns turn off the functional copy of these genes, thus causing the symptoms. Whether this region came from the sperm or the egg determines its pattern of methylation and consequently its pattern of gene expression (Fig. 17.6). About 75% of cases of these syndromes are due to deletions. Less often they result from point mutations, imprinting errors, or inheriting both copies of chromosome 15 from one parent.

Angelman's syndrome (AS) is due to the *UBE3A* gene. It encodes a ubiquitin ligase that is normally expressed in most tissues. However, only the maternal copy is expressed in nerve cells, and the paternal copy is inactivated by methylation. If the maternal copy is defective, mental retardation results.

Prader–Willi syndrome causes multiple defects in development, including that of the brain. It occurs when the paternal copy of 15q11-q13 is defective and the maternal copy is inactive due to methylation. Prader–Willi syndrome is more complex than Angelman syndrome and

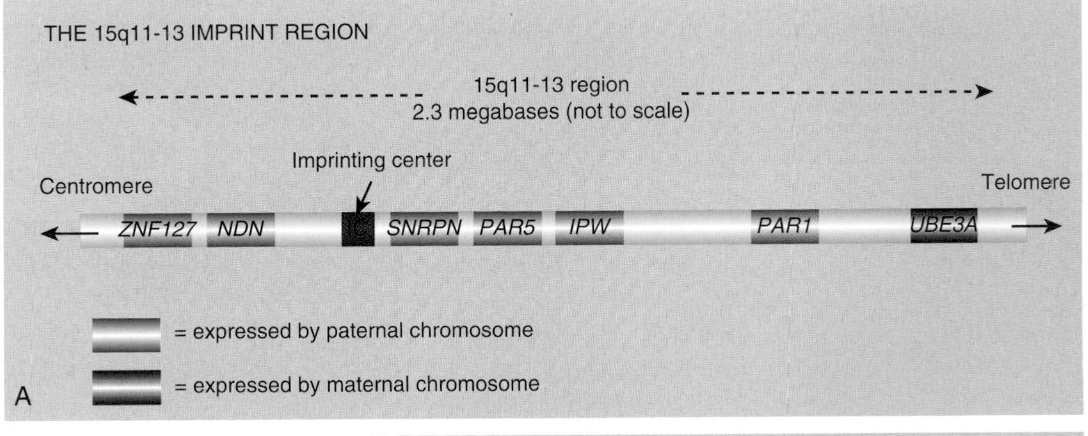

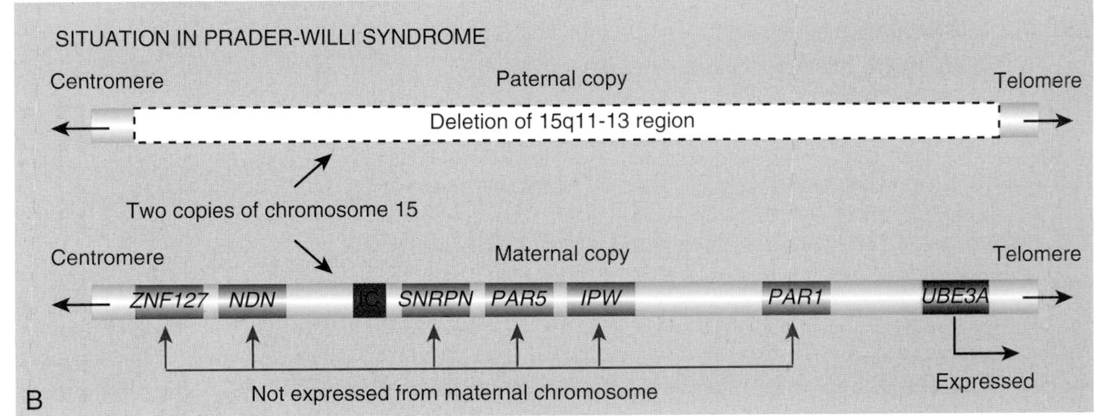

FIGURE 17.6 Prader–Willi and Angelman's Syndromes

(A) Chromosome 15 has a region between 15q11 and 15q13 that is imprinted. The paternal chromosome will express the genes shown in blue and turn off the genes in red due to imprinting. The maternal chromosome expresses the genes shown in red and methylates the genes in blue. In a normal diploid individual, a single copy of each gene is expressed either from the paternal or maternal chromosome. (B) In Prader–Willi syndrome, the paternal copy of chromosome 15 is deleted between region 15q11 and 15q13. Even though the other copy of chromosome 15 has the genes shown in blue, they are turned off because of imprinting, resulting in overall loss of function.

due to several closely linked genes, several with unknown functions. However, many of the symptoms seem due to defects in the *SNORD116* gene that encodes a small nucleolar RNA (snoRNA) gene.

No cure exists for imprinting problems, although supportive therapy helps in some cases. However, activating the intact but methylated copy of the needed gene provides a promising approach. For Angelman's syndrome, this has been achieved in mice, where topoisomerase inhibitors abolished silencing of the paternal Ube3a allele indirectly by reducing transcription of an imprinted antisense RNA. Topoisomerase inhibitors are used clinically to treat certain cancers, but they have some toxic side effects.

A few genes have methylation patterns that differ depending on whether they came from the male or female parent. Complex syndromes may result when one copy of such a gene is defective and the other is silenced by methylations.

MITOCHONDRIAL DEFECTS

The mitochondrial genome is a tiny fraction of an animal cell's genetic information. In terms of coding DNA, there are about 100 Mbp in the nuclear genome and only 15.4 Kbp in human mitochondrial DNA (mtDNA). Consequently, the vast majority of mutations affecting coding DNA occur in nuclear DNA. Furthermore, there are thousands of copies of mitochondrial DNA in most cells, so a mutation in one gene on one copy of mtDNA would have little effect on the whole organism. Because each original mutation must occur on a single molecule of mtDNA, the chances of its spreading in the mitochondrial population and becoming fixed seem highly unlikely. Nevertheless, the proportion of hereditary defects affecting mtDNA is surprisingly large.

One contributing factor is that mtDNA has a much higher mutation rate than nuclear DNA. Oxidation damage to DNA, due to reactive oxygen species generated by the respiratory chain, is much higher in mitochondria. In addition, there are fewer repair systems in mitochondria, and the mtDNA is not protected by histones. Moreover, mitochondria go through many more divisions than the cells that contain them. (Although animal mtDNA is evolving very fast, the situation in plants is quite different. Plant mtDNA is much larger [150 Kbp to 2.5 Mbp], contains introns, and evolves relatively slowly.)

Disorders due to mitochondrial mutations are maternally inherited. Sperm cells contribute nuclear DNA during fertilization but do not donate their mitochondria to the zygote. Consequently, all the mitochondria in a particular individual are derived from ancestors in the female egg cell, or **oocyte**, which has approximately 100,000 copies of mtDNA. Despite this, the rate at which mutations in mtDNA become established is some 10-fold greater than for nuclear DNA due to a bottleneck during the replication of mtDNA. When primordial germ cells develop into oocytes, only a few of their mitochondria divide and give rise to progeny (Fig. 17.7). This allows mutations to spread through the mitochondrial population in a specific cell line.

The population of mitochondria within an individual may all carry a mutation, a situation called **homoplasmy**, or there may be a mixture of mutant and normal mitochondria, called **heteroplasmy**. The severity of the defect will depend on the proportion of defective mitochondria, which may vary among individuals with the same mutation.

Mitochondria are highly specialized in generating energy via respiration, and the majority of mitochondrial genes affect the respiratory chain (Fig. 17.8). Consequently, almost

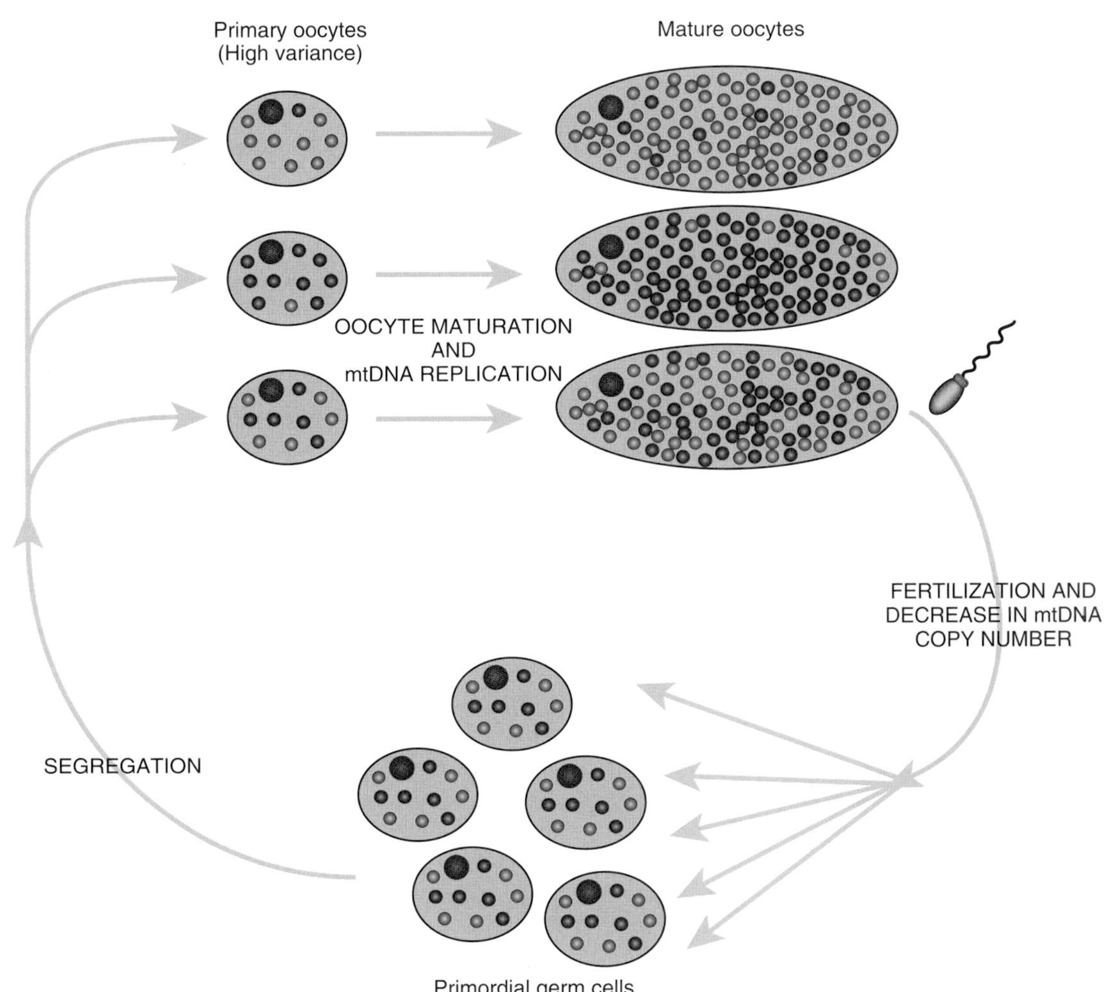

FIGURE 17.7 Mitochondrial DNA Bottleneck
In primordial germ cells *(bottom)*, about 50% of the mitochondria have a mutation (red). As they develop into primary oocytes, only a few mitochondria are passed on. Therefore, some oocytes will have just a few affected mitochondria *(top oocyte)*, some will have a majority of defective mitochondria *(middle)*, and others will have an equal proportion *(bottom)*. As the primary oocytes develop into mature eggs, the number of mitochondria increases, but the ratio of mutated to normal remains constant in each cell line.

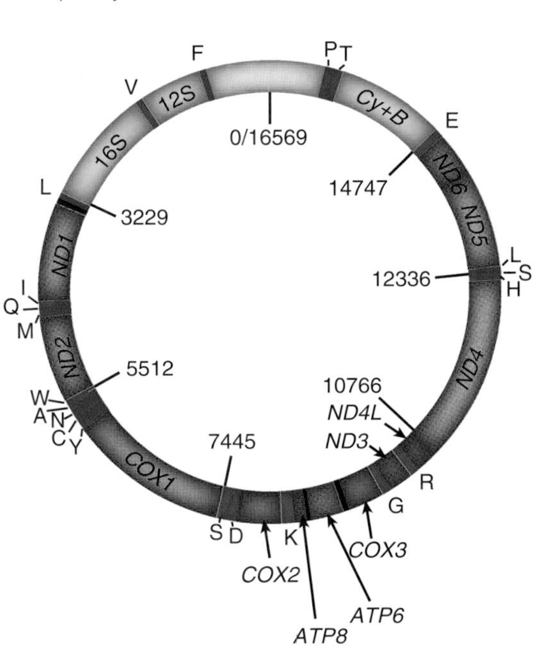

FIGURE 17.8 Map of Human Mitochondrial DNA
Mitochondrial DNA has genes that encode tRNA, rRNA, and proteins used in energy production. The tRNA genes are indicated by single-letter amino acid symbols. The two rRNA genes are 16S and 12S. The remaining genes are all for proteins in complexes I, III, IV, and V of the respiratory chain.

all genetic defects in mtDNA affect respiration and result in lowered energy production. Because different tissues of the body differ greatly in energy requirements, mitochondrial defects affect different tissues to very different extents. Brain and muscle are especially susceptible because of their high energy consumption.

Neuropathy, ataxia, and retinitis pigmentosa (NARP) syndrome is caused by base changes at position 8993 of mtDNA. These changes affect the *ATP6* gene, encoding a component of the **ATP synthetase**. This enzyme couples energy from the respiratory chain to the synthesis of ATP. Defects cause a lack of ATP to energize the cell. Nerve and muscle cells are most affected, and muscular weakness, dementia, seizures, and sensory malfunction are among the results. Symptoms occur when 70% to 90% of the mitochondria are affected. If more than 90% are affected, death in infancy is the usual outcome.

At present, no genetic therapies exist for mitochondrial defects. A major problem is getting gene therapy vectors into the mitochondria specifically.

Mutations in the mitochondrial genome also cause hereditary defects, mostly affecting energy generation. Unlike nuclear genes, mitochondrial mutations are inherited only via the maternal line.

IDENTIFICATION OF DEFECTIVE GENES

Although the DNA sequence of the whole human genome is available, connecting particular symptoms with a specific gene is not always straightforward. Here, we outline some general approaches to identifying the genes responsible for hereditary defects and finding their location on the human chromosomes. Confirmation of identity normally involves cloning and further genetic analysis.

One way to identify genes responsible for hereditary defects is to analyze the symptoms and then make an informed guess as to what kinds of proteins might be involved. Possible candidates are then chosen from the list of characterized genes and investigated further. Such **candidate cloning** is rarely successful because relatively few human genes have been functionally characterized. However, recent increases in proteomics research are revealing the functions of many mammalian genes, so this approach is becoming more realistic.

An improved variant of this approach comes from using model organisms— in particular, mice. The vast majority of human genes have homologs in mice. Moreover, unlike humans, mice may be directly used for genetic experimentation. As described in Chapter 16, it is possible to make mice in which both copies of any chosen gene have been artificially inactivated. Such **knockout mice** are then examined for symptoms. Several programs are systematically making mice with knockout mutations affecting every one of the 25,000 or so mammalian genes. This in turn should allow many human genes to be matched with possible symptoms. An alternative approach is to use RNAi (see Chapter 5) to knock down gene expression. This approach is especially useful when knockouts are lethal.

Functional cloning begins with a known protein that is suspected of involvement in a hereditary disorder. The amino acid sequence of the protein is determined. Nowadays this would most likely be done by mass spectrometry of peptide fragments generated from the protein by protease digestion. The protein sequence is then used to deduce the coding sequence of the gene. The gene is then located by examination of the human genome sequence.

Positional cloning is used when the nature of the gene product is unknown. In this case, the disease gene must be mapped at least approximately by a genetic approach before further DNA-based screening can proceed. The easiest cases are those in which there is a major chromosomal abnormality, such as a deletion, inversion, or translocation that may be visualized under the light microscope. This may localize the defect to a specific band on a particular chromosome. Alternatively, linkage studies on individuals from families afflicted by the inherited defect may locate the damaged gene close to other genetic markers. These other markers may be known genes, but more often they will be SNPs, RFLPs, VNTRs, or other sequence polymorphisms (see Figures 8.3 through 8.5 for details of these polymorphisms).

Such genetic mapping can localize a gene to around 1000 kb. This length of DNA may contain anywhere from 10 to 50 genes, depending on how crowded that region of the genome is. DNA from the suspect region is cloned. However, in positional cloning, we have no previously identified protein that can be used to check for the corresponding gene. Therefore, the hereditary defect must be identified at the DNA level. The suspect DNA may be scanned for the presence of functional genes by a variety of approaches:

(a) The presence of open reading frames indicates a possible coding sequence. Note that in higher organisms, the coding sequence will typically be fragmented into several exons separated by noncoding introns. These introns may be very long and frequently account for more of the overall length of the gene than the exons.

(b) CpG (or CG) islands are often found upstream of transcribed regions in vertebrate DNA. These are GC-rich regions that are often methylated for regulatory purposes.

(c) Coding DNA tends to evolve more slowly than noncoding DNA. Consequently, coding DNA from one animal will often hybridize to DNA from a range of related organisms while noncoding DNA does not. Zoo blots are often used to identify coding DNA.

(d) Tissues most severely affected by a genetic disease may contain significant levels of mRNA transcribed from the gene responsible for the defect. Hybridization can be used to see if candidate DNA sequences match those in the mRNA pool. This assumes that the gene in question is reasonably highly expressed. This will usually be true for genes encoding structural proteins and enzymes but not for those encoding regulatory proteins. Note that the mRNA must be isolated from a healthy person because the defective gene might not be transcribed in patients with the defect.

(e) Ultimately, sequencing of DNA from healthy and affected individuals should show a difference—if the suspected gene is truly responsible for the hereditary defect.

Various approaches are used to clone genes responsible for human hereditary defects. Some depend on the known location of the gene; others depend on gene function. Information from model organisms with equivalent defects has often been very useful.

GENETIC SCREENING AND COUNSELING

Close relatives are more likely to produce malformed children because of harmful recessive alleles coming together. It is also true that two people from otherwise unrelated families that both have a history of the same hereditary disease might be wise to avoid having children. Today, we can do more than give general advice. Testing of prospective parents for high-risk genetic diseases can now be done for an increasing number of genes. The typical approach is to use PCR analysis (see Chapter 4). This will reveal whether a mutant copy of the gene is present in a DNA sample from the person tested. If marital partners both test positive for a recessive defect, they will have to decide whether or not to take the risk of having children. For a recessive defect, in which both parents are carriers, one in four children will get the disease.

The basic problem with genetic screening is that our ability to detect genetic defects has far outrun our ability to cure them. Thus, screening may reveal a defect about which nothing can be done, and this may cause severe psychological distress to the individuals concerned. For example, Huntington's disease is an autosomal dominant condition that results in movement disorder, dementia, and ultimately death. The symptoms usually appear only in the 30s, although they may be delayed as late as the 60s in some patients. Genetic screening helps those who are afflicted decide whether or not to have children. It also provides them with the knowledge that they will eventually develop an incurable and fatal disease. Some cancers and coronary heart disease have a genetic component (often multigenic). Screening may allow affected individuals to avoid factors that tend to trigger the disease. On the other hand, especially if the treatment is complicated and/or only partially effective, such knowledge may be a burden.

Genetic screening sometimes allows successful intervention by modifying diet or lifestyle. The classic case is **phenylketonuria**. The absence of the enzyme **phenylalanine**

hydroxylase causes a buildup of phenylalanine and a deficiency of the product, tyrosine (Fig. 17.9). Excess phenylalanine is neurotoxic; if untreated, this condition results in severe damage to the central nervous system, and patients require permanent care for life. However, a diet low in phenylalanine avoids the buildup of this amino acid and largely alleviates the problems. Phenylketonuria occurs in 1 in 10,000 live births and is routinely screened for in developed nations using blood samples taken a week or so after birth. The diet may be relaxed somewhat later in life after development of the nervous system is largely complete.

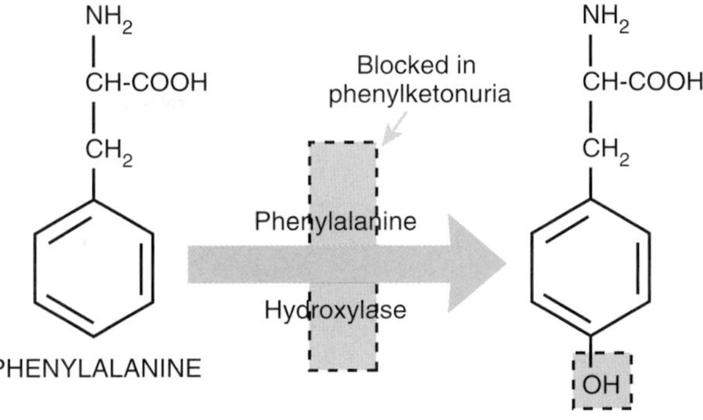

Given that a reliable test is available, it is generally agreed that neonatal screening is justified if the disease is reasonably frequent (say 1 in 10,000 or more), the disease is severely damaging, and treatment is available that significantly improves the condition.

FIGURE 17.9 Phenylketonuria
Phenylalanine hydroxylase catalyzes the conversion of phenylalanine to tyrosine. If the enzyme is absent, phenylalanine accumulates and is toxic to the nervous system.

It is also possible to examine embryos during early pregnancy by drawing samples of amniotic fluid that contain some cells from the fetus, a procedure called **amniocentesis**, or by examining extra-embryonic fetal tissue. These cells may be genetically screened for genetic defects. Embryos doomed to grow up with hereditary defects could then be aborted early if the parents wish to do so. It is also now possible to screen embryos obtained by *in vitro* fertilization before they are implanted. This avoids aborting an unwanted fetus should a serious genetic defect be discovered.

> Genetic screening is possible for a wide and growing number of hereditary defects. Unfortunately, most genetic defects cannot yet be cured.

GENERAL PRINCIPLES OF GENE THERAPY

The most straightforward use of gene therapy is to deal with a hereditary defect due to a single gene that occurs only when both copies of the gene are defective, that is, a recessive condition. Introducing a single good copy of the gene can then cure the defect. This treatment is sometimes known as **replacement gene therapy**. Furthermore, it would obviously simplify treatment if the disease mostly affects just one or a few organs. The main steps in replacement gene therapy are as follows:

- **(a)** Identification and characterization of gene
- **(b)** Cloning of gene
- **(c)** Choice of vector
- **(d)** Method of delivery
- **(e)** Expression of gene

The first step is to identify the genetic defect and to clone a good copy of the gene involved. The gene must then be delivered to the patient. Doing so involves choosing a vector together with a suitable method of delivery. In addition, the vector/gene construct must be designed to allow proper expression of the gene, once inside the patient. Delivery may be performed in a variety of ways. The vector/gene construct may be injected into the bloodstream or other tissue. It may be aerosolized and sprayed into the nose and airways. In some cases, cells are removed from the patient, engineered while growing in culture, and then returned to the patient. This approach is known as ***ex vivo* gene therapy** because the actual genetic engineering takes place outside the patient. (The direct delivery of the vector/gene construct to the patient is sometimes called *in vivo* gene therapy to contrast with this.)

In the laboratory, most manipulations are done with genes carried on bacterial plasmids. Although gene therapy has occasionally been performed directly with plasmid DNA carrying a therapeutic gene, more often specialized delivery systems are used. In most cases, a modified animal virus is used as the vector. Because viruses cause disease, they first need to be genetically disarmed in order to be used in gene therapy. About 70% of human gene therapy trials have used viral vectors (Table 17.2). Two main groups of viruses have been used: retroviruses and adenoviruses. In a smaller proportion of cases, DNA has been used directly or delivered inside liposomes.

Table 17.2 Delivery Systems Used in Gene Therapy Trials

	Trials	
Vector Used	Number	%
Retrovirus	455	22.1
Adenovirus	476	23.1
Other viruses	459	22.3
Naked DNA or RNA	359	17.4
Liposomes	112	5.4
Other categories	137	6.6
Unknown	65	3.2
Total	2063	100

Data as of 2013. Current data on the numbers and types of gene therapy trials are available from the website of the *Journal of Gene Medicine*: http://www.abedia.com/wiley/vectors.php

In gene replacement therapy, a functional copy of the gene responsible for the hereditary defect is inserted. The most popular approach is for the gene to be carried on an engineered virus vector.

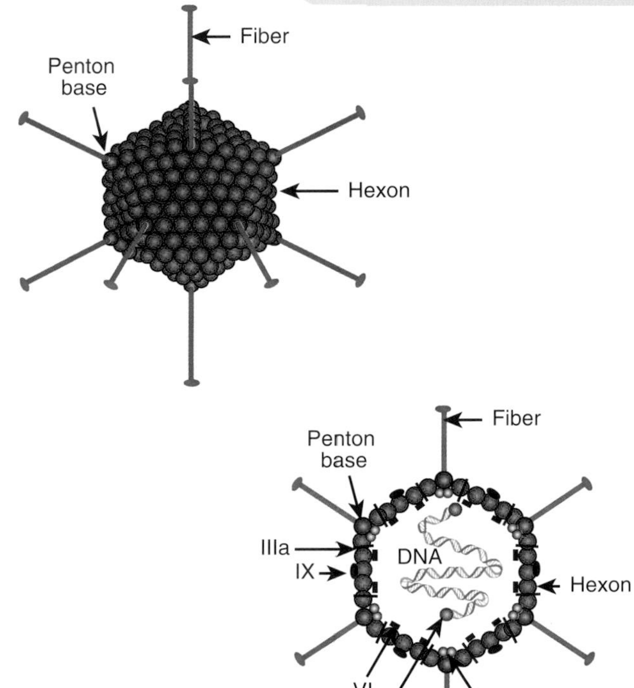

FIGURE 17.10
Adenovirus Structure
Adenoviruses have an icosahedral capsid with 20 faces and 12 vertices. Each face is composed of hexons (yellow balls). At each of the vertices are a penton base (red ball) and a fiber (black strand). Proteins IIIa, VI, VIII, and IX stabilize the capsid. A strand of double-stranded DNA is packaged inside the capsid, with two terminal proteins to protect the ends.

ADENOVIRUS VECTORS IN GENE THERAPY

Adenoviruses are relatively simple, double-stranded DNA viruses that infect humans and other vertebrates. The virus particle consists of a simple icosahedral shell, or capsid, containing a single linear dsDNA molecule of approximately 36,000 base pairs (Fig. 17.10). A terminal protein protects each end of the DNA. The capsid is made of 240 **hexons**, each of which is a trimer of the hexon protein. Hexons are named for their six-fold symmetry and are each surrounded by six neighboring hexons. The hexon protein has loops that project outward from the virus surface. Adenoviruses are subdivided into about 50 serotypes by their response to antibody binding. This variation is largely due to variation in the loops of the hexon protein.

Five faces of the virus particle converge at each vertex. Here is found a **penton**, which consists of a base (a pentamer) plus a fiber (a trimer). The fiber varies greatly in length in different subgroups of adenovirus, and its tip binds to the receptor on the host cell surface (Fig. 17.11).

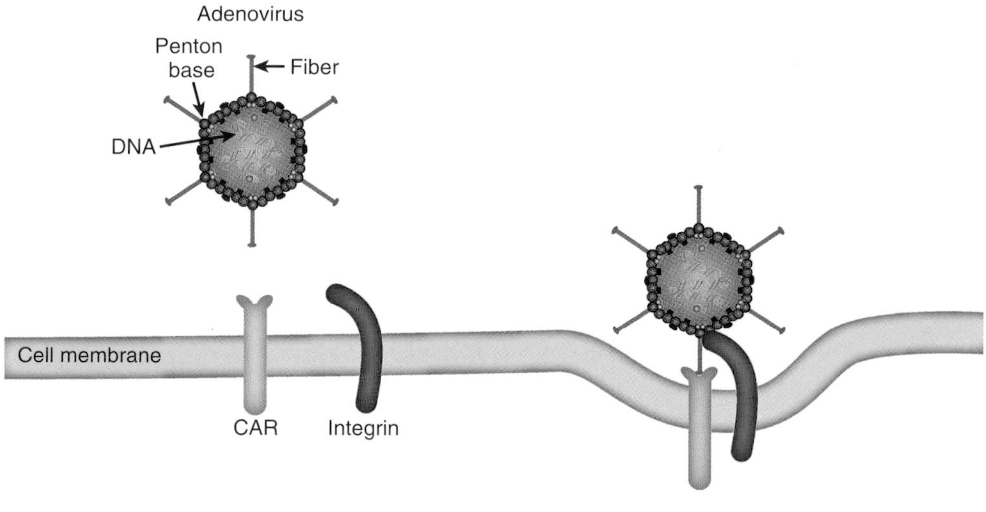

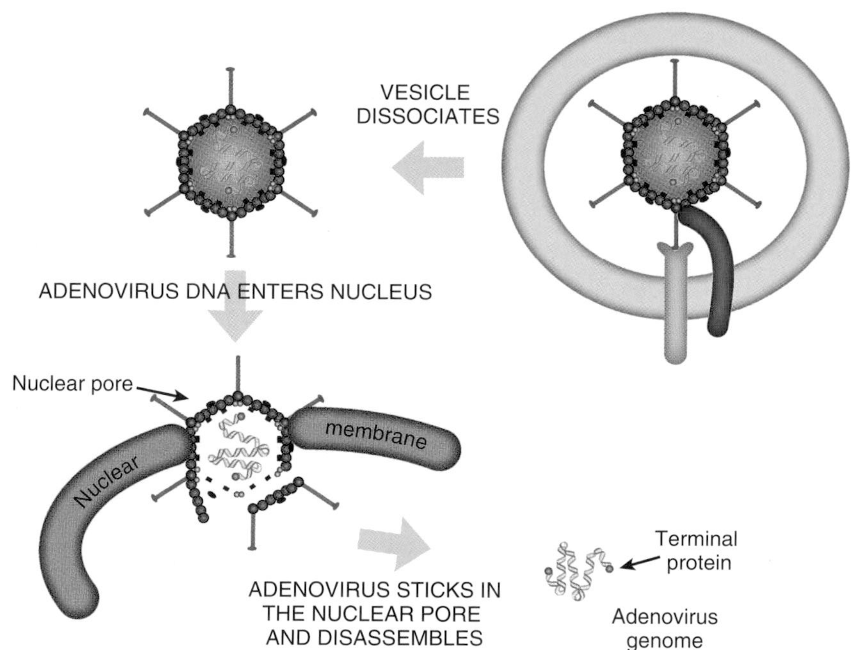

FIGURE 17.11 Adenovirus Entry

Adenoviruses enter human cells by recognizing two receptors: CAR and integrin. The virus is taken into the cell attached to the receptors and surrounded by a membrane vesicle that dissociates in the cytoplasm. The adenovirus injects its DNA into the nucleus through a nuclear pore.

The adenoviruses share the same receptor as B-group coxsackieviruses. This protein is therefore known as **coxsackievirus adenovirus receptor (CAR)**, and is a cellular adhesion protein found in several tissues.

After the fiber tip binds to CAR, the penton base binds to integrins on the host cell surface. (Integrins are transmembrane proteins involved in adhesion.) Next, the membrane puckers inward and forms a vesicle that takes the adenovirus inside the cell. The virus is then released into the cytoplasm and travels toward the nucleus. The virion is disassembled outside the nucleus, and only the DNA (with its terminal proteins) enters the nucleus.

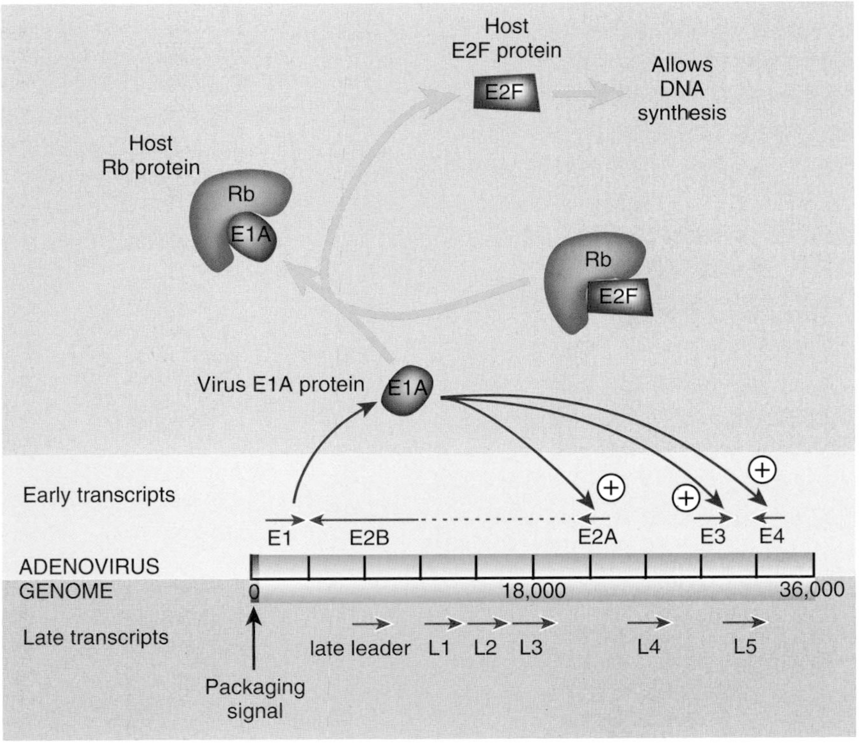

Adenoviruses were among the first viruses chosen as vectors for use in human gene therapy. Their advantages are as follows:

(a) Adenoviruses are relatively harmless. They cause mild infections of epithelial cells, especially those lining the respiratory or gastrointestinal tracts.
(b) Adenoviruses do not cause tumors.
(c) Adenoviruses are relatively easy to culture and to produce in large quantities.
(d) The life cycle and biology of adenovirus are well understood.
(e) The function of most adenovirus genes is known.
(f) The complete DNA sequence is available for several adenoviruses.

FIGURE 17.12
Role of Adenovirus E1A Protein
Eliminating E1A prevents adenovirus replication. E1A is a transcription factor that activates other adenovirus genes, such as E2A, E3, and E4. E1A also binds host cell Rb protein, thus releasing host E2F protein to activate DNA synthesis.

Although mild, adenoviruses do cause inflammation and can cause serious illness in patients with damaged immune systems. Therefore, when an adenovirus vector is designed for gene therapy, the virus needs to be disarmed by crippling its replication system. This is done by deleting the gene for **E1A protein**, a virus protein made immediately on infection. E1A has two functions (Fig. 17.12). First, it promotes transcription of other early virus genes. Second, it binds to host cell Rb protein, which normally prevents the cell from entering the S-phase. This prompts the host cell to express genes for DNA synthesis, which the virus utilizes for its own replication. In the lab, crippled adenovirus is grown in genetically modified host cells that have the viral E1A gene integrated into host cell DNA. The virus particles generated by this approach cannot replicate in normal animal cells.

The DNA of adenovirus is packaged by a headful mechanism. If the DNA is more than 5% shorter or longer than wild type, packaging fails (Fig. 17.13). Insertion of a therapeutic gene into an adenovirus will make the DNA longer. If the inserted gene is much longer than the deletion used to cripple virus replication, the virus will fail to assemble correctly. Deleting other nonessential virus DNA solves this problem.

Although genes carried on engineered adenovirus have been expressed successfully in both animal and human tissues, there are problems. The major difficulty is that adenovirus infections are short-lived. Thus, the therapeutic gene is expressed for only a few weeks before the immune system eliminates the virus. Furthermore, the patient develops immunity to the virus so that a second treatment with the same engineered virus will fail. Thus, adenovirus vectors cannot be used for long-term gene therapy for hereditary diseases.

Even with the limitations just described, adenovirus vectors may help deliver a deadly gene to cancer cells. Here, only a short period of expression should be needed. The CAR receptor is normally only expressed highly by epithelial cells, which limits adenovirus entry to these cell types. However, many cancers also express the CAR receptor at high levels. Consequently, the majority of gene therapy trials using adenovirus are now aimed at cancer cells (see later discussion).

Adenovirus has been widely used as a gene therapy vector. Eventually, the immune system eliminates the virus, restricting its use in long-term therapy for inherited conditions.

CYSTIC FIBROSIS

Cystic fibrosis is the most common single-gene defect in the Western world with about 1 in 2000 white children suffering from it. This condition is due to homozygous recessive mutations. In other words, two defective copies of the gene, one from each parent, must be inherited for the child to suffer from the disease. Humans with a single defective allele are carriers but do not show symptoms, as a single wild-type version of the gene is sufficient for normal health.

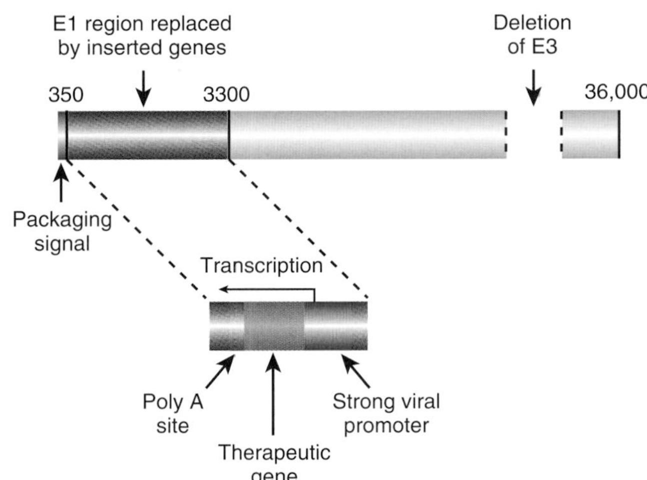

FIGURE 17.13 Engineered Adenovirus
The length of DNA in a virus particle must be very close to the natural adenovirus length of 36,000 base pairs for proper packaging. The therapeutic gene replaces the E1 region. If the gene is much longer than E1A, then a region that contains gene E3 is deleted to keep the overall length of the DNA constant.

The protein encoded by the cystic fibrosis gene is referred to as the **CFTR protein** (for cystic fibrosis transporter) and is found in the cell membrane, where it acts as a channel for chloride ions. In healthy people, this channel can be opened or shut as needed by the cell. The CFTR protein consists of a series of membrane-spanning segments with a central control module (Fig. 17.14). If the control module has a phosphate group attached, the channel is open, and when the phosphate group is removed, it shuts. The phosphate is derived from ATP, which promotes chloride transport.

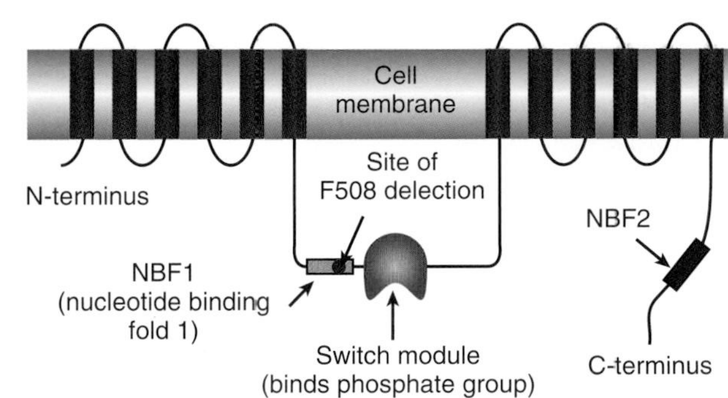

FIGURE 17.14 Cystic Fibrosis Protein Transports Chloride
The CFTR protein has 12 hydrophobic segments that span the cell membrane. Small loops connect the 12 transmembrane segments, except for segments 6 and 7, which are linked by a large protein loop. In the three-dimensional configuration, the 12 transmembrane segments organize to form a channel, and the large protein loop regulates whether the channel is open or shut.

In cystic fibrosis patients, control of the chloride channel is defective. This, in turn, affects a variety of other processes. The most harmful result is that the mucus lining the lungs is abnormally thick. Lack of chloride ions leads to lack of sufficient water in the mucus. This not only causes obstructions but also allows the growth of harmful microorganisms. Cells lining the airways of the lungs are killed and replaced with fibrous scar tissue—hence, the name of the disease. Eventually, the patient succumbs to respiratory failure.

The general location of the cystic fibrosis gene (to within 2 million base pairs) was found by screening large numbers of genetic markers (RFLPs) in members of families with a cystic fibrosis patient. The gene lies on the longer (q) arm of chromosome 7 between bands q21 and q31. After this region of the chromosome was subcloned in large segments, the cystic fibrosis gene was found to occupy 250,000 base pairs and have 24 exons, encoding a protein of 1480 amino acids (Fig. 17.15). Because 1480 amino acids need only 4440 base pairs to encode them, this means that scarcely 2% of the cystic fibrosis gene is actually coding DNA. The rest consists of introns.

About 70% of those with cystic fibrosis in North America share the same genetic defect. They all have a small deletion of three bases that code for the amino acid phenylalanine at position 508 of the normal protein. In Denmark, 90% of cystic fibrosis cases are due to this ΔF508 deletion (F = phenylalanine in single-letter code), whereas in the Middle East it accounts for only 30%. The other cystic fibrosis cases are the result of more than 500 different mutations, which makes genetic screening difficult.

The distribution and relatively high frequency of the ΔF508 deletion suggest that it originated in Western Europe, perhaps around 2000 years ago, and has been positively selected. Although two defective copies of the *CFTR* gene are highly detrimental, heterozygotes with one ΔF508 allele and one normal allele show only moderate reduction of ion flow and water movement. This does not cause cystic fibrosis but is sufficient to significantly reduce the loss of fluids during diarrhea. Such heterozygotes are more resistant to the effects of enteric infections such as typhoid and cholera that cause dehydration via extensive diarrhea.

> The primary defect in cystic fibrosis is in a chloride ion transport protein. Secondary effects include buildup of mucus in the lungs and bacterial infection. Heterozygotes, with only one defective *CFTR* gene, show resistance to infectious diseases that cause dehydration.

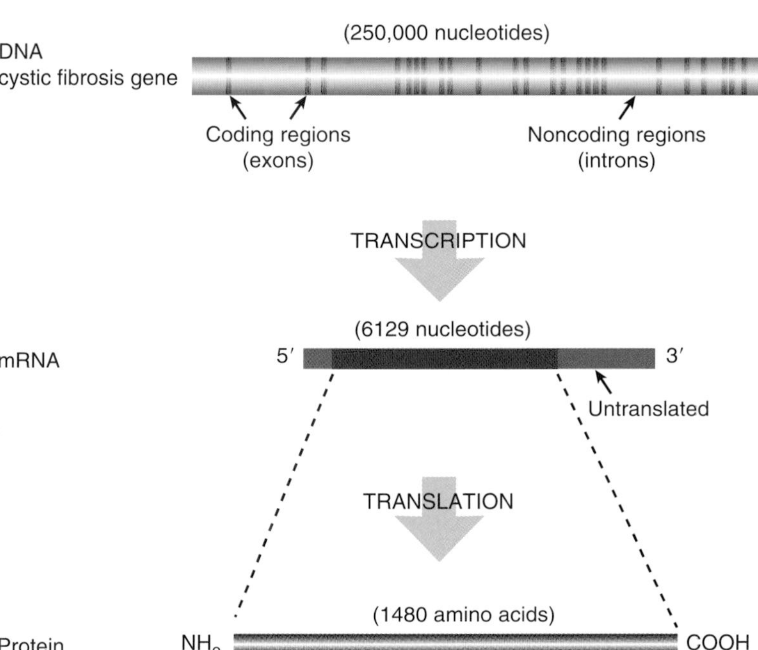

FIGURE 17.15
Cystic Fibrosis Gene
The cystic fibrosis gene covers an extremely large region of DNA (250,000 nucleotides) and has 24 exons. After transcription and splicing, the mRNA is only 6129 nucleotides; thus, the majority of the gene consists of introns.

CYSTIC FIBROSIS GENE THERAPY

Because the lungs are relatively accessible to viral infection, cystic fibrosis was an early candidate for gene therapy. The healthy version of the cystic fibrosis gene was cloned and inserted into a crippled adenovirus. Aerosols containing the engineered adenovirus with the cystic fibrosis gene were sprayed into the noses and lungs, first of rats, and then of humans. In some instances, the healthy cystic fibrosis gene was expressed and normal chloride ion movements were also restored. Unfortunately, expression fell off over a 30-day period, and repeated doses of virus had little effect, largely because of recognition and destruction of the virus by the immune system. This failure was a major early disappointment to the field of gene therapy. Nonetheless, gene therapy has proven effective in other cases (see later discussion) and, somewhat paradoxically, pharmaceuticals have been developed for cystic fibrosis based on detailed genetic analysis.

Novel drugs based on the molecular biology of the CFTR protein have been developed recently (Fig. 17.16). Ivacaftor (VX-770; trade name Kalydeco®) binds to the CFTR protein and promotes ATP-independent operation. Ivacaftor helps only patients with the G551D mutation. This results in a CFTR protein that is correctly inserted into membranes but fails to transport chloride. Only 4% to 5% of persons with cystic fibrosis have the G551D mutation. The FDA approved Ivacaftor in 2012, but it costs nearly $300,000 per year at present.

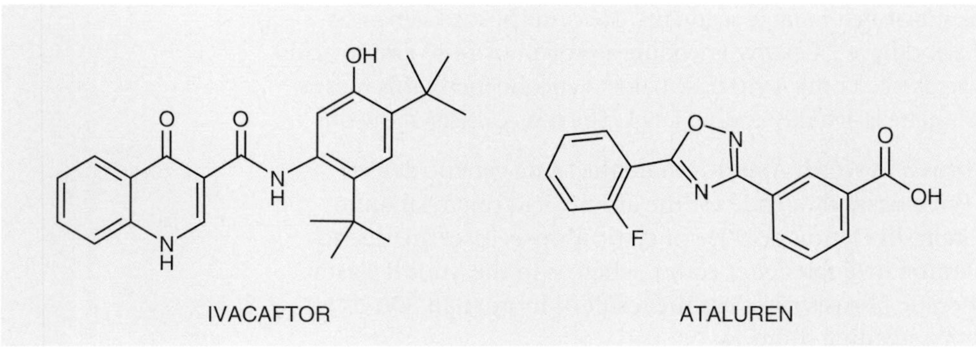

FIGURE 17.16
Ivacaftor and Ataluren
Ivacaftor binds to and reactivates CFTR protein with the G551D mutation. Ataluren suppresses nonsense mutations, thus preventing premature termination of protein synthesis.

Ataluren (PTC124; trade name Translana®) suppresses nonsense mutations that are responsible for premature termination of protein synthesis. Consequently, it can be used against several hereditary defects. However, for each condition only a small proportion of patients will have nonsense mutations, generally less than 10%. Ataluren has reached phase 3 clinical trials for cystic fibrosis.

Cystic fibrosis has been targeted for gene therapy because the lungs are readily accessible. Early attempts using adenovirus vectors were partially successful, but only for a brief period.

RETROVIRUS GENE THERAPY

Retroviruses infect many types of cells in mammals. They need dividing cells for successful infection and will not infect many tissues where host cell growth and division have come to a standstill. Moreover, the genetic material of retroviruses passes through both DNA and RNA stages. This means that introns must be removed from any therapeutic genes before they are cloned into a retrovirus. Despite these extra technical difficulties, a retrovirus has the distinction of carrying the first gene in successful human gene therapy (see later discussion).

The retrovirus particle has an inner nucleocapsid consisting of an RNA genome inside a protein shell and an outer envelope, derived from the cytoplasmic membrane of the previous host cell. The basic retrovirus genome consists of three genes (*gag, pol,* and *env*) enclosed between two **long terminal repeats (LTRs)**, although more complex retroviruses such as HIV have extra genes involved in regulation. The LTR sequences are needed for integration of the DNA version of the virus genome into the host cell DNA. Between the upstream or 5′ LTR and the *gag* gene is the **packaging signal** (Fig. 17.17), which is essential for packaging the RNA into the virus particle.

Vectors for gene therapy have been derived from the simpler retroviruses, especially **murine leukemia virus (MuLV)**. The vectors have all the retrovirus genes removed, and as a result, they are completely defective in replication. They retain only the packaging signal and the two LTRs (Fig. 17.17) and can carry approximately 6 to 8 kb of inserted DNA. A virus promoter in the 5′ LTR drives expression of the cloned gene. Because the vector lacks *gag, pol,* and *env,* it cannot make virus particles. Hence, these functions must be provided by a **packaging construct**, a defective provirus that is integrated into the DNA of the producer cell (see Fig. 17.17). The packaging construct lacks the packaging signal, so although it is responsible for manufacture of virus particles, it is not packaged itself. The virus particles generated contain only the retrovirus vector carrying the cloned gene.

After infection of the patient, the RNA inside the retroviral vector is reverse transcribed to give a DNA copy. (Although the retroviral vector does not carry a copy of the gene for reverse transcriptase, a few molecules of reverse transcriptase enzyme are packaged in retrovirus particles.) Ideally, the cloned gene, enclosed between the two LTR sequences, is then integrated into host cell DNA.

Because the retroviral vectors are completely devoid of genes for retrovirus proteins, they do not cause an immune response or significant inflammation. Furthermore, their ability to integrate into host cell DNA means that the therapeutic gene will become a permanent part of the host cell genome. In principle, the retrovirus could integrate into a host cell gene or regulatory sequence and cause harm. In practice, because most DNA in animal cells is noncoding, the chances are low, and only occasional cells would be damaged.

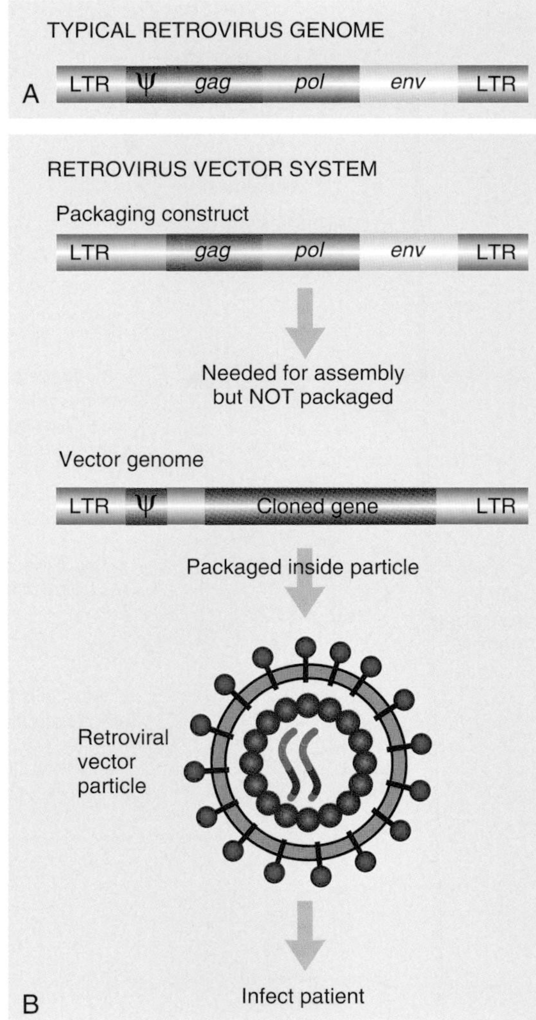

FIGURE 17.17 Retrovirus Genome and Vector System
(A) The retrovirus genome has a packaging signal (ψ) and the genes *gag, pol,* and *env* flanked by two direct repeats known as LTRs. (B) Retrovirus gene therapy uses two virus constructs. The therapeutic vector carries the cloned gene and packaging signal flanked by two LTRs. The packaging vector has the three genes necessary for virus assembly and packaging: *gag, pol,* and *env.* Because the packaging vector does not carry the packaging signal, it is never packaged and does not infect the patient. When both constructs are present, the therapeutic vector plus cloned gene is packaged into the capsid.

More serious problems are that retroviral vectors can carry only small amounts of DNA (about 8 kb) and cannot infect nondividing cells. However, the lentivirus family of retroviruses (to which HIV belongs) is unusual in being able to infect some nondividing cells. Naturally, using HIV itself is risky, but lentiviruses that infect other mammals, such as feline immunodeficiency virus (FIV), may be used to derive vectors.

Engineered retroviruses are the most frequently used viral vectors in gene therapy. Defective retrovirus vectors are grown in cells with an integrated helper virus to allow formation of virus particles.

RETROVIRUS GENE THERAPY FOR SCID

Severe combined immunodeficiency (SCID) occurs when both the B cells and T cells of the immune system are defective and results in an almost totally defective immune response. Children with SCID have to be shielded from all contact with other people and are kept inside special sterile plastic bubble chambers. Without immune protection, any disease, even a cold, could prove fatal. Several genetic defects are known that cause SCID. About 25% are due to mutations in the ***Ada* gene** that encodes the enzyme **adenosine deaminase**. This enzyme is needed for the metabolism of purine bases, and its absence prevents development of lymphocytes (white blood cells including both B cells and T cells).

The first successful instance of human gene therapy used a retroviral vector to provide a functional copy of the *Ada* gene to a child with SCID. The cells affected by SCID are the lymphocytes that circulate in the blood, where they carry out immune surveillance. Dividing cells in bone marrow generate these lymphocytes (Fig. 17.18). For gene therapy, bone marrow cells are removed from the patient and maintained in cell culture outside the body. Because bone marrow cells divide often to replenish the supply of blood cells, they are suitable for retroviral infection. The cultured bone marrow cells are infected with genetically engineered retrovirus carrying the *Ada* gene and are then returned to the body.

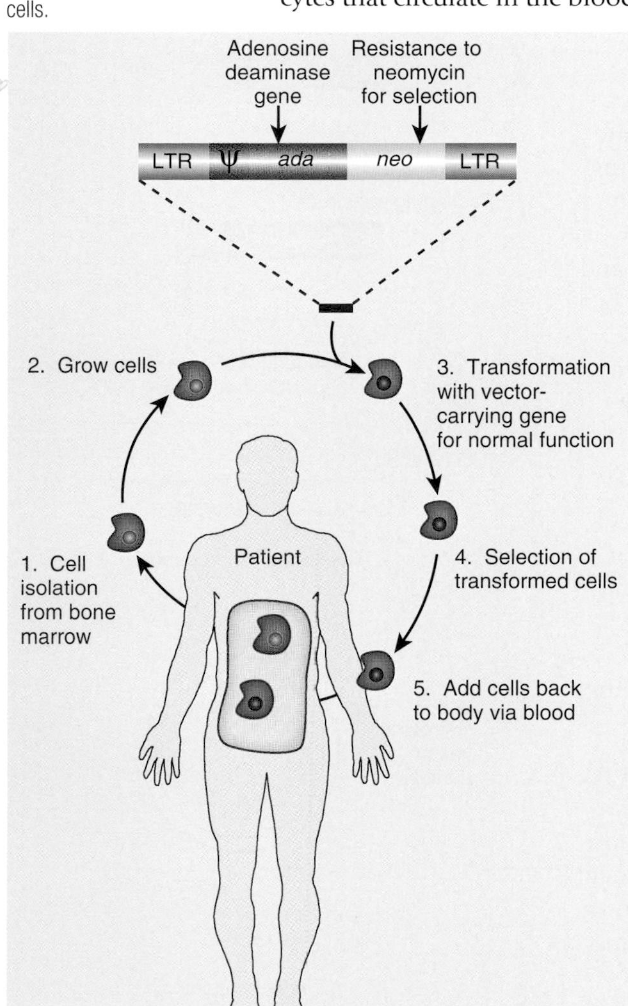

FIGURE 17.18
***Ex Vivo* Retroviral Gene Therapy for *Ada* Deficiency**
Gene therapy for SCID requires the removal of bone marrow cells from the patient. The cells are cultured, and the mutant *Ada* gene is replaced with a functional copy. The bone marrow culture is treated with neomycin to kill nontransformed cells. The bone marrow cells are then replaced in the patient and repopulate the blood supply with normal blood cells.

Since 1991, several children have been treated using this approach. However, because T cells live for only 6 to 12 months, the procedure must be repeated at intervals. This problem has been tackled by using blood **stem cells**, which divide to provide the precursors to all types of blood cells, including more stem cells. These cells are found in bone marrow but only in very small numbers. However, umbilical cord blood has a much higher proportion of stem cells. So in 1993, blood stem cells were obtained from the umbilical blood of several newborn babies who were known to carry homozygous defects in *Ada*. The *Ada* gene was introduced into the stem cells on a retroviral vector, which has resulted in a long-term supply of healthy T cells.

In the preceding cases, the patients have been receiving injections of purified ADA enzyme as well as gene therapy. It is therefore unclear how much of their improvement is due to the gene therapy, even though these patients now have functional T cells in their blood. However, a clear-cut result has been obtained more recently with another variant of SCID. This variant is due to defects in the receptor for several **interleukins**, including IL-7, a protein that promotes development of T cells from stem cells. Because B cells need helper

T cells in order to function, both B and T cells are inactive. In this case, the gene for the missing subunit of the IL-7 receptor was inserted into cultured stem cells using a retroviral vector.

Because purified protein is not used in treating this type of SCID, the patients had to depend on gene therapy alone. The patients did develop normal numbers of T cells and were successfully vaccinated against several infectious diseases, indicating a proper immune response. Consequently, they were able to leave their bubbles and enjoy normal lives. Unfortunately, 5 of 19 subjects developed leukemia, due to retrovirus insertion into other genes. Thus, the use of retrovirus for gene therapy needs further safeguards before it can be used routinely.

Newer vectors, with more effective safeguards, derived from the Lentivirus family of retroviruses are being tested. However, due to the tendency of retroviruses to insert into human chromosomes at multiple possible sites, these early problems have prompted the use of other virus vectors, especially adeno-associated virus.

> Some types of severe combined immunodeficiency have been successfully treated by gene therapy using a retrovirus vector.

ADENO-ASSOCIATED VIRUS

Because of the problems with using adenoviruses and retroviruses discussed in the preceding section, other DNA viruses have been considered as vectors. The most promising is **adeno-associated virus (AAV)**. Indeed, the majority of active clinical trials to cure genetic defects use AAV (as of 2013). AAV is a defective, or "satellite," virus that depends on adenovirus (or some herpes viruses) to supply some necessary functions. Consequently, it is usually found in cells that are infected with adenovirus. Unlike adenovirus, AAV seems to be entirely harmless. The benefits of using AAV are as follows:

(a) It does not stimulate inflammation in the host.
(b) It does not provoke antibody formation and can therefore be used for multiple treatments.
(c) It infects a wide range of animals, as long as an appropriate helper virus is also present. It can therefore be cultured in many types of animal cells, including those from mice or monkeys.
(d) It can enter nondividing cells of many different tissues, unlike adenovirus.
(e) The unusually small size of the virus particle allows it to penetrate many tissues of the body effectively.
(f) AAV integrates its DNA into a single site in the genome of animal cells (the *AAVS1* site on chromosome 19 in humans). This allows the therapeutic gene to be permanently integrated.
(g) Several different serotypes of AAV are available that show preference for infecting different types of tissue.

One drawback is that the AAV genome is small (4681 nucleotides of single-stranded DNA), and the virus can carry only a relatively short segment of DNA. (AAV is unusual in packaging both plus and minus strands into virus particles. Although each virus particle contains only one ssDNA molecule, a virus preparation contains a mixture of particles, half with plus and half with minus strands.) On entering a host cell, the DNA is converted to the double-stranded replicative form, or RF, which is used for both replication and transcription. In the absence of helper virus, AAV integrates into the host chromosome and becomes latent.

Clinical trials using AAV as vector have shown positive results for several conditions, in particular, Leber congenital amaurosis (LCA), which causes childhood blindness. Recessive defects in multiple genes affecting the structure and function of the retina may cause LCA. About 10% are due to defects in retinal pigment epithelium specific 65 kD protein (RPE65). Preliminary success with mice and dogs led to human trials. Within weeks, vision was improved, especially in younger patients. A contributing factor for this success is that the eye is easy to access when administering gene therapy vectors.

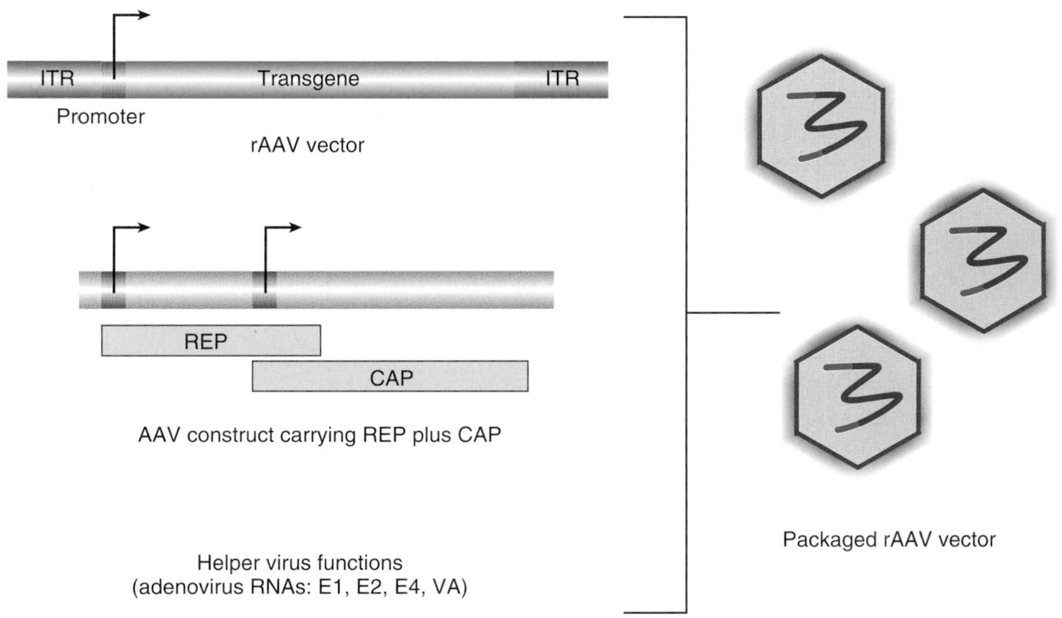

FIGURE 17.19 Generation of rAAV Vector

Two AAV-derived viruses are used, together with a cell line that provides helper functions originally contributed by adenovirus. The recombinant AAV (rAAV) vector carries the transgene plus a suitable promoter, enclosed by two ITRs (inverted terminal repeats) that are required for packaging of the construct. The second AAV construct supplies the REP and CAP proteins (expressed from overlapping genes). In reality, the REP and CAP genes possess multiple promoters and undergo alternative splicing (not shown here). Consequently, REP encodes four proteins involved in replication and packaging, and CAP encodes three structural proteins that form the capsid.

As for retrovirus vectors, two virus constructs are used for AAV: one carries the therapeutic gene, and the other supplies the virus replication and capsid proteins. In addition, a cell with helper virus functions from adenovirus is required (Fig. 17.19). The rAAV vector consists of the transgene (plus suitable promoter) flanked by two inverted terminal repeats (ITRs). These are required for a DNA segment to be recognized by the packaging system. The second AAV construct and the host cell (carrying adenovirus helper genes) provide all other needed components and functions.

AAV has provided most of the recent success stories in gene therapy. Clinical trials using AAV are ongoing for hemophilia B, some types of muscular dystrophy, lipoprotein lipase deficiency, Pompe disease (defect in glycogen breakdown), Sanfilippo syndrome (defect in heparin sulfate breakdown), Batten disease (defect in breakdown of lipoproteins), age-related macular degeneration, severe heart failure, Alzheimer's disease, Canavan disease (aminoacylase 2 deficiency), and Parkinson's disease. A current list of clinical trials is available at http://clinicaltrials.gov for the United States and at https://www.clinicaltrialsregister.eu for Europe plus Australia.

Adeno-associated virus (AAV) is being developed as a gene therapy vector. A major advantage of using it over adenovirus is that AAV does not provoke antibody formation and can be used for multiple treatments.

NONVIRAL DELIVERY IN GENE THERAPY

However sophisticated a viral delivery system may be, nonviral vectors are inherently safer. Nonetheless, they have been relatively neglected because viruses are more efficient. However, several unfortunate incidents have occurred with viral vectors, especially retroviruses, including the occasional appearance of leukemia-like disease. This has resulted in renewed interest in nonviral delivery systems.

About 70% of gene therapy trials have used viral vectors. Various alternative approaches have also been investigated, although few have been effective or widely used. They include

(a) Use of naked nucleic acid (DNA or less often RNA). Many animal cells can be transformed directly with purified DNA. The therapeutic gene may be inserted into a plasmid and the plasmid DNA used directly. Between 10% and 20% of gene therapy trials have used unprotected nucleic acid.

(b) Liposomes are spherical vesicles composed of phospholipid. They have been used in around 10% of gene therapy trials (see later discussion).

Assorted other methods, combined, account for 5% or less of trials:

(c) Particle bombardment. DNA is fired through the cell walls and membranes on metal particles. This method was developed to get DNA into plants and is therefore discussed in Chapter 15. However, it has also been used to make transgenic animals and is occasionally used for humans.

(d) Receptor-mediated uptake. DNA is attached to a protein that is recognized by a cell surface receptor. When the protein enters the cell, the DNA is taken in with it.

(e) Polymer-complexed DNA. Binding to a positively charged polymer, such as polyethyleneimine, protects the negatively charged DNA. Such complexes are often taken up by cells in culture and may be used for *ex vivo* gene therapy.

(f) Encapsulated cells. Whole cells engineered to express and secrete a needed protein may be encapsulated in a porous polymeric coat and injected locally. Foreign cells excreting nerve growth factor have been injected into the brains of aging rats. The rats showed some improvement in cognitive ability, suggesting that this approach may be of value in treating neurological conditions such as Alzheimer's disease.

A variety of approaches, other than viruses, can be used to get foreign DNA into target cells. These methods include using naked DNA or liposomes and, rarely, DNA bound to artificial polymers or proteins, particle bombardment, and encapsulated whole cells.

LIPOSOMES AND LIPOFECTION IN GENE THERAPY

About 10% of gene therapy trials have used **liposomes**. These hollow microscopic spheres of phospholipid can be filled with DNA or other molecules during assembly. The liposomes will merge with the membranes surrounding most animal cells, and the contents of the liposome end up inside the cell (Fig. 17.20), a process known as **lipofection**. Although lipofection works reasonably well, it is rather nonspecific because liposomes tend to merge with the membranes of any cell.

A formidable problem during anticancer gene therapy is how to get foreign DNA specifically into cancer cells (see the following section). Lipofection is a promising approach because "armed" liposomes can be injected directly into tumor tissue. In fact, liposomes are probably of more use in delivering proteins than DNA, something not feasible when using viruses as genetic engineering vectors. For example, toxic proteins such as **tumor necrosis factor (TNF)** can be packaged inside liposomes and

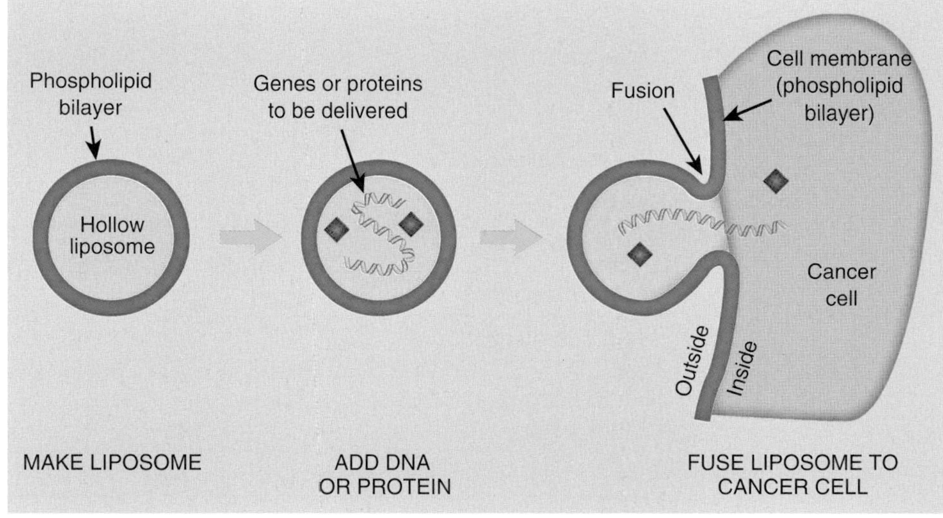

FIGURE 17.20 Delivery of Anticancer Agent by Lipofection

Liposomes are hollow spheres surrounded by a lipid bilayer. They can be filled with DNA or proteins such as TNF. Liposomes interact with cell membranes because both have hydrophobic lipid layers. Once in contact, the liposome merges with the cell membrane, delivering its payload into the cell.

injected into tumor tissue. The liposomes merge with the cancer cell membranes, and the lethal proteins are then released inside the cancer cells.

Liposomes are hollow microscopic spheres of phospholipid. They can be used to carry foreign DNA or proteins to a target tissue.

AGGRESSIVE GENE THERAPY FOR CANCER

The original idea behind gene therapy was curing hereditary defects by replacing defective genes. However, there is no inherent reason why gene therapy must only be "defensive" and suppress defects. We can go on the offensive and provide genes whose products may cure a disease even though the genes we use were not responsible for the problem in the first place. The best examples of such **aggressive gene therapy** are not in curing hereditary defects but in the treatment of cancer. Here, the objective is to kill cancer cells. In fact, as shown in Table 17.3, the majority of gene therapy trials are now directed against cancer.

Although most cancers are not inherited via the germline, cancer is nonetheless a genetic disease. In the case of hereditary disease, we may attempt to replace the defective component, thus preventing cell death. In contrast, when dealing with a cancer, we need to destroy the cancer cells, or at least inhibit their growth and division. Several strategies have been used and may be classified as follows:

(a) Gene replacement
(b) Direct attack
(c) Suicide
(d) Immune provocation

Gene replacement therapy for cancer is analogous to its use in correcting hereditary defects. The cancer is analyzed to identify the mutant gene(s) that are responsible. The wild-type version of the oncogene or tumor suppressor gene is then inserted into the cancer cells. For example, the wild-type version of the p53 gene has been delivered to p53-deficient cancer cells. The delivery method is usually via an adenovirus vector, but sometimes liposomes have been used.

In the direct plan of attack, a gene that helps kill cancer cells is used. For example, the ***TNF* gene** encodes tumor necrosis factor. This is produced by white blood cells known as **tumor-infiltrating lymphocytes (TILs)**. These cells normally infiltrate tumors where they release TNF, which is fairly effective at eradicating small cancers. To suppress a large cancer that is out of control, TNF production must be increased. First, the *TNF* gene is cloned.

Table 17.3 Target Conditions for Gene Therapy Trials

	Trials	
Disease Targeted	Number	%
Cancer	715	66.4
Single gene defects	95	8.8
Vascular diseases	92	8.6
Infectious diseases	72	6.7
Other diseases	33	3.1
Investigative	69	6.4
Total	1076	100

Then white blood cells are removed from the patient and cultured. Multiple copies of the *TNF* gene—or perhaps an improved *TNF* gene with enhanced activity—are introduced into the white cells. Then the white cells are injected back into the patient.

Although TNF is very effective in killing cancer cells, it is also toxic to other cells. Thus, high levels of TNF are dangerous to the patient. There are two sides to this problem. One is limiting TNF or other toxic agents to the cancer cells. The other is getting the toxic agent to the relatively inaccessible cells on the inside of a tumor. A variety of modifications are being tested to solve these problems—for example, putting the TNF gene under control of an inducible promoter and using adenovirus to transfer the gene into cancer cells. The chosen promoter is designed to be induced by agents already used in treating cancer cells, such as radiation or cisplatin.

The suicide strategy is actually a hybrid of anticancer drug therapy with gene transfer therapy. A harmless compound, or **prodrug**, is chosen that can be converted to a toxic anticancer drug by a specific enzyme. If the enzyme is present, the cell expressing it will commit suicide when the prodrug is available (Fig. 17.21). Consequently, an enzyme that is not present in normal human cells must be chosen for this approach. The gene encoding the suicide enzyme must be delivered to the target cancer cells, usually by a viral vector or in liposomes. If the enzyme is successfully expressed in the cancer tissue, then the toxic drug will be generated inside the cancer cells. Thus, the prodrug can be administered to the patient by normal means but is specifically lethal for the cancer cells.

In practice, two major suicide enzyme/prodrug combinations have been used. Gene therapy has been used to deliver the enzyme **thymidine kinase**, originally from herpes virus, to cancer cells. The nontoxic prodrug, the nucleoside analog **ganciclovir**, is converted to its monophosphate by thymidine kinase (hence, its clinical use in treating herpes virus infections). Because only the cancer cells have thymidine kinase, all the noncancerous cells are unaffected. Normal cellular enzymes then convert the monophosphate to ganciclovir triphosphate (GCV-TP). This acts as a **DNA chain terminator** (Fig. 17.22). DNA polymerase incorporates GCV-TP into growing strands of DNA. However, lack of a 3′-OH group prevents further elongation of the nucleic acid strand. DNA synthesis is thus inhibited and the cell is killed. A similar scheme involves the conversion of 5-fluorocytosine to 5-fluorouracil by **cytosine deaminase**, originally from bacteria. Again, cellular enzymes finish the job by making the phosphorylated nucleoside that actually inhibits DNA and RNA synthesis.

FIGURE 17.21
Suicide Gene Therapy
Suicide gene therapy begins by delivering a therapeutic gene into the cancer cells. The gene encodes an enzyme that will convert a nontoxic prodrug into a toxic compound. Because noncancerous cells do not have the suicide enzyme, they are not affected.

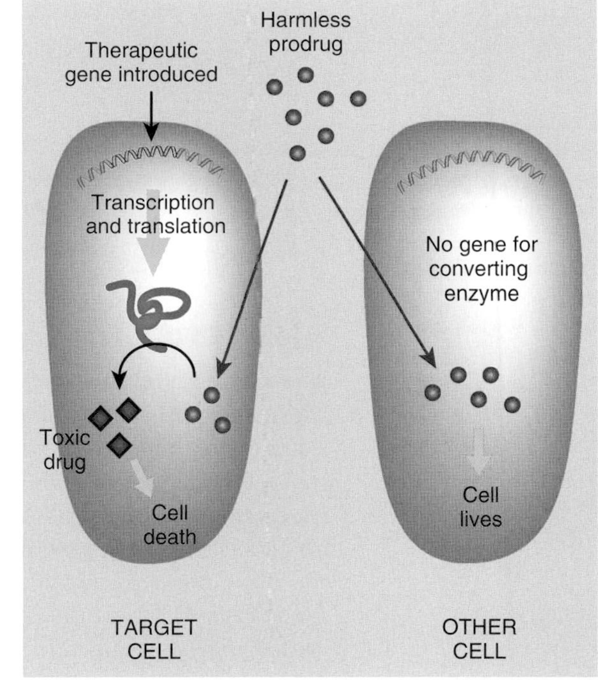

A more indirect approach relies on the body's natural defenses. Our immune systems are effective at killing cancers, provided they identify them while still small. To survive, a cancer has to evade the body's immune surveillance. In this approach, a gene that attracts the attention of the immune system to the tumor cells is inserted. For example, the HLA (= MHC) proteins are exposed on the surfaces of mammalian cells where they act in cell recognition. Different individuals have different combinations of **HLA genes**, which act as molecular identity tags so that cells of the body are recognized as "self." If HLA genes that are not originally present in a particular individual are inserted into cancer cells, the tumor appears alien, and the immune system will now mount an assault.

A related approach is to use **cytokines**. These short proteins attract immune cells and stimulate their division and development. The genes for several cytokines of the interleukin family (especially IL-2, IL-4, and IL-12) have been used to provoke immune attacks on cancer cells.

Aggressive gene therapy uses engineered genes to attack cancer rather than to replace defects. Sometimes therapeutic genes are used to increase the activity of the body's own anticancer systems. In other cases, the product of the therapeutic gene is used to specifically activate a toxic drug within cancer cells.

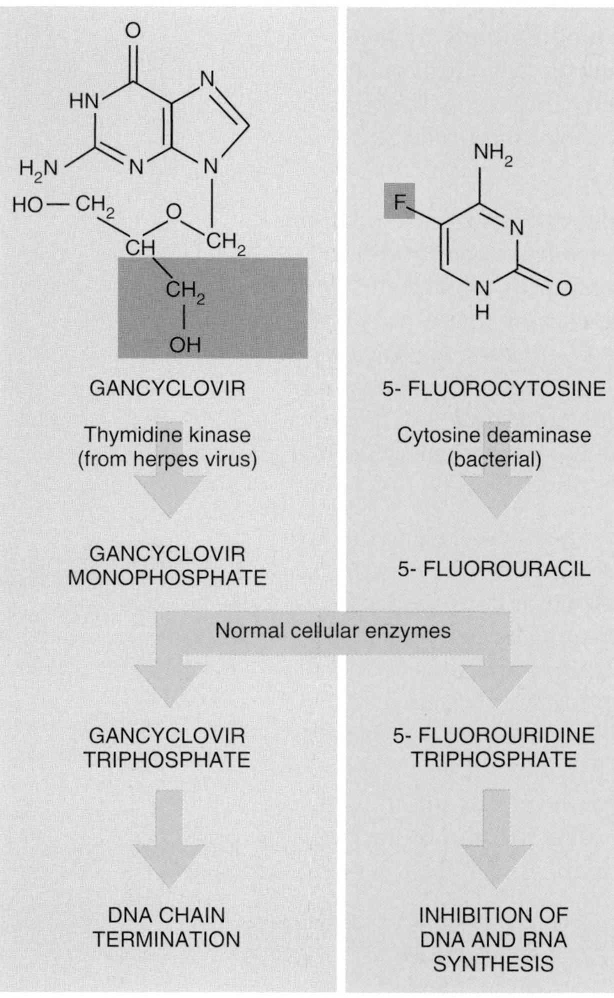

FIGURE 17.22 Ganciclovir and 5-Fluorocytosine
Ganciclovir is converted to a DNA chain terminator in cells containing thymidine kinase (from herpes virus). Similarly, 5-fluorocytosine is converted to a DNA and RNA synthesis inhibitor in cells that have cytosine deaminase (from bacteria). The groups highlighted in yellow prevent addition of further nucleotides. Genes for thymidine kinase or cytosine deaminase must be inserted into the cancer cells via gene therapy.

USING RNA IN THERAPY

The recent explosion of information about RNA has resulted in increased interest in using RNA clinically. The earliest attempts to use RNA therapeutically used antisense RNA to suppress gene expression, but this has largely fallen out of favor. Most present clinical uses of RNA focus on RNA interference or microRNA. Antisense RNA, artificial aptamers, and ribozymes are also being considered, although to a lesser extent.

Practical use of RNA faces two main technical problems. RNA oligonucleotides are taken up poorly by many cells and, in addition, are not very stable. Use of liposomes to deliver RNA is one way to overcome the uptake problem. Another approach is to provide a gene that encodes the therapeutic RNA and generate the RNA internally by transcription.

The issue of stability has led to the use of modified oligonucleotides. Segments made of DNA rather than RNA are just as effective, are easier to synthesize, and are more stable than RNA. Oligonucleotides (either RNA or DNA) with phosphorothioate linkages are more stable than those with phosphodiester linkages. Other modifications to stabilize RNA include blocking its 2′-OH with methyl or methoxy-ethyl groups or replacing its 2′-OH with fluorine. Locked nucleic acid (LNA) also protects the 2′-OH by linking it to ring carbon atom #4 (Fig. 17.23). Oligonucleotides made of peptide nucleic acid are extremely stable.

Therapeutic use of RNA oligonucleotides has increased recently. Uptake and stability are the two main practical problems. Various chemically modified oligonucleotides that are more stable than native RNA are often used.

ANTISENSE RNA AND OTHER OLIGONUCLEOTIDES

Antisense RNA binds to the corresponding messenger RNA and prevents its translation by the ribosome. Antisense RNA and related antisense nucleic acids are now being tested for possible therapeutic effects. In most cases, the objective is to inhibit the expression of a target gene. Two main alternatives exist when using antisense RNA. It is possible to use a full-length **anti-gene** that is transcribed to give a full-length antisense RNA. The anti-gene must be carried on a suitable vector and expressed in the target cells. Alternatively, much shorter artificial RNA oligonucleotides may be used. An antisense RNA of 15 to 20 nucleotides is often capable of binding specifically to part of the complementary mRNA and preventing translation. Such oligonucleotides may be designed to block the 5′-end of a transcript, thus preventing translation from the beginning. However, other sites within the mRNA are often just as effective.

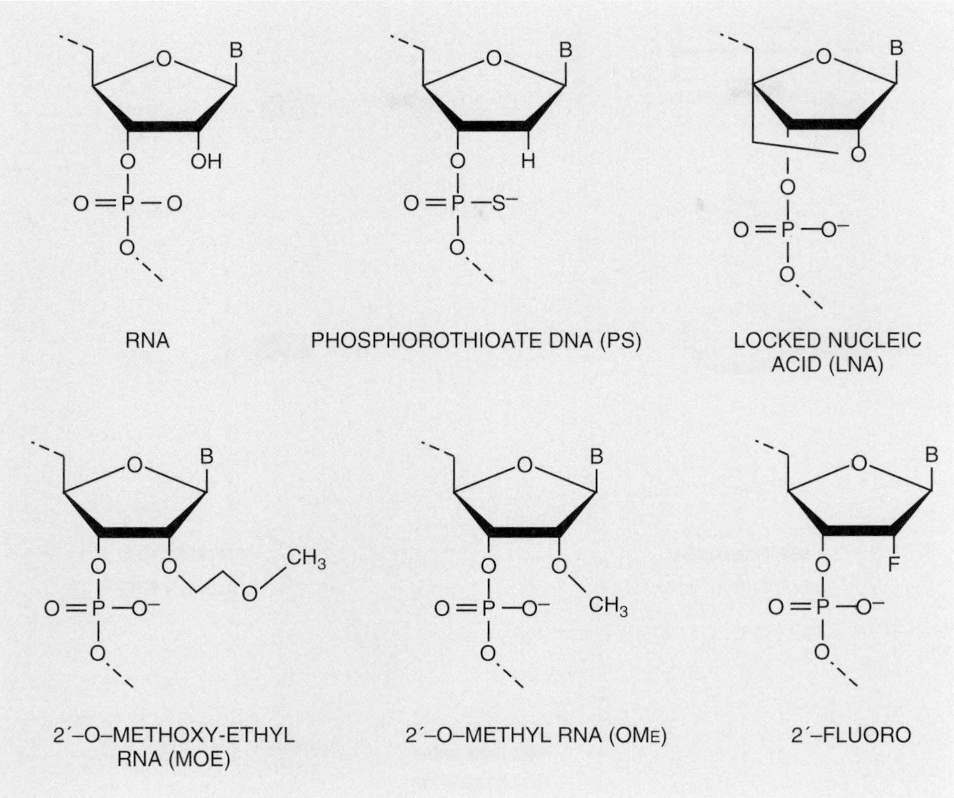

FIGURE 17.23
Modifications of Therapeutic Oligonucleotides
The unmodified RNA structure is shown at top left. The modifications include phosphorothioate DNA, locked nucleic acid (LNA), and three 2′ modifications of RNA. Modified from Burnett JC, Rossi JJ (2012). RNA-based therapeutics: current progress and future prospects. *Chem Biol* **19**, 60–71.

The use of a full-length anti-gene has been tested experimentally against cancer cells in culture. Malignant glioma is the most common form of human brain cancer. Here, the glial cells, which form the interstitial tissue between nerve cells, grow and divide in an uncontrolled manner. Often this behavior is due to overproduction of **insulin-like growth factor 1 (IGF1)**, which stimulates the growth and division of these cells. An *anti-IGF1* gene was constructed by reversing the orientation of the cDNA version of the ***IGF1* gene** (Fig. 17.24). The *anti-IGF1* gene was placed under control of the **metallothionein** promoter, which responds to trace levels of heavy metals. The construct was inserted into glioma cells in culture. Adding a low concentration of zinc sulfate turned on the *anti-IGF1* gene, and the cancer cells stopped dividing and no longer caused tumors if injected into rats.

Antisense RNA oligonucleotides are capable of preventing expression of specific mRNAs. Clinical use requires problems of stability and delivery to be overcome.

APTAMERS—BLOCKING PROTEINS WITH DNA OR RNA

Artificial oligonucleotides can be designed to fold up and fit into binding sites of any shape. Even if an enzyme does not normally bind nucleic acids, scientists could design an oligonucleotide that would block its active site. Such oligonucleotides do not correspond to natural nucleic acid sequences and are known as **aptamers**.

Aptamers may be made by using a combination of molecular selection and PCR amplification. Random oligonucleotides (of 50 or 60 bases) are artificially synthesized by solid-state synthesis, using a mixture of all four bases at each step. The pool of random sequences is then ligated at each end to primer binding sequences (approximately 20 bases long). The mixture is poured through a column to which the target protein is

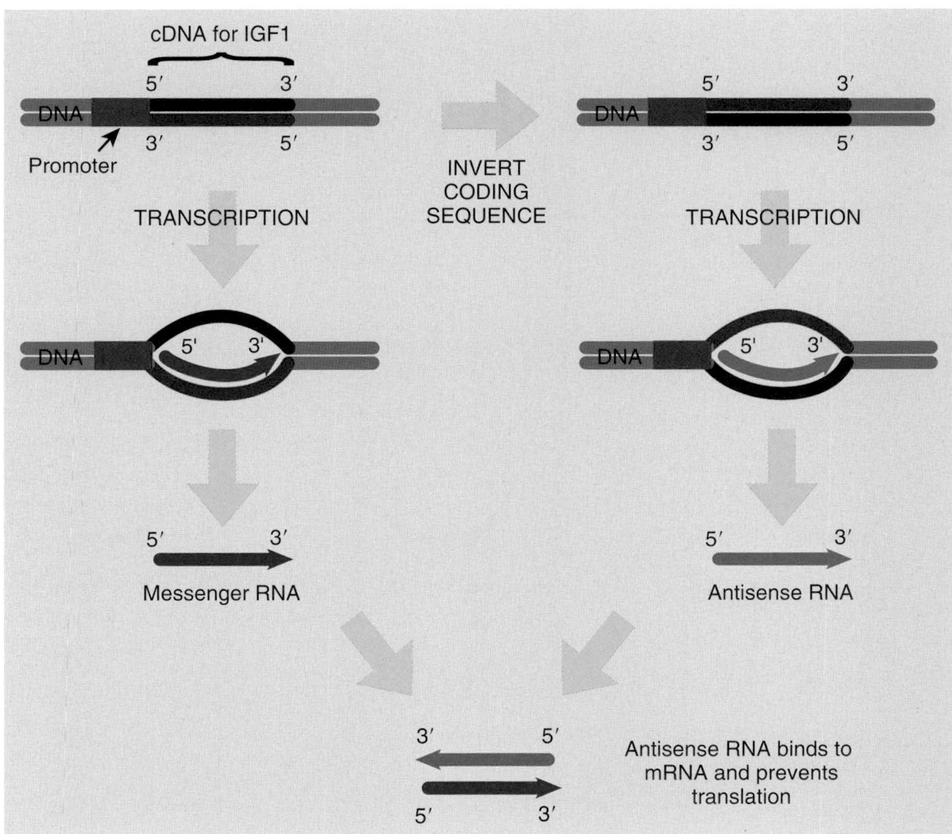

FIGURE 17.24
Full-Length Antisense Therapy for Cancer
The cDNA gene for IGF1 was inverted so that the RNA transcribed from it was complementary to the mRNA from the wild-type IGF1 gene. When this antisense construct was transformed into glial cells, the antisense RNA bound all the wild-type IGF1 mRNA, preventing production of IGF1 protein. The antisense IGF1 gene was controlled by a metallothionein promoter that turns on only in the presence of heavy metals.

attached (Fig. 17.25). Some oligonucleotides will bind to the protein and are therefore retained by the column. These are dissociated and then amplified by PCR. The bound oligonucleotides will be a mixture, some of which bind weakly and some strongly. Therefore, the binding and PCR steps are repeated two or three times to select oligonucleotides with high binding affinity. If generation of RNA versions of aptamers is desired, a T7-RNA polymerase promoter is included in one of the primer binding sequences.

This procedure has been demonstrated experimentally for the enzyme **thrombin**, a protease involved in the blood-clotting cascade. The antithrombin DNA aptamer is short-lived because it is degraded rapidly *in vivo*. It might perhaps be useful as an anticlotting agent, such as during bypass surgery. Several DNA and RNA aptamers against thrombin and other blood clotting factors are in clinical trials.

Recently, pegaptanib (Macugen; from Eyetech Pharmaceuticals, Pfizer, New York) has been approved for treatment of blindness due to age-related macular degeneration (Fig. 17.26). Pegaptanib is an RNA aptamer that inhibits the action of an isoform of vascular endothelial growth factor (VEGF) that is responsible for abnormal vascularization of the eyes. Binding of pegaptanib to VEGF prevents VEGF from binding to its receptor.

Short artificial oligonucleotides known as aptamers can be designed to block protein activity.

RIBOZYMES IN GENE THERAPY

RNA molecules with catalytic ability are known as **ribozymes**. Some ribozymes act by cleaving other RNA molecules. Such ribozymes have catalytic domains that carry out the reaction fused to domains that recognize the substrate RNA by base pairing (Fig. 17.27).

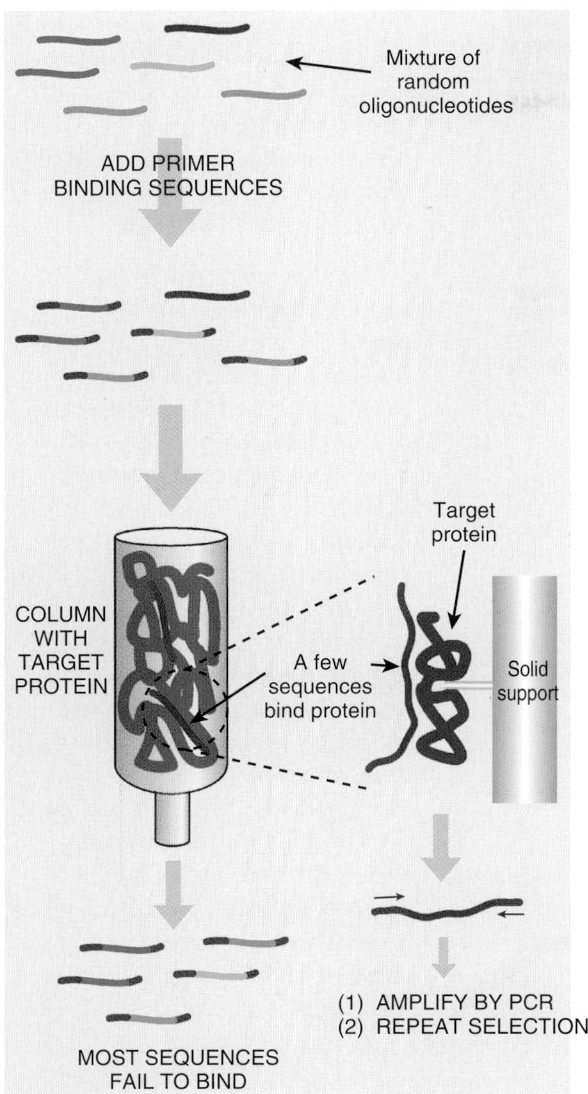

FIGURE 17.25
Selection of Aptamer against Enzyme
To find an oligonucleotide that binds to a specific enzyme, a pool of oligonucleotides with random sequences is synthesized. Short primer binding sequences are added to each end for later PCR amplification. The pool of random oligonucleotides is passed over a column with the target enzyme. Those oligonucleotides that recognize the enzyme stick to the column while everything else passes through. The sequences that bound to the column can be isolated and amplified by PCR.

FIGURE 17.26
Pegaptanib Aptamer against Enzyme
The RNA aptamer pegaptanib binds the VEGF growth factor and prevents VEGF from binding to its receptor. Pegaptanib is linked to a 40 kDa polyethylene glycol polymer at the 5′ end and has a 3′-dT cap. The ribose of the RNA backbone is modified at the color-coded positions by 2′-methoxy (green) or 2′-fluoro (orange) groups. Modified from Esposito CL et al. (2011). New insight into clinical development of nucleic acid aptamers. *Discov Med* **11**, 487–496.

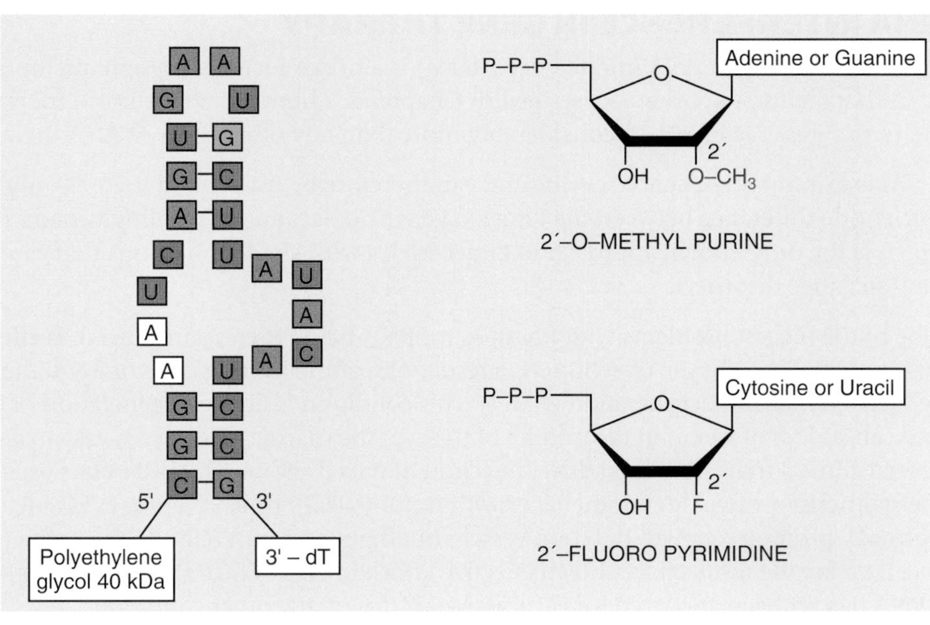

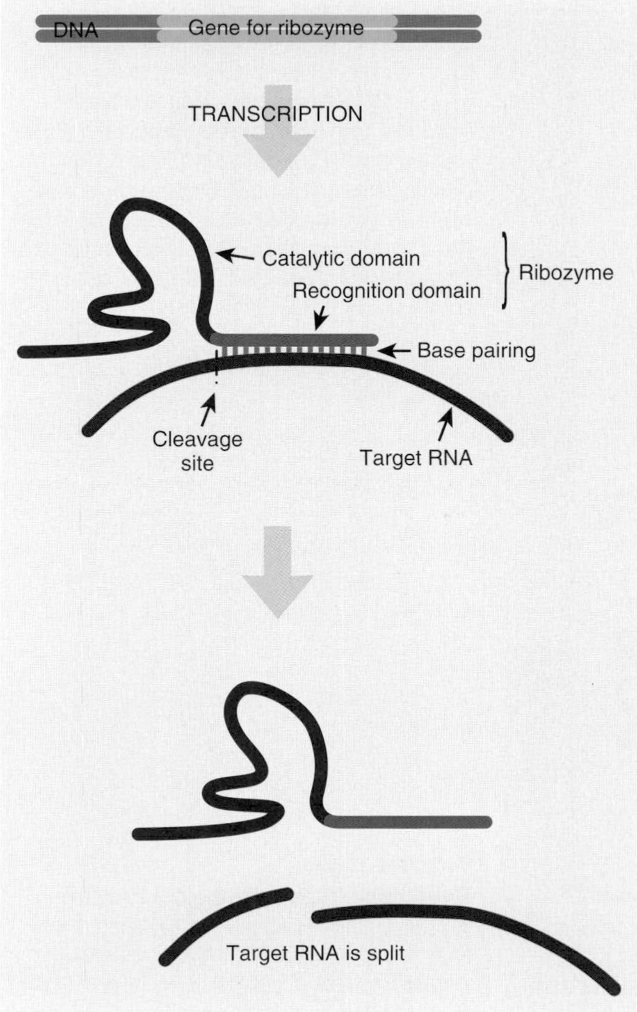

FIGURE 17.27
Targeted Ribozymes
Gene therapy can also be used to transfer a gene for a ribozyme into a defective cell. The ribozyme has two domains: the catalytic domain acts as an enzyme, and the recognition domain binds to a specific target RNA. Once bound, the enzyme domain cleaves the target RNA.

If the sequence of the substrate recognition domain is changed, the ribozyme will recognize a different target RNA molecule. Thus, ribozymes may be engineered to recognize and destroy any target messenger RNA molecule.

So far, the use of ribozymes is still experimental. Usually, a vector carrying a gene encoding the ribozyme is used for delivery to the target cell. Transcription of this gene generates the ribozyme RNA, which then binds and cleaves the target mRNA. All naturally occurring nucleic acid enzymes are made of RNA. However, there is no chemical reason why DNA cannot act catalytically. Indeed, **deoxyribozymes** may be synthesized artificially, using sequences equivalent to those of RNA ribozymes. The DNAzymes are less easily degraded and may be manufactured in multiple copies through use of a PCR-based technique. Most ribozymes presently under investigation are targeted at cancer cells or virus infections.

Ribozymes that act by cleaving RNA target molecules have been proposed as therapeutic agents.

RNA INTERFERENCE IN GENE THERAPY

As noted previously, RNA interference (RNAi) is a major focus of present attempts to create RNA therapeutics. RNAi was described in Chapter 5. About a dozen clinical trials using RNAi are in progress (as of 2012), considerably more than any other form of RNA therapy.

RNAi is extremely sequence specific and can therefore be used when there is only a single nucleotide difference between the targeted genetic defect and its healthy variant. For example, it is the only known approach to target alleles with single mutations that cause dominant negative disorders.

One of the biggest problems with RNAi, as for RNA-based therapy in general, is effective delivery of the RNA to the site of action. As a result, one of the leading cases of RNAi therapy is for age-related macular degeneration (AMD). This condition is due to degeneration of the retina and causes loss of vision in the middle of the eye (the macula). The eye is easy to access, and several clinical trials have used direct injection of naked siRNA. The siRNA has usually targeted the synthesis of vascular endothelial growth factor (VEGF) or its receptor (VEGF-R1). VEGF normally promotes growth of blood vessels, but high levels can damage the retina (or other tissues). So far, the results have shown effective knockdown of VEGF expression by siRNA, but no siRNA has yet been approved for clinical use for this or any other condition.

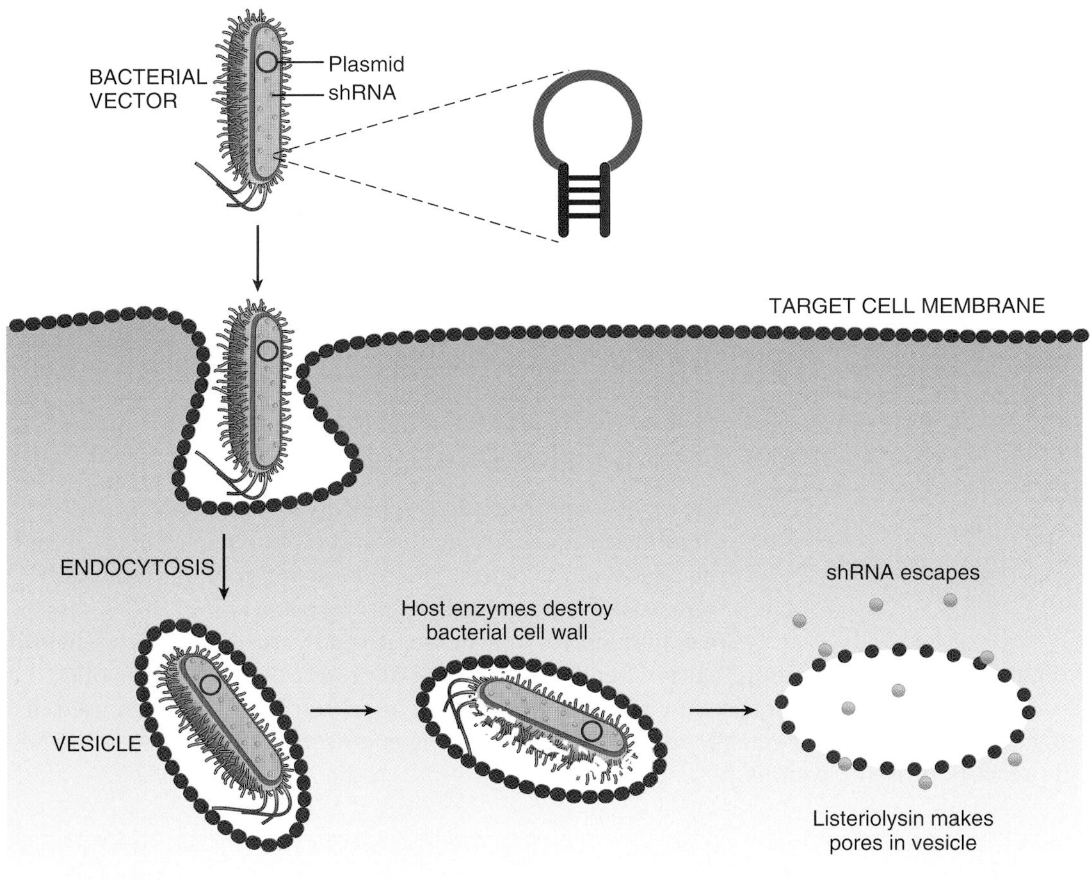

FIGURE 17.28 Trans-Kingdom RNAi Therapy
The bacterial vector contains a plasmid that drives synthesis of shRNA. The bacteria are taken up by endocytosis and then destroyed by host digestive enzymes. Listeriolysin released from the bacteria then makes pores in the host vesicle. This releases the shRNA.

Other approaches to delivering siRNA include the use of liposomes or even of whole living bacteria, by a process named transkingdom RNAi. This approach is being developed by Marina Biotech. Bacterial vectors that synthesize therapeutic RNA are taken by mouth in a capsule that releases them in the intestines, where intestinal cells take them up (Fig. 17.28). Nonpathogenic bacteria, such as harmless strains of *Escherichia coli*, are modified for use as follows:

(a) Providing a plasmid vector carrying the gene for the therapeutic RNA. Short hairpin RNA (shRNA) is a convenient form of RNA to induce RNA interference. The shRNA folds into a hairpin whose double-stranded region triggers the RNAi response.

(b) Inserting the gene for T7 RNA polymerase, which is needed to generate large amounts of the therapeutic RNA.

(c) Inserting the *inv* gene from *Yersinia tuberculosis*. The INV gene product, the invasin protein, allows entry into animal cells by binding the beta-integrin receptors on the cell surface. The bacteria end up inside vesicles in the cytoplasm of the target cells. Host cell enzymes destroy the bacteria, and the therapeutic RNA is released but still trapped inside the vesicles.

(d) Inserting the *hlyA* gene from *Listeria monocytogenes*. The HlyA gene product, listeriolysin O, makes pores in the vesicle membrane. This allows the therapeutic RNA to escape from the vesicles inside the target cell.

Both *Yersinia tuberculosis* and *Listeria monocytogenes* are moderate pathogens that possess the ability to enter animal cells. They are both well characterized and a convenient source of cell-entry genes that are used for engineering other bacteria.

RNA interference has been proposed as a mechanism for gene therapy. So far, the major problem is effective delivery of the therapeutic RNA.

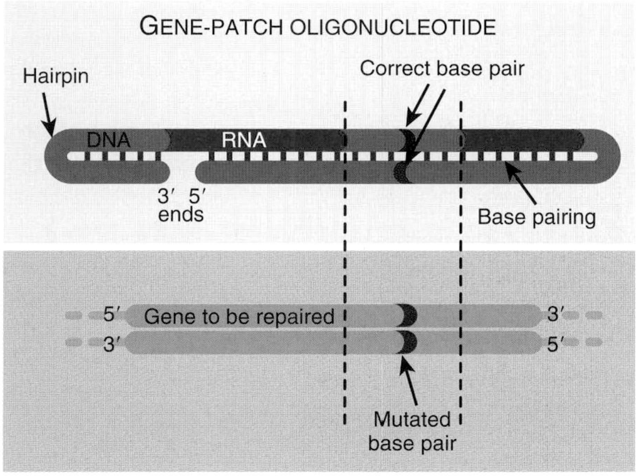

FIGURE 17.29 Patching Defective Gene by Oligonucleotide Crossover

A special gene-patch oligonucleotide can be synthesized to provide a corrected copy of a short specific region of a defective gene. The oligonucleotide is designed to promote a crossover in the defective region of the gene.

GENE EDITING WITH NUCLEASES

In gene therapy, we normally think of replacing the whole defective gene with a complete functional copy. However, some genetic defects consist of just a single base change, or perhaps a cluster of closely linked base alterations or a very short deletion. In such cases, the defective gene could be repaired (or "patched") rather than replaced. This approach is known as gene (or genome) editing.

The earliest approach used relatively short double-stranded oligonucleotides carrying the wild-type sequence that covered the region of the defect. The cells' own repair mechanisms were used to insert this by crossing over (Fig. 17.29). Crossover frequencies may be improved by using RNA–DNA hybrid oligonucleotides. In addition, hairpin bends may be used to protect the ends of the oligonucleotide and prevent degradation by exonucleases. This approach has never been used in practice and has largely been abandoned in favor of using engineered nucleases to edit DNA at precisely chosen locations.

Small, localized genetic defects could be repaired, or "patched," by using a variety of techniques. Early attempts used relatively short oligonucleotides and crossing over. More recent attempts use engineered nucleases.

GENOME EDITING WITH ENGINEERED NUCLEASES

Localized lesions in DNA can be mended by using engineered nucleases to precisely edit the DNA. Several kinds of nucleases have been considered for this technique. Three of the most versatile are zinc finger nucleases, transcription activator-like effector (TALE) nucleases, and homing endonucleases. The nucleases must be engineered to change their DNA recognition specificity to that of the chosen, target DNA. Indeed, the relative ease with which the DNA recognition site can be altered is the underlying reason these three nucleases were chosen for genome editing. Nucleases from the CRISPR defense system of bacteria have also been engineered. These, however, differ in using guide RNA for targeting and are therefore discussed separately.

As discussed in Chapter 11, zinc fingers are DNA-binding motifs that are found in many regulatory proteins. Each zinc finger (ZNF) motif consists of a Zn ion surrounded by 25 to 30 amino acids and specifically binds three bases in the DNA. Multiple zinc finger motifs can be joined end to end to generate larger domains with longer recognition sites. Typically, from 3 to 6 ZNF motifs are joined, hence recognizing from 9 to 18 bases. Such a recognition domain may be attached to a nuclease domain and will guide it to the chosen target sequence.

The best-known nuclease domain is from the *Fok*I restriction enzyme. This enzyme is conveniently divided into two protein domains, the recognition and nuclease domains, which can be easily cut apart. The *Fok*I nuclease domain must dimerize to cleave DNA. This means that two engineered nucleases must bind next to each other and in inverse orientations, at the recognition sequence, as shown in Figure 17.30. This greatly improves target specificity.

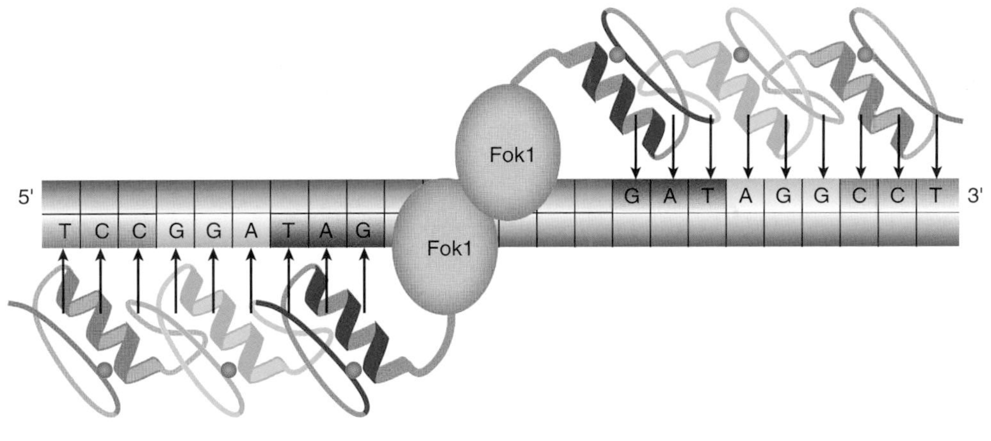

FIGURE 17.30
DNA Binding of Engineered *Fok*I Nuclease
A pair of engineered ZNF nucleases is shown binding to an inverted repeat sequence in the DNA. Each nuclease monomer has three zinc finger domains, which provide sequence recognition. In reality, the two *Fok*I catalytic domains are associated in three dimensions to form a dimer.

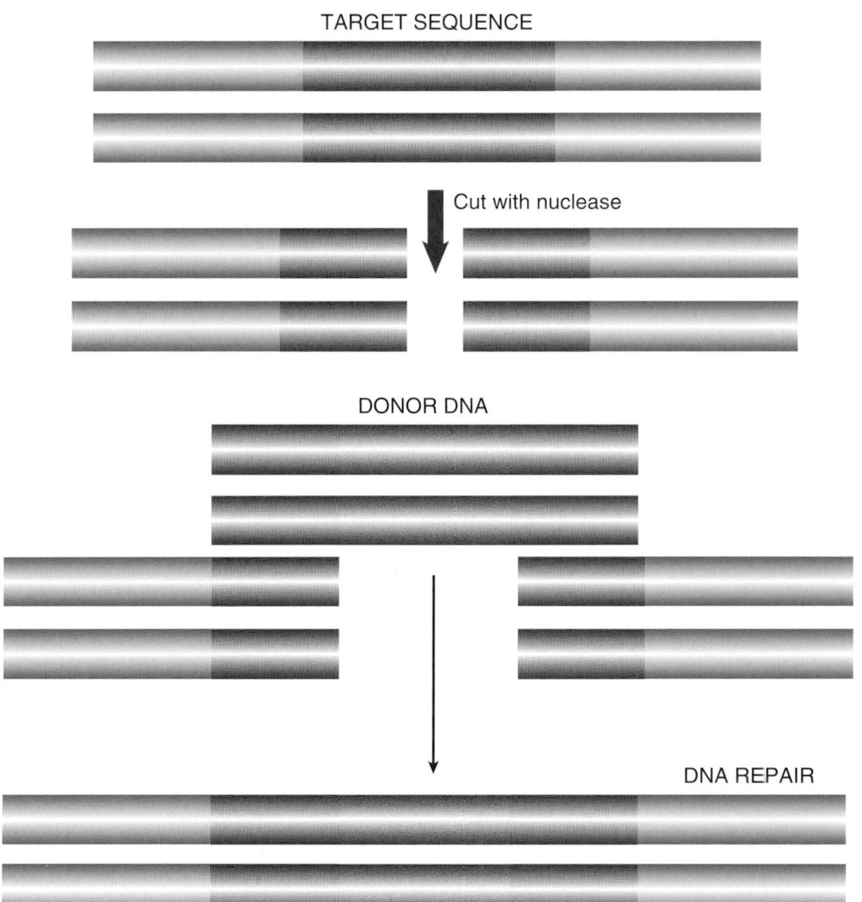

FIGURE 17.31
Genome Editing: Repair and Insertion
After the target site is cut with an engineered nuclease, genome editing relies on the host cell repair systems to insert the therapeutic correction. The corrected DNA sequence (green) is provided on a molecule of donor DNA that shares homology with the target site at each of its ends (red).

For genome editing, a double-stranded break is made in the chosen target sequence by the engineered nuclease. Modifications to the target gene are then inserted during DNA repair by host cell repair systems. Eukaryotic cells possess multiple systems that efficiently repair double-stranded breaks. The desired sequence correction may be introduced at the break site by providing short segments of DNA that are homologous at their ends to the target region, as shown in Figure 17.31. It is also possible to inactivate a target gene by providing donor DNA that contains a small deletion.

Correction of inherited defects has been successfully performed in cultured cells of both animal models and humans. So far, the only clinical trials using engineered ZNF nucleases for genome editing are destructive rather than corrective. They involve knocking out the CCR5 receptor protein that takes part in HIV infection, thus making human cells resistant to HIV.

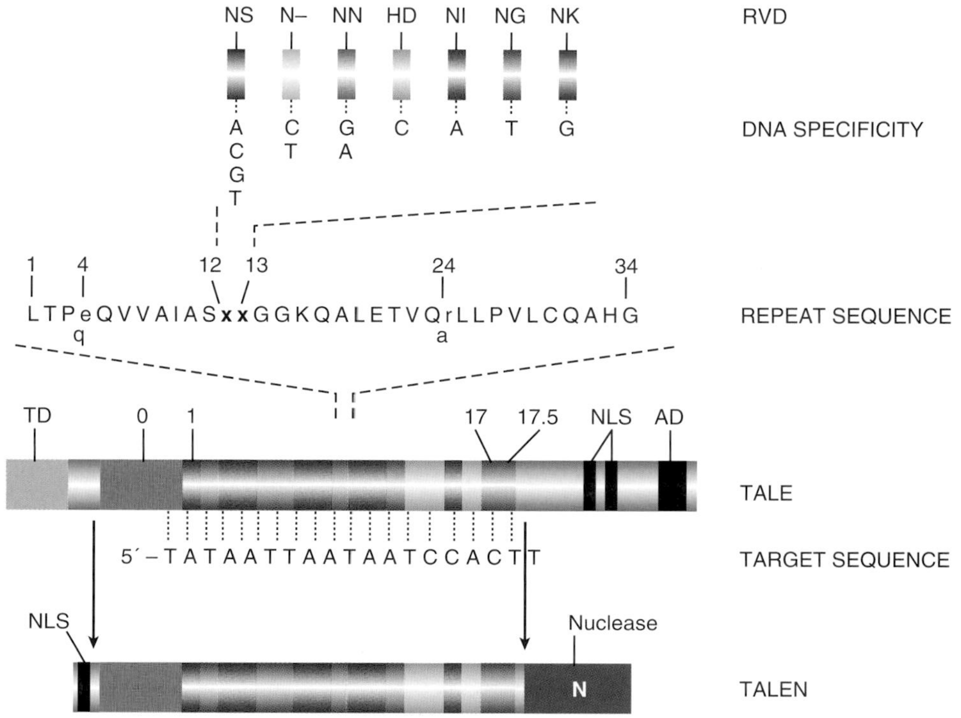

FIGURE 17.32 Structure of TALE Nuclease
A typical transcription activator-like effector (TALE) has an N-terminal translocation domain (TD), central repeat units (0–17.5) that bind DNA, and a C-terminal region with nuclear localization signals (NLS) and a transcriptional activation domain (AD). A single repeat unit is shown on top. The hypervariable residues in positions 12 and 13 (repeat-variable di-residue, RVD) dictate DNA specificity. (Amino acids in positions 4 and 24 show minor variability.) The central TALE repeat units can be fused to a nuclease domain to generate a sequence-specific TALE-nuclease. Modified from Mussolino C Cathomen T (2012). TALE nucleases: tailored genome engineering made easy. *Curr Opin Biotechnol* **23**, 644–650.

The engineered TALE nucleases comprise another class of nucleases useful for genome editing. Like the engineered ZNF nucleases, these nucleases also consist of a *Fok*I nuclease domain linked to a DNA recognition domain. However, in this case, the recognition domain comes from a TALE protein. Invading bacteria of the genus *Xanthomonas* inject these proteins into plant cells. Once inside the plant, TALE proteins enter the nucleus, where they bind DNA and activate host genes that are needed for growth and division of the bacteria.

TALE proteins have a central DNA-binding domain that may be cut out and fused to a nuclease domain. Moreover, the TALE DNA-binding domain has a simpler DNA recognition code than zinc fingers (Fig. 17.32) and hence can be engineered very easily to bind to novel DNA sequences. Work on TALE proteins is more recent than with Zn fingers, and engineered TALE nucleases are still in the experimental phase.

Homing endonucleases are responsible for the insertion of an unusual class of mobile elements into host genomes. In homing, a mobile element present in only one of a pair of alleles is copied and inserted into the other. The process involves specific cleavage within the target allele by the homing endonuclease, which is encoded by the mobile element. They are frequently located within self-splicing introns or inteins. Since insertion is specific for a single gene in a eukaryotic genome, the homing endonucleases recognize extremely long recognition sites of 12 to 40 bp.

Homing endonucleases are thus more specific in their sequence requirements than either ZNF or TALE nucleases. However, they are difficult to modify, as the nuclease and recognition functions do not form separate domains. At present, they seem an outside possibility for genome editing. Genome editing has even been suggested for whole chromosome defects, such as Down syndrome (see Box 17.2).

Box 17.2 Correcting Chromosomal Trisomy by Genome Editing

As mentioned early in this chapter, Down syndrome is due to having three copies of chromosome 21 (trisomy 21). The recent emergence of genome editing by engineered nucleases has suggested a possible approach to gene therapy for Down syndrome. The objective is to eliminate or silence one of the three copies of chromosome 21. So far, these objectives have been achieved using cultured stem cells. Although this approach is clearly impractical for the patient as a whole, it is possible that stem cells that give rise to rapidly dividing tissues (for example, bone marrow) could be replaced with cured versions. This might alleviate some of the symptoms of Down syndrome.

One alternative is to eliminate the third copy of chromosome 21. This was achieved by inserting a DNA cassette into one of the three copies of chromosome 21 using adeno-associated virus (AAV). The TkNeo cassette carries two transgenes: one conferring resistance to neomycin (Neo) and the other conferring sensitivity to ganciclovir (Tk). Sensitivity to ganciclovir is due to expression of the enzyme, thymidine kinase (as explained in Fig. 17.22). The resistance gene was used to select for insertion of the DNA cassette. Then cells were treated with ganciclovir, which selects against the presence of the cassette. Although some cells just lost the DNA cassette, others lost the whole third copy of chromosome 21 (Fig. B, part B).

The other alternative, selectively silencing one copy of chromosome 21, is more interesting. This approach relies on the fact that human females have two copies of the X chromosome, whereas males have only one. To avoid gene dosage problems, one copy of the two X chromosomes in females is normally silenced through a mechanism involving *Xist* RNA, a very long noncoding RNA of approximately 17 Kb. The *Xist* RNA is transcribed from the *Xist* gene and then spread over the chromosome where it is being expressed (Fig. B, part A). The *Xist* gene on the other copy of the X chromosome is inactivated by methylation of its regulatory region.

The *Xist* gene was transferred to chromosome 21 by genome editing; it is one of the longest segments of DNA inserted into a new location using this technique to date. The *Xist* gene was inserted into a 36 base-pair target sequence within an intron of the DYRK1A gene on chromosome 21 using an engineered *Fok*I Zn finger nuclease. When the *Xist* gene was inserted into one copy of chromosome 21 in cultured stem cells possessing three copies, it behaved the same as when located on the X chromosome. That is, *Xist* RNA was transcribed and spread over this particular copy of chromosome 21, causing its inactivation (Fig. B, part B). The other two copies of chromosome 21 were unaffected. The inactivated copy of chromosome 21 condensed into a so-called Barr body (as seen for inactive X chromosomes) massively enriched in heterochromatin.

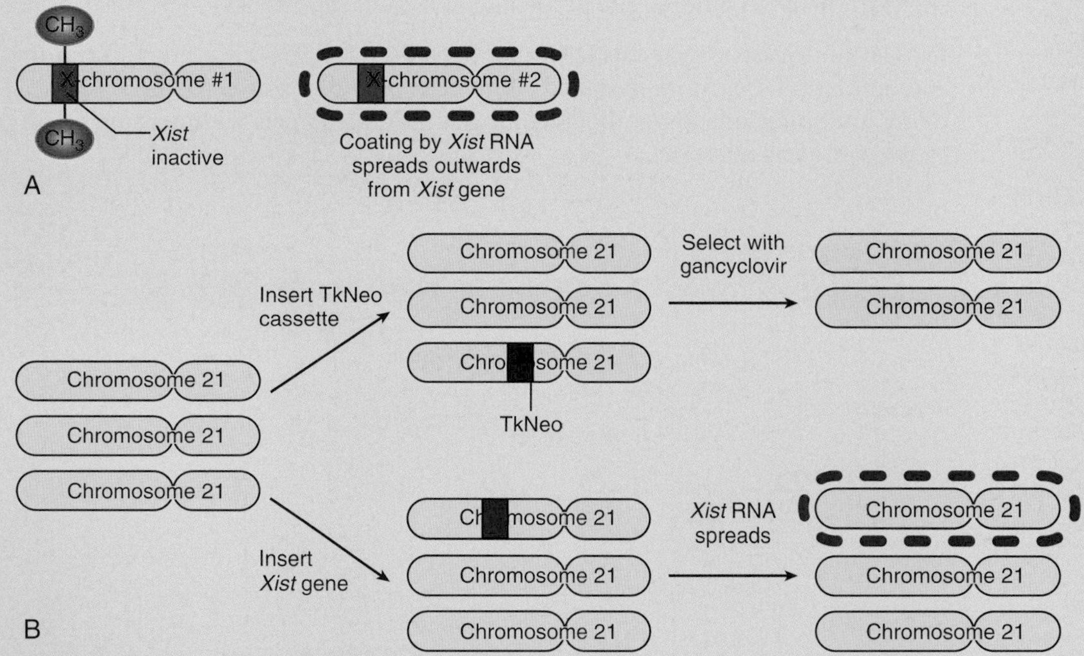

FIGURE B Suppressing Trisomy by Genome Engineering
(A) Mechanism of action of *Xist* RNA. (B) Suppression of trisomy 21 by insertion of a TkNeo cassette or by a copy of the *Xist* gene.

Various engineered nucleases have been tested for possible use in genome editing. Those based on zinc fingers have reached the clinic. TALE nucleases and homing endonucleases are still experimental.

GENOME EDITING WITH CRISPR NUCLEASES

The CRISPR system of bacteria defends against hostile intruders by degrading the nucleic acids of intruding viruses or other foreign genetic elements such as plasmids or transposons. Both foreign RNA and DNA are targeted. CRISPR stands for clustered regularly interspaced short palindromic repeats, which refers to the way foreign sequences are stored on the bacterial chromosome. For an overview, see Figure 5.3.

CRISPR is sometimes regarded as functionally equivalent to the RNA interference system of higher organisms. However, the mechanism and components of the CRISPR system are totally different to those of the RNAi system. RNAi is based on structure, whereas CRISPR is based on memory. RNAi targets double-stranded RNA rather than particular sequences. In contrast, the CRISPR system stores an array of short sequence fragments characteristic of foreign genetic elements (viruses, plasmids, and transposons). When nucleic acids appear whose sequences contain matches to those stored, they are destroyed. Furthermore, unlike RNAi, CRISPR destroys DNA, not just RNA.

Foreign sequences are stored by the CRISPR system on the bacterial chromosome as an array of short DNA segments separated by identical repeat sequences. They are transcribed to give a long RNA molecule that is cut into short segments of RNA (crRNA). The CRISPR associated (Cas) nucleases use these short crRNA molecules as guides. Thus, to use CRISPR for genome editing of mammalian cells, scientists need to construct specific guide RNAs rather than to engineer the DNA binding site of the nuclease itself (Fig. 17.33).

The Cas9 nuclease, which cuts DNA, has been used for genome editing in cultured mouse and human cells. Custom-designed chimeric guide RNA is used for targeting (see Fig. 17.33). Using multiple guide RNAs allows cutting at multiple targets within a single host cell genome by the same Cas nuclease.

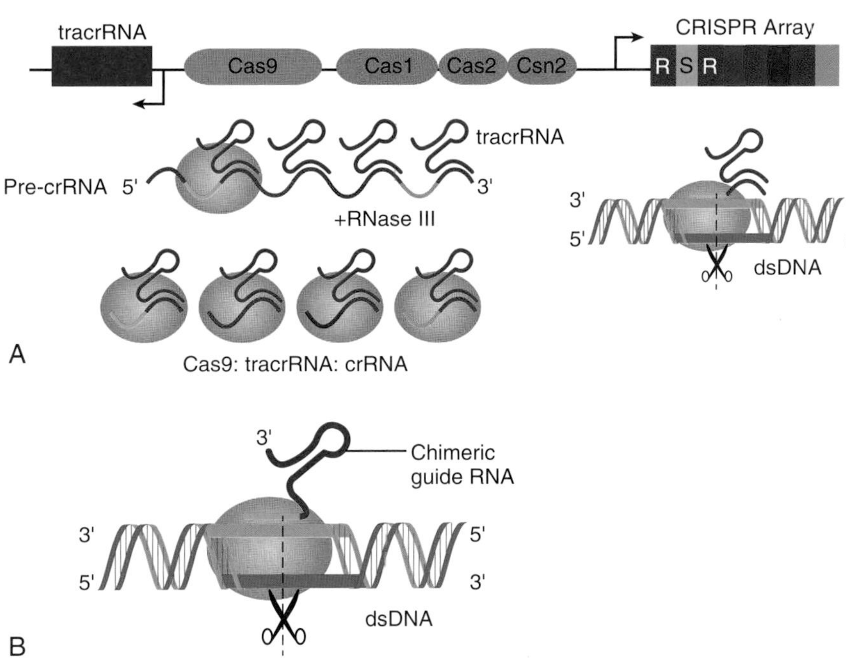

FIGURE 17.33
Using the CRISPR System in Genome Editing
(A) The type II CRISPR system includes the tracrRNA, located upstream from the DNA-cutting nuclease Cas9. The tracrRNA (blue) binds to the crRNA via the repeated sequences (purple) between the individual recognition sequences. The Cas9 nuclease binds the tracrRNA:crRNA complex and locates the target DNA, which is then cleaved. In the natural state, the target DNA would belong to an invading virus or plasmid. (B) Chimeric RNA consisting of tracrRNA joined at its 5′ end to a target-specific recognition sequence is used for genome editing. In this case, the target DNA is part of the animal genome being edited. Modified from Fineran PC, Dy RL (2012). Gene regulation by engineered CRISPR-Cas systems. *Curr Opin Microbiol* **18**, 83–89.

Summary

Various genetic defects affect the human population. Some are due to mutations in single genes and are reasonably well understood. Others involve multiple genes and are very complex. Many mutations have minor effects, whereas others are life threatening. At present, our ability to analyze hereditary defects has far outrun our ability to treat them. Genetic engineering can be used to combat hereditary defects. This approach is mostly still under development, but a few successes have been achieved. Modified viruses, including adenovirus and retroviruses, are the favorite vectors for inserting genes into human patients, although other vectors such as liposomes are being tested. At present, most successes have been achieved using adeno-associated virus (AAV). Aggressive gene therapy versus cancer uses genes and proteins for attack rather than replacement of genetic defects and has shown considerable promise. New approaches include the use of RNA interference and other RNA-based technologies. Most recently, genome editing has emerged as a significant possibility. Engineered nucleases are used to correct small segments of the genome, rather than replacing whole genes.

End-of-Chapter Questions

1. Why might a defective gene not have an affect on an organism's genotype?
    - **a.** because the organism is haploid and the other copy compensates for the defective copy
    - **b.** because the mutation is often lethal
    - **c.** because mutations can never be passed to progeny
    - **d.** because the organism is diploid and the other copy compensates for the defective copy
    - **e.** none of the above
2. Which of the following is the first molecular disease to be identified?
    - **a.** sickle cell anemia
    - **b.** phenylketonuria
    - **c.** hemophilia
    - **d.** cystic fibrosis
    - **e.** Duchenne's muscular dystrophy
3. What causes Down syndrome?
    - **a.** an extra Y chromosome
    - **b.** an extra X chromosome
    - **c.** an extra copy of chromosome 21
    - **d.** an extra copy of each chromosome
    - **e.** none of the above
4. What are dominant negative mutations?
    - **a.** a mutation that activates a gene when it would not normally be active
    - **b.** a mutant protein that loses its own function and also interferes with the function of another, non-mutated, protein
    - **c.** a mutation that causes a protein to lose function
    - **d.** a mutation that produces no detrimental effects on the protein
    - **e.** none of the above

(Continued)

5. What are dynamic mutations?
   a. tandem repeats of bases within the coding regions of DNA that increase in number each generation
   b. mutations that affect the coding sequence of the DNA
   c. mutations that cause defects in methylation of DNA
   d. mutations that affect mitochondrial DNA
   e. none of the above
6. When does imprinting occur?
   a. when DNA is methylated at every base
   b. when the enzymes in the DNA methylation pathways are mutated
   c. when the mitochondrial DNA is not methylated
   d. methylation patterns in gametes survive and affect gene expression in the new organism
   e. none of the above
7. Why are mitochondrial defects surprisingly prevalent?
   a. Damage to DNA due to reactive oxygen species is much higher.
   b. Mitochondrial DNA is not protected by histones.
   c. Mitochondria contain few repair systems.
   d. Mitochondrial DNA has a much higher mutation rate than nuclear DNA.
   e. all of the above
8. How are mitochondrial disorders inherited?
   a. maternally
   b. paternally
   c. during fetal development
   d. all of the above
   e. none of the above
9. In what way can the presence of a genetic defect on a segment of DNA be screened after positional cloning?
   a. sequencing of the DNA
   b. hybridization of mRNA extracted from the tissues most affected by the disease to suspect DNA
   c. by the presence of open reading frames
   d. identification of CpG islands
   e. all of the above
10. Which ion channel is defective in people with cystic fibrosis?
    a. sodium transporter
    b. potassium transporter
    c. chloride transporter
    d. calcium transporter
    e. zinc transporter
11. What advantage does a ΔF508 heterozygote have over a patient with cystic fibrosis (i.e., homozygous for ΔF508)?
    a. ΔF508 heterozygotes are more susceptible to malaria.
    b. They are more resistant to the effects of enteric diseases, such as typhoid and cholera.
    c. They have no advantage since only one defective copy is needed to cause cystic fibrosis.
    d. The ΔF508 mutation does not cause disease, so there is no advantage.
    e. none of the above

12. Which statement about CRISPR is false?
- **a.** CRISPR is a bacterial system that combats foreign DNA.
- **b.** CRISPR uses guide RNAs derived from foreign genetic material.
- **c.** Cas nuclease destroys foreign genetic material.
- **d.** CRISPR is exactly the same as RNAi.
- **e.** Short sequences of foreign DNA are stored within the host's own genome.

13. Of the following nucleases, which is currently in clinical trials for gene editing?
- **a.** Cas
- **b.** Zinc-finger
- **c.** Homing
- **d.** TALE
- **e.** CRISPR

14. Which of the following has been approved for age-related macular degeneration?
- **a.** CRISPR
- **b.** Antisense RNA
- **c.** RNAi
- **d.** Gene editing
- **e.** aptamers

15. Why is mating between close relatives not a good idea, genetically?
- **a.** Recessive genes within the family have a chance to come together and cause disease.
- **b.** Dominant alleles within the family have a chance to come together and cause disease.
- **c.** The mitochondrial DNA is not genetically diverse.
- **d.** DNA methylation patterns have more tendency to create imprinting.
- **e.** none of the above

16. What is a specific case in which genetic screening has saved and improved the quality of life for an individual through diet modification?
- **a.** Duchenne's muscular dystrophy
- **b.** cystic fibrosis
- **c.** Prader-Willi syndrome
- **d.** phenylketonuria
- **e.** Down syndrome

Further Reading

Antunes, M. S., Smith, J. J., Jantz, D., & Medford, J. I. (2012). Targeted DNA excision in Arabidopsis by a re-engineered homing endonuclease. *BMC Biotechnology, 12*, 86.

Burnett, J. C., & Rossi, J. J. (2012). RNA-based therapeutics: current progress and future prospects. *Chemistry & Biology, 19*(1), 60–71.

Cassidy, S. B., Schwartz, S., Miller, J. L., & Driscoll, D. J. (2012). Prader–Willi syndrome. *Genetics in Medicine: Official Journal of the American College of Medical Genetics, 14*(1), 10–26.

Colella, P., & Auricchio, A. (2012). Gene therapy of inherited retinopathies: a long and successful road from viral vectors to patients. *Human Gene Therapy, 23*(8), 796–807.

Hafez, M., & Hausner, G. (2012). Homing endonucleases: DNA scissors on a mission. *Genome, 55*(8), 553–569.

Keeling, K. M., & Bedwell, D. M. (2011). Suppression of nonsense mutations as a therapeutic approach to treat genetic diseases. *Wiley Interdiscip Rev RNA, 2*(6), 837–852.

Lage, H., & Krühn, A. (2010). Bacterial delivery of RNAi effectors: transkingdom RNAi. *Journal of Visual Experiment*, (42) pii: 2099.

Lei, Y., Lee, C. L., Joo, K. I., Zarzar, J., Liu, Y., Dai, B., Fox, V., & Wang, P. (2011). Gene editing of human embryonic stem cells via an engineered baculoviral vector carrying zinc-finger nucleases. *Molecular Therapy: The Journal of the American Society of Gene Therapy, 19*(5), 942–950.

Limberis, M. P. (2012). Phoenix rising: gene therapy makes a comeback. *Acta biochimica et biophysica Sinica, 44*(8), 632–640.

Merk, D., & Schubert-Zsilavecz, M. (2013). Repairing mutated proteins—development of small molecules targeting defects in the cystic fibrosis transmembrane conductance regulator. *Expert Opinion in Drug Discovery, 8*(6), 691–708.

Mussolino, C., & Cathomen, T. (2012). TALE nucleases: tailored genome engineering made easy. *Current Opinion in Biotechnology, 23*(5), 644–650.

Nelson, D. L., Orr, H. T., & Warren, S. T. (2013). The unstable repeats—three evolving faces of neurological disease. *Neuron, 77*(5), 825–843.

Papaioannou, I., Simons, J. P., & Owen, J. S. (2012). Oligonucleotide-directed gene-editing technology: mechanisms and future prospects. *Expert Opinion on Biological Therapy, 12*(3), 329–342.

Rettig, G. R., & Behlke, M. A. (2012). Progress toward *in vivo* use of siRNAs-II. *Molecular Therapy: The Journal of the American Society of Gene Therapy, 20*(3), 483–512.

Roberts, N. J., Vogelstein, J. T., Parmigiani, G., Kinzler, K. W., Vogelstein, B., & Velculescu, V. E. (2012). The predictive capacity of personal genome sequencing. *Science Translational Medicine, 4*(133) 133ra58.

Seyhan, A. A. (2011). RNAi: a potential new class of therapeutic for human genetic disease. *Human Genetics, 130*(5), 583–605.

Urnov, F. D., Rebar, E. J., Holmes, M. C., Zhang, H. S., & Gregory, P. D. (2010). Genome editing with engineered zinc finger nucleases. *Nature Reviews. Genetics, 11*(9), 636–646.

Wang, T., Bray, S. M., & Warren, S. T. (2012). New perspectives on the biology of fragile X syndrome. *Current Opinion in Genetics & Development, 22*(3), 256–263.

Zhang, P., Huang, A., Ferruzzi, J., Mecham, R. P., Starcher, B. C., Tellides, G., et al. (2012). Inhibition of microRNA-29 enhances elastin levels in cells haploinsufficient for elastin and in bioengineered vessels—brief report. *Arterioscler Thromb Vasc Biol, 32*(3), 756–759.

Cloning and Stem Cells

Biotechnology

http://dx.doi.org/10.1016/B978-0-12-385015-7.00018-1

INTRODUCTION

Stem cells have ignited controversy among groups of politicians, religious leaders, and many people in the public. Whether or not scientists should study stem cells is hotly debated in the United States, but other parts of the world are also creating rules and regulations as to the use and study of stem cells. Many of the reports seen in the media provide a very negative slant to the use of stem cells in scientific research, but most of the reports do not explain the reason stem cells are so controversial. In addition, most of the accounts do not explain what a stem cell is and why scientists want to study them. One of the key goals for this chapter is to provide the answers to these questions and to describe the importance these cells have in your tissues and organs. Because of the key role stem cells play in our body, the idea of using stem cells to cure diseases and disabilities has evolved. This chapter explains the motive and current state of research for the use of stem cells in some therapies. Most importantly, the chapter distinguishes between the types of stem cells that are the focus of the heated political debates and the other types of stem cells, and the continuum of characteristics that range between these types of cells.

Along the same tangent of political hotbed issues lies the idea of cloning. The media has sensationalized the idea of cloning animals, and this chapter will also present the scientific facts about cloning so that the reader will have a greater understanding of why cloning has been important to advancements in science. In fact, cloning was an early step in understanding that a differentiated nucleus from any cell can be stimulated to become any type of cell of the body, and even a whole new organism.

WHAT IS A STEM CELL?

A **stem cell** is a cell that has the ability to develop into different types of cells with specific functions during the growth of any multicellular animal (Fig. 18.1). In addition, stem cells are **undifferentiated** and retain the ability to divide or proliferate, unlike most cells, which lose the ability to divide after they specialize or **differentiate**. For example, not only does the red blood cell not divide, but it is so specialized that it does not even retain its nucleus. The stem cell's ability to become any cell in the body would provide a source for cell types like neurons or cardiac cells that could be used to cure diseases or fix disabilities. Stem cells could create differentiated cells that could substitute for an organ transplant, precluding the need for donor organs.

FIGURE 18.1
Stem Cells
Stem cells have three main characteristics: the ability to replicate to make more stem cells, to remain undifferentiated, and to produce daughter cells that differentiate into different cell types when put into the right environment.

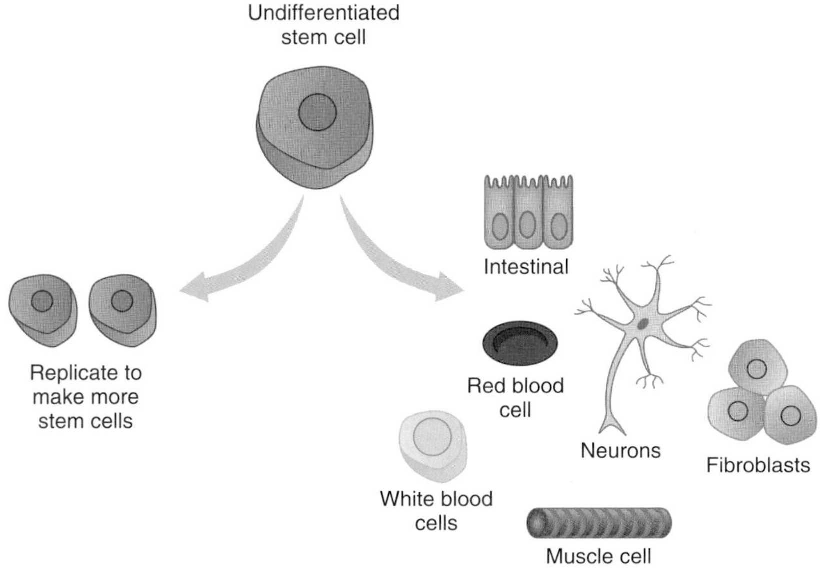

Stem cells replenish any cells in our body and the bodies of other animals on the planet after a certain amount of time. Older cells are unstable due to aging and may accumulate genetic mutations that could develop into a cancer or create excess free radicals that can damage the surrounding healthy cells. The removal of old or damaged cells is called apoptosis and is covered in detail in Chapter 20. The stem cells replenish the tissues when the aged cells are removed, keeping the tissues and organs functioning at their peak. Stem cells have to maintain themselves so that they can continue to make progeny to differentiate.

Since a stem cell can develop into different tissues, the ultimate stem cell is a fertilized egg, or **zygote**. When a sperm and egg join, the resulting cell develops into the entire organism, forming every single cell type possible. Stem cells or early embryonic cells from a zygote are **totipotent**; that is, these cells have the information to from any type of cell. During the early stages of zygotic development, each of the cells maintains the ability to form the entire animal. In fact, if the zygote splits into two cells, identical twins form. As the zygote develops, more and more cell divisions lead to a hollow ball of cells called a **blastula**. In humans, this stage of development occurs even before the embryo implants into the woman's uterus. These very early stages of development occur while the fertilized egg travels through the fallopian tubes. At one end of the hollow ball of cells, more cell division and cell migration lead to a build-up of cells called the **inner cell mass**. When cells from the inner cell mass are isolated and grown in a culture dish, they retain the ability to form every tissue of the organism. The resulting cell lines are called **embryonic stem cell lines** (Fig. 18.2). Since mice and humans have very similar early development, embryonic stem cell lines can be derived from a mouse inner cell mass also. Of course, mouse stem cells cannot form human neurons to fix spinal cord injuries, so researchers are also interested in using human embryonic stem cells, which is the cause for controversy. If life begins at conception, as many people believe, the destruction of embryos to isolate stem cells is like destroying a baby. Patients undergoing *in vitro* fertilization to conceive a baby routinely form multiple blastulas. The blastulas are frozen until the woman is ready to have them implanted in her uterus. More blastulas are made than actually are implanted, and often they are destroyed anyway. The ability to generate and use human embryonic stem cell lines is guided by government policy, which is evolving. For many years, the National Institutes of Health (NIH) of the United States was banned by presidential order from providing any federal funding for research on human embryonic stem cells. As of the time of writing, a different presidential order has lifted the restrictions, but research is still highly regulated. Depending on the politics of the seated president in the United States, the ability to study human-derived stem cells may or may not be restricted.

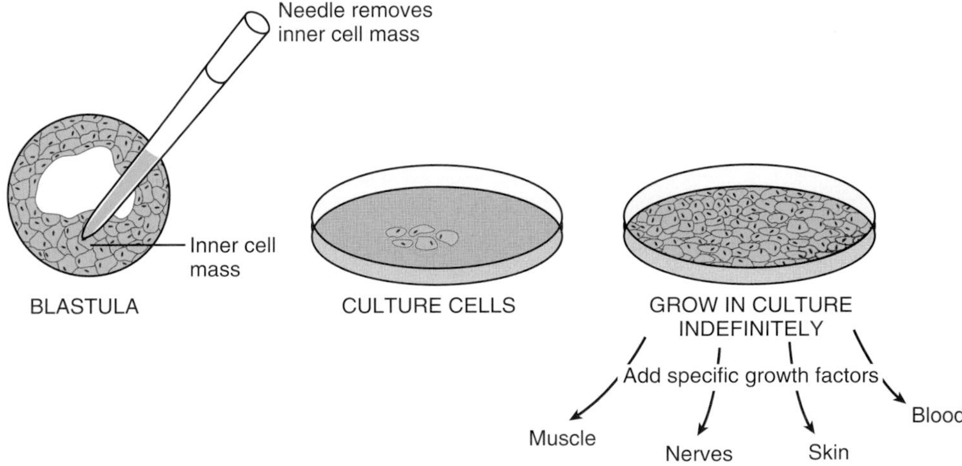

FIGURE 18.2 Creating an Embryonic Stem Cell Line
Cells of the inner cell mass from the blastula can be isolated and cultured in a dish where they can grow indefinitely. If the cells are treated with the proper growth factors, they can specialize into different cellular types such as blood, skin, nerves, or muscle.

Once inner cell mass cells begin growing in a cell culture dish, the cells must be assayed or tested for the three main stem cell characteristics: the ability to grow indefinitely, the ability to remain in an undifferentiated state, and the ability to produce daughter cells that can differentiate into many cell types. To assay for indefinite growth, these cell lines must grow and divide in a dish for many generations. To assay for their undifferentiated state, scientists check the cells by microscopy to see if they have the same shape and size. In addition, undifferentiated cells will express certain proteins called **transcription factors** unique to stem cells. In particular, OCT4 and NANOG, two transcription factors, are produced and function only in stem cells, so the cell lines must be tested periodically for their expression. The cell lines must also be tested for whether or not the cells retain the ability to differentiate. If they pass the previous tests, then some of the cells are isolated and stimulated to differentiate by treating with special growth factors, which cause the cells to ball up into **embryoid bodies** and then differentiate into neurons or muscle. Embryoid bodies can also spontaneously form different cell types (Fig. 18.3). The ability to control which types of cells that embryonic stem cells can differentiate into is one key goal of stem cell research. This ability could provide a source of cells that could regenerate human tissues that are damaged due to disease, injury, or even aging. In another test, the cell line in question can be injected into a mouse that has had its immune system

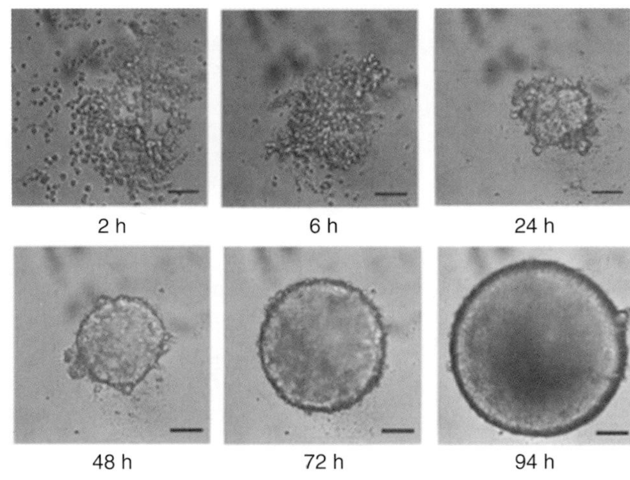

FIGURE 18.3 Embryoid Body Formation
Embryonic stem cells aggregate and form large spherical masses when they are unable to adhere to the bottom of the cell culture dish. Reprinted with permission from Kurosawa H (2007) Methods for inducing embryoid body formation: in vitro differentiation system of embryonic stem cells. *J Biosci Bioengineering* 103(5), 389–398.

suppressed. If the cells are stem cells, they will create a **teratoma**, which is a cancer that has cells of a type different from the surrounding tissue, usually more than one cell type also.

In addition to embryonic stem cells, other types of stem cells are also available for study. **Adult** or **somatic stem cells** exhibit the three same properties of stem cells. That is, they are able to divide and repopulate a tissue or organ, they are not specialized or differentiated, and they have the ability to specialize when put into the proper environment. The main difference between an adult stem cell and embryonic stem cell is the extent of differentiation. Adult stem cells are only **multipotent**; that is, they can only differentiate into a small subset of cells. In contrast, embryonic stem cells can differentiate into any cell type of the organism (totipotent). For example, hematopoietic stem cells can become any type of blood cell. There are stem cells in our muscles that regenerate muscle. Intestinal stem cells regenerate the cells that line the intestine so that the nutrients are absorbed from the food passing through our intestine. Each of these adult stem cells holds potential for potential stem cell therapies in its associated tissues.

Stem cells have three main characteristics: they maintain the ability to divide continually, they are undifferentiated, and they have the ability to differentiate into multiple cell types. Embryonic stem cells are totipotent, meaning they are able to produce any type of cell found in the organism. Adult or somatic stem cells are able to differentiate into different cell types but are multipotent; that is, they are restricted to the tissues in which they originate. Embryonic stem cell lines are created from the inner cell mass of the blastula stage of an embryo from many different mammals, including humans, mice, and primates. The cell lines can be induced to differentiate by forming embryoid bodies that ultimately differentiate into different cell types.

IDENTIFYING ADULT STEM CELLS

The stem cell must have the three main characteristics, but identifying exactly which cell has all three traits is difficult in adult tissues. The stem cell population is often a very small percentage of the tissue or organ. In some tissues, stem cells are scattered throughout the tissue, whereas in other locations they are found in a defined niche (see following discussion). The study of stem cells in our bodies is still in its infancy, and so the defining characteristics for these cells will become clearer as more research is performed.

There has not been one single substance or molecule that definitively identifies every adult stem cell; therefore, the leading theory is that each adult stem cell type is unique. There are some possible characteristics for identification of certain stem cells, but they are specific to the organism or tissue. One marker of *Drosophila* stem cells is clusters of endoplasmic reticulum–like vesicles called **spectrosomes** (Fig. 18.4). Germline stem cells of *Drosophila* ovaries contain

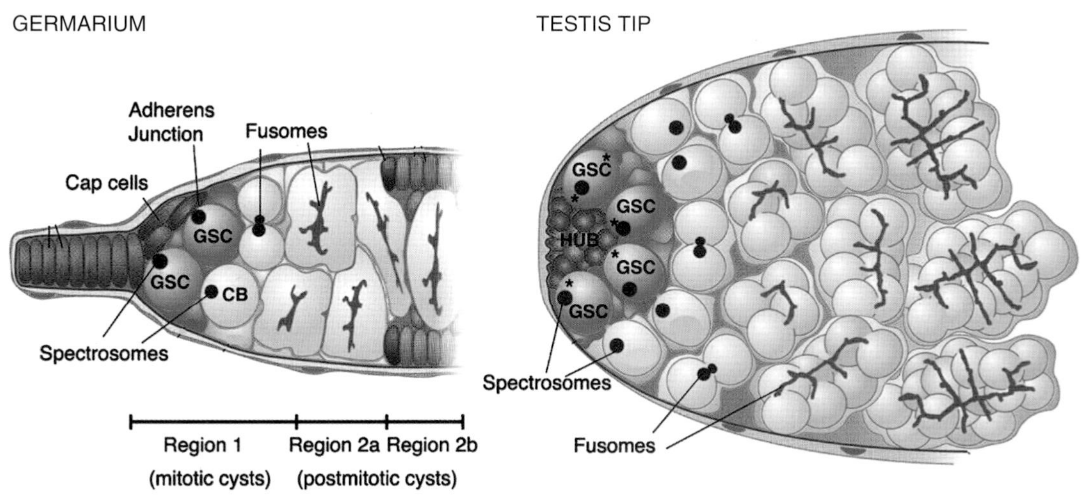

FIGURE 18.4
Germline Stem Cells of *Drosophila*
Germline stem cells of model organisms are used to study what molecules are markers of stem cells. The female ovary (A) and male testis tip (B) from *Drosophila* are shown. The spectrosomes (red dots) are found in the germline stem cells (GSC) and the first descendants of the stem cells. Reprinted with permission from Lighthouse DV, Buszczak M, Spradling AC (2008) New components of the Drosophila fusome suggest it plays novel roles in signaling and transport. *Dev Biol* 317(1), 59–71.

spectrosomes. These organelle-like complexes are associated with the mitotic spindle during cell division and are important for the asymmetric division of the stem cell. In the *Drosophila* ovary, stem cells divide asymmetrically so that one of the daughters remains as a stem cell, and the other daughter cell differentiates into a nurse or egg cell. The spectrosomes contain membrane skeleton proteins such as spectrin and ankyrin. Spectrosomes persist in the first cells after the stem cells divide, called the cystoblasts, but are then absent in the fully differentiated oocyte.

Apart from organelle-like structures, stem cells can be distinguished from other cells by various signal transduction proteins. For example, some stem cells express proteins that keep the cells from differentiating. Some candidate proteins include Notch, a key protein that keeps germline stem cells of *C. elegans* from differentiating into eggs. Notch signal transduction pathways are also important for many different adult stem cells, including those from the intestine, skin, lungs, blood, and muscle. Unfortunately, signal transduction proteins do not disappear immediately as the stem cell differentiates. In fact, the marker may persist even after a stem cell has fully differentiated and therefore may not be specific enough for identifying stem cells from their surrounding differentiated cells.

The best way to identify the adult stem cells is called **fate mapping** or **lineage analysis**, which involves labeling the potential stem cell with a molecular marker and then tracing where the marker is found after many generations (Fig. 18.5). If the marker is labeling a true stem cell, as the stem cell divides, the marker will pass to the daughter cells and remain in the stem cell. If the marked cell is not a stem cell, the marker will not be passed onto descendants since normal differentiated somatic cells do not divide by mitosis. Most adult differentiated cells do not divide at all and are considered postmitotic. Fate mapping was originally used to trace the development of organisms from early in embryogenesis to adulthood, which is analogous to the process of identifying the stem cells.

Finding a good marker for the potential stem cell is key to following the fate of any of its descendants. In small model organisms or in cultured cells, direct observation of cell division can provide helpful information. This method does not work for tissues within a larger organism or in humans. For these organisms, using a molecular marker such as β-galactosidase or green fluorescent protein (GFP) is a better option. β-Galactosidase can be visualized by exposing the tissues to X-gal; whereas GFP can be visualized directly with fluorescence microscopy. To determine the fate of lung stem cells, a retrovirus was engineered to include the gene for β-galactosidase. The recombinant virus was then used to infect human lung tissue that was collected during a lung transplant. The infected lung tissue was cultured with rat trachea so that it could coat the surface and then was transplanted back into the rat. Any human lung cell harboring the retrovirus/β-galactosidase fusion maintained the integrated viral genome in its own DNA and then passed it onto any descendants during mitosis. The results demonstrated that the human lung cells contained multipotent stem cells that could regrow complete submucosal glands, which are essential to lung function. These

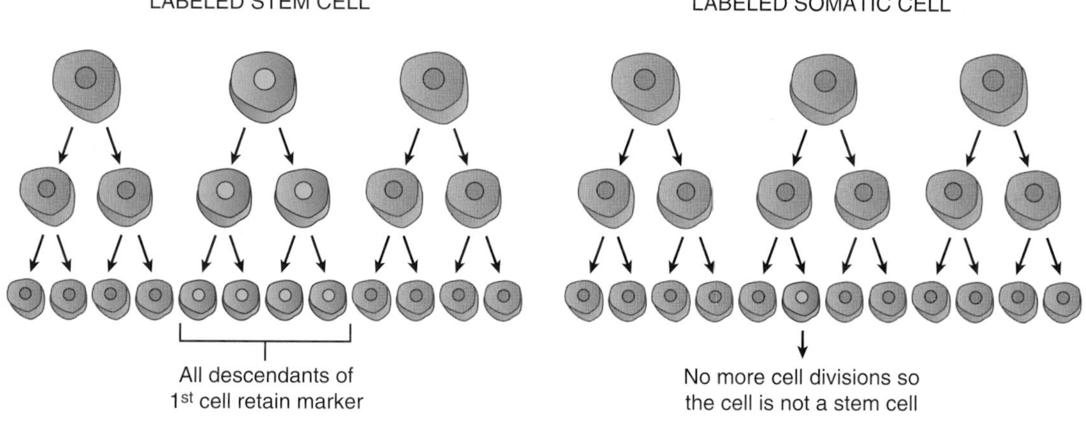

FIGURE 18.5
Lineage Analysis
Adult stem cells can be identified by fate mapping. First, a marker or label is added to cells from a tissue. If the cell that is labeled is a stem cell (left), then the descendants of that cell are also labeled. If the labeled cell is already differentiated, there will be no descendants with the label.

glands produce mucus that is used to trap any particles that are inhaled. The tissue isolated from the transplanted rat trachea had fully developed glands that turned blue upon exposure to β-galactosidase, meaning these were derived from the human tissue, not the rat.

One modification of this analysis is to isolate the cell populations at varying time points after infection with the retrovirus. Tissues or cells isolated shortly after infection will identify the stem cells, whereas isolating the cells at longer times postinfection will mark the stem cells and their descendants.

Another method of marking a stem cell is to create a genetic mosaic by **transplantation**. For example, quail cells can be transplanted into chicken embryos. The quail cells have a unique structure to their heterochromatin and therefore can be distinguished from the recipient chicken cells. Using microscopy, scientists were able to identify yolk cells as the primary source of blood progenitor cells. Instead of two different but related species, current transplantation studies move cells from transgenic organisms into normal organisms. For example, cells from a transgenic mouse that is expressing the gene for green fluorescent protein can be seeded into the cells of a normal animal. The green fluorescent cells can then be traced to determine if they are stem cells or just somatic cells by determining how long they persist and whether they have created GFP-labeled descendants. These studies have been done extensively in mice and a few studies in primates to identify stem cell populations (Fig. 18.6).

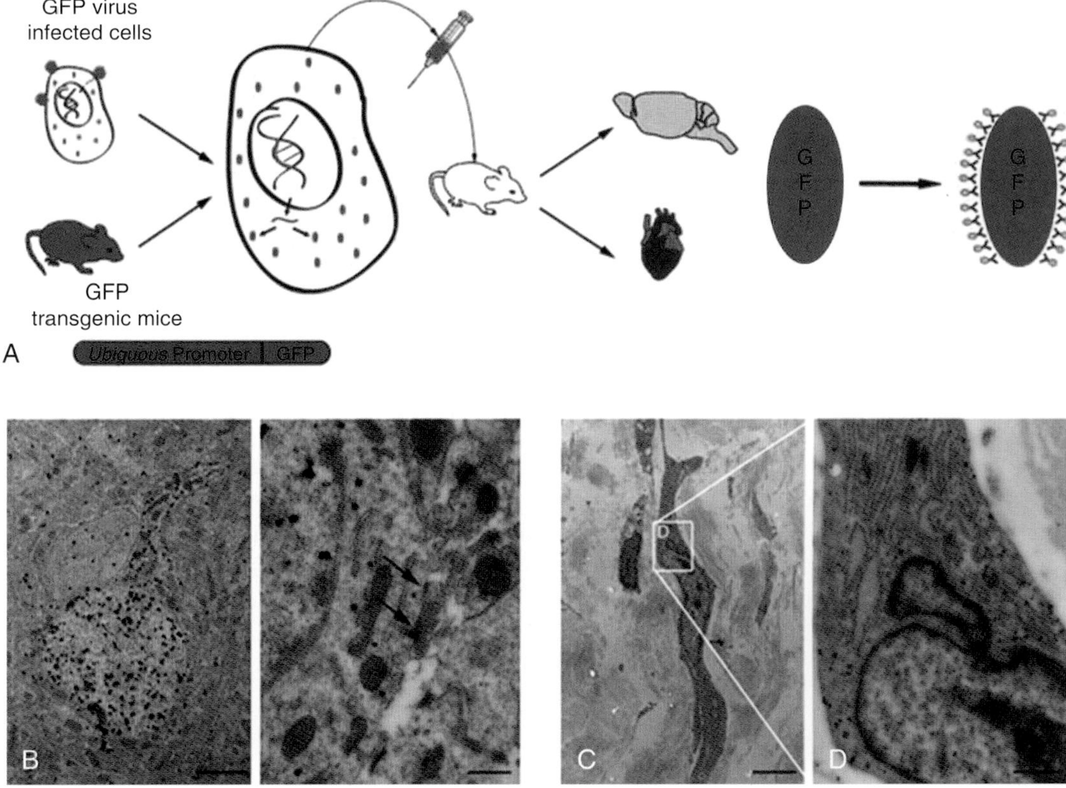

FIGURE 18.6 Cell Transplantation

(A) GFP marks cells from the donor population. The donor can be derived from transgenic mice or cultured cells expressing GFP (left). When the marked cells are transplanted into an unlabeled mouse, the tissues (either brain or heart) can be analyzed for GFP by labeling the protein with an antibody that has a detection system attached. For electron microscopy, the antibody is detected with a gold bead that appears as a black spot on micrographs, since the gold is electron dense. (B) On the left, a single neuron has multiple GFP-gold beads attached to the various locations within the cell. On the right, a magnified image shows the location of some GFP-beads within the cell. (C) GFP-positive stromal stem cells found in a population of heart fibroblast cells surrounded by collagen. The heart cells were damaged, and the transplanted cells were able to establish in the damaged area. (D) A magnified image of the fibroblast cell shows the individual GFP-beads inside the cell. Used with permission from Sirerol-Piquer MS, Cebrián-Silla A, Alfaro-Cervelló C, Gomez-Pinedo U, Soriano-Navarro M, Verdugo JM (2012). GFP immunogold staining, from light to electron microscopy, in mammalian cells. *Micron* 43(5), 589–599.

Identifying adult stem cells is more difficult because they are distributed around the body and comprise a very small fraction of the total cells in the tissues or organs. Scientists hope to identify markers or traits of stem cells that can distinguish them from surrounding cells. Some possible markers include organelle-like structures such as the spectrosome or signal transduction molecules like Notch, but these are not very specific to the stem cell population, and they are also not found in every stem cell throughout the body. Two of the best techniques to identify stem cells are analyzing lineage and creating genetic mosaics.

THE KEY FEATURES OF A STEM CELL NICHE

A **stem cell niche** is a local microenvironment that directly promotes or protects a population of stem cells. There are two key components to a stem cell niche, the microenvironment and the stem cell, and each of their functions relies on the other. The environment emits cues or signals to the stem cells that keep the cells from differentiating—that is, keeps the cells competent to develop into a variety of different cells. Some of these signals originate from cells that encase the niche, and in other cases, the extracellular matrix provide the cues. In addition, **autocrine signals** from the stem cell itself can keep the cells in their undifferentiated state (Fig. 18.7). The stem cells that stay encased in the niche provide a pool of cells that can be triggered to form the other cells of our body. If the stem cell niche runs out of cells, then there are no precursors for tissues to regenerate. In contrast, an extracellular environment that causes overproliferation, that is, creates too many stem cells, leads to tumors. Therefore, proper niche function requires a balance, and there cannot be too many stem cells or too few.

The cells and extracellular matrix that make up the environment depend on the type of stem cell that they house. For example, the model organism *C. elegans* has a stem cell population that can generate more oocytes or eggs for the worm (Fig. 18.8). These stem cells are kept in a niche that is created by one single cell, called the distal tip cell, which caps the area and sends long processes or fibrils that surround the area. This single cell emits a signal that prevents the stem cells from becoming an egg. If this single cell is destroyed in any way, the stem cells turn into eggs. If this single cell is moved to a different location, the nearby cells receiving its signals turn into stem cells.

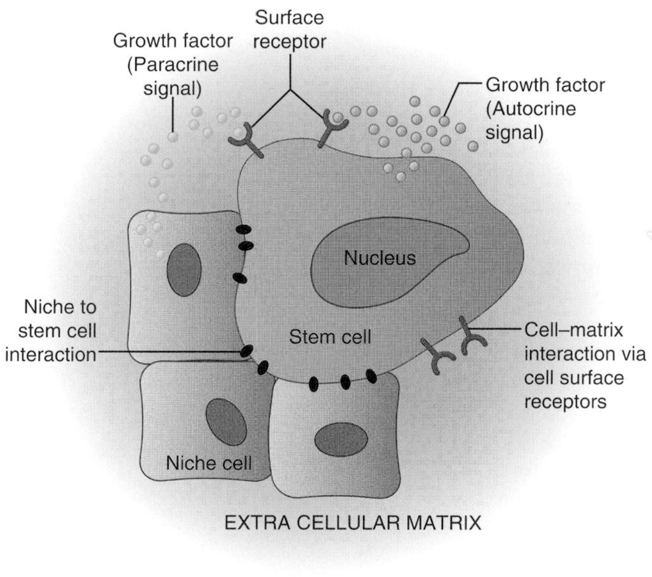

FIGURE 18.7
Stem Cell Niche
The stem cells found in a niche interact with the niche cells and the extracellular matrix. The stem cell can receive signals from both components via direct interactions (cell to cell or cell to matrix). Stem cells can also receive signals such as growth factors that are secreted from themselves (autocrine) or from the niche cells (paracrine). These signals bind to cell surface receptors and trigger the stem cell to stay undifferentiated.

Stem cells contained within a niche often undergo **asymmetric cell division** to produce one daughter cell that differentiates and another daughter cell to maintain the stem cell population. The signals that affect each of the two daughters are different, and can either arise from an **intrinsic asymmetry** between the two daughters or from an **extrinsic asymmetry** (Fig. 18.9). In an intrinsic asymmetry, one daughter cell receives a signaling molecule such as a protein, RNA, or macromolecule, and the other daughter cell does not. Depending on the role of the intrinsic signal, the daughter cell that receives the molecule either stays a stem cell or differentiates. External asymmetries also drive one daughter cell to differentiate and the other to stay a stem cell. For example, if the cell division pushes one of the daughters too far from the niche and the signals that keep the stem cell from differentiating, then this daughter will start to differentiate. Another possibility is that the daughter cell in the new environment may receive new signals that cause it to differentiate. There can also be **symmetrical renewal** where stem cell division creates two more stem cells. Additionally, a mitotic division can also produce two cells that differentiate. This process is termed **symmetrical differentiation** (Fig. 18.10). If all the stem cells in a niche had symmetrical differentiations,

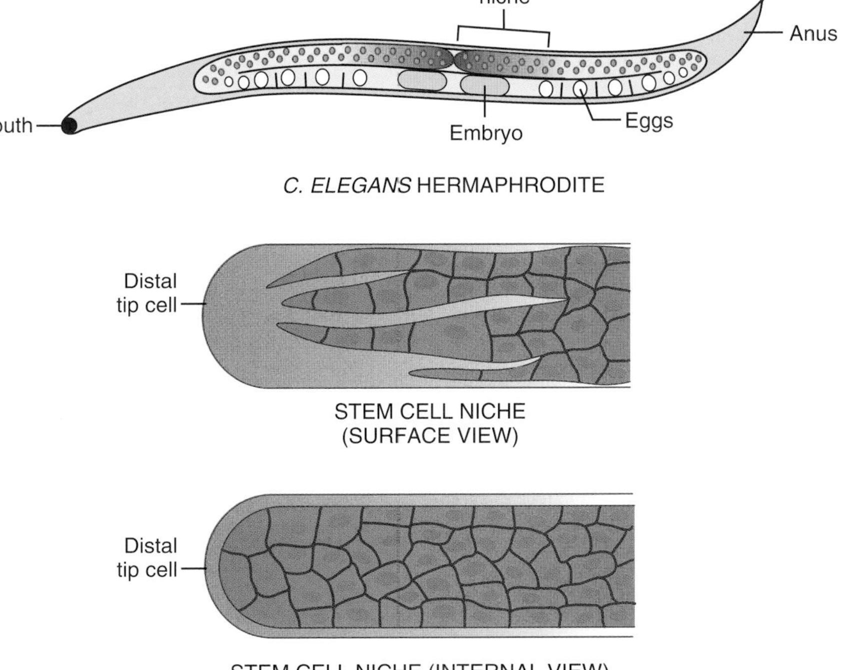

FIGURE 18.8
***C. elegans* Oocyte Stem Cells Are Found in a Defined Niche**
The stem cells responsible for creating more oocytes for the *C. elegans* hermaphrodite are found in a niche that is capped by a single cell called the distal tip cell. This cell emits signals that activate the Notch signal transduction pathway, which in turn keeps the stem cell from differentiation. As the stem cell moves away from the niche, the Notch signals fade, and then the cell enters into meiosis to form a haploid oocyte. As the stem cells divide, cells on the periphery of the niche are displaced and exit the range of the distal tip cell signals.

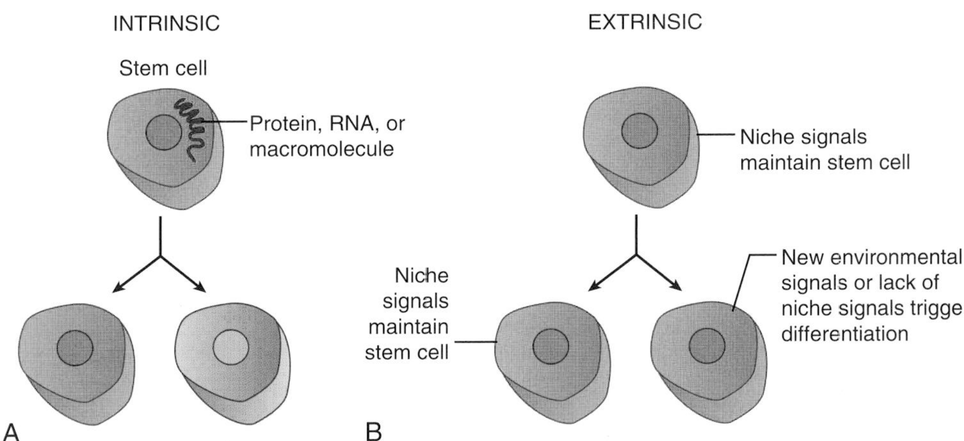

FIGURE 18.9
Asymmetrical Cell Divisions Are due to Intrinsic or Extrinsic Asymmetries
(A) Stem cells may have intrinsic factors such as proteins, RNA, or macromolecules that are sequestered unequally during cell division. The signaling factors can either induce the recipient cell to differentiate or may keep the cell from differentiating, depending on their function. (B) Extrinsic signals from the niche often maintain cells in their stem cell fate, and when the daughter moves to a new environment, the loss of the signal or the addition of a new signal triggers this daughter to differentiate. The other cell remains in the same environment and therefore remains a stem cell.

the organism would run out of stem cells. On the other hand, if the stem cell always divided with symmetrical renewal, the niche would have too many stem cells, which is a scenario that resembles a cancer.

One key component to keeping the stem cell in a niche is the adherence to the extracellular matrix (ECM). This involves anchoring the stem cell to the ECM through cell surface receptors such as E-cadherin. If this surface receptor is removed from germline or follicular stem cells, these stem cells rapidly leave the niche and differentiate. E-cadherin mediates stem cell adhesion in the *Drosophila* testes stem cell niche, in the neuronal stem cells of the brain region called the **subventricular zone**, and the stem cell niche for creating blood cells.

Stem cell niches are quite varied throughout the body and can be categorized by their make-up. **Open stem cell niches** have no geographical boundaries for the microenvironment and the stem cells. In the open variety of niche, interaction between neighboring niches is common, where stem cells from one region can migrate to another. Lineage analysis has determined that stem cells from one location have the ability to migrate from their niche and populate neighboring niches. An example of stem cells found in open environments is the seminiferous

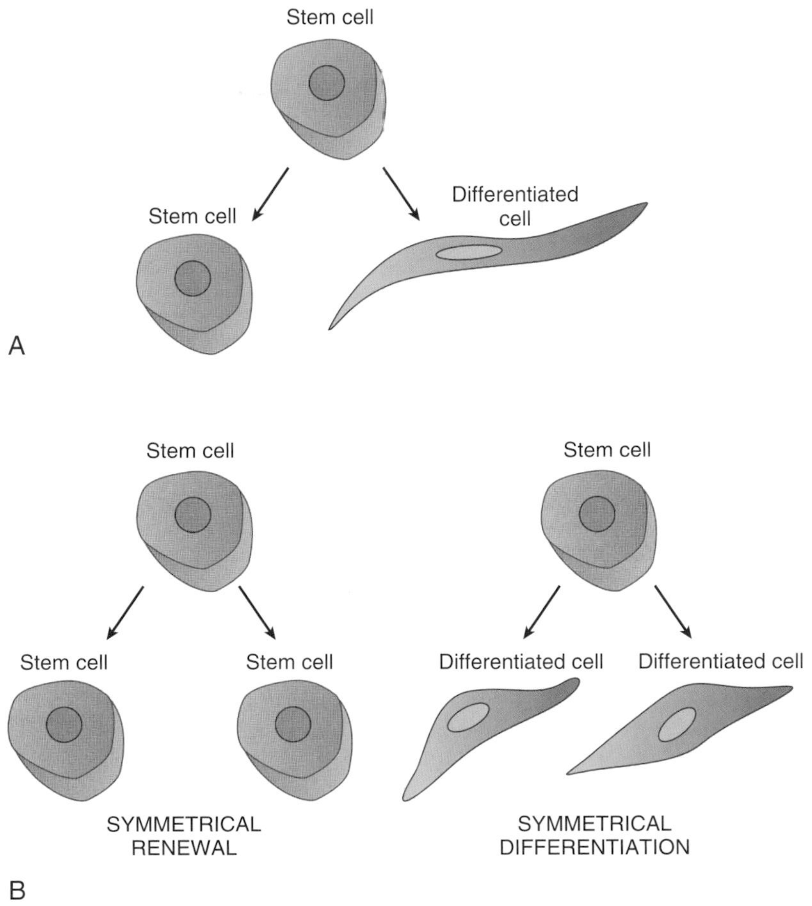

FIGURE 18.10 Stem Cell Divisions Can Be Symmetrical or Asymmetrical
(A) Stem cells divide asymmetrically supplying the niche with a replacement stem cell and the other cell that differentiates into a specialized cell. (B) Stem cells divide symmetrically, either providing two new stem cells (symmetrical renewal) or two cells to differentiate (symmetrical differentiation).

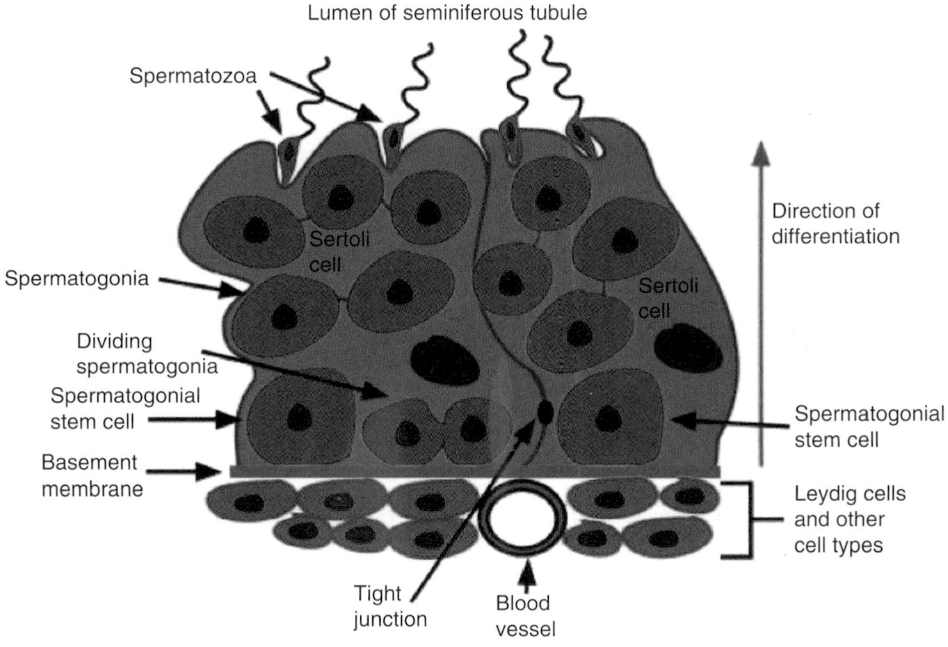

FIGURE 18.11 Open Stem Cell Niche of Seminiferous Tubules
The spermatogonial stem cells (green ovals with red dots) are found in an open niche situated between Sertoli cells (purple cells) and a basement membrane (blue line). Figure used with permission from L'Hernault SW (2013). Spermatogonia. In *Brenner's Encyclopedia of Genetics*, 2nd ed. (London, UK: Academic Press), pp. 533–535.

tubules in mammals (Fig. 18.11). In contrast, **closed stem cell niches** have fixed boundaries that enclose the environment and stem cells. A closed niche is surrounded by an extracellular matrix or capped by cells that prevent the stem cells from leaving the area. In *C. elegans*, the stem cell niche has a closed environment capped by a distal tip cell. Stem cells can leave the area in only one point, and differentiation begins upon exiting of the niche (see Fig. 18.8).

Stem cells reside in niches, or microenvironments that provide cues or signals to keep the stem cells from differentiating into adult cells. Autocrine signals from the stem cells themselves, signals from niche cells, and signals from the extracellular matrix act together to maintain a pool of stem cells to replenish the organ or tissue. The niche can be a closed area that is encased by a layer of cells or an open niche that is more open to the rest of the tissues. Stem cells divide asymmetrically, providing one daughter cell to maintain the stem cell pool and one daughter cell to differentiate. The signal to differentiate or stay a stem cell can originate from an intrinsic signal or extrinsic signal. Stem cells can also divide symmetrically, producing either two new stem cells or two daughter cells that differentiate.

HEMATOPOIETIC STEM CELLS IN THE BONE MARROW

Bone marrow harbors hematopoietic stem cells (HSCs) that replenish the blood supply of our bodies and the bodies of mammals. Over 7×10^9 red and white blood cells are needed for each kilogram of body weight each day for an average human. HSCs are found around the small blood vessels within the marrow and at the interface of the bone and marrow (Fig. 18.12; see gold-colored ovals). The region where the marrow and bone meet is called the **endosteum**, and the cells that are found there are referred to as endosteal cells. There is a complex mixture of cells within the endosteal niche. One set of cells includes osteoblasts and osteoclasts that are essential for making the bone. In addition, mesenchymal stromal stem cells are present, which differentiate into the bone and cartilage cells. Finally, Schwann cells encase and insulate autonomic neurons in the marrow (not shown). Another group of cells, the vascular cells, includes the small blood vessels (medullary vascular sinus in Fig. 18.12). These vessels are encased with a layer of **endothelial cells** (red) and adventitial reticular cells (brown), which provide another microenvironment for HSCs (gold).

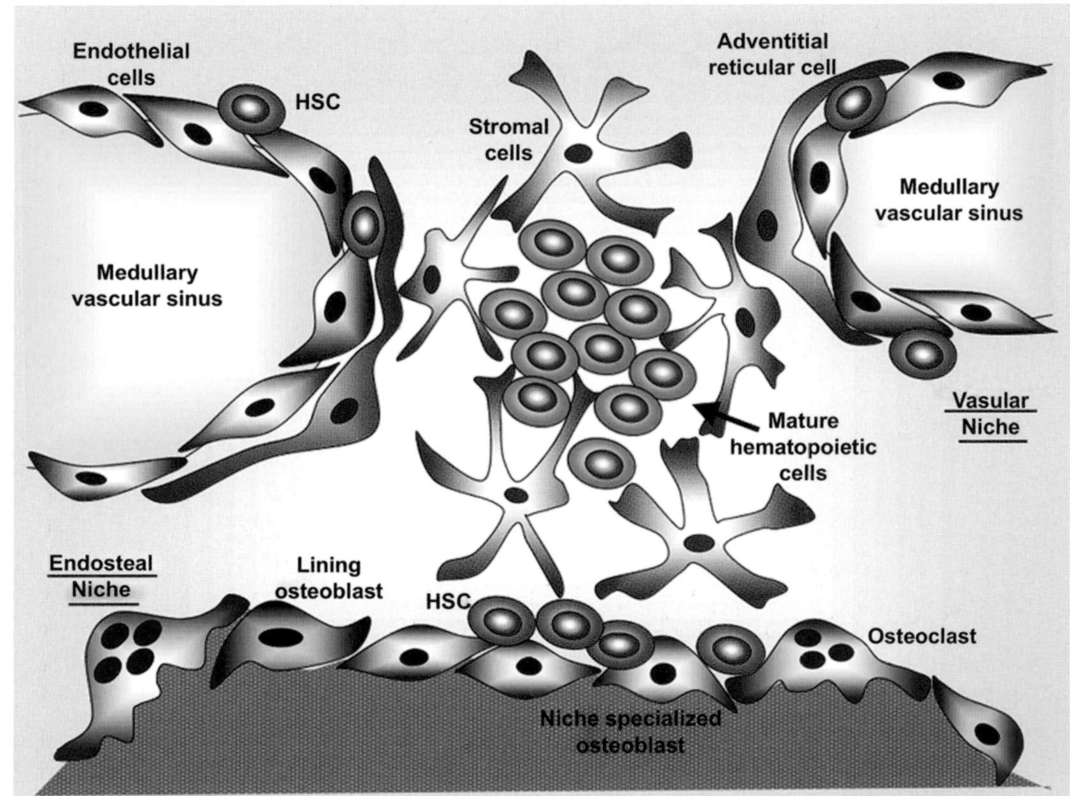

FIGURE 18.12 Hematopoietic Stem Cell Niche

Hematopoietic stem cells are located near the interface of the bone marrow and bone surface (the endosteal niche) and also near vascular sinuses (vascular niche). The HSCs in each niche divide to form mature hematopoietic cells that populate the marrow tissues. Used with permission from Di Maggio N, Piccinini E, Jaworski M, Trumpp A, Wendt DJ, Martin I (2011). Toward modeling the bone marrow niche using scaffold-based 3D culture systems. *Biomaterials* 32(2), 321–329.

The cells of this niche fall into three categories: vasculature, endosteal cells, and hematopoietic stem cell progeny. The vasculature includes arterioles (small arteries) that lie near the endosteal surface. These are associated with nerves found in the marrow. Endothelial cells form a barrier between the hematopoietic stem cell progeny and the bloodstream because they line the small blood vessels. Osteoblasts secrete factors that induce hematopoietic stem cell differentiation, which lie close to the endosteum.

Many different factors regulate hematopoietic stem cell maintenance and differentiation. Stem cell factor (SCF), a secreted protein found in the hematopoietic stem cell niche, keeps the stem cells in their undifferentiated state. Mice missing the gene for SCF die of severe anemia. Extracellular matrix proteins such as N-cadherin adhere the stem cells in the niche environment. Although most HSCs stay in the niche areas, some HSC also migrate out of the niche and into the bloodstream. After circulating through the body, the HSCs return to the niche, which may be an important behavior to repopulate an existing stem cell niche or establish new stem cell niches. Two key signal transduction pathways, Notch and Wnt, work in union to maintain the stem cell pool as undifferentiated cells; therefore, mutations in the Notch pathway components cause HSCs to differentiate into blood cells and eventual depletion of the stem cell niche.

Two different stem cell populations exist in the HSC niche. One set of stem cells transiently self-renews, whereas the other set maintains the stem cell state long term. During differentiation, the stem cell can become either a lymphoid progenitor or a myeloid progenitor (Fig. 18.13). The erythroid cell lineage develops into red blood cells from the myeloid progenitor, as well as macrophages, megakaryocytes, and neutrophils. The lymphoid progenitor divides to form T cells, B cells, and NK cells. During development along these two lines, various factors stimulate and modulate the development from stem cell into mature blood cell. For example, interleukin-7 (IL-7) triggers the formation of T lymphocytes or T cells. B lymphocytes or B cells also require IL-7, as well as other cytokines (not shown), to differentiate into these immune cells. The combination of hematopoietic growth factors, position in the body, and the early signals received in the bone marrow produces the large number of cells needed every day for proper circulatory and immune system function.

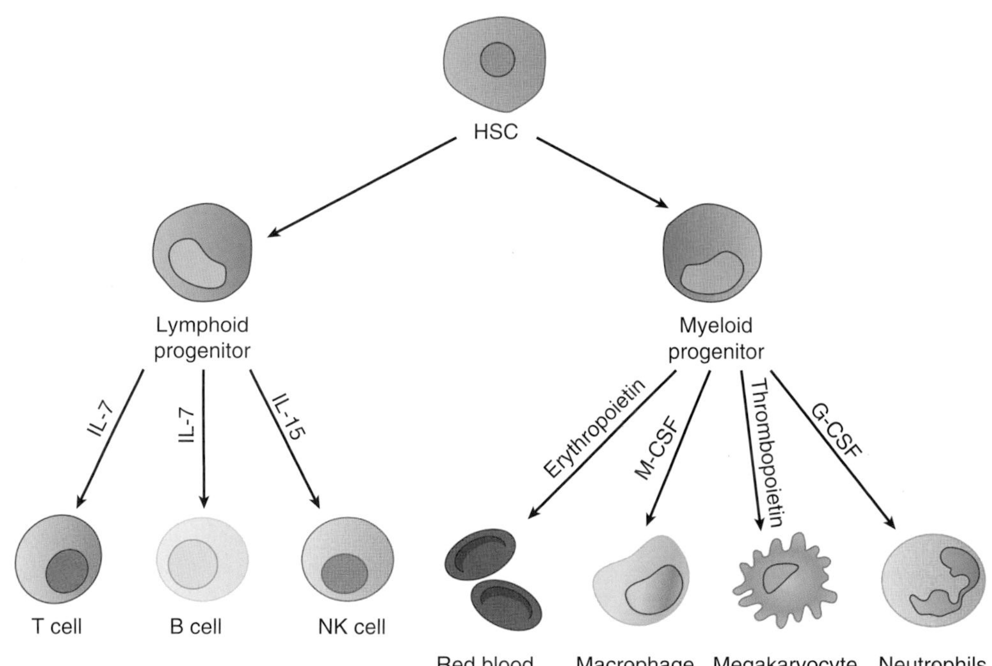

FIGURE 18.13 Factors Stimulate Blood Cell Development
Hematopoietic stem cells (HSCs) develop into lymphoid precursor cells that respond to interleukins such as IL-7 to form T cells and B cells or to IL-15 to form natural killer, or NK, cells. In addition to the lymphoid lineage, HSCs develop into myeloid progenitor cells that also respond to cytokines and growth factors. Erythropoietin triggers the myeloid precursor to become red blood cells. Macrophage-colony stimulating factor (M-CSF) triggers macrophage formation. Thrombopoietin stimulates megakaryocyte formation that then develops into platelets. The other cells such as neutrophils, eosinophils, and basophils develop in response to granulocyte-colony stimulating factor (G-CSF).

Hematopoietic stem cells are found within a niche situated in the bone marrow. The hematopoietic stem cell niche contains a variety of different cells and tissues, including blood vessel cells, bone cells, neural cells, and bone marrow cells. Extracellular matrix, signal transduction proteins, circulating signaling proteins, and signals from cells in the marrow stimulate stem cells to differentiate into circulating red and white blood cells.

INTESTINAL EPITHELIAL STEM CELLS

The tissue of the small intestine replenishes itself completely every 3–5 days in a mouse and even faster when damaged. Adult stem cells rapidly provide the various cell types found in the intestine. Simple columnar epithelial cells, called **enterocytes** or absorptive epithelial cells, line the entire intestine, coating the villi or small finger-like projections (Fig. 18.14).

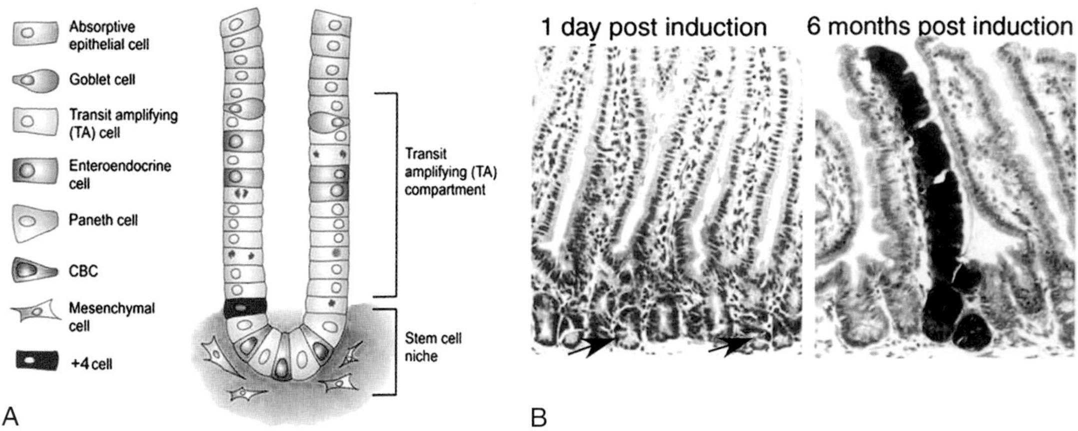

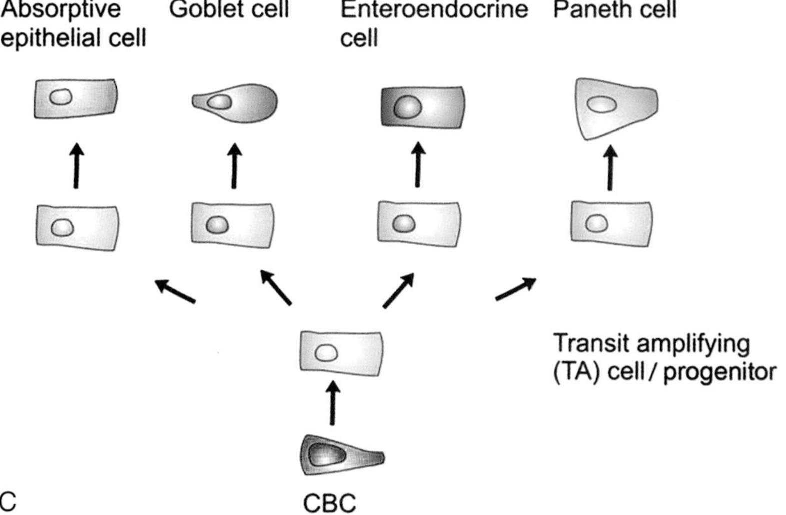

FIGURE 18.14 Intestinal Stem Cell Niche

(A) Crypts, or the valley between two different finger-like projections called villi, harbor a population of stem cells that replenish the villi with new absorptive epithelial cells, enteroendocrine cells, and goblet cells. The stem cell niche contains Paneth cells, which secrete antimicrobial peptides that regulate the type of bacteria found in the intestine. The stem cells are labeled CBC for crypt base columnar cells. (B) Lineage tracing of Lgr5-positive cells expressing GFP are colored blue. After cells are induced to make GFP protein for one day, only one CBC cell is blue in two different crypts. After 6 months, one entire villus has cells with GFP that they inherited from the original CBC cell, whereas the other labeled CBC stem cell did not create descendants that populated the villus with cells. (C) CBC, or intestinal stem cells, can differentiate into all four types of cells found in the villi of the intestine: absorptive epithelial cells, goblet cells, enteroendocrine cells, and Paneth cells. The differentiation occurs via a transit amplifying (TA) progenitor cells. Used with permission from Vries RG, Huch M, Clevers H (2010). Stem cells and cancer of the stomach and intestine *Mol Oncology* 4(5), 373–384.

These cells absorb nutrients from the intestinal lumen and pass them into the bloodstream. Scattered among the enterocytes, goblet cells and enteroendocrine cells secrete mucous and digestive hormones, respectively. Each of these cell types differentiates from stem cells found in the crypts between the villi. The crypts also contain Paneth cells that produce antimicrobial peptides and regulate the type of bacteria found in the intestine.

Intestinal stem cells (ISCs or CBCs for crypt base columnar cells) exist in a niche at the base of two different villi, sending cells up the sides of each adjacent villi. The CBCs can be identified because they express a specific marker called Lgr5, which is a membrane receptor for a key developmental signal transduction pathway. Lgr5 positive (Lgr5+) stem cells divide rapidly, sending cells up the sides of the villi. Initially, the stem cell daughter cells exist in an intermediate state of differentiation, called **transit amplifying cells**. The transit amplifying population retains the ability to divide by mitosis, but only a finite number of times. Fully differentiated intestinal cells cannot divide. The cells migrate up the villi until they reach the top, and then are shed from the villus into the lumen, where they are excreted with the waste. The continual movement of cells ensures that the nutrients are absorbed by young healthy cells.

Interestingly, lineage tracing of CBCs has established the mechanism for stem cell division and niche maintenance in adult intestinal tissues of mouse. Prevailing thought was that every stem cell divided to provide differentiated cells for the tissues, but not every CBC divides to repopulate the villi. In some cases, CBCs are lost, and in other cases, neighboring stem cells repopulate a stem cell niche. The results suggest that stem cells work in a process called **neutral drift**, where there are some stem cells in the crypt that are dividing to repopulate the stem cell pool, and where other stem cells in the crypt differentiate. Eventually, the crypt stem cells become descendent of only one of the stem cells. In other words, the stem cells in a crypt are dividing asymmetrically, where some cells divide into two stem cells, and other cells divide into two differentiated cells (see Fig. 18.10B). This behavior was identified by studying the amount of labeled progeny from stem cells containing the fluorescent gene such as GFP under the control of the Lgr5 promoter. The stem cell and its descendants harboring the GFP reporter stain blue glow green in a fluorescent microscope. Looking at intestine tissue at various times after turning on Lgr5-*GFP* fusion expression has shown this movement of descendants from the crypt to the tip of the villus. In some of the crypts, the entire villus was coated with green descendants from the stem cell, but in other crypts, none of the descendants were green, thus confirming neutral drift. Some stem cells divide and repopulate the entire villus, and in other villi, the labeled stem cell only produced differentiated progeny.

Intestinal stem cells are located at the base of the crypts between the villi of the small intestine. The cells have a marker called Lgr5, which is actually a key protein in a signal transduction pathway. The stem cells divide into transit amplifying cells, and then fully differentiate into absorptive epithelial cells, also called enterocytes, goblet cells, Paneth cells, and enteroendocrine cells. Intestinal stem cells divide asymmetrically and symmetrically, populating the villi with descendant cells in a pattern reminiscent of neutral drift. Symmetrical division result in some stem cells creating descendents that repopulate the entire crypt, whereas other stem cells differentiate and are lost from the crypt.

INDUCED PLURIPOTENT STEM CELLS

Until recently, differentiated cells were thought to be unchangeable. Two seminal discoveries have identified a mechanism that converts normal differentiated adult tissues into stem cells. In other words, a few small changes in gene expression in adult cells revert the cell into an undifferentiated pluripotent state. These **induced pluripotent stem cells (iPSCs)** divide

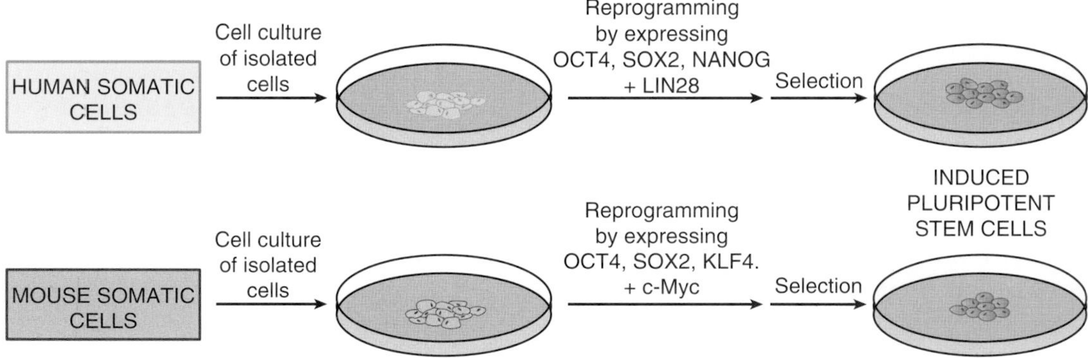

FIGURE 18.15
Induced Pluripotent Stem Cells (iPSCs)
Human differentiated somatic cells can be isolated and cultured in dishes. If the transcription factors OCT4, SOX2, NANOG, and LIN28 are activated, the cells return to their undifferentiated stem cell state. In a similar way, mouse somatic cells can be reprogrammed from the differentiated state to an undifferentiated stem cell fate by expressing four transcription factors: OCT4, SOX2, KLF4, and c-MYC.

indefinitely and remain undifferentiated unless treated with factors that reprogram them into a completely different cell type (Fig. 18.15). For example, a fibroblast can be reverted to a stem cell and then reprogrammed into a neuron or muscle cell. The first report of this appeared in 2006. Takahashi and Yamanaka converted mouse fibroblasts into embryonic stem cells (ESCs) by the expression of four different genes. Then in 2007, Yu and colleagues converted human cells into ESCs using four factors, although these four factors were different from those used by Takahashi and Yamanaka. These ESCs are now known as iPSCs.

Three major topics of scientific study on the regulation of cell differentiation merged into the discovery that normal somatic cells could be reprogrammed into iPSCs. The first key piece of research by John Gurdon in 1962 had shown that the nucleus of a somatic cell could be transplanted into an unfertilized frog egg, which stimulated the unfertilized egg to develop into a tadpole (see cloning sections later). The finding highlighted the idea that somatic nuclei retain all the information necessary to direct the synthesis of the entire organism. During the 1980s, researchers began looking for molecules that regulate gene transcription, especially for transcription factors that promote different cell fates. In this era, seminal work discovered transcription factors such as MyoD that can convert one cell type into another. MyoD can convert a fibroblast into a muscle cell. The third line of research that led to the discovery of induced pluripotent stem cells was the discovery and culturing of mouse ESCs from the inner cell mass of the blastula. These three lines of evidence led to the idea that perhaps key proteins regulate how and when a cell differentiates.

Ever since the discovery of iPSCs, many new lines of study have been developed. In fact, the use of somatic cells to create the undifferentiated cells for regenerative medicine has a lot of promise. Personalization of a treatment is one key advantage to using iPSCs for regenerative medicine. If a patient has a disease or injury that requires the replacement or regeneration of his or her organs, the idea of using one of the patient's somatic cells to regenerate the damaged organ would preclude the use of antirejection drugs. The regenerated organ or tissue would be derived from the patient's own body and therefore be immunologically compatible. Another advantage is the ability to circumvent the use of embryos for studying cellular differentiation. The iPSCs are similar to ESCs in many ways, although some researchers have found differences in gene expression and methylation patterns between the two. In fact, it is still undecided as to whether or not iPSCs and ESCs are identical, but the fact is that the iPSCs are very close to the same pluripotent stem cells derived from embryos and therefore will be less controversial to study. In addition, these types of cells could be used for drug discovery and toxicology. To directly test an experimental drug on human iPSCs would provide better evidence to the efficacy and safety of the new drug. Finally, this technology might lead to the discovery of new ways for cells in our body to be directly reprogrammed from one state to another; for example, this might lead to discoveries that could reprogram a cancer cell into a harmless somatic cell.

Making iPSCs involves expressing four different proteins. In mouse fibroblasts, the four different proteins are OCT4, SOX2, KLF4, and c-MYC. In human cells, c-MYC protein causes the cells to die, so in human cells, the four factors are OCT4, SOX2, NANOG, and LIN28. OCT4 is a transcription factor encoded by the gene called *Pou5f1*. This transcription factor is present in the zygote during the early cell divisions and is ultimately expressed in the inner cell mass of the blastula. In the adult, OCT4 protein is found only in germ cells, that is, the cells that develop into eggs or sperm. The transcription factor has a homeodomain, which has two helix–turn–helix folds for binding DNA, and is an important regulator for NANOG protein expression. In fact, many researchers believe this one factor may be the "master regulator" of embryonic development. Transgenic mice that do not make any OCT4 protein have been created by disrupting the gene for OCT4. The eggs produced without OCT4 do not create pluripotent inner cell mass cells. Instead, the inner cell mass develops with the outer part of the blastula, which forms extra-embryonic tissues such as the placenta. In support of the important role for OCT4 in development, if ESCs are genetically altered to not express OCT4, then the cells do not retain the pluripotent phenotype and form extra-embryonic tissues also.

The other genes involved in iPSC generation are important transcription factors for development also. SOX2 can form a heterodimer with OCT4 to activate transcription of key genes for the pluripotent phenotype, such as *Nanog*. SOX2 is not expressed as early in development as OCT4 and is preferentially found in early development of neural stem cells. KLF4 and c-MYC are other key molecules to create iPSCs, but KLF4 is dispensable since other members of the KLF family, KLF2 and KLF5, can replace KLF4 in the process. Additionally, c-MYC appears to be mouse specific since its overexpression in human cells is lethal. In human cells, NANOG and LIN28 are used. NANOG is also important for mouse iPSCs but is used to stabilize the pluripotent phenotype rather than establish it. The current view is that NANOG is not absolutely required to induce pluripotency. The final transcription factor, LIN28, was first identified in *C. elegans*, as a key regulator of developmental timings; that is, when the gene is defective, the worms do not molt to the next stage of development at the proper time. In addition, LIN28 has been found to regulate gene expression by binding to RNA and modulating its translation, which seems to be its key role to maintain a state of "stemness" in the iPSCs.

The ability to create iPSCs is an easy process that involves simply expressing the four different factors, but the reality is that only 1% of the cells actually reprogram fully. There is some sort of unknown regulatory element that prevents the majority of somatic cells from reprogramming. Although adult cells rarely revert to an undifferentiated state, the ability to create some of these cells has changed our view of cellular physiology. Adult cells of one tissue do retain the ability to become other types of cells. We just do not understand the complexity of the process yet.

Adult cells can be reprogrammed into undifferentiated stem cells by expressing four different transcription factors to form induced pluripotent stem cells. Mouse and human iPSCs are created using two different cocktails of transcriptions factors. The main regulator for cellular reprogramming appears to be OCT4.

STEM CELL THERAPY

The uses of stem cells as therapies for diseased tissues or injuries and even to stop the aging process are some of the potential benefits of stem cell research. Research into the potential use of embryonic stem cells, adult stem cells, as well as induced pluripotent stem cells is ongoing.

Using hematopoietic stem cells is probably the most widely used stem cell therapy to date. In 1945, people who were exposed to lethal doses of radiation died because they were unable to make blood. After reproducing the radiation sickness in mice, researchers discovered that transplanting healthy bone marrow could save mice. The bone marrow was a rich source of hematopoietic stem cells (HSCs) that were able to repopulate the marrow of irradiated mice.

Bone marrow transplants are widely used today to treat people with cancers of the white blood cells such as leukemia, lymphoma, or multiple myeloma. The HSCs in bone marrow can repopulate marrow from patients with aplastic anemia, sickle cell anemia, neutropenia, and other blood diseases. Other reasons include treatment for people who have had chemotherapy, which destroyed their bone marrow stem cells.

Donor bone marrow is derived from the patient (**autologous**) by harvesting his or her own stem cells before undergoing chemotherapy. These stem cells are then replaced when the chemotherapy is finished. This is also called a rescue transplant. For some patients, autologous transplants are not possible, and therefore, other people must donate HSCs. **Allogeneic** bone marrow has to match the person, so it is most likely to come from a family member (Fig. 18.16). Unrelated people can be a match, so even if a family member is not a match, there is still some possibility of finding a matching donor. Surprisingly, umbilical cord blood is another rich source of HSCs. These cells can be frozen and stored when a child is born so that there is always a rich source of HSCs for that person throughout his or her lifetime. The blood cells found in umbilical cord are even more immature than the cells of regular bone marrow; therefore, cord blood transplants do not have to be as closely matched as bone marrow transplants. Another source of HSCs is the peripheral, circulating blood. Remember that the stem cells often leave their niche area and migrate into the bloodstream. If the patient receives a dose of granulocyte-colony stimulating factor (G-CSF), a cytokine that induces the migration of stem cells out of the marrow and into the bloodstream, then in a few days, a large number of HSCs will be circulating in the bloodstream. The HSCs can be directly filtered from the blood by collecting the white blood cells with a cell surface protein called CD34. Of these cells, between 5% and 20% will be HSCs, which is double the amount found in the bone marrow itself. This way of collecting HSCs for transplant is by far the easiest and least painful; therefore, it has rapidly become the method of choice.

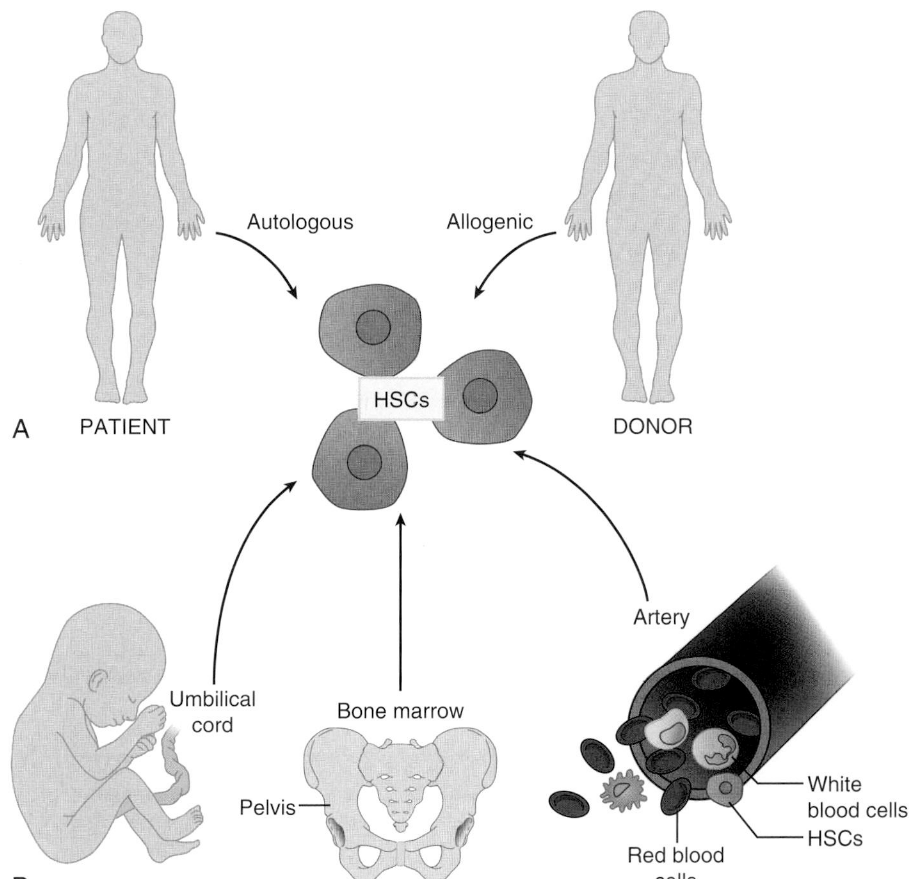

FIGURE 18.16 Sources of Hematopoietic Stem Cells for Transplants
(A) Autologous hematopoietic stem cell transplants use hematopoietic stem cells harvested directly from the patient before undergoing his or her treatment. Allogenic hematopoietic stem cell transplants use HSCs harvested from donors with the same blood type and similar HLA types. (B) The three main tissues that are rich in HSCs include umbilical cord blood, bone marrow, and peripheral blood after a patient is treated with G-CSF, which induces the HSCs to migrate out of the bone marrow and into the bloodstream.

In the case of leukemia, bone marrow transplants cure certain types of this disease. Leukemia can be broken into different categories depending on the onset and type of blood cell affected by the cancer. The two most common types of leukemia are chronic lymphocytic leukemia (CLL) and acute myeloid leukemia (AML). The overall survival rate for all leukemia has increased from 34.1% in 1975–1977 to 59.2% in 2003–2009. CLL has the highest survival rate at 83.1%. The use of stem cell therapies and the development of new drugs have significantly changed how this cancer is perceived and treated. Hodgkin's lymphoma is treated through a variety of methods but uses stem cell therapy-based treatments in cases in which standard chemotherapy does not destroy the cancer cells. There are so many treatment options that the survival rate is 93.5% for those younger than 45 years of age.

HSCs can also attack and destroy unrelated cancers. For example, metastatic renal cell (kidney) carcinoma tumors treated with isolated HSCs from peripheral blood shrunk because the stem cells stimulated the immune system to attack the tumors. Therefore, further research will determine if HSCs are capable of destroying other tumors.

Stem cell therapies hold lots of promise for treating diseases and disabilities such as Parkinson's disease, arthritis, diabetes, spinal cord injuries, and other neurodegenerative diseases such as amyotrophic lateral sclerosis. As of this writing, there are a few clinical trials on stem cell treatments that are derived from human embryonic stem cells. These treatments are expected to cure a retinal disease called Stargardt's Macular Dystrophy (SMD) and age-related macular degeneration. In 2012, researchers reported that hESC cells were differentiated into retinal pigment epithelial cells and transplanted into patients. The patients did not lose vision in subsequent months, and the treatment did not have any adverse side effects. As of this writing, the study is still recruiting patients for treatment, and no conclusive results have been obtained. Additional studies are trying to determine if a variety of different diseases, such as Alzheimer's, sickle cell anemia, heart disease, digestive system diseases, Parkinson's disease, and many others (see clinicaltrials.gov for a listing of trials using stem cells), can be cured with stem cells.

In late 2014, a new clinical trial for treating Type I diabetes was launched. Patients with Type I (formerly known as juvenile) diabetes do not produce insulin because their pancreatic β cells are destroyed by their immune system. The patients in the clinical trial will be treated with insulin-producing pancreatic β cells derived from stem cells. Doug Melton's group at Harvard University found a way to induce both human embryonic stem cells and induced pluripotent stem cells to differentiate into β cells of the pancreas. The differentiated cells produce insulin in response to glucose. The procedure involves the step-wise addition of 11 different compounds to induce the stem cells to fully differentiate. When these cells are transplanted into a mouse pancreas, the cells produce insulin in response to blood glucose levels, and the mouse model for diabetes is cured. The problem with transplanting these cells directly into human pancreas is that our immune system will attack and kill the cells. The company is working to develop a capsule that protects the cells from our immune system but allows the passage of glucagon and insulin in and out of the stem cell-derived β cells.

In addition to direct therapy for patients, cultured stem cells can be used to assess the efficacy and safety of newly developed drugs. The cultured stem cells treated with an experimental drug can also be assessed for adverse reactions. Scientists want to use induced pluripotent stem cells for this type of research since they can provide a source of different cell types. The drugs' effect on the specific cell type can then be evaluated in a dish rather than an animal model. In fact, most newly developed drugs are not brought to the market because of toxicity to the cardiovascular tissue. A source of differentiated cardiac muscle cells or cardiomyocytes from iPSCs would help eliminate bad drugs before they reach a patient and cause damage. Currently, a procedure has been developed that converts iPSCs into cardiac muscle cells. The differentiated cells include many different types of cells such as ventricular, atrial, and nodal cells; therefore, these cells mimic an animal model but do not require animal experimentation.

Newly designed drugs also can be tested on induced pluripotent stem cells from diseased tissue. A sample of tissue from the actual patient can be induced to form stem cells by expressing the four known transcription factors. For example, iPSCs have been created from motor neurons from patients with spinal muscular atrophy. These cells do not produce enough survival motor neuron 1 (SMN1) protein. These cells can be used to screen for drugs that enhance the amount of SMN1 and therefore provide a potential treatment for the disease.

Bone marrow transplants are one stem cell therapy that is widely used and is very effective to treat disorders and cancers of the blood. In other diseases, stem cells may be essential to regenerate tissues that have been destroyed by accidents or diseases. For example, stem cells have been differentiated to create retinal tissue and insulin-producing β cells. Stem cells can also provide a source of tissue to test newly discovered drugs for efficacy and safety.

SOMATIC CELL NUCLEAR TRANSFER

One of the basic discoveries leading to the development of induced pluripotent stem cells (iPSCs) was the discovery that the somatic nucleus had the information to create the entire organism. The nucleus just needed the correct environment. The major accomplishment for this field of research was the creation of Dolly the sheep in 1996. Although she created a major furor in the media, from a scientific viewpoint, her cloning was a relatively small step in a developing technology. Cloning animals relies on the technique of **nuclear transplantation**. This process actually dates back to 1952 when nuclei from early frog embryos were transplanted into eggs from which the nucleus had previously been removed. Some attempts gave rise to normal embryos. Nuclear transplantation can be used to generate a group of identical **cloned animals**. Several nuclei from the same donor are transplanted into a series of enucleated eggs. Since the 1980s, nuclear transfer in a variety of mammals has been performed successfully using nuclei from early embryos (morula or blastocyst stages). Fusing a somatic cell with an empty egg cell transfers the donor nucleus into a completely undifferentiated cytoplasm. A brief electrical pulse fuses the two cell membranes into one embryo. In 1995, nuclei from cultured embryonic cells of sheep were successfully transplanted. Two lambs, Megan and Morag, were produced by this technique at the Roslin Institute in Edinburgh, Scotland. In 1996 the same research group produced Dolly by nuclear transplantation from an adult cell line—the epithelial layer of the mammary gland. Thus, Dolly was the first mammal to be produced using a nucleus from a differentiated cell line (Fig. 18.17).

Mammals may be cloned by nuclear transplantation in which the nucleus from a somatic cell is inserted into an egg cell whose own nucleus has been removed.

DOLLY THE CLONED SHEEP

The cloning of Dolly the sheep showed that it is possible to reset the clock of an adult cell to zero and start development again, a process that is reproduced in making pluripotent stem cells. In Dolly's case, the trick was to starve cultured udder cells from the donor animal so that both the cell and the DNA stopped dividing (i.e., the cells entered the **G_0 stage** of the cell cycle; see Chapter 4). When the resting G_0 nucleus is placed in an egg cell whose

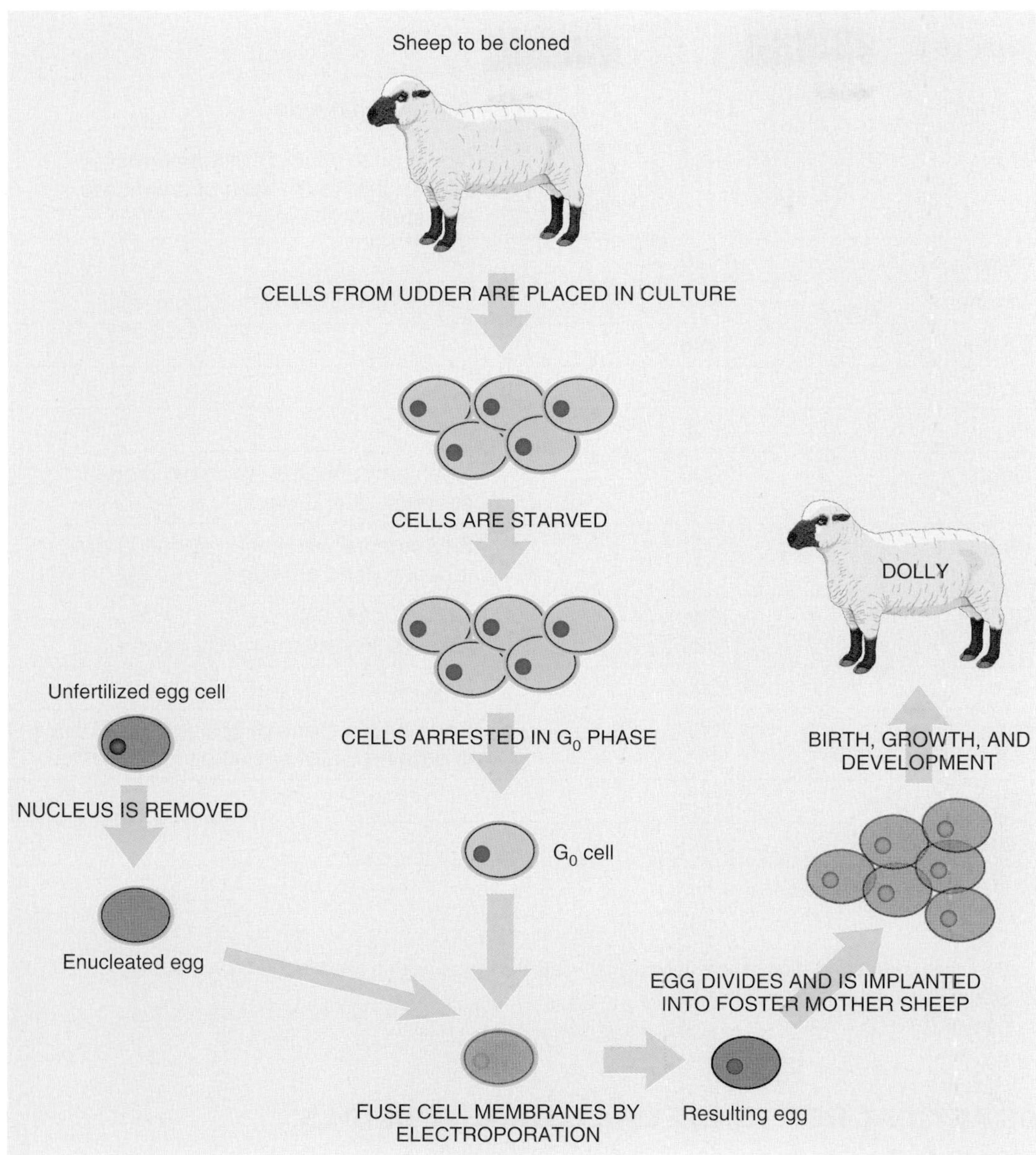

FIGURE 18.17
Scheme for Cloning Sheep
To clone a mammal such as a sheep, cells from the udder are isolated, grown in culture, and then starved in order to arrest them in G_0 of the cell cycle. Unfertilized egg cells from another sheep are also harvested, and the nucleus is removed. An electrical stimulus fuses the G_0 udder cell with the enucleated egg, thus placing a somatic cell nucleus into an undifferentiated cytoplasm. The eggs that result are put back into a foster mother, and the offspring are screened for DNA identical to the donor sheep.

own nucleus has been removed, it starts dividing again. The egg is then transplanted into a female animal, where it will develop into an embryo. If all goes well, a baby will be born (Fig. 18.17).

As mentioned previously, early in 1996, at the Roslin Institute in Scotland, the world's first cloned animal, Dolly the sheep, was born. The donor nucleus came from a mammary gland cell (also known as the udder) from a pregnant ewe. Since Dolly's birth, a variety of other animals, including cattle, pigs, goats, mice, and cats, have been cloned (Table 18.1). Dolly herself was mated and gave birth to a lamb of her own—named Bonnie—during Easter 1998.

Strictly speaking, Dolly was not a complete clone. In addition to the nucleus, which contains the majority of the genetic information, animal cells contain a few genes in their mitochondria. In Dolly's case, only the nuclear DNA was cloned. The mitochondrial DNA was provided by the egg cell that received the nucleus.

Table 18.1 Cloned Animals

Animal	Date	Name/Comments
Frog	1952	Nobel prize winning research that contributed to the idea of stem cells and stem cell research
Sheep	1996	Dolly
Mouse	1997	Cumulina (followed by 50 others!)
Cattle	1998	
Goat	1999	
Pig	2000	
Gaur	2000	Noah (endangered Asian ox; died of infection after 2 days)
Mouflon	2001	Endangered wild sheep (cloned from recently dead animal)
Cat	2001 2004	CopyCat (or CC) Little Nicky, commercially cloned cat
Deer	2003	Dewey
Dog	2005	Snuppy is a clone of Tei; cloning of dogs is commercially available in South Korea
Rhesus Monkey	2007	
Zebrafish	2009	

Following Dolly the cloned sheep, several mammals have now been cloned. However, the success rate is still low.

PRACTICAL REASONS FOR CLONING ANIMALS

Why clone sheep? Aside from showing that whole animals can be cloned, there are practical reasons. For millennia, humans have bred farm animals in attempts to improve them. Genetic duplication allows an improved animal to be widely distributed relatively quickly.

In addition, a flock or herd of genetically identical animals will give wool, milk, eggs, or meat of a more standardized quality. Conversely, genetically identical animals will all be susceptible to the same infections, and epidemics will spread faster and further.

Although cloning produces identical animals, the technique may paradoxically help protect genetic diversity. For example, in New Zealand, the last surviving cow of a rare breed was successfully cloned. Thus, cloning allows us to genetically rescue rare breeds of animals or endangered species but avoids mixing their genes with outsiders, as would happen in crossbreeding. Table 18.1, which lists animals cloned to date, includes some rarities, including the gaur and mouflon.

The most important use of animal cloning is in combination with transgenics. Previously created transgenic animals may be cloned for speedier distribution of the product. However, it is also possible to introduce transgenic DNA during the cloning process. The Roslin Institute, where Dolly was born, has since cloned sheep carrying the gene for **human factor IX**. The transgene was inserted into the nuclear donor cells while in culture. A variety of other transgenic animals carrying pharmaceutically important proteins have also been cloned.

The reason Dolly was cloned using a mammary gland cell should now be apparent. If a foreign protein is expressed in this tissue, it will be secreted into the milk and be easy to harvest commercially. Once a good transgenic cell line has been established in culture, nuclear transplantation can be used to generate several genetically identical animals for production purposes. Such transgenic cells can be stored over the long term by freezing in liquid nitrogen and have been referred to as *protoanimals.*

Cloned animals may be altered by transgenics, thus generating animals with useful properties more rapidly than traditional selective breeding.

IMPROVING LIVESTOCK BY PATHWAY ENGINEERING

It is possible to combine pathway engineering with cloning to produce improved livestock (as opposed to merely using transgenic animals as the source of a single useful protein). For example, the genes for the cysteine biosynthetic pathway are absent in mammals. Consequently, mammals, including sheep, cannot make the sulfur-containing amino acid cysteine and must receive it in their diet. Sometimes cysteine is limiting for wool growth. Adding extra cysteine to the diet works poorly because of its uptake and degradation by microorganisms in the sheep's intestine.

Many bacteria do synthesize cysteine. Two steps are needed, starting from the amino acid serine plus inorganic sulfide (Fig. 18.18). In enteric bacteria, these two key enzymes, **serine transacetylase** and **acetylserine sulfhydrylase**, are encoded by the *cysE* and *cysK* genes, respectively. These two bacterial genes have been cloned from *Salmonella* and placed under control of the mouse metallothionein promoter. The construct has been successfully integrated into transgenic mice, which expressed both enzymes. Furthermore, when these mice were deprived of the two sulfur-containing amino acids, cysteine and methionine, they remained healthy and made their own cysteine. The animals did need inorganic sulfide in their diet.

Although transgenic sheep carrying bacterial *cysE* and *cysK* genes have been made, high-level expression of these genes and synthesis of cysteine in the desired location (the epithelium of the rumen) have not yet been achieved. Several other schemes to improve livestock, including synthesis of other essential amino acids such as lysine and threonine, are in the early experimental stages.

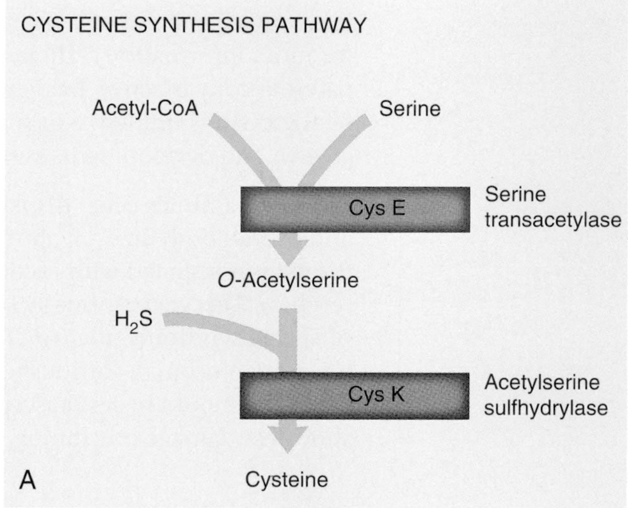

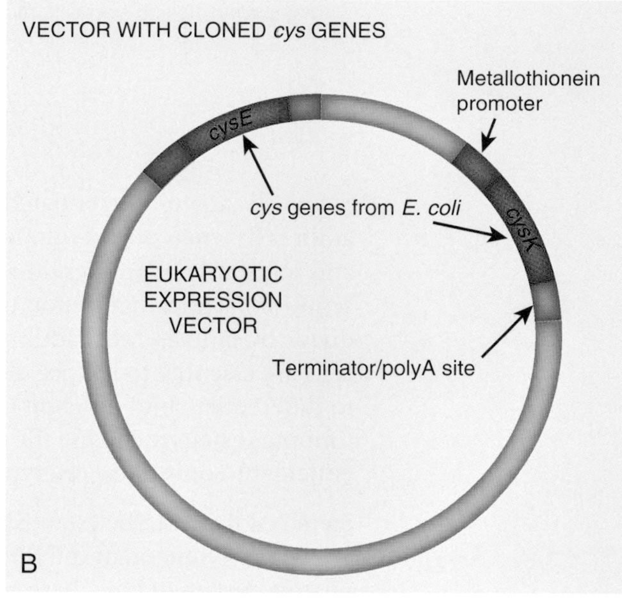

FIGURE 18.18 Biosynthesis of Cysteine from Serine plus Sulfide

(A) The cysteine biosynthetic pathway uses serine and acetyl-CoA. First, acetyl-CoA and serine are converted to *O*-acetylserine by serine transacetylase. Then acetylserine sulfhydrylase converts *O*-acetylserine to cysteine using hydrogen sulfide (H_2S). (B) The cysteine biosynthetic pathway has been cloned into a eukaryotic expression vector. The two *E. coli* genes are cloned behind the mammalian metallothionein promoter.

One way to significantly improve livestock would be to engineer pathways to synthesize essential amino acids that must presently be provided in the diet. So far, partial success has been achieved using mice as model organisms.

IMPRINTING AND DEVELOPMENTAL PROBLEMS IN CLONED ANIMALS

Even after the technological problems have been conquered, most attempts at cloning animals still fail. Successful nuclear transplantation involves reprogramming a nucleus from a differentiated cell. This is a complex process, and the low success rate is probably due to failure to properly reprogram. Recent investigations implicate DNA methylation as the major factor. Methylation patterns in cloned embryos are not identical to those in natural embryos. This is even true for most successfully cloned animals. However, Dolly and other cloned animals have given rise to genetically normal offspring, so cloning does not introduce permanent genetic alterations into an animal's germline.

Methylation generally shuts down eukaryotic genes that are not required in a particular tissue or at a particular stage in development. Methylation is also involved in **imprinting**, the regulatory mechanism that activates only the maternal or paternal copy of a gene (see Chapter 17 for more information). There are about 40 imprinted genes in humans. Usually, when the paternal copy is active, fetal growth is promoted; whereas when the maternal copy is active, fetal growth is limited. Not surprisingly, incorrect imprinting patterns cause aberrant fetal growth and development, as seen in many embryos derived from cloning.

Cloned mammals often display **large-offspring syndrome**. Their limbs, interior organs, and overall body size are abnormally large. The resulting animals are unhealthy. This syndrome is associated with incorrect imprinting of *IGF2R*, the gene for insulin growth factor 2 receptor. This gene normally shows maternal imprinting in mammals. In fetuses with large-offspring syndrome, the *IGF2R* gene showed altered methylation, and its expression was lower than normal. Curiously, the *IGF2R* gene is not imprinted in primates, so humans and monkeys should be less susceptible to large-offspring syndrome and might be expected to show less damage on cloning, at least from this particular cause.

Mammalian cloning suffers from a high failure rate, the cause of which is still not understood.

Summary

Stem cells are undifferentiated cells that have the potential to develop into different types of adult cells when placed into new environments. The process of changing their transcriptional and translational profiles during the process of differentiation is mediated by various transcription factors. In addition to the undifferentiated state, stem cells also retain the ability to divide by mitosis. Most adult or differentiated cells lose this potential, and therefore, stem cells are essential to replace old and worn out cells that are removed by apoptosis. The ability to differentiate into different cell types varies among different stem cells, with the ultimate totipotent stem cell being the newly formed zygote to multipotent stem cells that can differentiate into only a few cell types, such as intestinal stem cells.

Stem cell lines can be established and grown *in vitro* by isolating stem cells from embryos. These embryonic stem cell lines are controversial since the embryo is destroyed in this process, and therefore, these cell lines are strictly regulated. Stem cell lines from adult or multipotent stem cells can also be created, but the identification and isolation of stem cells from the complex tissue and/or organ varies from tissue to tissue. In many tissues, there has not been a definitive marker that identifies the stem cells. Other tissues have suspected markers such as spectrosomes found in *Drosophila* ovaries or signal transduction proteins such as Notch. The best way to identify a stem cell in adult tissues is by lineage tracing or fate mapping, where the descendants of a particular cell are labeled with a marker. Other methods include transplantation, where labeled cells from one organism are transplanted into embryos of an unlabeled organism.

Stem cells reside in a niche, or microenvironment, that keeps the stem cells in their undifferentiated state. Other cells within the niche, the extracellular matrix, and the stem cells themselves provide the signals that keep the stem cells within the niche undifferentiated. Niches can be open to the surrounding tissues or closed. Stem cells divide within the niche through a variety of mechanisms. Asymmetric division provides a replacement stem cell and a daughter cell that differentiates by either unequal partitioning of intrinsic factors or the movement from the niche. The external asymmetry from the one daughter cell leaving the niche can be sufficient to trigger the cell to differentiate. Symmetrical cell divisions of a stem cell produce either two more stem cells or two daughter cells that differentiate.

Two examples of adult stem cell niches are hematopoietic stem cell niches of the bone marrow and intestinal stem cell niches found at the base of the villi of the intestine. These stem cell niches create an environment that promotes stem cell survival, and when the cells leave the niche, they begin to differentiate. In intestinal cells, the first cells to leave the niche become transit amplifying cells, which are an intermediate state of differentiation. These cells are able to divide a few times before they fully differentiate into the different cells of the intestine.

Ground-breaking research on nuclear transplantation, transcription factors, and the creation of embryonic stem cell lines merged into the discovery that adult differentiated cells can be reprogrammed into undifferentiated cells by simply expressing a cocktail of four different transcription factors. The induced pluripotent stem cells (iPSCs) have similar characteristics as embryonic stem cells because they are able to grow and divide *in vitro*, they are undifferentiated, and when they are treated with different chemicals, the iPSCs can become differentiated cells.

Stem cells have a variety of potential applications that may provide cures or treatments for a variety of diseases. Treating leukemias and other blood-derived diseases with bone marrow transplants is one stem cell treatment that is used today. The blood stem cells can be isolated from the patient before he or she is treated (autologous), from a blood-type matched donor (allogeneic), or from circulating blood since blood stem cells leave their niche and travel through the bloodstream.

Stem cell research was based on early experiments of somatic cell nuclear transfer or cloning. This technique involves the transfer of a somatic cell nucleus into an enucleated egg. The somatic nucleus is stimulated with a mild electrical pulse, which stimulates the cell to develop into a blastula. When this is transplanted into a mother, the blastula forms a cloned animal. This technique has been used with a variety of different mammals, but the most famous cloned animal is Dolly the sheep. Cloning animals is useful for a variety of reasons, but the most important is to create genetically identical transgenic animals for research or production of biological drugs such as human factor IX. Cloning will also be useful for creating livestock that have new biochemical pathways.

Although cloning and stem cells can potentially solve a lot of agricultural problems, medical problems, and even environmental issues, the research is still very young. Potential problems exist with both technologies. For example, DNA methylation patterns that are established and altered in the embryo may not occur properly in a cloned animal. The epigenetic changes of stem cells are still not fully understood and may be among the reasons that only 1% of the cells treated with the four transcription factors fully develop into iPSCs, and why cloned animals tend to have problems such as large offspring syndrome.

End-of-Chapter Questions

1. Stem cells that have lost the ability to divide are called ______________.
   - a. totipotent
   - b. competent
   - c. undifferentiated
   - d. differentiated
   - e. transformed
2. Embryonic stems cells are derived from the ______________.
   - a. inner cell mass
   - b. embryoid bodies
   - c. zygote
   - d. implanted embryo
   - e. none of these
3. Oct4 and Nanog are transcription factors that are only expressed in ______________.
   - a. adult stem cells
   - b. differentiated embryonic stem cells
   - c. undifferentiated embryonic stem cells
   - d. the blastula
   - e. embryoid bodies
4. How are adult stem cells identified?
   - a. Lineage analysis of a marker within a labeled potential stem cell.
   - b. Identification of organelle-like structures called spectrosomes.
   - c. Genetic mosaicism by transplantation
   - d. Identification of signal transductions molecules specific to stem cells.
   - e. All of the above are ways in which adult stem cells can be identified.
5. When one daughter cell (from a stem cell that has asymmetrically divided) receives a signaling molecule that the other daughter cell does not, this is called ______________.
   - a. extrinsic asymmetry
   - b. intrinisic asymmetry
   - c. asymmetric cell division
   - d. symmetrical renewal
   - e. symmetrical differentiation
6. Concerning hematopoietic stem cells, which factor is mismatched with its function?
   - a. N-cadherin – adheres stem cells to the niche environment
   - b. Notch – maintain the stem cell pool as undifferentiated cells
   - c. SCF – maintains undifferentiated stem cells
   - d. Interleukin – trigger formation of T lymphocytes
   - e. Wnt – signals the stem cells to differentiate into blood cells
7. How are intestinal stem cells best identified?
   - a. location in the intestine
   - b. specific marker called Lgr5
   - c. presence of transit amplifying cells
   - d. proximity to Paneth and enteroendocrine cells
   - e. presence of neutral drift

8. An adult cell that reverts back into an undifferentiated state is called a/an ______________.

a. totipotent cell
b. induced pluripotent stem cell
c. untransformed cell
d. somatic cell
e. stem cell

9. Which protein is not part of the four needed in making iPSCs in humans?

a. OCT4
b. NANOG
c. SOX2
d. LIN28
e. C-MYC

10. What cytokine is given to patients to induce the migration of stem cells out of the marrow and into the blood stream?

a. human growth factor
b. granulocyte-colony stimulating factor
c. C-MYC
d. LIN28
e. none of these

11. All of the following diseases are targets of potential stem cell therapies discussed in the textbook except ______________.

a. Parkinson's disease
b. arthritis
c. spinal cord injuries
d. diabetes
e. All of the above are disease targets for therapy.

12. Two lambs, Megan and Morag, were produced ______________.

a. when the somatic nucleus was transplanted into an enucleated egg
b. when the nuclei from differentiated cell lines were transplanted into enucleated eggs
c. when the nuclei of cultured embryonic cells were transplanted into enucleated eggs
d. by fusing the nuclei of two egg cells
e. none of these

13. Why is Dolly not a complete clone?

a. Mitochondrial DNA was provided by the egg cell only.
b. The egg cell was not enucleated.
c. The mitochondria were also cloned into the egg cell from the mammary gland cell.
d. Only mitochondrial DNA was used.
e. There were two sources of nuclear DNA.

14. Which is not necessary when improving livestock by pathway engineering, specifically for cysteine biosynthesis?

a. serine transacetylase
b. acetyserine sulfhydrylase

(Continued)

   c. dietary inorganic sulfide
   d. metallothionein promoter
   e. *cysE* and *cysK* genes

15. All of the following are true statements about imprinting except ______________.
   a. methylation generally shuts down eukaryotic genes
   b. imprinting is a regulatory mechanism that activates only the maternal or paternal copy of a gene
   c. fetal growth is promoted when the maternal copy of a gene is active
   d. incorrect imprinting of IGF2R results in large-bodied cloned mammals
   e. when cloning animals, permanent genetic alterations are not introduced, as evidenced by the ability of a cloned animal to have genetically normal offspring

Further Reading

Adachi, K., Nikaido, I., Ohta, H., Ohtsuka, S., Ura, H., Kadota, M., et al. (2013). Context-dependent wiring of Sox2 regulatory networks for self-renewal of embryonic and trophoblast stem cells. *Molecular Cell, 52*, 380–392.

Barker, N., van Oudenaarden, A., & Clevers, H. (2012). Identifying the stem cell of the intestinal crypt: strategies and pitfalls. *Cell Stem Cell, 11*, 452–460.

Chen, S., Lewallen, M., & Xie, T. (2013). Adhesion in the stem cell niche: biological roles and regulation. *Development, 140*, 255–265.

Dalton, S. (2013). Signaling networks in human pluripotent stem cells. *Current Opinion in Cell Biology, 25*, 241–246.

Deng, W., & Lin, H. (1997). Spectrosomes and fusomes anchor mitotic spindles during asymmetric germ cell divisions and facilitate the formation of a polarized microtubule array for oocyte specification in *Drosophila*. *Development Biology, 189*, 79–94.

Green, R. M. (2014). Ethical considerations. In R. Lanza, & A. Atala (Eds.), *Essentials of Stem Cell Biology* (pp. 595–604). San Diego, CA, USA: Academic Press.

Gurdon, J. B. (1962). The developmental capacity of nuclei taken from intestinal epithelium cells of feeding tadpoles. *Journal of Embryology and Experimental Morphology, 10*, 622–640.

Hackett, J. A., & Surani, M. A. (2014). regulatory principles of pluripotency: from the ground state up. *Cell Stem Cell, 15*, 416–430.

Hsu, Y.-C., Li, L., & Fuchs, E. (2014). Transit-amplifying cells orchestrate stem cell activity and tissue regeneration. *Cell, 157*, 935–949.

Jiang, J., Chan, Y.-S., Loh, Y.-H., Cai, J., Tong, G.-Q., Lim, C.-A., et al. (2008). A core Klf circuitry regulates self-renewal of embryonic stem cells. *Nature Cell Biology, 10*, 353–360.

Jones, D. L., & Fuller, M. T. (2004). Stem cell niches. In R. Lanza, J. Gearhart, B. Hogan, D. Melton, R. Pedersen, J. Thomson, & M. West (Eds.), *Handbook of Stem Cells*, Vol. 2 (pp. 59–72). Academic Press.

Kurosawa, H. (2007). Methods for inducing embryoid body formation: *in vitro* differentiation system of embryonic stem cells. *Journal of Bioscience and Bioengineering, 103*, 389–398.

Morrison, S. J., & Spradling, A. C. (2008). Stem cells and niches: mechanisms that promote stem cell maintenance throughout life. *Cell, 132*, 598–611.

Nichols, J., Zevnik, B., Anastassiadis, K., Niwa, H., Klewe-Nebenius, D., Chambers, I., et al. (1998). Formation of pluripotent stem cells in the mammalian embryo depends on the POU transcription factor Oct4. *Cell, 95*, 379–391.

Okita, K., & Yamanaka, S. (2014). Induced pluripotent stem cells. In R. Lanza, & A. Atala (Eds.), *Essentials of Stem Cell Biology* (pp. 375–385). San Diego, CA, USA: Academic Press.

Pellegrini, G., & De Luca, M. (2014). Eyes on the prize: limbal stem cells and corneal restoration. *Cell Stem Cell, 15*, 121–122.

Perdigoto, C. N., & Bardin, A. J. (2013). Sending the right signal: notch and stem cells. *Biochimica et Biophysica Acta, 1830*, 2307–2322.

Schwarz, B. A., Bar-Nur, O., Silva, J. C. R., & Hochedlinger, K. (2014). Nanog is dispensable for the generation of induced pluripotent stem cells. *Current Biology, 24*, 347–350.

Shyh-Chang, N., & Daley, G. Q. (2013). Lin28: primal regulator of growth and metabolism in stem cells. *Cell Stem Cell, 12*, 395–406.

Tajbakhsh, S. (2014). Ballroom dancing with stem cells: placement and displacement in the intestinal crypt. *Cell Stem Cell, 14*, 271–273.

Takahashi, K., & Yamanaka, S. (2006). Induction of pluripotent stem cells from mouse embryonic and adult fibroblast cultures by defined factors. *Cell, 126*, 663–676.

Tan, D. W. -M., & Barker, N. (2014). Intestinal stem cells and their defining niche. *Current Topics in Developmental Biology, 107*, 77–107.

Welling, M., & Geijsen, N. (2013). Uncovering the true identity of naïve pluripotent stem cells. *Trends in Cell Biology, 23*, 442–448.

Yamanaka, S. (2012). Induced pluripotent stem cells: past, present, and future. *Cell Stem Cell, 10*, 678–684.

Yu, J., Vodyanik, M., Smuga-Otto, K., Antosiewicz-Bourget, J., Frane, J. L., Tian, S., et al. (2007). Induced pluripotent stem cell lines derived from human somatic cells. *Science, 318*, 1917–1920.

Cancer

Biotechnology

http://dx.doi.org/10.1016/B978-0-12-385015-7.00019-3

CANCER IS GENETIC IN ORIGIN

Both inherited diseases and cancers are genetic in origin. Inherited diseases are genetic defects that are passed on from one individual to another via the germline. Mutations must occur in the germline cells, which give rise to the eggs and sperm, to be passed on to the descendants of a multicellular organism. In contrast, each occurrence of cancer is limited to a single multicellular organism and is not passed on to the next generation. If a mutation does occur in somatic cells—those making up the rest of the body—a variety of possibilities may result (Fig. 19.1). A mutation that occurs early on in embryonic development may be highly detrimental, because each cell of the embryo gives rise to many cells during development. If the single precursor cell for a major organ or tissue suffers a serious mutation, the results may be serious or fatal. Most **somatic mutations** occurring later in development will affect only one or a few cells and will be of little major significance.

Some somatic mutations occurring after the organism has reached maturity are still dangerous. **Cancers** are the result of somatic mutations that damage the regulatory system controlling cell growth and division. Cancer often starts with a single cell starting to grow and divide again, long after it is supposed to have ceased (Fig. 19.1). Cancers develop in several stages and require multiple mutations. First, normal control of cell division is lost. Second, the abnormal (mutant) cell divides to form a microtumor. Many microtumors are formed throughout our bodies. Most of these are destroyed by the immune system, whereas others lie quiescent for many months to years.

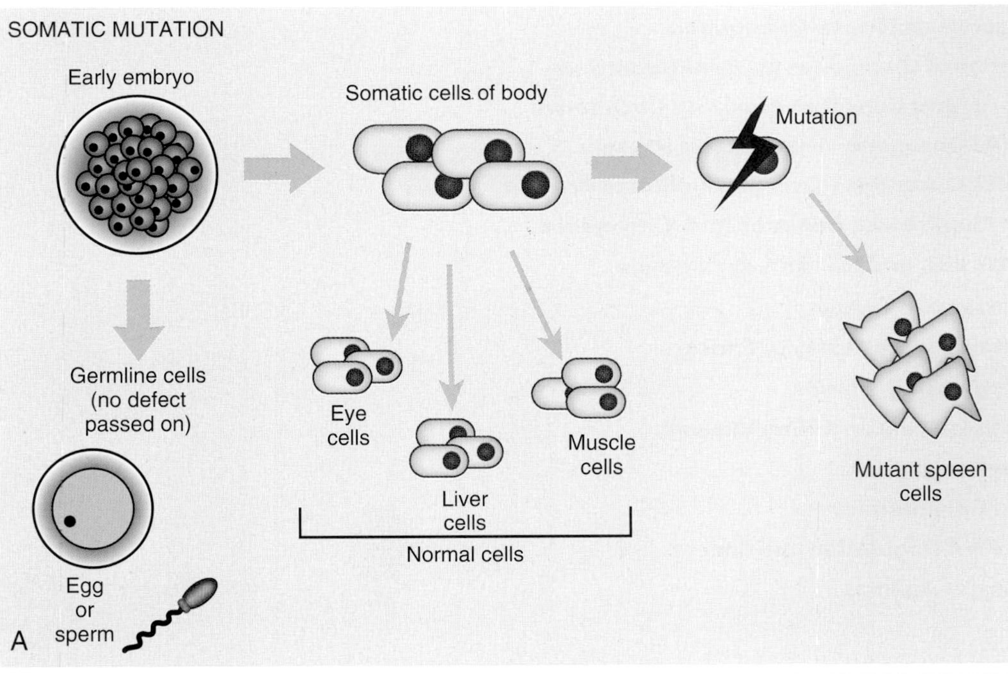

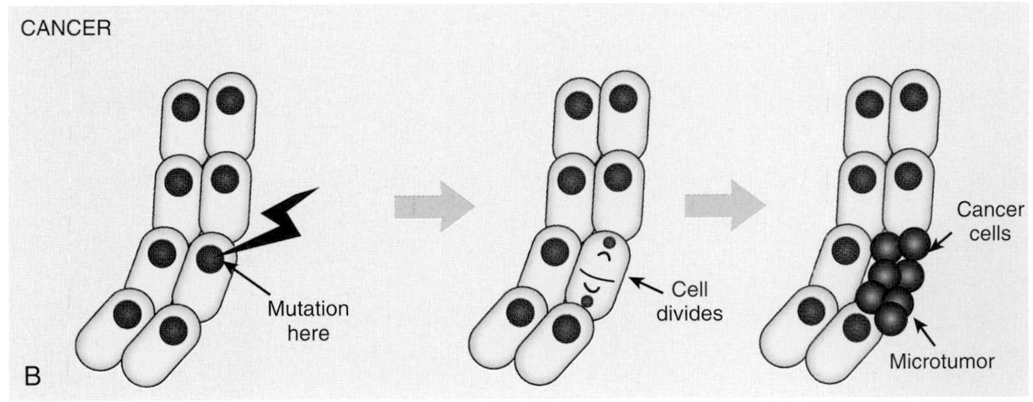

FIGURE 19.1
Cancer Is Caused by Somatic Mutations
(A) Early embryos have germline cells, which become eggs and sperm, and somatic cells that make up the rest of the organism. Depending on the timing, a somatic mutation usually affects only a small portion of the entire organism, such as the spleen. (B) Cancer occurs when a somatic cell mutates because of errors during DNA replication or exposure to a carcinogen. The mutation allows division in a cell that should no longer divide. The mutant cells keep dividing to form a microtumor.

Small microtumors of about a million cells lie dormant because they have no way of obtaining nutrients. But further mutations may allow the tumor to acquire a blood supply. The mutant cancer cells emit signaling molecules to surrounding tissues, attracting vascular endothelial cells to organize into blood vessels, a process called **angiogenesis**. Once the microtumor gets its own blood supply, it can continue to grow into a large mass. If this stays in one place, it is known as a benign tumor and can often be cut out by a surgeon, resulting in complete recovery. Finally, as a result of further mutations, a cancer may gain the ability to invade other tissues and form secondary tumors and is then said to be malignant. Cancers that have spread are much more difficult to cure.

Cancers are somatic mutations that disturb the control of cell division and cell death. Multiple somatic mutations are needed to form microtumors. Microtumors grow into large tumors when the tumor gains blood vessels.

ENVIRONMENTAL FACTORS AND CANCER

It is true that smoking cigarettes, harmful chemicals in our environment, and nuclear radiation all cause cancer. All of these factors act by causing mutations. Because cancers are primarily the result of mutations in somatic cells, those chemicals that cause mutations, known as mutagens, can cause cancer. Radiation that causes DNA damage similarly leads to mutations and cancers. Few mutations actually cause cancer. Most mutations do not even affect transcribed regions of DNA, and even if they do affect a particular gene, it is not likely to be involved in controlling cell division.

In practice, some cancer-causing mutations are due to environmental factors, whereas others occur spontaneously as a result of mistakes made during replication of the cell's DNA. Cancer-causing agents are often called **carcinogens**. Almost all carcinogens are also mutagens. Occasional discrepancies occur due to metabolism within the body, usually in the liver. Certain chemicals, which do not react with DNA themselves, may be altered by the body, giving rise to derivatives that do react with DNA. In this case, the original compound is by definition a carcinogen but not, strictly speaking, a mutagen.

Approximately 80% of cancers are derived from the epithelial cells that form the outer covering of tissues. Epithelial cells are the surface cells of the skin and are also found lining the intestines and the lungs. Because the outermost layers are constantly worn away, the underlying layers must keep dividing. Cells from tissues where cell division is rare only occasionally become cancerous. (Nerve and muscle cancers account for only 3% to 4% of the total.) In addition, surface cells are much more likely to be exposed to dangerous chemicals and harmful radiation.

Carcinogens are cancer-causing agents. Mutagens are agents that change the genetic code in a cell. Almost all carcinogens are mutagens.

NORMAL CELL DIVISION: THE CELL CYCLE

To understand further how cancer occurs, we must first consider the process of normal cell division. The eukaryotic cell cycle has four stages (see Chapter 4 and Fig. 4.9):

1. **G_1 phase**—The cell grows.
2. **S (synthesis) phase**—The DNA and chromosomes are duplicated.
3. **G_2 phase**—The cell grows and prepares to divide.
4. **M (mitosis) phase**—The cell and its nucleus divide.

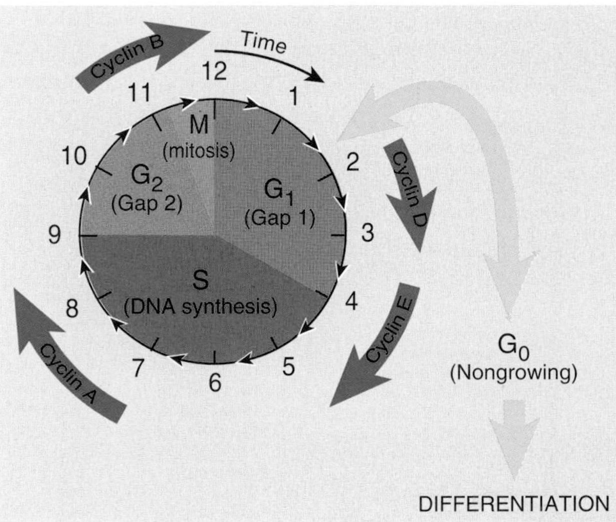

FIGURE 19.2
Eukaryotic Cell Cycle: Division versus Differentiation
The cell cycle normally consists of the four stages G_1, S, G_2, and M. However, if the conditions are right, rather than going from G_1 into the S phase, the cell may differentiate and enter G_0. If the cell does not differentiate, a signal is received from cyclin D and E, and the cell enters S phase and replicates its DNA. After about 5 hours, another signal from cyclin A triggers the cell to enter G_2. After cyclin B becomes active, the cell enters mitosis and divides.

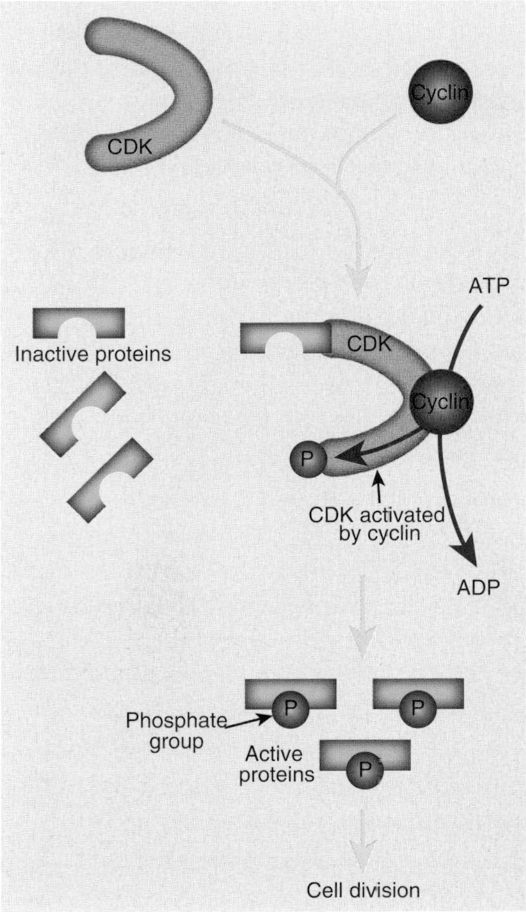

FIGURE 19.3
Cyclins Operate via Cyclin-Dependent Kinases
Before each cell can enter a new phase of the cell cycle, the cyclin must complex with a cyclin-dependent kinase (CDK). Addition of a phosphate from ATP activates the cyclin/CDK complex. The active complex then transfers the phosphate to other proteins that execute cell division.

In addition, cells may exit from the growth and division cycle into the **G_0 phase**. Most nondividing cells are in G_0. Many of these will differentiate and rarely divide again, under normal circumstances (Fig. 19.2).

To move from one stage to another requires the permission of proteins called **cyclins**, one for each major stage. The cyclins act as security checkpoints. They monitor the environment and also check to make sure that the previous stage of the cell cycle has been finished properly before moving on. The cyclins work in conjunction with the **cyclin-dependent kinases (CDKs)**. When the cyclin for a particular step in the cell cycle senses that conditions are correct, it binds to the appropriate CDK (Fig. 19.3). This process activates the CDK, which then adds phosphate groups to a series of other proteins. These are the enzymes and structural proteins that actually carry out the process of cell division. These proteins are on standby until the added phosphate group activates them. Several anti-oncogenes act by blocking the action of the cyclins (see later discussion).

Perhaps the most critical checkpoint is the transition between the G_1 and S phases, which is controlled by two transcription factors: **E2F** and p53 together with the **pRB** protein (product of the retinoblastoma gene, an anti-oncogene). E2F promotes the expression of several genes involved in DNA replication. It also increases synthesis of the cyclins E and A that control the cell cycle beyond G_1 (Fig. 19.4). Binding to pRB inactivates E2F until the cell receives a signal from the outside. External **growth factors** cause synthesis of cyclin D. This activates CDK4, which in turn phosphorylates pRB. Phosphorylated pRB releases E2F, which is then free to activate its target genes. The negative side of this regulation system is dominated by p53 protein (see later discussion).

Eukaryotic cells use cyclins to control the progression through the stages of cell division. At each stage, the cell reaches a checkpoint to ensure all the proteins and DNA are in the correct position before moving to the next stage. E2F and pRB control the checkpoint from G_1 to S.

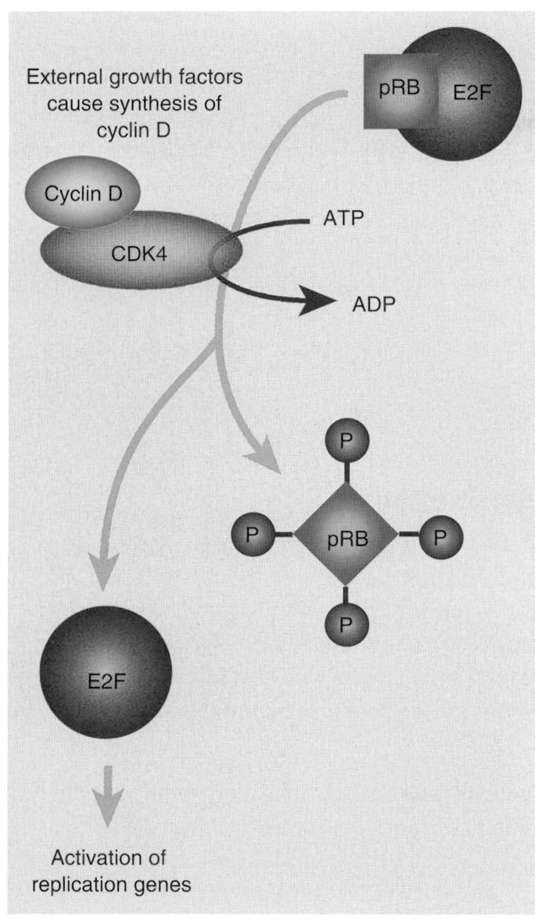

FIGURE 19.4 Control of G_1 to S Transition by pRB and E2F

At the end of G_1, cyclin D must be activated to initiate transition into S phase. This requires binding of cyclin D to its partner, CDK4. Once they are together, ATP is hydrolyzed and a phosphate is transferred to the cyclinD/CDK4 complex. This phosphate is then transferred to pRB, which releases E2F to transcribe genes needed to initiate DNA replication.

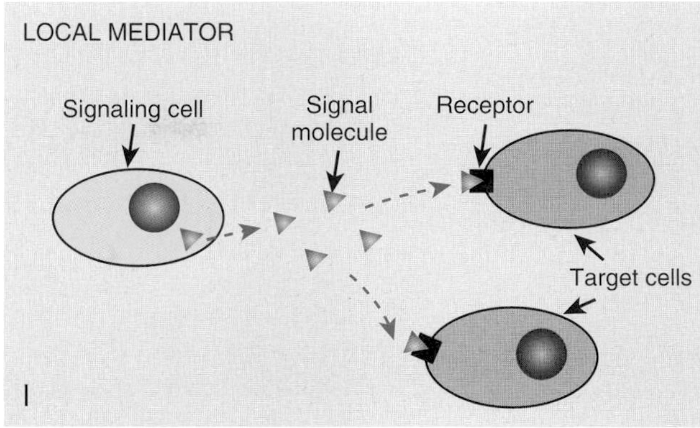

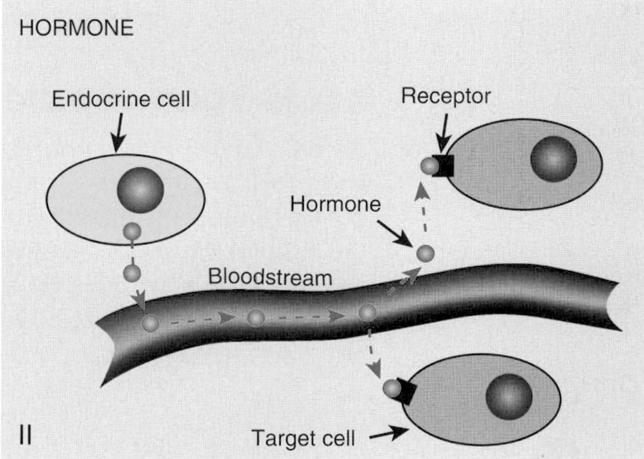

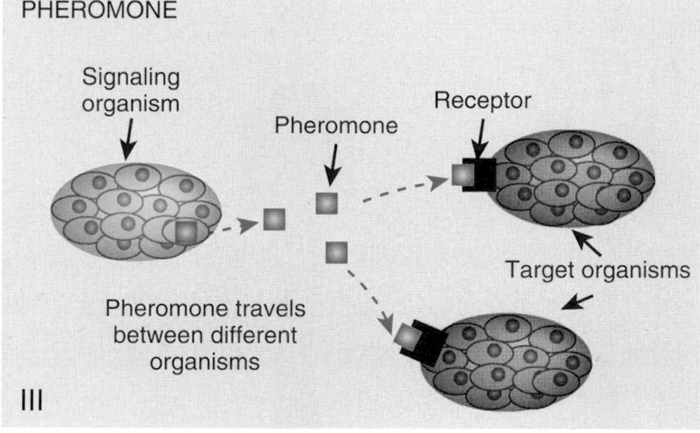

FIGURE 19.5 Local Mediators, Pheromones, and Hormones

Three different types of chemical signals are used to stimulate cells. Neighboring cells may communicate via local mediators. Endocrine cells secrete hormones that enter the bloodstream and carry signals to other cells far away from the original cell. Finally, single or multicellular organisms can emit pheromones that travel through the environment rather than within the organism.

CELLULAR COMMUNICATION

Multicellular organisms depend on coordinating the activities of many different cells, and this requires constant communication. Some signals are sent by cell-to-cell contact. More often, signals are sent by chemical means. **Local mediators** are molecules that carry signals between nearby cells, whereas **hormones** carry signals to remote tissues and organs in animals, plants, and fungi (Fig. 19.5). Those hormones and growth factors that affect cell division are especially important in cancer formation (see following discussion).

In addition to such internal signals, organisms may also send signals from one organism to another. Signal molecules sent from one organism to another are called **pheromones**. Even single-celled microorganisms such as yeast and bacteria communicate with each other through pheromones. Yeast secrete pheromones to signal readiness for mating. Note that pheromones are used for signaling between organisms (whether single or multicelled), whereas hormones circulate internally inside multicellular organisms.

In contrast to plants and fungi, multicellular animals also possess nervous systems to send messages. In this case, the signal travels in electrical mode along an extremely elongated cell, known as a neuron, until it reaches a junction, or synapse. Here, the signal must traverse the gap between two cells by chemical means. Each arriving electrical impulse stimulates the release of a pulse of neurotransmitter molecules from the nerve cell. This chemical signal diffuses across the narrow gap to the next cell. Nervous impulses travel much faster than hormones, and in addition, each neuron delivers its signal to only a single target cell. In contrast, hormones act on multiple recipient cells and travel relatively slowly.

Cellular communication is important in multicellular organisms. Local mediators send signals from one cell to its neighbors. Hormones trigger different cells at long distances because they travel in the bloodstream. Pheromones travel from one organism to the next via the air, water, or soil. Vertebrates have neurons that facilitate cellular communication by neurotransmitters.

RECEPTORS AND SIGNAL TRANSMISSION

A wide variety of molecules are used in signaling, but in all cases the recipient cell needs a **receptor**. Usually, the receptor is a protein situated in the cytoplasmic membrane with the site for binding the messenger facing the outside. When a signal molecule appears, it binds to the receptor, causing a conformational change. The receptor then passes the signal on to other proteins, known as **signal transduction (or transmission) proteins**. Both receptors and signal transduction proteins often dimerize or dissociate during the process of signal transmission. There are three main types of signal transmissions.

(a) **Phosphotransfer systems.** Activation of the receptor causes the phosphorylation of signal transduction proteins (Fig. 19.6). **Protein kinases** transfer phosphate groups from ATP to other proteins. The activated receptor may act as a protein kinase itself, or it may bind to and activate a separate protein kinase. Often several proteins take part in a phosphotransfer cascade that allows the signal to be amplified or modulated in a variety of ways.

(b) **Second messengers.** Activation of the receptor results in synthesis of a small intracellular signal molecule known as the second messenger (Fig. 19.7). This may directly activate or inhibit various enzymes and may also enter the nucleus and affect gene expression. Often a GTP-binding **G protein** links the receptor to the enzyme that makes the second messenger.

(c) **Ion channel activation.** In this case, the receptor itself acts as an ion channel (Fig. 19.8). On receiving the external message, it either opens or closes. The altered movement of ions through the channel then mediates further signaling.

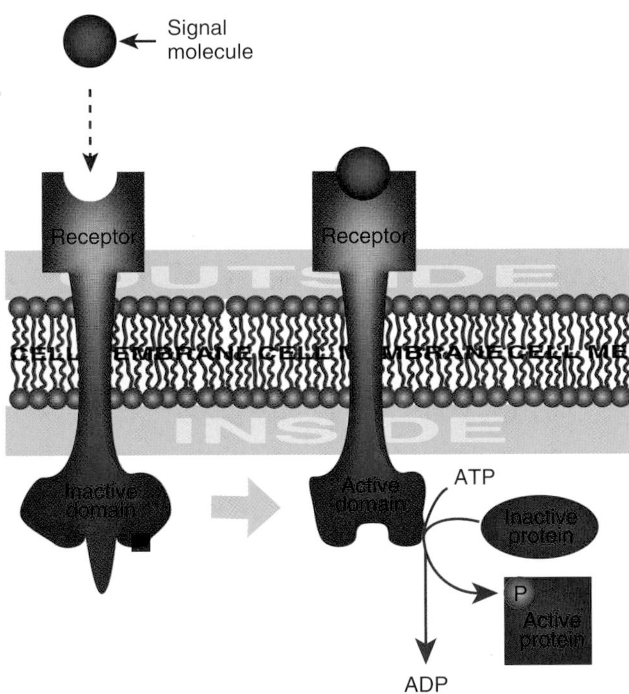

FIGURE 19.6
Phosphotransferase System
A signal molecule induces a conformational change in the receptor that induces self-phosphorylation of the receptor. The receptor then transfers the phosphate group to downstream proteins.

Signal transmission takes an external signal and transmits the information to the nucleus. Signal transmission occurs via phosphotransfer systems, second messengers, and ion channels.

CELL DIVISION RESPONDS TO EXTERNAL SIGNALS

A large number of extracellular growth factors and hormones exist. Many are specific for particular cells or tissues; for example, epidermal growth factor (EGF) stimulates epithelial

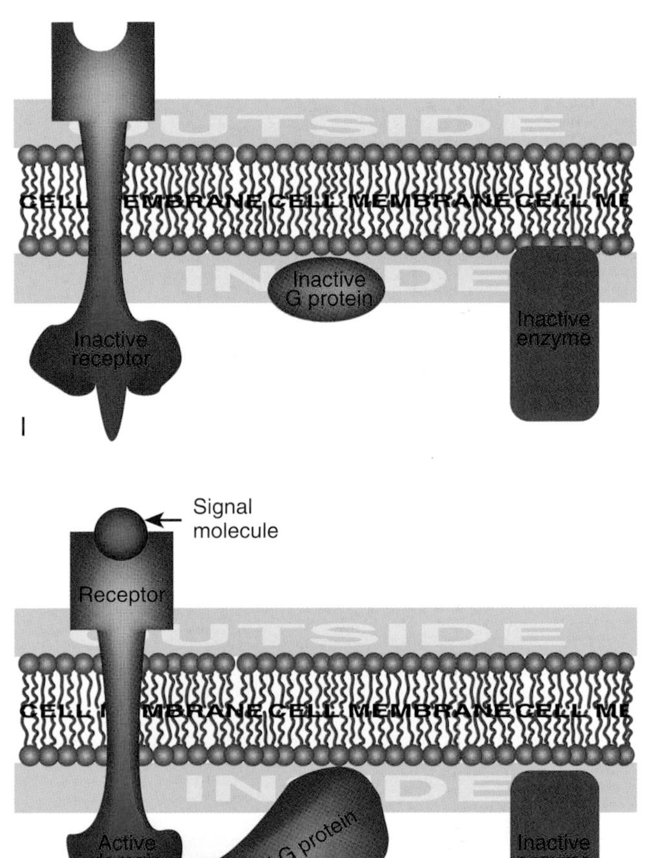

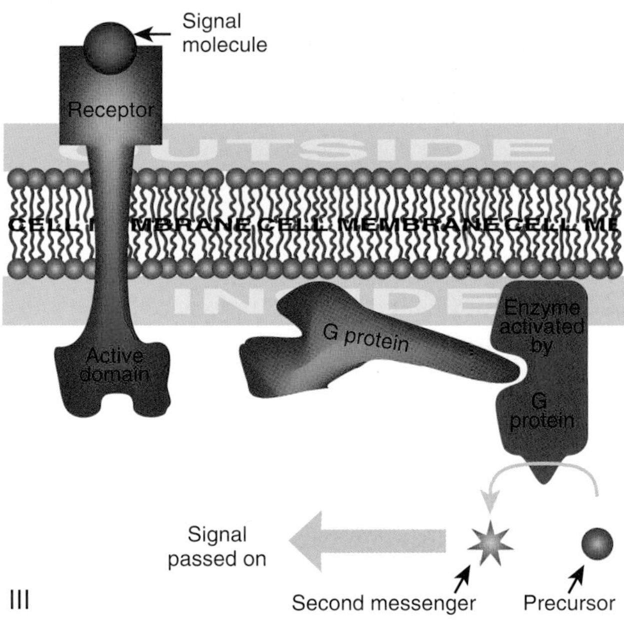

FIGURE 19.7 Second Messengers
Small intracellular signaling molecules called second messengers are produced in response to an activated receptor. The active receptor activates a G protein, which in turn activates a membrane-bound enzyme, which in turn synthesizes the second messenger.

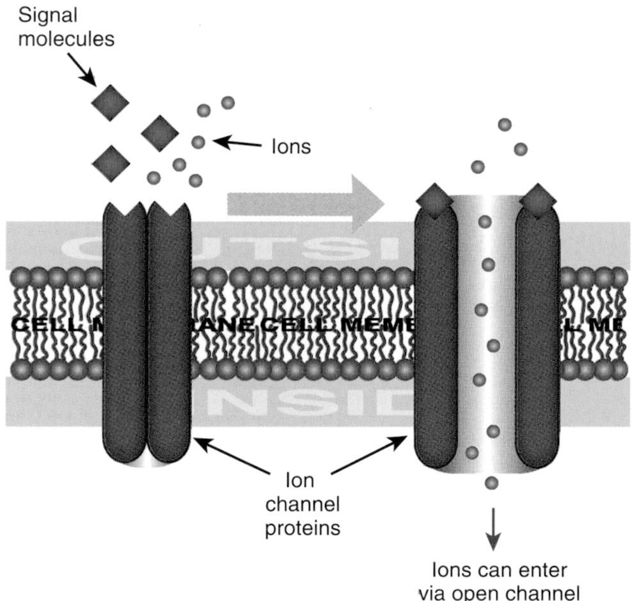

FIGURE 19.8 Ion Channel Activation
Binding of a signal molecule to an ion channel induces a conformational change that opens the channel. Ions may now pass through the open channel freely.

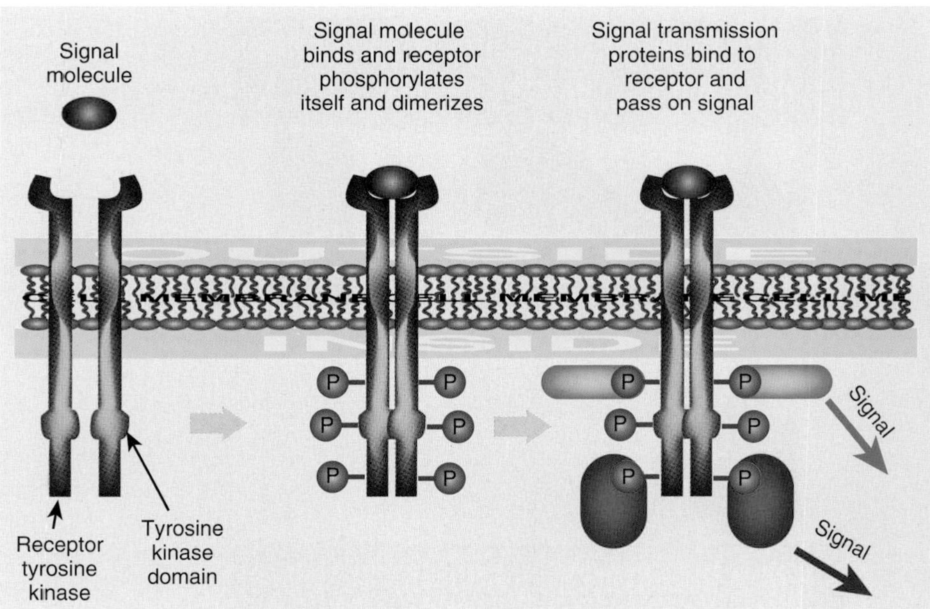

FIGURE 19.9 Activation of Growth Factor Receptor
Growth factor receptors have three domains: the extracellular binding site, the transmembrane region, and the intracellular tyrosine kinase domain. When a signal molecule binds to the extracellular domain, two receptor molecules bind together. This prompts the tyrosine kinase domains to activate each other by cross phosphorylation. The phosphates are then used to activate downstream signaling molecules, which in turn activate the transcription factors for cell growth.

cells, and fibroblast growth factor (FGF) stimulates muscle cells. Growth factors are bound by specific cell surface receptors. Binding activates the inner domain of the receptor, which is often a protein kinase itself or else activates an associated protein kinase (Fig. 19.9). Typically, the protein kinase activates a protein of the Ras family, and this in turn stimulates a phosphotransfer cascade. This cascade consists of three **mitogen-activated protein (MAP) kinases**. MAP kinase kinase kinase (MAPKKK) activates MAP kinase kinase (MAPKK), which activates MAP kinase (MAPK). Finally, MAPK phosphorylates transcription factors that activate genes needed early in cell division, including the gene for cyclin D.

Cell division is activated by growth factors, which send a signal from the outside of the cell to the nucleus via a cascade of protein kinases.

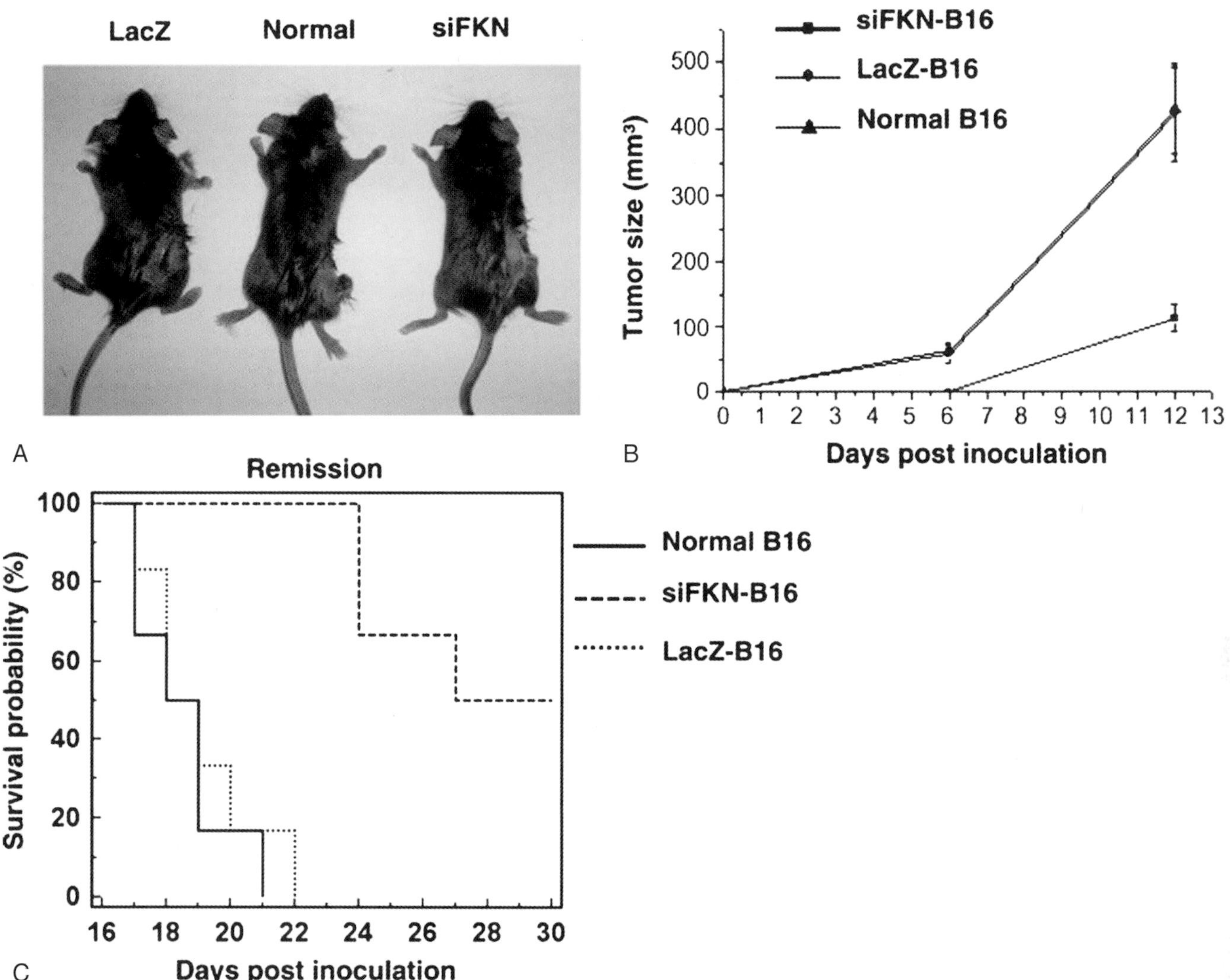

FIGURE 19.13 Human Cancer Cells Cause Tumors in Mice

Mouse melanoma cell line B16-F0 cells were grown *in vitro* with DMEM medium and fetal bovine serum, and then 1×10^6 live cells were injected into the right-side flank of C57BL/6 mice. Tumors with blood vessels developed within 3 weeks (middle mouse). In the control mouse on the left, the B16-F0 cells were transfected with a plasmid vector containing the gene for β-galactosidase before implantation in the mouse. In the mouse on the right, the B16-F0 cells were transfected with a plasmid vector containing a short-interfering RNA (siRNA) that blocks the expression of CX_3CL1/fractalkine, a cell adhesion molecule. The tumor is much smaller and has much less vascularization when the adhesion protein is blocked. From Ren T, Chen Q, Tian Z, Wei H (2007). Down-regulation of surface fractalkine by RNA interference in B16 melanoma reduced tumor growth in mice. *Biochem Biophys Res Commun* **364**, 978–984. Reprinted with permission.

2. Cell surface receptors: These proteins are found in the cell membrane, where they receive chemical messages from outside the cell. They pass the signal on, often by activating other proteins, such as G-proteins.
3. Signal transduction proteins: These proteins pass on the signal from outside the cell to proteins or genes involved in cell division. Many of these are protein kinases that activate or inactivate other proteins by addition of a phosphate group.
4. Transcription factors: These proteins bind to and switch on genes in the cell nucleus. This results in the synthesis of new proteins, as opposed to the activation of those already present.

Proto-oncogenes become oncogenes after a mutation makes the protein more active, the gene is duplicated, or a mutation in the promoter increases the expression of the protein.

Cell division is promoted by growth factors that bind and activate specific cell surface receptors. The internal portion of the receptor activates proteins that move to the nucleus, where they activate transcription factors that turn on genes for growth.

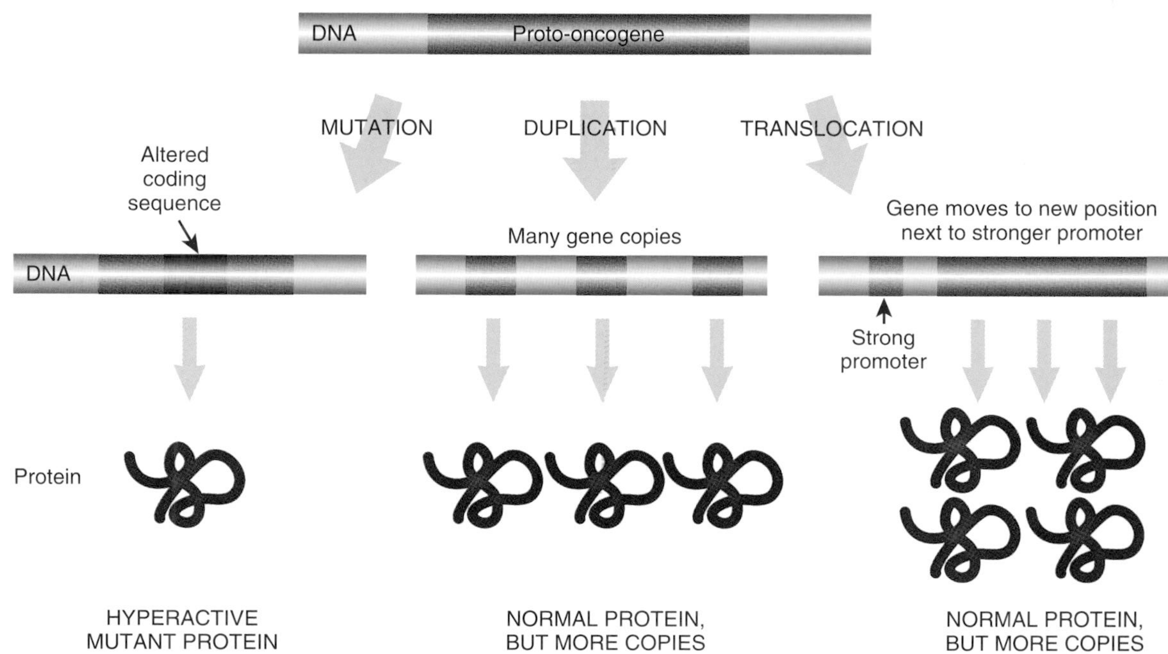

FIGURE 19.14 Oncogenic Mutations Result in Increased Activity
Three different possible mutations can cause a proto-oncogene to become oncogenic. First, the actual protein sequence may be altered by a mutation, resulting in a hyperactive protein. Second, the entire proto-oncogene may be duplicated one or more times. The extra copies cause an abnormally high amount of protein to be produced. Third, the proto-oncogene may be moved next to a strong promoter due to chromosomal rearrangements. The new promoter increases gene expression, and hence, the level of protein is increased.

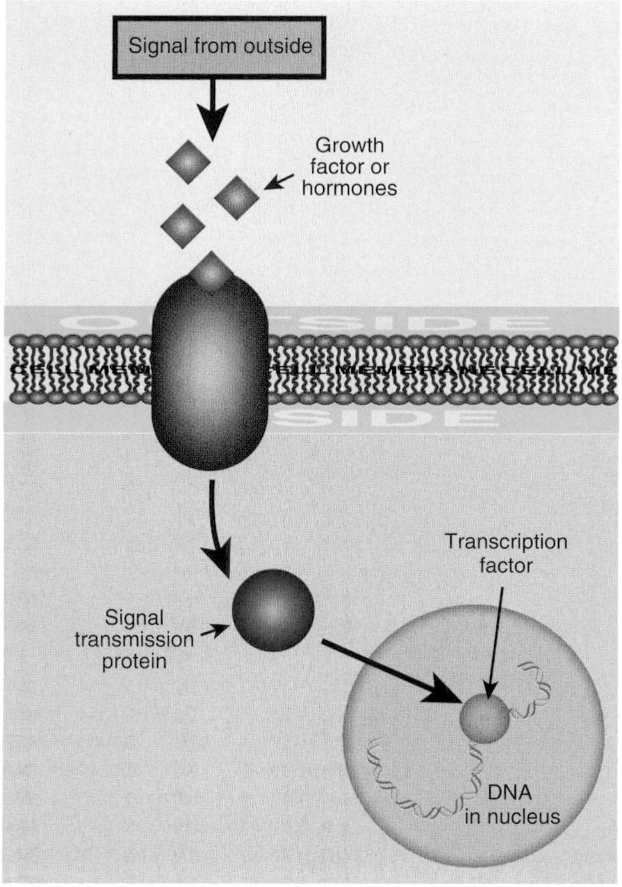

FIGURE 19.15 Components That Promote Cell Growth and Division
The execution of a cell growth signal involves a variety of steps. First, the signal to grow involves the production of growth factors. These bind and activate a cell surface receptor, which in turn activates a variety of intracellular proteins that transmit the signal from the cell membrane to the nucleus. Inside the nucleus, a transcription factor activates genes necessary for cell division and growth.

THE *RAS* ONCOGENE—HYPERACTIVE PROTEIN

Ras proteins are a closely related family of proteins that transmit signals for cell division in humans, flies, and even yeast cells (Fig. 19.16). Growth signals are received from outside the cell by receptors at the cell surface. The activated receptor transfers the signal to intracellular Ras protein. Ras is a G-protein, and after receiving a signal, normal Ras protein binds guanosine triphosphate (GTP) and enters signal-emitting mode. After transmitting a brief pulse of signals, Ras splits the GTP into guanosine diphosphate (GDP) plus phosphate and relapses into standby mode again. The cancer-causing form of the Ras protein is locked permanently into the signal-emitting mode and never splits the GTP. Therefore, it constantly floods the cell with signals urging cell division, even when none is received from outside.

The ***ras* oncogene** is the result of a single base change in the structural region of the gene. This causes an alteration in a

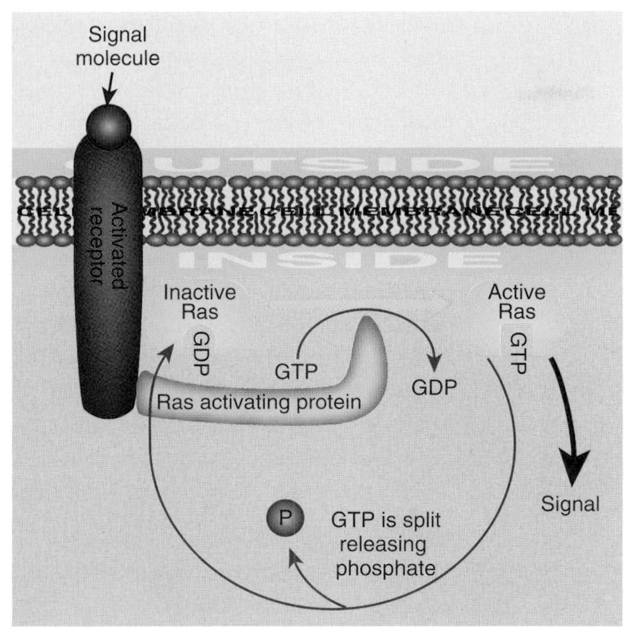

FIGURE 19.16 Function of Ras Protein
When a cell surface receptor receives a signal to grow, it transmits the signal to Ras. Activated Ras binds GTP and sends the growth signal to the nucleus. After the signal has been sent, GTP is hydrolyzed to GDP, and Ras becomes inactive once again.

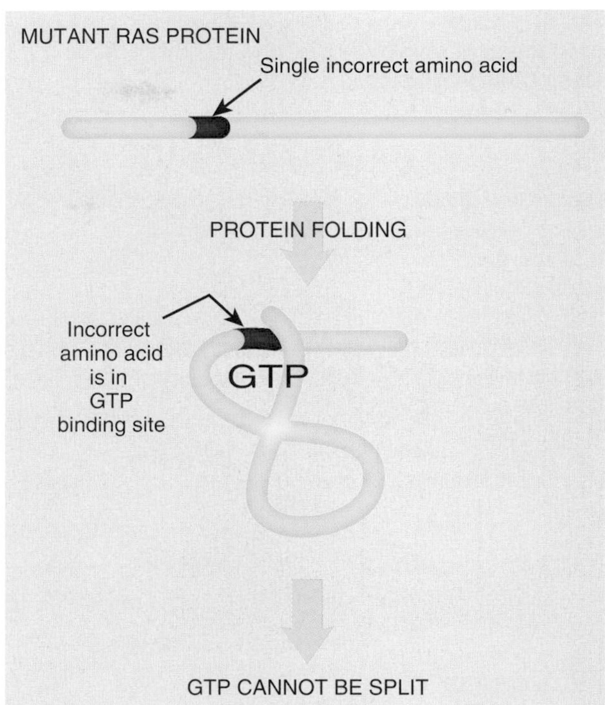

FIGURE 19.17 Mutations Creating the *ras* Oncogene Are Highly Specific
The transition from the normal proto-oncogene to the *ras* oncogene involves a single amino acid substitution. This amino acid change prevents the hydrolysis of GTP by Ras; therefore, the protein is always in its signal-emitting mode.

single amino acid in the encoded protein (Fig. 19.17). Most *ras* mutations alter the amino acid at position 12; others affect position 13 or 61. Only a few very specific mutations can create a *ras* oncogene from the proto-oncogene. The 3D structure of the Ras protein has been solved by X-ray crystallography. Those few amino acid residues that are changed by oncogenic mutations are all directly involved in the binding and splitting of the GTP. The consequence of hyperactivation of Ras is uncontrolled cell division and the beginnings of a possible cancer. Mutations of *ras* are found in about 25% of cancers and have been analyzed in detail in cancers of the lung, colon, pancreas, and thyroid.

No present therapies exist that are targeted to Ras protein. However, clinical trials are testing the use of RNA interference against the KRas protein in pancreatic cancers. The siRNA is released slowly from a polymer matrix and is specifically targeted against *KRas* genes with a G12D mutation (Gly to Asp at position 12).

Ras becomes an oncogene when a mutation alters amino acids at position 12, 13, or 61. The oncogenic form of Ras protein continually sends signals to the nucleus for growth. Normal Ras protein signals growth only when a growth factor binds to its cell surface receptor.

THE *MYC* ONCOGENE—OVERPRODUCTION OF PROTEIN

Some oncogenes are due to mutations that alter the structure of a protein such as Ras. Instead of a hyperactive mutant protein, other oncogenes suffer changes that vastly increase the amount of the protein formed, although the protein itself is unchanged.

A well-known example is the ***myc* oncogene**, which encodes a transcription factor that switches on several other genes needed for cell division. Myc is an abbreviation for

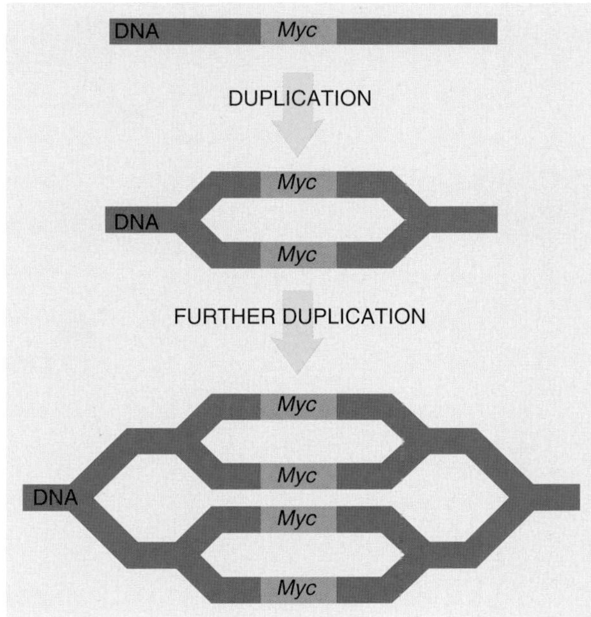

FIGURE 19.18 Overproduction of Myc by Gene Dosage Effect Normally, the gene for Myc is present in two copies. Because of aberrant chromosomal replication, the *myc* gene may be duplicated. Once this occurs, the number of *myc* oncogenes grows exponentially. Since all these copies are expressed, the amount of Myc protein increases and cancer results.

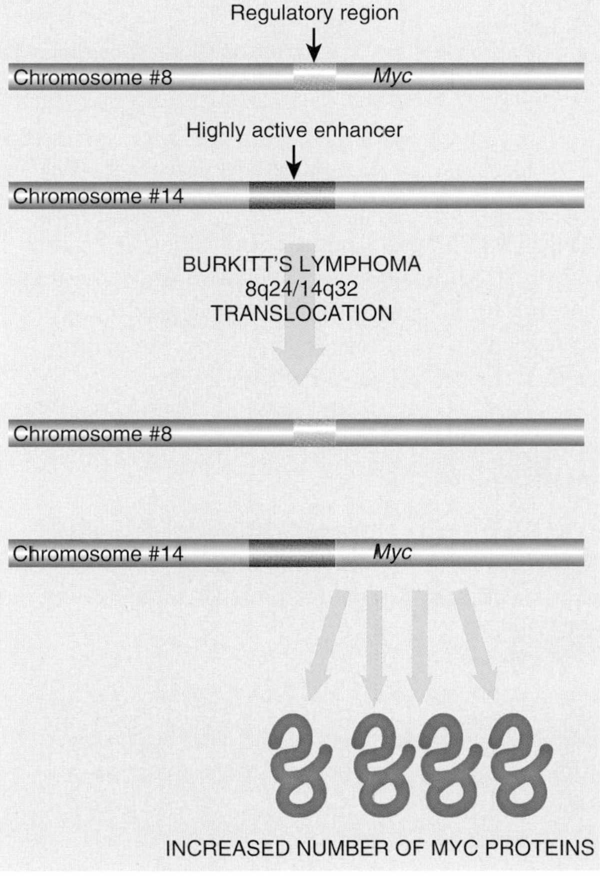

FIGURE 19.19 Overproduction of Myc by Changing Promoters Chromosomal translocation of the *myc* oncogene may enhance the amount of protein by fusing the structural portion of the gene to a highly active promoter.

myelocytoma, a cancer of the myelocytes, which are a type of bone marrow cell. Although *myc* was first found in the myelocytomas of chickens, it is very common in many different human cancers. A **Myc protein** overdose can occur in two ways. Some *myc*-dependent cancers result from chromosomal changes in which the *myc* gene is duplicated many times. Instead of the normal two copies, 50 to 100 copies may occur because of mistaken duplication of the segment of DNA carrying the *myc* gene (Fig. 19.18). The Myc protein will then be overproduced by 50- to 100-fold, too.

Alternatively, the number of copies of the *myc* gene may remain unchanged, but their regulation may be altered. In Burkitt's lymphoma, a rare translocation swaps segments of two unrelated chromosomes. This separates the *myc* structural gene from its own regulatory region and fuses it to the highly active promoter of another gene (Fig. 19.19). The Myc protein is now produced continuously in substantial amounts instead of being strictly regulated as before.

Mutations in *myc* are among the most common in mammalian cancers. Most types of human cancers show overexpression of Myc, although the frequency differs considerably from one type of tumor to another. In particular, mutations in *myc* tend to occur during the later progression of many cancers, including those of lung, breast, and ovary.

No present therapies exist that are targeted to Myc protein. However, expression of the *myc* gene is repressed by the microRNA miR-34c, which opens the possibility of future therapy using RNA interference.

Myc is a common oncogene. In some cancers, the number of *myc* genes increases due to aberrant DNA replication. In other cancers, the *myc* gene is translocated and linked to a strong promoter. Both mutations increase the amount of Myc protein, resulting in too much cell growth.

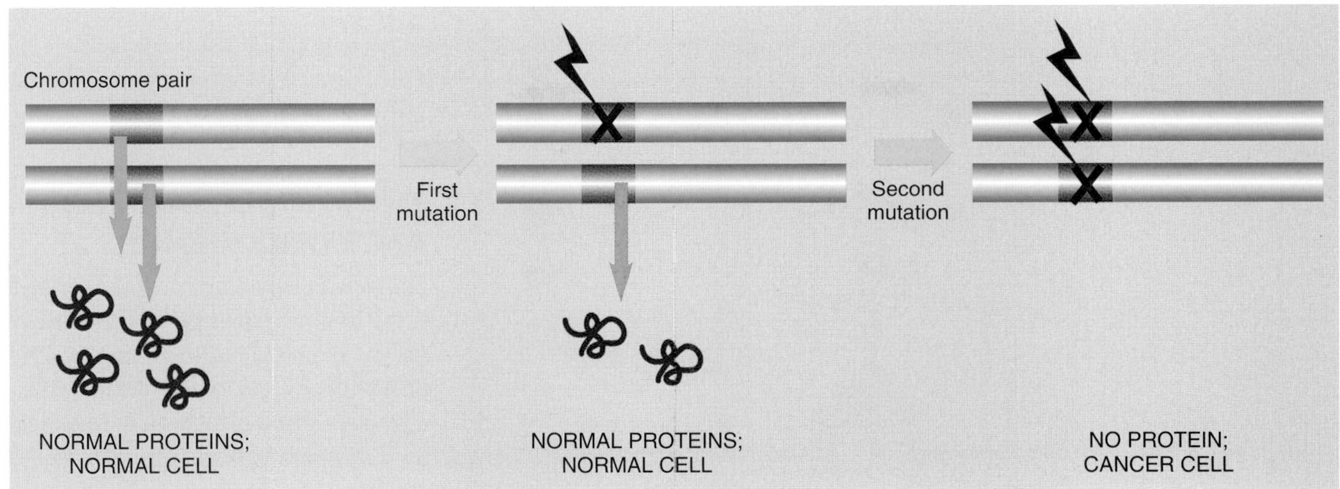

FIGURE 19.20 Mutations in Tumor-Suppressor Genes Are Recessive
In a normal situation, two wild-type anti-oncogenes are present. Two mutation events must occur to completely inactivate the anti-oncogene.

TUMOR-SUPPRESSOR GENES OR ANTI-ONCOGENES

Tumor-suppressor genes, or anti-oncogenes, normally suppress cell division. Consequently, it is necessary to inactivate both copies to initiate cancerous growth. A mutation in one copy of a tumor-suppressor gene has no effect; that is, these are recessive mutations. When both copies of a gene have been inactivated by null mutations, this is known as the **nullizygous** state. The two most common anti-oncogenes are Rb and p53 (see later discussion). Almost all human tumors inactivate either Rb or p53 and often both. Many of the proteins encoded by anti-oncogenes are DNA-binding proteins, often with zinc fingers.

There are two ways to end up with both copies of a gene inactivated (Fig. 19.20). First, during division of the cells that form the body, two successive somatic mutations may occur. First, one copy of the gene is inactivated, and then a second mutation strikes the second copy of the same gene. Although this seems highly unlikely, it will occasionally happen given enough time and a sufficiently high number of cells.

The second route is more frequent. The first mutation occurs in one copy of the gene in a germline cell of an ancestor. If this defective copy is passed on, a newborn individual will start life with one copy of the gene already inactivated in every cell. A somatic mutation that inactivates the second copy may then occur during cell division. In other words, one defective allele is inherited, and the other is acquired by somatic mutation. Individuals who inherit a single defective anti-oncogene do not always develop cancer. What they inherit is an increased chance of doing so.

This scheme applies to a dozen or more tumor-suppressor genes. Some of them are involved in a wide range of cancers, such as the **retinoblastoma (Rb) gene**, originally discovered and named for a rare cancer of the retina of the eye, but also involved in many other tumors. Other anti-oncogenes are very tissue specific, and inactivation of both copies triggers cancer of a particular organ, such as the Wilms' tumor gene responsible for a kidney cancer.

Anti-oncogenes encode proteins whose normal role is to inhibit growth or prevent cell division. Some signals that promote growth and division come from outside the cell—for example, hormones or growth factors circulating in the blood. Some proteins encoded by anti-oncogenes detect these external signals and block their effect on cell division (Fig. 19.21). An example is the PTEN protein that antagonizes insulin signaling and thus restricts cell growth and protein synthesis.

In addition to external control, many somatic cell lines are also preprogrammed. They have a preset number of allowed cell divisions. An internal "generation clock" counts off the number of permitted divisions left, and when it reaches zero, growth and division stop. Some anti-oncogenes take part in this system, and when they are defective, the cell fails to stop dividing.

Tumor-suppressor genes require a mutation in both copies in order to cause cancer. Sometimes both mutations occur in the same cell at different times. Other cancers form because the offspring received one mutated copy of the gene from a parent and therefore need only one mutation to occur.

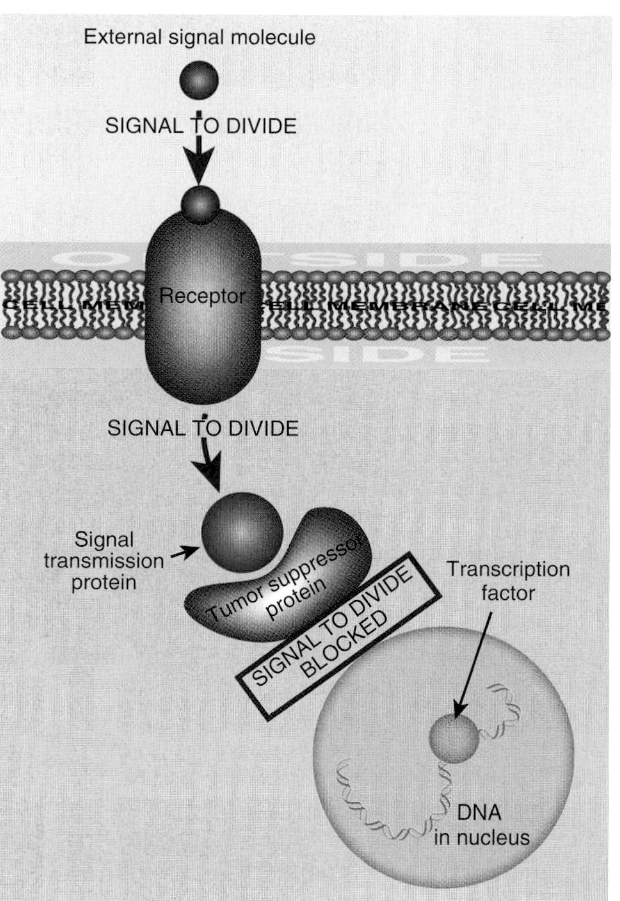

FIGURE 19.21 External Signal for Cell Division
Because growth-promoting signals often circulate in the blood, most cells in the body will be exposed to the growth cue. Only some cells divide, because proteins encoded by anti-oncogenes block transmission of the external stimulus. Anti-oncogene proteins may bind to signal transmission proteins and prevent them from activating transcription factors.

THE p16, p21, AND p53 ANTI-ONCOGENES

The best-known human anti-oncogene is the infamous ***p53* gene**, found on the short arm of chromosome 17. The **p53 protein** (also known as TP53) is a DNA-binding protein that, together with proteins p16 and p21, acts as an emergency braking system for the cell cycle. The *p53* gene is involved in a very large number of diverse cancers because its behavior differs from that of a standard anti-oncogene. Mutant alleles of typical anti-oncogenes are recessive, and a single mutation does not allow cell division. In contrast, a single defective p53 allele does show such effects, even in the presence of a second, normal copy of the gene. Thus, p53 mutations are dominant negatives. Well over half of all human cancers are defective in p53.

The reason for this aberrant behavior is the formation of mixed tetramers. The protein encoded by the *p53* gene assembles into groups of four (Fig. 19.22). When a cell has one good copy of the *p53* gene and one bad copy, it will produce a mixture of functional and defective p53 protein subunits. They will assemble into mixed tetramers, and even if a tetramer contains some good subunits, a single defective subunit will cripple the whole assembly. The likelihood of assembling a wholly good tetramer from such a 50:50 mixture of functional and defective subunits is $\{1/2\} \times \{1/2\} \times \{1/2\} \times \{1/2\} = 1/16$. Thus, only one-sixteenth of the p53 protein assemblies will function correctly even though half of the individual protein subunits are normal.

If a cell's DNA is damaged in any way, the normal p53 protein activates the gene for p21. The **p21 protein** then blocks the action of all of the cyclins and freezes the cell wherever it is in the cell cycle until the damage can be repaired (Fig. 19.23). The p16 protein acts similarly but just blocks cyclin E. It stops division at the critical point just before the cell enters the S-phase in which the DNA is replicated.

The p53 protein is not necessary for normal cell division. Mice with both copies of the *p53* gene knocked out grow normally to start with. However, they all die of cancer after 3 or 4 months. The role of p53 is to shut down cell division in DNA emergencies, such as ultraviolet radiation damage. The p53 pathway also responds to a shortage of nucleotides or a lack of oxygen. In severe cases, p53 may initiate programmed cell death, via Bax protein (see Chapter 20) rather than merely arresting cell division (Fig. 19.24). The p53 protein is inactive when first made and cannot bind to its DNA recognition sequence. It may be activated to form the tetramer by binding to single-stranded DNA or by phosphorylation.

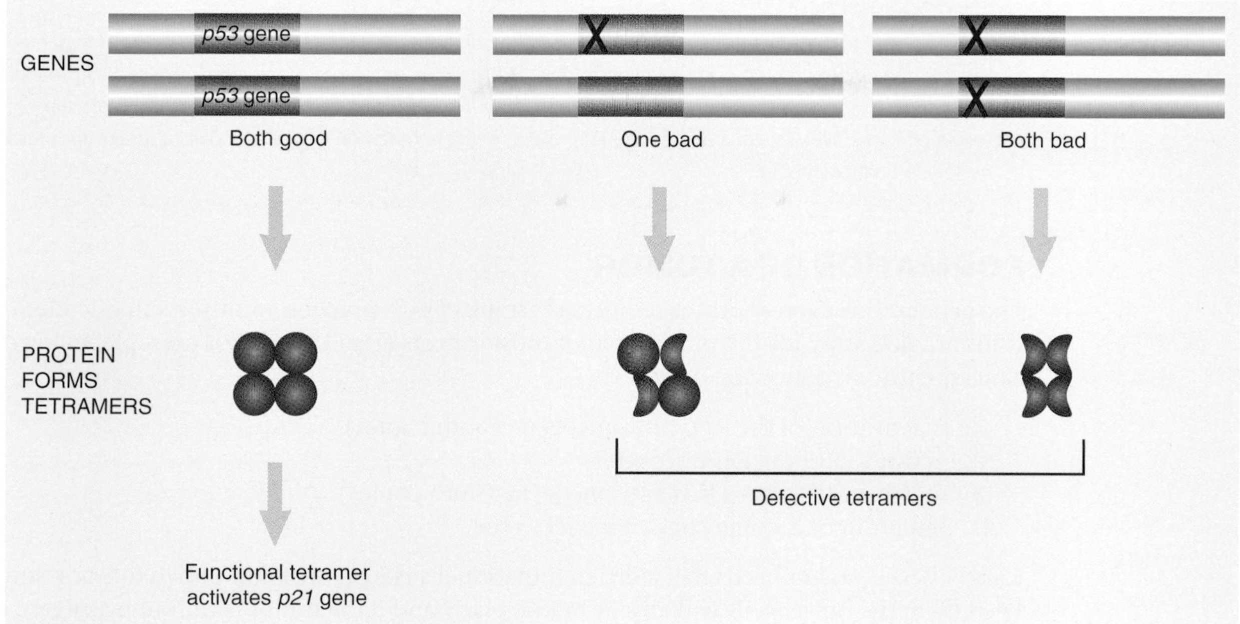

FIGURE 19.22 Mixed Tetramers of p53 Protein

Mutations in the *p53* gene are detrimental whether one or both copies are mutated. p53 protein works as a tetramer, so even if only one of the alleles has a mutation, most of the tetramers are defective. p53 functions correctly only if all four subunits of the tetramer are normal.

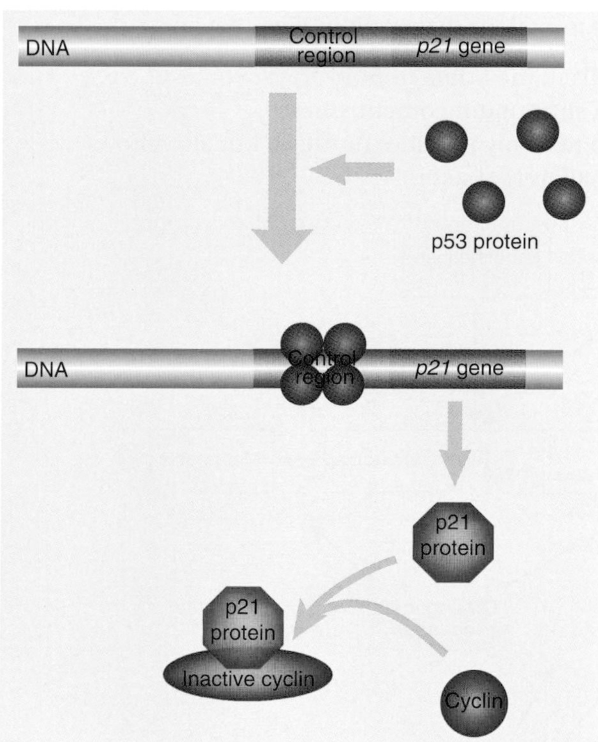

FIGURE 19.23 p53 and p21 Block the Cell Cycle

When a cell senses DNA damage, the p53 protein forms active tetramers that bind to the control region of the *p21* gene. p53 stimulates the transcription and translation of p21 protein. The p21 protein then binds to and inactivates the cyclins, preventing progression of the cell cycle.

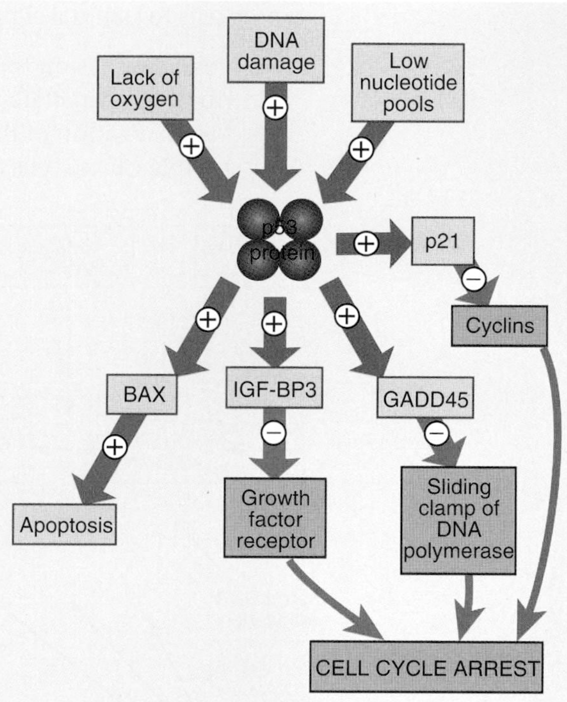

FIGURE 19.24 Central Role of p53

Three possible cues activate p53 protein: lack of oxygen, DNA damage, and low nucleotide pools. Activated p53 affects many different targets. If there is severe damage, p53 activates Bax protein and initiates programmed cell death. Otherwise, p53 activates cell cycle arrest by turning on p21, which in turn blocks the cyclins. Active p53 can also block synthesis by DNA polymerase through the action of GADD45. Active p53 also binds to growth factor receptors to block any further growth signals.

The p53 protein prevents defective cells from progressing through the cell cycle and, in extreme damage, triggers the cell to commit suicide via apoptosis. Both alleles of the *p53* gene are expressed as protein and can form mixed tetramers. Thus, even if only one allele has a mutation, most of the p53 protein tetramer is inactive. This is a rare example in which a single mutation in one of a pair of tumor-suppressor genes can cause cancer.

FORMATION OF A TUMOR

The generation of a real tumor requires several steps. In practice, multiple somatic mutations are necessary for the production of most cancers (Fig. 19.25). For example, many colon cancers carry the following defects:

1. Inactivation of the *APC* anti-oncogene (both copies)
2. Activation of the *Ras* oncogene
3. Inactivation of the *DCC* anti-oncogene (both copies)
4. Mutation of a single copy of the *p53* gene

Cancer development needs half a dozen mutational steps before a full-blown tumor results. Even then, the cancer cells will all stay in one place, and if the tumor is cut out by surgery, all may still be well. However, cancers do not stay put forever. Eventually, they bud off cancer cells that travel around the body, settle down in other tissues, and grow into secondary tumors. This is known as **metastasis**, and once things reach this stage, it is virtually impossible to find and remove all of the cancers. Other mutations, which aren't fully understood, are necessary for cancer cells to start traveling. These include mutations that result in the following:

1. Loss of adhesion to neighboring cells in the home tumor
2. Ability to penetrate the membranes surrounding other tissues
3. Vascularization of the tumor, which not only provides nutrients but also allows mobile cancer cells access to the circulatory system

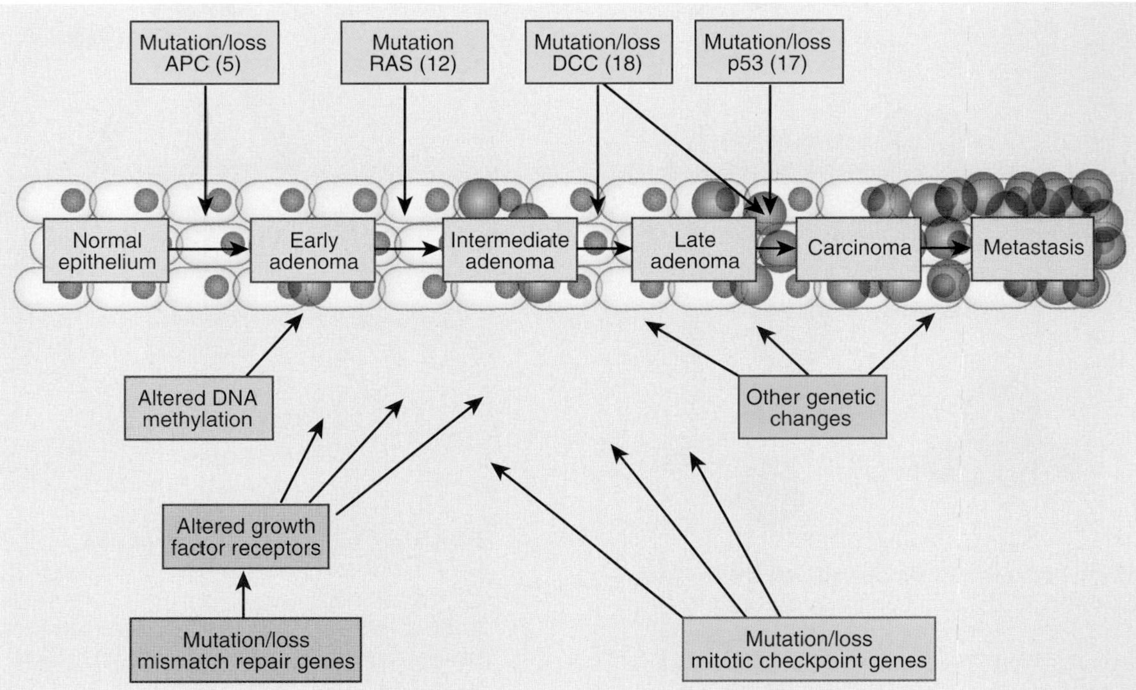

FIGURE 19.25 Steps in Tumor Development and Metastasis
The number of mutations required before a metastatic colon cancer emerges demonstrates the complexity of cancer. The loss of *APC* on chromosome 5 begins the process, followed by mutations in *ras*, inactivation of *DCC*, and finally a defect in *p53*. Chromosome locations of genes are shown in parentheses. Each mutation leads the cells further from normality, causing them to become more and more dysfunctional.

A cancerous tumor requires multiple genetic mutations to form. For it to become metastatic, even more mutations are required.

INHERITED SUSCEPTIBILITY TO CANCER

It is thought that 5% to 10% of cancers may be largely due to inherited defects. Many of the genes responsible for this are poorly understood, but they may be divided into three general categories.

First, as already noted, it is possible to inherit one defective copy of an anti-oncogene. In this case, every somatic cell starts life with one faulty copy, and only a single somatic mutation is needed to completely inactivate the pair of anti-oncogenes. (Note: Inheriting two defective copies of an anti-oncogene is normally lethal. Artificially engineered mice that are doubly negative for such genes generally die before birth.)

Second, there are indirect effects on cancer frequency due to genetic differences between races or within populations. For example, some skin cancers result from mutations caused by ultraviolet radiation from the sun. White people, especially those exposed to high levels of sunshine in the tropics or Australia, develop skin cancer much more often than darker-skinned people. The reason is obvious: the more pigment, the less UV radiation penetrating to your DNA.

Last but not least, mutations in **mutator genes** affect the rate at which further mutations occur during cell division. Mutator genes include genes that take part in DNA synthesis, such as the genes encoding DNA polymerase. Other mutator genes are involved in DNA repair. They have been analyzed in detail in bacteria, although less is known for humans. Nonetheless, it appears that certain inherited forms of colon cancer are due to defects in genes involved in mismatch repair. This, in turn, increases the mutation rate of all other genes including the tumor-suppressor genes. Defects in mutator gene are generally recessive, like those in typical tumor-suppressor genes.

Inherited breast cancer falls into this category. Inheriting a single defective copy of either the ***BRCA1*** or ***BRCA2*** **genes** (breast cancer A genes) predisposes women to breast cancer. About 0.5% of U.S. women carry mutations in *BRCA1*, which also predisposes to cancer of the ovary. As with other tumor-suppressor genes, the second copy must mutate during division of somatic cells for a cancer to arise. Individuals with two defective *BRCA* alleles die as embryos. Both BRCA1 and BRCA2 proteins are involved in DNA repair. Both bind to RAD51, which takes part in mending double-stranded DNA breaks (Fig. 19.26). In addition, BRCA1 has a dual role as a transcriptional regulator of other components needed for DNA repair.

The role of BRCA1 and BRCA2 in DNA repair allows a unique approach to therapy. If BRCA1 or BRCA2 is defective and the remaining

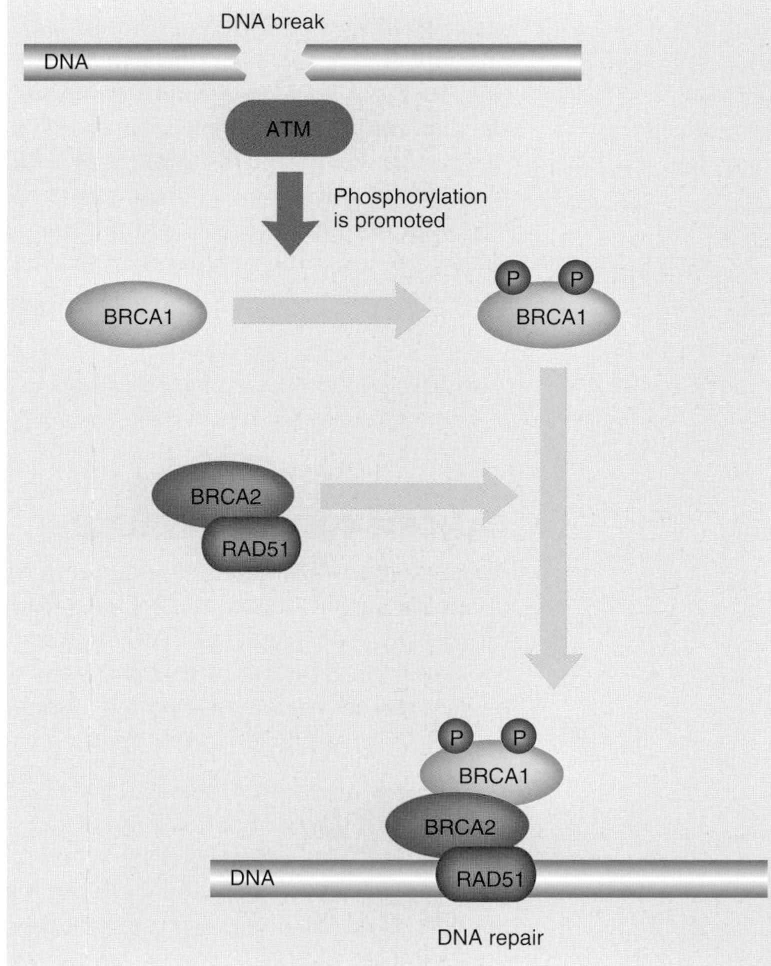

FIGURE 19.26
BRCA1 and BRCA2 Bind to RAD51
The presence of a double-stranded DNA break triggers the phosphorylation of BRCA1 via the ATM protein. BRCA2 plus RAD51 then bind to phospho-BRCA1 and take part in the DNA repair process.

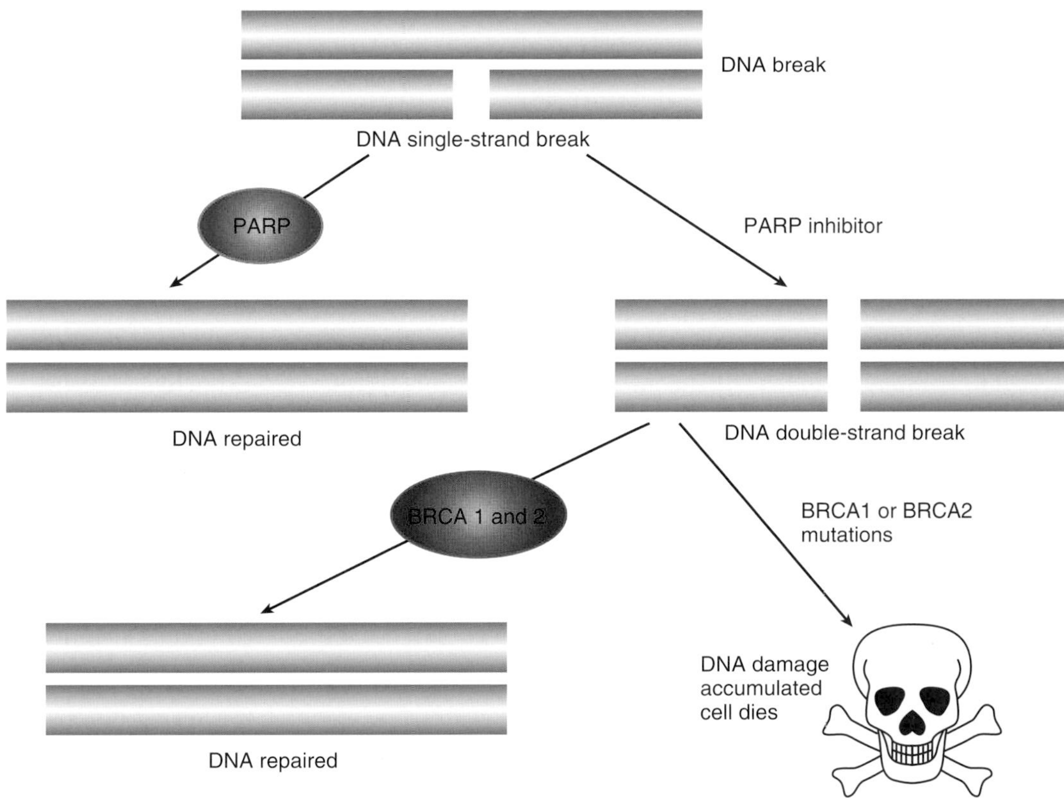

FIGURE 19.27 DNA Repair Needs PARP or BRCA1/2
PARP is needed to mend single-strand DNA breaks. Inhibition of PARP results in increased damage and the emergence of double-stranded DNA breaks. These could be mended if BRCA1/BRCA2 were functional. But if BRCA1 or BRCA2 is defective, the cell is doomed.

major DNA repair pathways are inhibited, then replicating cells will be unable to repair DNA properly. In practice, such cells will die due to the accumulation of lethal genetic errors (Fig. 19.27). The enzyme Poly(ADP-ribose) polymerase (PARP) is a sensor of DNA damage that binds to both single- and double-strand breaks. It adds poly(ADP-ribose) chains that recruit repair proteins to the breaks. In particular, PARP is essential for the base excision repair pathway that repairs single-strand breaks. If these breaks are left unrepaired, they will lead to double-strand breaks. Inhibition of PARP in tumors with BRCA1 or BRCA2 defects causes cell death due to accumulation of unrepaired DNA damage. Several PARP inhibitors were in clinical trials as of 2013, with mixed results.

Inherited defects in anti-oncogenes, components of DNA replication, and DNA repair systems can also predispose a person to cancer. Some genetic differences make people more susceptible to certain cancers.

CANCER-CAUSING VIRUSES

Although retroviruses that cause cancer in chickens and mice were important in the discovery of oncogenes, few human cancers are due to retroviruses. The most famous human retrovirus causes AIDS (see Chapter 21). Although some AIDS patients die of cancers, the AIDS virus does not cause cancer directly. Rather, it cripples the immune system that would otherwise kill off most cancer cells before they get too far out of control. Kaposi's sarcoma, often seen in AIDS patients, is caused by secondary infection with another virus, **human herpesvirus 8 (HHV8)**.

The first genuine cancer-causing retrovirus to be discovered was **Rous sarcoma virus (RSV)**. Long ago, the ancestor of RSV picked up a copy of the chicken *src* oncogene. The genome inside a retrovirus particle is RNA. When RSV infects chickens, the RNA is converted into a DNA copy by reverse transcriptase (see Chapter 21). The viral DNA is then integrated into the host cell DNA along with the *src* oncogene that it carries. Expression of the *src* oncogene causes muscle tumors, known as **sarcomas**.

Only about 15% of human cancers are due to viruses, and most of them are due to DNA viruses (papillomaviruses and herpesviruses). The major human cancer-causing viruses are listed in Table 19.1. However, simian virus 40 (SV40), which causes cancer in monkeys, has been studied most. SV40 acts by blocking the cell's tumor-suppressor genes. First, the virus integrates its DNA into the host cell genome. Second, it makes a virus protein, the T-antigen, which binds to the cell's Rb and p53 proteins (Fig. 19.28). This activates cell division and may lead to a tumor.

Human papillomaviruses (HPV) act in a similar manner but usually produce only benign growths, that is, warts. Most papillomaviruses rarely cause dangerous tumors. However, a subgroup of human papillomaviruses that are sexually transmitted are responsible for virtually all human cervical cancers and, occasionally, cancers of other organs. In these papillomaviruses, two virus proteins, E6 and E7, are responsible for inactivating p53 and pRb, respectively. Recently, a vaccine against the cancer-associated HPV genotypes (HPV16, HPV18) has been about 95% successful in preventing infection. This is the first vaccine to prevent a human cancer.

Viruses cause only 15% of human cancers. The majority of cancer-causing viruses are DNA viruses, especially papillomaviruses.

Table 19.1 Oncogenic Viruses of Humans

Virus	Examples of Cancer
DNA viruses	
Human papillomavirus	Cervical cancer, skin cancer
Epstein–Barr virus	Burkitt's lymphoma, Hodgkin's lymphoma
Human herpesvirus 8	Kaposi's sarcoma
RNA viruses	
HTLV-1	Adult T-cell leukemia
Hepatitis B virus	Liver cancer

ENGINEERED CANCER-KILLING VIRUSES

Lytic viruses infect and kill their target cells (as opposed to integrating or lying latent). If lytic viruses happen to infect cancer cells, then they kill these, too. If such viruses could be made specific for cancer cells, they could provide an effective therapy against cancer. Several viruses are being considered as the basis for such engineered **oncolytic viruses**. Those that have reached clinical trials include adenovirus, herpes simplex, measles, and vaccinia. At present, engineered oncolytic herpes virus seems to be closest to clinical use. However, we discuss engineered measles virus here, as it is only slightly further behind in development and illustrates the principles needed in an oncolytic virus with fewer complications.

Measles has several advantages as an anticancer virus. Most people in advanced nations are already immunized against measles, providing a level of protection against nonspecific tissue infection and other problems. Furthermore, certain lineages of measles virus have inherent strong antitumor activity. However, they are not cancer specific and infect other cell types also. To make them cancer specific, scientists must change their receptor specificity. The natural receptors for measles are the human cell surface proteins CD46 and SLAM (signaling lymphocyte activation molecule; CD150), which are recognized by the H-protein of measles virus (Fig. 19.29).

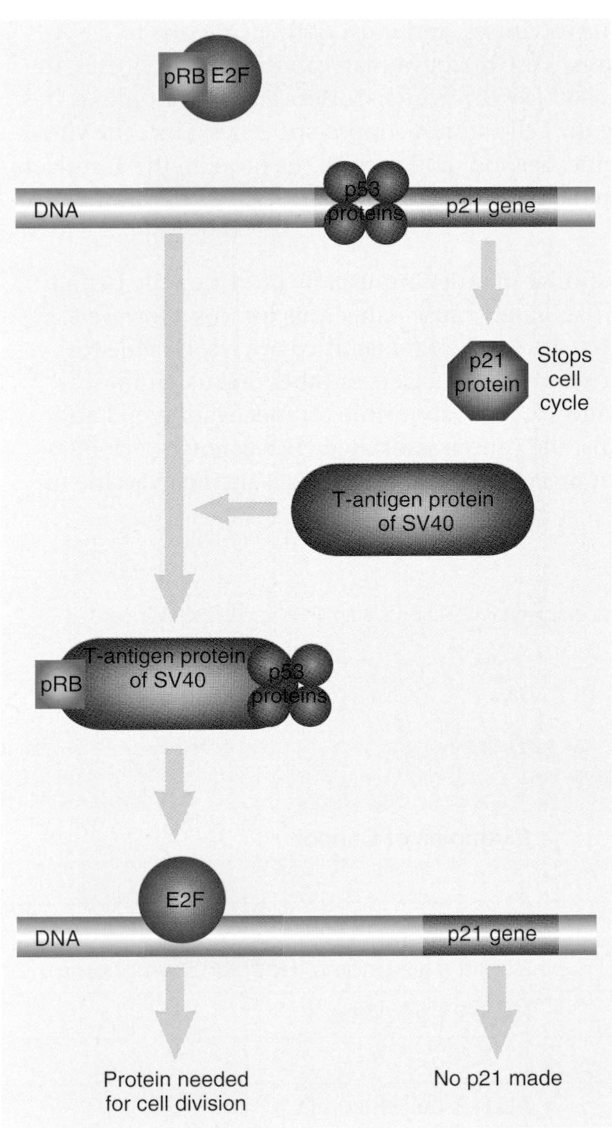

FIGURE 19.28 DNA Tumor Virus Inactivates pRb and p53
Normally, p53 and pRB act together to stop the cell cycle. If the cell is infected with SV40, the virus produces a protein called the T-antigen. This protein binds to pRB and p53, thus freeing E2F to stimulate cell division.

Unlike many other virus families, the paramyxoviruses, which include measles, use two separate virus proteins to recognize (protein H) and enter the host cell (protein F). Consequently, it is possible to radically alter the recognition protein, H, without crippling virus entry. Measles viruses with engineered H-proteins have been made. The modified H-proteins have mutations that prevent binding to CD46 or SLAM plus an extra C-terminal domain that recognizes other host cell surface proteins. The new recognition domain was generated as a single-chain antibody Fv fragment, or scFv (see Chapter 6 for structure and generation of scFv chains).

Retargeted measles viruses have been designed to bind to CD38 or EGFR (epidermal growth factor receptor), proteins often expressed at high levels by cancer cells (Fig. 19.30). Consequently, these engineered viruses were specific for cultured cancer cells with CD38 or EGFR on the cell surface. Better yet, when mice with implanted human tumors displaying CD38 or EGFR proteins were inoculated with the retargeted viruses, the tumors were efficiently destroyed. It is hoped that this approach will eventually work on human cancer patients.

Present opinion is that when they reach clinical deployment, oncolytic viruses should probably be used in combination with other anticancer agents.

Changing the viral protein that binds to the human host cell so that it recognizes only cancer cells makes a normal virus oncolytic; that is, the virus will infect and kill only cancer cells.

CANCER GENOMICS

The advent of high-throughput DNA sequencing means that it is possible to sequence the whole genome of the cells of a cancer and compare them to the genome of nonmutated cells from the same patient. This approach is recent and still largely experimental. However, it can reveal which particular oncogenic mutations are present in any given cancer, and this in turn may allow a superior choice of treatment tailored to the individual patient.

Many novel cancer genes have been revealed by large-scale sequencing of cancer genomes. They affect a wide variety of processes including cell cycle regulation, signal transmission, DNA methylation, histone modification, RNA splicing, protein degradation, telomere stability, and metabolism.

The situation is more complex than was originally thought, however, because many cancers contain mixtures of cells with different mutations. Indeed, if cancer cells carry mutations in DNA repair—as many do—then a wide range of mutations will emerge in the descendents of such cells. Many of these extra mutations will not be involved in the progression of the cancer. For example, breast cancer is often triggered by defects in the BRCA genes for DNA repair (see earlier discussion). In such cases, as many as 2,000 genes (around 10% of the total) may carry mutations. Of these, many, perhaps half, are silent mutations; nonetheless, identifying those mutations that affect cancer progression brings to mind the traditional vision of the needle in the haystack.

Worse still, some cancer cells contain major chromosomal aberrations. They include deletions and duplications that may cover hundreds of genes, although typically they include six to seven. This makes it extremely difficult to decide which gene was critical in cancer development. In addition, cancer cells often contain multiple rearrangements of chromosomal segments. Perhaps around 10% of cancers carry such major alterations, although the frequency varies greatly among different kinds of cancer, and in addition, the number of major rearrangements rises as cancers progress.

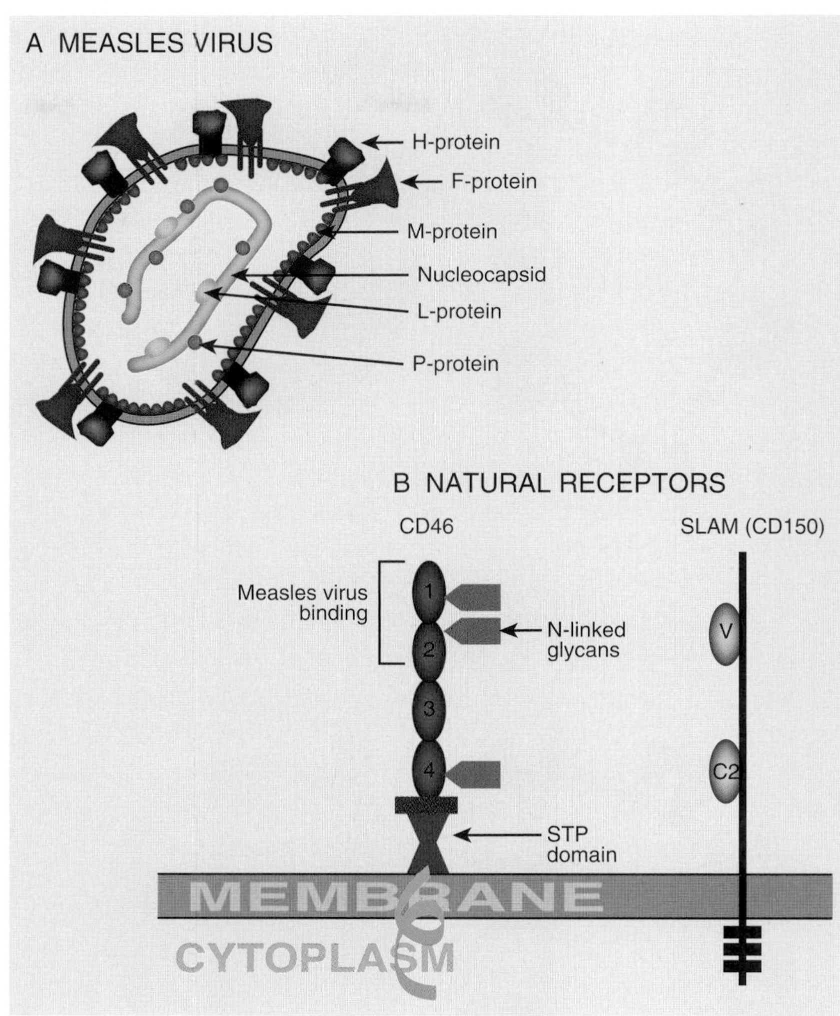

FIGURE 19.29
Natural Receptors for Measles Virus
Measles virus uses two cell surface proteins, CD46 and SLAM, as receptors. (A) Measles virus is an enveloped virus whose RNA genome is inside a nucleocapsid. The glycoproteins, H and F, are on the surface of the virus particle and are required for entry into the host cell. The internal L and P proteins take part in virus replication, whereas M is a structural protein. (B) Domain structures of CD46 and SLAM (signaling lymphocyte activation molecule).

An unforeseen benefit of cancer genomics is that certain anticancer drugs may fail in traditional clinical trials, wasting a great deal of time and money; nonetheless, such drugs might still be beneficial to a subset of patients who share mutations in certain genes. Sequencing of individual genomes and comparison to trial outcomes can reveal if a drug showing overall unfavorable results is actually useful for a restricted group of patients.

For example, some cancers that are resistant to standard treatments have mutations affecting the receptors for fibroblast growth factors (FGFs). The FGFs promote cell growth and division and the generation of a blood supply in certain tissues. Several experimental drugs (e.g., Dovitinib and Ponatinib, trade name Iclusig®) inhibit the function of FGF receptors and are therefore suitable for this subset of cancers, although they might well be harmful to patients lacking FGF receptor mutations.

There are significant disparities in both cancer frequency and outcome between different racial groups in the United States. Although socioeconomic factors are partly responsible, genomics data have revealed race-specific differences in the prevalence of certain oncogenic mutations. We discussed the role of the *RAS* oncogene family earlier. Mutations in *KRAS* occur in many lung cancers. However, their frequency among cancer patients varies greatly with race. Asians typically have few *KRAS* mutations, whereas Africans have most and whites are intermediate (Fig. 19.31). The opposite trend is seen in mutations in *EFGR*, encoding the epidermal growth factor receptor.

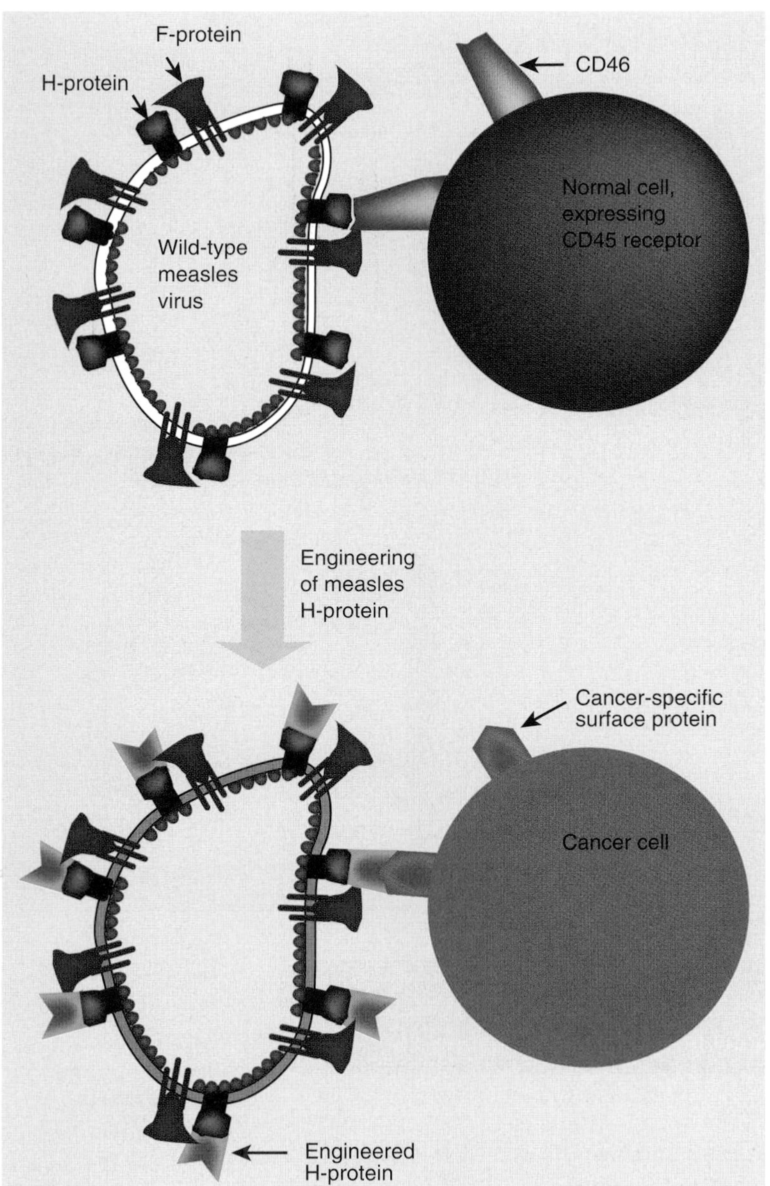

FIGURE 19.30 Cancer Retargeted Measles Virus
The external glycoprotein H may be engineered to recognize a different host cell protein. Instead of binding to CD46 or SLAM, it now binds to cancer-specific surface proteins such as CD38 or EGFR.

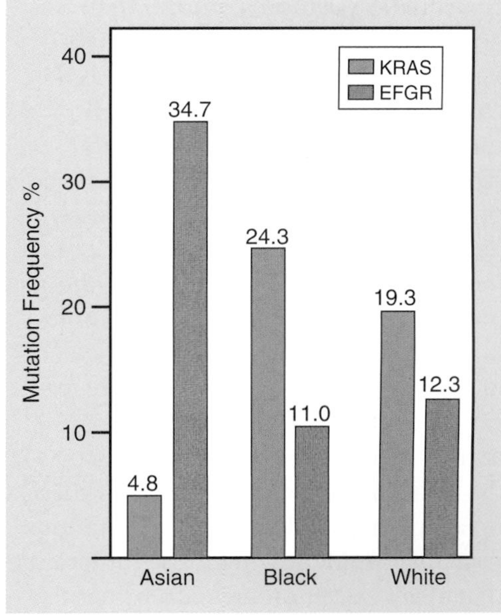

FIGURE 19.31 Racial Differences in Cancer Mutation Frequency
The spectrum of mutations in the *KRAS* and *EFGR* genes found in lung cancers varies considerably among different ethnic groups. Asians have more mutations in the *EFGR* gene, whereas blacks have more mutations in the *KRAS* gene. Whites are intermediate in both cases. Based on data taken from El-Telbany A, Ma PC (2012). Cancer genes in lung cancer: racial disparities: are there any? *Genes Cancer* **3**, 467–480.

Genomics is providing a vast amount of individualized information about the mutations in cancer cells, including the differences between different cells within the same cancer. In some cases, it is possible to suggest changes in therapy based on such data.

CANCER EPIGENOMICS

The blooming field of epigenomics is also relevant to cancer. In this case, we are concerned with alterations in cell regulation due to epigenetic changes to the DNA as opposed to changes in DNA sequence. The most frequent epigenetic changes are in DNA methylation and in histone modifications. Since these alterations affect the regulation of a wide range of cell operations, including growth and division, it is hardly surprising that damaging these forms of regulation may promote cancer development.

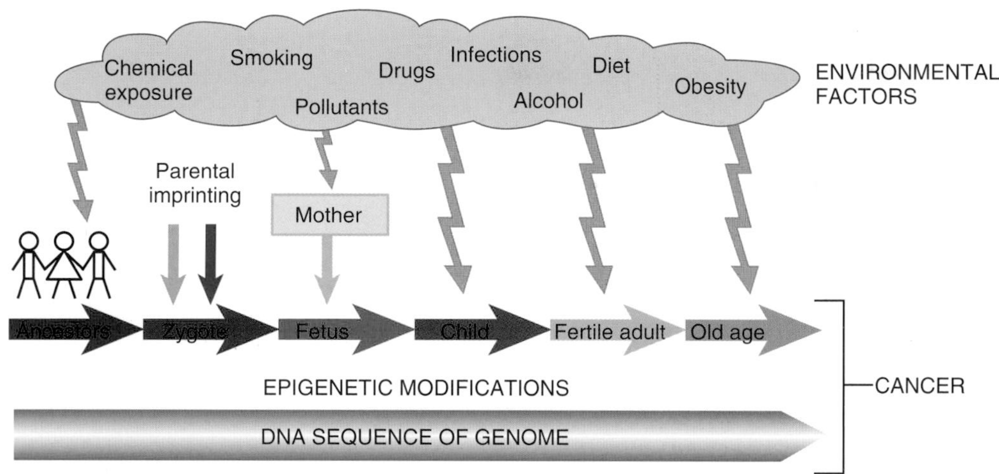

FIGURE 19.32 Epigenetic Timeline and Cancer
A variety of environmental factors affect successive stages of life and provoke epigenetic modifications that steadily accumulate. Even some epigenetic modifications that occurred in ancestors may persist over the generations. Imprinting is due to differ DNA methylation patterns in the male and female gametes that are maintained in the zygote (see Chapter 17 for imprinting). Some of these epigenetic modifications, together with mutations present in the DNA sequence of the genome, may promote cancer.

To qualify as epigenetic, changes in DNA methylation and in histone modification must be passed on from one generation of cells to the next. For multicellular animals, this may be divided into those changes passed on from one organism to its descendents and those passed on between successive generations of cells within the same organism. This is illustrated for the case of breast cancer in Figure 19.32. Despite the effects of the *BRCA* genes discussed previously, there is considerable difference between identical twins in the development of breast cancer. This is where epigenetics plays a role. Exposure of family ancestors to toxic chemicals, especially xenoestrogens, may cause epigenetic changes that are passed between generations. Exposure of the fetus while in the womb triggers other changes. Finally, epigenetic alterations that result from factors such as diet are passed on from one cell to its descendents within a particular individual.

Most epigenetic changes occur within the upstream regulatory regions of genes and so affect expression of the nearby gene. Generally, increased methylation decreases gene expression by blocking access of activators. Conversely, decreased methylation increases gene expression. The effects of histone modifications are more complex, as there are multiple sites for modification and, moreover, several chemical classes of modification exist—methylation, acetylation, and phosphorylation being the most common. Some of these epigenetic modifications are induced by compounds found in the diet (see Box 19.1).

In several known examples, the methylation level of the promoters of oncogenic genes correlates with cancer progression. However, it is often uncertain whether these are truly inherited epigenetic changes or whether they are the secondary result of other factors affecting DNA methylation. At present, epigenetic data is being gathered in large amounts but has yet to make a significant clinical impact.

Epigenomics is an active experimental field that is generating a flood of data. Hopefully, this will ultimately allow improved cancer treatment.

Box 19.1 Eat Your Veggies: Epigenetics versus Cancer

Epigenetic alterations are generated in response to multiple environmental factors such as diet. In some cases, the chemical components of food that provoke the epigenetic response have been identified. Many common fruits and vegetables contain compounds that affect DNA methylation or histone modifications, or both (Fig. A). Notable among these compounds are polyphenols from plants such as turmeric, soybeans, broccoli, and green tea. As indicated in Figure A, fruits and vegetables also contain a wide variety of other types of compounds with similar effects.

Many polyphenols act by inhibiting DNA methyltransferase (DNMT) and thus reducing DNA methylation levels. Examples of DNMT inhibitors include resveratrol, found in grapes and red wine, and epigallocatechin-3-gallate (EGCG) from green tea.

Other polyphenols affect histone modification. Although the range of possible histone modifications is wide, many polyphenols affect acetylation by inhibiting histone deacetylases (HDACs). Hence, the histone acetylation level increases. Examples of HDAC inhibitors include caffeic acid from coffee and curcumin from turmeric. Potent compounds that are not polyphenols include allyl derivatives from garlic and isothiocyanate derivatives from broccoli. They both act as potent HDAC inhibitors.

Inevitably, some plant compounds affect microRNA expression. Curcumin and isothiocyanates affect the level of multiple miRNAs in cultured cancer cells. For example, curcumin inhibits the expression of the microRNA, miR-21, which is known to promote cancer if overexpressed (see below).

Thus, the chance of developing cancer can be reduced by consuming a diet high in fruit and vegetables that promote "good" epigenetic alterations. Moreover, active compounds from plants are being tested on cancer cells in culture for possible future use as additives during cancer therapy.

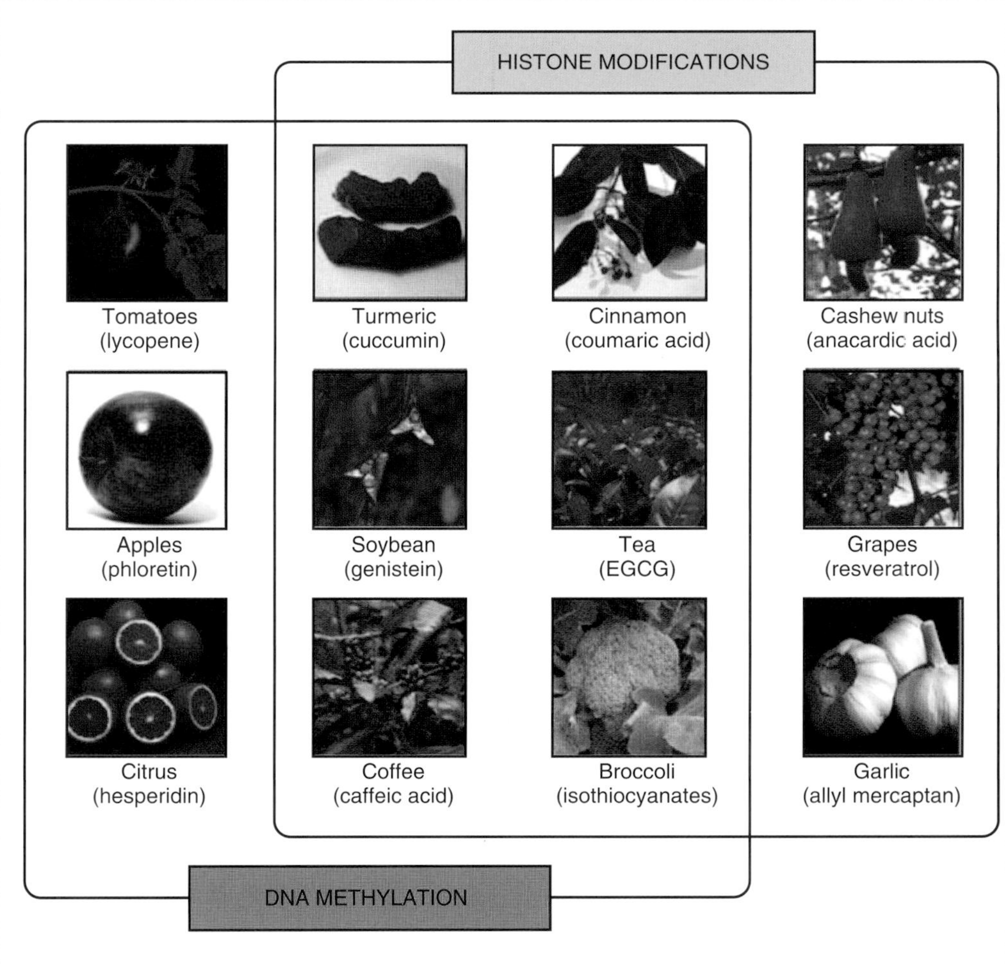

FIGURE A Epigenetic Modifiers from Plants
Several plant compounds are known that provoke epigenetic modifications. Some plant compounds affect histone modifications, whereas others DNA methylation. As indicated, many plants contain compounds that trigger both kinds of alteration. From Link A, Balaguer F, Goel A (2010). Cancer chemoprevention by dietary polyphenols: promising role for epigenetics. *Biochem Pharmacol* **80**, 1771–1792.

MICRO RNA REGULATION AND CANCER

Another relatively new area is the effect of RNA regulatory mechanisms on cancer. The principles of RNA-based regulation and the RNA interference defense system were covered in Chapter 5. In particular, both systems act on messenger RNA rather than on gene expression per se. Thus, microRNA acts by controlling the translation of mRNA, and RNA interference operates by triggering the destruction of mRNA from viruses or other hostile sources.

In humans, miRNAs are known to play a role in all aspects of cellular regulation, including cell growth and division. Rather than turn genes on or off completely, miRNA tends to modulate expression. Furthermore, many miRNAs regulate multiple genes. This results in some very complex regulatory networks. Probably more than half of all human genes are regulated by miRNAs, many of which exist as families of related miRNAs. Just as with regular protein-encoding genes, oncogenic miRNAs exist that promote cancer, whereas tumor suppressor miRNAs do the opposite. Moreover, miRNAs often interact with transcription factors such as Myc, E2F, or p53 that are known to play a major role in cancer.

The miR-34 family of miRNAs is an example of a tumor suppressor, whereas miR-21 is an oncogenic miRNA that regulates multiple target genes (Fig. 19.33). Typically, miR-21 acts by preventing translation of the mRNA of its target genes, although in a few instances it directly inhibits the activity of protein translation factors—for example, STAT3 of the Jak-Stat signaling pathway. In the case of tumor suppressors, proposed therapy consists of administering extra miRNA. This approach has suppressed tumors in experiments with mice. Conversely, when oncogenic miRNA is overexpressed, therapy consists of giving the corresponding antisense RNA. Again, experiments in mice have been successful. Clinical trials in humans to date consist of miRNA profiling rather than therapy.

Mutations affecting cancer also occur in regulatory RNA, especially microRNA. Experiments in mice suggest that miRNA-based therapy may be possible.

ANTICANCER AGENTS

Our ability to diagnose cancer has outrun our ability to treat it (see Box 19.2). So far, the most successful anticancer agents that have emerged from the application of molecular biology are monoclonal antibodies and small molecule inhibitors. The development and

Box 19.2 Cancer Revisionism and Lead Time Bias

Since the "War on Cancer" was started, most of the major advances have been in the diagnosis and understanding of cancer, rather than its cure. Because of the massive vested interests in the cancer research industry, cancer statistics are often presented in an overly optimistic way. One frequently used measure of misinformation is what is known as *lead time bias*.

Consider a patient who is diagnosed with a serious cancer and dies 5 years later. Now consider what happens if the ability to diagnose cancer improves and the same cancer is detected 5 years earlier. If treatment has not improved, the patient will die 10 years after diagnosis. In the second case, the patient survived for 10 years with cancer as opposed to only 5 years in the first case. However, no real increase has occurred in survival. The patient did not live any longer but was merely diagnosed earlier and had the burden of worrying about cancer for 5 years longer. Through use of this bogus measure of survival time, it is sometimes made to look as if great advances have been made in curing cancer. In fact (using data from 1950 to 1982), massive improvements in genuine life expectancy have been made in only a few relatively rare forms of cancer.

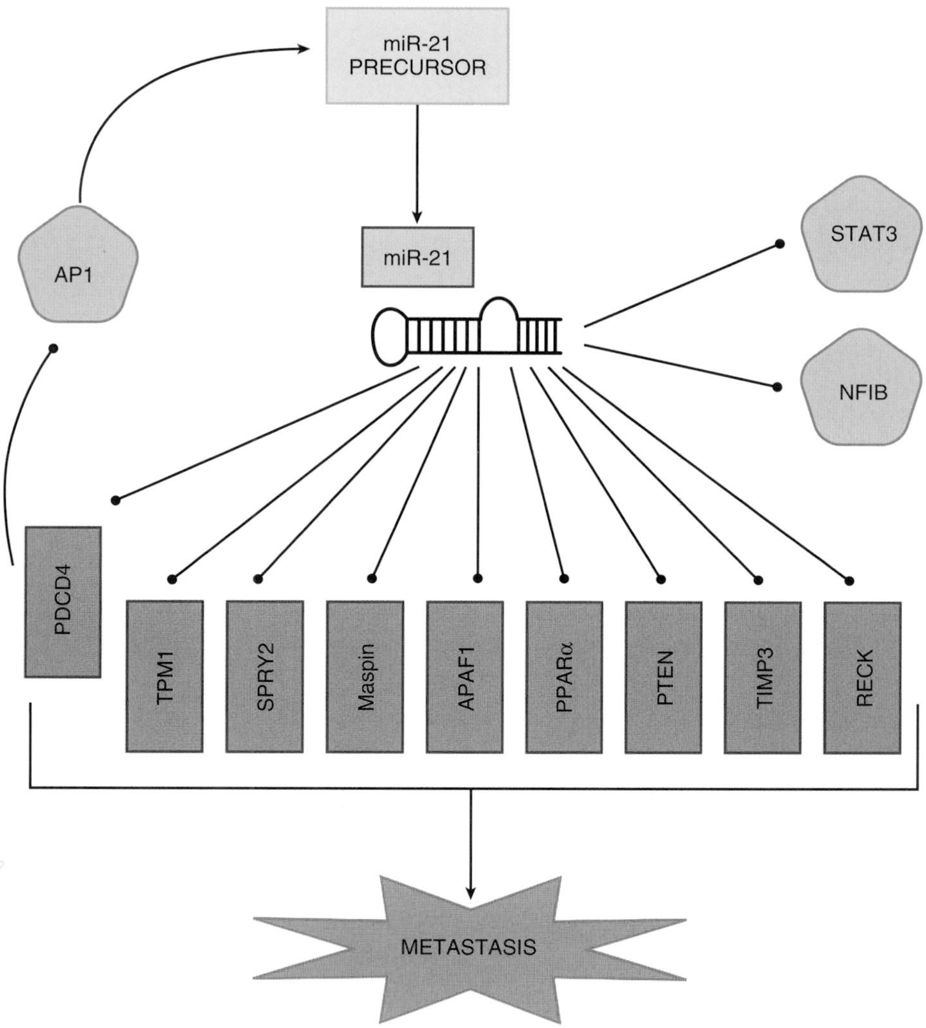

FIGURE 19.33 miR-21 Regulatory Network
The microRNA, miR-21, regulates an unusually high number of other genes, a small selection of which are shown here. miR-21 regulates genes in green rectangles by inhibiting translation of their mRNA. Purple pentagons represent protein translation factors that are directly inhibited. Note the feedback loop by which miR-21 promotes its own synthesis by inhibiting PDCD4, which inhibits the translation factor AP1, which in turn stimulates the synthesis of the precursor to miR-21. The net result of miR-21 overexpression is the promotion of metastatic cancer cell growth. Black lines ending in circles = inhibition; green arrows = activation.

properties of monoclonal antibodies were discussed in Chapter 6. Monoclonal antibodies used for cancer therapy are often targeted against extracellular growth factors or their receptors. For example, bevacizumab (marketed as Avastin®) binds to vascular endothelial growth factor and inhibits the development of blood vessels in tumors, whereas cetuximab binds to the receptor for epidermal growth factor and slows tumor growth and metastasis. The names of most monoclonal antibodies in clinical use end in "-ab." It has even been proposed that whole "armed" bacteria could be used to treat cancer (see Box 19.3).

The use of small molecule inhibitors against cancer has a long history. Some early anticancer agents, such as the platinum compound cis-platin, were discovered largely by chance. In contrast, systematic screening of large compound libraries has been used more recently. Such recent compounds often have names ending in "-ib." Targets include a variety of components that take part in cell division, for example:

1. DNA topoisomerases: These enzymes take part in uncoiling DNA during chromosome replication (see Chapter 4). Inhibitors include etoposide and anthracyclines such as daunorubicin.
2. Microtubules of the mitotic spindle: These are vital for cell division and chromosome partition. Inhibitors include colchicine and taxane.
3. Cyclin-dependent kinases: These play a major role in regulating the cell cycle (see above). At present, several inhibitors are in clinical trials, including alvocidib and seliciclib.

Box 19.3 Radioactive Bacterial Therapy Against Cancer

At the opposite extreme from small molecule inhibitors is the use of whole, living bacteria loaded with radioactive isotopes. This novel approach is in the experimental stage but has shown promising effects in mice with metastatic cancers.

The *Listeria monocytogenes* bacterium is a moderately virulent pathogen that enters human cells very efficiently (Fig. B) and possesses the unusual ability to transfer itself from cell to cell. *Listeria monocytogenes* is normally acquired from contaminated food and causes gastrointestinal distress in healthy people. However, it may be life threatening in immunocompromised patients. Because of its proficiency in entering animal cells, *Listeria* has previously been tested as a vector for the delivery of DNA or vaccines to cell interiors. As usual with vectors used in biotechnology, the bacteria are first modified to make them harmless by knocking out several virulence genes, and have proven to be safe and effective.

In the approach discussed here, the objective is to kill cancer cells by radiation, while sparing normal cells of the body. An attenuated strain of *Listeria* is loaded with a toxic radioisotope, Rhenium-188, that kills cells by radiation damage to their DNA. Rhenium-188 is already in use for treating some cancers. For bacterial delivery, the radioisotope is attached to the bacteria via antibodies specific for *Listeria*. Because the *Listeria* is attenuated, it cannot infect normal cells successfully. However, cancer cells are vulnerable due to their location inside the tumor where the immune system is suppressed. This approach was especially effective at killing metastatic tumors, which are secondary tumors that have just started to grow in new locations.

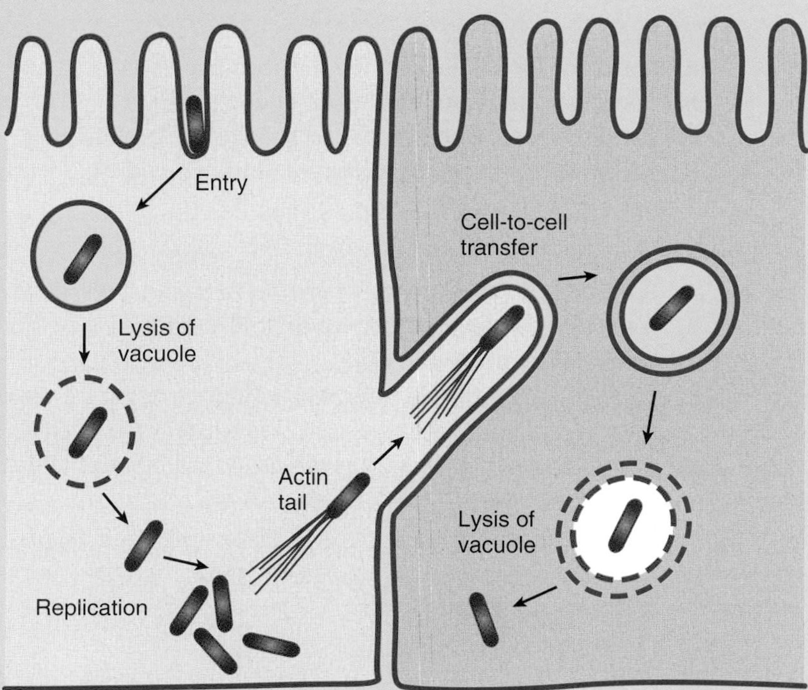

FIGURE B Cell Entry by Listeria
Listeria monocytogenes can enter epithelial host cells by using the two bacterial surface proteins InlA and InlB. This promotes uptake via phagocytosis, and the *Listeria* is temporarily trapped in a vesicle. It then uses phospholipases (PlcA and PlcB) plus a pore-forming protein toxin, LLO, to burst the vesicles and escape into the cytoplasm. *Listeria* also uses its ActA protein to take control of host actin protein. This is converted into "actin comet tails" that enable transfer into neighboring cells.

Summary

Cancer is a genetic disease; that is, a mutation in the DNA must occur for cancer to occur. In normal cells, growth begins when a growth factor binds to receptors on the cell surface. This activates a cascade of proteins that trigger the activation of transcription factors in the nucleus. This turns on genes that cause the cell to divide. Cell division is a highly regulated process, with multiple redundant safeguards. When the DNA of a normal cell is damaged by a carcinogen or mutagen, the cell cycle is halted until the damage is repaired. Cell cycle checkpoints are controlled by cyclins and transcription factors such as E2F and pRB. In cancers, mutations in the regulatory system allow defective cells to continue around the cycle and divide. In practice, cancer formation needs multiple mutations because there are so many internal controls. A large tumor emerges only after blood vessels form to provide nutrients to the internal cells. Without a blood supply, the cancer stays a microtumor. Finally, some tumor cells escape from the primary tumor and travel to different sites in the body, a process called metastasis.

A small proportion of cancers are due to virus infection. Ironically enough, certain viruses are being engineered to destroy cancer cells. Modern approaches to cancer that are still under development include genomic analysis and the use of RNA interference.

End-of-Chapter Questions

1. What causes cancer?
   - **a.** somatic mutations that disrupt normal cell division and death
   - **b.** germline mutations that result in cell death
   - **c.** somatic mutations that cause cells to die prematurely
   - **d.** germline mutations that disrupt normal cell division and death
   - **e.** none of the above
2. How might a carcinogen not be a mutagen as well?
   - **a.** A carcinogen is always a mutagen because it will always react with DNA.
   - **b.** A carcinogen may be converted in the body to a benign substance, which does not interact with DNA.
   - **c.** A chemical that does not react with DNA could be converted in the body to give rise to a derivative that does interact with DNA.
   - **d.** A carcinogen never reacts with DNA, but instead alters protein folding.
   - **e.** all of the above
3. Which protein regulates the cell cycle by monitoring the environment of each stage and making sure the previous cycle is complete?
   - **a.** cyclin synthases
   - **b.** cyclin-dependent kinases
   - **c.** E2F
   - **d.** cyclins
   - **e.** p53
4. What activates cell division?
   - **a.** Ras
   - **b.** E2F
   - **c.** growth factors
   - **d.** cell surface receptors
   - **e.** phosphotransferase
5. Which of the following statements about oncogenes and anti-oncogenes is not true?
   - **a.** Mutations in oncogenes are dominant.
   - **b.** Both copies of tumor-suppressor genes must be inactivated to give rise to cancer.

- **c.** Mutations in oncogenes are recessive and can often be overcome by the non-affected, wild-type allele.
- **d.** Tumor-suppressor genes normally protect cells from cancer.
- **e.** All of the above statements are true.

6. How do oncogenic viruses form?
- **a.** Upon excision from the genome, the virus "picks up" an oncogene from the host.
- **b.** A naturally occurring oncogene within a virus becomes mutated.
- **c.** A virus is exposed to carcinogens, which mutates DNA.
- **d.** Because viruses are haploid, they do not have a backup copy of the mutated oncogene, and thus become oncogenic.
- **e.** none of the above

7. How might oncogenes be detected?
- **a.** transduction of suspect DNA into animal cells using a virus
- **b.** transformation of suspect DNA into animal cells and observation of cell growth (i.e., lack of contact inhibition)
- **c.** using phage-display technology
- **d.** by cloning the suspect gene onto an expression vector and assaying for the presence of the defunct protein in bacterial cells
- **e.** none of the above

8. Which of the following does not cause a proto-oncogene to become an oncogene?
- **a.** a mutation in the coding region that changes the amino acid sequence to make the protein hyperactive
- **b.** a mutation within the regulatory region of the proto-oncogene that increases expression of that gene
- **c.** a duplication in the proto-oncogene
- **d.** a genetic rearrangement of the proto-oncogene that places the gene behind a strong promoter, thus increasing expression
- **e.** all of the above

9. How does *Ras* become an oncogene?
- **a.** by a mutation in the coding region of the gene that creates a loss of an important function in cell division
- **b.** by a mutation in the genetic code that alters one of three amino acid residues involved in the binding and splitting of GTP
- **c.** by a duplication of the *Ras* proto-oncogene that results in more protein expressed
- **d.** by a genetic rearrangement that places *Ras* proto-oncogene downstream of a strong promoter
- **e.** none of the above

10. A mutation in which oncogene tends to occur in the later progression of lung, breast, and ovarian cancers?
- **a.** *Ras*
- **b.** cyclin-dependent kinase
- **c.** *Rb*
- **d.** *Myc*
- **e.** none of the above

(Continued)

11. What is nullizygous?
- **a.** when both copies of an anti-oncogene have been inactivated by mutations
- **b.** when one copy of an anti-oncogene is inactivated but the other is still active
- **c.** when both copies of an anti-oncogene are not active, but not due to a mutation, rather the anti-oncogenes are not needed in the cell at that time
- **d.** when both copies of a anti-oncogene are active and overproducing protein
- **e.** none of the above

12. Why does a mutation in only one p53 anti-oncogene allele cause a defect in the tetrameric protein?
- **a.** A mutation in only one of the p53 anti-oncogene alleles does not cause a defect in the protein.
- **b.** Because p53 is expressed from both alleles and one mutant copy causes half of the tetramer to be defective.
- **c.** Because only one p53 allele is active at any given time.
- **d.** Because each allele is expressed, making both good and bad p53 proteins to assemble into mixed tetramers.
- **e.** none of the above

13. Which of the following is a mutation associated with colon cancer?
- **a.** inactivation of both copies of the *DCC* anti-oncogene
- **b.** activation of the *Ras* oncogene
- **c.** mutation in one copy of the *p53* gene
- **d.** inactivation of both copies of the *APC* anti-oncogene
- **e.** all of the above

14. Which of the following inherited defects predisposes a person to cancer?
- **a.** DNA repair systems
- **b.** anti-oncogenes
- **c.** components of DNA replication
- **d.** genetic differences between races (e.g., skin color and melanoma)
- **e.** all of the above

15. How is a virus made oncolytic?
- **a.** by integrating the genes for tumor-suppressor and necrosis factors into the viral genome under a constitutive promoter
- **b.** by changing the location of a viral DNA integration into the host cell so that it causes a gene knockout in the defective gene that is causing the cancer
- **c.** changing the viral protein that binds to the host cells so that it only recognizes and infects, then kills, cancer cells
- **d.** by cloning genes for apoptosis into the viral genome, which integrates into the host cancer cells and kills the cells
- **e.** none of the above

16. The result of inhibiting PARP is all of the following except ______________.
- **a.** an accumulation of double-strand breaks in the DNA
- **b.** inhibition of base excision repair pathways
- **c.** cell death
- **d.** activation of p53
- **e.** lack of poly(ADP-ribose) chains being added to single- and double-strand breaks

17. Which of the following is not true regarding cancer epigenetic changes?
- **a.** Changes that occur within the DNA sequence.
- **b.** DNA methylation influences gene expression.
- **c.** Changes that are passed from parent generation to daughter generations.
- **d.** Histone modification
- **e.** Methylation decreases gene expression by blocking the activator binding sequences.

18. The chemical component of grapes that may be involved in epigenetic changes is called _______________.
- **a.** caffeic acid
- **b.** resveratrol
- **c.** curcumin
- **d.** isothiocyanate

19. Which organism discussed in the textbook that offers a mechanism to deliver radioisotopes directly into cancer cells?
- **a.** *Escherichia coli*
- **b.** *Listeria monocytogenes*
- **c.** *Staphylococcus aureus*
- **d.** Lytic viruses
- **e.** Lysogenic viruses

20. Which of the following is not involved in cellular communication?
- **a.** pheromones
- **b.** local mediators
- **c.** hormones
- **d.** mRNA
- **e.** All of the above are involved in communication between cells.

21. How does signal transmission occur?
- **a.** phosphotransfer systems
- **b.** ion channel activation
- **c.** second messengers
- **d.** all of the above
- **e.** none of the above

Further Reading

Balmaña, J., Domchek, S. M., Tutt, A., & Garber, J. E. (2011). Stumbling blocks on the path to personalized medicine in breast cancer: the case of PARP inhibitors for BRCA1/2-associated cancers. *Cancer Discovery, 1*(1), 29–34.

Blanpain, C. (2013). Tracing the cellular origin of cancer. *Nature Cell Biology, 15*(2), 126–134.

Chen, P. S., Su, J. L., & Hung, M. C. (2012). Dysregulation of microRNAs in cancer. *Journal of Biomedical Science, 19*, 90.

Frazer, I. H., Leggatt, G. R., & Mattarollo, S. R. (2011). Prevention and treatment of papillomavirus-related cancers through immunization. *Annual Review of Immunology, 29*, 111–138.

Garraway, L. A., & Lander, E. S. (2013). Lessons from the cancer genome. *Cell, 153*(1), 17–37.

Hilton, J. F., Hadfield, M. J., Tran, M. T., & Shapiro, G. I. (2013). Poly(ADP-ribose) polymerase inhibitors as cancer therapy. *Frontiers in Bioscience: a Journal and Virtual Library, 18*, 1392–1406.

Manning, A. L., & Dyson, N. J. (2011). pRB, a tumor suppressor with a stabilizing presence. *Trends in Cell Biology, 21*(8), 433–441.

Msaouel, P., Iankov, I. D., Dispenzieri, A., & Galanis, E. (2012). Attenuated oncolytic measles virus strains as cancer therapeutics. *Current Pharmaceutical Biotechnology, 13*(9), 1732–1741.

Nana-Sinkam, S. P., & Croce, C. M. (2013). Clinical applications for microRNAs in cancer. *Clinical Pharmacology and Therapeutics, 93*(1), 98–104.

Quispe-Tintaya, W., Chandra, D., Jahangir, A., Harris, M., Casadevall, A., Dadachova, E., et al. (2013). Nontoxic radioactive *Listeria*^{at} is a highly effective therapy against metastatic pancreatic cancer. *Proceedings of the National Academy of Sciences of the United States of America, 110*(21), 8668–8673.

Russell, S. J., Peng, K. W., & Bell, J. C. (2012). Oncolytic virotherapy. *Nature Biotechnology, 30*(7), 658–670.

Sandoval, J., & Esteller, M. (2012). Cancer epigenomics: beyond genomics. *Current Opinion in Genetics & Development, 22*(1), 50–55.

Aging and Apoptosis

Biotechnology

http://dx.doi.org/10.1016/B978-0-12-385015-7.00020-X

INTRODUCTION

Aging is a process we all start at birth and continue until death. Some people seem to age very gracefully, whereas others show signs of aging very early. Even ignoring the effects of infection or accident, quite a large range of life spans occurs among humans. Interestingly, there is a dramatic difference in the average life span of similar-sized mammalian species. For example, mice live about 2 to 3 years, whereas other rodents of similar size average about 5 to 10 years. Therefore, age is a relative term with differences within and between species.

Until recently, there was no clear molecular reason for aging other than slow accumulation of wear and tear on our cells, tissues, organs, and ultimately on our bodies as a whole. Breakthroughs in our understanding of aging occurred when mammalian cells were grown in the laboratory and aging was studied in model organisms such as *Caenorhabditis elegans*. Many people try to offset the signs of aging and spend vast amounts of money on products that claim to repair or mask the visible signs. Not surprisingly, the biotechnology industry hopes to tap into this flow of money by finding biological solutions to aging.

Aging can be broken into two interconnected branches of alterations. The first branch includes genomic instabilities such as accumulation of genetic changes, telomere shortening, and epigenetic changes in DNA. The second branch deals with cellular dysfunctions in processes such as intercellular communication, nutrient sensing, protein function, and mitochondrial function. These processes occur in the normal aging process, but they can go awry. If the normal process accelerates due to mutation, the organism will age faster, a set of conditions called progeroid syndromes. In contrast, if the normal process slows, the organism should live longer. These damages, once reaching a certain threshold, result in cellular senescence or apoptosis, depending on the situation and extent of damage.

GENETIC PHENOMENA ASSOCIATED WITH AGING

Accumulation of Genetic Changes

Environmental assaults such as UV rays from the sun, chemicals, and biological agents are external sources of DNA damage. In addition, cells acquire mutations from internal assaults such as DNA replication mistakes and various hydrolytic reactions. These damages occur at random, and our body has multiple lines of defense to protect and repair our DNA (see Chapter 4). As we age, though, some changes are unrepaired and persist in the next generation of cells. **Somatic mosaicism**, that is, where some somatic cells have different genomic sequences than the organism as a whole, may be more common than originally thought. When blood cells from adults over the age of 70 had their whole genomes sequenced, the results showed that 1% to 2% of these cells had different genomic sequences than expected, probably due to mutations accumulating without repair. In contrast, the blood from adults younger than 50 had less than 1% of the cells with genomic differences. Some of the changes that occur with age include loss of chromosomes, copy number variations in genes, base damage that leads to missense mutations, and chromosomal translocations.

Oxidative stress is a common culprit for generating mutations and results in genetic alterations in both the nuclear and mitochondrial genomes. Because the majority of oxidizing radicals form during aerobic respiration, the mitochondria play a central role in this cause of aging. In brief, mitochondrial electron transport, which manufactures most of the ATP, uses about 85% of the oxygen. Partial reduction of the O_2 molecule creates highly **reactive oxygen species (ROS)**, including superoxide ions ($\bullet O_2^-$), peroxides (H_2O_2), and hydroxyl radicals ($OH\bullet$). These reactive oxygen metabolites damage protein, lipid, and DNA nonspecifically (Fig. 20.1).

Mitochondrial DNA (mtDNA) is a main target for ROS because of proximity, but also because the DNA is not wrapped around histones like its nuclear counterpart. These mutations often affect the genes that encode the electron transport chain, rendering these enzymes

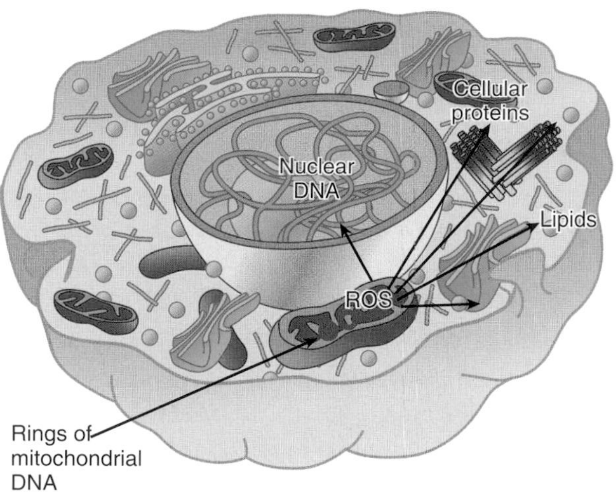

MUTATED mtDNA MAKE DEFECTIVE ELECTRON TRANSPORT PROTEINS

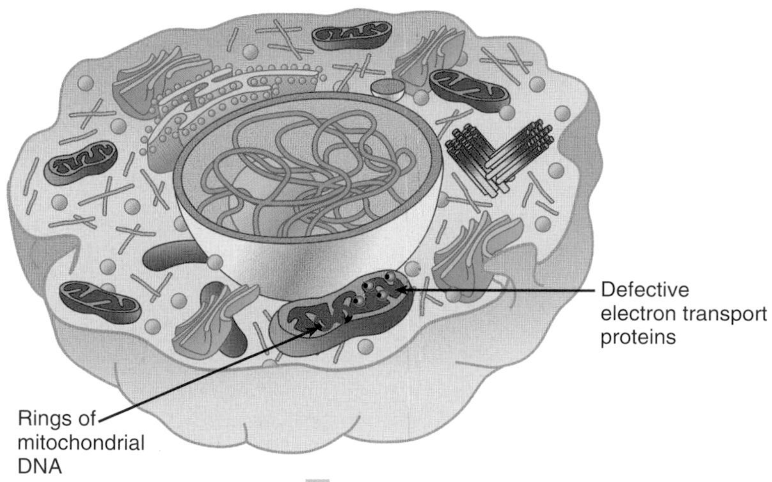

MUTATED ELECTRON TRANSPORT PROTEINS MAKE MORE ROS AND LESS ATP

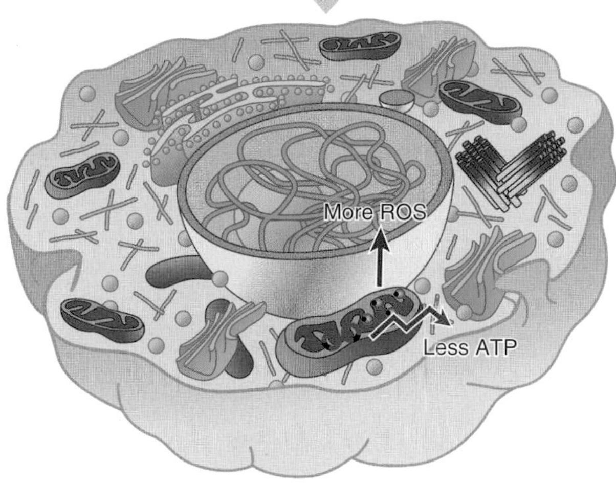

FIGURE 20.1 Oxidative Stress Damages Surrounding Molecules
Reactive oxygen species (ROS) attack mitochondrial DNA, cellular proteins, lipids, and nuclear DNA. Damaged mitochondrial DNA may produce mutant respiratory chain proteins, which in turn make more ROS and/or less ATP.

less efficient and more likely to produce more free radicals, which damages the mtDNA further, thus creating a vicious cycle. This type of damage accumulates until it reaches a threshold. After that point, the cells may initiate senescence or apoptosis, depending on the severity and extent of the damage (Fig. 20.2).

Genetic mutations occur throughout our lives, and most are repaired. As we age, unrepaired mutations accumulate. Many of these mutations occur due to the generation of reactive oxygen species (ROS), which can attack the mitochondrial DNA, nuclear DNA, proteins, and lipids.

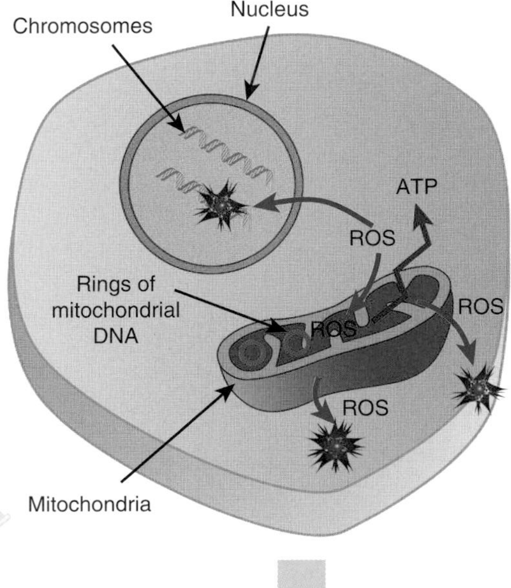

FIGURE 20.2 DNA Damage Activates Senescence
Mitochondrial respiration produces ATP but also generates reactive oxygen species that damage cell components. If the damage is extensive, the cell will senesce or die.

Telomere Shortening

In normal cells, the length of the **telomeres** controls whether or not cells continue to divide, or **senesce**, which is a state of quiescence where the cell remains metabolically active but no longer divides (see the later discussion of cellular senescence). Telomeres are highly repetitive sequences at the ends of linear eukaryotic chromosomes (see Chapter 4 for details). During the replication of chromosomes, DNA polymerase cannot replicate the DNA at the ends. Consequently, each chromosome shortens each time the cell divides (Fig. 20.3). When the telomeres reach a critically short length, the cell enters senescence.

In some respects, the telomeres can act as a molecular clock that determines how long a cell lives. Thus, normal human chromosomes have telomeres between 5 and 15 kb long, whereas senescent cells have telomeres 4 to 7 kb long. They are bound by a protein complex called **shelterin** that protect the ends of the DNA from double-stranded break repair, which would fuse the ends of different chromosomes together—a disastrous mistake (Fig. 20.4). The enzyme **telomerase**, which is particularly active in reproductive cells, repairs the shortened ends of the chromosome by adding more repeats to the ends (Fig. 20.5). Telomerase is a nontypical reverse transcriptase with two components: protein and RNA. The protein component called TERT catalyzes the synthesis of new DNA, and the RNA component called TERC acts as a template.

In most human cells, telomerase activity is very low, so these cells are highly susceptible to telomere shortening. In contrast, mice have significant telomerase activity in most cells and have much longer telomeres (approximately 40 to 60 kb). These differences are evident when cells are cultured. Mouse cell lines divide many more times before entering senescence and sometimes become immortal spontaneously. Mutant mice without the RNA component of telomerase appear normal for a few generations, although the telomeres shorten with each generation, because they can no longer be extended. At a critical point in telomere length, the mutant mice begin to show chromosomal fusions, higher cellular senescence, and increased apoptosis. Thus, the mouse cells come to resemble human cells (which normally have shorter telomeres).

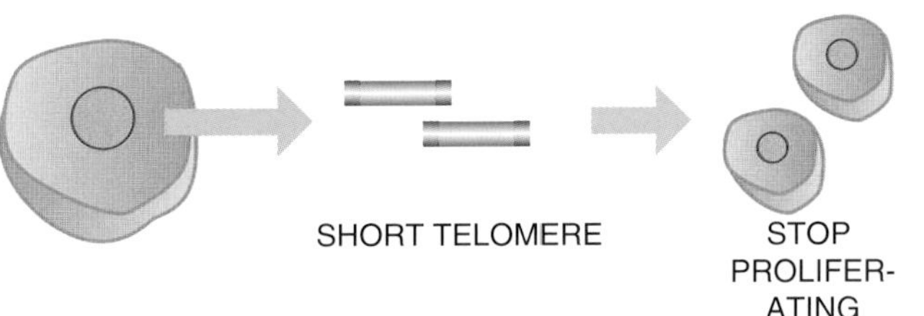

FIGURE 20.3 Telomere Shortening Causes Cellular Senescence
A cell with long telomeres is able to continually divide and multiply. A cell with short telomeres may divide only one more time before entering senescence.

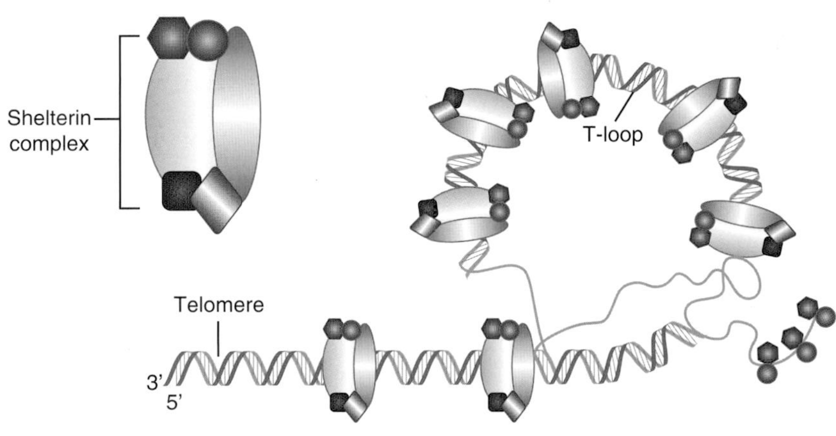

FIGURE 20.4
Shelterin Protects the Ends of DNA
Telomeres are structures found at the end of linear chromosomes that must be protected from double-stranded break repair. A complex of six proteins, called shelterin, coats the telomere and keeps the DNA looped to protect the ends.

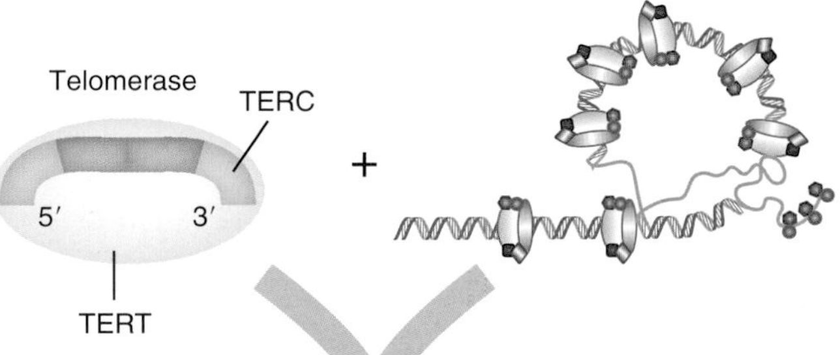

RNA TEMPLATE OF TELOMERASE RECOGNIZES TELOMERE ENDS

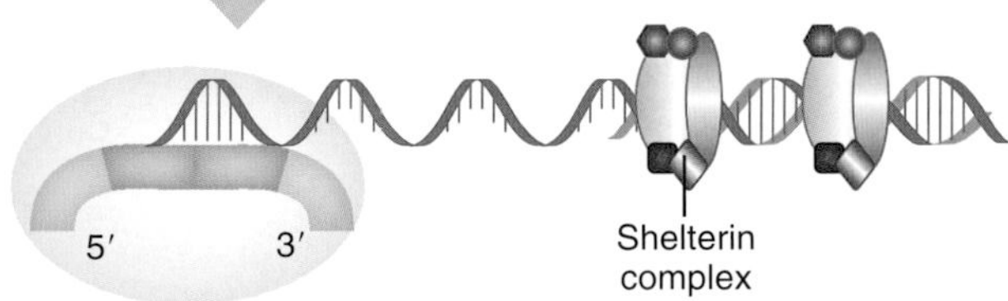

ENZYME PORTION ADDS SEQUENCES USING RNA TEMPLATE

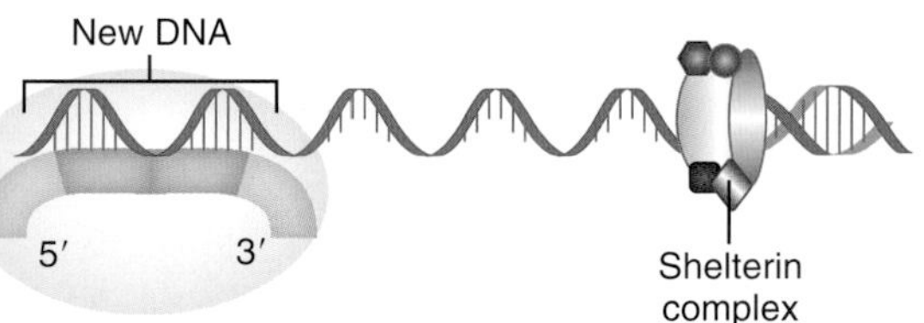

FIGURE 20.5
Telomerase Increases the Length of Telomeres
Telomerase consists of a protein plus an RNA component. The RNA recognizes the telomere repeats, and the protein elongates the telomeres.

The human inherited disease dyskeratosis congenita (DKC) is linked to lower telomerase activity (see Table 20.1). People with DKC have blocked tear ducts, learning difficulties, pulmonary disease, graying and loss of hair, and osteoporosis, to name just a few of the problems. Interestingly, many of these are also symptoms of aging. The gene responsible for DKC encodes a protein that processes rRNA precursors in the nucleolus. In DKC patients, the level of the RNA component of telomerase is much lower than normal, and the telomeres are unusually short. Some of the symptoms of DKC are probably due to improper telomere length.

DNA polymerase cannot synthesize the end of chromosomes, and therefore, telomere ends get shorter with each cell division. The enzyme telomerase can lengthen telomeres in some cells. Mutations in telomerase components have symptoms that recapitulate aging, suggesting the telomere length and stability are key genetic factors in aging.

Table 20.1 Diseases with Symptoms Associated with Aging

Disease	Gene	Frequency	Function	Consequence of the Mutation	Symptoms
Werner Syndrome	Werner syndrome gene *(WRN)*	1: 200,000 (US)	Werner protein repairs and maintains DNA	Defects in DNA replication and DNA repair	Cataracts, type 2 diabetes, osteoporosis, atherosclerosis, graying of hair, diminished fertility
Telomerase Deficiency (dyskeratosis congenita, or DKC)	Telomerase and shelterin gene mutations	1: 1 million	Protect and stabilize chromosome ends	Unable to protect DNA ends during replication	Premature development of pulmonary fibrosis, premature graying, osteoporosis
Hutchinson–Gilford progeria	Lamin A/C (*LMNA*)	1: 4 million	Nuclear lamina is the region on the inner side of the nuclear membrane	Lamins are implicated in organizing chromatin domains within the nucleus	Dwarfism, baldness, aged-looking skin, premature arteriosclerosis, and other severe cardiovascular diseases
Bloom Syndrome	*BLM* gene	unknown	RecQ helicase unwinds DNA during replication	Chromosome instability with gaps and breaks in DNA	Short stature, sun-sensitive skin, increased risk of cancer, diabetes, early menopause (women), no sperm (men)
Xeroderma pigmento-sum	Nucleotide Excision Repair	1: 1 million	Repairs DNA, especially damage from UV exposure	Buildup of unrepaired DNA	Sensitivity to UV light; dry skin, propensity to develop skin cancer; cataracts
Trichothiodystrophy	*ERCC2, ERCC3,* or *GTF2H5*	1: 1 million	DNA repair, basal transcription	Reduce the amount of TFIIH general transcription factor, which affects gene transcription and DNA repair	Osteoporosis, cataracts, fragile hair, neurodegeneration
Cockayne Syndrome	*ERCC6* or *ERCC8*	2 in 1 million	Transcription coupled DNA repair	Damaged DNA is not repaired	Thin hair, cachexia, retinal degeneration, hearing loss, neurodegeneration, cataracts, sensitivity to light

Epigenetic Changes

Unlike the other genetic changes associated with aging, epigenetic changes do not change the nucleotide sequence of the genome. These changes modify the DNA and the proteins associated with the DNA.

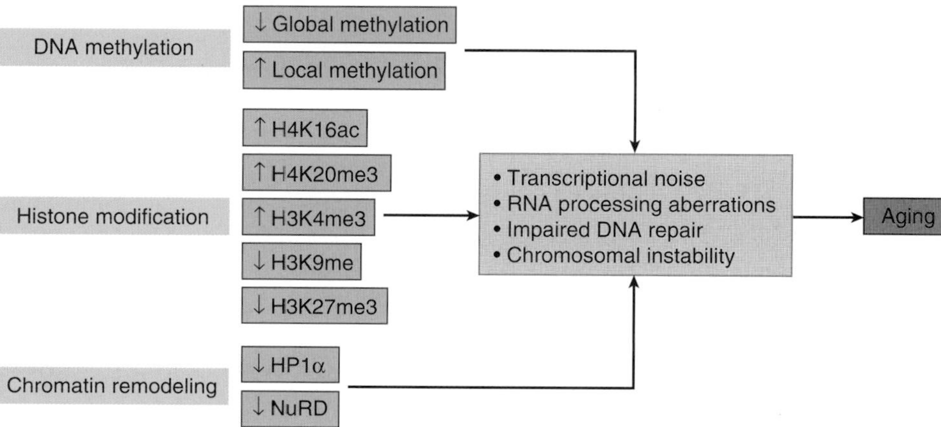

FIGURE 20.6 Epigenetic Changes Associated with Aging

DNA methylation, histone modification, and chromatin remodeling alter gene transcription, RNA processing, DNA repair and chromosome instability, which can then lead to aging. DNA methylation changes include decreased global methylation, and an increased local methylation of certain genes. Histone modifications include variations in histone 3 and 4 (H3 and H4) changes such as acetylation (ac) and methylation (me) on the different lysine residues (K16, K20, K4, K9, and K27). Chromatin remodeling includes lower expression of HP1α and NuRD complexes.

Epigenetic changes include DNA methylation, histone modification, and chromatin remodeling (see Chapter 2; also Fig. 20.6). First, **DNA methylation** adds methyl groups ($-CH_3$) to CG or CNG sequences on one strand of the chromosome. The methyl groups modulate gene transcription by compacting the chromatin DNA and preventing transcription factors from accessing the genes. Studies have demonstrated that DNA methylation decreases as we age, suggesting that aging increases gene transcription of genes that were originally quiescent. There are regions of the genome that do not follow this tenet, and become hypermethylated during aging, such as the lamin A/C gene. Lamins are proteins that line the nucleus to provide architecture around the genome. Although the gene is essential, the role it plays in aging was ambiguous until is was found to be the causative agent of Hutchinson–Gilford syndrome, otherwise known as progeria. This disease has many symptoms that appear to be premature aging. Since mutations of lamin A/C are associated with accelerated aging in progeria, the hypermethylation of this gene could cause some of the age-related decline symptoms seen in normal persons.

Another nonsequence change associated with DNA expression is **histone modification**. DNA wraps around histones in eukaryotes, and post-translational modifications such as methylation, phosphorylation, ubiquitylation, and acetylation of histones determine if other proteins can access the DNA. Histone modifications are reversible, since enzymes such as demethylases and deacetylases remove the added groups. The system allows cells to control what type of regulatory protein can access each particular gene (see Chapter 2). *C. elegans* and *Drosophila* have extended life spans when certain histone methylases are deleted, suggesting that the acquisition of methylated histones is another potential molecular cause of aging.

Sirtuins are another group of histone modification proteins associated with aging. These are NAD^+-dependent histone deacetylases (HDAC), and were first discovered in the budding yeast, *S. cerevisiae.* Despite being a single-celled eukaryote, yeast undergoes aging of two different types: **replicative aging** and **chronological aging**. Replicative aging applies to dividing cells and is measured by the number of daughter cells one mother can produce. The average number of replicative divisions is 20–25. Chronological aging is induced by nutrient deprivation. The length of time the cell can live without dividing determines its chronological age. Yeast cells that have excess Sir2, the founding member of the sirtuin family, have a longer replicative life span. Excess Sir2 removes the acetyl groups from histones so the DNA can wrap around the histone tightly, and therefore, homologous recombination is less likely to occur. In contrast, when the gene for Sir2 is mutated, and the enzyme cannot remove any acetyl groups from the histones, then the DNA cannot tightly associate with the histones. The loosely wrapped DNA is more likely to undergo homologous recombination, especially at the rRNA genes because these are highly repetitive DNA sequences. Recombination events in this area of the genome create extrachromosomal ribosomal DNA circles. In cells missing Sir2, circular ribosomal DNA accumulates and creates genomic instability, which causes the mutant cells to senesce earlier than normal. Homologs of Sir2 have the same effect on the life span of the mouse and in cultured human cells, suggesting that the function is conserved in evolution.

Chromatin structural changes are another feature of aging that contributes to some of the changes in our cellular functions. **Euchromatin** and **heterochromatin** are two forms of chromatin found in all cells. Euchromatin is the loosely packed chromatin that is found around genes that are actively being expressed. Heterochromatin is tightly packed DNA and histones (nucleosomes) that are usually found around repeated sequences, such as telomeres and centromeres. The amount of heterochromatin declines with age. When heterchromatic DNA relaxes, repetitive DNA regions can become transcribed into nonfunctional mRNA, which in turn destabilizes the genome. Some of the proteins associated with maintaining the amount and location of heterochromatin are heterochromatin protein 1α (HP1α), the Polycomb group proteins, and the NuRD complex. In *Drosophila*, extra HP1α expression increases the life span, whereas mutations that disrupt HP1α function decrease the life span. Although this applies to *Drosophila*, the results suggest that genomic instability may be another important aspect of aging.

DNA methylation, histone modification, and chromatin structural changes alter DNA expression and genomic stability. Sirtuins are deacetylases that remove acetyl groups from histones. When these proteins are overexpressed, yeast cells, cultured mammalian cells, and mice have longer life spans. HP1α also increases longevity in *Drosophila*, presumably by preventing heterochromatin from loosening.

CELLULAR DYSFUNCTION AND AGING

Nutrient Sensing

The role of metabolism and nutrition in aging has been revealed by studying the nematode worm *C. elegans*, which normally lives for 2 to 3 weeks. Several mutations are known that increase the life span and/or the worm's entry into a type of hibernation called **dauer** (Fig. 20.7). Normally, these worms go through four larval stages before they become adults. When there is not enough food or water, a worm goes into the dauer stage. It produces a thicker cuticle to prevent dehydration and closes off its mouth and anus so it cannot eat. It does not feed until stimulated with more food. Some mutations in the genes *daf-2* and *age-1* cause the worms to enter the dauer stage. Many of these mutations are **temperature sensitive**, that is the worm is normal at the permissive temperature, and when shifted to a higher or lower temperature, then the mutation becomes evident. Other *daf-2* or *age-1* mutations cause an increased life span, of up to 2 months, without inducing the dauer stage.

The DAF-2 and AGE-1 proteins both regulate metabolic rate (Fig. 20.8). According to sequence data, *Age-1* encodes a worm homolog to the catalytic subunit of phosphatidylinositol-3-kinase (PI3K), and *daf-2* encodes a homolog to the insulin receptor. The mammalian homologs of these two proteins are involved in insulin signaling, and are found in the **insulin and IGF-1 signal transduction pathway (IIS pathway)**. The pathway is responsible for coordinating growth, differentiation, and metabolism in response to the environment. When this pathway is impaired in humans, we develop type 2 diabetes (adult-onset), and some evidence suggests that the pathway may be involved in longevity. The worm mutations in *daf-2* and *age-1* block the signal transmitted by the insulin-like receptor. The normal DAF-2 and AGE-1 proteins affect the cell by deactivating the FOXO family of transcription factors and keeping them in the cytoplasm. Since these transcription factors activate genes associated with resistance to oxidative stress and decreased metabolism, these genes remain off when FOXO and DAF-16 are in the cytoplasm. So when signaling through the IIS pathway is decreased by mutations in *daf-2* and *age-1*,

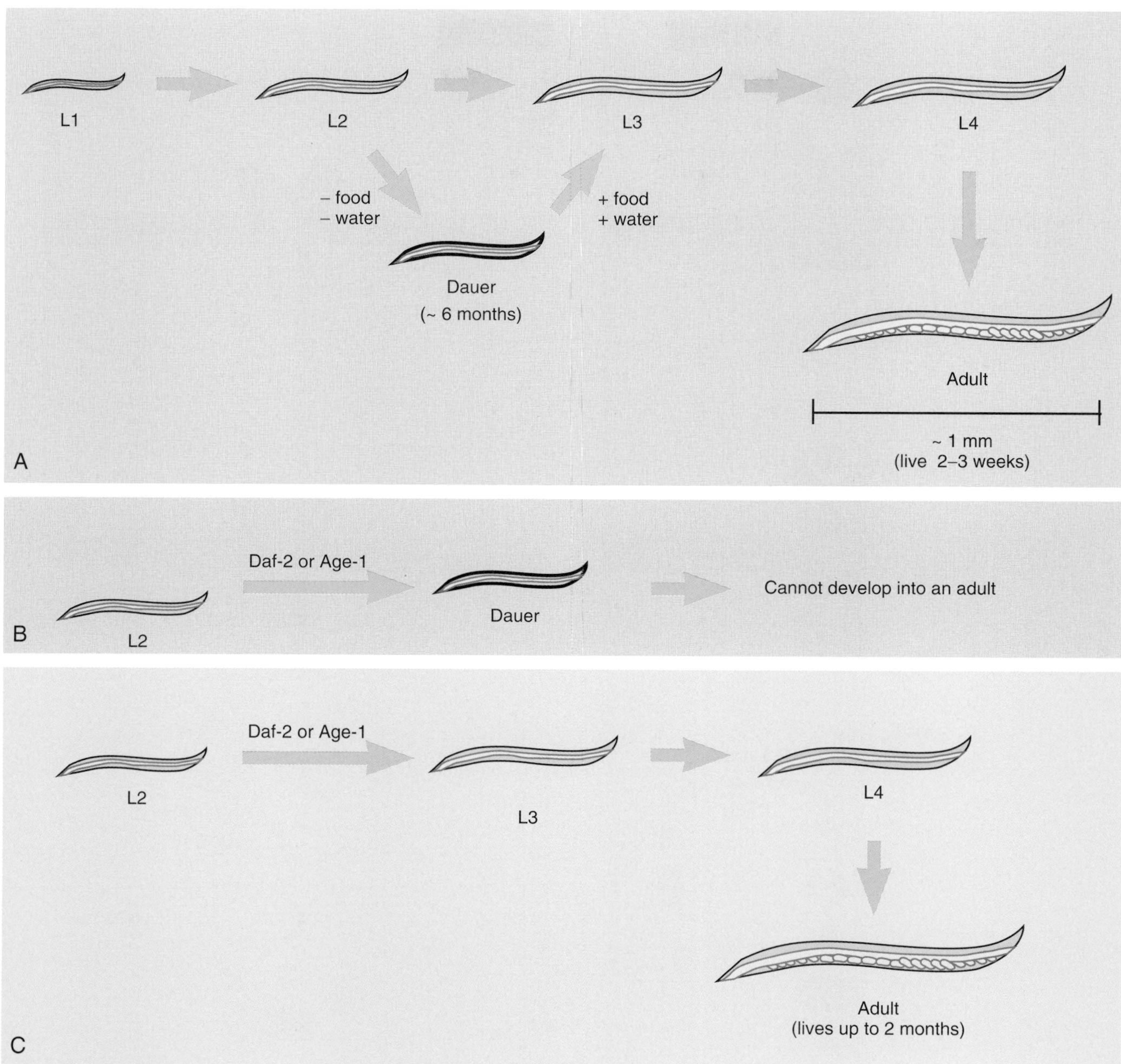

FIGURE 20.7 Effects of *daf-2* and *age-1* on *C. elegans* Life Span and Dauer Formation
(A) Normal worms go through four larval stages before becoming adults, which live about 2 to 3 weeks under normal conditions. If worms encounter starvation while in L2, they can hibernate as a dauer for up to 6 months. (B) Some *daf-2* or *age-1* mutations cause the worms to enter the dauer state no matter what their nutritional status. (C) Other *daf-2* or *age-1* mutations do not induce the dauer state but result in greatly increased life span.

FOXO and DAF-16 are free to enter the nucleus and activate stress-resistance and metabolism-lowering genes; therefore, worms show lower metabolic rates than wild-type and more resistance to oxidative damage.

Interestingly, **dietary** or **caloric restriction** decreases the activity of the IIS pathway and can extend the life span of many organisms, from yeast to nonhuman primates. Although a diet low in caloric intake can extend life span in these organisms, they must have complete nutrition for this effect. If, for some reason, there is a nutrient deprivation, then the life span increase is not seen. Although the effect of caloric restriction on extending life span has been known for many years, there is still a mystery as to all the molecular events involved.

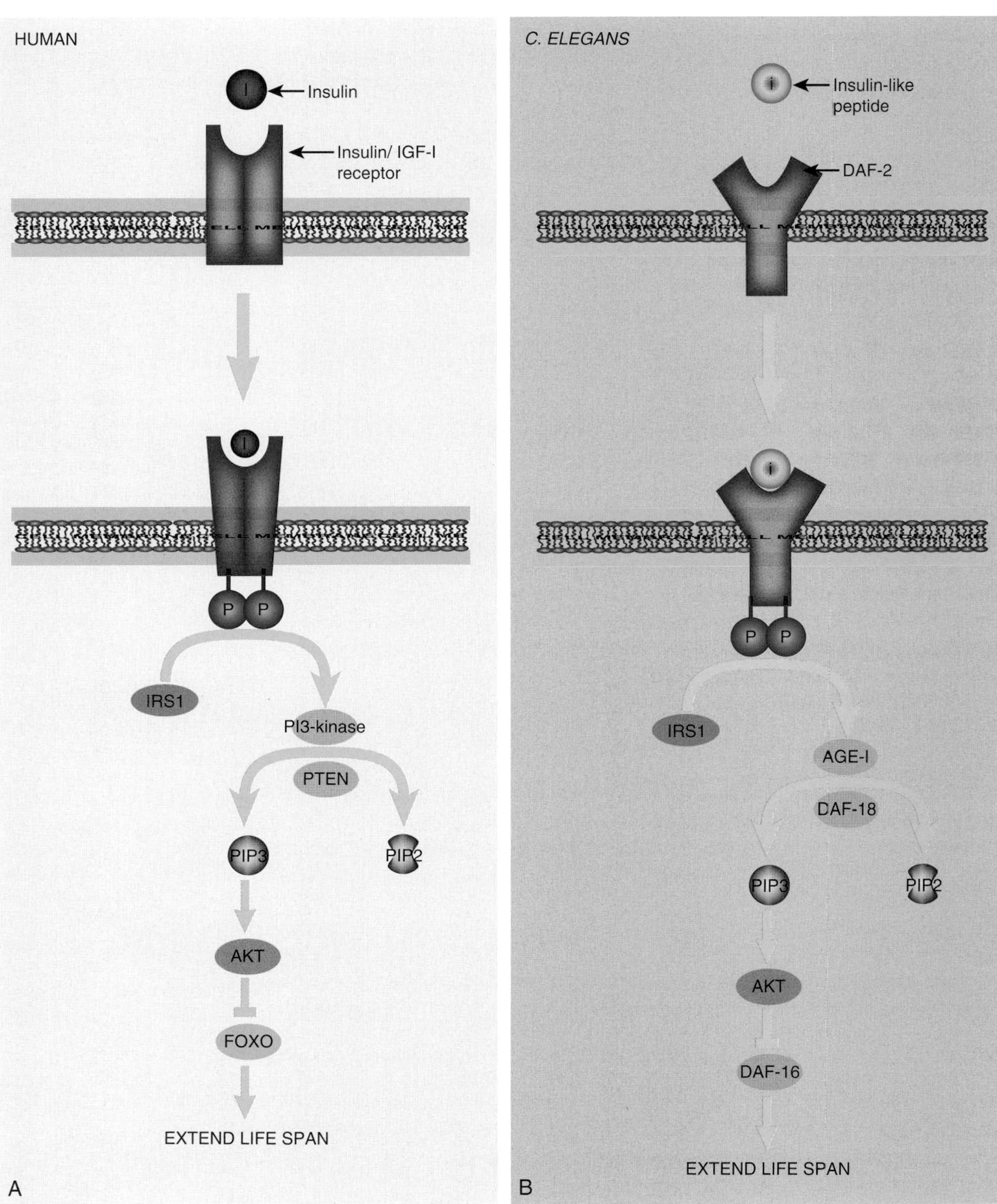

FIGURE 20.8 Insulin Signaling in Humans and *C. elegans*

(A) When insulin binds to the human insulin receptor, the intracellular domain is activated by phosphorylation. The phosphate is transferred via IRS-1 to PI3-kinase, which converts the lipid molecule phosphatidylinositol 3,4-bisphosphate (PIP_2) to phosphatidylinositol 3,4,5-triphosphate (PIP_3) by adding a phosphate. PIP_3 triggers the intracellular response to insulin by preventing the FOXO family of transcription factors from entering the nucleus. (B) In worms, Daf-2 protein binds an insulin-like molecule and activates AGE-1 protein. This protein is similar to the PI3-kinase of mammals, and phosphorylates PIP_3, which prevents the FOXO homolog, DAF-16, from entering the nucleus. When mutations or caloric restriction block signaling through these pathways, FOXO or DAF-16 can enter the nucleus and activate genes associated with reduced oxidative stress and decrease in metabolism, which can extend the life span of the organism.

also results in an accumulation of damaged proteins. Targeting this system may also slow the aging process; for example, in *C. elegans*, having more proteasome subunit RPN-6 extends life span of the worm.

The microautophagy-lysosome system and chaperone-mediated autophagy identify defective cytosolic proteins and move them into the lysosome for degradation. In the microautophagy system, the lysosome simply engulfs whole areas of the cytosol by wrapping its organelle membrane around the area. The method is essential but not very well understood in mammalian cells. This form of autophagy is also implicated in certain types of cell death (see the later section on necroptosis). In contrast, chaperone-mediated autophagy uses a chaperone to identify the defective protein and move it into the lysosome. The most common chaperone used for this system is called the heat shock chaperone of 70kD, or hsc70. Hsc70 recognizes a specific receptor called lysosomal-associated membrane protein type 2A (LAMP-2A). In aging tissues, both of these systems do not work as efficiently, but when artificially enhanced in model organisms, they extend life span. Protein quality control systems are essential for efficient cellular functions, and as cells age, the accumulation of these proteins adds stress to the cell and ultimately to the tissue and organism as a whole.

Proteins need to be degraded and recycled in order for a cell to function efficiently. Cells can try to fix proteins by refolding them with chaperones, but if they are not able to be fixed, then the cell simply degrades them. The ubiquitin-proteasome system, microautophagy, and chaperone-mediated autophagy degrade proteins when they are damaged beyond repair.

Mitochondria and Aging

Remember that each mitochondrion has multiple copies of its genome, each cell has many mitochondria, and each tissue has many cells. Damage from many events must accumulate for mitochondrial defects to be noticeable. Highly metabolically active cells tend to accumulate oxidative damage faster due to the accumulation of ROS from metabolic pathways. Neurons are particularly sensitive since they have high metabolic activity, few antioxidant enzymes, and they do not replicate. As described previously, metabolism (particularly in the mitochondria) creates ROS such as hydrogen peroxide (H_2O_2) and superoxide (O_2^-). Two other related categories of ROS include **reactive nitrogen species (RNS)**, such as nitric oxide (NO) and peroxynitrite ($ONOO^-$), and reactive lipid species (RLS) that accumulate in brains of people with Parkinson's or Alzheimer's disease. They form during incomplete reduction of oxygen during the mitochondrial electron transport chain and other associated energy-producing metabolisms.

Antioxidant enzymes can alleviate the effects of oxidant stress. There are several of these with varied roles. The two most familiar are **superoxide dismutase (SOD)** and **catalase**.

There are two forms of SOD with different metal ions at the active site. Cu/Zn SOD resides primarily in the cytoplasm, whereas Mn SOD concentrates in the mitochondria. SOD converts superoxide ions to oxygen plus hydrogen peroxide. Catalase converts hydrogen peroxide to oxygen plus water:

$$\text{Superoxide dismutase (SOD): } 2 \bullet O_2^- + 2H^+ \rightarrow O_2 + H_2O_2$$

$$\text{Catalase: } 2H_2O_2 \rightarrow O_2 + 2H_2O$$

The jury is still out as to the addition of antioxidant enzymes' ability to prolong the life span of the organism, but in general, there seems to be an increase in ROS with age, and a subsequent decrease in antioxidants, which may lead to the gradual decline called aging.

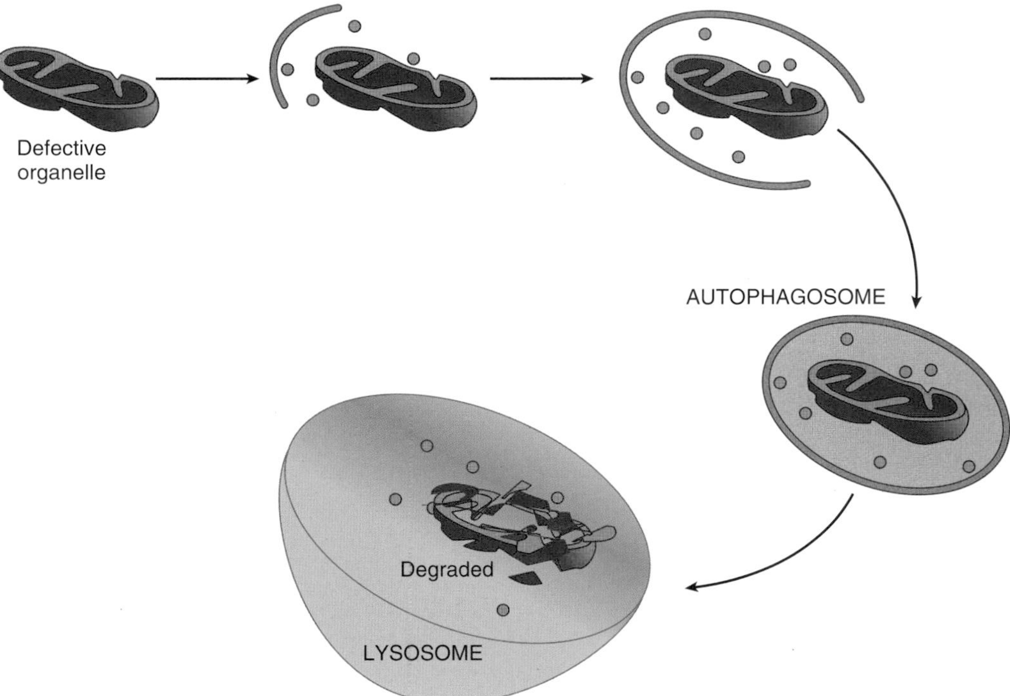

FIGURE 20.10
Macroautophagy
Defective organelles such as mitochondria can be engulfed by lysosomes and degraded. First, the whole region with the defective organelle is surrounded by an autophagosome, which then attaches to the lysosome. The entire contents of the autophagosome are then degraded within the lysosome.

Mitochondria turnover declines with age, leading to accumulation of damaged mitochondria within the cells. **Autophagy** is the mechanism that destroys and recycles defective cellular components; it is also sometimes referred to as **mitophagy** when it targets mitochondria (Fig. 20.10). There are three different types of autophagy: **macroautophagy** (also referred to as autophagy), microautophagy, and chaperone-mediated autophagy. The two latter deal with defective proteins and were described previously in the section "Protein Degradation During Aging." Macroautophagy or autophagy targets whole organelles such as mitochondria to be degraded. A double-membrane vesicle called an **autophagosome** forms around the defective organelle, fuses with the lysosome to become an **autophagolysosome**, and then the entire organelle is degraded by the lysosomal enzymes. This process increases when there is nutrient deprivation, and inhibition of autophagy induces changes in the tissue that resemble aging. These experimental findings suggest that autophagy of defective mitochondria are an effective way to prevent symptoms of aging.

Reactive oxygen species are produced from the partial reduction of oxygen, and they trigger oxidative stress. Mitochondria are a major source of ROS, and the amount of ROS increase with age. Cellular antioxidants work to remove ROS.

Defective mitochondria that produce more ROS can be removed by autophagy, a process that degrades the organelles within a double-membrane structure called an autophagosome that fuses with a lysosome to become an autophagolysosome.

CELLULAR SENESCENCE

Cellular damages, including both the cellular dysfunctions associated with nutrition and protein functions and the genetic mistakes that affect the DNA, accumulate in cells and work in overlapping and interrelated pathways to control whether a cell survives or not. **Senescence** prevents the damages from getting out of control. This metabolic state is activated when cells are damaged, and prevents the cell from creating new damaged descendents since senescent cells exit the cell cycle and simply stop dividing. Senescence was discovered when normal adult cells from a mouse or a human grew in laboratory dishes.

The cells divide for a certain number of generations and then stop dividing. Even the addition of growth inducers has no effect. However, the cells do not die as long as they are fed and maintained properly. This is also called cellular or replicative senescence. When Hayflick and Moorhead published this discovery, they noted that the number of replications may be "an expression of aging or senescence at the cellular level" (1961). Human fibroblasts from a fetus will divide 60 to 80 times in culture, whereas fibroblasts from an older person divide only 10 to 20 times. Replicative senescence depends on the number of cell divisions, not the calendar. The allowed number of divisions is programmed into the cell, rather than being controlled by circulating hormones or surrounding tissues. As described previously, cellular senescence occurs primarily when telomeres become short but can also be activated by other factors such as oncogene-induced senescence.

Some key characteristics are associated with cellular senescence (Fig. 20.11). First, senescent cells are cell cycle arrested (see Chapter 4). The cells are metabolically active; that is, they produce proteins, generate energy, and function in their normal capacity, yet the cells do not replicate their DNA or divide. Related to senescence is the physiological condition called **terminal differentiation**. These cells also arrest in G_1 but remain arrested in the cell cycle because of a physiological signal from their environment. In contrast, cells can senesce without changing their physiological role. Second, senescent cells have active tumor suppressor molecules. The tumor suppressors, RB and p53, are the two key regulators that control whether or not a cell enters senescence or apoptosis (cell death). Their activated forms accumulate in cells about to enter senescence. Third, senescent cells are metabolically distinct and produce molecules only associated with the growth-arrested state. These cells accumulate a substance called senescence-associated β-galactosidase. This enzyme works to degrade sugars in the lysosome of a cell, and the increased amount of the enzyme reflects an increase in the number of lysosomes of senescent cells. Senescence cells also have an altered chromatin structure with senescence-associated heterochromatic foci that are visible with DNA dyes. In contrast, DNA dyes of normal cells do not form foci and have a homogenous staining pattern. Finally, senescent cells secrete more factors, a phenomenon called senescence-associated secretory phenotype. Some secreted substances include proinflammatory cytokines and extracellular matrix factors.

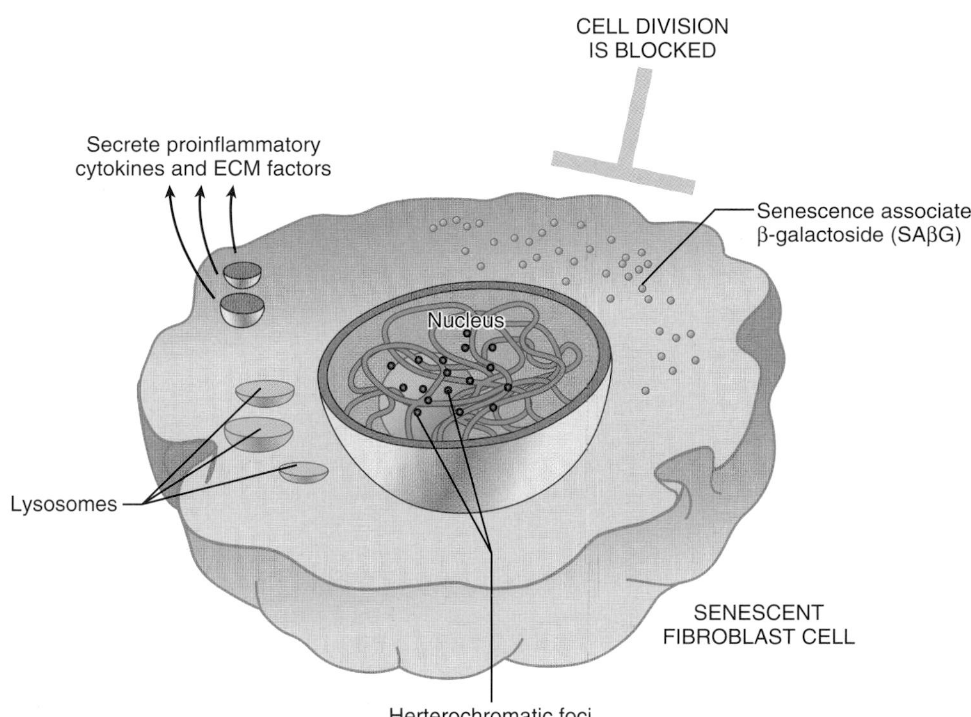

FIGURE 20.11 Phenotypes Associated with Senescence
Senescent cells undergo a series of physiological changes: cell division is blocked and molecules to keep the cell from dividing are produced; more senescence-associated β-galactosidase is produced; more lysosomes are created; and more proinflammatory cytokines and extracellular matrix factors are produced.

If all our cells entered a senescent state, then we would never be able to heal a wound or recover from damage due to bacterial or viral attack. Therefore, some of our cells never enter a senescent state. Obviously, during embryogenesis and development, very few cells are senescent. All multicellular organisms contain **stem cells**, immature or precursor cells from which differentiated cells originate (see Chapter 18). As we age, the number of stem cells decreases, but some remain to replenish various tissues, especially the skin, intestinal lining, immune system, and blood cells. **Immunosenescence** is one example of stem cell depletion that occurs with aging. The number of hematopoietic stem cells decline with age and result in fewer white blood cells circulating in the blood.

Some cells are not subject to senescence. Besides stem cells, germline cells do not enter senescence and always maintain the ability to divide when necessary. Tumor cells are another example of nonsenescent cells. Unlike stem cells and the germline, tumor cells have mutated in a way that overrides the entry into senescence.

Senescent cells are arrested in cell-cycle progression, have activated tumor suppressor proteins, express marker molecules such as senescent-associated β-galactosidase and senescent-associated heterochromatic foci, and secrete proinflammatory cytokines and extracellular matrix molecules.

PROGRAMMED CELL DEATH

If and when cellular damage accumulates beyond a point of repair, normal cells activate programs to commit suicide. Cells that initiate the selfless act of suicide can die in a controlled pathway called **programmed cell death**. The most studied programmed cell death is **apoptosis**, and is sometimes referred to as type I programmed cell death. When you get sick, your immune system makes new cells to fight off the infection. If these extra cells just burst and died when the battle was over, the body would have quite a mess to clean up! Instead, the cells calmly activate genes that initiate apoptosis, which occurs in an orderly manner with genes regulating each step of the process. Few to none of the cytoplasmic components are released, and neighboring cells engulf the neatly packaged debris. All the components are recycled. This form of cell death is essential during development to reshape and remodel tissues from the juvenile stages to adult.

During apoptosis, the dying cell undergoes a process that is morphologically distinct and recognizable by sight. This was first described in 1972, but not until the past 20 years has the process been analyzed genetically. First, the cell membrane starts to form blebs, or regions that balloon out (Fig. 20.12). The nucleus shrinks, condenses, and divides into smaller fragments. Finally, the entire cell shrinks and divides into large condensed fragments called **apoptotic bodies**. Other cells engulf the debris by phagocytosis, thus cleaning up the remnants. The proteins, lipids, and nucleic acids are digested and recycled.

Another method of cell death is autophagy or type II programmed cell death. As described in the preceding sections, autophagy consists of three distinct processes that converge on the lysosome, and each of them can be used to remove damaged cellular proteins or damaged mitochondria, but it can also degrade entire cells and leave a large number of autophagic vacuoles.

When cells are damaged from external injury, oxygen starvation, or energy depletion, they undergo **necrosis**. Necrotic cells swell because their osmotic balance is perturbed. Their proteins are denatured and degraded. Finally, the cell ruptures and dies, which triggers the immune system by releasing the cellular contents including proinflammatory cytokines. Until recently, necrosis was considered an unregulated form of cell death, but recent evidence

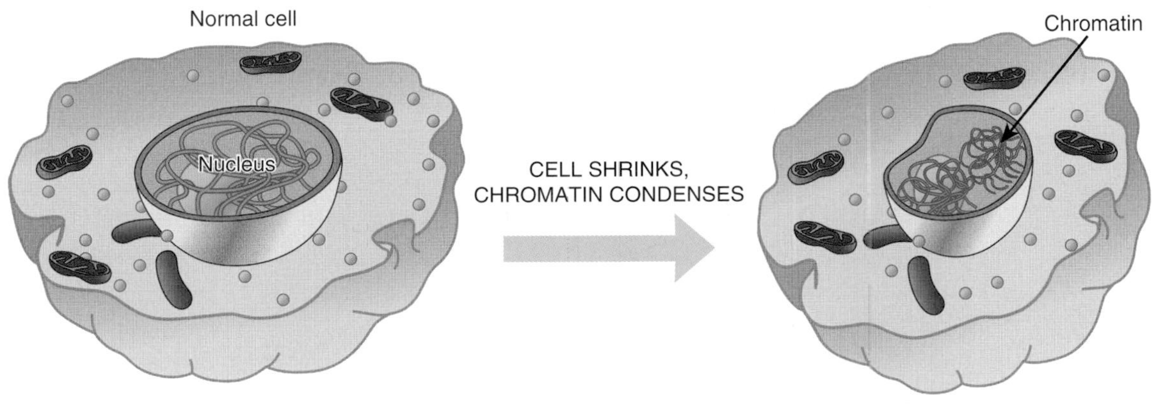

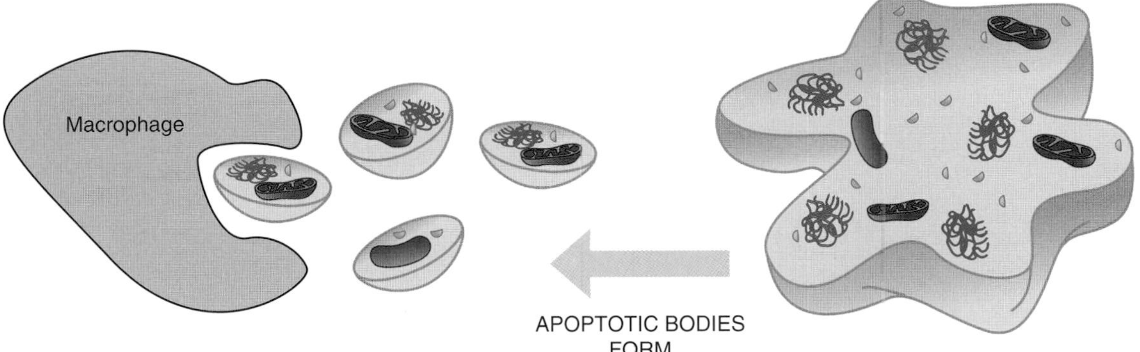

FIGURE 20.12 Apoptosis

The cell shrinks and chromatin condenses as apoptosis begins. Then the nucleus fragments and other cellular components are degraded by enzymes called caspases. Finally, the condensed apoptotic bodies are eaten by macrophages.

has shown that this form of cell death is also programmed genetically. Hence, a new term, **necroptosis**, or type III programmed cell death is now used to describe necrotic death initiated by molecular signals, specifically the signal from a serine/threonine kinase called RIP1. In comparison with apoptosis, necroptosis is similar since molecules and proteins trigger the cell to die this way, but it differs from apoptosis since this form of death elicits an immune response and apoptosis does not.

Three forms of programmed cell death are used to remove defective cells from tissues. Apoptosis, or type I programmed cell death, is a series of enzymatic steps that condense the cell components and nuclei and create apoptotic bodies that are removed by phagocytosis. Autophagy, or type II programmed cell death, is characterized by large autophagic vesicles within the dying cell. Necroptosis, or type III programmed cell death, is similar to apoptosis since it also is executed by a series of molecular signals but differs from apoptosis since the cells burst and release factors that elicit an immune response.

APOPTOSIS INVOLVES A PROTEOLYTIC CASCADE

The first well-documented account of apoptosis was seen in the nematode *C. elegans.* This small worm is normally found in soil and eats soil bacteria. The worm is used as a model organism for developmental genetics because it is easy to maintain, has

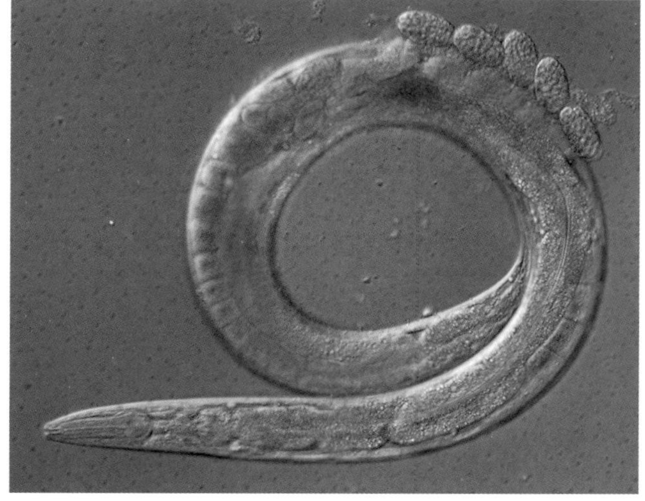

FIGURE 20.13 Nomarski Image of Adult *Caenorhabditis elegans*
Nomarski optics transforms differences in density into differences in height, giving an image that looks like a three-dimensional relief. The technique is useful in visualizing tissues that are several cells deep. Courtesy of Jill Bettinger, Virginia Commonwealth University, Richmond, VA.

four developmental stages, and is transparent (see Fig. 20.7). Through use of a special microscope, the **Nomarski** or **differential interference microscope**, every cell of *C. elegans* can be seen (Fig. 20.13). Using this technique, scientists have been able to trace each cell division from the single-celled egg through the entire adult. During development, *C. elegans* generates 1090 cells, but the final adult worm has only 959. The other 131 cells die via apoptosis, and the corpses of these dying cells are visible with Nomarski microscopy.

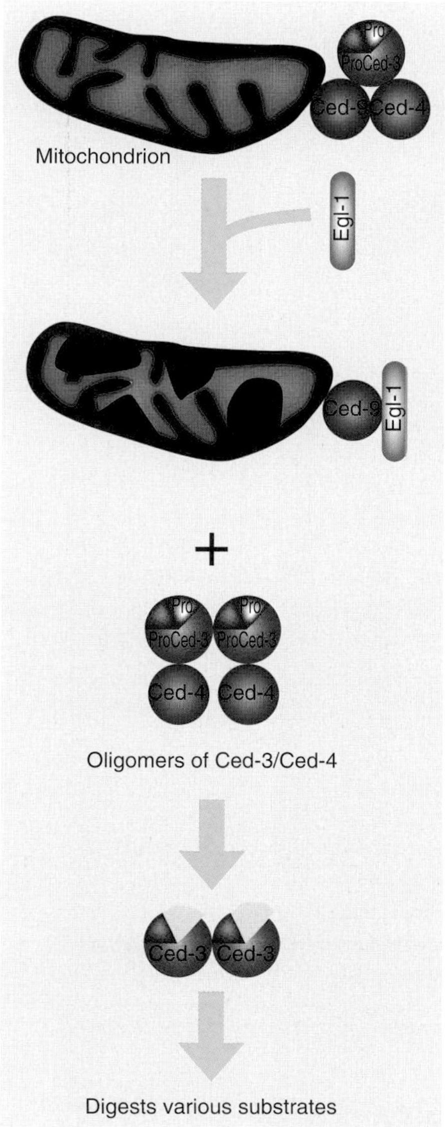

Many mutant worms were identified in which the number of corpses of dying cells were more or less than usual. Four genes involved in aberrant apoptosis were characterized initially. Three of these are *ced-3, ced-4,* and *ced-9* (ced = cell death abnormal), and the fourth is *egl-1* (for egg laying defective). When the *ced-3, ced-4,* or *egl-1* genes are defective, there are more than 959 cells in the adult worm, and fewer corpses form during development. Thus, these genes initiate or execute the apoptotic program. When *ced-9* is defective, there are fewer than 959 cells but more corpses, indicating more apoptosis than normal. Thus, CED-9 protein inhibits apoptosis.

FIGURE 20.14 Programmed Cell Death in *C. elegans*
In the membrane of *C. elegans* mitochondria, a complex of CED-9, proCED-3, and dimers of CED-4 remains inactive. When apoptosis is triggered, EGL-1 protein binds to CED-9, releasing CED-4 and proCED-3. Tetramers of CED-4 are formed and cleave the inhibitory domain from proCED-3. Activated CED-3 is the central enzyme that activates the remaining caspases and executes the apoptosis program.

The CED and EGL proteins work in a cascade that initiates cell death. That is, the action of one protein activates the next protein, which in turn activates further proteins. Genetic and biochemical experiments have given the following model for apoptosis in C. *elegans* (Fig. 20.14). The three CED proteins form an inactive complex in the membrane of the mitochondrion. A signal from surrounding cells activates the synthesis of EGL-1 protein. EGL-1 binds to CED-9 and removes it from the complex. This activates CED-4, which is a protease that specifically cleaves a small inhibitory domain from the end of CED-3. Activated CED-3 forms a heterotetramer of two small and two large domains. This in turn digests various cellular proteins by cutting after aspartic acid residues. This type of enzyme is known as a **caspase** (cellular aspartate-specific protease). Once CED-3 is active, it cleaves inhibitory domains off other proteases, nucleases, and other caspases, thus executing the apoptotic program.

During development, *C. elegans* generates more cells than are found in the adult worm. The extras die via apoptosis.

The CED and EGL proteins of *C. elegans* work in a cascade that initiates cell death. CED-3 and CED-4 are special proteases called caspases, which are key regulators to apoptosis.

MAMMALIAN APOPTOSIS

Comparison of the genome sequence of *C. elegans* with those of mice and humans has allowed the discovery of many mammalian genes that are homologs of worm apoptosis genes. Because apoptosis in humans occurs throughout development and adulthood, there are significantly more proteins and regulators associated with apoptosis than in nematode worms. There are more than 15 caspases with varied tissue specificity and functions. In addition, there are more than 20 mammalian proteins that are similar to CED-9. The precise roles of proteins involved in regulating mammalian apoptosis are still being elucidated.

Mammalian cells have two major pathways to trigger apoptosis (Fig. 20.15). These pathways eventually converge on the caspases. One pathway is designated the **death receptor pathway**, or **extrinsic pathway**, because the death signal originates at a cell surface receptor protein, or **death receptor**. An external signal molecule binds to the extracellular domain on the receptor. The receptor transmits the signal to its intracellular domain and then recruits a variety of internal proteins to the membrane. This ultimately activates the caspases. The other pathway is the **mitochondrial death pathway**, or **intrinsic pathway**. It is usually triggered by intracellular catastrophe, such as irreparable DNA damage. This activates proteins in the mitochondria, which release different effector molecules to activate the caspases.

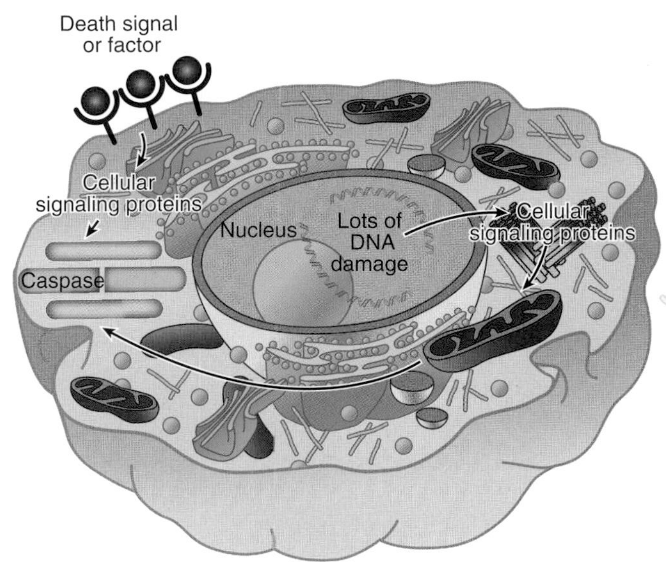

FIGURE 20.15 Activation of Apoptosis
Apoptosis in mammals has two different ways of being activated. An external death signal or factor can bind to a cell surface receptor and initiate an intracellular signal to active caspases. An internal signal such as massive DNA damage can also initiate a signal cascade in the mitochondria, which also activates the caspases.

The death receptor pathway can function independently of the mitochondrial pathway, although both pathways are often activated simultaneously. The signal molecules that bind to the death receptor are quite varied, and the biological significance of most is still under investigation since there is variation among cell types. One pathway is activated by Fas ligand, which binds to the CD95 receptor (Fig. 20.16). This highly glycosylated cell surface protein is found in immune cells. When CD95/Fas binds to its receptor on the target cell (the CD95 receptor—CD95R or Fas Receptor), it causes the receptor to trimerize. The intracellular domain of each receptor is activated by proximity to the others. Activated CD95R recruits a complex of proteins known as **death-inducing signaling complex (DISC)** to the cell membrane. One component of DISC is caspase-8, which cleaves its own prodomain, releasing the active form. The main target for caspase-8 is caspase-3, the direct homolog of CED-3 in *C. elegans.* Once caspase-3 is activated in mammalian cells, the cells die of apoptosis.

Another key apoptotic death receptor, tumor necrosis factor receptor 1 (TNFR1), also regulates inflammation in addition to apoptosis. First, the ligand, tumor necrosis factor α (TNFα), binds the receptor and activates the intracellular portion through a change in the shape (Fig. 20.17). The activated receptor recruits complex I to the membrane, and the entire set of proteins becomes internalized. Complex I has TRADD (TNF receptor–associated protein with a death domain), TRAF2 (TNF receptor–associated factor 2), and two serine/threonine kinases called RIP1 (receptor interacting proteins) and RIP3. The intracellular complex then associates with caspase-8 and

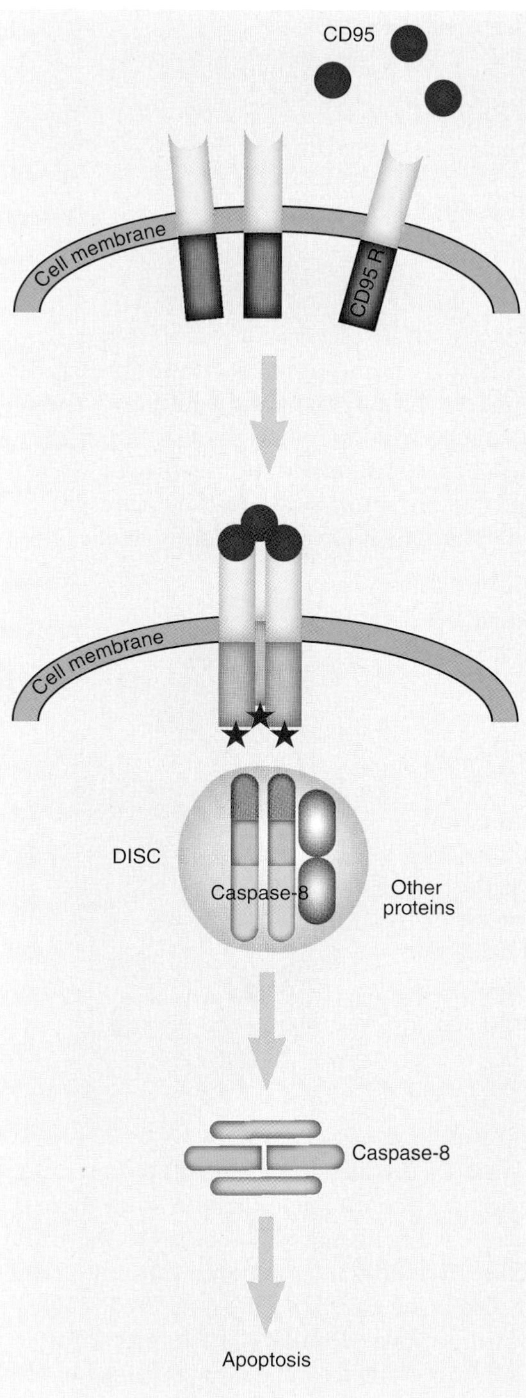

FIGURE 20.16
Death Receptor Pathway
An external signaling molecule called CD95 can initiate cell death. When three molecules of CD95 bind to the CD95 receptor, the three receptors come together to form one complex, which activates the intracellular domains. The DISC complex then binds to the receptor. The key component of DISC is caspase-8. When three molecules of caspase-8 come together, the prodomains are cleaved off and the remaining domains form a hetero-tetramer that activates caspase-3, which in turn activates the other caspases.

FADD (Fas-associated death domain) to become complex II. Activated caspase-8 initiates apoptosis via caspase-3 activation and activation of the mitochondrial pathway.

The mitochondrial pathway is usually activated by internal stimuli such as irreparable DNA damage. The signal converges on a family of mitochondrial proteins named after its founding member, **Bcl-2**, first identified in a B-cell lymphoma. Bcl-2 is homologous to the *C. elegans* CED-9 protein. Unlike *C. elegans*, which has only a single protein of this class, mammals have a family of many proteins, some of which induce apoptosis, whereas others protect cells from apoptosis. The members of the Bcl-2 family are active as dimers, and the relative number of each protein controls the death pathway (Fig. 20.18). Thus, if there are more dimers of Bcl-2, which is anti-apoptotic, then the cell will not die. If there are more dimers of Bax, a pro-apoptotic family member, then the cell dies.

If a pro-apoptotic signal is received, the pro-apoptotic members of the Bcl/Bax family allow **cytochrome *c*** (cyt *c*) to escape from the mitochondria through pores in its outer membrane (Fig. 20.19). Once released, cyt *c* induces formation of the **apoptosome**, a signaling complex containing cyt *c*, caspase-9, and Apaf-1. Apaf-1 is the mammalian homolog to CED-4 of *C. elegans*. However, whereas CED-4 is blocked by CED-9, Apaf-1 has an auto-inhibitory domain that is part of the Apaf-1 protein itself. Furthermore, CED-4 forms a tetramer when activated, but Apaf-1 forms a heptamer (Fig. 20.19). The Apaf-1 heptamer in the apoptosome activates caspase-9, which in turn activates caspase-3 as in the death receptor pathway.

Two different pathways activate apoptosis: (1) the death receptor, or extrinsic, pathway; and (2) the mitochondrial, or intrinsic, pathway.

In the death receptor pathway, CD95 or Fas ligand binds to its receptor, CD95R, which aggregates with two other activated CD95Rs. The intracellular domains activate DISC, which in turn activates caspase-3. This pathway can be activated by a variety of receptors including the tumor necrosis factor receptor.

The ratio of pro-apoptotic (Bax) and anti-apoptotic (Bcl-2) dimers controls whether cells die via the mitochondrial pathway. The pro-apoptotic signal triggers the release of cytochrome *c* from the mitochondria. Cytochrome *c* binds to Apaf-1 and caspase-9 to form the apoptosome.

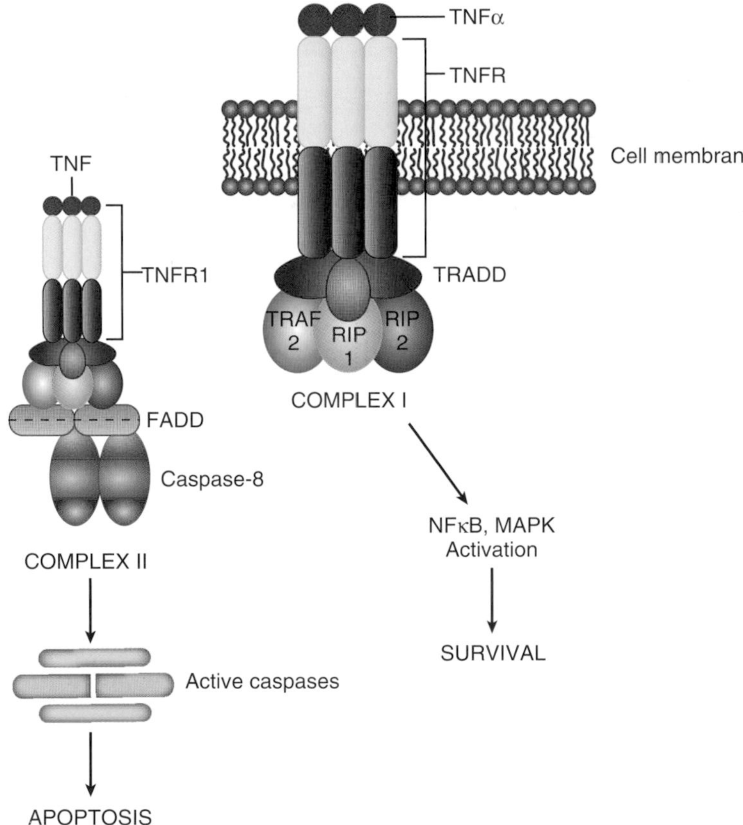

FIGURE 20.17 Tumor Necrosis Factor Induction of Apoptosis
Three TNFα molecules bind to three TNFRI receptors to aggregate the receptors into one complex. This recruits TRADD, TRAF2, RIP1, and RIP3 to the intracellular portion of the receptor to form Complex I. Complex I and the receptors are internalized and then associate with FADD and caspase-8 forming Complex II. Two molecules of caspase-8 cleave the inhibitory pro domains from the caspase to activate it and induce apoptosis.

FIGURE 20.18 Bax/Bcl-2 Control of Apoptosis
The ratio of pro-apoptotic (Bax) and anti-apoptotic (Bcl-2) dimers controls whether cells die via apoptosis.

CASPASES

Caspases are proteases that cleave their target proteins after certain specific aspartate residues. Specificity results from recognition of the four amino acid residues following the aspartate. There are many different caspases. They are highly conserved throughout evolution and are found in mammals, flies, nematodes, and even hydra. Scientists regard them as central executioners because they carry out the death sentence of apoptosis.

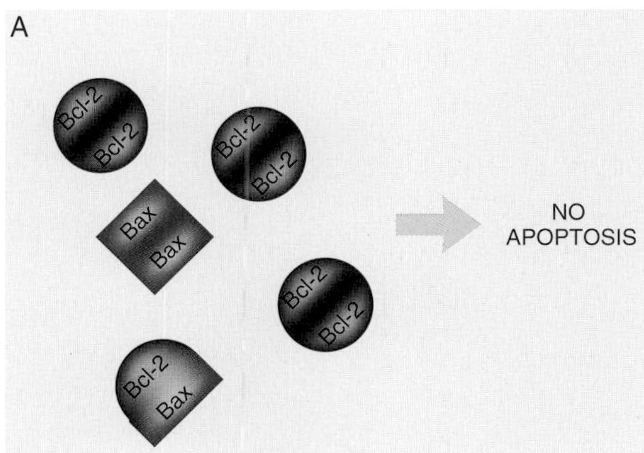

Caspases work as heterotetramers, that is, an assembly of four domains (Fig. 20.20). Most mammalian caspases have two p20 domains and two p10 domains. The p20 domain has the enzyme active site; therefore, each active caspase can cleave two different proteins at the same time. The inactive form of a caspase has three different domains: the prodomain, which blocks the active site; a p20 domain; and a p10 domain.

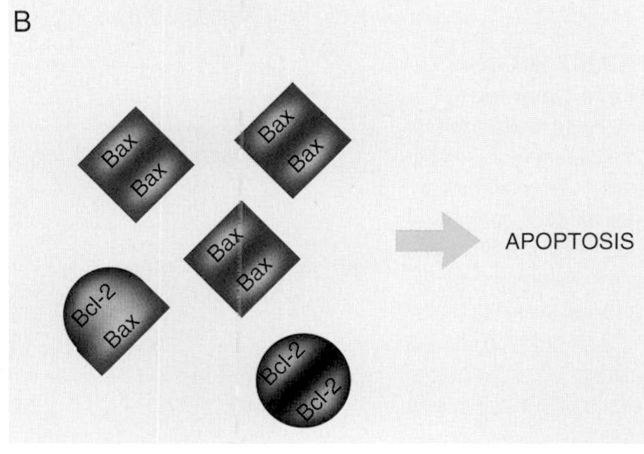

There are three ways to activate caspases. The first method requires another, previously activated caspase to cleave between the prodomain and p20, and between p20 and p10 (see Fig. 20.20). When two molecules have been digested, the p20 and p10 domains associate into their final structure. The second method of caspase activation occurs via self-association (Fig. 20.21). For example, when two molecules of caspase-8 trimerize, each caspase-8 cleaves its neighbor

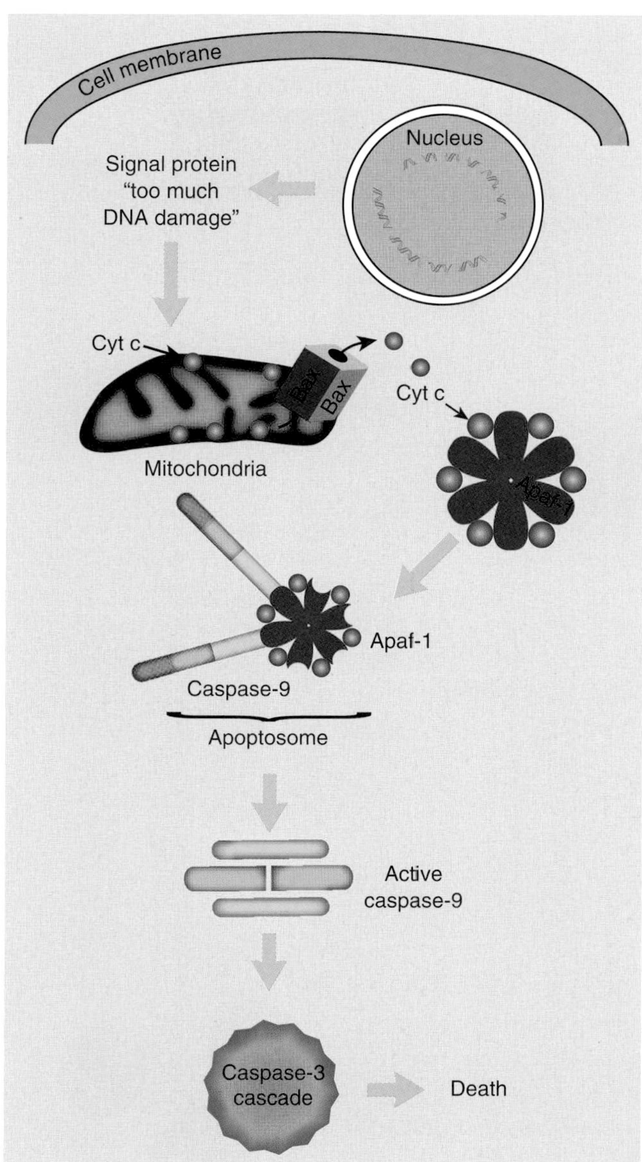

FIGURE 20.19 Mitochondrial Pathway of Apoptosis
Severe DNA damage signals for an increase in Bax/Bax dimers. These form a channel through which cytochrome *c* escapes from the mitochondria. Cytochrome *c* binds to Apaf-1 and induces a conformational change. Activated Apaf-1 assembles into a heptamer, which binds to pro-caspase-9 and cleaves off the prodomain. Activated caspase-9 activates caspase-3, and the rest of the caspases carry out the death sentence.

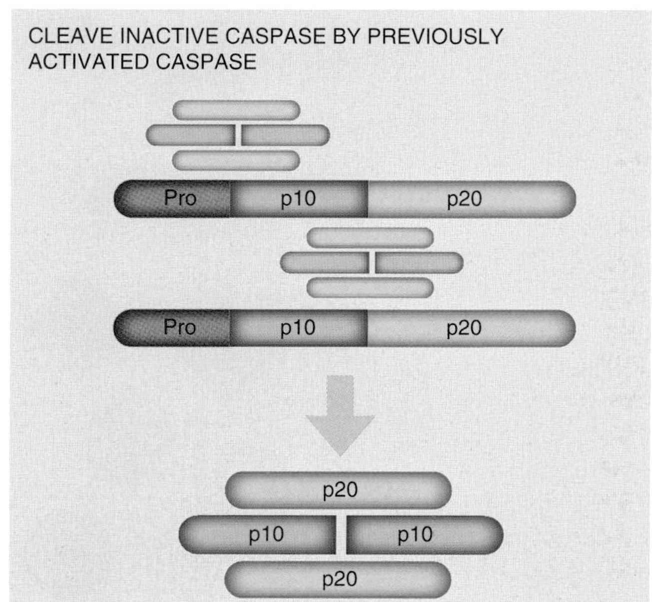

FIGURE 20.20 Active Caspases Are Heterotetramers
Caspases are usually found as inactive monomers with three domains: the prodomain, which inhibits activity, the p10 domain, and the p20 domain with the active site. When the prodomain is removed, two p10 domains and two p20 domains come together as a heterotetramer.

and forms the heterotetramer. This mechanism is used to activate the caspase cascade in the death receptor pathway of apoptosis. The third method involves specific regulatory proteins that are not caspases. For example, when the apoptosome forms in the mitochondrial pathway of apoptosis, two regulatory proteins, Apaf-1 and cytochrome *c*, bind to and initiate cleavage of caspase-9 (see Fig. 20.19).

There are more than a dozen different caspases in humans, each with its own exclusive set of targets. In some cases, digesting the target protein inactivates it. In some cases, cleavage releases part of the protein that has a new function. In other cases, the caspase cleaves an inhibitory domain from an inactive protein, releasing the active form.

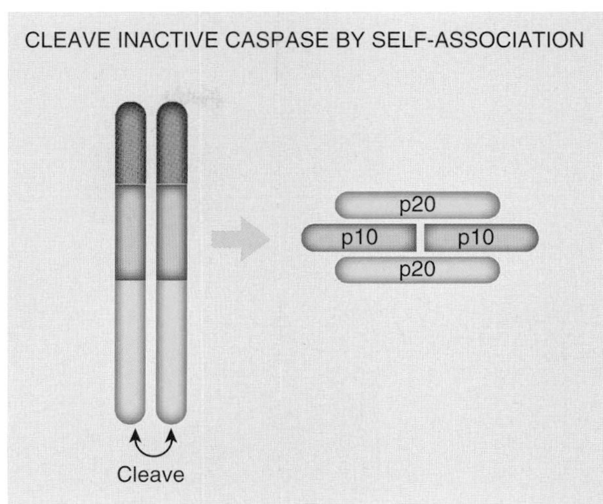

FIGURE 20.21 Self-Association Activates a Caspase
Association of two inactive caspase monomers triggers one enzyme to cleave its neighbors, resulting in two active p10 and p20 domains.

EXECUTION PHASE OF APOPTOSIS

Activating CED-3 in *C. elegans* and caspase-3 in humans triggers the cellular changes associated with apoptosis. Caspase-3 activates other caspases, and each digests different cellular substrates, such as **caspase-activated DNAse (CAD)**, a nuclease that cuts nuclear DNA between the nucleosomes (Fig. 20.22). The fragments of DNA are about 180 base pairs in length, which is a size of DNA fragment cut between nucleosomes. To determine if cells are in the process of apoptosis, scientists isolate the genomic DNA and look at the fragment sizes. If they are in multiples of approximately 200 base pairs, this implies the cells were going through apoptosis.

There are many other substrates of caspases. Proteins that maintain nuclear structure, such as lamins, are digested so that the nucleus shrinks and breaks into small pieces. The cytoskeleton is cleaved, and the cell architecture disintegrates and compacts. Other organelles are digested so they lose their structure, which creates small compact granules of cellular material called apoptotic bodies.

As a way to modulate caspases, there is a family of proteins called **Inhibitors of apoptosis (IAP)**. Each member of this family of proteins has three tandem repeats of highly conserved amino acid sequences called baculoviral IAP repeats (BIR). IAPs can inhibit caspase-8, caspase-9, and even the central executioner, caspase-3. One inhibitor of caspase-8, FLICE-like inhibitory protein (FLIP), has three alternately spliced forms. All three forms have the death domains, but only the long form has the caspase-like domain. When there is a lot of the long form in the cell, this can replace caspase-8 in complex II formation and inhibit the activation of caspase-8, thus blocking apoptosis.

CORPSE CLEARANCE IN APOPTOSIS

The apoptotic bodies are removed by phagocytosis. In *C. elegans,* neighboring cells take up the apoptotic bodies. In mammals, the situation is more complex. In a few areas of the body, neighboring cells engulf the apoptotic bodies. In most cases, though, macrophages engulf the apoptotic bodies and clean the area (Fig. 20.23). Macrophages are cells of the immune system whose primary role is to digest anything foreign, such as invading bacteria. Normally, macrophages recruit other immune cells to the site of infection. However, when dealing with apoptotic bodies, they do not recruit other immune cells. Thus, macrophages can distinguish an apoptotic body from a foreign invader.

How does the macrophage know that apoptotic bodies do not require an immune response? Normally, when macrophages ingest bacteria, they secrete soluble proteins to recruit other immune cells. Apparently, apoptotic bodies have molecules on the surface that trigger the macrophages to digest them but without secreting

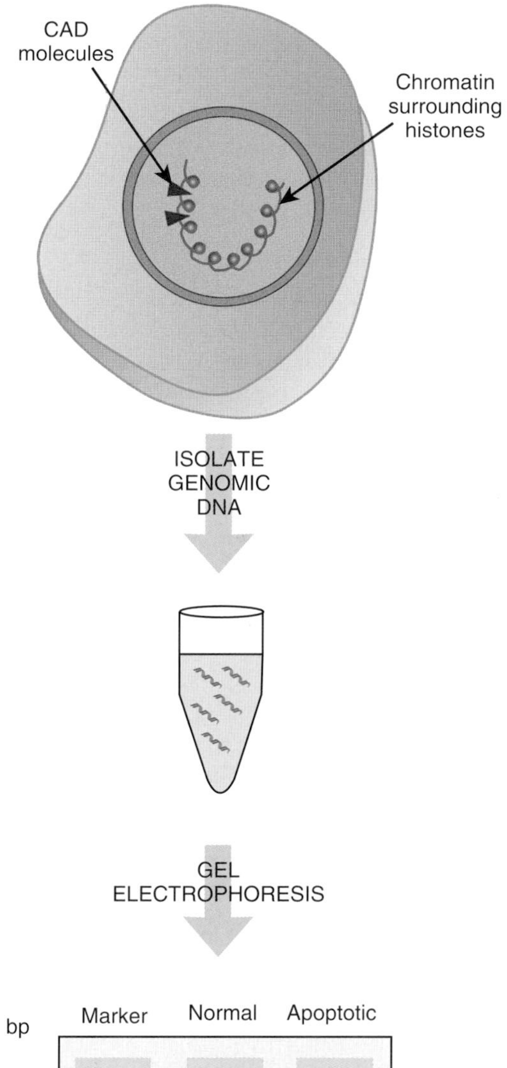

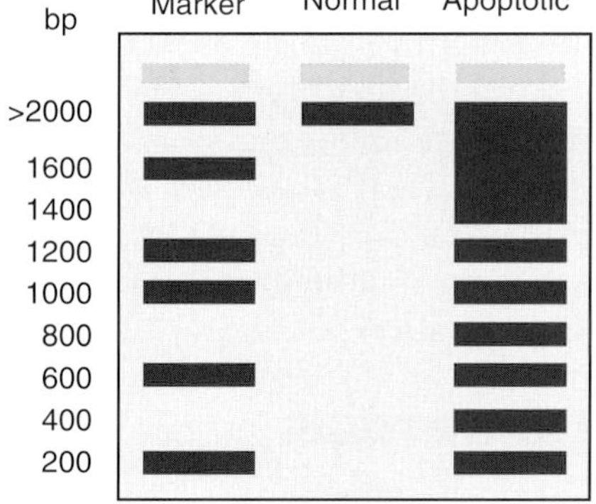

FIGURE 20.22 CAD Cleaves Nuclear DNA During Apoptosis
One of the targets for caspases is CAD, which digests the nuclear chromatin. After CAD is activated by the caspases, it cuts genomic DNA between the histone cores. Because about 180 base pairs of DNA are wound around each histone core, DNA is fragmented into pieces differing in size by around 200 base pairs. When genomic DNA is isolated from a cell in apoptosis and electrophoresed through an agarose gel, the DNA forms a ladder.

the immune signaling factors. The numbers and mechanisms of "eat me" receptors are quite complex since different mammalian tissues may have different mechanisms to activate the macrophage. One such molecule appears to be phosphatidylserine, a phospholipid normally found only in the inner leaflet of the cell membrane (Fig. 20.24). During apoptosis phosphatidylserine may be translocated to the outer leaflet. Macrophages have receptors for phosphatidylserine, allowing them to recognize the apoptotic body as "self."

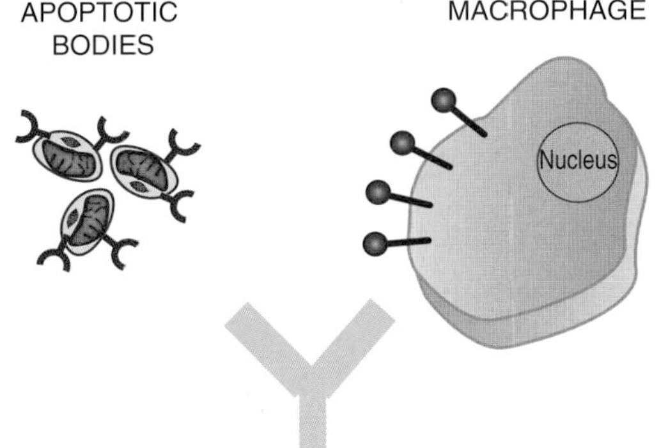

FIGURE 20.23
Removal of Apoptotic Bodies
In mammals, most apoptotic bodies are removed by macrophages. These recognize cell surface receptors on the apoptotic body and ingest the entire cellular fragment. Any remaining proteins are broken down and recycled by the macrophage.

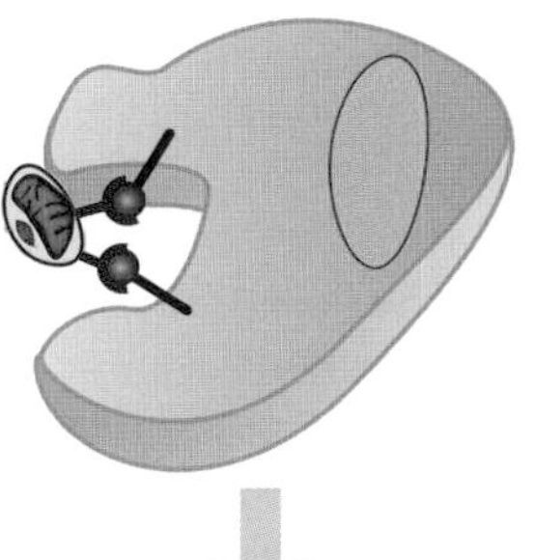

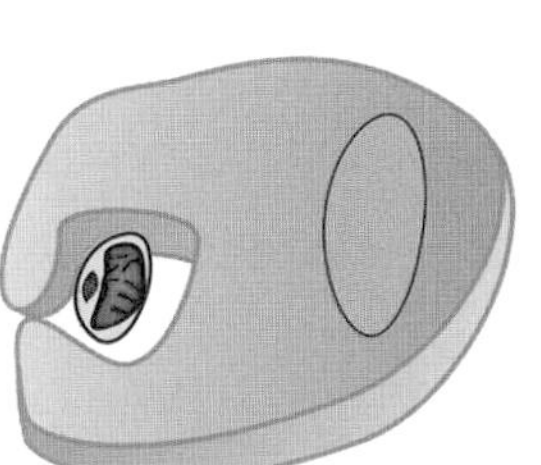

FIGURE 20.24
Apoptotic Bodies Do Not Trigger an Immune Response by Macrophages
(A) Normally, macrophages ingest invading pathogens such as bacteria. This triggers the macrophage to release factors that attract other immune cells. (B) In contrast, the cell surface signal on apoptotic bodies does not trigger the macrophage to release these factors, so other immune cells are not recruited to the site.

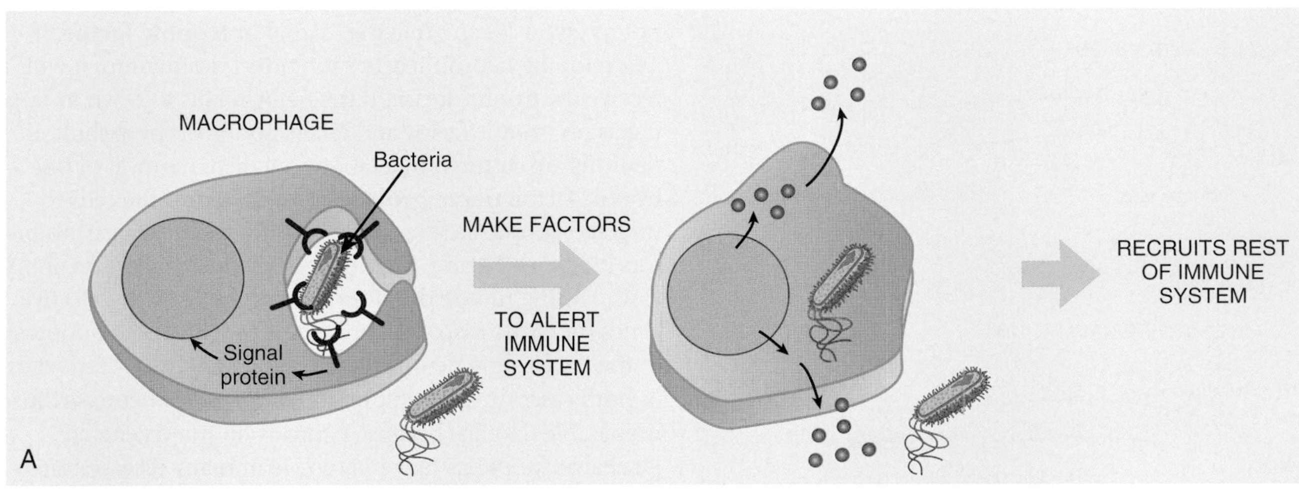

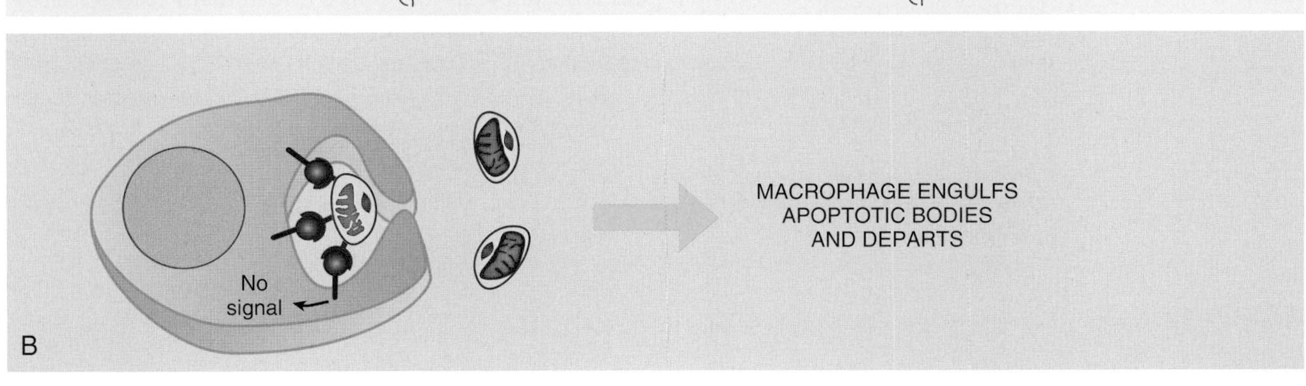

Macrophages digest apoptotic bodies and recycle the cellular proteins without triggering other immune cells.

CONTROL OF APOPTOTIC PATHWAYS IN DEVELOPMENT

Although controlling the onset of apoptosis is very complex, the ramifications of losing control are dire. Too much apoptosis or inappropriate activation of apoptosis can destroy fully functioning cells. Death of too much tissue can kill a developing organism. Not enough apoptosis, especially during development, can create surplus tissue and disrupt the normal operation of tissues and organisms. Many disease states arise from inappropriate or defective apoptosis, too.

In *C. elegans,* sex development depends on apoptosis. *C. elegans* comes in two sexes: males, which produce only sperm, and hermaphrodites, which produce both sperm and eggs. No true females are produced naturally. Part of the decision to become male or hermaphrodite hinges on apoptosis (Fig. 20.25). Two neurons control the muscles around the vulva so that eggs can be laid. If the neurons are present, the worm is a hermaphrodite. If the neurons are absent, the worm cannot lay eggs. The presence or absence of these hermaphrodite-specific neurons (HSN) depends on apoptosis. Defects in *elg-1* affect whether the worm becomes a fully functioning hermaphrodite. Without *elg-1,* too much apoptosis occurs; therefore, all worms are missing the HSN and no egg-laying worms are produced.

FIGURE 20.25 Apoptosis Controls *C. elegans* Sex Determination
C. elegans has two sexes: males that produce sperm and hermaphrodites that produce both sperm and eggs. One nerve, the HSN, determines whether the worm will lay eggs or not. The presence or absence of HSN depends on apoptosis in development. EGL-1 mutations induce too much apoptosis; therefore, the HSN nerve is killed during development, and *egl-1* hermaphrodites cannot lay eggs.

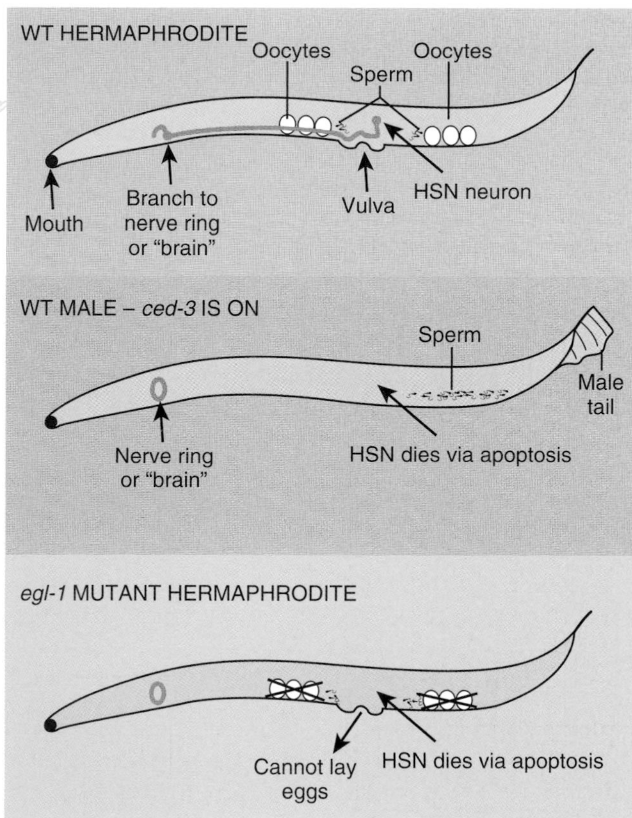

Apoptosis is tightly regulated during an immune response. During infection, the body responds to the attack by increasing the number of white cells of the immune system. When the infection is past, the body eliminates the surplus immune cells via apoptosis. Immune cells use the death receptor pathway to trigger apoptosis (see earlier discussion). Too much apoptosis would deplete our immune system of essential cells and disable the immune response.

During development, a large number of neurons undergo apoptosis. One theory suggests that neurons die if they do not receive a "keep on living" signal or **trophic factor**. If a developing neuron reaches its correct destination, it will receive the trophic factor. If the neuron fails to reach its target, it gets no trophic factor and enters apoptosis by default. If neurons are cultured in a laboratory dish, removal of one trophic factor, **nerve growth factor**, induces the cells to undergo apoptosis (Fig. 20.26). Addition of caspase inhibitors blocks cell death, proving the cells were dying via apoptosis. During mouse development, embryos with defective genes for either caspase-3 or caspase-9 die. Lack of apoptosis in the developing neural system is the main cause for death in both cases. In the adult brain, apoptosis of neurons causes irreparable damage because neurons do not regenerate. Extensive apoptosis may play a role in many diseases, such as Alzheimer's, Parkinson's disease, Huntington's disease, and amyotrophic lateral sclerosis (ALS). The exact role of apoptosis in these diseases is still being investigated.

Apoptosis controls the presence or absence of the HSN, which controls whether *C. elegans* can lay eggs via their vulva.

When neurons do not receive trophic factor signals, the cells die via apoptosis.

NECROPTOSIS

Necrosis, which was identified many years before the discovery of apoptosis, has one main characteristic: the early rupture of the cellular membrane that leaks the cellular contents into the intracellular area, which then elicits an immune response. Interestingly, most scientists believed necrosis to be unregulated, but various lines of evidence found this tenet not completely true. Some of the discoveries that led to the idea that necrosis is actually a controlled cellular mechanism include (1) the discovery that during development there are necrotic events not caused by pathological assaults; (2) necrosis can be induced by certain molecules binding to membrane receptors; and (3) necrosis can be regulated by genetics, epigenetics, and drugs. In particular, a necrotic signal transduction pathway centering on a serine/threonine kinase called **Receptor Interacting Protein 1**, or **RIP1,** is called **necroptosis** to designate that it can be controlled genetically.

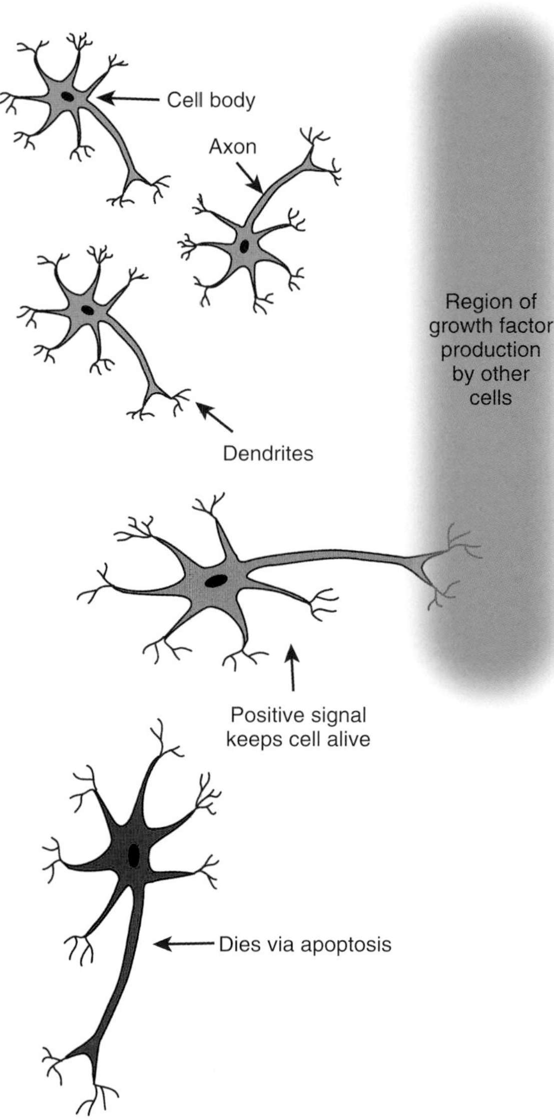

FIGURE 20.26 Growth Factors Keep Neurons Alive
Nerve cells undergo apoptosis if their dendrites do not receive trophic factors such as nerve growth factor.

Necroptosis signals overlap with the extrinsic pathway or death receptor pathway of apoptosis (Fig. 20.27). When tumor necrosis factor α (TNFα) binds to its death receptor called tumor necrosis factor receptor 1 (TNFR1), one of the three different pathways can be activated depending upon what molecules associate with the intracellular domain. If complex I forms, then NFκB activates cell survival genes. If complex II forms with dimers of caspase-8 then apoptosis is activated. If capase-8 associates with FLIP, caspase-8 cannot self-activate, and therefore, caspases are not activated. Instead, the complex activates necroptosis where the cell ruptures and leaks all of its intracellular contents into the surrounding tissue. Interestingly, RIP1 kinase activity is essential for necrosis, but not for apoptosis. RIP1 and RIP3 inhibitors are called necrostatins, and they inhibit RIP1 kinase activity, thus blocking necroptosis. The exact mechanisms for how RIP1 and RIP3 activate necroptosis are still under investigation and will provide a great insight as to how cells balance the decision of whether or not they die, and if so by what mechanism.

TNFα binding to TNFR1 can elicit either cell survival via NFκB, apoptosis via caspase activation, or necroptosis via RIP1 and RIP3 depending upon which complex forms after the receptor is activated. To activate necroptosis, caspase-8 associates with FLIP rather than another caspase-8, which then activates the RIP1 and RIP3 kinases, which then triggers cellular suicide that activates the immune system.

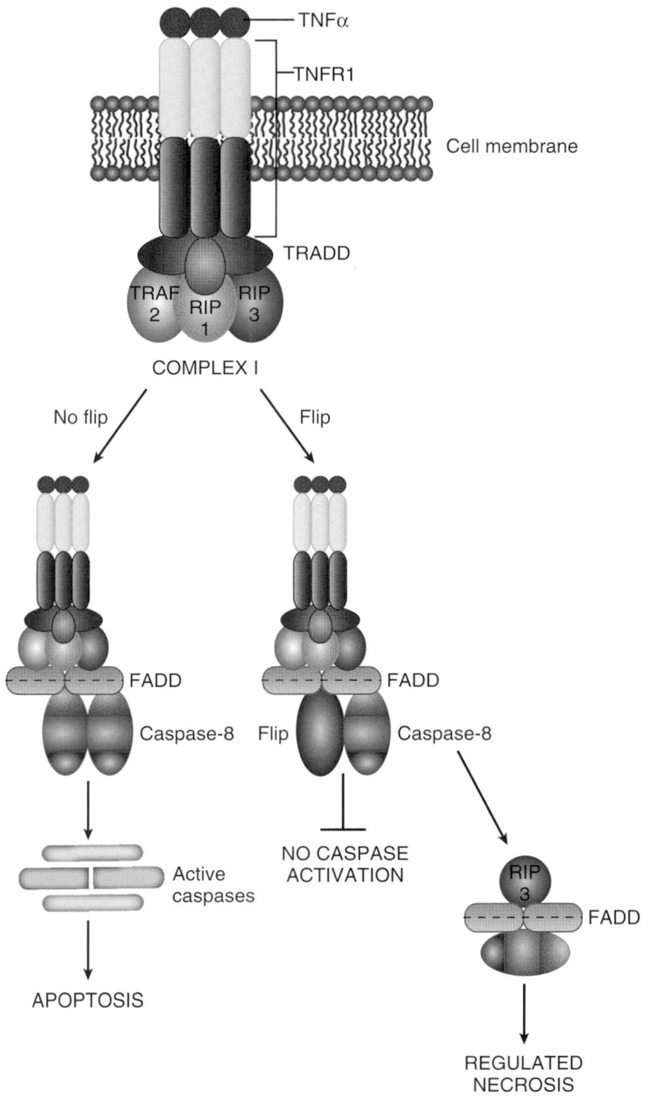

METABOLIC CONTROL OF CELL DEATH

As discussed in the section on aging, metabolic function has a major influence on life span and health span in organisms. Metabolism also has its control over certain molecules of apoptosis. As cells age, the accumulation of genetic alterations and cellular dysfunctions causes the cells to senesce (see the earlier section "Cellular Senescence"). To senesce, cells must exit the cell cycle and become quiescent. Cyclins and cyclin-dependent kinases (CDKs) control the timing for moving through the cell cycle, and the D-type cyclins control the G1 stage. The production of D-type cyclins is regulated by metabolism. An enzyme that converts fructose-6-phosphate to fructose-2,6-bisphosphate, which functions during glycolysis, regulates the expression of D-type cyclins, thus linking metabolic state with the exit from the cell cycle. In apoptosis, cytochrome *c* is released from the mitochondria to assemble the APAF1. The redox state of cytochrome *c* is essential for its ability to promote cell death. Reduced forms of cytochrome *c* cannot initiate apoptosis and therefore offer a way for the metabolic state of the cell to dictate whether or not the cell will die via apoptosis. Another example of metabolic control occurs with caspase-2, which cleaves pro-apoptotic BID (a molecule related to Bax), which in turn releases cytochrome *c* from the mitochondria. Caspase-2 is inhibited by glucose and NADPH of the pentose phosphate pathway, and so when these metabolic products are abundant, apoptosis via caspase-2 is inhibited. The metabolic control of many different points of apoptosis is essential to proper cellular function, and since this research is still in its infancy, many other controls are sure to be identified.

FIGURE 20.27
Necroptosis
An alternate cell death pathway called necroptosis is activated in cells that have the short version of FLIP and caspase-8.

Metabolic enzymes control many aspects of senescence and apoptosis. Cyclin D expression occurs when nutrients are available since the gene is turned on by enzymes from the glycolysis pathway. The redox state of cytochrome c dictates whether or not it can initiate apoptosis. And finally, caspase-2 is controlled by glucose and NADPH of the pentose phosphate pathway.

CANCER, AGING, AND PROGRAMMED CELL DEATH

The pathways of senescence and apoptosis work to protect the organism from damaged cells. Cancer cells have deleterious mutations in the genome but for some reason have suppressed the senescence and apoptosis pathways so that they continue to grow and divide. A damaged cell can evade internal programs for cell quiescence or death in many ways.

One major factor that triggers cellular senescence is the presence of oncogenic mutations, that is, mutations that promote cancer (see Chapter 19 for details). To defend the organism, tainted cells sacrifice themselves by starting the genetic program for senescence or apoptosis, thus entering **premature senescence**. This form of senescence occurs without the shortening of telomeres, a hallmark feature of regular senescence. One oncogenic mutation is in the **Ras gene**. One common mutation converts glycine at position 12 to valine (Ras^{G12V}). The mutation causes cells to enter senescence without the requisite telomere shortening or other genomic instability. The mutation is also found in many cancers, including pancreatic and lung. Two other key signal transduction pathways regulate oncogene-induced senescence: **p53 pathway** and **p16^{INK4A}-RB (retinoblastoma protein) pathway**, both of which are involved in control of the cell cycle.

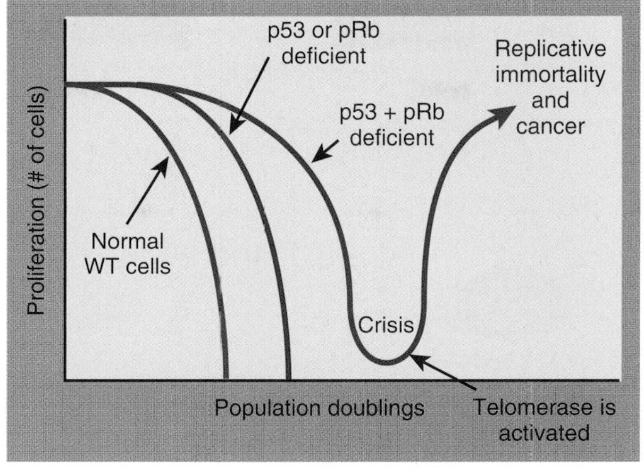

FIGURE 20.28
Defects in p53 and pRb Extend Cellular Life Span
Normal cells proliferate for a specific time and then stop dividing. Cells with mutations in p53 or pRb continue to divide and replicate longer than normal. If both p53 and pRb are mutated, the cells divide for even longer. Eventually, the mutants do stop dividing unless some further oncogenic mutation occurs.

If either p53 or pRb is defective, cells replicate for more cycles than usual (Fig. 20.28). If cells are defective for both p53 and pRb, the effect is additive, and the replicative life span is longer than with either mutation alone. Nonetheless, such double mutants do eventually enter senescence, after an extended number of cell divisions; thus, cells have other mechanisms to trigger senescence.

The normal p53 protein is a transcription factor that regulates its own gene as well as a variety of other genes in the suppression of tumors. In particular, p53 activates transcription of **p53-upregulated modulator of apoptosis (PUMA)**, a member of the Bcl-2 protein family that activates Bax, thus promoting apoptosis. Other important genes activated by p53 include p21, a cell cycle inhibitor that causes the cell to stop dividing. The role of p53 in the death or survival of a damaged cell is very complex and is still being investigated, but p53 mutations are found in up to 50% of some cancers.

When a cell becomes damaged, it enters senescence early. Severely damaged cells commit suicide via apoptosis. Damaged cells that bypass the senescence program and block apoptosis are called cancer. Ras, p53, and pRb are proteins that regulate senescence. Mutations in these genes make the cells resistant to senescence and often cause cancer.

PROGRAMMED CELL DEATH IN BACTERIA

Although apoptosis *per se* does not occur in single-celled organisms, a genetic system that kills *Escherichia coli* when under extreme stress does exist. Morphologically, the death does not resemble apoptosis, but like apoptosis, the death system is genetically encoded. In *E. coli*, an **addiction module** of two genes controls the death-inducing system (Fig. 20.29). One gene encodes a toxin, MazF, which is quite stable. The second gene encodes the antitoxin, MazE, which prevents the toxin from killing the bacteria. The antitoxin is unstable and degrades very fast after translation. If its transcription or translation is stopped or slowed in any way, the level of antitoxin plummets and the toxin kills the bacteria.

The MazF toxin is a specific endoribonuclease that degrades messenger RNA (mRNA). It recognizes the sequence ACA and cleaves to the 5′ side. Such enzymes have been named mRNA interferases and have now been found in a variety of bacteria. When bacteria do not produce the antitoxin, the MazF toxin is no longer degraded. The net result is the destruction of mRNA followed by a halt in protein synthesis. Cell death rapidly follows.

When bacteria are depleted of nutrients, transcription and translation slow down, and this may trigger the MazEF suicide system. Perhaps some *E. coli* commit suicide for the good of the rest, because the proteins, lipids, and nucleic acids of the dead cell could provide food

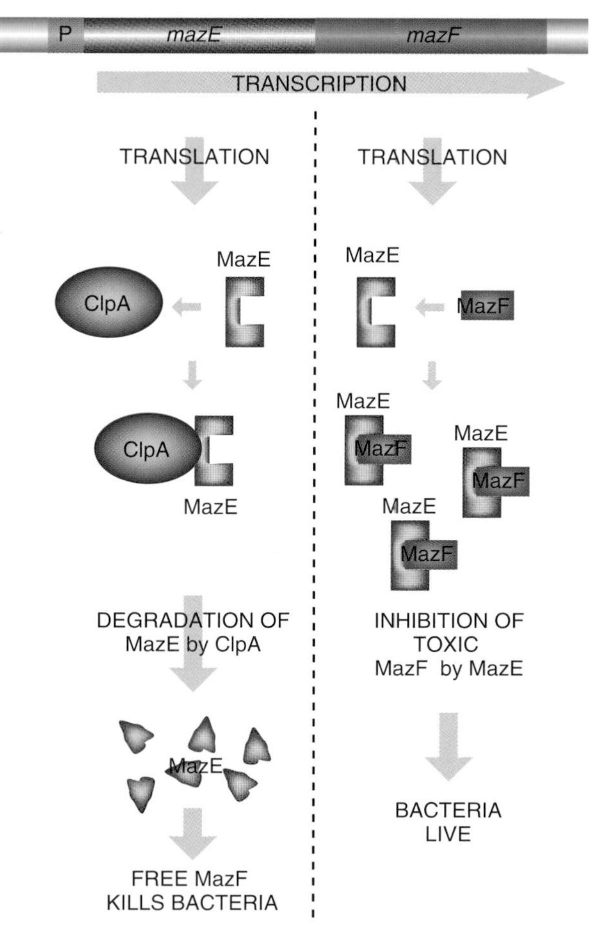

FIGURE 20.29 MazEF System of *E. coli*
Two genes, *mazE* and *mazF*, control whether or not *E. coli* self-destruct. MazF protein is a toxin that kills the bacteria by degrading mRNAs. It is very stable and is produced continually. The MazE antitoxin protects *E. coli* from death by inhibiting the MazF toxin. However, MazE is degraded very fast by the ClpA protease. Under normal conditions, *E. coli* continually makes the antitoxin, but if the bacteria encounter stress, they may halt protein synthesis. The antitoxin is no longer made, and the toxin is free to initiate the suicide program.

for nearby cells. This theory suggests that, despite being unicellular, bacteria have genetic programs for the good of the population. Another theory is that the MazEF suicide system is designed to limit the multiplication of bacterial viruses. Indeed, mutants of *E. coli* deleted for the whole *mazEF* operon give higher yields of bacteriophage when they burst. In wild-type cells, the MazEF system kills the cell before virus replication is complete and thus reduces the number of viruses produced.

Addiction modules also exist in bacteriophages such as P1 and lambda that are maintained in a lysogenic state. The toxin/antitoxin pair of proteins, called PhD and Doc, respectively, prevents the bacteriophage genome from being destroyed or lost during *E. coli* growth. The bacteriophage genome encodes both the toxin and the antitoxin. Just as in the MazEF system, the toxin protein is very stable, whereas the antitoxin degrades quickly and must be produced continually. If the P1 or lambda genome is lost, the stable toxin protein will kill the bacteria. Interestingly, the toxin produced by P1 does not kill *E. coli* directly but acts by activating the bacterium's own MazEF system (Fig. 20.30). The P1 toxin inhibits translation of the MazE antitoxin, which activates the MazF toxin, which in turn kills the cell.

E. coli have a programmed cell death system that is controlled by two proteins: MazE and MazF. MazE is the antitoxin and requires constant synthesis; MazF is a toxin that kills the cell when MazE is missing.

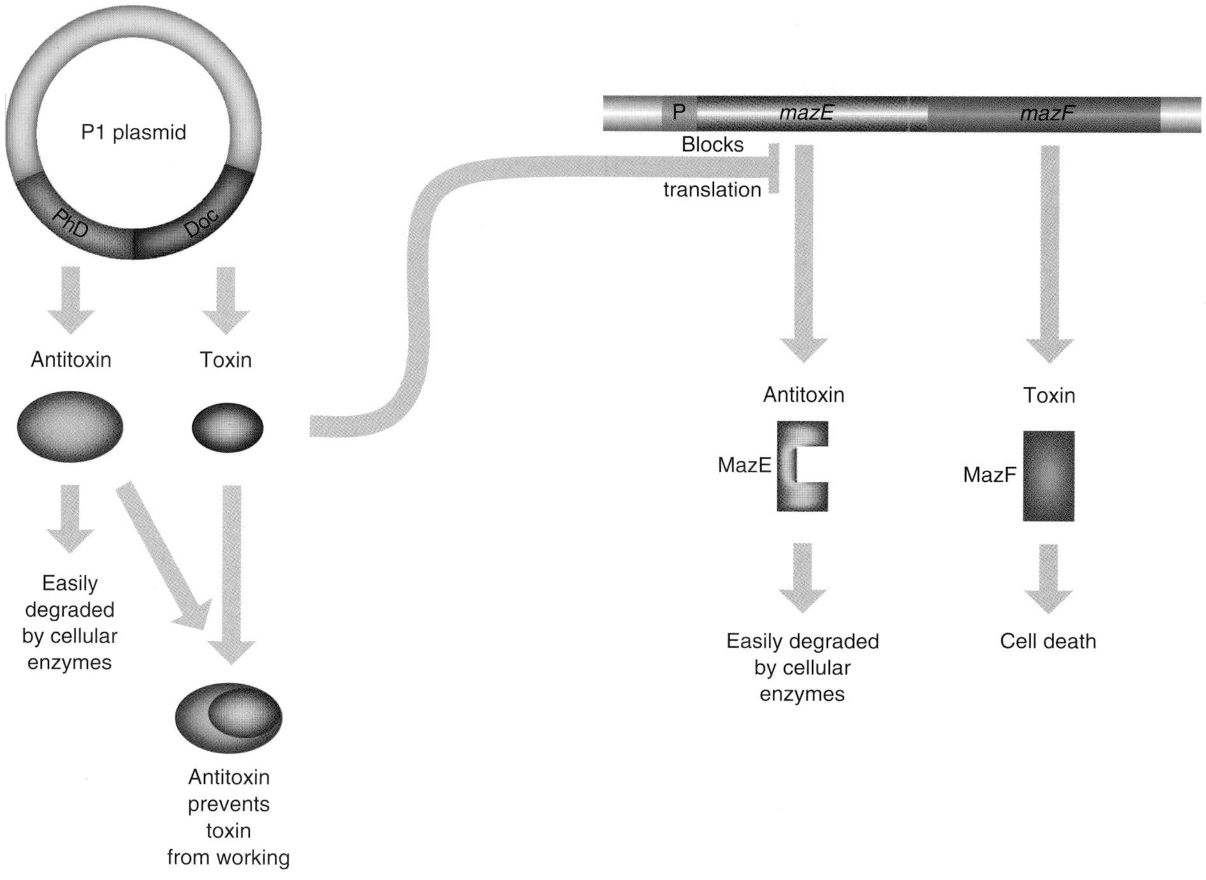

FIGURE 20.30 Phd/Doc System of P1
PhD/Doc is an addiction module carried by P1 that encodes an antitoxin/toxin pair. The antitoxin is unstable and must be synthesized continually, or else the toxin will become active. If, for some reason, the PhD protein is not produced, Doc inhibits translation of MazE. Without MazE, the MazF toxin activates the suicide program.

Summary

All organisms age, some more slowly than others, but every living organism slowly accumulates mutations and other genetic changes over time. There are two major branches of age-related decline, genetic alterations and cellular dysfunction, that overlap in many different ways. Some genetic alterations seen in aging include a general increase in mutations, telomere shortening, and epigenetic changes. These include changes in methylation patterns, alterations of post-translational modifications of histones, and decreases in the amount of heterochromatin. Any mutations in the genes can affect cellular function, and aging is associated with alterations in nutrient sensing, decreases in protein degradation, and an increase in mitochondrial dysfunction.

At a certain threshold, genetic and cellular mutations initiate senescence, a state of cellular physiology in which cells continue to function but do not divide. Senescent cells have activated tumor suppressor proteins, express marker molecules such as senescent-associated β-galactosidase and senescent-associated heterochromatic foci, and secrete proinflammatory cytokines and extracellular matrix molecules.

Instead of senescence, cells can activate programmed cell death. Three forms of programmed cell death are used to remove defective cells from tissues. Apoptosis, or type I programmed cell death, is a series of enzymatic steps that condense the cell components and nuclei and create apoptotic bodies that are removed by phagocytosis. Autophagy, or type II programmed cell death, is characterized by large autophagic vesicles within the dying cell. Necroptosis, or type III programmed cell death, is similar to apoptosis since it also is executed by a series of molecular signals, but differs from apoptosis since the cells burst and release factors that elicit an immune response.

Apoptosis is an important process to development of all organisms because apoptosis controls tissue remodeling and physiology. Apoptosis occurs via a cascade of different enzymes from two different pathways. In the death receptor pathway, or extrinsic pathway, a ligand binds a death receptor in order to activate an intracellular cascade of protein activations. The intracellular proteins converge on caspase-8, which activates caspase-3, and the remaining caspases to digest and destroy the cell. In the mitochondrial or intrinsic pathway, an internal signal such as DNA damage, triggers the ratio of Bcl-2 to Bax to have more pro-apoptotic Bax dimers. These dimers cause the release of cytochrome *c* from the mitochondria, which in turn activates the caspase cascade. Necroptosis is a recently identified programmed cell death pathway that is activated by some of the same death receptors as apoptosis. This pathway ends in the cell rupturing, leaking its intracellular contents, and eliciting an immune response.

Many organisms besides humans have apoptosis, including *E. coli*. The apoptosis in *E. coli* is not morphologically similar, but it is genetically encoded, and it kills the cell for the good of the population, much like apoptosis in eukaryotic organisms.

End-of-Chapter Questions

1. What is cellular senescence?
- **a.** when cells are alive but are no longer dividing
- **b.** when cells finally die
- **c.** when the cells are alive, but are no longer expressing genes
- **d.** when cells divide uncontrollably
- **e.** none of the above

2. Which enzyme lengthens the telomeres in some cells?
- **a.** DNA polymerase
- **b.** polyadenylate polymerase
- **c.** telomerase
- **d.** RNA polymerase
- **e.** none of the above because telomeres cannot be lengthened

3. Which two enzymes are responsible for decreasing the effects of oxidants on cellular components?
- **a.** catalase and telomerase
- **b.** superoxide dismutase and telomerase
- **c.** PUMA and telomerase
- **d.** catalase and superoxide dismutase
- **e.** catalase and PUMA

4. Besides *age-1*, a mutation in which gene of *C. elegans* has been shown to increase lifespan and induce hibernation?
- **a.** *daf-2*
- **b.** catalase
- **c.** superoxide dismutase
- **d.** telomerase
- **e.** peroxidase

5. Which enzymes are considered the key regulators of apoptosis?
   - a. caspases
   - b. telomerases
   - c. catalases
   - d. superoxide dismutases
   - e. none of the above
6. Which of the following apoptotic pathways is usually triggered by intracellular catastrophe, such as irreparable DNA damage?
   - a. death receptor pathway
   - b. apoptotic caspase activation pathway
   - c. death-inducing pathway
   - d. mitochondrial death pathway
   - e. none of the above
7. Which two proteins control programmed cell death in *E. coli?*
   - a. CED-3 and CED-2
   - b. MazE and MazF
   - c. caspase and catalase
   - d. Bax and Bcl-2
   - e. none of the above
8. Which protein complex binds to telomeres and protects the ends from double-strand break repair?
   - a. shelterin
   - b. telomerase
   - c. DNA polymerase
   - d. TERT
   - e. Ras
9. Which of the following is not a genetic phenomena associated with aging?
   - a. chromosomal translocations
   - b. copy number variations
   - c. histone modifications
   - d. shortening of telomeres
   - e. necroptosis
10. Why is mtDNA more susceptible to damage by ROS?
    - a. The location of mtDNA is far away from the protective confines of the organelle.
    - b. The mtDNA is not protected by histones.
    - c. ROS interfere with the histones bound to mtDNA.
    - d. The mtDNA contains more oncogenes than nuclear DNA.
    - e. Methylated mtDNA is more likely to be attacked by ROS.
11. What are sirtuins?
    - a. DNA methylases
    - b. NAD^+-dependent histone deacetylases
    - c. ubiquitylation proteins
    - d. proteases
    - e. cyclin-dependent kinases

(*Continued*)

12. Which gene(s) is/are hypermethylated in Hutchinson–Gluillford syndrome in humans?
    a. sirtuin
    b. p53 and pRb
    c. lamin A/C
    d. caspase-2
    e. telomerase
13. The mTOR system in mammals regulates lifespan by ______________.
    a. regulating catabolic processes
    b. sensing the amount of AMP and ADP as byproducts of ATP utilization
    c. modifying histones through the addition of acetyl groups
    d. methylating regions of DNA that need to be silenced, such as those involved in apoptosis
    e. sensing the availability of amino acids
14. Which of the following does not degrade proteins and cellular components?
    a. ubiquitin-proteasome
    b. microautophagy lysosome
    c. chaperone-mediated autophagy
    d. heat shock proteins
    e. all of the above degrade proteins.
15. All of the following are key characteristics of senescent cells except ___________.
    a. cell cycle arrested
    b. initiation of apoptosis
    c. metabolically distinct
    d. altered chromatin structure
    e. production/secretion of more factors, like cytokines
16. Which statement about apoptosis is not true?
    a. Apoptotic bodies are eliminated by macrophages and some nearby cells.
    b. Cancer cells often bypass apoptotic pathways.
    c. An immune response is activated when macrophages phagocytize apoptotic bodies.
    d. Apoptosis is type-I programmed cell death.
    e. Apoptosis is a quiet, well-regulated and controlled cell death.
17. When external signals activate mammalian apoptosis, this is called __________.
    a. death receptor pathway
    b. intrinsic pathway
    c. mitochondria death pathway
    d. caspase pathway
    e. tumor necrosis factor pathway
18. Necroptosis involves a serine/threonine kinase called ______________.
    a. RIP1
    b. TNF-α
    c. TNFR
    d. FLIP
    e. caspase-8

19. Metabolic control of cell death includes all of the following except ___________.
  a. regulation of D-type cyclins by a glycolytic enzyme
  b. reduced state of cytochrome c
  c. cleavage of BID by caspase-2
  d. concentration of glucose and NADPH in the pentose phosphate pathway
  e. production of oncogenes

Further Reading

Anisimov, V. N., Berstein, L. M., Popovich, I. G., Zabezhinski, M. A., Egormin, P. A., Piskunova, T. S., et al. (2011). If started early in life, metformin treatment increases life span and postpones tumors in female SHR mice. *Aging, 3*, 148–157.

Balaban, R. S., Nemoto, S., & Finkel, T. (2005). Mitochondria, oxidants, and aging. *Cell, 120*, 483–495.

Chung, J. H., Manganiello, V., & Dyck, J. R. B. (2012). Resveratrol as a calorie restriction mimetic: therapeutic implications. *Trends in Cell Biology, 22*, 546–554.

Cornu, M., Albert, V., & Hall, M. N. (2013). mTOR in aging, metabolism, and cancer. *Current Opinion in Genetics & Development, 23*, 53–62.

Galluzzi, L., & Kroemer, G. (2008). Necroptosis: a specialized pathway of programmed necrosis. *Cell, 135*, 1161–1163.

Hayflick, L., & Moorhead, P. S. (1961). The serial cultivation of human diploid cell strains. *Experimental Cell Research, 25*, 585–621.

Kaeberlein, M., McVey, M., & Guarente, L. (1999). The SIR2/3/4 complex and SIR2 alone promote longevity in Saccharomyces cerevisiae by two different mechanisms. *Genes & Development, 13*, 2570–2580.

Lee, J., Giordano, S., & Zhang, J. (2012). Autophagy, mitochondria and oxidative stress: cross-talk and redox signalling. *The Biochemical Journal, 441*, 523–540.

Moo-Young, M., Sauerwald, T. M., Lewis, A., Dorai, H., & Betenbaugh, M. J. (2011). Apoptosis. In M. Moo-Young (Ed.), *Comprehensive Biotechnology* (2nd ed.) (pp. 483–494). Amsterdam, The Netherlands: Academic Press.

Rahman, I., Bagchi, D., Zhu, H., & van der Harst, P. (2014). Telomere biology in senescence and aging. In I. Rahman & D. Bagchi (Eds.), *Inflammation, Advancing Age and Nutrition* (pp. 71–84). London, UK: Academic Press.

Storer, M., Mas, A., Robert-Moreno, A., Pecoraro, M., Ortells, M. C., Di Giacomo, V., et al. (2013). Senescence is a developmental mechanism that contributes to embryonic growth and patterning. *Cell, 155*, 1119–1130.

Wilkinson, J. E., Burmeister, L., Brooks, S. V., Chan, C.-C., Friedline, S., Harrison, D. E., et al. (2012). Rapamycin slows aging in mice. *Aging Cell, 11*, 675–682.

Viral and Prion Infections

Biotechnology

http://dx.doi.org/10.1016/B978-0-12-385015-7.00021-1

VIRAL INFECTIONS AND ANTIVIRAL AGENTS

Many human diseases are due to viruses. These agents consist of genomes of either DNA or RNA inside a protein shell. Despite this deceptive simplicity, virus infections are less well understood than bacterial diseases, largely because viruses cannot be grown alone in culture but depend on a host cell. Until recently, protection against virus diseases relied on public health measures and vaccination. Only since the late 1980s have a significant number of specific antiviral agents become available.

Pathogenic bacteria contain many unique components not found in eukaryotic cells, which can be targeted by antibiotics. In contrast, because viruses rely on the host cell for almost all of their metabolic reactions, they usually have few unique components apart from the structural proteins of the virus particle. Consequently, most chemical agents that prevent virus metabolism are also toxic to the host cells. Another problem is that viruses mutate rapidly and so develop resistance to antiviral agents relatively quickly. This is especially serious for RNA viruses, such as influenza or HIV, which have extremely high mutation rates.

Like pathogenic bacteria, viruses must also attach to and invade host cells. Recognition proteins on the surface of the virus capsid bind to specific receptors on the surface of the host cell. After entry, viral replication occurs at the expense of the host cell, which supplies not only raw material and energy, but also the ribosomes needed for synthesis of viral proteins and often many of the enzymes required for synthesis of viral nucleic acids as well. Finally, new virus particles are assembled and exit the cell. These stages, and some corresponding antiviral agents are shown in Figure 21.1 and listed in Table 21.1. Antiviral agents that combat HIV are discussed in the later section on AIDS. In addition to HIV, we have chosen to focus on influenza, as it is one of the most widespread human viruses and illustrates many facets of virus biology and of the use of biotechnology for virus control.

Finally, we discuss prions. These infectious agents were originally believed to be anomalous viruses, hence their inclusion here. However, they consist solely of protein, with no enclosed nucleic acid. Thus, they are definitely not viruses despite sharing the superficial properties of size and infectiousness. Indeed, recent work suggests prion disease is related to other neurological disorders, not normally regarded as infectious.

> Relatively few antiviral agents are available compared to the number of antibiotics for treating bacterial infections. Moreover, most antivirals have harmful side effects.

INTERFERONS COORDINATE THE ANTIVIRAL RESPONSE

Interferons are a class of proteins induced in animal cells in response to virus infection. Clinical treatment with interferons is used to treat viral infections in a few cases (e.g., against hepatitis B and hepatitis C infections). **Interferons α** and **β (INF α** and **INF β)** block the spread of viruses by interfering with virus replication. (**Interferon** γ is quite distinct and is not induced directly by virus infection. It responds to intracellular pathogens.) Double-stranded RNA, which is symptomatic of the replication of most RNA viruses, activates secretion of interferons α and β. They bind to the interferon receptors of both the infected cell itself and its neighbors. Locally, this triggers a phosphorelay signal pathway that activates several genes that combat virus infection (Fig. 21.2). Interferons also help activate immune system cells, such as NK cells, which selectively destroy virus-infected cells.

Antiviral proteins induced by interferon include oligoadenylate synthetase, which converts ATP into 2′-5′-linked poly(A). This removes the ATP required as an energy source for viral

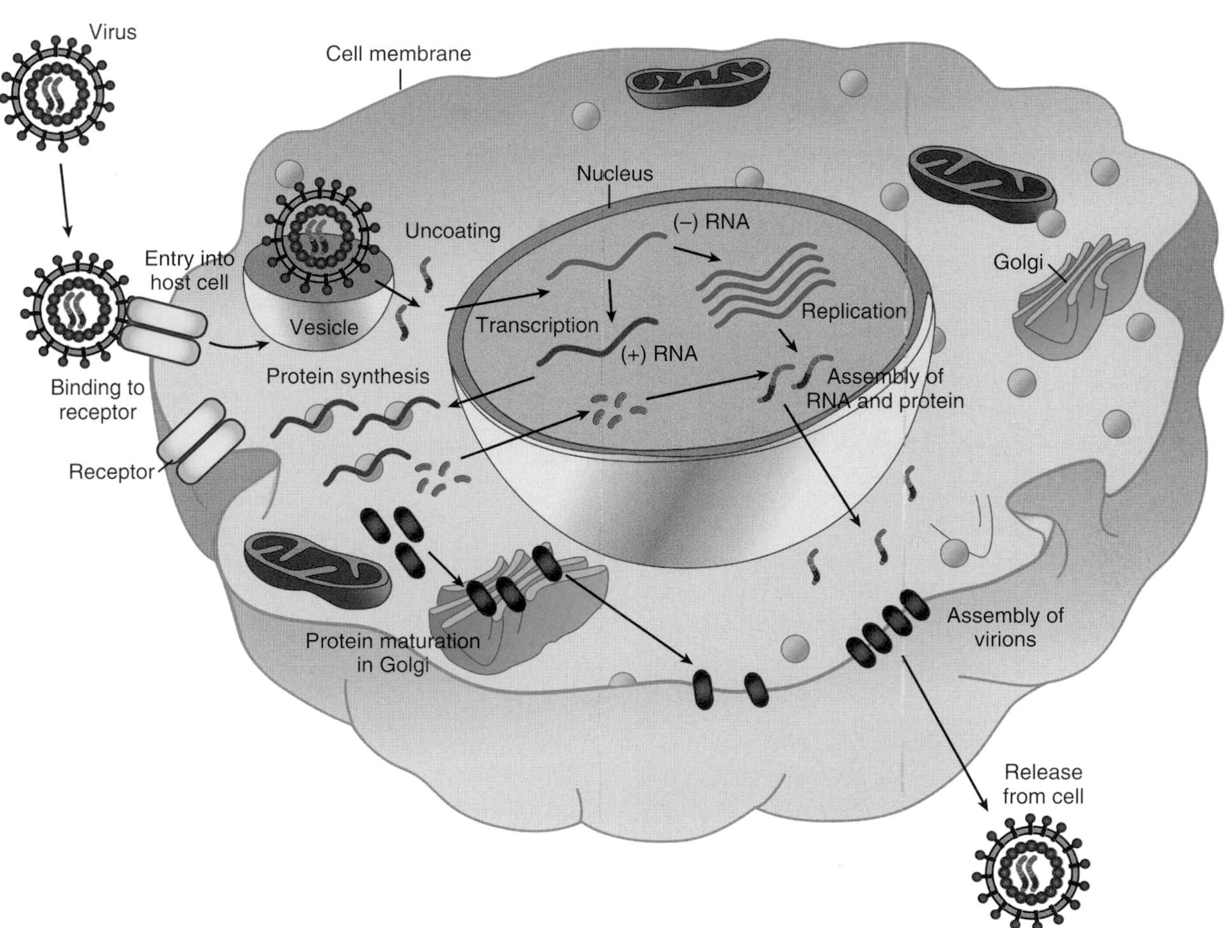

FIGURE 21.1 Virus Life Cycle with Antiviral Targets
The stages of virus life cycle provide several possible targets for antiviral agents. The virus shown here contains negative sense RNA (red). This is transcribed in the nucleus to give positive RNA (orange) that is translated by the ribosomes in the host cell cytoplasm to give viral proteins. Some proteins associate with the viral nucleic acid, whereas others move to the host cell membrane, via the Golgi apparatus, before virus assembly.

Table 21.1 Virus Life Cycle and Antagonists

Stage in Life Cycle	Possible Antiviral Agents
Binding to receptor	WIN compounds (picornavirus), zinc (rhinovirus)
Entry into host cell	Amantadine (influenza)
Gene expression	Interferons α and β
Reverse transcription (retroviruses only)	Reverse transcriptase inhibitors
Replication	Nucleoside analogs
Assembly of virions	Protease inhibitors
Release from cell	Neuraminidase inhibitors (influenza)

replication. In addition, 2′-5′-poly(A) activates an endonuclease that cleaves viral RNA. P1 kinase is also activated and phosphorylates initiation factor eIF2, halting protein synthesis. The **Mx proteins** are GTPases that interfere with the assembly of the RNA polymerase of negative-strand RNA viruses (e.g., influenza, parainfluenza). The Mx proteins form a ring that surrounds the viral RNA (Fig. 21.3), thus preventing the RNA polymerase from moving along and replicating the genome.

FIGURE 21.2 Interferons and Antiviral Proteins
The presence of dsRNA inside an infected cell triggers production of INFα and INFβ. These are secreted to neighboring cells, bind to the interferon receptor, and activate various antiviral proteins. P1 kinase blocks protein synthesis by phosphorylating eIF2 (an elongation factor). Oligo(A) synthetase converts ATP to 2′,5′-poly(A), which activates an endonuclease to digest dsRNA and depletes the ATP supply. Without ATP and protein synthesis, the virus cannot survive in the host cell.

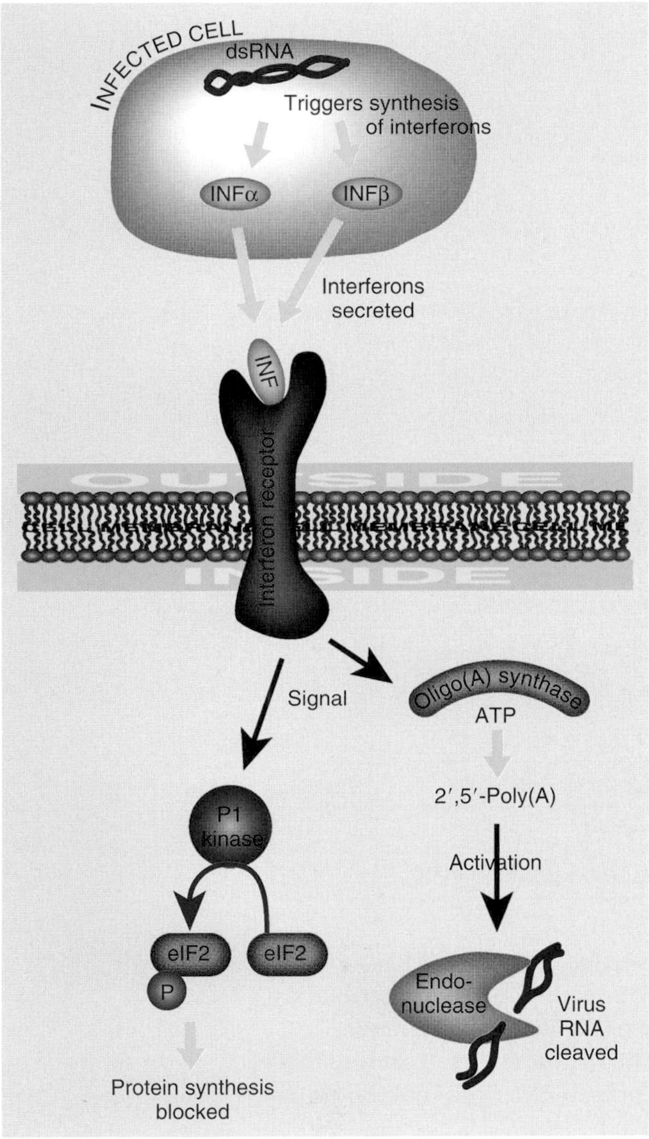

Different strains of influenza virus differ in their susceptibility to Mx proteins, and conversely, different animals have slightly different Mx proteins. These variants play a major role in determining both virulence and the transmission of virus between different animals.

Interferon alpha was one of the first mammalian proteins to be manufactured via genetic engineering. However, its clinical effects have been disappointing except in a few cases, such as treatment of hepatitis C. Recent attempts at antiviral therapy have moved away from interferons and focused on using the RNA interference system.

Interferons are animal proteins that promote the antiviral response by inducing synthesis of a range of enzymes with specific antiviral activities.

ANTIVIRAL THERAPY USING RNA INTERFERENCE

The basics of RNA technology were discussed in Chapter 5. Antisense RNA and ribozyme therapy have been proposed for antiviral therapy, but neither has proven effective so far. However, using RNA interference (RNAi) to treat virus infection looks promising.

RNA interference is a natural defense system used by cells to protect themselves against invasion by RNA viruses. RNAi targets double-stranded RNA (dsRNA) derived from RNA virus replication and destroys both the dsRNA and corresponding single-stranded RNA (in practice, this will usually be viral mRNA). RNAi is triggered by short dsRNA molecules of just over 20 nucleotides, known as short-interfering RNA (siRNA).

Not surprisingly, many viruses have evolved mechanisms to avoid destruction by RNAi. However, in mammals, administration of artificially synthesized siRNA around 17–21 nucleotides long provokes a strong RNAi response even against viruses with protection mechanisms. The sequence of the siRNA is designed to represent conserved regions of the RNA virus genome.

RNAi therapy is especially useful for viruses infecting the respiratory tract. The reason is that the siRNA can be administered easily by inhalation. RNAi is effective against respiratory syncytial virus, influenza, parainfluenza, measles, and several coronaviruses. The siRNA sequences can be screened for effectiveness in cell culture before being used on whole organisms. Phase II clinical trials using siRNA against respiratory syncytial virus are underway, and so far the results are promising.

RNAi can also protect plants against RNA viruses. To achieve this resistance, scientists engineer constructs that generate siRNA internally into transgenic plants (see Chapter 15 for details of plant genetic engineering). One common approach is to express RNA that folds into hairpin structures and hence includes a length of dsRNA. This triggers the synthesis of siRNA and RNAi, and the plants become resistant to the virus as a result (Fig. 21.4). RNAi is now being investigated for protecting crop plants, especially rice, against RNA viruses. Several RNA viruses of rice that are spread by insects may cause major crop losses.

RNAi is a natural form of defense against RNA viruses. It can be stimulated by administration of siRNA. Respiratory infections are especially easy to treat with RNAi because the siRNA can be inhaled as a spray.

INFLUENZA IS A NEGATIVE-STRAND RNA VIRUS

Influenza virus, an **orthomyxovirus**, is an example of a negative-strand single-stranded RNA virus. In other words, the virus genome is present in the virus particle as noncoding (= antisense = negative-strand) RNA. The flu virus particle contains a segmented genome consisting of eight separate pieces of single-stranded RNA ranging from 890 to 2341 nucleotides long. These pieces are each packed into an inner **nucleocapsid** and are surrounded by an outer envelope (Fig. 21.5). Although the outer membrane is derived from host-cell material, it contains virus-encoded proteins such as neuraminidase, hemagglutinin, and ion channels. These viral proteins are made on the ribosomes of the infected host cell and are involved in virus recognition and entry into successive host cells. The hemagglutinin (H) and neuraminidase (N) of influenza differ slightly but significantly between strains of flu. These variants are designated by H and N numbers. Thus, the Spanish flu of 1918 was H_1N_1, and the avian flu presently spreading worldwide is H_5N_1. The virulent outbreak of novel avian flu in China in 2013 was H_7N_9. This virus contains segments from several different avian flu strains. Genome analysis confirms increased virulence and implies resistance to amantadine (see following discussion).

When a flu virus comes in contact with an appropriate host cell, it is engulfed and ends up inside a vesicle. Both the vesicle and the outer coat of the virus particle are dissolved, releasing the nucleocapsids, which enter the nucleus. The nucleocapsids disassemble inside the nucleus, releasing the RNA molecules (Fig. 21.6).

FIGURE 21.3 Mechanism of Action of Mx Protein

The viral RNA polymerase moves along the viral genomic RNA to replicate it. The Mx protein assembles into ring structures that surround the viral RNA. This blocks the movement of the RNA polymerase and consequently prevents replication. From Gao S et al. (2011). Structure of myxovirus resistance protein a reveals intra- and intermolecular domain interactions required for the antiviral function. *Immunity* **35**, 514–525.

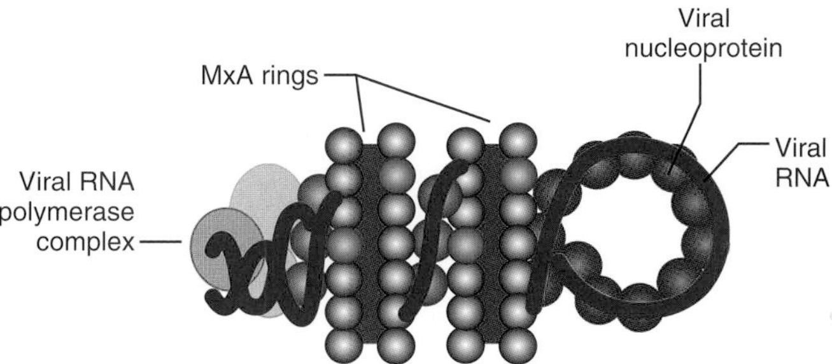

FIGURE 21.4 RNAi versus Plum Pox Virus

Transgenic *Nicotiana benthamiana* plants were constructed that expressed a hairpin RNA that triggers RNAi versus plum pox virus (the agent of "sharka disease"). *N. benthamiana* is a close relative of *N. tabacum*, the tobacco plant that grows in Australia. Wild-type and transgenic *Nicotiana benthamiana* plants were then tested against infection with plum pox virus. After 7 days, severe wilting was seen in the wild-type but not the transgenic plants. From Pandolfini T, et al. (2003). Expression of self-complementary hairpin RNA under the control of the *rolC* promoter confers systemic disease resistance to plum pox virus without preventing local infection. *BMC Biotechnol* **3**, 7.

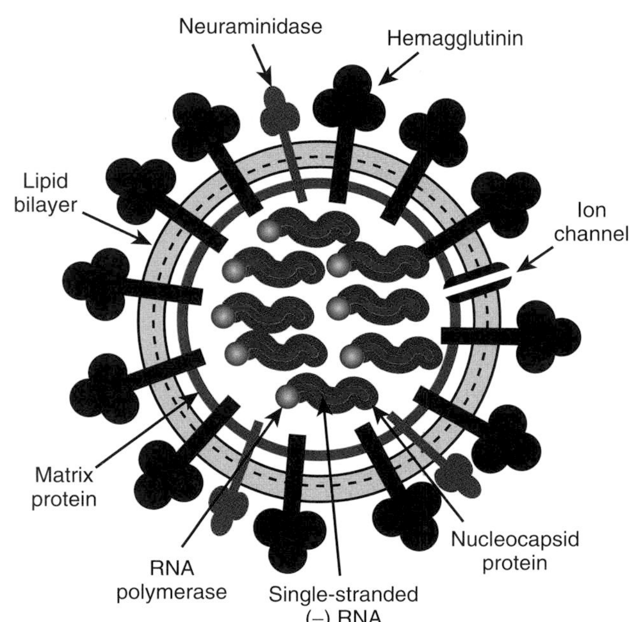

FIGURE 21.5
Structure of the Influenza Virus
The influenza virus has an outer envelope containing neuraminidase, hemagglutinin, and ion channels. Several individual negative-strand ssRNA molecules are packaged within the outer membrane. Each strand is coated with nucleocapsid proteins. An RNA replicase molecule is also included with each ssRNA strand to ensure expression.

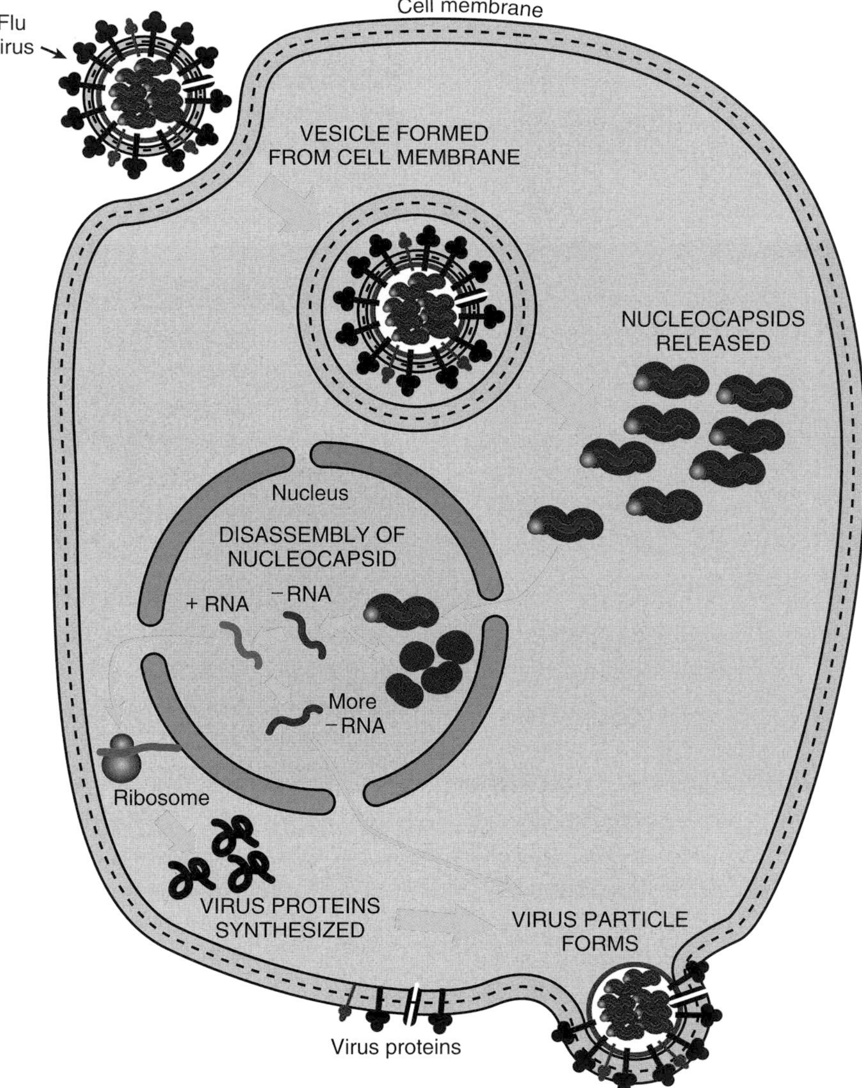

FIGURE 21.6
Life Cycle of the Influenza Virus
After entry into the host cell, the nucleocapsids enter the nucleus before disassembly. There the viral replicase makes positive RNA strands and more negative strands. The (+) RNA strands are exported to the ribosomes, where they act as mRNA and are translated. The resulting viral proteins are assembled into more virus particles, together with the (–) RNA strands.

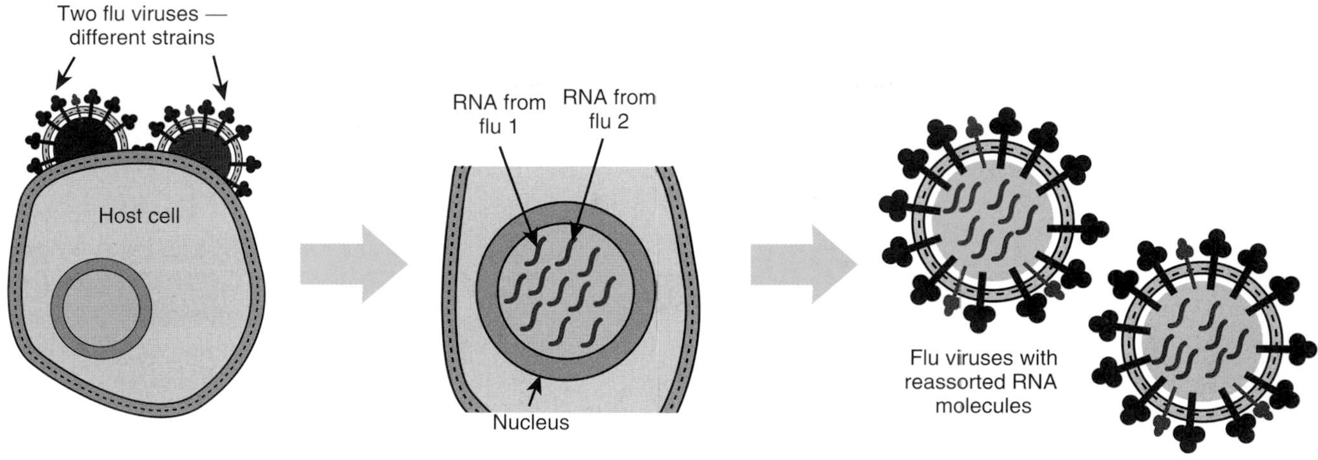

FIGURE 21.7 Influenza Viral Genomes Can Switch RNA Segments
If two different influenza strains infect the same host cell, the genomes of both will enter the nucleus. When new virus particles are formed, some nucleocapsids from strain 1 may be packaged with strain 2, and vice versa. Thus, complete ssRNA molecules from different influenza strains may be reshuffled to generate new assortments. Such reshuffling more often happens in pigs and birds than in human hosts.

Replication of the influenza RNA occurs in the nucleus. The viral mRNA exits the nucleus just like normal cellular mRNA and travels to the ribosomes in the cytoplasm. Here, the proteins for the new virus particles are made.

Because influenza virus has its genes scattered over eight separate molecules of RNA, different strains of flu can trade segments of RNA and form new genetic combinations (Fig. 21.7). In addition, mutations occur at a higher rate during RNA replication than in DNA. These two mechanisms result in a lot of genetic diversity. Consequently, different strains of flu emerge every couple of years. The changing surface antigens of the virus allow it to avoid immune recognition. These different flu strains vary greatly in their apparent virulence. However, this depends as much on the immune history of the human population as on genetic changes in the virus.

Influenza viruses fall into two major groups: influenza A and B. Mutation of both A and B causes annual epidemics due to slow antigenic drift. Influenza B is largely restricted to humans and has less genetic variation. Influenza A has a wider host range ("people, pigs, and poultry"). As a result, influenza A gives rise to severe but less common epidemics due to reassortment of viruses from different hosts during mixed infections.

The Spanish flu of 1918–1919 was the worst influenza A pandemic so far and is estimated to have killed around 50 million people (more than World War I). Will there be another major flu pandemic soon? The major threat seems to be the successive versions of avian flu emerging in Asia. Relatively few humans catch these viruses by direct transmission from birds. The real danger is that these avian viruses will mutate to become transmissible from person to person.

Amantadine is a tricyclic amine that binds to the M2 protein, one of the transmembrane ion channels found in the outer envelope of influenza A virus. M2 is not expressed by influenza B, and consequently, amantadine works only against type A influenza. Amantadine blocks the M2 ion channel, and this stops entry of protons, which prevents uncoating of the virus particle (Fig. 21.8). Thus, entry of the virus is prevented. Amantadine must be given very early in infection. Amantadine was the first specific antiviral agent to be discovered, although its mode of action was only elucidated later.

Influenza (both A and B) may also be treated with neuraminidase inhibitors, such as oseltamivir (=Tamiflu) or zanamivir. These inhibitors are analogs of *N*-acetylneuraminic acid. Neuraminidase normally cleaves this from the virus receptor, allowing progeny virus particles to be released. If neuraminidase is inhibited, progeny virus is trapped in infected cells. Resistance can arise due to mutations in the N protein; for example, H247Y (changing His247 to Tyr) results in resistance to oseltamivir but not to zanamivir.

Influenza is an extremely common viral infection of humans and some other animals. Its genome consists of eight pieces of RNA of negative complementarity. As a result, it shows a high rate of both mutation and recombination. Very few drugs are available to treat influenza.

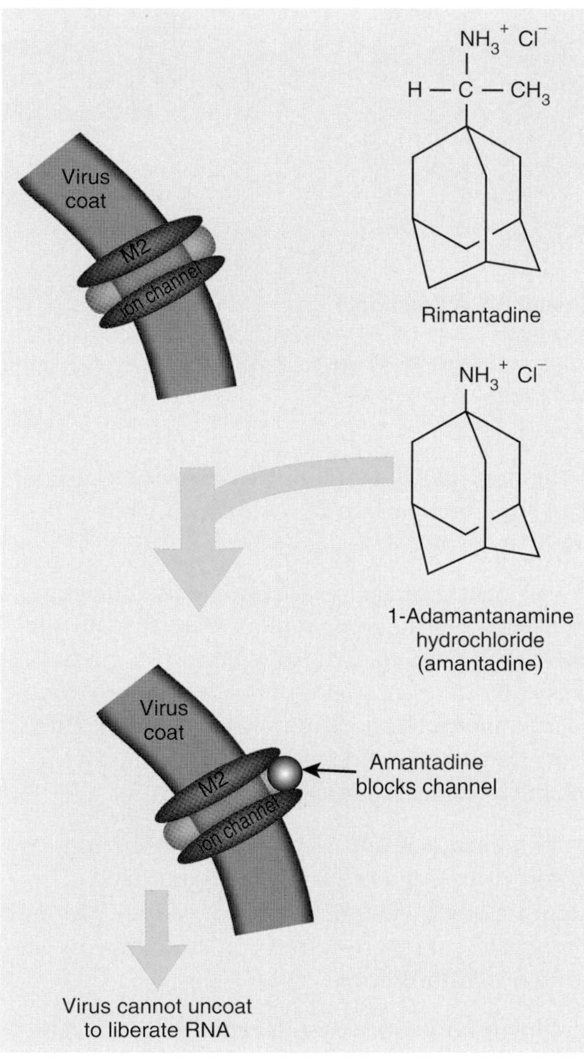

FIGURE 21.8 Amantadine Blocks M2 Ion Channel
The amantadine molecule blocks ions from passing though the M2 channel in the virus coat, thus preventing uncoating and RNA molecule release.

THE AIDS RETROVIRUS

Acquired immunodeficiency syndrome (AIDS) is caused by human immunodeficiency virus (HIV), which damages the immune system. Most AIDS patients die of **opportunistic infections**. These infections are seen only in patients with defective immune systems and are caused by assorted viruses, bacteria, protozoans, and fungi that are normally relatively harmless but may attack if host defenses are down. In addition, without immune surveillance, cancers caused by other viruses or somatic mutations often grow out of control.

HIV infects white blood cells belonging to the immune system, the **T cells**. The **CD4 protein** is found on the surface of many T cells, where it acts as an important receptor during the immune response (see Chapter 6). HIV also uses the CD4 protein as a receptor (Fig. 21.9). The **gp120** protein in the outer envelope of HIV is a glycoprotein with a molecular weight of 120 kDa. It recognizes and binds to CD4, which is needed for entry of the virus.

The CD4 protein is also found on the surface of some other immune system cells—the monocytes and macrophages. HIV does not seriously harm these two cell types, but the cells become reservoirs to spread the virus to more T cells. The damage to the T cells is most critical to immune function. Once HIV has entered the T cell, the DNA form of the retrovirus genome integrates into the host chromosome and begins to express virus genes. Viral proteins are manufactured on host ribosomes. In particular, the HIV envelope protein, gp120, is made in large amounts and inserts into the T-cell membrane. The gp120 on the surface of infected T cells binds to the CD4 protein on other T cells. Consequently, several T cells clump together and fuse (Fig. 21.10). The giant, multiple cell soon dies. About 70% of the body's T cells carry the CD4 receptor. As they gradually die off, the immune response fades away over a 5- to 10-year period.

AIDS is caused by a retrovirus that uses the CD4 protein on the surface of T cells as a receptor. Damage to T cells cripples the immune response, leaving the body open to other infections.

CHEMOKINE RECEPTORS ACT AS CO-RECEPTORS FOR HIV

The entry of HIV into T cells requires binding of virus to both the CD4 protein and one of several chemokine receptors, which act as **co-receptors**. The **chemokine receptors** are

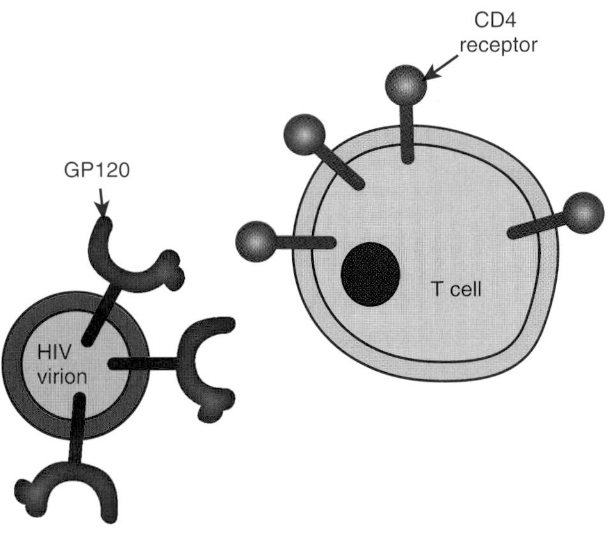

FIGURE 21.9 HIV Uses CD4 Protein as Receptor
HIV particles are coated with gp120, which recognizes the T cells of the immune system. The viral glycoprotein gp120 binds to protein CD4, on the surface of the T cell. The viral particle is then taken into the T cell, where it takes over the cellular machinery to produce more virus.

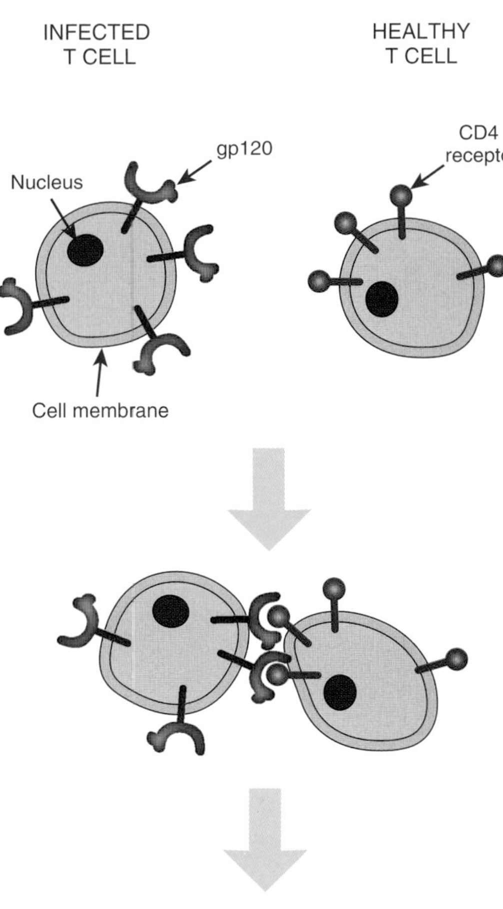

FIGURE 21.10 Fusion of Infected T Cells
Once HIV has entered the T cell, gp120 is made in large amounts and is inserted into the host cell membrane. T cells with gp120 in their membranes bind to other T cells via the CD4 receptor, which causes the cells to fuse. The process continues until large clumps of T cells form. These cells soon die, crippling the immune system.

membrane proteins with seven *trans*-membrane segments. They bind **chemokines**, a group of approximately 50 small messenger peptides that activate the white blood cells of the immune system and attract them to the site of infections. The most important chemokine receptors for HIV entry are **CCR5** and to a lesser extent CXCR4.

Mutations in CCR5 are largely responsible for the small proportion of the population who are naturally resistant to HIV infection. The *CCR5Δ32* allele has a deletion of 32 base pairs and results in non-functional CCR5 protein. Individuals homozygous for *CCR5Δ32* are vastly less susceptible to infection by HIV (although not totally resistant). In addition, if these individuals are infected, the disease progresses much more slowly. About 2% of Europeans are homozygous for *CCR5Δ32* and 14% are heterozygous. Heterozygotes are mildly protected and show slower progression, in accord with the lower levels of CCR5 protein on the surfaces of their T cells. The origin of the *CCR5Δ32* allele has been traced back to around 700 years ago in northwest Europe, at about the time of the Black Death. Conceivably, the defects in CCR5 were selected by providing resistance against the bubonic plague. Variations in susceptibility to AIDS also result from alterations in the DNA sequence of the promoter for the *CCR5* gene. Presumably, these alterations cause variations in the level of CCR5 protein expressed.

Receptors that take up important molecules into animal cells are often the targets for viruses. It is quite possible for the same host cell protein to be used as a receptor by unrelated infectious agents, including both viruses and bacteria. Thus, the myxoma poxvirus, which causes immune deficiency in rabbits, also uses the CCR5 and CXCR4 chemokine receptors. Which receptors are used by smallpox or other poxviruses is still unknown. Other pathogens, including the malaria parasite, also target chemokine receptors, although not CCR5 and CXCR4. Scientists are presently trying to identify the functions of the various receptors on immune cells in the hope of understanding how viruses exploit them for their own use.

Entry of HIV into target cells requires co-receptors. Natural resistance to AIDS results from defects in co-receptors, especially the CCR5 chemokine receptor.

TREATMENT OF THE AIDS RETROVIRUS

No complete cure or effective vaccine yet exists for AIDS, although several treatments are available that significantly extend patients' lives. About 50% of the antiviral drugs in clinical use are for AIDS. The fundamental problem with all anti-AIDS drugs is that HIV is an RNA virus and so has a relatively high mutation rate. HIV mutates at a rate of approximately one base per genome per cycle of replication. Even within a single patient, HIV exists as a swarm of closely related variants known as a **quasi-species**. Consequently, strains of HIV resistant to individual drugs appear at a relatively high frequency. Attempts to control AIDS (Fig. 21.11; Table 21.2), whether by using vaccines, protein processing inhibitors, or antisense

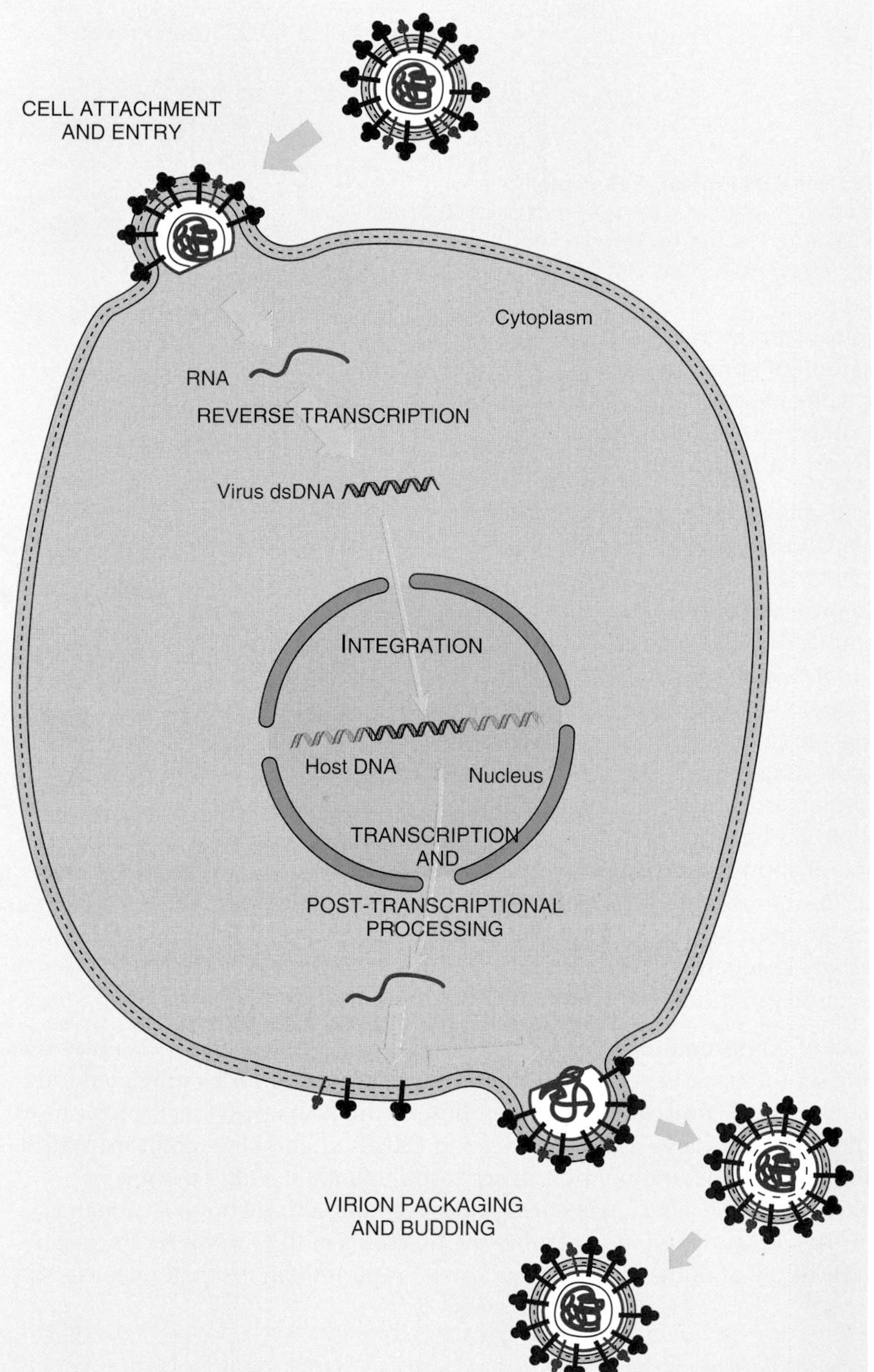

FIGURE 21.11 Possible Steps for HIV Inhibition
HIV infections could be stopped at the following steps: (1) at the cell surface, competing molecules could prevent virus attachment; (2) enzyme inhibitors may block the action of reverse transcriptase; (3) integration of the viral genome could be prevented; (4) transcription and translation could be blocked; (5) finally, blocking virion packaging and budding would protect other cells from becoming infected.

Table 21.2 HIV Antagonists

Stage in Life Cycle	Possible Antiviral Agents
Binding to receptor	CCR5 co-receptor inhibitor; Maraviroc
Membrane fusion	Fusion inhibitors
Reverse transcription	(a) Nucleoside analogs (chain terminators) (b) Non-nucleoside reverse transcriptase inhibitors (NNRTI)
Integration	(a) Integrase strand transfer inhibitor (INSTI) (b) Integrase LEDG inhibitors (LEDGIN)
Assembly of virions	Protease inhibitors

RNA, all face the same problem: HIV will mutate to produce resistant variants. In practice, this problem may be partially overcome by simultaneous treatment with several drugs that hit different targets.

Azidothymidine (**AZT**, or zidovudine) was one of the first drugs used against AIDS. It is an analog of thymidine that lacks the 3′-hydroxyl group. Various other **nucleoside analogs** that lack the 3′-hydroxyl group are also in use. AZT and other 3′-deoxy base analogs are converted to the 5′-triphosphate by the cell and then incorporated into the growing DNA chain during reverse transcription (Fig. 21.12). Because AZT lacks a 3′-hydroxyl group, the DNA chain cannot be extended. AZT is thus a DNA **chain terminator**. Although AZT is incorporated more readily by the viral reverse transcriptase than by most host-cell DNA polymerases, it is not completely specific. Thus, one major drawback is that AZT partially inhibits host DNA synthesis in uninfected cells of the body. In particular, it is toxic to bone marrow cells (B cells), which are another part of the immune system. Mutations in the HIV reverse transcriptase may cause resistance to base analogs. For example, Met41Leu (i.e., replacement of methionine at position 41 with leucine) increases resistance to AZT by 4-fold, and a second mutation of Thr215Tyr gives an overall 70-fold resistance.

Certain drugs that do not bind at the active site can also inhibit reverse transcriptase. They are referred to as **non-nucleoside reverse transcriptase inhibitors** (**NNRTI**; Fig. 21.13). They bind to the enzyme at a separate site, relatively close to the active site. This distorts the structure of reverse transcriptase and inhibits its activity. Unfortunately, mutations that alter the NNRTI binding site occur quite frequently, and they give rise to resistant reverse transcriptase enzyme. These drugs are therefore generally used in combination with nucleoside analogs.

Most individual HIV proteins are joined together as polyproteins when first made and must therefore be cut apart by HIV protease. For example, the *env* gene is transcribed and translated to give gp160, which is cleaved to gp41 and gp120. The *gag* gene encodes a polyprotein that includes the proteins of the virus core. Consequently, inhibition of polyprotein cleavage will prevent the assembly of the virus particle. The HIV protease recognizes and binds a stretch of seven amino acids around the cleavage site. This step may be blocked with **protease**

FIGURE 21.12 Nucleoside Analogs Act as Chain Terminators
Two examples of chain terminators are azidothymidine (AZT) and acyclovir, which replace thymine and guanine, respectively. AZT has an azido group on the 3′ position of the deoxyribose ring rather than a hydroxyl. The entire deoxyribose ring is altered in acyclovir. In both cases, the analogs are incorporated into DNA during the reverse transcriptase reaction. Once the analog has been inserted, reverse transcriptase cannot elongate the DNA chain any further because the analogs lack the 3′-OH group to which the next nucleotide would be added.

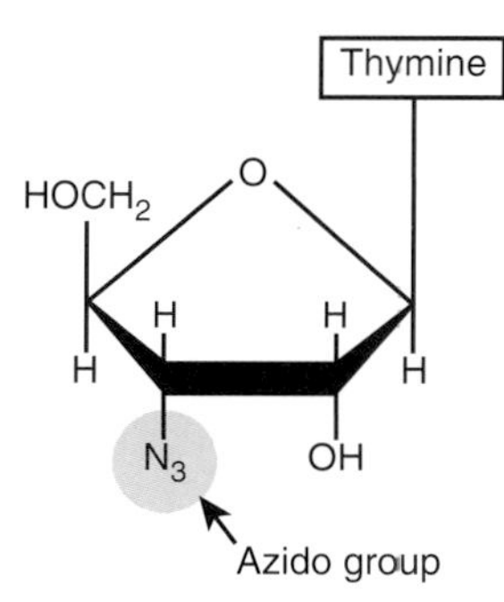

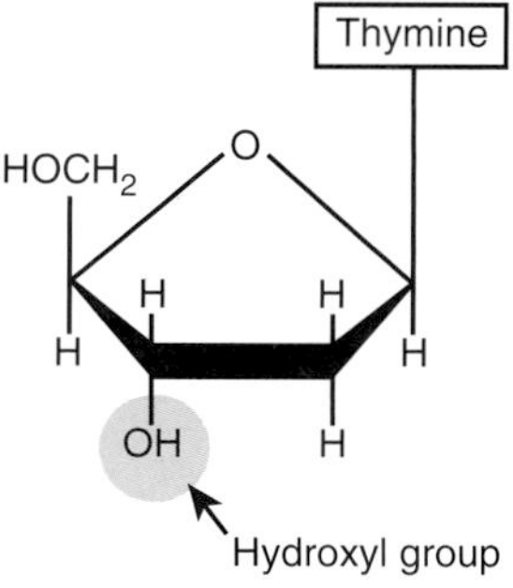

ACYCLOVIR

Guanine
HOCH2
O
H
H
"Acyclic sugar"
H
H

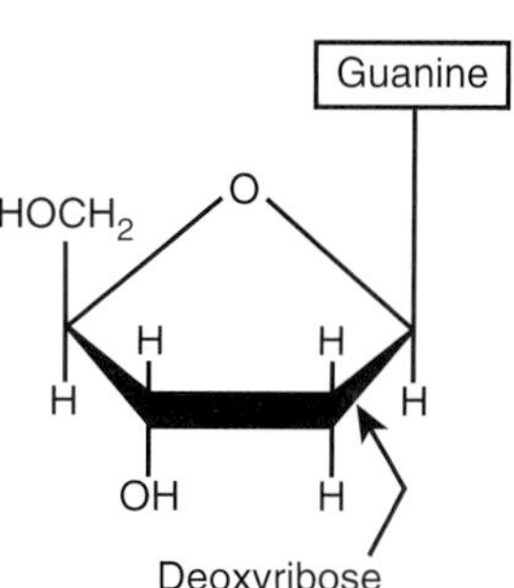

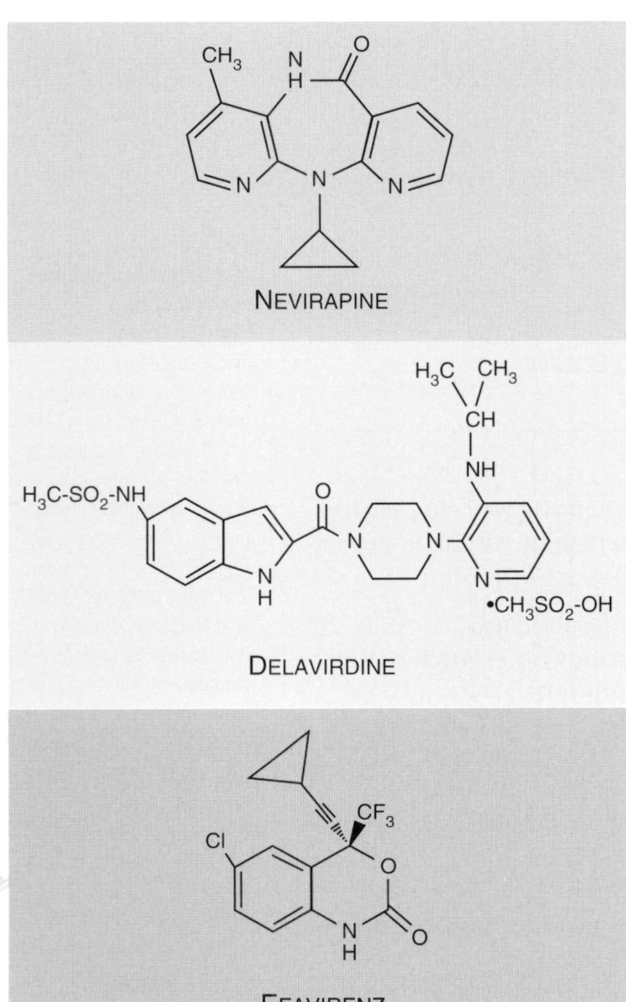

FIGURE 21.13 Non-Nucleoside Reverse Transcriptase Inhibitors

Chemical structures of three NNRTIs that are presently in use: nevirapine, delavirdine, and efavirenz. They are specific for HIV-1 and have no effect on HIV-2.

inhibitors that are analogs of several amino acid residues around the cut site (Fig. 21.14). For example, saquinavir is an analog of Asn-Tyr-Pro.

At present, the favored approach in AIDS therapy, referred to as Highly Active Anti Retroviral Therapy (HAART), is to use three or four drugs with different mechanisms in combination. Different drugs should *not* be used one after the other because this allows resistance to develop to each drug in turn. If several drugs are used simultaneously, emerging virus mutants that are resistant to one drug will be killed by the others. A typical cocktail consists of two chain termination inhibitors plus a non-nucleoside reverse transcriptase inhibitor or a protease inhibitor. Since the mid 1990s, deaths from AIDS have dropped 60% to 80% in those nations whose citizens can afford expensive long-term treatment with costly pharmaceuticals. In the United States, treatment with such a cocktail may cost from $800 to $1500 per month, although the cost keeps dropping.

Integrase inhibitors and fusion inhibitors are newer additions to the AIDS arsenal. Integrase inhibitors prevent integration of the HIV DNA into the host genome. Fusion inhibitors prevent the fusion of host and viral membranes that occurs during the uptake process, after receptor binding. CCR5 co-receptor inhibitors block binding to those HIV strains that use CCR5. Before use, the patient must be checked for the co-receptor specificity of the virus, an expensive process. These drugs are generally used when others fail due to resistance or harmful side effects.

Hydroxyurea was once used in cocktails to treat AIDS. Hydroxyurea inhibits enzymes of the human host cell that are needed for the AIDS virus to replicate. Because human genes encode these proteins, the virus cannot mutate to produce hydroxyurea-resistant enzymes. The advantage and the problem with hydroxyurea are that it also inhibits human cell DNA replication. Hydroxyurea is no longer used due to its toxicity.

Most recently developed antiviral agents were designed to treat AIDS. They include nucleoside analogs (chain terminators), non-nucleoside reverse transcriptase inhibitors, protease inhibitors, integrase inhibitors and fusion inhibitors.

INFECTIOUS PRION DISEASE

Prions are proteins with unique properties that are capable of causing inherited, spontaneous, or infectious disease. The **prion protein (PrP)** exists in two conformations, the normal harmless or "cellular" form (**PrP^c^**) and the pathogenic (**PrP^Sc^**) form, named after **scrapie**, a disease of sheep (Fig. 21.15). Rogue prion proteins bind to their normal relatives and induce them to refold into the disease-causing conformation. Thus, a small number of misfolded prions will eventually subvert the population of normal proteins. Over time, this leads to neural degeneration and eventually death.

Mutations within the ***Prnp* gene** that encodes the prion protein may result in prions with a greatly increased likelihood of misfolding. This causes hereditary prion disease. Several clinically different variants are known, depending on the precise location of the mutation within the prion protein and the nature of the amino acid alteration. The most common is

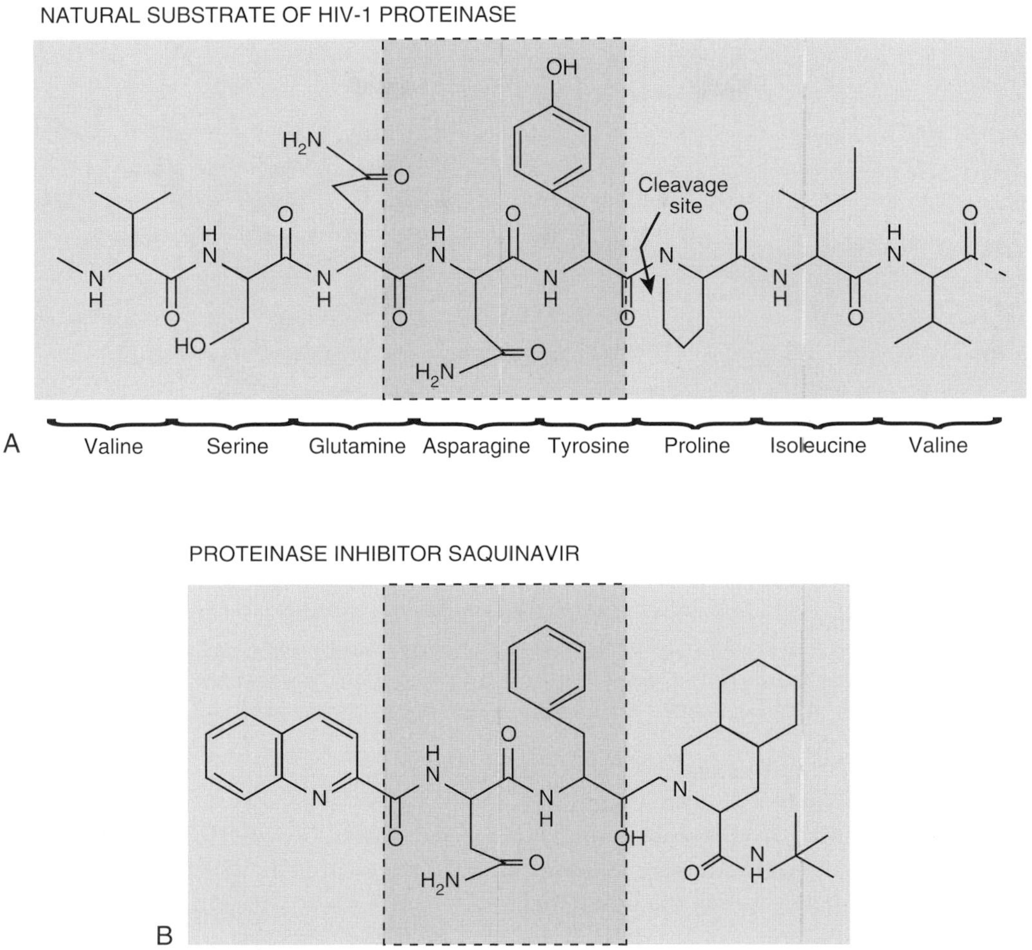

FIGURE 21.14 **Protease Inhibitors**

(A) HIV-1 protease recognizes Asn-Tyr-Pro, cleaving the protein between the tyrosine and proline. (B) Saquinavir has a structure that mimics these three amino acids. HIV-1 proteinase binds to saquinavir but cannot cleave or release it because the cleavage site is missing.

Creutzfeldt–Jakob disease (CJD). Even normal prions occasionally misfold. The result is spontaneous prion disease, which occurs at a rate of about one per million of the human population.

The pathogenic misfolded prions form insoluble aggregates known as **amyloids**. These are fibrils consisting of protein with a high beta-sheet content. The beta-sheets are short and form stacks that run sideways relative to the long axis of the fibers (Fig. 21.16).

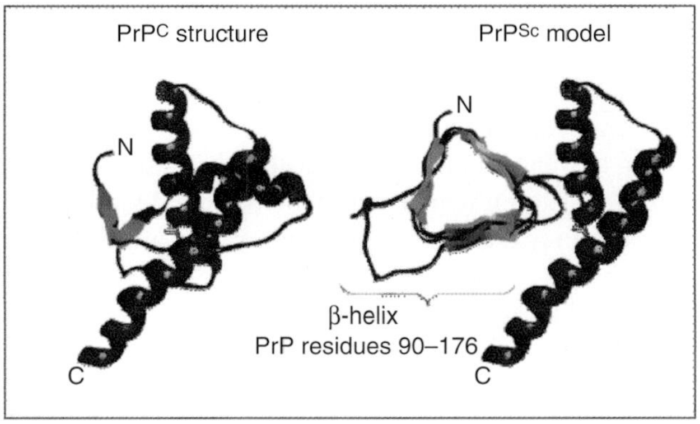

FIGURE 21.15 **Normal and Pathogenic Forms of the Prion Protein**

The PrP^c structure is on the left, and the PrP^{Sc} structure is on the right. Note the greatly increased proportion of beta-sheet in the PrP^{Sc} structure. From Eghiaian F (2005). Structuring the puzzle of prion propagation. *Curr Opin Struct Biol* **15**, 724–730. Reprinted with permission.

If misfolded prions are transmitted to another susceptible host, the result is infectious prion disease, also known as **transmissible spongiform encephalopathy (TSE)**. Such an infection can be passed from one cell to another and one animal to another by entry of the PrP^{Sc} form of the prion. The two individuals may be of the same or different species. Infection of a new victim by prions is relatively difficult. It requires uptake of rogue prion proteins from infected nervous tissue, especially brain, but the details of infection remain obscure. The best-known infectious prion diseases are

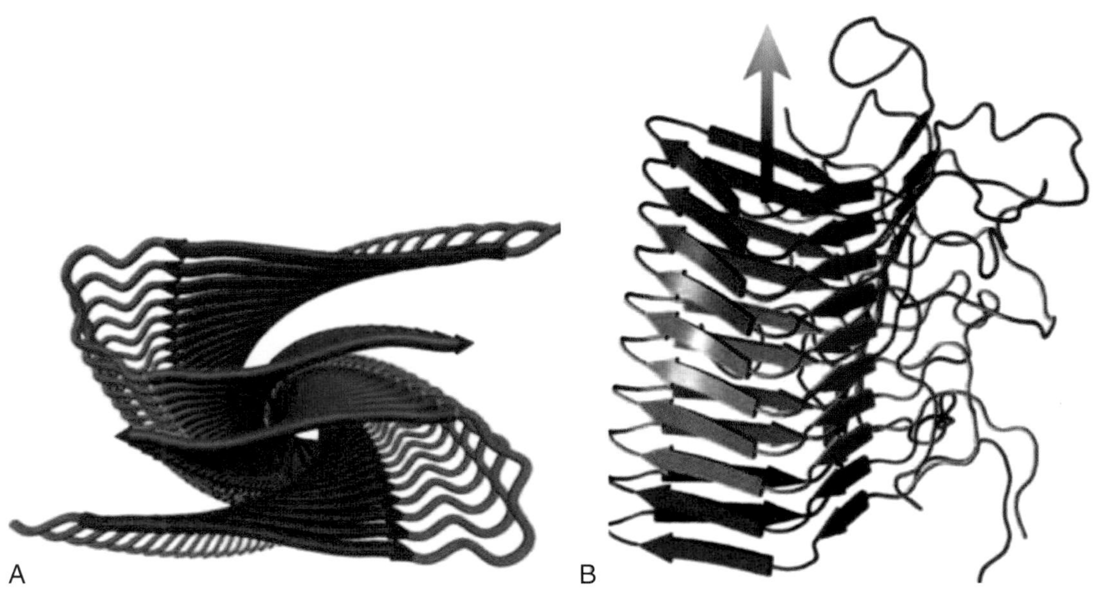

FIGURE 21.16 Amyloid Structure
Models of a typical amyloid structure found in prion proteins. (A) The $A\beta^{1-40}$ structure is parallel in-register with each peptide in a hairpin configuration and monomers stacked. Two stacks are aligned as shown. (B) The yeast HET-s prion domain (residues 218–289) forms a two-turn β-helix with partial directly repeated sequences in the peptide aligned. (From Wickner RB et al. (2011). Prion diseases of yeast: amyloid structure and biology. *Semin Cell Dev Biol* **22**, 469–475.)

1. Scrapie, a disease of sheep and goats
2. **Kuru**, a disease of cannibals
3. **Mad cow disease**, officially known as **bovine spongiform encephalopathy (BSE)**
4. **Chronic wasting disease (CWD)** of deer and elk, which can be transmitted via saliva, unlike the other TSEs

Scrapie is a disease of sheep and related animals that has been recorded going back several hundred years in Europe. The name comes from the behavior of infected sheep that constantly scrape themselves against fences, trees, or walls and often seriously injure themselves (Fig. 21.17). Only certain breeds of sheep are susceptible because of the slight differences in prion sequence between breeds. Dead and decomposing sheep may contaminate the grass of their fields with prion proteins. These proteins are unusually stable and long-lived and may be eaten by healthy sheep.

Kuru was transmitted by ritual cannibalism and used to be endemic among the Fore tribe of New Guinea. The women had the honor of preparing the brains of dead relatives and participating in their ritual consumption. As a result, 90% of the victims were women, together with younger children who accompanied them. Developing symptoms took from 10 to 20 years, but once they did, the progression from headaches to difficulty walking to death from neural degeneration took from 1 to 2 years. No one born since 1959, when cannibalism stopped, has developed kuru.

Brain degeneration, or spongiform encephalopathy, due to misfolded prions is possible in any mammalian species. In addition to scrapie, BSE, and CWD, a variety of less well-characterized prion diseases are known in other animals. In a way, they are really all the same disease because there is a single prion gene encoding a single prion protein that is found in the brain of all mammals. Symptoms vary slightly from species to species, but after a long incubation period, the result is degeneration and death of cells of the central nervous system. As the popular name *mad cow disease* indicates, progressive degeneration of the brain and nervous system causes the infected animals to behave bizarrely during later stages of the disease.

Mad cow disease was spread by overly intensive farming practices. Animal remains, including the brains, were ground up and incorporated into animal feed. Because sheep remains were included in feed for cows, the epidemic of mad cow disease, which began in England in 1986, was originally blamed on sheep with scrapie. However, people in England and other European countries have eaten sheep with scrapie since the 1700s without any noticeable ill effects. Nor have any other domestic animals, including cows, ever caught scrapie, despite sharing the same fields. Moreover, sheep prions are not infectious for cows. It is now thought that a random flip-flop event converted a normal prion into the rogue form inside a cow's brain somewhere in England in the late 1970s or early 1980s. The rogue cow prions were recycled in animal feed and spread, eventually causing an epidemic. After mad cow disease broke out in England, the recycling of animal remains in feed was prohibited and infected herds were destroyed.

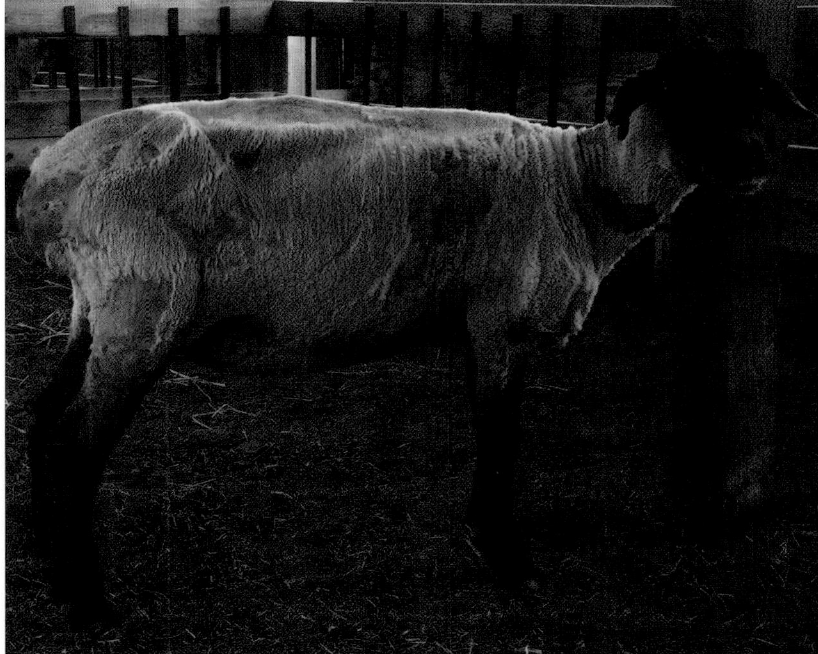

FIGURE 21.17
Sheep with Scrapie
This sheep with scrapie is from the Caine Veterinary Teaching Center, Caldwell, ID. Photograph provided by Sharon Sorenson.

Mad cow disease can be transmitted to humans, but the rate of infection is extremely low. The first human cases were confirmed in 1996 and were named **variant CJD** in an attempt to obscure their origin. However, when the rogue cow prion infects humans, the misfolded prions are characteristic of mad cow disease, not genuine CJD. In humans with CJD or kuru, the precise conformation of the misfolded prions is different. The human victims of mad cow disease are scattered randomly throughout the population, suggesting that relatively few humans are actually susceptible to infection. As of 2013, about 225 people, mostly in England, have come down with BSE. Calculations based on the history and age distribution of BSE in humans since the outbreak started suggest an average incubation period of about 15 years and that the total number of cases will be under 300. These estimates reflect the extremely low infectivity of prions when crossing from one species to another.

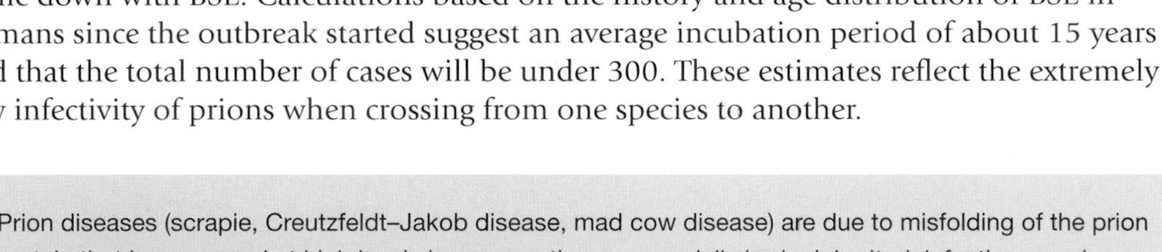

Prion diseases (scrapie, Creutzfeldt–Jakob disease, mad cow disease) are due to misfolding of the prion protein that is expressed at high levels in nervous tissue, especially brain. Inherited, infectious, and spontaneous variants of prion disease are found.

DETECTION OF PATHOGENIC PRIONS

The emergence of mad cow disease (BSE) has created the need to screen cows and their products for the presence of the pathogenic form of the prion protein (PrP^{Sc}). This screening is presently done by immunological detection. Early assays lacked separate antibodies specific to the normal (PrP^{c}) and pathogenic (PrP^{Sc}) isoforms. Consequently, because the pathogenic form of the prion is protease resistant, samples were first treated with protease to destroy the normal (PrP^{c}) form and then subjected to immunological testing by Western blotting (see Chapter 6). The overall procedure is tedious and only moderately sensitive. The development of isoform specific antibodies led to the **conformation-dependent immunoassay (CDI)** and was a major improvement.

Further improvements relied on amplification schemes. The first of these was the **protein misfolding cyclic amplification (PMCA)** procedure, which amplifies the levels of misfolded prion in a manner analogous to the use of PCR for amplifying DNA (Fig. 21.18). This allows greatly increased sensitivity of detection of PrP^{Sc} in clinical samples. Small samples suspected of containing PrP^{Sc} are mixed with normal brain homogenate containing a surplus of the normal PrP^{c}. The PrP^{c} is converted to PrP^{Sc} and incorporated into the growing PrP^{Sc} aggregates. The sample is then sonicated to break up the aggregates. This procedure is repeated for several cycles. Increases of around 60-fold over five cycles are typical.

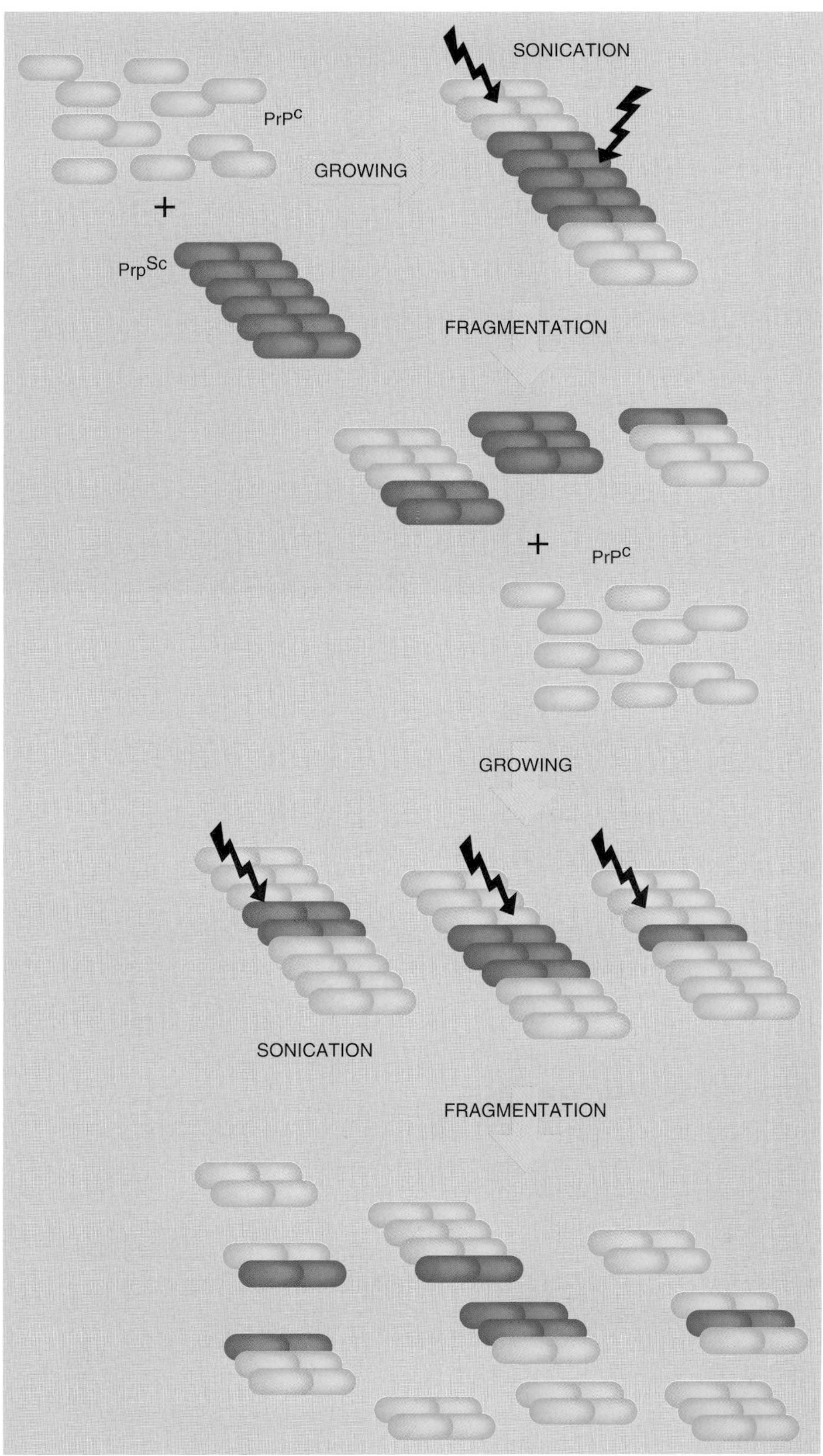

FIGURE 21.18 Protein Misfolding Cyclic Amplification (PMCA)
Amplification involves multiple cycles of incubating PrP^{Sc} in the presence of excess PrP^{c} followed by sonication. During the incubation periods, the size of PrP^{Sc} aggregates (purple) increases because of incorporation of normal prion protein (blue). During sonication, the aggregates are disrupted, producing more pathogenic conversion units.

Various modifications are increasing the sensitivity of the current methods for detecting PrP^{Sc}. Improvements include using recombinant prion protein (expressed by bacteria) and replacing sonication with shaking to give the quaking-induced conversion (QuIC) assay. This allows detection of subfemtogram amounts of PrP^{Sc} within 24 hours. Another subtle but effective improvement is based on the principle of real-time PCR, which uses fluorescence for increased speed and sensitivity (see Chapter 4). A similar real-time version of the quaking-induced conversion assay has been developed using the dye thioflavin T, which fluoresces upon binding to aggregated prion protein. A novel method to detect prions using nanopores is in development (see Box 21.1).

Detection of prions is technically difficult. Cyclic amplification of prions has greatly increased the sensitivity of detection.

APPROACHES TO TREATING PRION DISEASE

At present, there is no effective treatment for any of the prion diseases, although a variety of agents are being tested. Relatively few drugs cross the blood–brain barrier effectively. Nonetheless, random screening of those known to do so revealed that both quinacrine and chlorpromazine eliminate prions from infected animal brain cells in culture (Fig. 21.19). (Quinacrine is a rarely used antimalarial drug, and chlorpromazine is

Box 21.1 Nanopore Detection of Single Prion Protein Molecules

Nanopores may be used to capture and identify single protein molecules. (For background information on nanotechnology, see Chapter 7.) Both glass and protein nanopores have been shown to detect the prion protein. The nanopore is immersed in a conducting solution, and the current through the nanopore is monitored. When large molecules enter the nanopore, they partially block it, and consequently, there is a drop in the ionic current through the nanopore.

In the experiment featured here, a natural protein, α-hemolysin, is used as the nanopore. It is inserted in a lipid bilayer, as shown in Figure A. The prion protein is attracted into the nanopore by a transmembrane voltage that attracts the positively charged N-terminus of the prion protein. Both the normal and pathogenic forms of prion protein can be detected. More importantly, single molecules of the two prion forms can be distinguished by their ion-current signatures.

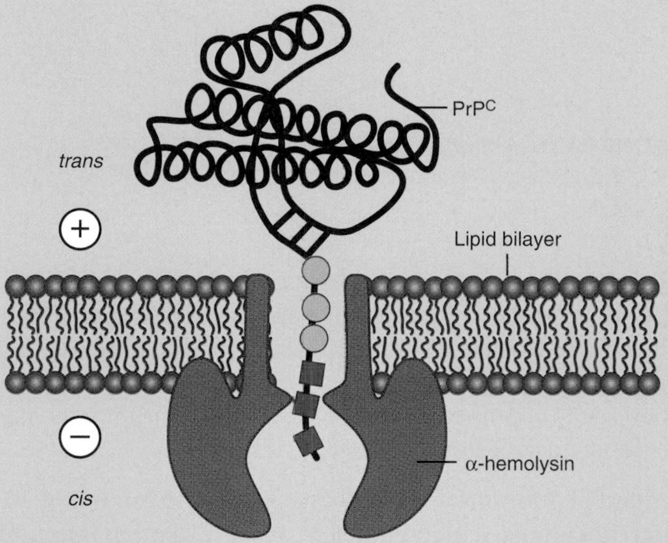

FIGURE A Nanopore Detection of Prion Protein
Capture of a single prion molecule by an alpha-hemolysin nanopore. From Jetha NN et al. (2013). Nanopore analysis of wild-type and mutant prion protein (PrP(C)): single molecule discrimination and PrP(C) kinetics. *PLoS One* **8**(2), e54982.

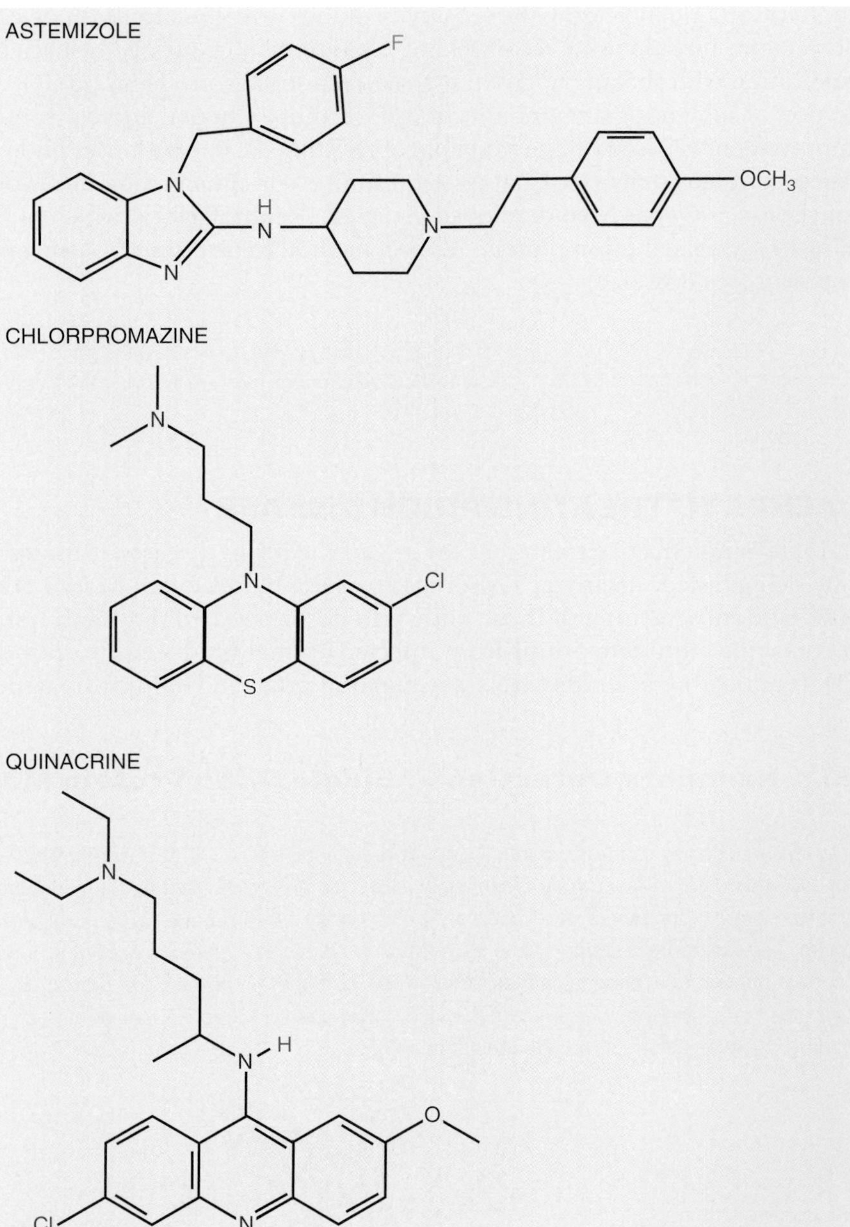

FIGURE 21.19 Antiprion Agents
Structures of chlorpromazine, quinacrine, and astemizole.

widely used to treat schizophrenia.) Quinacrine binds to PrP^{c} and probably helps to stabilize it against conformational change induced by PrP^{Sc}. Unfortunately, these compounds do not cure the disease in whole animals. A variety of structurally similar compounds are being screened. Some have antiprion activity, but none has yet progressed to clinical testing. One strong candidate is Astemizole, which has been previously used as an antihistamine.

Removal of prions from infective material is also important. In particular, blood transfusion using contaminated blood has been a source of prion infection. Filters have recently been developed that remove prions. Combinatorial libraries (see Chapter 11) were screened for ligands that bound prion protein. The ligands were then attached to resins and placed in columns for filtration of blood or other liquids that might contain active prions. When scrapie-infected hamster blood was filtered in this manner and then injected into hamsters, prion infection was prevented.

RNA interference (see Chapter 5) is widely used to suppress gene expression in laboratory studies. It is possible to generate siRNA that will suppress expression of the *Prnp* gene in mice. To generate the siRNA in prion-infected cells, scientists used a retrovirus vector that expresses short hairpin RNAs. These are processed in the target cells by Dicer to give the siRNA. This in turn triggers RNA interference directed against *Prnp* mRNA, which is degraded. Retroviral vectors were chosen because they can infect nongrowing cells, such as those of the nervous system. At least in mice, intracranial injection of the vector-siRNA construct reduced prion levels and prolonged survival.

Knocking out the prion gene in livestock is another approach to eliminating prion disease. Transgenic mice lacking both copies of the *Prnp* gene were engineered several years ago. They grow and develop normally; however, they are unable to make prion protein and are resistant to infection by pathogenic prions. This confirmed that the host cell is responsible for making new prion proteins. During infectious prion disease, these proteins change conformation. Although the prion gene is not needed for survival and its role is still unclear, it does appear to be involved in long-term memory and spatial learning.

Recently, cattle lacking both copies of the *Prnp* gene have been engineered and after 2 years are normal in growth and development. Brain cells from such animals are resistant to prion infection. *Prnp* knockout livestock could be used to provide prion-free products, if transgenic animals are approved as a source of human food.

There is presently no treatment for prion disease, although several lead compounds have been found with partial activity in cell culture.

PRIONS IN YEAST

For a long time, the strange behavior of the mammalian prion protein was thought to be unique. However, other proteins whose misfolded forms catalyze conversion of the normal protein into the misfolded version exist. Furthermore, the misfolded versions form insoluble amyloid aggregates. The first of these to be discovered were the **yeast prions**. More recently, it seems that prion-like behavior may be involved in certain neurological conditions, such as Alzheimer's, Huntington's, and Parkinson's diseases (see later discussion).

Yeast prions were discovered to be the cause behind some weird genetic behavior. The yeast prions are nonlethal and are soluble cytoplasmic proteins rather than membrane proteins. Nonetheless, yeast prions show nucleic acid-free inheritance, and their misfolded forms catalyze conversion of the normal protein into the misfolded version. Furthermore, the misfolded versions of yeast prions form insoluble amyloid aggregates, like those of mammals. However, although they display similar structural domains, the mammalian and yeast prion proteins show no sequence homology.

The two best-known yeast prions are [URE3] and [PSI+], which are the misfolded forms of the Ure2p and Sup35p proteins. Ure2p takes part in nitrogen regulation, and Sup35p is a translation termination factor. Yeast with the prion version of Sup35p show different colony morphology and can be monitored by using a variety of reporter constructs (Fig. 21.20). Surveys of yeast strains have revealed quite a few more prions, several with unknown roles. Around a third of several hundred wild strains of *Saccharomyces* contain prions.

As with mammalian prions, a given yeast prion may exist as a variety of strains with slightly different structures. Although yeast prions are generally nonlethal, some variants of the [PSI+] prion are severely detrimental or even lethal. Even the relatively harmless common forms of the well-known yeast prions convey a small growth disadvantage under most natural conditions. However, yeasts carrying prions have an advantage under specific conditions. The [URE3] prion allows yeast to better use poor nitrogen sources, while the [PSI+] prion acts as a nonsense suppressor and promotes better growth in strains with nonsense mutations.

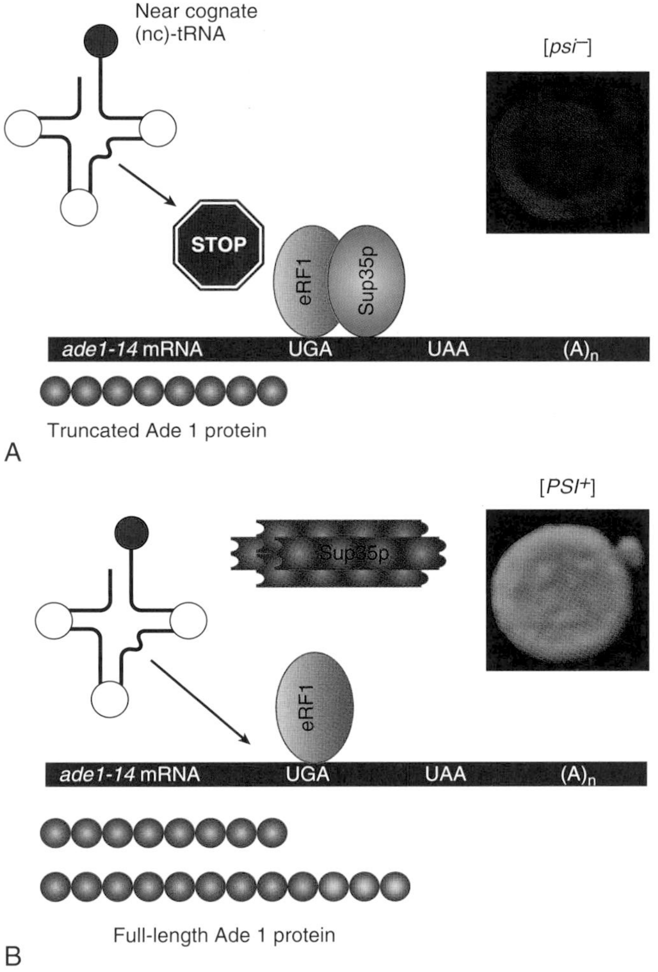

FIGURE 21.20 Yeast Prion Phenotype

The [PSI^+] prion of *S. cerevisiae* is an alternative form of the Sup35 protein that causes defective termination of translation. This may be monitored by using the *ade1–14* allele as reporter gene. This allele contains a premature UGA stop codon that blocks adenine synthesis and causes accumulation of a red pigment. (A) In prion-free [psi^-] cells, the Sup35 protein binds to eRF1, termination occurs, and red colonies are observed. (B) In [PSI^+] cells, most of the Sup35 takes part in the prion aggregates. Defective termination allows near-cognate tRNAs to translate the UGA codon. Consequently, functional *ADE1* gene product is made, and white colonies are observed. Modified from Tuite MF (2013). The natural history of yeast prions. *Adv Appl Microbiol* **84**, 85–137.

USING YEAST PRIONS AS MODELS

A new approach to screening compounds for use in prion therapy is based on the use of yeast prions. As mentioned previously, Sup35p is a translation termination factor. Its prion form, [PSI+], is insoluble and inactive. Conversion of Sup35p to [PSI+] causes increased read-through of stop codons. This forms the basis for a clever and quick genetic screening system for possible antiprion drugs.

Yeast mutants defective in adenine biosynthesis turn red because of accumulation of metabolic by-products. A yeast mutant with a nonsense mutation (i.e., a premature stop codon) in the ADE1 gene thus forms red colonies. If the Sup35p protein is in its prion form, read-through of stop codons occurs, and enough full-length protein is made to allow adenine synthesis; that is, the mutation is suppressed. Consequently, prion-positive strains form white colonies. If the prions are lost, the yeast goes back to forming red colonies. This allows rapid color-based screening of chemical compounds simply by adding them to the medium and looking for those causing a white-to-red color shift of the yeast colonies. Although a variety of compounds have been found that block prion replication, none have yet been clinically effective.

Prions have been discovered in yeast. This has allowed systematic screening for antiprion agents.

AMYLOID PROTEINS IN NEUROLOGICAL DISEASES

Several degenerative diseases of the nervous system are characterized by the buildup of insoluble amyloid protein aggregates in cells of the brain. They include Alzheimer's, Huntington's, and Parkinson's diseases. There is also some evidence that amyloid fibrils may be involved in non-neurological conditions, including atherosclerosis and arthritis.

The amyloid-beta protein of Alzheimer's and the alpha-synuclein protein of Parkinson's disease both demonstrate prion-like behavior. Both form insoluble aggregates that can self-propagate due to seeding by misfolded protein. In both cases, mutations are known that increase the formation of amyloid. Although most cases of Parkinson's disease are sporadic, around 10% of cases are inherited. Both point mutations and duplications in the gene for alpha-synuclein are associated with the familial forms of Parkinson's disease.

At least in animal models, both amyloid-beta and alpha-synuclein can be transmitted between cells. However, there is no evidence that either Alzheimer's or Parkinson's diseases can be transmitted from person to person.

Summary

Viruses rely on a host cell for growth and replication. As a consequence, it is often difficult to stop viral replication without damaging the host. The recent increased development of antiviral agents has largely been driven by the AIDS epidemic and, more recently, by the threat of pandemic flu. Agents have been developed that target most of the important steps in the virus life cycle. Recently, RNA interference has been used against viral infections, first in plants and then in people.

Prions are infectious proteins that cause neurodegenerative diseases in mammals. Fungi, including yeasts, also have prions, but they act as regulatory mechanisms for adjusting to changes of environment. The amyloid state that is typical of the aggregated pathogenic form of the prion protein is also found in proteins involved in other neurodegenerative diseases. So far, no effective therapy has been found to combat prion infection or other amyloid-related conditions.

End-of-Chapter Questions

1. Which of the following is a possible antiviral agent?
   - **a.** interferon β
   - **b.** amantadine
   - **c.** nucleoside analogs
   - **d.** protease inhibitors
   - **e.** all of the above
2. All of the following are potential outcomes of the presence of dsRNA inside a human cell except
   - **a.** activation of NK cells
   - **b.** interferon production
   - **c.** blocking of protein synthesis by P1 kinase
   - **d.** activation of an endonuclease
   - **e.** increase ATP supply
3. What prompts interferon β to be secreted?
   - **a.** ssDNA
   - **b.** dsRNA
   - **c.** ssRNA
   - **d.** dsDNA
   - **e.** mtDNA
4. Which of the following is not a component of influenza virus?
   - **a.** nucleocapsid
   - **b.** neuraminidase
   - **c.** negative ssRNA
   - **d.** outer envelope
   - **e.** caspase
5. Which influenza group causes the most severe outbreaks?
   - **a.** influenza A
   - **b.** avian influenza
   - **c.** influenza B
   - **d.** influenza AB
   - **e.** all of the above

(Continued)

6. What is the mode of action for the drug Tamiflu?
   a. hemagglutinin protease
   b. ion channel blocker
   c. neuraminidase inhibitor
   d. reverse transcriptase blocker
   e. none of the above
7. Which HIV protein binds to CD4?
   a. ganglioside GM_1
   b. neuraminidase
   c. hemagglutinin
   d. gp120
   e. chemokines
8. A mutation in which gene is responsible for natural resistance to HIV infection?
   a. CD4
   b. CCR6
   c. CCR5
   d. CXCR4
   e. gp120
9. What is the main problem with treatment of HIV/AIDS?
   a. availability of antiviral drugs
   b. the high mutation rate of the virus
   c. socioeconomic issues within populations
   d. adequate testing facilities for the disease
   e. adequate education about HIV/AIDS prevention

10. What is the mode of action for AZT?
    a. a DNA chain terminator
    b. reverse transcriptase inhibitor
    c. gp120 analog
    d. CCR5 analog
    e. none of the above
11. What is the favored method for HIV/AIDS therapy?
    a. treatment with a two-drug cocktail
    b. treatment with a three-drug cocktail
    c. treatment with a reverse transcriptase inhibitor only
    d. treatment with a DNA chain terminator
    e. treatment with a protease inhibitor only
12. How do prions cause disease?
    a. Prion proteins bind to cells and induce apoptosis.
    b. Prion proteins bind to DNA polymerase and prevent replication.
    c. Prion proteins induce normal cellular proteins to refold into the prion form.
    d. Prion proteins induce an immune response against the “self.”
    e. none of the above
13. Which of the following diseases is not caused by a prion?
    a. kuru
    b. BSE
    c. scrapie

d. HIV
e. CJD

14. Which method is used to identify prion infections?
a. PMCA
b. PCR
c. RT-PCR
d. brain biopsy
e. none of the above

15. How might prion diseases be treated?
a. RNAi
b. *Prnp* knockouts
c. removal of prions using a filter
d. quinacrine or chlorpromazine treatment
e. all of the above

Further Reading

Berkhout, B., Eggink, D., & Sanders, R. W. (2012). Is there a future for antiviral fusion inhibitors? *Current Opinion in Virology, 2*, 50–59.

Biasini, E., Turnbaugh, J. A., Unterberger, U., & Harris, D. A. (2012). Prion protein at the crossroads of physiology and disease. *Trends in Neurosciences, 35*, 92–103.

De Clercq, E. (2012). Human viral diseases: what is next for antiviral drug discovery? *Current Opinion in Virology, 2*, 572–579.

DeVincenzo, J. P. (2012). The promise, pitfalls and progress of RNA-interference-based antiviral therapy for respiratory viruses. *Antiviral Therapy, 17*, 213–225.

Eisenberg, D., & Jucker, M. (2012). The amyloid state of proteins in human diseases. *Cell, 148*, 1188–1203.

Gao, S., von der Malsburg, A., Dick, A., Faelber, K., Schröder, G. F., Haller, O., Kochs, G., & Daumke, O. (2011). Structure of myxovirus resistance protein a reveals intra- and intermolecular domain interactions required for the antiviral function. *Immunity, 2011*(35), 514–525.

Halfmann, R., Jarosz, D. F., Jones, S. K., Chang, A., Lancaster, A. K., & Lindquist, S. (2012). Prions are a common mechanism for phenotypic inheritance in wild yeasts. *Nature, 482*, 363–368.

Ison, M. G. (2011). Antivirals and resistance: influenza virus. *Current Opinion in Virology, 1*, 563–573.

Karapetyan, Y. E., Sferrazza, G. F., Zhou, M., Ottenberg, G., Spicer, T., et al. (2013). Unique drug screening approach for prion diseases identifies tacrolimus and astemizole as antiprion agents. *Proceedings of the National Academy of Sciences of the United States of America, 110*, 7044–7049.

McGlinchey, R. P., Kryndushkin, D., & Wickner, R. B. (2011). Suicidal [PSI+] is a lethal yeast prion. *Proceedings of the National Academy of Sciences of the United States of America, 108*, 5337–5341.

Polymenidou, M., & Cleveland, D. W. (2012). Prion-like spread of protein aggregates in neurodegeneration. *Journal of Experimental Medicine, 209*, 889–893.

Sano, K., Satoh, K., Atarashi, R., Takashima, H., Iwasaki, Y., Yoshida, M., et al. (2013). Early detection of abnormal prion protein in genetic human prion diseases now possible using real-time QUIC assay. *PLoS One, 8*(1), e54915.

Sierra-Aragón, S., & Walter, H. (2012). Targets for inhibition of HIV replication: entry, enzyme action, release and maturation. *Intervirology, 55*, 84–97.

Verhelst, J., Parthoens, E., Schepens, B., Fiers, W., & Saelens, X. (2012). Interferon-inducible protein Mx1 inhibits influenza virus by interfering with functional viral ribonucleoprotein complex assembly. *Journal of Virology, 86*, 13445–13455.

Zhou, Y., Yuan, Y., Yuan, F., Wang, M., Zhong, H., Gu, M., & Liang, G. (2012). RNAi-directed down-regulation of RSV results in increased resistance in rice (*Oryza sativa* L.). *Biotechnology Letters, 34*, 965–972.

Biological Warfare: Infectious Disease and Bioterrorism

Biotechnology

http://dx.doi.org/10.1016/B978-0-12-385015-7.00022-3

INTRODUCTION

The term *biological warfare* typically conjures images of medieval warriors tossing dead cattle over city walls or clandestine government agents secretly releasing mysterious microbes into enemy territory. Of course, biological warfare does encompass such activity, but the vast majority of what constitutes biological warfare is far more mundane. Ever since life evolved on Earth about 3.8 billion years ago, organisms have constantly devised new ways to kill each other. Any organism that makes use of toxins—from bacteria to snakes—is engaging in a form of biological warfare. Humans who engage in biological warfare do so by taking advantage of these toxin-producing organisms.

THE NATURAL HISTORY OF BIOLOGICAL WARFARE

An entire textbook could be filled with examples of organisms that employ toxins to kill other organisms. We therefore touch only briefly on the natural history of biological warfare.

Bacteria are particularly adept at biological warfare. While humanity finds antibiotics incredibly useful in our battle against infectious disease, bacteria did not create them for our benefit. Instead, they make antibiotics to kill off other bacteria that are competing for the same habitat or resources. Similarly, bacteria synthesize toxic proteins known as **bacteriocins** to kill their relatives because closely related strains of bacteria are likelier to compete with each other. For example, many strains of *Escherichia coli* deploy a wide variety of bacteriocins (referred to as colicins) intended to kill other strains of *E. coli*. The genes for colicins are normally carried on plasmids, and many of these plasmids are commonly used in molecular biology and genetic engineering (see Chapter 3). *Yersinia pestis,* the plague bacterium, also makes bacteriocins (called pesticins in this case) designed to kill competing strains of its own species (Fig. 22.1).

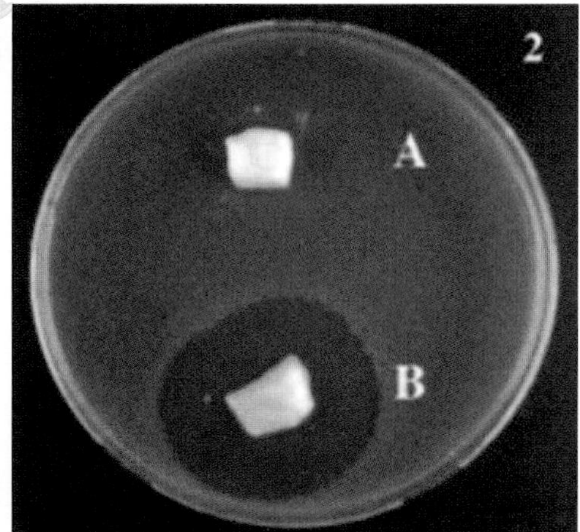

FIGURE 22.1 Bacteriocins Inhibit Other Bacteria

A bacteriocin-producing strain of *Lactococcus* in a piece of cheese can inhibit the growth of a related microorganism. From Garde S, et al. (2011). Outgrowth inhibition of *Clostridium beijerinkii* spores by a bacteriocin-producing lactic culture in ovine milk cheese. *Int. J Food Microbiol* **150**, 59–65.

A point of clarification: The distinction between *bacteriocin* and *toxin* has to do with the target. Bacteria deploy bacteriocins against their fellow—often closely related—bacteria with the deliberate intention of killing them. In contrast, proteins produced by bacteria that act against higher organisms are referred to as **toxins**. Perhaps counterintuitively, pathogenic bacteria do not usually "intend" to kill the organisms they infect. Rather, they want to manipulate them long enough to survive and reproduce. The longer the host stays alive, the longer it provides a home for the infecting bacteria. Just like antibiotics, some bacterial toxins are useful to humans. The bacterium *Bacillus thuringiensis* produces an insect-killing toxin that is harmless to vertebrates, and this "Bt toxin" has been used extensively in genetically modified crops. (See Chapter 15.)

Lower eukaryotes also regularly engage in biological warfare. *Paramecium,* a ciliated protozoan, carries symbiotic bacteria *(Caedibacter)* known as **kappa particles** that grow and divide inside the larger eukaryotic cell (Fig. 22.2).

Strains of *Paramecium* with kappa particles are known as killers and, due to unknown genetic factors and resistance mechanisms, are naturally tolerant of them. Killer strains release kappa particles into the environment, and if a sensitive *Paramecium* (i.e., one lacking the ability to harbor kappa particles) eats and digests just a single kappa particle, a protein toxin is released and kills the *Paramecium.* Interestingly, the toxin is not encoded by a gene on the bacterial chromosome, but on a plasmid derived from a defective bacteriophage. So a toxin encoded by a virus infecting the kappa particle bacterium has been commandeered for the purpose of killing other strains of *Paramecium.*

This phenomenon is not at all unusual. Many toxins used by pathogenic bacteria that infect humans are actually encoded by foreign DNA of nonchromosomal origin, such as viruses, plasmids, or transposons. These elements are often integrated into the chromosome of pathogenic strains of bacteria. For example, the only strains of *Corynebacterium diphtheriae*—the causative agent of diphtheria—that are dangerous to humans are the ones that carry a toxin-encoding virus.

Higher eukaryotes can either create their own toxins—such as the venom produced by snakes and scorpions—or expropriate toxins produced by other species. One species of caterpillar that feeds on tobacco plants can exhale noxious nicotine at spiders, chasing them away. Other insects rely on microbes to wage biological warfare. Certain parasitic wasps inject their eggs into the maggots (i.e., larvae) of plant-eating insects. After the eggs hatch, the newborn wasps eat the living maggots from the inside (Fig. 22.3).

The maggots are eventually killed, and a new generation of wasps is released. The secret to the wasp's success is the injection of an adenovirus along with the eggs. The virus targets the maggot's "fat body" (vaguely equivalent to the liver of higher animals) and cripples the maggot's developmental control system and immune system. The maggot loses its appetite for plants and is prevented from molting and turning into a pupa, the next stage in its life cycle.

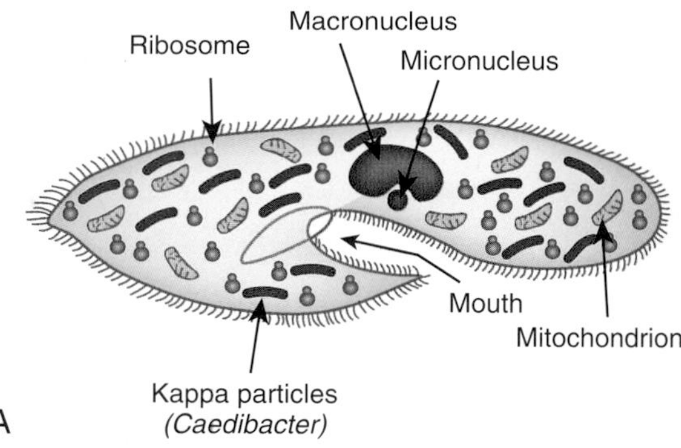

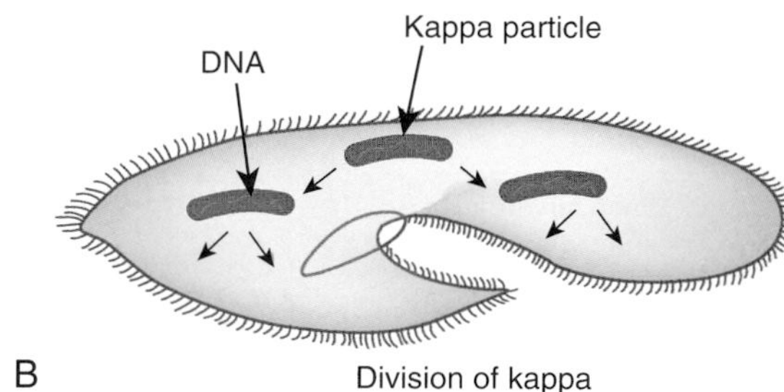

FIGURE 22.2 Killer *Paramecium* Uses a Bacterial Toxin
(A) The kappa particles are found in the cytoplasm of the *Paramecium*. (B) Kappa particles are symbiotic *Caedibacter* that are found in many strains of *Paramecium*, yet they have their own DNA and divide like typical bacteria.

> Many different kinds of organisms engage in biological warfare. Bacteria kill other bacteria with antibiotics or bacteriocins. They also make toxins that are targeted at higher organisms. Eukaryotes can either make their own toxins or commandeer those produced by lower organisms.

MICROBES VERSUS MAN: THE RISE OF ANTIBIOTIC RESISTANCE

Although we rarely perceive it this way, infectious disease is just another manifestation of biological warfare that is ubiquitous throughout life. The evolutionary relationship between hosts and pathogens is essentially a never-ending arms race. When a pathogen evolves a new toxin, the host evolves a response to it. Humanity has taken this arms race one step further by utilizing technology such as vaccines and industrial-scale manufacturing of antibiotics. However, the microbes are fighting back.

Perhaps the biggest problem plaguing medical microbiology today is the rise of antibiotic resistance. There are many reasons why bacteria have developed this resistance, but all of the explanations have one thing in common: the proliferation and misuse of antibiotics. For instance, medical doctors often prescribe antibiotics to patients who have an infection, even if it is unknown whether the disease is bacterial. Other times, the wrong antibiotic is prescribed. In many developing countries, antibiotics can be bought over the counter without a prescription. Compounding the dilemma, patients who receive antibiotics often do not comply with the recommended dose, ending treatment as soon as they feel better. This has the effect of selecting for the survival of the bacteria that have already developed a slight resistance to the drug. When the patient propagates the infection, he unintentionally passes on these toughened survivors. The widespread use of antibiotics in animal feed—which farmers use to fatten up livestock—is also a major contributor to the problem.

Today, many experts worry about "incurable" infections. Methicillin-resistant *Staphylococcus aureus* (MRSA) gets a lot of media attention, but it is not the only worrisome microbe.

There have been reports from around the world of totally drug-resistant tuberculosis, which as the name implies, appears to be resistant to all treatment. In a 2013 report, the Centers for Disease Control and Prevention (CDC) issued an urgent warning about infections from (1) *Clostridium difficile*, which causes diarrhea and is often acquired by patients in health-care settings who were treated with antibiotics for other infections; (2) Carbapenem-resistant Enterobacteriaceae (CRE), such as *Klebsiella* and *E. coli*, which also cause health-care-associated infections and may be resistant to all known antibiotics; and (3) *Neisseria gonorrhoeae*, the etiologic agent of gonorrhea, which is growing in resistance to several antibiotics.

While these developments are alarming, much research is being done to combat the rise of antibiotic resistance. Although microbes have responded to our antibiotic assault, we are developing some new weapons to regain the upper hand.

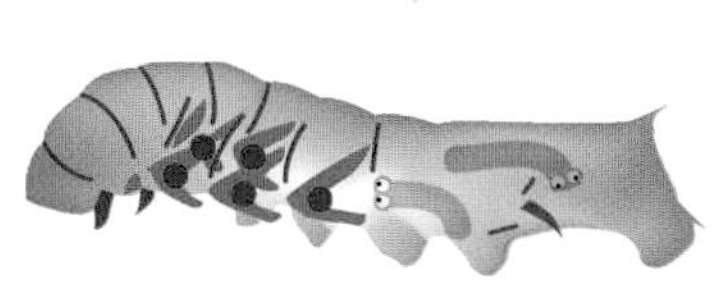

FIGURE 22.3 Wasps Use Viruses against Maggots
Certain types of wasps lay their eggs inside tobacco hornworm larvae. The wasp lands on the back of the larva and injects the eggs plus adenovirus into the maggot through the ovipositor. The adenovirus prevents the larva from eating and therefore developing into a pupa. When the eggs hatch, the young use the insides of the larva as a food source, to grow and develop into adult wasps.

Novel Targets for Antibiotics

Although there has been speculation of an inevitable "post-antibiotic era," there are still plenty of opportunities for the development of novel antibiotics.

One strategy is to attack previously unexploited vulnerable spots in a bacterium's metabolism or life cycle, preferably those that bacteria cannot easily defend by acquiring resistance. For instance, bacteria use iron chelators, known as **siderophores**, to bind iron and extract it from host proteins. Siderophores are excreted, bind iron, and are then taken back into bacteria by specialized transport systems. Absence of high-potency siderophores largely abolishes virulence in both plague and tuberculosis. Because mammals do not make siderophores, their unique biosynthetic pathways provide an attractive target for development of novel antibiotics. Yersiniabactin, the siderophore of several pathogenic *Yersinia* species, is capped by a salicyl group (Fig. 22.4).

The intermediate in the pathway, produced when ATP activates salicylate, is salicyl-AMP. A chemically synthesized analog of salicyl-AMP, called salicyl-AMS, replaces the phosphate with a sulfamoyl group. The compound is highly active and specifically inhibits siderophore synthesis. This prevents the growth of *Yersinia* under iron-limiting conditions, such as encountered in the human body.

Another strategy is to screen novel microbes for antibiotics. As discussed earlier, bacteria produce antibiotics for the explicit purpose of killing other bacteria. Since most microbes that exist in nature have neither been cultured nor identified, it is likely that many natural antibiotics have yet to be discovered. In 2013, a new antibiotic, called anthracimycin, was isolated from an Actinomycete that lives in the ocean. The new antibiotic is active against *Bacillus anthracis* and MRSA, and modifying it with chlorine groups expanded its spectrum of activity.

Yet another strategy is to identify and clone potential antimicrobial biosynthetic pathways. For example, based on its DNA sequence, one research group cloned a biosynthetic gene cluster from an Actinomycete called *Saccharomonospora* that was predicted to produce an antimicrobial lipopeptide. Expressing the gene cluster resulted in the discovery of a new antibiotic, taromycin A. The major advantage of this technique is that it can be applied to microbes that are difficult to culture in the laboratory.

A different approach is to disrupt existing antibiotic resistance, rather than developing new antibiotics. For example bacteriophage, such as those that live in the human gut, can shuttle antibiotic resistance genes between bacteria. Consequently, developing drugs that kill or disable bacteriophage is an innovative way to combat the spread of antibiotic resistance.

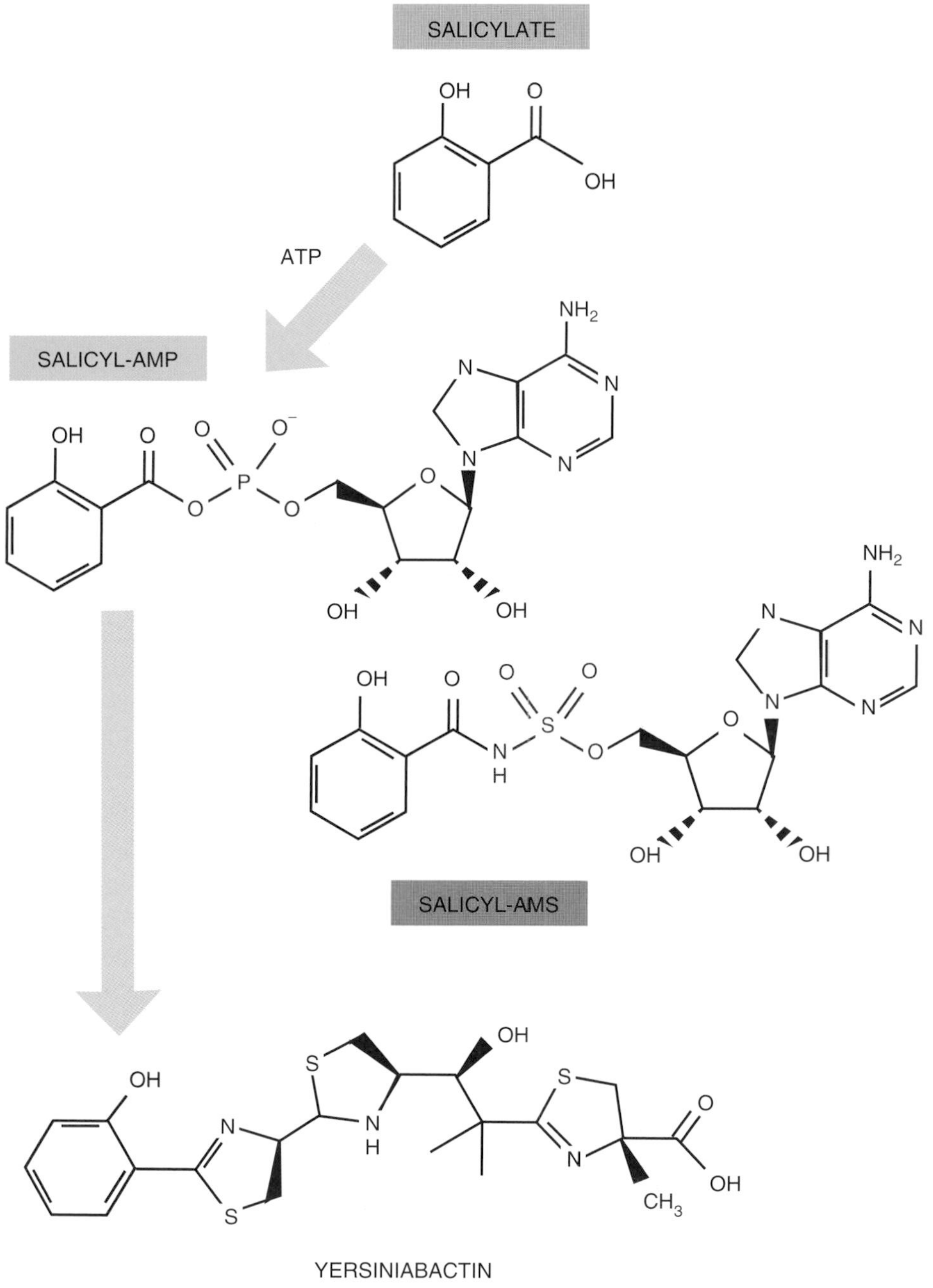

FIGURE 22.4
Salicyl-AMS Inhibits the Production of Yersiniabactin
The structure of yersiniabactin shows the salicyl group in red. The precursor, salicyl-AMP, is made by activating salicylate with ATP. The sulfamoyl analog, salicyl-AMS, inhibits the incorporation of the salicyl group into yersiniabactin.

Additionally, disrupting bacterial **quorum sensing** has been suggested. Bacteria use quorum sensing as a communication system in order to coordinate behavior (Fig. 22.5).

By releasing particular chemical compounds into the environment, bacteria can detect when a threshold population density, or "quorum," has been reached. Many pathogens construct antibiotic-resistant biofilms after the population has reached a particular density. Disrupting their communication system would cripple their ability to coordinate behavior and keep the bacteria more vulnerable to antibiotics.

Phage Therapy and Bacterial Predators

The history of **phage therapy**—that is, using bacteriophage (also called "phage") to treat bacterial infections—begins in France in 1921. That year, microbiologist Felix d'Hérelle used phage to treat patients suffering from dysentery (Fig. 22.6).

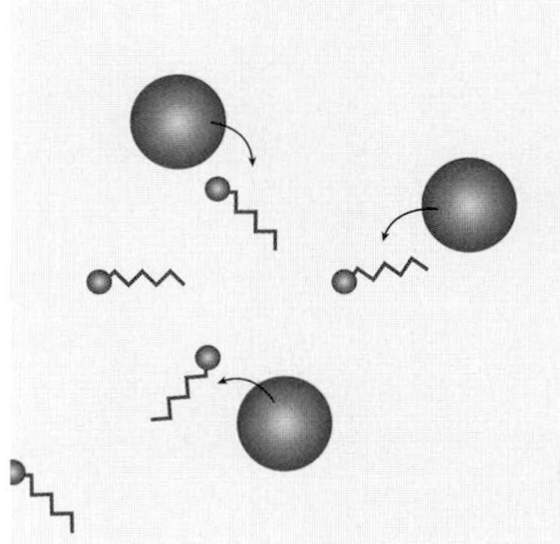

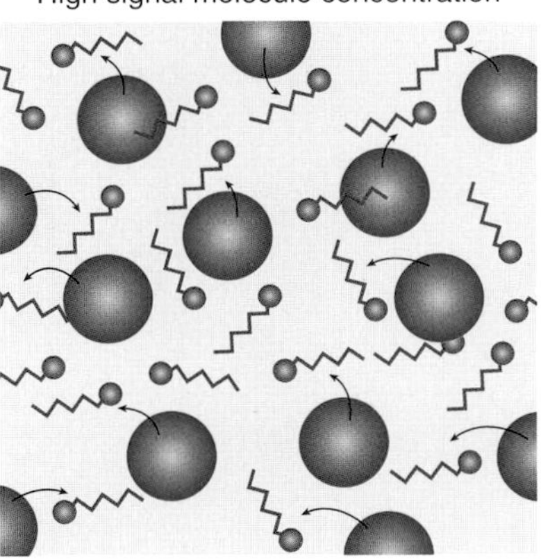

FIGURE 22.5
Quorum Sensing
Bacteria can coordinate behavior by detecting the presence of a signal molecule that indicates the density of the population. From Boyen F, et al. (2009). Quorum sensing in veterinary pathogens: mechanisms, clinical importance and future perspectives. *Vet. Microbiol* **135**, 187–195.

FIGURE 22.6
Felix d'Hérelle
Microbiologist Felix d'Hérelle helped pioneer phage therapy.

In 1927, he also used phage therapy to treat cholera victims in south Asia. Unfortunately, many other scientists in the United States and elsewhere were unable to replicate his work, and when the widespread production of antibiotics started in 1945, the scientific community mostly lost interest in phage therapy. The French, however, enthusiastically practiced phage therapy into the 1990s and, during those seven decades, there were reports of successful treatment of typhoid fever, colitis, septicemia, skin infections, and various other bacterial diseases. Other countries that embraced phage therapy include Poland, Russia, and Georgia. Today, patients there can receive phage therapy for chronic and antibiotic-resistant bacterial infections.

Since the 1990s, the Western scientific community has renewed its interest in phage therapy. One benefit of using phage, as opposed to antibiotics, is their specificity. Antibiotics kill many different types of bacteria—which is harmful if they destroy helpful gut bacteria—but individual phage species infect only a group of very closely related bacteria. Every bacterial infection could, in theory, be targeted by a highly specific phage.

As predicted, however, bacteria also can develop resistance to phage, mainly through thwarting viral attachment. Now, researchers are investigating the use of **lysins**, a class of toxins that phage use to dismantle bacterial cell walls as part of their lytic cycle (Fig. 22.7). Because lysins target conserved regions within peptidoglycan, it is believed that bacteria will be less able to develop resistance. Lysins work best against Gram-positive bacteria, but genetic engineering can expand the spectrum of activity to include Gram-negative bacteria also.

As an alternative to phage, it may be possible to deploy predatory bacteria against human pathogens. *Bdellovibrio*, which invades other bacteria rather like a virus, and *Micavibrio*, which attaches to bacterial cell surfaces, have been shown to kill antibiotic-resistant pathogenic bacteria *in vitro*.

Fighting Pathogens with Genetic Engineering

Because of a persistent fear that we will run out of novel antibiotics, many clever new technologies have been suggested to fight bacterial infections. Some of the most promising of these antibiotics utilize genetic engineering.

For example, many pathogenic *Escherichia coli* use the FimH adhesin to bind to mammalian cells via mannose residues on surface glycoproteins. Several alkyl- and aryl-mannose derivatives bind with extremely high affinity to the adhesin and block its attachment to the natural receptor. Such mannose derivatives, therefore, could serve as anti-adhesin drugs. However, manufacturing pharmaceuticals is quite expensive. It would be far cheaper to genetically engineer nonpathogenic strains of *E. coli* to express the mannose derivatives on their cell surfaces. Pathogenic bacteria would then bind to these decoys instead of to mammalian cells. This would also avoid the need for continuous administration of sugar derivatives because the decoy strains of *E. coli* would multiply naturally in the intestine. Alternatively, nonpathogenic strains of *E. coli* could be engineered with genes for adhesins that would allow them to compete with pathogens for mammalian cell receptors. (Such engineered strains would also have the advantage of being able to deliver protein pharmaceuticals or large segments of DNA for gene therapy into mammalian cells.)

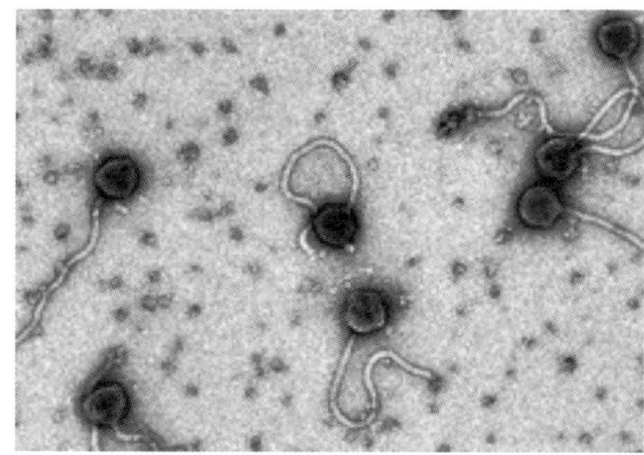

FIGURE 22.7
Bacteriophage Tsamsa Kills *Bacillus anthracis*
The lysin isolated from the bacteriophage Tsamsa kills *Bacillus anthracis* and other closely related species. From Ganz HH, et al. (2014). Novel giant Siphovirus from *Bacillus anthracis* features unusual genome characteristics. *PLoS One* **9**(1), e85972.

A different approach is to generate altered toxins that interfere with their natural analogs. Typical A-B bacterial toxins are made from a single "active" A subunit, which carries out a toxic enzymatic reaction inside a target cell, and often several "binding" B subunits, which serve as a delivery system by attaching to the cell surface. Because several properly functioning binding subunits are required to deliver the active subunit, one approach to antitoxin therapy relies on utilizing **dominant-negative mutations** in the binding subunit of the toxin. The mechanism involves the binding of a defective protein subunit to functional subunits resulting in a complex that is inactive overall. (The term *dominant-negative* refers to mutations in which an abnormal gene product sabotages the activity of the wild-type gene product. Consequently, most dominant-negative mutations affect proteins with multiple subunits.) Dominant-negative mutations have been deliberately isolated in the B protein (called the "protective antigen") of anthrax toxin. Mixing mutant subunits with wild-type ones resulted in the assembly of inactive heptamers that bind the A subunits (called "lethal factor" and "edema factor") of anthrax toxin. As a result, the toxic A subunits cannot be transported into target cells (Fig. 22.8). This technique has been shown to protect both cultured human cells and whole mice or rats from death by lethal levels of anthrax toxin.

Fighting Pathogens with Nanotechnology

Many of the advances in nanotechnology aimed at fighting pathogens involve the creation of bactericidal surfaces (see Chapter 7 for more on nanotechnology). Several metals are inherently antibacterial. For instance, silver ions kill bacteria through several mechanisms, such as generating reactive oxygen species and disrupting protein disulfide bonds. Surfaces coated with silver, selenium, and copper nanoparticles all show antimicrobial activity.

Metals are not the only option. A substance known as "black silicon" is made of tiny "nanopillars" that are able to physically destroy bacteria, including endospores, through mechanical stress (Fig. 22.9). Antimicrobial activity has also been demonstrated with stacked carbon nanotubes called nanocarpets (see Chapter 7). Additionally, polymers of esters and cyclic hydrocarbons reduce attachment of bacteria. Such discoveries could allow for improved sanitation in health-care settings and the manufacture of antimicrobial medical devices.

Antibiotic resistance is a growing concern, but contrary to popular reports, it is not necessarily an intractable problem. Novel targets for antibiotics, phage therapy, genetic engineering, and nanotechnology provide multiple possibilities for fighting antibiotic-resistant pathogens.

A BRIEF HISTORY OF HUMAN BIOLOGICAL WARFARE

Throughout history, humans have devised new and innovative ways to kill other humans. When technology was primitive, warriors used whatever nature provided. Burning crops was probably the easiest and earliest form of warfare aimed at undermining an enemy, as was poisoning a community's drinking water with dead or rotting animals.

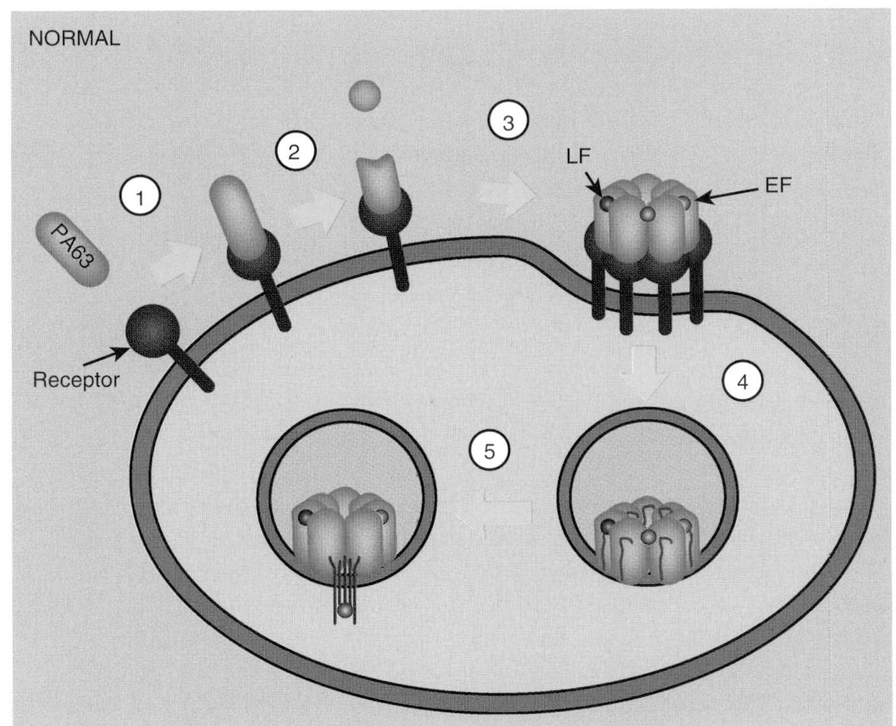

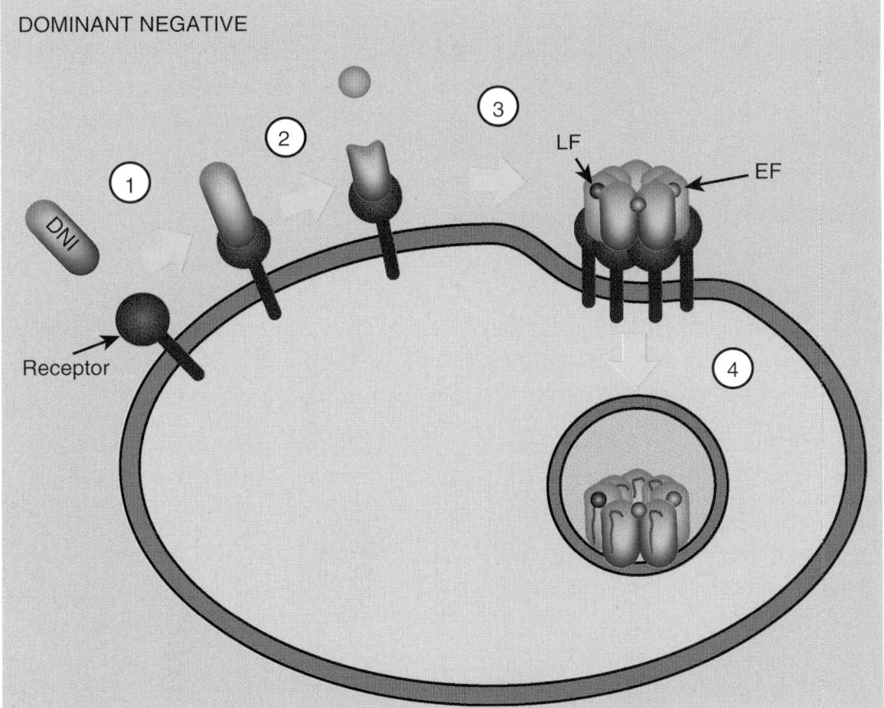

FIGURE 22.8 Dominant-Negative Mutations

For anthrax, the B subunit (called PA63 protein or "protective antigen") binds the A subunits (called lethal factor, LF, and edema factor, EF) and transports them into the target cell cytoplasm via an endocytic vesicle. The dominant-negative inhibitory (DNI) mutant of the PA63 protein (purple) assembles together with normal PA63 monomers (pink) to give an inactive complex that cannot release the LF and EF toxins from the vesicle into the cytoplasm.

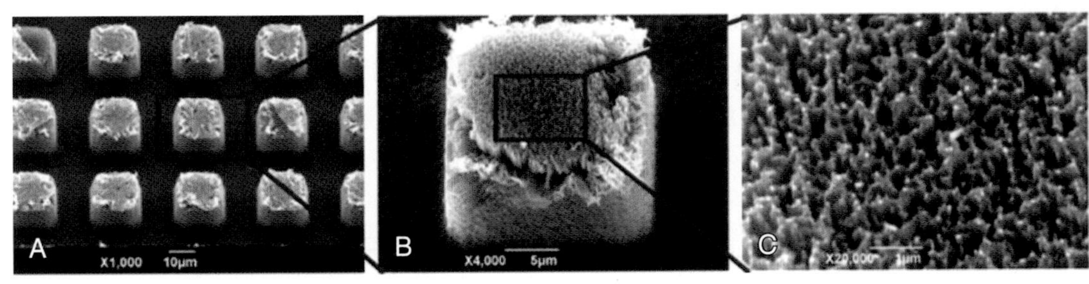

FIGURE 22.9 Nanostructures Can Kill Bacteria

Scanning electron micrograph of black silicon surface showing its hierarchical structures. (A) Periodically arranged micropillar arrays; (B) a micropillar with nanostructures; (C) nanostructures formed on the top of the micropillar. From He Y, et al. (2011). Superhydrophobic silicon surfaces with micro-nano hierarchical structures via deep reactive ion etching and galvanic etching. *J Colloid Interface Sci* **364**, 219–229.

Early Human Biological Warfare

Slightly more advanced forms of biological warfare emerged when soldiers began dipping spears in feces and throwing poisonous snakes. During the **Black Death** epidemic of the mid-1300s, the Tartars catapulted plague-ridden corpses over the walls into cities held by their European enemies. Although this is sometimes credited with spreading the plague, rats and their fleas were far more effective at spreading **bubonic plague** than contact with corpses (Fig. 22.10).

FIGURE 22.10
Bubonic Plague
This painting by Arnold Böcklin, simply titled *Plague*, depicts the fear that bubonic plague provoked in antiquity. From ET Rietschal, et al. (2004). How the mighty have fallen: fatal infectious diseases of divine composers. *Infect Dis Clin North Am* **18**, 311–339.

Given the state of hygiene in most medieval towns or castles, there was little need to provide an outside source of infection. With plague, typhoid, smallpox, dysentery, and diphtheria already around, all that was usually necessary was to let nature take its course. Similarly, a widespread myth exists that European settlers purposefully infected Native Americans with smallpox. While it is true that the British military attempted this strategy during the French and Indian War in the mid-1700s, the vast majority of Native American deaths—perhaps as much as 95% of the population—were due to inadvertent infection with smallpox and other diseases.

The truth is, until very recently, humans were not particularly hygienic. Consider, for instance, that antiseptic surgery—invented by Joseph Lister and now considered a mainstay of modern medicine—wasn't widely adopted until the late 1870s. Before then, armies and civilian populations were so dirty and disease-ridden that practicing germ warfare was like throwing mud on a pig. It is only in our modern hygienic age that biological warfare has become a more meaningful threat.

Modern Human Biological Warfare and Bioterrorism

Modern biological warfare began during World War I. Although the Germans refused to use biological agents against people, they did use them against animals, infecting Allied horses with glanders (*Burkholderia mallei*) and anthrax. The French also employed glanders against German horses. During World War II, the infamous Japanese Unit 731 experimentally infected Chinese prisoners of war with horrifying diseases, such as cholera, epidemic hemorrhagic fever, and venereal disease. It was also responsible for dropping plague-infected "flea bombs" on cities in China, although this likely had little effect partly because plague was already endemic to the region (Fig. 22.11).

After World War II, particularly during the Korean War, the United States ratcheted up its biological weapons program. Perhaps the most controversial aspect of the program was the purposeful release of biological agents, such as the relatively harmless *Serratia marcescens*, over American cities to study weapons dispersal. The military unintentionally infected 11 civilians, one of whom died. By 1969, the U.S. had weaponized anthrax and tularemia. However, in 1975, the U.S. renounced all biological weapons by signing the Biological Weapons Convention (BWC).

The Soviet Union also signed the BWC but then deceitfully enlarged its efforts. The scope of the Soviet program was astonishing. The Soviets manufactured several hundred tons of anthrax, and an accidental release in 1979 killed 66 people. The former USSR also made thousands of pounds of smallpox and plague, and in 1989, they supposedly managed to weaponize Marburg virus, which causes a deadly hemorrhagic fever similar to Ebola. These allegations remain unconfirmed. Finally, under President Boris Yeltsin in 1992, Russia ended its biological weapons program, but the fate of the weapons stockpiles remains unclear.

FIGURE 22.11
Unit 731
Japanese military Unit 731 killed thousands of Chinese people with experimental infections and biological warfare. Source: Figure 6 from: López-Muñoz F, et al. (2007). Psychiatry and political-institutional abuse from the historical perspective: the ethical lessons of the Nuremberg Trial on their 60th anniversary. *Prog Neuropsychopharmacol Biol Psychiatry* **31**, 791–780.

Today, biological warfare is feared less from nations and more from terrorist groups or "lone wolves." But there is disagreement over just how much of a threat this poses. Many believe that terrorists would be incapable of carrying out an effective, large-scale biological attack. For instance, in 1984, the Rajneesh cult gave food poisoning to about 750 citizens of a small Oregon town for political purposes by adding *Salmonella* to salad bars. Aum Shinrikyo, a Japanese cult that perpetrated a sarin gas attack in the Tokyo subway in 1995, experimented with biological weapons, but to no avail. The 2001 U.S. anthrax attack (discussed in more detail in the following section) killed only 5 people. Skeptics point to incidents like these as evidence that bioterrorists are incapable of inflicting widespread damage. Other analysts disagree (Fig. 22.12).

Some biological agents, such as anthrax, require little expertise to grow or weaponize. With microbiological information universally available on the Internet, some experts believe that it is just a matter of time before a large bioterrorist attack occurs. A small crop duster airplane loaded with anthrax and flown over a major city could potentially kill hundreds of thousands if not millions of people. Exacerbating the problem is the fact that a 2010 federal commission found the United States to be completely unprepared in the event of a bioterrorist attack.

Psychological Impact and Cost

During the Vietnam War, the Viet-Cong guerillas dug camouflaged pits as booby traps. Inside, they often positioned sharpened bamboo stakes or splinters smeared with human waste. Although it was possible to contract a nasty infection from these, the main purpose was psychological. The tactic worked. The response of American troops was to alter their movements in a way that was disproportionate to the actual threat. An analogous scenario played out following the 2001 anthrax attack in the United States in which there was a colossal disruption of postal services and massive new expenses. Yet, only 5 people died in the attack. (Compare that to the roughly 62,000 Americans who died from influenza and pneumonia that same year.)

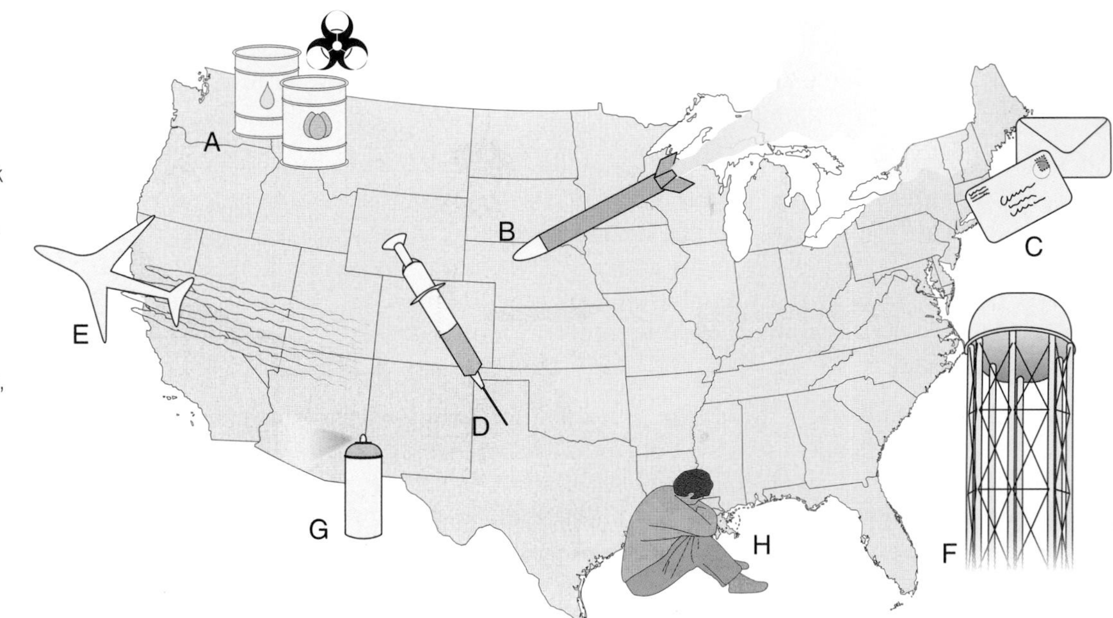

FIGURE 22.12
Bioterrorism
Some experts believe that a large-scale bioterrorist attack will occur in the not-too-distant future, but others say bioterrorism is an ineffective tactic. Attack methods include contamination of food and water supplies (A), bombs (B), using the mail (C), contamination of water (F), spraying aerosolized agents (E, G), direct injection (D), or the infiltration of "suicide infectees" (H). From Osterbauer PJ, Dobbs MR (2005). Neurobiological weapons. Neurol Clin **23**, 599–621.

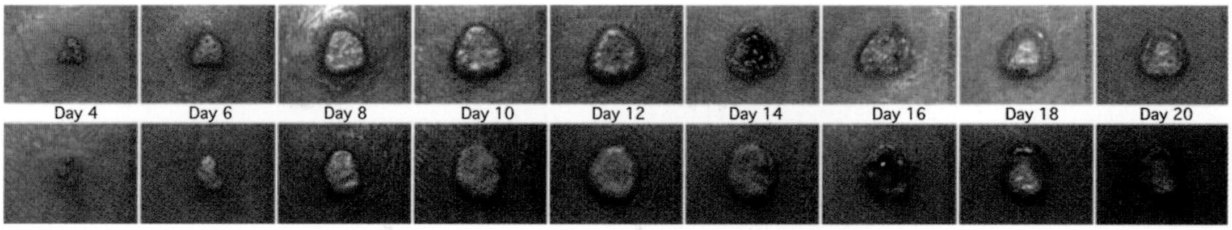

Day 4 Day 6 Day 8 Day 10 Day 12 Day 14 Day 16 Day 18 Day 20

FIGURE 22.13
Smallpox Vaccine
How the normal skin reaction to smallpox vaccination progresses in two patients. Source: Centers for Disease Control and Prevention.

Both of these examples serve to underscore two important points: First, biological warfare will almost certainly have a far greater psychological impact than direct impact; and second, protective measures against biological attacks are costly and inconvenient. For instance, giving soldiers vaccines against all possible biological agents would be impractical and possibly dangerous if they have been developed under emergency conditions without thorough testing. Also, vaccines have side effects. Consider the anthrax vaccine used by the U.S. army that was approved in 1971. Vaccination requires six inoculations plus annual boosters. It produces swelling and irritation at the site of injection in 5% to 8% and severe local reactions in about 1% of those inoculated, although major systemic reactions are rare. Although it works against "natural" exposure, it is uncertain whether it would protect against a concentrated aerosol of anthrax spores.

Or consider the smallpox vaccine (Fig. 22.13). For every 1 million people vaccinated, the CDC estimates that 1,000 people will have serious side effects, 14 to 52 people will have life-threatening side effects, and 1 or 2 people will die. Is it worth vaccinating an entire army or nation—knowing ahead of time that many will die or become sick—to protect them against an unlikely threat? From an epidemiological standpoint, the answer is clearly no, which explains why citizens do not receive smallpox vaccinations. The general rule in public health is to vaccinate only if the risk of the disease is greater than the risk of vaccination.

Even if widespread vaccination is forgone in favor of other measures, such as protective clothing or respirators, there is still the financial cost. A nation that invests heavily in bioterrorism preparedness could have spent that money in more productive ways. Dressing troops in special clothing and equipment could promote heat stress or make them easier targets for conventional weaponry. Additionally, medications taken prophylactically to prevent infectious diseases are expensive, rarely 100% effective, and may have long-term negative health consequences.

Biological warfare has been practised since ancient times. However, it has only rarely been effective. Naturally occurring infectious diseases have killed far more people. Still, bioterrorism may pose a serious threat today. Even if an attack kills relatively few people, the psychological impact could be enormous.

IDENTIFYING SUITABLE BIOLOGICAL WARFARE AGENTS

Biological warfare is used to kill, injure, and psychologically intimidate enemies. Many naturally occurring diseases are effective agents, although it might be possible to "improve" them with genetic engineering, as discussed later.

What makes for an effective biological agent? Five major factors need to be considered.

Preparation. Some pathogenic microorganisms are relatively easy to grow in culture, whereas others are extremely difficult or expensive to manufacture in sizeable quantities. Viruses, for instance, can grow only inside host cells, and culturing animal cells is more complex than growing bacteria. Similarly, pathogenic eukaryotes such as *Plasmodium* (malaria) or *Entamoeba* (amoebic dysentery) are difficult to culture on a large scale, although some pathogenic fungi can be grown relatively easily. Bacteria are generally the easiest to manufacture

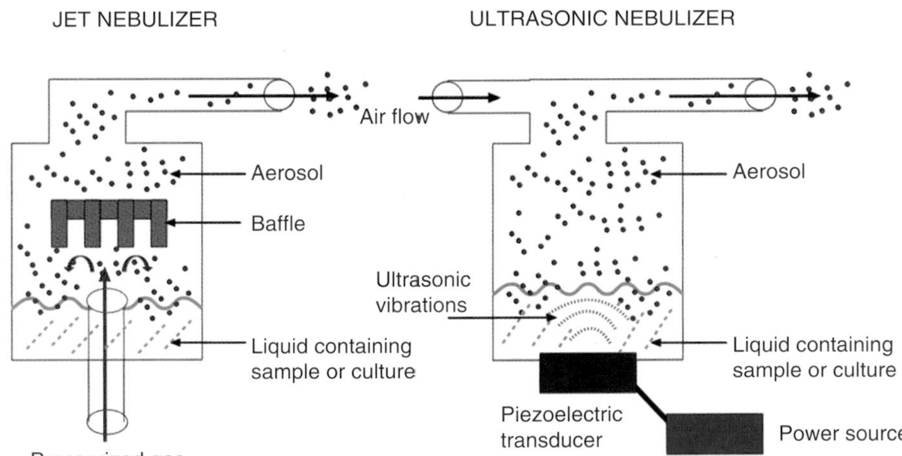

FIGURE 22.14
Nebulizer
A medical nebulizer could be used to aerosolize a biological agent for an indoor attack. Two general types of nebulizer are in use: the jet nebulizer that uses pressurized gas and the ultrasonic nebulizer that relies on ultrasonic vibrations.

on a large scale, but most bacterial infections can be cured with antibiotics. Viruses, though more difficult to grow, have the advantage of being largely incurable despite a small and growing range of specific antiviral agents.

Another factor is **weaponization**. The disease agent must be prepared in a manner that facilitates storage and dispersal. Because bacterial cells and spores tend to clump together spontaneously, they must be weaponized to allow effective delivery.

Dispersal. Dispersal is a particular challenge for biological weapons. The most likely option would be some form of airborne delivery. However, if applied outdoors, this tactic would be vulnerable to the whims of the weather. Not only is a pleasant breeze required, but also the wind needs to blow in the right direction! During the 1950s, the British government conducted field tests with harmless bacteria. When the wind blew them over farmland, many of the airborne bacteria survived the trip and reached the ground alive. In contrast, when the wind blew the bacteria over industrial areas, especially oil refineries or similar installations, the airborne bacteria were almost all killed. Ironically, air pollution may help protect an urban population from a bioterrorist attack. To aerosolize a biological agent for an indoor attack, a building's ventilation system or a medical nebulizer could be used (Fig. 22.14).

Persistence. Persistence may be the most difficult factor to consider. On the one hand, the biological agent should be able to persist in storage until it is ready to be deployed, and it must survive long enough in the environment to infect the enemy. On the other hand, it should not persist so long that the victor is unable to invade and conquer enemy territory.

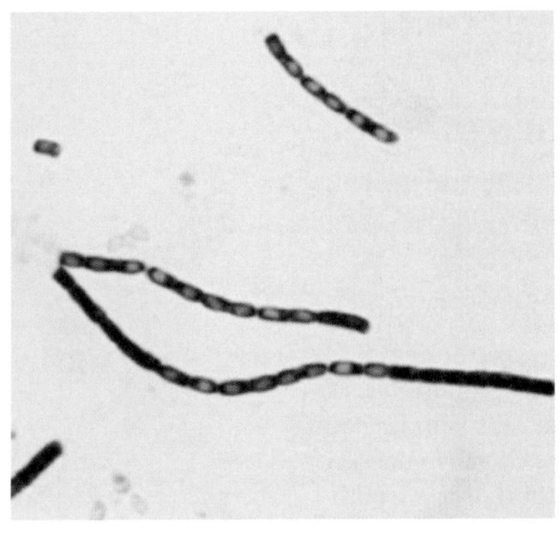

FIGURE 22.15
Spores of *Bacillus anthracis*
Anthrax spores, which are seen here forming inside bacterial cells, are difficult to destroy and last a very long time. From Ringertz SH, et al. (2000). Injectional anthrax in a heroin skin-popper. *Lancet* **356**, 1574–1575.

Many infectious agents are sensitive to desiccation and become inactive if exposed to air for significant periods of time. Moreover, natural UV radiation from the sun also inactivates many bacteria and viruses. Thus, most biological warfare agents must be protected from this "open air factor" before use and then dispersed as rapidly as possible. For instance, many viruses last only a few days, if even that, outside their animal or human hosts. (However, infections due to these agents may persist among the local population.)

Anthrax is often chosen as a biological weapon because of its ability to persist for long periods of time. The bacterium *Bacillus anthracis,* which causes the disease, spreads by forming spores that are tough and difficult to destroy (Fig. 22.15). When suitable conditions return, for example, inside the lungs of a human, the spores germinate and resume growth as normal bacterial cells, releasing life-threatening toxins.

Incubation time. A problem unique to biological warfare, compared to conventional weapons, is that death or incapacitation from infectious disease is a relatively slow process. Even the most virulent pathogens, such as Ebola virus or pneumonic plague, can take a few days to kill. An infected enemy would therefore still be capable of fighting for a significant period. Yet, a biological agent that kills too quickly may not have time to spread among the enemy population.

High-containment laboratories. High-containment laboratories are needed for research and development of infectious biological agents. Biological containment is rated on a scale with four levels. Biosafety level 1 (BSL-1) microbes are mostly harmless, such as nonpathogenic *E. coli*. BSL-2 organisms are human pathogens, but not easily transmitted in the laboratory, such as *Salmonella*. BSL-3 organisms are dangerous and often can be transmitted via aerosol, such as tuberculosis and SARS. BSL-4 laboratories are for extremely dangerous and easily transmissible microbes, such as Ebola.

The whole BSL-4 laboratory is sealed off and kept at a little under normal atmospheric pressure. In case of a leak, outside air will flow into the laboratory, helping ensure contaminated air will remain there instead of seeping out. Operations are conducted inside safety cabinets with glove ports. To enter a BSL-4 lab, a researcher must use an air lock and exchange outside clothes for a separate set of lab clothes, including a special "spacesuit" that is equipped with its own air supply (Fig. 22.16).

FIGURE 22.16
Biohazard Clothing, Then and Now
(A) Even during the bubonic plague, doctors wore protective clothing to prevent exposure to the deadly pathogens. The large beak was often stuffed with flowers and herbs to create a pleasant scent that was thought to keep away the plague, as illustrated in Bartholin, Thomas Hafniae, 1654–1661: *Historiarum anatomicarum.* Courtesy U.S. National Library of Medicine. (B) Today's suits are more scientific and streamlined, but serve the same purpose. Laboratory worker wearing BSL-4 protective gear. Courtesy of USAMRIID, DoD, and the NIAID Biodefense Image Library.

When finished, a scientist leaves behind his lab clothes and uses an exit equipped with disinfectant showers and ultraviolet lights. Some high-containment labs are designed so that the only exit is via total submersion in a pool of disinfectant. Ultraviolet lights are used to sterilize both the laboratories themselves and the air locks, especially when working with viruses.

Using high-containment facilities for research is expensive and time consuming. For manufacturing biological weapons on an industrial scale, the inconveniences are correspondingly worse. However, terrorist groups or rogue nations may only care about secrecy and may be willing to forgo biosafety considerations. The U.S. Army's criteria for a biowarfare agent are given in Box 22.1.

Five major factors that influence the use of a biological warfare agent include preparation, dispersal, persistence, incubation time, and the necessity of high-containment laboratories. A variety of viruses, bacteria, and toxins have been proposed as effective agents. These are classified into three categories by the Centers for Disease Control and Prevention (CDC) according to their level of risk.

Box 22.1 Requirements for Biological Warfare Agents

According to the U.S. Army, a biological warfare agent should fulfill the following requirements:

1. It should consistently produce death, disability, or damage.
2. It should be capable of being produced economically and in militarily adequate quantities from available materials.
3. It should be stable under production and storage conditions, in munitions, and in transportation.
4. It should be capable of being disseminated efficiently by existing techniques, equipment, or munitions.
5. It should be stable after dissemination from a military munition.

A CLOSER LOOK AT SELECT BIOLOGICAL WARFARE AGENTS

The Centers for Disease Control and Prevention (CDC) has classified biological warfare agents into three categories based on the potential level of threat they pose to society. These categories are summarized in Table 22.1. Anthrax was used in the 2001 bioterror attack in the USA (see Box 22.2).

Anthrax and Other Bacterial Agents

Anthrax is a virulent disease of cattle that infects humans quite easily. It is caused by the bacterium *Bacillus anthracis,* which is relatively easy to culture and forms spores, which can survive harsh conditions that would kill most bacteria. The spores may lie dormant in the soil for years and then germinate on contact with a suitable animal victim.

Three main forms of anthrax occur. Cutaneous anthrax, that is, infection of the skin, is rarely dangerous. Gastrointestinal anthrax occurs mostly in grazing animals and is relatively rare among humans, although it can occur via ingestion of bacteria or spores from contaminated meat. Inhalational anthrax, in which the spores enter via the lungs, gives a high death rate. In many ways, anthrax is the ideal biological weapon—lethal, highly infectious, and cheap to produce, with spores that store well.

The problem with anthrax, however, is that the spores are so tough and long-lived that getting rid of them after hostilities are over is nearly impossible. During World War II, the British tested anthrax (using sheep as the targets) on the tiny island of Gruinard, which lies off the coast of Scotland. Although it was firebombed and disinfected, the island remained uninhabitable for nearly 50 years because of anthrax spores still surviving in the soil. Finally, in 1990, the island was declared safe after it was treated with a solution of formaldehyde and seawater. The indestructibility of anthrax spores would thus be problematic for a military occupation, but it could be useful as a defense

Table 22.1 CDC-Listed Agents Relevant to Biological Warfare

CATEGORY A AGENTS INCLUDE ORGANISMS THAT POSE A RISK BECAUSE:	
▪ They can be easily disseminated or transmitted person-to-person ▪ They cause high mortality ▪ They might cause public panic and social disruption ▪ They require special action to protect public health	
Bacteria	
Anthrax	*Bacillus anthracis*
Plague	*Yersinia pestis*
Tularemia	*Francisella tularensis*
Viruses	
Smallpox	*Variola major*
Filoviruses	Ebola hemorrhagic fever
	Marburg hemorrhagic fever
Arenaviruses	Lassa fever
	Junin virus (Argentine hemorrhagic fever)
Toxins	
Botulinum toxin from *Clostridium botulinum*	
CATEGORY B AGENTS INCLUDE THOSE THAT:	
▪ Are moderately easy to disseminate ▪ Cause moderate morbidity and low mortality ▪ Require improved diagnostic capacity and enhanced surveillance	
Bacteria	
Brucellosis	*Brucella* (several species)
Glanders	*Burkholderia mallei*
Melioidosis	*Burkholderia pseudomallei*
Q fever	*Coxiella burnetti*
Several food- or waterborne enteric diseases, including	*Salmonella, Shigella dysenteriae, Vibrio cholerae*
Viruses	
Alphaviruses	Venezuelan encephalomyelitis
	Eastern and Western equine encephalomyelitis
Toxins	
Ricin toxin from *Ricinus communis* (castor bean)	
Epsilon toxin from *Clostridium perfringens*	
Enterotoxin B from *Staphylococcus*	
CATEGORY C AGENTS:	
Emerging pathogens that could possibly be engineered for mass dissemination in the future, such as Nipah virus, hantaviruses, flaviviruses (yellow fever, dengue fever), multidrug-resistant tuberculosis	

Box 22.2 The 2001 Anthrax Attack in the United States

Shortly after the terrorist attack on the World Trade Center in September 2001, anthrax spores were distributed via the U.S. Postal Service. The anthrax attack was notable in two respects. First, it killed only a small handful of victims, supporting the contention that biological warfare is not usually very effective in practice. Second, it generated a vastly disproportionate reaction, illustrating the importance of the psychological aspects of bioterrorism. Undoubtedly, governmental overreaction and public panic did far more damage than the anthrax attack itself.

An insider in America's own biodefense research establishment perpetrated the attack. Detectives believe the culprit was Bruce Ivins, an army scientist and anthrax expert, but he committed suicide in 2008 without ever being charged. The FBI officially closed the case in 2010. However, doubts surrounding the evidence against Ivins have led some observers to call for a new investigation.

The attacker used the Ames strain of *Bacillus anthracis,* which is widely used in laboratories across the United States. A major problem with tracing the origin of anthrax outbreaks is that all the various strains of *Bacillus anthracis* are closely related and difficult to tell apart. No differences in either 16S rRNA or 23S rRNA sequence occur between different strains. In practice, analysis is done using single-nucleotide polymorphisms (SNPs) or variable number tandem repeats (VNTRs; see Chapter 23). For example, the *vrrA* gene of *Bacillus anthracis* contains from two to six copies of the sequence CAATATCAACAA within the coding region for a protein of unknown function (Fig. A).

These repeats were probably originally generated by slippage of DNA polymerase during replication. The repeats do not alter the reading frame, but result in corresponding repeats of the four-amino-acid sequence Gln-Tyr-Gln-Gln within the encoded protein. Several other VNTRs are also now used, including some on the pOX1 virulence plasmid. The greatest diversity of *Bacillus anthracis* strains, as assessed by multiple VNTR analysis, comes from southern Africa, which is therefore regarded as the probable homeland of anthrax.

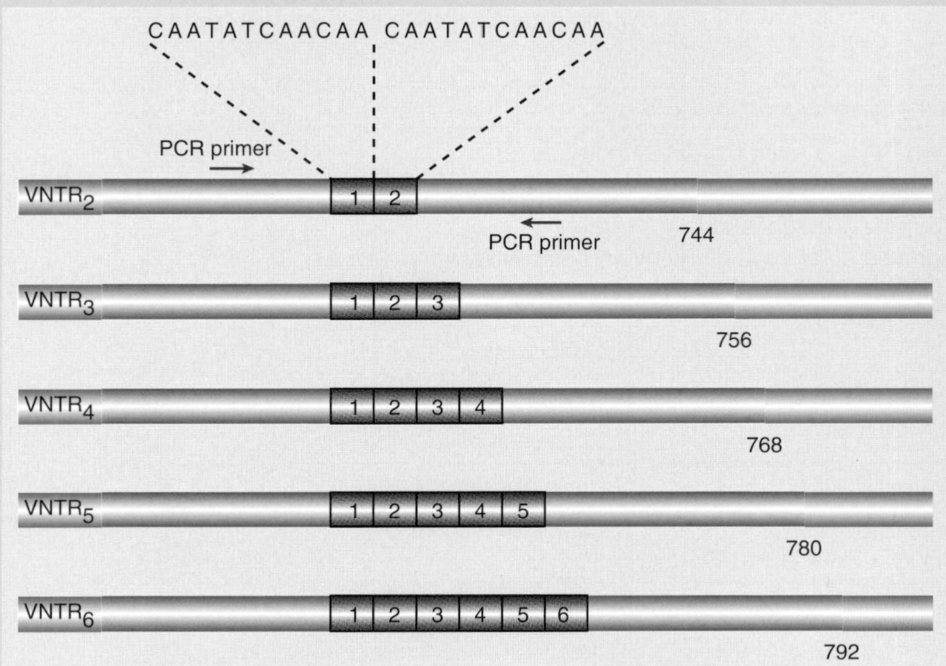

FIGURE A The *vrrA* VNTR of *Bacillus anthracis*
The *vrrA* gene of anthrax (blue) has a stretch of repeats in the coding region. Different strains of anthrax have different numbers of repeats (green) due to polymerase slippage and can therefore be traced by comparing the number of repeats. PCR is used to amplify the region containing the repeats. The length of the PCR product reveals the number of repeats.

mechanism. Anthrax spores seeded into the soil of sparsely populated land could serve to protect against foreign invaders.

Yersinia pestis, the causative agent of bubonic plague, was responsible for the notorious Black Death epidemics of the Middle Ages and is a current bioterrorism threat. Plague is typically spread by flea bites. However, aerosolized bacteria can cause the pneumonic form of plague. This form of the disease is highly infectious and has a mortality rate close to 100% if untreated. Besides Japan's use of plague as a biological weapon during World War II (discussed previously), the British biological warfare center at Porton Down maintained large-scale plague cultures for several years following the war. In the 1960s, the United States experimented with spreading plague among rodents in Vietnam, Laos, and Cambodia to little practical effect. Because it can be obtained relatively easily from nature, plague may be an attractive weapon for bioterrorists (Fig. 22.17).

FIGURE 22.17
Plague Reservoir
Rodents, such as squirrels and prairie dogs, serve as a natural reservoir for plague. A California ground squirrel is shown here. From Hayward P. (2013). Rare zoonoses in the USA. *Lancet Infect Dis* **13**, 740–741.

An unconfirmed report in 2009 indicated that 40 al-Qaeda terrorists in Algeria accidentally became infected with plague and died, presumably from a biological weapons experiment gone awry.

Other potential bacterial agents include the following:

- *Brucella.* **Brucellosis** is a disease of cattle, camels, goats, and related animals. Brucellosis was developed as a biological weapon by the United States from 1954 to 1969. In humans, it behaves erratically, both in the time for symptoms to emerge and the course of the disease. Although human victims often fall severely ill for several weeks, it is rarely fatal, even if untreated. It could be used as an incapacitating agent.
- *Francisella tularensis.* **Tularemia** is a disease of rodents or birds that has a death rate of 5% to 10% in humans if untreated. It is highly infectious and generally regarded as an incapacitating agent.
- *Burkholderia pseudomallei.* **Melioidosis** is related to glanders *(Burkholderia mallei)*, a disease of horses. Melioidosis is a rare disease of rodents from the Far East that is spread by rat fleas. Melioidosis is more virulent than glanders and, untreated, is fatal some 95% of the time in humans.

In many ways, anthrax is one of the best biological weapons. It is lethal, highly infectious, and easy to produce, and it has long-lasting spores. Plague is also lethal, and its pneumonic form can spread from person to person. Potential incapacitating agents include brucellosis and tularemia.

Smallpox and Other Viral Agents

Variola, the viral etiologic agent of **smallpox**, is a member of the poxvirus family. These large viruses contain double-stranded (ds) DNA (Fig. 22.18).

Poxviruses are the most complex animal viruses and are so large they may be seen with a light microscope. They measure approximately 0.4 by 0.2 microns, compared to 1.0 by 0.5 microns for bacteria such as *E. coli.* Unlike other animal DNA viruses, which replicate inside the cell nucleus, poxviruses replicate their dsDNA in the cytoplasm of the host cell. They build subcellular factories known as inclusion bodies, inside which virus particles are manufactured. Poxviruses have 185,000 nucleotides encoding 150 to 200 genes, about the same number as the T4 family of complex bacterial viruses.

Variola virus infects only humans, which allowed its eradication by the World Health Organization, a task completed by 1980. Smallpox is highly infectious and exists as two variants: *Variola major* with a fatality rate of 30% to 40% and *Variola minor* with a fatality rate of around 1% (Fig. 22.19). Their genome sequences differ by approximately 2%. *Vaccinia* virus is a related, mild poxvirus of unknown origin that is used as

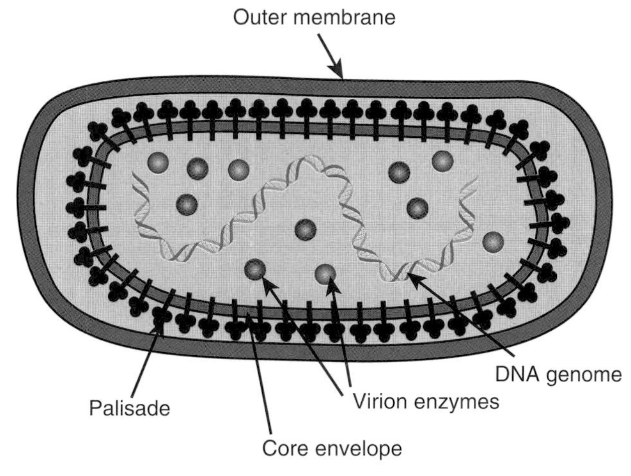

FIGURE 22.18
Variola Virus
Poxviruses, including smallpox, are closely related in structure and DNA sequence. They have genomes of dsDNA surrounded by two envelope layers. A protein layer, known as the palisade, is embedded within the core envelope. Premade viral enzymes are also packaged with the genome to allow replication immediately on infection. Poxviruses infect animals, and the outermost viral membrane is derived from the membrane of the previous host cell.

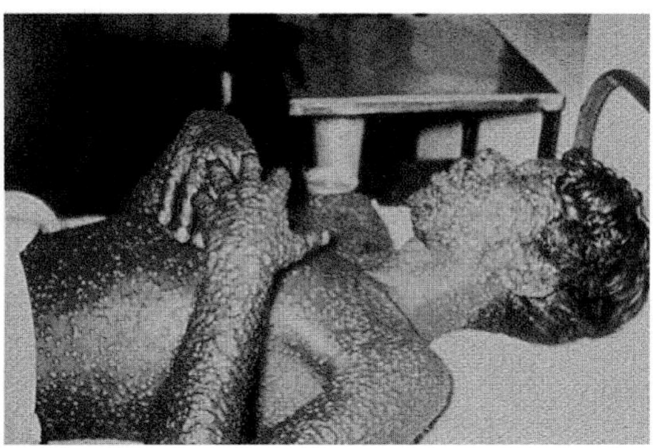

FIGURE 22.19
Smallpox
A person with smallpox develops a characteristic rash. Source: Centers for Disease Control and Prevention.

a live vaccine. Immunity is conferred against several closely related poxviruses including smallpox and monkeypox.

For governments, the preparation of large amounts of a virus whose particles are fairly stable and long-lived, such as smallpox, is feasible. It is believed that the former Soviet Union had done so, as discussed earlier. The virus could be delivered using a medical nebulizer (see Fig. 22.14) or through the use of suicidal volunteers who would deliberately infect themselves and then travel to densely populated target areas. They would mingle with as many people as possible, by attending large events and utilizing mass transit. However, transmission requires close contact, and when a person is most contagious, he may be feeling far too ill to actually walk around in public.

Though many viruses are difficult to culture in large amounts and are unstable during storage, several others have been considered as possible biological warfare agents:

- **Filovirus.** Ebola and Marburg viruses make up a family of negative single-stranded (ss) RNA viruses, known as the Filoviruses, which form long, thin filaments (Fig. 22.20). Patients vomit and ooze blood from various orifices, including their eyes and ears. Ebola outbreaks in Sudan and Zaire have had 80% to 90% fatality, but closely related strains exist that are not as virulent. For example, in 1989, an Ebola outbreak occurred among long-tailed macaques in a research facility in Reston, Virginia. However, this strain was not lethal to humans. Transmission generally requires substantial exposure to infected body fluids, so filoviruses are difficult to acquire by casual exposure. In practice, this makes them a relatively poor choice for use as a biological warfare agent.
- **Flavivirus.** Dengue fever and yellow fever are both caused by members of the Flavivirus family. Yellow fever is frequently lethal, whereas dengue is rarely fatal, but it is very painful and incapacitates its victims for several days. However, both are spread by insect bites, which would make their use as biological weapons difficult.
- **Arenavirus.** An Arenavirus that appeared in the Lassa River region of Nigeria in the late 1960s causes Lassa hemorrhagic fever, which symptomatically resembles an Ebola infection. This segmented ssRNA virus has extremely high mortality and is typically spread by rodents.

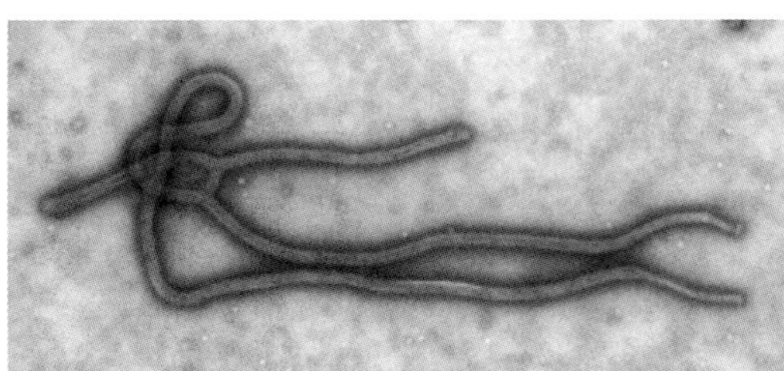

FIGURE 22.20
Ebola Virus
This electron micrograph depicts Ebola virus. From Moran GJ, Talan DA, Abrahamian FM (2008). Biological terrorism. *Infect Dis Clin North Am* **22**, 145–187. Courtesy of Centers for Disease Control and Prevention/Cynthia Goldsmith.

> Smallpox is the most likely virus to be used as a biological warfare agent. It is highly infectious, and the death rate may reach 30% to 40%. Other viruses with higher mortality rates may be prohibitively difficult to distribute.

Rust and Other Fungal Agents

Fungal agents perhaps may be most effectively used against staple crops, for instance, cereals and potatoes, which are an important part of the food supply. A wide variety of fungi exist that destroy these crops, such as rusts, smuts, and molds. Their spores are often highly infectious and easily dispersed by wind or rain, and in many cases there is no effective treatment.

Soybean rust and wheat stem rust are examples of pathogenic fungi that could destroy major crops. In addition to destruction of the crop, certain fungi may produce toxins. For example, when **ergot** grows on rye or other cereals, it produces a mixture of toxins that cause a syndrome referred to as ergotism, which can lead to convulsions, hallucinations, and even death

(Fig. 22.21). Some researchers believe that community ergot poisoning may have been responsible for the hysteria that led to the Salem witch trials.

FIGURE 22.21
Ergot on Quackgrass
Fungi such as *Claviceps purpurea* infect various grains such as wheat or rye as well as grasses such as quackgrass, shown here. The mature fungus forms purple to black bodies, called ergot bodies or sclerotia, where the grain would normally be positioned. Courtesy of David Barker, Department of Horticulture and Crop Science, Ohio State University, Columbus, OH.

Another potential fungal agent is *Aspergillus flavus*, which infects cereals and legumes and produces the carcinogenic **aflatoxin**. Acute aflatoxin poisoning can cause liver damage and death, and chronic exposure can cause cancer. Herculean efforts are made to keep the food supply free of aflatoxin.

From the perspective of biological warfare, there are many advantages to using a fungal agent against crops. First, an entire crop might have to be screened even if only a small part was infected, causing major disruptions and economic losses. Second, dispersal could be rather easily accomplished by spraying fungal spores with a crop duster airplane over farmland. Alternatively, seeds could be infected, especially since many of them are imported to the United States and may be more easily accessed for contamination. Third, modern agriculture is particularly vulnerable to infection because large acres of genetically identical cultivars are often planted in high density. This lack of genetic variability could allow for an infection to spread rapidly. Finally, fungal agents that attack crops pose little danger to those using them.

The spores of highly infectious fungi could be used as biological warfare agents to target staple crops.

Purified Toxins

Another approach to biological warfare is to use purified toxins rather than a living infectious agent. A variety of toxins are known that may be purified in substantial quantities. Bacteria, primitive eukaryotes such as algae or fungi, higher plants, and animals all make toxins (Table 22.2).

Botulinum toxin. The most toxic substance known is **botulinum toxin**. It is made by the anaerobic bacterium, *Clostridium botulinum*, and is the cause of botulism, a severe form of

Table 22.2 Toxins Relevant to Biowarfare

Toxin	LD50 (μg/kg)	Producer Organism
Botulinum toxin A	0.001	Bacterium (*Clostridium botulinum*)
Enterotoxin B	0.02	Bacterium (*Staphylococcus*)
Ciguatoxin P-CTX-1	0.2	Marine dinoflagellate
Batrachotoxin	2	Poison arrow frog
Ricin	3	Castor bean (*Ricinus communis*)
Tetrodotoxin	8	Pufferfish
VX	15	Synthetic nerve agent
Anthrax lethal toxin	50	Bacterium (*Bacillus anthracis*)
Aconitine	100	Plant (monkshood, a.k.a. wolf's bane)
Mycotoxin T-2	1200	Fungus (*Fusarium*)

food poisoning. It has been proposed as a biological warfare agent but has actually found its most frequent application in cosmetics, under the name Botox. It is also used to treat a few clinical conditions in which a muscle relaxant is needed.

Botulinum toxin is a neurotoxin that blocks transmission of signals from nerves to muscles, thus causing muscular paralysis. The incredible potency of botulinum toxin is due to its enzymatic activity. It is a zinc protease that cleaves SNARE proteins in the neuromuscular junction that are required for release of the neurotransmitter acetylcholine. Death is generally due to paralysis of the lungs and respiratory failure (Fig. 22.22).

C. botulinum almost never causes infections but will grow in improperly canned food. Proper canning uses a pressure cooker to destroy the hardy spores produced by *Clostridium*. If the spores are not destroyed, they can germinate. After the bacteria die, they release botulinum toxin, which accumulates in the food. Merely 50 ng of botulinum toxin is enough to kill the average human. The toxin can, however, be destroyed by heating.

Terrorists of the Japanese cult Aum Shinrikyo, discussed previously, have attempted to use botulinum toxin. Aerosols were dispersed at various sites in Tokyo and at U.S. military installations in Japan on several occasions between 1990 and 1995. The attacks failed, mainly because

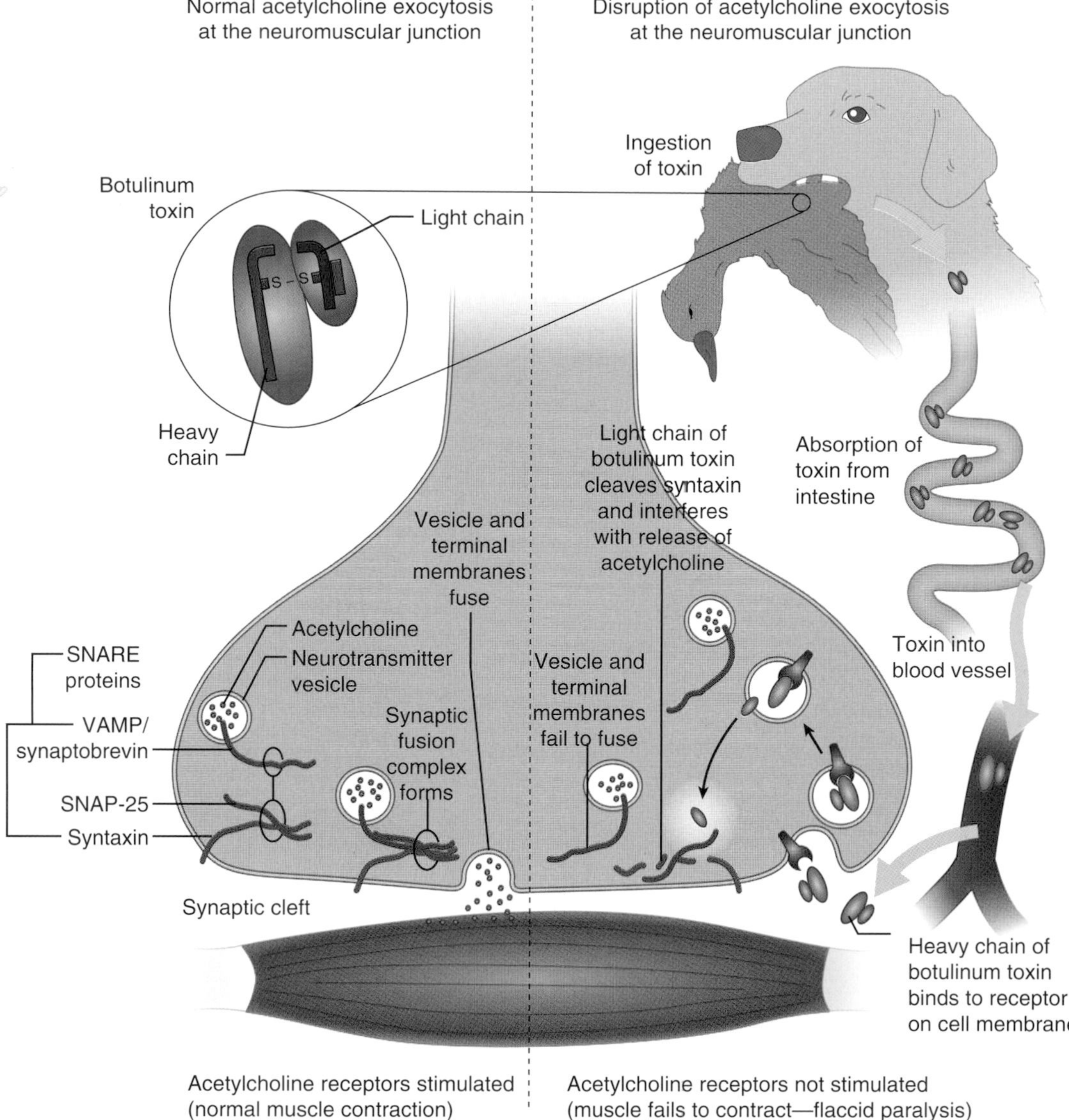

FIGURE 22.22 Mechanism of Botulinum Toxin
Botulinum toxin disrupts the normal functioning of a neuromuscular junction by inhibiting the release of acetylcholine. From Sykes JE (2014). Tetanus and botulism. In *Canine and Feline Infectious Diseases*, Ch. 54, pp. 520–530. Saunders/Elsevier, Philadelphia, PA, USA.

the cult used strains of *C. botulinum* that failed to produce toxin. On the other hand, millions of people have willingly had extremely dilute preparations of botulinum toxin (Botox) injected into their face to eliminate wrinkles. The procedure works because botulinum toxin inhibits the muscle contraction responsible for causing them.

FIGURE 22.23 Seeds of the Castor Bean, *Ricinus communis*
Courtesy of Dan Nickrent, Department of Plant Biology, Southern Illinois University, Carbondale, IL.

Ricin. Many higher plants make **ribosome-inactivating proteins (RIPs)**. These enzymes split the N-glycosidic bond between adenine and ribose from a specific sequence in the large-subunit ribosomal RNA. Clipping adenine from the rRNA totally inactivates the ribosome. A single RIP molecule is sufficient to inactivate all the ribosomes and kill a whole cell. Because RIPs are synthesized as precursor proteins that are fully processed only after exiting the plant cell's cytoplasm, the toxin does not kill the plant. Intact ribosomes from different types of organisms differ greatly in their sensitivity to RIPs. Mammalian ribosomes (which contain 28S rRNA) are by far the most sensitive. On the other hand, the activity of many RIPs against bacterial ribosomes (which contain 23S rRNA) is low or negligible, and this has allowed the genes for some RIPs to be cloned and expressed in *E. coli*.

Like many bacterial toxins, **ricin** is a typical A-B toxin in which the A chain exhibits toxic enzymatic activity and the B chain mediates entry into the target cell. Ricin is lethal at around 3 μg/kg body weight, meaning that 300 μg should kill a large human. Ricin is extracted from the seeds of the castor bean plant, *Ricinus communis* (Fig. 22.23). This plant is widely grown, both for ornamentation and on a large scale for castor oil production. Because of its widespread availability, high toxicity, stability, and lack of any antidote, there are several examples of the use of ricin as a biological weapon.

Ricin achieved international notoriety in 1978 when the Bulgarian defector Georgi Markov was assassinated in a London street by ricin. The communist assassin wielded a modified umbrella that injected a hollow 0.6-mm-diameter metal sphere, filled with ricin, into Markov's leg. In 1991, four members of the Patriots Council, an extremist group in Minnesota with an antigovernment and antitax ideology, purified ricin in a home laboratory. They were arrested for plotting to kill IRS and law enforcement agents with ricin. In late 2013, actress Shannon Richardson pleaded guilty for mailing letters containing ricin to President Barack Obama and New York City Mayor Michael Bloomberg in a scheme to frame her estranged husband. In a completely separate incident, in January 2014, James Dutschke also pleaded guilty to sending ricin to President Obama and other government officials with the intention of framing an Elvis Presley impersonator with whom he had a personal feud.

Abrin. Though less well known, **abrin**, which is also a ribosome-inactivating protein, is four times more toxic than ricin. Abrin is derived from the seeds of *Abrus precatorius*, commonly known as jequirity or rosary pea (Fig. 22.24).

The beautiful seeds are widely used in jewelry, particularly rosary beads. However, the seeds are so toxic that, if broken or damaged, a small prick in the skin is sufficient to absorb a lethal dose of abrin. There have been reports of abrin poisoning in jewelry makers, as well as in individuals who ingested seeds, but there are no known instances of abrin being used as a biological weapon.

FIGURE 22.24 Seeds of the Rosary Pea, *Abrus precatorius*
The beautiful seeds of the rosary pea are highly toxic when the coat is damaged. Courtesy of Kenneth R. Robertson, University of Illinois.

Conotoxin. Cone snails are predators that use a venom cocktail containing at least 100 different conotoxins to paralyze and kill their prey. The most dangerous cone snail to humans, *Conus*

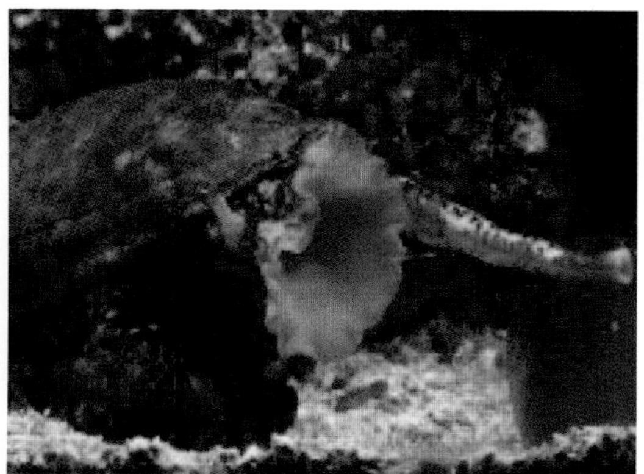

FIGURE 22.25
Cone snail, *Conus geographus*
The cone snail produces highly toxic venom. From Andreotti N, et al. (2010). Therapeutic value of peptides from animal venoms. In *Reference Module in Chemistry, Molecular Sciences and Chemical Engineering, Comprehensive Natural Products II. Chemistry and Biology.* Vol. 5 Ch. 10, pp. 287–302. Edited by Reedijk, J. Elsevier, Waltham, MA, USA.

geographus, stabs fish with a venom-filled "harpoon" located in its proboscis (Fig. 22.25).

Death from the sting of a cone snail is largely due to α-conotoxins, which cause muscle paralysis leading to respiratory arrest. Other toxins may trigger cardiovascular collapse. Symptomatically, α-conotoxins resemble botulinum toxin, although the mechanism of action is different. Because most conotoxins are short peptides 10–30 amino acids in length, the concern from a biological warfare perspective is not that a government or terrorist would harvest venom from cone snails but that the toxins would be chemically synthesized.

Purified toxins are possible biological warfare agents. Natural toxins can be isolated from bacteria, plants, and animals, but other toxins could be chemically synthesized. Botulinum toxin disrupts the neuromuscular junction and is the most potent toxin known. Ricin and abrin are ribosome-inactivating proteins that are made by certain plants.

ENHANCING BIOLOGICAL WARFARE AGENTS WITH BIOTECHNOLOGY

It is often suggested that genetic engineering could be used to create more dangerous versions of infectious agents. Although there is some truth to this assertion, consider the following:

Suppose a bioterrorist tries to genetically modify a harmless laboratory bacterium, such as *E. coli*. The bacteria could be engineered to go "under cover" when they enter the human body, hiding from the immune system. Additionally, the bacteria could be programmed to rebuff immune cells by injecting them with toxins, and other genes could be added for ripping vital supplies of iron away from blood cells. Finally, the bacteria could be modified to be highly infectious. Such a biological agent would make for a fearsome weapon.

Unfortunately, this bacterium already exists. It is called *Yersinia pestis*. It is the agent of bubonic plague and is still endemic in many parts of the world, including China, India, Madagascar, and the United States. Instead of devoting years to genetically engineer a lethal biological weapon, a bioterrorist could simply isolate one of Mother Nature's very own products. The "improvement" of infectious diseases by genetic engineering, therefore, is probably a minor threat.

Still, genetic engineering of biological warfare agents is theoretically possible, so we briefly consider the issue here.

Engineering Pathogens to Be More Lethal

The Soviet germ warfare facility is known to have modified smallpox virus and generated a variety of artificial mutants and hybrids. The details are largely unavailable. However, recent experiments with mousepox (*Ectromelia* virus) have given disturbing results. Mousepox is related to smallpox, but it only infects mice. Its virulence varies greatly depending on the strain of mouse. Genetically resistant mice rely on cell-mediated immunity, rather than antibodies. Natural killer (NK) cells and cytotoxic T cells destroy cells infected with mousepox virus, thus clearing the virus from the body.

Researchers modified mousepox virus by inserting the human gene for the cytokine interleukin-4 (IL-4). IL-4 is known to stimulate the division of B cells, which synthesize antibodies. The rationale for engineering the virus was that IL-4 would stimulate the production of antibodies and lead to an improved and more balanced immune response. What actually happened was the opposite of what was expected: the creation of a virus with vastly greater virulence. Not only did it kill all of the genetically resistant mice, but it also killed 50% of mice that had been vaccinated against mousepox. The expression of excess IL-4 suppressed the NK cells and cytotoxic T cells. Furthermore, it failed to increase the antibody response. The reasons are not fully understood, but they do serve as a reminder that the immune system is under extremely complex control.

Similar results have been seen with strains of *Vaccinia* virus, which is used for vaccination against smallpox. Whether insertion of IL-4 or other immune regulators into smallpox itself would lead to increased virulence by undermining the immune response is unknown. poxviruses already possess genes designed to protect the virus by interfering with the action of NK cells and cytotoxic T cells (Fig. 22.26). These are the cytokine response modifier (*crm*) genes, and they vary in effectiveness among different poxviruses. One reason smallpox is so virulent may be that it already subverts the body's cell-mediated immune response. In this case, adding IL-4 would not be expected to increase virulence.

Creating Camouflaged Viruses

With genetic engineering, it is also possible to hide a potentially dangerous virus inside a harmless bacterium. This strategy is already used in nature when bacteriophages insert their genomes into bacterial chromosomes or plasmids and later re-emerge to infect other hosts.

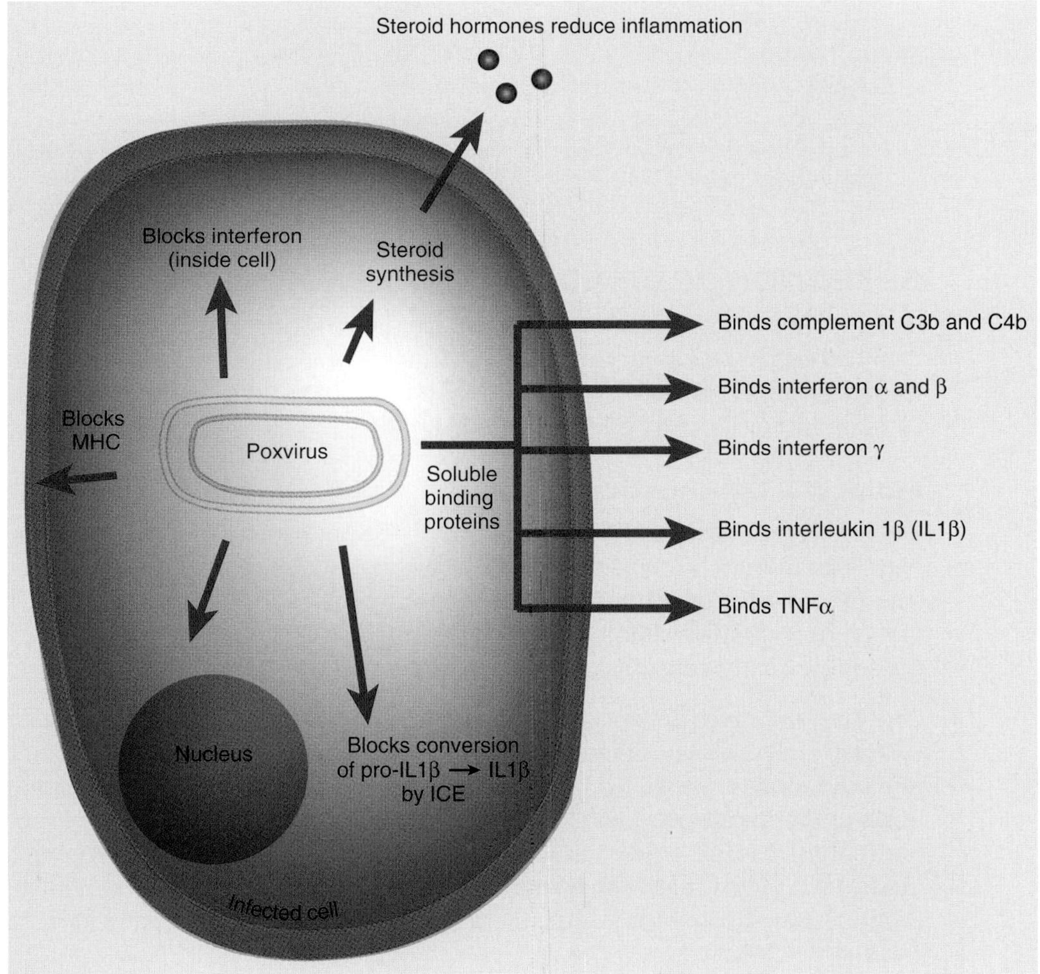

FIGURE 22.26 Poxvirus Immune Evasion
Poxvirus deploys many different proteins to prevent the infected cell from being attacked by the host's immune system.

Theoretically, cloning the entire genome of a small animal or plant virus into a bacterial plasmid could create a biological weapon. Larger viruses could be accommodated with bacterial or yeast artificial chromosomes. In the case of RNA viruses, a cDNA copy of the virus genome must first be generated by reverse transcriptase before cloning it into a bacterial vector. Any virus containing a **poison sequence**, a base sequence that is not stably maintained on bacterial plasmids, could perhaps be cloned as separate fragments. Such a strategy works for yellow fever virus, but a complete, functional cDNA requires ligation of the fragments *in vitro*.

Many cell types, both bacterial and eukaryotic, can take up DNA or RNA under certain circumstances by transformation. Consequently, the naked nucleic acid genomes of many viruses, both DNA and RNA, are infectious even in the absence of their protein capsids or envelopes. Thus, once a viral genome is cloned, the DNA molecule containing it may itself be infectious. Alternatively, the cDNA version of some RNA viruses can successfully infect host cells and give rise to a new crop of RNA-containing virus particles. This has been demonstrated for RNA viruses such as poliovirus, influenza, and coronavirus.

The cleverest strategy for generating an RNA virus is to clone the cDNA version of its genome onto a bacterial plasmid downstream of a strong promoter (Fig. 22.27). The natural RNA version of the viral genome will be generated by transcription. When induced, the bacterial cell would generate a large number of infectious viral particles. A dangerous human RNA virus loaded into a harmless intestinal bacterium under the control of a promoter designed to respond to conditions inside the intestine could pose a formidable threat.

Genetically engineering biological warfare agents to make them deadlier is a minor threat since many naturally occurring microbes are already very dangerous. However, certain poxviruses have been modified to become more virulent. Inserting viral DNA into plasmids carried by harmless bacteria could create camouflaged viruses.

DETECTION OF BIOLOGICAL WARFARE AGENTS

In the laboratory, some pathogenic bacteria grow slowly or not at all. This may be because the microbe has fastidious nutrient requirements or is otherwise difficult to culture outside its host organism. However, thanks to advances in biotechnology, infectious microbes can be identified using a variety of different techniques.

Molecular Diagnostics

Rather than attempting to grow and identify disease-causing agents using classical microbiological techniques, molecular diagnostics analyzes molecules; typically DNA, but RNA, proteins, and volatile organic compounds can also be used. (Other diagnostic methods involve the use of antibody technology and are discussed in Chapter 6.) Molecular techniques have the advantage of being quicker, more accurate, and more sensitive.

One diagnostic method is called fluorescent *in situ* hybridization (FISH; for details see Chapter 3). Biopsies or other patient samples are directly probed with fluorescent DNA oligonucleotides specific to a pathogen of interest. If the pathogen is present, the probe binds to the complementary DNA in its chromosome and the fluorescence can be visualized under a microscope. A new innovation, called peptide nucleic acid (PNA), replaces the negatively charged sugar-phosphate backbone of DNA with a neutral peptide backbone. Probes made of PNA bind complementary DNA more tightly and enter bacterial cells more easily (Fig. 22.28).

Most other methods based on DNA detection involve extracting DNA from a sample followed by amplification via PCR. Because primers can be designed to amplify DNA sequences unique

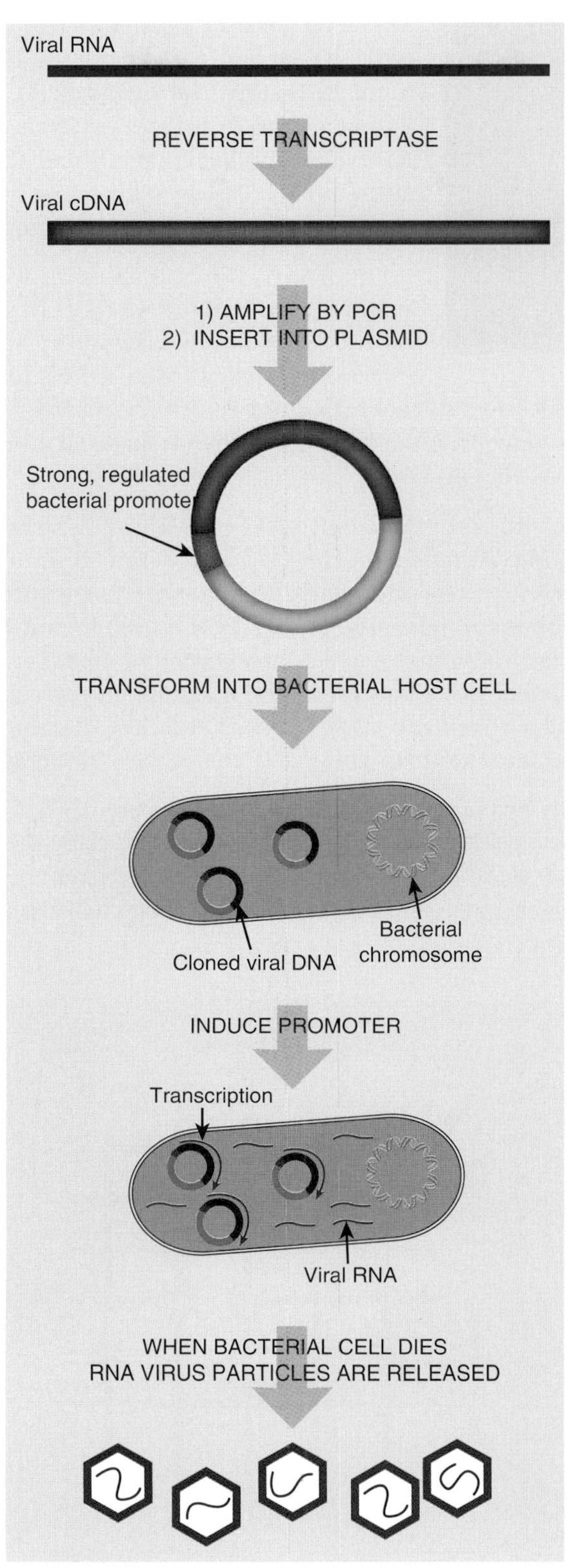

FIGURE 22.27
Expression of Cloned RNA Virus
Cloning an RNA virus requires making a double-stranded DNA copy using reverse transcriptase. The cDNA is inserted into an appropriate bacterial plasmid and transformed into bacterial cells. To control the expression of the viral DNA, a strong promoter is placed upstream of the viral cDNA. If the promoter is inducible, when the bacteria are given the appropriate stimulus, the viral cDNA will be expressed, resulting in production of viral particles that could infect many people.

FIGURE 22.28
PNA FISH Probe
The yeast *Candida albicans* is detected in a blood culture using fluorescent *in situ* hybridization (FISH) with a peptide nucleic acid (PNA) probe. Fluorescence microscopy; original magnification 500x. From Bravo LT, Procop GW (2009). Recent advances in diagnostic microbiology. *Semin Hematol* **46**, 248–258.

to a particular pathogen, PCR itself can serve as a diagnostic tool. The advantages of PCR are that it theoretically requires only a single molecule of target DNA and works on microbes that cannot be cultured in the laboratory. The downside is that PCR is susceptible to contamination and false positives. A variant of PCR (see Chapter 4), called randomly amplified polymorphic DNA (RAPD), can be used to distinguish different strains of the same bacterial species. This capability is useful in epidemiology for tracking the spread of infectious disease.

Additionally, every species of microorganism has a different **small-subunit ribosomal RNA** (SSU rRNA) sequence (16S rRNA in bacteria and 18S rRNA in eukaryotes). Thus, if a patient has an unknown infection, clinicians can use PCR to amplify the gene encoding the microbe's SSU rRNA. Primers that recognize the conserved region of SSU rRNA are used to amplify the gene. The PCR fragment is then sequenced and compared with a database of known DNA sequences.

Another technique that generally relies on SSU rRNA is **checkerboard hybridization**. This allows multiple bacteria to be detected and identified simultaneously in a single sample. A series of probes corresponding to different bacteria are applied in horizontal lines across a hybridization membrane (Fig. 22.29). PCR is used to amplify a portion of the SSU rRNA gene from clinical samples, which may contain a mixture of pathogens. The PCR fragments are then labeled with a fluorescent dye and applied vertically to the membrane. After denaturation and annealing to allow hybridization, the membrane is washed to remove unbound DNA. Those samples that hybridize to the probes appear as bright fluorescent spots.

A potentially revolutionary technology called PLEX-ID has been developed by Abbott Laboratories. It combines traditional PCR with mass spectrometry to identify unknown microbes in patient samples. DNA is extracted and many different sets of primers are used to amplify various target sequences. The fragments are then analyzed with a mass spectrometer to

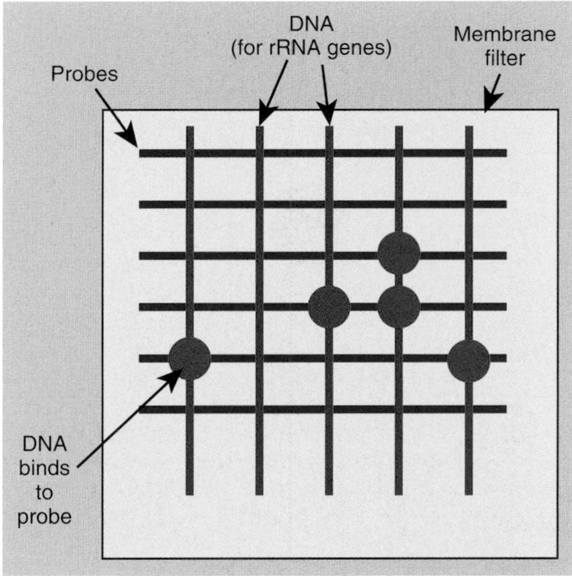

FIGURE 22.29 Checkerboard Hybridization
Probes corresponding to 16S rRNA for each candidate bacterium are attached to a membrane filter in long horizontal stripes (one candidate per stripe). DNA from patient samples is extracted and amplified by PCR using primers for 16S rRNA. The PCR fragments are tagged with a fluorescent dye and applied in vertical stripes. Each sample is thus exposed to each probe. Wherever a 16S PCR fragment matches a 16S probe, the two bind, forming a strong fluorescent signal where the two stripes intersect.

determine their mass. From this information, the DNA sequence can be deduced and the pathogen identified. PLEX-ID can make a diagnosis in 8 hours.

In the future, it may be possible to diagnose disease using an "electronic nose." As the name implies, the device detects volatile organic compounds that are released by pathogens or by the body in certain diseased conditions.

Biosensors

Biosensors are devices for the detection and measurement of reactions that rely on a biological mechanism (Fig. 22.30). Biosensors have been traditionally used in medical diagnostics and in food and environmental analysis. By far the biggest use has been the clinical monitoring

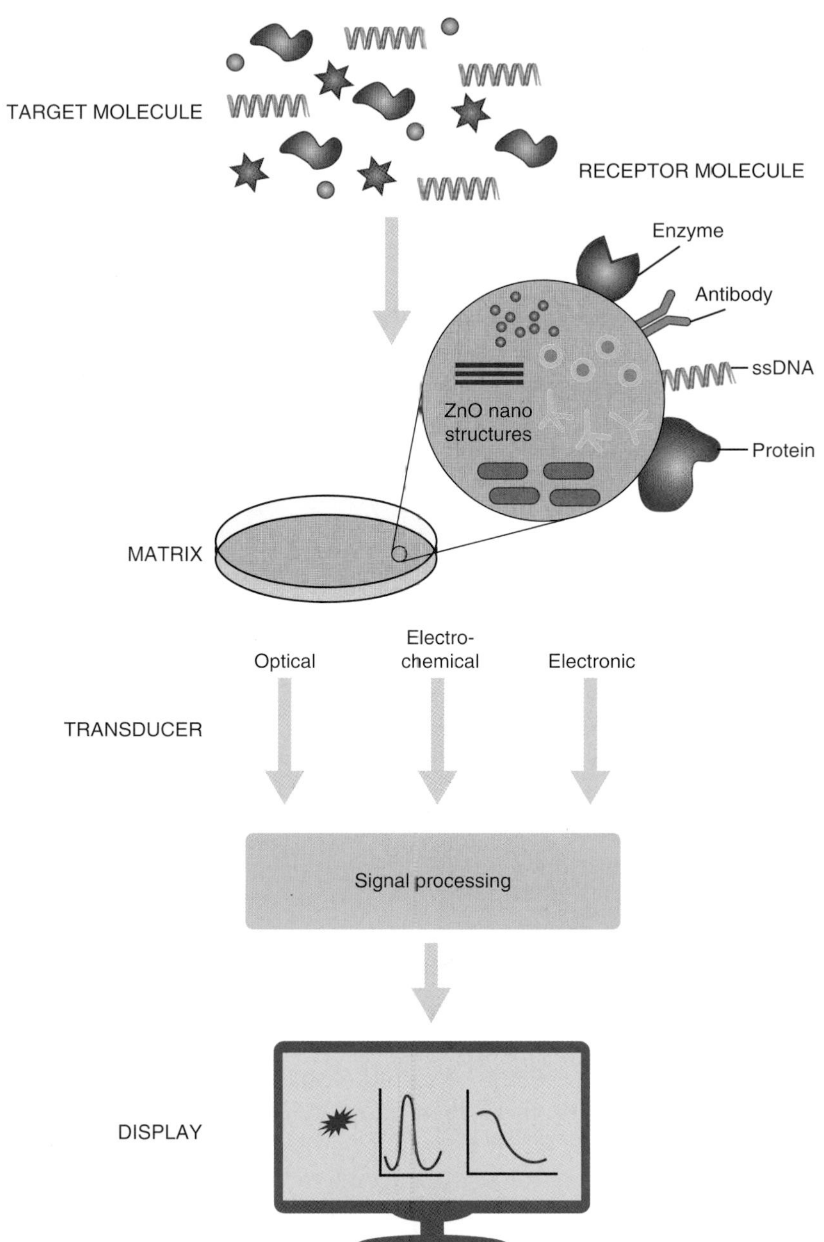

FIGURE 22.30 Biosensors

Biosensors, in general, share a common design. A highly specific biological receptor molecule detects or interacts with a target molecule of interest, for instance, a biological warfare agent. A signal is generated, processed, and displayed for the user. Modified from Arya SK, et al. (2012). Recent advances in ZnO nanostructures and thin films for biosensor applications: review. *Anal Chim Acta* **737**, 1–21.

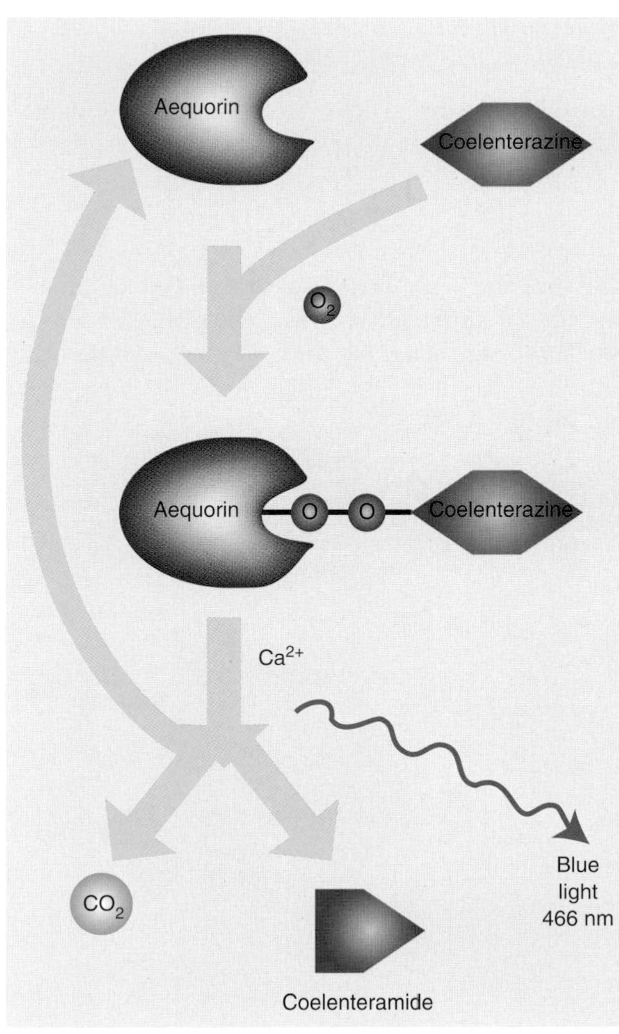

FIGURE 22.31 Light Emission by Aequorin
Aequorin, from *Aequorea victoria,* emits blue light when provided with its substrate, coelenterazine, plus oxygen and calcium. The enzyme binds to aequorin via the oxygen; and when calcium is present, the complex emits blue light, degrades the substrate to coelenteramide, and releases carbon dioxide.

of glucose levels in diabetics using the enzyme glucose oxidase.

There is growing interest today in using biosensors to detect biological warfare agents. Placing biosensors in high trafficked areas, such as in malls or subway stations, could allow for continuous surveillance. Additionally, handheld devices giving a rapid response at the site of a possible attack would be highly useful. Several proposals exist that would use specific antibodies or antibody fragments as detectors for biological warfare agents (see Chapter 6 for antibody engineering).

B cells carry antibodies specific for one antigen, so one proposal is to use whole B cells in a biosensor. When an antigen binds to the antibody on the surface of a B cell, it triggers a signal cascade. Engineered B cells have been made that express **aequorin**, a light-emitting protein from the luminescent jellyfish *Aequorea victoria.* Aequorin emits blue light when triggered by calcium ions (Fig. 22.31). Living jellyfish actually produce flashes of blue light, which are transduced to green by the famous green fluorescent protein (GFP).

In a biosensor, when a B cell detected a disease agent (or any specific antigen), calcium ions would flood into the cell due to activation of a signal cascade (Fig. 22.32). This in turn triggers light emission by aequorin. The light emitted is detected by a sensitive charge-coupled device (CCD) detector. This approach could detect 5 to 10 particles of a biological warfare agent. Approximately 10,000 B cells specific to different pathogens could be assembled in array fashion onto a chip placed inside the biosensor.

Another scheme developed by the Ambri Corporation of Australia uses antibody fragments mounted on an artificial biological membrane, which is attached to a solid support covered by a gold electrode layer. Channels for sodium ions are incorporated into the membrane. When the ion channels are open, sodium ions flow across the membrane and a current is generated in the gold electrode. The ion channels consist of two modules, each spanning half the membrane. When top and bottom modules are united, the ion channel is open. When the top module is pulled away, the ion channel cannot operate. Binding of biological warfare agents by the antibody fragments separates the two halves of the channels, which in turn affects the electrical signal (Fig. 22.33).

Diagnosing pathogenic bacteria with molecular techniques, particularly using the genes encoding ribosomal RNA sequences, is faster and more sensitive than traditional microbiological methods. Biosensors use biological components themselves to monitor for suspicious biomolecules.

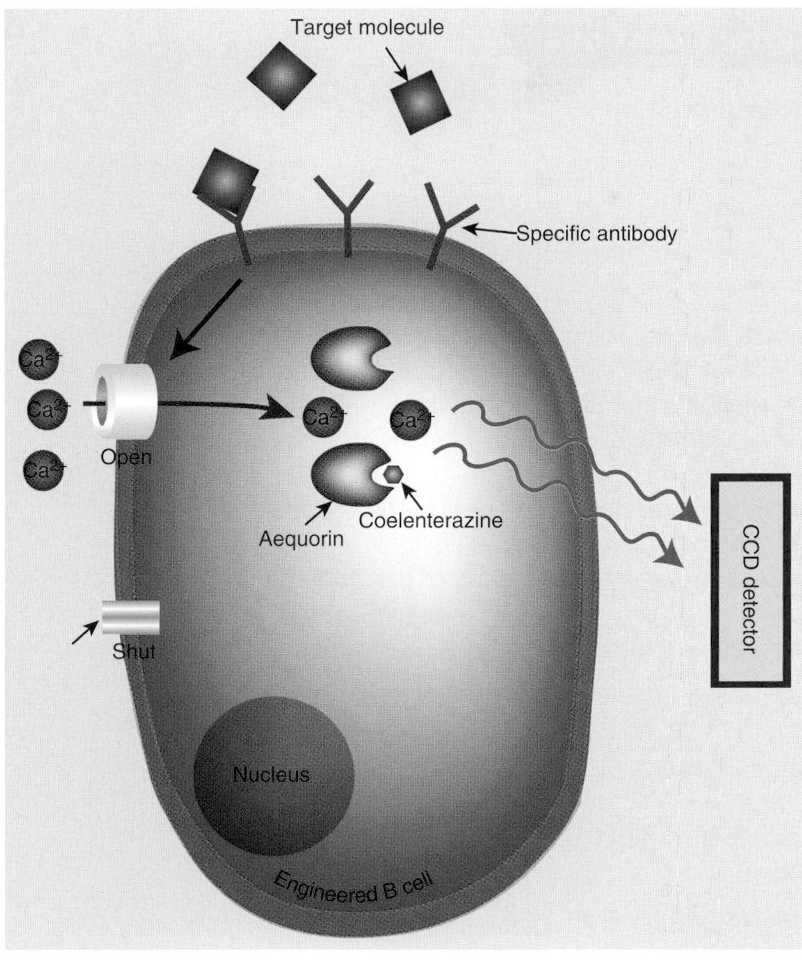

FIGURE 22.32
B-Cell Optical Biosensor
Expressing aequorin in a B cell would provide a detection system for B-cell activation. When a trigger molecule, such as a biological warfare agent, binds to receptors on the B cell, the calcium channels are opened and calcium floods the cell. The high calcium levels would activate aequorin to emit blue light. A charge-coupled device (CCD) would measure the photon emissions and warn the user of a biological agent.

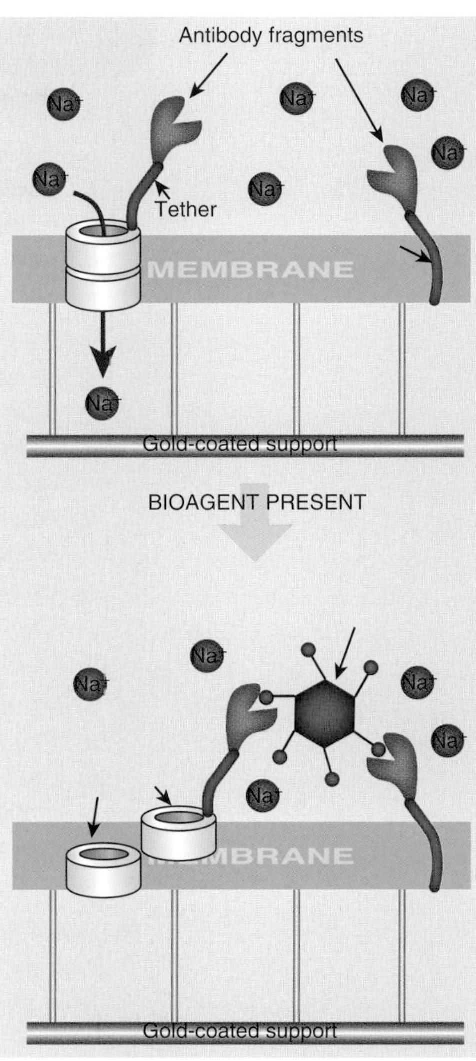

FIGURE 22.33
Antibody Ion-Channel Biosensor
Antibody fragments that bind specific biological agents can be engineered and tethered to a fixed location on an artificial membrane. Another molecule of the same antibody fragment is tethered to a sodium channel. The artificial membrane is carried on a gold-coated solid support that acts as an electrode. This detects sodium ions that pass through the ion channel. When a biological agent is present, the antibody fragments bind it, pulling the top half of the sodium channel out of alignment with the bottom half. Sodium ions no longer pass to the gold electrode, decreasing the signal.

Summary

Biological warfare has been around since life first evolved. Humans are most concerned about the biological warfare directed at us, including infectious diseases and the ability of microbes to evolve resistance to antibacterial agents. Although this development is worrisome, new strategies and technologies are being developed to fight back against the growing problem of antibiotic resistance.

Humans have often attempted to use biological agents in warfare, although with little overall success so far. Several highly virulent infectious agents including anthrax, plague, and smallpox, as well as certain biological toxins such as ricin and abrin, are regarded as likely biological warfare agents. Whether or not genetic engineering can create "improved" bioweapons is as yet uncertain. Developing quicker ways to detect and diagnose microbes is an active area of research.

End-of-Chapter Questions

1. What can bacterial toxins kill?
   a. insect cells
   b. human cells
   c. other bacterial cells
   d. protozoa
   e. all of the above

2. Which statement is true regarding novel antimicrobial strategies?
   a. Siderophores are a good target because they do not exist in humans, and thus the side effects would be diminished.
   b. Disruption of quorum sensing causes bacteria to produce more antibiotic-resistant biofilms.
   c. Alkyl- and aryl-mannose derivatives bind to FimH adhesins and enhance attachment to the natural receptor.
   d. Phage therapy increases the virulence of the bacteria due to transduction.
   e. Production of aflatoxin inhibits fungal agents from growing on cereals.

3. Which of the following is an important consideration of germ warfare?
   a. dispersal
   b. persistence of the agent
   c. incubation time
   d. storage and preparation of the agent
   e. all of the above

4. According to the U.S. Army, which of the following is a requirement for biological weapons?
   a. It should be able to be produced economically.
   b. It should consistently produce death, disability, or damage.
   c. It should be stable from production through delivery.
   d. It should be easy to disseminate quickly and effectively.
   e. All of the above are requirements for biological agents.

5. Which one of the following has rarely been considered as a possible biological warfare agent?
   a. viruses
   b. bacteria
   c. pathogenic eukaryotes
   d. pathogenic fungi
   e. none of the above

6. According to the text, which of the following is one of the best biological weapons?
   a. anthrax
   b. malaria
   c. amoebic dysentery
   d. smallpox
   e. none of the above

7. Since *B. anthracis* strains are closely related, how is it possible to differentiate between the strains?
   a. 16S rRNA sequencing
   b. VNTRs
   c. 23s rRNA sequencing
   d. gene expression profiles
   e. none of the above

8. How does bubonic plague spread?
   a. ticks
   b. person-to-person
   c. fleas
   d. rodents
   e. birds

9. Which of the following is used as a live vaccine for smallpox?
   a. *Variola* major
   b. monkeypox
   c. *Variola* minor
   d. *Vaccinia* virus
   e. none of the above

10. To what virus family do dengue fever and yellow fever belong?
   a. flaviviruses
   b. poxviruses
   c. filoviruses
   d. variola viruses
   e. arenaviruses

11. Which of these is a siderophore?
   a. aflatoxin
   b. PA63
   c. yersiniabactin
   d. anthracimycin
   e. lysin

12. What is the mode of action for ricin?
   a. inactivation of transcription
   b. inactivation of rRNA
   c. activation of the apoptosis pathway
   d. inactivation of the immune system
   e. creation of pores in cell walls

13. Which of the following could be used as a biological agent against agriculture crops?
   a. viruses
   b. bacteria
   c. pathogenic fungi spores
   d. prions
   e. none of the above

(*Continued*)

14. Which of the following gained virulence upon introduction of the IL-4 gene?
- **a.** monkeypox
- **b.** smallpox
- **c.** chickenpox
- **d.** mousepox
- **e.** camelpox

15. How are pathogens detected by using biosensors?
- **a.** by antibodies that are connected to components to give electrical signals or trigger light emission
- **b.** by isolating the pathogen directly from the sample
- **c.** biosensors detect antibodies against specific pathogens, similar to a Western blot
- **d.** by using PCR to amplify variable regions of the pathogen's genome
- **e.** none of the above

16. Which of the following has improved sanitation in healthcare settings by preventing the attachment of bacteria to surfaces?
- **a.** black silicon
- **b.** nanopillars
- **c.** nanocarpets
- **d.** polymers of esters and cyclic hydrocarbons
- **e.** all of the above

17. What is the mechanism of action for botulinum toxin?
- **a.** inhibits the release of acetylcholine
- **b.** activates the release of acetylcholine
- **c.** causes muscle contraction
- **d.** mimics the action of acetylcholine
- **e.** stimulates acetylcholine receptors

18. Which of the following techniques uses both PCR and mass spectrometry to identify a pathogen within eight hours?
- **a.** FISH
- **b.** PNA
- **c.** VNTR
- **d.** SNP analysis
- **e.** PLEX-ID

Further Reading

Abedon, S. T., Kuhl, S. J., Blasdel, B. G., & Kutter, E. M. (2011). Phage treatment of human infections. *Bacteriophage, 1*, 66–85.

Anderson, P. D., & Bokor, G. (2012). Conotoxins: potential weapons from the sea. *Journal of Bioterrorism and Biodefense, 3*, 120.

Centers for Disease Control and Prevention. (2013). *Antibiotic Resistance Threats in the United States* 2013.

Croddy, E. (2002). *Chemical and Biological Warfare: A Comprehensive Survey for the Concerned Citizen.* New York City, NY: Copernicus Books.

Domaradskij, I. V., & Orent, L. W. (2006). Achievements of the Soviet biological weapons programme and implications for the future. *Revue Scientifique et Technique (International Office of Epizootics), 25*, 153–161.

Dublanchet, A., & Bourne, S. (2007). The epic of phage therapy. *Canada Journal of Infectious Disease and Medical Microbiology, 18*, 15–18.

Fang, Y., Rowland, R. R., Roof, M., Lunney, J. K., Christopher-Hennings, J., & Nelson, E. A. (2006). A full-length cDNA infectious clone of North American type 1 porcine reproductive and respiratory syndrome virus: expression of green fluorescent protein in the Nsp2 region. *Journal of Virology, 80*, 11447–11455.

Halverson, K. M., Panchal, R. G., Nguyen, T. L., Gussio, R., Little, S. F., Misakian, M., Bavari, S., & Kasianowicz, J. J. (2005). Anthrax biosensor, protective antigen ion channel asymmetric blockade. *The Journal of Biological Chemistry, 280*, 34056–34062.

Jolley, K. A., Brehony, C., & Maiden, M. C. (2007). Molecular typing of meningococci: Recommendations for target choice and nomenclature. *FEMS Microbiology Reviews, 31*, 89–96.

Li, Y., Sherer, K., Cui, X., & Eichacker, P. Q. (2007). New insights into the pathogenesis and treatment of anthrax toxin-induced shock. *Expert Opinion on Biological Therapy, 7*, 843–854.

Muldrew, K. L. (2009). Molecular diagnostics of infectious diseases. *Current Opinion in Pediatrics, 21*, 102–111.

Neumann, G., & Kawaoka, Y. (2004). Reverse genetics systems for the generation of segmented negative-sense RNA viruses entirely from cloned cDNA. *Current Topics in Microbiology and Immunology, 283*, 43–60.

Osborne, S. L., Latham, C. F., Wen, P. J., Cavaignac, S., Fanning, J., Foran, P. G., & Meunier, F. A. (2007). The Janus faces of botulinum neurotoxin: sensational medicine and deadly biological weapon. *Journal of Neuroscience Research, 85*, 1149–1158.

Palomino, J. C. (2006). Newer diagnostics for tuberculosis and multi-drug resistant tuberculosis. *Current Opinion in Pulmonary Medicine, 12*, 172–178.

Pastagia, M., Schuch, R., Fischetti, V. A., & Huang, D. B. (2013). Lysins: the arrival of pathogen-directed anti-infectives. *Journal of Medical Microbiology, 62*, 1506–1516.

Paterson, R. R. (2006). Fungi and fungal toxins as weapons. *Mycological Research, 110*, 1003–1010.

Pinton, P., Rimessi, A., Romagnoli, A., Prandini, A., & Rizzuto, R. (2007). Biosensors for the detection of calcium and pH. *Methods in Cell Biology, 80*, 297–325.

Prentice, M. B., & Rahalison, L. (2007). Plague. *Lancet, 369*, 1196–1207.

Roxas-Duncan, V. I., & Smith, L. A. (2012). Of beans and beads: ricin and abrin in bioterrorism and biocrime. *Journal of Bioterrorism and Biodefense*, S7, 002.

Schrallhammer, M. (2010). The killer trait of *Paramecium* and its causative agents. *Palaeodiversity, 3*(Suppl.), 79–88.

Simon, J. D. (2011). Why the bioterrorism skeptics are wrong. *Journal of Bioterrorism and Biodefense*, S2, 001.

Sliva, K., & Schnierle, B. (2007). From actually toxic to highly specific—novel drugs against poxviruses. *Virology Journal, 4*, 8.

Stirpe, F. (2004). Ribosome-inactivating proteins. *Toxicon, 44*, 371–383.

Trull, M. C., du Laney, T. V., & Dibner, M. D. (2007). Turning biodefense dollars into products. *Nature Biotechnology, 25*, 179–184.

van Belkum, A. (2007). Tracing isolates of bacterial species by multilocus variable number of tandem repeat analysis (MLVA). *FEMS Immunology and Medical Microbiology, 49*, 22–27.

Yamanaka, K., et al. (2014). Direct cloning and refactoring of a silent lipopeptide biosynthetic gene cluster yields the antibiotic taromycin A. *PNAS, 11*, 1957–1962.

Forensic Molecular Biology

Biotechnology

http://dx.doi.org/10.1016/B978-0-12-385015-7.00023-5

THE GENETIC BASIS OF IDENTITY

DNA technology has many practical uses. Because every individual has a unique DNA sequence, DNA samples can be used for identification. The legal system is now using DNA evidence to determine guilt or innocence (see Box 23.1). The application of DNA technology began in Britain in the mid-1980s and appeared in America shortly afterward. Today many societies have reached the point of compiling DNA databases of known criminals—especially serious offenders. However, the most frequent use of DNA evidence is actually in cases of unknown or disputed paternity.

Identity can also be established in other ways. Just a casual glance reveals major differences among people. Geneticists refer to this outward appearance as the *phenotype.* Most physical differences between people are due to complex interactions of several genes during development. Some are obvious at a glance; others require close observation. Fingerprints are the classic example of a phenotype used in law enforcement. They are due to variations in the pattern of dermal ridges, small skin elevations on our fingers. Fingerprint patterns depend on more than one gene (i.e., they are **multigenic**). This creates the huge genetic diversity underlying this phenotype. Although you might expect the fingerprints of identical twins to be the same, they are not identical. Minor variations in fingerprint patterns occur as a result of environmental factors affecting development. Fingerprints were being used for identification by the late 1800s.

Retinal scans provide a more high-tech form of unique identification. These scans take advantage of the unique pattern of blood vessels on the retina at the back of the eyes. Scanning typically takes about a minute because several scans are needed. A subject must place his or her eye close to the scanner, keep his or her head still, and focus on a rotating green light. Infrared light is used for scanning because blood vessels on the retina absorb this light better than the surrounding tissues do. A computer algorithm is then used to convert the scan into digital data. There is about 10-fold more information in a retinal scan than in a fingerprint.

Previously, retinal scanning was mostly used in high-security situations—for example, by the CIA, FBI, NSA, and some prison systems. However, it is now being used for animal identification (Fig. 23.1). The Optibrand Corporation now makes a handheld scanner for livestock, especially cattle. The setup incorporates a Global Positioning System (GPS) time, date, and location stamp. The Optibrand system can identify and track individual animals through the food processing chain. This capability is of special relevance in tracing any animals suspected of being exposed to mad cow disease.

Box 23.1 Forensics Has Two Definitions

Originally, the word *forensics* referred to debating evidence in a courtroom. It derives from the Latin word *forensic*, referring to the forum, the public area where meetings and trials were held. Forensic science (or forensic studies) is the use of science in gathering evidence. Although the term *forensic science* is used most often in relation to the law, it is also used in fields such as history and archeology. Most recently, the term *forensics* has been used to mean the same as *forensic science*. Indeed, many dictionaries now accept that *forensics* can mean either *debating evidence* or *forensic science*.

Fingerprints and retinal scanning use highly detailed biological data for identification of individual humans or animals.

FIGURE 23.1 Optibrand Retinal Scanning of Cattle
Retinal scanning is now being used for cattle identification. Courtesy of Optibrand Ltd. LLC, Fort Collins, Colorado.

BLOOD, SWEAT, AND TEARS

All kinds of body tissues and fluids may be used to establish identity. Although blood analysis is most common, other body fluids such as sweat, tears, urine, saliva, and semen also have cells with surface proteins that can be analyzed.

The membranes of red blood cells contain several proteins and lipids with attached carbohydrate portions that are exposed on the outside of the cell. These are highly antigenic, and in **blood typing**, they are referred to as **blood antigens**. Binding of an antibody to the corresponding antigen is highly specific. Consequently, two related blood antigens with only relatively small differences in shape can be told apart because different antibodies bind them.

Several groups of blood antigens are used for identification. The best known is the **ABO blood group** system. Three different glycolipids, A, B, and O, are involved. They consist of different carbohydrate structures attached to the same lipid. The A antigen is made by adding *N*-acetyl-galactosamine to the end of the O antigen and the B antigen by adding galactose (Fig. 23.2). Antibodies are made against the A and B antigens, but the shorter O "antigen" is poorly antigenic and invokes little antibody production. The closely related enzymes that make the A and B antigens are encoded by different alleles of the same gene. Absence of this enzyme gives the O allele.

Moving to the molecular level, the proteins and polysaccharides made by all cells provide another set of identifying features. Good examples of individual differences that involve these features are the various blood types found in human populations. But for ultimate identification at the molecular level, we must examine the genes themselves to determine the genotype. This is what is meant by DNA typing or **DNA fingerprinting**, a technique described in a following section.

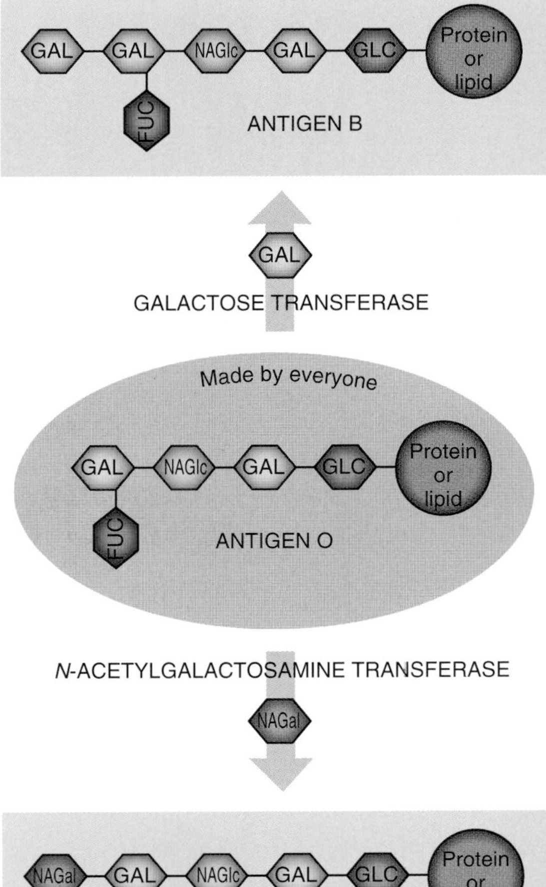

FIGURE 23.2 ABO Blood System Molecules
Antigenic glycolipids determine each individual's blood type. The enzyme *N*-acetyl-galactosamine transferase makes A antigen by adding *N*-acetyl-galactosamine to the end of the O antigen. Galactose transferase makes B antigen by adding galactose to the end of the O antigen. Abbreviations: NAGal, *N*-acetyl-galactosamine; GAL, galactose; NAGlc, *N*-acetyl-glucosamine; FUC, fucose; GLC, glucose.

Three alleles—A, B, and O—are present in the population. Because we all have two copies of each gene, we all have two alleles for the ABO system. They may be identical or different in any given person. The A and B alleles are both dominant; therefore, if an allele for either A or B is present, that antigen will be expressed (Table 23.1). A person with one A and one B allele will express both antigens on his or her blood cell surfaces (AB blood group).

People do not make antibodies against those antigens present on their own red blood cells. Each individual makes antibodies against foreign antigens, that is, those present only in other individuals. Consequently, people with type A blood express antibodies against the B antigen, whereas those with type B blood express antibodies to the A antigen. People with type O blood express antibodies to both A and B antigens, and those with type AB have neither antibody (see Table 23.1). If a person is given blood of the wrong type, that person's antibodies cause the foreign blood cells to clump together or agglutinate. People with type AB blood are considered universal acceptors because they do not react against any type of blood. Conversely, type O people are universal donors but can accept only type O themselves because they have antibodies to both the A and B antigens.

Table 23.1 ABO Blood System

Alleles Present	Blood Type Expressed	Antigens Made	Antibodies
A A	A	A	anti-B
A O	A	A	anti-B
B B	B	B	anti-A
B O	B	B	anti-A
A B	AB	A and B	neither
O O	O	neither	anti-A and anti-B

Because humans are diploid, a child may belong to a different blood type than either of his or her parents. For example, an AO mother will make A antigen, and a BO father will make B antigen. Despite this, these parents can have a type O child because the child may inherit the O gene on one chromosome from the heterozygous mother and another O gene from the heterozygous father (Fig. 23.3).

Usually, we know who the mother of a child is. Mostly, it is the father who may be difficult to identify. If a man is accused of being a father in a paternity suit, ABO typing will exclude him in only around 15% to 20% of cases, assuming he is innocent. Consider, for example, a mother who has blood type A with a daughter of type AB. We know the father must have contributed the B allele to the daughter. Therefore, the father must be BO, BB, or AB but cannot be AA, AO, or OO. Therefore, anyone accused of paternity but who has type A or type O blood is innocent. Individuals with type B or AB blood might be guilty. Because large numbers of people share each blood group, ABO blood typing alone cannot provide proof of guilt, although it can prove innocence.

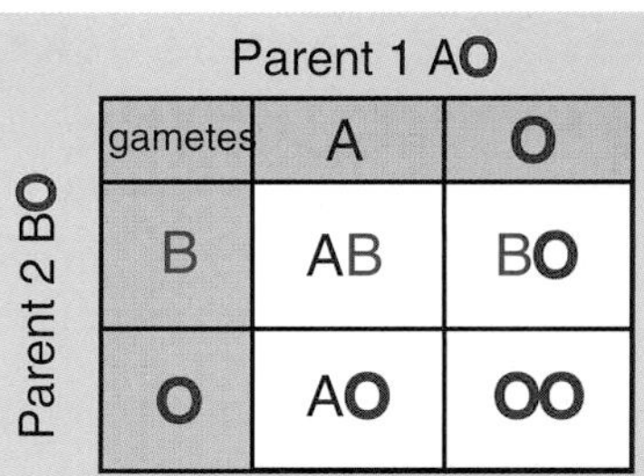

FIGURE 23.3
Inheritance of ABO Blood Type
If parents with two different blood types have children, the offspring may differ in blood type from either parent. In this example, parent 1 has the alleles for type O and A antigens. Parent 2 has the alleles for type B and O antigens. Mendelian analysis shows that the child could have type AB, type B, type A, or type O—that is, all the possibilities.

Several other blood and tissue antigen systems similar in principle to the ABO system are also used in forensic medicine. Using the HLA system of white blood cells, the chance of exclusion is over 90%. When the HLA and ABO systems are combined, the chances of exclusion are about 97%. Including the analysis of blood serum proteins with the others makes exclusion almost certain, if the accused is indeed innocent.

The following formula is used to determine the combined probability (P) of exclusion from multiple tests with individual probabilities P1, P2, P3, and so on:

$$P = 1 - (1 - P1)(1 - P2)(1 - P3)(1 - P4), \text{etc}.$$

Blood typing relies on identifying antigens present on the surface of blood cells. The widely used ABO system consists of three different alleles distributed among the human population.

FORENSIC DNA TESTING

Blood typing was once the primary forensic evidence used in many criminal cases. During the 1980s, the use of DNA in forensics became established. Today DNA evidence is regarded as superior to virtually every other type of forensic evidence in establishing identity. No two individuals (except identical twins) have the same DNA. During gamete development and fertilization, sets of individual chromosomes are distributed to offspring in so many possible combinations that it is incredibly unlikely that any two individuals will have the same DNA. Identical twins are the exception that proves the rule because they occur when the egg divides after fertilization has already happened. (Modern advances allow us to tell apart even identical twins, by using alterations to the DNA that occur after birth; see the following discussion).

DNA tests alone, without supportive evidence, have sometimes been sufficient for conviction when identity was the key issue (see Box 23.2). DNA evidence is now almost always sufficient for exoneration of misidentified individuals who were wrongly convicted of committing a crime. DNA evidence can be obtained from any body tissue or secretion that has cells with nuclei that contain DNA. Tissue samples taken from suspects are then compared with evidence obtained from the crime scene. Several types of analysis may be used to determine whether DNA found at the scene of a crime matches that of the suspect or the victim. The most widespread is known as **DNA fingerprinting**.

DNA evidence can provide virtually unambiguous identification of individuals if carried out properly.

Box 23.2 Heroic Cockatoo Provides DNA Evidence

In September 2002, a fascinating example of DNA evidence was presented to a grand jury in Dallas, Texas. A pet bird pecked and clawed intruders while trying to protect its owner. Blood found at the crime scene apparently came from a wound the bird pecked on the head of one of the suspects. The bird, a white-crested cockatoo, stood 18 inches tall and had a beak powerful enough to snap thin tree branches. Sadly, the intruders killed both the bird and its owner. Both the murdered owner and the bird were autopsied, and the bird's beak and claws were checked for blood in a search for additional DNA evidence. When confronted with the condemning DNA evidence, one of the suspects confessed—but blamed his partner for the actual killing.

DNA FINGERPRINTING

DNA fingerprinting relies on the unique pattern made by a series of DNA fragments after separating them according to length by gel electrophoresis. DNA samples from different suspects, the victim, and samples from the crime scene are first purified. The samples are then processed to generate a set of DNA fragments. When Alec Jeffreys invented DNA fingerprinting in 1985 in England, the DNA was cut with restriction enzymes to generate fragments because PCR had not yet been invented. In addition, early DNA fingerprinting used radioactivity for labeling the DNA fragments. Nowadays, DNA is prepared by PCR, and fluorescent dyes are used for labeling. In addition, modern DNA fingerprinting uses repeated sequences (short tandem repeats or STRs) for routine identification purposes, as described here.

The polymerase chain reaction (PCR) is a procedure for amplifying tiny amounts of DNA. The details of PCR are discussed in Chapter 4. Early DNA fingerprinting, using restriction enzymes, required relatively long strands of DNA. However, not only can PCR be used on short segments of DNA, but it also needs far smaller amounts of DNA. Thus, PCR can amplify a segment of DNA starting from only a picogram (10^{-12} g), although microgram (10^{-6} g) quantities or larger are better. In fact, PCR can be used successfully to analyze DNA from a single cell. In current practice, forensic DNA analysis is almost all done by PCR-based methodology.

For the first generation of DNA fingerprints, restriction enzymes were used to generate the variation in DNA fragment size between individuals. Variations in the DNA base sequence of restriction enzyme recognition sites result in differences in the size of the fragments. Such sequence differences are called *restriction fragment length polymorphisms* (*RFLPs*; see Chapter 3). There is approximately one difference in every 1000 nucleotides between nonrelated individuals. Many different restriction enzymes with distinct recognition sites exist, and in practice several such enzymes are used on each DNA sample. Even if mutations have changed a few bases of the target sequence around the cut site, there is usually still enough similarity for probes to bind. The entire process requires several weeks to finish.

The steps involved in RFLP-based DNA fingerprinting are as follows (Fig. 23.4):

1. The DNA is cut with a restriction enzyme.
2. The DNA fragments are separated by length or molecular weight by gel electrophoresis.
3. The fragments are visualized by Southern blotting. The separated fragments are transferred from the gel to nylon paper. Then a radioactively labeled DNA probe is added. The probe binds to those DNA fragments with complementary sequences.
4. The blot is covered with radiation-sensitive film to give an autoradiograph. This shows the location of those DNA fragments that reacted with the radioactive probe.

The final product of a DNA fingerprint is an autoradiograph that contains at least five essential lanes (Fig. 23.5). The markers are standardized DNA fragments of known size, which have been radioactively labeled. They help determine the size of the various fragments.

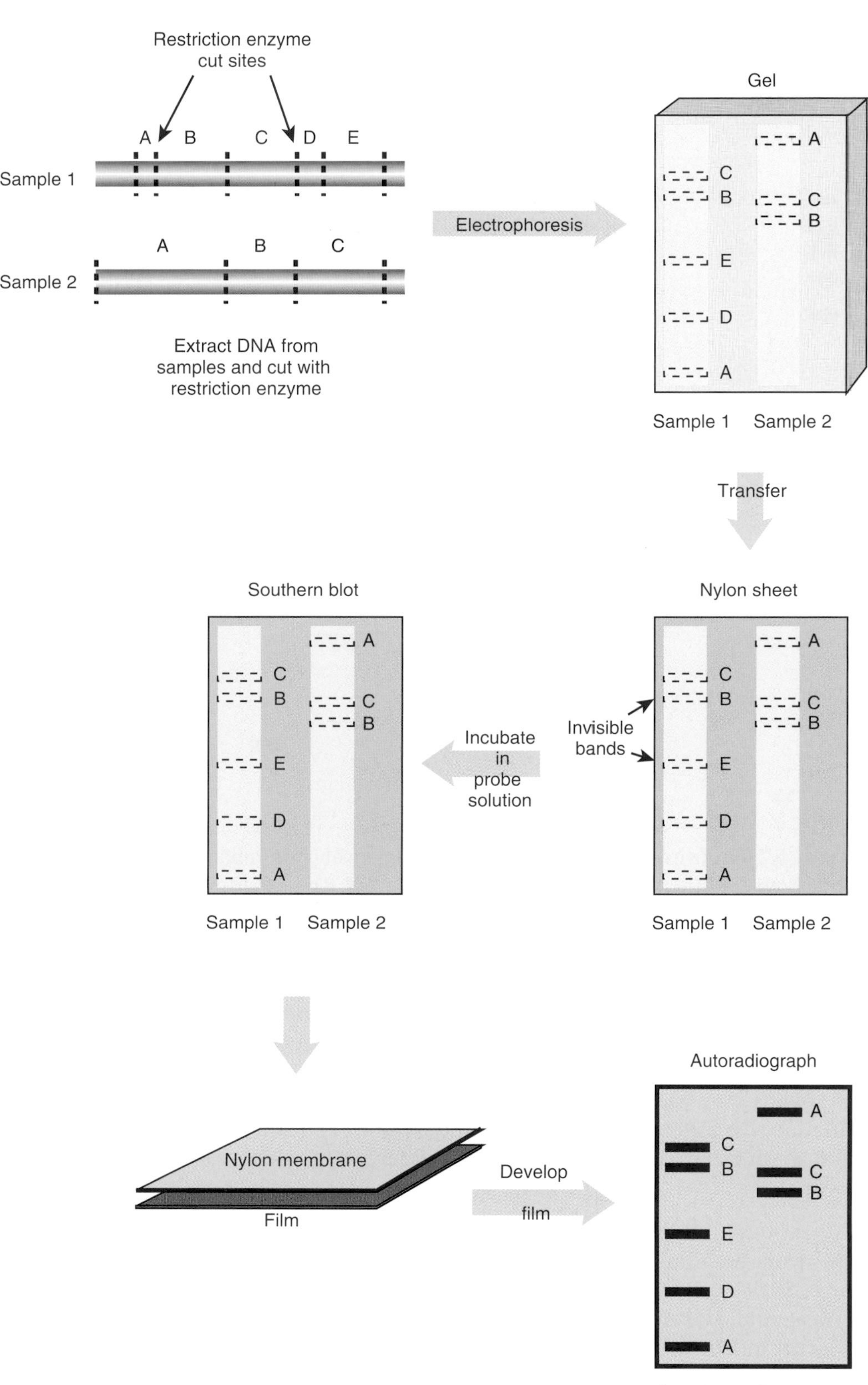

FIGURE 23.4 Outline of RFLP-Based DNA Fingerprinting

A pattern of different-sized DNA fragments is generated during DNA fingerprinting. The sequence of the DNA determines where the restriction enzyme cuts in the first step; thus, different people will have a different pattern of fragments for the same restriction enzyme. These differences may be used to identify people.

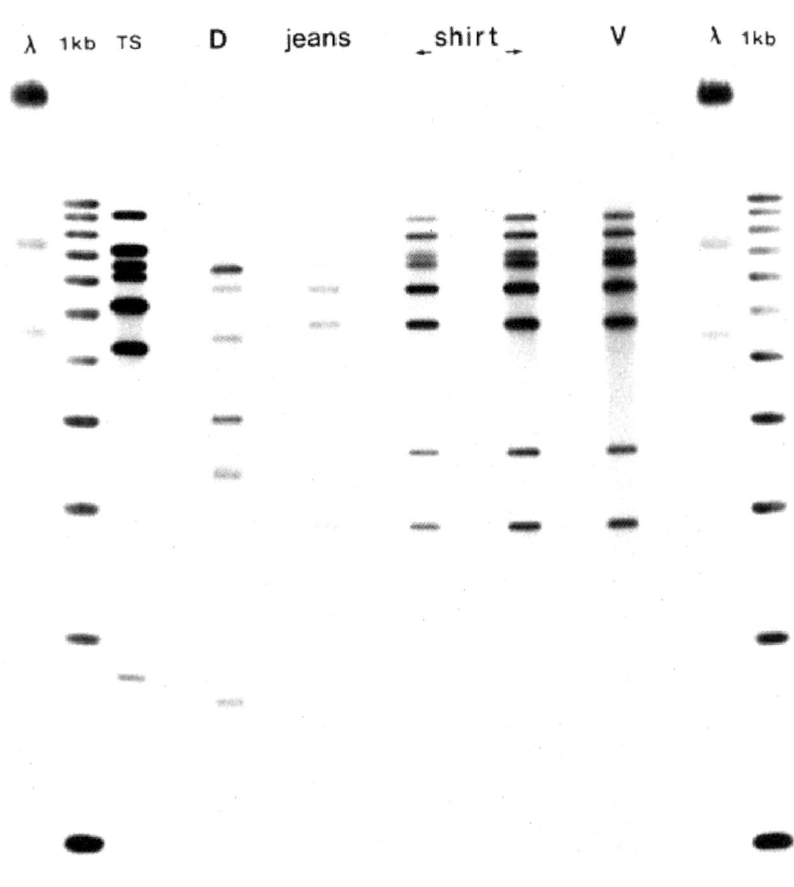

FIGURE 23.5 DNA Fingerprint
Actual DNA fingerprint showing that the pattern of DNA fragments of the victim (V) were found on the defendant's clothing (jeans/shirt). The first two and last two lanes are the standard size markers (labeled λ and 1 kb). The lane marked TS is a positive control showing that the fingerprint technique was successful. The lane marked D is the defendant's DNA pattern. Reprinted with permission from Quick Publishing, LC.

The "control" is DNA from a source known to react positively and reliably to the DNA probes and shows whether the test has worked as expected. The experimental lanes have samples from the victim, the defendant, and the crime scene. In this example, blood from the defendant's clothing was compared with his own blood and the victim's blood. The DNA from the clothing actually matches that of the victim.

Two variants of DNA fingerprinting have been used: **single-locus probing (SLP)** and **multiple-locus probing (MLP)**. In SLP, each probe is specific for a single site, that is, a single locus, in the genome. Because humans are diploid, each SLP probe normally gives rise to two bands from each person, provided that the chosen locus shows substantial allelic variation. Occasional persons will be homozygous and show only a single band. For full identification using SLPs, it is necessary to run several reactions, each using a different SLP probe.

Multilocus probes bind to multiple sites in the genome and consequently generate multiple bands from each individual. Because it is not known which particular band comes from which particular locus, interpretation is difficult. Furthermore, statistical analysis is impractical, and data cannot be reliably standardized for reference. In practice, fingerprints generated by MLP must be directly compared with others run on the same gel. Historically, MLP was used before SLP. However, SLP analyses use smaller amounts of material and are easier to interpret and compare. Statistical analysis and population frequencies are possible using SLP data. Consequently, MLP methods have largely been displaced by SLP analysis.

In addition to preparing DNA for fingerprinting, PCR may be used to generate DNA segments for direct comparison by DNA sequencing or hybridization. PCR is especially useful for amplifying small regions of DNA with high person-to-person variability. If the sequences of two samples match in several highly variable regions, they are probably from the same person.

In this approach, DNA from a forensic sample is amplified by PCR and compared with DNA from the suspect. For hybridization, spots of DNA samples are bound to a membrane and tested for binding a DNA probe that is tagged with a fluorescent dye (radioactive probes were used in early work). In such dot blots, the probe either binds or doesn't bind, so any spot is either positive or negative (Fig. 23.6). Alternatively, segments of DNA that have been amplified by PCR can be fully sequenced.

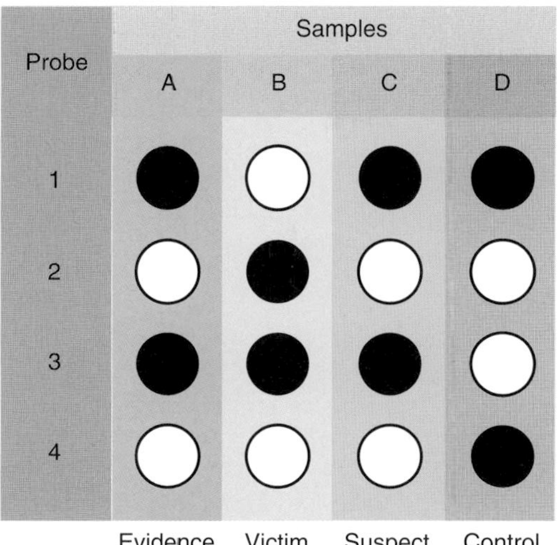

FIGURE 23.6 PCR plus Dot Blots for Identification
This example compares whether or not evidence from a crime scene matches the victim or suspect. DNA from the victim, the suspect, and the evidence was isolated. PCR was used to amplify a short segment of the DNA using primers flanking a highly variable region. The PCR products from the evidence, victim, suspect, and a control were each applied as four separate spots to a nylon membrane. The membrane was treated with four different probes, such that each probe was in contact with each of the different PCR products. Probes bound specifically to those spots that had sequences matching the probe. Notice that the pattern of the suspect and the evidence are identical. Therefore, the suspect's DNA was present at the scene of the crime.

DNA fingerprinting relies on the unique pattern found in different individuals when a series of DNA fragments is separated according to length. The polymerase chain reaction (PCR) is now used to generate the DNA fragments for DNA fingerprinting.

USING REPEATED SEQUENCES IN FINGERPRINTING

A variation of DNA fingerprinting is to look at regions of the DNA that contain **variable number tandem repeats (VNTRs)**. As discussed in Chapter 8, this means that sequences of DNA are repeated multiple times and that different people have different numbers of repeats. Repeat sequences vary greatly in length; however, for forensic purposes, relatively short repeated sequences are now generally used and are known as **short tandem repeats (STRs)**—see later discussion. VNTRs usually occur in noncoding regions of DNA. Hence, using VNTRs protects privacy in the sense that an individual's coding DNA is not revealed during forensic investigations. Originally, restriction enzymes were used to cut out the DNA segment containing the VNTR. Nowadays, DNA fragments for VNTR analysis are generated by PCR. As before, the DNA bands are separated by gel electrophoresis and visualized by Southern blotting. Figure 23.7 shows corresponding DNA fragments from three individuals who differ in the number of repeats in the same VNTR. Consequently, the length of the fragment differs from person to person.

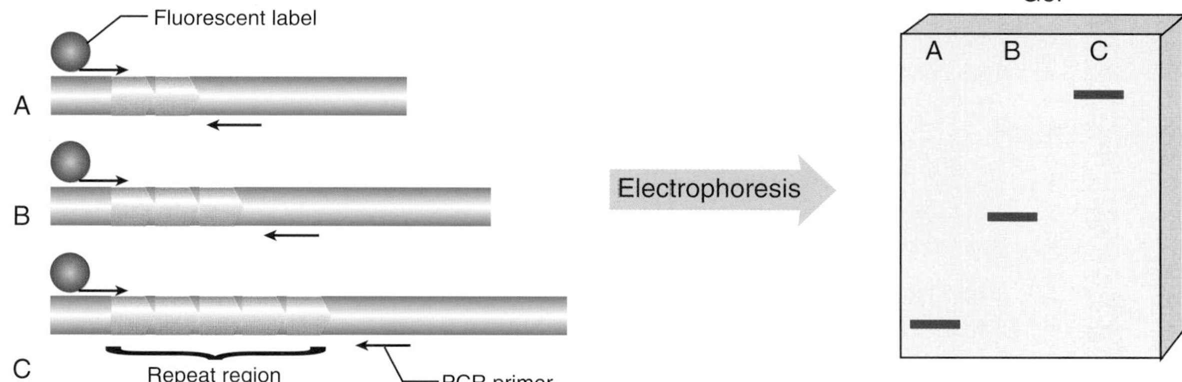

FIGURE 23.7 VNTR Fingerprinting
Genomic DNA has regions with repeated sequences. In each individual, the number of repeats varies, and therefore, the lengths of these regions can be compared to distinguish identities. The repeated region is isolated using PCR from three individuals marked A, B, and C. One of the PCR primers is labeled with a fluorescent dye. The fragments are run on agarose gels to compare the lengths.

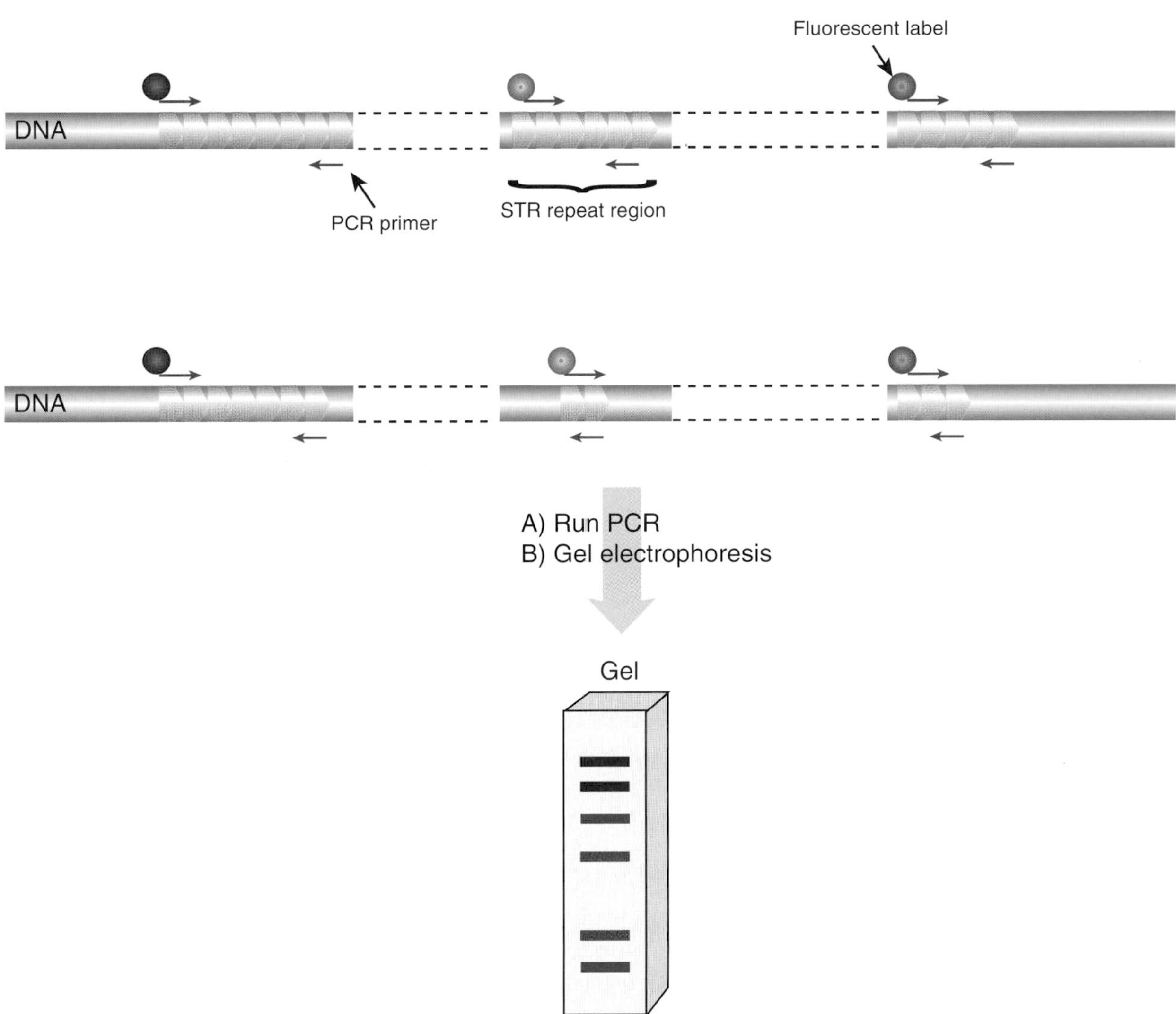

FIGURE 23.8 Multiplex STR Fingerprinting
Three different STR loci are amplified using PCR primers. Each set of primers is labeled with a different fluorescent label to distinguish each locus from the other. The PCR products are run on an agarose gel to determine the length of the fragment and therefore the number of repeats. This individual is heterozygous for each locus because there are two different-sized bands for each PCR primer set.

There is often an enormous variation between people in the number of repeats at any particular VNTR site. So there is a very low likelihood of any two people matching exactly. Some VNTRs have 100 to 200 different variants, making them very useful for forensic analysis. Indeed, the original DNA fingerprints used highly variable VNTRs with long repeat sequences. Although VNTRs are not genuine genes, their variants are regarded as alleles, and so VNTRs are considered to be "multiallelic" systems. One practical problem is that multiple tandem repeats give so many closely packed bands that standard agarose gel electrophoresis cannot discriminate the different fragments. Different types of gels such as polyacrylamide may resolve closely spaced bands. Alternatively, the fragments can be separated on a gradient gel.

The STR is a subcategory of VNTR in which the repeat is from two to six nucleotides long. Most STRs are not as variable in the number of repeats as VNTRs with longer unit sequences. Many STR sequences have only 10 to 20 alleles and hence cannot provide unique identification alone. However, many STR loci are available, and if several are analyzed simultaneously, this will provide enough data that the pattern would be unique for each individual.

It is possible to analyze several STR loci simultaneously by running multiple PCR reactions in the same tube using different primers **(multiplex PCR**; Fig. 23.8). This requires that the multiple sets of primers do not interfere with each other, which is often difficult to achieve when running six or more amplifications together. Nonetheless, commercial kits are now available that can run up to 13 STR analyses in one reaction tube. Such a multiplex analysis gives a gel track with up to 2*N* bands (where *N* is the number of loci analyzed). For this example, 13 STR loci would produce 26 bands. Fewer bands will be seen in individuals who are homozygous at any of the chosen loci. Despite the apparent complexity, the STRs that are used derive from known sequences at known chromosomal locations, and hence the individual bands can be identified and entered into computer databases. Multiplex STR analysis is the basis of the national database set up in the United Kingdom in 1995. Similar databases are now used in other European nations and, since 1998, in the United States.

Present-day DNA fingerprinting uses repeated sequences. In particular, STR sequences are used because they are convenient for analysis. Multiple standardized PCR reactions for several STRs can be run in the same tube.

PROBABILITY AND DNA TESTING

If two DNA samples are different, then they must have come from different people. Hence, DNA testing can readily exclude an individual from being suspected. But what if two DNA samples match for whatever tests we have run? Positive identification requires the use of probability. Inclusion depends on the assumption that it is highly improbable that the DNA of the suspect and the DNA from the evidence match merely by chance. Through use of DNA profiling, it is now possible to achieve probabilities of less than 1 in the total world population of a chance match. We should be cautious if close relatives are suspects in criminal proceedings because the probability of a match is obviously much greater than for the general population.

The following general steps are important when determining the probability of a match:

1. From the same population to which the suspect belongs, select a random sample of individuals.
2. Determine the genotype of these randomly selected individuals and estimate the frequency of the alleles at the loci used in DNA typing.
3. Calculate the probability of finding the genotype of the suspect by assuming that the suspect's alleles at each individual locus represent a random selection from the population in general. We also assume that the alleles tested are not linked but are independent of each other.
4. Multiply together the frequencies that are determined from the various loci. The number obtained represents the overall probability that the suspect's DNA would match the evidence by chance.

The details of population genetics used to establish probabilities for genetic screening, whether DNA or blood groups, are beyond the scope of this book. However, the probabilities from DNA testing are now sufficiently good in practice to make identification virtually certain, provided that the tests are carried out properly on reasonably good samples of DNA (see Box 23.3). Convictions have been obtained using DNA evidence where the probability of a chance match was 1 in 100, but with the addition of supporting evidence. However, in cases in which the evidence is primarily based on DNA, juries often expect astronomical odds such as 1 in several billion.

Modern DNA analyses using multiple STR sequences can provide almost total certainty of unique identification.

Box 23.3 Exoneration of Misidentified Individuals Using DNA

As already noted, if two DNA samples are different, they must be from different individuals. Consequently, DNA testing can demonstrate innocence in a very clear-cut manner. This approach has been used to exonerate a number of wrongly convicted individuals in cases in which identity was the key issue for conviction. Many of these overturned convictions occurred before DNA testing was routine. In such cases, samples of biological evidence that can provide DNA must have been stored and be available for modern DNA analysis.

As of early 2014, just over 300 people have been exonerated in the United States by the use of DNA. Figure A shows the number of such exonerations per year until 2010. The average time in prison served by those exonerated was around 14 years. About three-quarters of these wrongful convictions were due to misidentification by eyewitnesses. About half involved invalid forensics (such as overreliance on blood group analysis). In nearly half of the DNA exoneration cases, the real criminal was identified by the DNA testing.

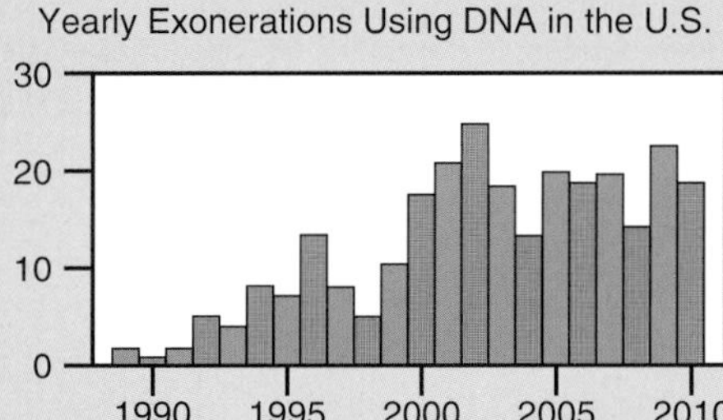

FIGURE A Exonerations by Year
Number of exonerations per year in the United States based on using DNA for identification. Data used are from The Innocence Project, http://www.innocenceproject.org/know/.

THE USE OF DNA EVIDENCE

The Frye rule (*Frye v. United States,* 1923) concerns the admissibility of scientific evidence. The principle states that new scientific tests must be generally accepted in appropriate scientific circles before evidence from them is admissible in courts. A "helpfulness" standard, which involves the use of expert witnesses to assist the court in interpreting scientific evidence, is also applied in some states. By 1996, DNA evidence had been admitted in more than 2500 criminal cases in the United States. There are relatively few U.S. government labs doing DNA testing, however. Accredited private labs perform most forensic work, and these services are available to both the prosecution and defense.

The main impact of DNA technology has been the far greater certainty with which individuals can be associated with or excluded from a particular crime than was possible with traditional blood tests. Experience has shown that if DNA testing is given as evidence, there is a higher probability of conviction than without DNA testing. DNA evidence is commonly used in cases of rape. However, in most cases of rape, the accused admits knowing the alleged victim and identity is not an issue. DNA testing can also be used by law enforcement to narrow the number of possible suspects, given that simple DNA testing protocols can determine sex and racial characteristics.

The British judicial system has led the way in DNA testing. In one early case, DNA testing was used to exclude the person first suspected of the sexual assault and murder of a young girl. But to find the real perpetrator, the police screened more than 5000 men in the village by blood testing (ABO and phosphoglucomutase), only to find no match with anyone. Ironically, the murderer was discovered because it was revealed that he had paid another man to give blood for testing. DNA testing subsequently confirmed, with high probability, that his DNA matched that of the semen sample taken from the victim. A conviction was finally obtained. Overall, there is considerable public support in Britain for maintaining DNA profiles on the entire population.

In the United States, the FBI presently maintains a national DNA data bank, and computer searches often screen it to find suspects. Although some worry about invasion of privacy, those who feel that liberty includes the freedom to walk down the street without being assaulted regard these developments favorably. Can DNA information be misused? The answer, of course, is that any information can be abused. People with different racial or genetic characteristics have been persecuted in the past. Then again, detailed DNA sequence information is hardly necessary for identifying people by race. In practice, the vastly increased accuracy of DNA testing compared to, say, blood group analysis means that unique individual identification is usually possible, and consequently, racial bias is largely excluded in DNA testing.

DNA testing has become widespread in industrial nations, and many countries are setting up national DNA data banks.

DNA IS ALSO USED TO IDENTIFY ANIMALS

Obviously, DNA can be used to identify animals as well as people. What happens when a sleazy operator mixes low-quality bonito in with premium tuna? The taste test may tell us something is wrong, but this is not sufficient for legal action. DNA analysis can reveal the species of fish present, even if mixed in with others. Sequence differences and profiles of PCR fragments generated from the cytochrome b gene, carried on mitochondrial DNA, can be used to distinguish multiple closely related fishes. The food regulatory agencies of the European Union are now using this procedure.

When American newspaper the *Boston Globe* investigated local restaurants in 2011 and sent samples for DNA testing, it found that nearly half (48%) of the fish were misidentified. Although deliberate fraud is sometimes to blame, a major problem is that several hundred species of fish are used as food, and many can be told apart only by experts—assuming that the fish are still whole and have not yet been processed into less recognizable forms. A more "scientific" study found that 75% of red snapper sold in the United States was mislabeled. As a result, the U.S. Food and Drug Administration is now implementing DNA testing procedures.

DNA testing is also used to monitor meat products. Early in 2013, a scandal broke out in England and Ireland when it was discovered through DNA analysis that some beef products contained around 25% horsemeat. The scandal then spread to most other European countries. Most of the horsemeat was eventually traced to poorer Eastern European countries where horses are sent for slaughter. Horsemeat (correctly identified) is acceptable for human consumption in several countries and also used in animal food.

A rather less tasteful use for the DNA-based identification of animals is in tracing unscooped dog poop. A few upscale apartment complexes mandate that dog owners must have their pets' DNA entered into a database. When owners fail to clean up after their pets, the offending deposit is sampled and analyzed to identify the dog responsible. The company providing this service is known as PooPrints™, and its website can be accessed at http://www.pooprints.com/. It operates at between 30 and 40 locations in the United States, with outliers in Israel and Singapore.

DNA analysis of animal droppings also can be put to more scientific use. Fecal samples contain DNA not only from the depositing animal but also from the animals and/or plants it has eaten. This analytic approach has allowed quantification of the diet of predators whose movements are difficult to follow in the wild. For example, snow leopards (*Panthera uncia*) in the mountains of Mongolia are often accused of killing livestock. Recent analyses of droppings showed that domestic goats and sheep comprised about 20% of the diet; and wild sheep and Siberian ibex, the other 80% (Fig. 23.9).

This approach has even been used on the fossilized feces (coprolites) of the extinct cave hyena (*Crocuta crocuta spelaea*). This hyena was much larger than its modern African relatives and hunted large prey, such as wild horses. The cave hyena became extinct around 13,000–14,000 years ago. DNA from samples of coprolites from French caves showed that a major component of the hyena's diet was the red deer (*Cervus elaphus*), an animal that still survives.

DNA sequences may be used to identify animals. DNA analysis can reveal wrongly labeled fish and meats in food. In addition, DNA analysis of animal droppings can be used to reveal their diets.

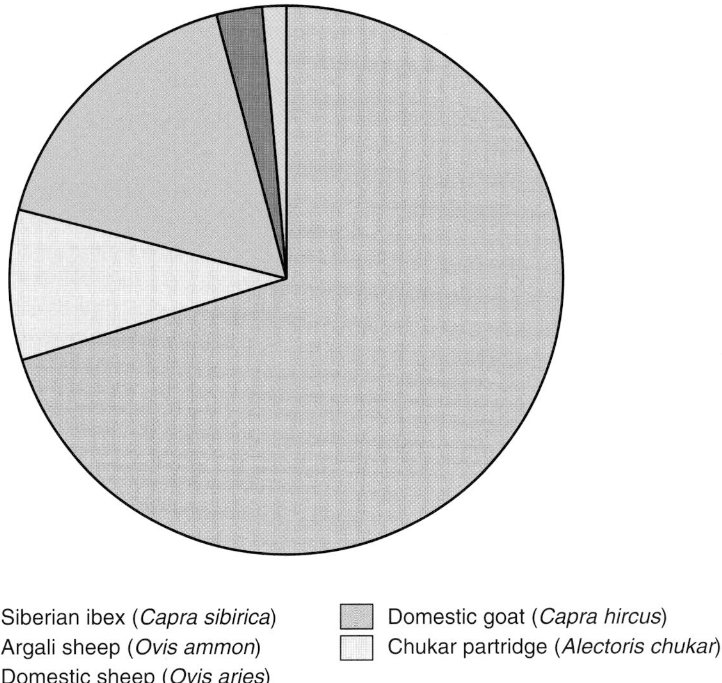

FIGURE 23.9
Composition of Snow Leopard Prey
Relative frequency of different species consumed by snow leopards from the Tost Mountains, South Gobi, Mongolia. The proportions are based on the analysis of DNA in the feces of the snow leopards. From Shehzad W, et al. (2012). Prey preference of snow leopard (*Panthera uncia*) in South Gobi, Mongolia. *PLoS One* **7**(2), e32104.

TRACING GENEALOGIES BY MITOCHONDRIAL DNA AND THE Y CHROMOSOME

Mitochondrial DNA sequences have been very useful in tracing the recent evolution of the human species at the molecular level. Analysis of mitochondrial DNA (mtDNA) can also be used in forensics. The main advantage is that mitochondrial DNA is present in multiple copies per cell and so is relatively easier to obtain in sufficient amounts for analysis. The sequence of mtDNA varies by 1% to 2% between unrelated individuals.

The major disadvantage is that mitochondrial DNA does not vary between closely related individuals. Mitochondria are inherited maternally, and mitochondrial DNA sequences are therefore shared among groups of people derived from the same maternal lineage. If two samples of DNA show different mitochondrial sequences, this indicates that they come from different people. However, the opposite is not true. Identical mitochondrial sequences are found in people related on the mother's side.

Mitochondrial DNA has been used to derive family ancestries. Indeed, it is now possible to submit personal samples of DNA for analysis to companies such as Oxford Ancestors. The company's MatriLine service allows persons of European descent to trace their maternal ancestry back to one of seven ancestral females (Fig. 23.10). Almost everyone in Europe, or whose maternal roots are in Europe, is descended from one of only seven women whose descendants make up well over 95% of modern Europeans. For genealogical purposes, each of these seven women may be regarded as the founder of a "maternal clan." For those whose maternal roots lie outside Europe, a similar analysis is available but is not yet so detailed.

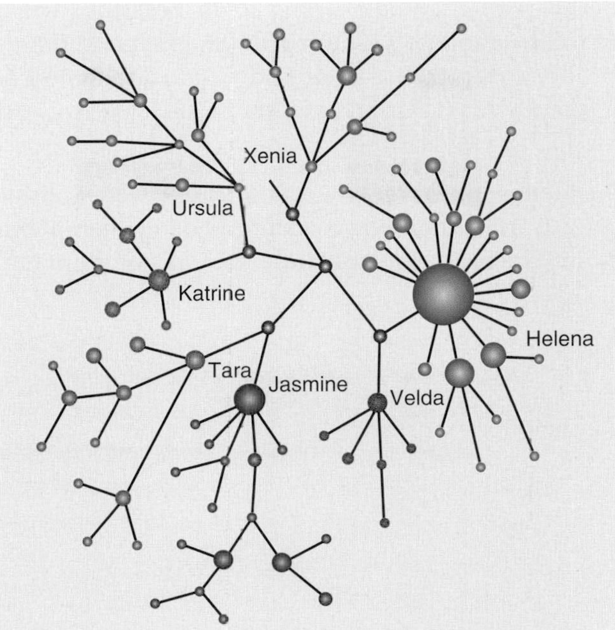

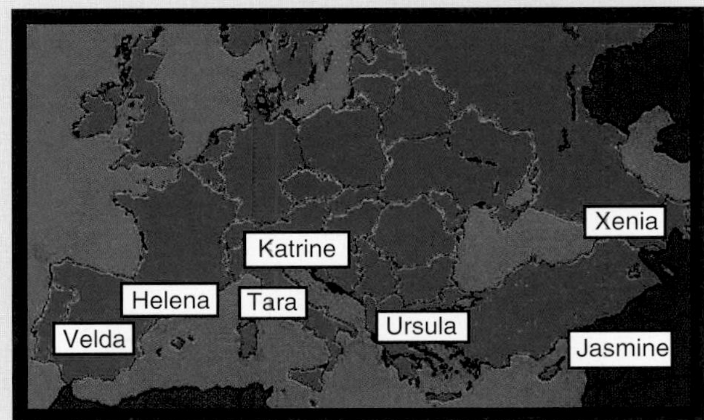

FIGURE 23.10 Seven Daughters of Eve
Oxford Ancestors refers to these so-called "Seven Daughters of Eve" as Ursula (Latin for "she-bear"), Xenia (Greek for "hospitable"), Helena (Greek for "light"), Velda (Scandinavian for "ruler"), Tara (Gaelic for "rock"), Katrine (Greek for "pure"), and Jasmine (Persian for "flower"). People of European descent can trace their lineage back to one of these seven women by comparing the mitochondrial DNA sequences. Used with permission from Oxford Ancestors (http://www.oxfordancestors.com).

In contrast to mitochondrial DNA, the Y chromosome follows a paternal pattern of inheritance. The Y chromosome contains many STR sequences in noncoding regions. However, most have few different alleles, and so only a few are suitable for forensic analysis. One advantage of using Y-linked STR loci is that any sequence specific to the Y chromosome must have come from a male. This is often useful in cases of sexual assault.

Genealogies are increasingly being traced by DNA sequencing. Mitochondrial DNA is maternally inherited and is often used to specifically trace maternal ancestry. Y chromosome sequences can be used to trace male ancestry.

IDENTIFYING THE REMAINS OF THE RUSSIAN IMPERIAL FAMILY

An interesting example of forensics concerns the identification of the remains of the Russian royal family. Analysis of both short tandem repeats in chromosomal DNA and sequencing of mitochondrial DNA was involved. The last Tsar of Russia, Nicholas Romanov II, was executed in July 1918, together with his family and a handful of servants. After execution by a firing squad of Bolshevik soldiers, the bodies were buried in a hidden mass grave. The burial site was rediscovered in 1989, and in 1991 nine skeletons were excavated. Thorough forensic analysis of the bones, clothing, and personal possessions from the grave provided evidence that some of the skeletons belonged to the tsar and his family. American and British teams, at the invitation of the Russian government, then carried out DNA testing.

Nuclear and mitochondrial DNA tests were performed on the nine bone samples. Five of the bodies were clearly related, as demonstrated by STR analysis at five different genetic loci (Fig. 23.11). These were Tsar Nicholas; his wife, the Tsarina Alexandra; and three of their four daughters. The fourth daughter and their son, Prince Alexei, the heir to the throne, were missing; their bodies had apparently been destroyed completely by burning before the mass burial. The other four remains were those of servants who were unrelated to the royal family.

The identity of the remains of the tsarina was confirmed by sequencing mitochondrial DNA. Tsarina Alexandra was the granddaughter of Queen Victoria of England. Alexandra's sister, Princess Victoria of Hesse, was the grandmother of Prince Philip, Duke of Edinburgh,

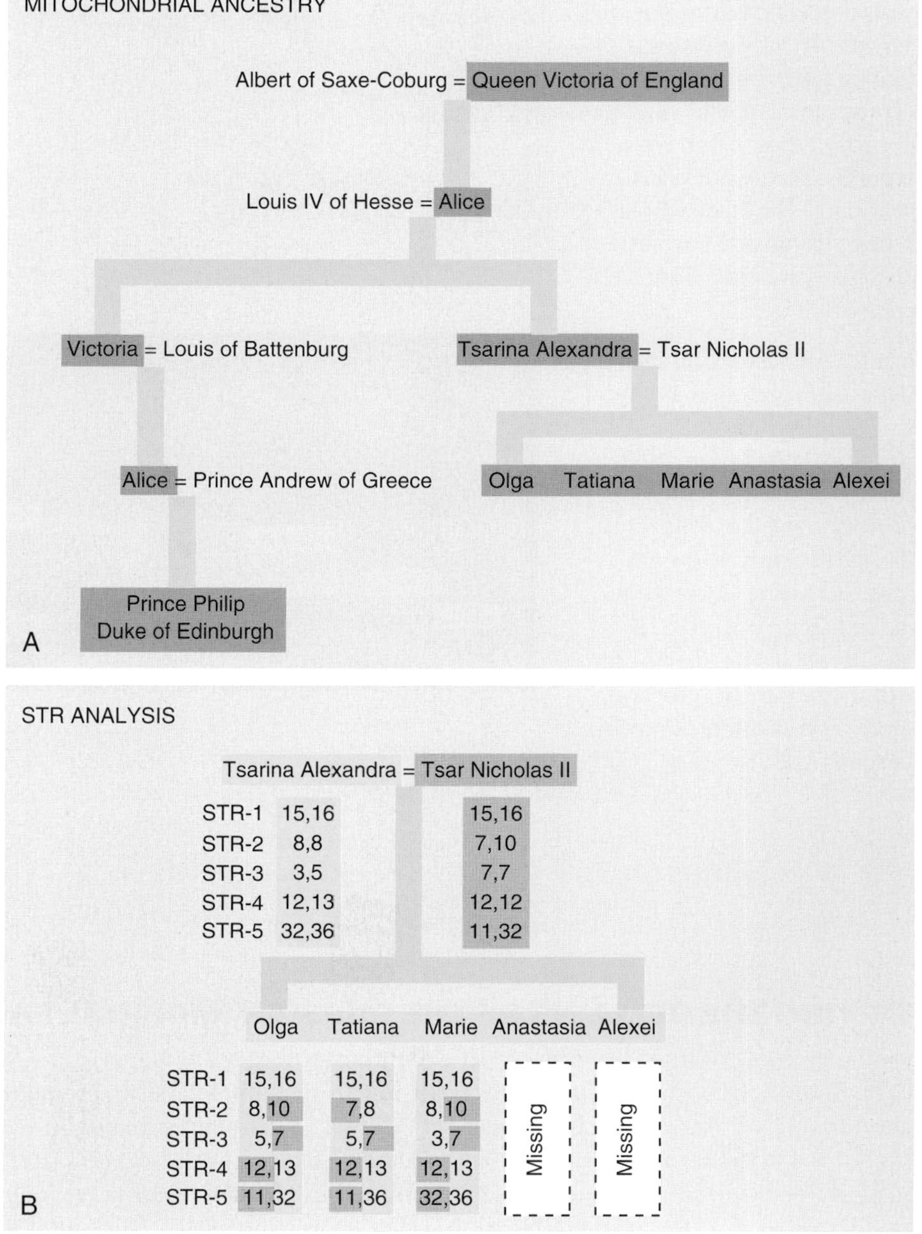

FIGURE 23.11 Russian Royal Family

(A) Family tree showing the ancestry of Tsarina Alexandra, Tsar Nicholas II, and their children. The yellow highlights show people with identical mitochondrial DNA. (B) STR analyses of the skeletal remains of Tsar Nicholas II and his family. The remains were examined with PCR primers for five different STRs (labeled 1–5). The three children had combinations of STR fragments found in either parent. The number of the STR is color coded, with red from Tsar Nicholas II and blue from Tsarina Alexandra. Two children's remains were missing from the grave.

husband of the present queen of England. A sample of blood provided by Prince Philip showed an mtDNA sequence that was identical to that of the remains presumed to belong to Tsarina Alexandra.

The mtDNA of the tsar himself proved more intriguing. Two distant maternal relatives of the tsar, Countess Xenia Sfiri and the Duke of Fife, contributed samples for comparison. Their mtDNA sequences were identical to each other. The tsar's mtDNA was identical to the relatives except at position 16169. Here, both relatives had T, but the tsar had a mixture, with 70% T and 30% C at position 16169. This suggested either that the sample was contaminated or that he had a rare condition known as **heteroplasmy**.

In a few individuals, there are two populations of mitochondria with slight differences in mtDNA sequence—that is, heteroplasmy. This condition is sometimes inherited via the maternal line. However, often the minority population of mitochondria is not present in all descendants. The matter was settled by exhumation of the body of Georgij Romanov, younger brother to the tsar, who died of tuberculosis in 1899. Georgij also showed heteroplasmy with the same mixture of T and C at position 16169 of his mtDNA. The rarity of heteroplasmy provides extremely high probabilities for correct identification when a match is found. In this case the likelihood ratio for authenticity was estimated at over 100 million!

On July 17, 1998, more than a million people attended the reburial of the last Imperial monarch of Russia, Tsar Nicholas II, together with his wife Tsarina Alexandra, and three of their five children, Olga, Tatiana, and Maria. The ceremony took place in the Peter and Paul Fortress in St. Petersburg (known as Leningrad during the communist period).

Analysis of both nuclear and mitochondrial DNA sequences successfully identified the remains of the Russian royal family.

GENE DOPING AND ATHLETICS

Many cases are known of athletes taking drugs such as steroids to enhance their performance. More recently, doping has expanded to include the use of recombinant proteins made by genetic engineering (see Chapter 10), such as human growth hormone (GH), together with related factors such as insulin-like growth factor-1 (IGF-1), which increases growth of muscle cells. The obvious next step in using advancing technology to gain an advantage in athletics is to use the genetic material itself.

The term *gene doping* refers to using the techniques used in gene therapy (see Chapter 17) to alter the expression of genes so as to promote physical superiority. Gene doping is banned, and so far no one is known to have practised gene doping. However, it is uncertain if the World Anti-Doping Agency (Wada) could actually detect it.

Consider the protein erythropoietin (EPO), which promotes the production of red blood cells and thus increases stamina and endurance (Fig. 23.12). Athletes, including riders in the Tour de France, have used this illegally since the late 1990s. The most famous was Lance Armstrong, who won the Tour de France seven years in a row (1999–2005) and was stripped of his titles for doping. Since EPO is a natural protein, it is extremely hard to detect whether athletes are taking extra EPO, although tests now exist. Excess EPO can be dangerous. Overproduction of red blood cells makes the blood more viscous. This increases stress on the heart and makes clotting more likely. This is especially dangerous if athletes suffer dehydration during long events.

But suppose that instead of injecting EPO protein an athlete inserted an extra EPO gene? Extra EPO protein would now be made internally but would be indistinguishable from

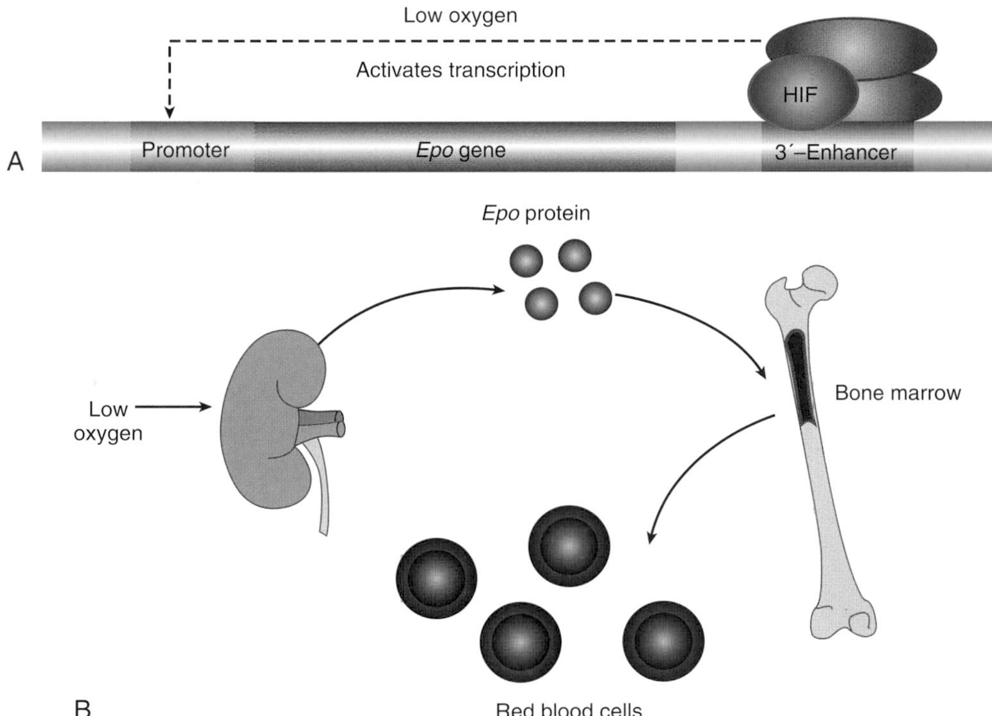

FIGURE 23.12 Mechanism of Erythropoietin Action
(A) The *Epo* gene is induced in response to low oxygen levels. When oxygen is low, the HIF transcription factor is converted from its inactive to its active form and binds to an enhancer sequence that lies beyond the 3′-end of the *Epo* gene. This stimulates transcription of the gene and leads to increased levels of EPO protein. (B) EPO is made in the kidney when oxygen is low. The EPO is released into the bloodstream and travels to bone marrow, where it stimulates the production of red blood cells.

the body's normal EPO. Two approaches for detection seem possible, although technically difficult. First, if the extra *Epo* gene is carried on a vector, or is inserted into the genome at a novel location, the flanking DNA would differ from that next to the natural copy of the gene. This could be detected. Second, the kidney normally makes most EPO protein (Fig. 23.12B). If the extra *Epo* gene was active in other locations, this might be noticeable.

EPO is one of the favored targets for gene doping. Others include vascular endothelial growth factor (VEGF, which also increases the supply of red blood cells), IGF-1 (see preceding discussion), and myostatin (which inhibits muscle development; hence in this case, gene expression would be decreased to increase muscle development).

Gene doping refers to the potential use of gene therapy techniques to enhance human athletic performance. Detection would be extremely difficult.

GENOMICS DRIVES ADVANCES IN FORENSICS

Molecular biology is still making major advances, some of which allow improved analysis of DNA or RNA. In this section, we survey some novel techniques that are still largely experimental but will probably enter the realm of forensics soon. Most of these new techniques were discussed in Chapter 8.

As noted in Chapter 8, it is now possible to detect single base changes (single nucleotide polymorphisms, SNPs) that differ between two or more human genomes. Large numbers of SNPs can be monitored simultaneously by using DNA microarrays. An increasing number of SNPs have been linked both to ancestry and to different alleles of genes responsible for known characteristics. Although most emphasis has been on medically relevant genes, a variety of other characteristics such as the color of skin, hair, and eyes have also been linked to certain SNPs. Such SNPs are located either in or near genes that affect such characters. Predictions based on SNP analysis are presently around 70%–80% accurate for eye color and 60%–70% accurate for hair color.

The use of STR loci for profiling depends on reference samples in the databases. But what happens if there is no match for DNA from a crime scene? For such cases, DNA arrays have recently come into use (see Chapter 8 for arrays). They allow monitoring for many DNA sequences simultaneously. Such DNA arrays include mitochondrial, X chromosome, and Y chromosome markers as well as large numbers of genome wide SNPs. For example, the Identitas v1 Forensic Chip carries over 200,000 probe sequences. It provides information on ethnic ancestry, sex, as well as hair and eye color. This chip is not accurate enough to be used as evidence in court, but it is useful as the basis for further investigation.

Although identical twins do indeed start out life with identical DNA sequences, it is now possible to tell them apart by their DNA. Two approaches can be used to tell apart identical twins, and both rely on changes to the DNA that occur during development. First, mutations occur over time in the somatic cells of the body. Since different mutations occur randomly, identical twins will steadily diverge in their spectrum of accumulated somatic mutations in different tissues. Detection of such differences requires massive ultra-deep sequencing (hundreds of millions of reads). Sperm are a good tissue for searching for somatic mutations since they are the end result of a high number of cell divisions. A pilot investigation using sperm found five SNPs that differed between a particular pair of identical twins. However, the time and cost of such an analysis limits the use of this method. In addition to single point mutations, copy number variations, such as deletions or duplications, are relatively frequent. They may be monitored in multiple tissues by using DNA arrays. Second, epigenetic modifications to DNA, especially DNA methylation (see Chapter 8), also occur during development. Again, there are differences between identical twins since many epigenetic DNA modifications respond to the environment.

Obviously, the base sequence of DNA does not reveal the age of an individual. Nonetheless, age may be determined from the analysis of DNA. As discussed in Chapter 8, DNA methylation is a major mechanism of epigenetic gene regulation. Methylation is reversible, as genes are switched on and off during development and aging. Nonetheless, as we age, the overall methylation of our DNA increases. In particular, the methylation of sites located within the promoter regions of certain genes shows an especially high correlation with aging. Through analysis of several such sites, it is possible to estimate age to within approximately 5 years. Forensic analysis can even reveal wealth, although not using DNA (see Box 23.4).

Forensic DNA analysis provides information on personal identity and characteristics. However, it does not reveal which tissues of the body the evidence sample comes from. Different tissues have different patterns of gene expression that is reflected in variation in their messenger RNA composition. RNA is less stable than DNA and is consequently technically more difficult to work with. Nonetheless, the mRNA composition of samples may be surveyed by using DNA microarrays. The mRNA hybridizes to the DNA probes in the array, and this allows identification of those genes that are highly expressed in the sample of interest (Fig. 23.13).

This approach, which is still experimental, can tell the difference between body fluids such as blood, saliva, vaginal secretions, or semen—an important consideration in cases such as sexual assault. As can be seen from Figure 23.13, many genes are expressed in multiple tissues, and those from blood and semen tend to overshadow the others. Nonetheless, each body fluid has an individual signature that allows it to be distinguished.

For routine screening, arrays are expensive and time consuming. Instead, the information from the array is used to choose just one or two genes that are highly and specifically expressed for each selected tissue. Then multiplex quantitative reverse transcriptase PCR (RT-PCR; see Chapter 4) is used to monitor expression of the chosen genes. Using this approach does indeed allow these four body fluids to be unambiguously identified.

The use of mRNA is compromised by its large size and poor stability, especially in old or degraded samples. It has been suggested that noncoding microRNA (miRNA), which is 20–25 bases in length, might be a superior replacement. Several miRNAs show tissue-specific expression and are present in sufficient amount for forensic screening.

Constant advances in the techniques of genomics generate corresponding advances in the amount and type of information that may be derived from the forensic analysis of both DNA and, recently, RNA.

Box 23.4 Forensic Wealth Analysis

At least so far, it is not possible to deduce wealth from DNA analysis! Nonetheless, through the use of more traditional chemical analyses, it is possible to differentiate between people based on socioeconomic status. HPLC and mass spectroscopy in particular are used to monitor the presence of trace metals and organic pollutants in human urine samples and other tissues.

With regard to trace metals, richer people have more mercury, arsenic, cesium, and thallium. This is most likely due to the rich consuming more salmon and sushi, as ocean fish accumulate these metals from seawater. Conversely, poorer people have more lead, cadmium, and antimony, which are probably derived from smoking, a habit more common among the poor.

Organic contaminants also differ. The rich contain more perfluoro-octanoic acid and perfluoro-nonanoic acid, possibly due to higher consumption of shellfish, and more benzophenone-3, an ingredient of sunscreen. Poorer people have more bisphenol A and phthalate derivatives. Bisphenol A is used in a variety of packaging such as cans and plastic containers for food. It is suspected of negative effects on health and may act as a xenoestrogen. Phthalates are widely used as plasticizers and also suspected of some negative health effects (Figure A).

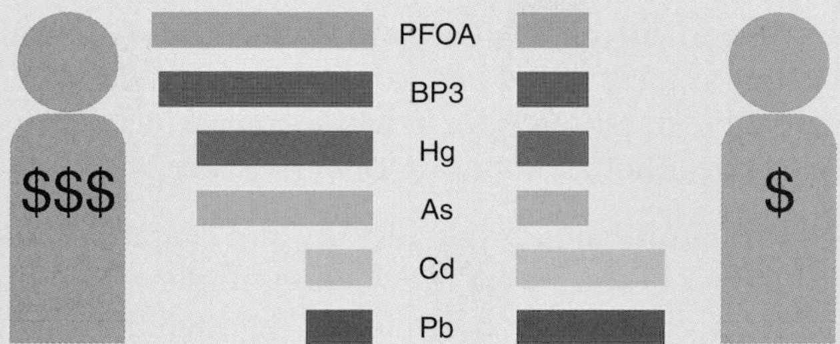

FIGURE A Rich versus Poor
Major differences in contaminating chemicals based on income. BP3 = benzophenone-3; PFOA = perfluoro-organic acids. Modified from Tyrrell J, et al. (2013). Associations between socioeconomic status and environmental toxicant concentrations in adults in the USA: NHANES 2001-2010. *Environ Int* **59**, 328–335.

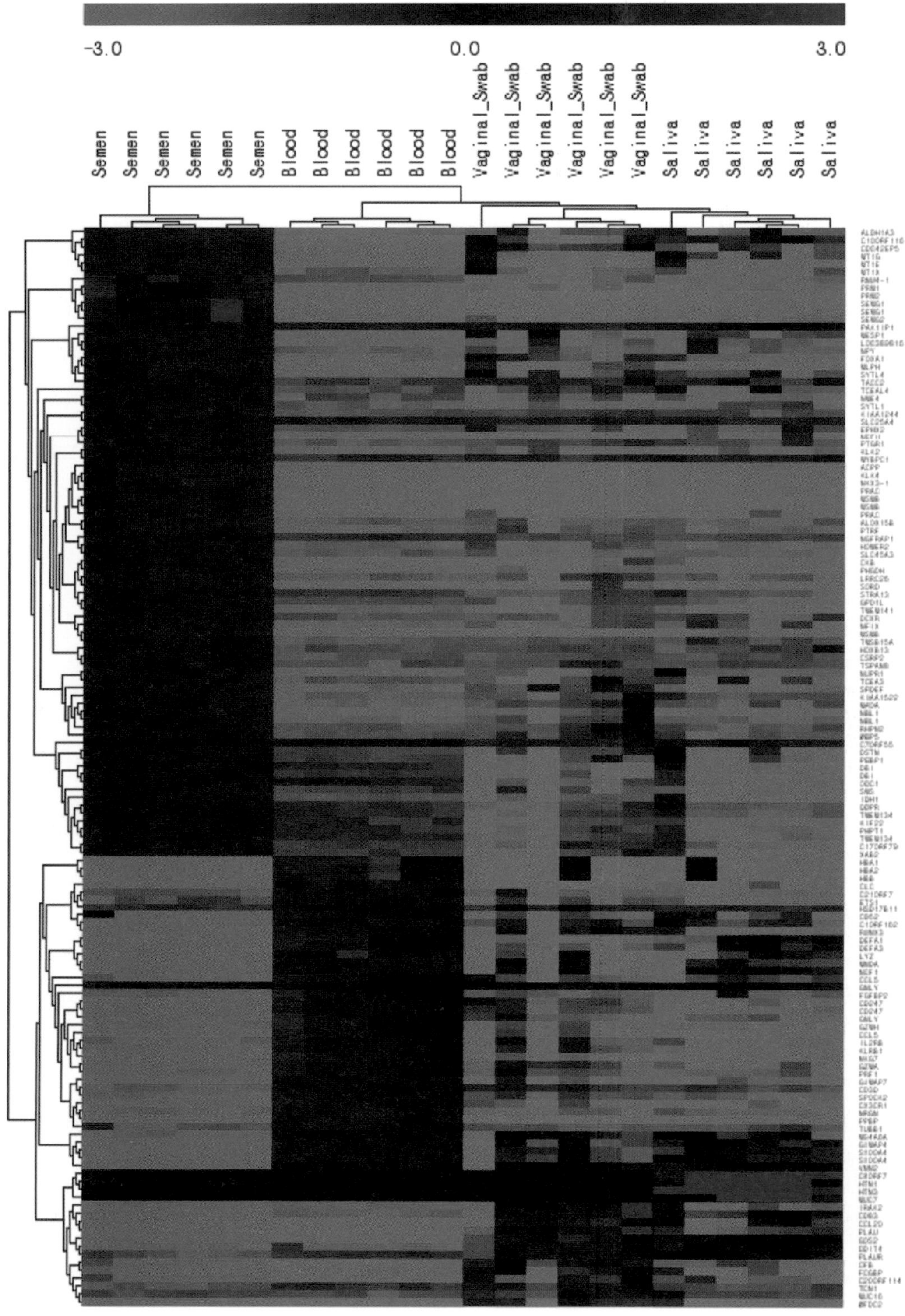

FIGURE 23.13 Tissue Identification via mRNA Analysis

A set of 137 tissue-specific candidate genes was selected for four different body fluids. An Illumina BeadChip Array platform was used to analyze the mRNA from each tissue by hybridization. In this experiment, mRNA was extracted from 24 Korean body fluid samples, 6 each from semen, blood, vaginal swab, and saliva. Unsupervised hierarchical clustering of 137 candidate genes was performed to give the diagram shown. Green = semen; red = blood; yellow = vaginal secretion; blue = saliva. Of these, 80 genes were semen specific, 40 were blood specific, 13 were vaginal secretion specific, and 4 were saliva specific. From Park SM, et al. (2013). Genome-wide mRNA profiling and multiplex quantitative RT-PCR for forensic body fluid identification. *Forensic Sci Int Genet* **7**, 143–150.

Summary

DNA sequences are unique to each individual, apart from identical twins (and even twins develop differences as they age). DNA samples can thus be used for personal identification. In particular, short tandem repeats (STRs) are used in forensic analysis to establish identity. They may be used to establish identity in criminal investigations, in cases of disputed paternity, or in historical and archaeological research. DNA fingerprinting has largely displaced older methods, such as blood typing, in these areas. Animals can also be identified by their DNA sequences. This analysis is now widely used to check for correct labeling of fish and meat for human consumption. A variety of novel analytic techniques will soon allow the determination of age, visible characteristics such as hair and eye color, and tissue derivation from forensic samples.

End-of-Chapter Questions

1. What technology can track cattle through the food processing chain?
- **a.** retinal scans
- **b.** fingerprints
- **c.** cattle color patterns
- **d.** DNA fingerprinting
- **e.** none of the above

2. Which type of body fluid can be analyzed?
- **a.** semen
- **b.** blood
- **c.** urine
- **d.** tears
- **e.** all of the above

3. In the ABO blood group, how many alleles are present in the population?
- **a.** 2
- **b.** 4
- **c.** 6
- **d.** 3
- **e.** 1

4. Which one of the following statements about ABO blood grouping is not correct?
- **a.** People with type A blood will express antibodies against type B blood.
- **b.** A person with AB blood type is the universal donor.
- **c.** A child may not have the same blood type as either one of his/her parents because humans are diploid.
- **d.** A person with O blood type is the universal donor.
- **e.** A type O person is only able to accept type O blood.

5. Which of the following techniques is not involved in the identification of DNA at crime scenes against possible suspects?
- **a.** PCR
- **b.** Western blot
- **c.** sequencing
- **d.** hybridization
- **e.** DNA fingerprinting

6. Which one of the following is not a step in DNA fingerprinting?
   a. DNA fragments are separated according to their molecular weights during gel electrophoresis.
   b. Autoradiography is used to identify the location of the radioactive-labeled probe after hybridization.
   c. A cDNA copy of the mRNA is made.
   d. Southern blotting is used to visualize the DNA fragments.
   e. The DNA is cut with restriction enzymes.
7. Why have MLP methods been displaced by SLP methods?
   a. Interpretation is difficult since it is unknown which band corresponds with which locus.
   b. In order to make comparisons, fingerprints from MLP must be run on the same gel.
   c. It is difficult to store the data from MLP methods in a database.
   d. Statistical analysis of MLP is difficult to obtain.
   e. All of the above are reasons SLP has replaced MLP.
8. What is a problem associated with using VNTRs to identify persons of interest in criminal cases?
   a. Multiple tandem repeats create densely packed bands during agarose gel electrophoresis, so other separation methods must be used.
   b. VNTRs are usually the same from person-to-person, so very little information is obtained from this technique.
   c. VNTRs occur within coding regions; thus privacy is not well protected.
   d. There have been no problems associated with using VNTRs to identify people.
   e. none of the above
9. What is the major difference between DNA fingerprinting and PCR?
   a. Both techniques provide adequate data for forensic science.
   b. DNA fingerprinting looks for differences in sizes, whereas PCR tests for the presence or absence of specific regions.
   c. DNA fingerprinting identifies regions that are absent or present, whereas PCR identifies different sized fragments.
   d. There are no major differences between the two techniques.
   e. none of the above
10. Which of the following terms describes running several PCR reactions simultaneously in one tube?
    a. DNA fingerprinting
    b. multiplex PCR
    c. VNTR
    d. RFLP
    e. STR PCR
11. When close relatives are suspects, what happens to the probability of a match during DNA testing?
    a. The probability increases.
    b. The probability decreases.
    c. The probability remains the same.
    d. The probability becomes 100%.
    e. The probability becomes 0.

(Continued)

12. What has been the main impact of DNA technology on the criminal justice system?
- **a.** uncertainty between the crime scene sample and the suspect's DNA
- **b.** providing loopholes in laws regarding DNA testing and admissibility in courts
- **c.** not much impact
- **d.** greater certainty when matching crime scene DNA with suspect DNA
- **e.** none of the above

13. What technique has been used to prevent low-quality bonito from being included with premium tuna?
- **a.** RFLP
- **b.** VNTR
- **c.** RT-PCR
- **d.** Western blot
- **e.** none of the above

14. How have mitochondrial DNA sequences been used?
- **a.** to identify differences between closely related people
- **b.** to roughly determine how many mitochondria are in one cell
- **c.** to derive family ancestries
- **d.** mitochondrial DNA sequences have not been used in forensics
- **e.** none of the above

15. What method was used to identify the remains of the Russian Imperial Family?
- **a.** PCR
- **b.** mitochondrial DNA sequencing
- **c.** VNTR
- **d.** RFLP
- **e.** Western blot

16. All of the following are factors in determining the probability of a genetic match except _______________.
- **a.** genotype determination
- **b.** allelic frequency
- **c.** random alleles
- **d.** linked alleles
- **e.** unlinked allele combinations

17. Which statement regarding SNPs and associated technology is incorrect?
- **a.** SNP analysis is accurate between 60–70% of the time for hair color.
- **b.** The DNA microarrays for STR loci profiling contain mitochondrial, X chromosome, Y chromosome, and various SNPs.
- **c.** SNP stands for small nuclear proteins.
- **d.** Ethnic ancestry, sex, hair and eye color can be determined using STR loci profiling.
- **e.** SNPs vary in identical twins.

18. All of the following are mechanisms of detection for gene doping with *Epo* except _______________.
- **a.** unnatural flanking regions surround *Epo* gene
- **b.** EPO produced in other tissues besides those associated with the kidney
- **c.** *Epo* gene on a vector

d. blood viscosity
e. *Epo* gene in a novel location

19. Chemical analysis of wealthy individuals' tissues have determined that all of the following trace metals are more abundant than in poorer individuals' tissues except ______________.
a. caesium
b. thallium
c. mercury
d. arsenic
e. lead

Further Reading

Bocklandt, S., Lin, W., Sehl, M. E., Sánchez, F. J., Sinsheimer, J. S., Horvath, S., & Vilain, E. (2011). Epigenetic predictor of age. *PLoS One, 6*(6), e14821.

Bon, C., Berthonaud, V., Maksud, F., Labadie, K., Poulain, J., Artiguenave, F., et al. (2012). Coprolites as a source of information on the genome and diet of the cave hyena. *Proceedings Biological Sciences/The Royal Society, 279*(1739), 2825–2830.

Butler, J. M. (2006). Genetics and genomics of core short tandem repeat loci used in human identity testing. *Journal of Forensic Sciences, 51*, 253–265.

Fortes, G. G., Speller, C. F., Hofreiter, M., & King, T. E. (2013). Phenotypes from ancient DNA: approaches, insights and prospects. *Bioessays, 35*(8), 690–695.

Gusmão, L., Butler, J. M., Carracedo, A., Gill, P., Kayser, M., Mayr, W. R., et al. (2006). DNA Commission of the International Society of Forensic Genetics (ISFG): an update of the recommendations on the use of Y-STRs in forensic analysis. *Forensic Science International, 157*, 187–197.

Keating, B., Bansal, A. T., Walsh, S., Millman, J., Newman, J., Kidd, K., & International Visible Trait Genetics (VisiGen) Consortium, et al. (2013). First all-in-one diagnostic tool for DNA intelligence: genome-wide inference of biogeographic ancestry, appearance, relatedness, and sex with the Identitas v1 Forensic Chip. *International Journal of Legal Medicine, 127*(3), 559–572.

Park, S. M., Park, S. Y., Kim, J. H., Kang, T. W., Park, J. L., Woo, K. M., et al. (2013). Genome-wide mRNA profiling and multiplex quantitative RT-PCR for forensic body fluid identification. *Forensic Science International Genitics, 7*(1), 143–150.

Pompanon, F., Deagle, B. E., Symondson, W. O., Brown, D. S., Jarman, S. N., & Taberlet, P. (2012). Who is eating what: diet assessment using next generation sequencing. *Molecular Ecology, 21*(8), 1931–1950.

Schneider, P. M. (2007). Scientific standards for studies in forensic genetics. *Forensic Science International, 165*, 238–243.

Shehzad, W., McCarthy, T. M., Pompanon, F., Purevjav, L., Coissac, E., Riaz, T., & Taberlet, P. (2012). Prey preference of snow leopard (*Panthera uncia*) in South Gobi, Mongolia. *PLoS One, 7*(2), e32104.

Sobrino, B., Brión, M., & Carracedo, A. (2005). SNPs in forensic genetics: a review on SNP typing methodologies. *Forensic Science International, 154*, 181–194.

van der Gronde, T., de Hon, O., Haisma, H. J., & Pieters, T. (2013). Gene doping: an overview and current implications for athletes. *British Journal of Sports Medicine, 47*(11), 670–678.

Varsha, A. (2006). DNA fingerprinting in the criminal justice system: An overview. *DNA Cell Biology, 25*, 181–188.

Bioethics in Biotechnology

Biotechnology

http://dx.doi.org/10.1016/B978-0-12-385015-7.00024-7

INTRODUCTION

As biotechnology continues its march forward, it will inevitably raise new moral and legal questions. Our expanding knowledge and capabilities—instead of resolving bioethical issues—will most likely create new challenges, and the line between ethical and unethical behavior will only become murkier. There are few "black and white" issues in bioethics, but instead varying shades of gray. Indeed, the complexity of bioethics is such that a student can now obtain a graduate-level degree in the field.

Much of what is regarded as "official" bioethics derives from the clinical arena, including the practice of medicine as well as clinical trials and experiments. In this chapter, however, we focus on issues whose scientific background has been described in previous chapters. Many of the questions raised are not unique to biotechnology. Who should control technology? What should be banned or permitted, and who should decide? Who should profit? Should access to novel and expensive technology be provided to those who cannot afford it? If so, who should pay? The answers to these questions differ according to personal beliefs and cultural outlook.

Therefore, we describe many ethical challenges, but we give few definite answers. The reason is that we feel moral decisions are for the reader to make, not the authors. Still, we have not attempted to artificially hide our own biases. We are in favor of the use of gene therapy and genetic engineering where they have been properly tested. Our inclination is to place more weight on genetics and less on environmental and cultural effects than most people nowadays. However, these attitudes may be changing, partly due to the increasing use of DNA in forensics and the accompanying greater public familiarity.

Finally, do keep in mind while pondering these ethical dilemmas that many of these issues ultimately arise from success. While the developing world still struggles with famine and disease, the developed world has largely solved these problems of yesteryear. However, instead of creating a utopia, we now have an entirely new set of troubles to contend with. Yet, just as technology helped solve those previous problems, we believe it also will play a major role in solving tomorrow's. Therefore, the issues discussed here should be viewed more as challenges than as forecasts of gloom and doom.

PRINCIPLES OF BIOETHICS

Ethics is defined as follows: "The philosophical study of the moral value of human conduct and of the rules and principles that ought to govern it; moral philosophy" (Collins online dictionary; http://www.collinsdictionary.com) or "the branch of knowledge that deals with moral principles" (Oxford online dictionary; http://www.oxforddictionaries.com/us/). Ethics then is, or should be, an objective attempt to clarify and assess ethical concepts and beliefs, but without necessarily agreeing or disagreeing with them. It points out hidden assumptions, non sequiturs, etc. It is "philosophy of morals" comparable to the philosophy of religion or philosophy of science. At this level, the ethicist does not provide answers to particular ethical dilemmas or provide moral guidance.

Unfortunately, the word *ethics* has become rather ambiguous. It is also used to refer to rules for behavior (such as *business ethics, medical ethics*) as opposed to the procedures for analyzing them. In response, the word *meta-ethics* is sometimes used for that part of philosophy that deals with moral principles.

In any case, decisions on real-life issues must be made. Our first problem in applying ethics to science is that philosophy itself is fragmented into several schools with different outlooks. Dealing with such differences is far beyond the scope of this book, but readers should realize that the scheme followed here is by no means universally accepted. The National Commission for the Protection of Human Subjects of Biomedical and Behavioral Research issued the Belmont Report in 1979. Although primarily aimed at biomedical research with human subjects, its basic principles—*autonomy, beneficence,* and *justice* Belmont Report in 1979—have often been applied to the broader areas of biotechnology. In 2005, the United Nations issued a Universal Declaration on Bioethics and Human Rights, including a much-expanded set of principles:

Universal Declaration on Bioethics and Human Rights

General Provisions

- Article 1 Scope
- Article 2 Aims

Principles

- Article 3 Human dignity and human rights
- Article 4 Benefit and harm
- Article 5 Autonomy and individual responsibility
- Article 6 Consent
- Article 7 Persons without the capacity to consent
- Article 8 Respect for human vulnerability and personal integrity
- Article 9 Privacy and confidentiality
- Article 10 Equality, justice and equity
- Article 11 Non-discrimination and non-stigmatization
- Article 12 Respect for cultural diversity and pluralism
- Article 13 Solidarity and cooperation
- Article 14 Social responsibility and health
- Article 15 Sharing of benefits
- Article 16 Protecting future generations
- Article 17 Protection of the environment, the biosphere and biodiversity

These principles may be viewed as expanding on the principles of autonomy (Articles 5–9), beneficence (Articles 3 and 4), and justice (Articles 10–15). Added to these is the protection of the biosphere for the benefit of future generations (Articles 16 and 17).

Autonomy, that is, informed consent and related issues, applies especially to medical research and clinical applications. In biotechnology, privacy and confidentiality (Article 9) receive greater emphasis. In particular, access to personal genome information has become a thorny problem.

Beneficence involves promoting benefit and avoiding harm to people (and animals)—physically, mentally, and to their rights. The concept of justice seems obvious, but in practice there are often conflicts, especially that between the interests of the individual and society.

The tricky part is applying these principles, to which there is widespread agreement—in principle—to particular issues. Not only do different individuals and different cultures vary in their outlooks, but also some scientific advances are new, and there may be insufficient data for accurate assessment of future effects and possible hazards. A brief scheme for making an ethical decision is shown in Figure 24.1.

Some questions that could be asked during the decision-making process are as follows:

What are the benefits of the proposed biotechnological advance?
Who specifically will benefit, and will the benefits be shared fairly?

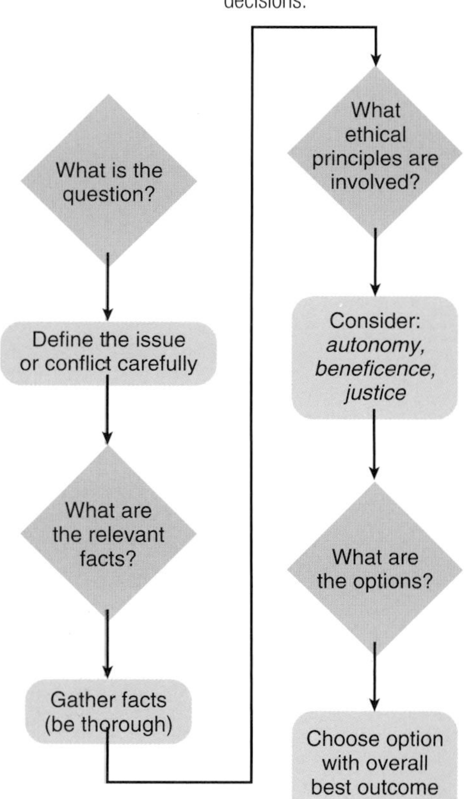

FIGURE 24.1
Ethical Decision Tree
A brief outline of a possible approach to making ethical decisions.

Is the proposed change cost effective, and does it use resources efficiently?
What are the possible hazards of the proposed biotechnological advance?
What is the likelihood that harm will result?
What precautions might be taken to minimize the harm?
Have those people been consulted who will be subjected to clinical tests, use novel products, or otherwise be affected?
What impact does the proposed change have on society as a whole?
What impact does the proposed change have on the environment?

USE OF THE PRECAUTIONARY PRINCIPLE

The **precautionary principle** states that if a proposed change has a possible risk of causing harm to people or to the environment, the burden of proving that it is safe (or very unlikely to cause harm) falls on those proposing the change. This approach is used when there is no overall scientific consensus. The extent and cost of possible damage versus the proposed benefits should be taken into account. The precautionary principle is often used in the realm of biology and medicine because changes cannot easily be reversed or usually even contained. Furthermore, modified organisms can replicate and spread. In biotechnology the precautionary principle has been applied to such issues as

(a) Spreading disease accidentally by using technology (such as the transmission of AIDS or hepatitis by blood transfusions or the emergence of mad cow disease by changing animal food processing).
(b) Introducing new pharmaceutical products, especially those generated by genetic engineering. Requiring pharmaceutical companies to perform clinical trials to show that new medications are safe is a well-established practice.
(c) Introducing genetically modified organisms into the environment.
(d) Creating artificial life.

In the physical arena, advances such as robots, 3D printers, and self-replicating nanotechnology are similar in involving novel risks that are hard to estimate. Whether assessing physical or biological risks, there are many difficulties. *Safety* is a relative term. The safety of a novel invention can only be assessed relative to other similar technology. Even if a new technology is not entirely safe, it may well be safer than the pre-existing technology it is intended to replace.

The European Union has applied the precautionary principle to legislation addressing those biological issues affecting agriculture and the environment. As a result, genetically modified organism (GMO) crops are not grown in most European countries. In contrast, GMO crops are widely planted in the United States. The reason is not that Europe officially regards GMOs as hazardous but that no clear benefit would result. Most GMO crops are variants of cotton, maize, and soybean that are pest resistant, which results in lower costs for farmers. Although grown widely in the United States, few of these crops are grown in Europe. In contrast, GMO food provides essentially no benefits for inhabitants of wealthy nations. Consequently, most Europeans support a "better safe than sorry" approach to agriculture. Note, however, that the European approach to regulating GMOs for medical use is quite different. In this case, many bioengineered products have been approved, even when there are significant risks, because many Europeans will benefit overall from improved or cheaper medications. Thus, the regulatory decisions that are made need not be the same in all nations. Note in this case that the application of the principle of beneficence gives quite different results when applied to the United States versus Europe.

Another question is how long safety testing should continue. The harmful effects of some novel drugs and chemicals are soon obvious. In other cases, the effects may be visible only after a prolonged period. A classic case is the early use of X-rays, even to examine the fetus during pregnancy. When this technique was first introduced, DNA was not even recognized as the genetic material, and the concept of damage to DNA-causing mutations was still way in the future. On the other hand, one cannot keep putting off decisions forever in case of future knowledge. A case in point is the approval of GMOs, including crops and livestock. Given that many people around the world are poorly fed, how long should testing be continued before releasing GMOs for use? (See the later section on GMOs.)

Finally, it is worth noting that people are notoriously poor at evaluating risk. A classic example is nuclear power versus coal. Most people view nuclear power as much more dangerous, yet despite several high-profile disasters, the nuclear power industry is responsible for far less loss of life than coal mining. While much of the public worries about the safety of vaccines and GMOs, the real risks in life are far more mundane, such as driving automobiles and smoking cigarettes.

QUESTIONS

- Should the precautionary principle—or some form of it—be codified into law? If not, how should the safety of new technologies be regulated?
- How should we approach matters when two ethical principles conflict?

THE POWER OF INFORMATION

For good or for worse, in the digital age, information is power. Data can provide people with the ability to make wiser decisions about their own lives. On the other hand, when information falls into the wrong hands—such as identity thieves and computer hackers—widespread havoc can result. How will biological data fit into our growing information economy? In the following sections, we consider the ethical implications of possessing, or being denied access to, certain types of biological information.

Privacy and Personal Genetic Information

The use of DNA for identification in both criminal investigations and civil cases (e.g., paternity suits) is now widely accepted. When properly done, forensic DNA analysis can help convict or exonerate defendants (see Chapter 23). Although using DNA evidence in the courtroom is no longer controversial, the creation of a national criminal DNA database is. Should every person who is arrested be required to provide a DNA sample? Also, it may be possible someday to deduce a person's physical and mental characteristics from DNA. Should detectives be allowed to build a criminal profile based on DNA from the crime scene, or should they be restricted to analyzing only regions of noncoding DNA that can be used to confirm identity?

Similarly, in the future, it may be possible to predict potential health problems by analyzing an individual's DNA. Currently, this can be done for a few inherited defects, such as Huntington's disease (see Chapter 17). Such knowledge could allow people to plan their lives accordingly. However, many people may prefer not to know, choosing blissful ignorance over worrisome foreknowledge, especially if no cure is available.

Personal genetic information is relevant not only to individuals, but also to health-care providers. Do health and life insurance companies have a right to know about potential health problems? Employers who provide health insurance can penalize smokers, and life insurance providers charge higher premiums based on age, weight, and cholesterol levels. Should insurance companies be allowed access to genetic information also? If a country has a single-payer health-care system, differential costs based on such factors are no longer relevant. Nonetheless, how much personal genetic information should the government be allowed to have? What would constitute an invasion of privacy? Should we all have a DNA-based identity card (Fig. 24.2) that provides not only unique biometric information but also health-care data useful in an emergency (e.g., allergies and blood type)?

Legislation addressing these issues has been passed in many industrial nations. For example, in the United States, the Genetic Information Nondiscrimination Act of 2008 prohibits health insurers from denying coverage or charging higher premiums based solely on genetic predisposition to a possible future disease. It also forbids employers from taking genetic

FIGURE 24.2
Genetic Identity Card
DNA fingerprinting that relies on a dozen STR sequences is the state-of-the-art method in forensic analysis. The amelogenin marker identifies the sex chromosomes. Such a fingerprint is sufficiently specific to distinguish more than 100 million million (10^{14}) individuals.

information into account for firing, hiring, or promotion. Support for passing the act was almost unanimous.

The question of privacy has also arisen in the process of basic research. Scientists may be able to identify volunteers who anonymously donated samples of their DNA. A minor controversy ensued after a team of researchers published the genome sequence of HeLa cells, a cell line derived from cancer cells taken from a patient named Henrietta Lacks who died in 1951. The Lacks family was upset that researchers did not seek their permission prior to publishing the information. The HeLa genome sequence may contain personal genetic information about Lacks's living descendants. On the other hand, any genome sequence that is made public with the owners consent will contain some information about their genetic relatives.

Another related concern involves whether or not it is appropriate to inform volunteers about disturbing genetic information if researchers accidentally discover it. For example, if scientists find out that a volunteer in a clinical trial has an incurable disease, should they tell him or her? What if they discover that a volunteer's father isn't whom the volunteer thinks it is?

QUESTIONS

- Should there be a national criminal DNA database? If so, who should be included?
- Would you prefer to know to which diseases you are susceptible? What if the disease is incurable or fatal?
- Do health-care systems or governments have a right to your genetic information? If yes, how much information?
- Suppose future advances in genome analysis reveal sensitivity to toxic metals or other workplace hazards. Should employers then be allowed to exclude those at risk from jobs involving such hazards?
- Were researchers wrong to publish the HeLa genome sequence without seeking the family's permission?
- Should genetic information be used to plan marriages and careers? Should couples simulate genetic combinations before deciding to have children?

The Problem of Dual-Use Research

The term *dual-use* refers to a useful invention that could be diverted to harmful ends. Thus, explosives developed to make mining safer can be used in armaments. In the biotechnology arena, the World Health Organization defines the term *dual use research of concern (DURC)* as "life sciences research that is intended for benefit, but which might easily be misapplied to do harm." In the United States, the National Science Advisory Board for Biosecurity (NSABB) considers the following factors about possible dual-use research:

- Increasing the harmful consequences of a biological agent or toxin.
- Disrupting immunity (or effective immunization) toward a biological agent or toxin.
- Making a biological agent or toxin resistant to useful preventative or therapeutic countermeasures.
- Improving the ability of a biological agent or toxin to evade detection.
- Improving the stability or transmissibility of a biological agent or toxin.
- Altering the host range (including tropism to particular tissues within the body) of a biological agent or toxin.
- Enhancing the susceptibility of a target population to a biological agent or toxin.
- Generating a novel pathogenic agent or toxin.
- Re-creating an extinct or eradicated pathogenic agent or toxin.

Dual-use research that is deemed dangerous or a security risk often generates a discussion within the scientific community about whether it should actually be published. In biotechnology, such research often involves deadly microbes (Fig. 24.3). On the one hand, investigating such microorganisms is necessary in order to understand them and to prevent disease; on the other hand, terrorists or rogue states could use the same information to genetically engineer even deadlier agents. A recent example was when researchers determined the mutations necessary to make a lethal H5N1 influenza strain more easily transmissible (Herfst et al., 2012). Another group, in an attempt to preempt such concerns, decided to withhold some information about a new type of botulism toxin that they discovered (Barash and Arnon 2014). In this case, the gene sequence for the novel toxin is being withheld until an antitoxin becomes available.

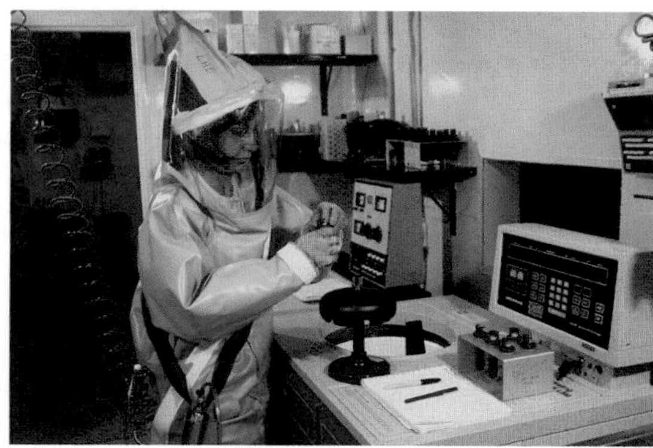

FIGURE 24.3
Dual-Use Research
Information about deadly microbes can be used for both benevolent and malevolent purposes. A CDC scientist wearing a protective suit with helmet and face mask is protected from pathogens as she conducts studies in the CDC BSL-4 laboratory in Atlanta, Georgia. Courtesy of Centers for Disease Control and Prevention; photo taken by Jim Gathany, 2002.

Are such actions necessary? Rogue states certainly might have the resources and technological ability to create more powerful biological weapons through genetic modification, but it is doubtful that terrorist groups would. The more realistic threat is terrorists acquiring biological weapons produced by a government.

But is it really worth a government's time and investment to genetically modify an infectious agent to become even deadlier? Remember, Mother Nature already produced smallpox, plague, Spanish flu, and antibiotic-resistant bacteria. It is difficult to imagine how humans could "improve" much on these microbes. From a military point of view, research on nuclear weapons has always yielded a greater return-on-investment, at least so far. (See further discussion on biological weapons later in this chapter.)

QUESTIONS

- Should dual-use research in biotechnology be published? If so, should access to the information be tightly controlled or widely available?
- If you were in charge of the military of a small rogue state, would you consider investing in biological weapons worthwhile?

Ownership of Genetic Information

The ownership of knowledge is an age-old question. It is generally accepted in the Western industrial nations that if you invent something, you are entitled to patent and profit from it. But how far do such rights extend? In the United States, products of nature cannot be patented, but until very recently, human gene sequences could.

The company Myriad Genetics patented the DNA sequences of two genes linked to breast cancer, *BRCA1* and *BRCA2*. The company also developed a diagnostic assay based on these gene sequences. Thus, the effect of the patents was to eliminate all competition from the market because no other company could create a DNA-based test for these breast cancer genes. Myriad Genetics was sued, and the U.S. Supreme Court resolved the case in 2013. The court ruled that patents could not be issued for "natural" DNA sequences, but they could be issued for any DNA that was manipulated in the lab—including complementary DNA (cDNA), which is DNA created from an mRNA transcript. In effect, it appears that the court is allowing human genes to be patented as long as the introns have been removed. Adding to the confusion is the uncertainty over how much manipulation is necessary for DNA to become patentable. Unfortunately, with so many unanswered questions, it is likely this issue will be debated for years to come.

The law is, however, much clearer on genetically modified organisms. Since these are considered new inventions that are not found in nature, they are regularly patented.

A related issue considers information in general. Should scientists working at universities with public funding be allowed to patent their discoveries and profit personally? In practice, most universities have schemes whereby profits are split, say 50:50, between investigator and institution. Critics claim that knowledge generated from taxpayer-funded research should be public property. A similar argument is used against publishers of scientific journals. Many feel that it is improper for scientists or citizens to have to pay to read scientific papers that were funded using public money. It is this reasoning, plus the available technology, that has largely provided the impetus to the growing "open access" movement.

QUESTIONS

- Should only modified gene sequences be eligible for patenting?
- How much manipulation is necessary for a gene sequence to be considered "modified?"
- Should information be copyrighted? What if that information was taxpayer-funded?

POSSIBLE DANGERS TO HEALTH FROM BIOTECHNOLOGY

Unintended consequences of new technology are, by definition, impossible to foresee. Most technologies have side effects that often cannot be predicted in advance. For example, technology that increased global life expectancy has created a society with a larger proportion of retired people who place a heavy burden on welfare and pension systems. Millions of old and sick people stress health-care systems. Combined with decreased infant mortality, a larger human population affects the environment by consuming scarce resources and causing pollution. Overcrowding promotes the emergence and spread of novel infectious diseases that can turn into pandemics.

As for the realm of genetic engineering, will a genetic construct—perhaps a reconstructed virus or some other GMO—one day escape from a laboratory and crossbreed with a wild organism, forming a fearsome hybrid monster (Fig. 24.4)? Much of the public is under the impression that scientists care little about the societal consequences of their research. But

FIGURE 24.4 Ancient Greek Chimera
The Chimera was a hybrid monster of ancient Greek legend that was thought to live in southwest Anatolia (present-day Turkey). It combined a lion, goat, and fire-breathing dragon. This modern representation was created by Rebecca Kemp, http://www.wildlife-fantasy.com/.

this is not true. Indeed, in 1975, during the early days of recombinant DNA research, the molecular biology community itself met at Asilomar, California, and called for a moratorium on those experiments that were seen as potentially hazardous. This respite allowed the National Institutes of Health (NIH) to generate guidelines to oversee recombinant DNA research. Recently, research involving the construction of deadly influenza viruses generated widespread controversy. The following sections discuss some of the realistic threats from biotechnology.

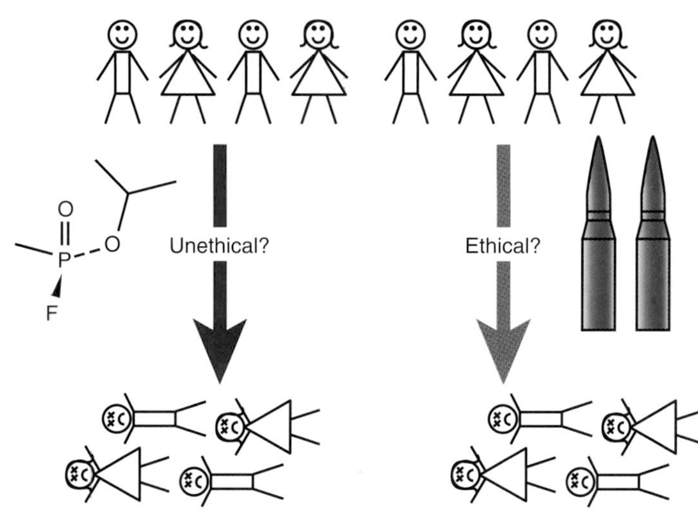

FIGURE 24.5 Syrian Civil War
The victims are just as dead whether killed "ethically" by bullets or "unethically" by sarin nerve gas.

Biological Weapons

Humanity considers some weapons so heinous that they should never be used, not even in wartime. Included among them are chemical and biological weapons. Most countries are signatories of the Chemical and Biological Weapons Conventions, which ban chemical and biological weapons, respectively. Why these weapons are considered particularly gruesome is a bit of a mystery. Conventional weapons, such as bullets and missiles, regularly kill far more people than chemical and biological weapons. For instance, in the Syrian civil war that began in 2011, the government of President Bashar al-Assad is believed to have killed approximately 1,000 people using sarin gas, but many times that number were killed with conventional weapons. Is it more ethical to kill a person with a bullet than with sarin gas (Fig. 24.5)?

Similarly, nuclear weapons are allowed under international law, although their proliferation is not. Given that the nuclear bombs of today are thousands of times more powerful than those dropped on Hiroshima and Nagasaki at the end of World War II, the destructive potential of these weapons is nearly beyond imagining. These weapons could obliterate millions of people within a matter of seconds, something that chemical and biological weapons could never do. Yet, nuclear weapons are legal, while chemical and biological weapons are not. Why?

09-11-01
This is next
Take Penacilin Now
Death to America
Death to Israel
Allah is great

Tom Brokaw
NBC TV
30 Rockefeller Plaza
New York NY 10112

FIGURE 24.6 Anthrax Letters
Envelopes containing anthrax spores were sent to political figures through the U.S. Postal Service in 2001. Source: Terrorism 2000/2001, Federal Bureau of Investigation, Counterterrorism Division.

In reality, a person is much likelier to survive a strike from a chemical or biological weapon than a nuclear weapon. However, chemical and biological agents are invisible and therefore "scarier" to most people than other weapons. This disproportionate fear was quite palpable in the days following the anthrax attacks in the United States in 2001, which occurred shortly after the terrorist destruction of the World Trade Center. Anthrax was sent as spores in envelopes to victims via the post office (Fig. 24.6). The number of casualties due to anthrax was very low, yet the associated fear was widespread and became a hot media topic.

Furthermore, studying highly pathogenic biological agents is itself controversial because—as we discussed earlier—it is a classical example of dual-use research: Knowing how to protect against an infectious disease inevitably provides information that would help in using the disease against an enemy. But this conundrum is true in other areas as well, such as nuclear technology, which can be used to build power plants or bombs.

Another issue is asymmetric warfare. Poor, developing nations are no match militarily against rich, developed ones. Germ warfare can be seen as a "poor man's nuclear weapon." Nations too poor to develop costly high-tech weapons could throw together crude biological weapons relatively easily and cheaply. Germ warfare thus represents a possible means by which developing nations could protect themselves against rich nations. A dictator might even be willing to release a biological agent within the borders of his own country and accept casualties to his own people, knowing that this would frighten off a rich invader. During World War I, typhus epidemics were common on the eastern front. The Serbians lost 150,000 men to typhus in the first 6 months of the war, including more than half of their 60,000 Austrian prisoners of war. Paradoxically, this actually aided the Serbs because the Austrians were so frightened by the typhus epidemic that they kept their armies out of Serbia for fear of infection. Would a despot purposefully infect his own people to fight off an enemy?

QUESTIONS

- Why do many people regard chemical or biological warfare as less ethical than the use of conventional weapons or even nuclear warfare?
- Is it more unethical for a microbiologist to design a lethal virus than for a physicist to design a better missile or tank?
- Is it unethical for poor countries to possess chemical and biological weapons but ethical for rich, advanced nations to possess nuclear weapons?

Antibiotic and Antiviral Resistance

Overuse and improper use of antibiotics have led to the spread of antibiotic resistance. It is consequently getting difficult to find effective antibiotics to treat certain infections that used to be susceptible. Some notable examples are the emergence of methicillin-resistant *Staphylococcus aureus* (MRSA), and the appearance of multiply resistant strains of tuberculosis and gonorrhea. The CDC reports that over 2 million Americans, 23,000 of whom subsequently die, become infected with antibiotic-resistant bacteria and fungi each year.

Many careless practices have led to the spread of antibiotic-resistant bacteria. Certain antibiotics are widely used in agriculture to promote growth of domestic animals and improve meat yields. The United States in 2009 consumed over 36 million pounds of antibiotics, only 20% of which were administered to humans. The other 80% were given to farm animals. (Unlike in the United States, in Europe this practice has been greatly restricted; consequently, the continent has seen a drop in antibiotic resistance.) Overprescription of antibiotics for minor ailments, even when bacteria may not cause them, is another factor. Worsening the problem is patient noncompliance, that is, patients feeling better and failing to finish the entire regimen of antibiotics. If the infecting bacteria are not totally destroyed, the survivors may gain resistance and spread. Surprisingly, antibiotics can even be purchased without a prescription over the counter in many countries. Poor countries additionally decrease the dosage and length of antibiotic treatment to save money.

Similar considerations apply to antiviral agents, such as those used to treat HIV/AIDS. The virus mutates so quickly (see Chapter 21) that treatment with just one antiviral agent will result in resistant HIV mutants appearing in most patients. In practice, HIV/AIDS patients are given cocktails of at least three antiviral agents so that the other antiviral agents will kill mutants resistant to one. Previously, patients had to take different pills at different times of the day, but newer combination therapies allow for a once-per-day treatment. However, poor people in developing countries may not have access to combination therapy and instead may rely on just a single antiviral.

QUESTIONS

- What policies, if any, should be implemented to prevent antibiotic and antiviral resistance? Should use of antibiotics in agriculture be forbidden? Should certain antibiotics be restricted to human use only? Should doctors' prescriptions be closely monitored?
- Should developed countries forbid the export of certain antibiotics and antivirals to poor countries in order to keep them safe from improper use?
- Should those unable to afford HIV combination therapy be forbidden to take a single antiviral agent because of the risk of producing a resistant virus?

GENETICALLY MODIFIED ORGANISMS

Modern biotechnology has made it possible to alter individual genes within particular species or even to move blocks of genetic information across major taxonomic boundaries (e.g., from bacteria to animals or plants), rather than merely selecting for desirable genetic variants within a population or by hybridizing closely related species. The technology involved was discussed in Chapter 15 on transgenic plants, Chapter 16 on transgenic animals, and Chapter 17 on human gene therapy.

Some people view the creation of transgenic animals and plants (genetically modified organisms, or GMOs) as interference with the natural world. To those with deeply held religious beliefs, such tinkering may be seen as interference with God's creation. Many within the scientific community, including the authors of this book, view genetic engineering as merely a faster and more efficient way to achieve what selective breeding has done historically. There is no compelling scientific evidence that GMOs cause adverse health effects in humans. On the other hand, their effects on the environment are less clear. In practice, many environmental and consumer groups remain adamantly opposed to GMOs.

Genetic engineering followed by cloning to distribute many identical animals or plants is sometimes seen as a threat to biodiversity. However, evidence examining the effect of GMOs on biodiversity has not shown this to be the case. Besides, this alleged threat is not unique to GMOs: humans have been replacing diverse natural habitats with artificial monoculture for millennia. Artificial environments have already replaced wild habitats all over the world. If a threat to biodiversity from GMOs does exist, it is miniscule compared to the systematic destruction of natural habitats, climate change, and the introduction of invasive species by more traditional human activities.

Although many of the issues are related, disapproval of genetic modification or cloning generally increases as we move from plants, to animals, to mammals, to animals kept as pets, and finally to humans. This is true for both the public and professional ethicists. Thus, the weighting given to different factors clearly changes as we move closer to ourselves.

Transgenic Plants

Despite public disapproval in the West, usage of GMO crops continues to grow, particularly in developing countries. According to the International Service for the Acquisition of Agribiotech Applications (ISAAA), 20 developing nations grow 52% of the world's GMOs, while 8 developed nations grow the remaining 48% (Fig. 24.7). In the United States, most major crops are GMOs (Table 24.1). Though many farmers have embraced biotechnology, the terminator seed controversy caused a rift between farmers and the companies that provide seeds. (See Box 24.1.)

Box 24.1 Terminator Genes in Seeds

One divisive aspect of the GMO controversy was the development of "terminator" technology. Crop plants were engineered so that their seeds would be sterile. The pretense was that this would prevent escape of GMO plants into the wild. The underlying motive was financial. Farmers would be forced to buy a new supply of seeds each year instead of planting seeds saved from the previous year's harvest. This would both increase the profits of the seed corporation and make farmers dependent on their seed suppliers. The attempted use of terminator technology caused a great deal of ill feeling.

The terminator scheme involves three transgenes:

(a) A gene for a toxin that is lethal only in developing seeds. The toxin gene is otherwise inactive due to a DNA spacer flanked by *loxP* sites inserted between the promoter and the coding sequence.

(b) A gene for Cre recombinase (see Chapter 14), which recognizes the *loxP* sites and recombines them, thus deleting the spacer sequence. This allows expression of the toxin gene.

(c) A gene encoding a variant of the TetR repressor (see Chapter 15) that prevents expression of the Cre recombinase gene.

Before sale, the seeds are soaked in a solution of tetracycline that binds to and inactivates the repressor. This allows the Cre recombinase to become active and remove the spacer sequence. The toxin gene is now expressed. Because the toxin does not harm the growing plant, except for the developing seeds, the crop grows normally except that the seeds are sterile.

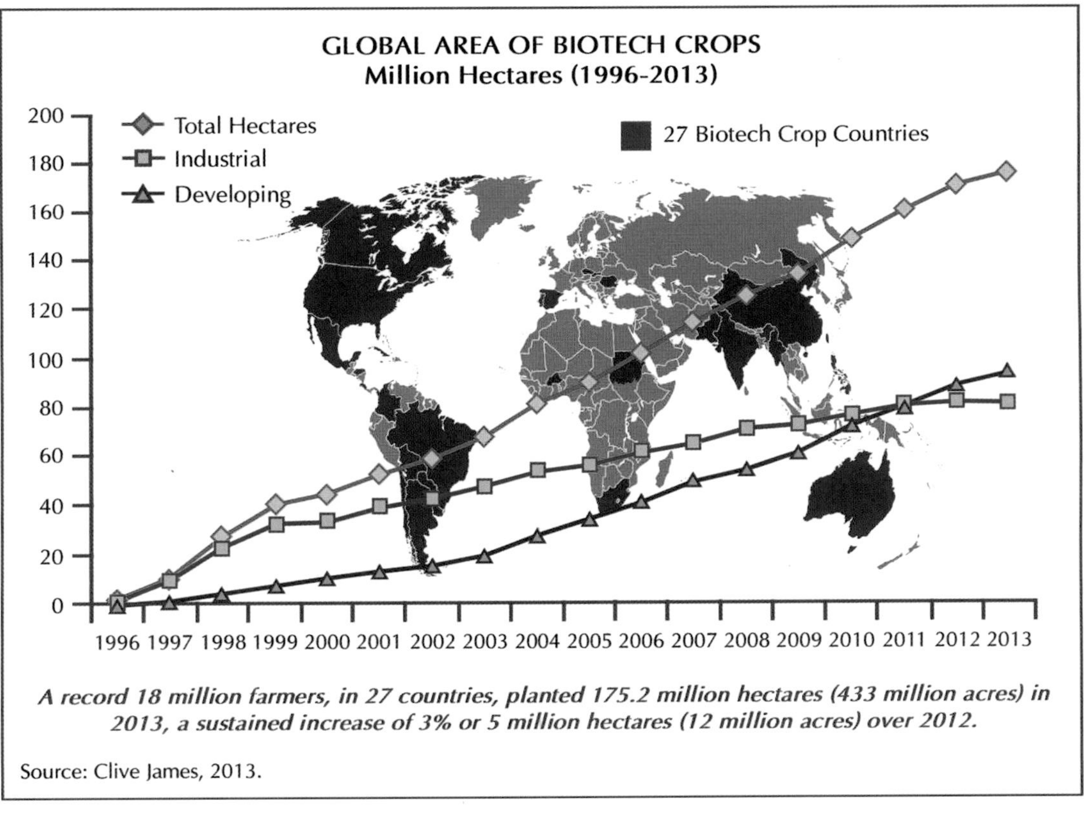

FIGURE 24.7 Transgenic Crops
Area of transgenic crops planted worldwide from 1996–2013. From James C (2013). *Global Status of Commercialized Biotech/GM Crops for 2013: ISAAA Brief 43*. ISAAA, Ithaca, NY. Courtesy of Clive James and the International Service for the Acquisition of Agri-biotech Applications.

Most transgenic animals can be contained fairly easily, but this is not true for transgenic crops. Plants with improved resistance to drought, disease, or insect pests would have an advantage in the wild, and thus, hybridization of transgenic crops with wild varieties would be expected to occur. Indeed, it already has. Wild maize (corn) in Mexico examined in 2001 contained transgenic DNA, even though planting transgenic corn

Table 24.1 Percentage of U.S. Crop That Is Genetically Modified

Soybeans (Herbicide-tolerant)	93%
Cotton (Herbicide-tolerant)	82%
Cotton (Bt)	75%
Corn (Herbicide-tolerant)	85%
Corn (Bt)	76%

Source: USDA, http://www.ers.usda.gov/data-products/adoption-of-genetically-engineered-crops-in-the-us/recent-trends-in-ge-adoption.aspx#.UnRjnZSienA

plants was stopped in 1998. Herbicide resistance has spread into wild canola plants, and herbicide-resistant "superweeds" already threaten farmers. A strain of herbicide-resistant wheat mysteriously appeared in a field in Oregon. Still, it should be remembered that hybridization and invasion by non-native species are a natural part of the evolutionary process. Therefore, it is currently unclear if the spread of transgenes in the environment represents an excessive risk to ecosystems and agriculture beyond what normally occurs in nature.

Complete containment of transgenic plants is unrealistic on an agricultural scale, but it can be reduced in various ways. For instance, planting GMOs 50 to 100 meters away from unmodified crops should keep cross-fertilization events below 0.5%. Still, even with mitigating tactics in place, seeds from GMO and non-GMO crops are impossible to keep wholly separate.

QUESTIONS

- Is genetic engineering of humans, animals, and plants wrong? Are there major ethical differences between altering humans versus animals and animals versus plants?
- Do the benefits of GMOs outweigh the potential risks?
- Who should get to decide if a farmer wants to plant GMOs?
- Should transgenic crops be banned until scientists are certain there are no environmental threats? How long is long enough?
- Should advanced countries avoid GMOs just to be safe? What about poor countries?
- Is it ethical for advanced countries to adopt policies that affect whether or not developing countries can plant GMOs?

Genetic Modification of Animals

A generation ago, Americans almost unanimously supported animal research. Due perhaps to the influence of animal rights groups, today only about half support it. Even though public support has fallen, some animal research will be necessary for the foreseeable future. It is simply not possible to understand many aspects of biology, such as physiology, toxicology, or immunology, without using live animals in basic research. On the other hand, animals are also used in large numbers for screening and testing products such as soaps and cosmetics. Much animal testing to ensure the safety of products used by humans was enforced by legislation a generation ago or more. At the time, the companies concerned often opposed such legislation for being unnecessary and expensive. Those in favor of the legislation argued that human health and safety outweighed other factors. Today, the moral tide has turned. Many now argue that routine animal testing of products such as cosmetics is not justified merely to avoid possible minor discomfort.

It is important to point out that scientists often look for alternatives to animals when possible, both for ethical and financial reasons. Tissue culture and computer models are common alternatives. The National Institutes of Health recently decided to retire most research chimpanzees, finding their use ethically questionable and scientifically unnecessary. According to the NIH, new scientific methods have made their use in research largely obsolete. From an ethical standpoint, Francis Collins, the Director of NIH, stated, "Their likeness to humans has made them uniquely valuable for certain types of research, but also demands greater justification for their use."

Humans have deliberately modified animals since time immemorial through selective breeding and hybridization. Additionally, human activity has led to inadvertent genetic modification of many organisms. Converting a patch of forest into farmland eliminates the natural inhabitants and selects for life forms adapted to croplands. We have undoubtedly selected for alterations in the mice that infest our fields and the insects that rely on human crops. If we did not grow so much corn, for instance, the European corn borer would likely be a rare insect. Whether we are aware of it or not, we routinely impose genetic selection on organisms around us, whatever we do.

Thus, the novelty of genetic engineering is not in *what* we are doing, but in *how* we do it. Today, we can transfer DNA from one organism to any other organism. Genes from bacteria and jellyfish can be snipped out and inserted into mice and rabbits. With our increasing knowledge and ability to manipulate organisms at the genetic level, we can create some truly bizarre creatures.

By manipulating the *Hox* genes, which control body plans and are important in embryonic development, scientists have created a fruit fly that grows legs on its head. More bizarrely, it has been proposed that manipulating *Hox* genes could "de-evolve" a chicken to a more ancestral form that somewhat resembles a dinosaur. It may even be possible to grow farm animals with multiple body parts to satisfy the world's demand for meat or lacking most of the brain to prevent suffering.

What seems to us a rather frivolous use of biotechnology is the creation of transgenic animals for artistic reasons, i.e., "transgenic art." The insertion of the gene that encodes green fluorescent protein (GFP) is now routine in genetic engineering. A rabbit named Alba, also called "GFP Bunny," was claimed as genuine transgenic art. Born in France in the spring of 2000, Alba was a transgenic albino rabbit expressing high levels of GFP. She was surrounded by controversy. An artist, Eduardo Kac, claimed that Alba was engineered at his request, but the scientists who created her refused to release the rabbit to the artist. According to Eduardo Kac, the GFP in GFP Bunny was intended to be a "social marker" and to raise questions about social integration. The artist's full viewpoint may be seen at

http://www.ekac.org/gfpbunny.html#gfpbunnyanchor

FIGURE 24.8
GloFish Tetra
You can purchase transgenic animals for your children's amusement and enlightenment! Courtesy of GloFish; http://www.glofish.com/meet-glofish/glofish-gallery/.

The GFP Bunny project was widely criticized as unethical for experimenting on animals for the sake of art. However, many breeds of dogs, as well as some birds and fish, have been historically bred for their appearance, with little ethical criticism.

Besides GFP, other fluorescent proteins with other colors are commonly used. Animals carrying such genes appear normal in daylight, but if illuminated in the dark with near-UV or blue light, they fluoresce. Fluorescent fish are available commercially under the trade name GloFish (Fig. 24.8), and they come in six amazing colors!

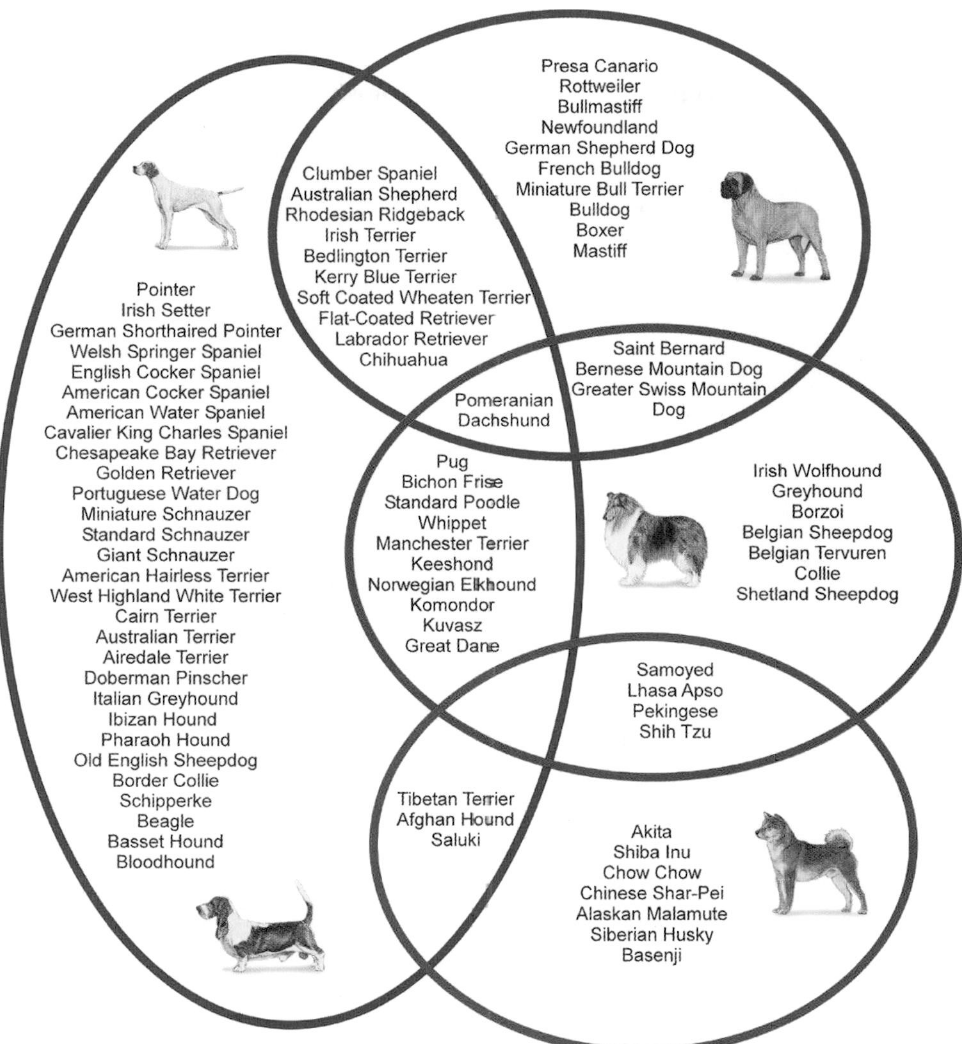

FIGURE 24.9 Population Structure of Dog Breeds

Five unrelated dogs from each of 85 breeds were genotyped using 96 (CA)n repeat-based microsatellites (VNTRs) that spanned the dog genome at an average density of 30 megabases. Four main genetically related clusters of breeds exist, as indicated by the colored circles, together with a variety of intermediate breeds. From Parker HG, Ostrander EA (2005). Canine genomics and genetics: running with the pack. *PLoS Genet* 1, e58. Public domain and courtesy of Elaine Ostrander.

Is it ethical to genetically modify animals for purely aesthetic reasons? Keep in mind that humans have done exactly that for a long time already. Pugs, a popular dog breed, have cute, flattened faces (and, consequently, breathing problems). Many other dog breeds and animals were bred mostly for show rather than for food or work (Fig. 24.9). Perhaps not surprisingly, such show breeds often have health problems. Transgenic technology has sped up this process and allowed more drastic alterations than the traditional crossbreeding methods.

QUESTIONS

- Is genetically modifying animals less ethical than modifying plants?
- Is it ethical to meddle with genes involved in embryonic development? What if it creates a grotesquely deformed animal?
- Should people be allowed to clone their pets? (See Box 24.2.)
- Is genetic engineering ethical if it does not introduce foreign DNA from another species?
- Would it be ethical to develop a chicken with, say, eight legs for food?
- Should "transgenic art" be allowed?

Box 24.2 Cloned Pets from Genetic Savings & Clone

Julie (last name withheld by request) of Texas became the first paying client to receive a pet clone, when the "twin" of her deceased cat Nicky—dubbed "Little Nicky"—was presented to her at a December 10 holiday party thrown by Genetic Savings & Clone (GSC) at a San Francisco restaurant.

As the first clone delivered to a paying client, Little Nicky made a huge splash in the media when GSC announced the delivery in December 2004. "He looks identical, his personality is extremely similar, they are very close," said Julie, an airline employee from Dallas who placed the order, during an interview on *Good Morning America* on December 23, 2004. Little Nicky, who was born October 17 in Austin, Texas, is a clone of Nicky, a Maine Coon who died in November 2003 at age 17.

However, in December 2006, Genetic Savings & Clone shut down due to lack of demand.

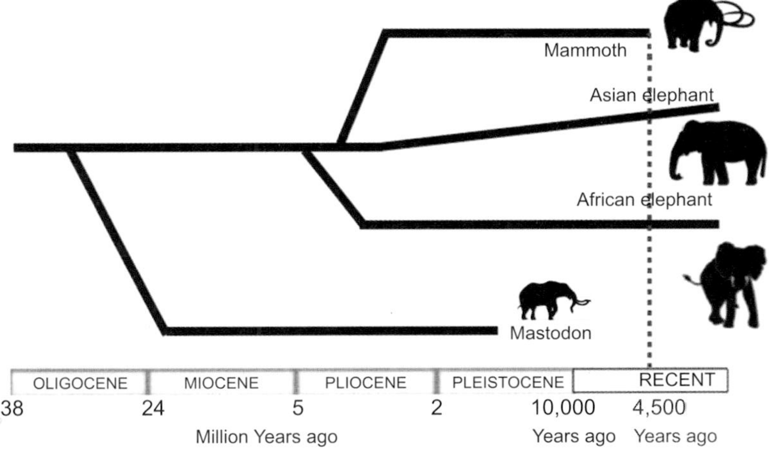

FIGURE 24.10 Mammoth and Elephant Evolution
The extinct mammoth is actually more closely related to the Asian elephant than the Asian and African elephants are to each other. Note also that modern-day elephants are larger than the mammoth, which in turn was larger than the mastodon.

De-Extinction: Resurrecting the Dead

Modern biotechnology has raised the intriguing possibility of bringing extinct species back to life. The animal that gets scientists and the public most excited is the woolly mammoth. In 2013, blood was recovered from the frozen body of a woolly mammoth in Siberia, raising the possibility of cloning a mammoth in the same manner as Dolly the sheep and several other animals. Elephants are the nearest living relatives of mammoths (Fig. 24.10). Theoretically, scientists could perform nuclear transplantation (also called somatic cell nuclear transfer) by transferring a mammoth cell nucleus into an enucleated elephant egg cell. The resulting embryo would gestate in an adult female elephant.

There are many problems with resurrecting extinct species, however. DNA has a chemical half-life of only 520 years or so under most conditions, even if frozen. Most mammoths went extinct about 10,000 years ago, which means the DNA in any frozen carcass would be mostly degraded. A population of mammoths survived on Wrangel Island in the Arctic Ocean until 1650 BC, but this length of time is still equivalent to seven half-lives of DNA. Therefore, a more promising approach may be to sequence as much mammoth DNA as possible, rebuild the original genome sequence, and clone it into artificial chromosomes. Even assuming that a woolly mammoth could be cloned, the animals would have little genetic diversity, so propagating the population would be difficult. Also, some people worry that bringing mammoths back solely for the purpose of putting them on display in a zoo is unethical, while others worry that de-extinction projects might undermine conservation efforts.

Resurrecting the woolly mammoth might be a long shot, but other extinct species may be more likely. The Long Now Foundation has listed 26 animals and plants that it believes are candidate species for de-extinction, including dodos, thylacines, and saber-toothed cats.

QUESTIONS

- Is resurrecting an extinct species ethical?
- What about resurrecting Neanderthals or other extinct human ancestors?
- If species are brought back from extinction, should they be only kept in captivity or released into the wild?

HUMAN ENHANCEMENT, CLONING, AND ENGINEERING

Few issues are more controversial than applying biotechnology to the human race. The debate boils down to two fundamental questions: Is it ethical to modify our own bodies? Is it ethical to modify future generations of human beings? Many believe the former is ethical, but not the latter. Keep these two questions in mind as you ponder the following issues.

Genetic Screening in Pregnancy and of Newborns

Genetic screening can come in two forms: screening the parents before conception to determine the risk of producing children with inherited genetic disorders and screening the fetus post-conception.

The former has an obvious benefit: Parents who are at high risk of producing children with severe genetic defects may decide to forgo having children and to adopt instead. On the other hand, parents may decide to accept the risk or overcome the risk by using cutting-edge reproductive technology. For example, women with mitochondrial disease may in the near future be able to acquire a healthy woman's eggs. Replacing the donated egg's nucleus with a nucleus from the mother will ensure that the mother's genes are passed on, with the exception of the defective genes in the mitochondria. (Recall that fertilized eggs acquire all their mitochondria from the female who produces the eggs.) The resulting "three-parent embryo" will contain nuclear DNA from the mother and father and mitochondrial DNA from an unrelated, healthy woman. This procedure is presently (2015) prohibited in the United States. However, in February 2015, the UK gave final approval. The UK is the first nation to allow this, as with the original *in vitro* fertilization procedure.

Genetic screening can also be performed *in utero*, or shortly after birth. This allows early treatment of newborn infants, the classic case being phenylketonuria (PKU). People with PKU lack the enzyme that converts phenylalanine to tyrosine, and excessive amounts of phenylalanine causes permanent brain damage. Newborn screening allows infants with PKU to be given a diet low in phenylalanine, greatly reducing the damage. More recently, it has become possible to screen developing fetuses for a variety of genetic defects long before birth. Analytical techniques are constantly advancing, and a growing list of inherited defects can be monitored at ever-earlier stages of development.

However, prenatal genetic screening, especially whole genome sequencing, could also be used to decide whether to abort a fetus that does not have traits desirable to the parents. As understanding of the human genome increases, it will become possible to deduce such things as the probable future height, eye color, IQ, and beauty of the developing fetus. Most parents would like to have smart, healthy, and attractive children, and the temptation to have abortions based on these characteristics will soon become a reality.

Currently, a major problem, most notably in Asia and Eastern Europe, is *gendercide*—the selective abortion of female fetuses. In some cultures around the world, having a male child is preferable to a female. As a result, female fetuses are disproportionately aborted. This has resulted in vastly skewed gender ratios. Normally, about 105 male infants are born for every 100 females, but in some regions, that ratio is 130 to 100. The long-term societal consequences of these decisions remain to be seen, but anecdotal evidence suggests that some men are having great difficulty in finding mates and building families.

Of course, lurking in the background of this entire discussion is the central question about abortion: "When does human life begin?" From a biological perspective, *life* does not begin at conception because it is a continuum. Sperm cells are alive, and so are the eggs they fertilize. Fusion of egg and sperm to create a zygote forms a new living individual with a unique genetic constitution. Thus, rather than the beginning of life, the real issue is personhood. At what point does a developing fetus become a "person"? Attempts have been made to answer this question based on when the fetus can feel pain or has developed some semblance of consciousness, but conclusive biological evidence has not been demonstrated.

Finally, who should decide if an abortion is to be performed? From a genetic viewpoint, both father and mother have an equal share in the new individual, except for the mitochondrial DNA that is maternal in origin. But, from the viewpoint of biological resources, the mother has more invested and has traditionally been allowed to make the decision. The father, therefore, has fewer legal rights over the children. Is that fair? Although this outlook was not deliberately based on evolutionary considerations, it does in fact coincide with Darwinian logic.

QUESTIONS

- When does "personhood" begin? When does a fetus deserve full human rights?
- Are "three-parent embryos" ethical? Does the "third parent" have any rights to the child?
- Should whole genome sequencing of a fetus be allowed?
- Should defective fetuses be terminated by abortion? If so, who should make the decision whether a defective fetus lives or dies?
- Should we enforce paternity tests to make sure that the true genetic parents of a child are notified of any decisions about the child's welfare?
- What if it becomes possible to detect genetic predisposition to undesirable characteristics, such as violent crime or low IQ, in developing fetuses? Should we allow selective abortion?

Embryonic Stem Cells and Stem Cell Therapy

Entangled with the abortion controversy is the issue of embryonic stem cell research. Stem cells are the precursors to the differentiated cells that make up the body. Embryonic stem cells are derived from an embryo's inner cell mass and retain the ability to develop into any body tissue. Adult stem cells, on the other hand, are partially differentiated and can develop into only a handful of related cell types. For instance, hematopoietic stem cells can develop into only red and white blood cells (see Chapter 18 for more details on stem cells).

While using adult stem cells is rarely considered controversial, using embryonic stem cells is. The reason is that deriving them generally destroys the embryo. Objectors believe that purposefully destroying an embryo for research purposes is unethical, whereas those in favor believe it is unethical to halt medical research that may yield life-changing treatments for patients. Currently, in the United States, researchers can obtain federal grant money only if they derive embryonic stem cells from "leftover" embryos created by *in vitro* fertilization for reproductive purposes, only after the parents have granted permission. Creating embryos for the sole purpose of deriving embryonic stem cells is not eligible for federal grant money.

Partially to get past this legal hurdle, scientists have created induced pluripotent stem cells (iPSCs). Somatic cells can be genetically or chemically manipulated to revert them to an embryonic-like state. However, much is still unknown about iPSCs, including whether or not they are an adequate substitute for embryonic stem cells. Research suggests that important differences exist between the two cell types.

Stem cells excite the scientific community because they have the potential to be the "holy grail" of regenerative medicine. One day, it may be possible to grow any organ that needs to be replaced. Simpler organs, such as tracheas and bladders, have already been grown and transplanted into patients using adult stem cells. Tiny livers have been grown from iPSCs. With further research, it is hoped that more complicated structures, such as hearts and lungs, will eventually be grown in the lab.

One advance that would make regenerative medicine more likely is *therapeutic cloning*. Using somatic cell nuclear transfer, a nucleus from a patient's cell could be transferred to an

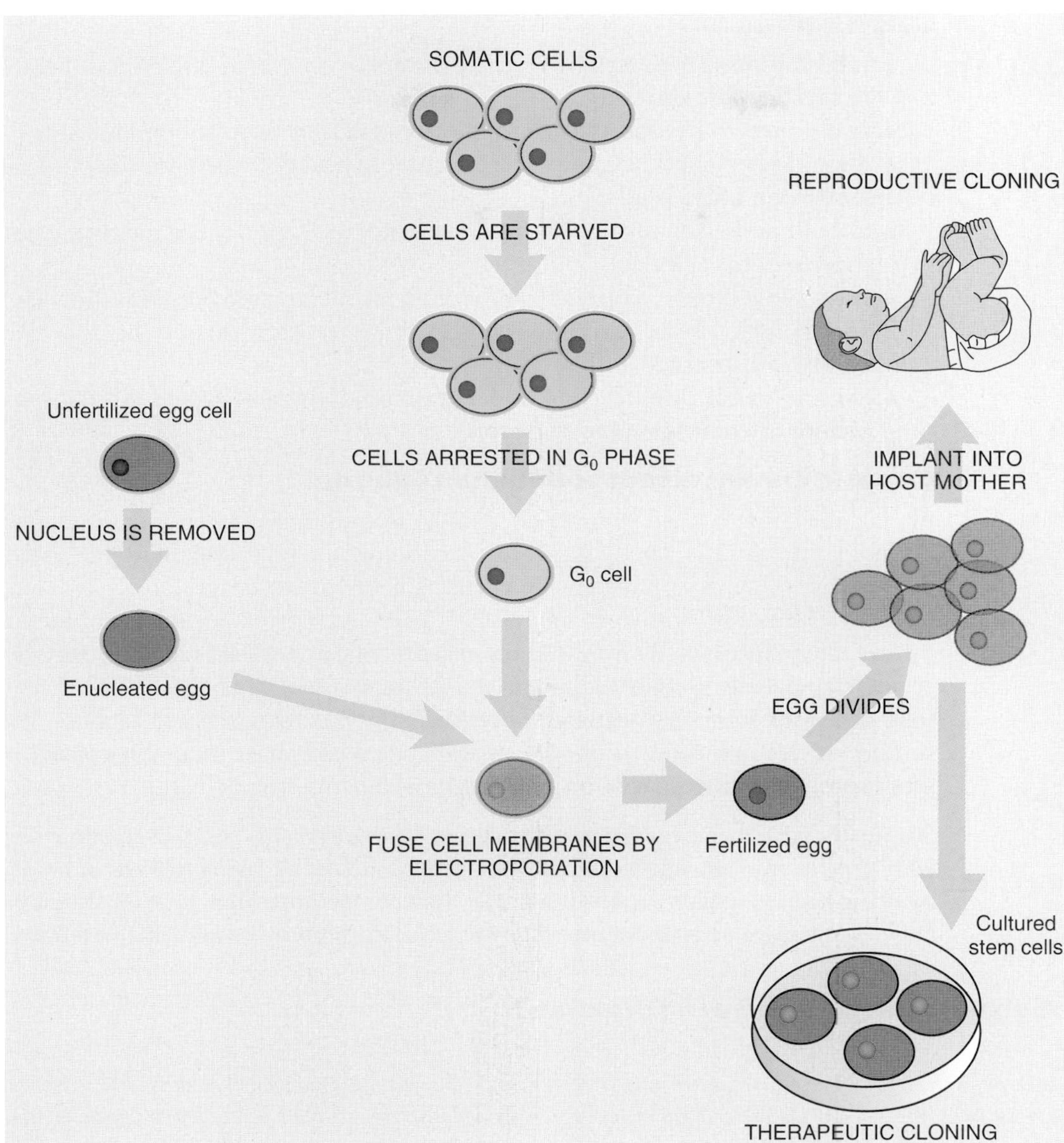

FIGURE 24.11 Therapeutic Cloning
Somatic cell nuclear transfer can be used to clone an entire human being or to derive embryonic stem cells for therapeutic purposes.

enucleated egg, which would then be coaxed into forming an embryo (Fig. 24.11). Harvesting the inner cell mass would derive embryonic stem cells that were 100% genetically identical to the patient (with the exception of mitochondrial DNA). Until recently, cloning primate embryonic stem cell lines has been fraught with technical difficulties. However, in 2007, cloned embryonic stem cells were derived from rhesus macaques, and in 2013, they were derived from humans.

Stem cell therapy is also an active area of investigation. In the future, it may be possible to heal spinal cord injuries, cure blindness, and regrow teeth using stem cells. This technology is still in its infancy, but unscrupulous practitioners have made outlandish claims of being able to cure almost any ailment known to man. These "treatments" are the 21st century equivalent of "snake oil," often involving techniques that are not supported by scientific evidence and may be harmful. One woman who had stem cells injected near her eye for cosmetic reasons instead grew bone under her skin, and a child who had stem cells injected into his brain grew a tumor.

QUESTIONS

- Is it ethical to destroy embryos to derive embryonic stem cells? Should embryonic stem cell research be allowed?
- Should researchers be allowed to harvest stem cells from aborted fetuses?
- Should researchers be allowed to create embryos specifically for the purpose of deriving embryonic stem cells?
- Should we personally benefit from technologies if we oppose the research methods used to develop them?
- If we allow human embryos to be cloned for therapeutic purposes, how far should such embryos be allowed to develop? Should we allow them to be used only to derive embryonic stem cells?
- Should people be allowed to undergo stem cell therapy, even if it is dangerous and unsupported by scientific evidence?

Genetic Enhancement and Human Cyborgs

By "genetic enhancement," we are referring to technologies that modify human somatic cells, but not germline cells. Therefore, in this section, we are excluding heritable changes. Gene therapy, by which we specifically mean curing genetic defects, has largely avoided raising moral concerns among the general public. Currently, most of the relevant bioethical issues involve safety since it is still mostly in the experimental stage (see Chapter 17 for a discussion of gene therapy). Early in its development, gene therapy suffered major setbacks when experimental therapies resulted in cases of death or serious injury. However, as sad as these cases are, it should be kept in mind that many gene therapy patients have incurable conditions or poor prognoses with little expectation of living long, healthy lives.

Eventually, gene therapy will give way to genetic enhancement. This will certainly be popular with athletes who already "dope" in various ways, such as by taking steroids or boosting their red blood cell count, to gain an advantage over their competitors. Therefore, we should fully expect athletes to engage in gene doping when the technology becomes available (see Box 24.3).

Box 24.3 Gene Doping—A Future Dilemma?

The advent of improved transgenic mice (e.g., Mighty Mouse, Marathon Mouse; see Chapter 16) has led to wishful thinking on the part of some people. Could the same improvements be applied to human athletes? Researchers involved in these areas have received quite a few inquiries from athletes looking for muscle enhancement.

The term *gene doping* refers to the provision of extra copies of genes that offer a competitive advantage (as opposed to doping with illegal steroids, for example). Greater muscle mass, increased endurance, or higher red cell counts are all possible. The extra proteins would be very similar or identical to the body's own. Hence, detection would be extremely difficult.

At the moment, gene doping is still a future worry. But given the pace of scientific development and the willingness of athletes to try performance-enhancing drugs, even if illegal, the issue may emerge into reality soon.

But gene doping raises an interesting conundrum. It is a fact that some people are born with superior genetics for athletic performance. Sports teams are already asking athletes to submit to genetic testing. One particularly interesting gene is ACTN3, which encodes a protein expressed in skeletal muscle cells. The R allele of the gene is commonly found in athletes who play power sports, such as sprinting and football, while the X allele is found in athletes who play endurance sports, such as long-distance running. Should athletes who are not genetically gifted be allowed to gene dope in order to "level the playing field?" Alternatively, should competitions be segregated along genetic lines so that athletes with inferior genes are not forced to compete against athletes with superior ones? And what of mechanical enhancements, such as "bionic" limbs (Fig. 24.12)? Oscar Pistorius is an Olympic sprinter whose amputated legs were replaced with carbon-fiber prostheses. Pistorius's prosthetic legs cause him to run differently than people with regular legs. Controversy ensued because some athletes felt he had an unfair advantage.

An issue related to genetic enhancement is the creation of human cyborgs. Cyborgs are organisms that are enhanced using electronic or mechanical technology (see Chapter 14 for cyborgs). For example, scientists are developing bionic arms and legs, which respond merely to thought, for patients who are missing limbs. Augmented reality glasses allow users to see the world enhanced with information pulled from the Internet. Strangely, not everybody is pleased with this technology. Professor (and cyborg) Steve Mann, who has augmented reality glasses permanently attached to his head, claims that angry employees assaulted him at a McDonald's in Paris. If his allegations are true, this could be the world's first anti-cyborg hate crime!

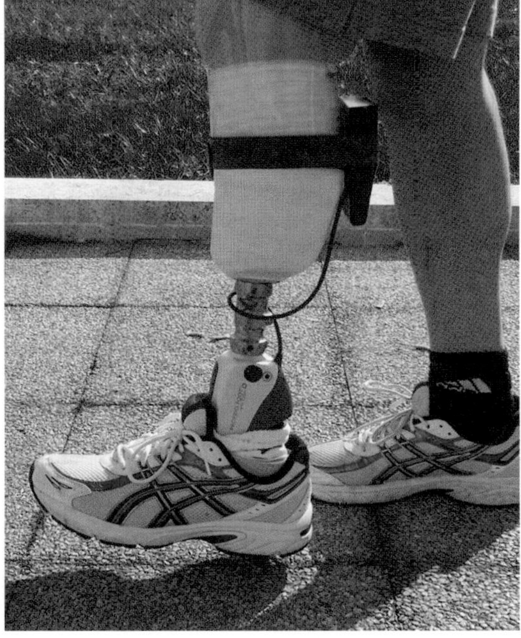

FIGURE 24.12
Bionic Foot
The Proprio-Foot®. During the swing phase, it automatically lifts the toe to reduce the risk of trips. Power comes from a battery on the rear side of the socket. From Delussu AS, et al. (2013). Assessment of the effects of carbon fiber and bionic foot during overground and treadmill walking in transtibial amputees. *Gait Posture* **38**, 876–882.

Humans aren't the only creatures that will become cyborgs. An educational company named Backyard Brains has developed a cyborg cockroach called RoboRoach (Fig. 24.13). Customers can order live cockroaches along with an electronic gadget that allows a person to control the cockroach's movement. The kit comes with instructions on how to perform surgery on the cockroach in order to install the device. Once set up, the user can command the cockroach to move left or right using a smartphone. Critics contend that RoboRoach will turn children into psychopaths by teaching them to treat living organisms as playthings.

QUESTIONS

- Is there an ethical difference between gene therapy and genetic enhancement? Should the former be allowed but not the latter?
- Should patients with serious illnesses be allowed to risk experimental gene therapy, even if it might kill them?
- Is gene doping cheating, or is it "leveling the playing field?" Should there be separate Olympic categories for those with different genetic makeups?
- Should people be allowed to turn themselves into cyborgs?
- Is creating a cyborg for the purpose of controlling an organism's mind or behavior unethical?

FIGURE 24.13 RoboRoach
Two views of RoboRoach, the cyborg cockroach from Backyard Brains. Courtesy of Backyard Brains; https://backyardbrains.com/products/roboroach.

Human Cloning

The same technique used for therapeutic cloning (discussed earlier) can be repurposed to clone entire human beings (see Fig. 24.11). Instead of harvesting the inner cell mass of an embryo to derive embryonic stem cells, scientists could implant the embryo into a surrogate mother who would give birth to a human clone. But why do this? There are possibly two reasons: creating "replacement" individuals and creating brainless humans for "spare parts."

Replacement clones might be desired by parents whose child died prematurely or by wealthy and powerful people who would like their legacy to live on in the form of a clone. Dictators might be interested in keeping a series of backup clones for emergencies. (It is rumored that various world leaders throughout history have had political decoys or body doubles.) However, cloned humans would have to grow and develop, just like any other human, and thus would not reach adulthood for many years. Furthermore, environmental influences would mean that although genetically identical, the clone would not be a behavioral replica. Identical twins are (mostly) genetically identical—i.e., natural clones—yet they still show considerable divergence in personality, behavior, and ability. From a practical viewpoint, there is little reason to clone human beings, but it is likely inevitable that some research group will attempt it.

Organ shortages are a common problem, particularly in the United States. Too few people sign up to be organ donors, resulting in a shortage. Organs are largely distributed on a first-come-first-served basis combined with urgency. Even here, the classic dilemma still exists: "If only one liver is available, should it be given to the alcoholic (who will abuse it), the aging textbook writer (whose best work is over), or a young child (whose future is unknown)?" Several solutions have been proposed to alleviate this crisis.

The first involves changing organ donation laws. From the 1980s onward, several European countries have introduced "presumed consent laws" for organ donation. Under this system, it is presumed that an individual is willing to contribute organs upon death, unless he has registered as a nondonor or there is other evidence to the contrary (e.g., from relatives). In most cases, this has greatly increased the supply of organs. But this approach does not solve the problem of organ rejection. Therefore, the second solution (discussed previously) is regenerative medicine, which is growing cloned organs in the laboratory. The third solution is far more gruesome: growing cloned, brainless humans as a source of "spare parts."

In addition to being ethically dubious, this solution would be technologically challenging because modifications would be needed to avoid developing complete individuals. In particular, it would be necessary to engineer a humanoid with no conscious mind or ability to feel pain, presumably by restricting brain development. In the distant future, will our world be one in which every person has a brainless backup clone in the basement to provide spare parts? Or will a central facility grow generic clones with elongated bodies, missing heads, and 20 kidneys? Thankfully, regenerative medicine will likely make this disturbing possibility an obsolete approach.

QUESTIONS

- Is human cloning ethical?
- Is there an ethical difference between cloning for replacement individuals as opposed to cloning brainless humans for spare parts?
- Should the United States switch to presumed consent for organ donation?

Human Genetic Engineering

By *genetic engineering*, we are specifically referring to the manipulation of the human germline. At present, this is illegal in most countries. Human genetic engineering is also not yet technically feasible, although this is likely to change fairly soon. Critics of genetically modified organisms, who claim that scientists are "playing God," object to human genetic engineering especially strongly. Most people who accept genetic engineering of animals and plants also oppose genetic engineering when applied to humans.

However, there are many potential benefits to human genetic engineering. This technology may allow humans to be engineered by choosing variants of existing human genes (e.g., resistance to disease, improved intelligence, or blue eyes). This can already be done in a primitive way by genetic screening followed by abortion of an undesirable fetus, as noted earlier. However, it may eventually be possible to deliberately engineer offspring for desired characteristics, perhaps by inserting DNA from other humans or even foreign DNA into the human genome, creating transgenic humans. We have previously discussed engineering animals to make their own essential amino acids, thus avoiding the need to provide these in the diet (see Chapter 16). What about combating malnutrition and vitamin deficiency in humans by inserting DNA encoding metabolic pathways for essential amino acids or vitamin C? Also, adding the gene for green fluorescent protein might make it easier to find wayward children in the dark!

In the first few decades of the 20th century, various countries around the world tried their hands at eugenics, the deliberate genetic improvement of the human species (see Box 24.4). The American eugenics movement promoted the sterilization of the mentally ill and violent criminals. Such schemes were based on selective breeding (as used on domestic animals over the centuries) rather than genetic engineering.

Box 24.4 Eugenics or Dysgenics?

The Napoleonic Wars provided a fascinating but accidental example of dysgenics (negative eugenics). Napoleon, who was himself short, deliberately recruited tall men into the French Imperial Army. Although Napoleon won many battles, his casualties were enormous, and even when victorious, he often lost more men than his enemies because he used columns many soldiers deep. Bullets and shells often passed through men in the front rows, killing those behind, too. The result of constantly selecting tall men and subjecting them to massive casualties was that the average height of the French nation decreased significantly in the generation that followed!

Since industrialization began, welfare programs of one kind or another have enabled the poor to raise children at the expense of social programs. On the other hand, overcrowding due to urbanization has greatly aided the spread of infectious disease, especially among the crowded and badly housed poor. Have we been selecting for or against the less able? No one really knows. Nonetheless, our actions are changing the human gene pool, whether we care to admit it or not. Today we possess better techniques not only for screening genetic defects but also for artificial manipulation of individual genes and whole organisms. Will we use this knowledge sensibly, or will this issue remain taboo while we continue to make changes to society that alter our genetic heritage in an ignorant and semi-random manner?

While eugenics is rejected today in theory, it still occurs in practice. The institutionalization of criminals and mental patients greatly reduces the number of their offspring. Aborting genetically defective embryos also affects the gene pool, as does providing welfare to those unable to feed their own children. Almost any major social change affects the human gene pool, including epidemics and government policies. Additionally, culture itself can change the gene pool. For example, the domestication of cattle resulted in the selection of lactase expression in adults. Lactase, the enzyme that breaks down the sugar lactose present in milk, is typically expressed only during childhood. Though most humans are lactose intolerant, a substantial proportion of people, mainly of European ancestry, can consume milk products their entire lives (Fig. 24.14). Whether we intend to or not and whether we are aware of the changes or not, humans are constantly modifying themselves genetically.

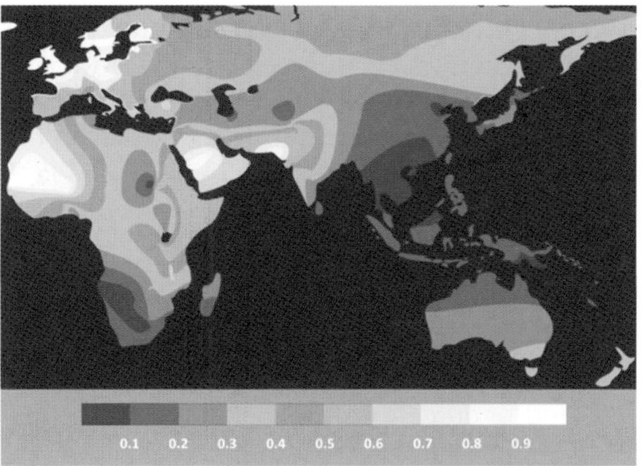

FIGURE 24.14
Lactose Tolerance around the World
About 30% of the adult human population makes lactase. Those who do not are lactose intolerant and have difficulty digesting milk and other dairy products. This proportion varies around the world; note that lactose tolerance is especially high in northwestern Europe. From Leonardi M et al. (2012). The evolution of lactase persistence in Europe. A synthesis of archaeological and genetic evidence. *Int Dairy J* **22**, 88–97.

Ultimately it may become possible not merely to fix specific genetic defects, but to enhance the human genome in various ways. Possibilities include improving resistance to assorted infectious diseases and to conditions such as cancer and heart disease and improvements to abilities such as vision, hearing, and intelligence. Perhaps most controversial is the suggestion of improving human beauty. At present, the public in Western nations is heavily opposed to human cloning and designer babies. However, the opposite is true in much of Asia, especially China and India.

QUESTIONS

- Is human genetic engineering "playing God?" Should it continue to be banned?
- Should violent criminals (e.g., rapists, serial killers) and the mentally ill be sterilized?
- Is eugenics inherently immoral? Or is there a benefit to improving society's gene pool?
- Is there an ethical way to engage in eugenics? For example, would individual decisions to have genetically modified offspring be acceptable?

ETHICS CHANGES OVER TIME

Advances in science and technology are often initially shunned due to their novelty. People often fear things of which they are ignorant, and many express those fears in terms of "ethics" and "morality." Biotechnology is very complex, and relatively few members of the general public understand the science behind it. Because of that, misinformation—both deliberate and accidental—becomes a real issue in the debate. One now-classic example concerns human cloning. In science fiction, clones typically emerge as full-grown copies equipped with the memories and skills of the original. In reality, if human clones were made, they would start out as embryos and need to acquire knowledge and skills as they grew up.

The other side of this phenomenon is that once a technology has become familiar, complaints against its supposed immorality fade away. Vaccines and rock music were both shunned by the Church when they were first introduced. Now, the Church gladly embraces both. The cloning of Dolly the sheep made enormous headlines and provoked a heated moralistic debate. Today, cloning a new animal species scarcely qualifies as a newsworthy event. When "test tube babies" (*in vitro* fertilization) first appeared, they were the subjects of intense ethical debate. Since Louise Brown, the world's first test tube baby, was born on July 25, 1978, there have been more than a million others. The "morality" of *in vitro* fertilization itself is rarely questioned any more. What is discussed today is the morality of modifications to the basic technology, such as its application in triparental babies or for prenatal genetic screening. The fact that many "fundamental ethical issues" fade away at much the same rate as fashions in clothing and music brings their deep significance into question.

Another factor that makes many bioethical decisions highly subjective is cultural viewpoint. Some religions and cultures embrace technology, whereas others reject it entirely. The Japanese have a fascination with robots, and it is conceivable that human cyborgs would be rather easily welcomed into their society in the future. But what if human cyborgs are rejected in other countries? Would this mean Japan is less ethical than other nations? How much of what we consider to be an issue of ethics is really just a difference in local customs?

As a general rule, we predict that many of the debates over the ethics of specific biotechnologies will fade away after society becomes more familiar with the benefits and risks of each technology. What will be debated in perpetuity, however, are the more basic underlying issues such as privacy, safety, fairness, and access. But do such underlying principles of ethics change over time and/or differ from one society to another? We leave that as a final unanswered question.

Summary

New advances in both knowledge and technical capability bring new ethical and regulatory problems. Closer inspection suggests that many apparently new bioethical issues are merely old issues in a new guise, such as privacy, safety, fairness, and access. Other issues do not seem to involve ethics so much as familiarity. Nonetheless, some aspects of biotechnology, such as the cloning and genetic engineering of humans, do pose legitimately new and difficult bioethical questions.

End-of-Chapter Questions

1. What percentage of the population is needed to grow food for everyone in advanced nations?
   - a. 5%
   - b. 10%
   - c. 90%
   - d. 2%
   - e. 50%
2. Which of the following was not a basic principle of the Belmont Report?
   - a. beneficence
   - b. autonomy
   - c. justice
   - d. privacy
   - e. All of the above are included in the Belmont Report.
3. Which of the following is an incidental side effect of technology?
   - a. Increased life-expectancy burdens the health care system.
   - b. Decreased infant mortality contributes to overcrowding and more use of finite resources.
   - c. Overcrowding because of increased life expectancy or decreased infant mortality promotes the emergence of infectious diseases.
   - d. Cars can enable a sick person to reach the hospital much faster than horses, but many people die in car accidents each year.
   - e. All of the above are side effects of technology.
4. What is one theory that describes why biological warfare creates more emotion than conventional warfare?
   - a. because of a fear of the invisible (i.e., microorganisms)
   - b. does not cause fear
   - c. because of multi-drug resistant bacteria
   - d. because of a lack in sufficient control mechanisms to contain an outbreak
   - e. none of the above
5. How has the supply of donated organs increased in European nations?
   - a. Upon death, people are assumed to be donors unless pre-registered as non-donors.
   - b. People are generally friendlier in European nations and are more willing to give.
   - c. Payment for the organs is allowed in European nations; therefore, people are willing to donate more.
   - d. The supply of donated organs has not increased in European nations.
   - e. none of the above
6. How has poverty contributed to the emergence of some antibiotic resistance bacteria?
   - a. The dosage and length of antibiotic treatment are decreased to save money.
   - b. Poverty has not contributed to the emergence because there are social programs in place that assure all individuals have proper medications.
   - c. People in impoverished countries have the same access to health care as rich individuals.

(Continued)

d. Antibiotics are more widely available in impoverished countries.
e. none of the above

7. If humans have been interfering with nature for thousands of years, why has genetic engineering been hotly debated?
   a. Genetic engineering is not hotly debated and has been widely accepted by most nations.
   b. Any interference, even by natural means, is considered "playing God."
   c. Genetic engineering allows us to make bigger changes much faster.
   d. Genetic engineering ensures that biodiversity is not compromised.
   e. none of the above

8. Which of the following is an example of how humans have genetically selected organisms without performing genetic engineering?
   a. Growing entire fields of crops eliminates the natural vegetation on the land.
   b. Selectively breeding animals to produce desired looks and features.
   c. Growing large amounts of a certain crop invites insect pests that would normally have not inhabited that land before the crop was planted.
   d. Selectively breeding crops for desired characteristics.
   e. All of the above are examples.

9. How might prenatal genetic screening be used now and in the future?
   a. to test for inherited defects in the fetus
   b. for early detection of diseases such as PKU
   c. to test for particular facial features
   d. to test for IQ
   e. all of the above

10. What would be the most practical use to cloning humans?
    a. to use as food
    b. to produce new organs or tissues for transplants
    c. to have a behavioral replica of one's self
    d. all of the above
    e. none of the above

11. The Universal Declaration on Bioethics and Human Rights expanded the Belmont Report. Which of these articles is not an example of the autonomy expansion?
    a. benefit and harm
    b. individual responsibility
    c. consent
    d. person without the capacity to consent
    e. human vulnerability and personal integrity

12. Who might be interested in a person's individual genetic information?
    a. military
    b. health care system
    c. insurance companies
    d. employers
    e. all of the above

13. What is a major ethical concern of DNA obtained during a criminal or civil case?
    a. privacy issues of DNA records
    b. if the procedures used to exonerate or implicate a suspect are accurate
    c. if the DNA obtained in the investigation matches the suspect's DNA
    d. entering DNA evidence into a trial is common practice and therefore, there are no ethical concerns
    e. none of the above

14. Why is genetic privacy a concern for the general public?
    a. In the future, it may be possible to determine health problems by analyzing DNA segments.
    b. Insurance companies could use that information to deny health benefits.
    c. Intelligence and mental capacity could be determined just by examining DNA segments.
    d. A DNA-based identity card may be required as a form of ID and health information.
    e. All of the above.

15. What is a debate surrounding genetic information ownership?
    a. Should the knowledge gained from taxpayer money be free information?
    b. Should researchers that are funded by taxpayers be able to patent and profit from their research?
    c. Should someone be able to patent preexisting information?
    d. If a researcher patented a particular apparatus, does he or she own the information obtained from the equipment?
    e. All of the above are debated.

16. Which is not a valid concern against dual-use research in terms of biosecurity?
    a. The stability and transmissibility of a biological agent or toxin could be improved.
    b. Immunity or effective immunizations could be disrupted.
    c. Ownership of the knowledge is always doubted.
    d. Increasing the pathogenesis or a biological agent.
    e. De-extinction.

Further Reading

Adcock, M. (2007). Intellectual property, genetically modified crops and bioethics. *Biotechnology Journal, 2*, 1088–1092.

Baoutina, A., Alexander, I. E., Rasko, J. E., & Emslie, K. R. (2007). Potential use of gene transfer in athletic performance enhancement. *Molecular Therapy: The Journal of the American Society of Gene Therapy, 15*, 1751–1766.

Barash, J. R., & Arnon, S. S. (2014). A novel strain of *Clostridium botulinum* that produces type B and type H botulinum toxins. *The Journal of Infectious Diseases, 209*, 183–191.

Barfoot, P., & Brookes, G. (2014). Key global environmental impacts of genetically modified (GM) crop use 1996–2012. *GM Crops Food, 5*(2), 149–160.

Carpenter, J. (2010). Peer-reviewed surveys indicate positive impact of commercialized GM crops. *Nature Biotechnology, 28*, 319–321.

Herfst, S., et al. (2012). Airborne transmission of influenza A/H5N1 virus between ferrets. *Science, 336*, 1534–1541.

Linster, M., et al. (2014). Identification, characterization, and natural selection of mutations driving airborne transmission of A/H5N1 virus. *Cell, 157*, 329–339.

Nicolia, A., Manzo, A., Veronesi, F., & Rosellini, D. (2014). An overview of the last 10 years of genetically engineered crop safety research. *Critical Reviews in Biotechnology, 34*, 77–88.

Paarlberg, R. (2010). GMO foods and crops: Africa's choice. *Immobilization Biotechnology, 27*, 609–613.

Roche, P. A., & Annas, G. J. (2006). DNA testing, banking, and genetic privacy. *The New England Journal of Medicine, 355*, 545–546.

Tachibana, M., et al. (2013). Human embryonic stem cells derived by somatic cell nuclear transfer. *Cell, 153*, 1228–1238.

Wolpert, L. (2007). The public's belief about biology. *Biochemical Society Transactions, 35*, 37–40.

Yearley, S. (2009). The ethical landscape: identifying the right way to think about the ethical and societal aspects of synthetic biology research and products. *Journal of the Royal Society, Interface/the Royal Society, 6*(Suppl. 4), S559–S564.

β3-adrenergic receptor Receptor found on fat cells that binds noradrenaline (norepinephrine)

β-galactosidase Enzyme that splits lactose and related disaccharides into simple sugars

β-lactamase Enzyme that degrades antibiotics of the β-lactam family

β-lactams Large family of antibiotics including both penicillins and cephalosporins

ΦC31 (phiC31) integrase A DNA integrase from bacteriophage phiC31 used to insert segments of foreign DNA at a specific site in genetic engineering

2D LC microcapillary column A column containing two different stationary phases that separates complex peptide fragment mixtures during liquid chromatography (LC)

2-micron circle (2μ circle) A small multicopy plasmid found in the yeast *Saccharomyces cerevisiae*, whose derivatives are widely used as vectors

3′ untranslated region (3′ UTR) Sequence at the 3′ end of mRNA, downstream of the final stop codon that is not translated into protein

30-nanometer fiber Chain of nucleosomes that is arranged helically, approximately 30 nm in diameter

30S initiation complex Initiation complex for translation that contains only the small subunit of the bacterial ribosome

4S pathway Pathway of four steps for removing sulfur from dibenzothiophene and related molecules

5′ untranslated region (5′ UTR) Region of an mRNA between the 5′ end and the translation start site

59-base element (59-be) (see also *attC*) A repeated genetic sequence that flanks an integron gene cassette and contains a seven-base-pair core that is highly conserved

5-bromo-2-deoxyuridine (BrdU) A base analog for thymine that is easily incorporated into DNA during replication

6HIS Six tandem histidine residues that are fused to proteins, thus allowing purification by binding to nickel ions that are attached to a solid support. Also known as polyhistidine tag

70S initiation complex Initiation complex for translation that contains both subunits of the bacterial ribosome

a factor Yeast mating factor that acts as a pheromone, attracting yeast cells of the opposite mating type

ABO blood group Set of blood antigens on surface of red cells consisting of three different but related glycolipids

abrin Plant toxin from seeds of the rosary pea (*Abrus precatorius*)

acceptor stem Base-paired stem of tRNA to which the amino acid is attached

acetylserine sulfhydrylase Enzyme that converts *O*-acetylserine plus hydrogen sulfide to cysteine

achondroplasia Inherited defect due to mutation of the *FGFR3* gene that encodes the fibroblast growth factor (FGF) receptor #3. Also known as short-limbed dwarfism

acquired immunity Type of immunity that responds to specific antigens over a course of days, has the ability to distinguish self from non-self, remembers past invading pathogens, and generates new antibodies

acquired immunodeficiency syndrome (AIDS) A disease caused by the HIV retrovirus, which slowly undermines the immune system by destroying helper T cells

activator proteins Protein that switches a gene on

***Ada* gene** Gene that encodes the enzyme adenosine deaminase

adaptive immune system Immunity that develops upon exposure to antigens that results in the production of antibodies. Can be passive (movement of antibodies from mother to fetus) or active (production of antibodies by B cells to attack a foreign bacteria or virus)

adaptors (also called linkers) A short double-stranded piece of DNA with know sequence that is added to the ends of DNA fragments

addiction module Set of genes that control cell death in bacteria such as *Escherichia coli*

adeno-associated virus (AAV) A defective or satellite virus that depends on adenovirus (or certain herpesviruses) to supply necessary functions

adenosine deaminase Enzyme involved in purine metabolism whose absence prevents development of B cells and T cells

adenovirus-36 Strain of adenovirus that causes obesity

adenoviruses Family of small, spherical, double-stranded DNA viruses that infect animals

adenylate cyclase Enzyme that synthesizes cyclic AMP

adherent cell lines Cultured cells that grow attached to the culture dish

adhesin Bacterial protein that binds to glycoproteins or glycolipids on animal cell surfaces

adjuvant (or carrier) An agent used to increase the antigenicity of a vaccine

ADP-ribose Molecular fragment consisting of adenosine diphosphate plus ribose that is normally obtained by cleavage of NAD

ADP-ribosylation Addition of ADP-ribose group to a protein, thereby altering its activity or completely inactivating it

adult stem cells A collection of cells isolated from adult somatic tissues that are able to divide indefinitely, remain undifferentiated, but have the ability to differentiate into specialized cells of the organ or tissue in which they are originally derived

adverse drug reaction (ADR) Term used to describe serious side effects or lack of drug efficacy in certain patients

Aedes aegypti A mosquito species that transmits yellow fever

aequorin A protein from the luminescent jellyfish *Aequorea victoria* that emits blue light when triggered by calcium ions

affinity A biochemical value (represented as an equilibrium constant, K_d) that describes how tightly a ligand and target bind to each other

affinity HPLC Chromatography technique in which the stationary phase has a binding site for a specific molecule (e.g., an antibody to a particular protein)

aflatoxin A naturally occurring toxic metabolites produced by Aspergillus species that are associated with various diseases and have been shown to cause cancer in animals

A-form An alternative form of the double helix, with 11 base pairs per turn, often found for double-stranded RNA, but rarely for DNA

agarose A polysaccharide from seaweed that is used to form gels for separating nucleic acids by electrophoresis

aggressive gene therapy Therapy of cancers or infections by introducing genes or gene products that act to kill cancer cells or infectious agents

agouti Brown coat color in mice due to a dominant allele

AIDS (acquired immunodeficiency syndrome) A disease caused by the HIV retrovirus, which slowly undermines the immune system by destroying helper T cells

aligned (In reference to DNA sequence comparison) Matching the most similar DNA sequences between two or more different genes

alkaline phosphatase An enzyme that cleaves phosphate groups from a wide range of molecules

allele One particular version of a gene, or more broadly, a particular version of any locus on a molecule of DNA

allogeneic Tissues or cells derived from an organism from the same species, but may not be identically matched

alpha (α) factor Yeast mating factor that acts as a pheromone, attracting yeast cells of the opposite mating type

alpha complementation Assembly of functional β-galactosidase from N-terminal alpha fragment plus rest of protein

alternative splicing library Library of gene or protein sequences generated by randomized inclusion or exclusion of exons from an original protein

Alu element An example of a SINE, a particular short DNA sequence found in many copies on the chromosomes of humans and other primates

amantadine Antiviral agent that blocks ion channels found in the outer envelope of influenza A virus

aminoacyl tRNA synthetases Enzyme that attaches an amino acid to tRNA

aminopeptidase A protease that removes amino acids from a polypeptide chain starting at the amino-terminal end

amniocentesis Procedure for drawing samples of amniotic fluid that contains some fetal cells

AMP-activated kinase (AMPK) Kinase that signals nutrient scarcity by activating cellular uptake of glucose with other energy producing processes and is implicated in aging

amyloid β A peptide fragment of a protein found in neuritic senile plaques

amyloid aggregate Insoluble clump of misfolded proteins found in several disease conditions including Alzheimer's and prion disease

amylopectin Branched component of starch

amylose Linear component of starch

analyte Mixture of molecules (such as proteins) that is being studied by chromatography

Angelman's syndrome Inherited defect resulting in loss of function of genes on the maternally derived copy of chromosome 15 that are subject to imprinting

angiogenesis The development of new blood vessels

Anopheles gambiae A mosquito species that transmits malaria

anthrax Virulent bacterial disease of cattle that readily infects humans and is caused by *Bacillus anthracis*

antibodies Proteins of the immune system that recognize and bind to foreign molecules (antigens)

anticodon Group of three complementary bases on tRNA that recognize and bind to a codon on the mRNA

antigen capture immunoassay Method of protein detection where an antibody is linked to a solid support and captures its cognate protein from a mixture

anti-gene Gene that gives a full-length antisense RNA when transcribed

antigens Molecules that cause an immune response and that are recognized and bound by antibodies

anti-oncogene Gene that acts to prevent unwanted cell division (same as tumor-suppressor gene)

antiparallel Parallel, but running in opposite directions

antisense (or noncoding) The strand of DNA that has complementary sequence to mRNA

antisense DNA Single-stranded DNA with a sequence complementary to a specific target molecule of DNA or RNA, usually mRNA; binding of mRNA by antisense DNA may block its translation

antisense DNA oligonucleotide Short antisense DNA

antisense RNA RNA complementary in sequence to messenger RNA and that therefore base-pairs with it

anti-Shine–Dalgarno sequence Sequence on 16S rRNA that is complementary to the Shine–Dalgarno sequence of mRNA

antiterminator proteins Protein that allows transcription to continue through a transcription terminator

AP-1 (activator protein-1) Eukaryotic transcription factor that activates a variety of different genes

apoptosis Genetic program that eliminates damaged cells or cells that are no longer needed without activating the immune system

apoptosome A complex of signaling proteins that activates mammalian caspase-3 during apoptosis

apoptotic bodies Dense granular fragments of an apoptotic cell

aptamer Oligonucleotide that binds to another molecule, often a protein, but is not encoded by a naturally occurring DNA sequence

Argonaut (AGO) family Enzymes found within the RISC complex that degrade any cut any cellular RNAs that are complementary to the guide strand of the siRNA

artificial chromosome A self-replicating element used to clone large fragments of DNA that has three main components: an origin of replication, a centromere, and telomeres

ascospores Type of spore made inside an ascus by fungi of the ascomycete group, including yeasts and molds

ascus Specialized spore-forming structure of ascomycete fungus

aspartic protease A class of protease that has two aspartic acid residues in its active site

assembler A general-purpose device for molecular manufacturing capable of guiding chemical reactions by positioning molecules

asymmetric cell division A cell division that produces two daughter cells that have different fates. In stem cell niches, one daughter cell retains the stem cell fate and the other daughter cell differentiates into a cell making up the surrounding tissue or organ

asymmetry Uneven in distribution of various factors during cell division

atomic force microscope (AFM) An instrument able to image surfaces to molecular accuracy by mechanically probing their surface contours

ATP synthetase Enzyme that synthesizes ATP using energy from the respiratory chain

attenuated vaccine A live pathogen that no longer causes a disease state but still stimulates the immune system to create antibodies

attenuation riboswitch Type of riboswitch that causes premature termination by inducing an alternative stem and loop structure in the leader region of the mRNA when triggered by a signal

autocrine signals A chemical signaling molecule that is produced from the same cell or same cell type in which it activates

autogenous regulation Self-regulation, i.e., when a DNA-binding protein regulates the expression of its own gene

autologous Tissues or cells derived from the same individual

automatic DNA sequencer Machine that separates a dye-labeled dideoxy sequencing reaction into increasing larger fragments, analyzes the final nucleotide of each fragment by recording the color, and prints the results in a graphical format to determine the sequence

autophagolysosome A digestive vacuole found in the cytoplasm that forms when an autophagosome and lysosome fuse

autophagosome A vesicle found in the cytoplasm that contains defective organelles or other cellular components

autophagy The process that degrades cellular components by the lysosomes

autoradiography A technique of allowing radioactive materials to take pictures of themselves by laying them flat on photographic film

avidin A protein from egg white that binds biotin very tightly

azidothymidine (AZT) Nucleoside analog that acts as a chain terminator. It is used against AIDS and is also known as zidovudine

B cell The type of immune system cell that produces antibodies

B-cell receptors Membrane bound form of an antibody that is found on the surface of B cells. Responsible for signaling the B cell the presence of a pathogen and for processing the antigens from the pathogen to activate other immune cells (i.e., T cells)

Bacillus anthracis Bacterium that causes anthrax and that readily infects humans

back-crossed Procedure in which pollen from the progeny of two parents is cross-pollinated onto one of the original parents

bacmid Hybrid cloning vector made from a baculovirus and a plasmid

bacteriocin Toxic protein made by bacteria to kill closely related bacteria

bacteriophage Virus that infects bacteria

bacteriophage vector A bacteriophage genome in which nonessential bacteriophage genes are replaced with a multiple cloning site where other DNA fragments can be cloned. The vector can be maintained within a bacterial cell since it contains all the necessary genes for replication, growth, and lysis

baculoviruses Family of DNA viruses that infect insects and related invertebrates and are widely used as vectors

bait The fusion between the DNA-binding domain of a transcriptional activator protein and another protein as used in two-hybrid screening

barcode sequence A unique DNA sequence added to the linker or adapter for next-generation sequence that is used to discern one genome from another

base Alkaline chemical substance; in molecular biology especially, refers to the cyclic nitrogen compounds found in DNA and RNA

base peak Peak for the most intense ion detected in a sample

base substitution Mutation in which one base is replaced by another

basic peptide A short length of basic amino acids that facilitate translocation across biological membranes. These are derived from known proteins or made synthetically

Bcl-2 Mitochondrial protein that helps control entry into apoptosis

beta-galactosidase or β-galactosidase (LacZ) Enzyme that splits lactose and related molecules to release galactose

B-form The normal form of the DNA double helix, as originally described by Watson and Crick

bifurcation Division of two different taxonomic groups from each other during evolution

bioaugmentation Adding degrading microorganisms (normal or engineered) plus nutrients to remove a contaminant from the environment

biohydrogen Hydrogen that is produced by organisms such as algae or bacteria

bioinformatics The computerized analysis of large amounts of biological sequence data

biomimetics The imitation of biological structures in engineering

biopanning Method of screening a phage display library for a desired protein by binding to a bait molecule attached to a solid support

bioreactors Large chambers or vats used to grow organisms such as yeast or bacteria so that the biotechnological product can be harvested on a large scale

bioremediation Using any microorganism, fungus, plant, or enzyme to clean up or restore the environment to its original condition

biosafety level Rating system (BL1–BL4) for high-containment laboratories

biosensor Detection or monitoring device that relies on a biological mechanism

biosensory system A genetic circuit that is controlled by a specific environmental cue

biostimulation Using different chemical compounds to stimulate naturally occurring microorganisms to remove a contaminant from its environment

Black Death Bubonic plague, or more specifically, the great plague epidemic of the mid-1300s

blastocyst A very early stage of the embryo

blastula In animals, the stage of development where a hollow ball of cells has a fluid filled cavity. Forms three to five days after fertilization in human development

blood antigens Glycoproteins and glycolipids on the surface of blood cells that are used for antigenic analysis

blood typing Use of differences in blood antigens to determine identity

blunt ends Ends of a double-stranded DNA molecule that are fully base-paired and have no unpaired single-stranded overhang

Bohr radius The natural preferred distance between positive and negative charges in a material

Botox Botulinum toxin, especially as used for cosmetic purposes

botulinum toxin Protein toxin made by *Clostridium botulinum* that blocks nerve transmission

botulinum toxin type A Specific type of botulinum toxin used clinically and for cosmetic purposes

bovine spongiform encephalopathy (BSE) Official name for mad cow disease, a brain degeneration disease of cattle caused by prions

BRCA1 Breast cancer A1 gene, involved in DNA repair. Defects in this gene predispose women to breast cancer

BRCA2 Breast cancer A2 gene, involved in DNA repair. Defects in this gene predispose women to breast cancer

bridge amplification A specialized type of PCR in which the forward and reverse primers are attached to the surface of a flow cell so the template form a U-shape on the surface. Used for Illumina-type of next-generation sequencing methodology

Bt toxin Toxin found in the soil bacterium *Bacillus thuringiensis*, which kills certain caterpillars that destroy crops

bubonic plague Virulent bacterial disease spread by fleas and primarily found in rodents

bud The new asexual daughter cell of a yeast that forms as a bulge on the surface of the mother cell

bZIP proteins Family of transcription factors that each have a leucine zipper domain

CAD (caspase-activated DNase) A nuclear DNase that cleaves DNA between histones; activated by caspases during apoptosis

callus An unorganized, proliferating mass of undifferentiated plant cells

callus culture Dedifferentiating a mass of plant tissue *in vitro* and then growing the undifferentiated tissue in a petri dish

calmodulin A small calcium-binding protein of animal cells

caloric restriction The theory that decreasing the overall number of calories without altering the nutrient intake will extend the lifespan

cancer Tissue or cluster of cells resulting from uncontrolled cell growth and division due to somatic mutations affecting cell division

candidate cloning Approach to cloning that involves making an informed guess as to what kind of protein is likely to be involved in a hereditary disorder

CAP (catabolite activator protein) A transcription enhancer protein that activates the transcription of sugar metabolizing operons when bound by cAMP (also CRP-cyclic AMP receptor protein)

cap Structure at the 5′ end of eukaryotic mRNA consisting of a methylated guanosine attached in reverse orientation

capsid Shell or protective layer that surrounds the DNA or RNA of a virus particle

CAR (coxsackievirus adenovirus receptor) Receptor on animal cells shared by adenoviruses and B-group coxsackieviruses

carboxypeptidase Enzyme that removes C-terminal amino acid residue from a protein

carboxypeptidase H A particular carboxypeptidase that is needed during the processing of insulin

carcinogen Cancer-causing agent

carrier (or adjuvant) An agent used to increase the antigenicity of a vaccine

caspase Enzyme that carries out the protein degradation characteristic of apoptosis; caspases cleave other proteins after aspartic acid residues

caspase-activated DNase (CAD) A nuclear DNase that cleaves DNA between histones; activated by caspases during apoptosis

catalase Enzyme that converts hydrogen peroxide to oxygen plus water

catecholamines Family of neurotransmitters including dopamine, adrenalin (= epinephrine), and noradrenalin (= norepinephrine)

catenated Structure in which two or more circles of DNA are interlocked

CCR5 A protein that acts as a coreceptor for entry of HIV into host cells. Its natural role is as a chemokine receptor

CD4 protein Protein found on the surface of many T cells that is used as a receptor by the AIDS virus

cell cycle Series of stages that a cell goes through from one cell division to the next

cell-mediated immunity Type of immunity that relies on T cells and cell–cell interactions (rather than on antibodies made by B cells)

cellular senescence Arrested cell division of cultured mammalian cells; same as replicative senescence

cellulose Structural polymer of β-1,4-linked glucose found in plant cell walls

central dogma Basic plan of genetic information flow in living cells that relates genes (DNA), message (RNA), and proteins

centromere Region of eukaryotic chromosome where the microtubules attach during mitosis and meiosis

centromere (Cen) sequence Sequence at centromere of eukaryotic chromosome that is needed for correct partition of chromosomes during cell division

cephalosporins Subfamily of antibiotics belonging to the β-lactam family of antibiotics

CFTR protein Cystic fibrosis transporter—protein encoded by the cystic fibrosis gene and found in the cell membrane where it acts as a channel for chloride ions

CG islands Region of DNA in eukaryotes that contains many clustered CG sequences that are used as targets for cytosine methylation

chain termination sequencing Method of sequencing DNA by using dideoxynucleotides to terminate synthesis of DNA chains. Same as dideoxy sequencing

chain terminator Nucleoside analog that is incorporated into a growing DNA chain but prevents further extension of the DNA chain

chaperone Protein that mediates the refolding of misfolded or unfolded proteins

chaperone-mediated autophagy A system of protein degradation in which a chaperone protein recognizes the protein through a specific amino acid sequence and transfers it to the lysosome for degradation

chaperone-mediated folding Mechanism of repairing misfolded proteins where a chaperone enzymatically unfolds and refolds the defective proteins

chaperonin Protein that oversees the correct folding of other proteins

checkerboard hybridization Technique for mass screening of DNA samples using probes to 16S rRNA genes corresponding to many different bacteria

chemokine receptor Cell surface protein required for uptake of chemokines into animal cells

chemokines Group of approximately 50 small messenger peptides that activate the white cells of the immune system

chimeric animal Animal in which different cells vary genetically

ChIP-chip Technique to identify the sequence of transcription factor binding sites by whole-genome arrays after the binding sites are cross-linked to the transcription factors and isolated by immunoprecipitation

chloroplast transit peptide A small peptide added to the N-terminus of a protein that directs the protein from the ribosome into the chloroplast

cholera toxin Protein toxin made by *Vibrio cholerae* that hyperactivates the adenylate cyclase of animal cells

cholesterol Sterol derivative found in humans and other animals that has a major effect on atherosclerosis

chromatin Complex of DNA plus protein that constitutes eukaryotic chromosomes

chromatin immunoprecipitation (ChIP) Immunoprecipitation of transcription factors cross-linked to their binding sites on DNA

chromatography Assorted techniques that separate a mixture of molecules by size or chemical properties

chromatography column A long tube that contains the stationary phase for chromatography

chromosome walking Method for cloning neighboring regions of a chromosome by successive cycles of hybridization using overlapping probes

chronic wasting disease (CWD) Prion disease of deer and elk in the northern United States

chronological aging Length of time the yeast cell can live without dividing

circum-sporozoite protein Protein on surface of sporozoite stage of malarial parasite

cistron Segment of DNA (or RNA) that encodes a single polypeptide chain

clade A group of organisms that share a common evolutionary ancestor

cladistics A system of biological classification based on using quantitative analysis of physical traits to construct evolutionary relationships

clamp-loading complex Group of proteins that loads the sliding clamp of DNA polymerase onto the DNA

class I major histocompatibility complex (class I MHC) MHC proteins consisting of two chains of unequal length and found on the surface of all cells. Their role is to display fragments of proteins originating inside the cell

class II major histocompatibility complexes (class II MHC) MHC proteins consisting of two chains of equal length and found only on the surface of certain immune cells. Their role is to display fragments of proteins from digested microorganisms

ClfA (clumping factor A) A protein found on the surface of *Staphylococcus aureus* that allows the bacteria to adhere to fibrinogen of the host cell

cloned animals Genetically identical animals derived from the same original cell line

cloning vectors Any molecule of DNA that can replicate itself inside a cell and is used for carrying cloned genes or segments of DNA. Usually a small multicopy plasmid or a modified virus

closed stem cell niches A tissue that has a defined geographical location for the stem cell population that is enclosed by a specific set of cells or extracellular matrix

Clostridium botulinum Bacterium that makes botulinum toxin

cluster of differentiation (CD) antigen Cell surface proteins found on leukocytes and used to classify the different types of leukocytes

coding strand (see also nontemplate or sense) The strand of DNA equivalent in sequence to the messenger RNA (same as plus strand)

codon Group of three RNA or DNA bases that encodes a single amino acid

codon bias Some organisms use only a subset of redundant codons for a particular amino acid, whereas other organisms will use a different subset. This bias can hinder expression of foreign proteins in genetic engineering

codon usage The favoring of different alternative codons for the same amino acid in different organisms

co-immunoprecipitation Method of identifying protein interactions by using antibodies to one of the proteins

"cold" Slang for nonradioactive

colicin Toxic protein or bacteriocin made by *Escherichia coli* to kill closely related bacteria

combinatorial library Large series of related molecules that have been systematically generated by combining chemical groups and/or molecular motifs

combinatorial screening Screening of a large series of related molecules for those with useful properties

comparative genomics Comparing genome sequences of different organisms, especially in attempts to determine potential functions for genes or proteins by comparison with characterized examples

competent Cell that is capable of taking up DNA from the surrounding medium

complementarity determining region (CDRs) Short segment that forms a loop on the surface of the variable region of an antibody, thus forming part of the antigen-binding site

complementary DNA (cDNA) DNA copy of a gene that lacks introns and therefore consists solely of the coding sequence. Made by reverse transcription of mRNA

complex transposon A transposon that moves by replicative transposition

c-onc Cellular version of an oncogene

conditional mutation Mutation whose phenotypic effects depend on environmental conditions such as temperature or pH

conformation-dependent immunoassay (CDI) Assay that uses specific binding of monoclonal antibodies to distinguish between different forms of the prion protein

conjugation Process in which genes are transferred by cell-to-cell contact in bacteria

conservative substitution Replacement of an amino acid with another that has similar chemical and physical properties

conservative transposition Same as cut-and-paste transposition

conserved noncoding elements (CNE) Areas of the human genome that are conserved in the various other animal genomes but do not encode protein-producing genes

constant region Region of an antibody protein chain that remains constant in sequence

constitutive Term used to describe genes that are expressed during all conditions

constitutive heterochromatin Regions of more or less permanent heterochromatin found on both copies of homologous chromosomes, especially around centromeres

constitutive promoter A promoter that functions in all tissues at all times

construct Term used to describe a cloning vector derived from various different genes and DNA segments assembled into one

contact inhibition Inhibition of cell division that occurs due to contact with neighboring cells

contig A stretch of known DNA sequence, built up from smaller cloned fragments, that is contiguous and lacks gaps

contig map A genome map based on contigs

controlled pore glass (CPG) Glass with pores of uniform sizes that is used as a solid support for chemical reactions such as artificial DNA synthesis

Coomassie Blue A blue dye used to stain proteins

copy number The number of copies of a gene or plasmid found within a single host cell

core enzyme The part of DNA or RNA polymerase that synthesizes new DNA or RNA (i.e., lacking the recognition and/or attachment subunits)

coreceptor Protein required for virus entry into host cell in addition to the main virus receptor

corepressor In prokaryotes—a small signal molecule needed for some repressor proteins to bind to DNA; in eukaryotes—an accessory protein, often a histone deacetylase, involved in gene repression

cos **sequences (lambda cohesive ends)** Complementary 12 base-pair-long overhangs found at each end of the linear form of the lambda genome

cosmid Small multicopy plasmid that carries lambda *cos* sites and can carry around 45 kb of cloned DNA

cosuppression An alternate term used to describe RNA interference-like phenomena

coxsackievirus adenovirus receptor (CAR) Receptor on animal cells shared by adenoviruses and B-group coxsackieviruses

C-peptide Connecting peptide that originally links the A- and B-chains of insulin but is absent from the final hormone

CRE (cyclic AMP response element) A specific DNA sequence found in front of genes that are activated by cyclic AMP in higher organisms

Cre A recombinase from bacteriophage P1 that directs recombination at specific sites (*loxP* sites)

Cre recombinase Enzyme encoded by bacterial virus P1 that catalyzes recombination between inverted repeats (*loxP* sites)

Cre/*loxP* A system allowing a specific gene to be added or removed during development by exploiting Cre recombinase to excise DNA sequences flanked by *loxP* recombination sites

CREB protein (cyclic AMP response element binding protein) Protein that regulates genes in response to cyclic AMP in animal cells by binding to CRE sequences in promoters

Creutzfeldt–Jakob disease (CJD) Inherited or spontaneous brain degeneration disease of humans caused by prions

crossover Structure formed when the strands of two DNA molecules are broken and rejoined with each other

cross-pollination Taking the pollen from one plant and placing it on the stigma of another in order to direct the exchange of genetic information

CRP protein (cyclic AMP receptor protein) Bacterial protein that binds cyclic AMP and then binds to DNA (see also CAP, catabolite activator protein)

Cry protein Crystalline protein found in *Bacillus* bacterial spores that breaks into delta endotoxin (Bt toxin)

CTXphi Filamentous bacteriophage that carries genes for cholera toxin and lysogenizes *Vibrio cholerae*

cut-and-paste transposition Type of transposition in which a transposon is completely excised from its original location and moves as a whole unit to another site

cycle sequencing Technique that combines PCR and chain termination sequencing to determine the sequence of a template DNA

cyclic AMP (cAMP) (cyclic adenosine monophosphate) A signal molecule used in global regulation (in bacteria) and as a second messenger (in higher organisms). A cyclic mononucleotide of adenosine that is formed from ATP

cyclic AMP response element (CRE) A specific DNA sequence found in front of genes that are activated by cyclic AMP in higher organisms

cyclic AMP response element binding protein (CREB protein) Protein that regulates genes in response to cyclic AMP in animal cells by binding to CRE sequences in promoters

cyclic GMP (cyclic guanosine monophosphate) A signal molecule used as a second messenger by eukaryotic cells

cyclic phosphodiesterase Enzyme that degrades cyclic nucleotides, including cyclic AMP and cyclic GMP

cyclin-dependent kinase (CDK) Specialized protein kinase that is activated by a cyclin and participates in control of cell division

cyclins Family of proteins that controls the cell cycle

cysteine protease A class of protease with cysteine in its active site

cystic fibrosis An inherited disease in which the major symptom is the accumulation of fibrous tissue in the lungs, ultimately due to defects in transmembrane chloride channel

cytochrome *c* Mitochondrial protein that functions in electron transport; released from mitochondria during apoptosis

cytogenetic map A visual chromosome map based on light and dark bands due to staining and seen under the light microscope

cytokine response modifier genes (*crm* genes) Genes of poxviruses that interfere with the action of NK cells and cytotoxic T cells

cytokines Short peptides that stimulate cell growth and division, especially of immune cells

cytosine deaminase Enzyme, usually from bacteria, that converts cytosine into uracil

data mining The use of computer analysis to find useful information by filtering or sifting through large amounts of data

dauer Stage of *C. elegans* life cycle initiated by low food and water; characterized by lack of movement and eating

***db* gene** Gene that encodes the receptor for the hormone leptin

death receptor A cell surface receptor that transmits an external signal to die to the intracellular proteins responsible for killing the cell

death receptor pathway Program of apoptosis that involves activating membrane-bound receptors with extrinsic factors

death-inducing signaling complex (DISC) A complex of proteins activated by binding to a trimerized death receptor; the proteins initiate apoptosis

defensin A Antibacterial peptide made by mosquito

degenerate primer Primer with several alternative bases at certain positions

degradome The complete set of proteases that are expressed at one time, under a defined set of conditions

dehydrogenase Type of enzyme that removes hydrogen atoms from its substrate

deletion Mutation in which one or more bases is lost from the DNA sequence

delta endotoxin Actual toxin released from the Cry protein into the gut of caterpillars that digest *Bacillus* bacterial spores

demethylases An enzyme that removes methyl groups

***de novo* methylase** An enzyme that adds methyl groups to wholly nonmethylated sites

***de novo* proteins** Proteins with new sequences not previously found that perform a new function

deoxyribose The sugar with five carbon atoms that is found in DNA

deoxyribozyme Artificial DNA molecule that acts as an enzyme

depth of coverage In reference to whole genome sequencing. The number of individual sequence reads for a particular region of a genome where the higher number reflects a greater accuracy

***DHFR* gene** Gene that encodes the enzyme dihydrofolate reductase

diabetes mellitus Group of related diseases causing inability to control level of blood sugar due to defect in insulin production

diabody Artificial antibody construct made of two single-chain Fv (scFv) fragments assembled together

dibenzothiophene (DBT) Thiophene fused to two benzene rings, widely regarded as a model compound for organic sulfur in coal and oil

Dicer An enzyme that cuts double-stranded RNA into small segments of 21 to 23 nucleotides in length (siRNA)

dicots (see also dicotyledonous plants) Plants with broad leaves with netlike veins, whose seedlings have two cotyledons, or seed leaves

dicotyledonous plants (see also dicots) Plants with broad leaves with netlike veins, whose seedlings have two cotyledons, or seed leaves

dideoxynucleotide Nucleotide whose sugar is dideoxyribose instead of ribose or deoxyribose

dietary restriction The theory that decreasing the overall number of calories without altering the nutrient intake will extend the lifespan

differential fluorescence induction (DFI) Method to identify genes active in infectious agents that might be used for potential vaccines. Fluorescently labeled library clones are used to determine which genes are active after the infectious agent enters the cell

differential gel electrophoresis (DIGE) Two dimensional gel electrophoresis of two or three differentially labeled protein samples on the same gel so as to compare samples within the same gel matrix

differential interference microscopy (see also Nomarski optics) Type of polarization microscopy that transforms differences in density into differences in height so that the image looks like a three-dimensional relief

differentiate During growth or development, the acquisition of cellular traits or behaviors that are specific to one type of tissue or organ

dihydrofolate reductase Enzyme that takes part in one carbon metabolism and is needed for the synthesis of thymine and adenine

dimethoxytrityl (DMT) group Group used for blocking the 5′ hydroxyl of nucleotides during artificial DNA synthesis

dioxygenase Enzyme that inserts two oxygen atoms into its substrate, thus yielding a diol

diphtheria toxin Protein toxin made by *Corynebacterium diphtheriae* that inactivates elongation factor EF-2 of animal cells

diploid Having two copies of each chromosome and hence of each gene

direct immunoassay (see also reverse-phase array) Method of protein detection in which the protein is bound to a solid support and an antibody with a detection system is added

directed evolution Technique for enhancing or altering the original activity of an enzyme by randomly mutating the gene and then screening for the new activity

directed mutagenesis Deliberate alteration of the DNA sequence of a gene by any of a variety of artificial techniques

DISC (death-inducing signaling complex) A complex of proteins activated by binding to a trimerized death receptor; the proteins initiate apoptosis

disulfide bonds A bond between two closely situated cysteine residues in a protein that stabilizes the tertiary structure

***DMD* gene** Gene located on the X chromosome that encodes dystrophin and is defective in Duchenne's muscular dystrophy

DNA (deoxyribonucleic acid) Nucleic acid polymer of which the genes are made

DNA adenine methylase (Dam) A bacterial enzyme that methylates adenine in the sequence GATC

DNA cassette Deliberately designed segment of DNA that is flanked by convenient restriction or recombinase sites

DNA chain terminator A nucleotide analog that is incorporated into a growing DNA chain and stops further elongation

DNA chip technology Method of hybridization in which a chip is used to simultaneously detect and identify many short DNA fragments. Also known as DNA array technology or oligonucleotide array technology

DNA cytosine methylase (Dcm) A bacterial enzyme that methylates cytosine in the sequences CCAGG and CCTGG

DNA fingerprinting Individual identification by means of differences in DNA sequence generally visualized as the pattern of DNA fragments after separation by gel electrophoresis

DNA gyrase An enzyme that introduces negative supercoils into DNA, a member of the type II topoisomerase family

DNA helicase Enzyme that unwinds double-helical DNA

DNA invertase An enzyme that recognizes specific sequences at the two ends of an invertible segment and inverts the DNA between them

DNA ligase Enzyme that joins DNA fragments covalently, end to end

DNA methylation Addition of methyl ($—CH_3$) groups onto DNA to regulated protein access to various regions

DNA microarray Chip used to simultaneously detect and identify many short DNA fragments by DNA–DNA hybridization. Also known as DNA array or oligonucleotide array detector, same as DNA array or DNA chip

DNA polymerase An enzyme that elongates strands of DNA, especially when chromosomes are being replicated

DNA polymerase III (Pol III) Enzyme that makes most of the DNA when bacterial chromosomes are replicated

DNA SELEX Method in which catalytically active DNA sequences are altered by mutations to use new substrates

DNA shuffling Method of artificial evolution in which genes are cut into segments that are mutagenized, mixed, shuffled, and rejoined

DNA synthesizer Machine that adds one nucleotide after another in a 3′ to 5′ direction to assemble a length of DNA

DNA vaccine Vaccine that consists of DNA that encodes a specific antigenic protein of the disease agent. After entering the host, the DNA is expressed, and the immune system reacts to the foreign protein

dominant-negative mutation Mutation giving a gene product that not only is functionally defective but also inactivates the wild-type gene product. Usually occurs with multi-subunit proteins

dot blot Hybridization technique in which various DNA samples are attached to a filter in small spot. The blot can then be probed with a gene of interest to find matching sequences

double helix Structure formed by twisting two strands of DNA spirally around each other

Down syndrome Defective development, including mental retardation, resulting from an extra copy of chromosome 21

driver Normal DNA used in suppressive subtraction hybridization that is used to "subtract" or remove the common sequences

***dszABC* operon** Group of genes encoding pathway for removing sulfur from dibenzothiophene

Duchenne's muscular dystrophy One particular form of muscular dystrophy, degenerative muscle disease

duplication Mutation in which a segment of DNA is duplicated

dynamic mutation Mutation consisting of multiple tandem repeats that increase in number over successive generations

dystrophin Protein encoded by the *DMD* gene whose malfunction causes muscular dystrophy. Dystrophin helps attach muscle fibrils to the membranes of muscle cells

E1A protein An adenovirus early protein that promotes transcription of other early virus genes and binds to host cell Rb protein

E2F Regulatory protein that controls synthesis of cyclins E and A

early genes Genes expressed early during virus infection and that mainly encode enzymes involved in virus DNA (or RNA) replication

ecdysone Steroid hormone from insects that is involved in molting

edema factor (EF) Protein toxin made by anthrax bacteria that acts as an adenylate cyclase

edible vaccine Vaccine that is expressed in the edible part of a plant. Eating an edible vaccine would confer resistance to the disease state

elastin Protein encoded by the *ELN* gene and found in the elastic tissues of skin, lung, and blood vessels

elastin-like polypeptide (ELP) A protein created in the laboratory that has elastin-like properties

electrochemical detector A device that detects changes in oxidation and reduction of proteins (or other molecules) in the mobile phase during chromatography

electrode-oxidizing organisms Microorganisms that use electrons from a cathode of a microbial fuel cell to reduce substances in the chamber, and establish a flow of electrons to form a current

electrode-reducing organisms Microorganisms that donate electrons to an anode of a microbial fuel cell, thus establishing a current flow

electroporation Technique that uses high-voltage discharge to make cells competent to take up DNA

electrospray ionization (ESI) Type of mass spectrometry in which gas-phase ions are generated from ions in solution

ELISA See enzyme-linked immunosorbent assay

elongation (in reference to protein translation) The process of adding amino acids onto a growing polypeptide chain

elongation factors Proteins that are required for the elongation of a growing polypeptide chain

elution Removing bound molecules from a chromatography column by passing through buffer

embryoid body A three-dimensional ball of embryonic stem cells that forms before the stem cells differentiate into specialized cell types

embryonic stem cell Stem cell derived from the blastocyst stage of the embryo

embryonic stem cell line A cell line that grows *in vitro* with the appropriate nutrients and factors that was derived from the inner cell mass of a blastula embryo and retains all the characteristics of a stem cell

emulsion PCR A PCR reaction that occurs in water droplets of an emulsion of water and oil that is used during 454 sequencing technology

endomesoderm Mesoderm that derives from the endoderm of a two-layered blastodisc

endopeptidase Protease that cuts within a polypeptide chain

endosteum A layer of connective tissue that lies between the bone and bone marrow in long bones

endothelial cells A thin flattened cell found on the surface of blood vessels and lymph vessels

endotoxin Lipopolysaccharide from the outer membrane of gram-negative bacteria

enhancer sequences Regulatory sequence outside, and often far away from, the promoter region that binds transcription factors

enterochelin (or enterobactin) A siderophore used by *E. coli* and many enteric bacteria

enterocytes Simple columnar epithelial cells that line the intestinal villi and absorb nutrients from the intestinal lumen and pass them into the blood stream

enterotoxin Protein toxin secreted by enteric bacteria

enteroviruses Group of small RNA viruses that infect animal intestines and includes poliovirus

Enviropig™ Transgenic pig with much less phosphorus in its waste

enzyme-linked immunosorbent assay (ELISA) Sensitive method to assay the amount of a specific protein in a sample using enzyme-linked antibodies

epigenetic change 1. Refers to inherited changes that are not due to alterations in the DNA sequence. 2. Changes in gene regulation that persist after tissue culture but are not inherited by any progeny of the plant

epigenetic inheritance Inherited genetic features such as methylation and histone modifications that affect gene expression but do not alter the DNA sequence

epigenome The total number of epigenetic changes that are found in a genome and are passed onto the organism's offspring

epitopes Localized regions of an antigen to which an antibody binds

EPSPS An enzyme essential for making amino acids of the aromatic family in plants and bacteria

ergot Fungus that grows on cereals, especially rye, and produces a mixture of toxins that cause a syndrome called ergotism

error-prone PCR Type of PCR in which mutations are introduced at random during the amplification steps

erythromycin An antibiotic of the macrolide family that inhibits protein synthesis

erythropoietin A protein required for proper development of red blood cells

Escherichia coli A species of bacterium commonly used in genetics and molecular biology

ethidium bromide A stain that specifically binds to DNA or RNA and appears orange if viewed under ultraviolet light

euchromatin Lightly packed form of chromatin that has more genes that are being actively transcribed

eugenics Deliberate improvement of the human race (or other species) by selective breeding

event A term used to distinguish different insertions of the same transgene. The transgene is identical, but the integration sites will vary

excisionase (Xis) Enzyme that reverses DNA integration by removing a segment of double-stranded DNA and resealing the gap. In particular, lambda excisionase removes integrated lambda DNA

***ex vivo* gene therapy** Gene therapy by removal, engineering, and return of cells derived from the same original patient

exon Segment of a gene that codes for protein and that is still present in the messenger RNA after processing is complete

exopeptidase A protease that cuts starting at either the carboxy- or amino-terminal end of the polypeptide chain

exotoxin Protein toxin secreted by bacteria

explant Tissue taken from an original site in a plant and transferred to artificial medium for growth and maintenance

expressed (as pertaining to genes) Converting a DNA region into RNA and/or protein to be used by the living cell

expressed sequence tag (EST) A special type of STS derived from a region of DNA that is expressed by transcription into RNA

expression library (see also expression vector) A library where each cloned piece of DNA is expressed into a protein by the vector

expression vector Vector designed to enhance gene expression, usually by providing a strong promoter that drives expression of the cloned gene

extrinsic asymmetry During stem cell division, one of the daughter cells leaves the environment or moves too far from an external signal that is keeping the stem cell in an undifferentiated state. Without the signal, the daughter cell differentiates

extrinsic pathway Pathway of initiating apoptosis that originates as an external signal such as a ligand binding to a cell surface receptor

Fab fragments Antigen-binding fragments of an antibody

facultative heterochromatin Heterochromatin that has the ability to return to normal euchromatin

family (in relationship to gene sequences) Group of closely related genes that arose by successive duplication and perform similar roles

fast neutron mutagenesis Mutagenesis technique in which seeds are exposed to fast neutrons that induce small deletions

fast neutrons Free neutrons with kinetic energy around 1 MeV that are used to create DNA deletions in plant seeds

fate mapping A method to trace a group of cells or individual cells through development or to trace cell divisions in a tissue or organ of an adult organism. The method can identify stem cell populations in adult tissues

Fc fragment The stem region of an antibody, i.e., the fragment that does not bind the antigen

ferritin An iron storage protein of animals

Fierce Mouse Transgenic mouse with abnormal aggression

filoviruses Family of negative-strand ssRNA viruses that includes Ebola virus and Marburg virus

fimbria (plural fimbriae) Thin helical protein filaments found on the surface of bacteria; same as pilus

FLAG tag A short peptide tag (AspTyrLysAspAspAspAspLys) that is bound by a specific anti-FLAG antibody and that may be attached to proteins

flight tube Tube or channel in which ions generated during mass spectroscopy are separated according to size or charge

flippase (same as Flp recombinase and Flp protein) Enzyme encoded by the 2-micron plasmid of yeast that catalyzes recombination between inverted repeats (FRT sites)

floral dip (also called *in planta transformation*) Method in which a transgene is carried by *Agrobacterium* and transformed into *Arabidopsis* by dipping a developing flower bud into a solution of *Agrobacterium* plus a surfactant

flow cytometry Analyzing (without sorting) different cells as they flow past a fluorescent detector by observing the attached fluorescent antibodies

Flp protein Enzyme encoded by the 2-micron plasmid of yeast that catalyzes recombination between inverted repeats (FRT sites)

Flp recombinase (same as flippase and Flp protein) Enzyme encoded by the 2-micron plasmid of yeast that catalyzes recombination between inverted repeats (FRT sites)

fluorescence-activated cell sorting (FACS) Technique that sorts cells (or chromosomes) based on fluorescent labeling

fluorescence detector A detector that records fluorescence emitted by the passing mobile phase after it has been excited by light

fluorescence *in situ* hybridization (FISH) Using a fluorescent probe to visualize a molecule of DNA or RNA in its natural location

***Fok*I** A particular type II restriction endonuclease with separate recognition and nuclease domains

foster mother (as used in genetics) Female animal that carries engineered embryos

founder animal Animal that is the original host for a transgene and maintains it stably

fragile X syndrome Inherited defect resulting in a fragile site within the long arm of the X chromosome that can be seen by microscopic observation

Francisella tularensis Bacterium that causes tularemia

FRT site Flp recombination target, the recognition site for Flp recombinase

functional cloning Approach to cloning that starts with a known protein that is suspected of involvement in a hereditary disorder

functional genomics The study of the whole genome and its expression

G_0 phase Resting phase in which eukaryotic cells no longer grow or divide

G_1 phase First stage in the eukaryotic cell cycle, in which cell growth occurs

G_2 phase Third stage in the eukaryotic cell cycle, in which the cell prepares for division

G-418 Aminoglycoside antibiotic that kills animal cells by blocking protein synthesis; also known as geneticin

Gal4 protein Transcriptional activator from yeast that has a DNA-binding domain and a transcription activating domain

ganciclovir A nucleoside analog that is converted by cells to a DNA chain terminator. Used against virus infections and in cancer therapy

ganglioside GM_1 Glycolipid found in eukaryotic cell membranes that is used as a receptor by cholera toxin

Gateway® cloning vector Series of cloning vectors that move the gene of interest from one vector to another using the lambda *attP*, *attB*, *attL*, and *attR* recognition sites and the two enzymes, integrase and exisionase

GC ratio The amount of G plus C relative to all four bases in a sample of DNA. The GC ratio is usually expressed as a percentage

gel electrophoresis Electrophoresis of charged molecules through a gel meshwork in order to sort them by size

gene cassette Simple genetic element that has one or two open reading frames flanked by repeats called 59-be or *attC* and inserts into integrons

gene creature Genetic entity that consists primarily of genetic information, sometimes with a protective covering, but without its own machinery to generate energy or replicate macromolecules

gene dosage compensation The phenomena in which the amount of proteins and RNA produced from sex chromosomes is equalized between males and females in diploid organisms

gene fusion Structure in which parts of two genes are joined together, in particular when the regulatory region of one gene is joined to the coding region of a reporter gene

GENE impedance (GENEi) An alternate term used to describe RNA interference. Proposed to encompass all the different phenomena that induce the degradation of a specific RNA transcript homologous to a short-interfering RNA

gene library Collection of cloned segments of DNA that is big enough to contain at least one copy of every gene from a particular organism. Same as DNA library

gene superfamily Group of related genes that arose by several stages of successive duplication. Members of a superfamily have often diverged so far that their ancestry may be difficult to recognize

gene testing Analyzing DNA sequences to identify potential disease states in a patient

gene therapy Medical treatment that involves use of recombinant DNA technology; originally referred to curing hereditary defects by introducing functional genes

general secretory system (Sec system) Standard system for exporting proteins across membranes that is found in most organisms

general transcription factors Proteins that work to enhance or repress gene expression for all genes

geneticin (G-418) Aminoglycoside antibiotic that kills animal cells by blocking protein synthesis

genetic changes Changes in plants after tissue culture that are inherited by the progeny, including alterations in chromosomes

genetic circuits Combinations of genes, promoters, enhancers, and repressors that control the output or expression of the final gene product

genetic mapping Determining the linear order of genetic markers determined by the frequency in which two markers stay together during mating

genetic maps Maps of genetic markers and/or genes ordered by using linkage information but without exact base-pair distances

genetic markers Physical landmarks used to construct genomic maps that are genetic in origin, such as genes, single nucleotide polymorphisms

genetic surgery Less common name for gene replacement therapy

genome browser A computer program that allows the scientist to look graphically at the entire genome of an organism to identify regions or genes of interest

genome map A graphical representation of an organism's genome

genome-wide association study (GWAS) Examining genetic variation among different individuals to see if the variation associates with a particular trait

germline Reproductive cells producing eggs or sperm that take part in forming the next generation (in eukaryotic organisms)

ghrelin Peptide hormone that signals hunger

glucocorticoid hormones Group of steroid hormones in mammals that are involved in water and ion balance

glutathione-*S*-transferase (GST) Enzyme that binds to the tripeptide glutathione; often used in making fusion proteins

glycogen Polymer of glucose linked by α-1,4 bonds and used as a storage polysaccharide by animals and bacteria

glycosylation The addition of sugar residues (usually several forming short chains) to proteins after translation

glyphosate A weed killer that inhibits synthesis of amino acids of the aromatic family in plants

gp120 Glycoprotein of 120 kDa found in outer envelope of HIV that binds to CD4 protein on host cell, resulting in entry of the virus

G-proteins Class of GTP-binding eukaryotic proteins involved in signal transmission

gratuitous inducer A molecule (usually artificial) that induces a gene but is not metabolized like the natural substrate; the best known example is the induction of the *lac* operon by IPTG

green fluorescent protein (GFP) A jellyfish protein that emits green fluorescence and is widely used in genetic analysis

growth factor Protein or other chemical messenger circulating in the blood that carries signals for promoting growth to the cell surface

guanylate cyclase Enzyme that synthesizes cyclic GMP

hairpin A double-stranded base-paired structure formed by folding a single strand of DNA or RNA back on itself

hairpin ribozyme Small catalytic RNA molecule with four helices around two internal loops

hammerhead ribozyme Small catalytic RNA containing three helices around one core loop

hapblocks Groups of common SNPs that are commonly associated with each other during analysis of the human genome

haploid spores A fungal sexual offspring that carries a single copy of all the genes (generally used of organisms that have two or more sets of each gene)

haploinsufficiency Situation in which a defect in one of the two copies of a diploid gene causes significant phenotypic defects

haplotype blocks (hapblocks) Groups of common SNPs that are commonly associated with each other during analysis of the human genome

heat-shock proteins Protein that mediates the refolding of misfolded or unfolded proteins

heat-stable oral vaccines Vaccines that are made by expressing a disease antigen in plant leaf tissue. Once the plant tissue is grown, it can be freeze-dried and encapsulated so that refrigeration is unnecessary

heavy-chain antibodies (hcAb) An antibody that is from camels but has only a single heavy chain molecule

heavy chains The longer of the two pairs of chains forming an antibody molecule

helper component proteinase (Hc-Pro) A polyviral protein that inhibits accumulation of plant siRNAs, but has no effect on the spread of RNA interference to other parts of the plant

helper virus A virus that provides essential functions for defective viruses, satellite viruses, and satellite RNA

hemicellulose Mixture of polymers found in plant cell walls that consist of several sugars, especially mannose, galactose, xylose, arabinose, and glucose

hemimethylated Methylated on only one strand

hemolysin Protein toxin that lyses red blood cells

hepatitis delta virus (HDV) A single-stranded RNA satellite virus that has ribozyme activity

Herceptin Trade name of trastuzumab, a monoclonal antibody to the HER2 receptor on metastatic breast cancer cells that is used as a therapeutic treatment to block growth and spread of the cancer cells

heritable A trait that is passed from the parent to the offspring

herpesviruses Family of DNA-containing viruses that cause a variety of diseases and sometimes cause tumors. They contain dsDNA and an outer envelope surrounding the nucleocapsid

heterochromatin Tightly packed form of chromatin that is found in centromeres and telomeres; histone 3 is di- and tri-methylated on lysine 9 (H3K9) in these regions of the genome

heterologous Derived from a different species

heteroplasmy Presence of different mitochondrial (or chloroplast) genomes in a single individual

hexon Protein subunit of virus capsid with six-fold symmetry that is therefore surrounded by six neighboring subunits

high-pressure liquid chromatography (HPLC) A type of chromatography in which a liquid phase containing a mixture of molecules is passed over a solid phase under high pressure

histone Special positively charged protein that binds to DNA and helps to maintain the structure of chromosomes in eukaryotes

histone acetyl transferase (HAT) Enzyme that adds acetyl groups to histones

histone deacetylases (HDACs) Enzyme that removes acetyl groups from histones

histone modification The addition of methyl or acetyl groups onto histones that modulates the access various transcription regulators has to genes

HIV (human immunodeficiency virus) The member of the retrovirus family that causes AIDS

HLA genes Family of genes for proteins found on the cell surface and acting in cell recognition

hole The absence of an electron from an atom. Used in combination with electrons to create current in a semiconductor

homologous cosuppression A type of RNAi in which multiple copies of a transgene decrease the expression of related host genes

homologous recombination Switching of DNA pieces between two strands of DNA that is initiated by two regions of the DNA with very high homology

homoplasmy Condition in which the population of mitochondria within an individual are all identical

hormone Molecule that carries signals inside multicellular organisms

Horseradish peroxidase (HRP) An enzyme extracted from horseradish root used to visualize the location of the attached molecule (such as DNA or antibody); the enzyme oxidizes various substrates, such as luminol that releases light when it reacts with HRP

"hot" Slang for radioactive

housekeeping genes Genes that are switched on all the time because they are needed for essential life functions

human chorionic gonadotropin (hGC) Hormone produced by the placenta during pregnancy

human factor IX One of the proteins involved in blood clotting

human herpesvirus 8 (HHV8) Virus of the herpesvirus family that causes Kaposi's sarcoma, often seen in AIDS patients

human immunodeficiency virus (HIV) The member of the retrovirus family that causes AIDS

human leukocyte antigens (HLAs) Another name for the proteins of the major histocompatibility complex (MHC) of humans, due to their presence on the surface of white blood cells

humanized (of an antibody) Replacing all of a protein, except the antigen-binding region, with human-encoded sequences

humoral immunity Type of immunity that relies on B cells to produce antibodies

Huntington's disease Inherited defect affecting nerve cells that results in loss of control of the limbs, impaired cognition, and dementia

hybrid dysgenesis A genetic mechanism causing abnormally low frequency of viable offspring in a mating between two parents

hybridization Pairing of single strands of DNA or RNA from two different (but related) sources to give a hybrid double helix

hybridoma A cell made by researchers in which an antibody-producing cell (B cell) is fused with a myeloma cell to form a self-proliferating cell that produces one specific monoclonal antibody

hydrolase Type of enzyme that degrades its substrate by hydrolysis

ICAM1 (intercellular adhesion molecule 1) Protein found on the surface of animal cells that is used as a receptor by many viruses of the picornavirus family

ice nucleation factor Protein that acts as a seed for ice crystals to form

***IGF1* gene** Gene that encodes insulin-like growth factor 1

immune memory Memory of antigens previously encountered by the immune system due to specialized memory B cells

immunity protein Protein that provides immunity. In particular, bacteriocin immunity proteins bind to the corresponding bacteriocins and render them harmless

immunocytochemistry Technique of visualizing specific cellular proteins by using antibodies. When the antibody binds the protein, the label reveals its position

immunoglobulin G (IgG) The class of antibody with a γ (gamma) heavy chain

immunoglobulins Another name for antibody proteins

immunohistochemistry Technique of visualizing specific proteins in a tissue section using labeled antibodies

immunosenescence As people age, the hematopoietic stem cells lose their ability to form new white blood cells, and result in an age-related decline in immune function

imprinting When the expression of a particular allele depends on whether it originally came from the father or the mother (imprinting is a rare exception to the normal rules of genetic dominance)

indexing Mixing multiple DNA samples together in one next-generation sequencing reaction. Each sample has a different barcode sequence in the adapters

induced pluripotent stem cells (iPSCs) Normal somatic cells from either mouse or human that can be induced to de-differentiate by expressing a cocktail of four different transcription factors that are only expressed in stem cells

***in planta Agrobacterium* transformation** (see also floral dip transformation) Method in which a transgene is carried by *Agrobacterium* and transformed into *Arabidopsis* by dipping a developing flower bud into a solution of *Agrobacterium* plus a surfactant

***in vivo* induced antigen technology (IVIAT)** Method used to identify genes active after an infectious agent enters the host cell. The technique identifies antibodies from the patient that react with intracellular proteins of the disease agent

***inaZ* gene** Gene encoding ice nucleation protein of *Pseudomonas syringae*

inclusion bodies Dense crystals of misfolded, nonfunctional proteins found in host cells that are expressing a foreign protein

indigo Bright blue pigment based on the indole ring system

indole Ring system containing nitrogen and found in tryptophan and indigo

inducer (signal molecule) Molecule that exerts a regulatory effect by binding to a regulatory protein

inducible promoter A promoter that functions only under special circumstances

influenza virus Member of the orthomyxovirus family with eight separate ssRNA molecules

inhalational anthrax Form of anthrax in which the spores of *Bacillus anthracis* enter via the lungs and the death rate is high

inhibitors of apoptosis (IAP) A family of proteins that inhibit cell death by modulating caspases during apoptosis

initiation factors Proteins that are required for the initiation of a new polypeptide chain

initiator box Sequence at the start of transcription of a eukaryotic gene

inner cell mass Thickened area of cells of a blastula that forms by cell division and cell migration. The cells have the ability to form every cellular type in the body and are used to make embryonic stem cell lines

insertion Mutation in which one or more extra bases are inserted into the DNA sequence

insertional inactivation Inactivation of a gene by inserting a foreign segment of DNA into the middle of the coding sequence

insulator binding protein (IBP) Protein that binds to insulator sequence and is necessary for the insulator to function

insulator sequence A DNA sequence that shields promoters from the action of enhancers and also prevents the spread of heterochromatin

insulators DNA sequences that shield promoters from the action of enhancers and also prevent the spread of heterochromatin

insulin receptor Protein on cell surface that acts as a receptor for insulin

insulin Small protein hormone made by the pancreas that controls the level of sugar in the blood

insulin and IGF-1 signal transduction pathway (IIS pathway) Signal transduction pathway that is activated by insulin and insulin-like growth factor-1 (IGF-1), which controls cellular growth and is implicated in aging

insulin-like growth factor 1 (IGF1) Peptide that stimulates the growth and division of certain animal cells

integrase (*intI*) Enzyme that inserts a segment of dsDNA into another DNA molecule at a specific recognition sequence. In particular, lambda integrase inserts lambda DNA into the chromosome of *E. coli*

integron analysis Identifying the genes embedded within integrons in order to identify useful new genes from the environment

integron Genetic element consisting of an integration site (for gene cassettes) plus a gene encoding an integrase

intein Self-splicing intervening sequence that is found in a protein

interferon γ (INF γ) Protein induced in animal cells in response to intracellular pathogens

interferons (INF) Family of proteins induced in animal cells in response to virus infection or intracellular pathogens

interferons α and β (INF α and INF β) Proteins induced in animal cells in response to virus infection and that induce antiviral responses

interleukin-4 (IL-4) A cytokine that (among other effects) stimulates the division of B cells, which synthesize antibodies

interleukins Subclass of cytokines involved in development of immune system cells

internal ribosomal entry site (IRES) Sequence allowing the translation of multiple coding sequences on the same message in a eukaryotic cell. IRES sequences are found on some animal viruses

intrinsic asymmetry The unequal inheritance of a specific factor such as a protein, RNA, or macromolecule that changes the fate of one daughter cell during stem cell division

intrinsic pathway Pathway of initiating apoptosis that originates from an internal event that activates the mitochondrial Bcl-2 family of proteins

intron Segment of a gene that does not code for protein but is transcribed and forms part of the primary transcript

invasin Bacterial protein that provokes an animal cell to swallow the bacteria

inverse PCR Method for using PCR to amplify unknown sequences by circularizing the template molecule

inversion Mutation in which a segment of DNA has its orientation reversed but remains at the same location

invertase (strictly, DNA invertase) An enzyme that recognizes specific sequences at the two ends of an invertible segment and inverts the DNA between them

ion-exchange HPLC Chromatography technique that separates noncharged from charged molecules. The stationary phase has a net charge that attracts charged molecules, but neutral molecules are not retained

IPTG (isopropyl-thiogalactoside) A gratuitous inducer of the *lac* operon

IRES Internal ribosomal entry site

isoelectric focusing Technique for separating proteins according to their charge by means of electrophoresis through a pH gradient

isoform One of several alternative forms of a protein that are encoded by the same gene. They differ because of alternative splicing of mRNA or alternative processing of precursor protein

JNK (Jun amino-terminal kinase) A eukaryotic protein that transfers a phosphate group from itself to AP-1 in order for AP-1 to bind and activate gene transcription

jumping gene Popular name for a transposable element

junk DNA Term used to describe defective selfish DNA that is of no use to the host cell it inhabits and that can no longer move or express its genes

kappa particles Symbiotic bacteria (*Caedibacter*) that grow inside killer *Paramecium* and make toxin

killer *Paramecium* *Paramecium* carrying kappa particles and therefore capable of killing sensitive paramecia

kinase Enzyme that catalyzes the addition of phosphate groups onto another protein

knockout mice Mice containing genes that have been inactivated by genetic engineering, usually by insertion of a DNA cassette to disrupt the coding sequence

kuru Brain degeneration disease of cannibals caused by prion

LacI repressor Repressor protein that controls *lac* operon

lactate dehydrogenase (LDH) Enzyme that interconverts pyruvate and lactate

lactoferrin An iron transport protein of animals

lactose acetylase Protein product of the LacA gene that has unknown function in lactose metabolism

lactose permease (LacY) The transport protein for lactose

***lacZ* gene** Gene encoding β-galactosidase; widely used as a reporter gene

lagging strand The new strand of DNA that is synthesized in short pieces during replication and then joined later

large-offspring syndrome Imprinting defect that causes abnormal body size

late genes Genes expressed later in virus infection and that mainly encode enzymes involved in virus particle assembly

latency State in which a virus replicates its genome in step with the host cell without making virus particles or destroying the host cell. Same as lysogeny, but generally used to describe animal viruses

lawn A uniform growth of bacteria that coats the entire surface of the growth medium. Single colonies of bacteria are not visible

LCR (same as locus control region) Regulatory sequence in eukaryotes found in front of a cluster of genes that it controls

leading strand The new strand of DNA that is synthesized continuously during replication

lectin Plant protein that specifically binds carbohydrates

leptin A protein hormone that controls the appetite and the burning of fat by the body

leptin receptor Receptor for leptin, encoded by the *db* gene

lethal factor (LF) Protein toxin made by anthrax bacteria that is a protease and cleaves several host cell mitogen-activated protein kinase kinases (MAPKKs)

ligated In biotechnology, joining up DNA fragments end to end using an enzyme such as DNA ligase

light chains The shorter of the two pairs of chains forming an antibody molecule

lignin Insoluble polymer of cross-linked aromatic residues found in plant cell walls

lineage analysis A method to trace a group of cells or individual cells through development or to trace cell divisions in a tissue or organ of an adult organism. The method can identify stem cell populations in adult tissues

linkage Two markers are linked when they are inherited together more often than would be expected by chance, usually because the two markers are found close to one another on the same chromosome

linkage mapping Determining the linear order of genetic markers determined by the frequency in which two markers stay together during mating

linkers (also called adapters) A short double-stranded piece of DNA with known sequence that is added to the ends of DNA fragments

lipofection Use of liposomes to transfer DNA or proteins into a target cell

liposomes Vesicles with an aqueous core surrounded by a phospholipid shell that may be used to deliver oligonucleotides, drugs, or other molecules across the cell membrane

live cell microarrays A method used to analyze siRNA library clones in which the library DNA is spotted onto a glass slide and eukaryotic cells are grown over the slide. The siRNA library clones are taken up, and the cells are assessed for physical differences

local mediator Molecule that carries signals between nearby cells

locus control region Regulatory sequence in eukaryotes found in front of a cluster of genes that it controls

long interspersed element (LINE) Long sequence found in multiple copies that makes up much of the moderately repetitive DNA of mammals

long terminal repeats (LTRs) Direct repeats of several hundred base pairs found at the ends of retroviruses and some other retroelements required for insertion into host DNA

***loxP* site** Specific sequence that is recognized by Cre recombinase

***luc* gene** Gene encoding luciferase from eukaryotes

luciferase Enzyme that emits light when provided with a substrate known as luciferin

luciferin Chemical substrate used by luciferase to emit light

Lumi-Phos An artificial substrate that is split by alkaline phosphatase, releasing an unstable molecule that emits light

***lux* gene** Gene encoding luciferase from bacteria

lysins Proteins made by bacterial viruses that destroy the cell membrane to release newly made virus particles

lysogeny Type of virus infection in which the virus becomes largely quiescent, makes no new virus particles, and duplicates its genome in step with the host cell. Same as latency but used of bacterial viruses

lysozyme An enzyme found in many bodily fluids that degrades the peptidoglycan of bacterial cell walls

lytic phase Type of growth in which a virus generates many virus particles and destroys the cell

M (mitosis) phase Fourth phase of the eukaryotic cell cycle, in which the cell divides; also known as mitosis

macroalgae Larger multicellular photosynthetic plant-like organisms commonly called seaweed

macroautophagy The process that degrades cellular components like organelles inside a double-membrane structure

macrolides Class of antibiotics derived from the polyketide pathway

mad cow disease Brain degeneration disease of cattle caused by prions, also known as bovine spongiform encephalopathy (BSE)

magnetosomes Prokaryotic organelle that contains mineralized magnetic crystals of Fe_3O_4 or Fe_3S_4 surrounded by a protein layer

maintenance methylase Enzyme that adds a second methyl group to the other DNA strand of half-methylated sites

MalE protein Carrier protein for maltose found in the periplasmic space of *E. coli*

maltose-binding protein (MBP) Protein of *E. coli* that binds maltose during transport; often used in making fusion proteins

mammalian Target of Rapamycin (mTOR) A kinase that regulates cell growth, cell proliferation, cell survival, protein synthesis, and transcription

***MAO-A* gene** Gene that encodes monoamine oxidase A

MAR proteins Proteins that form the connection between the chromosomes and the nuclear matrix

Marathon Mouse Transgenic mouse that runs farther than a normal mouse

massively parallel sequencing Sequencing technologies that use miniaturized platforms for sequencing millions of DNA fragments in parallel

MAT* locus** Chromosomal locus in yeast that controls the mating type and exists as two alternative forms: *MAT**a or *MAT*α

matrix attachment regions (MAR) Site on eukaryotic DNA that binds to proteins of the nuclear matrix or of the chromosomal scaffold—same as SAR sites

matrix-assisted laser desorption-ionization (MALDI) Type of mass spectrometry in which gas-phase ions are generated from a solid sample by a pulsed laser

mediator complex A protein complex that transmits the signal from transcription factors to the RNA polymerase in eukaryotic cells

melanocortin receptor Receptor for one of several hormones of the melanocortin family

melanocortins Family of peptide hormones all derived from the same precursor protein: pro-opiomelanocortin (POMC)

melt When DNA is used, refers to its separation into two strands as a result of heating

melting temperature, Tm Temperature at which 50% of a protein is denatured

memory cells Specialized B cells that wait for possible infections instead of manufacturing antibodies

Mendelian disease A disorder that is inherited from one generation to the next in a similar pattern as a trait studied by Mendel which is usually attributed to a mutation to a single gene such that one allele is dominant over another allele

metabolic fingerprinting Characterizing all the metabolites that are present under a certain set of conditions or at a certain time

metabolome The total complement of small molecules and metabolic intermediates of a cell or organism

metagenomic library A library of DNA sequences isolated from an environmental sample. Contains gene sequences from many different organisms and genetic elements

metagenomics The study of all the genomes in a particular environment

metalloprotease A protease that uses a metal cofactor to facilitate protein digestion

metallothionein A metal-binding protein whose synthesis is induced only when certain heavy metals are present

metallothionein promoter Promoter of gene encoding metallothionein; used in genetic engineering because it is very strong and is induced by traces of zinc or other metals

metastasis Process in which cancer cells from a primary tumor move around the body and form secondary cancers

methionine sulfoximine Toxic analog of methionine

methotrexate Antibiotic that inhibits the enzyme dihydrofolate reductase in animal cells

methyl *tert*-butyl ether (MTBE) A fuel additive that oxygenates gasoline so the engine runs with less knocking and the gasoline burns more completely

methylcytosine binding proteins Type of protein in eukaryotes that recognizes methylated CG islands

microalgae Prokaryotic or eukaryotic photosynthetic microorganisms that are unicellular or simple multicellular structures that live in terrestrial or aquatic environments

microautophagy-lysosome system A system of protein degradation where the proteins are taken into the lysosome by direct transfer

microbial fuel cells Device that uses microorganisms to obtain electrons for creating electrical currents

microfluidics Manipulation of fluids on a micrometer scale

microRNAs (miRNAs) Small regulatory RNA molecules of eukaryotic cells

microsatellite polymorphisms Genetic markers that consist of very short repeated sequences of 2 to 5 nucleotides in length

mifepristone A synthetic steroid that is used in high concentrations to induce abortions and, in low concentrations, to induce genes artificially

Mighty Mouse Transgenic mouse with colossal muscle development

mini-gene Miniature gene encoding part of a protein, usually made by artificial DNA synthesis

minisatellite Another term for variable number tandem repeat, a cluster of tandemly repeated sequences in DNA whose number of repeats differs from one individual to another

Minos A transposon originally found in insects

minus (–) strand The noncoding strand of RNA or DNA

mismatch repair system DNA repair system that recognizes mispaired bases and cuts out part of the DNA strand containing the wrong base

missense mutation Mutation in which a single codon is altered so that one amino acid in a protein is replaced with a different amino acid

mitochondrial death pathway Program of apoptosis that involves activating mitochondrial proteins to kill the cell; often activated by internal factors such as DNA damage

mitogen-activated protein (MAP) kinases Family of signal transmission proteins that form part of a phosphotransfer cascade in animal cells

mitophagy Degradation of mitochondria by autophagy or engulfment and degradation by a lysosome

mobile DNA Segment of DNA that moves from site to site within or between other molecules of DNA

mobile phase The liquid or solvent containing the mixture of molecules that moves over the stationary phase in column chromatography

modification enzyme Enzyme that binds to the DNA at the same recognition site as the corresponding restriction enzyme but methylates the DNA

modular design Taking various useful domains of different proteins and combining them into a new engineered protein

molar absorptivity The absorbance of a 1 molar solution of pure solute at a given wavelength. The higher it is, the more light is absorbed

molecular chaperone Protein that oversees the correct folding of other proteins

molecular phylogenetics Study of evolutionary relationships using DNA or protein sequences

molecular weight standards Mixture of varying sized DNA or proteins of known size used to compare to unknown proteins

monoamine oxidase (MAO) Enzyme involved in degradation of neurotransmitters of the catecholamine family

monocistronic mRNA mRNA carrying the information of a single cistron that is a coding sequence for only a single protein

monoclonal antibody A pure antibody with a unique sequence that recognizes only a single antigen and that is made by a cell line derived from a single B cell

morpholino-antisense oligonucleotides Synthetic oligonucleotides with morpholino rings instead of ribose and phosphorodiamidate linkages between nucleotides

mousepox (*Ectromelia* virus) Relative of smallpox virus that infects mice

mRNA or messenger RNA The class of RNA molecule that carries genetic information from the genes to the rest of the cell

multidimensional protein identification technique (MuDPIT) A method of analyzing complex mixtures of proteins with mass spectroscopy that separates the proteins using a liquid chromatography microcapillary column (2D LC microcapillary column)

multigenic Due to interaction of several genes

multiple cloning site (MCS) (see also polylinker) A stretch of artificially synthesized DNA that contains cut sites for seven or eight widely used restriction enzymes

multiple locus probing (MLP) Variant of DNA fingerprinting in which multiple probes are used

multiple nuclear polyhedrosis virus (MNPV) A particular baculovirus widely used as a cloning vector

multiplexing Mixing multiple DNA samples together in one next-generation sequencing reaction. Each sample has a different barcode sequence in the adapters

multiplex PCR Running several PCR reactions with different primers in the same tube

multipotent A term that describes a cell that can develop into a limited number of specialized cell types

multivalent vaccine A vaccine prepared from two or more related pathogen species or multiple antigens from a single pathogen

murine leukemia virus (MuLV) A simple retrovirus frequently used to construct vectors for gene therapy

muscular dystrophy Several diseases that result in the wasting away of muscle tissue and cause premature death

mutation An alteration in the DNA (or RNA) that composes the genetic information

mutation breeding Using mutagens to induce genetic changes in plants in order to make or increase a desirable trait

mutation hot spot Region of DNA where alterations in DNA due to mutation are common

mutator gene Gene that affects the rate at which mutations occur

Mx proteins Antiviral proteins of animal cells that interfere with RNA polymerase of negative strand RNA viruses

myc A small DNA segment that encodes for a peptide epitope that can be recognized by antibodies. The tag is used to mark uncharacterized proteins for analysis

***myc* oncogene** Oncogenic version of gene encoding Myc protein

Myc protein Transcription factor involved in switching on several genes involved in cell division

myeloma cells Cancer cells derived from B cells, which therefore express immunoglobulin genes

myelomas Naturally occurring cancers derived from B cells that express antibodies, which are used to create hybridomas

NAD Nicotinamide adenine dinucleotide; a cofactor that carries reducing equivalents during dehydrogenations; NAD usually acts in degradative pathways

NADP Nicotinamide adenine dinucleotide phosphate; a cofactor that carries reducing equivalents during dehydrogenations; NADP usually acts in biosynthetic pathways

nano- A prefix meaning one billionth (1/1,000,000,000)

nanobody (Nb) Recombinant antibody fragment that contains only the variable region of the single heavy chain of the Camelidae family

nanocarpets Structure formed by stacking a large number of nanotubes together, with their cylindrical axes aligned vertically. The carpet has antibacterial qualities and the ability to change color

nanoparticles Particles of submicron scale—in practice from 100 nm down to 5 nm in size—that can be constructed in a variety of shapes

nanorods Nanoparticles that have a long cylindrical shape. Only the diameter must be in the nanoscale range

nanoscale ion channel Small nanoscale channel created using biological molecules that puncture small holes in the membrane and allow passage of ions under controlled conditions

nanoshells Hollow nanosized particles that can hold different materials

nanotechnology Control of the structure of matter based on molecule-by-molecule control of products and by-products; the products and processes of molecular manufacturing, including molecular machinery

nanotubes Cylinders made of pure carbon with diameters of 1 to 50 nanometers that have novel properties that make them potentially useful in a wide variety of applications

nanowires Wires of dimensions in the order of a nanometer (10^{-9} meters) range, which can be metallic, semiconducting, and insulating. They can be designed of repeating organic units such as DNA

naphthalene oxygenase Enzyme that carries out the first step in naphthalene breakdown by inserting oxygen into the aromatic ring

necroptosis Type III programmed cell death pathway that is triggered by TNF-alpha binding to the TNF receptor in the presence of caspase inhibitors; cells death is disordered but triggers an immune response

necrosis Death of a cell characterized by cellular swelling and rupture; elicits an immune response

negative (–) strand The noncoding strand of RNA or DNA

negative regulation Regulatory mode in which a repressor keeps a gene switched off until the repressor is removed

neomycin phosphotransferase Enzyme that confers resistance to antibiotics such as kanamycin and neomycin

nerve growth factor A soluble trophic factor that is required to keep neurons alive

neurofibrillary tangles Intracellular clumps of protein found in neurons of Alzheimer's patients

neuron Nerve cell

neuropeptide Y (NPY) Peptide in brain that increases feeding and so makes animals fatter

neurotoxin Toxin that attacks nerve cells

neurotransmitter Molecule that carries signals across synapses between cells within the nervous system

neutral drift A population of stem cells will divide and differentiate such that over time, the entire population of descendant cells will be derived from only one of the stem cells. The other stem cells are lost over time, and do not produce descendants that populate the tissue

next-generation sequencing Sequencing technologies that use miniaturized platforms for sequencing millions of DNA fragments in parallel

***N*-formylmethionine (fMet)** Modified methionine used as the first amino acid during protein synthesis in bacteria

nick A break in the backbone of a DNA or RNA molecule (but where no bases are missing)

nicotine Alkaloid in tobacco that raises levels of uncoupling protein 1 in brown fatty tissue, hence promoting fat metabolism

nitric oxide (NO) Gaseous molecule used in signaling by animal cells

nitric oxide synthase (NO synthase) Enzyme that synthesizes nitric oxide

nitrocellulose A pulpy or cotton-like substance formed into sheets and used to attach proteins in Western blotting

NO synthase (nitric oxide synthase) Enzyme that synthesizes nitric oxide

Nomarski optics Type of polarization microscopy that transforms differences in density into differences in height so that the image looks like a three-dimensional relief

noncoding (or antisense) The strand of DNA that has complementary sequence to mRNA

nonhomologous end joining (NHEJ) DNA repair system found in eukaryotes that mends double-stranded breaks, but can introduce deletions or insertions

non-natural amino acids Modified amino acids that have functional groups that are useful for biochemical analysis or protein engineering

non-nucleoside reverse transcriptase inhibitor (NNRTI) Antiviral agent that inhibits the reverse transcriptase of viruses such as HIV but is not a nucleoside analog

nonsense mutation Mutation due to changing the codon for an amino acid to a stop codon

nontarget organism Any organism exposed to a specific insecticide, herbicide, or transgenic plant which that product was not intended to harm

nontemplate The strand of DNA equivalent in sequence to the messenger RNA (same as plus strand)

noradrenalin Neurotransmitter of the catecholamine family (also known as norepinephrine)

norepinephrine Neurotransmitter of the catecholamine family (also known as noradrenalin)

Northern blots Hybridization technique in which a DNA probe binds to an RNA target molecule

nosocomial infections Those infections acquired in a hospital setting

npt The gene that encodes neomycin phosphotransferase

phenotype catalogue A listing of proteins that cause the same phenotype when their function is disrupted

phenotypic signature A set of physical features that are categorized together to classify a particular cell function, such as adherence, motility, or cell division

phenylalanine hydroxylase The enzyme that converts the amino acid phenylalanine into tyrosine

phenyl-boronate Resin used to bind beta-lactamases

phenylketonuria Inherited defect causing lack of the enzyme phenylalanine hydroxylase and hence a buildup of phenylalanine

pheromone Molecule that carries signals between organisms

***phoA* gene** Gene encoding alkaline phosphatase; widely used as a reporter gene

phosphate group Group of four oxygen atoms surrounding a central phosphorus atom found in the backbone of DNA and RNA

phosphodiester bond The linkage between nucleotides in a nucleic acid that consists of a central phosphate group esterified to sugar hydroxyl groups on either side

phosphodiesterase 5 (PDE5) One particular cyclic phosphodiesterase of animal cells that degrades cyclic GMP

phospholipase Enzyme that degrades phospholipids

phosphoramidite A monoamide of a phosphite diester found on nucleoside phosphoramidites which are used during chemical synthesis of DNA

phosphorelay Term to describe transferring one phosphate from one location to another to successively activate different proteins

phosphorodiamidate An uncharged version of the phosphodiester group linking nucleotides in which one of the oxygen atoms is replaced with an amidate group

phosphorothioate A phosphate group in which one of the four oxygen atoms around the central phosphorus is replaced by sulfur

phosphorothioate oligonucleotide A synthetic oligonucleotide with a sulfur replacing one of the four oxygen atoms in the phosphodiester linkage between nucleotides

phosphotransferase system System found in many bacteria that transports sugars and regulates metabolism. It operates by transferring phosphate groups

photodetector Sensitive instrument that detects an optical signal and converts it into an electrical signal

photolithography Method used to synthesize oligonucleotides directly on a DNA chip where light is passed through a mask to selectively activate some regions and keep other regions blocked

photolyase Enzyme that catalyzes the repair of DNA thymine dimers in response to blue light

phylogenetic tree A diagram showing the evolutionary relationship of different organisms

physical mapping Determining the actual distance between genetic markers by physically associating a marker with a location along the chromosome

physical maps Genetic maps that give physical DNA base-pair lengths between features

phytoremediation Using plants to sequester various water or soil contaminants so that it cannot effect other wildlife

picornaviruses Virus family that includes enteroviruses (such as poliovirus) and rhinoviruses (which cause the common cold)

piezoelectric ceramics Materials that change shape in response to an applied voltage

piggyBac A transposon originally found in insects

pilin Protein that makes up the main part of a pilus

pilus (plural pili) Thin helical protein filaments found on the surface of bacteria; same as fimbria

piwi-interacting RNAs A class of small non-coding RNAs that are encoded by genes in the organism's genome, and are used to inhibit transposon movement; specific to germ cells

plaques (when referring to viruses) A clear zone caused by virus destruction in a layer of cultured cells or a lawn of bacteria

plasmid incompatibility The inability of two plasmids of the same family to coexist in the same host cell

plasmid Self-replicating genetic elements that are sometimes found in both prokaryotic and eukaryotic cells. They are not chromosomes or part of the host cell's permanent genome. Most plasmids are circular molecules of double-stranded DNA, although rare linear plasmids and RNA plasmids are known

Plasmodium falciparum Protozoan parasite that causes the malignant form of malaria

plus (+) strand The coding strand of RNA or DNA

poison sequence Base sequence, often found in virus genomes, that is not stably maintained or replicated on plasmids in bacterial hosts

poly(A) tail A stretch of multiple adenosine residues found at the 3′ end of mRNA

polyacrylamide gel electrophoresis (PAGE) Technique for separating proteins by electrophoresis on a gel made from polyacrylamide

polyadenylation complex Protein complex that adds the poly(A) tail to eukaryotic mRNA

polycistronic mRNA mRNA carrying multiple coding sequences (cistrons) that may be translated to give several different protein molecules; only found in prokaryotic (bacterial) cells

polyclonal antibody Natural antibody that actually consists of a mixture of different antibody proteins that all bind to the same antigen

polygenic A particular phenotype that is influenced by more than one gene

polyglutamine tract Run of multiple glutamine residues in a protein

polyhedrin Protein that comprises polyhedron structure of baculoviruses

polyhedron In reference to viruses, the packages of virus particles embedded in a protein matrix that are formed by baculoviruses

polyhistidine tag (His6 tag) Six tandem histidine residues that are fused to proteins, thus allowing purification by binding to nickel ions that are attached to a solid support

polyhydroxyalkanoate (PHA) Type of bioplastic polymer made by certain bacteria from hydroxyacid subunits

polyhydroxybutyrate (PHB) Bioplastic polymer made by certain bacteria from hydroxybutyrate subunits

polyketide Class of natural linear polymers whose backbone consists of two carbon repeats with keto groups on every other carbon (when first synthesized)

polylinker (see also multiple cloning site [MCS]) A stretch of artificially synthesized DNA that contains cut sites for seven or eight widely used restriction enzymes

polymerase chain reaction (PCR) Amplification of a DNA sequence by repeated cycles of strand separation and replication

polymorphism A difference in DNA sequence between two related individual organisms

polyploid Having more than one set of chromosomes per cell

polysome Group of ribosomes bound to and translating the same mRNA

polytene chromosomes Giant chromosomes found in cells that replicate their DNA without dividing into separate cells. Found in *Drosophila* salivary gland cells

polyvalent inhibitor Inhibitor consisting of several linked inhibitor molecules that consequently binds multiple copies of its target, giving a very high overall binding affinity

position effect variegation (PEV) The effect of chromosomal position on the expression of a particular gene. For example, genes embedded within heterochromatin are not expressed, but if positioned in euchromatin, the same gene would be expressed

positional cloning Approach that attempts to locate a gene by its position on a chromosome and is used when the nature of the gene product is unknown

positive (+) strand The coding strand of RNA or DNA

positive regulation Control by an activator that promotes gene expression when it binds

post-transcriptional gene silencing (PTGS) Plant version of RNA interference

poxviruses Family of large animal viruses with dsDNA and approximately 150 to 200 genes

Prader–Willi syndrome Inherited defect resulting in loss of function of genes on the paternally derived copy of chromosome 15 that are subject to imprinting

pRB (retinoblastoma protein) Protein associated with a malignant cancer of the retina; involved with cellular senescence

precautionary principle The principle that the burden of proving that a proposed change is safe falls on those proposing the change

premature senescence Entering senescence before the normal number of cell divisions

pre-microRNAs Longer precursor molecules that are converted into microRNAs

preproinsulin Insulin as first synthesized, with both a signal sequence and the connecting peptide

presenilins Transmembrane proteins found in the Golgi apparatus and endoplasmic reticulum; associated with Alzheimer's disease

prey The fusion between the activator domain of a transcriptional activator protein and another protein as used in two-hybrid screening

PriA Protein of the primosome that helps primase bind

primary antibody First antibody to the protein of interest used to identify a protein in Western blotting

primary transcript The original RNA molecule obtained by transcription from a DNA template, before any processing or modification has occurred

primase Enzyme that starts a new strand of DNA by making an RNA primer

principle of independent assortment Alleles of a gene sort independently during formation of a gamete

principle of segregation Alleles of a gene segregate or separate into different gametes during gamete formation, but then reunite during fertilization

prion protein (PrP) Brain protein that may exist in two forms, one of which is pathogenic and may cause transmissible prion disease

Prnp Gene that encodes the prion protein

probe molecule Molecule that is tagged in some way (usually radioactive or fluorescent) and is used to bind to and detect another molecule

prodrug A harmless compound that is converted to an active drug by a specific enzyme

programmed cell death Genetic program that eliminates damaged cells or cells that are no longer needed without activating the immune system

proinsulin Precursor to insulin that contains both the A and B chains plus the connecting peptide

promoter Region of DNA in front of a gene that binds RNA polymerase and so promotes gene expression

pronuclei The parental male and female nuclei in a fertilized egg just before nuclear fusion

prophage Bacteriophage genome that is integrated into the DNA of the bacterial host cell

protease inhibitor Inhibitor of protease enzymes, in particular antiviral agent that inhibits the protease of viruses such as HIV

protease Same as proteinase; an enzyme that degrades polypeptides by hydrolysis

proteasome Multisubunit complex that degrades proteins labeled with ubiquitin

protective antigen (PA) Protein that acts as the delivery system for both anthrax toxins

protein Polymer made from amino acids; may consist of several polypeptide chains

protein A Antibody binding protein from *Staphylococcus* that is often used in making fusion proteins

proteinase Same as protease; an enzyme that degrades polypeptides by hydrolysis

protein engineering Altering the sequence of a protein by genetic engineering of the DNA that encodes it

protein fusion Hybrid protein made by joining the coding sequences of two proteins together in a frame

protein fusion vector Vector designed for fusing cloned proteins to a carrier protein to help expression and/or export

protein interactome The total of all the protein–protein interactions in a particular cell or organism

protein kinase Enzyme that transfers phosphate groups to other proteins, thus controlling their activity

protein kinase A (PKA) One particular protein kinase of animal cells that is activated by cyclic AMP

protein misfolding cyclic amplification (PMCA) Protocol that amplifies misfolded prions in a manner analogous to PCR

proteome The total set of proteins encoded by a genome or the total protein complement of an organism

proteomics Study or analysis of an organism's complete protein complement

proto-oncogene Original, unmutated wild-type allele of an oncogene

protoplasts Plant cells that have been dissociated and whose cell walls are removed

provirus Virus genome that is integrated into the host cell DNA

PrP^c Normal, "cellular" form of the prion protein

PrP^{Sc} Pathogenic, "scrapie" form of the prion protein

pseudogenes Defective copies of a genuine gene

pulsed field gel electrophoresis (PFGE) Type of gel electrophoresis used for analysis of very large DNA molecules and that uses an electric field of "pulses" delivered from a hexagonal array of electrodes

PUMA (p53-upregulated modulator of apoptosis) A Bcl-2 protein family member that activates Bax and promotes apoptosis

purine Type of nitrogenous base with a double ring found in DNA and RNA

pyrimidine Type of nitrogenous base with a single ring found in DNA and RNA

pyrrolysine Amino acid resembling lysine but has a pyrroline ring attached to the end of the R-group

quantum confinement The phenomenon seen in nanoscale structures where an electron-hole pair is kept within a structure that is near its natural Bohr radius. Energy states are discrete in these structures

quantum dots An alternative name for fluorescent nanocrystals that are small enough to show quantum confinement and are used in biological labeling

quantum yield The ratio of photons absorbed to photons emitted during fluorescence

quasi-species Group of related RNA-based genomes that differ slightly in sequence but arose from the same parental RNA molecule

quelling Fungal version of RNA interference

quorum sensing Regulatory system used by bacteria to coordinate their response to population density

radiation hybrid mapping Mapping technique that uses cells (usually from a rodent) that contain fragments of chromosomes (generated by irradiation) from another species

radical replacement Replacement of an amino acid with another that has different chemical and physical properties

radiochemical detector A device that detects radioactively labeled molecules, for example, in the mobile phase during chromatography

random shuffling library Library of gene or protein sequences generated by randomized shuffling and linking of short segments

randomly amplified polymorphic DNA (RAPD) Method for testing genetic relatedness using PCR to amplify arbitrarily chosen sequences

Ras gene Founding member of a family of related proteins that are small GTPases used for cell signal transduction of growth pathways

***ras* oncogene** Oncogenic version of gene encoding Ras protein

Ras protein GTP binding protein involved in transmitting signals concerning cell division in animals

raster-scan An image displayed line by line

rBST Acronym for recombinant bovine somatotropin

reactive nitrogen species (RNS) Nitrogen containing molecules formed by the incomplete reduction of oxygen during oxidative metabolism

reactive oxygen species (ROS) Molecules that have chemically reactive forms of oxygen such as oxygen ions and peroxides

read The individual sequence of each fragment of DNA as determined by next-generation sequencing

read depth The number of independent sequences determined by next-generation sequence for a specific region of a genome

reanneal Renaturation of single-stranded DNA into double-stranded DNA

receptor Molecule that binds another molecule, such as a hormone or a nutrient. In particular, receptors are often proteins that participate in signaling and are situated in the cell membrane facing outward

Receptor Interacting Protein (RIP1) Serine threonine kinase essential to activating necroptosis

recombinant bovine somatotropin (rBST) Bovine growth hormone produced in another organism

recombinant human somatotropin (rHST) Human growth hormone produced in another organism

recombinant plasmids Plasmids that contain segments of DNA not originally found in the plasmid, most likely from another organism

recombinant protein Protein expressed from recombinant DNA gene

recombinant tissue plasminogen activator (rTPA) Tissue plasminogen activator produced in another organism

recombinase Enzyme that catalyzes recombination between inverted repeats

recombineering Technique that uses homologous recombination to insert foreign DNA into a DNA vector

refractive index detector A device that records changes in the velocity of light as it passes through liquid such as the eluate of a column

release factor Protein that recognizes a stop codon and brings about the release of a finished polypeptide chain from the ribosome

replacement gene therapy Curing inherited defects by introducing functional copies of the defective gene

replication fork Region where the enzymes replicating a DNA molecule are bound to untwisted, single-stranded DNA

replicative aging The number of daughter cells a yeast cell can produce in its lifetime

replicative form (RF) Double-stranded form of the genome of a single-stranded DNA (or RNA) virus. The RF first replicates itself and is then used to generate the ssDNA (or ssRNA) to pack into the virus particles

replicative senescence (cellular senescence) Arrested cell division of cultured mammalian cells

replicative transposition Type of transposition in which two copies of the transposon are generated: one in the original site and another at a new location

replicator A system able to build copies of itself when provided with raw materials and energy

replicon Molecule of DNA or RNA that contains an origin of replication and can self-replicate

replisome Assemblage of proteins (including primase, DNA polymerase, helicase, SSB protein) that replicates DNA

reporter genes Any gene whose product is easy to see or assay that is often fused to a regulator region of another gene. The fusion can then be identified easily after expression in another organism

repressor Regulatory protein that prevents a gene from being transcribed

resolution The sharpness of a peak after separation by chromatography

restriction endonuclease Type of endonuclease that cuts double-stranded DNA at a specific sequence of bases, the recognition site

restriction enzyme-generated siRNA (REGS) A method to generate an RNAi library that uses restriction enzymes to create random double-stranded 21- to 23-nucleotide-long segments of genes

restriction enzyme Type of endonuclease that cuts double-stranded DNA at a specific sequence of bases, the recognition site

restriction fragment length polymorphism (RFLP) A difference in restriction sites between two related DNA molecules that results in production of restriction fragments of different lengths

retinal scan Scan of the unique pattern of blood vessels on the retina

retinoblastoma (*Rb*) gene Anti-oncogene that is responsible for a rare cancer of the retina of the eye

retinoblastoma protein (pRB) Protein associated with a malignant cancer of the retina; involved with cellular senescence

retroviruses A family of animal viruses with single-stranded RNA inside two protein shells surrounded by an outer envelope. Once inside the host cell, they use reverse transcriptase to convert the RNA version of the genome to a DNA copy

reverse-phase array (see also direct immunoassay) Method of protein detection in which the protein is bound to a solid support and an antibody with a detection system is added

reverse-phase HPLC Chromatography technique that passes a solution of proteins over a nonpolar stationary phase under high pressure. Hydrophobic molecules attach to the stationary phase and hydrophilic molecules elute from the column

reverse transcriptase An enzyme that uses single-stranded RNA as a template for making double-stranded DNA

reverse transcriptase PCR (RT-PCR) Variant of PCR that allows genes to be amplified and cloned as intron-free DNA copies by starting with mRNA and using reverse transcriptase

reverse vaccinology Approach to making new vaccines that uses genomics to identify new epitopes or proteins from infectious agents that are highly antigenic without causing disease

reverse-phase assay (or direct immunoassay) Method of protein quantification and detection, where the protein is bound to a solid support, and an antibody with a detection system is added

RFLP (restriction fragment length polymorphism) A difference in restriction sites between two related DNA molecules that results in production of restriction fragments of different lengths

rhinoviruses Group of small RNA viruses that cause a major percentage of cases of the common cold

Rho (ϱ) protein Protein factor needed for successful termination at certain transcriptional terminators

Rho-dependent terminator Transcriptional terminator that depends on Rho protein

Rho-independent terminator Transcriptional terminator that does not need Rho protein

rHST Acronym for recombinant human somatotropin

ribonuclease (or RNase) Enzyme that cuts or degrades RNA

ribonucleoprotein A protein that has RNA associated with it

ribose The five-carbon sugar found in RNA

ribosomal RNA or rRNA Class of RNA molecule that makes up part of the structure of a ribosome

ribosome The cell's machinery for making proteins

ribosome binding site (RBS) Sequence close to the front of mRNA that is recognized by the ribosome; found only in prokaryotic cells

ribosome-inactivating protein (RIP) Toxic protein that inactivates ribosomes by releasing adenine from a specific site in large-subunit rRNA

riboswitches Domains of messenger RNA that directly sense a signal and control translation by alternating between two structures

ribotyping Identification of bacteria or other living organisms based on the sequence of their small-subunit ribosomal RNA

ribozyme RNA molecule that acts as an enzyme

ricin Highly toxic ribosome-inactivating protein from the castor bean

right-handed helix In a right-handed helix, as the observer looks down, the helix axis (in either direction), each strand turns clockwise as it moves away from the observer

RNA (ribonucleic acid) Nucleic acid that differs from DNA in having ribose in place of deoxyribose

RNA interference (RNAi) Response that is triggered by the presence of double-stranded RNA and results in the degradation of mRNA or other RNA transcripts homologous to the inducing dsRNA

RNA polymerase Enzyme that synthesizes RNA

RNA polymerase I Eukaryotic RNA polymerase that transcribes the genes for the large ribosomal RNAs

RNA polymerase II Eukaryotic RNA polymerase that transcribes the genes encoding proteins

RNA polymerase III Eukaryotic RNA polymerase that transcribes the genes for 5S ribosomal RNA and transfer RNA

RNA SELEX Method in which catalytically active RNA sequences are altered by mutation to identify new ligands or substrates for their activity

RNA-dependent RNA polymerase (RdRP) RNA polymerase that uses RNA as a template and is involved in the amplification of the RNAi response

RNAi library A library that expresses double-stranded RNA to activate RNA interference. Each library clone inactivates a gene from the organism of interest

RNA-induced silencing complex (RISC) Protein complex induced by siRNA that degrades single-stranded RNA corresponding in sequence to the siRNA

RNA-SIP Method used to enrich the RNA in an environmental sample. A stable isotope is mixed with an environmental sample and becomes incorporated into any life form that is using RNA for protein expression

rolling circle replication Mechanism of replicating double-stranded circular DNA that starts by nicking and unrolling one strand and using the other, still circular, strand as a template for DNA synthesis. Used by some plasmids and viruses

ROM (reactive oxygen metabolites) Highly reactive oxygen-derived molecules or ions with extra electrons (especially superoxide ions, peroxides, and hydroxyl radicals)

***rosy* gene** Gene of *Drosophila* that affects eye color

Rous sarcoma virus (RSV) Cancer-causing retrovirus of chickens

rTPA Acronym for recombinant tissue plasminogen activator

RU486 Progesterone antagonist; ingredient of abortion pill

RXR protein Nuclear protein that forms mixed dimers with receptors for androgens, vitamin D, thyroxine, and retinoic acid

S phase Second stage ("synthesis" phase) in the cell cycle, in which chromosomes are duplicated

sarcoma Cancer originating from muscle cells

satellite viruses A defective virus that needs an unrelated helper virus to infect the same host cell in order to provide essential functions

scanning tunneling microscope (STM) An instrument able to image conducting surfaces to atomic accuracy; has been used to pin molecules to a surface

scFv fragment (single-chain antibody) Engineered antibody with VH and VL domains linked together by a short peptide chain

SCID (severe combined immunodeficiency) Immune defect due to lack of both T cells and B cells. About 25% of inherited SCID cases are due to adenosine deaminase deficiency

scintillation counting Detection and counting of individual microscopic pulses of light

scrape-loading A method of getting oligonucleotides into cultured cells by gently scraping cells growing on the surface of a dish. The mechanical scraping introduces small breaks in the cell membrane that allow oligonucleotides into the cytoplasm

scrapie Brain degeneration disease of sheep and goats caused by prion

second messenger Intracellular signal molecule that is made when a cell surface receptor receives a message

secondary antibody Antibody to the primary antibody that has a detection system attached. Used in Western blotting to identify a protein of interest

selective pressure (in data analysis) Continued pressure or refinement of a particular data set based on set criteria

selenocysteine Amino acid resembling cysteine but containing selenium instead of sulfur

semiconductors Materials that are between a conductor and insulator in electrical conductivity. The electrical current travels via electron-hole pairs

semiconservative replication Mode of DNA replication in which each daughter molecule gets one of the two original strands and one new complementary strand

senesce The act of entering a quiescent cellular state where the cell functions and is living, but does not divide

senescence A quiescent cellular state where the cell functions and is living, but does not divide

senile neuritic plaques Aggregates of degenerating neurons in brains of Alzheimer's patients

sense strand The strand of DNA equivalent in sequence to the messenger RNA (same as plus strand)

sensor kinase A protein that phosphorylates itself when it senses a specific signal (often an environmental stimulus, but sometimes an internal signal)

sequence tagged site (STS) A short sequence (usually 100–500 bp) that is unique within the genome and can be easily detected, usually by PCR

serine protease A protease that has a serine residue in its active site

serine transacetylase Enzyme that converts acetyl-CoA plus serine to *O*-acetylserine

severe combined immunodeficiency (SCID) Immune defect due to lack of both T cells and B cells. About 25% of inherited SCID cases are due to adenosine deaminase deficiency

shelterin Complex of six proteins that protect DNA telomeres from being recognized as broken DNA and marked for repair

Shine–Dalgarno sequence Same as RBS; sequence close to the front of mRNA that is recognized by the ribosome; found only in prokaryotic cells

short-hairpin RNA (shRNA) Genetically engineered RNA with complementary sequences that fold to become double-stranded. Used to activate RNA interference

short-interfering RNA (siRNA) Double-stranded RNA with 21 to 22 nucleotides involved in triggering RNA interference in eukaryotes

short interspersed element (SINE) Short repeated sequence that makes up a major fraction of the moderately or highly repetitive DNA of mammals

short tandem repeats (STR) Subclass of VNTR with short repeated sequences

shotgun sequencing Approach in which the genome is broken into many random short fragments for sequencing. The complete genome sequence is assembled by computerized searching for overlaps between individual sequences

shuttle vector A vector that can survive in and be moved between more than one type of host cell

siderophore Iron chelator used by microorganisms to extract and bind iron from their environments

sigma subunit Subunit of bacterial RNA polymerase that recognizes and binds to the promoter sequence

signal molecule (see also inducer) Molecule that exerts a regulatory effect by binding to a regulatory protein

signal transmission protein Protein involved in transmitting signals, often from cell surface receptor to gene regulators

sildenafil Generic name for Viagra

silence In genetic terminology, refers to switching off genes in a relatively nonspecific manner

silver stain A sensitive dye used to stain proteins

simian virus 40 (SV40) Cancer-causing virus of monkeys

similarity search Comparing newly sequenced DNA with other known DNA sequences to determine its identity

single-chain antibody (scFv fragment) Engineered antibody with VH and VL domains linked together by a short peptide chain

single-chain Fv (scFv) Fv fragment of an antibody engineered so that the VH and VL domains are linked together by a short peptide chain

single-locus probing (SLP) Variant of DNA fingerprinting in which a single probe is used

single nucleotide polymorphism (SNP) A difference in DNA sequence of a single base change between two individuals

single-stranded binding protein A protein that keeps separated strands of DNA apart

sirtuins A family of histone modifying enzymes that are associated with aging

size exclusion chromatography Chromatography technique that separates on the basis of size

sliding clamp Subunit of DNA polymerase that encircles the DNA, thereby holding the core enzyme onto the DNA

small subunit ribosomal RNA (SSU rRNA) The 16S rRNA in prokaryotes or 18S rRNA in eukaryotes; target RNA that is isolated from the environment to identify and catalogue all the different organisms

smallpox Virulent virus disease that infects only humans

Smart Mouse Transgenic mouse that has improved learning and memory

snRNA Small nuclear RNA; involved in splicing introns from eukaryotic primary RNAs

sodium bisulfite Chemical that deaminates cytosine to uracil

soft robotics Robotics modeled on soft biological structures

sodium dodecyl sulfate (SDS) A detergent widely used to denature and solubilize proteins before separation by electrophoresis

somatic cell Cell making up the body, as opposed to the germline

somatic mosaicism The genomic sequence for some somatic cells changes due to unfixed mutations and errors, resulting in a subset of cells with a slightly different genomic sequence than the rest of the body

somatic mutation Mutation that occurs in somatic cells and is not passed on to the next generation via the germline

somatic stem cells A collection of cells isolated from adult somatic tissues that are able to divide indefinitely, remain undifferentiated, but have the ability to differentiate into specialized cells of the organ or tissue in which they are originally derived

somatotropin A polypeptide hormone that controls cell growth and reproduction in humans and other animals

Southern blots A method to detect single-stranded DNA that has been transferred to nylon paper by using a probe that binds DNA

specific immunity Type of immunity that responds to specific antigens over a course of days, has the ability to distinguish self from nonself, remembers past invading pathogens, and generates new antibodies

specific transcription factors Regulatory proteins that exert their effect on a single gene or operon or on a very small number of related genes

spectrosomes An organelle-like structure found in the male and female germlines of *Drosophila* that contains a lot of membrane skeletal proteins that functions to orient the mitotic spindle in the germline stem cell so that one daughter differentiates and the other daughter remains in the stem cell niche

spindle Microtubule-based structure in which chromosomes attach and separate during mitosis or meiosis

spliceosome Complex of proteins and small nuclear RNA molecules that removes introns during the processing of messenger RNA

splicing factors Molecules that remove intervening sequences and rejoin the ends of a molecule; usually refers to removal of introns from RNA

***src* oncogene** Oncogene carried by Rous sarcoma virus and originally derived from chicken cells

Stable Isotope Labeling by Amino acids in Cell culture (SILAC) A technique used to discern two different protein samples using heavy isotopes such as ^{13}C and ^{15}N during mass spectroscopy

stable isotope probing (SIP) Method for enriching the DNA in an environmental sample. The sample is incubated with a stable isotope that is incorporated into any actively dividing life form, and then the labeled DNA is isolated

stanol Sterol derivative whose double bond has been reduced

starch Polymer of glucose linked by α-1,4 bonds and used as a storage polysaccharide in plants

Starlink Controversial transgenic corn that inadvertently entered the human food supply before being fully evaluated by the government

stationary phase The material packed into a column whose physical or chemical properties separate a mixture of molecules such as proteins into separate fractions

stem cell A cell that can divide indefinitely, that is undifferentiated, and that has the ability to develop into different cell types such as muscle, nerve, intestine, etc.

stem cell niche A local microenvironment that directly promotes or protects a population of stem cells within an organism

steroid 1. Type of polycyclic lipophilic molecule that includes cholesterol, sex hormones, and corticosterols. 2. Sterol derivative having keto rather than hydroxyl groups

steroid hormones Hormones with a steroid structure, such as the sex hormones

steroid receptor Dual-function protein that acts as receptor for steroid hormone and as transcription factor

sterols Class of lipophilic biological compounds with four fused nonaromatic rings

sticky ends Ends of a double-stranded DNA molecule that have unpaired single-stranded overhangs, generated by a staggered cut

STR (short tandem repeats) Subclass of VNTR with short repeated sequences

strand slippage During replication, DNA polymerase may slip along the template strand at a repeat and reattach at a different site. The daughter strand will have a different number of repeats from the parental DNA strand

streptavidin Enzyme from *Streptomyces avidinii* that binds to biotin with a very high affinity

streptolysin O A toxin from *Streptococci pyogenes* that attaches to cholesterol in cellular membranes and aggregates into a circular structure, forming a pore for molecules to enter the cell

structural gene Sequence of DNA (or RNA) that codes for a protein or for an untranslated RNA molecule

subunit vaccines Vaccines that contain a single polypeptide from the disease agent. The immune system creates antibodies to the single polypeptide and inhibits the infectious agent by attacking via the single target protein

subventricular zone A region of the brain situated near the ventricles or fluid filled regions (near the connections of the left and right hemispheres) that serves as a source of neural stem cells

subviral agents Infectious agents that are more primitive than viruses and encode fewer of their own functions

supercoiling Higher level coiling of DNA that is already a double helix

superoxide dismutase (SOD) Enzyme that converts superoxide to oxygen plus hydrogen peroxide

suppressive subtraction hybridization (SSH) Culture enrichment technique for metagenomics that identifies differences in DNA content between two different environments

surface-enhanced laser desorption-ionization (SELDI) A form of mass spectroscopy in which the protein of interest is in a liquid phase that is placed on a solid surface and then ionized with a laser

suspension cells Cultured cells that grow in a liquid nutrient media, not attached to any surface

suspension culture Dedifferentiating a mass of plant tissue and then growing the cells in a liquid medium

symbiotic theory Theory that the organelles of eukaryotic cells are derived from symbiotic prokaryotes

symmetrical differentiation A cell division in a stem cell niche that produces two differentiated daughter cells rather than retaining the stem cell fate

symmetrical renewal A cell division in a stem cell niche that produces two more stem cells

synapse Junction between cells, across which signals are carried by chemical molecules known as neurotransmitters

syncytium Giant cell with many nuclei

synthetic biology The term used to describe the field of scientific study that uses biological components such as genes, promoters, and enhancers to create new biological pathways, devices, or systems

T cells Cells of the immune system that remove virally infected cells and secrete soluble factors that activate other cells in the immune system, particularly B cells. Responsible for cell-mediated immunity; make T-cell receptors rather than antibodies

T lymphocytes (also T cells) Type of immune system cell that is responsible for cell-mediated immunity and that makes T-cell receptors instead of antibodies. These mature in the thymus

TA cloning Procedure that uses *Taq* polymerase to generate single 3′-A overhangs on the ends of DNA segments that are used to clone DNA into a vector with matching 3′-T overhangs

***tac* promoter** Hybrid promoter containing the ribosome-binding site from the *trp* promoter and the operator sequence from the *lac* promoter

TALE nucleases (TALENs) Transcription activator-like effector nucleases that are a fusion of the *FokI* restriction enzyme DNA cleavage domains with a DNA binding domain from TALE proteins

tandem mass spectroscopy Mass spectroscopy technique with multiple steps of selection and analysis that use successively smaller fragments for starting material

***Taq* DNA polymerase** Heat-resistant DNA polymerase from *Thermus aquaticus* that is used for PCR

target DNA DNA that is the target for binding by a probe during hybridization or the target for amplification by PCR

target sequence 1. Sequence on host DNA molecule into which a transposon inserts itself. 2. Sequence within the original DNA template that is amplified in a PCR reaction

targeting vector Vector designed to promote the integration of a transgene into a specific location

targeting-induced local lesions in genomes (TILLING) Mutagenesis technique in which plant seeds are soaked in a chemical mutagen to induce point mutations and the genomic DNA is analyzed by PCR and hybridization

TATA binding factor or TATA box factor Transcription factor that recognizes the TATA box

TATA binding protein Transcription factor that recognizes the TATA box

TATA box Binding site for a transcription factor that guides RNA polymerase II to the promoter in eukaryotes

TATA box protein Transcription factor that recognizes the TATA box

tau Protein normally associated with microtubules that clumps and forms helical aggregates in Alzheimer's patients

taxonomy Scientific classification of organisms based on physical or genetic relatedness

telomerase Enzyme made of RNA plus protein that re-elongates telomeres by adding DNA to the end of a eukaryotic chromosome

telomeres Special repeated sequences that cap the ends of linear eukaryotic chromosomes

temperature-sensitive mutation Mutation that shows a different phenotype at different temperatures

template Strand of DNA used as a guide for synthesizing a new strand by complementary base-pairing

teratoma A tumor that usually develops in the gonads which is made of tissues or differentiated cells not normally found in that tissue

terminal differentiation Expression of final phenotype for a cell type or tissue

tester DNA from the experimental sample that has sequences unique to that environment. Used in suppressive subtraction hybridization

***tet* operon** Cluster of bacterial genes that confers resistance to the antibiotic tetracycline

***tetO* operator** Site in front of *tet* operon where repressor binds

TetR repressor Repressor protein that controls *tet* operon

tetracyclines Family of antibiotics with four fused rings derived from the polyketide pathway

therapeutic cloning Cloning to obtain tissue for transplantation as opposed to generating a new individual

thermocycler Machine used to rapidly shift samples between several temperatures in a preset order (for PCR)

thermogenin Alternative name for uncoupling protein 1 (UCP1)

thermostable Able to withstand high temperatures without loss of function

theta-replication Mode of replication in which two replication forks go in opposite directions around a circular molecule of DNA

thiophene Five-membered aromatic ring containing sulfur and four carbon atoms

threonine protease A protease that has an active-site threonine

thrombin A protease involved in the blood-clotting cascade

thymidine kinase Enzyme that converts thymidine and related nucleosides to their monophosphate derivatives

thyroid hormone (thyroxine) Hormone made by thyroid gland

thyroxine Hormone made by thyroid gland

time-of-flight (TOF) Type of mass spectrometry detector that measures the time for an ion to fly from the ion source to the detector

Tm (see also melting temperature, Tm) Temperature at which 50% of a protein is denatured

***TNF* gene** Gene encoding tumor necrosis factor

TOL plasmid (pTOL) Plasmid-carrying pathway for degradation of toluene

topoisomerase I Enzyme that alters the level of supercoiling or catenation of DNA (i.e., changes the topological conformation)

totipotent A term that describes a cell that can develop into all the different types of cells of the body

toxin Poisonous molecule, often a protein, of biological origin; in particular, refers to proteins made by pathogenic bacteria

transcription Process by which information from DNA is converted into its RNA equivalent

transcription bubble Region where DNA double helix is temporarily opened up, thus allowing transcription to occur

transcription factor Protein that regulates gene expression by binding to DNA in the control region of the gene

transcription start site Starting point where a gene is converted into its RNA copy

transcriptional gene silencing An alternate term used to describe RNA interference-like phenomena

transcriptome The total sum of the RNA transcripts found in a cell, under any particular set of conditions

transfer RNA (tRNA) RNA molecules that carry amino acids to the ribosome

transferrin An iron transport protein of animals

transformation 1. (as used in bacterial genetics) Process in which genes are transferred into a cell as free molecules of DNA. 2. (in cancer biology) Conversion of a normal cell into a cancer cell

transgene A foreign gene that is inserted into an organism using genetic engineering

transgenic An organism with a foreign piece of DNA stably integrated into its genome

transgenic art Art form that involves transgenic animals and plants

transgenic plant A plant containing a gene (transgene) from a different plant or other organism

transit amplifying cells A population of cells in a state of development between the stem cell fate to a fully differentiated state

transition Mutation in which a pyrimidine is replaced by another pyrimidine or a purine is replaced by another purine

translation Making a protein using the information provided by messenger RNA

translational expression vector Vector designed to enhance gene expression at the level of translation

translatome The total set of proteins that have actually been translated and are present in a cell under any particular set of conditions

translocation 1. Transport of a newly made protein across a membrane by means of a translocase. 2. Sideways movement of the ribosome on mRNA during translation. 3. Removal of a segment of DNA from its original location and its reinsertion in a different place

transmissible spongiform encephalopathy (TSE) Infectious form of prion disease

transplantation The transfer of specially marked cells from one organism to another to determine if the marked cells are stem cells competent to repopulate and divide in the adult tissue in which they originally derived

transposable element A mobile segment of DNA that is always inserted in another host molecule of DNA. It has no origin of replication of its own and relies on the host DNA molecule for replication. Includes both DNA-based transposons and retrotransposons

transposase Enzyme responsible for moving a transposon

transposition The process by which a transposon moves from one host DNA molecule to another

transposon A mobile segment of DNA that is always inserted in another host molecule of DNA. It has no origin of replication of its own and relies on the host DNA molecule for replication. Includes both DNA-based transposons and retrotransposons

transversion Mutation in which a pyrimidine is replaced by a purine or vice versa

trehalase Enzyme that degrades trehalose into two glucoses

trehalose A nonreducing storage sugar of plants that protects against dehydration

trehalose-phosphate synthase Enzyme that combines UDP-glucose and glucose 6-phosphate into trehalose 6-phosphate

trehalose 6-phosphate phosphatase Enzyme that removes a phosphate from trehalose 6-phosphate

triplets (see also codons) Group of three RNA or DNA bases that encodes a single amino acid

tRNA$_i$ (initiator tRNA) RNA molecule that carries the first amino acid of a protein to the ribosome

tRNA$_i^{fMet}$ Designation for a tRNA charged with *N*-formyl-methionine, which is used as the first amino acid in a prokaryotic protein

trophic factors Soluble molecules that tell neurons, or other cells, to continue to live

tularemia Bacterial disease of rodents or birds that has a death rate of 5% to 10% in humans

tumor-infiltrating lymphocyte (TIL) White blood cell that secretes TNF

tumor necrosis factor (TNF) Protein that kills cancer cells and is produced by tumor-infiltrating lymphocytes

tumor-suppressor gene Gene that acts to prevent unwanted cell division (same as anti-oncogene)

two-component regulatory system A regulatory system consisting of two proteins:a sensor kinase and a DNA-binding regulator

two-dimensional polyacrylamide gel electrophoresis (2D-PAGE) A technique used to separate proteins first by size and then by isoelectric focusing (i.e., by electrical charge)

two-hybrid system Method of screening for protein–protein interactions that uses fusions of the proteins being investigated to the two separate domains of a transcriptional activator protein

type I restriction enzyme Type of restriction enzyme that cuts the DNA a thousand or more base pairs away from the recognition site

type I secretory system Specialized export system that spans both inner and outer membranes of gram-negative bacteria such as *E. coli*

type I toxin Toxin that triggers a harmful (internal) response by binding to a cell surface receptor

type II restriction enzyme Type of restriction enzyme that cuts a fixed number of bases away from its recognition site

type II secretory system Specialized export system that spans the outer membrane only

type II toxin Toxin that damages the cell membrane

type III toxin Toxin consisting of a toxic factor (A protein) that enters the target cell together with a delivery system (B protein)

tyrosyl-tRNA synthetase Enzyme that charges tRNA with tyrosine

ubiquitin A small 8 kD protein that self-conjugates to form long polyubiquitin chains; it is used to mark proteins for degradation by the proteasome

ubiquitin-proteasome system Protein complex that degrades proteins that have been tagged with ubiquitin chains

ultraviolet detector A device that records changes in ultraviolet absorption due to solutes in the mobile phase

uncoupling (of mitochondria) Dissociation of the operation of the respiratory chain from ATP synthesis, which results in the energy being wasted as heat

uncoupling protein (UCP) Protein that uncouples the respiratory chain of mitochondria and releases heat

undifferentiated A state of cellular development where a cell does not perform any specific function within an organ or tissue

universal genetic code Version of the genetic code used by almost all organisms

vaccination Artificial induction of the immune response by injecting foreign proteins or other antigens

vaccines A solution containing various derivatives or parts of an infectious agent that do not cause disease, but induce the body to make memory B cells

variable number tandem repeat (VNTR) Cluster of tandemly repeated sequences in DNA, whose number of repeats differs from one individual to another

variable region Region of an antibody whose sequence is varied by gene segment shuffling in order to provide many alternative antigen-binding sites

variant CJD (variant Creutzfeldt–Jakob disease) Name used for human cases of mad cow disease

variant Creutzfeldt–Jakob disease (variant CJD) Name used for human cases of mad cow disease

Varkud satellite ribozyme A small ribozyme that initiates self-cleavage and replication of the Varkud satellite virus RNA found in the mitochondria of *Neurospora*

V(D)J recombination Mechanism of creating diversity in antibody structure that involves recombination of different segments, alternate splicing of the mRNA, and finally producing an antibody with one V, one D, and J segment linked to a constant region

vector vaccine A vaccine in which a nonpathogenic virus or bacterium (the vector) is engineered to express a protein or peptide from the disease agent on its surface. The immune system creates antibodies to the expressed protein, which confers immunity to the disease agent

VHH Designation for the variable region of a single heavy chain antibody found in members of the Camelidae family, which is engineered to form nanobodies

Viagra (sildenafil) Drug used to treat male erectile dysfunction; acts by inhibiting phosphodiesterase 5 and keeping cyclic GMP levels up

virion Virus particle

virosphere All the viruses found within our biosphere

virulence factors Inherited properties that allow pathogenic microorganisms to successfully infect their hosts

virulence plasmid Plasmid that carries genes involved in virulence and pathogenicity

virus-induced silencing An alternate term used to describe RNA interference-like phenomena

VNTR (variable number tandem repeats) Cluster of tandemly repeated sequences in the DNA, whose number of repeats differs from one individual to another

v-onc Virus-borne version of an oncogene

VP16 activator Gene activator protein from herpes simplex virus

weaponization Conversion of an infectious biowarfare agent to a physical form that can be easily spread

weaponized Refers to agents that have been physically prepared for use as biological weapons, such as by making a stable powder

West Nile virus Virus of the flavivirus family originally from the Middle East/Africa that is now spreading in North America

Western blotting Detection technique in which a probe, usually an antibody, binds to a protein target molecule

whole exome sequencing Sequencing only the exon sequences of the human genome rather than the entire human genome using specific PCR primers to amplify all the exons during Next-generation sequencing

WIN compound Antiviral agent that prevents the attachment of many picornaviruses

wobble Less rigid base-pairing but only for codon/anticodon pairing during translation

xenobiotic Chemical compound that possesses significant biological activity, but is foreign to the environment

xenoestrogen Foreign polycyclic molecule that binds to steroid receptor and mimics the action of estrogens

X-inactivation The condensation and complete shutting down of gene expression of one of the two X chromosomes in cells of female mammals

X-Phos 5-bromo-4-chloro-3-indolyl phosphate, an artificial substrate that is split by alkaline phosphatase, releasing a blue dye

xylose Five-carbon sugar that is a major component of various hemicellulose polysaccharides

yeast artificial chromosome (YAC) Single-copy vector based on yeast chromosome that can carry very long inserts of DNA. Widely used in the Human Genome Project

yeast prion Yeast protein that behaves in some ways like the prion protein of mammals

Yersinia pestis Bacterium that causes both the bubonic and pneumonic forms of plague

Z-form An alternative form of double helix with left-handed turns and 12 base pairs per turn. Both DNA and dsRNA may be found in the Z-form

zinc finger domain A protein domain of 25 to 30 amino acids containing cysteines and histidines that bind to zinc

zinc finger nucleases (ZFNs) An artificial nuclease that is created by protein engineering a zinc finger domain to a DNA recognition domain from the enzyme *FokI*

zoo blot Comparative Southern blotting using DNA target molecules from several different animals to test whether the probe DNA is from a coding region

zygote The cell that arises from the fusion of a sperm and egg, which is also called a fertilized egg

INDEX

Note: Page numbers followed by "b", "f" and "t" indicate boxes, figures and tables respectively.

C

F

G

L

M

N

Q

R

W

X

Y

Z